THE PRINCIPLES OF INSECT PHYSIOLOGY

THE PRINCIPLES OF INSECT PHYSIOLOGY

by

V. B. WIGGLESWORTH

C.B.E., M.D., F.R.S.

EMERITUS PROFESSOR OF BIOLOGY IN THE UNIVERSITY OF CAMBRIDGE; FORMERLY DIRECTOR, AGRICULTURAL RESEARCH COUNCIL UNIT OF INSECT PHYSIOLOGY

Seventh Edition

With 412 illustrations

LONDON
CHAPMAN AND HALL

First published in 1939
by Methuen and Co. Ltd
Second edition 1944
Third edition 1947
Fourth edition, revised, 1951
Fifth edition, with addenda, 1953
Sixth edition, revised, 1965
Reprinted once
Seventh edition 1972
published by Chapman and Hall Ltd
11 *New Fetter Lane London EC4P 4EE*
Reprinted 1974

Printed in Great Britain by
Butler & Tanner Ltd, Frome

SBN 412 11490 9

Published in the U.S.A.
by Halsted Press, a Division
of John Wiley & Sons, Inc.
New York

PREFACE

INSECTS PROVIDE an ideal medium in which to study all the problems of physiology. But if this medium is to be used to the best advantage, the principles and peculiarities of the insect's organization must be first appreciated. It is the purpose of this book to set forth these principles so far as they are understood at the present day. There exist already many excellent text-books of general entomology; notably those of Imms, Weber, and Snodgrass, to mention only the more recent. But these authors have necessarily been preoccupied chiefly with describing the diversity of form among insects; discussions on function being correspondingly condensed. In the present work the emphasis is reversed. Structure is described only to an extent sufficient to make the physiological argument intelligible. Every anatomical peculiarity, every ecological specialization, has indeed its physiological counterpart. In that sense, anatomy, physiology and ecology are not separable. But regarded from the standpoint from which the present work is written, the endless modifications that are met with among insects are but illustrations of the general principles of their physiology, which it is the aim of this book to set forth. Completeness in such a work is not possible, or desirable; but an endeavour has been made to illustrate each physiological characteristic by a few concrete examples, and to include sufficient references to guide the student to the more important sources.

The physiology of insects is to some the handmaid of Economic Entomology. For although it is not the purpose of physiology to furnish directly the means of controlling insect pests, yet the rational application of measures of control—whether these be insecticides of one sort or another, or artificial interferences with the insect's environment—is often dependent on a knowledge of the physiology of the insect in question. Physiology may thus serve to rationalize existing procedures, or to discover the weak spots in the ecological armour of a species. A knowledge of the ecology of a species is always necessary to its effective control; its ecology can be properly understood only when its physiology is known.

1939

PREFACE TO FOURTH EDITION

DESPITE THE turmoil of the intervening years there have been substantial advances in our knowledge of insect physiology since this book was published in 1939. Some of these represent new principles; others illuminate established notions. In this revised edition an attempt has been made to incorporate this new material into the old framework. It is the aim of such a book as this to provide a framework sufficiently well founded to support the new knowledge but not so rigid as to impede its progress.

1950

PREFACE TO THE FIFTH EDITION

IN THIS edition a few corrections and alterations have been made in the main text. At the end of each chapter will be found a number of Addenda, which include the most recent advances with cross-references to the pages which they supplement.

1953

PREFACE TO THE SIXTH EDITION

DURING THE twelve years that have elapsed since the last revision of this book the study of insect physiology has been transformed. New techniques in electron microscopy, neurophysiology, and many sides of biochemistry have led to spectacular advances. The rate at which new material is being published is rising steeply.

A complete revision of the text with the incorporation of this new knowledge has been undertaken, and some fifty new illustrations have been introduced. As in earlier editions the bibliography has been restricted—so far as is compatible with providing authority for each statement. That has entailed the addition of more than two thousand new references.

It goes without saying, however, that to master the advances that are being made on so many fronts, and to sift the essential discoveries from the non-essential, is becoming beyond the range of a single author. It is only the conviction that the physiology of any group of animals must be treated as a unity that has encouraged the present writer to make the attempt.

1965

PREFACE TO THE SEVENTH EDITION

THIS EDITION is a reprint of the sixth edition supplemented by inclusions, amounting to some ten per cent. of the whole text, in which the major advances of the past seven years have been incorporated, together with a few new illustrations. In order to hold down the cost of reprinting, page references to the new material are given in square brackets at the end of the relevant paragraphs; but all the new entries are printed together, under appropriate captions, in an addenda section at the end of each chapter, where the many new references to the literature will be found.

1972

CONTENTS

Chapter I

Development in the Egg

Cuticular membranes of the egg: the chorion—Most insects begin their independent life within an egg. When the egg is laid it is enclosed by two envelopes, the chorion and the vitelline membrane. The egg-shell or chorion is the product of the follicular cells in the ovary of the mother (p. 704). It has been studied in great detail in *Rhodnius*[6] where it consists of some seven layers of proteinaceous material modified in various ways by tanning (p. 34), by association with lipids, and perhaps by the formation of sulphur linkages (Fig. 1). Two layers compose the outer part or 'exochorion', the substance of which, termed 'chorionin'[79] resembles the 'cuticulin' of the epicuticle (p. 35). The inner part or 'endochorion' is composed of five layers, the innermost (like the outermost layer of the cuticle) being rich in polyphenols.

The egg is covered with a layer of cement secreted by the colleterial glands of the female. This secretion secures the eggs to the surface on which they are laid; or it may serve to bind them together into a compact capsule or oötheca, as in Blattidae and Mantidae (p. 706). In the familiar egg rafts of *Culex*, on the other hand, the eggs are held together not by any cement, but only by the surface tension of the water on which they rest.[18] The chorion may be thin and flexible, as in the eggs of such insects as *Tenebrio*, which are laid in protected situations; usually it is more or less rigid, as in the exposed eggs of Lepidoptera, Hemiptera, &c. In Phasmidae, and likely enough in other groups, it is strengthened by the incorporation of lime in its substance.[50, 48]

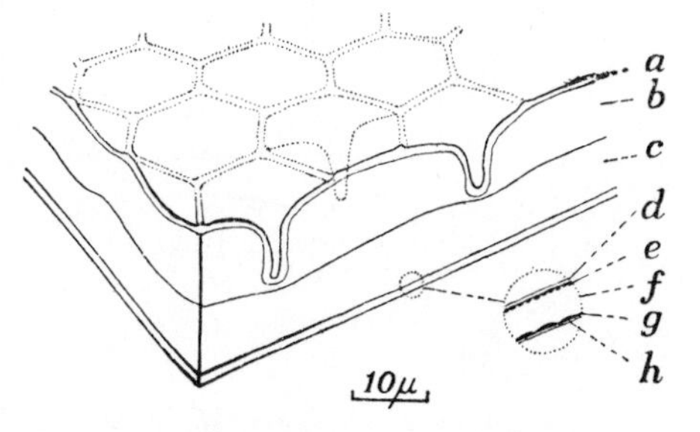

FIG. 1.—Section of chorion in egg of *Rhodnius* (schematic) (*after* BEAMENT)

a, resistant exochorion; *b*, soft exochorion (both of the lipoprotein 'chorionin'); *c*, soft endochorion; *d*, amber layer of lipidized tanned protein; *e*, outer polyphenol layer; *f*, resistant layer of tanned protein; *g*, inner polyphenol layer; *h*, primary wax layer.

The coverings of the egg must protect it also from the loss of water by evaporation. The copious colleterial secretion or 'spumaline' over the eggs of *Malacosoma* (Lep.) is hygroscopic and probably helps to conserve moisture.[33] The chorion itself may assist in the retention of water, particularly if its substance is kept dry,[20] but it is not usually responsible for the protection of those eggs which are highly resistant to desiccation. In the egg of *Rhodnius*, for example, all the layers of the chorion are freely permeable to water. Waterproofing is due to a layer of wax, perhaps 0·5μ thick, which is laid down by the developing oöcyte before ovulation and is firmly attached to the inside of the chorion. The egg is therefore waterproofed before laying, the process being independent of fertilization.[6] This wax resembles in its properties the waterproofing waxes of the cuticle (p. 35). Likewise in the egg of *Bombyx*, a wax layer below the endochorion is the chief barrier to transpiration.[130] In the *Culex* egg an oily layer seems to exist between the exochorion and the endochorion;[18] in *Lucilia* the waterproofing lipoid is laid down by the oöcyte between the chorion and the chorionic vitelline membrane,[20] and the same applies to *Aedes hexodontus*.[95] In the grasshopper *Melanoplus* a hard white wax in a layer about 0·1μ thick is deposited, as in *Rhodnius*, beneath the chorion,[71] and a secondary wax layer on the surface of the serosal cuticle (p. 4).[45, 125] The egg is not, of course, completely impermeable to water, for insect eggs kept under too dry conditions may fail to hatch, in some cases because the embryo within is desiccated, in others because the chorion itself becomes too hard for the young insect to compass its escape.[14, 19]

Respiratory mechanisms—Respiration takes place through the general surface of the chorion when this is thin; but some eggs show structural adapta-

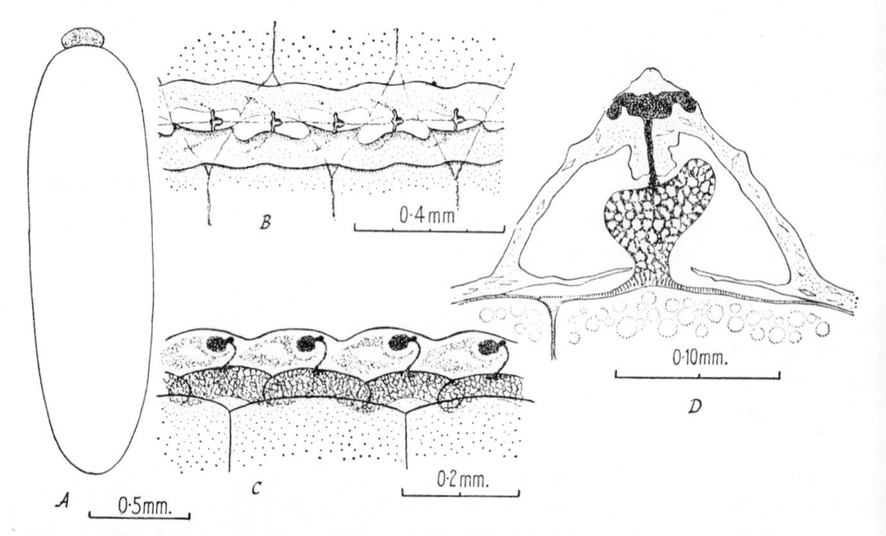

FIG. 2.—Respiratory system in the egg of *Blattella* (*after* WIGGLESWORTH and BEAMENT)

A, dorsal view of single egg showing the curved respiratory process at the anterior pole. *B*, crest of the oötheca seen from above; overlying the respiratory process on each egg is a small T-shaped chamber filled with air. *C*, crest of the oötheca seen from the side; the air chamber has been injected and the black contents extend along a curved duct to the respiratory process. *D*, transverse section of the crest of the oötheca after injection showing how the air (replaced by the black injection material) extends from the exterior to the respiratory process and so to the thin chorion below.

tions which combine rigidity with the needs of respiration. It was shown by Leuckart (1855) that in many eggs the shell is lined by a porous air-filled layer which communicates with the environment so that diffusion of the respiratory gases can proceed in a gaseous phase to all parts of the egg surface. The filaments at the anterior pole of the eggs of *Nepa* and *Ranatra* are the openings of this respiratory system. Around the rim of the cap on the egg of *Rhodnius* are numerous 'pseudo-micropyles' which serve the same purpose;[6, 87] if these are blocked the egg is asphyxiated.[80] The 'sperm cups' of the Lygaeid *Oncopeltus* have a central canal which serves as a micropyle (p. 705) and spongy air-filled walls which serve for respiration.[87] Elaborate arrangements of the same general type occur in the eggs of Lepidoptera,[87] of *Dixippus*,[87] *Syrphus*, *Drosophila*,[87] *Gasterophilus*,[131] and other flies,[49] and in the oötheca and eggs of *Blattella*.[87] The passage of oxygen through the oötheca in cockroaches is made possible by the moulding of ducts, when the oötheca is formed, which lead from the air-filled chorion to the outside air (Fig. 2).[87] The system fills with air while the egg is still bathed in fluid in the oviduct.[49] The mechanism of filling presents the same problems as the filling of the tracheal system in submerged eggs or larvae (p. 364). It is presumably the result of the development of hydrophobe properties in the porous substance by tanning or wax secretion.[87] [See p. 20.]

The horns on the eggs of Diptera are well adapted to meet the demands of environments that may be either dry or flooded: they can obtain air from solution in water by means of the air-containing horns, acting as a gill or 'plastron' (p. 383) which is highly resistant to the entry of water; but in dry air the loss of moisture will be limited to the small cross-sectional area at the base of the horn.[108]

The vitelline membrane originates from the cell wall of the ovum. In grasshoppers[70] and many other insects it remains a delicate structure and disappears as soon as development begins. In Tachinidae and other Diptera it is probably of twofold origin—an inner oöplasmic zone, and an outer chorial zone, which is at first thick and viscous but later condenses to form a tough membrane;[49] so that the egg of *Lucilia*, for example, is enclosed, beneath the true chorion, in a relatively thick tough 'chorionic vitelline membrane';[20] in *Drosophila* this layer containing protein, lipid, and polysaccharide, is certainly a product of the follicular cells of the mother.[114] Muscid eggs in moist conditions will continue their development within this sheath after the chorion has been stripped off.[86] In *Dytiscus*, also, the vitelline membrane is a very definite structure.[7]

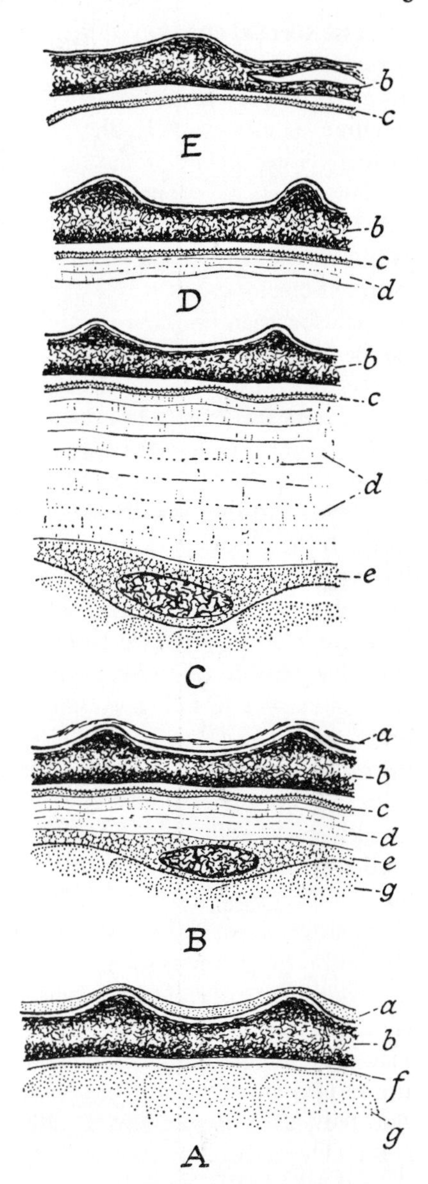

FIG. 3.—Egg membranes in *Melanoplus differentialis* (*after* SLIFER)

A, egg newly laid; *B*, egg incubated 11 days at 25° C. (serosa has laid down the yellow cuticle and part of white cuticle); *C*, egg incubated 21 days (serosal cuticle fully formed); *D*, egg 3 days before hatching (serosal cuticle much reduced); *E*, egg ready to hatch (nothing remains but chorion and the yellow part of the serosal cuticle); *a*, temporary secretory layer over chorion; *b*, chorion; *c*, yellow cuticle; *d*, white cuticle; *e*, serosa; *f*, vitelline membrane; *g*, yolk.

The serosal cuticle—Later in development a layer of epithelium, the serosa, is formed beneath the vitelline membrane; and this may lay down another cuticle. In *Dytiscus* it is a stout laminated structure which has been called the 'secondary vitelline membrane'.[7] In the grasshopper *Melanoplus* it closely resembles the ordinary integument of the insect (p. 28), consisting of two layers, (i) a thin 'yellow cuticle', responsible for the impermeability of the egg membranes for salt ions,[39] secreted at 6 or 7 days after laying (at 25° C.), and (ii) a laminated white cuticle, composed largely of chitin (p. 29), apparently free from protein, secreted during the next week (Fig. 3). When complete this inner layer is generally thicker than the chorion and is responsible for the toughness of the grasshopper's egg.[70] A similar serosal cuticle occurs in *Pteronarcys* (Plecoptera).[47] In the egg of *Locusta* if the fluid which later fills the space between the serosa and the serosal cuticle is exposed to the air it quickly forms a tanned protein-lipid membrane (p. 35);[111] when the eggs of grasshoppers swell, any damage to the serosal cuticle is repaired in this way.[106]

Absorption of water—Eggs laid in moist surroundings may absorb water through their cuticular membranes. This was first noted by Réaumur (1740) in the eggs of sawflies and ants and later by Rathke (1844) in the eggs of *Gryllotalpa* from damp soil, and in the aquatic Trichoptera; it has since been observed in many other insects.[14,60] In *Dytiscus* the egg, at first easily deformed by pressure, becomes tense and hard, increasing in diameter from 1·2 to 2·25 mm.; the chorion splits and falls away and the egg is enclosed only in the vitelline membrane and serosal cuticle.[7] The egg of *Locusta migratoria*, which will develop normally only in moist soil, increases in weight from 6·3 mg. to 14 mg., the water content increasing from 52 to 82 per cent.[60] As the egg swells the chorion usually cracks and peels away; but if kept moist it stretches and remains intact until hatching.[70] In the Capsid *Notostira*, water is absorbed from the tissues of the plant in which the egg is laid. Absorption begins about 55 hours after laying, at the same time as the formation of tissue becomes visible, and it reaches its maximum (a 75 per cent. increase) in about 160 hours (Fig. 4). In this case the chorion is resistant to distension; the swelling forces off the preformed egg-cap, and the egg is prevented from bursting only by the formation over the yolk of a cuticular plug laid down by the serosa and probably comparable with the serosal cuticle of other insects.[40] The eggs of Aleurodids are implanted into the leaves of plants by a stalk which bears a thin-walled terminal bladder; this is able to extract water from the plant and so make up for loss of water from the egg surface[83] (Fig. 5).

Special structures for the absorption of water probably occur in other insects also. Thus at the posterior pole of the egg of the grasshopper *Melanoplus* there is a small circular area in the 'yellow cuticle' secreted by a group of enlarged and modified serosal cells. The chorion over this region is more permeable than elsewhere. The whole structure is termed the 'hydropyle', since it is responsible for the uptake of water which begins as soon as the yellow cuticle is formed.[70,128] In *Locustana* the inner layer (the white cuticle) in this region has very fine pore canals, apparently containing cytoplasmic filaments from the hydropyle cells, which link up with larger canals in the thickened outer layer (the yellow cuticle).[45] In all cases the absorption of water is probably an active process, and not due simply to osmosis;[14] in *Melanoplus*, for example, the uptake varies with metabolic activity irrespective of the membranes present.[76] In the eggs of *Phyllopertha*

(Scarabaeidae), which absorb twice their own weight of water in the early stages of incubation, this is a process of controlled osmosis brought to an end by re-waterproofing of the shell by the embryo.[117] In the cricket *Acheta*[118, 99] water is likewise absorbed over the entire surface of the shell; uptake ceases when the outermost layer of the serosal cutical undergoes phenolic tanning. [See p. 20.]

In the serosal cuticle of *Pteronarcys* there is a thickened circular area apparently corresponding with the hydropyle of Orthoptera.[47] In the floating rafts of *Culex* eggs there is a specialized area at the upper, that is, the posterior extremity of each egg where a droplet of water invariably condenses and is probably absorbed as hatching approaches.[18] Embryos of Collembola and *Campodea* possess a remarkable 'dorsal organ' which consists of filamentous processes arising from a pit in the anterior dorsal region and spreading over the entire surface of the embryo beneath the chorion. It is suggested that these may be concerned in the absorption of water.[77] In the eggs of endoparasitic Hymenoptera, laid in the body

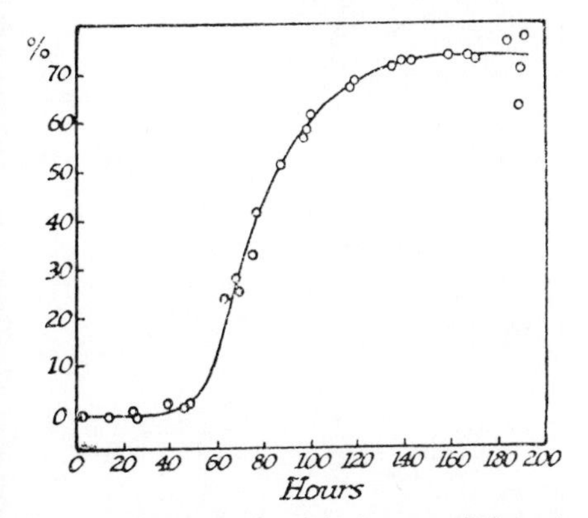

FIG. 4.—Increase of weight in eggs of *Notostira erratica* at 28° C. in contact with neutral water (*after* JOHNSON)

Ordinate: percentage increase in wet weight. Abscissa: hours after oviposition.

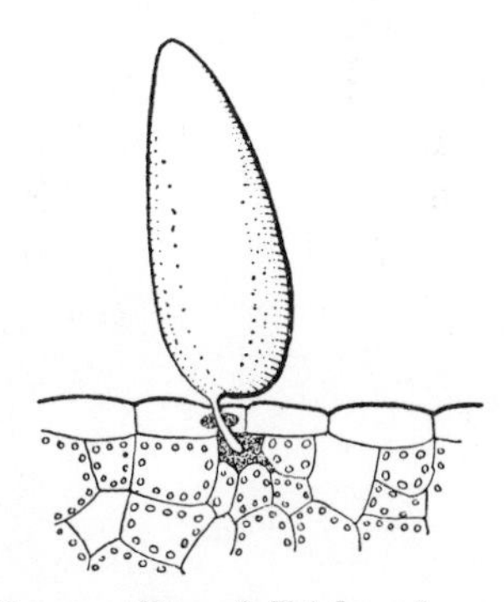

FIG. 5.—Egg of *Trialeurodes vaporarium* with the process, through which water is absorbed, inserted into a leaf and surrounded by cement (*after* WEBER)

fluids of other insects, the egg-shell is reduced to an excessively delicate envelope which must be permeable to many substances in the blood of its host. These eggs are poor in yolk and increase in volume enormously in the course of incubation. The egg of the Braconid *Dinocampus*, for example, may increase in cubic content over 1,200 times.[38]

Early development—Fertilization of the egg usually takes place about the time of laying (p. 716). The fertilized nucleus lies embedded in the yolk, surrounded by cytoplasm which ramifies throughout the egg enclosing in its meshes the yolk spheres. At the surface of the egg there is a zone where the cytoplasm is rather denser and is free from yolk; this is the cortical layer, the 'Keimhautblastem' of Weismann.[86] Apart from those exceptional insects in which the egg is deficient in food reserves[100] the fertilized nucleus divides without segmentation of the plasma; the daughter nuclei are surrounded by islands of cytoplasm connected to one another and to the cortical layer by cytoplasmic strands. At first these cleavage divisions take place simultaneously throughout the egg; but after a definite number of mitoses, peculiar to each species, the nuclei begin to divide independently. In *Drosophila*, for example, there are twelve synchronous cleavages.[129] The cleavage nuclei arrange themselves in a single layer enclosing a spherical or pyriform space in the centre of the egg (Fig. 6, A). As they divide they move towards the periphery of the egg and the space which they surround expands and changes its form and becomes poorer in cytoplasm than the yolk elsewhere in the egg (Fig. 6, B). The advancing cells leave behind some of their number to form vitellophags, or yolk nuclei, and in the honey-bee further yolk cells are later budded off from the blastoderm.[121] Ultimately the nuclei reach the cortical layer (Fig. 6, C). The cytoplasm, previously trailing like a tail behind each nucleus, is drawn in; the lateral partitions between the cells now appear; then their inner limits are formed, a basement membrane is laid down, and the formation of the blastoderm is complete. In a limited ventral region of the egg

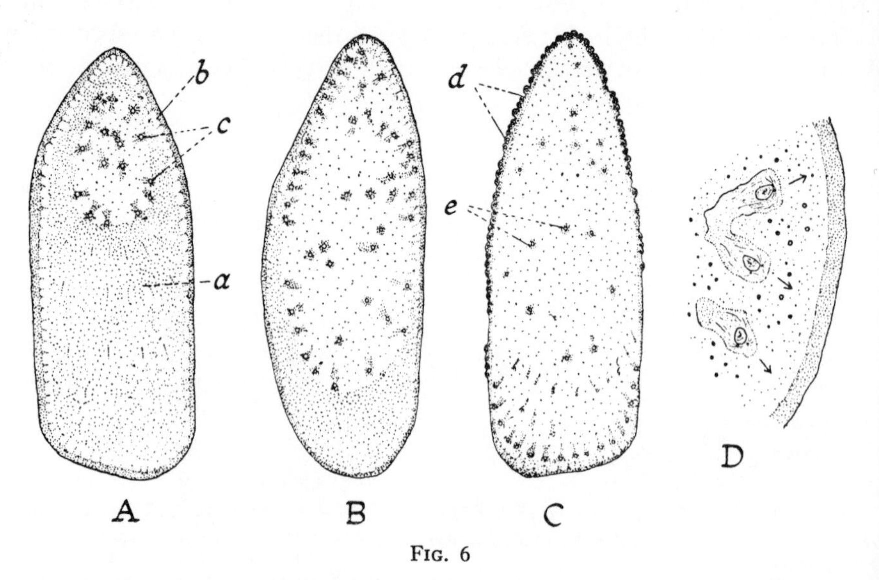

FIG. 6

A, B, and *C,* longitudinal sections of eggs of *Pieris,* showing migration of cleavage nuclei into the cortical layer (*after* EASTHAM). *A,* 6 hours after laying; *B,* 8 hours; *C,* 10 hours, nuclei of anterior pole have entered the cortical layer to form the blastoderm. *a,* yolk; *b,* cortical layer; *c,* cleavage nuclei; *d,* blastoderm; *e,* yolk cells. *D,* cleavage nuclei in *Calliphora* approaching the cortical layer (*after* STRASBURGER).

the cells of the blastoderm become higher and form a cylindrical epithelium; this is the germ band which is later differentiated to form the embryo; the remainder of the blastoderm, made up of flattened cells, forms the serosa.[23, 59]

It is not known what mechanism ensures the simultaneous mitosis of the nuclei in the early stages of segmentation, which seems to persist as long as protoplasmic continuity remains between them.[57] We shall see the same phenomenon in the healing of wounds (p. 106). Nor is it known whether the nuclei migrate actively towards the surface or whether they are carried along passively by an outward streaming of the plasma.[59] In *Calliphora* the nuclei are always orientated so that the centriole is directed towards the surface of the egg, and after each cell division the inner daughter nucleus swings round so that the centriole looks outwards (Fig. 6, D); here the cells or nuclei are actively concerned.[74] But there is also evidence of an outward movement of the cytoplasm independent of the nuclei. In *Pieris* there are elongated flecks radiating in the untraversed regions of the yolk;[23] and in *Tenebrio*, granules in the yolk plasma wander through the reticulum ahead of the nuclei to reach the cortical layer. During the first four hours after the egg of *Tenebrio* is laid there is an outward movement of the plasma which causes a gradual thickening of the cortical layer, especially in the region of the presumptive germ band, ventrally and laterally at the posterior pole; so that in this insect the future position of the germ band can be made visible by staining *before* the entry of nuclei.[26]

Organization—The nuclei in the egg contain the chromosomes bearing the genes which are responsible for the hereditary characters (p. 88); but the general form and organization of the embryo is controlled by agents located in the plasma at the periphery of the egg, quite remote from the dividing nuclei. The organism seems from the outset to be something to which the constituent nuclei are subservient, their function determined by their position in relation to the whole. On the other hand, the constitution of the cytoplasm reflects the earlier action of the genes, and the functional integrity of the cytoplasm depends on the presence of the nucleus even in these earliest stages of development. In *Drosophila*, for example, deficiency of the X-chromosome may result in failure to produce a blastoderm and all subsequent morphological changes are lacking.[55] [See p. 20.]

Three centres control the beginnings of organization in the insect egg.

(i) The initial stimulus comes from a *cleavage centre*, situated at the level of the future head rudiment. Here cleavage and migration of the nuclei begin.[42, 115, 116] In the egg of *Drosophila* 1–2 hours after laying, that is, between the first and fourth cleavage, the region most sensitive to exposure to X-rays is that occupied by the 'cleavage centre'.[93]

(ii) At the posterior pole of the egg there is an *activation centre* (Bildungszentrum) which is brought into activity by interaction with the cleavage nuclei. When these arrive physico-chemical changes occur in the plasma of this region[65] which thus influences the peripheral zone of the egg in such a way that it can proceed to the subsequent stages of development. It does this probably by giving off a material substance which permeates the egg from behind forward. If this centre is eliminated at a very early stage by excision or exposure to ultra-violet light, cleavage and migration of nuclei occur as usual, but the resulting blastoderm is solely of the extra-embryonic type; no germ band is formed.[65] Elimination of other parts of the egg have not this effect; if the anterior parts are removed the embryo is merely smaller and the germ band displaced backwards (Fig. 8, B).[58] As the influence of this centre spreads forwards, during the early hours of development, increasingly large areas around the posterior pole must be excised or burned if the formation of the embryo is to be prevented. Within a few hours, although there is still no visible differentiation of any sort within the egg, the process of activation is complete. A centre of this kind has been demonstrated in the dragon-fly *Platycnemis* (Fig. 7),[65] in the ant *Camponotus*,[58] in the weevil *Sitona*,[58] the bean beetle *Bruchus*,[11] the mealworm *Tenebrio*,[26] and in *Culex*.[110] No such centre has been detected in *Leptinotarsa*,[105] *Melasoma*,[113] or *Calliphora*.[122] In *Acheta* the activation centre is located, not at the posterior pole but in the presumptive thorax, close to the 'differentiation centre'.[120]

The most direct demonstration of an activation centre has been obtained in the leafhopper *Euscelis*. The centre in question is closely associated with the mass of symbionts (p. 522) at the posterior pole of the egg. If this mass is displaced into the anterior half of the egg a complete embryo is induced in this region (Fig. 9).[126] An induction of a different kind can be demonstrated in the egg of *Chironomus*.

The cytoplasm at the anterior pole tends to induce head development, that at the posterior pole to induce abdominal development. If the egg at the stage of

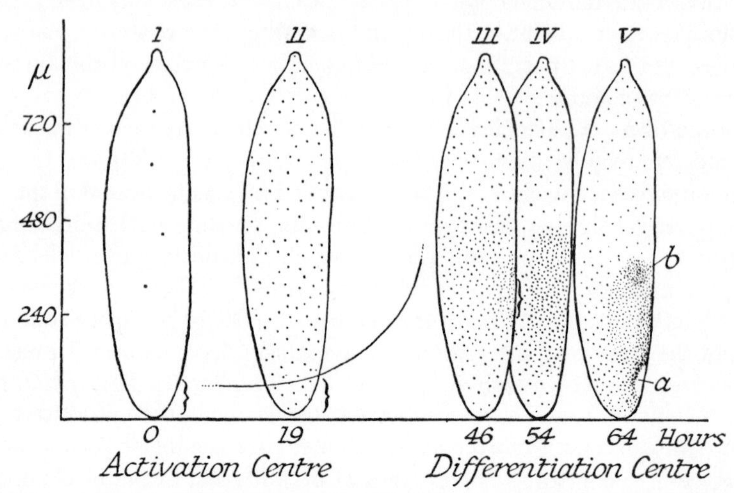

FIG. 7.—Organization in egg of *Platycnemis pennipes* (Odonata) (*after* SEIDEL)

Ordinate: length of egg in μ. Abscissa: time of development in hours at 21·5° C. (i) 0 hours, stage with 4 nuclei (position of activation centre indicated by bracket). (ii) 19 hours, stage with 256 nuclei. Influence from activation centre spreads forward as indicated by the curved line; (iii) 46 hours, cells accumulating first in region of differentiation centre indicated by bracket; (iv) 54 hours, germ band forming; (v) 64 hours, germ band fully formed: **a**, point where germ band sinks into the yolk; ***b***, cephalic lobes forming.

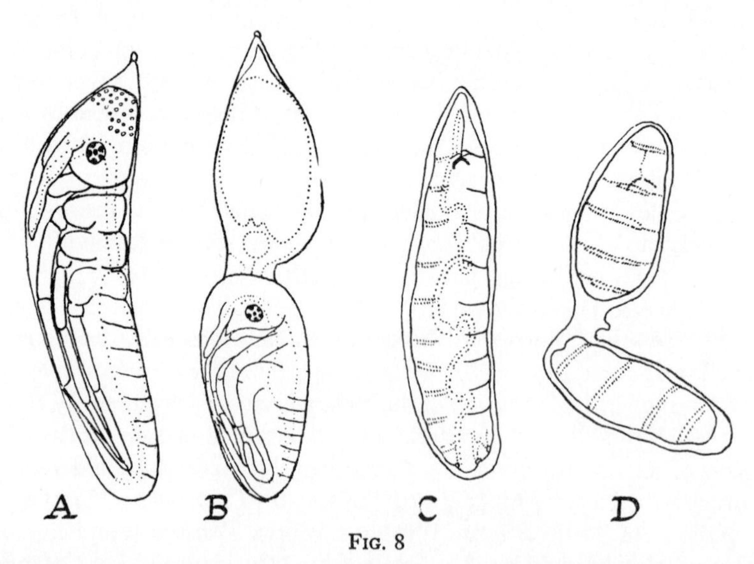

FIG. 8

A, *Platycnemis pennipes*; normal embryo shortly before hatching; *B*, dwarf embryo of the same age produced by ligaturing the egg in the middle at the stage with 4 cleavage nuclei (*after* SEIDEL); *C*, Muscid larva, normal embryo shortly before hatching; *D*, embryo of *Calliphora* ligatured at the stage with 16 nuclei, showing mosaic development (*from* SEIDEL, *after* REITH and PAULI).

nuclear division is centrifuged towards the posterior pole, a larva with a head at each end is formed. If centrifuged towards the anterior pole, a larva with an abdomen at each end is developed (Fig. 10).[133] [See p. 20.]

(iii) Towards the middle of the presumptive germ band, in a position corresponding with the future thorax of the embryo, there is a *differentiation centre.* As soon as this centre has been induced to begin its activity, under the influence of the activation centre, it provides the focus from which all subsequent processes of development spread forwards and backwards. It is a part of the cortical plasma zone with no visible difference from any other part. But when the cleavage nuclei reach the surface and form the blastoderm they accumulate first in the region of this centre. Later, the blastoderm is thickest here; cell divisions are most numer-

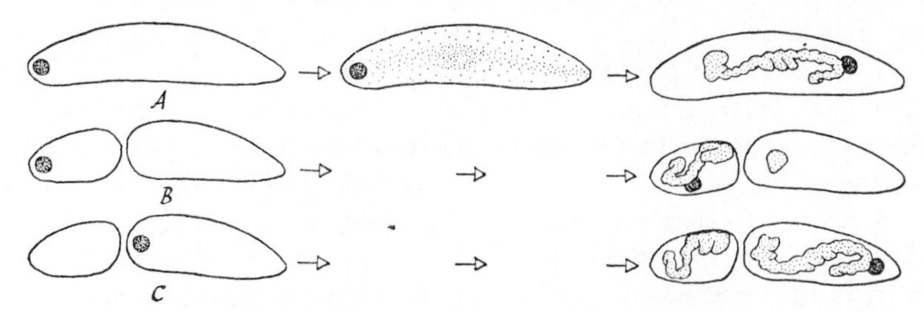

FIG. 9.—The effect of posterior pole material on the induction of embryonic development in the egg of *Euscelis* (*after* SANDER)

A, normal development; the dark sphere represents the ball of symbionts. *B*, effect of ligature at an early stage; an embryo is developed only in the posterior fragment. *C*, the ball of symbionts with associated cytoplasm, is displaced forwards before ligature; an embryo is developed in both fragments.

ous; and differentiation of body segments and appendages extends forwards and backwards from this point. The differentiation centre is in fact the place where visible differentiation of form begins in point of time and at which, later, the degree of differentiation at any moment is greatest. A centre of this type, which determines the future function of the nuclei that come into its sphere of influence, has been demonstrated in Orthoptera (*Acheta*[120]), Hemiptera (*Pyrrhocoris*[65]),

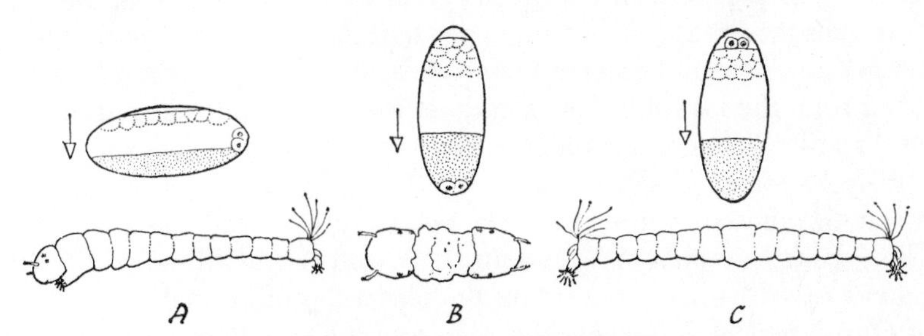

FIG. 10.—The effect of centrifugation on development of the embryo in the egg of *Chironomus* (*after* YAJIMA)

A, lateral centrifugation; normal development. *B*, centrifugation towards the pole cells at the posterior end; an embryo with two heads develops. *C*, centrifugation towards the anterior end; an embryo with two abdomens develops.

Odonata (*Platycnemis*[65]) (Fig. 7), Neuroptera (*Chrysopa*[8]), Hymenoptera (*Camponotus*[58] and *Apis*[63]), Lepidoptera (*Ephestia*[64]), Coleoptera (*Tenebrio*,[26] *Leptinotarsa*[105]) and Diptera (*Culex*[110]). It is said to act, not by giving out a diffusible substance, but by initiating a contraction wave in the yolk system, spreading forwards and backwards and forcing the blastoderm cells, hitherto distributed uniformly, to fill the space that is left. Indeed, artificial depressions in the yolk system caused by heat or by ultra-violet rays will bring about this directed movement of blastoderm cells to any desired part of the egg.[65] Likewise in the egg of *Acheta* the aggregation of blastoderm cells at the differentiation centre, as well as the subsequent movements of blastokinesis, are all the result of active contraction, first of the yolk plasma and then of the embryonic membranes. The surface of the yolk serves as a 'conveyor belt' for the blastoderm cells.[120]

Determination—The cleavage nuclei are 'isopotent'; they are capable of forming any part of the future embryo, depending upon which region of the cortical plasma they eventually reach. The future gonads, for example, take their origin from the 'pole cells'—cleavage cells which enter the posterior extremity of the egg. It was shown by Hegner (1908) that elimination of a limited area of cytoplasm at the posterior pole of the egg of *Leptinotarsa*, a region containing the so-called 'germ cell determinants', prevents the formation of the 'pole cells' and of the gonads. In *Drosophila* a cleavage nucleus may divide so that one daughter nucleus lies in the polar plasma, the other outside it. Then the former will become a germ cell, the latter an ordinary blastoderm cell.[3, 54, 56] Irradiation of the posterior pole plasma in *Culex* results in complete sterility in both sexes: no sperm are formed in the testes and neither oöcytes nor nurse cells in the ovaries.[123]

The entire cortical cytoplasm of the egg is rich in nucleoprotein, but the relative importance of protein and nucleic acid in the early stages of determination is uncertain.[98] Ribonucleic acid is particularly evident in the cytoplasm of the posterior pole; and in the unfertilized egg of *Apanteles* it is a prominent site of alkaline phosphatase activity.[132] In the ant *Formica*, eggs with a large accumulation of nucleic acids at the hind pole develop into females; those with a smaller accumulation into workers.[96] In *Drosophila* the polar granules are made up of dense masses of ribosomes[119] which are later included within the pole cells,[101, 102] and there is some evidence that these granules are indeed required for germ cell differentiation. In Cecidomyid midges the nuclei of the germ line are sharply distinguished from the future somatic nuclei by the retention of their full complement of chromosomes. In *Wachtliella persicariae* there are 40 chromosomes in the germ cells, 8 in the soma of the female, 6 in the soma of the male. The single mother germ cell when it enters the nucleic acid-rich zone of the pole plasma is protected from the loss of chromosomes. If the nucleoproteins are removed the germ cells suffer chromosome reduction and the ovaries do not develop properly. Conversely, exposure of somatic nuclei to the nucleoprotein of the pole plasma will arrest chromosome elimination.[103] [See p. 20.]

The egg of *Drosophila* may be centrifuged and the distribution of the migrating nuclei upset without preventing the formation of normal insects.[35] In the egg of *Platycnemis*, at the stage with two or four nuclei, one nucleus may be killed by focussing a pencil of ultra-violet rays on it and yet a normal embryo is developed.[65] Even when they reach the surface (at the 256 nuclear stage in *Platycnemis*) defects can still be made good. Thus, until the differentiation centre has performed its function, the egg is capable of extensive 'regulation'; that is, complete formation from reduced material. If the egg of *Platycnemis* is ligated in the middle soon after laying a dwarf embryo is developed (Fig. 8, B). If the ventral plasma is divided longitudinally by means of a cautery, twin embryos are formed, the smaller included sometimes within the body of the larger.[65] Similar doubling has been produced in *Bruchus* by exposure to hydrogen cyanide, which causes a physiological separation by the differential inhibition of oxygen uptake,[12] and in *Leptinotarsa* by dividing the germ longitudinally.[105]

In the egg of the honey-bee 12 hours after laying, the 'potency' to form all parts of the body is still so concentrated around the differentiation centre that one-fifth of the entire egg can be removed at the anterior end and development of a complete dwarf embryo follows. As development proceeds, the 'potency' to form the different regions spreads outward over the egg; and in the bee, by 24 hours after laying, the prospective functions of all the main parts are finally determined.[63] The egg is then a 'mosaic' egg. If any part is eliminated, a corresponding region of the completed embryo is missing.

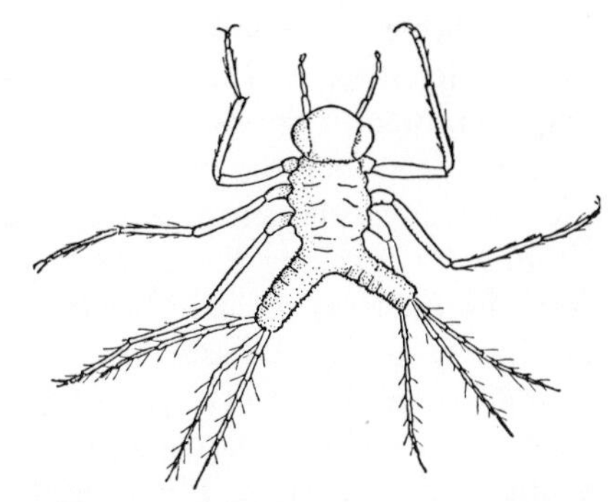

FIG. 11.—*Platycnemis pennipes*, showing doubling of the posterior part of the body as the result of dividing the prospective germ band by ultra-violet irradiation at the stage when the cells were beginning to collect around the differentiation centre (Fig. 7 (iii)) (*from a photograph by* SEIDEL)

In the greenhouse cricket *Tachycines*, as in *Platycnemis* (Fig. 11), the embryonic germ is still capable of well-marked regulation at the time the egg is laid; and if the germ band is divided with a needle, many types of doubling, at first of the whole body and later of individual organs, can be produced.[41, 42] The converse process, the fusion of eggs to form a single embryo, is seen in Phasmids. Senile *Carausius* females will lay eggs containing anything up to 10 oöcytes which fuse to form a compound egg. If fusion takes place early in development, regulation occurs and a normal insect results; if fusion takes place later, double monsters are produced, doubling being most frequent at the anterior end where differentiation first occurs.[16] Other Phasmid eggs may have two micropyles; in them there is no fusion where the embryos come in contact; each develops as far as it is able in the space available (Fig. 12).[16] Two or more embryos may occur occasionally in the eggs of the grasshoppers *Dociostaurus* and *Melanoplus*.[72]

Determination is completed at different times in different insects. In the bee, as we have seen, it is complete in 24 hours after laying;[63] in the bean beetle *Bruchus* in 6½ hours;[11] while in Muscidae[52, 58] and in *Drosophila*[29] the egg has already reached the 'mosaic' state by the time it is laid. The process of determination, like the changes which precede it, is likewise independent of the nuclei in the egg. In *Platycnemis*, regulation is still possible in the late blastoderm stage.[65] In *Sialis* determination occurs between the fourth and fifth cleavages, at the moment when the blastema is formed by peripheral migration of the cytoplasm.[21] In *Sitona*, the parts are fully determined by the time the blastoderm is formed.[58] In *Bruchus*, the posterior cytoplasmic regions of the egg are determined before the entrance of the cleavage nuclei, the more anterior parts shortly after the cleavage cells arrive.[11] Whereas in Muscid eggs, determination in the cortical plasma is complete before cleavage has begun (Fig. 8, D).[58] In all cases, determination of the main outline of the body takes place before the individual organs. When the process is complete it is possible to map

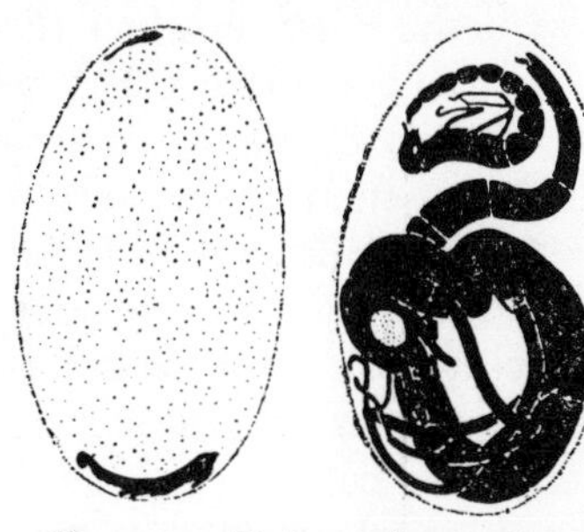

FIG. 12.—Twin embryos in eggs of the Phasmid *Clonopsis* at different stages of development (semi-schematic) (*after* CAPPE DE BAILLON and PILLAULT)

out the prospective embryo by destroying localized spots on the egg surface with ultra-violet rays. In *Platycnemis* it has been found that at first the embryonic map delimited in this way covers almost the entire egg; as the cells congregate to form the embryonic rudiment the various zones become concentrated around the differentiation centre towards the posterior half of the egg.[65]

Broadly speaking there are two egg types.[41, 42, 116] In the one the relative amount of plasma is small, the visible germ band is small, shows no differentiation at the outset and has a great capacity for regulation (*Platycnemis* and *Tachycines* are examples). In the other type the amount of plasma is relatively large, the germ band is long, occupying the greater part of the egg and it often shows some visible differentiation. These eggs are mosaic at the time of laying (*Musca*, *Drosophila*). Between the 'regulation' and 'mosaic' types there is a transitional group comprising Hymenoptera, Megaloptera, Coleoptera[73] and Lepidoptera[43, 44] in which the capacity for regulation is reduced in varying degrees.

Determination of imaginal structure in the egg—In *Drosophila*, as in the Muscidae, the egg at the time of laying is a mosaic egg. Local injuries effected by ultra-violet light during the first 4 hours after laying cause local defects in the resulting larva. But the imaginal characters are unaffected; in respect to imaginal characters the egg is still capable of 'regulation'. But if the egg is irradiated similarly 7 hours after laying or later (between 4 and 7 hours the egg is so sensitive to treatment with ultra-violet light that experiments cannot be made during that period) localized defects are produced in the corresponding region of the adult epidermis without any visible effect during larval development.[29] Similar results can be obtained by puncturing the egg at different levels with a needle.[35] Indeed it can be shown that during this second developmental period a wave of 'determination' in respect of imaginal characters spreads backwards from the thorax until the egg becomes a mosaic egg in respect to adult structure also—just like the determinative change that spreads from the differentiation centre of *Platycnemis* and other 'regulation' eggs. Similarly in *Tineola*, purely larval or purely imaginal defects can be obtained by ultra-violet irradiation of the egg at suitable times.[43] Metamorphosis in fact may be regarded as a kind of repeated embryonic development.[32]

Later development—The later stages of embryonic development, the formation of the germ layers, the sinking of the embryo into the yolk and its subsequent revolution around the egg (blastokinesis), in the course of which the invaginated part of the blastoderm (the amnion) and the superficial part (the serosa) are absorbed into the body of the embryo, the growth of the embryo around the yolk, the formation and histological differentiation of the various organs and tissues, and the variations in these processes which occur in different groups of insects are fully described in the text-books. Germ layer formation follows the primary segmentation and it likewise starts at the level of the differentiation centre and spreads forwards and backwards. It begins with the sinking in of the middle plate and ends with the formation of the coelom epithelium. The formation of organs begins as a cellular differentiation which spreads, in all layers in each segment, from the middle laterally. All the organ systems appear first in the germ band on the ventral side of the egg; later they are carried sideways and dorsally by dorsal closure.[8, 66] (Fig. 13.) [See p. 20.]

During this late process of differentiation, which has been studied in great detail in *Chrysopa*, the ectoderm appears to be a **self** differentiating system, independent of the mesoderm. On the other hand, the differentiation of mesodermal structures in *Chrysopa* does depend upon the overlying ectoderm, which seems to induce the corresponding mesodermal organs in much the same way as the mesoderm in Amphibia induces differentiation in the ectoderm. If localized

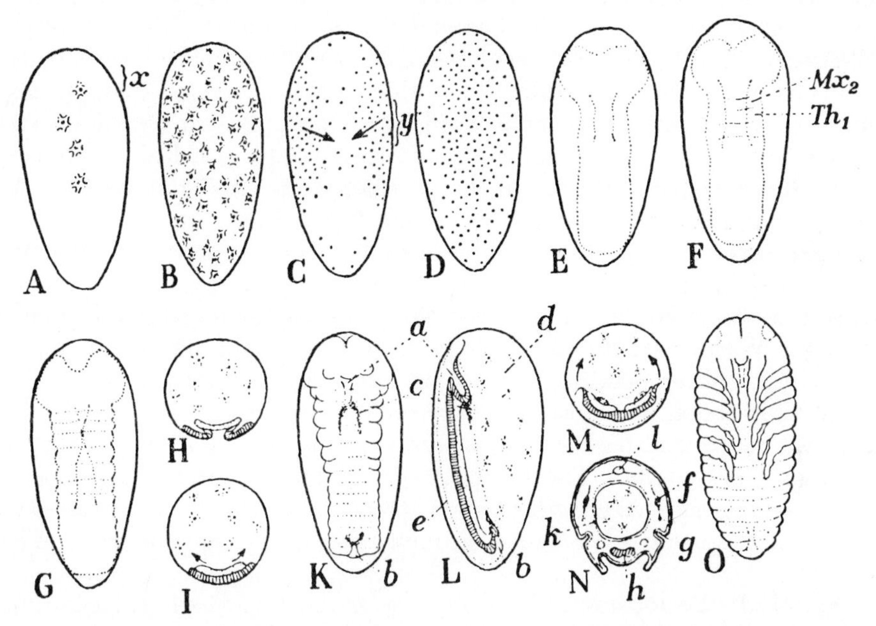

FIG. 13.—Embryonic development of an insect (schematic) (*after* SEIDEL)

A, four cell stage; *x*, marks position of cleavage centre. *B*, cleavage cells uniformly dispersed below surface of egg; stage with 128 nuclei. *C*, blastoderm cells massing at level of the differentiation centre (*y*). *D*, ventral germ band formed by fusion of the cell masses from either side. *E*, head lobes formed and primitive groove (middle plate) appearing in region of differentiation centre. *F*, rudiments of segments appearing: second maxilla and first thoracic. *G*, extension of primitive groove and segmentation forwards and backwards from differentiation centre; middle plate sinking in to form mesoderm at level of differentiation centre. *H*, cross section showing the sinking in of middle plate at stage reached in *F*. *I*, the same at stage reached at most advanced point (the differentiation centre) in *G*: mesoderm and ectoderm formed. *K*, appendages appear first in second maxillary and first thoracic segments; stomodoeum (*a*) and proctodoeum (*b*) formed by invagination; endoderm (*c*) growing inwards from these invaginations. *L*, longitudinal section at this stage; *d*, yolk; *e*, amnion. *M*, transverse section at same stage; the three germ layers are growing in the direction of the arrows to enclose the yolk. *N*, dorsal closure complete; *f*, gonads; *g*, tracheae; *h*, ventral nerve cord; *k*, gut; *l*, heart. *O*, ventral view at this stage.

regions of the ectoderm are killed, the underlying structures in the mesoderm fail to develop.[8] This same induction of mesoderm by the overlying ectoderm has been confirmed in *Leptinotarsa*; furthermore, in this insect, differentiation of the endoderm is due to induction by contact between splanchnic mesoderm and endoblast, itself derived wholly from stomodoeal and proctodoeal ingrowths.[105] In the embryo of *Carausius* there is a striking reciprocal influence between the developing compound eye and optic lobes.[53]

Factors controlling the rate of development—The physiological control of the separate processes of later development has not been studied; but much information exists about the factors that determine the rate of development as a whole. The most important factor is temperature (p. 684): within the vital limits of each species development is accelerated by high temperatures and retarded by low, so that below a given temperature, peculiar to each species, no development takes place; growth is restored when the temperature is raised.

In many insects development is delayed by low humidity in the surrounding air.[14, 62] At 20° C. the egg of *Sitona* (Col.) has an incubation period of 10½ days at a relative humidity of 100 per cent., 21 days at a humidity of 62 per cent.,[4] and in *Lucilia* there is a linear relation between saturation deficiency and the duration of egg development.[25] Presumably the delay under dry conditions is due to lack of water in the egg. For example, at the time of laying the egg of *Notostira* (Capsidae) does not contain enough water for complete development. If it is not given the opportunity of absorbing water (p. 4) embryonic development ceases, recommencing only when water is supplied.[40] Development is delayed from the same cause in the eggs of Aleurodids (p. 4) if laid on drying or wilting leaves,[83] and in the egg of the grasshopper if the 'hydropyle' (p. 4) is covered with material impermeable to water.[70]

Eggs of some insects can remain dormant for long periods in a desiccated state. Those of *Sminthurus* (Collembola), which hatch in moist conditions in 8 or 10 days, will shrivel and collapse if dried. They have been kept in this state for 271 days. When wetted they quickly become spherical again and hatch in 12 days or so.[34] Eggs of the South African locust, *Locustana pardalina*, which hatch in moist soil in about two weeks, have survived desiccation for 3½ years.[14] In such cases it is only at a restricted and usually an early stage of development that the eggs show this resistance. The undeposited eggs of many parasitic Hymenoptera are in a state of arrested growth so long as they are stored in the oviducts—since further development is dependent on immersion in the nutrient fluids of the host.[27]

Arrested development—This type of arrested development, controlled directly by external conditions and brought to an end when the temperature or water relations are favourable again, is sometimes termed 'quiescence'. But many insects, particularly those which pass the winter in the egg, undergo an arrest of development which may persist even though environmental conditions are favourable. For this state of dormancy Henneguy (1904) adopted the term 'diapause'—a word originally coined by Wheeler (1893) to specify that stage of development at which the embryo ceases to move tail first over the yolk, but remains stationary for a time before beginning the final movements of blastokinesis. We shall discuss later the occurrence of this state in larval, pupal, and adult life (pp. 111, 728); here we shall consider diapause only in relation to embryonic development. The arrest takes place at a different stage in different groups. Among Lepidoptera, almost no development occurs until the spring in *Zephyrus betulae*, &c., whereas the larva of *Argynnis paphia*, &c., is fully developed in the autumn;[81] *Orgyia* eggs cease developing when the germ band has formed and is beginning to give rise to mesoderm.[17]

Diapause in Bombyx mori—The classic example of diapause in the egg is afforded by the silkworm. Silkworm eggs laid in the autumn will not develop immediately even if kept warm; growth is completely arrested at an early stage. As was shown originally by Duclaux,[22] they will not hatch even in the spring if they have been kept warm (15–20° C.) throughout the winter; they will complete their development only if they have been exposed to a temperature around 0° C. for several months. Some races of *Bombyx mori* are single brooded or 'univoltine', so that every generation shows a prolonged period of arrest during embryonic development; other races are 'bivoltine' or even 'tetravoltine'; in these there are one or more uninterrupted generations during the summer before the winter generation of diapause eggs is produced. The diapause eggs are characterized by having the serosa laden with pigment.[78]

The mechanism of diapause and the mode of action of low temperature in bringing it to an end (this 'reactivation' by cold is a characteristic of diapause at all stages of growth) will be discussed in conjunction with post-embryonic development (p. 108). But some indication of the kind of physiological factors that are at work is given by the study of voltinism in the silkworm. Voltinism is to some extent hereditary; but when the races are crossed, clear-cut segregation does not occur. For the voltinism of the offspring is influenced by the temperature at which eggs of the preceding generation were incubated (eggs incubated at 25° C. tend to produce moths laying hibernating eggs; those incubated below 15° C. tend to produce non-hibernating eggs) and by the effect of temperature on the larva. But even more important than temperature is the effect of the length of day to which the eggs are exposed during incubation. If the eggs or young larvae are exposed to 13 hours of daylight, or less, the resulting moths lay non-hibernating unpigmented eggs; exposed to a day-length of 14 hours or more they lay dark hibernating eggs (Fig. 87).[92]

The voltinism of the eggs is determined by some influence from the somatic cells of the mother. For a batch of eggs from a single female is generally uniform as regards voltinism; and if the ovaries of one race are transplanted into another race during the larval stage, the eggs from these ovaries always show the voltinism of their new host.[28, 82] The suboesophageal ganglion of the female silkworm moth secretes a 'hibernation substance' or 'diapause hormone' which leads to the production of diapause eggs. But the secretion of this factor is under the control of the brain, acting by way of the oesophageal connectives. The brain normally inhibits the secretion of the diapause hormone; if the brain is removed immediately after pupation in moths which would otherwise have laid non-diapause eggs, they proceed to lay eggs which enter diapause.[90, 107] [See p. 20.]

The influence of the female parent on the diapause of the offspring is seen also in the beetle *Timarcha violaceo-nigra*. In the autumn the female lays eggs which go into diapause at an early stage of development. But exposure to low temperature causes her to lay eggs which hatch without any arrest.[2] [See also p. 113.]

Diapause in Orthoptera—There are all degrees of dormancy between a simple quiescence and a true diapause. The silkworm is an extreme example of the latter type; the grasshopper *Melanoplus differentialis* may be taken as an example of arrested development which is much more readily influenced by environmental factors.

In the winter generation of *Melanoplus* growth ceases when the embryo is fully differentiated but quite small.[13, 69] It remains in this state without further cell divisions until the spring. Then growth is vigorously resumed; the embryo undergoes blastokinesis, moving around the lower end of the egg, rotating on its long axis, and finally growing dorsally to enclose the remaining mass of yolk (Fig. 14).[69] These changes in growth activity are marked by changes in respiration (p. 633): there is a peak of oxygen uptake during the initial stage of growth (which occupies about three weeks at 25° C.), a long period of very low oxygen uptake during the diapause, and a rapid rise when growth is resumed (Fig. 15, B).

This rhythm of development and arrest in the *Melanoplus* egg is not so

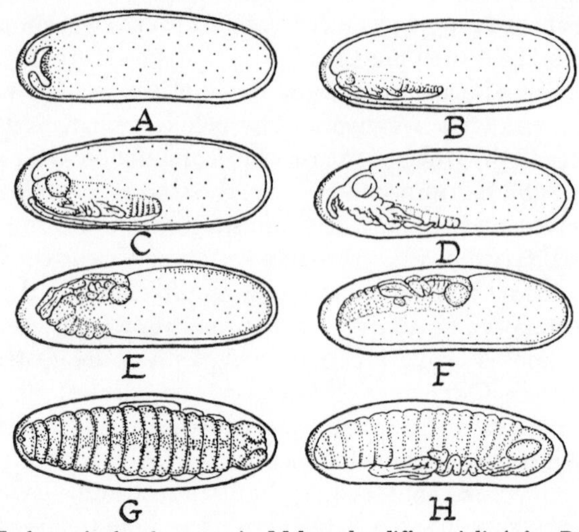

FIG. 14.—Embryonic development in *Melanoplus differentialis* (*after* BURKHOLDER)

A, appearance of germ band (5th day); *B*, embryo sunk into the yolk (10th day); the peak of oxygen consumption occurs at this stage; *C*, embryo in diapause; *D*, embryo beginning to revolve; amnioserosal membranes have ruptured; *E*, embryo half revolved; *F*, early revolution completed; *G*, dorsal view, and *H*, lateral view after the dorsal line has fused and lateral rotation has taken place.

fixed as that in *Bombyx mori*; for at any time during the winter a constant high temperature will initiate development, and the diapause can be prevented altogether in some of the eggs if they are kept at 25° C. from the time of laying. But an inborn rhythm probably exists in this insect also; for some eggs have been kept at 25° C. for over a year without starting to grow and yet have remained viable;[10] the longer the egg is exposed to cold in the winter, the greater is the rate of development upon returning to 25° C.; and once development has been resumed it cannot be arrested again by cold.[9] Moreover, *Melanoplus* eggs kept throughout at 25° C., which may appear to have no diapause, still show an initial peak of oxygen uptake succeeded by a fall (Fig. 15, A); whereas in the eggs of *Chortophaga* (a grasshopper which spends the winter as a nymph) the oxygen uptake rises steadily as development proceeds.[9] In *Melanoplus* the condition is readily brought to an end by treating the eggs with xylol or other lipid solvents. This acts upon the hydropyle, rendering it permeable to water and stainable by aqueous solutions of dyes. One factor in the development of diapause may be the secretion of a substance with wax-like properties at the hydropyle.[71] Indeed, with the electron microscope it is possible to see a dense homogeneous layer on the outer surface of the hydropyle, which becomes discontinuous when diapause is ended.[128] In the egg of *Locustana* there are two types of arrest: a temporary type resulting from lack of moisture (p. 14) and a more permanent type unaffected by

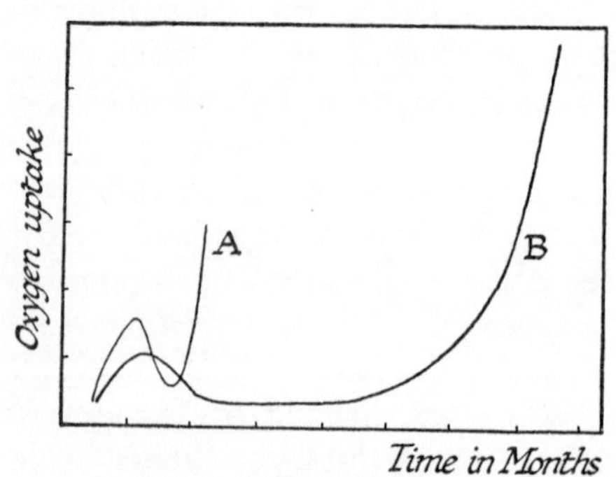

FIG. 15.—Oxygen uptake in eggs of *Melanoplus differentialis*

A, eggs at constant temperature of 27° C.; *B*, eggs kept at outdoor temperature and oxygen uptake estimated at that temperature (*after* BODINE)

favourable conditions. In the egg of this locust the water-proofing of the hydropyle region appears to be due not to a wax layer but to a proteinaceous material which becomes impermeable on drying.[45]

The optimum temperature for bringing diapause to an end is usually well above 0° C., but in *Acheta* the time required for this process was shortened from 17 days at 5° C. to 8 hours at −5° C. and to 20 minutes at −16·5° C. The same result is brought about by applications of urea or ammonia:[109] so perhaps this is an injury effect. We shall discuss further the physiological nature of diapause in later chapters (p. 109).

Dormancy in mature embryo—Some insects complete their development in the egg before becoming dormant. That is so in the beetle *Timarcha tenebricosa.* The egg containing the fully developed larva may be kept at 15° C. for 6 months without hatching, and if the larva is extracted it soon dies. Whereas if kept at 5° C. for some time the larvae will survive extraction; and if exposed to severe frost for several months they all hatch simultaneously a few days after return to room temperature.[1] Mosquitos belonging to the genus *Aëdes* are another example of this type. The eggs containing fully formed larvae will survive in a dry state for several months (eggs of *Aëdes aegypti* have hatched after keeping for 40 days at 28° C. in air dried with sulphuric acid[14]); upon immersion in water many of them hatch within a few minutes. There is no suggestion in this case that reactivation by cold is necessary (although that appears to be the case in *Aëdes hexodontus*[95]). It seems that some specific stimulus to hatching is needed. In the case of *Aëdes vexans* this stimulus appears to be a reduction of the dissolved oxygen in the medium—a reduction brought about by the presence of micro-organisms and various organic substances.[31] The action seems to be exerted on the central nervous system of the larva in the egg.[112] Hatching may even be induced by crowding of the eggs.[75]

Pleuropodia—In the mature embryo a pair of glandular organs is often present on the first abdominal segment. These were named 'pleuropodia' by Wheeler, who showed that they are homologous with the appendages of the first abdominal segment. They appear to be best developed in the less specialized groups (Blattidae, Mantidae and other Orthoptera, some Coleoptera); less so in other Coleoptera and Hemiptera; while in Hymenoptera and Lepidoptera they are vestigial or wanting.[36] Sometimes, as in *Belostoma* and *Ranatra* (Hem.),[36] they sink into the body wall, their distal ends prolonged into long thread-like structures forming a tuft projecting from the orifice (Fig. 16); sometimes, as in Orthoptera, they are stalked bodies projecting from the surface, made up of a single layer of very large cells.[61] They attain their greatest size just before the insect hatches, and then degenerate.

The function of these organs may

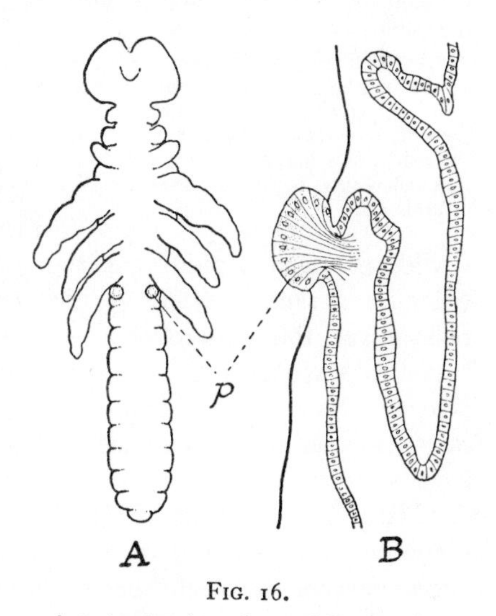

FIG. 16.

A, isolated early embryo of *Ranatra fusca* to show position of pleuropodia; *B*, detail of pleuropodia as seen in sagittal section. *p*, pleuropodia (*after* HUSSEY).

vary from one insect to another. In the viviparous Polyctenid *Hesperoctenes* their processes form a nutritive sheath which later secretes a protective cuticle (p. 732); in *Rhodnius* they are perhaps responsible for producing the iridescent cuticle which appears over embryo and serosa during blastokinesis;[46] and other secretory and excretory functions have been ascribed to them.[36, 61] In *Melanoplus* the inner white layer of the serosal cuticle (p. 4) is digested and dissolved shortly before hatching, the outer yellow cuticle being unchanged (Fig. 3, D, E). The enzymes responsible for this digestion are secreted by the pleuropodia; for if these are excised before digestion has occurred, the embryo develops normally but is unable to break through the cuticle, which remains thick and tough (Fig. 17).[70] The serosal cuticle in *Pteronarcys* is similarly digested, apparently by the secretion of the pleuropodia,[47] and the same function has been ascribed to the pleuropodia in *Pediculus*.[94] The process is analogous to what happens in the insect cuticle during moulting (p. 43). In the eggs of the stick insect, in which calcium carbonate forms a white crystalline layer on the inner surface of the shell, there is a considerable reduction in the ash content of the shell at the time of hatching.[48]

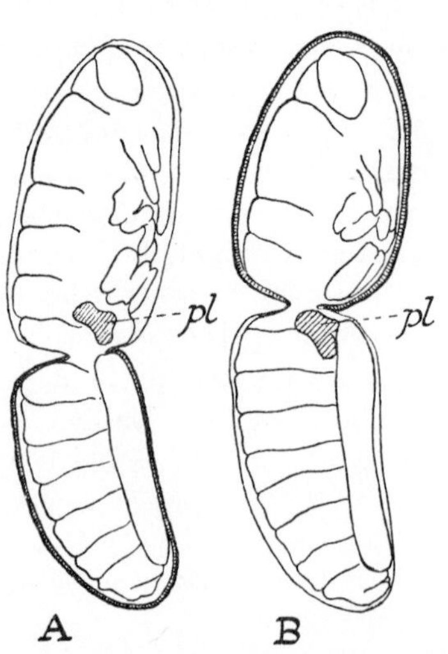

FIG. 17.—Diagram of an experiment to demonstrate the origin of the hatching enzyme in *Melanoplus* from the pleuropodia (*after* SLIFER)

Eggs with chorion removed are ligatured: *A*, behind pleuropodia; *B*, in front of pleuropodia. The white inner layer of the serosal cuticle is dissolved in the half containing the pleuropodia; the thin yellow outer layer is not attacked. *Pl*, pleuropodia.

Hatching—When its development is complete the embryo is faced with the problem of breaking the membranes and escaping from the egg. The main obstacles are the chorion, the vitelline membrane sometimes (in Muscids), and the serosal cuticle (in *Pteronarcys*, *Dytiscus*, Acridiids, and perhaps in other insects). In addition to these, many insects lay down a provisional cuticle which is later shed and replaced by a more substantial cuticle before hatching takes place. This so-called 'embryonic cuticle' is present in most hemimetabolous insects (Orthoptera, Hemiptera, &c.) and in some holometabolous forms such as *Tenebrio* and *Dytiscus* (Col.); it is widespread in Lepidoptera, though commonly overlooked because the cuticular membrane is eaten by the larva before hatching, along with the yolk residue and the amnion.[124] A protective role is commonly ascribed to the embryonic cuticle; as in grasshopper larvae which work their way up to the soil surface after hatching.[97] Or this embryonic moult may permit the establishment of new muscular attachments to the cuticle.[127] If this is indeed the main function it is to be expected that embryonic cuticles will be found in all insects.

Many insects are provided with cuticular structures, spines or blades, hard plates or eversible bladders, which are used to cut through the membranes of the egg or to force off a preformed egg-cap. These 'hatching spines' are developed on the embryonic cuticle (in Hemiptera, Neuroptera, Anoplura, &c.), or upon the true skin of the larva and are thus retained throughout the first instar (in Nematocera, Aphaniptera, Carabids, Dytiscids, Chrysomelids, &c.).[24, 67]

The mechanism of hatching is pretty constant. The first sign that it is imminent is the appearance of pumping movements in the pharynx indicating that the insect is swallowing the amniotic fluid. This has been observed in *Agrion* (Odonata) and *Hydrobius* (Col.)[5] and in many other insects.[67] Swallowing continues until all the fluid is absorbed and the insect fills the shell completely. It may then rupture the chorion and the other membranes by simple muscular force (we have seen that the serosal cuticle may first be weakened by digestion of the inner parts (p. 18)), or it may contract the posterior parts of its body and, by driving the blood forwards to the head, cause this to burst the shell. In the eggs of *Calliphora* and other Muscids the chorion contains a strongly hydrophilic protein which swells in a humid atmosphere and sets up strains that aid the larva in

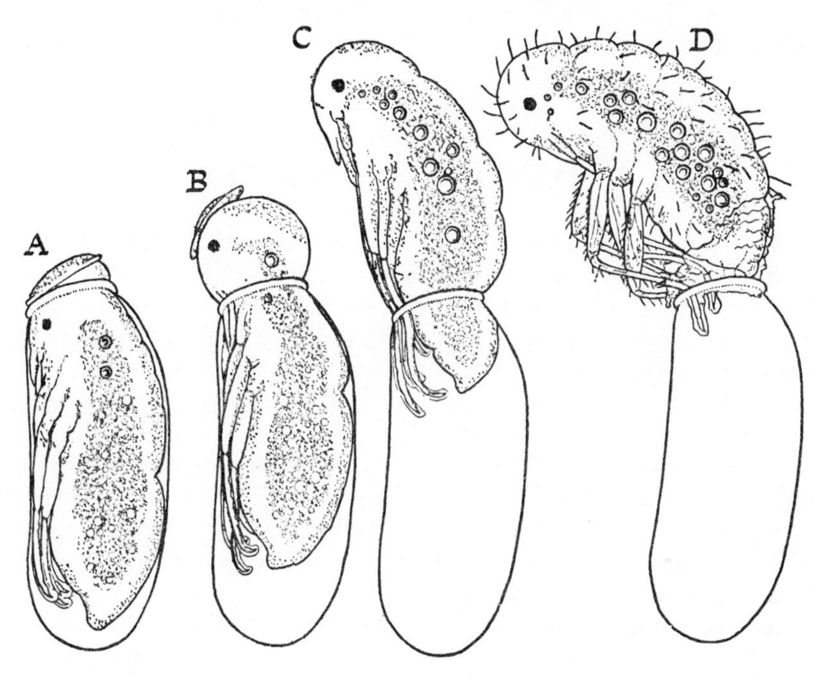

FIG. 18.—Hatching of *Cimex* from the egg (*after* SIKES and WIGGLESWORTH)

A, cap of the egg being forced off; gut and abdomen showing waves of peristalsis towards the head; a few air bubbles in the gut. *B*, active peristalsis continues; head distended and bulging from the egg. *C*, larva, almost free from the egg, has started to swallow air vigorously. *D*, embryonic cuticle has split and slipped backwards, allowing spines to stand erect; tracheae have filled with air; larva swallowing air.

rupturing the shell.[89] The hatching spines may serve to concentrate this pressure at one point of the egg (often the egg-cap, as in Pentatomidae) or they may be used deliberately to cut an opening in the chorion (as in Aphaniptera). The Lepidoptera simply gnaw their way out with the mandibles; the Muscidae tear a hole in the vitelline membrane with their mouth hooks.[67]

Often, when the amniotic fluid has been swallowed, air diffuses through the chorion and the insect may increase its bulk by swallowing this. Where the insect has to pass the narrow orifice that was closed by the egg-cap, it does so by peristaltic waves of muscular contraction passing forwards over the body, squeezing out the head as a tense little vesicle (Fig. 18). The embryonic cuticle of *Pediculus* and other Anoplura carries a tooth-bearing crown which not only serves to aid in hatching[67, 85] but provides a channel for the conduction of air to the mouth.[51]

It has been claimed that the swallowed air is ejected from the anus and serves to expel the larva from the egg; but that is not so; peristalsis of the body wall provides the force for hatching.[67] During this process the embryonic cuticle may split and remain as a frill round the mouth of the shell (Anoplura; Hemiptera) or the insect may leave the egg completely with this cuticle intact (Odonata; aquatic Hemiptera; Orthoptera). Thereupon it swallows air again, blowing itself up until this cuticle is forced to split; and it may continue the process or, in aquatic forms, it may swallow water, until its skin has stretched and it has reached a size far greater than the egg from which it came.[67]

A peculiar method of hatching is described in the uterine egg of *Glossina*. The larva bursts the chorion by means of an oral egg-tooth. The chorion is then apparently seized by a cellular organ, the 'choriothete', on the floor of the uterus and stripped off the young larva.[37]

ADDENDA TO CHAPTER I

p. 3. **Respiratory mechanisms**—The 'resistant protein' layer of the chorion (Fig. 1) appears to carry the air-filled meshwork in *Rhodnius*; in some Heteroptera the meshwork is so fine that it can be detected only with the electron microscope.[135] In many eggs subject to flooding the external openings of the air-filled meshwork are widely spread to form a 'plastron' which can serve as a physical gill for the extraction of dissolved oxygen (see p. 383). That is well seen in the eggs of many Muscid Diptera.[138]

p. 5. **Absorption of water**—The cycle of water uptake may thus be secondary to changes in the plasticity of the chorion and serosal cuticle rather than changes in the active control of water absorption.[134] The same applies to the Gryllid *Scapsipedus*.[137]

p. 7. **Organization**—Synthesis of RNA does not begin until the migrating cleavage nuclei make contact with the differentiated regions of the cortical plasma.[140] Transplantation of cleavage or blastoderm nuclei into denucleated eggs of *Leptinotarsa* can result in normal embryos.[145]

p. 8. The two types of duplication were produced irrespective of the orientation of the centrifuged egg.[142] In the Chironomid *Smittia* as well as *Chironomus* a similar replacement of the head and thorax by a series of abdominal segments can be induced by ultra-violet irradiation. The effect is attributed to a disturbance of the normal gradients in the oöplasm which control the genome of the blastoderm cells.[139, 146] In *Chironomus* larvae with 'double abdomen' the gonad develops only in the one abdomen which contains the pole cells.[146]

p. 10. **Determination**—The substance of the 'oösome', the RNA-containing material of the posterior pole, has been traced back to the chromidial granules which form on the nuclear membrane of the oöcyte. The oösome material accumulates around these granules.[141]

p. 12. **Anatrepsis**—In insects generally the 'yolk system' seems to play a decisive role in the invagination and migration of the germ band. In *Euscelis* contractile structures in the yolk system are responsible.[144]

p. 15. **Diapause in Bombyx**—The actual source of the diapause factor in the *Bombyx* female is probably a pair of neurosecretory cells lying just behind

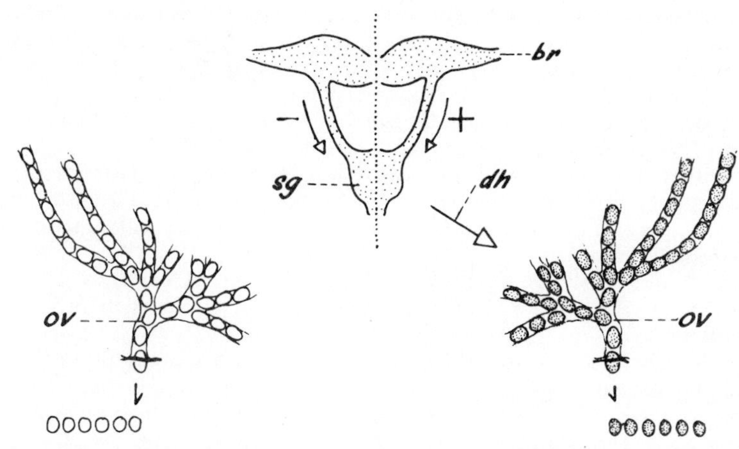

FIG. 18*a*.—The role of the brain and suboesophageal ganglion in the control of diapause in the silkworm egg (*after* FUKUDA).

To the left: inhibition by the brain leads to production of non-diapause eggs. To the right: stimulation by the brain leads to production of diapause eggs. *br*, brain; *dh*, diapause hormone; *ov*, ovary; *sg*, suboesophageal ganglion.

the centre of the suboesophageal ganglion (Fig. 18*a*). In pupae destined to give rise to non-diapause eggs these cells are laden with neurosecretory material which is stored and not released. In pupae destined to give diapause eggs the secretion is released and is virtually absent from the cytoplasm.[136, 143]

REFERENCES

[1] ABELOOS, M. *C.R. Acad. Sci.*, **200** (1935), 2112–14 (diapause in mature embryo: *Timarcha*, Col.).

[2] —— *C.R. Acad. Sci.*, **212** (1941), 722–4 (maternal influence on diapause in egg: *Timarcha*, Col.).

[3] ABOÏM, A. N. *Rev. Suisse Zool.*, **52** (1945), 53–154 (pole cells: *Drosophila*).

[4] ANDERSEN, K. T. *Z. Morph. Oekol. Tiere*, **17** (1930), 649–76 (effect of humidity and temperature on the egg: *Sitona*, Col.).

[5] BALFOUR-BROWN, F. *Trans. R. Soc. Edin.*, **47** (1910), 317–40; *Nature*, **92** (1913), 20–24 (hatching of eggs of water beetles).

[6] BEAMENT, J. W. L. *Quart. J. Micr. Sci.*, **87** (1946), 393–439; *Proc. Roy. Soc.*, B, **133** (1946), 407–18; *J. Exp. Biol.*, **23** (1947), 213–33; *Bull. Ent. Res.*, **39** (1948), 359–83; **39** (1949), 467–88 (egg-shell: *Rhodnius*, Hem.).

[7] BLUNCK, H. *Z. wiss. Zool.*, **111** (1914), 76–157 (development in the egg: *Dytiscus*, Col.).

[8] BOCK, E. *Z. Morph. Oekol. Tiere*, **35** (1939), 615–702; *Arch. Entw. Mech.*, **141** (1941), 159–247 (determination in egg of *Chrysopa*).

[9] BODINE, J. H. *Physiol. Zool.*, **2** (1929), 459–82 (diapause in egg of *Melanoplus*, Orth.).

[10] BODINE, J. H., and ROBBIE, W. A. *Physiol. Zool.*, **13** (1940), 391–7; **16** (1943), 279–87 (diapause in egg of *Melanoplus*).

[11] BRAUER, A., and TAYLOR, A. C. *J. Exp. Zool.*, **73** (1936), 127–52 (organization in the egg: *Bruchus*, Col.).

[12] BRAUER, A. *Physiol. Zool.*, **11** (1938), 249–66; *J. Exp. Zool.*, **112** (1949), 165–93 (reduplication in egg: *Bruchus*, Col.).

[13] BURKHOLDER, J. R. *Physiol. Zool.*, **7** (1934), 247–70 (respiration during diapause: *Melanoplus*, Orth.).

[14] BUXTON, P. A. *Biol. Rev.*, **7** (1932), 275–320 (insects and humidity: review).

[15] CAPPE DE BAILLON, P. *La Cellule*, **31** (1920), 1–245 (eggs and hatching in Tettigoniids).

[16] —— *C.R. Acad. Sci.*, **180** (1925), 1371–3; **181** (1925), 479–81; *Bull. Biol. Fr. Belg.*, **62** (1928), 378–87; **67** (1933), 1–33; *Ann. Sci. Nat. Zool.* (sér. 10), **20** (1937), 169–88 (embryology of double monsters: Phasmidae).

[17] CHRISTENSEN, P. J. H. *Zool. Jarhb., Abt. Anat.*, **62** (1937), 567–82 (histology of winter eggs of *Orgyia*, Lep.).

[18] CHRISTOPHERS, S. R. *Trans. Roy. Ent. Soc. Lond.*, **95** (1945), 25–34 (structure of egg and egg-raft: *Culex*, Dipt.).

[19] CLARK, N. *J. Anim. Ecol.*, **4** (1935), 82–7 (effect of temperature and humidity on eggs of *Rhodnius*, Hem.).

[20] DAVIES, L. *J. Exp. Biol.*, **25** (1948), 71–85 (egg-shell: *Lucilia*, Dipt.).

[21] DU BOIS, A. M. *Rev. Suisse Zool.*, **45** (1938), 1–92 (determination in egg: *Sialis*, Megalopt.).

[22] DUCLAUX, E. *C.R. Acad. Sci.*, **69** (1869), 1021–2; **83** (1876), 1049–51 (effect of the winter cold on silkworm eggs).

[23] EASTHAM, L. *Quart. J. Micr. Sci.*, **71** (1927), 353–94 (embryology of *Pieris*, Lep.).

[24] v. EMDEN, F. *Z. wiss. Zool.*, **126** (1925), 622–54; *Proc. Roy. Ent. Soc. Lond.*, A., **21** (1946), 89–97 (hatching spines).

[25] EVANS, A. C. *Parasitology*, **26** (1934), 366–77 (effect of humidity and temperature on the egg: *Lucilia*, Dipt.).

[26] EWEST, A. *Arch. Entw. Mech.*, **135** (1937), 689–752 (determination, &c., in egg: *Tenebrio*, Col.).

[27] FLANDERS, S. E. *Ann. Ent. Soc. Amer.*, **35** (1942), 251–66 (physiological egg types in parasitic Hymenoptera).

[28] FUKUDA, S. *Zool. Mag.* (Japan), **52** (1940), 415–29 (voltinism in silkworm).

[29] GEIGY, R. *Arch. Entw. Mech.*, **125** (1931), 406–47; *Rev. Suisse Zool.*, **48** (1941), 483–94 (determination in egg of *Drosophila*, Dipt).

[30] GIMINGHAM, C. T. *Trans. Ent. Soc. Lond.*, 1926, 585–90 (egg burster in Aphididae).

[31] GJULLIN, C. M., *et al. J. Cell. Comp. Physiol.*, **17** (1941), 193–202 (hatching of eggs: *Aedes*, Dipt.).

[32] HENSON, H. *Biol. Rev.*, **21** (1946), 1–14 (embryology and metamorphosis).

[33] HODSON, A. C., and WEINMAN, C. J. *Tech. Bull. Minn. Agric. Exp. Sta.*, No. 170 (1945), 1–31 (diapause in eggs of *Malacosoma*, Lep.).

[34] HOLDAWAY, F. G. *Commonw. Austral. Counc. Sci. Med. Res.* Pamphlet No. 4 (1927), (bionomics of *Sminthurus*, Collemb.).

[35] HOWLAND, R. B., *et al. J. Exp. Zool.*, **70** (1935), 415–24 ; **73** (1936), 109–52; *Proc. Amer. Phil. Soc.*, **84** (1941), 605–16 (determination in egg of *Drosophila*, Dipt.).

[36] HUSSEY, P. B. *Entom. Amer.*, **7** (1926), 1–80 (pleuropodia).

[37] JACKSON, C. H. N. *Proc. Roy. Ent. Soc. Lond.*, **23** (1948), 36–8 (hatching of eggs: *Glossina*, Dipt.).

[38] JACKSON, D. J. *Proc. Zool. Soc. Lond.*, 1928, 597–630 (eggs of *Dinocampus*, Braconidae).

[39] JAHN, T. L. *Proc. Soc. Exp. Biol. Med.*, **33** (1935), 159–63; *J. Cell. Comp. Physiol.*, **7** (1935), 23–46; **8** (1936), 289–300 (permeability of egg membranes: *Melanoplus*, Orth.).

[40] JOHNSON, C. G. *Trans. Soc. Brit. Entom.*, **1** (1934), 1–32; *J. Exp. Biol.*, **14** (1937), 413–21 (egg membranes and water absorption: *Notostira*, Hem.).

[41] KRAUSE, G. *Arch. Entw. Mech.*, **132** (1934), 115–205; *Z. Morph. Oekol. Tiere*, **34** (1938), 499–564; *Arch. Entw. Mech.*, **139** (1939), 639–723; **146** (1953), 275–370; *Zool. Jahrb., Anat.*; **75** (1957), 481–550 (organization in the egg: *Tachycines*, Orth.).

[42] —— *Verhandl. vii internat. Kongr. f. Entom.*, 1938, *Berlin*, **2** (1939), 772–9; *Biol. Zbl.*, **59** (1939), 495–536 (types of determination in insect eggs).

[43] LÜSCHER, M. *Rev. Suisse Zool.*, **51** (1944), 531–627 (determination in egg: *Tineola*, Lep.).

[44] MASCHLANKA, H. *Arch. Entw. Mech.*, **137** (1938), 714–72 (determination in egg: *Ephestia*, Lep.).

[45] MATTHEE, J. J. *Nature*, **162** (1948), 226–7; *Union of S. Africa Dept. Agric. Sci. Bull.* No. **316** (1951), 83 pp. (physiology of egg membranes: *Locustana*, Orth.).

[46] MELLANBY, H. *Quart. J. Micr. Sci.*, **79** (1936), 1–42 (embryology of *Rhodnius*, Hem.)

[47] MILLER, A. *Ann. Ent. Soc. Amer.*, **33** (1940), 437–77 (egg membranes: *Pteronarcys*, Plecopt.).

[48] MOSCONA, A. *Quart. J. Mior. Sci.*, **91** (1950), 183–93 (lime in egg-shell: *Bacillus*, Orth.).

[49] PANTEL, J. *La Cellule*, **29** (1913), 1–289 (egg membranes in parasitic Diptera).

[50] —— *C.R. Acad. Sci.*, **168** (1919), 127–9 (calcium in egg shell: Phasmidae).

[51] PATAY, R. *Bull. Soc. Zool. France* **66** (1941), 182–9 (respiration in egg: *Pediculus*).

[52] PAULI, M. E. *Z. wiss. Zool.*, **129** (1927), 482–540 (determination in eggs of *Musca* and *Calliphora*, Dipt.).

[53] PFLUGFELDER, O. *Biol. Zbl.*, **66** (1947), 372–87 (determination of compound eye, &c.: *Carausius*, Orth.).

[54] POULSON, D. F. *The Embryonic Development of Drosophila melanogaster*, Actualités scientifiques et industrielles, No. 498. Hermann et Cie, Paris, 1937.

[55] —— *J. Exp. Zool.*, **83** (1940), 273–325; *Amer. Nat.*, **79** (1945), 340–63 (gene action in early development: *Drosophila*, Dipt.).

[56] —— *Proc. Nat. Acad. Sci. U.S.A.*, **33** (1947), 182–4 (pole cells: *Drosophila*, Dipt.).

[57] RABINOWITZ, M. *J. Morph.*, **69** (1941), 1–49 (early embryology: *Drosophila*, Dipt.).

[58] REITH, F. *Z. wiss. Zool.*, **139** (1931), 664–734 (determination in egg: *Camponotus*, Hym.); *ibid.*, **147** (1935), 77–100 (ditto: *Sitona*, Col.).

[59] RICHARDS, A. G., and MILLER, A. *J.N.Y. Ent. Soc.*, **45** (1937), 1–16 (physiology of egg development: review).

[60] ROONWAL, M. L. *Bull. Ent. Res.*, **27** (1936), 1–14 (egg structure: *Locusta*, Orth.).

[61] ROONWAL, M. L. *Phil. Trans. Roy. Soc.*, B., **227** (1937), 175–244 (embryology of *Locusta*, Orth.).

[62] SCHIPPER, A. L. *Physiol. Zool.*, **11** (1938), 40–53 (desiccation and development: *Melanoplus* eggs).

[63] SCHNETTER, M. *Z. Morph. Oekol. Tiere*, **29** (1934), 114–95; *Arch. Entw. Mech.*, **131** (1934), 285–323 (differentiation centre in egg of honey-bee).

[64] SEHL, A. *Z. Morph. Oekol. Tiere*, **20** (1931), 533–98 (segmentation and germ-band formation: *Ephestia*, Lep.).

[65] SEIDEL, F. Organization in the insect egg. *Z. Morph. Oekel. Tiere*, **1** (1924), (*Pyrrhocoris*, Hem.); *Biol. Zbl.*, **48** (1928), 230–51; *Arch. Entw. Mech.*, **119** (1929), 322–440; *Arch. Entw. Mech.*, **126** (1932), 213–76; **131** (1934), 135–87; **132** (1935), 671–751 (*Platycnemis*, Odonata); *Verh. deutsche Zool. Ges.*, **38** (1936), 291–336 (review).

[66] SEIDEL, F., BOCK, E., and KRAUSE, G. *Naturwiss.*, **28** (1940), 433–46 (determination in insect egg: review).

[67] SIKES, E. K., and WIGGLESWORTH, V. B. *Quart. J. Micr., Sci.*, **74** (1931), 165–92 (hatching from the egg).

[68] SLIFER, E. H. *J. Morph.*, **51** (1931), 613–18 (diapause in embryo of *Melanoplus*, Orth.).

[69] —— *J. Morph.*, **53** (1932), 1–22 (growth of embryo of *Melanoplus*, Orth.).

[70] —— *Quart. J. Micr. Sci.*, **79** (1937), 493–506; **80** (1938), 437–57; *J. Morph.*, **63** (1938), 181–205; (egg membranes in *Melanoplus*, Orth.).

[71] —— *J. Exp. Zool.*, **102** (1946), 333–56; **110** (1949) 183–204; *Discussions of Faraday Soc.*, **3** (1948), 182–7; *Science*, **107** (1948), 152; *Ann. Ent. Soc. Amer.*, **42** (1949), 134–40; *J. Morph.*, **87** (1950), 239–74; *J. Exp. Zool.*, **138** (1958), 259–82 (wax and diapause: *Melanoplus* egg).

[72] SLIFER, E. H. and SHULOW, A. *Ann. Ent. Soc. Amer.*, **40** (1947), 652–5 (polyembryony in grasshopper eggs).

[73] SMRECZYŃSKI, S. *Zool. Jahrb.*; *Abt. Physiol.*, **59** (1938), 1–64 (determination in egg: *Agelastica*, Col.).

[74] STRASBURGER, E. H. *Z. wiss. Zool.*, **145** (1934), 621–41 (cell movements during cleavage: *Calliphora*, Dipt.).

[75] THOMAS, H. D. *J. Parasit.*, **29** (1944), 324–8 (hatching of eggs: *Aëdes*, Dipt.).

[76] THOMPSON, V., and BODINE, J. H. *Physiol. Zool.*, **9** (1936), 455–70 (dehydration and oxygen consumption: eggs of *Melanoplus*, Orth.).

[77] TIEGS, O. W. *Quart. J. Micr. Sci.*, **83** (1942), 153–69; **84** (1942), 35–47 ('dorsal organ': Collembola and *Campodea*, Thysan.).

[78] TIRELLI, M. *Physiol. Zool.*, **14** (1941), 70–7 (diapause eggs of *Bombyx mori*).

[79] TOMITA, M. *Biochem. Z.*, **116** (1921), 40–7 (composition of chorion: silkworm).
[80] TUFT, P. H. *J. Exp. Biol.*, **26** (1950) 327–34 (respiration in the egg: *Rhodnius*, Hem.).
[81] TUTT, J. W. *British Lepidoptera*, **1**, London, 1899.
[82] UMEYA, Y. *Bull. Sericult. Exp. Sta. Chosen*, No. **1** (1926), 1–26 (alternation of voltinism in the silkworm).
[83] WEBER, H. *Z. Morph. Oekol. Tiere*, **23** (1931), 575–753 (biology of *Trialeurodes*, Hem.).
[84] —— *Lehrbuch der Entomologie*, Jena, 1933.
[85] WEBER, H. *Biol. Zbl.*, **59** (1939), 98–109; 397–409 (hatching spines: *Haematomyzus*).
[86] WEISMANN, A. *Z. wiss. Zool.*, **13** (1863), 159–220 (development in the egg: *Calliphora* Dipt.).
[87] WIGGLESWORTH, V. B., and BEAMENT, J. W. L. *Quart. J. Micr. Sci.*, **91** (1950), 429–52; *J. Ins. Physiol.*, **4** (1960), 184–9 (respiration of insect eggs).

SUPPLEMENTARY REFERENCES (A)

[88] BANKS, C. J. *J. Exp. Biol.*, **26** (1949), 131–6 (absorption of water by eggs of *Corixa*, Hem.).
[89] DAVIES, L. *J. Exp. Biol.*, **27** (1950), 437–45 (hatching of eggs in Muscidae).
[90] FUKUDA, S. *Zool. Mag. (Japan)*, **60** (1951), 121; *Proc. Japan. Acad.*, **27** (1951), 582–6 (determination of diapause in eggs of silkworm).
[91] GANDER, R. *Rev. Suisse Zool.*, **58** (1951), 215–78 (hatching of eggs of *Aedes aegypti*).
[92] KOGURE, M. *J. Dept. Agric. Kyushu Imp. Univ.*, **4** (1933), 1–93 (effect of light and temperature in determination of diapause in eggs of silkworm).
[93] ULRICH, H. *Naturwissenschaften*, **38** (1951), 121; *Biol. Zbl.*, **70** (1951), 274–85 (effect of local application of X-rays to egg of *Drosophila*).

SUPPLEMENTARY REFERENCES (B)

[94] BAUDISCH, K. *Z. Morph. Oekol. Tiere*, **47** (1958), 436–88 (pleuropodia in *Pediculus*).
[95] BECKEL, W. E. *Can. J. Zool.*, **36** (1958), 541–54 (hatching of eggs in *Aëdes hexodontus*).
[96] BIER, K. *Biol. Zbl.*, **73** (1954), 170–90 (nucleic acids and caste determination in *Formica*).
[97] BLACKITH, R. E. *Comp. Biochem. Physiol.*, **1** (1961), 99–107 (water reserves of hatchling locusts).
[98] v. BORSTEL, R. C. in *The Beginnings of Embryonic Development*, Washington, Amer. Assoc. Adv. Sci. 1957, 175–99 (nucleus and cytoplasm in early insect development).
[99] BROWNING, T. O. *et al. J. Exp. Biol.*, **30** (1953), 104–15; **37** (1960), 213–17 (water uptake in egg of *Acheta*).
[100] BUCHNER, P. *Z. Morph. Oekol. Tiere*, **46** (1957), 481–528 (yolk-free eggs in insects: review).
[101] COUNCE, S. J. *Ann. Rev. Entom.*, **6** (1961), 295–312 (insect embryogenesis: review).
[102] —— *J. Morph.*, **112** (1963), 129–45 (polar granules in *Drosophila* egg).
[103] GEYER-DUSZYŃSKA, I. *J. Exp. Zool.*, **141** (1959), 391–447 (chromosome elimination in Cecidomyidae).
[104] GOUIN, F. J. *Fortschr. Zool.*, **14** (1962), 87–114 (experimental embryology of insects: review).
[105] HAGET, A. *Bull. Biol.*, **87** (1953), 124–217 (experimental embryology of *Leptinotarsa*).
[106] HARTLEY, J. C. *Quart. J. micr. Sci.*, **102** (1961), 249–55 (egg shell in *Locusta*).
[107] HASEGAWA, K. *J. Fac. Agric. Tottori Univ.*, **1** (1952), 84–124; *Nature, Lond.*, **179** (1957), 1300–1; *J. Exp. Biol.*, **40** (1963), 517–29 (diapause hormone in *Bombyx*).
[108] HINTON, H. E. *Quart. J. Micr. Sci.*, **101** (1960), 313–32; *Phil. Trans. Roy. Soc.*, B **243** (1960), 45–73; *J. Ins. Physiol.*, **7** (1961), 224–57; *Sci. Prog.*, **100** (1962), 97–112 (respiratory systems of insect egg-shells).
[109] HOGAN, T. W. *Nature*, **186** (1960), 98; *Aust. J. Biol. Sci.*, **13** (1960), 14–29; 527–40; **14** (1961), 419–26; **15** (1962), 538–42 (diapause in egg of *Acheta*).
[110] IDRIS, B. E. M. *Z. Morph. Oekol. Tiere*, **49** (1960), 387–429; *Arch. Entw. Mech.*, **152** (1960), 230–62 (experimental embryology of *Culex*).

[111] JONES, B. M. *Proc. Roy. Soc.*, B **148** (1958), 263–77 (membrane formation by the amniotic fluid in *Locusta*).

[112] JUDSON, C. L. *J. Ins. Physiol.*, **9** (1963), 787–92 (nervous system and hatching of *Aëdes aegypti*).

[113] JURA, C. *Zool. Polon.*, **8** (1957), 177–99 (experimental embryology of *Melasoma*).

[114] KING, R. C., and KOCH. E. A. *Quart. J. micr. Sci.*, **104** (1963), 297–320 (formation of chorion and vitelline membrane of *Drosophila*).

[115] KRAUSE, G. *Ergebn. Biol.*, **20** (1958), 159–98; *Symposium: Inst. Intern. Embryol.* 1961, 302–37 (experimental embryology: review).

[116] KRAUSE, G., and SANDER, K. *Adv. in Morphogenesis*, **2** (1962), 259–303 (experimental embryology in insects: review).

[117] LAUGHLIN, R. *J. Exp. Biol.*, **34** (1957), 226–36 (water absorption by egg of *Phylloper-tha*).

[118] MCFARLANE, J. E. *Can. J. Zool.*, **38** (1960), 232–41; 1038–9 (water absorption by egg of *Acheta*).

[119] MAHOWALD, A. P. *J. Exp. Zool.*, **151** (1962), 201–16 (fine structure of pole cells, &c., in *Drosophila*).

[120] MAHR, E. *Z. Morph. Oekol. Tiere*, **49** (1960), 263–311; *Roux Arch. Entw. Mech.*, **152** (1960), 263–302; **152** (1961), 663–724 (experimental embryology of *Acheta domestica*).

[121] MÜLLER, M. *Zool. Jahrb., Physiol.*, **67** (1957), 111–50 (origin of yolk cells in *Apis*).

[122] NITSCHMANN, J. *Verh. Dtsch. Zool. Ges. Graz 1957* (1958), 533; *Ibid. Frankfurt 1958*, (1959), 370–7 (experimental embryology of *Calliphora*).

[123] OELHAFEN, F. *Roux Arch. Entw. Mech.*, **153** (1961), 120–57 (experimental embryo-genesis in *Culex*).

[124] OKADA, M. *Jap. J. Appl. Ent. Zool.*, **2** (1958), 295–6 (embryonic cuticle in *Chilo*).

[125] SALT, R. W. *Can. J. Zool.*, **30** (1952), 55–82 (water absorption in eggs of *Melanoplus*).

[126] SANDER, K. *Roux Arch. Entw. Mech.*, **151** (1959), 430–97; **151** (1960), 660–707; *Verh. Dtsch. Zool. Ges., Saarbrücken 1961* (1962), 315–22 (experimental embryology in *Euscelis*).

[127] SHARAN, R. K. *et al. Ann. Zool.*, **3** (1958), 1–8; *Ann. ent. Soc. Amer.*, **53** (1960), 538–41 (embryonic cuticle in *Locustana* and *Dysdercus*).

[128] SLIFER, E. H., and SEKHON, S. S. *Quart. J. Micr. Sci.*, **104** (1963), 321–34 (hydropyle in *Melanoplus*, fine structure).

[129] SONNENBLICK, P. in *Biology of 'Drosophila'*, (M. Demerec, Ed.), Wiley, New York, 62–167 (early embryology of *Drosophila*).

[130] TAKAHASHI, Y. *Jap. J. Appl. Ent. Zool.*, **3** (1959), 80–5 (permeability of chorion in *Bombyx*).

[131] TATCHELL, R. J. *Parasitology*, **51** (1961), 385–94 (respiration in egg of *Gasterophilus*).

[132] TAWFIK, M. F. S. *J. Ins. Physiol.*, **1** (1957), 286–91 (alkaline phosphatase in pole plasma of *Apanteles* egg).

[133] YAJIMA, H. *J. Embryol. Exp. morph.*, **8** (1960), 198–215 (embryonic determination in *Chironomus*).

SUPPLEMENTARY REFERENCES (C)

[134] BROWNING, T. O. In *Insects & Physiology*, Beament & Treherne (Ed.), London, Oliver and Boyd, 1967, 315–28; *J. Exp. Biol.*, **51** (1969), 99–105 (uptake of water in egg of *Locusta* and *Teleogryllus*).

[135] COBBEN, R. H. *Med. No. 151: Lab. Entom. Agric. Univ. Wageningen* 1968, 475 pp. (egg shell structure in Heteroptera).

[136] FUKUDA, S. and TAKEUCHI, S. *Embryologia*, **9** (1967), 333–53 (diapause factor in suboesophageal ganglion of *Bombyx*).

[137] GRELLET, P. *J. Insect Physiol.*, **17** (1971), 1533–53 (water uptake in egg of *Scapsipedus*, Gryllidae).

[138] HINTON, H. E. *J. Insect Physiol.*, **13** (1967), 647–51; *Ann. Rev. Ent.*, **14** (1969), 343–68 (respiratory systems of insect egg shells).

[139] KALTHOFF, K. and SANDER, K. *Wilhelm Roux Arch. Entw. Mech. Org.*, **161** (1968), 129–146 (production of 'double abdomen' in *Smittia*, Chironomidae).
[140] LOCKSHIN, R. A. *Science*, **154** (1966), 775–6 (nucleic acid synthesis in early development).
[141] MENG, C. *Wilhelm Roux Arch. Entw. Mech. Org.*, **161** (1968), 162–208 (nature of the oösome in *Pimpla*, Ichneumonidae).
[142] OVERTON, J. and RAAB, M. *Devl. Biol.*, **15** (1967), 271–87 (development in centrifuged eggs of *Chironomus*).
[143] PARK, K. E. and YOSHITAKE, N. *J. Insect Physiol.*, **17** (1971), 1305–13 (fine structure of neurosecretory cells in suboesophageal ganglion of *Bombyx*).
[144] SANDER, K. *Verh. dtsch. zool. Ges.*, 1967, 81–9 (mechanism of Anatrepsis in *Euscelis*).
[145] SCHNETTER, W. *Verh. dtsch. zool. Ges.*, **30** (1966), 494–9 (transplantation of cleavage nuclei in egg of *Leptinotarsa*).
[146] YAJIMA, H. *J. Embryol. exp. Morph.*, **12** (1964), 89–100; **24** (1970), 287–303 (irradiation and 'double abdomen' &c. in *Chironomus*).

Chapter II

The Integument

THE PHYSIOLOGY of growth in insects is so profoundly influenced by the properties of their integument that we must devote a chapter to this subject before returning to the further course of their development. As in all Arthropods the integument consists of a single layer of epidermal cells (in insects often called the 'hypodermis'), which secretes over the surface of the animal, and over all those invaginations of the ectoderm that arise from it—the buccal cavity and fore-gut, the tracheae, the lower genital ducts and the multifarious glands that open upon the surface—a more or less inert cuticular membrane.

PROPERTIES OF THE CUTICLE

Epidermis and basement membrane—The epidermis is seen in its full development only when the new cuticle is being laid down; at other times it is

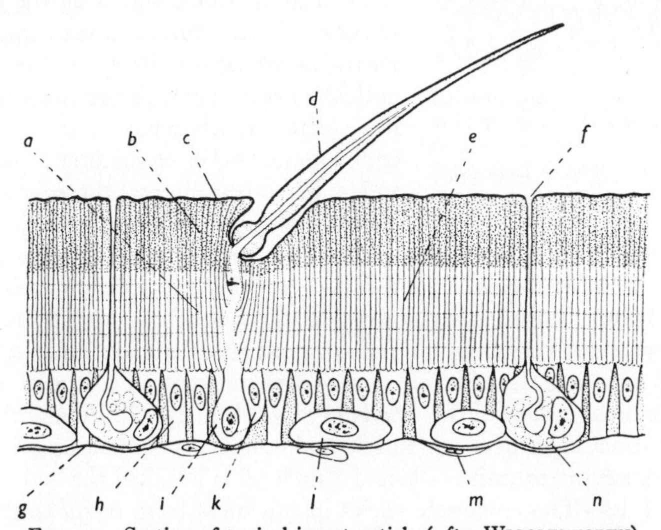

FIG. 19.—Section of typical insect cuticle (*after* WIGGLESWORTH)

a, laminated endocuticle; *b*, exocuticle; *c*, epicuticle; *d*, bristle; *e*, pore canals; *f*, duct of dermal gland; *g*, basement membrane; *h*, epidermal cell; *i*, trichogen cell; *k*, tormogen cell; *l*, oenocyte; *m*, haemocyte adherent to basement membrane; *n*, dermal gland.

exceedingly attenuated. When it is fully active the apical part of the epidermal cell is striated, with extensions into the vertical filaments (the pore-canals) of the cuticle. Differentiated from the epidermis are the dermal glands, which send fine ducts through the cuticle, and the oenocytes (p. 447) which may remain applied to its lower surface[114, 118, 119] or become incorporated in the fat body.

The epidermis rests upon a basement membrane probably formed from the con densed processes of epidermal cells,[120] with blood cells or haemocytes adherent to its lower surface and contributing to its substance[192] (p. 438), (Fig. 19).

Structure of cuticle—The 'cuticle' covering the epidermis may be an ex cessively delicate pellicle, as in the terminal branches of the tracheal system an the intracellular cavities of certain glands, or a thick, dense, horny armour suc as that which encases the thorax and appendages of many beetles. The charac teristic structure of the cuticle is best seen in the plates or sclerites on the dorsun of the abdomen as described, for example, in the larva[75] or adult[119] of *Tenebrio* in *Gryllotalpa*,[36] in aquatic Hemiptera,[76] in queen termites,[2] and in the Reduvii bug *Rhodnius*.[114, 118] It consists of three main layers (Fig. 19): (i) an outer thi refractile membrane, a few microns at most in thickness, sometimes darkly pig mented, but often apparently colourless, called the '*epicuticle*' (the 'Grenzlamelle or 'Grenzsaum' of German writers). (ii) Belov this is a rigid layer, usually amber coloured but sometimes almost black, which may com pose anything from one-twelfth to one-hal of the total thickness of the cuticle. Thi is the '*exocuticle*' (or 'primary cuticula',[10] 'Emailschicht',[11] 'Lackschicht',[86] or 'Pigment schicht'[51]). (iii) Finally there is a thick colour less elastic layer, the '*endocuticle*' (o 'secondary cuticula',[102] 'Hauptlage' o 'Innenlage'), which makes up the greater par of the entire structure. It is convenient to grou all the layers below the epicuticle as the 'pro cuticle', parts of which are hardened to forr the exocuticle, while parts remain as colourles endocuticle.[126, 177] Sometimes, as in *Blatt* and many other insects, there is a layer wit distinctive staining properties between the exo and endocuticle; this has been termed the 'mesocuticle';[180] sometimes, as i *Aphodius* larvae, it forms a zone some 5μ thick immediately below the epi cuticle.[168] It may be that in regions where an exocuticle appears to be absent, it i in fact present but very thin and indistinguishable from an 'inner epicuticle', a in caterpillars where the exuvium contains 12 per cent. of chitin,[106] and perhap in the abdominal cuticle of the *Rhodnius* larva (Fig. 21, C). Finally there is com monly a distinct layer between cuticle and epidermis containing a mucoid com ponent and serving to unite cells and cuticle; this is called the 'subcuticle'.[182]

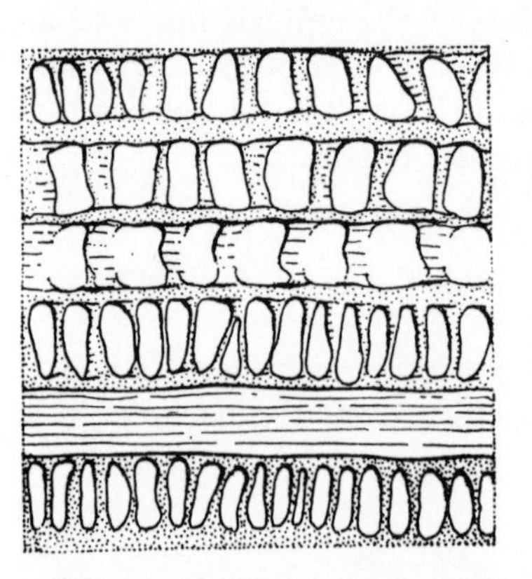

FIG. 20.—Section through cuticle of *Lucanus cervus*, showing the fibre bundles ('Balken') cut through at different angles (*from a photograph by* STEGEMANN)

Epicuticle—The epicuticle varies in thickness from 0·03μ in the mosquit larva to 2μ in *Periplaneta*[83] or 4μ in the larva of *Sarcophaga*.[23] It is a comple structure made up of several layers. Histologically it is sometimes possible t recognize two layers,[23, 83, 157] but by various experimental means and by follow ing the deposition of the epicuticle at the time of moulting, it has been show to be composed of at least three or four layers. In *Rhodnius* or *Tenebrio*,[118, 11] for example, there is (*a*) a layer of variable thickness, the 'cuticulin' layer whic is the refractile epicuticle as seen in histological sections, (*b*) a layer of wa perhaps 0·25μ thick, and (*c*) a very thin covering layer of 'cement' which may b detached by treatment with wax solvents. In the cockroach, *Periplaneta*, &c., th ement layer is a diffuse structure which is permeated by the waterproofing ipid;[50, 133] indeed, such interpenetration is probably general and the laminar tructure of the epicuticle is not be to interpreted too schematically.[195] The epi- uticle is variously folded and may bear minute little spicules or microtrichia, idiocuticular structures'.[56] It extends inwards to line the ducts and intracellular avities of the dermal glands.[85, 114]

Endocuticle and exocuticle and their modifications—The endocuticle s usually made up of obvious horizontal lamellae which are commonly regarded s providing for flexibility and stretching by sliding over one another.[2] But they nay in fact have no other significance than that they represent the layers of uticle deposited on successive days—comparable with the annual rings of a ree.[172] Movements take place articularly at the folds or con- unctivae between the segments f the body or appendages. At hese points the cuticle may be hinner than elsewhere (10μ as compared with 40μ on the stern- tes of *Tenebrio*[119]); often it is omewhat thicker (50μ as com- ared with 30μ in the tergites of he *Rhodnius* adult[114]); but an xocuticle is wanting (Fig. 21, B) r, as in the larva of *Tenebrio* Fig. 21, A)[75] and in *Liogryllus*,[36] t is broken up into little wedge- haped blocks. Similar wedges or ones of 'mesocuticle' occur in he flexible wing bases of Lepi- loptera.[185] Where great disten- ion has to be provided for, as in he pleural membrane of blood- ucking insects, or the general urface of the abdomen in the *Rhodnius* larva (Fig. 21, C) or he queen termite, an exocuticle s absent and the epicuticle in the unstretched state is thrown into deep folds.

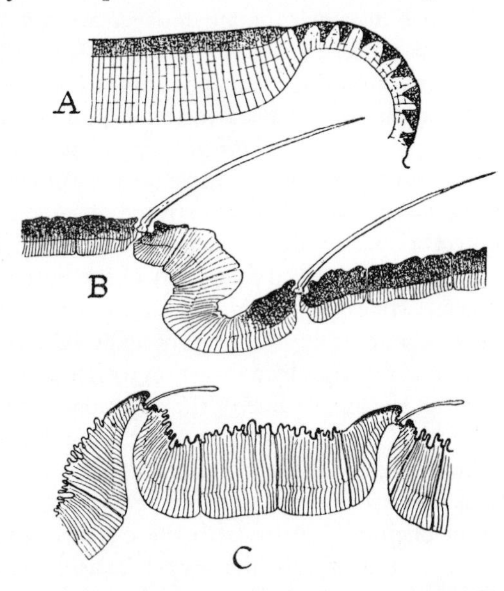

FIG. 21.—Examples of flexible and extensible cuticles

A, intersegmental membrane in *Tenebrio*; exocuticle broken up into wedge-shaped pieces (*after* PLOTNIKOW). *B*, intersegmental membrane in *Rhodnius* adult; exo- cuticle wanting. *C*, extensible cuticle of *Rhodnius* larva; epicuticle folded, exocuticle absent except around bristles (*after* WIGGLESWORTH).

Some insects can control the mechanical properties of the endocuticle. In the *Rhodnius* larva, for example, which takes a large distending meal of blood, some component in the haemolymph, liberated under the action of the nerves,[174] acts n the epidermal cells, which then in some manner render the cuticle less rigid, o that it is easily stretched;[136] they seem to cause a decrease in the cohesion of he chitin micelles.[174] [See p. 49.]

Rigidity in the cuticle is commonly supplied by the hardness of the exo- uticle; but in the Coleoptera the endocuticle has a characteristic structure which erhaps contributes to the firmness of the cuticle. As was first described by Meyer (1842) in the elytra of the stag beetle *Lucanus*, the lamellae of the endo- uticle are made up of parallel or anastomosing strands compressed laterally (the eams or 'Balken'[44]), embedded in a homogeneous matrix. The strands in successive layers run at an angle of about 60° to those above or below (Fig. 20) so that where they cross they give rise to star-like figures when seen in surface view. The strands are described by some as made up of bundles of fine fibrillae;[1] others regard them as having an alveolar or foam-like structure.[44] The former view agrees best with what is known of the micellar structure of the cuticle (p. 33). This arrangement of fibre bundles running in different directions in each superimposed lamella, is common in the skeletal structures of animals (in the placoid scales of Selachians, in the cornea and sclerotic of the eye[11]); in insects it is obvious only in Coleoptera; but it may really be more general, the separate elements being usually so small that they escape detection; for example, such a structure is described in *Gryllotalpa.*[36]

Pupae in which the appendages are free from the body, developing in protected situations, generally have a uniform thin cuticle. In pupae from exposed situations, the wings and appendages are often firmly stuck to the surface of the body by means of a fluid secretion poured over them at the time the last larval skin is shed. In *Leptinotarsa* (Chrysomelidae, Col.) this secretion is readily dissolved in water or alcohol in the early hours after pupation, so that the appendages become separated.[101] The extreme development along these lines is seen in the 'obtect' pupae of higher Lepidoptera, in which those surfaces of the appendages which are destined to form part of the outer sheath have a well-developed exocuticle, while the unexposed surfaces remain excessively delicate. At the last larval moult, when the pupa is exposed, the parts which form the sheath are cemented together by a secretion which soon hardens on exposure to the air, and only the surface outline of the appendages can be seen. Another modification of the cuticle for the protection of the pupa occurs in the higher Diptera. When their larvae are full-grown, the epidermis brings about a sclerotization of the thick soft endocuticle, transforming it into a hard dark shell or puparium, identical in chemical nature with the exocuticle of other insects, which serves as a permanent sheath for the very delicate pupa within.

There are many other modifications of the cuticle to meet special needs. It forms the cuticular spines, each attached to the margins of a socket by a thin flexible membrane. It is modified to provide other receptor organs (p. 256). Where it lines the larger cylindrical ducts, as in the salivary glands or the tracheae the cuticle is thrown into folds which commonly run a spiral course round the tube. Often the folds become filled with cuticular substance to give rise to a spiral thread or 'taenidium'.

Pore Canals—The endo- and exocuticle are traversed by numerous vertical lines. When seen in surface view these appear as minute dots, often arranged in polygonal fields separated by clear boundaries, corresponding to the limits of the epidermal cells.[36, 44, 75] On focussing the microscope up and down they appear to rotate, indicating that they run a spiral course.[36, 44, 75] These vertical lines were called 'pore canals' by Leydig (1855), and were regarded by him as filamentous processes of cytoplasm from the epidermal cells, around which the non-living substance of the cuticle was secreted. That interpretation is probably correct. It is supported by observations on *Leptinotarsa,*[102] *Gryllotalpa* and other insects,[36] aquatic Hemiptera,[76] *Rhodnius, Tenebrio,* and *Periplaneta.*[11] As several authors have pointed out, it brings the structure of the arthropod cuticle into line with that of bone and dentine in vertebrates.

In the larva of *Sarcophaga*[23] the pore canals are relatively coarse structures

1μ in diameter, 50–70 arising from each cell, about 15,000 canals per sq. mm. In this larva, their innermost section is straight, they have an intermediate section roughly coiled and a superficial branched section where they show the same tufted form described in larvae of Syrphidae and in the silkworm.[75] In the cockroach they are much finer structures.[83] What appear under the light microscope as straight parallel threads 1μ in diameter are seen with the electron microscope to consist of hollow tubes with an average diameter of $0{\cdot}15\mu$, thrown into a tight helix with a straight section near the epidermal cells. There are about 200 arising from each of the small epidermal cells, some 1,200,000 per sq. mm.

In some insects (the *Rhodnius* larva,[114, 118] *Periplaneta*,[83] the young larva of *Sarcophaga*[23]) the cytoplasmic contents of the canals apparently remain. When sections of the fresh cuticle, cut with the freezing microtome, are dried in air their contents shrink and on mounting the sections the air-filled canals appear as black threads. But in the mature larva of *Sarcophaga* they come to contain threads of chitin with an annular space around.[23] In the hard exocuticle of *Tenebrio* the canals contain solid sclerotized material free from chitin;[119] while in the outer part of the endocuticle of the caterpillar *Diataraxia*[106] they are filled with sclerotized material containing chitin.

When the epicuticle is exceptionally thick, as in the Mallophagan *Eomenacanthus*,[107] the pore canals can be seen to penetrate the cuticulin layer of the epicuticle. As a rule, when observed with the light microscope, they appear to end below the epicuticle; but there is physiological evidence that they do pierce the cuticulin layer (p. 40) and the structural arrangements will be described later in connection with wax secretion (p. 41).

'Sekretschicht'—As will be seen later (p. 41) the 'cement layer' over the epicuticle is the product of dermal glands and is discharged after the cuticle is complete. It has likewise been claimed that most of the Carabidae and Cicindelidae (with the exception of the Mantichorini group) differ from other beetles in having a thick 'Sekretschicht' or 'Sekretrelief' on the surface of the cuticle, that this layer is responsible for much of the pigmentation and that it is poured out from dermal glands. The 'Sekretschicht', it is said, may make up as much as one-third of the total thickness of the elytra.[86] It may be wholly responsible for the pigmentation or may be superimposed upon the usual exocuticle or 'Pigmentschicht' from which it differs in being slowly dissolved in 8 per cent. caustic potash, leaving the usual layers of the cuticle exposed (Fig. 22).[92, 93]

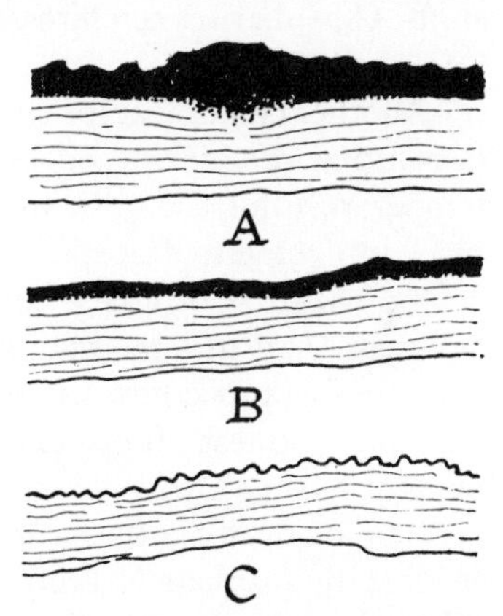

FIG. 22.—Cuticle of *Cicindela hybrida* showing the secretory layer (*after photographs by* STEGEMANN)

A, normal cuticle; B, after treating with 8% KOH at 60° C. for 36 hours (secretory layer partially dissolved); after 72 hours (secretory layer has disappeared).

Chemical composition: chitin—The best-known constituent of the cuticle is the nitrogenous polysaccharide named 'chitin' by Odier (1823) and later proved to be identical with the fungine which forms the cell wall of fungi. Chitin has the empirical formula $(C_8H_{13}O_5N)_x$; it is probably made up of long chains of acetylated glucosamine residues linked as shown on the following page.[7, 68]

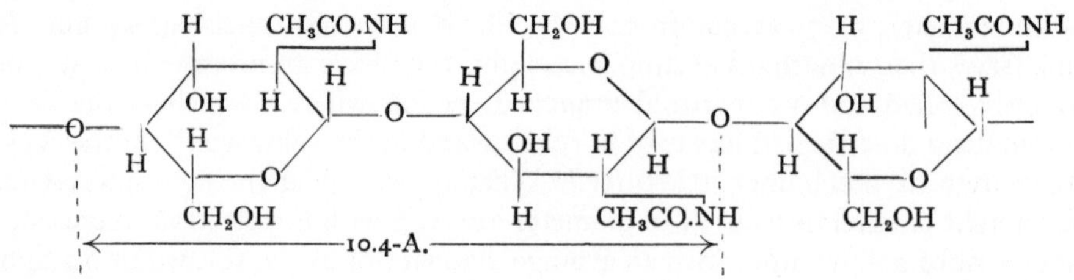

H $CH_3CO.NH$ CH_2OH H $CH_3CO.NH$
H H O H
OH H H OH
—O— H O O
H OH H H
H H H H O
CH_2OH H $CH_3CO.NH$ CH_2OH
←——10.4-A.——→

Chitin is insoluble in water, alcohol, ether and other organic solvents, dilute acids, dilute and concentrated alkalis. It is soluble in concentrated mineral acids, being hydrolysed to lower saccharides, glucosamine, acetic acid, &c. When treated with caustic alkalis at a high temperature the acetyl groups are detached and the polysaccharide 'chitosan' is produced. Chitosan gives a deep violet colour with iodine, and this provides the best test for chitin, van Wisselingh's test.[20, 111] It is performed as follows: the material is placed in a saturated solution of potassium hydroxide in a small tube, and this, together with a thermometer reading to 200° C., is immersed in a beaker of glycerine or paraffin, which is then heated to 150°–160° C. for 20 minutes (or longer if the material is not completely decolorized). The fragment is then transferred to a glass slide, washed in turn with 90 per cent. alcohol, 50 per cent. alcohol and water, and then flooded with 0·2 per cent. iodine in 1 per cent. sulphuric acid. An intense rose-violet colour is obtained. Chitosan prepared in this way is soluble in 3 per cent. acetic acid. It may be dissolved by warming in 10 per cent. sulphuric acid, and if left to stand at 70° C., spherites of chitosan-sulphate separate out.[56] Other tests are given by Campbell.[20]

Another colour test for chitin consists in treating the material with 'diaphanol' (a solution of chlorine dioxide in 50 per cent. acetic acid) in the dark at room temperature for a day or more. Chitin, even the most encrusted chitin in the black integument of beetles, will become soft and colourless and will give a violet colour with the ordinary zinc chloride and iodine reagent used in testing for cellulose.[47, 87] As a method of softening and bleaching hard pieces of cuticle this is a valuable procedure but as a test for chitin it is not entirely reliable.[20, 56] For example, the test often fails with the cuticle of caterpillars,[87] where the chitin content is very low.

The mode of synthesis of chitin is not fully known, but there is evidence that uridine diphosphate N-acetyl-glucosamine is the immediate precursor.[139, 140, 158] There is evidence also that chitin may contain small amounts of glucosamine residues in addition to the usual N-acetyl-glucosamine.[188] It seems that that may be the case in particular regions of the cuticle. Thus fully acetylated chitin is negative to periodic acid-Schiff (PAS) staining; but in *Rhodnius* the flexible membrane of the neck and the conjunctivae of the limbs are strongly PAS-positive,[192] and the same is true of the endocuticle of the wing bases in Lepidoptera.[185]

Chitin is usually the main constituent of the endocuticle, of which it forms about 60 per cent. of the dry weight in *Periplaneta*. It is present also in the exocuticle to the extent of 22 per cent. in *Periplaneta*; but it is completely absent from the epicuticle.[20] The entire abdominal sclerites of *Periplaneta* contain from 35 to 38 per cent. of chitin,[20, 47] the wings 18 to 20 per cent.[97] An average value for the entire cuticle of a wide variety of insects is 33 per cent. of the dry weight—ranging from 25 per cent. in the hemielytra of *Cercopis* to 55 per cent. in the larva of *Calliphora*.[59] In the scales of butterflies chitin seems to be absent.[82]

Protein—The bulk of the non-chitinous material in the cuticle is protein, which varies from 25 to 37 per cent. of the dry weight.[59] The inner parts of the cuticle readily give the protein colour tests, the biuret, Millon's and xanthoproteic reactions.[20, 56, 114] The extractable protein of the cuticle has been named 'arthropodin'.[28] [See p. 49.]

Arthropodin is very similar in general character in different insects;[160] but it is certainly a mixture of several proteins.[153] The nature of the union between protein and chitin in the insect cuticle is uncertain. It seems that the many protein fractions present are bound to chitin by bonds of varying type and varying strength.[154, 155] It is probable that chitin does not occur naturally uncombined with protein.[153, 179] Arthropodin has certain constant and peculiar characters. It is highly soluble in hot water, and after precipitation in 10 per cent. trichloroacetic acid it redissolves on heating.[28] In general properties it somewhat resembles sericin or silk gelatin.[96] In the larval cuticle of *Sarcophaga* and *Sphinx* a carbohydrate-containing protein, extractable in 5 per cent. sodium hydroxide, is present in small amounts.[96]

Another protein of a very different kind is a constant component of the cuticle. This is the rubber-like protein 'resilin', in which the protein chains are bound together in a uniform three-dimensional network to form a perfect rubber. The giant molecule has no limitations as to size or shape. In a few places, as in some of the tendons of the wing, it is present in pure masses of protein; but typically, in the elastic regions of the thorax, it occurs in continuous layers (2–5μ thick) separated by thin (0·2μ) chitinous lamellae. In ordinary cuticle it is present in smaller amounts between the chitin layers.[129, 189] Resilin provides the elasticity of the thorax of flying insects,[189] the elastic recoil of the salivary pump in Hemiptera[149] and is doubtless present in many other elastic structures. It is estimated that there are about 60 amino acid residues between two junction points of the protein chains. These are cross-linked through fluorescent aromatic amino acids of a new type, di- or tri-amino compounds with their amino groups built into the peptide chains.[127] [See p. 49.]

Micellar structure of the cuticle—The cuticle has a fibrillar structure. Chitin, like cellulose, is composed of long-chain polymers held together by hydrogen bonding over part of their length to form elongated particles or micellae.[144] Three distinct forms of chitin, α, β and γ, which differ in the folding and mutual association of the polysaccharide chains, have been distinguished crystallographically.[179] The micellae are orientated in the long axis of the criss-crossing 'Balken' of beetles, which are strongly birefringent as a result and are brought out very clearly in polarized light.[11, 32] In the endocuticle of other insects the micellae tend to lie at random in planes parallel to the surface. In the larva of *Calliphora*, however, they can be orientated in any desired direction by compression or extension of the cuticle. Thus when the larva rounds up before forming the puparium there is a 12 per cent. increase in its circumference. The consequent stretching brings about a transverse orientation of the crystallites, and this in turn is reflected in the tendency for the fully formed puparium to split in this direction.[28] It is probable that in these micellae, chitin is intimately associated, perhaps chemically united, with protein;[28] it has been suggested that the molecular chains of protein are orientated at right angles to the chitin chains to form a cross-grid structure lying in planes parallel to the cuticle surface.[178] In the bristles arising from the cuticle, the micellae are orientated in the long axis;[61] these hairs show 'form birefringence' with the greatest index of refraction parallel to the axis.

Sclerotin and the hardening of the cuticle—As was clearly recognized by Odier (1823), chitin is not responsible for the horny character of the exocuticle. The hardest or most highly 'sclerotized' cuticles often contain less chitin than the soft.[120] A brown or amber substance binds the chitin micellae together like cellulose with lignin, filling the intermicellar spaces and so masking the birefringence in the exocuticle.[78] This insoluble impregnating material is tanned protein, termed 'sclerotin'. In the oötheca of the cockroach it exists in the absence of chitin.[78] In *Periplaneta* and *Blatta* the left colleterial gland secretes the protein, a polyphenol oxidase, and a β-glucoside of protocatechuic acid; the right gland secretes a β-glucosidase. When the two secretions are mixed, protochatechuic acid is liberated and the protein becomes hard and dark: the protochatechuic acid is converted by the polyphenol oxidase into the corresponding quinone which tans the protein, linking adjacent polypeptide chains, blocking the reactive amino or imino groups and converting a soft and soluble protein into a hard, tough, dark, insoluble material:[78, 138, 190]

O O | COOH | Protein A | Protein B → OH OH | NH NH | Protein A | Protein B

In the oötheca of *Locusta* the phenol precursor is the glucoside of 3,4-dihydroxyphenylacetic acid.[187] [See p. 49.]

In the cuticle the same tanning process goes forward in the mixture of chitin and protein of the exocuticle, various phenolic substances derived from tyrosine being concerned.[33, 78, 137, 145, 146] The process has been studied in most detail in the 'puparium' of *Calliphora*. About 80 per cent. of the free tyrosine in the larva enters the cuticle.[161] It is hydroxylated to dihydroxyphenylalanine ('dopa') which is decarboxylated to dihydroxyphenylethylamine ('dopamine') and then acetylated to N-acetyldopamine which is the phenolic precursor of the tanning quinone.[161, 183] In the pupating silkworm the chief phenolic precursor seems to be protocatechuic acid.[162] In *Tribolium* catechol is present.[147] [See p. 49.]

The dark colour of hard cuticles is largely due to the tanned protein;[78] but some undeaminated tryosine is probably present and is utilized for melanin formation.[96, 28] Tyrosine labelled with ^{14}C is incorporated into the cuticle of *Schistocerca*, first into the sclerotizing agents, and much later into the patches of melanin. Tryptophane metabolites do not take part in either process.[181] The mandibles of beetles which can bite through copper, silver, zinc and tin, have a maximum hardness of about '3' on Mohs' scale of mineral hardness; that is, they are slightly harder than calcite. Thus the incorporation of calcite into the cuticle (as in Crustacea) does not increase its hardness.[130] Some hard cuticles are almost colourless; it is still uncertain whether these also are the result of tanning, or whether some other method of hardening is employed.[154] [See p. 49.]

Lime in the cuticle—In a few aquatic insects lime is deposited in the cuticle. In some *Pericoma* (Psychodidae, Dipt.) larvae, when they occur in water rich in lime, this is deposited on cuticular processes on the body surface.[52] In the larva of *Sargus* and other Stratiomyids, the lime is in the form of warts laid down in shallow pits at the time of moulting, and intimately mixed with the organic substance of the cuticle. It may form about 75 per cent. of the integument and is probably derived from the Malpighian tubes (p. 572).[69] In the puparium of the cherry fly *Rhagoletis cerasi* more than half the weight of the cuticle is made up from granules of calcium carbonate in the outer part of the endocuticle.[112] In the celery fly *Acidia heraclei* lime is poured out from the Malpighian tubes (p. 560) below the surface of the puparium where it forms an inner shell.[120] In these Trypetidae the puparium, although hard, is quite white. Impregnation with lime has taken the place of phenolic tanning, and if the lime is removed with acid the cuticle becomes limp and transparent.[78] [See p. 50.]

Composition of the epicuticle: cuticulin, wax and cement—The innermost layer of the epicuticle is a refractile amber-coloured membrane composed of a substance termed 'cuticulin'. Like the sclerotin of the exocuticle, cuticulin is highly resistant to solution in mineral acids and organic solvents. It differs from sclerotin in containing a large amount of fatty material which is set free and appears in the form of oily droplets when it is heated in concentrated nitric acid saturated with potassium chlorate.[56, 114, 118] Cuticulin is believed to consist of lipoprotein, which is perhaps denatured and condensed and is finally tanned along with the other proteins of the outer layers.[118, 119]

On the surface of the cuticulin, in some insects perhaps permeating that substance, there is a layer of waxy material. This varies in character from a soft grease in the cockroach, to a wax which is pale yellow and without crystalline form in the larvae of the sawfly *Nematus* or the cabbage caterpillar *Pieris*, hard, white, and crystalline in *Tenebrio* and *Rhodnius*.[117] The wax layer ranges in thickness from 0·1 to 0·4μ.[167] Although varying widely in physical properties, these waxes are of the same general type as beeswax (p. 606).[5, 80, 117] The wax has been studied in some detail in the silkworm where it makes up 4–4·5 per cent. of the cast skin. It is considered to be a mixture of paraffins of the probable order C_{25}–C_{31} and esters of *n*-alcohols and acids, both saturated and unsaturated, of the probable order C_{26}–C_{30}.[8] In the outer part of the epicuticle in the larva of *Sarcophaga* sterols appear to be present.[23]

The lipid material extractable from the cast cuticle of the Mormon cricket *Anabrus* contains hydrocarbons (48–58 per cent.), free fatty acids (15–18 per cent.), esters (9–11 per cent.), free cholesterol (2–3 per cent.), and acidic resins (12–14 per cent.). One resinous fraction appears to be identical with shellac.[131] It has indeed been suggested that the 'cement' layer (see below) which in the cockroach is permeated by the soft grease is composed of shellac.[132] *Periplaneta* lipid is similar to that of the Mormon cricket. It is composed of 75 per cent. hydrocarbons (chiefly an unsaturated C_{27} compound) and over 8 per cent. of free aldehydes; unlike the plant waxes, this material contains no alcohols—apart from 0·7 per cent. of sterol.[152] [See p. 50.]

Covering the wax layer is the protective layer of cement. The chemical nature of this is uncertain; the suggestion that it may be identical with shellac has been mentioned; and there is some evidence that it consists of phenol-tanned protein associated with lipid in some form; that is, a material of the same general type as cuticulin.[118, 119] In some insects, such as the caterpillar *Diataraxia*, the cement layer stains readily with the fat stain Sudan B.[106] In the cockroach the soft grease permeates this layer and is freely exposed on the surface.[50] [See p. 50.]

The cuticle lining the ectodermal invaginations has the same composition as that on the surface of the body. The larger branches of the tracheae contain chitin incorporated in the non-chitinous constituents;[20] the unicellular glands are lined only by cuticulin;[114] the finer branches of the tracheae and tracheoles contain no chitin; their composition is unknown.

Permeability of the cuticle—The most important physical properties of the cuticle are its rigidity and its impermeability. Enough has been said about its structure to indicate how rigidity and capacity for stretching are varied according to the habits of the insect or the part of the body. The permeability of the cuticle to water is equally varied, and bears little relation to thickness or to rigidity. As

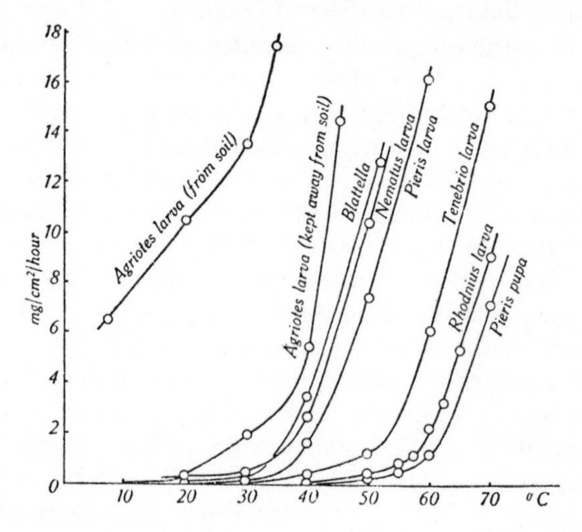

FIG. 23—Rate of transpiration from dead insects of different species in relation to temperature (*after* WIGGLESWORTH)

a rule it is almost impermeable, and it is this which makes possible the terrestrial life of such small creatures.

It is the epicuticle, and above all the wax layer of the epicuticle, which is responsible for this property;[56] and thus it is that such soft-skinned insects as the larvae of the clothes moth *Tineola* may be as resistant to desiccation as the heavily armoured forms.[65] If the wax layer is dissolved with chloroform or with suitable detergents, or if the cement and wax are abraded by fine mineral dusts, the rate of transpiration is enormously increased.[117] The rapid loss of water by insect larvae from the soil (such as *Tipula*, *Agriotes*, *Hepialus* or *Phyllopertha*) is the result of scratching of the cuticle by soil particles. If *Agriotes* is allowed to moult out of contact with the soil, its cuticle is impermeable to water like that of other insects.[117] If the insect is warmed, there is a critical temperature, characteristic of each species, above which the rate of transpiration increases very rapidly (Fig. 23). This may occur at a temperature at which the insect is still alive (about 30° C. in *Periplaneta*,[80] 37° C. in the larva of *Pieris*[117]); or it may not occur until the insect has been killed by the heat (49° C. in *Tenebrio* larva, 57° C. in *Rhodnius*, 58° C. in the pupa of *Pieris*).[117] The wax can be extracted from the cuticle and spread as a thin film upon some suitable membrane which is thus rendered waterproof. These artificial membranes show the same critical temperatures as the insects from which the waxes are obtained. [See p. 50.]

The curves shown in Fig. 23 were obtained by measuring the transpiration at constant relative humidity (0 per cent.) and variable air temperatures. A more exact measure of the critical temperature is given by carrying out the experiments at constant saturation deficiencies (that is, in air of constant drying power) and measuring the temperature at the surface of the cuticle itself. When this is done the increase in transpiration at the critical, or transition, temperature is exceedingly abrupt (Fig. 24), but the temperature at which it occurs agrees with that inferred from the results shown in Fig. 23.[133, 134]

There is good evidence that in many insects, notably in the cockroach and in various aquatic species, the most effective part of the wax in preventing water loss is an orientated 'monolayer' of the elongated lipid molecules which are held in contact with the water-containing cuticle by their polar groups. The critical temperature is some 5°–10° C. below the melting point of the wax; it is probably the temperature at which the intermolecular spacings between the wax molecules increase through molecular agitation and permit the water molecules to escape.[5, 134, 141] Such orientated monolayers have remarkable electrical properties which may well be concerned in the active transport of water and of ions to or from the body of the insect.[135]

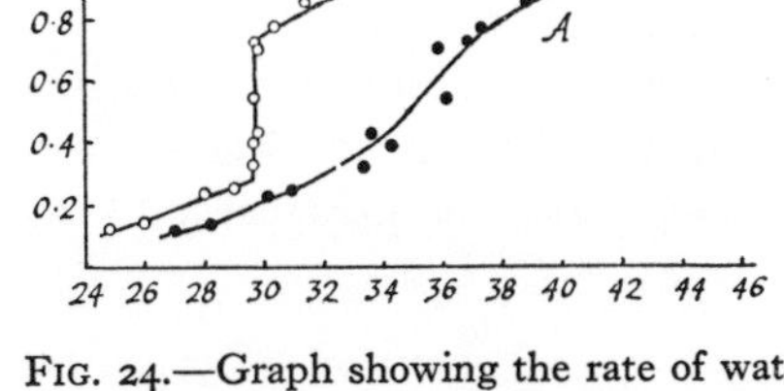

FIG. 24.—Graph showing the rate of water transpiration from a *Periplaneta* larva at constant saturation deficiency. *A*, in relation to air temperature. *B*, in relation to cuticle temperature (*after* BEAMENT)

Ordinate; loss of water in mg./mm. Hg/animal/hour. Abscissa; temperature (° C.).

In some insects, such as *Rhodnius* and *Tenebrio*, a second wax layer may form on the outer surface of the cement layer, and this leads to the appearance of a second transition point which occurs at a progressively higher temperature as the insect ages.[133]

In the cockroach,[117] in termites, and some other insects the waterproofing material can be removed by simple contact with sorptive dusts.[148] Electron diffraction studies of the pupal exuvium of *Calliphora* have provided direct evidence for the existence of a wax or lipid layer consisting of micro-crystals oriented perpendicular to the cuticle surface. When heated in the electron beam these waxes undergo irreversible disorientation corresponding with the irreversible increase in permeability to water after heating.[124] A somewhat different conception has been proposed for the waterproofing structure of the Muscid larva; this is pictured as consisting of alternating lipo-protein and protein zones, the whole being impregnated with lipids extractable by fat solvents.[43]

Both the amount of wax and its melting point may be influenced by the temperature to which the insect has been exposed. For example, the pupa of the moth *Dictyoploca japonica* under warm treatment had an average of 29·1 mg. of wax with a melting point of 60·3° C.; under cold treatment there was 12·0 mg. of wax with a melting point of 39·9° C.[163]

The tanning of the exocuticle, by reducing the hydrophilic properties of the protein, contributes to the impermeability of the cuticle, particularly if its surface is well dried.[117] The passage of water through the cuticle takes place much more readily from without inwards—a physical result of the asymmetrical structure.[6] Some insects can take up water from the air even when it is well below saturation (p. 666). In aquatic insects there are specialized regions, such as the anal papillae of mosquito larvae, where water and salts are normally absorbed.[115] The same applies to the respiratory organs of some other aquatic insects[48] and to the cuticle covering the rectal epithelium.[14, 113] In some species of Chironomid larvae the general cuticle may be permeable even to the more diffusible dyes.[3] Permeability to water in aquatic larvae is inversely proportional to the degree of development of the epicuticle.[184] Many aquatic adult insects, such as water-beetles, which leave the water on occasions, are as waterproof as the more permeable sorts of terrestrial insects, but aquatic larvae are commonly very permeable to water.[133] Indeed, in aquatic insects the wax layer is more important in securing unwettability of the cuticle than in waterproofing (p. 379),[134] and in *Podura aquatica*, which has a strongly hydrofuge cuticle, transpiration is high at all temperatures.[173] In *Sialis* larvae no evidence could be obtained of any sudden increase in transpiration at any given temperature.[186] It may be that different components in the epicuticle are responsible for its impermeability to different substances (water, oxygen, ions, &c.).[176]

Metallic colours of the cuticle—In many insects the surface of the cuticle itself, or of the flattened scales which are articulated to it, show metallic or iridescent colours. The physical basis of these colours is quite different in different

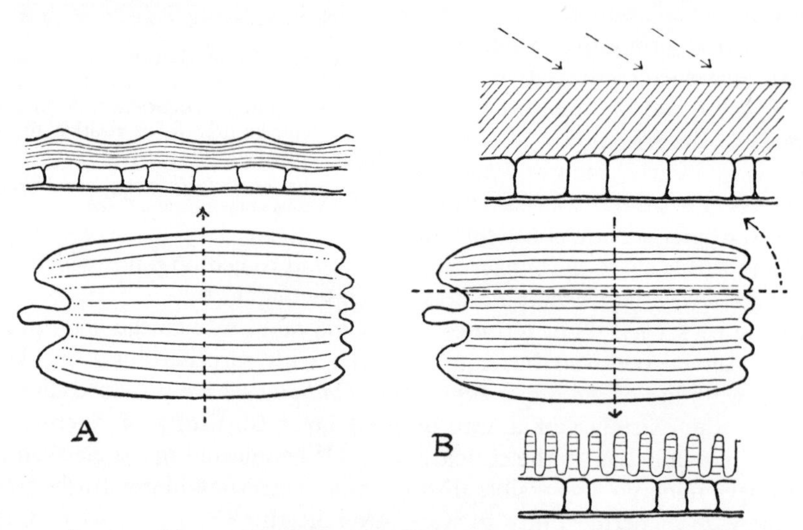

FIG. 25.—Two types of scales showing interference colours (schematic)

A, *Urania* type; transverse section above shows horizontal periodic structure in the upper lamina. *B*, *Morpho* type; longitudinal section above shows oblique periodic structure in the vertical ridges. Arrows above show direction of light to give interference. Below, transverse section showing ridges.

cases; in many it is unexplained. The metallic colours of Buprestid beetles are explained according to Walter's theory by the presence of granules of material in the surface secretion ('Sekretschicht') of the elytra; granules which, like such dyes as eosin, both absorb and reflect, very powerfully, light of a particular wave length; metallic green elytra appearing red-brown by transmitted light.[37] But in most cases the colours result from some periodic structure in the cuticle which gives rise to interference colours. In at least one case, that of the Lamellicorn *Serica*, the faint iridescence is due to diffraction by fine striae, 0·5–1·0μ deep and 1–1·5μ apart, evenly spread across the wing case; collodion impressions of the wing, bearing a cast of this structure, show the same iridescence as the wing itself.[64] In the majority of cases iridescence is due to interference by multiple thin films separated by material of slightly different refractive index.[12] Thin plates occur parallel to the surface in the Pentatomid *Calliphara*.[30] In the wing scales of Lepidoptera two main types are known: (i) The 'Urania type' occurring in Zygaenids, Papilios, Lycaenids, in which either the upper or lower lamella of the scale is much thickened and made up of numerous thin superimposed plates (Fig. 25, A).[64, 70, 95] (ii) The 'Morpho type', in which there is an oblique periodic lamination in the glass-like vertical ridges of the scales, so that the colour is visible only in the long axis of the scale and changes with the angle of vision (Fig. 25, B).[95] Visual demonstration of these structures has been made possible by the use of the electron microscope.[1, 31, 46, 165a] In the scales of the Diamond beetles *Entimus*, &c., oblique lamellae fill the interior of the scale, but are differently inclined in sharply defined areas, giving corresponding patches of colour.[64] The surface layer of the cuticle in the highly iridescent Chrysididae (Hym.) is composed of little vertical columns each made up of superimposed lamellae which are probably responsible for the colours.[29] In *Lucilia* and *Phormia*, desiccation at 100° C. causes the colour to change from metallic green to dark blue, suggesting an effect on the separation of thin plates.[59] [See p. 50.]

The periodic structures responsible for these colours presumably arises spontaneously by crystallization within the structure of the cuticle. For example, the puparium of blowflies viewed from within shows such interference colours;[78] and ordinary Lepidopterous scales show periodic projections apparently homologous to those which produce the colours in *Morpho*.[72] In the giant scales formed by polyploid epidermal cells the details of structure, the ribs, granules, &c., retain their normal size and spacing—their fine structure remains unchanged.[55] [See p. 50.]

FORMATION AND SHEDDING OF THE CUTICLE

Since the cuticle is incapable of growth and, in the more rigid parts such as the head capsule or appendages, is incapable even of being stretched, it must be shed from time to time as the insect grows, and a new and larger cuticle laid down in its place. In the next chapter we shall consider the physiological factors which regulate this process of growth, but here we may describe those aspects of the subject which bear upon the properties of the integument.

Formation of new cuticle—In the process of moulting or ecdysis the epidermal cells, which become greatly enlarged at this time,[18, 102] separate themselves from the old cuticle, the cytoplasmic processes being withdrawn, it is said, from the pore canals,[102] dispose themselves to provide the proper form for the next instar, and proceed to lay down the new integument.

They first secrete the **cuticulin layer** of the epicuticle. This appears as a delicate smooth membrane continuous over the whole surface of the body (Fig. 26, A). It then seems to expand so that it is thrown into the folds characteristic of the ensuing instar (Fig. 26, B).[102, 114] Or the extensive folding which allows for growth during the instar may be secured by some cells being long and others short so as to form tongue-like projections or false villi.[18] Immediately before the deposition of the cuticulin layer the oenocytes become enormously enlarged and lobulated (p. 449); they are composed largely of lipoprotein with the same stain-

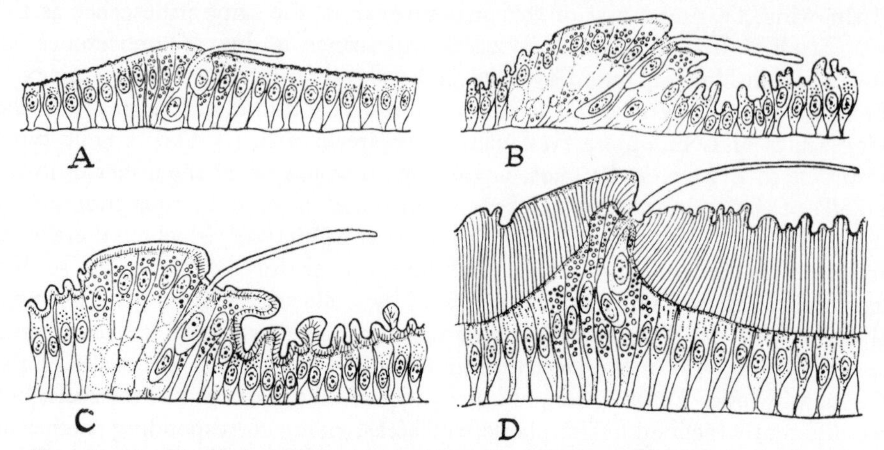

FIG. 26.—Longitudinal sections showing formation of new cuticle during moulting in 4th-stage larva of *Rhodnius* (*after* WIGGLESWORTH)

A, 10 days after feeding; epicuticle just laid down. *B*, 11 days; epicuticle thrown into folds. *C*, 12 days; endocuticle being laid down. *D*, 15 days; just before moulting.

ing reactions as the newly formed cuticulin, and as the cuticulin is laid down they rapidly decrease in size.[114, 118, 119] It seems probable that the oenocytes are concerned in the synthesis of the lipoprotein. In the larva of *Calliphora* the epicuticle arises by the coalescence of droplets aligned on the cell surface to form a continuous folded lipoprotein layer, and the inner epicuticle of cuticulin is added below this. Droplets with similar staining properties pass from the oenocytes through cytoplasmic filaments to the epidermal cells, and the same cycle of secretion is seen in the oenocytes during puparium formation.[196] [See p. 51.]

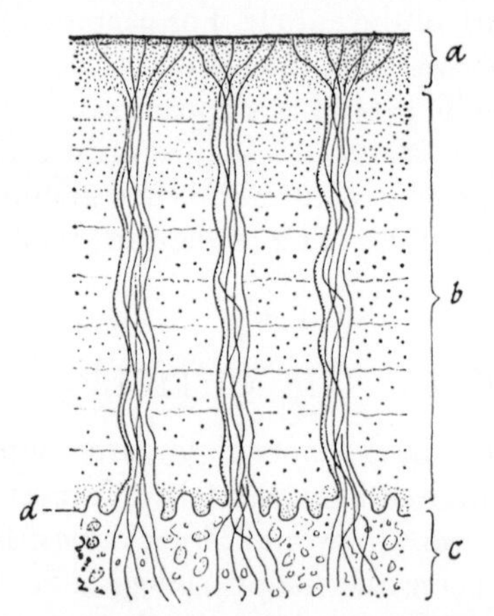

FIG. 27.—Semi-schematic figure showing the canals in the cuticle extending from the plasma membrane (*d*), through the exo- and endocuticle (*b*), to the inner limit of the epicuticle (*a*): and the wax canal filaments, collected within the pore canals, and extending from the cytoplasm (*c*) to the outer surface of the epicuticle (*after* LOCKE)

The subsequent changes in the epicuticle take place while the chitinous layers below are being deposited. The cuticulin layer is traversed by filamentous extensions from the pore canals (Fig. 27). Minute silver staining droplets, perhaps polyphenols bound to protein, [117, 118, 119] perhaps protein with reducing properties from some other cause,[146] are discharged from the tips of the canals and gradually fuse to form a more or less continuous layer termed the **polyphenol layer.** [See p. 51.]

Within a few hours before moulting the **wax layer** is laid down. The wax is likewise secreted by way of the pore canals and crystallizes on the surface, so that the insect is rendered more or less waterproof before moulting. The pore canals contain bundles of exceedingly fine filaments, so-called 'wax canal filaments' (Fig. 27) about 80Å thick, which are visible only with the electron microscope. At the outer limit of the pore canal just below the epicuticle the wax canal filaments spread out fanwise and pass right through the epicuticle. At their inner extremities they can be seen in whorled masses in the epidermal cells. They are particularly numerous in the cells of the wax gland of the honey-bee and in the wax-secreting areas of the cuticle of the Hesperid caterpillar *Calpodes*. At their outer extremities they contain esterase. They are probably concerned in the transmission of wax or wax precursors from the epidermal cells to the surface of the cuticle.[166] [See p. 51.]

Some evidence exists that waxes may be solubilized and secreted in association with protein; but in *Periplaneta* the soft wax (m.p. 34° C.) secreted throughout life on the surface of the cuticle becomes converted to a hard white wax (m.p. 55–60° C.) on exposure to the air. This may be the result of evaporation of a volatile solvent[132] or possibly of polymerization of some sort.[152] [See p. 51.]

The **cement layer** is discharged from the dermal glands and poured over the surface of the wax just before or just after moulting. At this time the dermal glands are distended with secretion; an hour or so later they are empty and the wax is covered.[118, 119] In most insects these cement-producing dermal glands are scattered all over the surface. In *Tenebrio* a single type containing silver reducing secretion is present.[119] In *Rhodnius* there are two types (recalling the dual origin of the oöthecal sclerotin in the cockroach). In caterpillars the secretion comes from the large paired segmental glands of Verson.[106] In the larvae of various Lepidoptera and Hymenoptera there is a gland in the main trunk of each trachea.[84] In Diptera no dermal glands have been reported. In the cockroach there is a thick cement layer, but the waterproofing grease permeates it and is freely exposed.[50]

While the formation of the epicuticle is being completed the **procuticle** is being laid down (Fig. 26, B, C). The precise way in which this is achieved has excited much controversy. It doubtless varies in different insects, but a very common mechanism is that originally proposed by Leydig and accepted by many later authors,[9, 15, 36, 76, 114] according to which the cuticular substance is secreted around the filiform outgrowths from the epidermal cells which later constitute the pore canals. There is evidence that these vertical filaments extend into the substance of the epidermal cells;[56, 114] there is therefore very little difference from a physiological point of view between the secretion of cuticular material outside the cell and a transformation of cellular substance into cuticle.[11, 104] Doubtless both processes occur. Where the extracellular secretion is discharged in fluid form, there will be no indication of cell boundaries in the resulting cuticle, as in *Rhodnius*;[114] where the transformation takes place within the body of the cell, the cuticle will bear this impress and will be divided into little columns, as in the exocuticle of the elytra in many beetles.[56] In the Collembolan *Podura* electron-microscopic studies have shown quite clearly that the cuticle is formed extracellularly, that is, beyond the limits of the plasma membrane.[173]

The question whether the contents of the pore canals are later converted into cuticle has already been discussed (p. 30). It is certain that in many insects with fully hardened cuticles cytoplasmic filaments from the cell still extend right up to the wax layer. If the cement and wax are removed by very gentle abrasion the cells below the cuticle react as though they had been wounded and they secrete more wax, through the substance of the cuticle, to repair the injury.[117] In such insects the cuticle must be regarded as 'alive' almost to the surface.[120] In fact, it seems more reasonable to regard the surface of the cuticulin layer, rather than the junction between the endocuticle and the main body of the cell, as the cell boundary.

A clear example of the formation of cuticle by the transformation of cell substance is seen in the formation of hairs and bristles. The microtrichia which are common on the epicuticle arise as little filaments from the surface of the epidermal cells.[100] Each large bristle arises by the outgrowth of a long cytoplasmic process from a large 'trichogen' cell. This process passes through an annular 'tormogen' cell that gives rise to the socket in which the bristle is articulated (Fig. 28).[34, 42, 62, 114] In the later stages of its development chitin and sclerotin are laid down in the substance of this outgrowth, the form of which is very largely influenced by the fine structure and properties of the materials of which it is composed.[61] [See p. 51.]

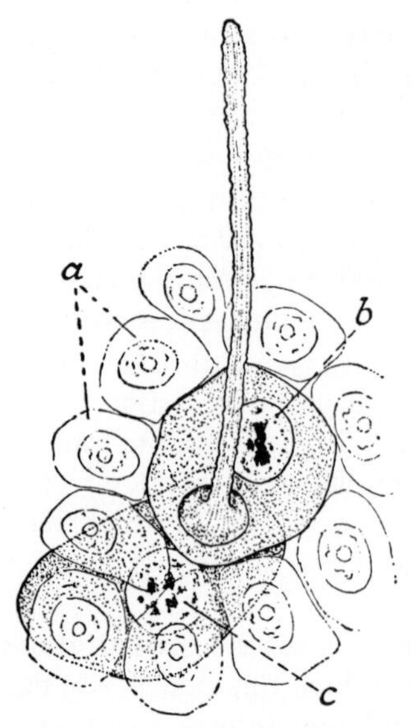

FIG. 28.—Early stage in formation of bristle during moulting of *Rhodnius* (*after* WIGGLESWORTH)

a, epidermal cells seen in surface view; **b**, tormogen or socket-forming cell; **c**, trichogen or hair-forming cell with process invested by the tormogen (*cf*. Fig. 26).

Very elaborate changes take place in the formation of scales, although in principle the mechanism of development is the same.[81, 94] The scale is the complicated cell wall of an exceedingly enlarged epithelial cell. The formative cell gives out a process which becomes club-shaped, and then flattens. Finally the plasma breaks down and air penetrates through openings in the cuticular wall into the interior of the scale.[53]

A rather special problem is presented by the criss-crossing strands or 'Balken' in the endocuticle of beetles. It is uncertain whether the adjacent epidermal cells co-operate in moulding them;[44] or whether they arise by the action of external stresses upon a homogeneous secretion from the cells.[11] It is worth noting that an orientation of this kind, on a submicroscopic scale, occurs in the chitinous cocoon of *Donacia* which is formed in the absence of any cells.[73]

In grasshoppers, cockroaches, dragonflies and other insects the successive laminae of the procuticle represent the amount of substance laid down in each succeeding 24 hours.[172] The cross-linking, or tanning to form the exocuticle is a secondary process that takes place after moulting (p. 34); but the cross-linking of the rubber-like cuticle 'resilin' (p. 33) occurs continuously as it is being laid down. At moulting in the locust there is a 6-hour pause in chitin formation, but resilin formation continues without interruption.[172] This may contribute to the boundary between layers deposited before and after moulting.[114] [See p. 51.]

The exocuticle alone of the inner (procuticle) layers is laid down before moulting; most of the cuticle is formed during the next few days (Fig. 34). Its deposition may be greatly influenced by nutrition;[120] in *Rhodnius*, for example, it varied from 8μ to 20μ with the amount of food taken.[116] It is at this time that the transfer of reserve glycogen[71, 118] and protein[118] from the fat body to the epidermis takes place. As the endocuticle is formed the reserves disappear.

Moulting fluid—When the epidermal cells separate from the old cuticle and begin to secrete the new, the space between the two cuticles is occupied by a thin plasma. In the later stages of moulting this space is filled by an abundant fluid, the moulting or ecdysial fluid, first clearly demonstrated by Newport. There can be little doubt that this fluid, which extends also throughout the tracheal system, arises mainly by exudation from the epidermal cells.[13, 18, 41, 75, 76, 88, 122] To what extent the dermal glands contribute to its secretion at an earlier stage in moulting is uncertain; their primary function, as we have seen, is the production of the cement layer. An occasional source of the moulting fluid is the secretion from the Malpighian tubes (p. 576). In the silkworm, the fluid contains crystals of oxalates and urates which come from these tubes and pass under the discarded cuticle of the rectum to reach the surface of the body.[75] These crystals are excluded if a ligature is tied round the anus,[90] an experiment which disproves the contention that during moulting the Verson glands assume an excretory function.[103]

The chief function of the moulting fluid is to digest and dissolve the inner layers of the old cuticle.[75, 102] It is a neutral salt-free fluid with proteins in solution which contains a protease and a chitinase.[114] During the early stages of moulting in the pupa of the Cecropia silkworm the exuvial space is filled with a transparent protein gel. In the later stages this is converted into a copious sol rich in proteinase and chitinase.[125] In *Podura* (Collembola), as studied with the electron microscope, the old cuticle is lifted away from the cells by a foam-like secretion, followed by electron-dense granules which contain the precursors of the chitinase and protease that become active only when the epicuticular layer is complete.[173] [See p. 52.]

Chitinase has been demonstrated in the moulting fluid of the silkworm,[156] which contains N-acetylglucosamine, glucosamine, and most of the natural amino acids,[197] and it probably occurs in all insects.[159] Complete hydrolysis of chitin is effected by two enzymes acting successively: a polysaccharidase ('chitinase') and an oligosaccharidase ('chitobiase').[159] Although most of the chitinase activity in *Periplaneta* is contained in the integument, the enzyme is distributed throughout the body, including the gut and haemolymph.[188] During digestion of the endocuticle and of the corresponding layer of the tracheae the cuticle becomes PAS-positive (cf. p. 32).[192]

The moulting fluid attacks only the endocuticle, which as a rule is completely broken down; the exocuticle is unaffected and so are the delicate epicuticular linings of the dermal glands.[114] By this process 86 per cent. of the abdominal cuticle in *Rhodnius* is absorbed and what is finally shed is a thin membrane consisting of little but the epicuticle (Fig. 29);[114] 83–85 per cent. is absorbed in the larva of *Tenebrio*;[25] 80–90 per cent. in the silkworm.[8, 58] As was suggested originally in *Leptinotarsa*,[102] the enzymes are probably secreted by the epidermal cells.[118] If a partial localized pupation is induced experimentally in caterpillars (p. 69), solution of the endocuticle by the exuvial fluid takes place, although no dermal glands are present;[54] and implanted pieces of integument which moult in the abdomen of caterpillars likewise show digestion and absorption of the endocuticle.[74] That the epidermis can dissolve the endocuticle is seen in the ptilinum of Muscid flies. When these emerge the endocuticle of the ptilinum is as thick as in other parts of the head capsule; but 14 days later nothing remains of it but the epicuticle with its spicules (Fig. 30).[60]

It has been commonly believed that the primary function of the moulting

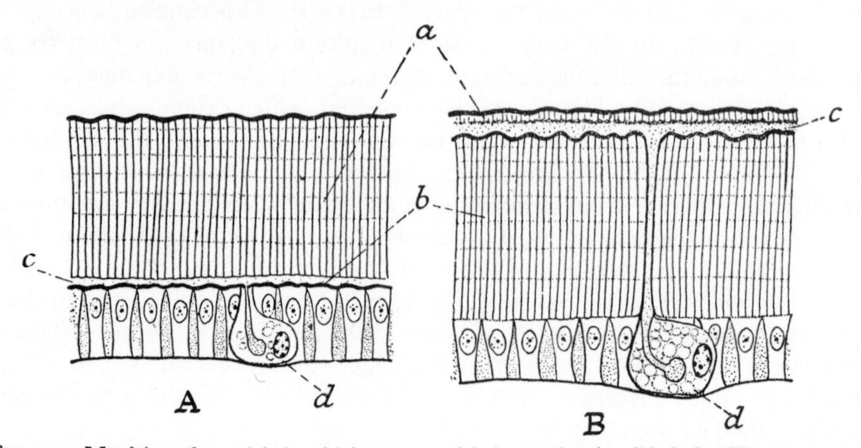

FIG. 29.—Moulting of a cuticle in which an exocuticle is wanting (*modified after* WIGGLESWORTH)

A, new epicuticle formed; digestion of old endocuticle scarcely begun. *B*, digestion and absorption of old endocuticle almost complete. *a*, old cuticle; *b*, new cuticle; *c*, moulting fluid; *d*, dermal gland.

fluid is to serve as a lubricant when the insect slides out of its skin. This moist layer will certainly keep the old cuticle soft and supple;[13] but by the time it is cast off almost all the moulting fluid has disappeared; the surface of the new skin is dry, the old cuticle very nearly so. This cannot, therefore, be regarded as its most important function.

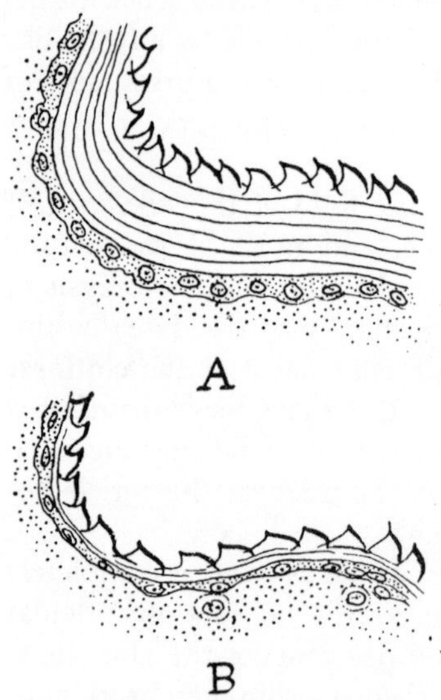

FIG. 30.—Section through ptilinum of *Calliphora*

A, newly emerged imago; *B*, 3 days after emergence, showing absorption of the inner layers of the cuticle (*after* LAING).

Almost all the moulting fluid, and the products of digestion from the cuticle, are in fact absorbed into the body of the insect before moulting takes place. Sugars labelled with ^{14}C and incorporated into the larval cuticle of *Hyalophora cecropia* appear to be specifically re-used for the synthesis of the pupal cuticle.[128] In the aquatic pupae of Nematocera and the nymphs of May-flies a film of air appears beneath the cuticle just before emergence,[22] probably as the result of absorption of the moulting fluid. In the silkworm much of the fluid is swallowed by the mouth;[105] but in most insects it is probably taken up through the general surface of the body. As can be proved by injecting vital dyes between the two cuticles, that is certainly so in *Rhodnius*.[114] Evidently the cuticle is still permeable to water at this stage; indeed, if the old cuticle is stripped off, the insect rapidly dries up; and if adult flies, *Calliphora*, are removed from the pupa one day before emergence and immersed in water they swell by osmosis and the wings expand.[26] It is only during the last hour or two before the old skin is shed that the wax is secreted and the cuticle made waterproof. At the moulting of the adult *Calliphora* the water in the moulting fluid is absorbed, leaving on the surface of the cuticle a denatured, hydrophobic, protein-lipid film which restricts water loss until waxy materials appear on the surface during the hardening process.[196] [See p. 52.]

In some insects, notably in Lepidoptera,[125, 168, 175] a so-called 'ecdysial membrane' appears between the old cuticle and the new. The nature and origin of this membrane, which resists digestion by the moulting fluid, is uncertain; perhaps the most likely suggestion is that it represents the innermost sheet of the old endocuticle modified by leakage of sclerotizing compounds from the new cuticle.[170, 175] [See p. 52.]

Mechanism of moulting—It is often claimed that moulting is a process of excretion which serves to eliminate waste products. But the greater part of the cuticle is reabsorbed into the body; only the indigestible sclerotin and cuticulin are shed. The digestion of the endocuticle, however, has another purpose besides the conservation of its substance. For in the young stages of all insects there is an 'ecdysial line' or 'Häutungsnaht' along which the cuticle splits most readily at moulting. It is usually a T-shaped line on head and thorax[24, 39] which appears white in pigmented insects because here the exocuticle is wanting and the endocuticle extends up to the epicuticle (Fig. 31). Consequently, when the endocuticle has been dissolved the ecdysial line constitutes a line of weakness where the slightest pressure will cause the cuticle to rupture.

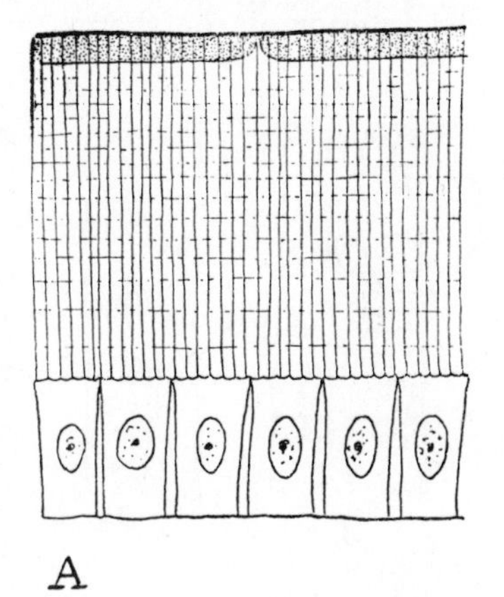

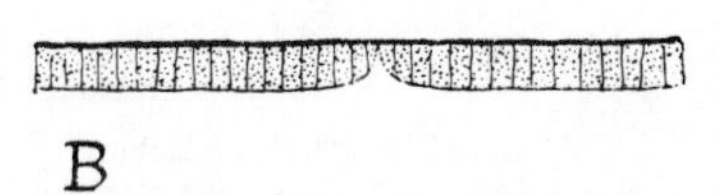

FIG. 31
A, transverse section of cuticle on dorsum of thorax of *Tenebrio* larva, showing the line of weakness or ecdysial line in the exocuticle (*after* PLOTNIKOW); *B*, the same after digestion of the endocuticle.

The mechanism by which the insect escapes from its cuticle and expands its wings and body has often been described. A very full and clear description of the process during the final moult of an Agrionid dragon-fly, which illustrates the principles common to most insects, is given by Brocher.[16] Like the insect hatching from the egg, the moulting insect contracts the abdomen and drives the body fluid into the head and thorax. It commonly assists this process, as was first shown by Jousset de Bellesme, by swallowing air[49] or, in aquatic insects, by swallowing water; if the crop of the cockroach is opened to the air the insect cannot moult, though it may remain alive a week or more.[24] The pressure so created splits the cuticle along the ecdysial line and the insect slowly draws itself out, often aided by gravity—for many insects hang head downwards while they moult. The linings of the ducts and glands[75] and of the tracheal branches down to a diameter of 1–2μ[13] come away with the old skin (p. 364).

The dorso-ventral and intersegmental muscles of the abdomen provide the pressure needed for moulting. In *Rhodnius* the intersegmental muscles are fully developed only at the time of moulting or hatching from the egg. Within 3 or 4 days after moulting the fibrils have disappeared; only the nuclei and a little cytoplasm persist within the innervated muscle sheaths and serve to rebuild the muscle at the time of the next moult (Fig. 32).[194] Similar cycles of muscle development and involution probably occur in many insects; for example, in Acrididae.[191]

In Saturniid moths, the longitudinal intersegmental muscles degenerate at metamorphosis in most segments of the body.[150] [See p. 52.]

Sometimes there are special structures to aid the splitting of the skin, analogous to those concerned in hatching from the egg: Acridiids have an extrusible bladder in the neck;[57] the aquatic Hemipteron *Limnotrechus* has a similar blood-filled vesicle which is extruded beneath the scutellum.[24] Mosquitoes have a row of rigid hairs on the mesothorax.[17] The most elaborate is the ptilinum of Muscid flies; this consists of a bladder attached to the margins of the frontal cleft, everted by the blood pressure, as suggested by Réaumur (1738), and retracted by a special temporary musculature.[60] Pupae of *Glossina* kept in dry surroundings lose too much water and are unable to escape from the puparium unless the top is cracked.[66] The evagination of the head from the interior of the thorax in the pupa of Cyclorrhaphous flies is similarly

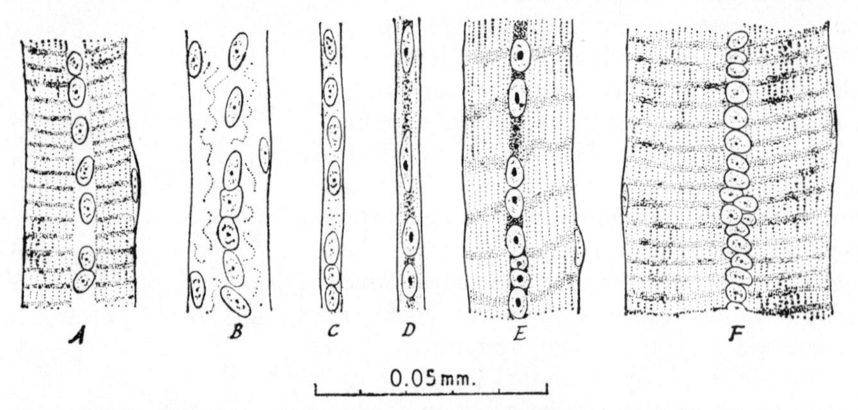

FIG. 32.—The cycle of involution and new formation of the ventral abdominal muscles in the 1st-stage larva of *Rhodnius* (*after* WIGGLESWORTH)

A, immediately after hatching. *B*, 4 days after hatching. *C*, immediately after feeding (14 days after hatching). *D*, 4 days after feeding, new muscle fibrils beginning to appear. *E*, 7 days after feeding. *F*, 11 days after feeding (newly moulted 2nd-stage larva).

effected by contractions of the abdominal muscles pressing the body fluids forwards.[27, 110] In adult *Calliphora* the muscles concerned in maintaining pressure degenerate within a few days after emergence.[143]

Expansion of new cuticle—Escaped from the old skin, the insect, whose new cuticle is still soft and pliant, swallows air (or water) vigorously again and increases its bulk like the insect newly hatched from the egg. The muscles of its body wall remain in a state of continuous contraction so that the pressure of the blood, which is normally not much removed from atmospheric pressure, is kept at a high level. In the newly moulted imago of *Aeschna* it is kept at about 75 mm. of water for several hours[89] until the cuticle is hard. If at this stage the gut is punctured, the insect collapses like a pricked balloon.[24] The pressure so maintained serves to expand the wings and other parts that were crumpled under the old skin. If the wings are punctured blood drips from them and they cannot expand.[16] Nor will they expand if the insect is narcotized with ether; though they will do so if the pressure is artificially increased by compression with the finger.[16] Perhaps the eating of the cast skin, a common habit among insects, helps to maintain the increased bulk.[24] Fraenkel[26] has described some ingenious experiments which demonstrate this mechanism of wing expansion. The blow-fly

Calliphora increases its volume 128 per cent. by swallowing air. The hydrostatic pressure of the blood is raised to 60–120 mm. Hg.[143] If it is allowed to swallow a little air and is then placed in carbon dioxide, this diffuses in with great rapidity (see p. 375) and raises the gaseous pressure so that the wings expand in a few seconds. The same instantaneous expansion takes place if the fly, after swallowing a little air, is exposed to a vacuum of 200 mm. of mercury.[26] The quantity of fluid in these insects is also important: the size of the body and wings in *Lucilia* is largely controlled by the volume of the blood.[67]

Hardening and darkening of cuticle—The cuticle of the newly moulted insect is generally colourless, always quite soft. During the next hour or so it hardens and darkens. This is no simple effect of contact with the air, for it fails to occur if the insect is killed immediately after moulting;[45] and if a part of the new cuticle is exposed by removal of the old cuticle 24 or 48 hours before moulting, although it appears fully formed it neither hardens nor darkens until moulting has taken place.[114] It has been suggested that the neurosecretory cells in the central nervous system may be responsible for the hormonal stimulus inducing these changes.[193] In *Calliphora*, expansion of the wings is generally complete in 20 to 30 minutes after emergence, darkening and hardening in 1 to 2 hours. But flies forced to keep digging through the soil for as long as 7 hours still remain pale and soft. Inflation, hardening and darkening occur as usual when they become free. Clearly these chemical processes are initiated by some nervous mechanism.[26] The brain is concerned in the release to the blood of some peptide hormone which acts on the epidermal cells and leads to the normal darkening and hardening of the cuticle.[142] The source of this hormone appears to be the median neurosecretory cells of the pars intercerebralis of the brain (p. 198).[151] As studied in different species of Diptera, hardening and darkening appear to be separate processes.[169] [See p. 52.]

Hardening and darkening result, as we have seen, from the tanning of the proteins to form sclerotin and cuticulin. The phenolic substances concerned are derived from tyrosine.[137] In the larva of *Sarcophaga* tyrosine accumulates in the cuticle[23] and diminishes in the blood[28, 96] during the hardening of the puparium. The phenolic derivatives not only form the polyphenol layer on the surface of the cuticulin (p. 40) but permeate the exocuticular parts which will become sclerotized.[78, 119] In the presence of oxygen, which is necessary for the hardening of the cuticle, and the enzyme polyphenol oxidase, which is plentiful in the cuticle of arthropods,[10] the quinones which tan the proteins are produced. These changes may occur in an epicuticle separated from the epidermal cells by a considerable thickness of endocuticle. It is doubtless one of the functions of the pore canals to enable the cells to exert this action at a distance.[114] In rubber-like cuticle (resilin) in which tanning does not occur, pore canals are absent.[172] In the larva of *Sarcophaga*, the polyphenol oxidase is concentrated in the inner part of the epicuticle; hardening of the endocuticle takes place from without inwards. In this insect the inner part of the endocuticle contains reducing substances which inhibit the formation of quinones and so prevent hardening.[23] The simultaneous darkening of the cuticle is due largely to the tanned protein,[78] partly to associated melanin.[28, 96] In addition to sclerotization it may be that the orientation and close packing of micelles is an essential factor in hardening—as it is in the hardening of silk and cellulose.[165]

Loss of water during moulting—Since the water-proofing wax is not laid down until shortly before the old skin is cast, the question arises whether moulting involves a serious loss of water. What data exist suggest that it does not. The dragon-fly *Aeschna* may show a 50 per cent. loss of weight during the

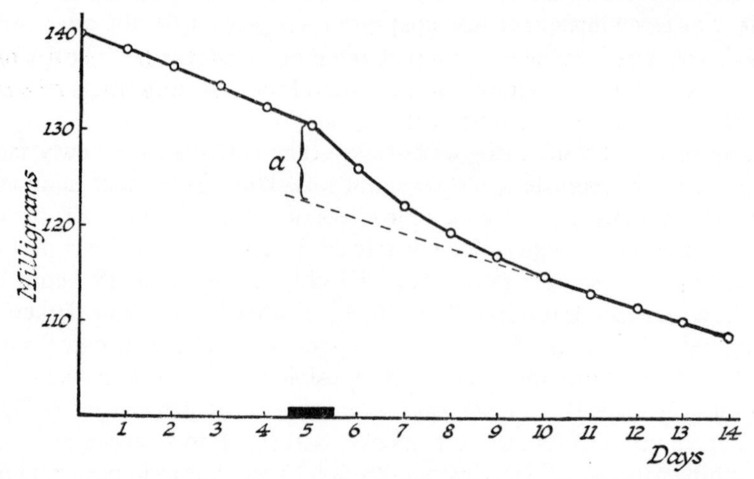

FIG. 33.—Loss of water during the moulting of *Rhodnius* 5th instar larva (*after* WIGGLESWORTH and GILLETT)

Ordinate: body weight. Abscissa: days. The thickening of the base line indicates the day of moulting. **a**, shows the extra loss of water due to moulting. (The dry weight of the cast cuticle has been added to the body weight after moulting.)

first 24 hours after emergence;[89] but dragon-flies are known to discharge from the anus many clear drops of water as soon as expansion is complete;[16, 35] they probably eliminate water swallowed by the nymph before moulting. In the meal-

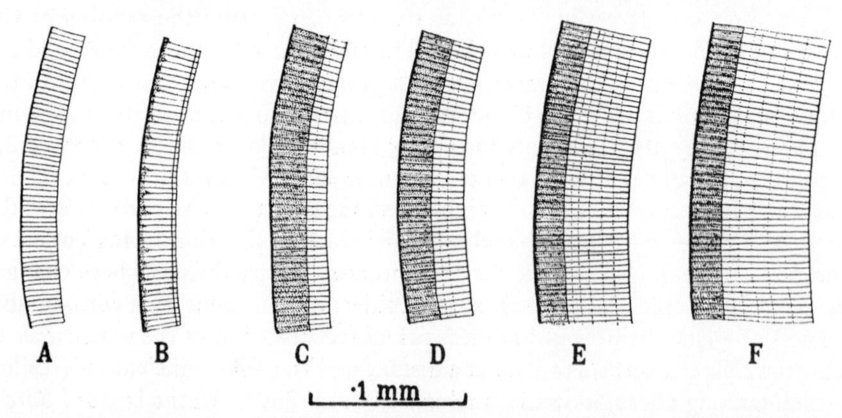

FIG. 34.—Deposition of cuticle after moulting. Transverse sections of cuticle of head of adult *Rhodnius*

Shading indicates sclerotin. *A*, immediately after moulting; *B*, 24 hours after moulting; *C*, 2 days; *D*, 4 days; *E*, 10 days; *F*, 1 month after moulting (*after* WIGGLESWORTH and GILLETT).

worm the conservation of water is so complete that moulting makes no measurable difference to the rate of evaporation;[19] in this insect there is no sign of any moulting fluid when the skin is cast.[122] In the *Tenebrio* adult the loss of water by transpiration during the first day after moulting is about four to six times the normal. This agrees with the observations which show that the polyphenol is almost but not quite covered by the wax when moulting takes place.[119] A similar increased loss occurs in *Carausius.*[91] In *Rhodnius*, weighing about 112 mg., the water lost with the old skin is about 1·1 mg., the skin itself about 1·8 mg. During the next 48 hours, especially during the first 12 hours after moulting, the rate of water loss is increased above the normal; but the total extra loss associated with moulting is equivalent to no more than the normal insect loses in about 4 days (Fig. 33).[121] On the other hand, at the time of emergence from the pupa the skin of the beetle *Popillia* is quite moist and the insect loses at this time about one-third of its weight.[63]

Thus the impermeability of the cuticle to water in these terrestrial insects is almost fully established by the time the skin is cast. But in some aquatic insects it seems to remain permeable longer. For the increase in volume of aquatic Hemiptera after hatching can be prevented by immersion in 0·8 per cent. sodium chloride, and this procedure will even restore to its initial volume a nymph hatching into pure water;[76] and besides the water that is swallowed after ecdysis in *Notonecta*, *Gerris*, &c., much seems to be taken up through the cuticle by osmosis.[98]

There is no increased evaporation before moulting, although the old cuticle may be becoming excessively thin. This demonstrates again that it is the outermost layers of the cuticle which are responsible for its impermeability. The same can be shown after moulting. For although the impermeability of the new cuticle is established almost at once, nearly the whole of the endocuticle is yet to be laid down. In *Rhodnius* this process requires at least 3 weeks (Fig. 34);[121] and in the elytra of Carabids the laminated structure is not complete 14 days after emergence.[92]

ADDENDA TO CHAPTER II

p. 29. **Plasticization of cuticle**—The temporary increase in plasticity of the cuticle of *Rhodnius* during feeding seems to be brought about by neurosecretory axons ending in the abdominal wall.[223]

p. 33. **Chitin and protein**—The microfibrils of chitin 25–65 Å in diameter are embedded in a homogeneous organic cement composed of cross-linked protein to give a tough material analogous to 'fibre glass'.[226, 239]

Resilin is an essential component in the cibarial pump of the tsetse-fly and other Diptera.[232]

p. 34. **Sclerotin and cuticle hardening**—The protein contained in the left colleterial gland secretion of *Periplaneta* has a high content of tyrosine, which may well be involved in the formation of the cross-links.[230]

An inactive precursor of the enzyme phenoloxidase is present in the haemolymph of *Calliphora* larvae; and a special activation enzyme is located in the cuticle.[234] In *Calpodes*, *Sarcophaga*, &c., the epicuticle appears to contain a prophenolase which is activated on injury to cause hardening and darkening of abrasions.[217]

p. 34. **Other methods of cuticle hardening**—Cystine occurs in all arthropod cuticles and disulphide bonds (as in keratin) probably afford a further mechanism for stabilizing protein, notably in *Machilis* where no evidence of diphenols can be obtained,[216] in *Smynthurus*,[216] and in the mesocuticle of *Lucilia* larvae.[212]

Stabilized lipid (cuticulin) combined with protein serves to stiffen the cuticle before sclerotization occurs. It seems always to be present in exocuticle that will later be tanned to form sclerotin, and in the untanned 'mesocuticle', including the tracheal taenidia.[241] Hydrogen bond formation, involving both protein and chitin, is another important factor in the consolidation of the cuticle.[219]

In the hard cuticle of *Schistocerca* and other insects much of the sclerotized protein appears to be cross-linked through the β-carbon atom in the aliphatic side-chain of dopamine or some metabolite of dopamine, such as acetyl dopamine, thus:

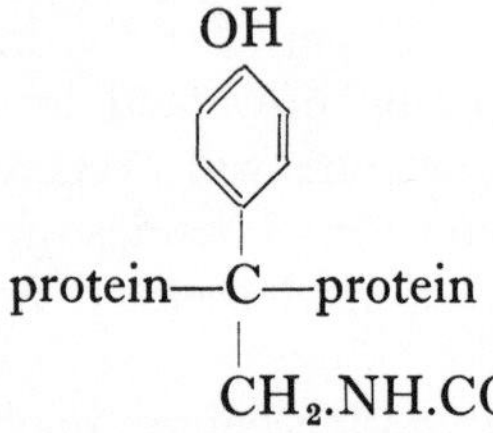

OH

protein—C—protein

$CH_2.NH.COCH_3$

This product is pale coloured and gives rise to ketocatechol on acid hydrolysis; it has been found in all insect species studied, but not in other arthropods.[198]

p. 35. **Lime in cuticle**—In the puparium of *Musca autumnalis*, also, calcification takes the place of tanning; like tanning it is controlled by ecdysone (p. 71).[211]

p. 35. **Extractable lipids**—It would seem that the soft cuticular waxes and greases of insects have a high content of hydrocarbons and a low content of long-chain alcohols. In *Tenebrio*, with a wax of intermediate hardness the hydrocarbons are all saturated and make up 10 per cent. of the cuticle wax; diols (notably pentacosane-8,9-diol) comprise 55 per cent.; free fatty acids 5 per cent.[204] In the caterpillar of the silkmoth *Samia cynthia* the white crystalline powder that accumulates on the surface of the cuticle is composed almost wholly of the straight chain saturated alcohol *n*-triacontanol ($C_{30}H_{62}O$).[203]

p. 36. **Cement layer**—Among termites, the species from dry environments have a relatively thick cement layer overlying the cuticle, whereas in species from damp environments this layer is very thin and delicate. It may be a significant factor in preserving the integrity of the water-proofing wax.[206]

p. 37. **Transpiration**—The fire-brat *Thermobia*, like the cockroach, has a mobile epicuticular grease as a water-proofing layer with a critical temperature just below 30° C. and well below the optimum temperature of this insect.[200]

p. 39. **Metallic colours**—Iridescence due to diffraction gratings is widely spread in Scarabaeidae and in some Carabidae;[213] and the brilliant colours produced by the stridulating files of Mutillid wasps result from diffraction of light by the ridges of the files.[213]

The pupa of *Danaus* has metallic gold spots. There are about four times as many lamellae per unit thickness at a gold spot as there are in the non-metallic areas of cuticle—where the lamellae are too widely spaced to produce colour by interference.[236]

In certain Scarabaeid beetles the metallic colours of the elytra are associated with the property of reflecting circularly polarized light in a manner analogous to cholesteric liquid crystals, but here determined by the helicoidal pitch of the cuticular microfibrils as their orientation changes systematically in successive microlamellae.[228] In *Plusiotis* and some other genera the incorporation of uric acid into the reflecting surface enhances the effect.[205]

p. 40. **Formation of epicuticle**—The first layer of the epicuticle to be laid down is a characteristic triple membrane, some 105 Å in total thickness, which first makes its appearance as a series of discrete plaques forming over the individual microvilli of the epidermal cells and eventually fusing to give a continuous layer.[207, 209, 220] This is the 'outer epicuticle', conspicuous in electron microscope sections, which is exceedingly resistant to breakdown by organic solvents and mineral acids, sometimes called the 'paraffin-containing' layer.[146] Some authors apply the term 'cuticulin' to this layer alone;[220, 233] and use the term 'protein layer'[220] for the main or inner layer of the epicuticle (originally called the 'cuticulin layer'[118]) which likewise is rich in bound lipid.[241] There are, indeed, substantial amounts of bound lipid throughout the substance of the cuticle; the material tanned to form sclerotin may well be a lipoprotein complex.[146, 241]

p. 40. **So-called 'polyphenol layer'**—As has been pointed out,[146, 242] the silver-binding material poured out from the tips of the pore canals is not the dihydroxy-phenol responsible for hardening. The nature of this material is still uncertain; it may be a protein or lipoprotein 'carrier' for the exuded wax.

p. 41. **Wax secretion**—The tergal wax glands in *Dacus* show similar characters to the wax-secreting epithelium of *Apis*, with bundles of hollow wax filaments converging on a perforated cuticular end organ.[208]

The major component of the cuticle grease in *Periplaneta* is the unsaturated hydrocarbon *cis-cis*-6,9-heptacosadiene;[201] on exposure to air it undergoes autoxidation to give stearal and stearic acid and perhaps polymers formed by free radicals. Free protocatechuic acid in the cuticle (the precursor of the *o*-quinone involved in tanning) serves also as an antioxidant which normally prevents the autoxidation of the cuticle lipid.[199]

p. 42. **Formation of bristles and scales**—In the formation of both bristles and scales, bundles of fibrils about 60 Å in diameter are arranged at regular intervals below the surface of the outgrowing process; and longitudinally oriented microtubules are distributed throughout its substance. These structures are believed to play a major role in establishing the form of the outgrowth.[221, 229, 231] They are rich in bound or structural lipid ('cuticulin').[242]

p. 42. **Lamination of cuticle**—In the locust *Schistocerca* during the day the cells lay down 'lamellate cuticle', in which the direction of orientation of the chitin fibrils changes every half hour or so to give rise to successive microlamellae. Whereas during the night the chitin fibrils retain a constant orientation so that the resulting cuticle is non-lamellate.[227] The apparent spiralization of the pore canals (see p. 31) is often a direct consequence of the laminar structure of cuticle: the canals are often flattened into ribbons by neighbouring microfibrils; these ribbons run straight when traversing layers with constant microfibril orientation, but rotate in phase to form twisted ribbons as they traverse lamellate layers in which the microfibrils change direction in each lamella.[220, 227]

In cuticle of the plywood type (Fig. 20), made up of successive layers each with the fibrils oriented in a preferred direction, as in adult *Lethocerus*, the change in angle at the point of transition is not abrupt but consists of a very thin layer of micro-lamellate cuticle.[226, 227] The arrangement of chitin microfibrils in the form of lamellate cuticle is probably a spontaneous process of self-assembly; a crystallization of an asymmetrical unit; whereas non-lamellate cuticle with a constant preferred orientation requires specialized control by the cells.[228]

p. 43. **Moulting fluid**—In the pupa of *Antheraea* the moulting gel contains its full complement of proteinases in inactive form; they are activated when the gel becomes diluted to form the moulting fluid about a week before adult emergence. Esterases, of unknown function, are present throughout.[215] In *Bombyx* the moulting fluid contains active enzymes from the time of its formation.[214]

p. 45. In *Hyalophora* and in the pupa of *Apis* most of the absorption of macromolecules in the moulting fluid is said to take place through the points of attachment of the muscles to the new cuticle.[218]

p. 45. **Ecdysial membrane**—In the *Tenebrio* pupa the 'ecdysial membrane' appears to be the insoluble residue of lipoprotein stabilizing-material added to the cuticle early in the pupal stage[207] and in *Rhodnius* it has a similar origin.[242]

p. 46. **Eclosion hormone**—In the silkmoth *Antheraea* there is a well defined 'eclosion hormone' liberated from neurosecretory cells in the brain, which switches the entire behaviour from pupal to adult type. This includes release of the labial gland secretion (p. 528) and all other acts involved in escaping from the pupal cuticle and cocoon, expansion of the wings, and lysis of intersegmental muscles.[237] These observations require correlating with the secretion of factor 'B' in *Rhodnius* at ecdysis, which induces breakdown of the prothoracic gland,[240] the tanning hormone 'bursicon' (p. 47), and the hormonal induction of muscle breakdown in *Hyalophora.*[222] During adult development in the pupa of *Hyalophora*, cytolytic enzymes, notably cathepsin and acid phosphatase, are synthesized and sequestered in lysosome particles in readiness for muscle breakdown.[222] The same applies to the reversible degeneration of flight muscles in *Leptinotarsa* during diapause.[235]

p. 47. **'Bursicon'**—The hormone which initiates melanogenesis, sometimes called 'bursicon',[211] also initiates endocuticle formation in the adult *Sarcophaga.*[210] A similar hormone is recorded in Orthoptera, Hemiptera, Coleoptera and Lepidoptera.[211] It has a molecular weight around 40,000.[211] In *Periplaneta* it appears to be present throughout the central nervous system but to be liberated only from the terminal abdominal ganglion.[224] It is believed to act upon certain of the blood cells, causing them to synthesize and release N-acetyl dopamine.[225] But some authors consider that the phenoloxidase concerned in tanning the cuticle is firmly bound to the cuticle structure and is quite distinct in properties from the phenoloxidase in the haemolymph.[243] The cells responsible for laying down hard cuticle in *Calpodes* larvae contain peroxidase localized in granules identified in the electron microscope as 'multi-vesicular bodies'.[221] Bursicon released from the terminal abdominal ganglion of *Locusta* controls darkening, deposition of endocuticle, water loss and tracheal emptying (see p. 364) and it may influence cuticle dehydration and so be involved in changes in cuticle plasticization.[238]

The unusual amino acid β-alanine is commonly present in cuticle. In *Sarcophaga* it is in the form of the peptide β-alanyl-tyrosine ('sarcophagine') which seems to serve as a carrier for the tyrosine used for tanning the puparium.[202]

[1] ANDERSON, T. F., and RICHARDS, A. G. *J. Appl. Phys.*, **13** (1942), 748–58 (structural colours studied with the electron microscope).
[2] AHRENS, W. *Jena Z. Naturw.*, **64** (1930), 449–530 (cuticle of termites).
[3] ALEXANDROV, W. F. *Acta Zool.*, **16** (1935), 1–19 (permeability of cuticle: Nematocera larvae).
[4] ARONSSOHN, F. *C. R. Soc. Biol.*, **68** (1910), 1111–13 (nature of epicuticle: *Apis* larva).
[5] BEAMENT, J. W. L. *J. Exp. Biol.*, **21** (1945), 115–31 (cuticular waxes).
[6] —— *Discussions of the Faraday Society*, No. 3 (1948), 221–2 (asymmetrical diffusion through cuticle).
[7] BERGMANN, M., *et al. Ber. deutsch. Chem. Ges.*, **64** (1931), 2436–40 (composition of chitin).
[8] BERGMANN, W. *Ann. Ent. Soc. Amer.*, **31** (1938), 315–21 (cuticular waxes: *Bombyx mori*, Lep.).
[9] BERLESE, A. *Gli Insetti*, Vol. 1, Milan, 1909.
[10] BHAGVAT, K., and RICHTER, D. *Biochem. J.*, **32** (1938), 1397–1406 (phenols in insect cuticle).
[11] BIEDERMANN, W. *Z. allg. Physiol.*, **2** (1903), 395–480 (structure and formation of cuticle).
[12] —— *Winterstein's Handb. d. vergl. Physiol.*, **3** (i) (1910) (colours of insects).
[13] BLUNCK, H. *Z. wiss. Zool.*, **121** (1923), 171–391 (metamorphosis in *Dytiscus*).
[14] BONÉ, G., and KOCH, H. J. *Ann. Soc. Zool. Belg.*, **72** (1942), 73–87 (permeability of hind-gut cuticle).
[15] BRAUN, M. *Arb. Zool.-Zootom. Inst. Würzburg*, **2** (1875), 121–66 (histological changes in moulting: *Astacus*).
[16] BROCHER, F. *Ann. Biol. Lac.*, **9** (1919), 183–200 (mechanism of moulting: *Agrionidae*).
[17] BRUMPT, E. *Ann. Parasit. Hum. Comp.*, **18** (1941), 75–94 (mechanism of emergence of adult mosquito).
[18] V. BUDDENBROCK, W. *Z. Morph. Oekol. Tiere*, **18** (1930), 700–25 (moulting in caterpillars; histology of integument).
[19] BUXTON, P. A. *Proc. Roy. Soc.*, B, **106** (1931), 560–77 (evaporation of water from the mealworm).
[20] CAMPBELL, F. L. *Ann. Ent. Soc. Amer.*, **22** (1929), 401–26 (tests for chitin).
[21] CASTLE, E. S. *J. Gen. Physiol.*, **19** (1936), 797–806 (double refraction of chitin)
[22] CAUSARD, M. *Bull. Soc. Ent. Fr.*, 1898, 258–61 (air beneath cuticle in aquatic insects at moulting).
[23] DENNELL, R. *Proc. Roy. Soc.*, B, **133** (1946), 348–73; **134** (1947), 79–110; 485–503 (cuticle of larva and puparium: *Sarcophaga*, Dipt.).
[24] EIDMANN, H. *Z. Morph. Oekol. Tiere*, **2** (1924), 567–610 (mechanism of moulting).
[25] EVANS, A. C. *Proc. Roy. Ent. Soc. Lond.*, A, **13** (1938), 107–10 (absorption of cuticle: *Tenebrio*, Col.).
[26] FRAENKEL, G. *Proc. Zool. Soc. Lond.*, 1935, 893–904 (emergence of *Calliphora*, Dipt., from the pupa).
[27] —— *Proc. Roy. Ent. Soc. Lond.*, A, **13** (1938), 137–9 (evagination of head: *Calliphora* pupa).
[28] FRAENKEL, G., and RUDALL, K. M. *Proc. Roy. Soc.*, B, **129** (1940), 1–35; **134** (1947), 111–43 (physical and chemical properties of cuticle: *Calliphora* larva and puparium, Dipt.).
[29] FREY, W. *Z. Morph. Oekol. Tiere*, **31** (1936), 443–89 (iridescent colours of Chrysididae, Hym.).
[30] GENTIL, K. *Ent. Rundschau*, **50** (1933), 105–110 (metallic colours: *Calliphara*, Pentatomidae).
[31] —— *Z. Morph. Oekol. Tiere*, **37** (1941), 590–612; **38** (1942), 344–55 (fine structure of iridescent scales: Lepidoptera).
[32] GONELL, H. W. *Z. physiol. Chem.*, **152** (1926), 18–30 (X-ray diffraction of chitin).
[33] HACKMAN, R. H., PRYOR, M. G. M., and TODD, A. R. *Biochem. J.*, **43** (1948), 474–7; **55** (1953), 631–7 (phenols in insects).

[34] HAFFER, O. *Arch. Naturgesch.*, *Abt.* A, **87** (1921), 110–66 (development of hairs: *Saturnia*, Lep.).

[35] HAMMOND, C. O. *Entomologist*, **61** (1928), 54–5 (emergence of *Erythromma*, Odonata).

[36] HASS, W. *Arch. Anat. Physiol.*, 1916, 295–338 (structure of cuticle).

[37] —— *Sitzungsb. Ges. Naturforsch. Fr. Berlin*, 1916, 332–43 (metallic colours of Buprestids, Col.).

[38] HENKE, K. *Z. vergl. Physiol.*, **1** (1924), 297–499 (pigment formation: *Pyrrhocoris*, Hem.).

[39] HENRIKSEN, K. L. *Not. Entom.*, **11** (1932), 103–27 (mechanism of moulting).

[40] HOLMGREN, N. *Anat. Anz.*, **20** (1902), 480–8; **21** (1902), 373–8 (structure of cuticle).

[41] HOOP, M. *Zool. Jarhb.*, *Anat.*, **57** (1933), 433–64 (moulting glands).

[42] HUFNAGEL, A. *Arch. Zool. exp. et gén.*, **57** (1918), 47–202 (metamorphosis in *Hyponomeuta*, Lep.: histology).

[43] HURST, H. *Brit. Med. Bull.*, **3** (1945), 132–7; *Discussions of the Faraday Society*, No. 3 (1948), 193–210 (permeability of cuticle).

[44] KAPZOV, S. *Z. wiss. Zool.*, **98** (1911), 297–337 (structure of cuticle).

[45] KEMPER, H. *Z. Morph. Oekol. Tiere*, **22** (1931), 53–109 (moulting: *Cimex*, Hem.).

[46] KINDER, E., and SÜFFERT, F. *Biol. Zbl.*, **63** (1943), 268–88 (fine structure of scales: *Morpho*, Lep.).

[47] KOCH, C. *Z. Morph. Oekol. Tiere*, **25** (1932), 730–56 (tests for chitin).

[48] KOCH, H. *Ann. Soc. Sci. Bruxelles*, **54** (1934), 346–61 (permeability of 'gills' of aquatic insects).

[49] KNAB, F. *Proc. Ent. Soc. Washington*, **11** (1909), 68–73; **13** (1911), 32–42 (mechanism of moulting).

[50] KRAMER, S., and WIGGLESWORTH, V. B. *Quart. J. Micr. Sci.* **91** (1950), 63–72.

[51] KREMER, J. *Zool. Jahrb.*, *Anat.*, **41** (1920), 175–272 (structure of elytra: Coleoptera).

[52] KRÜPER, F. *Arch. Hydrobiol.* **22** (1930), 185–220 (lime in cuticle: Diptera larvae).

[53] KÜHN, A., and AN, M. *Biol. Zbl.*, **65** (1946), 30–40 (fine structure of scales: Lepidoptera).

[54] KÜHN, A., and PIEPHO, H. *Biol. Zbl.*, **58** (1938), 12–51 (changes in epidermis and Verson's glands at pupation: *Ephestia*, Lep.).

[55] —— —— *Biol. Zbl.*, **60** (1940), 1–22 (structure of giant scales: *Ephestia*, Lep.).

[56] KÜHNELT, W. *Zool. Jahrb.*, *Anat.*, **50** (1928), 219–78 (structure and composition of cuticle).

[57] KUNKEL D'HERCULAIS, J. *C.R. Acad. Sci.*, **110** (1890), 807–9 (mechanism of moulting: Acrididae).

[58] KUWANA, Z. *Proc. Imp. Acad. Tokyo*, **19** (1933), 280–3 (deposition of cuticle: *Bombyx mori*, Lep.).

[59] LAFON, M. *Ann. Sci. nat. Zool.* (sér. 11), **5** (1943), 113–46 (chemistry of cuticle).

[60] LAING, J. *Quart. J. Micr. Sci.*, **77** (1935), 497–521 (ptilinum of *Calliphora*, Dipt.).

[61] LEES, A. D., and PICKEN, L. E. R. *Proc. Roy. Soc.*, B, **132** (1945) 396–423 (fine structure of bristles: *Drosophila*, Dipt.).

[62] LEES, A. D., and WADDINGTON, C. H. *Proc. Roy. Soc.*, B, **131** (1942), 87–110 (development of bristles: *Drosophila*, Dipt.).

[63] LUDWIG, D. *J. Exp. Zool.*, **60** (1931), 309–23 (emergence of *Popillia*, Col., from the pupa).

[64] MASON, C. W. *J. Phys. Chem.*, **30** (1926), 383–95; **31** (1927), 321–54, 1856–72 (structural colours).

[65] MELLANBY, K. *Ann. Appl. Biol.*, **21** (1934), 476–82 (evaporation from larva of *Tineola*, Lep.).

[66] —— *Bull. Ent. Res.*, **27** (1936) 611–32 (water and emergence: *Glossina*, Dipt.).

[67] —— *Parasitology*, **30** (1938), 392–402 (water and expansion of cuticle: *Lucilia*, Dipt.).

[68] MEYER, K. H., and MARK, H. *Ber. deutsch. chem. Ges.*, **61** (1928), 1936–9 (composition of chitin).

[69] MÜLLER, G. W. *Z. Morph. Oekol. Tiere*, **3** (1925), 542–66 (lime in cuticle: Stratiomyidae, Dipt.).

[70] ONSLOW, H. *Phil. Trans. Roy. Soc.*, B, **211** (1921), 1–74 (iridescent colours).

[71] PAILLOT, A. *C.R. Soc. Biol.*, **127** (1938), 1502–4 (glycogen in epidermis: *Bombyx mori*).

[72] PICKEN, L. E. R. *Phil. Trans. Roy. Soc.*, B, **234** (1949), 1–28 (fine structure of scales in Lepidoptera).

[73] PICKEN, L. E. R., *et al. Nature*, **159** (1947), 434 (fine structure of *Donacia* cocoon, Col.).

[74] PIEPHO, H. *Biol. Zbl.*, **58** (1938), 481–95 (absorption of old cuticle: *Galleria*, Lep.).

[75] PLOTNIKOW, W. *Z. wiss. Zool.*, **76** (1904), 333–66 (cuticle and moulting).

[76] POISSON, R. *Biol. Bull. Fr. Belg.*, **58** (1924), 49–204 (cuticle and moulting: aquatic Hemiptera).

[77] POYARKOFF, E. *Arch. Anat. Microsc.*, **12** (1910), 333–474 (moulting and metamorphosis: *Galerucella*, Col.).

[78] PRYOR, M. G. M. *Proc. Roy. Soc.*, B, **128** (1940), 378–93; 393–407; *Thesis, Cambridge* (1940), (mechanism of hardening in oötheca of *Blatta*, Orth. and insect cuticle).

[79] PRYOR, M. G. M., RUSSELL, P. B. and TODD, A. R. *Biochem. J.*, **40** (1946), 627–8; *Nature*, **159** (1947), 399–400 (phenolic substances and hardening of cuticle).

[80] RAMSAY, J. A. *J. Exp Biol.*, **12** (1935), 373–83 (evaporation of water from the cockroach).

[81] REICHELT, M. *Z. Morph Oekol. Tiere*, **3** (1925), 477–525 (development of scales: Lepidoptera).

[82] RICHARDS, A. G. *Ann. Ent. Soc. Amer.*, **40** (1947), 227–40 (chitin in scales: Lepidoptera).

[83] RICHARDS, A. G., and ANDERSON, T. F. *J. Morph.*, **71** (1942), 135–83 (structure of cuticle: *Periplaneta*, Orth.).

[84] SAKURAI, M. *C.R. Acad. Sci.*, **187** (1928), 614–15 (tracheal glands and moulting).

[85] SAINT-HILAIRE, K. *Z. Zellforsch. Mikr. Anat.*, **5** (1927), 449–94 (structure of salivary glands: Tenthredinidae larvae).

[86] SCHULZE, P. *Deutsche Zool. Ges. Verhandl.*, 1913, 165–95 (structure of elytra: Coleoptera).

[87] —— *Biol. Zbl.*, **42** (1923), 388–94; *Z. Morph. Oekol. Tiere*, **2** (1924), 643–66 (tests for chitin).

[88] SCHÜRFELD, W. *Arch. Entw. Mech.*, **133** (1935), 728–59 (Verson's glands and moulting fluid).

[89] SHAFER, G. D. *Stanford Univ. Publ. Biol. Sci.*, **3** (1923), 307–37 (growth and moulting of Odonata nymphs).

[90] SHIMIZU, S. *Proc. Imp. Acad. (Japan)*, **7** (1931), 361–2 (exuvial fluid of silkworm).

[91] SMALLMAN, B. N. *Proc. Roy. Soc. Edin.*, B, **61** (1942), 167–85 (loss of water at moulting: *Carausius*, Orth.).

[92] SPRUNG, F. *Z. Morph. Oekol. Tiere*, **24** (1932), 435–90 (structure of elytra in Carabidae).

[93] STEGEMANN, F. *Zool. Jahrb., Anat.*, **50** (1929), 571–80; *Z. Morph. Oekol. Tiere*, **18** (1930), 1–73 (structure of elytra: Cicindelidae, Col.).

[94] STOSSBERG, M. *Z. Morph. Oekol. Tiere*, **34** (1938), 173–206 (development of scales: *Ephestia*, Lep.).

[95] SÜFFERT, F. *Z. Morph. Oekol. Tiere*, **1** (1924), 171–306 (iridescent colours of scales: Lepidoptera).

[96] TRIM, A. R. *Biochem. J.*, **35** (1941), 1088–98 (proteins in insect cuticle).

[97] TAUBER, O. E. *J. Morph.*, **56** (1934), 51–8 (distribution of chitin: *Periplaneta*).

[98] TEISSIER, G. *Trav. Stat. Biol. Roscoff*, **9** (1931), 29–238 (growth and moulting).

[99] THEODOR, O. *Bull. Ent. Res.*, **27** (1936), 653–71 (water relations of larva of *Phlebotomus*, Dipt.).

[100] TIEGS, O. W. *Trans. Roy. Soc. S. Australia*, **46** (1922), 319–527 (metamorphosis in *Nasonia*, Chalcid. Hym.).

[101] TOWER, W. L. *Zool. Anz.*, **25** (1902), 466–72 (moulting fluid: *Leptinotarsa*, Col.).

[102] —— *Biol. Bull.*, **10** (1906), 176–92 (histological changes in integument during moulting: *Leptinotarsa*, Col.).

[103] VERSON, E. *Zool. Anz.*, **25** (1902), 652–54; *Z. wiss. Zool.*, **97** (1911), 457–80 (moulting glands in *Bombyx mori*).

[104] VIGNON, P. *Arch. Zool. exp. et gén.* (sér. 3), **9** (1901), 371–715 (formation of cuticle).

[105] WACHTER, S. *Ann. Ent. Soc. Amer.*, **23** (1930), 381–91 (moulting: silkworm).

[106] WAY, M. J. *Quart. J. Micr. Sci.*, **91** (1950), 145–82 (cuticle of *Diataraxia* larva, Lep.).

[107] WEBB, J. E. *Parasitology*, **38** (1947), 70–1 (structure of cuticle: *Eomenacanthus*, Mallophaga)
[108] WEBER, H. *Z. Morph. Oekol. Tiere*, **23** (1931) 575–573 (biology of *Trialeurodes*, Hem.).
[109] —— *Lehrbuch der Entomologie*, Jena, 1933.
[110] WEISMANN, A. *Z. wiss. Zool.*, **14** (1865), 187–336 (post-embryonic development: *Calliphora*, Dipt.).
[111] WESTER, D. H. *Zool. Jarhb., Syst.*, **28** (1910), 531–57 (tests for chitin).
[112] WIESMANN, R. *Vjschr. naturf. Ges. Zürich*, **83** (1938), 127–36 (lime in cuticle of puparium: *Rhagoletis*, Dipt.).
[113] WIGGLESWORTH, V. B. *Quart. J. Micr. Sci.*, **75** (1932), 131–50 (permeability of cuticle in hind-gut).
[114] —— *Quart. J. Micr. Sci.*, **76** (1933), 270–318 (cuticle and moulting: *Rhodnius*, Hem.).
[115] —— *J. Exp. Biol.*, **10** (1933), 1–37 (anal papillae of mosquito larvae).
[116] —— *Bull. Ent. Res.*, **3** (1942), 205–18 (entry of substances through cuticle: *Rhodnius*, Hem.).
[117] —— *J. Exp. Biol.*, **21** (1945), 97–114; *Proc. Roy. Ent. Soc. Lond.*, A, **22** (1947), 65–9 (transpiration through cuticle).
[118] —— *Proc. Roy. Soc.*, B, **134** (1947), 163–81 (structure and deposition of epicuticle: *Rhodnius*: Hem.).
[119] —— *Quart. J. Micr. Sci.*, **89** (1948), 197–217 (cuticle of adult *Tenebrio*, Col.).
[120] —— *Biol. Rev.*, **23** (1948), 408–51 (cuticle: review).
[121] WIGGLESWORTH, V. B., and GILLETT, J. D. *Proc. Roy. Ent. Soc. Lond.*, A, **11** (1936), 104–7 (loss of water at moulting: *Rhodnius*, Hem.).
[122] WILLERS, W. *Z. wiss. Zool.*, **116** (1916), 43–74 (histology of epidermis during moulting).

SUPPLEMENTARY REFERENCES (A)

[123] BEAMENT, J. W. L. *Nature*, **167** (1951), 652 (wax secretion in insects).
[124] HURST, H. *J. Exp. Biol.*, **27** (1950), 238–52 (crystalline structure of lipids on cuticle of *Calliphora* pupa).
[125] PASSONNEAU, J. V., and WILLIAMS, C. M. *J. Exp. Zool.*, **30** (1953), 545–60 (moulting fluid in pupa of *Hyalophora*, Lep.).
[126] RICHARDS, A. G. *The Integument of Anthropods*, Minneapolis, 1951.

SUPPLEMENTARY REFERENCES (B)

[127] ANDERSEN, S. O. *Biochim. Biophys. Acta*, **69** (1963), 249–62; *Adv. Ins. Physiol.*, **2** (1964), 1–66 (structure of resilin).
[128] BADE, K. L., and WYATT, G. R. *Biochem. J.*, **83** (1962), 470–8 (metabolic conversions at pupation in *Hyalophora*).
[129] BAILEY, K., and WEIS-FOGH, T. *Biochim. Biophys. Acta*, **48** (1961), 453–9 (composition of resilin).
[130] BAILEY, S. W. *Nature*, **173** (1954), 503 (hardness of arthropod mouth-parts).
[131] BAKER, G., PEPPER, J. H. *et al. J. Ins. Physiol.*, **5** (1960), 47–60 (composition of cuticular wax in *Anabrus*).
[132] BEAMENT, J. W. L. *Nature*, **167** (1951), 652; *J. Exp. Biol.*, **32** (1955), 514–38 (wax secretion in *Periplaneta*).
[133] —— *J. Exp. Biol.*, **35** (1958), 494–519 (temperature and waterproofing: *Periplaneta*); **36** (1959), 391–422 (various terrestrial insects); **38** (1961), 277–90 (aquatic insects).
[134] —— *Biol. Rev.*, **36** (1961), 281–320 (water relations of insect cuticle: review).
[135] —— *Nature*, **191** (1961), 217–21 (electrical properties of oriented lipid); *Adv. Ins. Physiol.*, **2** (1964), 67–129 (transport of water through cuticle).
[136] BENNET-CLARK, H. C. *J. Ins. Physiol.*, **8** (1962), 627–33 (control of mechanical properties of cuticle in *Rhodnius*).
[137] BRUNET, P. C. J. *Ann. N.Y. Acad. Sci.*, **100** (1963), 1020–34 (tyrosine metabolism in insects).

[138] BRUNET, P. C. J., and KENT, P. W. *Quart. J. Micr. Sci.*, **92** (1951), 113–27; **93** (1952), 47–69; *Proc. Roy. Soc.*, B, **144** (1955), 259–74 (formation and hardening of oötheca in *Periplaneta*).

[139] CANDY, D. J., and KILBY, B. A. *J. Exp. Biol.*, **39** (1962), 129–40 (chitin synthesis in *Schistocerca*).

[140] CAREY, F. G., and WYATT, G. R. *Biochim. Biophys. Acta*, **41** (1960), 178–9 (uridine diphosphate and chitin synthesis).

[141] CHEFURKA, W., and PEPPER, J. H. *Canad. Ent.*, **87** (1955), 145–71 (permeability properties of cuticle in *Melanoplus*).

[142] COTTRELL, C. B. *Thesis, Cambridge* (1960); *J. Exp. Biol.*, **39** (1962), 365–430 (control of hardening and darkening in adult blow-flies).

[143] —— *J. Exp. Biol.*, **39** (1962), 431–58; *Trans. R. Ent. Soc. Lond.*, **114** (1962), 317–33 (mechanism of moulting in adult blow-flies).

[144] DARMON, S. E., and RUDALL, K. M. *Faraday Soc. Discussion*, **9** (1950), 251–60 (infra-red and X-ray studies of chitin).

[145] DENNELL, R. *Biol. Rev.*, **33** (1958), 178–96 (hardening of insect cuticle: review).

[146] DENNELL, R., and MALEK, S. R. A. *Proc. Roy. Soc.*, B, **143** (1954), 126–36; **143** (1955), 239–57, 414–34; **144** (1955), 545–6; **145** (1956), 249–58 (cuticle of *Periplaneta*).

[147] DEVI, A., *et al. Experientia*, **19** (1963), 404–5 (catechol as tanning precursor in *Tribolium*).

[148] EBELING, W., and WAGNER, R. E. *J. Econ. Ent.*, **52** (1959), 190–212 (sorptive dusts and desiccation of termites, &c.).

[149] EDWARDS, J. S. *XI Int. Kongr. Entom. Wien 1960*, **3** (1960), 259–63 (elastic cuticle of salivary pump in *Platymeris*).

[150] FINLAYSON, L. H., and MOWAT, D. J. *Quart. J. Micr. Sci.*, **104** (1963), 243–51 (muscle breakdown in Saturniidae at metamorphosis).

[151] FRAENKEL, G., and HSIAO, C. *Science*, **138** (1962), 27–29; **141** (1963), 1057–8 (hormonal and nervous control of tanning in *Phormia*).

[152] GILBY, A. R., and COX, M. E. *Nature*, **195** (1962), 729; *J. Ins. Physiol.*, **9** (1963), 671–81 (cuticular lipids in *Periplaneta*).

[153] HACKMAN, R. H. *Biochem. J.*, **54** (1953), 362–77; *Austral. J. Biol. Sci.*, **8** (1955), 86–96; 530–6; **13** (1960), 568–77 (association of chitin with protein).

[154] HACKMAN, R. H. In *Biochemistry of Insects*, London: Pergamon Press 1959, 48–62 (biochemistry of insect cuticle: review).

[155] HACKMAN, R. H., and GOLDBERG, M. *J. Ins. Physiol.*, **2** (1958), 221–31 (association of chitin with protein).

[156] HAMAMURA, Y., and KANEHARA, Y. *J. Agr. Chem. Soc. Japan*, **16** (1940), 907 (chitinase in moulting fluid).

[157] ITO, T. *Bull. Sericult. Exp. Sta.*, **14** (1954), 249–52 (formation of cuticle in *B. mori*).

[158] JAWORSKI, E., WANG, L., and MARCO, G. *Nature*, **198** (1963), 790 (chitin synthesis in cell-free extracts of *Prodenia*).

[159] JEUNIAUX, C. *et al. Mém. Soc. Roy. Ent. Belgique*, **27** (1955), 312–19; *Arch. Int. Physiol. Biochim.*, **63** (1955), 94–103; **66** (1958), 121; **68** (1960), 411–12; *Bull. Soc. Zool. Fr.*, **86** (1961), 590–9 (chitinase in moulting fluid).

[160] JOHNSON, L. H., PEPPER, J. H., *et al. Physiol. Zool.*, **25** (1952), 250–8 (cuticle protein in *Anabrus*).

[161] KARLSON, P., *et al. Hoppe-Seyl Z.*, **318** (1960), 194–200; **327** (1962), 86–94; **330** (1963), 161–8; *Nature*, **195** (1962), 183–4 (chemistry of cuticular hardening).

[162] KAWASE, S. *Nature*, **181** (1958), 1350–1; *J. Sericult. Sci. Japan*, **29** (1960), 412–14 (chemistry of hardening and darkening of cuticle).

[163] KOIDSUMI, K. *Annot. Zool. Jap.*, **26** (1953), 162–75 (environment and cuticular lipids in insects).

[164] KRAMER, S., and WIGGLESWORTH, V. B. *Quart. J. Micr. Sci.*, **91** (1950), 63–72 (epicuticle of *Periplaneta*).

[165] KROON, D. B., VEERKAMP, T. A., and LOEVEN, W. A. *Koninkl. Nederl. Akad. Wetenschappen*, Proc. Ser. C, **55** (1952), 209–14 (extension and hardening of wings in Lepidoptera).

[165a] LIPPERT, W., and GENTIL, K. *Z. Morph. Oekol. Tiere*, **48** (1959), 115–22 (fine structure of iridescent scales).

[166] LOCKE, M. *Quart. J. Micr. Sci.*, **101** (1960), 333–8; *J. Biophys. Biochem, Cytol.*, **1**
(1961), 589–618 (wax secretion in insect cuticle).
[167] LOCKEY, K. H. *J. Exp. Biol.*, **37** (1960), 316–29 (thickness of wax layer of epicuticle)
[168] LOWER, H. F. *J. Morph.*, **101** (1957), 166–96 (cuticle of *Aphodius*); *Zool. Jahrb., Anat.*
76 (1957), 166–98 (cuticle of *Persectaria*, Lep.).
[169] MCLINTOCK, J. *Nature*, **201** (1964), 1245 (puparium formation in Diptera).
[170] MALEK, S. R. A. *Nature*, **178** (1956), 1185–6; *J. Ins. Physiol.* **2** (1958), 298–313 (ecdysia
membrane in *Schistocerca*).
[171] —— *Comp. Biochem. Physiol.*, **2** (1961), 35–50 (tanning of cuticle in *Schistocerca*).
[172] NEVILLE, A. C. *J. Ins. Physiol.*, **9** (1963), 177–86 (daily growth layers in cuticle o
Schistocerca); 265–78 (deposition of resilin and chitin in *Schistocerca*).
[173] NOBLE-NESBITT, J. *J. Exp. Biol.*, **40** (1963), 681–700; *Quart. J. Micr. Sci.*, **104** (1963)
253–70; 369–91 (cuticle of *Podura aquatica*).
[174] NUÑEZ, J. A. *Nature*, **199** (1963), 621–2 (nervous control of cuticle properties i
Rhodnius).
[175] RICHARDS, A. G. *J. Morph.*, **96** (1955), 537–64; *Z. Naturforsch.*, **13b** (1958), 811–13
(ecdysial membrane).
[176] —— *J. Ins. Physiol.*, **1** (1957), 23–39 (permeability of cuticle).
[177] —— *Ergebn. Biol.*, **20** (1958), 1–26 (cuticle of Arthropods: review).
[178] RICHARDS, A. G., and PIPA, R. L. *Smithson. Misc. Coll.*, **137** (1959), 247–62 (molecula
organization of cuticle).
[179] RUDALL, K. M. *Adv. Ins. Physiol.*, **1** (1963), 257–313 (chitin-protein complexes i
cuticle).
[180] SCHATZ, L. *Ann. Ent. Soc. Amer.*, **45** (1952), 678–85 (subdivisions of the cuticle).
[181] SCHLOSSBERGER-RAECKE, I., and KARLSON, P. *J. Ins. Physiol.*, **10** (1964), 261–6 (in
corporation of tyrosine into the cuticle of *Schistocerca*).
[182] SCHMIDT, E. L. *J. Morph.*, **99** (1956), 211–31 (subcuticular layer of the integument)
[183] SEKERIS, C. E. *Hoppe-Seyl. Z.*, **332** (1963), 70–8 (dopadecarboxylase in *Calliphor*
integument).
[184] SEMENOVA, L. M., and RODIONOVA, A. N. *J. Gen. Biol., Moscow*, **22** (1961), 128–3
(permeability of the cuticle to water in aquatic insects).
[185] SHARPLIN, J. *Canad. Ent.*, **95** (1963), 96–100 (flexible cuticle in wing bases of Lepi
doptera).
[186] SHAW, J. *J. Exp. Biol.*, **32** (1955), 330–52 (permeability of cuticle in *Sialis* larva).
[187] TOMINO, S. *Experientia*, **19** (1963), 231–2 (tanning agent in oötheca of *Locusta*).
[188] WATERHOUSE, D. F., HACKMAN, R. H., and MCKELLAR, J. W. *J. Ins. Physiol.*, **6** (1961)
96–112; 185–95 (chitinase in *Periplaneta*).
[189] WEIS-FOGH, T. *J. Exp. Biol.*, **37** (1960), 889–907; *J. Mol. Biol.*, **3** (1961), 520–31
648–67; *Adv. Ins. Physiol.*, **2** (1964), 1–66 (the rubber-like protein 'resilin' o
insect cuticle).
[190] WHITEHEAD, D. L., BRUNET, P. C. J., and KENT, P. W. *Nature*, **185** (1960), 610 (phenol
oxidase in *Periplaneta* oötheca).
[191] WIESEND, P. *Z. Morph. Oekol. Tiere*, **46** (1957), 529–70 (reduction of muscles in adul
grasshoppers).
[192] WIGGLESWORTH, V. B. *Quart. J. Micr. Sci.*, **97** (1956), 89–98 (PAS staining of cuticle i
Rhodnius).
[193] —— *J. Exp. Biol.*, **32** (1955), 485–91 (control of cuticular changes after moulting)
[194] —— *Quart. J. Micr. Sci.*, **97** (1956), 465–80 (involution of muscles after moulting i
Rhodnius, &c.).
[195] —— *Ann. Rev. Ent.*, **2** (1957), 37–54 (physiology of insect cuticle: review).
[196] WOLFE, L. S. *Quart. J. Micr. Sci.*, **95** (1954), 49–78; **96** (1955), 181–91 (cuticle o
Calliphora).
[197] ZIELINSKA, Z. M., and LASKOWSKA, T. *Acta. Biol. Exp.*, **18** (1958), 209–19 (amino acids
&c., in moulting fluid of *Bombyx*).

SUPPLEMENTARY REFERENCES (C)

[198] ANDERSEN, S. O. *et al. J. Insect Physiol.*, **16** (1970), 1951–9; **17** (1971), 69–83 (keto-catechols and sclerotization of cuticle).

[199] ATKINSON, P. W. and GILBY, A. R. *Science, N.Y.*, **168** (1970), 992 (autoxidation of cuticular lipid in *Periplaneta*).

[200] BEAMENT, J. W. L., NOBLE-NESBITT, J. and WATSON, J. A. L. *J. exp. Biol.*, **41** (1964), 323–30 (waterproofing of cuticle in *Thermobia*).

[201] BEATTY, I. M. and GILBY, A. R. *Naturwissenschaften*, **56** (1969), 373 (unsaturated hydrocarbon in cuticular wax of *Periplaneta*).

[202] BODNARYK, R. P. *J. Insect Physiol.*, **17** (1971), 1201–10 (β-alanine in puparium of *Sarcophaga*).

[203] BOWERS, W. S. and THOMPSON, M. J. *J. Insect Physiol.*, **11** (1965), 1003–11 (cuticular powder in *Samia* larva).

[204] BURSELL, E. and CLEMENTS, A. N. *J. Insect Physiol.*, **13** (1967), 1671–8 (cuticular lipids in *Tenebrio*).

[205] CAVENEY, S. *Proc. Roy. Soc. Lond.*, B, **178** (1971), 205–25 (uric acid in metallic cuticle of Scarabaeidae).

[206] COLLINS, M. S. and RICHARDS, A. G. *Ecology*, **47** (1966), 328–31 (cement layer and water loss in termites).

[207] DELACHAMBRE, J. *Z. Zellforsch.*, **81** (1967) 114–34; **108** (1970), 380–96; **112** (1971), 97–119 (epicuticle, &c., in *Tenebrio*).

[208] EVANS, J. J. T. *Z. Zellforsch.*, **81** (1967), 34–48 (wax secreting epidermis in *Dacus*).

[209] FILSHIE, B. K. and WATERHOUSE, D. F. *Tissue & Cell*, **1** (1969), 367–85 (formation of epicuticle in *Nezara*).

[210] FOGAL, W. and FRAENKEL, G. *J. Insect Physiol.*, **15** (1969), 1235–47 ('bursicon' and endocuticle formation in *Sarcophaga*).

[211] FRAENKEL, G., HSIAO, C., *et al. J. Insect Physiol.*, **11** (1965), 513–56; **13** (1967), 1387–94; *Science*, **151** (1966), 91–3 ('bursicon' in *Musca* and *Periplaneta*).

[212] HACKMAN, R. H. *et al. J. Insect Physiol.*, **17** (1971), 335–47; 1065–71 (disulphide and other bonds in cuticle of *Lucilia*).

[213] HINTON, H. E., GIBBS, D. F. *et al. J. Insect Physiol.*, **15** (1969), 549–52; 959–62 (diffraction gratings in Mutillidae and Carabidae).

[214] JEUNIAUX, C., *Chitine et Chitinolyse* Masson, Paris, 1963, 181 pp.

[215] KATZENELLENBOGEN, B. S. and KAFATOS, F. C. *J. Insect Physiol.*, **16** (1970), 2241–56; **17** (1971), 775–800; 823–32; 1139–51 (properties of moulting fluid in *Antheraea*).

[216] KRISHNAN, G. *et al. Current Sci.*, **33** (1964), 639–40; *Acta histochem.*, **34** (1969), 212–28 (disulphide bonding in cuticle of *Machilis* and Collembola).

[217] LAI-FOOK, J. *J. Insect Physiol.*, **12** (1966), 195–226 (phenolic tanning in wound repair).

[218] LENSKY, Y. *et al. Biol Bull.* **139** (1970), 277–95; (absorption of proteins in moulting fluid).

[219] LIPKE, H. and GEOGHEGAN, T. *J. Insect Physiol.*, **17** (1971), 415–25 (hydrogen bonding in cuticle).

[220] LOCKE, M. *Physiology of Insecta* **3,** Rockstein (Ed.), Academic Press, New York, 1964, 379–470; *J. Morph.* **118** (1966), 461–94; *Adv. Morphogenesis*, **6** (1967), 33–88 (epicuticle formation and general cuticle structure).

[221] LOCKE, M. *Tissue & Cell*, **1** (1969), 555–74; *J. Morph.*, **127** (1969), 7–40; (peroxidase in cuticle formation; pinocytosis in epidermis).

[222] LOCKSHIN, R. A. and WILLIAMS, C. M. *J. Insect Physiol.*, **10** (1964), 643–9; **11** (1965), 123–33; 601–10 (endocrine influence on muscle breakdown in Saturniids).

[223] MADDRELL, S. H. P. *J. exp. Biol.*, **44** (1966), 59–68 (control of mechanical properties of *Rhodnius* cuticle).

[224] MILLS, R. R. *et al. J. Insect Physiol.*, **11** (1965), 1047–53; 1269–75; **12** (1966), 275–80: 1395–1401; **13** (1967), 815–20 (tanning of cuticle in *Periplaneta*).

[225] MILLS, R. R. and WHITEHEAD, D. L. *J. Insect Physiol.*, **16** (1970), 331–40 (action of bursicon on blood cells in *Periplaneta*).

[226] NEVILLE, A. C. *Symp. R. ent. Soc. London*, **5** (1970), 17–39 (cuticle ultra-structure).

[227] NEVILLE, A. C. *et al. Quart. J. Micr. Sci.*, **106** (1965), 269–86; 315–25; *Adv. Insect Physiol.*, **4** (1967), 213–86 (review); *Tissue & Cell*, **1** (1969), 189–200; 355–66; 689–707 (nature and development of cuticular lamellae).

[228] NEVILLE, A. C. and CAVENEY, S. *Biol. Rev.*, **44** (1969), 531–62 (helical structure of cuticle of Scarabaeidae and reflection of circularly polarized light).

[229] OVERTON, J. *J. Cell Biol.*, **29** (1966), 293–305; *J. Morph.*, **122** (1967) 367–80 (fine structure of developing scales in *Ephestia* and bristles in *Drosophila*).

[230] PAU, R. N., BRUNET, P. C. J. and WILLIAMS, M. J. *Proc. Roy. Soc. Lond.*, B, **177** (1971), 565–79 (protein of colleterial gland in *Periplaneta*).

[231] PAWELETZ, N. and SCHLOTE, F. W. *Z. Zellforsch.*, **63** (1964), 840–70 (development of wing scales in *Ephestia*).

[232] RICE, M. J. *Nature*, **228** (1970), 1337–8 (role of resilin in feeding of *Glossina*).

[233] RINTERKNECHT, E. and LEVI, P. *Z. Zellforsch.*, **72** (1966), 390–407 (fine structure of cuticle of *Locusta*).

[234] SEKERIS, C. E. and MERGENHAGEN, D. *Science*, **145** (1964), 68–9 (phenoloxidase system in *Calliphora*).

[235] STEGWEE, D. *et al. J. Cell Biol.*, **19** (1963), 519–27 (lysosomes and muscle breakdown in *Leptinotarsa*).

[236] TAYLOR, R. L. *Ent. News*, **75** (1964), 253–6 (metallic spots on pupa of *Danaus*).

[237] TRUMAN, J. W. *et al. Science*, **167** (1970), 1624–6; *J. exp. Biol.*, **54** (1971), 805–14 (neuroendocrine control of eclosion in Saturniidae).

[238] VINCENT, J. F. V. *J. Insect Physiol.*, **17** (1971), 625–36 (role of bursicon at ecdysis in *Locusta*).

[239] WEIS-FOGH, T. *Symp. Roy. ent. Soc. Lond.*, **5** (1970), 165–85 (structure and formation of insect cuticle).

[240] WIGGLESWORTH, V. B. *J. exp. Biol.*, **32** (1955), 485–91 (breakdown of prothoracic gland in *Rhodnius*).

[241] WIGGLESWORTH, V. B. *Tissue & Cell*, **2** (1970), 155–79 (structural lipid in insect cuticle).

[242] WIGGLESWORTH, V. B. Unpublished observations.

[243] YAMASAKI, H. I. *J. Insect Physiol.*, **15** (1969), 2203–11 (cuticular phenoloxidase in *Drosophila*).

Chapter III

Growth

MOULTING

GROWTH IN the animal body is always more or less cyclical, periods of comparative rest alternating with periods of activity. But in no group is this so evident as in the insects, in which development is punctuated by a series of moults or ecdyses, each preceded by a period of active growth and followed by a period in which true growth may be entirely absent.

Growth ratios—Moulting is primarily a mechanism of growth, conditioned by the properties of the cuticle. In many insects the amount of growth which is achieved at each moult is predictable from certain empirical laws. It was shown by Dyar many years ago that the head capsule of caterpillars grows in geometrical progression, increasing in width at each moult by a ratio (usually about 1·4) which is constant for a given species. This rule applies to many parts of the body, so that when the number of the instar is plotted against the logarithm of some measurement on the insect, a straight line is generally obtained (Fig. 35).[283] It is sometimes possible, by the help of this rule, to deduce from incomplete series of cast skins the actual number of ecdyses—as in *Haematopota* (Tabanidae), the pharynx of which shows a very constant ratio of growth of 1·29.[52] In other species, such as *Heliothis* (Lep.)[103] and *Popillia* (Col.)[2] there are so many departures from Dyar's rule that it cannot be used for corroborating the number of instars. If extra ecdyses are produced in *Blattella germanica* by a less nutritious diet the growth ratio is diminished.[261] In *Bombyx mori* the number of moults may vary from 4 to 5 but the final head size is the same.[38]

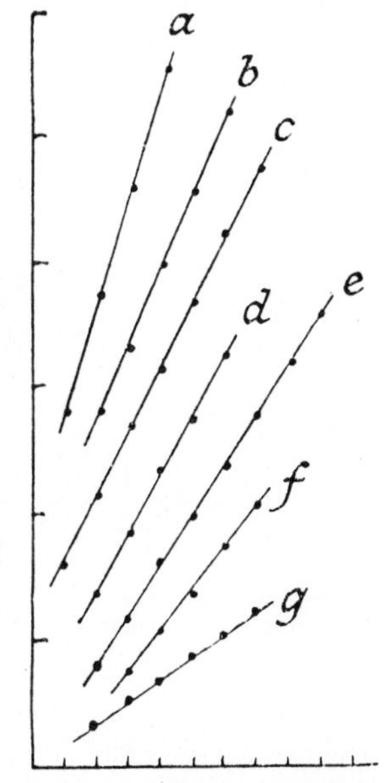

FIG. 35.—Chart illustrating Dyar's law (*after* TEISSIER)

Ordinate: logarithm of linear measurement. Abscissa: number of instar. a, *Bombyx mori*, ratio (r) = 1·91; b, *Philosamia ricini*, r = 1·52; c, *Crambus mutabilis*, r = 1·44; d, *Chaetopletus vestitus*, r = 1·40; e, *Sphodromantis bioculata*, r = 1·30; f, *Cladius isomerus*, r = 1·50; g, *Tenebrio molitor*, r = 1·12.

Another empirical law of growth is known as Przibram's rule. According to this the weight is doubled during each instar, and at each moult all linear dimensions are increased by the ratio 1·26 or $\sqrt[3]{2}$. This rule was deduced from measurements on *Sphodromantis*, which agree fairly well with these figures;[228] and the hypothetical explanation of the rule, that during each stage every cell in the body divides once and grows to its original size, is supported by the observation that in this insect there are in every stage an equal number of nucle per unit area of cuticle.[281] *Carausius* also moults when its weight is about double (Fig. 36, A).[82] In various other insects, while growth in weight is more or less continuous, growth in length of the rigid parts of the cuticle is discontinuous these often show a 25 per cent. increase at each moult, in conformity witl Przibram's rule.[282] A peculiar form of discontinuous growth is shown by aquati insects such as *Notonecta*, in which the weight may be abruptly doubled a moulting by the water swallowed or absorbed through the integumen (Fig. 30, B).[282]

On the other hand, agreement with this rule is often so inexact that i becomes of no practical value, as in *Lymantria*,[117] *Melanoplus*,[207] *Locusta*,[15] *Tenodera* (Mantidae),[227] *Popillia*.[185] It is of doubtful value, also, where th weight increases several times at each moult; in which case it is supposed tha such increase is always by some power of two.[20, 228] And the simple histologica

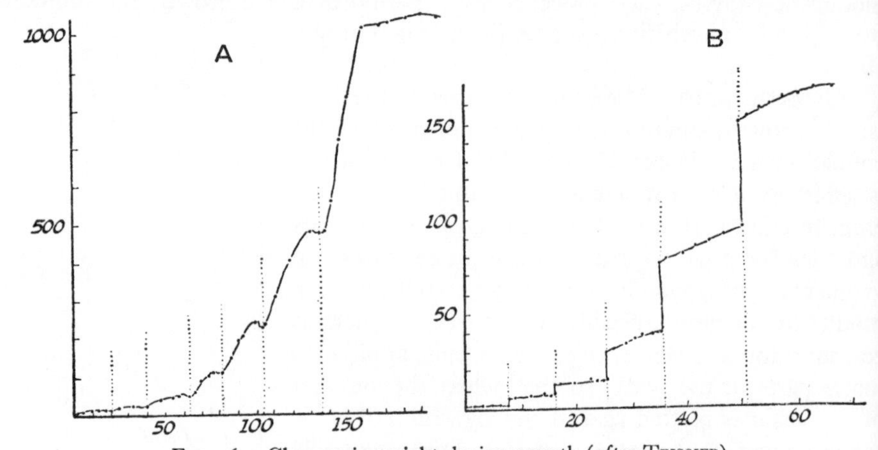

FIG. 36.—Changes in weight during growth (*after* TEISSIER)

A, Carausius morosus; B, Notonecta glauca. Ordinates: weight in milligrams. Abscissae: time in day The vertical dotted lines show the times when moulting occurred.

basis of this rule has had to be abandoned, because in Muscidae and other flies the rigid parts of which show fair agreement with the numerical rule,[4] the larva cells do not divide after the embryonic period; they grow in size only;[207] and i. other insects, as we shall see, there is much cellular breakdown and reconstruc tion at each moult, no simple dichotomous division. In *Rhodnius*, for example it is the increase in area of the integument which determines the increase in th number of cells; the cell population is adjusted to give a constant density pe unit area.[309, 311]

Allometric or Heterogonic growth—That the dimensions of a part of th body should increase at each moult by the same ratio as the body as a whole which is what Przibram's rule implies, presupposes the occurrence of 'harmoni growth'. Whereas growth in insects, as in so many animals, is generally 'dishar monic', 'heterogonic' or 'allometric'—the parts growing at rates peculiar to them selves, higher or lower than the growth rate of the body as a whole.[147, 513] Th law is expressed by the formula $x = ky^a$ (where x is the dimension of the whole, the dimension of the part, a the 'growth coefficient', and k another constant That is, the logarithm of the dimension of the part is proportional to the logarith of the dimension of the whole, so that when these measurements are plotted on a double logarithmic grid, a straight line is obtained (Fig. 37).

This type of growth may supervene late in development; in many insects it occurs chiefly during metamorphosis: the intermediate steps escape detection and it is the limitation of the total amount of growth achieved which is regulated in accordance with the formula. Moreover, this excess capacity for growth or 'growth potential' is often distributed unequally throughout an organ, and falls off in each direction from a centre of maximum growth. Thus in stag beetles (Lucanidae) there seems to be a gradient of growth intensity with its centre in the mandibles, falling away posteriorly.[147] In *Pediculus* there is a regular allometric growth of the various parts of the body throughout larval life; but at the last moult when the insect reaches maturity and suffers, in fact, a very mild degree of 'metamorphosis', the ratios of increase are changed.[50] Obviously, where growth rates in the body are regulated by such complex laws they cannot be expected to conform to Przibram's simple rule.

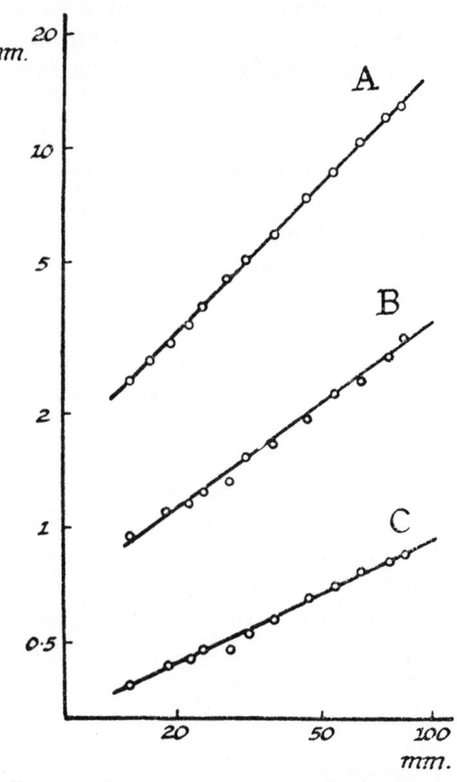

FIG. 37.—Heterogonic growth in *Carausius morosus*. Logarithmic co-ordinates

Length of posterior segment of prothorax (*A*), width of head (*B*), and diameter of eye (*C*) plotted against the total length of the body (*after* TEISSIER).

Number of instars—Simple rules which seek to relate amount of growth with moulting break down also where the number of instars is inconstant. Such inconstancy is very common.[145, 153, 291] Variations from 10–14 occur in the small wax moth *Achroea grisella*,[36] from 5–27 in the May-fly *Baëtis*. Often the female tends to have more moults than the male, as in *Sphodromantis*,[228] *Dermestes*,[170] *Tineola* and many other insects.[291] Where extra instars occur, the condition may be hereditary; different races of *Bombyx mori* have the number of ecdyses determined by mendelian factors;[202] and the occurrence of an extra instar in females of *Locusta* may be inherited.[154] Or the number may be determined by external conditions; raised temperature increases the number of moults from 9–11 in *Sphodromantis*,[228] from 4–5 and 5–6 in males and females respectively of *Dermestes*,[170] 11–15 at 25° C. and 15–23 at 30° C. in *Tenebrio*,[455] from an average of 4 at 18° C. to an average of 5 at 25° C. in *Ephestia kühniella*;[107] it has the reverse effect in *Melanoplus mexicanus* and in *Pieris brassicae*:[145] the larva of *Pieris* has 5 moults at 14–15° C., 4 at 15–20° C., 3 at 22–27° C.[158]

Inadequate nutrition, causing a prolonged larval period, may increase the number of moults enormously, especially in insects living in dry foodstuffs: by keeping the clothes moth *Tineola* on a rich or a poor medium the larval perioc could be varied from 26 days with 4 moults to 900 days with 40 moults.[291] Ir such cases moulting can take place without growth; indeed the insect may grov gradually smaller. In most insects moulting ceases when the body is fully grown but in Apterygota moulting may continue in the adult without change of size o organization.[229, 280] *Thermobia* may pass through 45–60 instars before death.[28]

Small larvae of locusts, derived from small eggs, undergo an extra moult t compensate for their smaller size at birth.[355] Caterpillars reared under crowdec conditions not only become more active and much darker in colour (p. 620) bu they feed more, and the duration of larval development may be shortened to 8c per cent. of that of solitary larvae and the number of instars is reduced.[454]

Causation of moulting—There are two aspects to the problem of moulting (i) the initiation of the process of growth and cuticle formation, and (ii) the determination of those changes in form that occur during ecdysis. Changes in form we shall consider later; here we are concerned only with the first aspect. And from this point of view there is no need to separate moulting from pupation; the essential phenomena of growth and differentiation and cuticle formation are common to both.

Moulting involves all parts of the body, whether supplied with nerves o not, at the same time. If spines or appendages are transplanted from one cater-pillar of *Vanessa* to another, they moult synchronously with their new host; limbs of *V. io* and *V. urticae* may be interchanged; or limbs of a 4th instar may be implanted on a 3rd; and always the time of moulting and the number o moults is determined by the host.[23] The same is found with transplants in *Carausius*[190] and *Galleria*;[210] and internal transplants (ovaries) in *Celerio* develop synchronously with their host.[51] These experiments suggest that moulting is determined by some factor outside the epidermis, a hormone circulating in the blood.

"Moulting Hormones"—The first experimental proof that moulting is brought about by a circulating hormone was given by Kopec[165] who produced

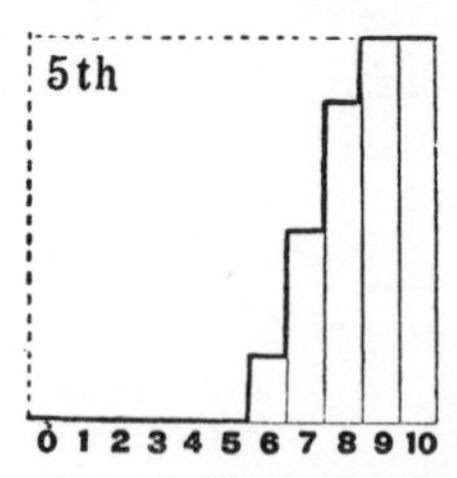

FIG. 38.—Chart showing the proportion of *Rhodnius* 5th-stage larvae which moulted, out o batches decapitated at different times after feeding

Ordinate: percentage moulting. Abscissa: days after feeding (*after* WIGGLESWORTH).

evidence that the pupal moult in Lepidoptera is induced by a hormone secreted in the region of the brain. If the larva of *Lymantria* is ligated and cut in two at days after the last larval moult, the anterior half pupates some 3 days later; the posterior half shows no change, though it survives for many weeks. In more recent years this conclusion has been confirmed in many groups of insects. Ir *Rhodnius* larvae, removal of the head during the first few days after feeding will prevent moulting (Fig. 38) although the headless insects may sometimes remain alive for longer than a year.[308] There is, however, a critical period a few days after feeding, after which moulting is no longer prevented by decapitation.[308] In Diptera (*Calliphora*,[96] *Drosophila*[25]) separation of the posterior half of the body from the region of the brain by a ligature will prevent the formation of the puparium (Fig. 39). That a stimulus from somewhere in the region of the brain is necessary for the pupal moult in Lepidoptera has been confirmed in *Sphinx* and *Celerio*,[216] in *Ephestia*,[174] in *Galleria* and *Bombyx*.[38] The same is true in the sawfly *Trypoxylon*,[257] and in many other insects.[524]

This stimulus to moulting is certainly hormonal in nature. Blood transfused through a capillary tube from a larva of *Rhodnius* fed a week or ten days before, will induce moulting in another larva decapitated the day after feeding (Fig. 40).[308] Haemolymph from the anterior half of a ligatured *Calliphora* larva will induce puparium formation in the posterior half.[96] The same applies to pupation

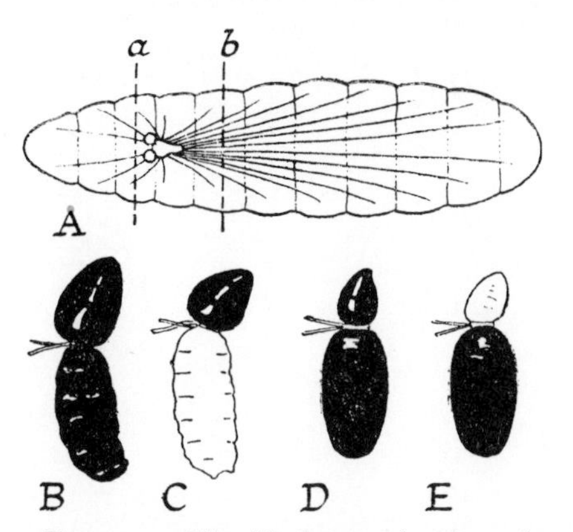

FIG. 39.—Muscid larva, showing the ganglion and the nerves, and the positions a, and b, at which ligatures are applied

B, larva ligatured at level b after the critical period: posterior fragment paralysed; both fragments have pupated. C, larva ligatured at same level before the critical period; pupation only of the anterior fragment containing the ganglion. D, larva ligatured at level a, after the critical period; both parts pupated; posterior part not paralysed. E, larva ligatured at same level before the critical period; pupation in posterior fragment only (*after* FRAENKEL).

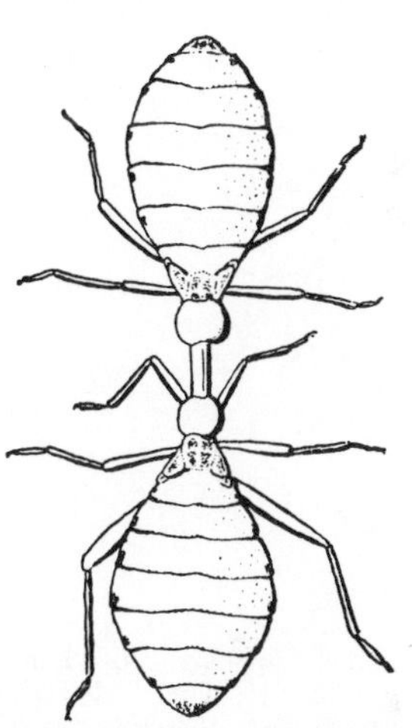

FIG. 40.—Two 4th instar larvae of *Rhodnius* decapitated and connected by a capillary tube sealed into the neck with paraffin wax (*after* WIGGLESWORTH)

in Hymenoptera.[257] Segments of limbs or fragments of integument implanted into the abdomen of *Rhodnius*[308] or in caterpillars,[210] will heal up so as to form closed cysts lined with cuticle; growth and moulting take place in these implants simultaneously with the host. If a piece of integument from a full-grown larva of *Galleria* is implanted into a very young larva it can be caused to undergo at least five extra moults. If the dormant cysts of the pupal testis of *Hyalophora* are immersed in the blood of a resting pupa (in diapause) they show no development; whereas in the blood of a pupa in which adult development has begun the cells undergo meiosis and differentiation into normal spermatids.[345, 352]

The moulting hormone is non-specific. Blood from *Rhodnius* will induce moulting in *Triatoma*, another genus, and in *Cimex*, which belongs to another family.[308] Implanted ring glands will induce pupation in the larvae of other species of *Drosophila*;[126] fragments of integument from the caterpillar *Ptychopoda* transplanted into *Galleria* are caused to moult;[210] and the isolated active substance to be mentioned later will induce moulting in insects of many sorts.

Brain hormone and thoracic gland hormone—It is now recognized that the moulting hormone is the product of endocrine glands which in the course of development have been budded off from the ectoderm associated with the mouthparts and carried backwards to the lower part of the head ('ventral glands') or to the fore-part of the thorax ('thoracic' or 'prothoracic glands') (p. 201). But these glands secrete their hormone only when activated by a second hormone liberated from the brain. This two-stage process seems to be general throughout the insects.[469, 475, 524, 531]

(i) **The neuro-endocrine system.** The system associated with the brain is termed the 'neuro-endocrine system'. It consists of nerve cells which have acquired a secretory function (neurosecretory cells) and a neurohaemal organ (the

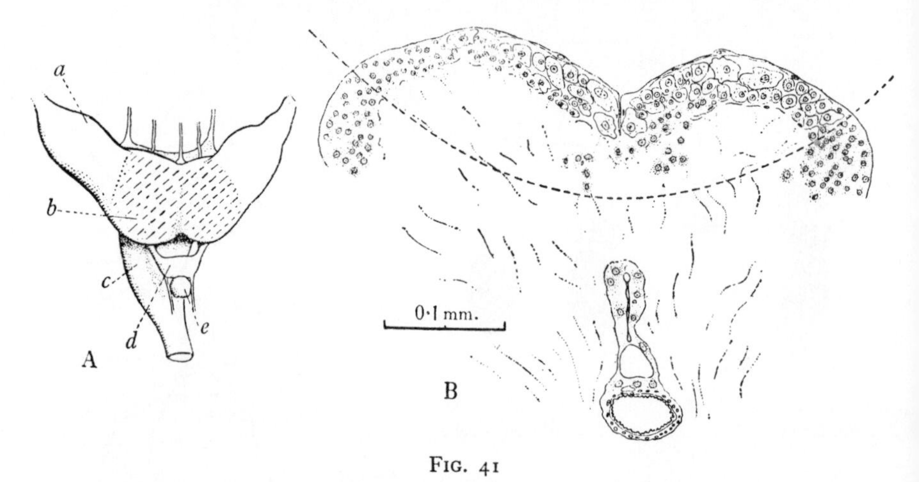

FIG. 41

A, brain of *Rhodnius*. *a*, optic lobes; *b*, protocerebrum; *c*, subœsophageal ganglion; *d*, corpus cardiacum; *e*, corpus allatum. The shaded area of the protocerebrum is the part which induces moulting when implanted. *B*, vertical section through the posterior part of the protocerebrum. In the central region are the large neurosecretory cells with fuchsinophil inclusions (*after* WIGGLESWORTH).

corpus cardiacum) through which their secretion is discharged into the circulating haemolymph. The neurosecretory cells responsible are found in the 'pars intercerebralis', or median dorsal region of the brain.[129] They are of several types distinguished by their staining properties.[427] The most characteristic type stains conspicuously with 'chromehaematoxylin' and with 'paraldehyde-fuchsin'.

These cells and the axons coming from them have a luminous blue appearance (Tindall's blue) when examined under dark-ground illumination.[510] This derives from the fact that their secretory product is always in the form of minute spheres of about 1,000–3,000 Å. in diameter.[493] The axons from the medial neurosecretory cells cross over in the brain and form the medial nerves to the corpus cardiacum (p. 199) where the axons end in bulbous extremities from which the hormone is believed to escape into the haemolymph.[492, 493] The transit of the neurosecretory product along the axons is very evident in the Blattid *Leucophaea*;[492] but it has been observed in many other insects: the silkworm,[357] the larvae of *Leptinotarsa*,[358] *Calliphora*,[510] *Rhodnius*,[525] *Schistocerca*,[416] &c.[531]

The region of the brain which contains the large neurosecretory cells[129] was shown by implantation experiments in *Rhodnius* to be the source of the factor that initiates moulting (Fig. 41).[308] This has been confirmed in the giant silk moth *Hyalophora* (*Platysamia*) *cecropia*,[314] in *Calliphora*,[479] in *Iphita* (Hemipt.),[466] and in other insects. In *Hyalophora* there is evidence that secretion from a lateral group of neurosecretory cells may also be necessary.[512] The hormone secreted by the neurosecretory cells in Lepidoptera is non-specific: implants of the brain from *Bombyx mori* will activate the prothoracic gland and initiate growth in other Lepidoptera,[329] and two brains of the very small Apterygote *Ctenolepisma* will activate the thoracic glands of the large pupa of *Philosamia*.[539]

(ii) **The thoracic gland system.** As was first clearly demonstrated in the silkworm *Bombyx*, the 'thoracic gland' is the immediate source of the hormone which induces larval moulting, pupation, and imaginal development;[100] but the hormone from the brain is necessary to activate the thoracic gland. This was proved experimentally in the over-wintering pupae of *Hyalophora* and *Telea*,[314] and has been confirmed in *Rhodnius*,[351, 521] *Sialis*,[482] *Calliphora*,[343, 479] the Wheat-stem Sawfly *Cephus*[383a] and other insects.[514] In the larva of *Tenebrio* the brain and

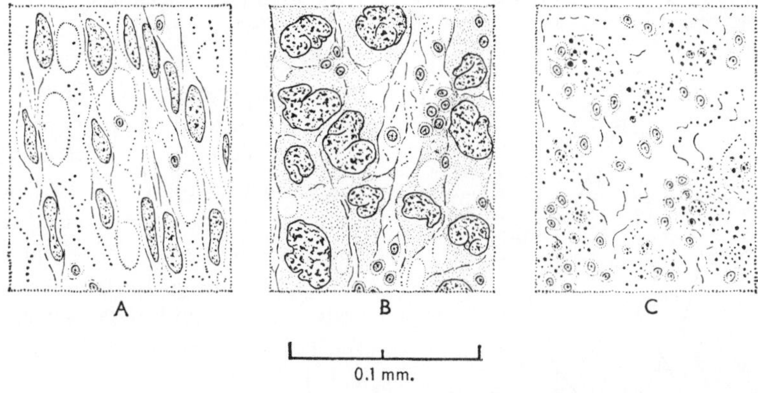

FIG. 42.—Changes in the thoracic gland during moulting in the 5th-stage larva of *Rhodnius*

A, unfed larva showing inactive gland cells and a few haemocytes. *B*, at 10 days after feeding; gland cells with dense cytoplasm and large lobulated nuclei; many more haemocytes present. *C*, at one day after moulting to the adult stage; numerous haemocytes around the degenerating nuclei of the gland cells (*after* WIGGLESWORTH).

the thoracic glands act successively as usual; the imaginal moult is controlled in the same way, but the critical period follows only a few hours after that of the pupal moult, which occurs in the prepupa, the two moults being closely integrated.[506] In the silkworm *Bombyx* the situation is similar: removal of the brain shows that the critical period for the pupal moult is at the beginning of the period of facultative feeding, and that for the imaginal moult a few days later, at the beginning of cocoon spinning.[373]

The thoracic glands go through a cycle of activity in which they show the greatest enlargement during the height of hormone secretion (Fig. 42).[521] This cycle is repeated at each moult until the insect becomes adult and then the cells break down and disappear, so that the insect is unable to moult again.[209, 521] But in the Thysanura (*Lepisma*, *Thermobia*, &c.), which continue to moult in the reproductive adult stage, the corresponding glands, the 'ventral glands', persist.[400] Extirpation of the thoracic or ventral glands will suppress moulting in *Locusta*,[430] in *Aeschna*,[491] and other insects; while moulting can be induced by the implantation of active glands in *Bombyx*,[100] *Hyalophora*,[314] the earwig *Anisolabis*,[473]

Aeschna,[491] *Rhodnius*,[521] &c. In *Gryllus* the brain is necessary for moulting,[347] but the histological changes in the prothoracic glands show that these also are probably concerned.[347] In *Periplaneta* the immediate source of the 'moulting hormone' is the prothoracic gland.[320]

In the larvae of higher Diptera, *Drosophila*, *Calliphora*, &c., the corpus cardiacum, the corpus allatum (p. 178) and the thoracic glands form a 'ring gland' (Weismann's ring) around the aorta just above the brain (Fig. 43). The large lateral cells of the ring, sometimes called the 'peritracheal gland'[479] (Fig. 126), form the homologue of the thoracic gland. The hormone which initiates puparium formation in these flies is secreted by these cells;[47, 126] the same hormone leads to histolysis of the larval organs and development of the imaginal discs;[27, 298] and it is responsible also for larval moulting; for if the isolated head segments of young *Drosophila* larvae are implanted into the abdomen of adult flies, they will moult only if the larval ring gland is implanted at the same time.[27] [See p. 119.]

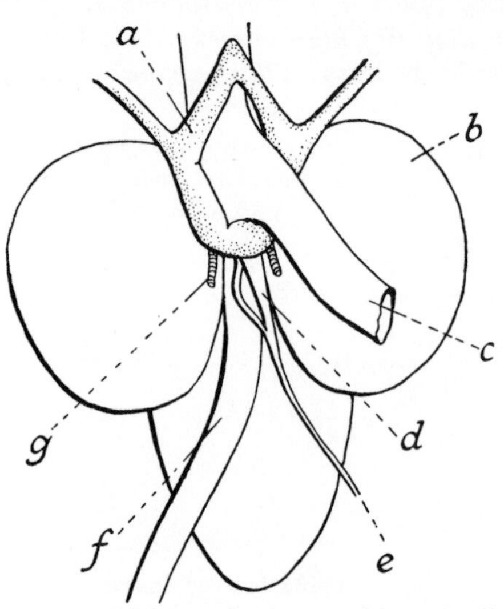

FIG. 43.—Weismann's ring or 'ring gland' in *Calliphora* (*after* BURTT)

a, Weismann's ring; *b*, brain; *c*, aorta; *d*, branch from recurrent nerve to Weismann's ring; *e*, recurrent nerve; *f*, gut; *g*, trachea to Weismann's ring.

Stimulus to secretion of the moulting hormone—Different insects are certainly caused to moult by different stimuli. The blood-sucking bugs *Cimex*[153] and *Rhodnius*[308] will not moult if they are starved, and if given small meals they will live indefinitely without moulting. In them it is not the state of nutrition but the distension of the abdomen by the meal which provides the stimulus to moulting. This stimulus is probably carried to the brain by the nerves, for section of the nerve cord prevents secretion of the moulting hormone.[308]

Stretch receptors have been demonstrated in the abdomen of *Rhodnius* (one on each side of each abdominal segment) which lead to the appearance of impulses in the nerves to the corpus cardiacum. They continue to discharge as long as the abdomen is distended.[511, 512] In *Locusta* it would seem that the hormone is being secreted continuously and that the stimulation of stretch receptors in the wall of the pharynx during feeding produces nerve impulses which travel via the frontal ganglion to the brain and bring about release of the hormone.[384] If several *Ptinus tectus* larvae are reared in one container, growth and moulting are greatly delayed.[124] Perhaps this points to a nervous inhibition of hormone secretion. Likewise the larva of the Wheatstem Sawfly *Cephus cinctus* is reluctant to resume development if removed from the stub of the wheat straw;[383a] and in the larva of the honey-bee contact with the silk cocoon provides a necessary stimulus to the brain for the initiation of pupation.[399] Similar effects are recorded in *Lucilia*.[194] And Uraniid larvae (*Chrysiridia*) may be caused to suffer a precocious metamorphosis and produce adults of half the usual size, apparently as the result of mechanical shock.[56] [See p. 119.]

Mode of action of the hormones from the brain and thoracic glands— The hormone from the neurosecretory cells of the brain activates the thoracic glands and causes them to secrete the moulting hormone. This is not a simple 'triggering' effect: the thoracic glands require the continued presence of the brain hormone until the 'critical period', when mitosis has begun and there is an adequate supply of moulting hormone in the blood.[308, 383a] In the full-grown larva of *Ephestia* the critical period coincides with a backward wave of mitoses throughout the epidermis; if the head and prothorax are tied off at this time, varying degrees of 'partial pupation' occur in the form of flecks or patches in those regions where mitoses have occurred.[174] The nature of the action of the brain hormone is unknown; one suggestion has been that the neurosecretory product of the brain may provide the labile raw material from which hormones manufactured by different parts of the endocrine system may be derived[531]—but there is no real experimental evidence to support this idea. In *Rhodnius* the haemocytes appear to play some undefined part in the initiation of moulting: if they are blocked with foreign matter, such as Indian ink, taken up by phagocytosis before the liberation of the thoracic gland hormone, moulting is delayed.[525]

The hormone of the thoracic gland is commonly called the 'moulting hormone'; it initiates all those changes in the epidermis and elsewhere which lead to the renewal of growth and the deposition of the new cuticle. Its action is most evident in insects such as *Rhodnius*[308] or the over-wintering pupa of *Hyalophora*,[314] in which the epidermal cells are in an inactive or dormant state until exposed to this hormone. In the embryo, of course, growth takes place before any endocrine glands have been developed; irradiation of the head region in the embryo of *Culex*, at the appropriate stage, results in complete lack of brain and corpus allatum complex, and yet the abdomen develops normally.[471] They are needed only to complete the moulting process that is ended by hatching;[431] and in the adult, in which thoracic glands are absent, the growth process associated with the healing of wounds can take place; these processes are not dependent on the moulting hormone. But although this hormone is primarily concerned in the induction of moulting, it may be responsible, when present in quite small amounts, for bringing about other changes which precede moulting in certain insects, notably the sclerotization and puparium formation in the larval cuticle of higher flies,[96] and the colour changes in the caterpillar of *Dicranura vinula* before it prepares its cocoon.[378]

The earliest visible effect of the hormone on the epidermal and other cells is described as 'activation': an enlargement of the nucleolus, the appearance of much ribonucleoprotein in the cytoplasm, and a great increase in the number of mitochondria.[526] These changes indicate an active renewal of protein synthesis, notably a renewal of enzyme synthesis. They are associated with a great increase in the respiratory enzymes (notably cytochrome *c*);[411, 445, 502] but the hypothesis that the synthesis of cytochrome *c* is the essential change that restores growth in the dormant insect has been abandoned.[410, 501] The moulting hormone will accelerate the incorporation of isotopically labelled amino acid into the proteins of the growing tissues,[509] and these syntheses entail a rise in respiratory metabolism.[541] But the same results follow injury to a non-moulting insect.[412, 508] [See p. 119.]

The characteristic of the moulting hormone is that it initiates these changes throughout the body in those tissues responsible for growth and moulting.[526, 531] The mechanism of this action is not known. It has been suggested that it might be influencing permeability relations within the cells.[509, 526] But there is an increasing tendency to think of these hormones as acting directly upon the nuclear genes, and bringing into operation those components of the gene pool that are needed to produce the enzymic syntheses necessary for growth and moulting. Evidence in support of this view consists in the visible enlargement of specific regions of the chromosomes (the so-called 'puffs' in the polytene chromosomes of *Drosophila* and *Chironomus*) of all the cells concerned. These changes may appear within 15 min. of exposure to the hormone in some gene loci; within an hour or two in others[364, 385] (Fig. 44). [See p. 120.]

Rhodnius larvae deprived of the actinomyces which normally develop in the gut will not moult as a rule beyond the fourth or fifth stage. The vitamins furnished by these organisms (p. 523) constitute a limiting factor in the formation or action of the moulting hormone.[42] In *Ephestia*, the formation of the pupation hormone is inhibited at low temperature (6–9° C.), although at this temperature the tissues are apparently still able to react to the growth stimulus if the hormone is present.[55, 525] On the other hand, absolute starvation, at least

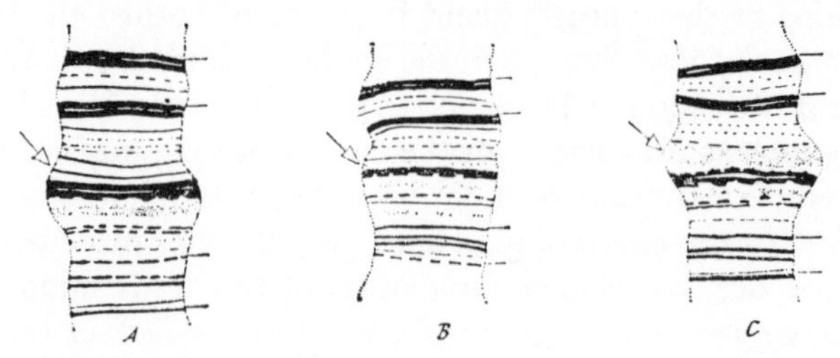

FIG. 44.—The 'puffing' of the locus 18-C on the chromosome I in the salivary glands of *Chironomus tentans* after the injection of ecdysone (*after* CLEVER)

A, untreated control. *B*, 30 minutes after injection. *C*, 2 hours after injection.

during the middle part of the last larval stage,[38] will usually precipitate pupation in caterpillars (pupae of *Galleria* one-tenth the normal weight have been produced[195]) while underfeeding or intermittent starvation cause pupation to be deferred[165, 206, 223] and may even result in some insects (*Lymantria*) attaining a larger size at pupation than the continuously fed controls.[166a] The same is true of mosquito larvae.[307] Larvae of *Tenebrio* and some other insects will moult several times although starved; starvation seems even to stimulate the process sometimes, as in the clothes moth *Tineola*, which may develop a sort of 'moulting frenzy' and cast the skin as many as eight times in two or three days without change of size.[291]

Regeneration of appendages tends to increase the number of moults in *Sphodromantis* (see p. 109)[228] and *Blattella*[317] and to cause moulting in the adults of *Machilis* (Thysanura).[229] This suggests a relation between the factors responsible for wound healing and for moulting. The healing of wounds (a form of localized moulting) can take place in the adult (p. 106). The absence of moulting in most adult insects is probably due to failure of the moulting hormone; for imaginal organs of the earwig *Anisolabis*[102] or the cricket *Acheta*[263] transplanted to the nymph will grow on and will moult when their host moults; and adult bed-bugs *Cimex* may be caused to moult by transfusing them with blood from moulting larvae of *Rhodnius*,[309] as also can the adults of *Rhodnius* itself.[308]

Chemical nature of the hormones from the brain and thoracic gland—The submicroscopic spherules, which are the main product of the neurosecretory cells in the brain, do not contain lipids but consist of a protein rich in sulphydryl groups.[359] However, it has been found that an oily extract from the brain of the silkworm will activate the thoracic glands, and it has been claimed that the active principle is cholesterol.[440, 442] In most vertebrate animals the active product of the neurosecretory cells consists of small polypeptides associated with inactive large protein molecules, and it has been claimed by other workers that preparations can be obtained from *Bombyx mori*, by extraction with saline, which are much more potent than the oily extracts in stimulating the thoracic gland, and in which the active principle is a protein.[421] [See p. 120].

The chemical investigation of the thoracic gland hormone started with the isolation of the factor which induces puparium formation in the posterior fragments of larvae of *Calliphora* ligated just before the critical period (Fig. 39). Apparently identical material was isolated from pupae of *Bombyx.mori* (injection into the posterior fragments of *Calliphora* being used as the method of assay) and has been obtained in crystalline form.[381] This purified moulting hormone is named 'ecdyson'. It has a molecular weight of 464 and the empirical formula $C_{27}H_{44}O_6$.[436] It appears to be a steroid, probably derived from cholesterol; indeed cholesterol labelled with tritium will give rise to radioactive ecdyson when injected into *Calliphora* larvae. This would explain the efficiency of cholesterol in favouring activity of the thoracic glands.[435]

Ecdyson exists in several slightly different molecular forms.[380, 434] It is non-specific in its action; ecdyson from different sources will induce puparium formation in *Calliphora*, renewed development in the diapause pupa of *Hyalophora*,[381] moulting in *Rhodnius*[525] and in other insects. A dose of about 10 μg/g. is needed to bring about complete development in these insects. [See p. 120.]

METAMORPHOSIS

Definition of metamorphosis—While growth in size is one purpose of moulting, change of form is another. Since the outward form is determined by the cuticle and can change only when the cuticle is shed, change of form is even more strikingly discontinuous than change in size. The intermediate stages of development become apparent only under special circumstances. There are all degrees of such discontinuity. In the Apterygota the gradations are extremely slight; the insect becomes sexually mature at some indefinite point, and even then moulting may continue.[229] Such insects are termed Ametabola: attaining the adult stage without a metamorphosis. In all other insects, although the change of form is generally slight at the earlier moults, there are conspicuous changes in the final stages. The adult stage is generally winged; its sexual appendages suddenly become prominent; and if its mode of life is changed it may suffer a dramatic transformation.

From the standpoint of phylogenetic classification, great importance is attached to whether the wing buds in the early stages are visible externally (Exopterygota) or concealed beneath the cuticle until metamorphosis begins (Endopterygota); and whether the active immature larva is transformed directly into the adult (Hemimetabola, showing 'incomplete metamorphosis') or whether a pupal stage, incapable of feeding and often comparatively inactive, is interposed between the larva and the adult so that metamorphosis takes place in two stages (Holometabola, showing 'complete metamorphosis'). From the standpoint of physiology, although it is convenient to use these terms, such distinctions are untenable. The histological processes of development are of the same type in all, though there are differences in degree. We shall therefore follow Handlirsch[128] in applying the term 'metamorphosis' to the more or less marked change of form which occurs when the insect becomes adult, whether this change requires two moults and a pupal stage or a single final moult. By the implantation of fragments of pupal integument of *Galleria* into pupating larvae they may be made to undergo a second pupal moult; and the imaginal integument will moult again.[210]

Where a pupa exists its formation marks the occurrence of metamorphosis. The pupa is best regarded as an intermediate instar which serves to bridge the morphological gap between a larva and an adult which have become adapted to widely different types of existence.[524] This intermediate stage serves another useful purpose in allowing two moulting steps for the development of the changed muscular system of the adult;[140] for the tendons and 'tonofibrillae' of the muscles are attached by way of the epidermal cells and endocuticle to the epicuticle, and therefore can be established only during moulting.[361]

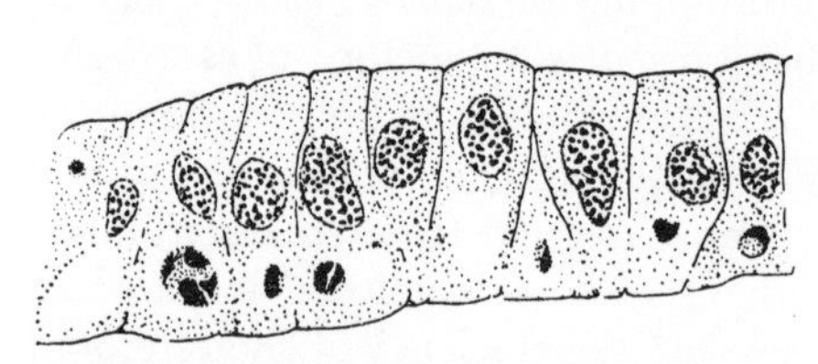

FIG. 45.—Epidermis of young pupa of *Calandra oryzae* showing appearance of chromatic globules (*after* MURRAY and TIEGS)

Histological changes—During moulting, even when unaccompanied by any striking change of form, there is much reconstruction in the epidermis. New cell divisions and the death and dissolution of unwanted nuclei take place simultaneously. The 'chromatic droplets', which appear as a result, have been described in *Galerucella* (Col.),[220] *Hyponomeuta* (Lep.),[146] and aquatic Hemiptera,[217] in *Calandra* (Col.) (Fig. 45),[197] *Tenebrio*,[311] and *Rhodnius* (Fig. 46).[311] They are not an essential phenomenon of metamorphosis; they do not signify 'dedifferentiation' or 'rejuvenation' among the cells, but merely serve to show that growth and cellular breakdown, 'histogenesis' and 'histolysis' go forward together.[311] [See p. 121.]

In those parts of the body and in those insects where the more extreme type of metamorphosis occurs, the rudiments or 'Anlagen' of the adult organs are already present in the early stages as clusters of undifferentiated embryonic cells, the 'imaginal discs', the significance of which was first appreciated by Weismann. These arise as thickenings of the epidermis, often with loose accumulations of mesenchyme cells below them. In Hymenoptera and higher Diptera the entire epidermis of the adult is derived from embryonic cells set aside for this purpose. But in most parts of the body in Lepidoptera and Coleoptera, and in all parts of the body in Hemimetabola, the adult cuticle is laid down by the same cells or by the daughter cells of those which previously laid down the larval cuticle. The chief function of the imaginal discs is to free certain ectodermal cells from the task of cuticle formation; the cell nests can thus grow more or less continuously and so permit the evolution of a highly specialized larval form which is uninfluenced by the form of the future adult (Fig. 47).[531]

The legs of caterpillars are sometimes said to contain at their base an imaginal disc which gives rise to the adult leg.[24] But that is not the case: the whole epidermis of the larval leg grows and differentiates to form the leg of the adult; the entire leg functions as an imaginal bud.[439, 446]

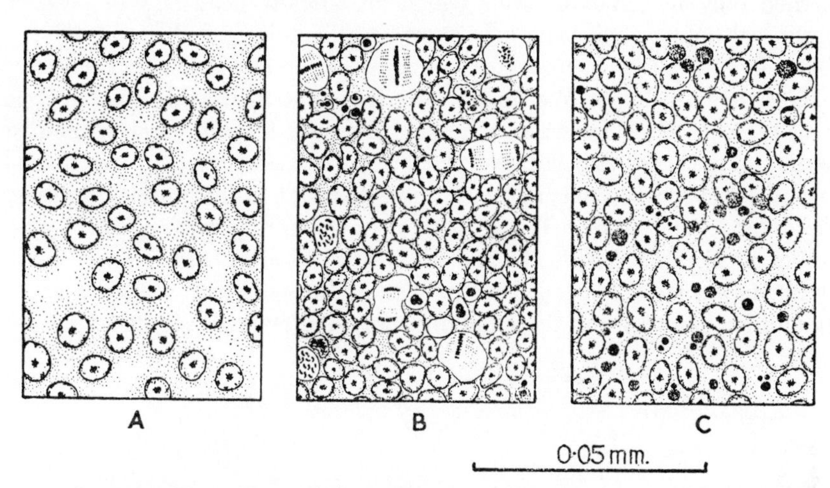

Fig. 46.—Corresponding regions of the epidermis of the abdomen in *Rhodnius* fifth-stage larva, seen in surface view at different times after feeding

A, 3 days, cell division not begun; *B*, 8 days, active mitosis, cells densely packed, nuclear breakdown beginning; *C*, 13 days, cells less dense, abundant chromatic droplets resulting from the breakdown of excess nuclei (*after* Wigglesworth)

The difference between metamorphosis in Hemimetabola and Holometabola is one only of degree. Even in the extreme metamorphosis of Diptera or Hymenoptera many of the larval organs and cells do not succumb but are reconstituted to form the adult body. There is never demolition and then reconstruction, but

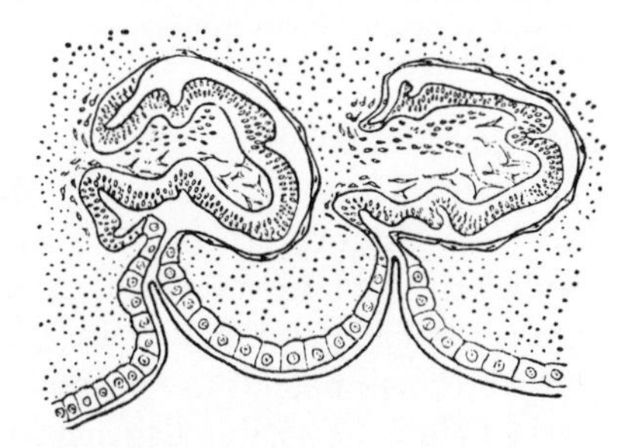

Fig. 47.—Longitudinal section through the imaginal discs of the wings in the full-grown larva of the ant *Formica* (*after* Pérez)

progressive substitution. The most specialized larval structures disappear, the most specialized parts of the imago are built anew from the imaginal discs; but many organs, such as the Malpighian tubes of Diptera and Coleoptera, though more or less reconstructed, are common to both larva and imago. Among the muscles of Muscidae and Hymenoptera there are all grades between survival and replacement.[207]

Histolysis—About the mechanism of histolysis there has been much controversy; particularly as to the part played by the phagocytes in the blood. At the present time it is agreed that there are great differences in different insects or in different organs of the same insect; and it is agreed that the phagocytes are concerned only in removing cells which are already dead and in process of autolysis—although such cells may show little histological change.[5, 207, 290] In the blow-fly *Calliphora*, haemocytes stuffed with granules of disintegrating tissue, the 'Körnchenkugeln' of Weismann, are abundant in the blood of the pupa (Fig. 48).[169, 207, 232] Whereas in the allied blow-fly *Lucilia*, autolysis and fragmentation of the muscles occur spontaneously; the haemocytes appear late on the scene and are concerned only in the ultimate digestion of the muscle fragments.[89] A like diversity is found in Hymenoptera: in the ant *Formica*, salivary glands, Malpighian tubes, fat body cells and many muscles are phagocytosed;[207] in the Chalcid *Mormoniella* (*Nasonia*), many of the tissues dissolve without the intervention of haemocytes, but these fall upon any undissolved fragments present in

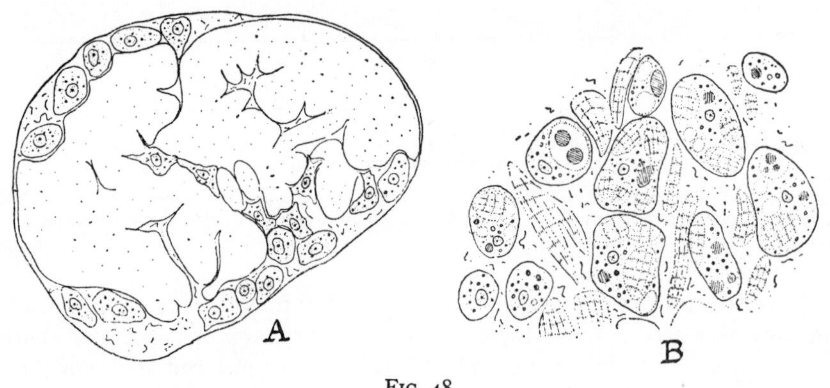

FIG. 48

A, transverse section of larval muscle of *Calliphora* 7 hours after pupation, showing the penetration of haemocytes beneath the sarcolemma and deep into the muscle; *B*, advanced stage in dissolution of the muscle (10 hours after pupation). Haemocytes stuffed with débris to form 'granular spheres'; some muscle fragments (sarcolytes) are still free (*after* PÉREZ).

the blood, and some tissues are attacked when they show little visible change; here chemical autolysis and phagocytosis take place side by side;[290] while in the honey-bee there is no evidence of phagocytosis in any organ, whether muscle, fat body, silk glands or gut wall.[201, 284] In Lepidoptera, extensive phagocytosis of dead cells may occur in the nerve ganglia[250] and muscles.[18] In *Calandra* (Col). some tissues are phagocytosed, others not;[197] in *Galerucella* (Col.) phagocytosis is general; many cells attacked show no visible degeneration.[220] [See p. 121.]

Metamorphosis and embryonic development—Although the mechanisms of histolysis and histogenesis vary greatly in detail in different insects, they always serve a common purpose: the replacement of the larval organism by the adult. The insect may be looked upon as made up of two organisms, larva and imago, existing in a single individual. As we saw in discussing early embryonic development (p. 7), organization is not merely a co-operation among cells; the 'organism' appears to be a differentiated continuum whose division into cells is a late occurrence in development. The 'individuation field' which moulds the body form is something superior to the cells.[313]

This idea of a dual determination in a single body is supported by those experiments on *Drosophila* and *Tineola* (p. 12) which show a wave of imaginal determination spreading throughout the egg, quite separate in point of time from the larval determination, and occurring at a stage of development when histological differentiation of both larval and imaginal structures is entirely wanting. One might therefore regard metamorphosis as the progression of this imaginal determination to the stage of visible differentiation—thereby superseding the larval differentiation which preceded it. This conception would make metamorphosis, what many have considered it, a return to embryonic development.[139] It consists in the realization of the latent imaginal characters.[125, 308]

The nature of metamorphosis—There are two ways of regarding metamorphosis.

(i) It may be supposed that the insect is undergoing a progressive development towards the adult form; that it is subject during its larval stages to some inhibition which prevents the completion of this development until it is fully grown; that the restraint is then removed and differentiation is completed.

(ii) Alternatively, it may be supposed that metamorphosis involves a switch into a line of development which is qualitatively different from that of the larva. In other words, that a new system of genes, previously latent in the chromosomes, is brought into action at a given point in development, and it is this system that is responsible for the control of the adult form. According to this conception,[522, 524, 531] the origin of metamorphosis has consisted in the independent evolution of two genetic systems, one controlling larval form and the other controlling adult form. The successive stages of the insect are thus comparable with the different forms in a polymorphic species (p. 98), or with the different forms of the parts of the body, which are likewise believed to result from the activation of different sets of genes.

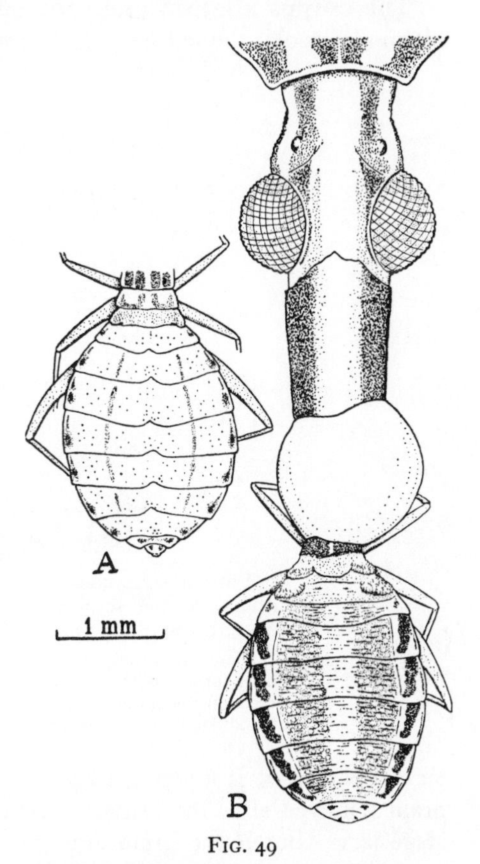

FIG. 49

A, thorax and abdomen of normal 2nd instar larva of *Rhodnius*; *B*, precocious 'adult' produced from a 1st instar larva by joining it with paraffin wax to the tip of the head of a moulting 5th instar larva (*after* WIGGLESWORTH).

Hormonal differences certainly exist in the blood at metamorphosis, and these are responsible for the morphological changes. If young larvae of *Rhodnius* are decapitated and transfused with the blood of 5th-stage larvae, which are in process of moulting to become adult, they undergo a precocious metamorphosis. Even 1st-stage larvae recently emerged from the egg, if treated in this way, will develop the cuticle and pigmentation rudimentary wings, abdominal structure and genitalia characteristic of the adult (Fig. 49).[308] Pieces of the integument from larvae of *Galleria* (Lep.) implanted into pupating larvae of the same species, or another species such as *Achroea*, moult to form pupal cuticle synchronously with their new host; further development of these implants to form imaginal cuticle takes place when the host pupa completes its metamorphosis.[210] By implantation of isolated fragments in the mature larva, the integument of newly hatched *Galleria* larvae can be caused to become pupal and then imaginal.[210]

The insect seems to be capable of developing its imaginal characters, that is, of undergoing metamorphosis at any stage—provided that the appropriate hormones are circulating in the blood.

The corpus allatum and the control of metamorphosis—The corpus allatum (p. 200) is the source of the hormone which maintains the development of

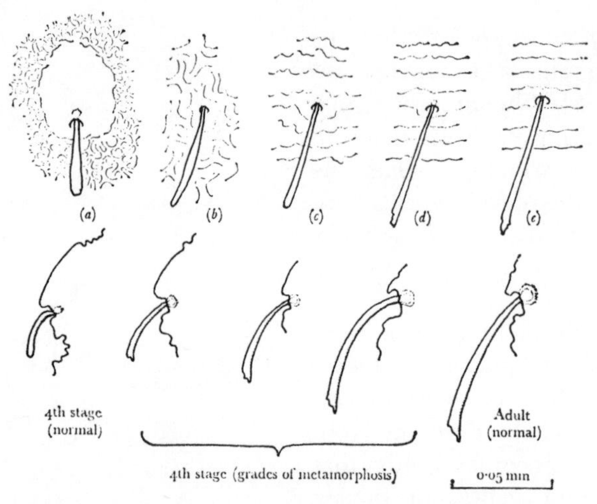

FIG. 50.—Precocious metamorphosis, and intermediate stages, in the sensory hairs of *Rhodnius*, produced by decapitation of the 3rd-stage larva

a, normal 4th-stage larva; *b*, *c* and *d*, intermediate and imaginal hairs formed in the decapitated insects; *e*, normal adult. (Above, hairs on the surface of the abdomen; below, from the margin of the abdomen in the same insects.)

larval characters. If a young stage (say the 3rd-stage larva) of *Rhodnius* has the brain removed after the critical period it completes its development to a 4th-stage larva. But if the brain and corpus allatum are removed it will develop partially adult characters (Fig. 50).[308] Conversely, if the corpus allatum from 3rd- or 4th-stage larvae in *Rhodnius* is implanted in the abdomen of 5th-stage larvae, these develop into giant larvae (Fig. 51, C); or, in less successful experiments, into forms intermediate between larva and adult (Fig. 51, D); or, sometimes, into adults with a small patch of cuticle of larval type immediately over the site of implantation (Fig. 52).[308] Some of the giant 6th-stage larvae have given rise to a partially larval 7th stage; others have developed into giant adults.[308]

In the silkworm, removal of the corpora allata from young larvae is followed by pupation. These larvae normally moult in the 5th stage to produce pupae weighing about 1·25 gm.; but larvae from which the corpora allata have been excised in the 2nd stage have been caused to turn directly into diminutive pupae weighing as little as 0·025 gm. (Fig. 53).[38] A similar though less dramatic effect is seen in the Phasmid *Carausius*. This usually becomes sexually mature in the 8th instar. If the corpora allata are removed in the 3rd instar, the insect becomes sexually mature in the 6th instar, having omitted the last two ecdyses.[209]

Carausius is an unusual species in that the females begin to reproduce par-

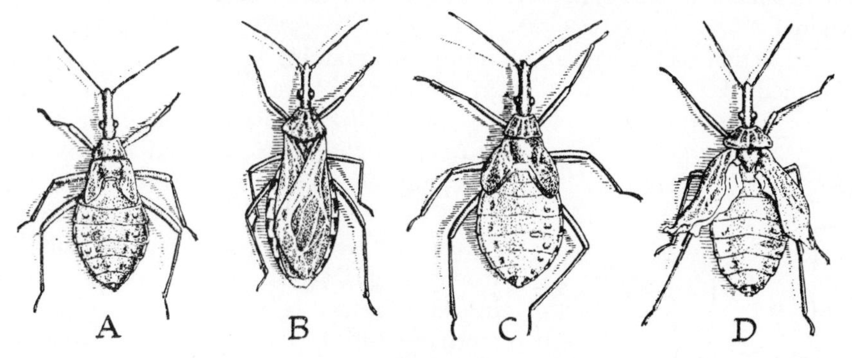

FIG. 51

A, normal 5th-stage larva of *Rhodnius*; *B*, normal adult *Rhodnius*; *C*, '6th-stage' larva produced by implanting the corpus allatum from 4th-stage larva into abdomen of 5th-stage; *D*, '6th stage' similarly produced, showing larval cuticle all over abdomen, wings, thorax, &c., intermediate between larva and adult (*from photographs by* WIGGLESWORTH).

thenogenetically (p. 725) before they are fully adult; they are in fact 'neotenic' insects[149] (p. 732). If five or six corpora allata are implanted from young larvae into these semi-mature adults they can be induced to make two further moults

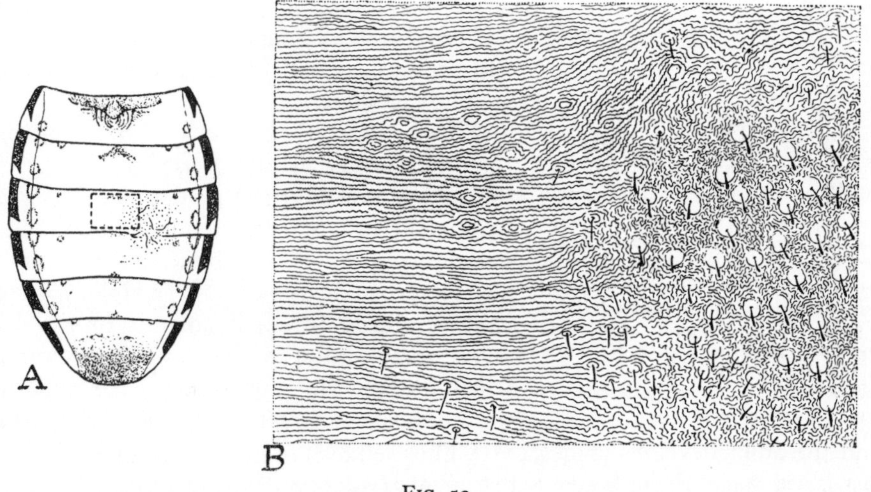

FIG. 52

A, dorsal view of abdomen of *Rhodnius* adult showing area of larval cuticle around site of implantation of corpus allatum of 4th instar larva; *B*, detail from *A* showing larval cuticle on right merging into adult cuticle on left. Note vestiges of sockets without bristles in the intermediate zone (*after* WIGGLESWORTH).

and become giant insects. After implantations into 6th-stage larvae as many as four extra moults have been made and giants up to 15·1 cm. in length, or twice the normal size, have been produced.[209] These giant forms show partial development of a median dorsal ocellus on the head—an imaginal character which does not normally appear in this species.[209] In the bisexual Phasmid *Bacillus* the males likewise become functional adults in the 5th instar if the corpora allata are removed in the 4th stage.[91] Even the normal adult *Periplaneta* seems to be slightly 'neotenic', or juvenile; for if it is induced experimentally to moult again, the external genitalia differentiate beyond the normal adult stage, they become 'superimaginal'.[366]

The action of the corpus allatum in maintaining larval characters has been confirmed (i) in the Orthoptera: in *Melanoplus*;[208] in the Blattid *Leucophaea*,[253] and in *Gryllus*.[218] (ii) In other Lepidoptera: in *Ephestia* and *Galleria*[211] and *Hyalophora*.[314] Not only does removal of the corpora allata from the young larvae of *Galleria* result in precocious metamorphosis, but implantations from young larvae into full-grown larvae result in extra instars leading to giant pupae and moths.[211] (iii) In Coleoptera: implantation of young corpora allata early in the last larval stage of *Tenebrio* has led to six extra moults and the production of very large larvae.[230] The role of the corpus allatum in maintaining larval characters is less easy to demonstrate in the Diptera. But the central part of the ring gland, which is homologous with the corpus allatum (p. 201), when implanted into the abdomen of *Drosophila*, will induce the local formation of larval integument in place of adult cuticle;[298] and the corpus allatum of the larva of *Calliphora* will produce a small patch of larval cuticle in an adult *Rhodnius*.[523]

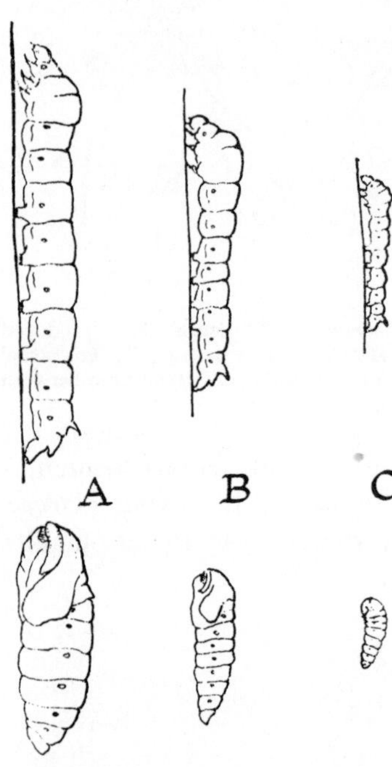

FIG. 53.—Precocious metamorphosis in the silkworm (*after* BOUNHIOL)

A, larva 7 days after final moult (at the stage when it is just large enough for pupation to be possible in the normal insect); below, the pupa resulting from starvation of such a larva. *B*, larva 4–5 days after third moult; below, pupa resulting from ablation of corpora allata at this stage. *C*, larva 3–4 days after second moult; below, pupa after ablation of corpora allata.

From these observations it is concluded that the corpus allatum hormone is secreted during the first four larval stages in *Rhodnius*, &c.; in the fifth stage secretion is inhibited by nervous control from the brain and then adult characters can develop.[308, 492] In Holometabola it is produced in small amount in the last larval stage and it is this low concentration which leads to the appearance of the pupa; in the pupal stage secretion is absent altogether and the adult develops (Fig. 54).[476] Thus removal of the corpus allatum in the last larval stage of the honey-bee[490] or of *Hyalophora*[535] results in partial omission of the pupal stage with the development of forms intermediate between pupa and adult. [See p. 121.]

In the *Thysanura*, in which the adults alternately moult and reproduce, the corpus allatum hormone is present, but does not control metamorphosis: no unduly small adults are produced if the gland is extirpated.[485] In the adults of all insects the corpus allatum again becomes active and its hormone plays an important part in reproduction (p. 724).

The juvenile hormone—The hormone of the corpus allatum is commonly called the 'juvenile hormone' or 'neotenin'. Like the hormones of the brain and thoracic glands it is non-specific: the blood of young *Rhodnius* larvae will prevent metamorphosis in *Triatoma* and in *Cimex*;[308] the corpus allatum of *Periplaneta* will maintain larval characters in *Oncopeltus* (Hem.)[339] or *Rhodnius*;[351] metamorphosis in *Galleria* can be prevented by the hormone from *Ephestia*, *Bombyx*, *Tenebrio* and *Carausius*,[341] and the corpus allatum of *Ctenolepisma* will prevent metamorphosis in *Philosamia*.[539]

An important effect of the juvenile hormone is the maintenance of the thoracic gland. In each larval stage this gland goes through a cycle of secretion (p. 67); but in the adult *Rhodnius* it breaks down within twenty-four hours after moulting.[351] Its breakdown is prevented if it has been exposed to the juvenile hormone.[521] In some insects it breaks down less readily: if the corpus allatum is removed from young *Carausius*[209] or in the cockroach *Leucophaea*[253] the insect makes several moults before metamorphosis occurs. Perhaps the hormone persists for some time in the blood.

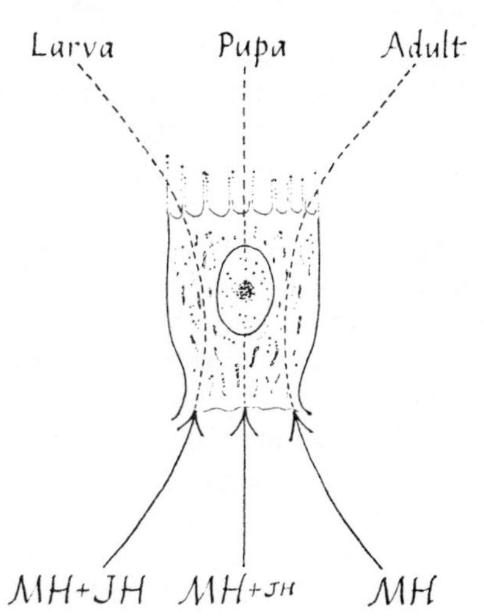

FIG. 54.—Hormonal control of metamorphosis in epidermal cell acted upon by moulting hormone (M.H.) and juvenile hormone (J.H.) in different proportions

The juvenile hormone acts directly upon the epidermal cells and ensures that they retain their larval characters, presumably by activating the requisite elements in the gene system. Even after metamorphosis this larval system is still capable of renewed activity in some of the cells. Adult *Rhodnius* caused to moult again after the implantation of juvenile corpora allata, lay down cuticle showing some larval characters; that is, a partial reversal of metamorphosis is possible.[308] By implanting small fragments of pupal or adult integument of *Galleria* into the abdomen of very young larvae it can be made to revert, in the course of several moults, to the larval type.[210, 342] Reversal has also been produced in the thorax of the earwig *Anisolabis*[473] and in the alimentary canal of *Galleria*.[477] [See p. 121.]

In certain insects the behaviour of the full-grown larva preparing for the pupal moult is quite different from that of the younger stages preparing for a larval moult. This behaviour is controlled by the juvenile hormone. If active corpora allata are implanted into the last stage larvae of *Galleria* it spins a slight silken web instead of a tough cocoon. Likewise, the Sphingid larva *Mimas tiliae* can be prevented from migrating down the tree trunk to the soil.[340]

Apart from normal metamorphosis, it seems that the juvenile hormone is concerned in some examples of polymorphism; such as the soldier caste of termites (p. 101),[432, 456] the solitary phase of locusts (p. 105),[429] and the wingless form of parthenogenetic Aphids (p. 104).[391, 451] Finally, it is possible that the juvenile hormone may be involved in the control of 'hypermetamorphosis' in *Meloë* and other insects.[139]

Chemical nature of juvenile hormone—The juvenile hormone seems to be identical with the hormone produced in the adult insect and concerned in reproduction (p. 707), for the hormone from the adult will prevent metamorphosis in the young stages, and the corpus allatum of the young stages will induce yolk formation in the adult female.[312] Active preparations of the hormone were first obtained by extracting with ether the lipid contents of the adult male *Hyalophora*.[534] By selective fractionation highly concentrated extracts have been prepared which reproduce all the activities of the juvenile hormone[404] including the 'reversal' of metamorphosis as in *Rhodnius*.[527] There is great variation in the susceptibility of different tissues to the effects of the juvenile hormone; tissues showing the most active cell division are the most resistant, that is, the most difficult to prevent from undergoing metamorphosis.[531]

Simple methods for assaying extracts for juvenile hormone activity have been devised,[404, 495, 527] and active preparations have been obtained from the most diverse sources: insects, in which only small amounts are usually stored,[527] crustacea, and many other invertebrates,[495] the suprarenal glands of cattle,[403] other organs of vertebrates, including man,[534] and even plants, bacteria, and yeasts.[496] The active substance responsible in the excreta of *Tenebrio*, in the suprarenal gland, and in yeasts, proved to be a mixture of the isoprenoid alcohol farnesol and its aldehyde farnesal;[494] and these materials and derivatives of them, when suitably formulated and administered, proved highly active in preventing and reversing metamorphosis in *Rhodnius*, and in inducing yolk formation.[528] The active material in extracts from *Hyalophora* and other silk moths likewise contains a mixture of farnesol and farnesal, and these materials can be synthesized by the insect from isotopically labelled mevalonic acid (the normal precursor of farnesol in living organisms) to yield radioactive farnesol.[494]

These results suggested that the natural juvenile hormone would probably prove to be an isoprenoid related to farnesol, which is well known as an intermediate in the biosynthesis of cholesterol and of carotenoids. Many derivatives of farnesenic acid or farnesol show high juvenile hormone activity, notably farnesyl methyl ether,[494] esters of epoxy-farnesenic acid[550] and other aromatic terpenoid ethers.[550, 646] The isolation and synthesis of the natural hormone in *Hyalophora* has borne out this conclusion. [See p. 121.]

Abnormal metamorphosis: prothetely and metathetely—Abnormal metamorphosis may occur in insects belonging to both hemimetabolous and holometabolous groups, and results in the production of forms intermediate between larva and adult or between larva and pupa.[287] This condition was named 'prothetely' by Kolbe on the assumption that these insects were larvae which had suffered a partial metamorphosis before their time. In larvae of *Simulium* (Nematocera) infested by the Nematode parasite *Mermis* a condition occurs in which development of the imaginal discs and sexual organs is inhibited although the larva continues to grow. For this abnormally retarded condition Strickland[277] proposed the term 'metathetely'. Recent authors have inclined to regard most examples of the so-called prothetely as being of this type: a form of 'neoteny' or retention of larval characters by insects which should have turned into pupae or adults—as in *Tenebrio*[180, 222] and *Tribolium*[58, 199] larvae, which occasionally develop wing pads (Fig. 55). Larvae showing only a slight degree of abnormality may continue to grow and moult, though they are apt to produce imperfect adults.[32, 180] [See p. 122.]

Where the insects in question have an indefinite number of instars and are prone to pupate over a wide range of size, there is little real distinction between prothetely and metathetely. On the other hand, where an insect like *Locusta*, which normally moults five times, develops incompletely adult characters in the 4th instar and becomes sexually active when still so small that the eggs cannot pass down the vagina,[155] this is a clear example of prothetely; and where adult insects, capable of reproduction, still retain a number of nymphal or infantile characters, as observed in *Gryllus campestris*,[70] this is an example of metathetely or neoteny.

These intermediate forms may be produced by abnormal conditions of many sorts. By high temperatures (29·5° C.), in *Tenebrio*;[222] by exposing the nearly mature larva to cold, in *Tenebrio*[180] and *Tribolium*;[199] by exposing larvae of *Tribolium* to an irritant gas emitted by the adult beetles[58] or to acetic acid vapour;[240] by keeping *Sialis* (Neuropt.) larvae under unusual conditions in the laboratory.[32] Their development is most satisfactorily explained by the production, too early or too late, in too high a concentration or too low, of the chemical factors which regulate metamorphosis.[32, 180, 277] For the intermediate state affects all parts of the body, including the gonads (Fig. 55).[113] This is certainly the explanation of the prothetely and metathetely produced experimentally in *Rhodnius* (p. 69); the former by providing young larvae with the blood of 5th-stage larvae, the latter by implanting into 5th-stage larvae the corpus allatum of some younger stage.[308] And *Galleria* pupae[211] or *Tenebrio* pupae[230] developed after the implantation of corpora allata, exactly resemble the naturally occurring abnormalities. When intermediates between larvae and pupae of *Galleria* are produced by implantation of young corpora allata into the last stage larvae, these produce a cocoon with characters intermediate between that prepared for pupation and the delicate structure prepared for larval moulting.[340]

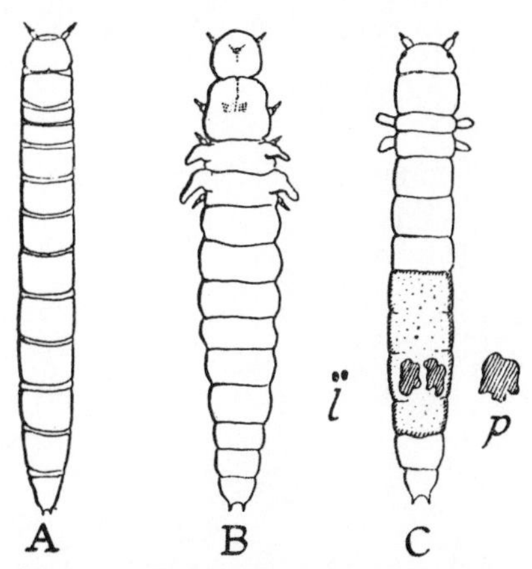

Fig. 55.—Prothetely in *Tenebrio molitor*

A, normal larva; *B*, form intermediate between larva and pupa; *C*, larva-like form with wing lobes, dissected to expose testes. *i*, testes of normal larva drawn to same scale; *p*, testis of normal pupa drawn to same scale (*after* v. Lengerken).

Prothetely in larvae of *Lymantria* is regarded by Goldschmidt[113] as an example of those developmental anomalies explicable by disturbances in the *rates* of various developmental processes; in this case an upset in the time relations of metamorphosis and evagination of imaginal discs. This idea may be extended to explain the normal restraint of metamorphosis in the young insect, if it be supposed that two processes are set in motion at each moult: (i) differentiation of the adult form and (ii) deposition of the new cuticle. If the second of these processes supervenes immediately after moulting has started, there will be little morphological change; that is, metamorphosis will be inhibited. If the second process is delayed, time will be allowed for the differentiation of adult characters, and metamorphosis will occur. Prothetely or metathetely would represent intermediate stages in a process of which only the end result would ordinarily become apparent.

This interpretation may be illustrated from the bed-bug *Cimex*. The homologies of the external organs of the male *Cimex* were elucidated by following the development of the epidermal rudiments.[64] If metamorphosis is partially 'inhibited' in this insect by joining it to a young larva of *Rhodnius*, it is caused to lay down cuticle before its time, and the genital appendages appear in an intermediate stage of development, so that the homologies become evident in the cuticular structure (Fig. 56).[308] Upsets in the normal rates of developmental processes are particularly liable to occur in hybrids (p. 95). It is noteworthy that the same applies to prothetely in Lepidoptera, which seems to be most frequent in hybrids, where a disturbance in the normal balance and timing of hormone activity might be expected to occur.[68] In some cases it may be a change in the amount and the timing of juvenile hormone secretion that is responsible; but similar abnormalities can be produced by excessive supply of the moulting hormone.[409, 521, 531]

Pupal development—In holometabolous insects one instar, the pupa, is normally interposed between the larva and the imago, and the formation of the

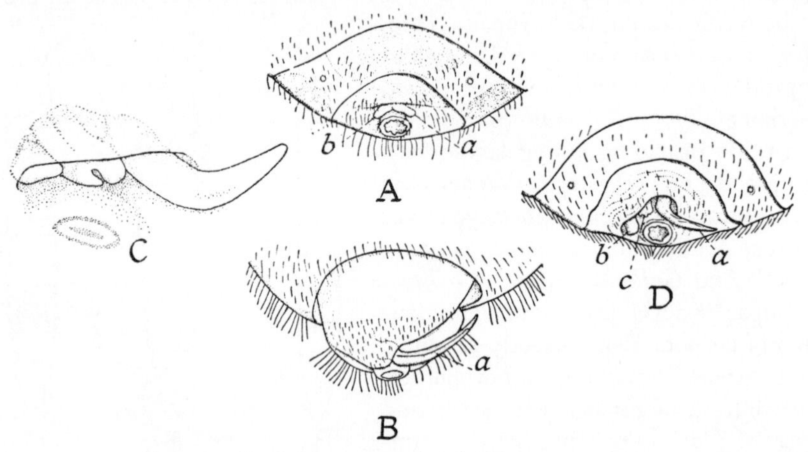

FIG. 56.—Genital segments of male *Cimex*, ventral view

A, normal 5th instar larva; *B*, normal adult; *C*, epidermal rudiments of the claspers, &c., in a male larva approaching the adult stage (*after* CHRISTOPHERS and CRAGG); *D*, a stage with characters intermediate between 5th instar and adult produced by transfusing a decapitated 5th instar larva with blood from a moulting 3rd instar larva of *Rhodnius* (*after* WIGGLESWORTH). *a*, left clasper; *b*, right clasper; *c*, aedeagus.

pupa marks the commencement of metamorphosis. There are, however, departures from this rule. In the higher Diptera metamorphosis begins with the formation of the puparium by the epidermis of the 3rd instar larva; but a delicate cuticle of larval type is moulted within the puparium before the pupa is formed (Fig. 57).[7, 239, 270] In *Calliphora* this 'pre-pupal cuticle' is formed only over the abdomen and is attached in front round the margin where the 3rd thoracic joins the 1st abdominal segment.[96a] It has been claimed that the so-called prepupal cuticle in *Drosophila*, *Calliphora*, and other flies is merely the unsclerotized inner part of the larval endocuticle added after the hardening of the puparium.[520]

In some sawflies there is a 'prepupal' instar, during which no feeding takes place and the salivary glands become modified for spinning the cocoon, before moulting to the pupa occurs.[83] In the males of some Coccids there are two immobile non-feeding stages before the adult is set free.

The physiological control of these more complex changes has not been studied; but here we may consider to what extent the continued development of the pupa is dependent on the presence of growth hormones. Conditions seem to vary in different insects. Larvae of *Lymantria* in which the brain is removed after the critical period (p. 64) so that they duly pupate, continue their development up to the formation of the imago; the prothoracic glands remaining, the brain is not needed to initiate pupal development.[165] On the other hand in the pupae of *Hyalophora* and *Telea*, which have an obligatory pupal diapause (p. 116) the brain is needed to activate the prothoracic glands.[314] If the pupa of *Drosophila* is extracted from the puparium and ligated in the middle between 1–24 hours after pupation, the posterior half does not develop; if the ligature is applied 36–48 hours after pupation, complete development of the posterior half occurs;[25, 27] developmental factors appear to come from the anterior half, presumably from the ring gland. Similar experiments on *Vanessa*[125] and *Phryganidia* (Lep.)[27] likewise show that if the abdomen of the pupa is separated from the thorax at an early stage, development of the adult abdomen does not occur; presumably because it is deprived of the secretion from the prothoracic glands. Transplantation experiments show that development of the ovaries in the pupa of *Celerio* (Lep.) is clearly dependent on hormones in the blo od.[51] (In *Phryganidia* the ovaries will develop normally even when separated from the thorax; so that they are not dependent on the centre which is necessary for the continued development of the ectoderm.[27])

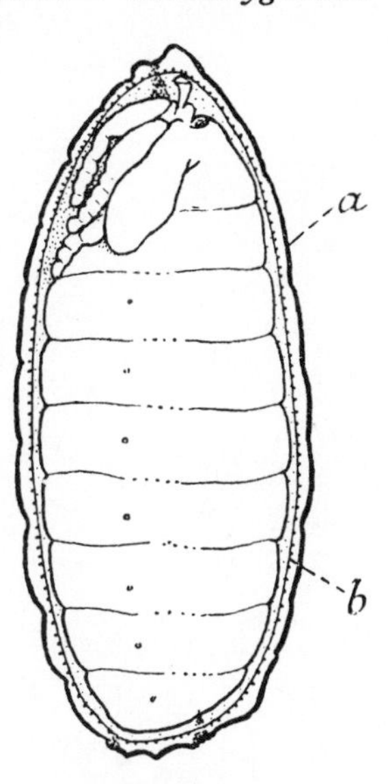

FIG. 57.—*Rhagoletis pomonella*

Pupa enclosed in two envelopes, a, the puparial shell, b, the 'pre-pupal skin'. The pupal head not yet evaginated from the thorax (*after* SNODGRASS).

The determination of structural characters during post-embryonic development will be considered in the next section, but one special case may be mentioned here. During pupal development in *Drosophila* the testis grows and acquires a degree of spiral coiling which differs in each species. This growth change seems to be induced by contact with the vas deferens—perhaps by means of a growth-promoting substance. If the testis is implanted into another species and establishes contact with the vas deferens, the type of coiling characteristic of the host species is induced.[274] If female genital discs are implanted into male hosts the resulting oviducts may become attached to the testis; when such contact occurs the testis degenerates.[29] Conversely, the pigmentation of the vasa efferentia which characterizes most forms of *D. melanogaster* is dependent on contact with the testis—perhaps merely as the result of migration of external covering cells from the testis.[274] [See p. 122.]

DETERMINATION OF CHARACTERS DURING POST-EMBRYONIC DEVELOPMENT

Differentiation in the epidermis—The control of form in the insect is mainly centred in the epidermis; changes in form result from local changes in the intensity of epidermal cell division.[530] Furthermore, the epidermis of the insect retains an embryonic character throughout post-embryonic development: it can differentiate into new organs. Thus the ordinary epidermal cells in the abdomen of the *Rhodnius* larva can retain this character, or differentiate to give rise to oenocytes, to dermal glands, or to sensilla of several kinds each furnished with a sense cell which gives off an axon that grows inwards to become a part of the central nervous system (p. 178), (Fig. 58).[522]

Each existing sensillum clearly inhibits the differentiation of a new sensillum for some distance around it, perhaps by the absorption of some essential nutrient or inductor substance;[309] and perhaps the same substance at a lower level

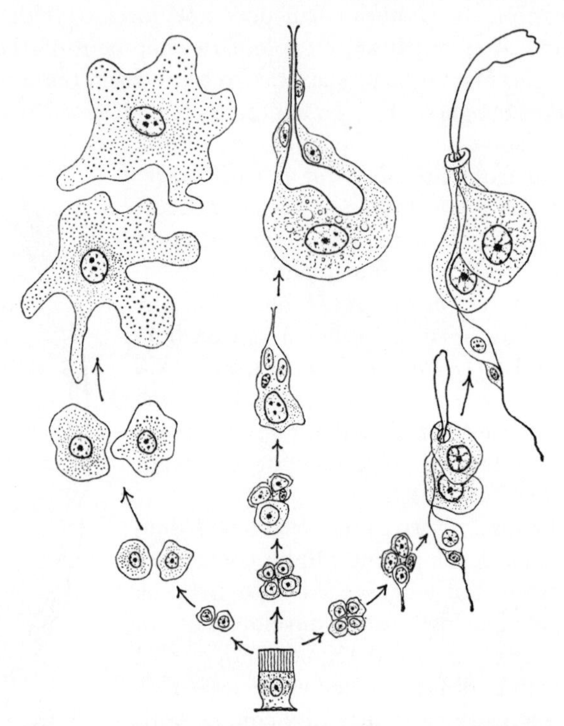

FIG. 58.—Differentiation in the epidermis in *Rhodnius*

To the left, formation of oenocytes: in the centre, formation of dermal gland: to the right, formation of tactile sense organ.

of concentration may be necessary for the differentiation of dermal glands.[522] In *Drosophila* the arrangement of the large bristles, or chaetae, and the form of these bristles, is dependent on an invisible 'prepattern' of determination established at an early stage of development and controlled genetically.[507] But even in *Drosophila* the pattern of some bristles can be modified by the mutual separation of the cells at quite a late stage of development.[504]

The eye of *Drosophila* provides a model system in which these problems of differentiation can be studied.[367] In the development of the compound eye in *Aëdes aegypti* the epidermis is induced to grow presumably by the moulting hormone, but the realization of the eye structure appears to depend on an intercellular determining factor diffusing from a differentiation centre which lies at the posterior margin of the epidermis of the prospective eye.[519] [See p. 122.]

Determination of appendages—As in the egg the general pattern of the organism is mapped out or 'determined' before there is any visible structure (p. 7), so throughout post-embryonic growth the structural details are determined well in advance of their development. The first stage in this process is reached when a given group of tissues is capable of giving rise to an organ with parts duly developed in harmony with one another, although the finer distinctions in the parts are still not fixed. This state is called primary organization ('primäres Organsystem').[24] It is best illustrated from observations on the appendages of *Vanessa* (Lep.).[24] The thickening of the epidermis of the limbs does not begin to appear until the end of the 4th larval stage; but already by the 3rd stage the distinctive characters of the adult fore limb and hind limb have been determined. If at this stage a part of the fore limb of the larva is transplanted on to the stump of the hind limb, the two parts work together harmoniously to produce a

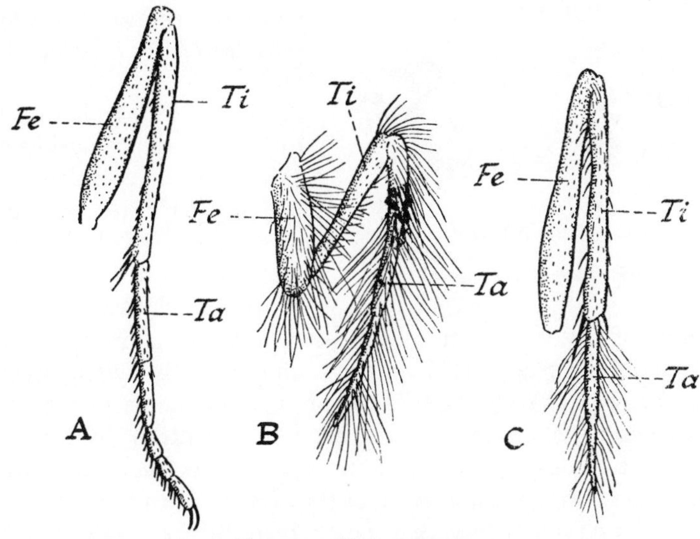

FIG. 59.—*Vanessa urticae*

A, normal hind leg; *B*, normal fore leg; *C*, leg produced by implanting larval fore leg on to stump of hind leg. The tarsus is chimaerical, combining the scale characters of both fore and hind legs. *Fe*, femur; *Ta*, tarsus; *Ti*, tibia (*from photograph by* BODENSTEIN).

limb with the normal arrangement of segments (the parts regulate themselves to a state of primary organization), but the characters are chimaerical, combining features of fore and hind limbs (Fig. 59). The same applies to the pupa. The first appendage normally lays down a part of the protective cuticle of the pupal sheath; the third appendage does not contribute to this sheath. This type of cuticle is, however, laid down by a first appendage transplanted to the position of a third in the 3rd larval stage: the character has already been determined though the time when it becomes manifest is dependent on the action of the hormones initiating metamorphosis in the host. Injury to the imaginal discs of eye or leg early in the last larval stage in *Drosophila* results in the formation of multiple organs. But during the first half of this larval stage determination becomes complete and this degree of 'regulation' is no longer possible.[299]

Determination of wing structure and pattern—All through development the characters become determined in an orderly sequence. Each part proceeds from the stage when it is still capable of 'regulation', to the 'mosaic' state (p. 11) long before the growth and movements of the cells put the plan into effect. The process has been followed in most detail in the wings.[54] The system of veins in the wings of the moth *Philosamia* becomes finally determined in the pupa several days old;[136] the venation in *Anax* (Odonata) is partly though not fully determined in the next to last nymphal stage.[203] Distortions produced experimentally in the developing wing of *Drosophila* may result in the displacement of the upper surface in relation to the lower; but the material of the upper epithelium which is destined to develop into a vein is able, by some means, to induce the corresponding material in the lower epithelium to form a vein surface.[178]

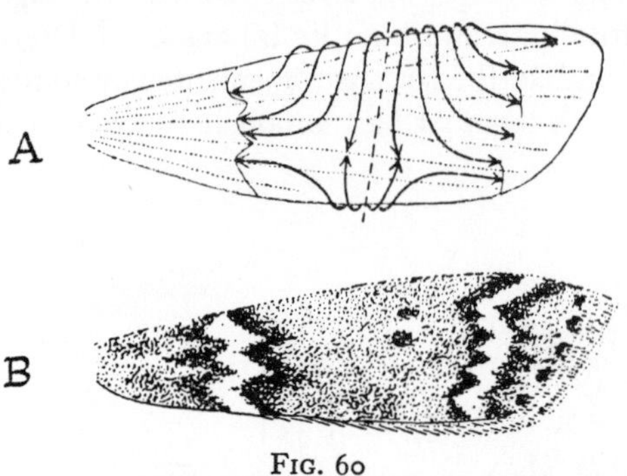

FIG. 60

A, course of spread of the influence determining the symmetrical wing pattern in *Ephestia* (*after* KÜHN); *B*, normal wing pattern of *Ephestia*.

In *Philosamia* a certain area of epithelium which will afterwards form the ocellus of the wing is already determined in the late caterpillar and after removal of this area no ocellus is formed. If some material is left it will organize itself to form an entire ocellus and if divided several complete ocelli may be formed. Clearly, the arrangement of the zones of the ocellus is still undetermined.[136]

A very clear description of the determination of the wing pattern has been worked out in *Ephestia*.[173] The wing of this moth bears a pattern which is approximately symmetrical about a transverse axis (Fig. 60). On the first day of pupal life no part of the upper surface of the wing is determined in respect to this pattern; by the fourth day determination is complete; and thereafter epithelial defects caused by burning simply obliterate corresponding areas of the mosaic. During these four days the influence (whatever its nature may be), which brings about determination, spreads outwards from the transverse axis of the wing. The course along which the determinative stream advances was worked out by burning small patches of the epithelium; the defects so produced create an obstruction to the spread so that it is possible to tell where it has reached at a given time. Similar results were obtained in *Abraxas grossulariata*.[172] (According to an alternative interpretation of this type of wing pattern, the determinative stream is thought to spread from the wing base in a succession of waves; these give rise to transverse lines, from each of which a second stream then spreads outwards in either direction and so determines the width of the bands to be produced.[179]) Observations on the circulation of the blood in the pupal wing show that the body fluids do in fact follow approximately a course spreading outwards and inwards from the middle of the wing, like that deduced for the spread of the symmetrical pattern (Fig. 61).[272] Supernumerary wing veins in *Ephestia* result in the formation of extra marginal spots. Perhaps these are likewise determined by the underlying blood lacunae.[159] The spreading of the transverse bands and the influence of the longitudinal veins are both apparent in the determination of the pattern.[179] Similar modifications of pattern have been observed in the related moth *Plodia*

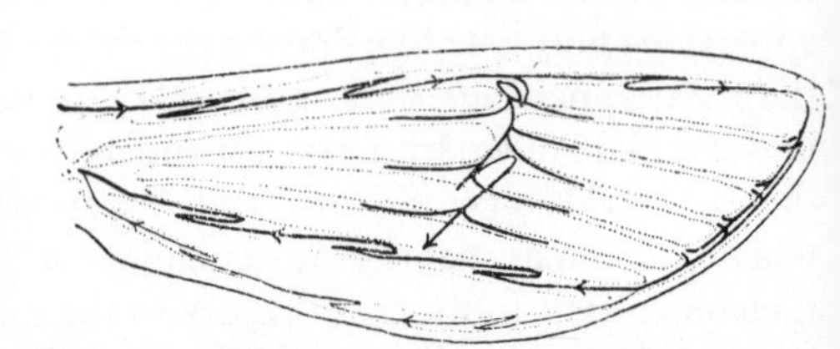

FIG. 61.—Circulation of haemolymph in pupa wing of *Ephestia*, schematic (*after* STEHR)

interpunctella, but without any clear evidence of a 'determination' stream,[517] although the same process may well occur in this insect also.[499]

The concentric expansion of the pattern zones of the wing, both in this symmetrical system and in the eye spots, as in *Philosamia*,[136] suggest a parallel with the periodic rings of precipitation (Liesegang rings) which may appear spontaneously in colloidal solutions.[15, 236] But what the nature of the determining influence may be is not known. It does not merely provide different quantities of pigment in the scales of the different zones; it brings about also morphological differences in the wing scales.[171] Partial transection of the wing with a hot lancet early in the pupal stage causes arrested development of scales and sockets distal to the defect.[186] But the two processes are independent, for in *Ephestia* and *Vanessa* the determination of pattern is complete at a time when the form of the scales is still undetermined.[162] The striking variations in the size of wing scales and their sockets in the different elements of the wing pattern probably result from the production of polyploid nuclei[137] such as are common in the healing of wounds[309] and during the growth of various organs in many insects.[48, 106, 139] In *Philosamia* the details of the colour pattern are determined later than the main outlines of the pattern as a whole, and the factors responsible for this determination, at least so far as they concern scale formation, seem to reside in the blood stream.[136] In the pupal wing of *Ephestia*, mitoses occur most abundantly in those regions which will later be darkly pigmented and along which the wing is folded transversely towards the end of pupal development; and they appear precisely at the time when experiment has shown that determination is complete.[41, 160] There must presumably be some connection between these phenomena. But the successive differentiation of wing scales during the development of the wing pattern,[413, 414] is exceedingly complex, and is rendered more complicated still by the action of various wing-scale mutations.[444]

'Critical periods' in wing development—It has long been known that the wing patterns of butterflies can be altered by exposing the pupae to cold or to heat during a certain critical period. Indeed, this is one of the factors which determine the seasonal forms of certain butterflies (p. 105). This critical or susceptible period is the time during which determination is taking place; by exposing pupae of *Ephestia*[92] or *Vanessa*[161] to high temperature at different times, it has been shown that there is a succession of such periods during the pupal stage, each relating to a different element in the structure or pattern of the wing. Thus in *Vanessa* pupae exposed for short periods to temperatures of 45–46·5° C., the wing pattern can be modified only during the first 48 hours after pupation; but this period is subdivisible into a great number of susceptible periods for the individual elements of the pattern. From 48–90 hours there is a susceptible period for loss of scales; from 90–102 hours one in which the form of the scales is affected, and so on.[161] In *Pyrrhocoris* (Hem.) the different elements in the wing pattern react differently towards abnormal temperatures.[135]

According to Goldschmidt's hypothesis[113, 116] the colour patterns of the wings of Lepidoptera are caused by different areas of the wing developing at different rates, so that at a given moment only certain areas are prepared to take up from the blood stream those metabolic products which furnish the various pigments. There are thus two processes at work: the development of the wing scales and the production of substances necessary for pigment formation; if the rate of either process is upset—whether by abnormal temperature, by local interference with the blood supply, or by the action of genes—abnormalities in the colour pattern result. But during the whole process of scale formation, the developmental stages run parallel in the scales of all parts of the wing.[162] In *Ephestia* the differences in pattern are associated with differences in the intensity of cell division (see above); there is some doubt as to whether there are differences in the *rate* of development;[41] and if warm or cold tubes are applied to parts of the wing in pupae of *Vanessa*, they will cause local delay or acceleration in development but do not affect the ultimate colour pattern.[108] So that if two competing processes are in fact at work they must be processes occurring at an earlier stage than visible differentiation of scale structure.

The effect of relative changes in the rate of developmental processes is seen also when the eye discs of *Drosophila* are transplanted into other larvae: the number of facets which appear in the eye is determined by the length of time development continues under the influence of the pupal differentiation factors.[28]

Action of genes in developmental physiology: rate genes—The characters of the individual are formed in the last analysis by the action of the genes in its chromosomes. In a few cases there is some evidence of the mechanism by which the genes exert their action. One mechanism is that which forms the central point of the theory of heredity elaborated by Goldschmidt[114] according to which the genes influence development by accelerating or retarding definite chains of reaction. There is believed to be a 'genic balance'; the final characters being the resultant of reactions set in motion by genes of different or opposite tendencies. For example, in the wing of *Drosophila* the same alterations of form can be brought about by the action of certain genes or by exposure of the larva to heat during a definite critical period. In either case the result is attributed, in terms of this theory, to a disturbance in the relative velocity of the different processes concerned in determination.[118] If the time of development of the *vestigial* mutant in *Drosophila* is prolonged by partial starvation, the size of the wings is increased (and this effect, curiously enough, may persist in the next generation).[62]

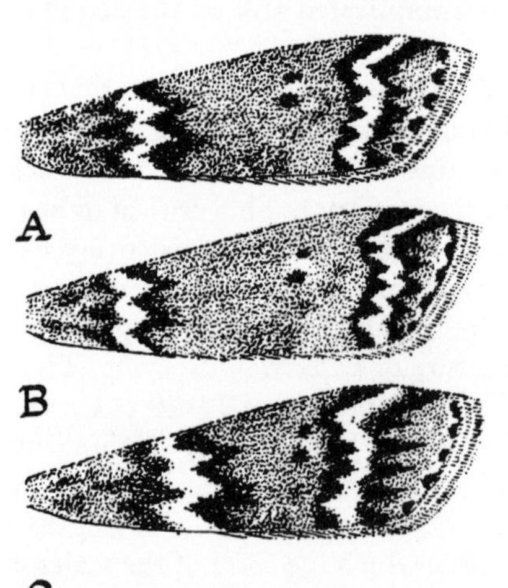

FIG. 62.—Forewings of *Ephestia kühniella*

A, wild type (*sy sy syb syb*). *B*, mutation with widened area between the symmetrical bands (*sy sy Syb Syb*); the same appearance can be produced by exposure of pupae of wild type to high temperature from 0–48 hours. *C*, mutation with narrow area between bands (*Sy sy syb syb*); the same can be produced by high temperature in pupae 48–72 hours old (*after* KÜHN).

Likewise in *Ephestia*, we have seen how the 'influence' which determines the symmetrical pattern of the wing spreads outwards from the axis of symmetry during the four days following pupation. If the pupa is subjected to heat (45° C. for 45 minutes) during the period 12–36 hours after pupation, the symmetrical area in the resultant moth is widened; a result which corresponds to the action of the mutant gene *Syb* (Fig. 62, B). Whereas pupae exposed to the same treatment 36–72 hours after pupation have the symmetrical area narrowed, as it is by the mutant gene *Sy* (Fig. 62, C). Now we have seen that in the course of determination all the insects pass through a stage with narrow pattern: the gene *Sy* merely has the effect of delaying the determinative stream in its early stages, and arresting it while the pattern is still narrow. Thus the mutant gene seems simply to have influenced in some way the velocity of a developmental process.[173] It is worth noting that the distribution of mitoses in the developing wing of *Ephestia* (p. 89) is shifted by the factor *Sy* and by heat just like the final pigment pattern.[41]

We have seen that prothetely or neoteny in insects is to be explained by some disturbance in the rate of developmental processes (p. 81). In the mealworm this results solely from environmental action and is not hereditary.[180] In *Gryllus campestris* its hereditary character shows that gene action is involved.[70] In *Drosophila melanogaster* there is a so-called 'giant' stock which affords an example of extension of development due to gene action: the larva continues feeding beyond the normal time and perhaps has an extra instar.[81] Similarly, genes controlling the relative size of parts in accordance with the heterogonic law (p. 62) must presumably act by influencing the *rate* of growth processes.[147]

Autonomous gene action—In general two modes of action by genes are recognized, (i) those whose action is confined to the cells in which they occur: autonomous gene action, and (ii) those which cause certain of the cells to liberate chemical substances and so exert their action at a distance and determine the characters of other tissues.

As an example of the first type may be cited the black-scaled mutant of *Ephestia, bb*. The local action of this gene is well seen in 'mosaics' in which 'somatic mutation' (through gene alteration or chromosome loss) has taken place and discrete patches of the recessive black pigment occur in the normal wing (Fig. 63).[171] There is a mutation in *Ephestia* which causes irregular chromosomal divisions in the earliest cleavages of the embryo; it results in recessive genes being separated from their alleles and irregularly distributed, so that mosaics of all kinds develop.[176]

In males of *Habrobracon* (Hym.) individuals mosaic for various characters may arise from binucleate eggs, the two oötids having been fertilized by spermatozoa bearing different genes.[122] The mosaic patches may be quite distinct and unaffected by the genes in neighbouring patches. But in males which have eyes mosaic for the two recessive whites (*wh* and ivory), genetically different though phenotypically similar, a line of the normal dark pigmentation may occur along the border of the ivory region. The dominant allelomorph to 'ivory' present in the 'white' region apparently causes the production of some substance that diffuses through and reacts with the dominant allelomorph to 'white', present in the 'ivory' region.[305] This is an example on the border line between the two types of gene action.

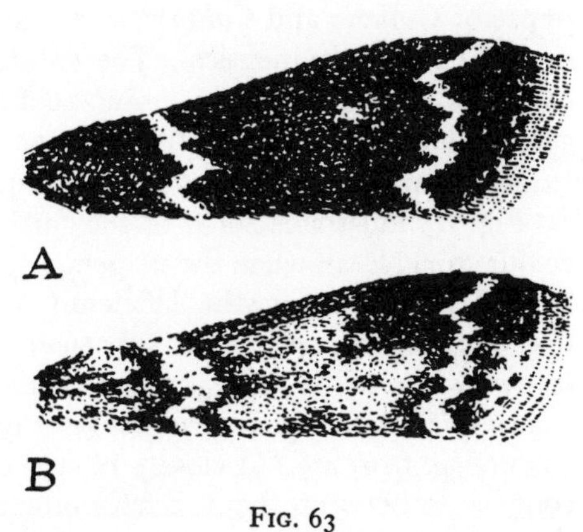

FIG. 63

A, forewing of black mutant (*bb*) of *Ephestia kühniella*; B, wing of the heterozygous form (*Bb*) with a black patch in the pattern due to somatic mutation (*after* KÜHN).

Genes producing diffusible substances—As an example of gene action of the second type may be noted the genes *A*, a^k, *a* in *Ephestia*, which form a graded series in respect to pigment characters; *AA* being black-eyed, *aa* red-eyed.[53] These genes affect the pigmentation of the larval epidermis, eyes, and testis, the pigmentation of the imaginal eyes and brain, the rate of development and the general 'vitality'. They demonstrate how manifold may be the actions of single genes. Transplantation experiments have shown that the appearance of the various *A* pigment characters is brought about in cells of *aa* constitution by a diffusible substance which is produced in the cells of testis, ovary and brain through the action of the gene *A* and given off into the blood. It is interesting to note that transplantation of an *Aa* ovary is as effective as an *AA* ovary; that is, the production of a definite quantity of '*A*' substance is not a direct effect of the *A* gene; production is preceded by a primary reaction in the plasma of the 'hormone'-forming cell. It is in this reaction that *A* is dominant over *a*.[175, 213] The '*A*' substance can also be supplied to *aa* individuals of *Ephestia* by implanting testes from various other Lepidoptera (*Plodia*, *Galleria*, *Carpocapsa*, &c.); a gene with the same properties is common to many species.[213]

Similar results have been obtained with the eyes of *Drosophila*. Most of the eye colours of *Drosophila* are genetically autonomous; but there are several exceptions, one of which is vermilion. Vermilion eye colour fails to develop in one eye of a gynandromorph (p. 74) if the other eye is of the wild type;[79] its colour is influenced by the genetic constitution of the other parts of the body. In such cases if the imaginal discs of the eyes are transplanted from one full-grown larva to another, although they are determined in the embryological sense and develop into eyes, their pigmentation may be influenced by 'diffusible factors' produced in the tissues of their new host: vermilion eyes assume the wild type pigmentation when transplanted to a wild host.[86] The v^+ substance responsible for this effect, which is lacking in the individuals of 'vermilion' constitution, is liberated from various tissues and circulates in the blood of the pupae 3–80 hours after pupation; it is effective if administered to larvae in the food,[14] but it can be taken up by the eyes only during a restricted period 65–70 hours after pupation. Like the '*A*' substance of *Ephestia*, it is non-specific; it is present in the blood of pupae of *Galleria* and *Calliphora*.[86] The v^+ substance and the '*A*' substance are in fact chemically identical. The substance is *l*-kynurenine, a product of the oxidative breakdown of the amino-acid tryptophane; *a* and *v* being homologous genes which intervene in tryptophane metabolism by producing the specific oxidizing enzymes necessary for the formation of the eye pigments (p. 617).[86, 214]

Fig. 64 illustrates some of the further complications of the vermilion eye-colour story. Even when the v^+ gene is present and kynurenine (v^+ substance) is formed, eye colour may be deficient ('cinnabar' eye) if the cn^+ gene is absent, so that hydroxykynurenine (cn^+ substace) is absent; or if the wa^+ gene has mutated to *wa* (white-eyed) so that the protein granules which carry the pigment are wanting.[443] Although pteridines and ommochromes which make up the eye colours (p. 616) are not closely related chemically there is some developmental connection between them. Such mutants as *w*, *rb*, *bw*, &c., besides controlling the amount of brown eye pigment also affect the quantity of pteridines; perhaps the two pigments compete for reactive sites in the protein granules of the pigment-forming cells.[408, 443]

Tryptophane metabolism seems to be similarly impaired in the green-eyed forms of *Musca* and *Phormia*: these develop the normal reddish-brown eyes after injection of 3-hydroxykynurenine.[370] Of the serosal pigments of the silkworm, some are autonomous, others are controlled by diffusable substances and are therefore influenced by the genetic constitution of the other tissues.[101]

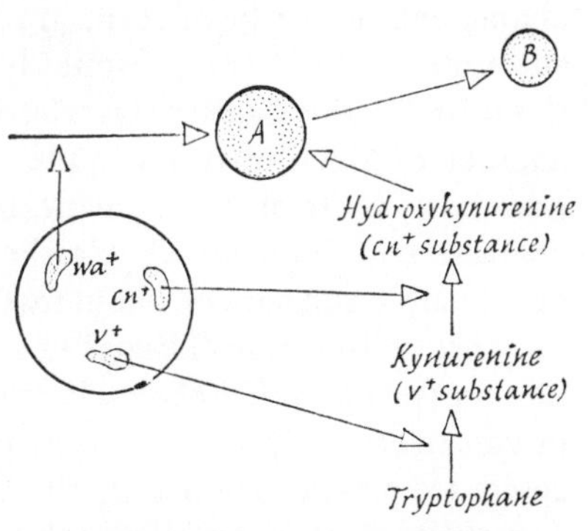

FIG. 64.—Greatly simplified scheme showing the action of some eye-colour genes (*after* KÜHN)

A, protein carrier for pigment: B, pigment. Further explanation in the text.

A comparable effect is that of the mutant *bar* in *Drosophila*, which results from the doubling of a segment of the X chromosome and leads to a great reduction in the number of facets in the compound eye (from a normal 600–650 to 40–60 in the female fly). This effect is greatly diminished if the mutant larvae are fed with an extract of *Calliphora* pupae.[87] This result may be reproduced with a great variety of imidazol derivatives, such as histidine or creatinine, the most active being methyl-hydantoine.[156] Acetamide and other amides are particularly active and may result in an eye of 'bar' constitution with even more facets than in the wild type insect.[433] The natural substance is of unknown composition but is widely distributed in insects.[49]

The influence of environment on gene action: phenocopies—The effect of a mutant gene may be greatly modified by the environment, notably by temperature and nutrition. Food containing breakdown products of tryptophane affects the colour of *vermilion Drosophila.*[86] Changes in temperature influence the number of facets in the *bar* mutant—probably by increasing the quantity of the active substance.[59] The exhibition of the gene *antenna-less* in *D. melanogaster* is influenced by nutrition: the effect is not manifest if the mutant larva is given an abundant supply of riboflavin.[120] And the manifestation or 'penetrance' of many other gene actions varies with the temperature (cf. p. 111).[297]

A striking example of a phenocopy is afforded by the effect of exposing the egg of *Drosophila* to ether vapour during a period of 15 minutes at 3 hours after fertilization. Fifty per cent. of the resulting flies have the metathorax more or less completely changed into a mesothorax, as brought about by the mutant gene 'bithorax'.[325, 328] This same result can be produced by a temperature shock during the same sensitive period.[338] Presumably this is the time at which the corresponding gene exerts its action.

As already mentioned, so-called 'phenocopies' resembling most of the known types of mutant characters can be produced in normal *Drosophila* by brief exposure to high temperature at different periods of development.[63, 130, 136, 138] And reversal of dominance can be obtained by this means. The developmental processes have presumably been thrown out of step with one another because of their different temperature coefficients. Consequently, the degree of dominance of a gene is a function of the environment within which the developmental process takes place. Goldschmidt[118] suggests that it is only such morphogenetic changes as result from differential effects on integrated reaction velocities which can be reproduced as phenocopies. But since genes act mainly through the synthesis of specific enzymes, substances which act as enzyme poisons will give rise to patterns of damage which reproduce the effects of mutant genes.[407]

Effects of the soma of the mother on the offspring—Most genetic characters, as we have seen, are autonomous and not influenced by the constitution of other tissues. Similarly, they are not influenced by the constitution of the soma of the mother. Ovaries transplanted from one strain of *Lymantria* to another produce offspring quite unaffected by the somatic characters of their new host; there is no 'somatic induction';[157] and the same is true of many characters in *Drosophila.*[65] On the other hand, where genes cause the liberation of diffusible substances, these may influence the characters of the offspring. We saw that voltinism in the eggs of silkworms is determined by the voltinism of the mother (p. 15); and *Ephestia* larvae of the constitution *aa* derived from a mother of constitution *Aa* show *A* type pigmentation in their early stages: their characters are pre-determined by the diffusible substance from the mother (Fig. 65). Later this effect wears off and they show *aa* characters.[175] The same effect in the offspring can be procured by implanting into *aa* mothers the testes or brain from *Ephestia* of *A* type or from *Acidalia* (*Ptychopoda*), *Plusia* or other Lepidoptera.[175] There are genes affecting the size of the testis in the adult *Drosophila pseudobscura* which do so by influencing the cytoplasm of the unfertilized egg: the series of processes between the gene and its effect may be a long one.[274] And there are other examples of this so-called 'maternal inheritance'.[212]

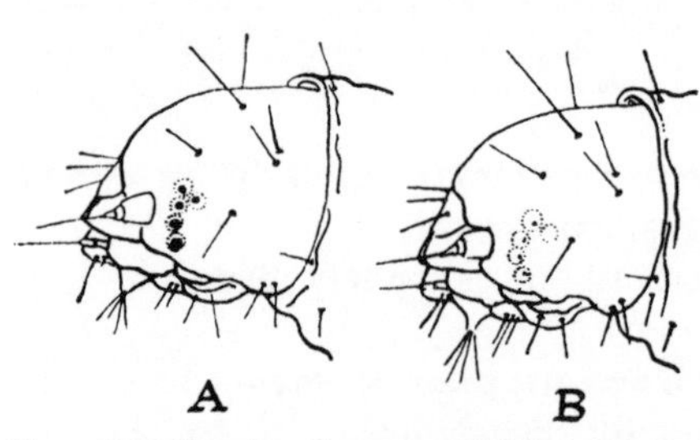

FIG. 65.—Heads of young *Ephestia* larvae both of the genetic constitution *aa*

A, derived from mother of constitution *Aa*, shows *A* type of eye pigmentation; *B*, derived from mother of constitution *aa*, shows *a* type of eye colour (*after* KÜHN).

Genes and the production of growth hormones—Genes may be responsible for the production of the growth-promoting hormones, and by causing the failure of these at various periods of growth they can exert a lethal action which may supervene quite late in development. The strain of *Drosophila* known as 'lethal giant' produces large bloated larvae which fail to form the puparium; but if the ring gland which secretes the pupation hormone (p. 68) is implanted into them from a normal larva, pupation will take place;[126] the ring gland and its constituent cells being considerably smaller in larvae of the *lgl* strain than in the normal larva.[254] And if the Sphingids *Celerio gallii* ♂ and *C. euphorbiae* ♀ are crossed, the female pupae fail to produce imagines; but the male pupae complete their development. In the female pupae it appears to be the growth hormones which are lacking; for wing germs and ovaries from the hybrid ♀♀ will develop normally if transplanted into the hybrid ♂♂;[51] and if the haemolymph of the female hybrid is replaced by haemolymph from the male hybrid it will develop normally to maturity.[462]

Most genes are pleiotropic—exerting multiple effects upon growth and metabolism. Such pleiotropy may be primary, or 'mosaic', in character, that is, having an autonomous effect in the different tissues; or it may be secondary, or 'correlative', that is, the gene affects a given organ, perhaps an organ of internal secretion, and by this means exerts its action from a distance upon various remote tissues.[127]

Determination of sexual characters—Determination of sex is a special case of the action of genes in development. It is generally agreed that sex is primarily determined by the chromosomes of the fertilized egg. One chromosome, the *X*-chromosome, of one sex is either unpaired (*XO*) or paired with a chromosome visibly different from itself (*XY*). Such an individual is heterogametic, for at meiosis when the chromosomal partners are separated, gametes of two sorts (*X* and *O* or *X* and *Y*) will be produced. The opposite sex has two *X*-chromosomes and will therefore produce gametes of one sort only. In most insects (e.g. *Drosophila*) the male is the heterogametic sex; the Lepidoptera are exceptional in showing female heterogamety. We shall discuss in a later chapter (p. 738) those variations in the relative numbers of the sexes that are determined by the behaviour of the male and female gametes and their chromosomes during maturation and fertilization. Here we shall consider only the physiological mechanisms by which the sexual characters are controlled in the course of development.

Gynandromorphs—Sex determination is an example of the first type of gene action. With few exceptions, to be noted later, the effect on the development of both primary and secondary sexual characters is exerted apparently within every cell in the body without the intermediation of hormones circulating in the blood. The most conclusive evidence of this is afforded by the occurrence of gynandromorphs, or sexual mosaics. These are individuals in which the chromosomal combination in the cells varies in different parts of the body; a state of affairs which may be brought about in several ways. In *Drosophila* they mostly result from the elimination of one of the *X*-chromosomes during early cleavage in a female. Thus a female starts with the constitution *XX*; but some of its cells become *XO* and therefore male.[196] In many Lepidoptera they result from the fertilization of binucleate ova, one nucleus having the *X*, the other the *Y*-chromosome.[66, 181] In certain strains of *Bombyx mori* there is an inherited tendency to produce binucleate ova and this is associated with gynandromorphism.[119, 150] In *Habrobracon* (Hym.) they arise from binucleate eggs of which only one oötid is fertilized;* females kept at a high temperature of 35–37° C. give rise to more binucleate eggs and more gynandromorphs.[122]

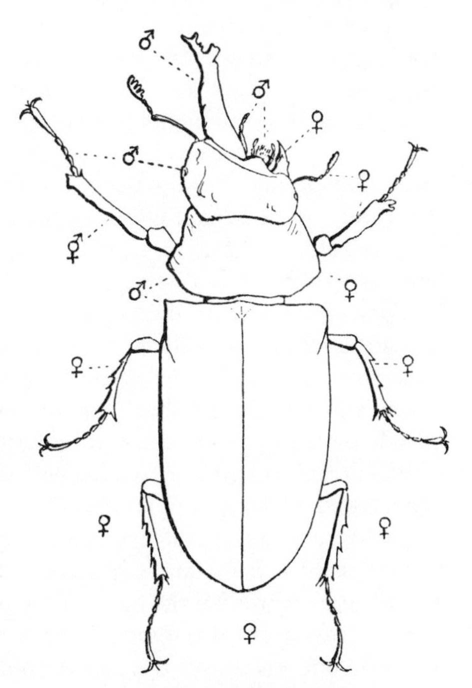

FIG. 66.—Antero-posterior gynandromorph of *Lucanus cervus*

The signs indicate the distribution of male and female characters (*after* V. LENGERKEN).

* In Hymenoptera, fertilized eggs give rise to females, unfertilized eggs to males (p. 740).

In the honey-bee both the fusion nucleus and a sperm nucleus may undergo cleavage in the same egg and so give rise to a mosaic individual;[486] but other modes of origin are possible in the bee.[487] In the Chalcid *Oöencyrtus* sex is determined by the temperature to which the parent female is exposed during development and during adult life; lower temperatures produce female progeny, higher temperatures male progeny; gynandromorphs appear when the parent female, at a critical stage of egg development, is exposed successively to lower and higher temperatures.[537] Likewise, the eggs of *Carausius* (which are normally all female) if incubated at a high temperature show partially male characters in the offspring: some parts male, some female, and some intermediate.[365] [See p. 122.]

The result in all cases is that some of the cleavage nuclei are male and some female. When they reach the cortical zone of the egg they become determined for a given part of the body (p. 10) and patches of one sex or the other are produced, depending on the make-up of the cells from which they happen to be formed. Sometimes the two lateral halves of the body are of opposite sexes; occasionally the partition is transverse; often the male and female characters are scattered over the body as an irregular mosaic (Fig. 66). In *Drosophila*, in which the germ plasma arises from a single cell, the gonads are always of the same sex in a given individual;[196] in some insects male and female gonads may be mixed.

In gynandromorphs, as in non-sexual mosaics (p. 89), it is evident that the sexual characters cannot be determined by circulating hormones. There is one case, however, in which diffusible factors seem to be concerned in sex determination. Male mosaics of *Habrobracon* (Hym.) may occasionally have feminized genitalia. This phenomenon is perhaps analogous to the production of dark pigment where two recessive white patches come in contact on the eye (p. 89). Such males may perhaps be mosaics of two types of tissue each recessive for a different sex factor, the dominant allelomorphs of both of which are necessary for femaleness; substances diffusing from one region to the other may interact to produce femaleness.[305]

Secondary sexual characters—At no stage of development do the gonads appear to exert any influence on the sexual characters elsewhere in the body. Castration of *Lymantria*[204] or *Bombyx mori* larvae[152] and of the cricket *Gryllus*[233] has no effect on sexual characters or behaviour. Nor has the implantation of the gonads of the opposite sex into castrated *Lymantria* larvae[193] any effect, even when carried out in very young larvae.[164] The sexual characters of antennae regenerated by such individuals are unaffected by the presence of the implanted gonads;[164] and the characters of wings are unchanged when the germs are transplanted in *Lymantria* larvae from one sex to the other.[164] Complete castration in *Drosophila*, secured by the ultra-violet irradiation of the pole cells (p. 10) in the egg, has no effect on the sexual characters of the resulting insect.[104]

Only one exception to such results has been reported. If the wing disc of a female *Orgyia* (Lep.) larva in the 4th instar is transplanted to a male and then extirpated, it regenerates as a male wing; whereas a male wing transplanted to a female regenerates as a male. Unless there is some other explanation for these results, the sex in the former case seems to have been reversed by the host.[205] Another special case, that of *Chironomus*, is discussed later (p. 97). [See p. 122.]

Ovaries implanted in a castrated male *Lymantria* may connect up with the cut end of the vas deferens to form a continuous duct;[164] but testes in gynandromorphs of *Drosophila* differentiate only when they are in contact with vasa deferentia; testes attached to oviducts degenerate. Perhaps diffusible substances from the genital ducts interfere with the development of the gonads of the opposite sex (cf. p. 83).[80]

Genic balance: Intersexes—Although these observations prove that sexual characters in insects are usually autonomous and not influenced by circulating hormones, yet it is possible even in insects for partial or complete reversal of sex to occur. The most familiar example of this is to be seen in the intersexual forms of the Gipsy Moth (*Lymantria dispar*) as studied in great detail by Goldschmidt.[115, 116] As in all Lepidoptera, the female is the heterogametic sex (*XY*) producing gametes of two sorts, *X* and *Y*. Yet it is possible to obtain normally functioning males with this genetic constitution (both secondary and primary sexual characters being reversed) which give rise to the two kinds of gametes; or to obtain individuals with the outward appearance of females but with the *XX* constitution. Between these extremes all intermediate grades of 'intersexuality' occur.

Goldschmidt explains these results by the theory of 'genic balance' already mentioned. Associated with the sex-chromosomes are factors determining sex: male-determining genes in the *X*-chromosomes, female in the *Y*. Further, the female-determining factor in the *Y*-chromosome is believed to transmit its influence to the substance of the gamete before the formation of the polar body, so that both sorts of egg, *X* and *Y*, will contain the female-determiner. The male-determining factor being located in the *X*-chromosome occurs only in the *X*-bearing gamete. Thus, if the male-determining factor is indicated by the symbol M, and the female factor by F, then the two kinds of gametes produced by the female, *X* and *Y*, will have the constitution MF and F, and the zygotes resulting from their fertilization will be MMF and MF. Now the sexual characters of the individual are supposed to result from the balance between these M and F determiners, each tending to produce different reactions. Normally in the MMF individual the male factors predominate and male characters appear; in the MF individual the female-determining influence is in excess. The 'strength' of the determiners, however, differs in different races of *Lymantria*, and if a 'strong' female is crossed with a 'weak' male, a single F determiner may outweigh two M determiners so that the genetic male *XX* (i.e. MMF) will develop all the characters of a female. The intermediate grades of intersexuality are explained by supposing that if in such an MMF individual the F determiner is only slightly in excess, the insect may start developing as a male and then at a given 'turning point' (and the greater the relative 'strength' of the F determiner, the earlier will this happen) it will switch over and complete its development as a female. It is but another example of characters being determined by the relative velocity of two opposing reactions that influence development (p. 88). We saw that in the case of wing patterns these genetical effects could be imitated by exposing the insect to abnormal temperatures, which are supposed to influence the two competing reactions to a different extent. And so it is with sexual characters; intersexes can be produced by exposing to extremes of temperature pure races of *Lymantria* with normally balanced determiners.[116, 168]

Intersexes and abnormal ratios between the numbers of the two sexes may also appear sometimes when two species are hybridized. Thus the cross between the Lepidoptera *Lycia hirtaria* ♂ × *Poecilopsis rachelae* ♀ produces males and intersexes but no females; and the cross *Lycia hirtaria* ♂ × *Nyssia zonaria* ♀ produces all males. As in the *Lymantria* crosses, this is doubtless due to differences in strength or 'valency' of the *X*-chromosome in the two species.[67] In the human louse *Pediculus*, intersexes are very common when the head louse *capitis* and the body louse *corporis* are crossed. In some matings they may amount to as much as 20 per cent. of the offspring; the percentage of females decreases, the intersexes being mostly masculinized females (Fig. 67).[151] And intersexes have resulted from the crossing of *Glossina* species.[295]

The normal determination of sex in *Drosophila* is explained by Bridges[44] by a theory somewhat analogous to that of Goldschmidt, according to which there is a balance between female tendency genes and male tendency genes scattered irregularly over all the chromosomes, sex chromosomes and autosomes alike. Intersexes arise from disturbances in this balance resulting, for instance, from the presence of an extra set of autosomes whose effect outweighs that of the sex chromosomes. In later views Goldschmidt[406] recognized that the autosomes in *Lymantria*, also, have a predominantly male-determining influence; and the same is true of the silkworm.[518]

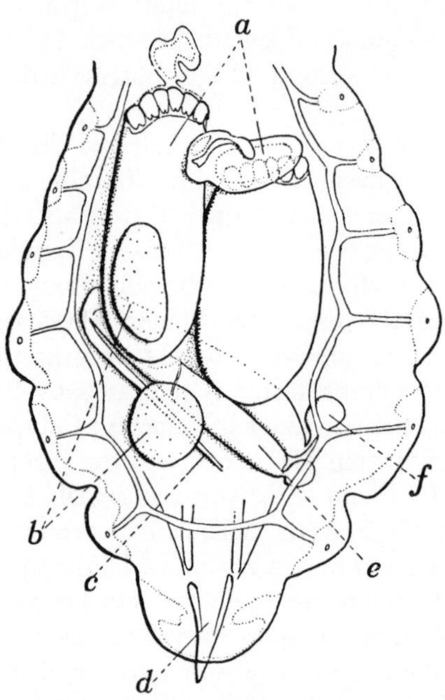

FIG. 67.—Intersex in *Pediculus* (*after* KEILIN and NUTTALL)

a, ovaries; *b*, testes; *c*, ejaculatory duct; *d*, penis; *e*, vesicula seminalis; *f*, male accessory gland.

These theories of a genic balance controlling sex provide a means of describing satisfactorily most of the observed facts. Argument has turned chiefly upon the nature of the turning point. Goldschmidt supposed that this occurred at a definite time; organs determined before that time were of one sex, those determined later were of the other. But there are many exceptions to this rule; intersexes of a given grade do not always show the same distribution of male and female characters; and intersexes resulting from crossing the parthenogenetic Psychid *Solenobia triquetella* with males of the bisexual race of the same species have their sexual arrangement very patchy and asymmetrical.[262] Therefore some authors favour the view that there is no switch-over but that the disturbance of genic balance results in an intermediate type of development throughout the period of growth;[11, 200] that normal sex determination occurs when F or M predominates; if the excess of one of these factors reaches a certain minimum, sex determination is purely genotypic; but if this 'epistatic minimum' is not attained, so that F and M are more or less in balance, sex determination is in part or wholly phenotypic and intersexes result.[419, 500]

Others consider that a switch-over does exist, but is not restricted to one point of time; it may occur at different times in different cells, even in different parts of the gonads.[262] [See p. 123.]

Parasitic reversal of sex—Another example of sex-reversal is seen in Hymenoptera parasitized by *Stylops*. In *Andrena* invaded by *Stylops* the yellow markings on the face of the male are lessened; they are caused to appear on the face of the female. As in the intersexes of *Lymantria*, those characters which are differentiated latest in development tend to be most readily affected by the parasite.[245] The effect seems to be related to the abstraction of nutriment from the host; for several parasites have more effect than one, the large female *Stylops* has more effect than the male, and no effect at all is produced in such hosts as *Polistes* and *Vespa*, which are fed by their parents according to their need, but only in such forms as *Andrena*, *Odynerus* (Fig. 68), *Sphex*, which are provided with a fixed ration by the egg-laying female. The stylopized specimens may be looked upon as intersexes in which the 'turning point' has occurred earlier in the most heavily parasitized individuals. According to this hypothesis, the sex determiner (often thought of as an 'intracellular hormone'), which would normally have the upper hand, is weakened or inhibited by lack of nourishment, so that the latent determiner of the opposite sex is enabled to exert its action and bring about a partial reversal of sex.[245]

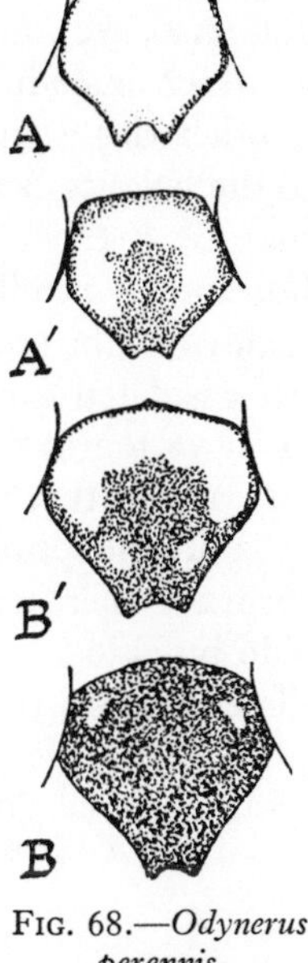

Fig. 68.—*Odynerus perennis*

A, clypeus of normal male; *A′*, stylopized male; *B′*, stylopized female; *B*, normal female (*after* Salt).

There is a similar partial reversal of sex in the Membracid *Thelia bimaculata* parasitized by the polyembryonic wasp *Aphelopus theliae*. Striking changes appear in the secondary sexual characters, the males assuming the external characters of females; the gonads are affected in some degree but even in completely feminized males the testis can still show spermatogenesis.[163] In male Delphacids (Homoptera) parasitized by *Elenchus* (Strepsiptera) both secondary sexual characters and the external genitalia are affected. In parasitized females no changes appear, and among the males it is only those specimens which hibernate as nymphs and so allow the triungulins sufficient time for feeding which become feminized.[131]

Fig. 69.—Effect of *Mermis* on *Chironomus*

A, normal female. *B*, parasitized female showing partially male reproductive organs (intersex); *C*, normal male (*after* Rempel).

It is interesting to note that among several Hymenopterous parasites (*Pimpla*;[60, 264] *Alysia*[143]) males preponderate amongst those emerging from small hosts. But whether their sex is determined by the failure of fertilization in the egg-laying female (p. 740) or by lack of nourishment during growth is uncertain. If several individuals of the egg parasite *Trichogramma* develop in a single host an increased proportion of males results; but this is attributed to the males competing more successfully under these conditions.[246]

In *Chironomus* females parasitized by the nematode *Mermis* the ovaries are destroyed in the last larval stage. The secondary sexual characters, the fore legs, antennae, &c., are not affected, but an unusual type of intersex is developed in which male internal and external organs of sex are formed in an insect that is otherwise a female (Fig. 69). This suggests that in *Chironomus*, in addition to the balance between male-determining and female-determining factors, there may be a third 'epigenetic' sex differentiating factor represented by a sex hormone controlling the development of the external genitalia.[235] Intersexual males of Chironomids can also be produced.[538] Oestrin-like substances have been isolated from the ovaries of *Attacus atlas* in quantities of the same order as in vertebrates;[183] but they have not yet been shown to exercise any physiological function in the insect.

Polymorphism—In many insects which occur naturally in more than one form the characters are controlled by genetic factors.[95] The different forms exist side by side in the natural environment, the relative abundance of each form depending on the pressure of natural selection to which each is exposed. This phenomenon is called 'balanced polymorphism'.[396] It is seen in the so-called 'chromosomal polymorphism' in *Drosophila*, where the different forms are not

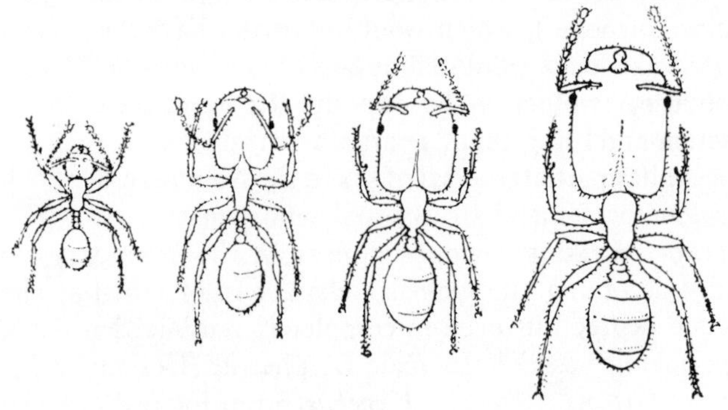

Fig. 70.—Heterogony as a cause of polymorphism. Neuters of *Pheidole instabilis* showing increase in the relative size of the head with absolute size of the body (*after* Wheeler)

visibly distinguishable but presumably possess small physiological differences;[39] and it is seen in the different forms of mimetic butterflies, notably in *Papilio dardanus*, where the relative numbers of the different forms varies in different localities in accordance with the abundance of the distasteful model species which they mimic.[503]

There are other insects in which the different forms occur in individuals of constant genetic constitution. These forms, likewise, of course, are produced by the action of the genes; but the gene function is 'switched', or latent genes are brought into action, under the impact of some change in the environment.[52] Some examples may be considered here, since they provide further examples of the mechanisms of determination.

When an insect shows well-marked heterogony (allometry) (p. 62), polymorphism may be simply a result of absolute size—variations in the quantity of food having caused the larvae to pupate early or late in development. This is well seen in the mandibles of Lucanid beetles and in the head types of polymorphic ants; in the large individuals the mandibles and head become disproportionately large (Fig. 70).[147, 304] It has been suggested that a special growth factor, Vitamin T, present in certain insects and lower fungi, when combined with a rich protein diet, will cause excessive growth with enlargement of the head in ants, cockroaches, and other insects.[111] In Hymenoptera diploid bi-parental males can be produced in addition to the normal impaternate haploid males (p. 740). Haploidy may be associated with a small size in the body cells and this, with or without starvation, may result, in *Habrobracon* for example, in dwarfism.[327]

Polymorphism in Hymenoptera—In the honey-bee the determination of characters in the females is again effected by nutrition, but here there are qualitative differences in the food.[540] Larvae fed throughout with 'royal jelly' from the salivary glands of the nurses become queens; those fed after about the middle of the third day with honey and pollen become workers; while intermediates appear if the diet of royal jelly is resumed in the late stages of larval growth.[182, 304, 516] The nature of the determining factor in the royal jelly is not known. Royal jelly has a most complex composition (p. 501) and that given to worker larvae is said to differ from that given to larvae destined to become queens.[331] It is maintained by some authors that there is no particular change in the composition of the diet of worker larvae or queen larvae; until about the third day all the larvae receive massive feeding; thereafter the worker larvae are undernourished; these are sterilized by partial inanition whereas the queen larvae suffer no underfeeding. This contention is supported by the observation that in the stingless bees (Meliponinae) the formation of queens is determined by the size of the food store provided for the larva before sealing down.[132] In the case of the honey-bee, the general consensus of opinion seems to be that both quality and quantity of food are important,[516] that nutrient balance is significant, but no single constituent determined the development of either caste.[393, 484] [See p. 123.]

In wasps the dimorphism of the female is much less striking. In *Vespa* the queen is larger than the worker but shows little difference otherwise, and when fully nourished the workers will produce eggs. In *Polistes* there is no morphological difference at all; the workers merely enter a temporary reproductive arrest, or diapause, apparently a consequence of the temperature during rearing.[392]

In ants such as *Pheidole* the distinction between female and worker is thought by some to be predetermined genetically in the egg;[294] but their form is very labile and susceptible to far-reaching modification by subsequent events. There is a tendency nowadays to reconcile the 'trophogenic' and 'blastogenic' hypotheses and to believe that both nutritional factors and genetic constitution contribute to caste determination.[94] This applies both to the production of sterile or fertile females and to the differentiation of sterile females as soldiers or workers.[182] Simple lack of food, as we have seen, may affect the proportions of the body by its effects on allometric growth. It may be that dimorphism in ants has resulted from the segregation of extreme forms within an allometric series originally tied to individual size.[536] In the ant as in other insects (p. 727) the yolk of the egg is reabsorbed if the oöcyte remains in the ovary and this appears to influence the caste of the offspring. Thus, if the rate of egg deposition is high, normal males and queens are produced, whereas if the rate is low the nutrient content of the eggs is reduced and the sterile female caste is the result.[94] In a new Aculeate colony with a single queen, the first brood contains only small workers; succeeding broods increase in size; and only after the largest workers have appeared are queens and males produced.[303] A colony of ants (*Leptothorax*) abundantly fed on emergence from hibernation produced 94 per cent. queens; a colony given the same food but in small amounts only, produced 30 per cent queens.[302] In *Pheidole* the production of soldiers occurs only when feeding on

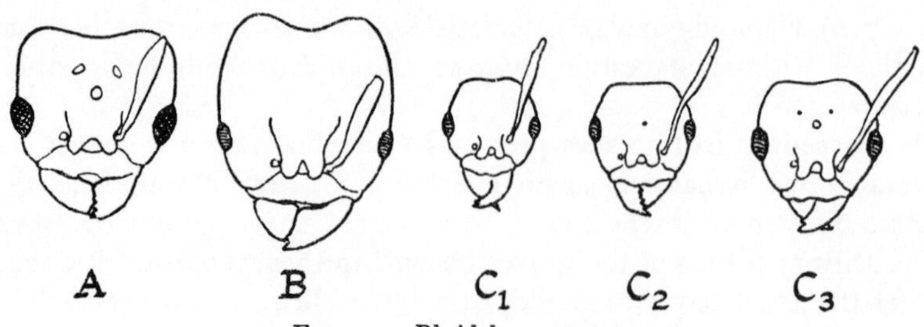

FIG. 71.—*Pheidole commutata*

A, head of queen; *B* head of soldier; C_1, head of normal worker; C_2 and C_3, heads of workers parasitized by *Mermis* ('mermithergates') (*after* WHEELER).

flesh coincides with a definite larval period of limited duration. If the larvae obtain concentrated protein during this short period they are able to grow very rapidly; they then develop into soldiers, otherwise they become workers.[110] In *Myrmica rubra* multiple factors operate in the control of queen production.[375] Ovaries from worker-presumptive third instar larvae of ants transplanted to queen-presumptive larvae develop in the manner of the ovaries of their host.[375] In certain genera of ants, if the nest is infested by Staphylinid beetles (*Lomochusa*) the workers serve the beetle larvae and neglect the larvae of the colony, resulting in the appearance of anomalous females called 'pseudogynes'.[182]

Invasion by parasites, notably the Nematode *Mermis*, may have the same result as malnutrition; but in addition it seems to have a markedly feminizing action, workers or soldiers assuming the characters of females ('ergatogynes') (Fig. 71).[304] These 'intercastes', like the 'intersexes' produced by stylopization in Hymenoptera, are thought to arise through the parasite changing the relative valency of the hereditary genes and rendering dominant (more or less) those which determine the female caste and which are normally inactive in the neuters.[294] If the intermediate forms between different castes in ants are to be regarded as 'mosaics' (p. 87), this strongly favours the blastogenic hypothesis. But if, as seems more probable, they are in reality 'caste intergrades' they can be readily accounted for by the trophogenic theory.[306]

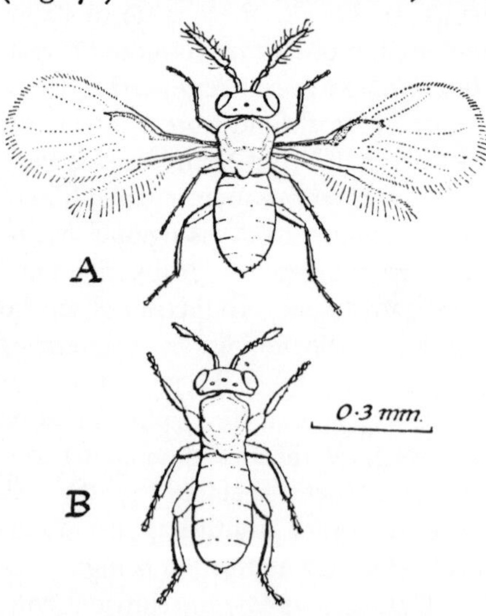

FIG. 72.—Dimorphism in the male of *Trichogramma semblidis* (*after* SALT)

A, winged male reared in eggs of Lepidoptera; *B*, apterous male reared in eggs of *Sialis*.

Effects of quantitative and qualitative differences in diet are seen also in parasitic Hymenoptera. Great differences in the size of Hymenopterous parasites can be brought about by the size of their hosts, and the small individuals are often apterous or brachypterous.[248] The Chalcid *Melittobia* develops in *Sceliphron* and other rge hosts which will provide sustenance for 500–800 larvae. The first 20 or o to develop are brachypterous females and eyeless males. All the rest, receiving erhaps different food, develop into adults of normal form. Similarly, the imorphism in the cocoons of the Ichneumonid *Sphecophaga* seems to depend pon whether they are spun by the early or late larvae feeding in a given ost.[256] The Ichneumonid *Pezomachus flavicinctus* is large and fully winged when reared from large hosts, tiny and apterous from small hosts;[288] and males of *Trichogramma semblidis* are apterous when reared from the eggs of *Sialis*, but winged and with striking differences in legs and antennae when reared from ggs of *Ephestia* and other Lepidoptera (Fig. 72).[247] There is some evidence that he controlling factor in this case is the oxygen supply; and the genetic basis of he differences is supported by the fact that in Belgium there is one strain of *T. emblidis* which gives exclusively apterous males, and one strain which gives about o per cent. of winged males.[372] The Ichneumonid *Gelis* has two forms of male, nacropterous and micropterous, with associated differences in the structure of horax, ocelli, and other parts. The micropterous form develops regularly in mall hosts, the macropterous form in large hosts, and intermediates do not occur.[488] Perhaps the dimorphism in the agamic and gamic generations of female Cynipids (p. 726) is due to their developing in different parts of the host plant.[247]

Polymorphism in termites—The immature stages of termites are highly plastic. In such primitive forms as *Zootermopsis* this plasticity extends beyond he 6th instar. The terminal types, alates, soldiers, workers or neotenics, are controlled by genetic patterns present in all individuals; which pattern is finally realized is determined by varying factors in the environment.[182] According to this view all the larvae have equal potentialities for all castes; when the factors for one caste are brought into operation by environmental conditions, the factors for the other castes are rendered ineffective—or partially ineffective, and then intercastes of all sorts occur.[3]

Among the lower termites the presence of the royal pair prevents the appearance of the reproductive castes. On removing the queen in *Zootermopsis* a new neotenic queen appears within 35–40 days: larvae or nymphs begin to reproduce while still retaining in part their juvenile body characters. Orphan colonies of *Zootermopsis* or *Reticulotermes* may contain some hundreds of neotenic sexuals. Likewise, soldiers in these genera[85] or in *Calotermes*[121] tend to inhibit the development of other soldiers; if the soldiers are removed the nymphs become transformed. It has therefore been suggested that the reproductive castes and the soldiers give off exudates containing substances which regulate growth and that the proportional representation of the castes in the population may be controlled by this hormone mechanism. It has been claimed that alcoholic and ether extracts of the queens will likewise delay the production of the new queen.[85] Young colonies do not produce winged forms. Perhaps these are inhibited by 'hormones' from the royal pair. But as the colony increases in numbers the inhibiting influence on each individual is lessened and alates appear. The females apparently prevent the occurrence of females and the males suppress males.[121, 182, 453] In *Calotermes* there is no permanent worker caste, but certain larvae may remain for long periods without moulting and thus constitute 'false workers' or 'pseudo-ergates'. Certain of these arise from the moulting of 7th-stage nymphs from which the wing lobes and eye discs disappear (Fig. 73).[326] This regressive moult is comparable with the rejuvenation or partial reversal of metamorphosis that can be induced experimentally by means of the juvenile hormone (p. 79). A similar regression occurs in *Reticulotermes*.[322]

In the establishment of neotenic sexual forms in *Calotermes* two processes are at work: (*a*) the determination of larvae, which must have reached at least the 7th instar, to produce replacement forms, and (*b*) the elimination of all but one sexual pair from the colony. If larvae in the reactive stage are present, removal of one sexual form from the colony for 24 hours may be sufficient to induce determination of neotenics. Such determination is normally inhibited by intimate contact or by substances given out by sexual forms: antennal contact through a wire gauze partition is insufficient to prevent it. On the other hand, antennal contact suffices to cause the elimination of the excess sexual forms that appear under these conditions.[337] [See p. 123.]

The 'pheromones' or 'social hormones' derived from the excrement of the royal pair are taken into the gut of the later stage larvae, or nymphs; as in *Reticu-*

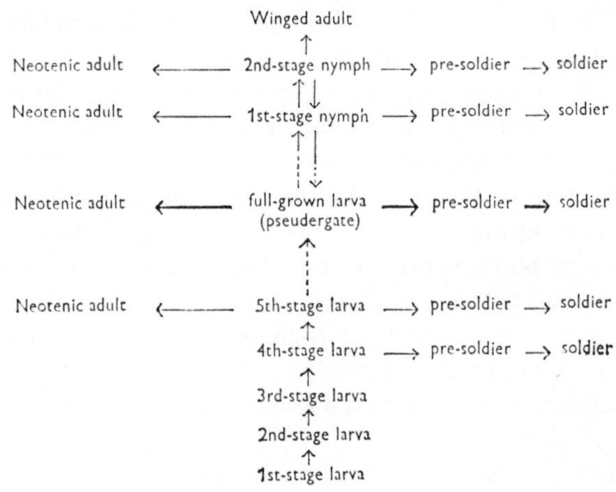

FIG. 73.—Course of development and caste formation in *Calotermes* (*after* (LÜSCHER)

Explanation in the text. The interrupted lines indicate the occurrence of several moults.

lotermes, substances from both male and female are necessary for complete suppression of supplementary reproductives. It may be that these pheromones circulating in the colony influence the endocrine system of the larvae and nymphs which receive them. The different castes (larva, nymph, adult, soldier, replacement sexual form) are probably controlled chiefly by a mechanism of differential timing and intensity of hormone secretion, notably a suitable relation between juvenile hormone and moulting hormone.[457] Soldiers, as we have seen (p. 79), can be produced by implantation of additional corpora allata. The general similarity of caste differentiation and metamorphosis (p. 75) is evident.[337] [See p. 123.]

In the higher, fungus-growing termites (Termitidae) the only method of producing substitute sexual insects is the laying of eggs by winged insects that have not left the nest. In the most primitive types of Termites certain workers can develop mature gonads with little outward change.[467]

Polymorphism in Aphids—Polymorphism is particularly evident among Aphids, in which there are differences on the one hand between the parthenogenetic and gamic generations, and on the other between alate and apterous forms among the parthenogenetic individuals (Fig. 74). These differences are certainly the result of environmental factors, but the nature of these doubtless varies from one species to another. Most Aphids begin to produce sexual forms in the autumn; but *Aphis fabae* has been reared parthenogenetically for nearly 3 years by extending the hours of artificial lighting during the winter months; whereas sexual females were made to appear in June of the first year by limiting the exposure to daylight to 8 hours.[73] And *A. forbesi*, given a $7\frac{1}{2}$-hour day, produced oviparous females in May, while if given a longer day, viviparous reproduction was still occurring in December.[189] In some cases this effect of illumination may in fact be one of nutrition, the plants being affected by the increased light; for the same effect is produced in root-feeding Aphids away from the direct action of the light.[189] But in *Megoura viciae* the day-length unquestionably acts directly upon the insect.[450, 451] It may be that physiological differences between

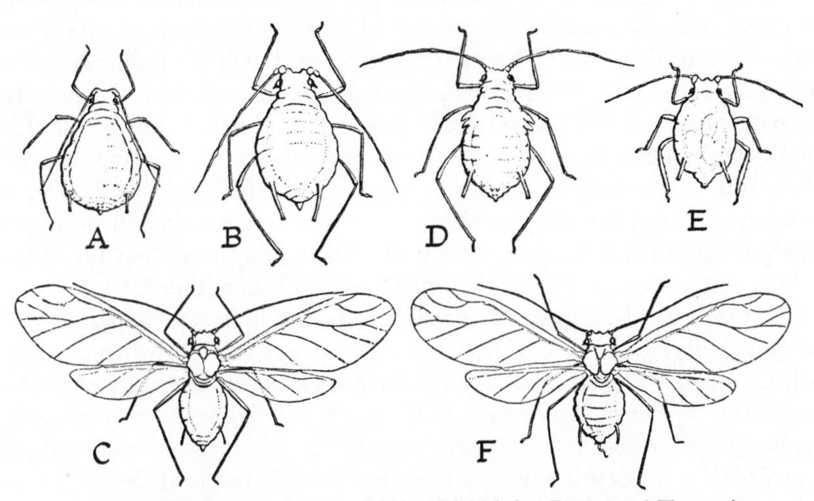

FIG. 74.—Polymorphism in *Aphis malifolii* (*after* BAKER and TURNER)

A, fundatrix; *B*, wingless parthenogenetic female; *C*, winged sexuparous parthenogenetic female; *D*, intermediate form; *E*, amphigonous or sexual female; *F*, male.

races influence the responsiveness to these external factors of light and temperature.[34]

In *Megoura viciae* the differentiation of the female embryos as oviparae or virginoparae is wholly governed by photoperiod and temperature acting upon the insect itself: high temperatures (exceeding 15° C.) and long photoperiods (exceeding $14\frac{1}{2}$ hours) promote the production of virginoparae.[450] But this applies only to older clones of virginoparae. Young clones (that is, the immediate offspring of sexual forms) when exposed to a short (12 hour) photoperiod produce only virginoparae; the capacity to produce sexual forms (males and oviparae) appears later as the clone ages. Indeed a definite time interval must elapse between the birth of the larvae which start the new parthenogenetic clone and the appearance of the capacity to produce sexual forms in response to a short day. It is remarkable that this time interval is independent of the number of parthenogenetic generations.

Light received through the eyes alone will not give these responses; the light which passes through the cuticle of the head acts upon some centre, perhaps a group of neurosecretory cells, in the dorsum of the brain which then presumably liberates a hormone acting upon the developing embryos.[450] It is remarkable that the mother is capable of influencing her developing offspring in this way when she herself is still a developed embryo in the ovary of the grandmother, where she responds to light transmitted through the abdominal wall.[450]

If females of *Macrosiphum solanifolii* which are producing gamic daughters are transferred from a low to a high temperature, they soon begin to give rise to intermediate forms, and within 10–14 days all the offspring are of the parthenogenetic type. The reverse change is brought about by exposing parthenogenetic forms to cold. The occurrence of intermediates during the transition periods suggests a switch over from one type to the other in the course of development. But it is interesting to note that during the change from gamic to parthenogenic forms the antennae and hind tibiae are first affected in the offspring; whereas during the reverse change from parthenogenic to gamic the body colour and reproductive system are affected first. Such differences in the order of reversal cannot easily be explained on the hypothesis of Goldschmidt that those parts determined latest in development will be influenced first in intermediate forms.[265, 266] The germaria of the ovaries are the first structures to show the change; gamic females have large germaria secreting much yolk. It may be that it is the germarium which controls the subsequent development of the ovariole and so, to some extent, the type of adult which results.[177] The early generations, derived from the fundatrices, are much more difficult to transform into oviparous females than are later generations. It appears that the offspring have been subject to some somatic influence from the mother.[316]

Alate forms can be caused to appear among the apterous viviparous Aphids by various external factors. *Viteus* produces winged forms when fed on dry or wilting plants.[265] In the Pea Aphid *Macrosiphum pisi* the chief factor is the concentration of the body contents of the parent.[251] Lack of water seems to be the primary factor also in *Toxoptera aurantii*; in this insect the critical period during which wing development is most readily affected is at the beginning of larval development.[238] There is much evidence to suggest that the immediate effect of many of the factors which lead to wing production in the offspring is to cause partial starvation in the mother. Perhaps that is the controlling factor.[269] Wingless females of *Aphis craccivora* produce winged offspring if kept at a low temperature on the mature leaves of their host plant, but wingless or intermediate offspring when reared on seedlings.[428]

In the Cabbage Aphis, *Brevicoryne brassicae*, the production of winged forms decreases as the protein content in the leaves of the food-plant rises.[90] But an important element in the determination of wing production in parthenogenetic females of Aphids is referred to as the 'group effect'—the simple presence of other Aphids.[321] Isolated wingless females of *Megoura* confined in an empty vial (2 × 1 inch) for 24 hours give birth solely to apterous offspring when they are fed; similar females confined under the same conditions in a group of ten, or even fewer, give birth to alate offspring.[451] Inherent factors imposed by the mother may also be involved, alate forms tending to produce apterous offspring; but intermediate forms frequently appear if the inherent and environmental factors conflict, or if conditions are changed in the course of development (Fig. 74, D).[72, 238] Such intermediates recall the phenomena of prothetely, &c. (p. 80), and they have again been attributed to disturbances in the rates of the developmental processes involved, such as those which fix the time of determination of characters and those which control the secretion of growth-producing hormones.[267] Histological observations on intermediate-winged forms of *Macrosiphum* support, in general, this 'time-of-determination' theory.[275] The course of wing development is probably controlled by hormones. Indeed, the application of juvenile hormone to the surface of *Megoura* larvae, which would have been winged, leads to the formation of the pattern and coloration of the apterous adult, which normally shows some retention of larval characters.[451] [See p. 123.]

Seasonal dimorphism in other insects—The spring and summer forms of the Nymphalid butterfly *Araschnia levana* are controlled to some extent by temperature;[278] but they are induced primarily by the photoperiod acting during larval development. Short days (13–15 hours of light) provoke diapausing pupae giving the spring form *levana*; long days (17–20 hours of light) produce non-diapause pupae giving the summer form *prorsa*.[463] In the froghopper *Euscelis* the seasonal characters are controlled solely by photoperiod and not at all by temperature.[464] In the Aleurodid *Aleurochiton complanatus* there are two seasonal types of puparia: a soft and almost colourless summer form which develops without delay, and a dark, sclerotized, winter form with white patterns of wax on its surface which has an obligatory phase of dormancy. The winter form results from exposure to short days during embryonic or early larval development. In these insects probably the physiological changes are primary and the morphological secondary; but in *Euscelis* the morphological differences persist without diapause or other detectable physiological effects.[465] Likewise, in the pupa of *Papilio xuthus* there is a partial correlation between diapause and orange colour.[432]

'Phase change' in Orthoptera, Lepidoptera, &c.—Locusts show a striking colour change, with conspicuous colour patterns, when they enter the gregarious migratory phase of their existence. This change is associated with intense muscular activity: hoppers of the gregarious type kept singly and undisturbed revert to the solitary phase; whereas if they are continually agitated they retain, at least in part, their 'gregaria' type of colouring. Perhaps under these conditions some diffusible substance is produced which controls the pigmentation. If so it seems to be passed on by the female to her offspring, for the 1st instar hoppers of 'gregaria' show their characteristic coloration within an hour after hatching.[395, 420] It is not certain, however, whether it is the agitation which induces the change of phase or whether this results from the sight or contact of their neighbours.[383]

The phases of locusts differ not only in colour pattern but in morphology, behaviour, and reproductive activity. In many ways the solitary phase can be regarded as a more juvenile, or neotenic, form;[437] and this is borne out by the fact that the ventral head glands (the homologue of the thoracic glands) persist in adult solitary locusts,[382] and that some of the features of the solitary phase, both of colour and of form, can be produced by supplying the gregaria form with active corpora allata. But phase change is not due solely to the level of corpus allatum activity.[429] Rearing conditions of parents and grandparents will influence the expression of phase characters in the offspring. [See p. 123.]

Similar effects of crowding are seen in Phasmida.[438] In *Gryllus brimaculatus* crowding intensifies macropterism and isolation favours brachypterism, but, somewhat unusually, it is isolation that leads to dark coloration.[398] This group effect depends on tactile, visual, olfactory, and vibratory stimuli.[452] A similar correlation (recalling that in Aphids) of low density and favourable food supply with brachypterous forms, and overcrowding, wilting of the host-plant, &c., with macropterous forms, is seen also in females of the plant hopper *Nilaparvata*.[441]

Phases like those of locusts occur in the African army worms *Laphygma exigua* and *L. exempta*[395] and in *Spodoptera*.[461] Reared in a crowd, these caterpillars are active, dark, and velvety ('gregaria'); reared in isolation they vary in colour from green to pinkish grey (solitaria').[461] Indeed, in many species of Lepidoptera the larvae become dark in colour and develop more rapidly when reared in crowds.[425, 454] [See p. 124.]

REGENERATION

Healing of wounds—Insects at all ages can repair injuries to their integument. In the early stages of healing the clotting of the blood (in some insects)

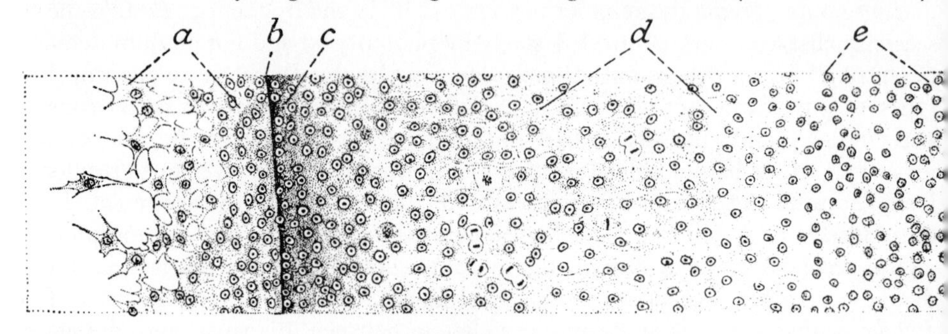

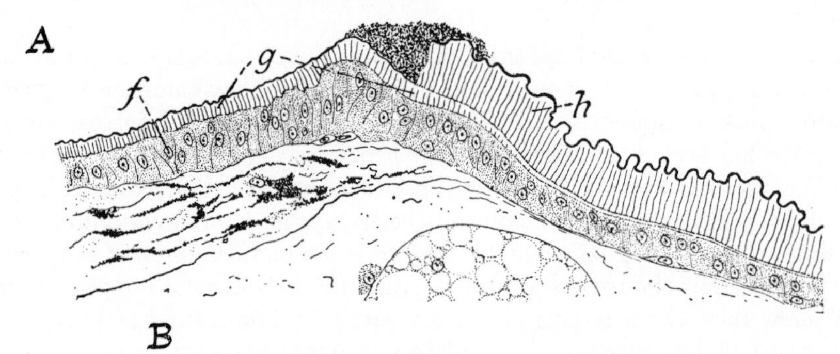

FIG. 75.—Wound healing in *Rhodnius* adult (*after* WIGGLESWORTH)

A, surface view of epidermis four days after a piece of cuticle had been excised. *a*, epidermal cells spreading over the gap; *b*, margin of excision; *c*, cells heaped up along margin of wound; *d*, sparse zone depleted of cells by migration to the wound, cell divisions occurring; *e*, unchanged epidermal cells. *B*, longitudinal section through the margin of such a wound after three weeks. *f*, epidermis established under excised region; *g*, new cuticle extending outwards under old cuticle; *h*, old cuticle at margin of wound.

and the accumulation of blood cells, play a part in the process (p. 436); but the essential reactions are those manifested in the epidermis. These reactions have been studied most closely in *Rhodnius*.[309] It appears that the dead or injured cells in the course of their breakdown give rise to substances, products perhaps of the hydrolysis of proteins, which exert an attraction upon the surrounding cells so that these migrate to the wound and congregate thickly around its margin, leaving a peripheral zone where the epidermal cells are very sparse (Fig. 75). If a piece of the integument has been removed, these aggregated cells spread across the wound, make good the defect, and lay down a new cuticle composed as usual of sclerotin and chitin. Meanwhile cell divisions take place in the sparse peripheral zone and continue until the normal density of the cells has been restored. The process of healing follows the same course in larvae and adults. Similar responses occur in *Ephestia* larvae.[374] These cellular migrations result in displacements of the cuticle pattern: existing larval patterns and the latent imaginal pattern which appears at metamorphosis are both displaced.[309] After the healing of burns in the caterpillar *Ptychopoda* there is the same displacement of the cuticular pattern as has been observed in *Rhodnius*.[309] In both these insects polyploid cells are common in the healed zone and in *Ptychopoda* these give rise to giant scales in the resulting adult.[334]

The cytological changes in the epidermal cells during wound-healing are the same as those during growth and moulting,[309, 526] but they take place also in the adult insect which, lacking thoracic glands, is not known to secrete the moulting hormone. In some insects, notably the pupae of the giant silk moth *Hyalophora*, injury causes a general increase in metabolism, apparently in response to some liberated 'injury factor',[501] but no general renewal of growth.[411]

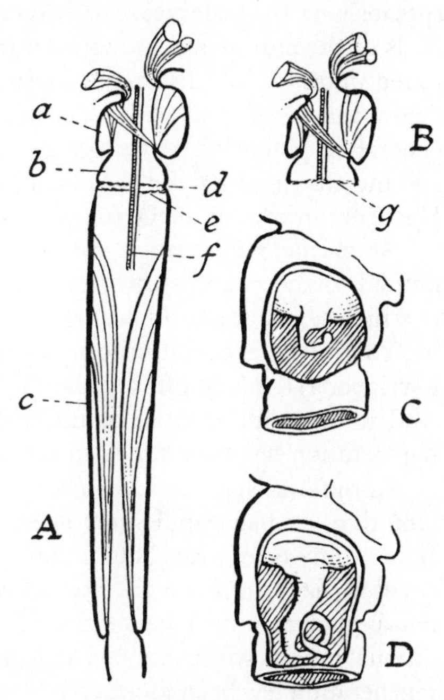

FIG. 76.—Autotomy and regeneration in Phasmids (*after* BORDAGE)

A, upper part of leg showing the junction between trochanter and femur where autotomy occurs; *B*, stump of leg after autotomy, sealed below by the upper half of the haemostatic membrane; *C*, *D*, stages in regeneration of limb within the coxal stump. *a*, coxa; *b*, trochanter; *c*, femur; *d*, groove where the break occurs; *e*, haemostatic membrane; *f*, trachea and nerve; *g*, upper half of haemostatic membrane.

The epidermis will re-establish its continuity when pupae of Saturniid moths, even different species, are cut into segments and joined together.[71] The same happens when the bugs *Rhodnius* and *Cimex* are decapitated and joined;[308] and when appendages are transplanted from one caterpillar to another.[23] In all these cases no other tissues besides the epidermis grow together; but in *Carausius* the head may be removed and replaced and both integument and gut (though not the nerve cord) are said to grow together again.[187] In many insects, on the other hand, no reunion is obtained on replacement of the head.[19]

Regeneration—We have seen that during both embryonic (p. 10) and post-embryonic development (p. 83) the parts of the organism, at first indeterminate, become progressively 'determined' or committed to the formation of particular organs. In many cases this determining influence, the so-called 'individuation field' which controls the type of organ that is to be produced, seems to persist in or around the organ long after it has appeared. In such an organ determination is not complete; developmental potencies still remain; and if it is removed it is capable of regeneration. Such regeneration may be regarded as the recapitulation of the ontogenetic process.[26]

It has long been known that the appendages of growing insects will reform at later moults if they are removed.[236] Many insects (Phasmids[109] and other

Orthoptera,[192] Agrionids,[61] &c.) are capable of detaching an injured limb, usually at the level of the trochanter. This phenomenon of 'autotomy' is particularly well marked in Phasmids. In these the articulation between trochanter and femur has become fixed and no muscles run across the joint (Fig. 76). In Phasmids a special muscle breaks off the leg at a predetermined level;[480] and in *Acheta*, also, there are three small muscles running from the coxa into the trochanter, whose sole function is to detach the limb at the point of weakness.[376] A two-layered haemostatic membrane crosses the limb at this point; the break occurs between the two layers of this membrane, and the upper portion walls off the little coxal cavity within which regeneration begins. The epidermis first spreads over the terminal scar. It then separates from this and from the cuticular walls of the stump and grows outwards in the form of a papilla which becomes rolled upon itself.[37] But regeneration will also take place after section above or below this level, and in insects not capable of autotomy.[192] In *Rhodnius*, regeneration potencies in the legs appear only at the level of the femoro-tibial joint and increase gradually in the distal direction. Unlike Phasmids and Mantids this Hemipteron cannot effect complete regeneration of a lost appendage.[336]

As a rule regeneration can take place only at moulting, and is therefore limited to the young stages; but in Apterygota (*Machilis*, *Lepisma*, *Thermobia*) in which sexual maturity occurs at some indefinite stage, and growth and moulting continue, regeneration can occur in the adult (as it does in Crustacea and Myriapoda)—injury often providing the stimulus to renewed moulting.[229, 280] And in an adult mantis, moulting has been observed after injury, and some regeneration has been noted in the adult in the absence of a moult.[225]

In the first-stage larva of *Blattella*, if the limb is autotomized shortly before moulting, a small papilla only is produced and the insect moults without delay. If autotomy takes place before the 'critical period' of the moult, a complete limb, equal in size to that on the normal side, is regenerated and moulting is delayed ten days or so until regeneration is complete. Repeated regeneration causes additional moults without any change in the ultimate size of the adult.[472] This rapid regeneration has been ascribed to a sudden discharge of moulting hormone from the thoracic glands;[472] and in *Periplaneta* there is experimental evidence that the thoracic gland hormone is necessary for regeneration.[368] But in the last-stage larva of *Ephestia* the regeneration of wing rudiments likewise causes a delay in moulting; and here it has been suggested that the activity of the thoracic gland is inhibited and it is for this reason that moulting is arrested until regeneration is complete. Indeed, regeneration in *Ephestia* will take place in the absence of the brain and thoracic gland.[478] [See p. 124.]

Determination of regenerating organs—The 'individuation field' may extend far beyond the visible organ—as in the imaginal discs of Lepidopterous larvae. If the wing disc of a young larva of *Lymantria* is removed, regeneration and normal wing development take place unless a considerable area of the surrounding epidermis is excised;[25, 292] under the influence of the individuation field, cells are capable of forming the wing when they would not normally do so. In the same way the cells around the base of the spines of *Vanessa* larvae act as an inductor of the regenerated spine;[21] but this effect is inhibited if the epidermis from another region is implanted on the wound.[22] When an organ, such as a leg, is transplanted to an abnormal site, its individuation field goes with it; and if amputated in its new position it regenerates as a leg.[21] Sometimes it is possible, up to a certain stage of growth, to divide the individuation field: if the imaginal discs of the wings in Lepidoptera are halved in the later larval stages both halves will give rise to a wing.[186]

The capacity for regeneration and the moulding of the outward form seem to reside exclusively in the epidermis. In *Sphodromantis*, antennae and legs of normal form will regenerate after removal of the corresponding ganglia. But they contain no nerve or muscle; only an amorphous granular mass.[279] In *Lymantria*, legs, antennae with normal sensilla, even the epidermal parts of the eye, will develop normally in the pupa after removal of the corresponding ganglia from the larva, though nerves and muscles are wanting;[166] and this has been confirmed in the flight muscles of the large Saturniid moth *Telea*.[470] In the course of normal regeneration the new muscles are perhaps budded off from the epidermal plate which forms over the wound;[98] perhaps they develop from mesodermal wandering cells (myoblasts).[209] The musculature of regenerating limbs in *Periplaneta* arises from blastema cells derived from haemocytes, and not from the epidermis.[474] The nerves arise by outgrowths from the nerves existing in the stump. The development of the muscles seems dependent on this nerve supply.[98] In *Blattella*, if regeneration requires no muscular reorganization it does not cause any delay in moulting.[472] [See p. 124.]

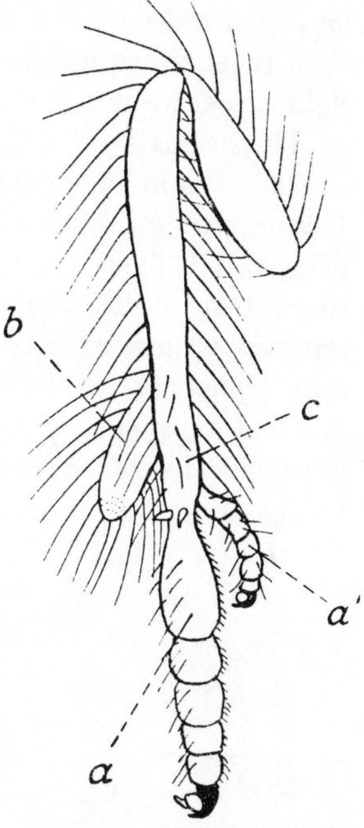

FIG. 77.—Triplicated limb in *Pyrameis cardui* produced by transplanting the two distal segments of the hind leg (in the last larval stage) on to the fore leg cut down to the second segment; the transplant being rotated 180°

Two of the branches *a* and *a'* are derived from the transplant; one branch *b* is derived by regeneration from the fore leg. The tibial region *c* is chimaerical, combining characters of both limbs (*after* BODENSTEIN).

Dissociation of larval and imaginal fields—If the limbs of certain beetles, such as *Tenebrio*, are amputated in the young larva, they regenerate in the course of larval life;[191] but in other beetles (*Hydrophilus*, Dytiscids,[191] *Timarcha*[1]) the wound simply heals over and no regeneration takes place until pupation, when a normally organized limb appears. In the latter case the individuation field for the larval organ seems to have been removed with the limb; morphogenetic activity is resumed only when the determination of the adult form is completed at metamorphosis. Amputation must be extended into the thorax in order to prevent this delayed regeneration.[39]

Abnormalities of regeneration: reduplication—The most frequent abnormality in regeneration is for the organ to be duplicated or even triplicated at the tip. When this occurs the arrangement always conforms to Bateson's law: the parts lie in one plane; the inner branch forms a mirror image of the normal; the third or outer forms a mirror image of the inner.[167] In *Mantis* this seems to result from some artificial interference with the wound surface.[226] In Lepidoptera, triplication has been induced by amputating the limb, rotating it 90° or 180° and allowing it to reunite. In these cases, by the use of transplanted appendages with recognizable characters, it has been shown that always two branches are derived from the transplant and one from the host (Fig. 77).[26, 360] Duplication may occur in the valves of the external genital organs of certain intersexes of *Lymantria* without preceding injury, and seem to be due to an abnormal degree of separation of parts which ought to be in contact.[112]

Thus reduplication appears to result from some upset in the spatial relations of the individuation field. In the case of amputated limbs replaced on the stump, if the orientation is normal, the two parts work together to form an organized whole; if inverted they work independently. The basal fragment then regenerates terminal parts; the distal fragment, influenced perhaps by the field or centre in the stump, also regenerates terminal parts.[26]

Homoeösis—Another abnormality is 'heteromorphous' regeneration: the replacement of an appendage by one belonging to another region of the body. In *Carausius* a fore limb may be regenerated in the place of an amputated antenna (Fig. 78).[43] This is particularly liable to happen if the antenna has been cut through near the base; but there seems to be no causal relation between the removal or retention of Johnston's organ (p. 262) and the type of regeneration,[99] such as has been claimed.[35] In *Cimbex* (Tenthredinidae), if the larva just before pupation receives a slight burn to the antenna, an antenna is developed in the adult; after a severe burn a leg-like appendage arises.[224]

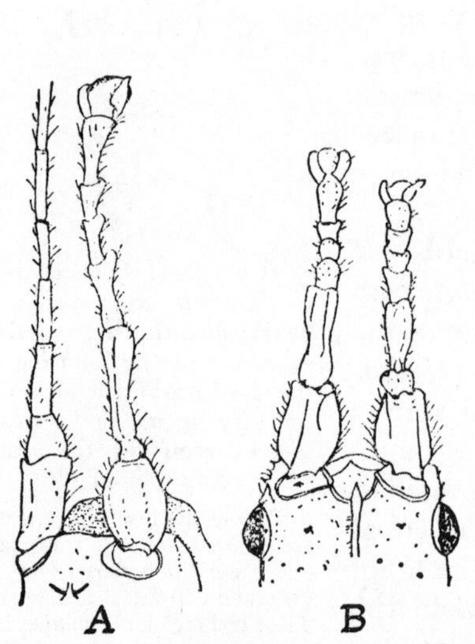

FIG. 78.—*Carausius morosus*, anterior part of head

A, right antenna regenerated as a leg; *B*, both antennae regenerated as legs (*from* KORSCHELT *after* CUENOT).

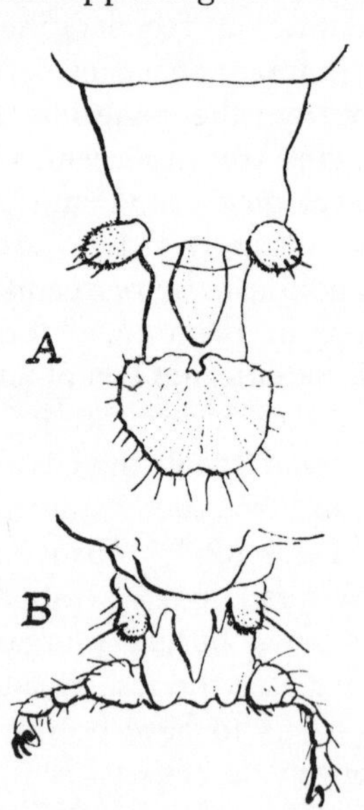

FIG. 79

A, normal mouth parts of *Drosophila*; *B*, mouth parts of *Drosophila* showing proboscipedia (*after* BRIDGES and DOBZHANSKY).

According to Przibram's explanation of this curious anomaly, the regenerating organ contains the potentiality to form both leg and antenna. The reactions which lead to the realization of these alternatives proceed at different velocities, and whichever process is the more rapid becomes dominant. Sometimes the potentiality for forming the normal appendage is temporarily masked: if the antenna in *Mantis* is cut off near the base in the early nymphal stages, a foot-like termination first appears, but is later replaced by a normal antenna.[226] The same idea may be expressed somewhat differently. According to the views of Child, metabolic activity, and with it regenerative potency, shows an axial gradient in the animal body, being most intense at the anterior extremity.[293] If it be supposed that the axial gradient becomes less steep with age, then the potency of differentiation at a given body level may be altered to a potency originally characteristic of a more posterior body level.[147] It must, however, be emphasized once more that the physiological mechanism in the whole process of determination is quite obscure.

However, there does seem to be a certain parallel between heteromorphous regeneration and those developmental anomalies caused by upsets in the rate of the component processes (p. 80). And just as effects of this type induced by abnormal temperatures can be copied by the action of certain genes, so various examples of 'hereditary homoeösis' are known which simulate heteromorphous regeneration. In *Drosophila*, for instance, the labella of the proboscis may be replaced by legs ('proboscipedia') (Fig. 79),[45] or the arista of the antenna may develop as a leg ('aristopedia'). The gene responsible for the latter condition is regarded as accelerating leg formation and so enabling the leg differentiation to dominate the antennal.[10] The expression of the aristopedia gene is favoured by cold; the expression of the proboscipedia gene is favoured by heat; and these results are attributed to a differential effect of temperature on the maturation of the imaginal disc on the one hand and on the production of the 'evocator' substances responsible for determining the character of the appendage on the other.[297, 299] Chemical inhibition of growth by means of colchicine may have the same effect,[299] and all grades of aristopedia can be produced in *Drosophila* by exposure of the larvae to a 'nitrogen mustard'.[371] It is interesting to note that genes which affect the tarsus but have no effect on the arista, do affect the tarsal component of the aristopedia organ.[300]

DIAPAUSE

Types of arrest—We have already discussed the arrest of development in the egg (p. 14) and in connection with reproduction we shall consider later the occurrence of diapause in the adult (p. 728). Growth may be arrested similarly at any stage of development and, as in the egg, the type of arrest varies in different species and even among individuals of a single species. Broadly speaking, insects fall into two groups: (i) those in which a continuous succession of generations occurs so long as conditions are favourable and in which growth is arrested only by the direct action of adverse circumstances, such as cold, drought, or starvation. These are sometimes termed 'homodynamic' insects,[241] and the arrest that occurs is called 'quiescence'. (ii) Those which show at some stage of their life history a prolonged arrest of growth which supervenes irrespective of the environment—'heterodynamic' insects,[241] with a true 'diapause'. Sometimes a heterodynamic generation may alternate with one or more homodynamic generations.

But these two classes represent only the extremes in a graded series. The kind of differences which exist can be best illustrated by examples. The meal-moth *Ephestia kühniella* will breed all the year round if kept sufficiently warm; but even at the same temperature the duration of development increases during the winter;[107] there is here a hint of an arrest related with the season. The blow-fly *Lucilia* will breed continuously if conditions are uniformly favourable; diapause in the larva is brought on by such adverse conditions as poor food, drought, cold,[69, 194] or excessive moisture.[88] But development does not proceed automatically upon return of the larva to a good environment; the arrest may be prolonged for weeks. Syrphid larvae hibernate and aestivate in a desiccated condition; development is quickly resumed if they are placed in contact with droplets of water;[33] there are varying degrees of arrest in different species, indicated by varying degrees of suppression in the maturation of the imaginal discs, the growth of which is favoured by exposure to low temperature.[258] The hibernation of *Phlebotomus papatasii* (Dipt. Nematocera) is primarily caused by low temperature, but inborn cyclical factors also exist to modify this effect. There is a latent inclination to diapause throughout the year: during the summer there are always some 2–12 per cent. of resting larvae in otherwise active broods; and this number is much increased by exposure to cold; but even when the larvae are kept at 30° C. the proportion which enters this resting state increases during the autumn, and hibernation continues in resting larvae transferred to this temperature.[242, 285] Arrest of growth may likewise supervene in the larva of *Anopheles plumbeus* at any season of the year, though far more frequent in the winter even under the same conditions of temperature and nourishment: the life cycle averages 22–29 days for eggs laid in May or June; 135–271 days for eggs laid in September.[133, 244] In this case the duration of daylight, acting either directly or by its effect on the fauna and flora of the water, may have influenced the results.[9] In *Pieris rapae* a diapause is not obligatory, but under natural conditions in England an increasing percentage of the pupae enter a dormant state as the season advances.[237]

The stage at which arrest takes place may be fixed in a species; *Carabus coriaceus* always hibernates as a 1st instar larva whether it derives from eggs laid in August or in October;[84] many Lepidoptera spend the winter only as full-grown larvae within the cocoon (e.g. *Pionea* and other Pyralids), the majority only as pupae. Or the arrest may occur at varying times: *Melolontha* may hibernate in several larval instars and as an immature adult; *Popillia japonica* (Col.) larvae show a resting period which occurs at different stages of development at different temperatures;[184] in *Reduvius personatus* terms of dormancy may supervene irregularly without any obvious relation to season or to environmental conditions (though they may be brought to an end by heat).[231]

Genetics and diapause—The number of generations which may take place before diapause supervenes, in other words the voltinism, may vary within a single species. This is well seen in the moth *Telea polyphemus*, the voltinism of which varies in different regions of North America. Voltinism in this species is regulated in large measure by environmental factors (exposure of the last larval stage to a falling temperature for about a week induces the pupa to become dormant), but the capacity to respond seems, within certain limits, to be dependent on the genetic constitution of the stock—many stocks being heterozygous in their genetic constitution relating to voltinism.[74] In *Locusta migratoria gallica* two strains have been isolated one with uniform, obligatory diapause, and the other entirely free of diapause.[448] The European Corn Borer *Pyrausta nubilalis* has both univoltine and bivoltine races with a different distribution in the United States.[8] The European Spruce Sawfly *Gilpinia polytoma* overwinters as a larva in the cocoon; diapause may occur at different stages in the development of the prepupal stage, and some of the resting larvae may lie over for seven years. After a rest at a low temperature, contact with water provides the stimulus for development. Here again there are univoltine and bivoltine races with different geographical distribution.[221] Similar genetic differences have been described in *Celerio*;[134] and they may be frequent, for it is a common experience during the rearing of a brood of caterpillars in the autumn for a part to pupate and produce another generation and a part to enter a winter diapause.

Factors inducing diapause—In general it seems that many of the cyclical diapause phenomena of insects are in fact induced by seasonal changes of one sort or another. But the arrest of development may be determined long before it becomes apparent and so the active factor is liable to be overlooked and a false impression created that there is an internal rhythm. For instance, in *Loxostege sticticalis*, in which the duration of the larval phase after the cessation of feeding may vary from 3 to 180 days, diapause may be induced by unfavourable nutrition for a restricted period in the quite early stages of larval growth; though in this particular case diapause can be prevented at all seasons of the year, whatever the diet, by keeping at a temperature of 32° C.[273] We saw that the voltinism of silkworm eggs can be determined in the larval stages of their mother (p. 14); similarly, in some Hymenopterous parasites (*Spalangia* and *Cryptus*) diapause in the larva is determined by the diet of the mother or by her age at the time of oviposition. In *Cryptus*, for example, in one experiment none of the larvae from the eggs laid by females in the first 5 days of adult life went into diapause; whereas 95 per cent. of those from eggs laid 16–20 days after emergence entered diapause. Even when a complete diapause did not supervene, the duration of development in the larvae increased with the age of the parent.[268] Larvae of the Pteromalid *Mormoniella* enter diapause at the end of the last larval stage if the mother is exposed to low temperatures during oögenesis;[497] and as the females become older a greater proportion of the progeny enter diapause at this stage, even at normal temperatures; in some females the change over is extremely sudden.[489] In some cases, slow deposition of the eggs, associated probably with partial reabsorption of the yolk (p. 727), may be responsible for causing diapause in the larval stage.[93]

The following examples illustrate some of the differences that exist in the factors producing diapause. Diapause in the Pink Bollworm *Platyedra gossypiella* is induced by lack of moisture in the food; the condition may persist two years or more; but pupation occurs within a few days after moistening.[271] In the pupa of the Corn Earworm *Heliothis armigera* diapause is caused by low temperature during the larval period; water content plays no part.[78] In the rice green caterpillar *Naranga aenescens* diapause supervenes in crowded cultures as the result of mutual stimulation.[424] The Wheat Stem Sawfly *Cephus cinctus* is a univoltine species with an obligatory diapause in the mature larva; after the ending of this diapause in the spring, it may be reinduced by partial desiccation.[249] Diapause in the young larvae of *Euproctis phaeorrhoea* is determined by nutrition: they do not enter diapause if fed exclusively on young foliage.[123] Diapause in the adult *Eurydema* (Hem.) can be induced by a brief fall in temperature during the third nymphal instar.[33] [See p. 124.]

The most important factor in the induction of diapause is the length of day.[532] Some Lepidoptera (*Phalera*, *Spilosoma*) have a fixed monovoltine rhythm uninfluenced by light; others require a certain critical length of day during larval life if immediate pupal development is to occur. In *Acronycta rumicis*[323] or *Diataraxia oleracea*[349] this light period must be longer than 16 hours; in *Pieris brassicae*[323] and *Grapholitha molesta*[324] 13–14 hours is sufficient. As the day length is shortened to perhaps 6 hours or less, diapause is again partially eliminated. In some insects, such as *G. molesta*,[324] all the pupae develop without diapause if the larvae have been kept in total darkness; in *A. rumicis*[323] about 80 per cent. develop after this treatment, in *D. oleracea*[349] about 20 per cent. In *Polychrosis botrana* the eggs and recently hatched larvae are the most sensitive stages; the later instars are almost insensitive (Fig. 80).[333] In *Pyrausta nubilalis* diapause is induced in the larva over a narrow range of photoperiods of 10–14 hours of light per day; a corresponding period of 12 hours of darkness is equally necessary.[362] Indeed, in many insects the length of the dark period is more important than the duration of the light: 'photoperiod' is really a misnomer.[449] In the Mymarid *Cataphractus* which parasitizes the eggs of *Dytiscus* and has several

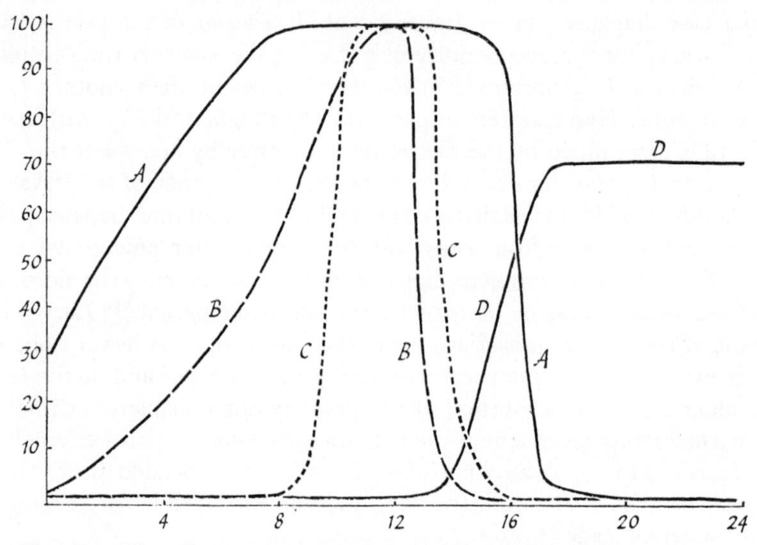

FIG. 80.—The effect of photoperiod on the incidence of diapause in some species of Lepidoptera

Ordinate; percentage of individuals entering diapause. Abscissa; hours of light in 24 hours. A, *Acronycta rumicis* (*after* DANILYEVSKY). B, *Grapholitha molesta* (*after* DICKSON). C, *Pyrausta nubilalis* (*after* BECK). D, *Bombyx mori*, bivoltine race (*after* KOGURE).

generations a year, diapause in the full-grown larva is induced in the late autumn by a short photoperiod of 7½ hours.[426] In a few cases, *Leptinotarsa* and *Euproctis similis* for example, diapause may increase again under constant illumination.[532] Some insects, such as *Dendrolimus*, require 4 weeks' exposure to a 9-hour photoperiod to induce diapause,[402] whereas diapause in *Diataraxia* can be prevented by a single photoperiod of suitable length and timing.[349] In all these insects the effect of the photoperiod is influenced to a varying extent by temperature, and by changed nutrition (as in *Carpocapsa pomonella*[401]) the result sometimes of changes in day-length operating on the host-plant (as in the Cabbage Root-fly *Erioischia brassicae*).[418] [See p. 124.]

In many of these insects, such as *Pieris*, *Pyrausta*, *Acronycta*, and doubtless many more, there are geographical races which are adapted to go into diapause at day-lengths appropriate to the latitudes at which they occur. Thus in *Acronycta* the limiting photoperiod ranges from nearly 20 hours at Leningrad in the north to $14\frac{1}{2}$ hours on the Black Sea coast in the south.[389] In the cabbage moth *Barathra brassicae* there is a transient summer diapause induced by a long photoperiod and a long persisting winter diapause induced by a short photoperiod.[459]

In *Bombyx mori* the effect of light is reversed: exposure of eggs to a long day induces moths of the next generation to lay diapause eggs (p. 15) (Fig. 80).[332]

Most insects respond to a defined photoperiod, but the last larval stage of the dragonfly *Anax* is induced to go into diapause in the late summer equally well by decreasing photoperiods on successive days. This diapause soon comes to an end and the insect is then held 'quiescent' by low temperature until the spring.[387] The threshold of sensitivity is usually adapted so that the light of the full moon (0·01–0·05 foot-candles) is ineffective.[449] The light is usually said not to be received through the eyes,[449] but in *Dendrolimus pini* the photoperiodic stimuli seem to be perceived through the ocelli of the caterpillar.[402] The egg of *Bombyx* first becomes responsive to photoperiod quite late in embryogenesis when all organ systems are differentiated.[332] [See p. 124.]

Diapause in parasitic insects—The development of endoparasitic insects is often conditioned by the continued development of the host. If *Plodia* larvae are kept under desiccating conditions (29° C., 20 per cent. relative humidity) the larvae of *Nemeritis* within them remain in the first stage; whereas if the host is kept at 26° C. and 80 per cent. R.H. they develop normally.[93] The larvae of the Tachinid *Erynnia* which overwinters in adults of *Galerucella* (Col.) cannot complete its development until the beetles begin feeding in the spring.[93] The Braconid *Apanteles glomeratus* normally forms clusters of cocoons over the pupating caterpillar of its *Pieris* host and overwinters as a prepupa; but if the *Pieris* sp. hibernates as a larva, the *Apanteles* larvae remain inside.[248] When *Trichogramma cacoeciae* is reared in eggs of *Cacoecia rosana* while they are in diapause, its development is inhibited for a period of 7 months; but if it is reared in eggs of *Ephestia* or *Mamestra* or in eggs of *Cacoecia* that are not in diapause, it develops rapidly. It seems to be the diapause of the host that retards the development of the parasite; active development of the host is not necessary, for if the egg is killed by heat and then parasitized, development occurs normally.[188] The Braconid *Chelonus annulipes* shows two types of arrest. If its host *Pyrausta nubilalis* goes into diapause, the parasitic larva does likewise; it produces a generation in 2 months in the summer host, in 10 months in the winter host. And the egg of *Chelonus* develops step by step with that of the host. If embryonic development of the host is delayed so is that of the parasite; and if the host egg is infertile development of the parasite ceases.[40] There is evidence that the effect of diapause of the host on the parasite can even be communicated to a hyperparasite.[248]

There is close hormonal coupling between the Hymenopterus parasites of Syrphids and their hosts. The larva of *Epistrophe bifasciata* (Dipt.) normally goes into diapause in the autumn, but if it is invaded by *Diplazon pectoratorius* (Hym.) puparium formation and moulting are evoked by secretions from the moulting parasite.[346] The synchronization of development in the pupa of *Bupalus piniarius* (Lep.) and its Tachinid parasite *Eucarcelia* seems to depend on the great sensitivity of the parasite to the thoracic gland hormone of the host;[498] and the rhythm of development in *Microgaster globatus* seems to depend on humoral factors associated with organ development in the host.[447] *Apanteles glomeratus*, when it hibernates in diapausing caterpillars of *Aporia crataegi*, appears to be influenced

by the physiological changes in the host. But the same species in *Pieris brassic*
is largely independent of the host: its diapause is determined by the direct effe
of day-length and temperature while it is still within the host larva.[460] *Pteromal*
puparium likewise responds directly to temperature and photoperiod and is influenced to a much lesser extent by the physiological condition of the host.[460]

Physiology of diapause—There are many limiting factors for growth i
insects. Mosquito larvae, *Aëdes*, which are not permitted to fill their trache
system cease growing in the 2nd instar for lack of *oxygen*;[310] and *sodium chlorid*
may likewise be a limiting factor in the growth of mosquito larvae.[310] *Rhodniu*
larvae deprived of their symbiotic Actinomyces cease growing in the 4th or 5t
instar for lack of *vitamins*.[42] *Rhodnius* larvae receiving a number of small meal
in place of one large one[308] or *Ephestia* larvae kept at too low a temperature[5]
fail to grow because they do not secrete the *moulting hormone*. We have alread
noted many examples of lack of *water* arresting growth. It is likely enough tha
many different factors are concerned in natural diapause. Two main hypothese
have been put forward. According to one the arrest is regarded as due to th
temporary absence of the hormones necessary to maintain growth.[308] Accordin
to the other, growth is thought to be inhibited by some chemical constituen
accumulating in the body.[13] [See p. 125.]

That the immediate cause of arrested growth is a failure to secrete the necessary hormones was suggested by the similarity between insects deprived of the
brain and insects in natural diapause (p. 64).[308] Diapause in *Lucilia* larvae
induced by desiccation can be terminated and the larvae induced to pupate
simply by placing them in empty glass vials out of contact with sand. Under
these conditions the ring gland shows histological signs of renewed activity; the
effect is attributed to a nervous stimulus to secretion of the pupation hormone.[19]
In the overwintering pupae of the giant silkmoths *Hyalophora* and *Telea* there is
conclusive evidence that the immediate cause of diapause is the absence of
growth-promoting factors the production of which is initiated in the brain; there
is no evidence of an inhibiting substance. Pupae without the brain may survive
in diapause for two years; growth in normal pupae is restored by chilling, as a
result of which the brain is induced to secrete a growth factor that activates the
prothoracic gland and, as in larval moulting, it is the secretion of this which
restores growth and differentiation.[314]

The arrest of growth by inhibitory substances is an idea developed particularly by Roubaud,[241] who, by analogy with muscular fatigue, regards diapause
as a kind of developmental fatigue ('asthenobiosis') which occurs cyclically after
a number of active generations and results from an intoxication of the tissues by a
surcharge of excretory products which may be transmitted from one generation
to the next. It is supposed that this intoxication disappears during a prolonged
resting stage at low temperature ('athermobiosis') during which metabolism is
much reduced while excretion proceeds. This conception, which has been much
elaborated in later publications,[243] is based entirely on hypotheses. The same
idea in less specific form has been used to explain the differences in voltinism
in different races of a given species by supposing that these are due to the production in different quantities, or at varying rates, of an inhibitory substance
('Latenzstoff'). When this substance reaches a sufficient concentration a latent
period or diapause supervenes; in some races this may happen in the first
generation, in others in the third.[114] It has further been suggested that the

lifferent length of hibernation period in related species may be due to differences
n the rate of destruction of 'Latenzstoffe' during the pupal rest, the 'Latenz-
toffe' being here identified, as in Roubaud's conception, with metabolic pro-
lucts accumulating early in metamorphosis.[51] And diapause in the egg of
Melanoplus (p. 15) is pictured as due to a hypothetical 'diapause factor' which
ncreases in amount and inhibits growth, but is gradually destroyed or itself
nhibited during exposure to low temperature, thus liberating the inherent
levelopmental factors.[30] One of the important changes in the egg of *Melanoplus*
vhen development is resumed is the activation of the pro-tyrosinase. It has been
uggested that the inhibitory 'diapause factor' may be some other protein
nolecule which can be adsorbed upon the activator surface and so cut it off
rom the pro-enzyme.[31] In the case of *Trichogramma* in the eggs of *Cacoecia*
quoted above, it certainly appears as though some chemical inhibitor is respon-
ible for diapause in the host and that it is ingested and acts on the parasite.[248]
n contrast with this is the case of *Chelonus* where growth ceases if the host egg
s infertile.[248]

These two hypotheses are not, of course, mutually exclusive: it is possible to
magine a 'diapause factor' as inhibiting the secretion of growth hormones. We
ave discussed already the 'diapause hormone' of the silkworm female which acts
n the eggs and determines the incidence of diapause in the eggs of the ensuing
generation (p. 15). It has been suggested that the fundamental factor in diapause
nay be a diminution in 'metabolic rate'.[268] In the resting pupa of *Hyalophora*
ytochrome *c* is virtually absent and metabolism is at a very low level.[315] Meta-
olism can be increased fourteen-fold by extensive injury—but this does not
ring diapause to an end and lead to the initiation of adult development.[412, 501]

It is generally agreed that in the post-embryonic stages of insects the im-
nediate cause of diapause is the arrest of secretion of growth hormones. That was
learly established in the pupa of *Hyalophora*,[314] and has been confirmed in many
nsects including the larva of the Wheat Stem Sawfly *Cephus cinctus*,[383a] the pupa
f *Luehdorphia japonica*,[422] the larva of *Pyrausta nubilalis*,[386] the pupa of the
phingid *Mimas tiliae*,[415] the larva of *Lucilia caesar*,[397] and many more.[531, 532]
The arrest may affect the neurosecretory cells in the brain or the thoracic glands
n different species.[533] In the *Hyalophora* pupa the brain in diapause seems to be
lectrically inexcitable; cholinesterase is indetectable and acetylcholine accumu-
ates. When electrical activity and cholinesterase reappear, the neurosecretory
ells release their hormone and diapause comes to an end.[511] But these changes
re not general in all insects.[498] It has been suggested by several authors that in
ome insects a secretion from the corpus allatum keeps the neurosecretory cells
nactive; but the evidence for this view is inconclusive.[531] In *Pyrausta* (*Ostrinia*)
ubilalis specialized cells of the hind-gut appear to set free into the blood a hor-
none ('proctodone') which is necessary to activate the neurosecretory cells of the
rain and thus bring diapause to an end.[363] Larvae of the Codling Moth *Cydia*
omonella are brought out of diapause if they are caused repeatedly to spin a fresh
ocoon; perhaps this is an effect on the relative water content.[286] Diapause may
lso be brought to an end by singeing or pricking the larvae, as in *Lucilia*;[241] by
he oviposition of Hymenopterous parasites, as in *Lucilia* attacked by *Alysia*;[142]
r by the development of parasites within the larva, as in *Lipara* (Chloropidae)
nd *Urophora* (Trypetidae).[296] [See p. 125.]

The nature of diapause: 'diapause development'—Diapause is essentially a physiological mechanism for survival during an adverse season, most commonly the winter cold. The mechanism seems to consist in the separation of physiological processes, some of which require a low temperature for their completion, while others require a high temperature. It seems that certain processes, called 'latent'[354] or 'diapause'[356] development, have become adapted by selection to proceed only in some low temperature range characteristic of the species; while the main process of development will go forward only in the upper range of temperature. Each of these two processes has a temperature curve of the usual type (p. 685) but with widely separated optima. In Fig. 81 this is illustrated in the egg of the Australian grasshopper *Austroicetes*.[356] When the temperature curves for the two processes do not overlap (as happens in the egg of the silkworm (p. 14)) development never takes place at an intermediate temperature; and when the two curves come close together the existence of diapause may be difficult to detect. This varying overlap is well seen among the Saturniid species *Philosamia ricini*, *P. cynthia*, *Antheraea pernyi* and *Saturnia pavonia*.[388, 449] [See p. 125.]

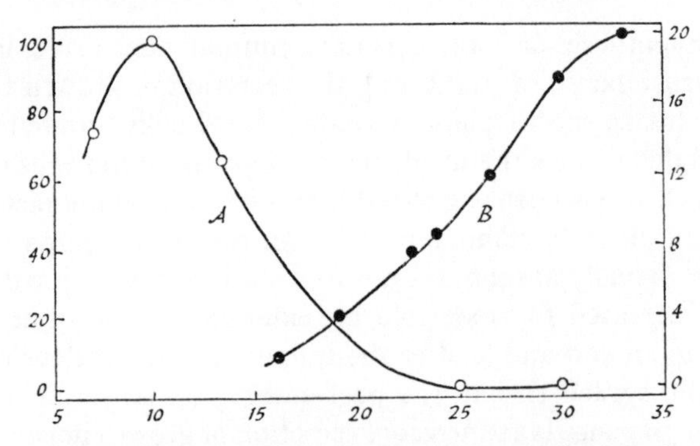

FIG. 81.—The influence of temperature on diapause development (curve A) and post-diapause development (curve B) in the eggs of *Austroicetes cruciata* (*after* ANDREWARTHA)

Ordinate; to the left, percentage of eggs to complete diapause development during 60 days at the temperature specified; to the right, relative rates of development of post-diapause embryos at different temperatures. Abscissa; temperature in ° C.

That is the usual state of affairs; but when diapause occurs in the hot dry season, 'diapause' or 'latent' development may require a high temperature for its completion. That is so in the moth *Diparopsis castanea* from tropical Africa where the optimum temperature for diapause development is 28° C.[449] In *Barathra brassicae* in Japan the summer dormancy is completed at higher temperatures than that in winter.[459] In *Pyrausta nubilalis*[458] and in *Pectinophora gossypiella*[35] no chilling is needed; diapause is ended by reversal of the photoperiod.

The optimum temperature for 'diapause development' is related with the temperature normally experienced by the species;[388, 449] e.g. 10 to 20° C. for subtropical species, 0 to −10° C. in temperate regions. Thus diapause development can occur at −5° C. in eggs of *Malacosoma disstria*,[417] at −10° C. in larvae of *Gilpinia polytoma*,[481] at −15° C. in pupae of *Saturnia pavonia*.[388] In some sawflies the diapausing eonymph needs at least 4 months at 0° C. and will lie over for a second winter if the duration of cooling is insufficient.[468] Pupae of the moth *Biston* have been known to pass through 7 years before resuming their development.[84] In the egg of the Wheat Bulb Fly *Leptohylemyia coarctata* it seems tha different temperatures are required for different stages of the diapause process—but a temperature of $-24°$ C. is highly effective.[515]

There is no reason to suppose that the same physiological processes are always involved in 'diapause development'. In post-embryonic diapause it seems usually to be some process in the brain which restores competence to the neurosecretory cells. In hibernating larvae of *Mormoniella* it looks as though at low temperatures (5° C.) the brain hormone accumulates, at high temperatures (25° C.) it may be broken down again.[497] This same hormone inactivation at high temperature (35° C.) is seen in *Cephus* larvae coming out of diapause,[383a] and in *Rhodnius* exposed to temperatures (35° C.) above the normal physiological range.[525] It must be something different in diapausing embryos of *Bombyx* or *Melanoplus* (p. 14) in which neither nervous system nor endocrine organs have been differentiated. It may even take place in fragments of such embryos.[377]

ADDENDA TO CHAPTER III

p. 68. **Prothoracic gland secretion**—During the prepupal stage of *Drosophila* the cells of the prothoracic gland give off numerous surface droplets, presumably containing ecdysone (p. 71), by exocytosis.[543] In full-grown larvae of *Calpodes* (Lep., Hesperidae) implanted prothoracic glands will induce pupation in the headless larva but will not induce pupation in the isolated thorax, although injected ecdysone will do this. It has therefore been suggested that the prothoracic glands act upon some tissue in the abdomen and induce this to liberate ecdysone.[641]

p. 68. **Stimulus to brain hormone secretion**—The histological changes in the neurosecretory cells of *Rhodnius* show that secretion begins immediately on feeding and lasts for 6–8 days; and a second phase of release accompanies ecdysis.[632] In cockroaches (and perhaps other insects) the initiation of a new moulting cycle after ecdysis requires a short period of feeding (some 12 hours). By using this response it is possible to get synchronous development of colonies of cockroaches.[586] Abnormal stimuli of many kinds can arrest growth; for example, mechanical constraint in *Galleria* larvae—even when all other conditions are favourable;[561, 625] in *Sarcophaga perigrina* pupation is delayed by contact with water, but hormone secretion is restored within a few hours in dry surroundings.[610]

p. 69. **Mode of action of hormones**—During larval-adult transformation in *Lucilia* the prothoracic gland hormone (ecdysone) shows a series of peaks which precede those periods when pronounced structural changes occur.[546] In order to exert its normal effect on growth and moulting the hormones must be present in the correct amounts. Too much hormone (as after injection of excessive quantities of ecdysone (p. 71) leads to precocious 'apolysis'[579] and rapid deposition of thin and amorphous cuticle.[548]

It may well be that the hormone or hormones liberated from the *pars intercerebralis* exert a direct effect on the tissues. That appears to be the case, for example, in the growth of the compound eye in larvae of *Aeschna*.[621] In *Calliphora* any excess of ecdysone introduced into the blood is rapidly inactivated by enzymes in the fat body.[582] In *Sarcophaga* the tissues are normally furnished with a continuous very small supply.[611]

p. 70. **'Puffs' on Giant Chromosomes**—Actinomycin D which specifically blocks DNA-dependent RNA synthesis causes puffs in giant chromosomes to disappear. It seems that the puffs are indeed the site of formation of messenger RNA.[556, 585, 588] On the major salivary gland autosomes of *Drosophila* 108 loci form puffs at some time during development.[545] The fact that the same puffs occur in salivary glands and mid-gut cells suggests that they are concerned in some general aspects of metabolism.[549] It may be that the salivary glands at metamorphosis are engaged in the production of substances to be utilized during histogenesis of the adult.[545] In the pulvilli of *Sarcophaga* the large number of 'puffs' in the giant polytene nuclei occur in an ordered sequence which is related to the development of the cuticular structure; this supports the belief that 'puffing' represents the visible activation of gene loci concerned in growth. They are not necessarily under direct hormonal control.[551, 557, 571, 644]

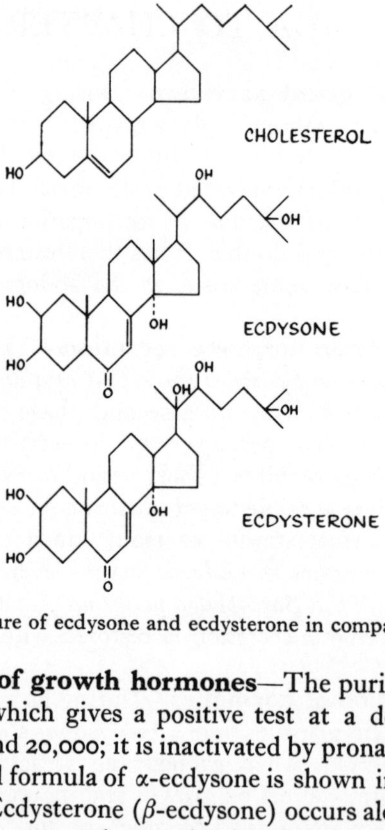

CHOLESTEROL

ECDYSONE

ECDYSTERONE

FIG. 81*a*.—The structure of ecdysone and ecdysterone in comparison with cholesterol.

p. 71. **Chemistry of growth hormones**—The purified peptide brain hormone from *Bombyx*, which gives a positive test at a dose of 0·02 μg, has a molecular weight around 20,000; it is inactivated by pronase but not by trypsin.[648]

p. 71. The chemical formula of α-ecdysone is shown in Fig. 81*a* in comparison with cholesterol. Ecdysterone (β-ecdysone) occurs along with α-ecdysone in many insects. These compounds are closely related to cholesterol but have a very large number of hydroxyl groups in the molecule which renders them soluble in water and led to their steroid nature being overlooked.[583] Ecdysone is readily converted to ecdysterone by the tissues of *Bombyx* and *Calliphora*, etc.[581, 606] In *Calliphora* ecdysterone is 3–5 times as active as ecdysone[575] and is indeed the main hormone present.[568] Ecdysone, ecdysterone and other variants of these steroids with moulting hormone activity have been isolated in large quantities from plants: *Taxus*, *Podocarpus*, *Polypodium*, etc.[578, 631]

p. 72. **Controlled lysis of cells**—We saw (p. 45) that extensive breakdown of muscles commonly occurs after moulting and metamorphosis. This is an orderly process involving the activation of lysis in the cells concerned, perhaps through induced enzyme synthesis.[598]

p. 74. **Histolysis**—In *Calliphora* the invasion of muscles by phagocytes is under hormonal control; a change in the haemocyte population is one of the first indications of the onset of metamorphosis.[559]

p. 78. **Corpus allatum and control of metamorphosis**—Pupal-adult intermediates have been produced in *Sarcophaga* by the application of juvenile hormone to the cuticle of young pupae.[630]

In *Rhodnius*, by exposing the last larval stage to juvenile hormone at progressively later times in development, graded intermediate forms are produced.[308] In *Galleria* this procedure has been shown to affect also the internal organs.[624]

p. 79. **Juvenile hormone action**—Reversal of metamorphosis has been effected also in the integument of *Oncopeltus*[592] and *Tenebrio*.[554] Experiments on *Tenebrio* suggest that the control of gene expression mediated by the juvenile hormone is at the translational level and involves the appearance of new transfer RNA's and their activating enzymes.[577] Characteristic changes in the 'puffing' pattern in *Chironomus* can be induced by compounds with juvenile hormone activity, suggesting specific alterations in RNA synthesis.[589, 596]

The juvenile hormone and various more or less related chemicals with juvenile hormone activity, when applied to the egg may cause disturbances in metamorphosis during post-embryonic development, or may disrupt development of the embryo.[615] These results have led to the suggestion that chemicals of this type might be of value as insecticides.

p. 80. **Chemical nature of juvenile hormone**—Fig. 81*b* shows skeleton formulae of some compounds with juvenile hormone activity: I, farnesol; II, farnesyl methly ether;[494] III, methyl ester of epoxy-farnesenic acid;[550] IV, the major natural juvenile hormone in *Hyalophora*;[617] V, a second, less abundant and slightly less active compound associated with IV,[604] VI, the methyl ester

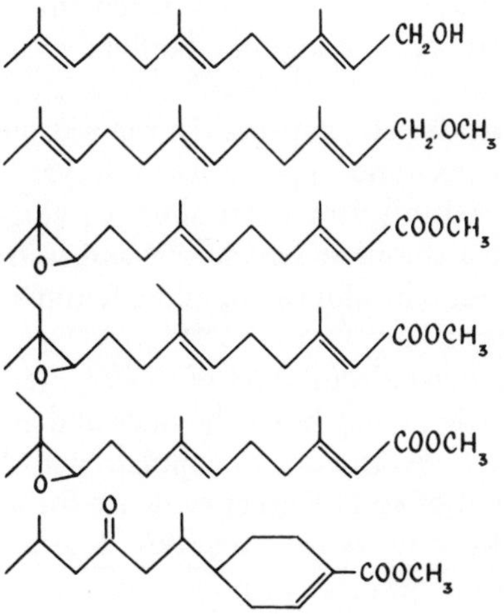

FIG. 81*b*.—Formulae of some compounds with juvenile hormone activity. Explanation in the text.

of todomatuic acid, a naturally occurring terpenoid of different structure in th resin of balsam fir, with very high juvenile hormone activity in Pyrrhocori bugs but not in other insects.[550, 628]

When isolated brains of *Hyalophora* plus corpora cardiaca and corpora allat are maintained *in vitro* they produce detectable amounts of juvenile hormone.[61

p. 80. **Prothetely and metathetely**—Infective organisms in insects ma liberate sufficient material with juvenile hormone activity to interfere wit normal metamorphosis; there is a partial retention of larval characters in Satur niid pupae[564] and in *Tribolium*[565] infected with Nosema. The converse effec is seen in Lepidoptera larvae infected with nuclear polyhedrosis virus, whic show precocious development of adult characters in antennae, mouth parts eyes, etc.[607]

p. 83. **Hormonal control of ecdysis**—The term 'ecdysis', originally equiva lent to 'moulting', is sometimes reserved for the final act of the moulting cycle the casting of the old skin. In silkmoths emerging from the pupa this ac depends on the clock-controlled release of a hormone from the brain which act on the central nervous system to set in motion the specific behaviour patter which culminates in ecdysis.[637]

p. 84. **Pattern and gradients**—One factor in the control of pattern in th body surface of insects is the existence of 'gradients', for example in the abdo minal segments or in segments of the legs.[597] Some gradients, for example i *Oncopeltus*, are responsible for determining the polarity, or direction of out growth, of the setae.[591] The commitment to form a given element in a patter is not absolute; 'transdetermination' is possible; and under experimental condi tions one cell type can be transformed into another.[567, 572, 593, 623] Trans determination is commonly believed to consist in the activation of new sets o genes. Such changed activities may result from the position of the cell in 'gradient', the type of determination being decided perhaps by the level of con centration of some chemical evocator.[591, 593, 634] Transdetermination is seen of course, at metamorphosis. In Hemimetabola, such as *Rhodnius*, the epiderma cells which lay down one characteristic component of the larval pattern in th presence of juvenile hormone, lay down a quite different type of cuticle formin a different element in the pattern of the adult, when the juvenile hormone i absent.[308, 645] Homoeösis (p. 110) is another familiar example of transdeter mination.[569]

p. 94. **Temperature and feminization**—In the mosquito *Aëdes* (*Ochlero tatus*) *stimulans* the normal sex ratio appears if the larvae are reared at a lo temperature; but larvae reared in warm water show varying degrees of femini zation extending up to almost complete females;[576] and in triploid intersexes o *Drosophila* it is possible to obtain almost complete feminization of the gonad by exposure of the segmenting eggs to 30° C.[590]

Among antero-posterior gynandromorphs of *Aëdes*, individuals with femal heads will take blood meals which may burst the male abdomen; and individual with male heads and female body attempt to copulate with females.[614]

p. 94. **Hormonal control of sex**—Direct evidence for a hormonal effect or sex has been obtained in the glow-worm *Lampyris*. The androgenic hormone which will cause total masculinization of both primary and secondary sexua characters in female larvae, is secreted in the mesodermic apical tissue, present in all the testicular follicles, and characteristic of the testes of Lampyrids. Th formation of the tissue is under the control of neurosecretory hormones from the brain: masculinization in the female can be induced by implantation of the brain (plus corpus allatum and corpus cardiacum) from a male larva. The interpretation of these results is made difficult by the fact that the female *Lampyris* is a neotenic form which retains many larval characters: prothoracic glands degenerate in the pupa of the male but persist in the adult female. In most insects sex is clearly determined at eclosion from the egg; in *Lampyris* it is only at the fourth larval stage that testes and ovaries become distinguishable.[608]

p. 96. **Intersexes**—In *Solenobia* it is established that sex of the soma and sex of the germ cells are determined by sex chromosomes. It seems likely that this is true also of *Lymantria*.[626] In 2X3A intersexes of *Drosophila* there may be a fine mosaic of male and female pigmentation of the abdominal tergites (as in *Solenobia*, p. 740). This phenomenon differs from typical gynandromorphism only in the finer character of the mosaic. Genetically, *Solenobia* and *Lymantria* intersexes may be regarded as uniform, in contrast to true, gynanders.[633]

p. 99. **Royal jelly**—One rather broad fraction of the dialysable portion of royal jelly is necessary for the development of queens. Though rapidly inactivated by storage, royal jelly freeze-dried and kept under nitrogen at −20° C retains its full activity for more than a year.[640]

p. 102. **Termite castes**—Elimination of supernumerary replacement reproductives is initiated by fighting between reproductives; injured reproductives are then destroyed by larvae and nymphs.[619]

There is some evidence that in field colonies of *Calotermes* the juvenile hormone may not only be produced by the corpus allatum of the individual termites but may be transported from one individual to another as a pheromone.[599]

p. 105. **Polymorphism in Aphids**—In *Aphis craccivora*[580] wing development can be influenced by nutrition, by temperature and by the photoperiod to which the mother is exposed, and even by brief contact, lasting no more than one minute, between adult females; in *Acyrthosiphon pisum*[635] tactile stimuli among crowded individuals and the nutritional state of the host plant can both control wing production among the offspring of treated females; in *Dysaphis devecta*[566] nutrition will determine the production of sexual forms as well as the appearance of apterae and alatae. It would appear that each aphid species has its characteristic responses.[594, 636] Even the removal of a single amino acid (isoleucine) from a synthetic diet may cause *Myzus persicae* larvae to switch over from alate to aptera production.[605] In *Brevicoryne* the loss of wing-buds in embryos appears to depend on maternal corpus allatum activity, whereas in the postnatal determination of apterae the corpus allatum activity of the individual aphid is the controlling factor.[643] In *Eriosoma pyricola* sexuparae are produced only in the roots of pear trees treated to induce cessation of shoot growth.[627] The action spectrum of the photoperiodic control of the type of progeny in *Megoura* shows the highest sensitivity at 450–470 mμ, in the blue region of the spectrum. The position of this absorption maximum suggests that the chromophore may be a carotinoid hydrocarbon perhaps complexed with protein.[595]

p. 105. **Phase change in locusts**—Isolated locusts reared in laboratories containing large cultures tend to develop 'gregarious' coloration: perhaps an airborne pheromone is involved.[563] This 'gregarization' pheromone is secreted with the faeces and is said to be received through the spiracles.[609] Implantation of active corpora allata into *Locusts migratoria* causes a shift in characters towards the solitary phase in maturing locusts and in their offspring: fertility is increased, more egg pods and pods containing more eggs are produced and there are fewer black larvae. Crowding or removal of corpora allata have the opposite effect.[553]

p. 106. In the cricket *Gryllodes sigillatus* larvae exposed to a photoperiod of 14 hours show maximal wing development; whereas a photoperiod of 10 hours suppresses wing formation completely; and adults exposed to a 14 hour photoperiod produce a greater proportion of winged progeny.[603]

p. 108. **Regeneration**—The two types of response seen in *Blattella*, which depend on whether amputation is done before or after a certain 'critical period', occur in other insects. In *Tachycines* amputation of the cercus late in the moulting cycle may delay regeneration for two moults; if the leg, including the femur, is removed, no regeneration takes place.[584] In *Galleria* (unlike *Ephestia*, p. 108) regeneration of imaginal discs requires the presence of the prothoracic gland or of injected ecdysone. The hormone appears to act directly on the imaginal discs themselves.[600] Experiments on *Periplaneta*[613] and experiments with regenerating tissues of *Leucophaea, in vitro*, have led to similar conclusions.[602]

p. 109. In *Periplaneta* regeneration takes place independently of the nervous system, although growth of the regenerated leg is delayed after denervation and the muscles are poorly developed and of course immobile.[613] Larvae of *Acheta* can regenerate a compound eye destroyed by cautery, with the formation of normal ommatidia and re-establishment of central connexions.[562]

p. 113. **Factors inducing diapause**—Many of the insects quoted above (p. 113), notably *Heliothis*,[642] are also influenced by photoperiodic conditions acting at various times in the life cycle.

p. 114. In *Pieris brassicae* the photoperiod acts directly on the brain itself to control secretion of the brain hormone and so regulate the onset of diapause. The young larval stages, in which the head capsule is entirely black, are insensitive. At the third moult a yellow triangle on the clypeus admits light to the brain and the larva becomes sensitive to photoperiod. Hormone secretion by such a brain transplanted to the abdomen of another larva can likewise be controlled by illumination.[555]

p. 115. In the foregoing account (pp. 113–15) it has been assumed that the insect is responding to the actual duration of the 'photophase' and 'skotophase' of the light regime, by reference to an innate response to a 'critical photoperiod'. But an alternative explanation has been put forward[552] according to which photoperiodic induction is related to an endogenous circa-dian rhythm which determines the receptivity of the central nervous system to the 'on' or 'off' switch of light. In the fruit-tree leaf-roller *Adoxophyes* in which diapause is induced by photoperiods below 16 hours the effect of short photoperiods of variable duration is destroyed by a light break of 2 minutes or even a photoflash of 10 msec. applied precisely 16 hours after the beginning of the photophase.[544] Likewise in *Pieris brassicae*, in which the 'light break' technique was first systematically applied to insect material, there are striking changes in the light sensitivity when the dark period is interrupted at different points.[552] The same applies to the induction of diapause in the larva of *Nasonia*.[620] Such results are more readily explained on the basis of a circa-dian rhythm. But this is not a general phenomenon.[560] For example, in the diapause of *Leptinotarsa*[647] or in the determination of embryos in *Megoura* (p. 105)[594] such interruptions of the skotophase are without effect. Further understanding of the physiology of time measurement in living cells will be needed to resolve this discrepancy.

p. 116. **Arrest at high temperature**—Growth and moulting in *Rhodnius* is arrested at 36° C although all other bodily functions continue apparently unchanged; growth is restored when the temperature is lowered.[525] In this case the high temperature interferes mainly with protein synthesis and this is reflected in the inhibition of cell division, impairment of differentiation, reduction in mid-gut protease, and the arrest of brain hormone secretion.[612] Just as *Anax* larvae show only a brief autumn diapause and are then arrested in quiescence by low temperatures throughout the winter,[558] so pupae of the butterfly *Luehdorfia* are arrested in quiescence by high temperature during the summer, complete their development to the adult in the autumn, and the fully developed 'pharate' adult then remains dormant until the spring when it emerges.[574]

p. 117. **Brain cholinesterase in diapause**—The changes in cholinesterase activity in the neuropile of insects seems not to be associated causally with the termination of diapause, but to be a feature of development of the nervous system that occurs equally in non-diapausing species.[629, 638]

p. 118. **'Diapause development'**—The embryo of *Aeschna mixta* shows a period of typical 'diapause development' which is optimal at 5–10° C., but there is much individual variation in the time required for its completion which ranges from 5–15 weeks.[622] In Australia *Sminthurus viridis* females produce diapause eggs under the influence of maturing food plants in the spring. These eggs require a period of 2–3 months of 'diapause development' in the summer before they will resume embryonic growth in moist and warm conditions.[639] Diapause in the pupa of *Antheraea* can likewise be terminated without chilling. A translucent area of cuticle on the head favours the penetration of light to the brain. Short day conditions (12 hour photoperiod) inhibit the ending of diapause, long day conditions (17 hours) promote termination. The light stimulus acts directly on the brain; probably on the neurosecretory cells. The effective wavelength extends from 398–509 mμ, the same spectral region as that of carotenoids (cf. p. 123) and an intensity of less than one foot-candle will activate the system.[542, 648] Diapause in *Antheraea* may be terminated also by high temperature alone.[601] In *Ostrinia* specialized cells in the hind-gut appear to set free into the blood a hormone ('proctodone') which is necessary to activate the neurosecretory cells of the brain and thus bring diapause to an end.[547]

REFERENCES

1 ABELOOS, M. *C.R. Soc. Biol.*, **113** (1933), 17–19 (regeneration: *Timarcha*, Col.).

2 ABERCROMBIE, W. F. *J. Morph.*, **59** (1936), 91–112 (growth of *Popillia*, Col.).

3 ADAMSON, A. M. *Proc. Roy. Soc.*, B, **129** (1940), 35–53 (termite castes).

4 ALPATOV, W. W. *J. Exp. Zool.*, **52** (1929), 407–32 (growth measurements: *Drosophila*).

5 ANGLAS, J. *Bull. Sci. Fr. Belg.*, **34** (1900), 1–111 (metamorphosis of Hymenoptera: histology).

6 —— *Bull. Soc. Ent. Fr.*, 1901, 104–7 (control of metamorphosis).

7 ASHBY, D. G., and WRIGHT, D. W. *Trans. Roy. Ent. Soc. Lond.*, **97** (1946), 355–79 (larva and puparium: *Psila*, Dipt.).

8 BABCOCK, K. W. *J. Econ. Ent.*, **17** (1924), 120–5 (voltinism in *Pyrausta*, Lep.).

9 BAKER, F. C. *Canadian Entom.*, **67** (1935), 149–53 (diapause in mosquito larvae).

10 BALKASHINA, E. I. *Arch. Entw. Mech.*, **115** (1929), 448–63 (hereditary homoeösis: 'aristopedia' in *Drosophila*).

[11] BALTZER, F. *Arch. Entw. Mech.*, **136** (1937), 1–43 (mechanism of production of intersexes: *Lymantria*, Lep.).

[12] BATAILLON, E. *C.R. Acad. Sci.*, **115** (1892); *Bull. Soc. Fr. Belg.*, **25** (1893), (cause of metamorphosis: *Bombyx mori*).

[13] BAUMBERGER, J. P. *Ann. Ent. Soc. Amer.*, **10** (1917), 179–86 (hibernation and diapause).

[14] BEADLE, G. W., and LAW, L. W. *Proc. Soc. Exp. Biol. Med.*, **37** (1938), 621–23 (control of eye colour in *Drosophila* by diffusible substances in the food).

[15] BECKER, E. *Z. vergl. Physiol.*, **24** (1937), 305–18 (Liesegang rings in pigmentation of *Vespa*).

[16] BECKER, E., *et al. Biol. Zbl.*, **59** (1939), 326–41; **61** (1941) 360–88 (extraction of puparium-forming substance: *Calliphora*, Dipt.).

[17] BECKER, G. *Biol. Zbl.*, **67** (1948), 407–44 (caste determination in termites).

[18] BLAUSTEIN, W. *Z. Morph. Oekol. Tiere*, **30** (1935), 333–54 (histology of metamorphosis: *Ephestia*, Lep.).

[19] BLUNCK, H., and SPEYER, W. *Z. wiss. Zool.*, **123** (1924), 156–208 (failure of reunion of severed head and neck: Coleoptera).

[20] BODENHEIMER, F. S. *Quart. Rev. Biol.*, **8** (1933), 92–95 (growth measurements: review).

[21] BODENSTEIN, D. *Z. wiss. Insektenbiol.*, **25** (1930), 23–35; *Arch. Entw. Mech.*, **128** (1933), 564–83; **130** (1933), 747–70; **133** (1935), 156–92 (transplantation and regeneration of limbs and spines in larva of *Vanessa*, Lep.).

[22] —— *Zool. Anz.*, **103** (1933), 209–13 (suppression of limb-regeneration by implantation of epidermis).

[23] —— *Arch. Entw. Mech.*, **128** (1933), 564–83 (moulting of transplanted limbs: Lepidoptera larvae).

[24] —— *Arch. Entw. Mech.*, **133** (1935), 156–92 (organization in limbs of Lepidoptera).

[25] —— *Ergebn. Biol.*, **13** (1936), 174–234 (regulation of post-embryonic growth: review).

[26] BODENSTEIN, D. *Arch. Entw. Mech.*, **136** (1937), 745–85; *J. Exp. Zool.*, **87** (1941), 31–53; (experimental reduplication of limbs: Lep.).

[27] —— *Arch. Entw. Mech.*, **137** (1938), 474–505 (control of metamorphosis: *Drosophila*); *Ibid.*, 636–60; *Biol. Zbl.*, **58** (1938), 329–32; *J. Exp. Zool.*, **82** (1939), 329–56; *Bull. Biol.*, **84** (1943), 13–33; 34–58; **86** (1944), 113–24 (ditto: *Phryganidia*, Lep.).

[28] —— *J. Exp. Zool.*, **86** (1941), 87–111 (determination of facet number: *Drosophila*, Dipt.).

[29] —— *Biol. Bull.*, **91** (1946), 288–94 (regulation of testis growth: *Drosophila*).

[30] BODINE, J. H. *Physiol. Zool.*, **5** (1932), 549–54 (mechanism of diapause).

[31] —— *Amer. Nat.*, **75** (1941), 97–106 (diapause factor and tyrosinase: *Melanoplus*).

[32] DU BOIS, A.-M., and GEIGY, R. *Rev. Suisse Zool.*, **42** (1938), 169–248 (metamorphosis, &c.: *Sialis*, Neuropt.).

[33] BONNEMAISON, L. *Ann. Epiphyt.*, **11** (1945), 19–56 (diapause: review) *C.R. Acad. Sci.*, **227** (1948) 985–7; 1052–4 (diapause: *Eurydema*, Hem.).

[34] —— *C.R. Acad. Sci.*, **226** (1948), 2093–4 (appearance of sexual forms: Aphidae).

[35] BORCHARDT, E. *Arch. Entw. Mech.*, **110** (1927), 366–94 (heteromorphous regeneration: *Dixippus*).

[36] BORCHERT, A. *Zool. Jahrb., Anat.*, **61** (1936), 99–106 (growth measurements and number of instars: *Achroea*, Lep.).

[37] BORDAGE, E. *Bull. Sci. Fr. Belg.*, **39** (1905), 307–54 (autotomy and regeneration: Orthoptera).

[38] BOUNHIOL, J.-J. *C.R. Acad. Sci.*, **203** (1936), 388–9; **205** (1937), 175–7; *C.R. Soc. Biol.*, **126** (1937), 1189–91; *Bull. Biol.*, Suppl. **24** (1938), 1–199 (brain and corpora allata in control of metamorphosis: *Bombyx mori*).

[39] BOURDON, J. *C.R. Soc. Biol.*, **124** (1932), 872–4 ; *Bull. Biol. Fr. Belg.*, **71** (1937), 466–99 (regeneration in *Timarcha*, Col.).

[40] BRADLEY, W. G., and ARBUTHNOT, K. D. *Ann. Ent. Soc. Amer.*, **31** (1938), 359–65 (effect of host (*Pyrausta*) on diapause in *Chelonus*, Braconidae).

[41] BRAUN, W. *Arch. Entw. Mech.*, **135** (1936), 494–520; *Cytologia*, **10** (1939), 40–3; *Biol. Bull.*, **76** (1939), 226–40 (cell divisions in relation to wing pattern: *Ephestia*, Lep.).

[42] BRECHER, G., and WIGGLESWORTH, V. B. *Parasitology*, **35** (1944), 220–4 (actinomyces and moulting: *Rhodnius*).

[43] BRECHER, L. *Arch. Mikr. Anat. u. Entw. Mech.*, **102** (1924), 549–72 (heteromorphous regeneration: *Dixippus*).

[44] BRIDGES, C. B. *Amer. Nat.*, **59** (1925), 127–37 (sex in relation to chromosomes: *Drosophila*).

[45] BRIDGES, C. B., and DOBZHANSKY, T. *Arch. Entw. Mech.*, **127** (1933), 575–90 (hereditary homoeösis: 'proboscipedia' in *Drosophila*).

[46] v. BUDDENBROCK, W. *Z. Morph. Oekol. Tiere*, **18** (1930), 700–25; *Z. vergl. Physiol.*, **14** (1931), 415–28 (Verson's glands and moulting hormone: Lep.).

[47] BURTT, E. T. *Proc. Roy. Soc.*, B, **124** (1937), 13–23; **126** (1938), 210–23 (corpora allata of Diptera).

[48] BUSHNELL, R. J. *J. Morph.*, **60** (1936), 221–41 (polyploidy in mid-gut epithelium *Acanthoscelides* Col.).

[49] BUTENANDT, A., *et al. Biol. Zbl.*, **65** (1946), 41–51 ('anti-bar substance': *Drosophila*, Dipt.).

[50] BUXTON, P. A. *Parasitology*, **30** (1938), 65–84 (allometric growth: *Pediculus*, Anopl.).

[51] BYTINSKI-SALZ, H. *Arch. Entw. Mech.*, **129** (1933), 356–78 (humoral control of growth in hybrids: *Celerio*, Lep.).

[52] CAMERON, A. E. *Trans. Roy. Soc. Edin.*, **58** (1934), 211–50 (life history of *Haematopota*, Dipt.).

[53] CASPARI, E. *Arch. Entw. Mech.*, **130** (1933), 353–81 (gene controlling eye colour in *Ephestia*, Lep.).

[54] —— *Quart. Rev. Biol.*, **16** (1941), 249–73 (development of colour patterns: Lepidoptera, review).

[55] —— *J. Exp. Zool.*, **86** (1941), 321–31 (temperature and pupation: *Ephestia*, Lep.).

[56] CATALA, R. *C.R. Acad. Sci.*, **208** (1939), 1349–51 (mechanical shocks and metamorphosis: *Chrysiridia*, Lep.).

[57] CAZAL, P. *Bull. Biol. Fr. Belg.*, Suppl. **32** (1948), 1–227 (retrocerebral endocrine glands in insects).

[58] CHAPMAN, R. N. *J. Exp. Zool.*, **45** (1926), 292–9 (prothetely: *Tribolium*, Col.).

[59] CHEVAIS, S. *Bull. Biol. Fr. Belg.*, **77** (1943), 1–108 (diffusible substance causing 'bar' eye: *Drosophila*).

[60] CHEWYREUV, I. *C.R. Soc. Biol.*, **74** (1913), 695–9 (relation between size of host and sex in Ichneumonidae).

[61] CHILD, C. M., and YOUNG, A. N. *Arch. Entw. Mech.*, **15** (1903), 543–602 (regeneration of limbs: Agrionidae).

[62] CHILD, G. *Biol. Bull.*, **77** (1939), 432–42 (rate of development and gene action: *Drosophila*).

[63] CHILD, G. P., *et al. Physiol. Zool.*, **13** (1940), 56–64 (phenocopies: *Drosophila*).

[64] CHRISTOPHERS, S. R., and CRAGG, F. W. *Ind. J. Med. Res.*, **9** (1922), 445–63 (homologies of external genitalia in *Cimex*, Hem.).

[65] CLANCY, C. W., and BEADLE, G. W. *Biol. Bull.*, **72** (1937), 47–56 (transplantation of ovaries in different genetic strains: *Drosophila*).

[66] COCKAYNE, E. A. *Trans. Roy. Ent. Soc. Lond.*, **83** (1935), 509–21 (origin of gynandromorphs: Lepidoptera).

[67] —— *Biol. Rev.*, **13** (1938), 107–32; *Proc. Roy. Ent. Soc. Lond.*, C, **9** (1945), 48–55 (genetics of sex in Lepidoptera: review).

[68] —— *Proc. Roy. Ent. Soc. Lond.*, A, **16** (1941), 55–9 (prothetely in *Smerinthus* hybrid, Lep.).

[69] COUSIN, G. *Bull. Biol. Fr. Belg.*, Suppl. **15** (1932), 341 pp. (diapause: *Lucilia*, Dipt.).

[70] —— *C.R. Acad. Sci.*, **200** (1935), 970–2; *Bull. Biol. Fr. Belg.*, **72** (1938), 79–118 (neoteny in *Gryllus*, Orth.).

[71] CRAMPTON, H. E. *Arch. Entw. Mech.*, **9** (1899), 293–318 (grafting together of pupae of Lepidoptera).

[72] DAVIDSON, J. *J. Linn. Soc. Zool.*, **36** (1927), 467–77 (formation of intermediates in *Aphis rumicis*).

[73] —— *Ann. Appl. Biol.*, **16** (1929), 104–34 (effect of environment in determining polymorphism of *Aphis rumicis*).

[74] DAWSON, R. W. *J. Exp. Zool.*, **59** (1931), 87–131 (voltinism and diapause: *Telea*, Lep.).
[75] DAY, M. F. *Nature*, **145** (1940), 264 (neurosecretory cells: Lepidoptera).
[76] DEWITZ, J. *Zool. Anz.*, **47** (1916), 123–4 (role of tyrosinase in metamorphosis: Diptera).
[77] DICKSON, R. C., and SAUNDERS, E. *J. Econ. Ent.*, **38** (1946), 605 (diapause: *Grapholitha molesta*, Lep.).
[78] DITMAN, L. P., *et al. J. Econ. Ent.*, **33** (1940), 281–5 (diapause: *Heliothis*, Lep.).
[79] DOBZHANSKY, T. *Arch. Entw. Mech.*, **123** (1931), 719–46 (interaction of genes in gynandromorphs of *Drosophila*).
[80] —— *Arch. Entw. Mech.*, **123** (1931), 719–46 (interaction between male and female ducts in gynandromorphs: *Drosophila*).
[81] DOBZHANSKY, T., and DUNCAN, F. N. *Arch. Entw. Mech.*, **130** (1933), 109–30 (genes affecting development: *Drosophila*).
[82] EIDMANN, H. *Z. Morph. Oekol. Tiere*, **2** (1924), 567–610 (mechanism of moulting).
[83] ELEISCU, G. *Z. angew. Ent.*, **19** (1932), 191 ('prepupal' instar: Ichneumonidae).
[84] V. EMDEN, F. *V^e. Congr. Internat. Ent. Paris 1932* (1933), 813–22 (diapause).
[85] EMERSON, A. E. *Ecol. Monographs*, **9** (1939), 287–300; *Amer. Mid. Nat.*, **21** (1939), 182–206 (determination of reproductive castes in termites).
[86] EPHRUSSI, B., and BEADLE, G. W. *Bull. Biol. Fr. Belg.*, **71** (1937), 54–74; *Genetics*, **22** (1937), 76–86; *Quart. Rev. Biol.*, **17** (1942), 326–38 (diffusible substances and development of eye colours in *Drosophila*: review).
[87] EPHRUSSI, B., KHOUVINE, Y., and CHEVAIS, S. *Nature*, **14** (1938), 203 (morphogenetic substance and 'bar' eye: *Drosophila*).
[88] EVANS, A. C. *Parasitology*, **27** (1935), 291–8 (diapause, &c., in *Lucilia* larvae, Dipt.).
[89] ——*Proc. Roy. Ent. Soc. Lond.*, A, **11** (1936), 52–4 (histolysis of muscle in pupa of *Lucilia*, Dipt.).
[90] —— *Ann. Appl. Biol.*, **25** (1938), 558–72 (effect of food on production of winged Aphids).
[91] FAVRELLE, M. *C.R. Acad. Sci.*, **216** (1943), 215–6 (corpora allata and metamorphosis: *Bacillus*, Orth.).
[92] FELDOTTO, W. *Arch. Entw. Mech.*, **128** (1933), 299–341 (critical periods in determination of wing pattern: *Ephestia*, Lep.).
[93] FLANDERS, S. E. *J. Econ. Ent.*, **35** (1942), 607; **37** (1944), 408–11 (diapause in parasitic insects).
[94] —— *Science*, **101** (1945), 245–6 (caste determination in ants).
[95] FORD, E. B. *Biol. Rev.*, **20** (1945), 73–88 (polymorphism).
[96] FRAENKEL, G. *Proc. Roy. Soc.*, B, **118** (1935), 1–12 (hormone controlling pupation *Calliphora*, Dipt.).
[96a] —— *Proc. Roy. Ent. Soc. Lond.*, A, **13** (1938), 158–60 (prepupal cuticle: *Calliphora* Dipt.).
[97] FREW, J. G. H. *Brit. J. Exp. Biol.*, **6** (1928), 1–11 (cultivation of insect tissues).
[98] FRIEDRICH, H. *Z. wiss. Zool.*, **137** (1930), 578–605 (regeneration: *Dixippus*).
[99] FRIZA, F., and PRZIBRAM, H. *Biol. Zbl.*, **53** (1933), 516–30 (Johnston's organ and 'aristopedia': *Sphodromantis* and *Drosophila*).
[100] FUKUDA, S. *Proc. Imp. Acad. Tokyo*, **16** (1940), 414–16; 417–20; *Zool. Mag.*, **53** (1941), 582–4; *Annot. Zool. Japon.*, **20** (1941), 9–13; *J. Fac. Sci. Tokyo Imp. Univ.* (iv) **6** (1944), 477–532 (prothoracic gland and hormone secretion: *Bombyx mori*).
[101] —— *Japanese J. Genetics*, **17** (1941), 97–108 (genes and serosal pigments: *Bombyx mori*).
[102] FURUKAWA, H. *Proc. Imp. Acad.*, **11** (1935), 158–60 (moulting of adult organs Dermaptera).
[103] GAINES, J. C., and CAMPBELL, F. L. *Ann. Ent. Soc. Amer.*, **28** (1935), 445–61 (Dyar's rule: *Heliothis*, Lep.).
[104] GEIGY, R. *Rev. Suisse Zool.*, **38** (1931), 187–288 (castration in *Drosophila* by irradiation of pole cells).
[105] GEIGY, R., and OCHSÉ, W. *Rev. Suisse Zool.*, **47** (1940), 225–41 (control of pupation *Sialis*, Megalopt.).
[106] GEITLER, L. *Biol. Zbl.*, **58** (1938), 153–79 (polyploidy in insects).
[107] V. GIERKE, E. *Arch. Entw. Mech.*, **127** (1932), 387–410 (number of instars *Ephestia*).

[8] GIERSBERG, H. *Z. vergl. Physiol.*, **9** (1929), 523–52 (determination of colour patterns: Lepidoptera).

[9] GODELMANN, R. *Arch. Entw. Mech.*, **12** (1901), 265–301 (autotomy and regeneration: *Bacillus*, Orth.).

[10] GOETSCH, W. *Naturwiss.*, **25** (1937), 803–8 (determination of soldiers among ants).

[11] —— *Oesterr. Zool. Z.*, **1** (1947), 193–274 ('vitamin T' and insect form).

[12] GOLDSCHMIDT, R. *Arch. Entw. Mech.*, **47** (1921), 654–67 (reduplication in genitalia of *Lymantria*).

[13] —— *Arch. Mikr. Anat.*, **98** (1923), 292–313 (prothetely and determination of wing patterns: Lepidoptera).

[14] —— *Physiologische Theorie der Vererbung*, Springer, Berlin, 1927.

[15] —— *Quart. Rev. Biol.*, **6** (1931), 125–42 (intersexuality in *Lymantria*).

[16] —— *Biol. Bull.*, **63** (1932), 337–56 (genetics and development: review).

[17] —— *Arch. Entw. Mech.*, **130** (1933), 266–339 (growth measurements: *Lymantria*, Lep.).

[18] —— *Biol. Zbl.*, **55** (1935) 535–54; *J. Exp. Zool.*, **100** (1945), 193–201 (control of form by 'rate genes': *Drosophila*).

[19] GOLDSCHMIDT, R., and KATSUKI, K. *Biol. Zbl.*, **51** (1931), 58–74 (hereditary gynandromorphism in *Bombyx mori*).

[20] GORDON, C., and SANG, J. H. *Proc. Roy. Soc.*, B, **130** (1941), 151–84 (nutrition and gene action: *Drosophila*).

[21] GRASSÉ, P. P., *et al. C.R. Acad. Sci.*, **223** (1946), 869–71; 929–31; *Rev. Suisse Zool.*, **53** (1946), 432–41 (caste determination: termites).

[22] GREB, R. J. *Biol. Bull.*, **65** (1933), 179–86 (mosaics in *Habrobracon*, Hym.).

[23] GRISON, P. *C.R. Acad. Sci.*, **225** (1947), 1089–90 (diapause: *Euproctis*, Lep.).

[24] GUNN, D. L., and KNIGHT, R. H. *J. Exp. Biol.*, **21** (1945), 132–43 ('group effect' on rate of growth: *Ptinus*, Col.).

[125] HACHLOW, V. *Arch. Entw. Mech.*, **125** (1931), 26–49 (control of pupal development: Lepidoptera).

[126] HADORN, E., *et al. Naturwiss.*, **25** (1937), 681–2; *Proc. Nat. Acad. Sci.*, **23** (1937), 478–84; *Arch. Entw. Mech.*, **138** (1938), 281–304; *Rev. Suisse Zool.*, **48** (1941), 495–509 (ring gland and pupation hormone in *Drosophila*).

[127] —— *Rev. Suisse Zool.*, **49** (1942), 228–36; *Arch. Julius Klaus-Stiftung Vererbforsch.*, **20** (1945), 82–95; *Symposia Soc. Exp. Biol.*, **2** (1948), 177–95 (pleiotropic gene action: *Drosophila*).

[128] HANDLIRSCH, A. *Schröders Handb. d. Entomologie*, **1** (1926), 1115 (post-embryonic development).

[129] HANSTRÖM, B. *Hormones in Invertebrates*, Oxford, 1939; *Biol. Generalis*, **15** (1942), 485–531 (organs of internal secretion in insects).

[130] HARNLY, M. H. *Biol. Bull.*, **82** (1942), 215–32 ('phenocopies': *Drosophila*).

[131] HASSAN, A. I. *Trans. Roy. Ent. Soc. Lond.*, **89** (1939), 345–84 (parasitic sex reversal: Delphacidae, Homopt.).

[132] HAYDAK, M. H. *J. Econ. Ent.*, **36** (1943), 778–92 (food and caste determination in honey-bee).

[133] HECHT, O. *Arch. Schiffs. Tropenhyg.*, **37** (1933), Beiht. 3, 87 pp. (diapause in mosquitos: review).

[134] HELLER, J. *Biochem. Z.*, **169** (1926), 208–34 ('latent' development or diapause: *Celerio* pupae, Lep.).

[135] HENKE, K. *Z. vergl. Physiol.*, **1** (1924), 297–499 (pigment formation in *Pyrrhocoris*, Hem.).

[136] —— *Naturwiss.*, **21** (1933), 633; 654; 665; 683; **34** (1947), 149–57; 180–6; **35** (1948), 176–81; 203–11; 239–46; *Rev. Suisse Zool.*, **55** (1948), 319–37 (determination of colour patterns).

[137] HENKE, K., *et al. Biol. Zbl.*, **61** (1941), 40–64; *Nachr. Akad. Wiss. Göttingen, Math. Phys. Kl.*, 1945, 20–38; *Biol. Zbl.*, **65** (1946), 120–35 (polyploid cells and size of wing scales: *Ephestia*, Lep.).

[138] —— *Z. indukt. Abstam. Vererbl.*, **79** (1941), 267–316 ('phenocopies': *Drosophila*).

[139] HENSON, H. *Biol. Rev.*, **21** (1946), 1–14 (insect metamorphosis).

[140] HINTON, H. E. *Trans. Roy. Ent. Soc. Lond.*, **99** (1948), 395–409; *Proc. Roy. Ent. Soc. Lond.*, C, **38** (1963), 77–85 (function of pupal stage).
[141] HODGE, C. *Physiol. Zool.*, **6** (1933), 306–28 (growth measurements: *Melanoplus*, Orth.).
[142] HOLDAWAY, F. G., and EVANS, A. C. *Nature*, 1930, Apr. 19th (parasitism by *Alysia* as stimulus to pupation in *Lucilia*).
[143] HOLDAWAY, F. G., and SMITH, H. F. *Australian J. Exp. Biol. Med. Sci.*, **10** (1932), 247–59 (relation between size of host and sex: *Alysia*, Ichneum.).
[144] HOOP, M. *Zool. Jahrb., Anat.*, **57** (1933), 433–64 (moulting glands).
[145] HOSKINS, W. M., and CRAIG, R. *Physiol. Rev.*, **15** (1935), 525–96 (insect physiology: review).
[146] HUFNAGEL, A. *Arch. Zool.*, **57** (1918), 47–202 (histology of metamorphosis: *Hyponomeuta*, Lep.).
[147] HUXLEY, J. S. *Problems of relative growth*, Methuen, London, 1932.
[148] ITO, H. *Bull. Imp. Tokyo Sericult, Coll.*, **1** (1918), 63–103 (corpora allata in Lepidoptera).
[149] JOLY, P. *Année Biol.*, **21** (1945), 1–34 (insect hormones: review).
[150] KATSUKI, K. *Biol. Zbl.*, **55** (1935), 361–83 (hereditary mosaic formation: *Bombyx mori*).
[151] KEILIN, D., and NUTTALL, G. H. F. *Parasitology*, **11** (1919), 279–328 (intersexes in *Pediculus*).
[152] KELLOGG, V. L. *J. Exp. Zool.*, **1** (1904), 601–5 (effect of gonads on sexual characters: *Bombyx mori*).
[153] KEMPER, H. *Z. Morph. Oekol. Tiere*, **22** (1931), 53–109 (moulting: *Cimex*).
[154] KEY, K. H. L. *Bull. Ent. Res.*, **27** (1936), 77–85 (growth measurements and extra instars: *Locusta*, Orth.).
[155] KEY, K. H. L., and EDNEY, E. B. *Proc. Roy. Ent. Soc. Lond.*, **11** (1936), 55–8 (precocious metamorphosis in *Locusta*, Orth.).
[156] KHOUVINE, Y., *et al. C.R. Acad. Sci.*, **217** (1943), 161–3 ('bar' substance in *Drosophila*).
[157] KLATT, B. *Z. indukt. Abstam. Vererbsl.*, **22** (1919), 1–50 (transplantation of gonads in different genetic strains: *Lymantria*, Lep.).
[158] KLEIN, H. Z. *Z. angew. Entom.*, **19** (1932), 395–448 (temperature and number of instars: *Pieris*, Lep.).
[159] KÖHLER, W. *Biol. Zbl.*, **60** (1940), 348–68 (effect of veins on wing pattern: *Ephestia*, Lep.).
[160] —— *Z. Morph. Oekol. Tiere*, **24** (1932), 582–681 (development of wing and wing pattern: *Ephestia*, Lep.).
[161] KÖHLER, W., and FELDOTTO, W. *Arch. Julius Klaus-Stiftung für Vererbungsforschung, Socialanthropologie und Rassenhygiene*, **10** (1935), 315–453 (critical periods in wing development: *Vanessa*, Lep.).
[162] —— —— *Arch. Entw. Mech.*, **136** (1937), 313–99 (determination of scale structure and wing pattern: *Vanessa*, Lep.).
[163] KOLLER, G. *Biol. Rev.*, **4** (1929), 269–306; *Hormone bei wirbellosen Tieren*, Leipzig, 1938 (internal secretions: review).
[164] KOPEĆ, S. *Arch. Entw. Mech.*, **33** (1911), 1–116; *Zool. Anz.*, **43** (1913), 65–74; *J. Exp. Zool.*, **36** (1922), 469–75 (effect of gonads on sexual characters: *Lymantria*, Lep.).
[165] —— *Biol. Bull.*, **42** (1922), 322–42; **46** (1924), 1–21; *Biol. Generalis*, **3** (1927), 375–84 (brain and metamorphosis: *Lymantria*, Lep.).
[166] —— *J. Exp. Zool.*, **37** (1923), 15–25 (nervous system and regeneration: *Lymantria*, Lep.).
[166a] —— *Z. vergl. Physiol.*, **26** (1938), 102–6 (effect of starvation on growth: Lepidoptera).
[167] KORSCHELT, E. *Regeneration und Transplantation*, Berlin, 1927–1931.
[168] KOSMINSKY, P. *Zool. Jahrb., Physiol.*, **30** (1911), 321–38 (external factors influencing sexual characters: Lepidoptera).
[169] KOWALEVSKY, A. *Z. wiss. Zool.*, **45** (1887) (post-embryonic development of Muscids).
[170] KREYENBERG, J. *Z. angew. Entom.*, **14** (1929), 140–88 (number of instars: *Dermestes*, Col.).
[171] KÜHN, A. *C.R. 12e Congr. Intern. Zool. Lisbon*, 1935, 1–18; *Naturwiss*, **24** (1936), 1–10 (action of genes in determination of wing patterns: *Ephestia*, Lep.).
[172] KÜHN, A., and V. ENGELHARDT, M. *Nachr. Ges. Wiss. Göttingen, Nachr. Biol.*, **2** (1936), 171–99 (determination of wing pattern: *Abraxas*, Lep.).

[173] KÜHN, A., and HENKE, K. *Abhandl. Ges. Wiss. Göttingen, Math.-phys. Kl.*, **15** (1929–36), 1–272 (determination of wing pattern in *Ephestia*, Lep.).

[174] KÜHN, A., and PIEPHO, H. *Ges. Wiss. Göttingen, Nachr. Biol.*, **2** (1936), 141–54; *Biol. Zbl.*, **58** (1938), 12–51 (hormones and pupation: *Ephestia*, Lep.).

[175] KÜHN, A., PLAGGE, E., *et al. Z. Ges. Wiss. Göttingen, Nachr. Biol.*, **2** (1935), 1–29, 251–6; *Z. Indukt. Abst. Vererbsl.*, **72** (1936), 127–36; *Biol. Zbl.*, **56** (1936), 406–9; **57** (1937), 114–26 (genes controlling eye colour through the production of diffusible substance: *Ephestia*, Lep.).

[176] KÜHN, A., and WOYWOD, D. *Z. Naturforsch.*, **1** (1946), 39–44 (mosaicism and gene action: *Ephestia*, Lep.).

[177] LAWSON, C. A. *Biol. Bull.*, **77** (1939), 135–45; *Amer. Nat.*, **73** (1939), 69–82 (determination of gamic females: Aphidae).

[178] LEES, A. D. *J. Genetics*, **42** (1941), 115–42 (determination of wing veins, &c.; *Drosophila*).

[179] LEMCHE, H. *Zool. Jahrb., Anat.*, **63** (1937), 125–288; *Ent. Medd.*, **24** (1945), 305–60 (wing patterns of Lepidoptera).

[180] v. LENGERKEN, H. *Zool. Anz.*, **58** (1924), 179–85; **59** (1924), 323–30; *Jena Z. Naturwiss.*, **67** (1932), 260–73 (prothetely, &c.: *Tenebrio* Col.).

[181] —— *Biol. Zbl.*, **48** (1928), 475–509 (origin of gynandromorphs).

[182] LIGHT, S. F. *Quart. Rev. Biol.*, **17** (1942), 312–26; **18** (1943), 46–63 (determination of castes in social insects).

[183] LOEWE, S., *et al. Biochem. Z.*, **244** (1932), 347–56 (oestrogens in *Lepidoptera*).

[184] LUDWIG, D. *Physiol. Zool.*, **5** (1932), 431–47 (diapause in *Popillia*, Col.).

[185] —— *Ent. News*, **45** (1934), 141–52 (growth measurements: *Popillia*, Col.).

[186] MAGNUSSEN, K. *Arch. Entw. Mech.*, **128** (1933), 447–79 (regulation of wing structure and pattern: Lepidoptera).

[187] MALABOTTI, A. *Akad. Anz. Wien*, No. 18, 1934 (reunion of severed head and neck: *Dixippus*, Orth.).

[188] MARCHAL, P. *Ann. Épiphyt. Phytogén.*, **2** (1936), 447–550 (diapause in host and parasite: *Trichogramma*).

[189] MARCOVITCH, S. *J. Agric. Res.*, **27** (1924), 513–22 (effect of length of day on appearance of sexual forms in Aphids).

[190] MAUSER, F. *Akad. Anz. Wien*, No. 17, 1934 (moulting of transplanted limbs: *Dixippus*, Orth.).

[191] MEGUŠAR, F. *Arch. Entw. Mech.*, **25** (1907), 148–234 (regeneration: Coleoptera).

[192] —— *Arch. Entw. Mech.*, **29** (1910), 499–586 (regeneration of limbs: Orthoptera).

[193] MEISENHEIMER, J. *Zool. Anz.*, **32** (1907), 393–400; **35** (1910), 446–50 (effect of gonads on sexual characters: *Lymantria*, Lep.).

[194] MELLANBY, K. *Parasitology*, **30** (1938), 392–402 (diapause: *Lucilia*, Dipt.).

[195] METALNIKOV, S. *Arch. Zool.*, **8** (1908), 489–588 (growth, &c., in *Galleria*, Lep.).

[196] MORGAN, T. H., and BRIDGES, C. B. *Carnegie Inst. Wash.* Publ. No. **278** (1919), 1–122 (origin of gynandromorphs: *Drosophila*).

[197] MURRAY, F. V., and TIEGS, O. W. *Quart. J. Micr. Sci.*, **77** (1935), 405–95 (metamorphosis of *Calandra*, Col.; histology).

[198] NABERT, A. *Z. wiss. Zool.*, **104** (1913), 181–358 (corpora allata).

[199] NAGEL, R. H. *Ann. Ent. Soc. Amer.*, **27** (1934), 425–8 (metathetely: *Tribolium*, Col.).

[200] NÜESCH, H. *Arch. Klaus-Stift. Vererbforsch.*, **16** (1941), 373–468; **22** (1947), 221–93; (intersexes in *Solenobia*).

[201] OERTEL, E. *J. Morph.*, **50** (1930), 295–340 (metamorphosis in honey-bee).

[202] OGURA, S. *Z. Indukt. Abstammungs-Vererb.*, **64** (1933), 205–68 (number of instars: *Bombyx mori*).

[203] OKA, H., and FURUKAWA, H. *Biol. Zbl.*, **55** (1935), 245–50 (determination of wing veins: Odonata).

[204] OUDEMANS, J. T. *Zool. Jahrb., Syst.*, **12** (1899), 71–85 (effect of castration on sexual characters: *Lymantria*, Lep.).

[205] PAUL, H. *Arch. Entw. Mech.*, **136** (1937), 64–111 (effect of gonads on sex of regenerated organs: *Orgyia*, Lep.).

[206] PEREDELSKY, A. A., and PASTUCHOVA, A. *Biol. Gen.*, **6** (1930), 327–52 (effect of starvation on metamorphosis).

[207] PEREZ, C. *Bull. Sci. Fr. Belg.*, **37** (1902), 195–427; *Arch. Zool.*, **4** (1910), 1–274; *Mem. Cl. Sci. Acad. Roy. Belg.*, **3** (1911), 1–103 (histological changes in metamorphosis: *Formica* (Hym.); *Calliphora* (Dipt.); *Polistes* (Hym.)).

[208] PFEIFFER, I. W. *J. Exp. Zool.*, **82** (1939), 439–61; *Trans. Conn. Acad. Arts. Sci.*, **36** (1945), 489–515 (corpus allatum and c. cardiacum in growth of *Melanoplus*, Orth.).

[209] PFLUGFELDER, O. *Z. wiss. Zool.*, **149** (1937), 477–512; **151** (1938), 149–91; **152** (1939), 159–84; 384–408; **153** (1940), 108–35; *Biol. Zbl.*, **66** (1947), 170–8; 211–35 (corpora allata, &c., and growth in *Dixippus*, Orth.).

[210] PIEPHO, H. *Naturwiss.*, **26** (1938), 383; 841–2; **27** (1939), 301–2; *Biol. Zbl.*, **58** (1938), 356–66; 481–95; **59** (1939), 314–26 (metamorphosis of implanted cuticle: *Galleria*, Lep.).

[211] —— *Biol. Zbl.*, **60** (1940), 367–93; *Arch. Entw. Mech.* **141** (1942), 500–83; *Naturwiss.*, **31** (1943), 329–35 (corpus allatum and control of metamorphosis: Lepidoptera, &c.).

[212] PLAGGE, E. *Naturwiss.*, **26** (1938), 4–11 ('maternal inheritance').

[213] —— *Arch. Entw. Mech.*, **132** (1935), 648; *Nachr. ges. wiss. Göttingen, Nachr. a.d. Biol.*, **2** (1936), 251–6 (gene controlling eye colour in *Ephestia*, Lep.).

[214] PLAGGE, E., and BECKER, E. *Biol. Zbl.*, **58** (1938), 231–42 (homology between eye pigment genes of *Ephestia* (Lep.) and *Drosophila* (Dipt.)).

[215] —— —— *Naturwiss.*, **26** (1938), 430–1 (extracts of pupation hormone).

[216] PLAGGE, E., *et al. Naturwiss.*, **23** (1935), 751–2; *Biol. Zbl.*, **58** (1938), 1–11 (source of pupation hormone: *Deilephila*, Lep.).

[217] POISSON, R. *Bull. Biol. Fr. Belg.*, **58** (1924), 49–204 (biology of aquatic Hemiptera; histology of moulting, &c.).

[218] POISSON, R., and SELLIER, R. *C.R. Acad. Sci.*, **224** (1947), 1074–5 (corpus allatum and metamorphosis: *Gryllus*, Orth.).

[219] POULSON, D. F. *Trans. Conn. Acad. Arts. Sci.*, **36** (1945), 449–87 (nature of ring gland: Diptera).

[220] POYARKOFF, E. *Arch. Anat. Microsc.*, **12** (1910), 333–474 (histology of moulting and metamorphosis: *Galerucella*, Col.).

[221] PREBBLE, M. L. *Canad. J. Res.*, **19** (1941), 295–454 (diapause: *Gilpinia*, Hym.).

[222] PRUTHI, H. S. *J. Camb. Phil. Soc.*, **1** (1924); *Nature*, Mar. 12, 1927 (prothetely: *Tenebrio*, Col.).

[223] ——*C .R Soc. Biol.*, **92** (1925), 76–7; *Brit. J. Exp. Biol.*, **3** (1925), 1–8 (effect of injury and starvation on pupation in Lepidoptera).

[224] PRZIBRAM, H. *Arch. Entw. Mech.*, **45** (1919), 69–80 (regeneration: *Cimbex*, Hym.).

[225] —— *Biol. Generalis*, **11** (1935), 189–203 (regeneration and moulting in adult Mantidae).

[226] —— *Brit. J. Exp. Biol.*, **3** (1920), 313–30 (transplantation and regeneration: review).

[227] PRZIBRAM, H., and BRECHER, L. *Arch. Entw. Mech.*, **122** (1930), 251–79 (growth measurements: *Tenodera*, Orth.).

[228] PRZIBRAM, H., and MEGUŠAR, F. *Arch. Entw. Mech.*, **34** (1912), 680–741 (growth measurements: *Sphodromantis*, Orth.).

[229] PRZIBRAM, H., and WERBER, E. I. *Arch. Entw. Mech* , **23** (1907), 615–31 (regeneration and moulting of adults: Thysanura).

[230] RADTKE, A. *Naturwiss.*, **30** (1942), 451–2 (corpus allatum and metamorphosis: *Tenebrio*, Col.).

[231] READIO, P. A. *Ann. Ent. Soc. Amer.*, **24** (1931), 19–39 (dormancy in *Reduvius*, Hem.).

[232] v. REES, J. *Zool. Jahrb., Anat.*, **3** (1888) (histology during metamorphosis: *Musca*).

[233] REGEN, J. *Zool. Anz.*, **35** (1910), 427–32 (effect of castration on *Gryllus*).

[234] REICHELT, M. *Z. Morph. Oekol. Tiere*, **3** (1925), 477–525 (scale and pigment formation in wings of *Lymantria*, Lep.).

[235] REMPEL, J. G. *J. Exp. Zool.*, **84** (1940), 261–89 (intersexes: Chironomidae).

[236] RICHARDS, A. G. *J. N.Y. Ent. Soc.*, **45** (1937), 149–210 (regulation of post-embryonic development: review).

[237] RICHARDS, O. W. *J. Anim. Ecol.*, **9** (1940), 243–88 (diapause: *Pieris*, Lep.).

[238] RIVNAY, E. *Bull. Ent. Res.*, **28** (1937), 173–9 (moisture affecting wing development in *Toxoptera*, Aphidae).

[239] ROBERTSON, C. W. *J. Morph.* **59** (1936), 351–98 (metamorphosis of *Drosophila*).

[40] ROTH, L. M., *et al. Ann. Ent. Soc. Amer.*, **34** (1941), 151–72; **37** (1944), 234–54 (abnor mal growth due to toxic gases: *Tribolium*, Col.).

[41] ROUBAUD, E. *Bull. Biol. Fr. Belg.*, **56** (1919), 455–544 (diapause: Muscidae, Dipt.).

[42] —— *Bull. Soc. Ent. Fr.*, 1927, 61–4 (types of cyclical arrest of growth).

[43] —— *Ann. Sci. Nat., Zool.*, **18** (1935), 39–52 (diapause).

[44] ROUBAUD, E., and COLAS-BELCOUR, J. *Bull. Soc. Path. Exot.*, **26** (1933), 965–72; **27** (1934), 546–51 (diapause in *Anopheles plumbeus*).

[45] SALT, G. *J. Exp. Zool.*, **48** (1927), 223–331; **59** (1931), 133–66 (stylopization and sexual characters).

[46] —— *J. Exp. Biol.*, **13** (1936), 363–75 (effect of superparasitism on sex ratio: *Trichogramma*, Hym.).

[47] —— *Parasitology*, **29** (1937), 539–53; **30** (1938), 511–22 (dimorphism of parasite determined by host: *Trichogramma*, Hym.).

[48] —— *Biol. Rev.*, **16** (1941), 239–64 (effects of hosts on parasites).

[49] SALT, R. W. *Canad. J. Res.*, D, **25** (1947), 66–86 (diapause: *Cephus*, Hym.).

[50] SÁNCHEZ, D. *Trav. Lab. Rec. Biol. Univ. Madrid*, **25** (1927), 1–39 (histolysis in nervous system: Lepidoptera).

[51] SCHAEFFER, C. W. *J. Agric. Res.*, **57** (1938), 825–41 (wing production in Aphids).

[52] SCHARRER, B. *Physiol. Rev.*, **21** (1941), 383–409; *The Hormones* (Pincus and Thimann (Edd.), New York), **1** (1948), 121–58 (insect hormones: review).

[53] —— *Endocrinology*, **38** (1946), 35–45; *Biol. Bull.*, **95** (1948), 186–98 (corpus allatum and prothoracic glands: *Leucophaea*, Orth.).

[54] SCHARRER, B., and HADORN, E. *Proc. Nat. Acad. Sci.*, **24** (1938), 236–42 (ring gland in *Drosophila* of 'lethal giant' strain).

[55] SCHARRER, B., and SCHARRER, E. *Biol. Bull.*, **87** (1944), 242–51; *Physiol. Rev.*, **25** (1945), 171–81 (neurosecretion in insects).

[56] SCHMIEDER, R. G. *Ent. News*, **50** (1939), 91–7; 125–31 (dimorphism: *Sphecophaga*, Hym.).

[57] —— *Anat. Rec.*, **84** (1942), 514 (control of metamorphosis: *Trypoxylon*, Hym.).

[58] SCHNEIDER F. *Mitt. Schweiz. ent. Ges.*, **21** (1948), 248–85 (diapause: Syrphidae, Dipt.).

[59] SCHRADER, K. *Biol. Zbl.*, **58** (1938), 52–90 (the brain and metamorphosis: *Ephestia*, Lep.).

[60] SCHÜRFELD, W. *Arch. Entw. Mech.*, **133** (1935), 728–59 (Verson's glands: Lepidoptera).

[61] SEAMANS, L., and WOODRUFF, L. C. *J. Kansas Ent. Soc.*, **12** (1939), 73–6 (number of instars: *Blattella*, Orth.).

[62] SEILER, J. *Arch. Entw. Mech.*, **119** (1929), 543–76; *Verhandl. deutsch. Zool. Ges.*, 1936, 147–50; *Rev. Suisse Zool.*, **44** (1937), 283–307; *Arch. Klaus-Stift. Vererbforsch.*, **20** (1945), 217–35 (intersexes resulting from racial crosses in *Solenobia*, Lep.).

[63] SELLIER, R. *C.R. Soc. Biol.*, **140** (1946), 965–6 (moulting of adult: *Acheta*, Orth.).

[64] SEYRIG, A. *Bull. Soc. Ent. Fr.*, **40** (1935), 67–70 (effect of size of host on sex: Ichneumonidae).

[65] SHULL, A. F. *Biol. Rev.*, **4** (1929), 218–48 (determination of types of individuals in Aphids).

[66] —— *Z. Indukt. Abstamm.-Vererbsl.*, **55** (1930), 108–26; **57** (1930), 92–101 (determination of type of reproduction in *Microsiphum*, Aphidae).

[67] —— *Biol. Bull.*, **68** (1935), 35–50; *Amer. Nat.*, **73** (1939), 256–69; *J. Exp. Zool.*, **89** (1942), 183–95 (factors affecting wing production in Aphids).

[68] SIMMONDS, F. J. *Bull. Ent. Res.*, **37** (1946), 95–97; *Phil. Trans. Roy. Soc.*, B, **233** (1948), 385–414 (diapause in Hymenopterous parasites).

[69] SMITH, L. M. *J. Agric. Res.*, **54** (1937), 345–64 (wing development in Aphids).

[70] SNODGRASS, R. E. *J. Agric. Res.*, **28** (1924), 1–36 (metamorphosis of *Rhagoletis*, Dipt.).

[71] SQUIRE, F. A. *Bull. Ent. Res.*, **30** (1940), 475–81; **31** (1940), 1–6 (diapause: *Platyedra*, Lep.).

[72] STEHR, G. *Rev. Suisse Zool.*, **54** (1947), 573–608 (wing pattern and circulation: *Ephestia*, Lep.).

[73] STEINBERG, D. M., and KAMENSKY, S. A. *Bull. Biol. Fr. Belg.*, **70** (1936), 145–87 (diapause in *Loxostege*, Lep.).

[74] STERN, C., *et al. Genetics*, **24** (1939), 162–79; *Growth*, **4** (1940), 377–82; *J. Exp. Zool.*, **87** (1941), 113–58; 159–80 (determination of testis characters: *Drosophila*).

[75] STILES, K. A. *Biol. Bull.*, **74** (1938), 430–60 (intermediate-winged Aphids).

[276] STRALLA, D. D. *C.R. Acad. Sci.*, **227** (1948), 1277–8 (ventral glands: Odonata).

[277] STRICKLAND, E. H. *Biol. Bull.*, **21** (1911), 302–38 (parasites affecting metamorphosis: *Simulium*, Dipt.).

[278] SÜFFERT, F. *Biol. Zbl.*, **44** (1924), 173–88 (seasonal dimorphism: *Araschnia*, Lep.)

[279] SUSTER, P. M. *Zool. Jahrb., Physiol.*, **53** (1933), 41–66 (nervous system and regeneration: Mantidae).

[280] SWEETMAN, H. L. *Bull. Brooklyn Ent. Soc.*, **29** (1934), 158–61; *Ecol. Monographs*, **8** (1938), 285–311 (regeneration and moulting of adults: Thysanura).

[281] SZTERN, H. *Arch. Entw. Mech.*, **40** (1914), 429–95 (growth measurements: *Sphodromantis*).

[282] TEISSIER, G. *Trav. Stat. Biol. Roscoff*, **9** (1931), 29–238 (growth measurements).

[283] —— *Livre Jubilaire E. L. Bouvier*, 1936, 334–42 (Dyar's law).

[284] TERRE, L. *Bull. Soc. Ent. Fr.*, 1899, 351–2 (histolysis of muscle in honey-bee).

[285] THEODOR, O. *Bull. Ent. Res.*, **25** (1934), 459–72 (hibernation: *Phlebotomus*, Dipt.).

[286] THERON, P. P. A. *J. Ent. Soc. Sthn. Afr.*, **6** (1943), 114–23 (diapause: *Cydia*, Lep.).

[287] THOMAS, C. A. *Ent. News*, **44** (1932), 91–6 (prothetely in Coleoptera and review).

[288] THOMPSON, W. R. *Bull. Soc. Ent. Fr.*, 1923, 40–2 (dimorphism of parasite determined by host: *Pezomachus*, Ichneumonidae).

[289] THOMSEN, E. *Naturwiss.*, **29** (1941) 605–6; *Vidensk. Medd. Dansk Naturhist. Foren.*, **106** (1942), 319–405 (ring gland: *Calliphora*, Dipt.).

[290] TIEGS, O. W. *Trans. R. Soc. S. Australia*, **46** (1922), 319–527 (histology during metamorphosis: *Nasonia*, Hym.).

[291] TITSCHACK, E. *Z. wiss. Zool.*, **128** (1926), 509–69 (growth and moulting: *Tineola*, Lep.).

[292] V. UBISCH, L. *Arch. Entw. Mech.*, **31** (1911), 637–53 (wing regeneration: *Lymantria*, Lep.).

[293] —— *Arch. Entw. Mech.*, **41** (1915), 237–50 (regeneration: Ephemeridae).

[294] VANDEL, A. *Bull. Biol. Fr. Belg.*, **64** (1930), 457–94 (intercastes in *Pheidole* (Formicidae) due to *Mermis* parasites).

[295] VANDERPLANK, F. L. *Proc. Roy. Ent. Soc. Lond.*, A, **20** (1945), 105–6 (intersexual *Glossina*, Dipt.).

[296] VARLEY, G. C., and BUTLER, C. G. *Parasitology*, **25** (1933), 263–8 (diapause and parasitism).

[297] VILLEE, C. A. *J. Exp. Zool.*, **93** (1943), 75–98; **96** (1944), 85–102 (effects of temperature on 'aristopedia' and 'proboscipedia': *Drosophila*).

[298] VOGT, M. *Biol. Zbl.*, **61** (1941), 148–58; 242–52; **62** (1942), 149–54; **63** (1943), 56–71; 395–446; *Naturwiss.*, **32** (1944), 37–9; *Arch. Entw. Mech.*, **142** (1942), 129–82; *Nature*, **157** (1946), 512 (ring gland and control of growth and metamorphosis: *Drosophila*).

[299] —— *Biol. Zbl.*, **65** (1946), 223–38; 238–54; **66** (1947), 81–105; *Z. Naturforsch.*, **1** (1946), 469–75; *Experientia*, **3** (1947), 156 (determination of imaginal discs: *Drosophila*).

[300] WADDINGTON, C. H. *Growth (Supplement)*, 1940, 37–44 (aristopedia: *Drosophila*).

[301] WEED, I. G. *Proc. Soc. Exp. Biol. Med.*, **34** (1936), 885–6 (corpora allata and moulting: *Melanoplus*, Orth.).

[302] WESSON, L. G. *Psyche*, **47** (1940), 105–111 (caste determination: *Leptothorax*, Hym.).

[303] WHEELER, W. M. *Social Life Among Insects*, London, Constable, 1923.

[304] —— *J. Exp. Zool.*, **8** (1910), 377–438; **50** (1928), 165–237 (parasitic castration; *Mermis* parasitism and intercastes among ants).

[305] WHITING, P. W., *et al. Biol. Bull.*, **65** (1933), 357–8; **66** (1934), 152–65 (sex intergrades in *Habrobracon*, Hym.).

[306] WHITING, P. W. *J. Heredity*, **29** (1938), 188–93 (caste determination in ants).

[307] WIGGLESWORTH, V. B. Nature, **123** (1929), 17 (starvation and delayed metamorphosis: Culicidae, Dipt.).

[308] —— *Quart. J. Micr. Sci.*, **76** (1933), 269–318; **77** (1934), 191–222; **79** (1936), 91–121; *J. Exp. Biol.*, **17** (1940), 201–22 (hormones controlling moulting and metamorphosis: *Rhodnius*, Hem.).

[309] —— *J. Exp. Biol.*, **14** (1937), 364–81; *J. Exp. Biol.*, **17** (1940), 180–200 (wound healing and moulting of adults: Hemiptera).

[310] —— *J. Exp. Biol.*, **15** (1938), 235–54 (chloride and oxygen as limiting factors in growth: *Aedes* larvae, Dipt.).

[311] WIGGLESWORTH, V. B., *Quart. J. Micr. Sci.*, **83** (1942), 141–52; **89** (1948), 197–217 (chromatic droplets: *Rhodnius*, Hem.; *Tenebrio*, Col.).
[312] —— *J. Exp. Biol.*, **25** (1948), 1–14 (corpus allatum: *Rhodnius*, Hem.).
[313] —— *Essays on Growth and Form*, Oxford, 1945, 23–40; *Symposia Soc. Exp. Biol.*, **2** (1948), 1–16 (role of cell in determination: *Rhodnius*).
[314] WILLIAMS, C. M. *Biol. Bull.*, **82** (1942), 347–55; **90** (1946), 234–43; **93** (1947), 89–98; **94** (1948), 60–5 (diapause; *Hyalophora*, &c., Lep.).
[315] WILLIAMS, C. M., and SANBORN, R. C. *Biol. Bull.*, **95** (1948), 282–3 (cytochrome system in diapause: *Hyalophora*, Lep.).
[316] WILSON, F. *Trans. Roy. Ent. Soc. Lond.*, **87** (1938), 165–80 (determination of polymorphism in Aphids).
[317] WOODRUFF, L. C. and SEAMANS, L. *Ann. Ent. Soc. Amer.*, **32** (1939), 589–99 (regeneration: *Blattella*, Orth.).
[318] YOKOYAMA, T. *Proc. Roy. Ent. Soc. Lond.*, A, **11** (1936), 35–44 (histology in a non-moulting strain of silkworm).

SUPPLEMENTARY REFERENCES (A)

[319] ANDREWARTHA, H. G. *Biol. Rev.*, **27** (1952), 50–107 (diapause: review).
[320] BODENSTEIN, D. *Proc. IXth Internat. Congr. Entom., Amsterdam 1951* (prothoracic gland and moulting: *Periplaneta*).
[321] BONNEMAISON, L. *Contribution à l'étude des facteurs provoquant l'apparition des formes ailées et sexuées chez les Aphidinae*. Thesis, Paris, 1951.
[322] BUCHLI, H. *C.R. Acad. Sci.*, **233** (1951), 206–8 (caste determination in *Reticulotermes*).
[323] DANILYEVSKY, A. S., *et al. Dokl. Akad. Nauk SSSR*, **59** (1948), 337–40; **60** (1948), 481–4; **68** (1949), 785–8; **71** (1950), 963 (photoperiod and diapause in Lepidoptera).
[324] DICKSON, R. C. *Ann. Ent. Soc. Amer.*, **42** (1949), 511–37 (photoperiod and diapause in *Grapholitha*, Lep.).
[325] GLOOR, H. *Rev. Suisse Zool.*, **54** (1947), 637–712 (ether and phenocopies in *Drosophila*).
[326] GRASSÉ, P. P., NOIROT, C., *et al. C.R. Acad. Sci.*, **224** (1947), 219–21; **230** (1950), 892–5 (caste determination in *Calotermes*).
[327] GROSCH, D. S. *J. Exp. Zool.*, **107** (1948), 289–313 (haploidy and dwarfism: *Habrobracon*, Hym.).
[328] HADORN, E. *Folia Biotheoretica*, No. 3 (1948), 109–26 (phenocopies in *Drosophila*).
[329] ICHIKAWA, M., and NISHIITSUTSUJI, J. *Annot. Zool. Jap.*, **24** (1951), 205–11 (hormone secretion in brain of Lepidoptera).
[330] KARLSON, P., and HANSER, G. *Z. Naturforsch.*, **7b** (1952), 80–3 ('pupation hormone' in *Drosophila*, etc.).
[331] v. RHEIN, W. *Arch. Entw. Mech.*, **129** (1933), 602–65; *Verh. dtsch. zool. Ges.* (1951), 99–101 (control of dimorphism in female honey bee).
[332] KOGURE, M. *J. Dept. Agric. Kyushu Imp. Univ.*, **4** (1933), 1–93 (determination of diapause in egg of silkworm).
[333] KOMAROV, O. S. *Dokl. Akad. Nauk SSSR*, **68** (1949), 789–92 (photoperiod and diapause: *Polychrosis*, Lep.).
[334] KÜHN, A. *Z. Naturforsch.*, **4b** (1949), 104–8 (wound-healing and polyploidy in *Ptychopoda*, Lep.).
[335] de LERMA, B. *Boll. Zool.* (suppl.), **17** (1950), 67–190 (insect endocrinology: review).
[336] LÜSCHER, M. *J. Exp. Zool.*, **25** (1948), 334–43 (regeneration of limbs in *Rhodnius*).
[337] —— *Rev. Suisse Zool.*, **58** (1951), 404–8; *Biol. Zbl.*, **71** (1952), 529–43; *Insectes Sociaux*, **3** (1956), 119–28; *Ann. N.Y. Acad. Sci.*, **89** (1960), 549–63; *Symp. Roy. Ent. Soc. Lond.*, **1** (1961), 57–67 (determination of neotenic reproductives in *Calotermes*).
[338] MAAS, A. H. *Arch. Entw. Mech.*, **143** (1948) 515–72 (phenocopies in *Drosophila*).
[339] NOVAK, V. J. A. *Nature*, **167** (1951), 132–3 (control of metamorphosis).
[340] PIEPHO, H., *et al. Z. Tierpsychol.*, **7** (1950), 424–34; **17** (1960), 261–9 (prothetely and cocoon formation: *Galleria*).
[341] —— *Biol. Zbl.*, **69** (1950), 1–10; 261–71 (corpus allatum and control of metamorphosis: *Galleria*).

[342] PIEPHO, H. and MEYER, H. *Biol. Zbl.*, **70** (1951), 252–60 (moulting of adult cuticle in *Galleria*, Lep.).

[343] POSSOMPÈS, B. *C.R. Acad. Sci.*, **231** (1950), 594–6 (role of brain in metamorphosis of *Calliphora*).

[344] RISLER, H. *Z. Naturforsch.*, **3b** (1948), 129–31 (size of nuclei during larval growth: *Ptychopoda*, Lep.).

[345] SCHMIDT, E. L., and WILLIAMS, C. M. *Anat. Rec.* (Suppl.), **105** (1949), No. 70; **111** (1951), No. 162 (growth hormone and differentiation of spermatocytes: Lepidoptera).

[346] SCHNEIDER, F. *Mitt. schweiz. ent. Ges.*, **23** (1950), 156–94; *Z. angew. Ent.*, **33** (1951), 150–62 (diapause in Syrphids and their Hymenopterous parasites).

[347] SELLIER, R. *C.R. Acad. Sci.*, **228** (1949), 2055–6; *Arch. Zool. exp. gén., Notes et Rev.* **88** (1951), 61–72 (brain and prothoracic gland in growth of *Gryllus*).

[348] WAGNER, G. *Z. Naturforsch.*, **6b** (1951), 86–90 (size of nuclei during growth in *Calliphora* larvae).

[349] WAY, M. J., *et al. Nature*, **164** (1949), 615; *J. Exp. Biol.*, **27** (1950), 365–76 (photoperiod and diapause in Lepidoptera).

[350] WIGGLESWORTH, V. B. *Proc. Roy. Ent. Soc. London*, Ser. C, **15** (1951), 78–81 (metamorphosis in insects: review).

[351] —— *Nature*, **168** (1951), 558; *J. Exp. Biol.* **29** (1952), 561–70 (thoracic gland and moulting in *Rhodnius*).

[352] WILLIAMS, C. M. *Growth Symposium*, **12** (1948), 61–74; *Fed. Proc.*, **10** (1951), 546–52 (biochemical mechanisms in insect growth).

SUPPLEMENTARY REFERENCES (B)

[353] ADKISSON, P. L., BELL, R.A., and WELLSO, S. G. *J. Ins. Physiol.*, **9** (1963), 299–310 (induction of diapause in *Pectinophora*).

[354] AGRELL, I. *Ark. Zool.*, **39A** (1947), No. 10, 1–48 ('latent', or 'diapause', development).

[355] ALBRECHT, F. O. *J. Agric. Trop. Bot. Appl.*, **2** (1955), 110–92 (compensatory growth in locusts).

[356] ANDREWARTHA, H. B. *Bull. Ent. Res.*, **34** (1943), 1–17; *Biol. Rev.*, **27** (1952), 50–107 ('diapause development').

[357] ARVY, L., BOUNHIOL, J. J., and GABE, M. *C. R. Acad. Sci.*, **236** (1953), 627–9 (neurosecretion in *Bombyx*).

[358] ARVY, L., and GABE, M. *79e Congr. Soc. Savantes*, (1954), 189–96 (neurosecretion in *Leptinotarsa*).

[359] —— *Mem. Soc. Endocr.*, **12** (1962), 331–44 (histochemistry of the neurosecretory product).

[360] BALAZUC, J. *Mem. Mus. Nat. Hist. Naturelle*, **25** (1948), 1–293 (teratology in *Tenebrio*).

[361] BARTH, R. *Zool. Jahrb., Anat.*, **69** (1945), 405–34 (tonofibillae in caterpillars).

[362] BECK, S. D. *Bull. Ent. Soc. Amer.*, **9** (1963), 8–16 (photoperiodism and diapause in *Pyrausta*).

[363] BECK, S. D., and ALEXANDER, N. *Science*, **143** (1964), 478–9 (hormonal activation of the brain in *Pyrausta*).

[364] BECKER, H. J. *Chromosoma*, **13** (1962), 341–84 ('puffs' in *Drosophila* chromosomes exposed to the moulting hormone).

[365] BERGERARD, J. *Biol. Exp.*, **246** (1958), 1930–3 (gynandromorphs in *Carausius*).

[366] BODENSTEIN, D. *J. Exp. Zool.*, **123** (1953), 189–232, 413–33; **124**, 105–15 (hormonal control of growth in *Periplaneta*).

[367] —— *Insect Physiology*, Roeder, K. D., Ed. Wiley: New York 1953, 822–65 (analysis of eye development in *Drosophila*).

[368] —— *J. Exp. Zool.*, **129** (1955), 209–24 (regeneration in *Periplaneta*).

[369] —— in *Recent Advances in Invertebrate Physiology*, B. T. Scheer, Ed., Oregon Univ. 1957, 197–211 (hormones and insect growth, review).

[370] —— *Smithson. Misc. Coll.*, **137** (1959), 23–41 (genetics of eye colour in *Phormia* and *Musca*).

[371] BODENSTEIN, D., and ABDEL-MALEK, A. *J. Exp. Zool.*, **111** (1949), 93–115 (induced aristapedia in *Drosophila*).

[372] BOUILLON, A. *La Cellule*, **53** (1949), 35–95 (dimorphism in *Trichogramma semblidis*).
[373] BOUNHIOL, J. J. *C.R. Acad. Sci.*, **235** (1952), 671–2, 747–8 ('critical periods' of moulting in *Bombyx*).
[374] BRAEMER, H. *Arch. Entw. Mech.*, **148** (1956), 362–90 (wound healing in *Ephestia*).
[375] BRIAN, M. V. *Insectes Sociaux*, **1** (1954), 101–22; **2** (1955), 1–34, 85–114; *Symp. Genet. Biol. Ital.*, **10** (1961), 170–2 (caste differentiation in *Myrmica*).
[376] BROUSSE-GAURY, P. *Bull. Biol. Fr. Belg.*, **92** (1958), 55–85 (autotomy in *Acheta*).
[377] BUCKLIN, D. H. *Anat. Rec.*, **117** (1953), 539 (ending of diapause in *Melanoplus* embryo).
[378] BÜCKMANN, D. *J. Ins. Physiol.*, **3** (1959), 159–89 (moulting hormone and colour change in *Dicranura* larva).
[379] —— *Fortschritte Zool.*, (1962), 165–237 (developmental physiology in arthropods: review).
[380] BURDETTE, W. C., and BULLOCK, M. W. *Science*, **140** (1963), 1311 (five active fractions in ecdyson).
[381] BUTENANDT, A., and KARLSON, P. *Z. Naturf.*, **9b** (1954), 389–91 (isolation of moulting hormone, ecdyson).
[382] CARLISLE, D. B., and ELLIS, P. E. *C.R. Acad. Sci.*, **249** (1959), 1059–60 (persistence of ventral glands in solitary adult locusts).
[383] CHAUVIN, R. See ref. 52, Chap. XIII.
[383a] CHURCH, N. S. *Canad. J. Zool.*, **33** (1955), 339–69 (hormones and diapause in *Cephus*).
[384] CLARKE, K. U., and LANGLEY, P. A. *J. Ins. Physiol.*, **9** (1963), 287–92, 363–74, 411–22, 423–30 (control of moulting in *Locusta*).
[385] CLEVER, U. *Chromosoma*, **12** (1961), 607–75; **13** (1962), 385–436; *J. Ins. Physiol.*, **8** (1962), 357–76 (ecdyson and gene activity in *Chironomus*).
[386] CLOUTIER, E. J., BECK, S. D., *et al. Nature*, **195** (1962), 1222–4 (brain and diapause in *Pyrausta*).
[387] CORBET, P. S. *J. Exp. Biol.*, **33** (1956), 1–14 (induction of diapause in *Anax* larvae).
[388] DANILYEVSKY, A. S. *Ent. Oboz.*, **30** (1949), 194–205 (temperature and diapause in Saturniids from different latitudes).
[389] —— *Ent. Oboz.*, **36** (1957), 5–27 (geographical races of insects with differing responses to photoperiod).
[390] —— *Photoperiodism and seasonal development of insects*, Oliver and Boyd, London, 1965.
[391] v. DEHN, M. *Naturwissenschaften*, **50** (1963), 578–9 (farnesol and wing development in *Aphis*).
[392] DELEURANCE, E. P. *Coll. Int. C.N.R.S.*, **34** (1952), 141–55 (social polymorphism in *Polistes*).
[393] DIXON, S. E., and SHUEL, R. W. *Canad. J. Zool.*, **41** (1963), 753–9 (royal jelly and caste determination).
[394] DOBZHANSKY, T. In *Insect Polymorphism*, Kennedy, J. E., Ed. *Symp. Roy. Ent. Soc. Lond.*, **1** (1961), 30–42 (chromosomal polymorphism in *Drosophila*).
[395] FAURE, J. C. See refs. 90 and 91, Chap. XIII.
[396] FORD, E. B. In *Insect Polymorphism*, Kennedy, J. S., Ed. *Symp. Roy. Ent. Soc. Lond.*, **1** (1961), 11–19 (balanced polymorphism).
[397] FRASER, A. *Proc. Roy. Soc. Edinb.*, B, **67** (1960), 127–40 (humoral control of diapause in *Lucilia* larva).
[398] FUZEAU-BRAESCH, S. *Bull. Biol. Fr. Belg.*, **94** (1960), 525–625 (polymorphism in *Gryllus*).
[399] FYG, W. *Mitt. Schwiz. Ent. Ges.*, **29** (1956), 404–16 (nervous stimuli and pupation in *Apis*).
[400] GABE, M. *Bull. Soc. Zool. Fr.*, **78** (1953), 117–93 (ventral glands in Thysanura).
[401] GAMBORO, P. *Boll. Zool.*, **21** (1954), 163–9; *Arch. Zool. Ital.*, **48** (1957), 207–19; *Mem. Accad. Patavine*, **72** (1960), 3–5 (nutrition and photoperiod in diapause in *Carpocapsa*).
[402] GEYSPITZ, K. F. *Dokl. Akad. Nauk SSSR*, **36** (1957), 548–60 (perception of photoperiodic stimulus in *Dendrolimus* larva).
[403] GILBERT, L. I., and SCHNEIDERMAN, H. A. *Science*, **128** (1958), 844 (juvenile hormone activity in adrenal cortex).
[404] —— *Trans. Amer. Micr. Soc.*, **79** (1960), 38–67 (bioassay for juvenile hormone).

[405] GILBERT, L. I., and SCHNEIDERMAN, H. A. *Gen. Comp. Endrocr.* **1** (1961), 453–72 (changes in juvenile hormone content in Lepidoptera during development).

[406] GOLDSCHMIDT, R. *Theoretical Genetics*, 1955, Univ. California Press (autosomes and sex determination).

[407] HADORN, E. *Advances in Genetics*, **4** (1951), 53–85; in *Chemical basis of development*, McElroy, W. D., and Glass, B., Ed. Johns Hopkins Press 1958, 76–82 (role of genes in development).

[408] HADORN, E. *Proc. X Int. Congr. Genetics*, **1** (1960), 338–53 (genetics and insect pigments).

[409] HALBWACHS, M. C., JOLY, L., and JOLY, P. *J. Ins. Physiol.*, **1** (1957), 143–9 (moulting hormone controlling body form in *Locusta*).

[410] HARVEY, W. R. *Ann. Rev. Ent.*, **7** (1962), 57–80 (metabolism in diapause: review).

[411] HARVEY, W. R., and WILLIAMS, C. M. *Biol. Bull.*, **114** (1958), 36–53 (cytochrome system in *Hyalophora* pupa during diapause).

[412] —— *J. Ins. Physiol.*, **7** (1961), 81–99 (injury metabolism in *Hyalophora* pupa).

[413] HENKE, K. *J. Embryol. Exp. Morph.*, **1** (1953), 217–26 (differentiation among insect wing scales).

[414] HENKE, K., and POHLEY, H. J. *Z. Naturforsch.*, **7b** (1952), 65–79 (differentiation in wing scales of *Ephestia*).

[415] HIGHNAM, K. C. *Quart. J. Micr. Sci.*, **99** (1958), 73–88 (hormones and diapause in *Mimas tiliae*).

[416] —— *Quart. J. Micr. Sci.*, **102** (1961), 27–38 (histology of neurosecretory system in *Schistocerca*).

[417] HODSON, A. C., and WEINMAN, C. J. *Tech. Bull. Minn. Agric. Exp. Sta.* No. 170 (1945), 31 pp. (diapause in eggs of *Malacosoma*).

[418] HUGHES, R. D. *J. Exp. Biol.*, **37** (1960), 218–23 (diapause in *Erioischia* (Dipt.) pupa).

[419] HUMBEL, E. *Rev. Suisse Zool.*, **57** (1950), 155–235 (intersexes in *Solenobia*).

[420] HUSAIN, M. A., *et al. Ind. J. Agric. Res.*, **6** (1936), 586–664; 1005–30 (phase changes in *Schistocerca*, Orth.).

[421] ICHIKAWA, M., and ISHIZAKI, H. *Nature*, **191** (1961), 933–4; **198** (1963), 308–9 (extracts of the brain hormone).

[422] ICHIKAWA, M., and NISHIITSUTSUJI, J. *Mem. Coll. Sci. Univ. Kyoto*, B, **22** (1955), 11–15 (pupal diapause in *Luehdorfia* (Lep.).

[423] ISHIZAKI, H. *Physiol. and Ecol.*, **8** (1958), 32–5 (colour and diapaus in pupa of *Papilio xuthus*).

[424] IWAO, S. *Physiol. and Ecol.*, **7** (1956), 28–38 (diapause in *Naranga* (Lep.) larva).

[425] —— *Mem. Coll. Agr. Kyoto*, **84** (1962), 1–77 (phase variation in caterpillars).

[426] JACKSON, D. J. *Nature*, **192** (1961), 823–4 (diapause in larva of *Cataphractus* (Hym.)).

[427] JOHANSSON, A. S. *Nyt. Mag. Zool.*, **7** (1958), 5–132 (neurosecretory system in *Oncopeltus*).

[428] JOHNSON, B., and BIRKS, P. R. *Ent. Exp. et Appl.*, **3** (1960), 327–39 (wing polymorphism in Aphids).

[429] JOLY, P. *Insectes Sociaux*, **3** (1956), 17–24; *Colloq. Int. Cent. Nat. Recherche Sci.*, **114** (1962), 77–88 (corpora allata and phase change in *Locusta*).

[430] JOLY, P., JOLY, L. and HALBWACHS, M. *Ann. Sci. Nat. Zool.* **18** (1956), 256–61 (ventral glands and moulting in *Locusta*).

[431] JONES, B. M. *J. Exp. Biol.*, **33** (1956), 174–85 (endocrine activity in embryo of *Locusta*).

[432] KAISER, P. *Naturwissenschaften*, **42** (1955), 303–4 (caste determination in termites).

[433] KAJI, S. *Annot. Zool. Japon.*, **26** (1954), 194–200 (factors affecting bar eye in *Drosophila*).

[434] KARLSON, P. *Vitam. and Horm.*, **14** (1956), 227–66; *Proc. IVth Int. Congr. Biochem.*, **12** (Biochemistry of Insects) (1959), 37–47; *Angew. Chem.*, **2** (1963), 175–82 (chemistry and biochemistry of insect hormones).

[435] KARLSON, P., and HOFFMEISTER, H. *Hoppe-Seyl. Z.*, **331** (1963), 298–300 (cholesterol as precursor of ecdyson).

[436] KARLSON, P., HOFFMEISTER, H., HOPPE, W., and HUBER, R. *Liebig's Ann. Chem.*, **662** (1963), 1–20 (chemistry of ecdyson).

[437] KENNEDY, J. S. *Biol. Rev.*, **31** (1956), 349–70; *Insect Polymorphism*, Symposium No. **1** (1961), Roy. Ent. Soc. Lond., 80–90 (polymorphism in locusts).

[438] KEY, K. H. L. *Austral. J. Zool.*, **5** (1957), 247–83 (phase change in Phasmids).

[439] KIM, CHANG-WHAN, *J. Embryol. Exp. Morph.*, **7** (1959), 572–82; *Proc. R. Ent. Soc. Lond.*, A **35** (1960), 61–4 (imaginal buds of legs in Lepidoptera).

[440] KIRIMURA, J., SAITO, M., and KOBAYASHI, M. *Nature*, **195** (1962), 729–30 (steroid hormone in *Bombyx*).

[441] KISIMOTO, R. *Nature*, **178** (1956), 641–2; *Jap. J. Ecol.*, **9** (1959), 94–7 (crowding and polymorphism in *Nilaparvata* (Homopt.)).

[442] KOBAYASHI, M., and KIRIMURA, J. *Nature*, **181** (1958), 1217 (brain hormone of *Bombyx*).

[443] KÜHN, A., *et. al. Naturwissenschaften*, **43** (1956), 25–8; *Biol. Zbl.*, **81** (1962) 5–17 (gene action and pigment formation).

[444] KÜHN, A., and MERKEL, A. *Biol. Zbl.*, **74** (1955), 113–45 (wing scale mutations in *Ephestia*).

[445] KURLAND, C. G., and SCHNEIDERMAN, H. A. *Biol. Bull.*, **116** (1959), 136–61 (cytochrome system in diapausing pupa of *Hyalophora*).

[446] KUSKE, G., *et al. Biol. Zbl.*, **8** (1961), 347–51; *Arch. Entw. Mech.*, **154** (1963), 354–77 (metamorphosis of legs in Lepidoptera).

[447] LABEYRIE, V. *C.R. Acad. Sci.*, **246** (1958), 2179–81 (diapause in *Microgaster*).

[448] LE BERRE, J. R. *Bull. Biol. Fr. Belg.*, **87** (1953), 227–73 (diapause race in *Locusta migratoria*).

[449] LEES, A. D. *The physiology of diapause in Arthropods*, 1955, Cambridge Univ. Press.

[450] —— *J. Ins. Physiol.*, **3** (1959), 92–117; **4** (1960), 154–75; **9** (1963), 153–64; *Cold Spring Harbor Symp. Quant. Biol.*, **25** (1960), 261–8; *J. Exp. Biol.* **41** (1964), 119–33 (photoperiod and polymorphism in *Megoura*).

[451] —— In *Insect Polymorphism*, Kennedy, J. S., Ed. *Symp. Roy. Ent. Soc. Lond.*, **1** (1961), 68–79 (clonal polymorphism in Aphids).

[452] LEVITA, B. *Bull. Soc. Zool. Fr.*, **137** (1962), 197–221 (group effect and polymorphism in *Gryllus*).

[453] LIGHT, S. F., and WEESNER, F. M. *J. Exp. Zool.*, **117** (1951), 397–414 (produtcion of supplementary reproductives in *Zootermopsis*).

[454] LONG, D. B. *Trans. R. Ent. Soc. Lond.*, **104** (1953), 543–85 (crowding and colour change in larvae of Lepidoptera).

[455] LUDWIG, D. *Ann. Ent. Soc. Amer.*, **49** (1956), 12–15 (temperature and number of moults in *Tenebrio*).

[456] LÜSCHER, M. *Rev. Suisse Zool.*, **65** (1958), 372–7 (production of soldiers in *Kalotermes*).

[457] LÜSCHER, M., and SPRINGHETTI, A. *J. Ins. Physiol.*, **5** (1960), 190–212; *Proc. XVIth Int. Congr. Zool. Washington* **4** (1963), 244–50 (corpus allatum and caste formation in *Kalotermes*).

[458] MCLEOD, D. G. R., and BECK, S. D. *Biol. Bull.*, **124** (1963), 84–96 (photoperiod and ending of diapause in *Pyrausta*).

[459] MASAKI, S. *Bull. Fac. Agr. Mie Univ.* No. **13** (1956), 26–46 (summer and winter diapause in *Barathra brassicae*).

[460] MASHENNIKOVA, V. A. *Rev. Ent. U.R.S.S.*, **37** (1958), 538–45 (diapause in *Apanteles* influenced by the host).

[461] MATTHÉE, J. J. *Bull. Ent. Res.*, **36** (1945), 343–71; *J. Ent. Soc. S. Africa*, **10** (1947), 16–23 (phase characters in locusts and Noctuids).

[462] MEYER, J. H. *Z. Wien. Ent. Ges.*, **38** (1953), 41–80 (blood transfusion in hybrid Sphingids inducing ovarial development).

[463] MÜLLER, H. J. *Naturwissenschaften*, **42** (1955), 134–5 (seasonal dimorphism in *Araschnia*).

[464] —— *Zool. Jb., Syst.*, **85** (1957), 317–430 (cyclical polymorphism in *Euscelis*).

[465] —— *Z. Morph. Oekol. Tiere*, **51** (1962), 345–74; 575–610; *Ent. Exp. Appl.*, **5** (1962), 125–38 (seasonal dimorphism in *Aleurochiton*).

[466] NAYAR, K. K. *Current Sci.*, **25** (1956), 192–3 (neurosecretory cells and moulting in *Iphita*).

[467] NOIROT, C. *Insectes Sociaux*, **3** (1956), 145–58 (replacement sexual forms in termites).

[468] NOVÁK, V. *Acta Soc. Ent. Cechosloven.*, **54** (1957), 269–76 (diapause stages in the sawfly *Cephaleia*).

[469] —— *Insektenhormone*, 283 pp. 1959. Verlag d. tschechoslow, Acad. d. Wissenschaft.

[470] NÜESCH, H. *Rev. Suisse Zool.*, **59** (1952), 294–301; *Verh. Naturf. Ges. Basel*, **68** (1957), 194–216 (effects of nerves on muscle development in *Telea*).

[471] OELHAFEN, F. *Arch. Entw. Mech.*, **153** (1961), 120–57 (embryonic development in *Culex* without endocrine glands).

[472] O'FARRELL, A. F., and STOCK, A. *Austral. J. Biol. Sci.*, **6** (1953), 485–500; **7** (1954), 302–7, 525–36; **9** (1956), 406–22 (regeneration and moulting in *Blattella*).

[473] OZEKI, K. *Sci. Papers Coll. gen. Educ. Univ. Tokyo*, **9** (1959), 256–62; **10** (1960) 88–97 (control of moulting and reversal of metamorphosis in *Anisolabis*).

[474] PENZLIN, H. *Arch. Entw. Mech.*, **154** (1963), 434–65 (leg regeneration in cockroach).

[475] PFLUGFELDER, O. *Entwicklungsphysiologie der Insekten*, 1958, 2nd. ed. Stuttgart.

[476] PIEPHO, H. *Verh. dtsch. Zool. Ges.* (Wilhelmshaven) (1951), 62–75 (hormonal control of metamorphosis).

[477] PIEPHO, H., and HOLZ, I. *Biol. Zbl.*, **78** (1959), 417–24 (rejuvenation in mid-gut of Lepidoptera).

[478] POHLEY, H. J. *Arch. Entw. Mech.*, **152** (1960), 182–203; **153** (1961), 443–58; **153** (1962), 492–503 (interactions between moulting and regeneration).

[479] POSSOMPÈS, B. *Arch. Zool. Exp. Gén.* **89** (1953), 203–364 (control of pupation and metamorphosis in *Calliphora*).

[480] —— *C.R. Acad. Sci.*, **253** (1961), 3089–91; **254** (1962), 574–6 (mechanism of autotomy in Phasmids).

[481] PREBBLE, M. L. *Canad. J. Res.*, D, **19** (1941), 323–46 (low temperature and ending of diapause in *Gilpinia*).

[482] RAHM, U. H. *Rev. Suisse Zool.*, **59** (1952), 173–237 (hormones and diapause in *Sialis*).

[483] REHM, M. *Arch. Entw. Mech.*, **145** (1951), 205–48; *Z. Zellforsch.* **42** (1955), 19–58 (neurosecretion and metamorphosis in *Ephestia*).

[484] REMBOLD, H. *Umschau*, (1961), 488–91, 524–5 (determination of queens in *Apis*).

[485] RICHTER, A. *Arch. Entw. Mech.*, **154** (1962), 1–28 (hormonal control of growth in *Lepisma*).

[486] ROTHENBUHLER, W. L., GOWEN, J. W., and PARK, O. *Science*, **115** (1952), 637–8 (mosaics in *Apis*).

[487] RUTTNER, F., and MACKENSEN, O. *Bee World*, **33** (1952), 71–9 (mosaics in *Apis*).

[488] SALT, G. *Quart. J. Micr. Sci.*, **93** (1952), 453–74 (polymorphism in the Ichneumonid *Gelis*).

[489] SAUNDERS, D. S. *J. Ins. Physiol.*, **8** (1962), 309–18 (parental age and diapause in *Nasonia*).

[490] SCHALLER, F. *Bull. Soc. Zool. Fr.*, **77** (1952), 195–205 (imaginal characters in *Apis* pupa after ligature behind the head).

[491] —— *Ann. Sci. Nat. Zool.*, **12** (1960), 755–868; *Bull. Soc. Zool. Fr.*, **87** (1962), 582–600 (control of metamorphosis in *Aeschna*).

[492] SCHARRER, B. *Biol. Bull.*, **102** (1952), 261–72 (the flow of neurosecretion to the corpora cardiaca in *Leucophaea*).

[493] —— *Mem. Soc. Endocr.*, **12** (1962), 89–97; *Z. Zellforsch.*, **60** (1963), 761–96; **62** (1964), 125–48 (fine structure of corpus allatum and corpus cardiacum in *Leucophaea*).

[494] SCHMIALEK, P. *Z. Naturforsch.*, **16b** (1961), 461–4; **18b** (1963), 462–5; 513–15; 516–19 (juvenile hormone activity of farnesol, &c., and its synthesis in the insect).

[495] SCHNEIDERMAN, H. A., and GILBERT, L. I. *Biol. Bull.*, **115** (1958), 530–5 (substances with juvenile hormone activity in invertebrates).

[496] SCHNEIDERMAN, H. A., GILBERT, L. I., and WEINSTEIN, M. J. *Nature*, **188** (1960), 1041–2 (juvenile hormone activity in micro-organisms and plants).

[497] SCHNEIDERMAN, H. A., and HOROWITZ, J. *J. Exp. Biol.*, **35** (1958), 520–51 (diapause in *Mormoniella* and *Tritneptis*).

[498] SCHOONHOVEN, L. M. *Arch. Neér. Zool.*, **15** (1962), 111–74; *Science*, **141** (1963), 173–4 (electrical activity in brain of diapusing insects).

[499] SCHWARZ, V. *Biol. Zbl.*, **81** (1962), 19–44 (determination of symmetrical wing pattern in *Plodia*).

[500] SEILER, J. *Experientia*, **5** (1949), 425–38 (intersexes; review).

[501] SHAPPIRIO, D. G. *Ann. N.Y. Acad. Sci.*, **89** (1960), 537–47 (injury metabolism in *Hyalophora* pupa).

[502] SHAPPIRIO, D. G., and WILLIAMS, C. M. *Proc. Roy. Soc.*, B, **147** (1957), 218–46 (cytochrome system in *Hyalophora*).

[503] SHEPPARD, P. M. In *Insect Polymorphism*, Kennedy, J. S., Ed. *Symp. Roy. Ent. Soc. Lond.*, **1** (1961), 20–9 (polymorphism in *Papilio dardanus*).

[504] SPICKETT, S. G. *Nature*, **199** (1963), 870–3 (genetic and local factors in determination of hairs in *Drosophila*).

[505] STEINBERG, D. M. *Proc. Xth Int. Congr. Ent. Montreal, 1956*, **2** (1958), 261–6 (differentiation and regeneration of wings in Lepidoptera).

[506] STELLWAGG-KITTLER, F. *Biol. Zbl.*, **73** (1954), 12–49 (critical periods in pupation and adult development in *Tenebrio*).

[507] STERN, C., *et al. Arch. Entw. Mech.*, **149** (1956), 1–25; *Cytologia* (Suppl.): Proc. Int. Genetics Symp. 1956 (1957), 70–2; *J. Fac. Sci. Hokkaido Univ.*, **13** (1957), 303–7 (factors in determination of bristles in *Drosophila*).

[508] STEVENSON, E., and WYATT, G. R. *Arch. Biochem. Biophys.*, **99** (1962), 65–71 (incorporation of leucine into protein in *Hyalophora*).

[509] TELFER, W. H., and WILLIAMS, C. M. *J. Ins. Physiol.*, **5** (1960), 61–72 (effects of diapause, development, and injury on incorporation of glycine into protein in *Hyalophora*).

[510] THOMSEN, E. *J. Exp. Biol.*, **31** (1954), 322–30 (transport of neurosecretory material in *Calliphora*).

[511] VAN DER KLOOT, W. G. *Anat. Rec.*, **120** (1954), 46; *Biol. Bull.*, **109** (1955), 276–94; *Amer. Zool.*, **1** (1961), 3–9 (electrical activity and cholinesterase of brain in diapause).

[512] —— *Ann. Rev. Ent.*, **5** (1960), 35–52 (neurosecretion: review).

[513] VOY, A. *Bull. Biol. Fr. Belg.*, **85** (1951), 237–66 (growth in *Blatta* and *Carausius*).

[514] WATSON, J. A. L. *J. Ins. Physiol.*, **10** (1964), 305–17 (control of the moulting cycle in *Thermobia*).

[515] WAY, M. J. *Trans. Roy. Ent. Soc. Lond.*, **111** (1959), 351–64; *J. Ins. Physiol.*, **4** (1960), 92–101 (diapause in eggs of *Leptohylemyia*).

[516] WEAVER, N. *Ann. Ent. Soc. Amer.*, **50** (1957), 283–94 (differentiation of queens in *Apis*).

[517] WEHRMAKER, A. *Zool. Jahrb., Physiol.*, **68** (1959), 425–96 (determination of wing pattern in Plodia).

[518] WHITE, M. J. D. In *Insect Reproduction*, Highnam, K. C., Ed. *Symp. Roy. Ent. Soc. Lond.*, **2** (1964), 1–12 (autosomes in sex determination).

[519] WHITE, R. H. *J. Exp. Zool.*, **148** (1961), 223–39; **152** (1962), 139–43 (differentiation of compound eye in *Aëdes*).

[520] WHITTEN, J. M. *Quart. J. Micr. Sci.*, **98** (1957), 245–50 ('prepupal' cuticle of cyclorrhaphous Diptera).

[521] WIGGLESWORTH, V. B. *J. Exp. Biol.*, **29** (1952), 561–70 (thoracic gland in *Rhodnius*); 620–31 (hormone balance in *Rhodnius*); **32** (1955), 485–91 (breakdown of thoracic gland).

[522] —— *Quart. J. Micr. Sci.*, **94** (1953), 93–112; *The control of growth and form*, Ithaca: Cornell Univ. Press 1959, 140 pp.; in *The cell and the organism* (J. A. Ramsay and V. B. Wigglesworth, Edd.), Cambridge Univ. Press, 1961, 127–43 (differentiation of sensilla, glands, &c., in the epidermis).

[523] —— *Nature*, **174** (1954), 556 (juvenile hormone in *Calliphora*).

[524] —— *The physiology of insect metamorphosis*, Cambridge Univ. Press, 1954, 152 pp.

[525] —— *J. Exp. Biol.*, **32** (1955), 649–63; *Quart. J. Micr. Sci.*, **97** (1956), 89–98 (haemocytes and neurosecretion, moulting hormone, &c.).

[526] —— *Symp. Soc. Exp. Biol.*, **11** (1957), 204–27; *J. Exp. Biol.*, **40** (1963), 231–45 (action of growth hormones in *Rhodnius*).

[527] —— *J. Ins. Physiol.*, **2** (1958), 73–84 (assay methods for juvenile hormone).

[528] —— *J. Ins. Physiol.*, **7** (1961), 73–8; **9** (1963), 105–19 (juvenile hormone effect of farnesol, &c.).

[529] —— In *Insect Polymorphism*, Kennedy, J. S., Ed. *Symp. Roy. Ent. Soc. Lond.*, **1** (1961), 103–13 (survey of insect polymorphism).

[530] —— *Symp. Soc. Exp. Biol.*, **18** (1964), 265–81 (homeostasis in insect growth).

[531] WIGGLESWORTH, V. B. *Adv. Insect Physiol.* **2** (1964), 243–332, (hormones and insect growth: review).

[532] de WILDE. J. *Ann. Rev. Ent.*, **7** (1962) 1–26 (photoperiodism in insects: review).

[533] WILLIAMS, C. M. *Biol. Bull.*, **103** (1952), 120–38 (brain and prothoracic glands in diapause).

[534] —— *et al. Nature*, **178** (1956), 212–13; **183** (1959), 405 (active extracts of juvenile hormone).

[535] —— *Biol. Bull.*, **121** (1961), 572–84; **124** (1963), 355–67 (role of juvenile hormone during moulting, pupation, and adult development in *Hyalophora*).

[536] WILSON, E. O. *Quart. Rev. Biol.*, **28** (1953), 136–56 (polymorphism in ants).

[537] WILSON, F., and WOOLCOCK, L. T. *Nature*, **186** (1960), 99–100; *Austral. J. Zool.*, **8** (1960), 153–69 (sex determination in *Oöencyrtus*).

[538] WÜLKER, W. *Arch. Hydrobiol.*, Suppl., **25** (1961), 127–81 *Mermis* and intersexes in Chironomidae).

[539] YASHIKA, K. *Mem. Coll. Sci., Kyoto,* **27** B (1960), 83–8 (corpus allatum and juvenile hormone in *Ctenolepisma*).

[540] ZANDER, E., and BECKER, F. *Erlanger Jahrb. Bienenk.*, **3** (1925), 161–246 (nutrition and queen determination in *Apis*).

[541] ZWICKY, K., and WIGGLESWORTH, V. B. *Proc. Roy. Ent. Soc. Lond.*, A, **31** (1956), 153–60 (oxygen consumption during the moulting cycle in *Rhodnius*).

SUPPLEMENTARY REFERENCES (C)

[542] ADKISSON, P. L. *Am. Nat.*, **98** (1964), 357–74 (carotenoids and response to photoperiod in *Antheraea*).

[543] AGGARWAL, S. K. and KING, R. C. *J. Morph.*, **129** (1969), 171–200 (ring gland function in *Drosophila*).

[544] ANKERSMIT, G. W. *Ent. exp. & appl.*, **11** (1968), 231–240 (photoperiod and diapause control in *Adoxophyes*, Tortricidae).

[545] ASHBURNER, M. *Chromosoma*, **21** (1967) 398–428; *Adv. Insect Physiol.*, **7** (1970), 1–75 (structure and function of polytene chromosomes in development).

[546] BARRITT, L. C. and BIRT, L. M. *J. Insect Physiol.*, **16** (1970), 671–7 (course of ecdysone secretion in *Lucilia*).

[547] BECK, S. D. *et al. Biol. Bull. mar. biol. Lab., Woods Hole*, **126** (1964), 185–98; **128** (1965), 177–88; *J. Insect Physiol.*, **11** (1965), 297–303 ('proctodone' and diapause in *Ostrinia*).

[548] BECK, S. D. and SHANE, J. L. *J. Insect Physiol.*, **15** (1969), 721–30 (effect of high dosage of ecdysone on *Ostrinia*).

[549] BERENDES, H. D. *Devl. Biol.*, **11** (1965), 371–84; *J. Exp. Zool.*, **162** (1966), 209–18 (puffing of polytene chromosomes in different cell types in *Drosophila*).

[550] BOWERS, W. S. *et al. Science*, **142** (1963), 1469–70; **154** (1966), 1020–1; **164** (1969), 323–5; *Life Sciences*, **4** (1965), 2323–31 (chemicals with juvenile hormone activity).

[551] BULTMANN, H. and CLEVER, U. *Chromosoma*, **28** (1969), 120–35 (control of chromosomal puffing in foot-pads of *Sarcophaga*).

[552] BÜNNING, E. and JOERRENS, G. *Z. Naturforsch.*, **15b** (1960), 205–13; **17b** (1962), 57–64 (circadian rhythms and time measurement in the photoperiodic response).

[553] CASSIER, P. *Insectes Soc.*, **12** (1965), 71–80; *Bull. Soc. Zool. Fr.*, **91** (1966), 125–48 (hormonal control of phase characters in *Locusta*).

[554] CAVENEY, S. *J. Insect Physiol.*, **16** (1970), 1087–1107 (effect of juvenile hormone on fine structure of cuticle in *Tenebrio*).

[555] CLARET, J. *Annls. Endocr.*, **27** (1966), 311–20; *C.R. hebd. Séanc. Acad. Sci. Paris*, **266** (1968), 1156–9 (pigmentation of head capsule and photosensitivity in *Pieris*).

[556] CLEVER, U. *Naturwissenschaften*, **19** (1964), 449–59; *Expl. Cell Res.*, **55** (1969), 317–22 ('puffing' and cell function in polytenic cells).

[557] CLEVER, U. *Am. Zoologist*, **86** (1966), 33–41 (gene activity and cell differentiation).

[558] CORBET, P. S. *J. Exp. Biol.*, **33** (1956), 1–14 (diapause in *Anax*, Aeschnidae).

[559] CROSSLEY, A. C. S. *J. Insect Physiol.*, **14** (1968), 1389–1407 (humoral control of haemocytes in *Calliphora*).

[560] DANILYEVSKY, A. S. *et al. Ann. Rev. Ent.*, **15** (1970), 201–44 (day length and diapause).

[561] EDWARDS, J. S. *J. Insect Physiol.*, **12** (1966), 1423–33 (neural control of metamorphosis in *Galleria*).

[562] EDWARDS, J. S. *et al.* In *Insects & Physiology*, Beament, J. W. L. and Treherne, J. E. (Eds.), Edinburgh, Oliver & Boyd, 1967, pp. 163–73; *Am. Zoologist*, **8** (1968), 221 (regeneration in compound eye of *Acheta*).

[563] ELLIS, P. E. and GILLETT, S. *Colloqu. int. C.N.R.S.* No. **173** (1968), 173–82 (aggregation of locusts to plant odours).

[564] FINLAYSON, L. H. and WALTERS, V. A. *Nature*, **180** (1957), 713–14 (*Nosema* causing metathetely in Saturniids).

[565] FISHER, F. M. and SANBORN, R. C. *Biol. Bull. mar. biol. Lab., Woods Hole*, **126** (1964), 235–52 (*Nosema* as source of juvenile hormone in *Tribolium*).

[566] FORREST, J. M. S. *J. Insect Physiol.*, **16** (1970), 2281–92 (factors in morph determination in *Dysaphis*).

[567] GARCIA-BELLIDO, A. *Exp. Cell Res.*, **44** (1966), 382–92 (transdetermination in disc cells of *Drosophila*).

[568] GALBRAITH, M. N. *et al. J. Insect Physiol.*, **15** (1969), 1225–33 (ecdysterone in *Calliphora*).

[569] GEHRING, W. *J. Embryol. exp. Morph.*, **15** (1966), 77–111 (homoeotic transdetermination in *Drosophila*).

[570] GILBERT, L. I. *The Hormones*, vol. **4** (1964), 67–134, Academic Press, New York (hormones and insect growth: review).

[571] GOLDBERG, E., WHITTEN, J. and GILBERT, L. I. *J. Insect Physiol.*, **15** (1969), 409–20 (soluble proteins correlated with chromosomal puffing in *Sarcophaga*).

[572] HADORN, E. *Devl. Biol.*, **13** (1966), 424–509 (transdetermination in imaginal discs of *Drosophila*).

[573] HERMAN, W. S. *Int. Rev. Cytol.*, **22** (1967), 269–347 (prothoracic glands, &c.: review).

[574] HIDAKA, T., ISHIZUKA, Y. and SAKAGAMI, Y. *J. Insect Physiol.*, **17** (1971), 197–203 (diapause and quiescence in *Luehdorfia* pupa).

[575] HOFFMEISTER, H. *Angew. Chem.*, **5** (1966), 248–9 (ecdysone and ecdysterone in *Calliphora*).

[576] HORSFALL, W. R. and ANDERSON, J. F. *J. exp. Zool.*, **156** (1964), 61–90 (temperature and sex determination in mosquitos).

[577] ILAN, J., ILAN, JUDITH and PATEL, N. *J. biol. Chem.*, **245** (1970), 1275–81 (mechanism of gene expression in *Tenebrio*).

[578] IMAI, S. *et al. Steroids*, **10** (1967), 557–65 (ecdysone-like steroids from plants).

[579] JENKIN, P. M. and HINTON, H. E. *Nature*, **211** (1966), 871 ('apolysis' in moulting cycle).

[580] JOHNSON, B. *Ent. exp. & appl.*, **8** (1965), 49–64; **9** (1966), 213–22; 301–13 (environmental control of wing polymorphism in *Aphis*).

[581] KAPLANIS, J. N., ROBBINS, W. E. and THOMPSON, M. J. *Science*, **166** (1969), 1540–1 (conversion of ecdysone to ecdysterone in *Manduca*, Sphingidae).

[582] KARLSON, P. and BODE, C. *J. Insect Physiol.*, **15** (1969), 111–18 (inactivation of ecdysone by fat body of *Calliphora*).

[583] KARLSON, P., HOFFMEISTER, H., HOPPE, W. and HUBER, R. *Liebigs Ann. Chem.*, **662** (1963), 1–20 (chemistry of ecdysone).

[584] KRAUSE, G. and GEISLER, M. *Roux Arch. EntwMech. Org.*, **160** (1968), 76–111 (regeneration of limbs and cerci in *Tachycines*).

[585] KROEGER, H. and LEZZI, L. *Ann. Rev. Ent.*, **11** (1966), 1–22 (regulation of gene action in insect development).

[586] KUNKEL, J. G. *J. Insect Physiol.*, **12** (1966), 227–35 (synchronized moulting in *Blattella*).

[587] LAI-FOOK, J. *J. Morph.*, **124** (1968), 37–78 (fine structure of wound repair in *Rhodnius*).

[588] LAUFER, H. *Am. Zoologist*, **8** (1968), 257–71 (factors regulating puffing in salivary chromosomes).

[589] LAUFER, H. and HOLT, T. R. H. *J. exp. Zool.*, **173** (1970), 341–52 (effects of juvenile hormone or chromosomal puffing in *Chironomus*).

[590] LAUGÉ, G. *C.R. Acad. Sci. Paris*, **265** (1967), 767–70; *Bull. Soc. zool. Fr.*, **94** (1969), 341–62 (high temperature and feminization in *Drosophila*).

[591] LAWRENCE, P. A. *J. Cell. Sci.*, **1** (1966), 475–98; *J. exp. Biol.*, **44** (1966), 607–20 (gradients in determination in *Oncopeltus*).

[592] LAWRENCE, P. A. *J. exp. Biol.*, **44** (1966), 507–22; *Devl. Biol.*, **19** (1969), 12–40 (hormonal control of development of hairs in *Oncopeltus*).

[593] LAWRENCE, P. A. *Adv. Insect Physiol.*, **7** (1970), 197–266 (polarity and patterns in insect development: review).

[594] LEES, A. D. *Adv. Insect Physiol.*, **3** (1966), 207–77 (control of polymorphism in Aphids: review).

[595] LEES, A. D. *Nature*, **210** (1966), 986–89 (photoperiodic timing mechanisms in insects).

[596] LEZZI, M. and GILBERT, L. I. *Proc. Nat. Acad. Sci.*, **64** (1969), 498–503 (juvenile hormone effects on 'puffing' pattern in *Chironomus*).

[597] LOCKE, M. *Adv. Morphogenesis*, **6** (1967), 33–88 (patterns and gradients in insect integument: review).

[598] LOCKSHIN, R. A. and WILLIAMS, C. M. *J. Insect Physiol.*, **10** (1964), 683–9; **15** (1969), 1505–16 (enzyme synthesis and lysis of muscles in Saturniids).

[599] LÜSCHER, M. *Proc. VI Int. Congr. Ent. IUSSI*, 1969, 165–70 (juvenile hormone as a pheromone in *Calotermes*).

[600] MADHAVAN, K. and SCHNEIDERMAN, H. A. *Biol. Bull. mar. biol. Lab., Woods Hole*, **137** (1969), 321–31 (hormones and regeneration in *Galleria*).

[601] MANSINGH, A. and SMALLMAN, B. N. *J. Insect Physiol.*, **17** (1971), 1735–9 (temperature and ending of diapause in *Antheraea*).

[602] MARKS, E. P. *et al. Gen. comp. Endocrinol.*, **11** (1968), 31–42; *Science*, **167** (1970), 61–2 (hormones and regeneration of *Leucophaea* tissues *in vitro*).

[603] MATHAD, S. B. and MCFARLANE, J. E. *Can. J. Zool.*, **46** (1968), 57–60 (photoperiod and wing development in *Gryllodes*).

[604] MEYER, A. D. *et al. Proc. Nat. Acad. Sci.*, **60** (1968), 853–60; *Mitt. Schweiz. Ent. Ges.*, **44** (1971), 37–63 (juvenile hormones of *Hyalophora*).

[605] MITTLER, T. E. and DADD, R. H. *Ann ent. Soc. Am.*, **59** (1966), 1162–6 (nutrition and wing determination in *Myzus*).

[606] MORIYAMA, H., NAKANISHI, K. *et al. Gen. comp. Endocrin.*, **15** (1970), 80–7 (metabolism of ecdysone in insects).

[607] MORRIS, O. N. *J. Invert. Path.*, **16** (1970), 173–9 (prothetely in virus-infected Lepidoptera).

[608] NAISSE, J. *Archs. Biol. Liège*, **77** (1966), 139–201; *Gen. Comp. Endocrinol.*, **7** (1966), 85–110; *J. Insect Physiol.*, **15** (1969), 877–92 (hormonal control of sex in *Lampyris*).

[609] NOLTE, D. J., MAY, I. R. and THOMAS, B. M. *Chromosoma*, **19** (1970), 462–73 (gregarization pheromone of locusts).

[610] OHTAKI, T. *Japan J. med. Sci. Biol.*, **19** (1966), 97–104 (arrest of pupation in *Sarcophaga*).

[611] OHTAKI, T., MILKMAN, R. D. and WILLIAMS, C. M. *Biol. Bull. mar. biol. Lab., Woods Hole*, **135** (1968), 322–34 (quantitative study of ecdysone in *Sarcophaga*).

[612] OKASHA, A. Y. K. *J. exp. Biol.*, **48** (1968), 455–86; *J. Insect Physiol.*, **14** (1968), 1621–34; **16** (1970), 545–53 (high temperature and arrest of growth in *Rhodnius*).

[613] PENZLIN, H. *Wilhelm Roux Arch. EntwMech. Org.* **155** (1964), 152–61 (nervous system and regeneration in insects).

[614] RAI, K. S. and CRAIG, G. B. *Proc. XI Int. Congr. Genetics*, 1963, **1**, 10.9 (gynandromorphs in *Aëdes*).

[615] RIDDIFORD, L. M. *et al. Proc. Nat. Acad. Sci.*, **57** (1967), 595–601; *Science*, **167** (1970), 287–8; *Devl. Biol.*, **22** (1970), 249–63 (juvenile hormone and anologues disrupting embryonic development).

[616] RINTERKNECHT, E. *Bull. Soc. zool. Fr.*, **91** (1966), 645–54; 789-800 (wound healing in *Locusta*).

[617] RÖLLER, H. *et al. J. Insect Physiol.*, **11** (1965), 1185–97; **15** (1969), 379–89; *Recent Progress in Hormone Research*, **24** (1968), 651–80 (chemistry of juvenile hormone).

[618] RÖLLER, H. and DAHM, K. H. *Naturwissenschaften*, **9** (1970), 454–5 (juvenile hormone secretion by corpus allatum *in vitro*).

[619] RUPPLI, E. and LÜSCHER, M. *Rev. Suisse Zool.*, **71** (1964), 627–32 (elimination of replacement reproductives in *Calotermes*).

[620] SAUNDERS, D. S. *et al. J. Insect Physiol.*, **14** (1968), 433–40; *Nature*, **221** (1969), 559–61 (circadian rhythms and photoperiodic clock in *Nasonia*).

[621] SCHALLER, F. *Ann. Endocrinol.*, **25** (1964), 122–7 (action of neuro-secretion from brain on egg development in *Aeschna*).

[622] SCHALLER, F. *J. Insect Physiol.*, **14** (1968), 1477–83 (temperature and embryonic diapause in *Aeschna*).

[623] SCHNEIDERMAN, H. A. *et al. Biology and the Physical Sciences* 1969, 186–208; *Nature*, **234** (1971), 187–94 (control systems in insect development).

[624] SEHNAL, F. *J. Insect Physiol.*, **14** (1968), 73–85 (corpus allatum and development of internal organs in *Galleria*).

[625] SEHNAL, F. and EDWARDS, J. S. *Biol. Bull. mar. biol. Lab., Woods Hole*, **137** (1969), 352–8 (body constraint and arrested development in *Galleria*).

[626] SEILER, J. *Monitore Zool. Ital.* **3** (1969), 185–212 (intersexes in *Solenobia* and *Lymantria*).

[627] SETHI, S. L. and SWENSON, K. G. *Ent. exp. & appl.*, **10** (1967), 97–102 (polymorphism in *Eriosoma*, Aphididae).

[628] SLÁMA, K. and WILLIAMS, C. M. *Biol. Bull. mar. biol. Lab., Woods Hole*, **130** (1966), 235–46 ('paper factor' and metamorphosis in *Pyrrhocoris*).

[629] SMALLMAN, B. N. and MANSINGH, A. *Ann. Rev. Ent.*, **14** (1969), 387–408 (cholinergic system in insect development: review).

[630] SRIVASTAVA, U. S. and GILBERT, L. I. *J. Insect Physiol.*, **15** (1969), 177–89 (juvenile hormone and metamorphosis in *Sarcophaga*).

[631] STAAL, G. B. *Koninkl. Nederl. Akad. Wetensch.*, **70** (1967), 409–18 (plants as source of insect hormones).

[632] STEEL, C. G. H. and HARMSEN, R. *Gen. comp. Endocrinol.*, **17** (1971), 125–42 (neurosecretion in brain of *Rhodnius*).

[633] STERN, C. *Rev. Suisse Zool.*, **73** (1966), 339–55 (intersexes and gynanders in *Drosophila*).

[634] STUMPF, H. F. *J. exp. Biol.*, **49** (1968), 49–60 (epidermal gradients and differentiation in *Galleria* pupa).

[635] SUTHERLAND, O. R. W. *J. Insect Physiol.*, **15** (1969), 1385–1410; 2179–201 (polymorphism in *Acyrthosiphon*, Aphididae).

[636] TOBA, H. H. and PASCHKE, J. D. *J. Insect Physiol.*, **13** (1967), 381–96 (polymorphism in *Therioaphis*, Aphididae).

[637] TRUMAN, J. W. and RIDDIFORD, L. M. *Science*, **167** (1970), 1624–26 (neuroendocrine control of ecdysis in Saturniids).

[638] TYSHTCHENKO, V. P. and MANDELSTAM, J. E. *J. Insect Physiol.*, **11** (1965), 1233–9 (electrical activity and cholinesterase in pupal diapause of *Antheraea*).

[639] WALLACE, M. M. H. *Aust. J. Zool.*, **16** (1968), 871–83 (diapause in *Sminthurus*).

[640] WEAVER, N. *Ann. Rev. Entom.*, **11** (1966), 79–102 (physiology of caste determination: review).

[641] WEIR, S. B. *Nature*, **228** (1970), 580–1 (control of moulting in *Calpodes*, Lep.).

[642] WELLSO, S. G. and ADKISSON, P. L. *J. Insect Physiol.*, **12** (1966), 1455–65 (photoperiod and diapause in *Heliothis*, Lep.).

[643] WHITE, D. F. *J. Insect Physiol.*, **17** (1971), 761–73 (corpus allatum and wingbud formation in *Brevicoryne*, Aphididae).

[644] WHITTEN, J. M. *Chromosoma*, **26** (1969), 215–44; *J. Morph.*, **127** (1969), 73–104 (puffing patterns and foot-pad development in *Sarcophaga*).

[645] WIGGLESWORTH, V. B. in *Cell Differentiation and Morphogenesis* (W. Beerman *et al.*) North Holland Publ. Co., Amsterdam, 1966, 180–209; *Insect Hormones*, Oliver & Boyd, Edinburgh, 1970, 159 pp. (hormonal regulation and differentiation in insects).

[646] WIGGLESWORTH, V. B. *J. Insect Physiol.*, **15** (1969), 73–94 (chemical structure and juvenile hormone activity).

[647] de WILDE, J. *Arch. Anat. Micr. Morph. Exp.*, **54** (1965), 547–64; *Mem. Soc. Endocrinol. No.* **18**, (1970), 487–514 (hormones and photoperiod in insect diapause).

[648] WILLIAMS, C. M., ADKISSON, P. L. *et al. Biol. Bull. mar. biol. Lab., Woods Hole*, **127** (1964), 511–25; **128** (1965), 497–507 (diapause in *Antheraea* pupa).

[649] YAMASAKI, M. and KOBAYASHI, M. *Bull Sericult. Expt. Sta., Tokyo*, **24** (1971), 523 (purification of brain hormone of *Bombyx*).

Chapter IV

Muscular System and Locomotion

ANATOMY AND HISTOLOGY

Anatomy—The muscles of insects fall into two groups; the skeletal muscle forming bands stretched across the articulations of the body wall, which serve to move one segment on another, and the visceral muscles which invest the internal organs. The *skeletal muscles* are made up of elongated contractile fibres lying parallel with one another or converging upon the point of insertion. They are often exceedingly numerous (Lyonet described 1,647 muscles in the goat-moth caterpillar as compared with 529 in man) and their arrangement differs greatly from one group of insects to another.[71] The *visceral muscles* may form a regular lattice of longitudinal and circular fibres, as around the gut of some insects, the circular fibres being exaggerated at places to form occlusive sphincters; or they may form an irregular feltwork of branching and anastomozing fibres, as in the wall of the crop (Fig. 82), in the ventral diaphragm or in the 'peritoneal' covering of the ovaries, where these fibres appear to merge imperceptibly into strands of connective tissue. In some insects, such as the bee, branching and anastomozing fibres of this kind may compose some of the muscle coats of the stomach wall.[45]

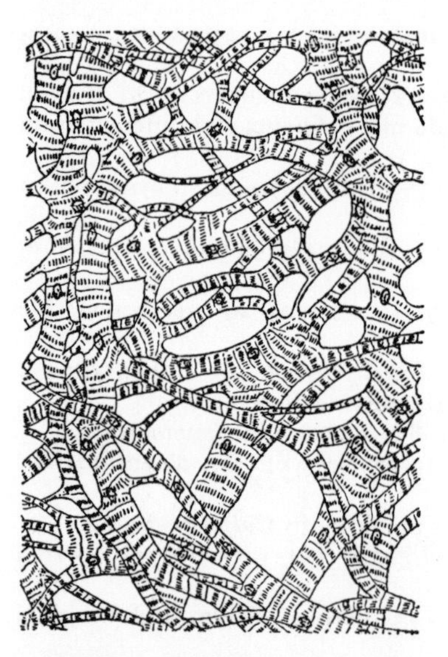

FIG. 82.—Musculature of the crop of *Calliphora*, Dipt., showing the branching and anastomozing fibres (*after* GRAHAM-SMITH)

Histology—All insect muscles seem to be made up of striated fibres; although in some of the visceral muscles the striations may be difficult to detect. Each fibre consists always of a number of parallel fibrillae or sarcostyles laid down in a nucleated plasma or sarcoplasm more or less laden with glycogen (Fig. 341, p. 596). But the degree of differentiation of the fibrillae, and their arrangement within the fibre, differs widely in various insects and in the different muscles of the same insect.[137]

In some of the muscles of the honey-bee larva[62] and in the larvae of many Diptera[71] the fibrils are minute threads with little visible differentiation; and they are invested by a thick layer of superficial plasma devoid of fibrillae (Fig. 83, A).

In adult insects three principal types of skeletal muscle are recognized.

(i) In the adults of all the higher Hymenoptera and Diptera,[62] and in adult

Dytiscus,[34] the nuclei are arranged in a row through an axial core of sarcoplasm extending the entire length of the fibre, and the fibrillae, as seen in transverse section, are arranged in flat bundles or lamellae radiating from the centre. These are sometimes called 'tubular muscles' or 'lamellar muscles' (Fig. 83, C).[45, 74] The flight muscles of Odonata, Blattidae and Mantidae are of this type.

(ii) 'Microfibrillar muscles' with closely packed cylindrical fibrils of 1–1·5μ diamater. Each fibre is ensheathed in a relatively tough structureless membrane, the sarcolemma, and the nuclei of the sarcoplasm are either scattered throughout the substance of the fibre or disposed immediately beneath the sarcolemma (Fig. 83, B).

(iii) The indirect flight muscles in the thorax of *Apis*, *Vespa*, &c., and many Diptera and Coleoptera, and in the tymbal muscles of Cicadas,[100] are sometimes called 'fibrous' or 'fibrillar muscles' (Fig. 83, D). Unlike the preceding types, which are always white in colour, these are yellowish or brownish. They consist

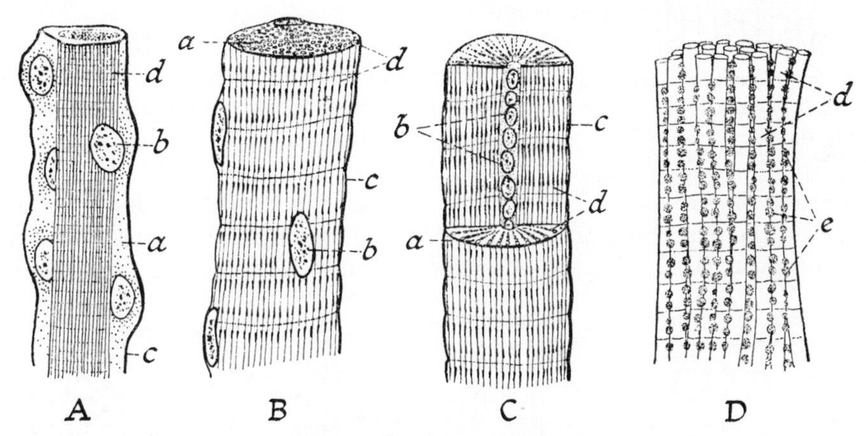

FIG. 83.—Types of skeletal muscle fibres in insects

A, from larva of honey-bee; *B*, from leg muscles of *Melolontha*; *C*, from leg muscles of honey-bee ('tubular muscles'); *D*, indirect flight muscles of honey-bee (a group of sarcostyles from 'fibrillar muscle') (*modified after* SNODGRASS). *a*, sarcoplasm; *b*, nuclei; *c*, sarcolemma; *d*, fibrils or sarcostyles; *e*, sarcosomes.

of bundles of very large fibrils, 2·5–3μ in diameter, which presumably correspond with the sarcostyles of the other types. In the wing muscles of the bee, which are of this type, there is no sarcolemma; the fibrils are loosely bound together by the tracheal endings. The entire fibre bundle is polygonal in cross section and 90–160μ in diameter.[62]

In all these types each sarcostyle or myofibril consists of alternating isotropic and anisotropic segments; these more or less correspond with the pale and dark-staining discs visible in the fixed tissue. In a given fibre these discs are at approximately the same level in neighbouring fibrils, so that the entire fibre has a banded or striated appearance. The details of this striation vary in complexity in different muscles. In its most elaborate form (Fig. 84) the light disc is traversed by a membrane, the Z line, (telophragma or Krause's membrane) attached all round the fibre to the sarcolemma, the compartment between adjacent membranes being termed a sarcomere. In the light disc, on either side of the telophragma, there may be a narrow row of dark dots, the accessory disc. And in the middle of the main dark disc there may be a pale stripe, the median disc or Henson's line, traversed occasionally by a second delicate membrane, the mesophragma. The same sequence of striations has been demonstrated with the electron microscope.[53]

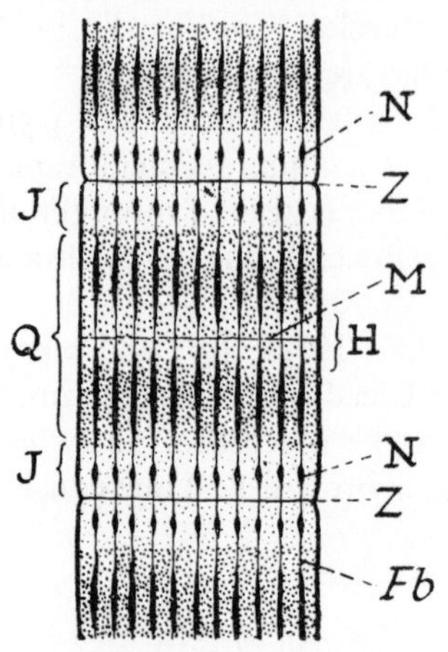

FIG. 84.—Diagram of striated muscle fibre

Fb, fibrilla or sarcostyle; *H*, median disc; *J*, light disc; *M*, mesophragma; *N*, accessory disc; *Q*, dark disc; *Z*, telophragma.

In the fibrillar type of wing muscles the striation is usually rather less complex than this. But telophragmata are generally present and at this level there is a concentration of endoplasmic reticulum linking adjacent fibrils together.[99] In all cases the telophragma seems to be a relatively tough membrane; for under abnormal conditions, when the fibres become swollen, they bulge in the middle of the sarcomeres and are constricted at the telophragmata. This happens both in the types of fibre provided with a sarcolemma and in the individual sarcostyles of the fibrillar wing muscles.[45]

All these types of muscle are furnished with tracheoles and with mitochondria. In the white 'microfibrillar muscles' of type (ii), the mitochondria are relatively inconspicuous; they lie in double rows at the level of the Z lines, and they are scattered between the fibrils;[32, 143] but the fibrils make up the main bulk of the muscle; mitochondria and tracheoles are much less conspicuous (Fig. 85, A)[91, 99] In the flight muscles of locusts, Hymenoptera,

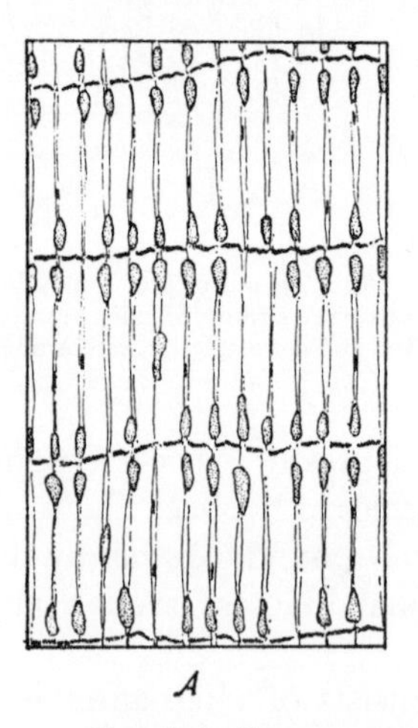

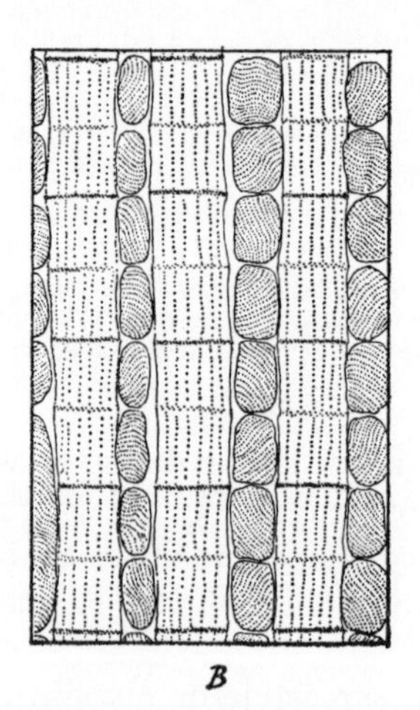

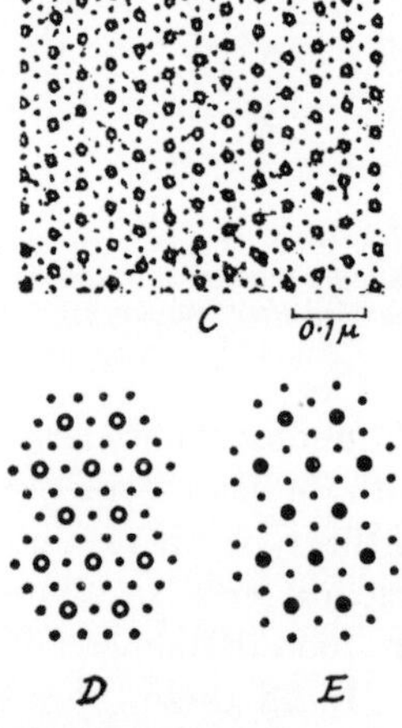

FIG. 85.—Fine structure of insect muscles as seen with the electron microscope

A, longitudinal section of coxal muscle of *Apis* showing mitochondria concentrated on each side of the Z line (*after* D. S. SMITH). *B*, longitudinal section of fibrillar flight muscle of *Polistes* with very large mitochondria (sarcosomes) between the fibrils (*after* D. S. SMITH). *C*, transverse section of fibrillar muscle of *Calliphora*, showing thick, apparently tubular, myosin filaments and fine actin filaments (*after* AUBER and COUTEAUX). *D*, Diagram of arrangement of filaments in *Calliphora* fibrillar muscle. *E*, the same in mammalian muscle for comparison (*after* AUBER and COUTEAUX).

&c. (Fig. 85, B) the mitochondria make up 40 per cent. of the total volume and they are furnished with a network of tracheoles.[91]

In the tubular flight muscles of *Aeschna* the mitochondria are conspicuous lab-like structures which lie opposite the radially arranged lamellar myofibrils.[134] 3ut it is in the fibrillar muscles that the mitochondria, commonly called 'sarco- omes', reach their greatest development.[32, 45] The staining properties of these odies, their relative contents of protein and lipids, the fact that cytochrome, ytochrome oxidase, and a whole series of dehydrogenases present in the muscle re confined to the sarcosomes is strong evidence for regarding them as giant nitochondria.[82] Under the electron microscope they show the bounding double nembrane and cristae characteristic of mitochondria.[134] During the first week of dult life in *Drosophila* and *Phormia* there is a progressive increase in wing-beat requency, and this is correlated with a great increase in size of the sarcosomes nd a threefold increase in their content of cytochrome *c*.[77, 115] In the mature nuscle nearly 40 per cent. of the wet weight is contributed by the sarcosomes.[116] f the wasp *Vespa* flies to exhaustion the sarcosomes in the wing muscles are wollen and disrupted with their contents extruded and the myofibrils dis- rganized by the accumulation of acid metabolites. After resting for ½–1 hour nany of the mitochondria have regained their normal form and structure.[107] imilar changes are seen after exhaustion in *Locusta* and after chill-coma in *'espa*.[132] In addition, the fibrillar flight muscles, for example in *Hydrophilus*[99] ave an exceedingly rich intracellular tracheole system closely linked with the uge mitochondria ($2{\cdot}2 \times 1{\cdot}5\mu$). The invading tracheoles draw with them a heath of plasma membrane from the surface of the fibre deeply in among the ndividual fibrils.[134]

During moderate contraction the anisotropic dark-staining segment is dimin- shed in extent, the pale segment increases somewhat;[31] but in strong contractions oth layers become shorter, and then their staining reactions change, so that the ark substance appears to separate and move to opposite ends of the sarcomere, rowding against the accessory discs and telophragmata.[32, 67] In this way the ark bands from adjacent sarcomeres are approximated and form a new dark isc, the contraction disc, so that the striation has the appearance of being eversed.[32, 62] The greater the contractility of the muscles the more frequent is he cross striation and the more nearly equal do the isotropic and anisotropic ands become.[68]

Studies of the myofibrils in the flight muscles of *Calliphora* with the electron nicroscope suggest that the general organization of the fibril is the same as in abbit muscle, and that contraction takes place by the sliding of the fine threads f actin in between the stouter array of myosin fibres concentrated in the A- ands (the anisotropic dark discs);[106] passive stretch affects only the I-bands (the sotropic pale discs) while the A-bands remain at constant length.[140] The arrange- nent of the large (primary) and small (secondary) filaments agrees in principle vith that in vertebrates, but the pattern of filaments is somewhat different. In ection the primary filaments appear tubular with an axial filament. (Fig. 85, C,) and E.)[86] [See p. 167.]

Muscle insertions—The muscles are united to the cuticle in many different vays. They may end directly on its inner surface, or they may be attached to the ase of the epidermal cells and joined to the cuticle by the plasma bodies of these, r by the intermediation of fibrillae passing through them. These fibrillae ('tono- brillae') can often be followed into the substance of the cuticle, and then appear o be chitinous; but some doubt exists as to whether they are derived from the pidermal cells upon which the muscles are exerting tension, or whether they are continuations of the myofibrillae. If the tonofibrillae are very elongated they ma form tendons of insertion; or the muscle may draw the epidermal cells inward so that they lay down a chitinous stalk which serves as a tendon. In its extrem form this process results in the development of apodemes and internal skeleta structures.[35, 71] Close study shows that as a rule the tonofibrils are inserted o the epicuticle.[87, 125] In *Calliphora* flight muscles the tubular tonofibrillae running from the cuticle to the sarcolemma are clearly the product of the epiderma cells.[85] [See p. 168.]

Nerve endings in muscle—The motor nerve axon is enclosed in a nucleated sheath made up of 'lemnoblasts', or Schwann cells. The axons are suspende in 'mesaxons' formed by complex infoldings of the plasma membrane of th Schwann cell. These infoldings often become loosely wrapped around the axons which are then said to be 'tunicated' (Fig. 112).[100] When the nerve reaches th muscle the basement membrane of the Schwann cells becomes continuous wit

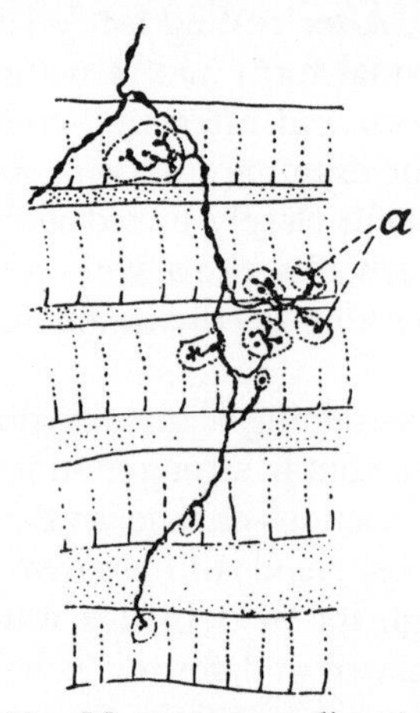

FIG. 86.—Motor nerve endings in oesophageal muscles of *Oryctes* larva a, Doyère's hillocks (*after* ORLOV).

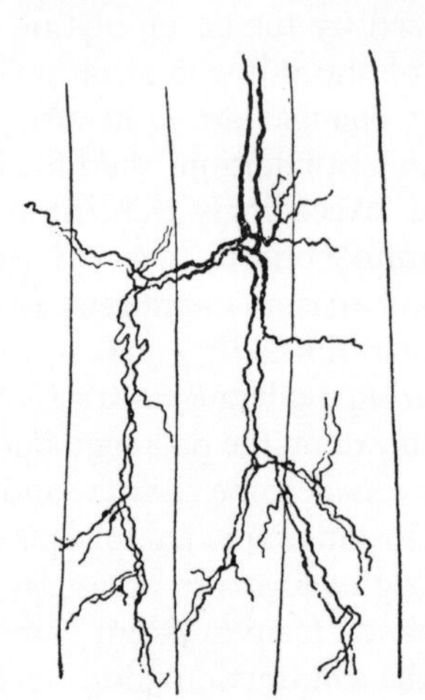

FIG. 87.—Double innervation in the leg muscle of *Decticus* (*after* MANGOLD)

the muscle sheath, or sarcolemma; at this point it encloses a group of sheat cells, containing numerous mitochondria, which form the Doyère's ending, o 'endplate'.[46, 137, 145] The axon ending contains abundant so-called 'presynapti vesicles', some 200–600 Å in diameter. These are believed to liberate a 'chemic transmitter substance' which passes to the muscle and causes the depolarizatio of the muscle bounding membrane which initiates the end-plate potentials th are followed by contraction. Along the terminal part of the axon are repeate points of close contact between the axon and the muscle, which are believed t be the actual sites of stimulation.[100, 134, 135] It is suggested that the endoplasmi reticulum at the level of the Z line is an invagination of the cell membran which serves to conduct impulses to the contractile fibrils within the musc fibre.[128, 133, 134]

In vertebrates the transmitter substance is probably acetylcholine; and th end-plate is rich in cholinesterase by which it is rapidly inactivated. In insec the nature of the transmitter substance is unknown. Cholinesterase is abser from the myosynapses of *Rhodnius*.[144] [See p. 168.]

The skeletal muscles of insects have a double innervation; two nerves are distributed side by side throughout the muscle, a given fibre receiving a suppl from both nerves (Fig. 87).[40, 84] Occasionally a third axon may occur.[111] That is so in certain of the fibres of the jumping muscle in the grasshopper *Romalea* which receive a 'fast' axon, a 'slow' axon exciting the muscle by depolarization, and an 'inhibitor' axon which produces hyperpolarization.[138] That is a state of affairs that has long been known in Crustacea. In some insects it is uncertain whether separate slow and fast muscle fibres exist.[89] [See p. 168.]

PHYSIOLOGICAL PROPERTIES OF INSECT MUSCLES

The movements of insects may be very slow, as in the crawling of some larvae, more rapid, as in the running of adult beetles, or excessively quick, as in the wing beats of Diptera or Hymenoptera, which may make more than 500 complete movements per second. We must now consider whether there are any peculiarities about the physiology of insect muscles which will account for these performances.

Absolute muscular power—Some insects are able to lift weights of greater mass than their own bodies, and leaping insects can project themselves great distances through the air. But these achievements are a simple result of their body size. For the power of a muscle varies with its cross section, that is, with the square of one linear dimension; while the volume or mass of the body varies with the cube of the linear dimensions. Consequently, as the body becomes smaller the muscles become relatively more powerful.

The absolute power of a muscle is defined by the maximum load it can raise per square centimetre of cross section. When expressed in this way there is no great difference between the muscles of insects and of vertebrates. Thus the value for man is 6–10 kg. per sq. cm., for the frog 3 kg., for the mandibular muscles of insects 3·6–6·9 kg.,[14] for the hind legs of *Tettigonia* 4·7 kg.[33] and for the flexor tibiae of *Decticus* 5·9 kg.[64] At ordinary body temperatures the tension and work developed by locust flight muscle are of the same order of magnitude as in the skeletal muscle of vertebrates.[140]

Properties of isolated muscles—The properties of the isolated muscles of insects are very similar to those of the skeletal muscles of vertebrates. In response to a single electrical stimulus the muscle makes a simple contraction or twitch (Fig. 88). Between the application of the stimulus and the commencement of

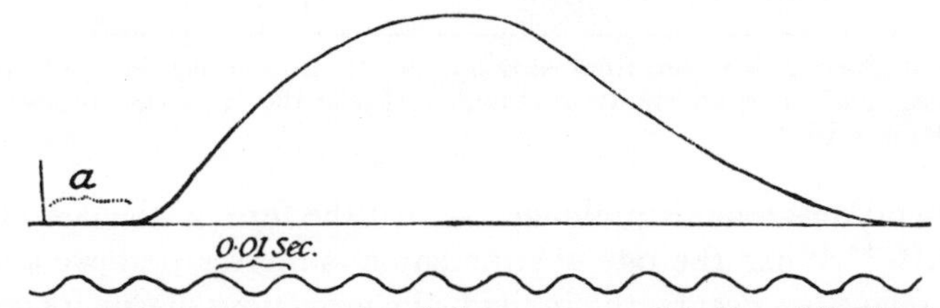

FIG. 88.—Kymographic record of simple twitch in extensor of hind leg of *Tettigonia* (*after* KAHN) Latent period (a), 0·012 sec.; total duration of twitch, 0·102 sec.

visible contraction there is a 'latent period' of variable duration; this is followed by a phase of contraction and then of relaxation. Since the contraction is usually recorded mechanically by causing the muscle to raise a lever which writes on a moving drum, the precise form of the twitch is greatly influenced by the mechanical arrangements.

If a second stimulus is applied during the latent period of the first, or during the early part of contraction, 'summation' occurs and the height of the contraction is increased (Fig. 89).[33] Summation is often very striking in insect muscles, particularly in the more slowly contracting types (it is much more evident in the muscles of *Hydrophilus* than in *Dytiscus* (*cf.* Table 1)).[57] It is well seen in the thoracic muscles of *Aeschna* (Fig. 90); stimuli of subthreshold intensity at the rate of 20 per second have no effect at first, but gradually they produce a contraction which mounts higher with each successive stimulus.[27] Locusts move

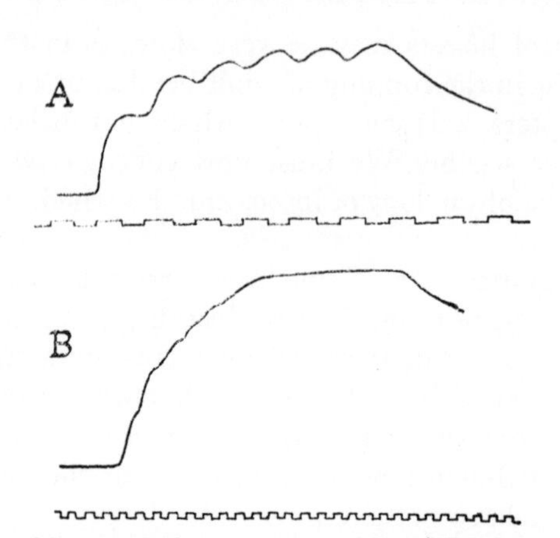

FIG. 89.—Kymographic records from extensor tibiae of *Decticus* at 5° C. (*after* SOLF)

A, stimulated at $\frac{1}{2}$-second intervals, as indicated by the time marker, showing summation; *B*, stimulated at $\frac{1}{5}$-second intervals, showing complete tetanus.

their wings by ordinary striated muscle with the usual one-to-one relation between nerve impulse and twitch contraction;[79] they may vibrate the wings at a rate of 18–20 wing strokes per second;[102] but the force of the contractions can be increased by 'double firing', that is, by two nerve impulses in close succession (4–8 milliseconds apart).[123]

At high temperatures the twitch becomes more rapid; at low temperatures

FIG. 90.—Kymographic record from wing muscles of *Aeschna coerulea*, given 20 stimuli per second, showing increasing sensitivity to stimuli which at the outset are below the threshold level (*after* HEIDERMANNS)

it is slower and relaxation is prolonged so that the form of the contraction wave is changed.[16, 36] When the rate of stimulation is increased above a certain frequency no relaxation occurs; the individual contractions are no longer apparent and a steady state of contraction or 'tetanus' persists (Fig. 89, B).

A single muscle in an insect contains relatively few fibres; in order to obtain smooth contractions the insect must employ different methods from those of vertebrates. There are two major differences. In vertebrates each muscle fibre receives the nervous stimulus at a single point, from which the excitation is propagated along the muscle fibre. This has been reported to occur in the leg muscles of *Periplaneta*[152]; but most authors consider that in the muscle fibres of insects here are only local motor end-plate potentials and no propagated action poten-ials; each fibre has multiple nerve endings each provoking a purely local con-raction. As in Crustacean muscle conduction of impulses is carried out solely y the motor nerve fibres: if the nerve is allowed to degenerate the muscle be-omes completely inexcitable to all forms of electrical stimulation.[80]

The second major difference from vertebrate muscle lies in the dual innerva-ion.[48, 96] The 'slow' nerve fibres produce a much smaller potential change in the nuscle. This is agreed by all authors to be a purely local phenomenon which is ot propagated along the fibre.[152] In *Carausius*, also, the 'fast' reaction is twitch-ike with an electrical response resembling the action potential in vertebrates; he 'slow' reaction is readily facilitating with an electrical response resembling nd-plate potential changes; it gives rise to slow smooth contractions and the naintenance of tonus (p. 189).[154]

The jumping leg-muscle of *Schistocerca* receives three motor nerve fibres, wo large and one small. One large axon is the 'fast' fibre, providing powerful 'all r nothing' twitches at a tension of 20,000 g./cm.^{2} of the cross-section. (This igh value is achieved by the 'pinnate' arrangement of the fibres converging upon central tendon.) The other large axon is the 'slow' fibre; it comes into action nly when impulses in excess of 15 per second pass down it; it is responsible for ll other types of muscle movement, from prolonged tonic contraction to quite apid walking movements.[110] In this instance it is claimed that the same muscle fibres are utilized for either the fast or slow contractions and both axons share the same end-plates.[89, 110, 111] The function of the third (small) axon may per-haps prove to be inhibitory, as in *Romalea*.[138] During natural movements there are long trains of 'slow'-fibre activity with occasional bursts of 'fast'-fibre acti-vity.

Insect muscles are much more tolerant to extreme variations in ionic con-centrations than are those of vertebrates. They remain active in media which completely paralyse vertebrate muscles. This difference is probably associated with the multiple innervation of insect muscle.[109] [See p. 169.]

Speed of response in insect muscle—Earlier records of the latent period between stimulus and contraction gave values ranging from 12–47 msec.[33, 57, 64] But the application of modern methods for recording the electrical and mech-anical events in contracting muscles has shown that in the thoracic muscles of *Periplaneta* the latent period is only 3·0 msec.[80] It was formerly suggested that even in the dragon-fly *Aeschna*, which makes only 28 wing-beats per second, maximum contraction and relaxation was impossible, and that during flight the muscles of these insects carry out movements corresponding to an incomplete tetanus.[27] But in the flight muscles of *Periplaneta* the contractions begin to fuse at rates of stimulus above 40–50 per second. This is well above the wing-beat frequency of this insect during flight.[80] Moreover the thoracic muscles normally operate at a temperature above 35° C. (p. 166); under these conditions the flight muscles of *Locusta* and *Schistocerca* can contract and relax without fusion at a frequency comparable to the normal wing-beat (20–28 cycles per second).[95, 102]

Records of the potential changes in the nerves supplying the wing muscles during flight show that these occur at the same frequency as the wing movements in *Periplaneta* (25–30 per second) and in *Agrotis* (Lep.) (35–40 per second). The relation between nerve stimulus and flight muscle contraction in these insects is therefore of the conventional type: one stimulus to one contraction.[79] But in

Diptera (*Calliphora*, *Lucilia*, *Eristalis*) and Hymenoptera (*Vespa*), in which the rate of wing-beat may go up to several hundreds per second, this relation is changed and the muscular contractions occur at a far higher rate than the changes in electrical potential.[50, 79]

Chronaxie and oscillatory contractions—An electrical stimulus of infinite duration must be of a certain threshold strength in order to produce a response in nerve or muscle. If the strength of stimulus is increased its duration may be shortened. The 'chronaxie' of Lapique is the duration of the stimulus, expressed in one-hundredths of a second (σ), which is necessary to produce a response when the strength of stimulus is double the threshold. It provides perhaps the most convenient measure of the rapidity of response in muscles and nerves.

The chronaxie in the human biceps is $0{\cdot}15\sigma$; in the gastrocnemius of the frog $0{\cdot}3\sigma$; in the slowly contracting muscles of *Chironomus* larvae it is 10σ, while in the imago it falls below 1σ[4]; in the leg muscles of *Hydrophilus* it is $0{\cdot}42\sigma$; in the wing muscles of the dragon-fly *Calopteryx* $0{\cdot}48$–$0{\cdot}72\sigma$, of *Bombus* $1{\cdot}84\sigma$ and of *Calliphora* $0{\cdot}8$–$1{\cdot}12\sigma$.[22] There seems to be a tendency for the chronaxie to be diminished as the wing surface is reduced.[37] Thus when measured in this way the flight muscles of insects appear less rapid than the gastrocnemius of the frog. It was therefore suggested that perhaps the wing muscles did not act directly on the wing, but served to set in vibration some elastic intermediary structure.[22, 27]

Nowadays the flight apparatus is regarded very largely as a mechanical resonant system which works under forced vibration near its natural frequency.[129] The isolated flight muscle of *Bombus* goes into a smooth tetanus with stimuli at 40–60 per second. In its normal state the shortening of the fibres is limited mechanically to about 1 per cent. of their length.[118] It is only when the muscle contracts against an inertial load and is suddenly released after rapid shortening and then suddenly stretched that its special properties appear.[92, 93]

In both the timbal muscle of the cicada, and the flight muscles of many insects, the muscles operate a 'click' mechanism, such that when they have contracted up to the critical point of the mechanism (p. 163) they are suddenly released. This sudden release seems to 'deactivate' the contractor muscle and at the same instant the sudden stretching of the opposing muscle excites it to contract.[126, 127] The result is a rapid oscillatory contraction of the opposing muscles, the rate of oscillation being mainly determined by the mechanical and elastic properties of the wall of the thorax. [See p. 169.]

This is essentially a 'myogenic' system: the muscles stimulate themselves or one another. In *Calliphora* lightly anaesthetized, a motor nerve impulse frequency of about 3 per second accompanies a regular wing-beat frequency of 120 per second.[50] The main function of the occasional nerve impulses is to maintain the myofibrils in a persistent 'active state'; yet an increase in frequency of the nerve stimulation does occur in flies whenever the wing-beat become more rapid.[151] In the beetle *Oryctes* the frequency of the oscillatory contractions is determined by the mechanical frequency of the system and is not influenced by the rate of nervous stimulation; but if the stimulus frequency is reduced the amplitude of the oscillation may fall. There are no potential changes in the surface membrane of the muscle synchronous with the oscillatory contraction.[119] This oscillatory response is peculiar to the fibrillar muscles of insects; it is not shown by any other known type of muscle. [See p. 169.]

Muscle tonus—In addition to their alternate contractions and relaxations the muscles may enter a state of prolonged steady contracture or tonus. In this state they may support the insect in some characteristic attitude and may have to bear a considerable weight. But the maintenance of this type of permanent contraction appears to demand no measurable increase in metabolism. *Carausius* shows no difference in oxygen consumption whether it is lying at rest on its back or in a tonic condition with its weight supported by the legs.[9] As we shall see later, this state is dependent on the integrity of the nerve supply (p. 189) and may be greatly influenced by the stimuli from certain sense organs (p. 312). In the muscles themselves (in *Hydrophilus*, flies, &c.) it is associated with electrical waves of small amplitude, present in only a few of the fibres in a given muscle and quite distinct from the electrical responses of much greater amplitude and frequency which accompany voluntary or reflex contraction.[54]

In the cockroach in the standing position there is a steady discharge of impulses passing down the slow nerve fibres to the depressor muscles of the leg, and producing the tonic contraction which raises the animal off the ground. Microscopic examination of a cockroach muscle in this state shows that the contraction is produced by many fibres responding in turn. On this background are superimposed short high-frequency bursts of impulses in the quick nerve fibres, producing vigorous contractions in the muscles and moving the animal rapidly over the ground. Slow movements are achieved entirely by means of the slow fibres. It would thus appear that there is a gradual transition from tonic contraction to slow active contraction and sudden outbursts of rapid movement.[48]

Chemical changes in muscle—The biochemistry of muscular contraction in insects agrees in general with that in vertebrates, but there are some striking variations which reflect differences in physiology. It is instructive to compare the leaping muscle of *Locusta* with the flight muscles of the same insect.[91] The *leaping muscle* resembles that of the vertebrate, with fibrils greatly developed but mitochondria and tracheoles relatively few. The locust is exhausted after 10–15 leaps: glycogen is broken down by anaerobic glycolysis to lactic acid and this is then oxidized by the lactic-dehydrogenase system as in mammalian muscle; after just two leaps lactic acid and pyruvic acid show a sixfold increase; at exhaustion a tenfold increase. A long rest is then required to remove and reconvert the lactic acid to glycogen. Glycerophosphate is formed in small amounts but glycerophosphate dehydrogenase is not very active.[155]

The *flight muscle* is adapted to show a continuous high rate of aerobic activity. Mitochondria and tracheoles are abundant (p. 148). There is no provision for anaerobic energy metabolism. Lactic dehydrogenase shows only one hundredth the activity of that in the leg muscles. After flying for 90 seconds in the absence of oxygen *Locusta* wing muscles show a 15–20 fold increase in glycerophosphate and pyruvic acid but lactic acid shows little change, and associated with these properties there is an exceedingly active glycerophosphate dehydrogenase. There is, in fact, a great development of the 'glycerophosphate cycle' which yields hydrogen for energy production by way of the cytochrome chain (p. 623)—a cycle which exists in vertebrate muscles but is of relatively small importance.[155]

A major source of muscular energy is activated acetic acid entering the citric acid cycle; this may come from pyruvic acid and so from carbohydrates; or it may come from fat which is the chief fuel used for flight in the locust. In the flight muscles this acetic acid is all consumed directly to carbon dioxide and water, and this is reflected in the extraordinarily high activity of the so-calle 'condensing enzyme' which catalyses the union of activated acetic acid wit oxaloacetic acid to yield citric acid for the citric acid cycle (p. 623). The relativ activities of this enzyme are: *Locusta* flight muscle, 131; *Locusta* leg muscle, 10 rat leg muscle, 1.[155]

In the flight muscles of *Musca* and other Diptera where carbohydrates are th chief fuel used in flight (p. 601), glycerophosphate derived from the glycolysis o carbohydrate is again the chief substrate for energy production.[130, 153] Glycoge and, to a less extent, fat are stored in the sarcoplasm of the muscle (p. 596) Lipase,[104] trehalase,[157] the glycolytic enzymes, succinic dehydrogenase, cyto chromes and cytochrome oxidase, and the numerous enzymes of the citric aci cycle, are concentrated in the mitochondria, and these enzymes are very muc more active in the flight muscles.[82, 90, 116, 143] In the *Periplaneta* male, which i capable of flight, the fresh wing muscles are opaque and pink in colour; in th female they are hyaline and white. At emergence the muscles of the male ar like those of the female, but the increase in the enzymes in the mitochondria particularly cytochrome and cytochrome oxidase, leads to the appearance of th red colour.[94, 114]

The oxidative processes in the mitochondria are coupled with the synthesi of adenosine triphosphate (ATP) which is later hydrolysed to adenosine di phosphate by ATP-ase, to yield the phosphate bond energy for muscular con traction and flight. In vertebrates the ATP-ase seems to be localized in th muscle fibril. In insects it occurs both in the sarcosomes[106] and in the fibrils;[15 in the honey-bee and other insects, as in vertebrates, actomyosin has a powerfu ATP-ase action.[120] In the flight muscles of *Dytiscus*,[76] and in *Phormia*,[82] sarco somes seem to be concerned in this process: isolated fibrils do not contract in th presence of adenosine triphosphate unless the sarcosomes remain adherent t them.[76] [See p. 169.]

As in most invertebrates, the 'phosphagen' in *Aeschna*[60] and in the blow flies *Lucilia* and *Calliphora*,[2] which provides a second reserve of phosphate bon energy, is arginine phosphoric acid (in place of the creatine phosphoric acid o vertebrates). But it is usually present in much smaller quantities, particularly i the flight muscles, than its equivalent in vertebrate muscle. It is produced out side the sarcosomes.[117] [See p. 169.]

LOCOMOTION

Walking—As was shown by Johannes Müller, the insect rests during walkin on a supporting triangle, formed by the anterior and posterior limb on on side and the middle limb on the other, while it carries forward the othe three legs. The fore leg acts as a tractor; the middle leg serves for suppor lifting the body on its own side and together with the hind leg raising th posterior part of the body; the hind leg acts as a propulsor and also turns th body in the horizontal plane. As the result of all these actions the centr of gravity of the insect falling within the supporting triangle of limbs carried forwards and outwards towards the apex of the triangle until it fal outside this base and its support is taken over by the other triangle of leg hence the body zig-zags slightly from right to left as it advances (Fig. 91). the insect is caused to walk on smoked paper it is found that the three legs o ach side come to lie successively on the same spot. The details of these movements and the mechanics f the muscles which bring them about were fully vorked out by Graber.[47]

When the walking insect is studied by cinematographic methods, however, it can be seen that the three legs forming the tripod for the time being are not lifted simultaneously but in the order: (i) foreleg, (ii) opposite middle leg, (iii) hind-leg. As a result, in some insects there may be wide departures from the classical description: three, four or ive legs may be on the ground at the same time; and the course followed in walking may be practically straight.[30] [See p. 169.]

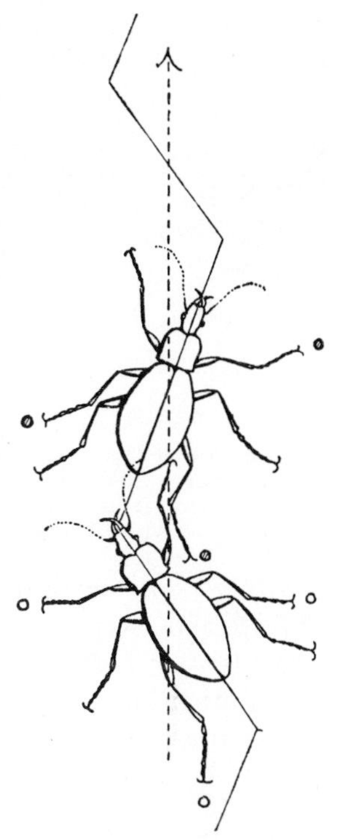

FIG. 91.—*Carabus* walking in the direction of the broken line (*after* v. LENGERKEN)

The continuous line shows the zig-zag course followed by the insect, considerably exaggerated; the circles indicate the legs resting on the ground in the phases represented.

Crawling of larvae—Some larvae walk by means of their thoracic legs in the same manner as adult orms; but they are often aided by provisional modifications of the abdomen. Thus Carabid, Chrysomelid and other beetle larvae propel themselves forwards by means of eversible 'pygopodia' arising from the terminal segment of the abdomen.[7] Similar structures occur on various Diptera-Nematocera and Siphonaptera larvae. Tenthredinid larvae and the caterpillars of Lepidoptera have well developed prolegs on many of the abdominal segments, and movement then consists of a wave of thickening and shortening running forwards along the body. The anal pair of legs is first carried forwards; the other pairs are moved onwards as the wave reaches them; and finally the thoracic appendages are similarly advanced to take up their new position and catch hold.[71]

A large part of the musculature of the body wall in caterpillars consists of small bands running across the many folds in the skin, their sole function being to maintain a steady internal pressure ('turgor muscles'). If the larva is punctured it shrinks and contracts under the action of these muscles; apparently the fall in pressure provides the stimulus to their contraction. In addition there are the true

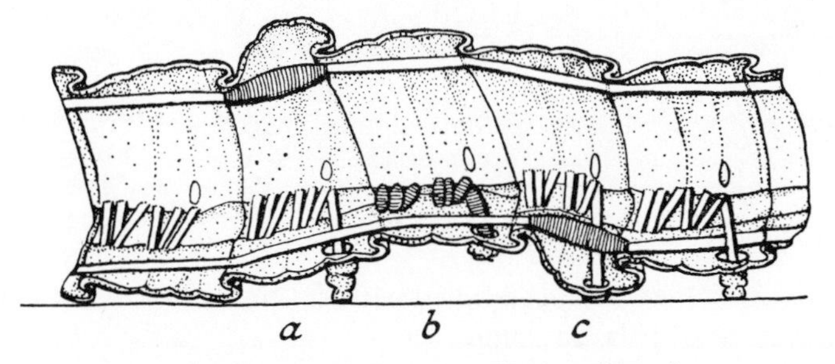

FIG. 92.—Schematic longitudinal section of a caterpillar to show the sequence of muscular contraction during the peristaltic wave

Three segments are active: in *a*, the dorsal longitudinal muscles are contracted; in *b*, the dorso-ventral muscles are contracted; in *c*, the ventral longitudinal muscles are contracted (*after* BARTH).

'locomotor muscles' (transverse, longitudinal and dorso-ventral muscles o greater length) which do not contract in the punctured larva but are throwr passively into folds. When the caterpilla crawls, the dorsal longitudinal muscle in one segment contract simultaneousl with the vertical muscles which lift th proleg in the segment behind, and wit the ventral longitudinal muscles in th segment behind that (Fig. 92). Thi mechanism is variously modified for th looping progression of Geometrids anc for the assumption of special attitudes In all these movements there is neces sarily a co-ordination between the turgo and locomotor musculature. The forme is highly important in movement; for the extension of any part of the body i brought about through the relaxation of its own muscles while the general in ternal tension is maintained by the turgor muscles elsewhere.[3]

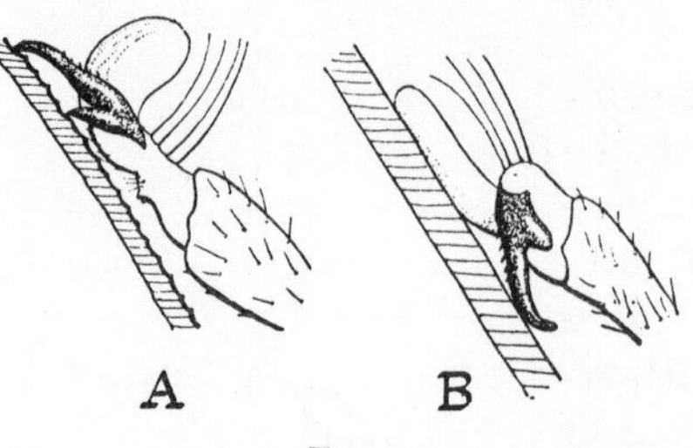

FIG. 93

A, foot of honey-bee applied to rough surface; it holds by the claws. *B*, applied to smooth surface; it holds by means of the adhesive organ (arolium) (*after* CHESHIRE).

In other larvae, such as Muscidae, and in motile pupae, legs are entirel absent and progression is effected by peristaltic movements or lateral twistin

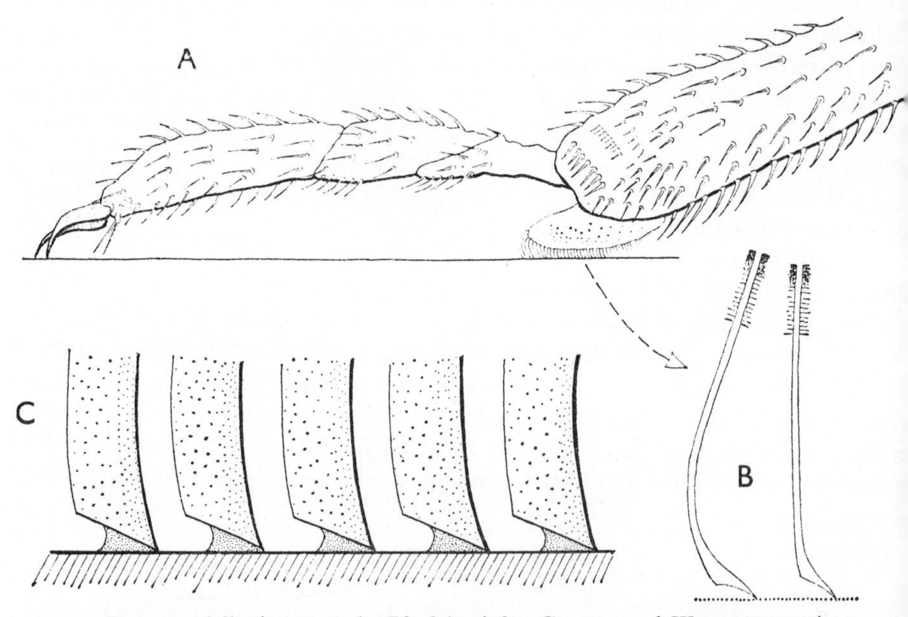

FIG. 94.—Adhesive organ in *Rhodnius* (*after* GILLETT and WIGGLESWORTH)

A, fore-leg of *Rhodnius* adult showing the adhesive pad at the lower end of the tibia. *B*, detail of tw hairs from the pad. *C*, diagram to show the mechanism of the adhesive organ. There is an oily secretio at the oblique tips of the hairs. These slide forward (to the left) easily, being lubricated. When move backwards (to the right) seizure occurs and the organ adheres to the surface.

movements of the body wall combined with friction against the surface due t backwardly directed hairs or spines.

Adhesive organs—Provided the surface on which the insect walks is suf ficiently rough, it can hold to it by means of its tarsal claws. If the surface i too smooth for the claws to grip, it makes use of 'adhesive organs' on the pulvilli empodia or tarsal or tibial pads (Fig. 93). These organs generally consist of dense collections of tubular hairs ('tenent hairs') with delicate expanded tips, moistened by some glandular secretion. By means of these, many insects can climb on perfectly clean glass; but the mechanism by which they do so has led to much controversy. By some authors they have been considered to act as suckers (as the funnel-shaped organs on the tarsus of the male *Dytiscus*[61, 71] certainly do), being held to the surface by the *atmospheric pressure*, the small amount of fluid present serving merely to make the union between the tubes and the surface airtight.[61, 72] Others have believed the force concerned to be the *surface tension* of the fluid secretion around the margins of the hairs.[58] Others again have supposed that the secretion is a sticky fluid and that it is the *cohesion* of this fluid which holds the insect to the surface.[19] At the present time it seems probable, at least in many cases (*Apis*,[1] *Rhodnius*[25]) that the extremities of the tenent hairs, being very soft and delicate, can be applied so closely to the surface, that in the presence of a small amount of fluid, seizure or *adhesion* takes place and the insect is held by surface molecular forces.[1, 17, 25] (Fig. 94) Such organs may be of value, of course, not only in clinging to smooth surfaces but for gripping smooth objects such as the bodies of other insects.[25] The Reduviid *Platymeris* can hold powerful beetles, such as *Oryctes*, with a polished surface, by means of the tibial adhesive pad; it can in fact support a tension of more than 50 gm.; an adult locust *Schistocerca* with its tarsal pads can support 5 gm.[101] [See p. 169.]

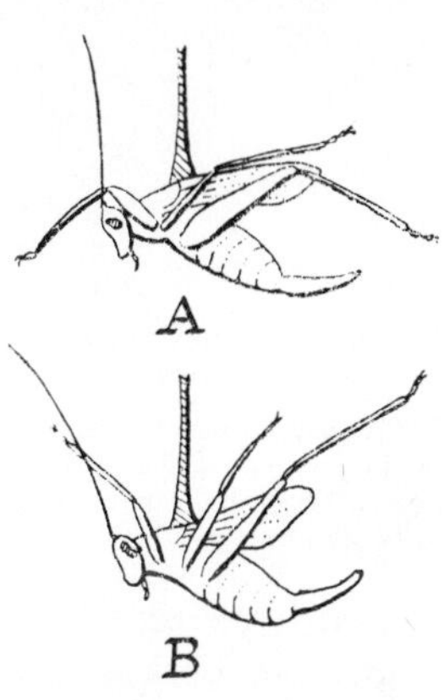

FIG. 95.—Falling reflex in *Meconema varians* (*after* VON BUDDENBROCK and FRIEDRICH)

A, normal attitude of insect supported from the thorax; *B*, attitude adopted when stream of air is blown from below.

Leaping is usually effected by the sudden combined extension of the tibiae of the hind legs, as in Acridiids, Siphonaptera, *Haltica*, &c. Occasionally it is brought about by a spring and catch mechanism. In Collembola or 'spring-tails' the anal fork is engaged below the 'retinaculum' of the 4th abdominal segment; when the anal fork is forcibly extended it slips out from the retaining catch and projects the insect into the air. A similar mechanism occurs in Elaterid beetles, in which the point of the prosternum is engaged in the fossette of the mesosternum; and in larvae of the 'cheese skipper' *Piophila*, in which the edge of the anal extremity is seized by the mouth hooks and suddenly released when the larva is in a state of tension. [See p. 169.]

Many insects (*Meconema* (Tettigoniidae), *Carausius*, *Forficula*, &c.) when allowed to fall 20 cm. or more through the air, almost always land on their feet. This results from the reflex assumption of a particular attitude. Legs and antennae are spread out and raised above the body, and the abdomen is curled upwards (Fig. 95). The reflex is induced by the removal of all contact stimuli (*cf.* p. 313) and by the effect of air currents on the antennae and tarsi. Once the appropriate attitude has been assumed the insect is turned into the dorso-ventral position mechanically by the resistance of the air.[12] The larvae of the Arctiid *Euchaetias* have their abundant hair tufts so arranged that they likewise land on their feet when falling, and they release themselves in such a way as to take advantage of this property.[23]

Locomotion on the surface of water—As a rule the cuticle is not readily wetted by water. Consequently, when the insect stands on water the surface tension acts in the opposite direction to gravity. Now the surface area of a body varies with the square of its radius, whereas its volume or mass varies with the cube of its radius; hence the ratio of surface to mass becomes progressively greater as the size of the body diminishes, and the forces resident in surfaces become relatively stronger. Many insects are in fact sufficiently small for surface forces to support the whole weight of the body, so that the insect on the water surface is held up as though by an elastic membrane (Fig. 96). In this it is doubtless assisted, in some cases at least, by glands in the tarsi producing a fatty secretion which enhances the 'hydrophobe' properties of the cuticle and its hairs. In *Podura aquatica* (Fig. 96)[124] and in *Anurida maritima*[88] the claws are wetted by water, and on the level surface the insect can walk as rapidly as on land. It anchors itself to the surface with its ventral tube. It can also raise a small hillock in the water surface by pressing down with head and tail and can then slide sideways down the slope. In a comparable manner *Aepophilus* (Hemiptera), resting on the surface, can deform the meniscus by muscular energy so as to be drawn up the slope at the margin of a pool of water.[88] The 'pond skaters', or Gerridae, spend the greater part of their life resting on the water surface in this way or rowing themselves along by simultaneous backward movements of the hind limbs. Gyrinid or 'whirligig' beetles lie in the water surface with only the polished hydrophobe scutum and elytra exposed; the ventral surface and the highly modified appendages with which they propel themselves along are completely immersed.

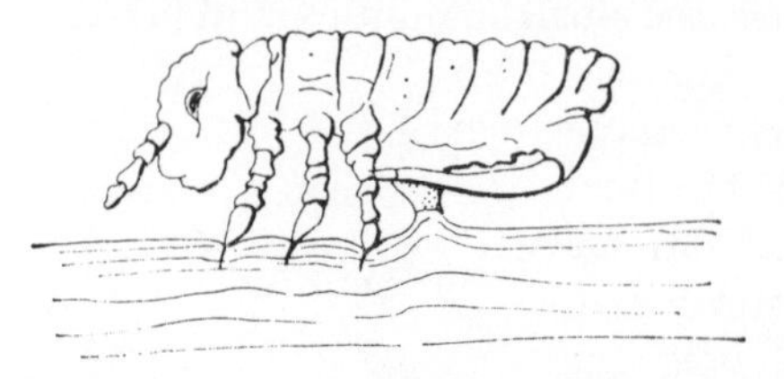

FIG. 96.—*Podura aquatica* standing on the surface of water (*modified after* BROCHER)

The claws are hydrophile, the legs are hydrophobe and depress the water surface. The ventral tube is hydrophile and by means of this the insect can anchor itself in the surface film.

Beetles of the genus *Stenus*, when they fall on water, utilize surface forces in a different way—by expelling from the anal glands[113] a substance which lowers the surface tension, so that they are drawn rapidly forwards like a toy boat propelled by camphor. If deprived of the tip of the abdomen they can no longer move in this manner.[6]

Swimming—Many different mechanisms are used by aquatic insects which swim beneath the water surface. Nematocerous larvae move by lateral flexions of the body, aided in the case of mosquito larvae by 'tail fins' composed of rows of fine bristles. Those larvae which breathe atmospheric air are sometimes raised by the buoyancy of the air they carry; in others (*Corethra* and *Mochlonyx*) the tracheae are modified to form hydrostatic organs by means of which the buoyancy can be adjusted, and they are thus able to maintain the body at any required level (p. 384). The nymphs of Anisopterous dragon-flies propel themselves forwards by forcibly ejecting water from the rectal gill chamber through the anus. In *Anax* larvae the pressure rises to about 30 cm. of water within 0·03 sec. and the insect is propelled at a speed of 30–50 cm./sec. for a short distance.[112] The Hymenopterous egg parasites, *Polynema*, &c., are exceptional in swimming under water by means of their wings. The aquatic Coleoptera and Hemiptera swim by movements of the legs. Hemiptera use their hind-legs simultaneously ke oars; and Dytiscids use the middle pair in addition, to a lesser extent, in the ame way; so that the sequence of leg movements is completely altered when he insect walks on land; whereas *Hydrophilus* and its allies swim with their nd and 3rd pair of legs by alternating movements.

The front of the body of the water beetle *Acilius* is very avourably shaped to reduce resistance; the flow of water is aminar over the anterior part of the body, but at the widest oint it changes into turbulence.[121]

The air stores which these insects carry for purposes of espiration (p. 381) play an important part, also, in the naintenance of equilibrium: if *Dytiscus* is deprived of its air tore it goes to the floor of the vessel to seek for air, and if larmed it swims upwards instead of down.[10]

Flight—In the more primitive insects, Orthoptera, Neuroptera, Isoptera, Odonata, the fore- and hind-wings are noved independently during flight; in *Agrion*, for example, vhen the fore-wing is depressed the hind-wing is elevated nd vice versa (Fig. 97).[70, 122] Whereas in Hymenoptera, Trichoptera, many Lepidoptera, Hemiptera, &c., fore- and nind-wings are united by various mechanisms to make a unctional unit. In Coleoptera, in which the fore-wings simply orm protective sheaths, they are sometimes held aloft during light, sometimes, as in *Cetonia*, &c., kept folded over the bdomen. In Diptera the hind-wings, and in male Strepsiptera he fore-wings have become reduced to richly innervated lub-shaped structures, the halteres, whose function in the egulation of flight will be considered later (p. 268). The nodification of the wings to give a single functional unit voids the disadvantage of the second pair of wings working n a region of turbulence produced by the first pair. This happens in Orthoptera, nd lower Neuroptera where the upstroke and downstroke of the fore-wings is n advance of the hind-wings. The dragon-flies (Odonata) have got over this

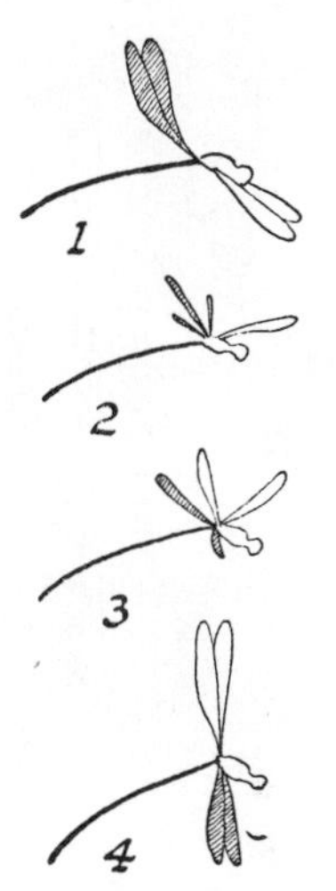

FIG. 97.—Successive positions of the wings of an Agrionid dragon-fly during flight

Fore-wings plain, hind-wings shaded (*from* v. BUDDENBROCK, *after cinematograph figures by* VOSS).

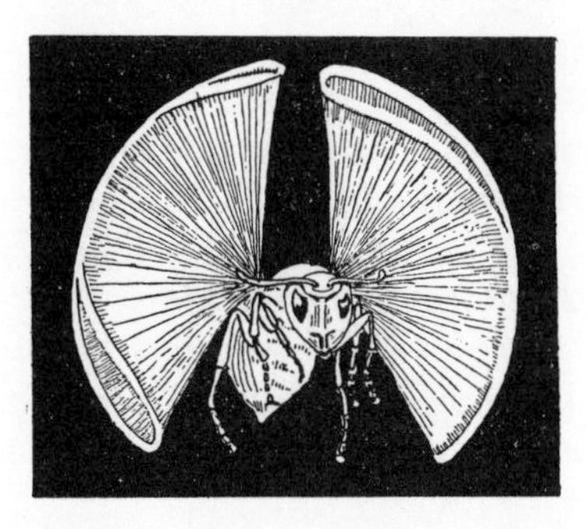

FIG. 98.—Wasp with gilded wing tips hovering in the sun; to show the extent of the excursions of the wings (*after* MAREY)

difficulty and developed a relatively efficient type of flight by reversing the order of the wing beat; the hind-wing beats first and thus meets the oncoming air before it is troubled by the passage of the fore-wing.[15] [See p. 169.]

In the Blattoidea the wings are operated predominantly by 'direct' flight muscles acting upon the base of the wing; these muscles remain important in all Orthoptera, Coleoptera and Odonata; but in all other groups, flight is effected chiefly by indirect muscles, vertical and longitudinal columns which deform the thoracic capsule by their contraction, the vertical columns elevating the wing, the longitudinal columns depressing it. The complex articulation of the wing gives a wide amplitude to the small and powerful movement imparted to the wing base: in the wasp, for example, the wings vibrate through a sector of 150° (Fig. 98).[42] In addition, there are direct muscles inserted into the wing itself. Of these the most important are the anterior and posterior pleural muscles which pull on the wing base before and behind the pleural fulcrum and so rotate the wing around its long axis (Fig. 99).[55, 62] They are concerned with the extension and flexion of the wings and with adjustments of the wing stroke.

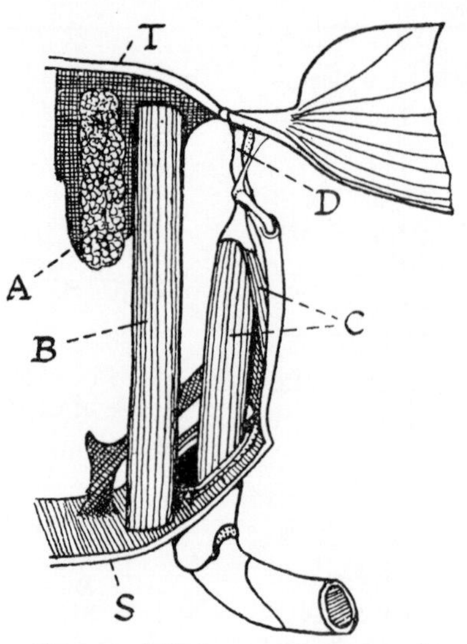

FIG. 99.—Diagrammatic cross-section of a winged segment, anterior view (*after* SNODGRASS)

A, longitudinal dorsal muscle, the indirecs depressor of the wing; *B*, tergo-sternal muscle, the indirect elevator of the wing; *C*, anterior pleural muscle which deflects the wing forwards; *D*, the pleural process on which the wing can rotate forward or backward; *S*, sternum, *T*, tergum.

The arrangement of the basal articulations in Diptera or Hymenoptera ensures not only that the force of contraction of the indirect muscles is transmitted to the wing but also that the twisting movements of pronation and supination occur at the correct phase of the cycle to produce lift and propulsion. The Coleoptera occupy a position intermediate between Diptera and Hymenoptera on the one hand and Orthoptera and Odonata on the other, in that certain of the direct muscles, besides twisting the wing, also assist in wing depression.[12]

In the relatively weak flying Orthopteron *Oedipoda* the flight muscles comprise only 8 per cent. of the total body weight, but in strong fliers they make up a far greater proportion: *Musca* 11 per cent., *Apis* 13 per cent., *Macroglossa* 14 per cent., *Aeschna* 24 per cent.[39]

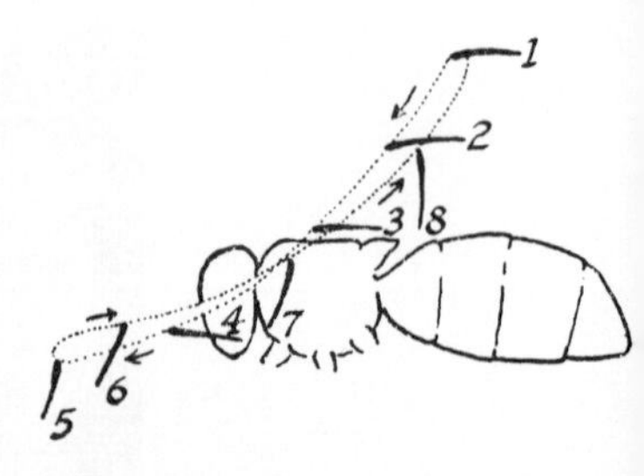

FIG. 100.—Course of the wing tip of a stationary *Volucella* (Dipt.), showing the inclination of the wing at successive points (*after* MAGNAN)

The movements of the wings during flight have been studied by gilding the wing tips so as to make their trajectory visible,[42] by cinematograph,[13, 38, 70] by stroboscopic methods,[15] and by observing the positions of the wings assumed naturally in the dead insect.[66] In the hovering stationary insect the wing tips trace in the air an elongated figure of 8, oblique to the long axis of the body (Fig. 100).[38] In some cases, such as *Eristalis*, the downstroke is performed more rapidly than the upstroke.[70] But when the fly is in motion the wings (e.g. in the blow-fly) describe a series of open loops going downwards and forwards and from below upwards and backwards (Fig. 101).[55]

In the course of this movement the wing rotates on its long axis. During the downstroke the surface of the wing looks downwards, whereas it cuts edgeways through the air to some extent during the upstroke, the posterior area being deflected. This twisting movement is due in part to the structure of the wing, which is rigid in front and flexible behind: the wing of a dragon-fly vibrated in a vacuum jar takes on the rotary movement automatically when air is admitted;[13] but it is chiefly effected by muscular action; for the wing is concave downwards as it descends, and in *Tipula* the same changes in inclin-

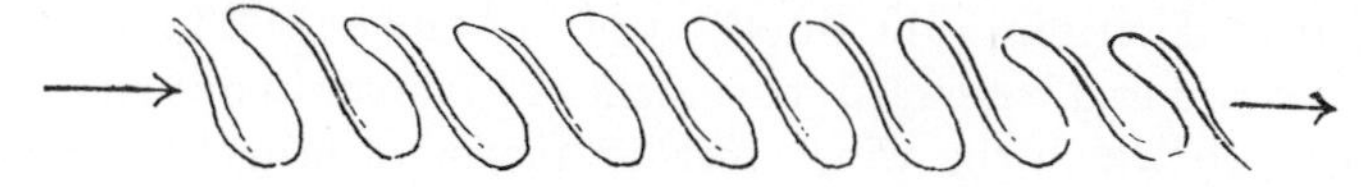

FIG. 101.—The course traced by the wing tip of *Calliphora* during flight (*after* RITTER)

ation during the upstroke and downstroke still take place when the wing is reduced to a short stump.[13]

This reversal of inclination during the elevation and depression of the wing produces the same mechanical effect as the revolution of a propeller blade.[63] Each wing, in fact, acts as a propeller which in the stationary insect (as has been demonstrated with Sphingid moths) draws air from above and in front and drives it backwards in a narrow concentrated stream (Fig. 102). The flying insect thus creates a zone of low pressure above and in front, and a zone of high pressure directly behind it.[18] In large Tabanids it is calculated that nearly 2 litres of air per second are passed through the wings.[105a] The wing-beat cycle is remarkably

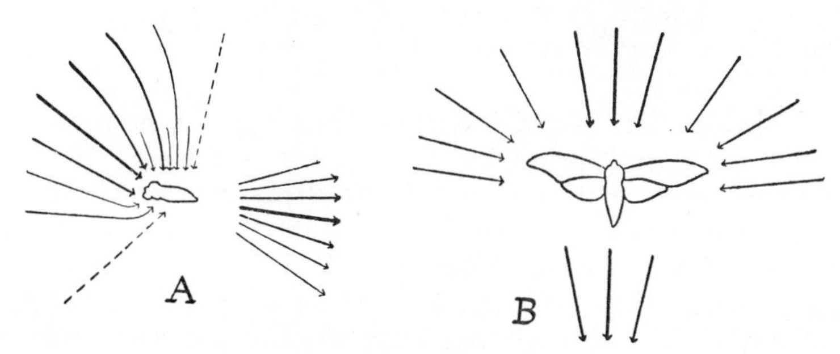

FIG. 102.—Air currents set up by the vibrating wings of a suspended Sphingid *A*, side view; *B*, dorsal view (*after* DEMOLL).

efficient: the flying insect produces a polarized flow of air from front to rear during approximately 85 per cent. of the cycle.[83]

The force of contraction of the flight muscles is largely taken up by the elasticity of the walls of the thorax and the elasticity of the antagonistic muscles, and much of this stored energy becomes available again for moving the wings in the opposite direction.[141] In *Calliphora*, *Sarcophaga*, Coccinellidae and other insects the stiffness of the thorax is reinforced by pleurosternal muscles and may reach the point at which a click mechanism develops: as the wings, erected above the back, are depressed they meet with increasing resistance until the stored elastic energy reaches a certain level; they then click over abruptly into the depressed position as the elastic energy is released. The reverse process occurs as the wings are raised.[92] The manner in which this mechanism evokes the oscillatory contractions of the flight muscles is described on p. 154. [See p. 169.]

Good fliers such as Diptera, Hymenoptera and Lepidoptera can also steer accurately; they can hover, go sideways or backwards, or rotate around the head or the tip of the abdomen. Steering is effected by unequal activity of the wings on the two sides: the fore-wings of *Pieris*, besides their up and down movements, can move forwards and backwards and around their long axis.[69] *Agrion* appears able to modify the inclination of the wing at will; when going upwards and backwards it lowers the body so that the wings move horizontally, and the inclination of the wing then determines the direction of movement.[13] The bee appears to regulate its forward and backward movement or to hover in one spot by varying the plane through which the wing vibrates: the more nearly horizontal is the plane of vibration, the greater is the upward drive and the smaller the forward drive (Fig. 103, B–D). It appears to bring about lateral movements by varying the amplitude of vibration on the two sides. The frequency of vibration always remains the same, but the amplitude on one side may diminish until it ceases altogether. Sideways movements are effected in this way.[66] [See p. 170.]

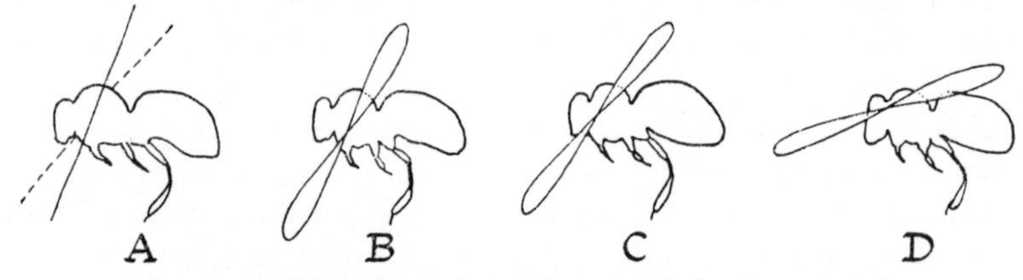

FIG. 103.—Flight in the honey-bee *Apis* (*after* STELLWAAG)

A, a bee turning: the lines show the plane of vibration of the wings on the two sides; *B*, plane of vibration during forward flight; *C*, during hovering; *D*, during backward flight.

Aerodynamics and flight control—Insect flight is probably based on conventional aerodynamic principles even in quite small insects. That has been proved by measurements in the locust *Schistocerca* which is dependent for lift entirely on the wings: changes in inclination of the body have no influence—in contrast to the effect of pitch on an aircraft.[142] Variations in lift are controlled by differences in wing twisting; the characters of the wing strokes, that is, the frequency of beat and the total angle over which the wing is moved, remain remarkably constant. But the power can be controlled by the number of muscle fibres that are brought into action.[150] The hind-wings are responsible for 70 per cent. of the total lift and thrust; about 80 per cent. of the lift being produced during the downstroke.[152] It is remarkable that certain of the wing muscles serve alternatively as leg muscles.[148] The vibrating stimulus of the moving air to the hairs on the front of the head causes the appearance of the characteristic rhythmic pattern of flight movements,[139] which are further modified by feed-back stimuli from a wide variety of sense organs in the wing hinge and on the wing surface.[14] Stretch receptors in the thorax of *Schistocerca*, which are stimulated by flight movements, provide nerve impulses which serve to increase the wing stroke frequency: if all the thoracic stretch receptors are destroyed, wing-stroke frequency is reduced to about half its normal value.[105, 149]

Although in the majority of insects the wings act after the manner of aerofoils, this is possible only if the Reynolds number is greater than 100. In the smallest insect wings the Reynolds number is so reduced that they cannot serve as aerofoils. These very small wings are made of bristles instead of a continuous membrane; they have probably abandoned altogether an aerofoil action and literally 'swim' through the air. This system, which must be wasteful of energy, is used by Mymaridae when they 'fly' under water moving their wings at a frequency of about two beats per second.[108]

Frequency of wing beat—The rate of vibration of the wings varies enormously in different groups. The earliest estimates were derived from the pitch of the sound produced; such values were at one time thought to be unduly high, it being suggested that each wing beat gives rise to two pressure waves;[26] but if overtones are excluded the pitch does give a fine measure of the rate of vibration.[65] The rate has been estimated also by tracing with the wing tip on a revolving kymograph, by the cinematograph or by stroboscopic methods. By adjusting the flash-frequency of an intermittent source of light until it is synchronous with the wing beat, the wings appear to stand still.[15, 73] The following are some of the values obtained, in wing beats per second. HYMENOPTERA: *Apis*, 190,[42] 180–203,[70] 250;[38] *Bombus*, 130,[38] 240;[42, 70] *Vespa*, 110.[42] DIPTERA: Tipulids, 48,[38] 44–73;[70] *Culex*, 278–307;[70] *Tabanus*, 96;[70] *Musca*, 190,[38] 180–97,[70] 330;[42] *Muscina*, 115–220;[29] LEPIDOPTERA: *Pieris*, 9,[70] 12;[38] *Colias*, 8;[51] *Saturnia*, 8;[38] *Macroglossa*, 72,[70] 85;[38] *Acidalia*, 32.[70] COLEOPTERA: *Melolontha*, 46;[38] *Coccinella*, 75–91;[70] *Rhagonycha*, 69–87.[70] ODONATA: *Libellula*, 20;[38] *Aeschna*, 22,[51] 28.[42] The wing-beat frequencies of a great number of insects have been deduced from the auditory tone during flight. The values range from 5–9 cycles per second in large butterflies (*Papilio*) to 587 in the male mosquito (*Aëdes*) and 988–1047 in the Ceratopogonid *Forcipomyia*, in which the flight muscles are enormously developed and the thorax very large in proportion to the rest of the body.[65] It is worth noting that the highest value among birds is in the humming bird, 30–50, which is the range found in most Noctuid moths.

Many factors, such as age, season, sex, humidity, &c., influence the frequency of wing beat. In female Culicids it ranges from 150–350, in males from 450 to nearly 600.[65] In *Drosophila melanogaster* it was 150 per second at 10° C., 250 per second at 37° C., the rate falling with fatigue.[15, 73] The rate is markedly affected by the load; for example, in different races of *Drosophila* species, when the volume of the thoracic muscles remains constant, the wing-beat frequency increases uniformly as the wing size decreases.[52] The resistance of the air plays a considerable part in some insects in determining the amplitude and the frequency of wing beat.[75] But experiments in which pieces are cut from various parts of the wings, or different parts are weighted by the addition of collodion, have shown that the inertia of the wing is often of equal or greater importance.[81] The normal *Tipula* with wings of 22 mm., with wings cut down to 10 mm., and with wings cut down to 5 mm., shows a ratio in frequency of 9 : 12 : 20.[8] If only one side is shortened, the rate of both is determined by the normal wing.[56] (*cf.*[65]) In *Forcipomyia* with the wings cut short and exposed to a high temperature, the wing-stroke frequency reached a value of 2218 per second![136]

It has already been pointed out that these very high rates of movement result from a special type of alternating contraction.[50] The wing-beat frequency in Diptera, for example, is controlled by the moment of inertia of the wings and the basic tension of the indirect muscles. During flight, muscle tension can be varied within certain limits and hence the frequency of the wing beat can also be varied.

There is no intrinsic resonance-frequency for the thorax itself. The halteres likewise have a frequency which depends on the moment of inertia.[97]

It is perhaps significant that some of the more powerful fliers such as Sphingids are unable to take flight without a period of fluttering during which the muscles of the thorax are warmed up above 30° C. (p. 670), whereas if they have been kept in an incubator at 34° C. they can fly straight away.[20] The power required to raise the weight of an insect off the ground rises at the 3·5th power of a linear dimension, whereas the wing area will increase only as the square. A heavier moth will therefore require more work from the flight muscles, giving higher frequency and amplitude of stroke and greater heat production. In *Daphnis nerii* the frequency of wing beat is determined mainly by the wing loading, and the temperature in the thorax increases linearly with the frequency. During the warming-up period before flight there is a steady rise in wing-beat frequency.[98] [See p. 170.]

Very diverse estimates have been made of the velocity of flight achieved by insects; the following are a few of the values which have been obtained. They are expressed in metres per second (13·4 m. per sec. = 30 miles per hour). HYMENOPTERA: *Apis*, 2·5,[38] 2·5–3·7,[18] 6;[24] *Bombus*, 3,[38] 3–5.[18] DIPTERA: *Tabanus*, 4,[38] 14[18]; *Musca* 2;[38] LEPIDOPTERA: *Pieris*, 1·8–2·3,[18] 2·5;[38] *Macroglossa*, 5,[38] Sphingids, 15.[18] COLEOPTERA: *Melolontha*, 2·2–3,[38] 2·5.[38] ODONATA and NEUROPTERA: *Anax*, 8,[38] *Libellula*, 4–10,[18] *Chrysopa*, 0·6.[18] Among these, Sphingids, many butterflies, and locusts can travel several hundred miles without a break.

Equilibrium during flight—Equilibrium in the flying insect has been studied only in the Diptera. The lower Diptera, such as Tipulids, have long thin abdomens and long legs which probably ensure their stability during flight; but

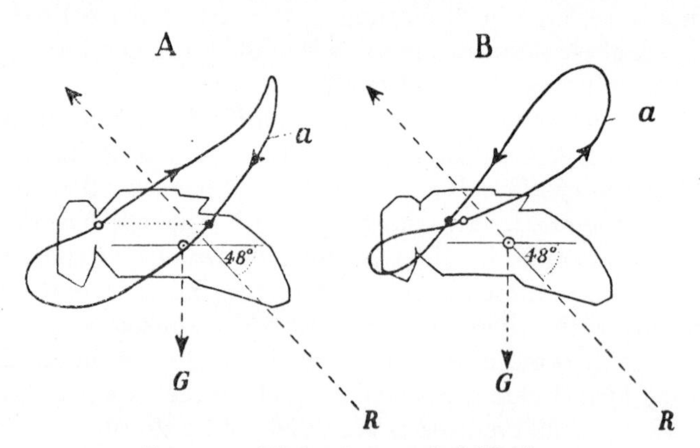

FIG. 104.—Flight of suspended *Muscina*

A, in still air; *B*, in air moving at 140 cm. per sec.; *G*, centre of gravity of the insect; *R*, the resultant of the forces exerted by the beating wings; a, the path travelled by the left wing tip (*after* HOLLICK).

it is otherwise in the more active higher forms such as Muscids. If a fly (*Muscina*) is suspended in still air its wing-beats create a resultant force which is equal to or greater than the weight of the insect. On an average this resultant is inclined forwards and upwards at an angle of 48° to the body axis of the insect; but it usually passes behind the centre of gravity (Fig. 104, A). The insect is therefore unstable and if suddenly released it dives to the ground. The wing tip in still air follows an elliptical course; but on exposure to an air current, which stimulates certain receptor organs in the antennae (p. 263) the wing tip assumes a figure of eight course (Fig. 104, B); the resultant now passes through or in front of the centre of gravity, and if the insect is released it will fly forwards.[29] Thus when the fly is moving through the air it is inherently stable in the antero-posterior or 'pitching' plane, and perhaps in the 'rolling' plane, but it is quite unstable in the horizontal or 'yawing' plane.[49] In this plane the fly maintains its equilibrium actively as the result of impressions received from its sense organs. Of these the most important are the eyes and the halteres. Many flies deprived of their halteres lose their equilibrium and are unable to fly; but the loss may be made good by attaching a thread of cotton to the tip of the abdomen.[21] The halteres serve as gyroscopic sense organs (p. 268) adapted to perceive deviations from the plane of their vibration. From physical considerations it would appear that they are most likely to be stimulated by deviations in the 'yawing' plane where natural stability is least; and flash photographs of the fly deprived of its halteres show that it does in fact follow a spiral course such as would be expected in that case.[49] But there is some evidence that the halteres can aid dynamic equilibrium in other planes also; and they are assisted in this by the eyes[103] and by sense organs responding to air currents.[131]

In the large dragon-fly *Anax* the flying insect maintains its equilibrium by means of three distinct reactions: (i) a dorsal light response (p. 325), (ii) a well marked optomotor reaction to the general visual pattern (p. 322), and (iii) the stimulation of receptors in the neck (the 'dynamic organ') by the inertia of the head (p. 268).[78]

ADDENDA TO CHAPTER IV

p. 149. **Muscle structure**—Comparative studies in *Periplaneta* show that the fast contracting muscles with the greatest endurance have more sarcosomes, more glycogen, a better oxygen supply and higher activity of oxidative enzymes; there are also more post-synaptic vesicles at the neuromuscular junctions.[205] Tonic and phasic muscle fibres from the hind femur of *Schistocerca* have a similar organization at the ultra-structural level, but they differ in the surface area of the sarcoplasmic reticulum and the volume of mitochondria.[171] The access of haemolymph to the muscle fibrils of the flight muscles of *Phormia* by way of deep invaginations of the surface plasma membrane (corresponding to the T-system of other fibres) has been described in detail.[207] In the segmental muscles of *Calliphora*[173] and *Phormia*[196] and in some visceral muscles of *Glossina*[201] the Z-disks are perforated, and during extreme contraction the myosin filaments pass through these holes and overlap the filaments in adjacent sarcomeres. This form of 'supercontraction' allows the muscles to shorten by 76 per cent., much more than the maximum of 50 per cent. attainable by vertebrate muscle.[196] Another interpretation of supercontraction (in *Lethocerus*) amounting to 60 per cent. of the resting length, assumes that the thick filaments decrease in length and increase in diameter.[218] The arrangement of myosin and actin filaments has been studied in detail in flight muscles of *Antheraea*.[164] In the striated visceral muscles of *Ephestia*, *Periplaneta* and *Carausius* each thick (myosin) filament is surrounded by twelve thin (actin) filaments rather than six, as in insect flight muscles and in vertebrate skeletal muscle.[206] In the ventral abdominal muscles of *Rhodnius* there is a constant ratio of thin to thick filaments of 6 : 1.[209] A high ratio of thin to thick myofilaments is said to be correlated with a slow work rhythm.[159] Intersegmental muscles which are of the slow variety have a ratio of 9 or 12 : 1;[173] but this distinction does not seem to be general.[180]

Muscle development—As seen during metamorphosis of *Calliphora*[173] or *Antheraea*[195] myoblasts originating from histoblasts, or from degenerating larval muscle, form the muscle Anlagen or invade the surviving larval muscles. They multiply by mitosis and fuse into long strings which are joined by nerve fibres from the larval muscles. Nerve supply seems necessary for full development and maintenance.[200]

p. 150. **Muscle insertion**—The fine structure of muscle attachments to the cuticle has been studied in a number of insects. The tonofibrillae within the epidermal cytoplasm are composed of fields of microtubules which run to 'conical hemidesmosomes' each situated opposite a pore canal, and from within each cone a 'muscle attachment fibre' extends up the pore canal to gain insertion on the epicuticle.[169, 186]

p. 150. **Transmitter substances**—There is accumulating evidence that the neuromuscular transmitter in insect muscle is L-glutamate, which will induce transient depolarization when applied iontophoretically to certain points on the surface of locust muscles.[163, 211, 212] A factor can be extracted from the nerves innervating the fore-gut and hind-gut of *Periplaneta* which will induce the same slow graded contractions as does stimulation of the nerves.[165] 5-Hydroxytryptamine (serotonin) and noradrenaline (and other catecholamines) affect visceral muscle but not somatic muscle; glutamic acid and Factor S (a paralyzing substance liberated from the nervous system during stress[172]) act only on somatic muscles.[172] A primary catecholamine has been localized in the ingluvial ganglion of *Blaberus*;[170] and four biologically active substances have been isolated from the hind-gut of *Leucophaea* and *Periplaneta*, two of which have been identified as L-glutamic and L-aspartic acids.[179]

p. 151. **Innervation of muscles**—Insect muscles do not show a typical action potential spreading in an all-or-none fashion over the fibre, as in vertebrates. Their electrically excited response is *graded*: it varies in magnitude with the electrical stimulus and it propagates with decrement. The multiterminal or distributed innervation ensures that the entire membrane of insect muscles is depolarized to more or less the same extent during neural stimulation. Na and K ions are not always essential for the electrical response; in *Carausius* and other insects, and in the heart muscle of *Hyalophora*, Mg may be involved in electrogenesis in the fibres.[188, 212]

In *Locusta* and *Schistocerca* a 'common inhibitory neurone' has been discovered which sends branches to most of the nerve trunks that innervate synergic and antagonistic muscles; it may promote more rapid leg movement in walking by facilitating relaxation of the antagonist muscles.[198] In *Periplaneta* the rate of relaxation of the slow muscles is certainly increased by activation of the inhibitory axons supplying them. This effect may well be helpful in the relaxation of slow muscles used in rapid movements.[181]

Insect muscle fibres receive separate axons which do not form synapses in the muscle but contain neurosecretory material (p. 198). These are thought to have some 'trophic' function. In *Phormia* larvae the motor terminals themselves appear to be neurosecretory.[197]

p. 153. **Ions and contraction**—Rigorous removal of calcium ions from the bathing medium abolishes the contraction of locust muscle: Ca ions are involved in the excitation-contraction coupling process[158] presumably in the activation of the adenosine triphosphatase of the actinomyosin system.

p. 154. **Oscillatory contractions of fibrillar muscle**—The fibrillar muscles owe their unusual behaviour to the fact that, unlike vertebrate skeletal muscle, shortening is followed by a delayed fall in tension, and elongation is followed by a delayed rise in tension.[183]

p. 154. **Contraction of visceral muscles**—The contractions of the striated visceral muscles of insects, as in the hind-gut of *Periplaneta*, are also essentially myogenic. Stretching results in depolarization which evokes repetitive action potentials whose frequency and strength depend on the degree of stretching. Depolarization and contraction can be induced also by nerve stimuli ('presynaptic potentials').[193]

p. 156. **Biochemistry of muscle**—During development of the adult *Lucilia* in the pupa there is a progressive increase in the flight muscle mitochondria which is completed after emergence. This goes parallel with the formation of dehydrogenase and oxidase enzymes and in the final state the mitochondrial cristae are covered with the so-called 'elementary particles'.[215] Likewise in the honeybee, there is a period of about 9 days after emergence, before flight is possible, during which the flight muscles and their enzymes develop; indeed sustained flight is not possible until the citric acid cycle (p. 623) is fully functional at 16–20 days after emergence.[160] In the flight muscles of *Calliphora* the specific ATP-ase is confined to the thick filaments in the A bands, that is, presumably the myosin component.[208] But in *Phormia* the splitting of ATP still occurs in the Z-discs after the quantitative extraction of myosin.[217]

Experiments on isolated myofibrils of *Locusta* suggest that granules, capable of acting as a relaxing factor, occur in insect muscle, as they do in vertebrate muscle.[210]

p. 156. The sole phosphagen in various orders and families of insects has proved to be phosphoarginine.[182]

p. 157. **Walking**—In *Periplaneta* the alternating tripod gait is essentially the same over the entire range of speeds of locomotion, except the slowest. As shown by amputations, the timing of leg movements is strongly influenced by feed back from peripheral receptors.[174]

p. 159. **Adhesive organs**—Adhesive organs are of course used for other purposes such as holding on to prey, or the opposite sex in mating. In fleas the inner surfaces of the antennae of the male are adapted to grip the female during pairing.[203]

p. 159. **Leaping**—In Siphonaptera, also, jumping is effected by a spring and catch mechanism, the energy being stored in an elastic pad of resilin (p. 33) between the notum and pleuron; the release of the spring being effected by a second muscle.[161] The resilin ligament in question is derived from the wing-hinge ligament of flying insects.[194] Locusts use a similar mechanism in which the energy is stored in the elastic components of the extensor muscle system.[166]

p. 161. **Flight**—In Coleoptera, *Melolontha* for example, under optimal conditions the elytra may serve to carry 10 per cent. of the body weight.[189]

p. 164. In the flight of *Calliphora* the downstroke of the wing generates most of the lift and only during the mid-part of the stroke does it generate any thrust. During the first half of the upstroke much thrust is generated together with a small amount of lift. The second half produces no useful forces.[190] In Calliphorid flies the flight motor of myogenic muscles is regulated externally by the various direct muscles which modify the linkages between motor and wing. At take-off the lateral muscle stiffens the thorax; the myogenic muscles are activated by nervous stimuli; the 'starter' muscle causes a jump from the ground and begins the oscillation of the thorax. Then the wings are drawn forwards and the full wing-beat develops within the first six strokes. The entire initiation process requires from 30–60 msec.[192]

In *Drosophila* the flight machinery is considerably simpler; the number of variable parameters is greatly reduced. Lift varies directly with the body angle; lift and thrust are adjusted by shifting the horizontal component of the lower extreme wing position; the upper wing position remaining constant.[213] The forward distance travelled during one full wing cycle, just over twice the total span of both wings, is approximately equal to the performance of a conventional propeller.[199] In locusts the effect of controlling inputs is to cause slow changes in the average excitation of individual motor neurons: that is, slow control over average performance, rather than fast control over performance in a particular cycle.[214]

In *Aeschna* the elastic structure used for storage of flight energy is chiefly in the wing muscles; less than 25 per cent. is in the thoracic box and the wing hinges. In *Sphinx* it is mainly in the hard cuticular box, elastic ligaments being small or absent.[216]

p. 164. **Steering in flight**—Turning (or yawing) in *Calliphora* is brought about by the non-fibrillar basalar muscles controlling abduction and adduction of the wings. In the flying insect during yawing these muscles fire at a rate equal to the wing-beat frequency, at which rate they must be in a smooth tetanus.[178] A similar mechanism operates in the Sphingid *Manduca*.[185] In the rhinoceros beetle *Oryctes* during yawing rotations, an increase in the frequency of nervous input to the flight muscles of the appropriate side brings about a greater amplitude of oscillation of these muscles and so leads to a unilateral increase in amplitude of the wing stroke.[167] In *Schistocerca* turning is effected by changing the inclination of the wing on one side;[175] the campaniform sensilla (p. 259) on the wings being necessary for the regulation of this wing twisting.[176] Locusts sense the direction of the wind by tactile receptors on the face (p. 258). Changes in 'wind direction' evoke rotation of the head about the long axis of the body and the resulting stimulation of cervical hair receptors (p. 264) leads to rudder-like movements of the abdomen and legs.[168] Wind-sensitive receptors on the antennae also control flight movements.[177] A great increase in lift is produced if the wing, during the downstroke, shows a rapid change in the angle of incidence. That is particularly the case in the higher Diptera.[162]

Gliding flight is common in *Schistocerca*, usually for a few seconds up to two minutes, but sometimes for prolonged periods.[202] Some butterflies are expert gliders and have adopted the ideal gliding angles, the scales on the wings improving their aerodynamic properties.[191]

p. 166. **Wing muscles and heat production**—During wing vibration in the Sphingid *Mimas* and in Saturniids, some flight muscles and their antagonists are active simultaneously so that much heat is generated with only a small amplitude of wing movement.[184] In *Acilius* this process can go forward with the wings folded.[187]

[1] ARNHARDT, L. *Arch. Bienenkunde*, **5** (1923), (adhesive organs in honey-bee).
[2] BALDWIN, E., and NEEDHAM, D. M. *J. Physiol.*, **80** (1933), 221–37 (phosphorus in resting fly muscle).
[3] BARTH, R. *Zool. Jahrb., Anat.*, **62** (1937), 507–66 (muscles and mechanism of walking in caterpillars).
[4] BENJAMIN, O. J. *C.R. Soc. Biol.*, **116** (1934), 390–1 (chronaxie in *Chironomus*, Dipt. larva).
[5] BHAGVAT, K., and SARMA, P. S. *Ind. J. Med. Res.*, **31** (1943), 173–81 (pyruvic acid after vitamin B_1-deficient diet: *Corcyra*, Lep.).
[6] BILLARD, G., and BRUYANT, G. *C.R. Soc. Biol.*, 1905, 102–3 (locomotion on water surface in *Stenus*, Col.).
[7] BRASS, P. *Zool. Jahrb., Syst.*, **37** (1914), 65–122 (last abdominal segment in Coleoptera larvae).
[8] V. BUDDENBROCK, W. *Arch. ges. Physiol.*, **175** (1919), 125–64 (halteres and rate of wing beat).
[9] —— *Arch. ges. Physiol.*, **185** (1920), 1–6 (tonic contraction of muscles: *Dixippus*, Phasmida).
[10] —— *Handb. norm. path. Physiol.*, **15**, 1 (1930), 88–96 (attitudes of the body in insects, &c.).
[11] —— *Handb. norm. path. Physiol.*, **15**, 1 (1930), 349–61 (insect flight: review).
[12] V. BUDDENBROCK, W., and FRIEDRICH, H. *Zool. Jahrb., Physiol.*, **51** (1932), 131–48 (falling reflex in insects).
[13] BULL, L. *C.R. Acad. Sci.*, **138** (1904), 590–2, 755–7; **149** (1909), 942–4; **150** (1910), 129–31 (insect flight).
[14] CAMERANO, G. *Arch. Ital. Biol.*, **18** (1893), 149 (absolute power of insect muscle).
[15] CHADWICK, L. E. *Physiol. Zool.*, **12** (1939), 151–60; *Psyche*, **46** (1939), 1–18; *Bull. Brooklyn Ent. Soc.*, **35** (1940), 109–12 (wing motion in flight: *Drosophila* and dragonfly).
[16] CREMER, E. *Zool. Jahrb., Physiol.*, **54** (1934), 191–223 (stimulatory physiology: wing muscles of Odonata).
[17] DAHL, F. *Zool. Anz.*, **7** (1884), 38–41; *Arch. Naturgesch.*, **50** (1881), 146–93; *Arch. Mikr. Anat.*, **25** (1885), 236–63 (adhesive organs of insects).
[18] DEMOLL, R. *Der Flug der Insekten und der Vögel*, Jena, 1918.
[19] DEWITZ, H. *Arch. ges. Physiol.*, **33** (1884), 440–81; *Zool. Anz.*, **7** (1884), 400–5 (adhesive organs of insects).
[20] DOTTERWEICH, H. *Zool. Jahrb., Physiol.*, **44** (1928), 399–425 (fluttering before flight in Lepidoptera).
[21] FRAENKEL, G. *Proc. Zool. Soc. Lond.* A, **109** (1939), 69–78 (function of halteres).
[22] FREDERICQ, H. *Arch. internat. Physiol.*, **30** (1928), 300–5 (chronaxie of insect muscles).
[23] FRINGS, H. *Amer. Mid. Nat.*, **34** (1945), 662–72 (falling reflex in caterpillars).
[24] V. FRISCH, K. *Aus dem Leben der Bienen*, Springer, Berlin, 1927.
[25] GILLETT, J. D., and WIGGLESWORTH, V. B. *Proc. Roy. Soc.*, B, **111** (1932), 364–75 (adhesive organ on tibia of *Rhodnius*, Hem.).
[26] HANNES, F. *Biol. Zbl.*, **46** (1926), 128–42 (musical tone and rate of wing beat in honey-bee).
[27] HEIDERMANNS, C. *Zool. Jahrb., Physiol.*, **50** (1931), 1–31 (stimulatory physiology: flight muscles of *Aeschna*, Odonata).
[28] HELLER, J. *C.R. Soc. Biol.*, **121** (1936), 414–16 (phosphorus compounds in *Deilephila*, Lep., pupa and adult).
[29] HOLLICK, F. S. J. *Phil. Trans. Roy. Soc.*, B, **230** (1940), 357–90 (equilibrium in flight: *Muscina*, Dipt.).
[30] HUGHES, G. M. *J. Exp. Biol.*, **29** (1952), 267–84 (co-ordination of locomotion in insects).
[31] HÜRTHLE, K. *Arch. ges. Physiol.*, **126** (1909), 1–164 (structure of striated muscle: *Hydrophilus*).
[32] JORDAN, H. E. *Anat. Rec.*, **16** (1919), 217–45; **19** (1920), 97–124; *Amer. J. Anat.*, **27** (1920), 1–67 (structure of striated muscle in insects).
[33] KAHN, R. H. *Arch. ges. Physiol.*, **165** (1916), 285–336 (physiology of insect muscle).

[34] KIELICH, J. *Zool. Jahrb., Anat.*, **40** (1918), 515–36 (structure of insect muscle).
[35] KORSCHELT, E. *Z. wiss. Zool.*, **150** (1938), 494–526 (muscle insertions).
[36] KRAEMER, F. K. *Zool. Jahrb., Physiol.*, **52** (1932), 86–117 (effect of temperature on form of muscle twitch: *Dytiscus, Lucanus*, Col.).
[37] LAPIQUE, M. *C.R. Soc. Biol.*, **136** (1942), 298–300 (chronaxie of insect muscles).
[38] MAGNAN, A. *Le vol des Insectes*, Hermann et Cie, Paris, 1934.
[39] MAGNAN, A., *et al. C.R. Acad. Sci.*, **195** (1932), 559–61 (relative weight of wing muscles in different insects).
[40] MANGOLD, E. *Z. allg. Physiol.*, **5** (1905), 135–205 (innervation of insect muscle).
[41] MARCHAL, P. *Richet's 'Dictionaire de Physiologie'*, **9** (1910), 273–386 (physiology of insects).
[42] MAREY, U. *La locomotion animale*, Paris, 1901.
[43] MELIN, D. *Uppsala Univ. Årssk.*, 1941, No. 4, 1–247 (insect flight, halteres, &c.).
[44] MONTALENTI, G. *Boll. Inst. Zool. Univ. Roma*, **4** (1927), 133–50 (nerve endings in muscle).
[45] MORISON, G. D. *Quart. J. Micr. Sci.*, **71** (1927), 395–463; 563–651; **72** (1928), 511–26 (muscles of honey-bee).
[46] ORLOV, J. *Z. wiss. Zool.*, **122** (1924), 425–502 (innervation of gut: Lamellicorn larvae).
[47] PACKARD, A. S. *Textbook of Entomology*, Macmillan, London, 1903.
[48] PRINGLE, J. W. S. *J. Exp. Biol.*, **16** (1939), 220–31 (motor mechanism of insect leg).
[49] —— *Phil. Trans. Roy. Soc.*, B, **233** (1948), 347–84 (mechanism of halteres).
[50] —— *J. Physiol.*, **108** (1949) 229–32 (excitation of flight muscles, Diptera).
[51] PROCHNOW, O. *Schröder's Handbuch d. Entomologie*, **1** (1920), 534–69 (mechanism of insect flight).
[52] REED, S. C., WILLIAMS, C. M., and CHADWICK, L. E. *Genetics*, **27** (1942), 349–61 (wing beat in *Drosophila* races).
[53] RICHARDS, A. G., *et al. Proc. Soc. Exp. Biol. Med.*, **51** (1942), 148–52 (striation of insect muscle).
[54] RIJLANT, P. *C.R. Soc. Biol.*, **111** (1932), 631–5 (electrical changes during tonic and voluntary contractions: *Hydrophilus*, Col.).
[55] RITTER, W. *Smiths Misc. Coll.*, **56** No. 12 (1911), 1–76 (mechanism of flight: *Calliphoro*)
[56] ROCH, F. *Biol. Zbl.*, **42** (1922), 359–64 (effect of load on rate of wing beat: *Diptera*).
[57] ROLLET, A. *Sitzgsber. Akad. Wiss. Wien, Math. Nat. Kl.*, **89** (1884), 436; *Denkschr. Akad. Wiss. Wien, Math. Nat. Kl.*, **53** (1887), 193 (physiology of insect muscle).
[58] ROMBOUTS, J. E. *Zool. Anz.*, **7** (1884), 619–23 (adhesive organs of flies).
[59] SARMA, P. S. *Ind. J. Med. Res.*, **31** (1943), 161–3 (pyruvic acid formation after exercise: *Corcyra*, Lep.).
[60] SCHÜTZE, W. *Zool. Jahrb., Physiol.*, **51** (1932), 505–46 (chemical composition of insect muscle).
[61] SIMMERMACHER, G. *Z. wiss. Zool.*, **40** (1884), 481–556 (adhesive organs on tarsal segments of insects).
[62] SNODGRASS, R. E. *Anatomy and Physiology of the Honey-bee*, New York, 1925.
[63] —— *Smithsonian Rept. for 1929*, 1930, 383–421 (How insects fly).
[64] SOLF, V. *Zool. Jahrb., Physiol.*, **50** (1931), 175–264 (stimulatory physiology: Orthoptera muscle).
[65] SOTAVALTA, O. *Acta. Entom. Fenn.*, **4** (1947), 1–177 (wing-beat frequency).
[66] STELLWAAG, F. *Z. wiss. Zool.*, **95** (1910), 518–50; *Biol. Zbl.*, **36** (1916), 30–44 (mechanism of flight and steering in honey-bee).
[67] V. STUDNITZ, G. *Z. Zellforsch. Mikr. Anat.*, **23** (1935), 1–12 (structure of striated muscle in insects).
[68] V. STUDNITZ, G., and BRENNER, H. *Verhandl. vii internat. Kongr. Entom. 1938 Berlin*, **2** (1939), 927–32 (striation of insect muscle).
[69] V. UEXKÜLL, J. *Arch. ges. Physiol.*, **202** (1924), 259–64 (wing movements in *Pieris*, Lep.).
[70] VOSS, F. *Verhandl. deutsch. Zool. Ges.*, **23** (1913), 118–42; **24** (1914), 59–90 (mechanisms of flight in insects).
[71] WEBER, H. *Lehrbuch der Entomologie*, Jena, 1933.

[72] WEST, T. *Trans. Linn. Soc. Lond.*, **23** (1862), 393–421 (adhesive organs on foot of fly, &c.).
[73] WILLIAMS, C. M., and CHADWICK, L. E. *Science*, **98** (1943), 522–4 (stroboscopic studies of insect flight).
[74] WILLIAMS, C. M., and WILLIAMS, M. V. *J. Morph.*, **72** (1943), 589–99 (flight muscles *Drosophila*).

SUPPLEMENTARY REFERENCES (A)

[75] CHADWICK, L. E., *et al. Biol. Bull.*, **97** (1949), 115–37; **100** (1951), 15–27 (air density and wing movements in *Drosophila*).
[76] HANSON, J. *Nature*, **169** (1952), 530–3 (sarcosomes and muscular contraction).
[77] LEVENBOOK, L., and WILLIAMS, C. M. *Anat. Rec.* (suppl.), **111** (1951), No. 158 (mitochondrial cytochrome C and wing-beat frequency of *Phormia*, Dipt., etc.).
[78] MITTELSTAEDT, H. *Z. vergl. Physiol.*, **32** (1950), 422–63 (equilibrium during flight: *Anax*, Odonata).
[79] ROEDER, K. D. *Biol. Bull.*, **100** (1951), 95–106 (potential changes in flight muscles, and wing movements in insects).
[80] ROEDER, K. D., and WEIANT, E. A. *J. Exp. Biol.*, **27** (1950), 1–13 (neuro-muscular transmission in *Periplaneta*).
[81] SOTAVALTA, O. *Nature*, **170** (1952), 1057–8; *Ann. Zool. Soc. Vanamo*, **15** (1952), 1–67 (effect of mutilation and loading of wings on wing-stroke frequency in insects).
[82] WATANABE, M. I., and WILLIAMS, C. M. *J. Gen. Physiol.*, **34** (1951), 675–89; **37** (1953), 71–90 (sarcosomes as mitochondria; enzyme content, &c.).
[83] WILLIAMS, C. M., and GALAMBOS, R. *Biol. Bull.*, **99** (1950), 300–7 (flight sounds and wing-beat frequency in *Drosophila*).

SUPPLEMENTARY REFERENCES (B)

[84] AUBER, J. *Z. Zellforsch.*, **51** (1960), 705–24 (dual innervation of insect muscles).
[85] —— *J. Microsc.*, **2** (1963), 325–36 (fine structure of muscle insertion).
[86] AUBER, J., and COUTEAUX, R. *J. Microsc.*, **2** (1963), 309–24 (ultra-structure of fibrillar muscle in Diptera).
[87] BARTH, R. *Zool. Jahrb., Anat.*, **69** (1945), 405–34 (insertion of tonofibrillae into the epicuticle in caterpillars).
[88] BAUDOIN, R. *Bull. Biol. Fr. Belg.*, **89** (1955), 16–164 (surface phenomena and insects living on the water surface).
[89] BECHT, G., HOYLE, G., and USHERWOOD, P. N. R. *J. Ins. Physiol.*, **4** (1960), 191–210 (neuro-muscular transmission in *Periplaneta*).
[90] BHAKTHAN, N. M. G., *et al. J. Anim. Morph. Physiol.*, **9** (1962), 143–51; **10** (1963), 56–62 (histochemistry of enzymes in insect muscle).
[91] BISHAI, F. R., and ZEBE, E. *Zool. Anz.*, **23** (1959), Suppl. 314–19 (enzymes and metabolism in muscles of *Locusta*).
[92] BOETTIGER, E. G. In *Recent Advances in Invertebrate Physiology*, Symposium, Oregon Univ. (B. T. Scheer, Ed.) 1957, 117–42; *Ann. Rev. Ent.*, **5** (1960), 53–68 (insect flight muscles and flight mechanism).
[93] BOETTIGER, E. G., and FURSHPAN, E. *Biol. Bull.*, **102** (1952), 200–11 (mechanics of flight in Diptera).
[94] BROOKS, M. A. *Ann. Ent. Soc. Amer.*, **50** (1957), 122–5 (succinoxidase in muscles of *Periplaneta*).
[95] BUCHTHAL, F., WEIS-FOGH, T., and ROSENFALCK, P. *Acta Physiol. Scand.*, **39** (1957), 246–76 (contractions in isolated flight muscles of locusts).
[96] CERF, J. A., GRUNDFEST, H., *et al. J. Gen. Physiol.*, **43** (1959), 377–95 (fast and slow contractions in muscles of *Romalea*).
[97] DANZER, A. *Z. vergl. Physiol.*, **38** (1956), 259–83 (role of resonance in flight mechanism of Diptera).
[98] DORSETT, D. A. *J. Exp. Biol.*, **39** (1962), 579–88 (preparation for flight in Sphingidae).
[99] EDWARDS, G. A., and RUSKA, H. *Quart. J. Micr. Sci.*, **96** (1955), 151–9 (fine structure and metabolism of insect muscle).
[100] EDWARDS, G. A., RUSKA, H., and HARVEN, E. *J. Biophys. Biochem. Cytol.*, **4** (1958), 251–6 (neuromuscular junctions in fibrillar muscles of the cicada).

[101] EDWARDS, J. S. *Proc. Roy. Ent. Soc. Lond.*, A, **37** (1962), 89–98 (adhesive organs of *Platymeris*).

[102] EWER, D. W., and RIPLEY, S. H. *J. Exp. Biol.*, **30** (1953), 170–7 (properties of flight muscles of *Locusta*).

[103] FAUST, R. *Zool. Jahrb., Physiol.*, **63** (1952), 326–66 (function of halteres).

[104] GEORGE, J. C., *et al. Experientia*, **14** (1958), 250; *J. Exp. Biol.*, **37** (1960), 308–15; *Nature*, **192** (1961), 356; *J. Anim. Morph. Physiol.*, **7** (1960), 141–9; **10** (1963), 47–62 (lipase and other enzymes in insect muscle).

[105] GETTRUP, E. *J. Exp. Biol.*, **40** (1963), 323–33 (stretch receptors and locust flight).

[105a] HOCKING, B. *Trans. R. Ent. Soc. Lond.*, **104** (1953), 223–345 (range and speed of flight in insects).

[106] HANSON, J. *Nature*, **169** (1952), 530–3; *Biochim. Biophys. Acta*, **20** (1956), 289–92 (fine structure and contractile mechanism of insect muscle).

[107] HOFFMEISTER, H. *Z. Zellforsh.*, **54** (1961), 402–20; **56** (1962), 809–18 (changes in fine structure of mitochondria in exhausted flight muscles of *Vespa*).

[108] HORRIDGE, G. A. *Nature*, **178** (1956), 1334–5 (flight of very small insects).

[109] HOYLE, G. *J. Physiol.*, **127** (1955), 90–103 (effects of ions on neuromuscular transmission in *Locusta*).

[110] —— *Proc. Roy. Soc.*, B, **143** (1955), 281–92; 343–67 (neuromuscular mechanisms in *Locusta*).

[111] —— *Comparative physiology of the nervous control of muscular contraction*. Cambridge Univ. Press, 1957.

[112] HUGHES, G. M. *J. Exp. Biol.*, **35** (1958), 567–83 (co-ordination of swimming in *Dytiscus*, *Hydrophilus*, &c.).

[113] JENKINS, M. F. *Trans. Roy. Ent. Soc. Lond.*, **110** (1958), 287–301 (anal glands of *Dianous* (Staphylinidae)).

[114] KRAMER, S. *Proc. Xth Int. Cong. Ent.*, **1** (1958), 569–79 (sexual difference in flight muscles of *Periplaneta*).

[115] LEVENBOOK, L. *J. Histochem. Cytochem.*, **1** (1953), 242–7 (mitochondria in insect flight muscle).

[116] LEVENBOOK, L., and WILLIAMS, C. M. *J. Gen. Physiol.*, **39** (1956), 497–512 (cytochrome in mitochondria of flight muscles in *Phormia*).

[117] LEWIS, S. E., and FOWLER, K. S. *Nature*, **194** (1962), 1178–9 (phosphoarginine in flight muscle of *Calliphora*).

[118] MCENROE, W. Quoted by Boettiger (1960).

[119] MACHIN, K. E., and PRINGLE, J. W. S. *Proc. Roy. Soc.*, B, **151** (1959), 204–25; **152** (1960), 311–30 (fibrillar flight muscle in *Oryctes*, Col.).

[120] MARUYAMA, K., *et al. Z. vergl. Physiol.*, **39** (1956), 21–4; **40** (1957), 451–3, 543–8; *J. Cell. Comp. Physiol.*, **51** (1958), 173–87; *J. Ins. Physiol.*, **3** (1959), 271–92 (insect actomyosin and ATP).

[121] NACHTIGALL, W. *Nature*, **190** (1961), 224–5 (hydrodynamics of *Acilius* Col. during swimming).

[122] NEVILLE, A. C. *J. Exp. Biol.*, **37** (1960), 631–56 (flight mechanics of Anisoptera).

[123] NEVILLE, A. C., and WEIS-FOGH, T. *J. Exp. Biol.*, **40** (1963), 111–21 (physiology of flight muscles in *Schistocerca*).

[124] NOBLE-NESBITT, J. *J. Exp. Biol.*, **40** (1963), 681–700 (wetting properties of cuticle in *Podura aquatica*).

[125] —— *Quart. J. Micr. Sci.*, **104** (1963), 369–91 (fine structure of muscle insertions in *Podura*).

[126] PRINGLE, J. W. S. *J. Physiol.*, **124** (1954), 269–91; in *Recent advances in invertebrate physiology* (B. T. Scheer, Ed.), Univ. Oregon, 1957, 99–115 (myogenic rhythms in fibrillar muscle).

[127] —— *Insect flight*, Cambridge Univ. Press, 1957.

[128] RUSKA, H., EDWARDS, G. A., and CAESAR, R. *Experientia*, **14** (1958), 117–20 (transmission of excitation in muscle).

[129] RUSSENBERGER, H. and M. *Mitt. Naturf. Ges. Schaffhausen*, **27** (1960), 2–88 (flight apparatus of *Aeschna* as a mechanical resonant system).

[130] SACKTOR, B. *Proc. IVth Int. Congr. Biochem.*, **12** (1959), (biochemistry of insects) 138–52; *Ann. Rev. Ent.*, **6** (1961), 103–30 (biochemistry of insect flight).

[131] SCHNEIDER, G. *Verh. dtsch. Zool. Ges.*, (1951), 195–9; *Z. vergl. Physiol.*, **35** (1953), 416–58 (function of halteres in *Calliphora*).

[132] SCHWALBACH, G., and AGOSTINI, B. *Z. Zellforsch.*, **61** (1964), 855–70; **62** (1964), 113–20 (swelling of mitochondria in exhaustion and chill-coma).

[133] SHAFIQ, S. A. *Quart. J. Micr. Sci.*, **105** (1964), 1–6 (fine structure of nerve endings in flight muscle of *Drosophila*).

[134] SMITH, D. S. *J. Biophys. Biochem. Cytol.*, **8** (1960), 447–66 (innervation of fibrillar muscle in *Tenebrio*); **10** (1961), 123–58 (structure of fibrillar muscle); **11** (1961), 119–45 (flight muscle in *Aeschna*); *Rev. Canad. Biol.*, **21** (1962), 279–301 (sarcoplasmic reticulum in insect muscle); *J. Cell. Biol.*, **19** (1963), 115–38 (flight muscle sarcosomes in *Calliphora*).

[135] SMITH, D. S., and TREHERNE, J. E. *Adv. Ins. Physiol.*, **1** (1963), 401–84 (neuromuscular junction in insects: review).

[136] SOTAVALTA, O. *Biol. Bull.*, **104** (1953), 439–44 (wing-stroke frequency in *Forcipomyia*).

[137] TIEGS, O. W. *Phil. Trans. Roy. Soc.*, B, **238** (1955), 221–359 (histology of flight muscles).

[138] USHERWOOD, P. N. S., and GRUNDFEST, H. *Science*, **143** (1964), 817–18 (inhibitory nerve axons to grasshopper muscle).

[139] WEIS-FOGH, T. *Nature*, **164** (1949), 873 (aerodynamic sense organ in locusts).

[140] —— *J. Exp. Biol.*, **33** (1956), 668–86 (tetanic contraction and passive stretching in flight muscles of *Schistocerca*).

[141] —— *Proc. XVth Int. Congr. Zool.*, 1958 (1959), 383–5 (elasticity of the thorax in the flying locust).

[142] WEIS-FOGH, T., and JENSEN, M. *Phil. Trans. Roy. Soc.*, B, **239** (1956), 415–584 (biology and physics of locust flight).

[143] WIGGLESWORTH, V. B. *Quart. J. Micr. Sci.*, **97** (1956), 465–80 (cycles of muscle development and breakdown in *Rhodnius*).

[144] —— *Quart. J. Micr. Sci.*, **99** (1958), 441–50 (cholinesterase in *Rhodnius*).

[145] —— *Quart. J. Micr. Sci.*, **100** (1959), 285–98 (peripheral nervous system in *Rhodnius*).

[146] —— *New Scientist*, **8** (1960), 101–4 (fuel and power in flying insects).

[147] WILSON, D. M. *J. Exp. Biol.*, **38** (1961), 471–90 (nervous control of flight in locusts).

[148] —— *J. Exp. Biol.*, **39** (1962), 667–77 (combined leg and wing muscles in locusts).

[149] WILSON, D. M., and GETTRUP, E. *J. Exp. Biol.*, **40** (1963), 171–85 (stretch reflex controlling wing-beat in *Schistocerca*).

[150] WILSON, D. M., and WEIS-FOGH, T. *J. Exp. Biol.*, **39** (1962), 643–67 (control of wing-beat in *Schistocerca*).

[151] WILSON, D. M., and WYMAN, R. J. *J. Ins. Physiol.*, **9** (1963), 859–65 (nervous control of wing-beat frequency in Diptera).

[152] WILSON, V. J. *J. Exp. Biol.*, **31** (1954), 280–90 (slow and fast responses in *Periplaneta* leg muscle).

[153] WINTERINGHAM, F. P. W., *et al. Biochem. J.*, **59** (1955), 13–21 (energy sources in flight muscle of *Musca*).

[154] WOOD, D. W. *J. Exp. Biol.*, **35** (1958), 850–61 (electrical responses in muscles of *Carausius*).

[155] ZEBE, E. *Biochem. Z.*, **332** (1960), 328–32; *Umschau*, 1960, 40–43; *Ergebn. Biol.*, **24** (1961), 248–86 (energy metabolism in insect muscle).

[156] ZEBE, E., and FALK, H. *Exp. Cell. Res.*, **31** (1963), 340–4; *Z. Naturforsch.*, **18b** (1963), 502–3 (localization of apyrase in insect flight muscle).

[157] ZEBE, E., and MCSHAN, W. H. *J. Cell. Comp. Physiol.*, **53** (1959), 21–9 (trehalase in wing muscles of *Leucophaea*).

SUPPLEMENTARY REFERENCES (C)

[158] AIDLEY, D. J. *J. Physiol.*, **117** (1965), 94–102 (calcium ions and contraction of locust muscle).

[159] AUBER, J. *J. Microscopie*, **5** (1966), 28; *Am. Zool.* **7** (1967), 451–6 (fine structure of myofilaments in insects).

[160] BALBONI, E. R. *J. Insect Physiol.*, **13** (1967), 1849–56 (maturation of flight muscles in honey-bee).

[161] BENNET-CLARK, H. C. and LUCEY, E. C. A. *J. exp. Biol.*, **47** (1967), 59–76 (mechanism of jumping in the flea).

[162] BENNETT, L. *Science*, **167** (1970), 177–9 (achievement of lift in *Schistocerca* flight).

[163] BERÁNEK, R. and MILLER, P. L. *J. exp. Biol.*, **49** (1968), 83–93 (L-glutamate as neuromuscular transmitter).

[164] BIENZ-ISLER, G. *Acta anat.*, **70** (1968), 416–33; 524–53 (fine structure of fibrils in flight muscle of *Antheraea*).

[165] BROWN, B. E. *Science*, **155** (1967), 595–7 (neuromuscular transmitter in visceral muscle of *Periplaneta*).

[166] BROWN, R. H. J. *Nature*, **214** (1967), 939 (mechanism of locust jumping).

[167] BURTON, A. J. *Nature*, **204** (1964), 1333; *J. exp. Biol.*, **54** (1971), 575–85 (control of turning by *Oryctes* in flight).

[168] CAMHI, J. M. *J. exp. Biol.* **52** (1970), 519–31 (correction of yaw by postural changes in flying locusts).

[169] CAVENEY, S. *J. Cell Sci.*, **4** (1969), 541–59 (muscle attachment to cuticle in Apterygota).

[170] CHANUSSOT, B. *et al. C.R. Acad. Sc. Paris*, **268** (1969), 2101–4 (catecholamine in ingluvial ganglion of *Blaberus*).

[171] COCHRANE, D. G., ELDER, H. Y. and USHERWOOD, P. N. R. *J. Physiol.*, **200** (1968), 68–9 (fine structure of tonic and phasic muscles in *Schistocerca*).

[172] COOK, B. J. *et al. Biol. Bull. mar. biol. Lab., Woods Hole*, **133** (1967), 526–38; *J. Insect Physiol.*, **15** (1969), 445–55; 963–75 (Factor S and biologically active amines in *Periplaneta* and *Blaberus*).

[173] CROSSLEY, A. C. *J. Insect Physiol.*, **14** (1968), 1389–1407 (fine structure of intersegmental muscles in *Calliphora* larva).

[174] DELCOMYN, F. *J. exp. Biol.*, **54** (1971), 443–69 (peripheral receptors and walking in *Periplaneta*).

[175] DUGARD, J. J. *J. Insect Physiol.*, **13** (1967), 1055–63 (directional change in flying locusts).

[176] GETTRUP, E. *J. exp. Biol.*, **44** (1966), 1–16 (sensory regulation of wing twisting in locusts).

[177] GEWECKE, M. *Nature*, **225** (1970), 1263–4 (antennae as wind-sensitive receptors in locusts).

[178] HEIDE, G. *Z. vergl. Physiol.*, **59** (1968), 456–60 (steering in flight of *Calliphora*).

[179] HOLMAN, G. M. and COOK, B. J. *J. Insect Physiol.*, **16** (1970), 1891–1907 (neuromuscular transmitters from hind-gut of *Leucophaea*).

[180] HUDDART, H. and OATES, K. *J. Insect Physiol.*, **16** (1970), 1467–83 (fine structure of insect muscle).

[181] ILES, J. F. and PEARSON, K. G. *J. exp. Biol.*, **55** (1971), 151–64 (peripheral inhibition in *Periplaneta*).

[182] JESO, F. di, *et al. C.R. Soc. Biol.*, **159** (1965), 809; 1112 (phosphagen in insect muscle).

[183] JEWELL, B. R. and RÜEGG, J. C. *Proc. R. Soc. Lond.* (*B*) **164** (1965), 428–59 (oscillatory contraction of fibrillar muscle).

[184] KAMMER, A. E. *J. exp. Biol.*, **48** (1968), 89–109 (motor patterns during flight and warm-up in Lepidoptera).

[185] KAMMER, A. E. *J. Insect Physiol.*, **17** (1971), 1073–86 (control of turning in *Manduca* during flight).

[186] LAI-FOOK, J. *J. Morph.*, **123** (1967), 503–28 (fine structure of muscle insertions in *Rhodnius* and *Calpodes*).

[187] LESTON, D. PRINGLE, J. W. S., and WHITE, D. C. S. *J. exp. Biol.*, **42** (1965), 409–14 (muscle activity during warm-up in *Acilius*, Col.).

[188] McCANN, F. V. *Comp. Biochem. Physiol.*, **12** (1964), 117–23 (conduction in myocardium of *Hyalophora*).

[189] NACHTIGALL, W. *Verhdl. dtsch. Zool. Ges., Kiel* 1964, 319–26 (aerodynamics of flight in Coleoptera).

[190] NACHTIGALL, W. *Z. vergl. Physiol.*, **52** (1966), 155–211 (biophysics of flight in *Calliphora*).

[191] NACHTIGALL, W. *Z. vergl. Physiol.*, **54** (1967), 210–31 (gliding flight in butterflies).

[192] NACHTINGALL, W. *et al. Z. vergl. Physiol.*, **47** (1967), 77–97; **61** (1968), 1–20 (neuromuscular control of flight in *Calliphora*).

[193] NAGAI, T. *et al. J. Insect Physiol.*, **15** (1969), 2151–67; **16** (1970), 437–48 (contraction of visceral muscles in *Periplaneta*).

[194] NEVILLE, A. C. and ROTHSCHILD, M. *Proc. Roy. ent. Soc. Lond. (C)*, **32** (1967), 9–10 (jumping mechanism in the flea).

[195] NUESCH, H. *Ann. Rev. Ent.*, **13** (1968), 27–43 (muscle development and regeneration: review).

[196] OSBORNE, M. P. *J. Insect Physiol.*, **13** (1967), 1471–82 (super-contraction in muscles of *Phormia*).

[197] OSBORNE, M. P. *Symp. Roy. ent. Soc. Lond.*, **5** (1970), 77–100; *Z. Zellforsch.*, 116 (1971), 391–404 (neuromuscular junctions and stretch receptors: review).

[198] PEARSON, K. G. and BERGMANN, S. J. *J. exp. Biol.*, **50** (1969), 445–71 (inhibitory motor neurones in *Periplaneta* and *Locusta*).

[199] PRINGLE, J. W. S. *Adv. Insect Physiol.*, **5** (1968), 163–227 (comparative physiology of the flight motor; review).

[200] RANDALL, W. C. *J. Insect Physiol.*, **16** (1970), 1927–43 (nerve supply and muscle regeneration in *Galleria*).

[201] RICE, M. J. *J. Insect Physiol.*, **16** (1970), 1109–22 (super-contraction in visceral muscles of *Glossina*).

[202] ROFFEY, J. *Anim. Behav. London*, **11** (1963), 359–66 (gliding flight in locusts).

[203] ROTHSCHILD, M. and HINTON, H. E. *Proc. Roy. ent. Soc. Lond. (A)* **43** (1968), 105–7 (holding organs on antennae of male fleas).

[204] SACKTOR, B. *Adv. Insect Physiol.*, **7** (1970), 267–347 (control mechanisms in insect flight muscle: review).

[205] SMIT, W. A., BECHT, G. and BEENAKKERS, A. M. T. *J. Insect Physiol.*, **13** (1967), 1857–68 (physiology of fast contracting muscles of *Periplaneta*).

[206] SMITH, D. S., GUPTA, B. L. and SMITH, U. *J. Cell Sci.*, **1** (1966), 49–57 (fine structure of insect visceral muscle).

[207] SMITH, D. S. and SACKTOR, B. *Tissue & Cell*, **2** (1970), 355–74 (plasma membrane in flight muscle of *Phormia*).

[208] TICE, L. W. and SMITH, D. S. *J. Cell Biol.*, **25** (1965), 121–35 (localization of ATP-ase in flight muscles of *Calliphora*).

[209] TOSELLI, P. A. and PEPE, F. A. *J. Cell Biol.*, **37** (1968), 445–81 (fine structure and development of intersegmental muscles in *Rhodnius*).

[210] TSUKAMUTO, M., NAGAI, Y. *et al. Comp. Biochem. Physiol.*, **17** (1966), 569–81 (relaxing granules in muscles of *Locusta*).

[211] USHERWOOD, P. N. R. *et al. Nature*, **210** (1966), 634–6; **219**, (1968), 1169–72; **223** (1969), 411–13; *J. exp. Biol.*, **49** (1968), 341–61 (glutamate as neuromuscular transmitter in insects).

[212] USHERWOOD, P. N. R. *Adv. Insect Physiol.*, **6** (1969), 205–75 (electrochemistry of insect muscle: review).

[213] VOGEL, S. *J. exp. Biol.*, **44** (1966), 567–78; **46** (1967), 383–92 (flight in *Drosophila*).

[214] WALDRON, I. *J. exp. Biol.*, **47** (1967), 201–28 (control of flight muscle activity in locusts).

[215] WALKER, A. C. and BIRT, L. M. *J. Insect Physiol.*, **15** (1969), 305–17; 519–27 (development of respiratory enzymes and mitochondria in flight muscle of *Lucilia*).

[216] WEIS-FOGH, T. Proc. XII int. Congr. Ent., 1965, 186–8 (elasticity and wing movements in insects).

[217] ZEBE, E. *et al. Histochemie*, **4** (1964), 161–80; *Experientia*, **22** (1966), 96 (sites of splitting of ATP in flight muscle of *Phormia*).

[218] ZEBE, E., MEINRENKEN, W. and RÜEGG, J. C. *Z. Zellforsch.*, **87** (1968), 603–21 (super-contraction of flight muscles of *Lethocerus*, Hemipt.).

Chapter V

Nervous and Endocrine Systems

THE NERVOUS system is composed of excessively elongated cells which transmit electrical disturbances, or impulses, from one part of the body to another. These nerve cells, or 'neurones', are derived in the course of development from the ectoderm. Each consists of a nucleated cell body and a long filament, or 'axon' (Fig. 105). The axon generally gives off a lateral branch or 'collateral' near its origin and both axon and collateral end in fine branching fibrils, the 'terminal arborization'. Similar fibrils arising from the nerve cell body form the 'dendrites'. Where the bundles of axon filaments run freely through the body they constitute the nerves. But the greater part of the nerve cells and their processes are massed in a series of segmental ganglia, united by longitudinal connectives, which constitute the central nervous system. Sensory or afferent neurones convey impulses inwards from the sense organs; motor or efferent neurones convey impulses outwards to the muscles, glands, &c.; and association neurones link the sensory and motor neurones together within the central nervous system.

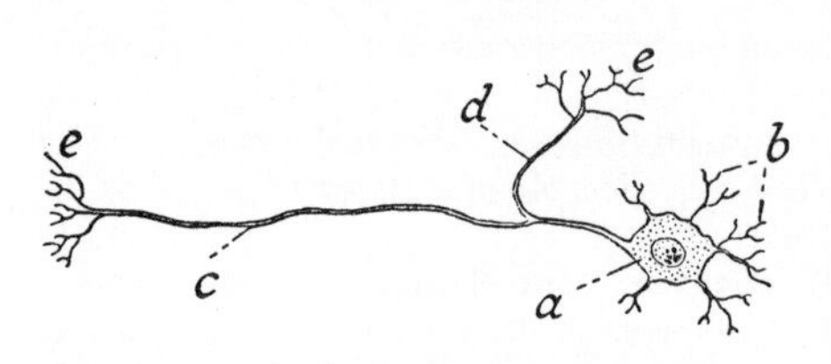

FIG. 105.—Diagram of neurone (*after* SNODGRASS)

a, nerve cell body or neurocyte; *b*, dendrites; *c*, axon; *d*, collateral; *e*, terminal arborizations.

Neurones of the nervous system—The *sensory neurones* have their cell bodies situated near the periphery of the body. They are generally bipolar, a distal process, or dendrite, running to a sense organ adapted to receive some particular type of stimulus (p. 256), and a proximal process, or axon, running to the central nervous system (Fig. 106, *b*). In *Rhodnius*[186] and in *Limnephilus* (Trichopt.)[166] the sense cells, along with the other cells which make up the sensilla (p. 84), are differentiated from the ordinary epidermal cells during post-embryonic development. The sense cell itself gives off an axon process which grows inwards along with some existing nerve to make connection with the central nervous system.[186] If these axons are cut they will regenerate, and sometimes, when they fail to meet another nerve, they may continue growing in the form of a 'circular nerve' with no connection to the central nervous system.[186] (Fig. 107) The proper development of the nerve centres in the brain is dependent on the ingrowing processes from the antennal sense cells and compound eyes.[156, 15] During post-embryonic growth in *Oncopeltus* the 'neuropile' (the central mass of the ganglion, devoid of nerve cell bodies) (Fig. 108, *e*) increases in size 24-fold whereas the cellular cortex increases only sixfold;[144] and the same change is seen during pupal development in *Drosophila*.[160] Perhaps this results from the in growth of immense numbers of new sensory axons. In the antennae there is extensive fusion of the sensory axons as they enter the antennal nerve: a fusion of perhaps 15 : 1 in the antenna of *Rhodnius*,[188] many hundredfold in the antennal flagellum of *Phormia*,[127] and 4 : 1 in the labella of *Calliphora* and *Phormia*;[176] but there is no fusion in the sensory nerves from the limbs.[188] [See p. 202.]

There are also sensory neurones of a second type. The cell bodies of these are located on the inner surface of the body wall and on the wall of the alimentary canal. They are either bipolar or multipolar and their distal processes, which are often numerous and finely branching, run to the epidermis and somatic and visceral muscles, while their central axons run to the ganglia of the central nervous system (Fig. 106, *c*). Neurones of this type are particularly abundant in soft-skinned larvae, where they anastomose to form a rich nerve plexus beneath the epidermis, the terminal branches of which seem to end freely on the basement membrane.

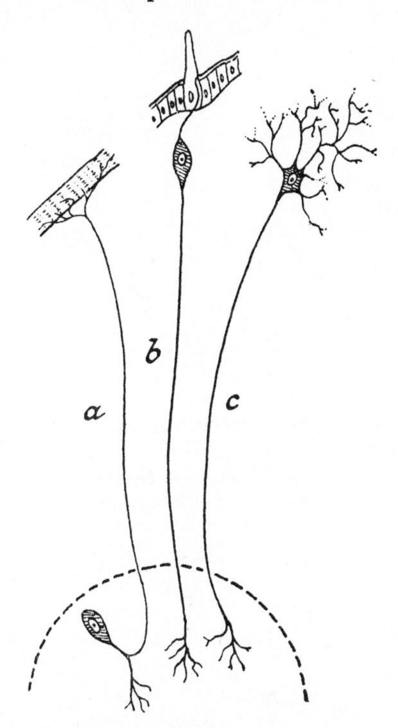

FIG. 106.—Diagram showing the types of neurones which compose the peripheral nerves

The broken line indicates the limit of the central nervous system. **a**, motor neurone supplying muscle fibre, with unipolar cell body in the central ganglion; **b**, sensory neurone type i, with bipolar cell body sending an unbranched process to sense organ; **c**, sensory neurone type ii, with multipolar cell body giving off branched processes to various tissues.

The cell bodies of the *motor neurones* are situated in the central nervous system. They are unipolar, and consist of a large pyriform cell body, devoid of dendrites, located in the peripheral part of the ganglion and connected by a stalk to the 'neuropile' (Fig. 106, *a*). The stalk divides into a collateral and an axon filament. The terminal arborizations of the collateral are connected with those of association neurones or sensory neurones. The bundles of axon filaments constitute the motor nerves. These, as we have seen (p. 150), branch around the muscle fibres and end either in Doyère's hillocks or in fine superficial ramifications. As a rule, a given nerve will contain both sensory and motor fibres.

The *association neurones* are similarly placed in the outer parts of the ganglia. Most of them have small cell bodies with nuclei rich in chromatin. Their massed nuclei form the 'globuli' of the outer layers of the ganglia. Some, like the motor neurones, are very large and are connected to 'giant axons' which may be as much as 4·5μ in diameter, and which may run the whole length of the nerve cord (p. 188). It has been suggested that in *Periplaneta* the giant ascending nerve fibres may be composite in origin and made up of fused axons from cell bodies in different ganglia.[138, 153]

Fig. 108 shows schematically the chief elements in a typical ganglion as studied in the nymph of *Aeschna*.[94] Each ganglion contains: (i) Motor neurones, with their cell bodies, collaterals, and axons; (ii) sensory axons from the periphery; (iii) association neurones of the transverse commissure communicating across the nerve cord or uniting the motor and sensory elements on their own side; (iv) association neurones of the longitudinal connectives, with cell bodies in the lateral part of the ganglion and processes running through the connectives to neighbouring ganglia; (v) neurones of the unpaired ventral nerve. [See p. 202.]

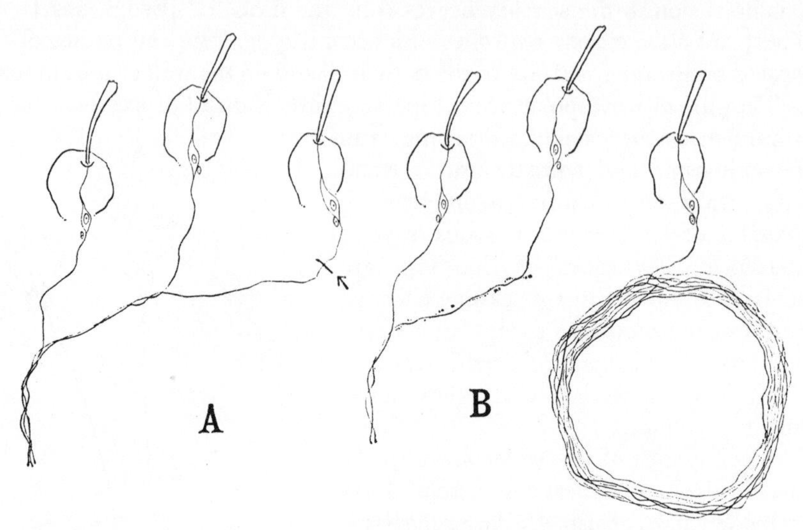

FIG. 107.—Schematic figure showing regeneration of a sensory nerve axon (*after* WIGGLESWORTH)

A, three sensory hairs with their axons forming a small nerve. The arrow shows point of interruption of one axon by a burn. *B*, the same after repair. The outgrowing axon has formed an annular nerve with no connexion with the central nervous system.

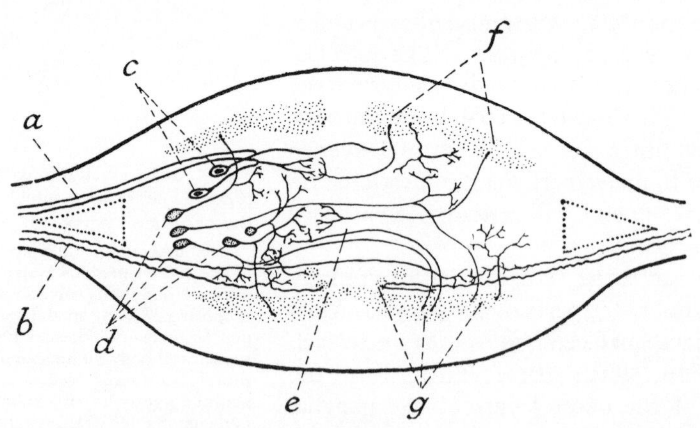

FIG. 108.—Diagrammatic cross-section of abdominal ganglion of an *Aeschna* nymph (*after* ZAWARZIN)

a, dorsal root carrying motor axons; *b*, ventral root carrying sensory axons; *c*, cell bodies of motor-neurones; *d*, cell bodies of association neurones; *e*, central neuropile; *f*, dorsal fibre tracts of longitudinal commissures; *g*, ventral fibre tracts of longitudinal commissures.

Each nerve arises by two roots, one conveying the thicker fibres to the dorsal part of the ganglion, the other carrying the finer fibres to the ventral parts. The dorsal root appears to contain the motor fibres: pressure on the dorsum of the ganglion in *Dytiscus*, or pricking in this region, causes motor paralysis without anaesthesia, while the same treatment of the ventral parts causes anaesthesia without loss of movement.[12] And in beetles with immovable wings (*Blaps*, *Timarcha*, *Carabus auratus*) the alar nerve from the second thoracic ganglion is reduced, and only the ventral, presumably sensory, root persists.[12] Electrical recordings in *Anax* have confirmed that the dorsal roots are motor and the ventral sensory.[130]

When nerves are cut the nerve fibres regenerate from both the cut ends: motor nerves from the proximal stump, sensory nerves from the distal stump; so that within a few weeks both motor and sensory functions are restored.[119, 189] Indeed, regeneration can take place within the brain and lead to the recovery of quite complicated responses previously eliminated by cuts through the ganglion.[128]

Anatomical arrangement of ganglia—Typically, each body segment has a pair of ganglia connected across the mid-line and joined to the ganglia of

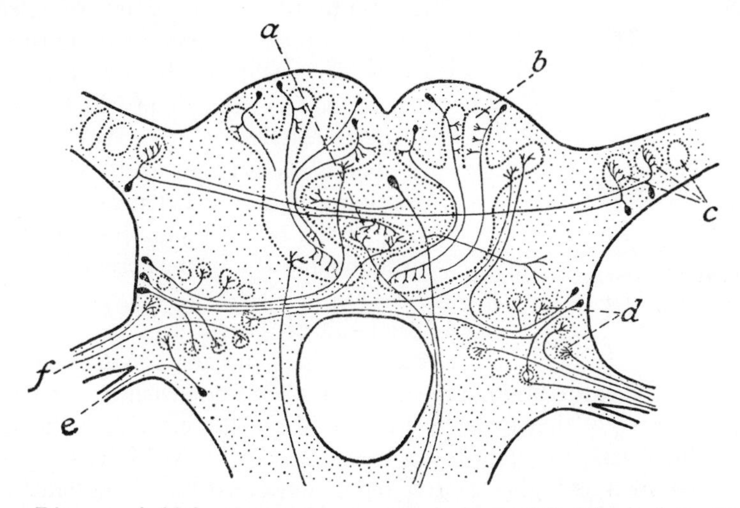

FIG. 109.—Diagram of chief centres and nerve tracts in brain of *Periplaneta* (*after* HANSTRÖM)

a, central body; *b*, mushroom bodies or corpora pedunculata; *c*, optic centres; *d*, antennal centres or glomeruli; *e*, motor nerve of antenna; *f*, sensory nerve of antenna.

adjacent segments by paired connectives containing no cell bodies. But there is always some degree of fusion between successive ganglia. In primitive insects such as *Machilis* there are 3 discrete pairs in the thorax and 8 in the abdomen, the last abdominal ganglion being composite. In higher forms the fusion of ganglia is more extensive. The abdominal ganglia are fewer; the first abdominal pair often fuses with that of the metathorax; in higher Diptera the thoracic ganglia become united into one; in Hemiptera all the thoracic and abdominal ganglia have coalesced to form a single mass.

In the head, the ganglia of the mandibular, maxillary and labial segments have always coalesced to form a suboesophageal ganglion which gives off nerves to these appendages. It is united by a pair of stout oesophageal connectives with the supraoesophageal ganglion or brain, which lies dorsal to the alimentary tract.

The brain consists of three parts, always fused into a single mass: (i) the 'protocerebrum', which forms the greater part, represents the united ganglia of the optic segment, and innervates the compound eyes and ocelli; (ii) the 'deutocerebrum', is derived from the fused ganglia of the antennary segment; and (iii) the 'tritocerebrum', which is formed by the ganglia of the third or intercalary segment of the head, consists of two small widely separated lobes attached to the dorsal lobes of the deutocerebrum; it innervates the labrum and the anterior part of the gut.

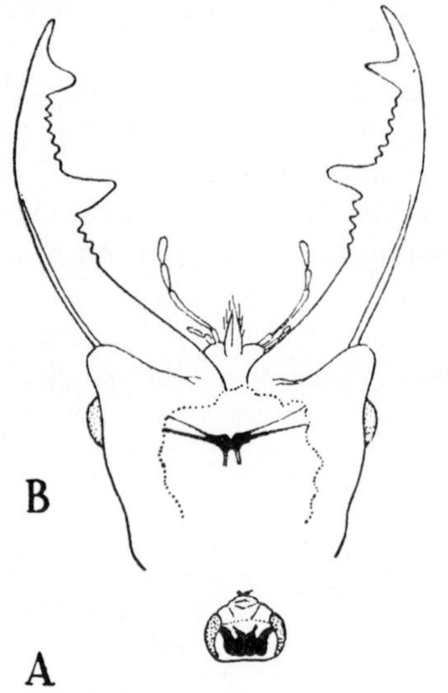

Apart from a few motor elements which innervate the antennae and perhaps others which control the pigment migrations in the eye, motor neurones are almost completely absent from the brain. The nerves it receives are all from the great sense organs of the head; and the nerve cells it contains are nearly all association neurones. As in the other ganglia, these are massed in the cortex: and in addition, in the protocerebrum, there are centrally placed groups of association cells or of their fibres forming the pedunculate or mushroom bodies, central body, ventral bodies, &c., upon which fibres from all parts converge (Fig. 109).

FIG. 110.—A, head of *Apis* with brain exposed; B, the same in *Lucanus*, drawn to the same scale (*after* JAWLOWSKI)

The brain is largest in those insects with the most complex behaviour: it composes $\frac{1}{4200}$ of the body volume in *Dytiscus*, $\frac{1}{400}$ in an ichneumon, $\frac{1}{280}$ in *Formica*, $\frac{1}{174}$ in the bee (Fig. 110).[33, 43] The central groups of ganglionic cells in the protocerebrum are largest in the social insects: the mushroom bodies comprise $\frac{1}{5}$ of the entire brain in the bee; in the ant *Formica* they are enormous in the workers, in which they make up $\frac{1}{2}$ the brain, smaller in the females and much smaller in the males, although these greatly exceed the workers in body size.[43] They are regarded as higher centres regulating behaviour. The females of the weevil *Deporaus betulae* and its allies have a larger number of globuli cells in the protocerebrum than the males; this is probably correlated with the complex leaf-rolling behaviour of the female.[167] There are visible differences also in the olfactory lobes: these are larger in the male, in which the sense of smell is more highly developed.[34]

Histology and histochemistry of the nervous system—The diameter of the axons in the leg nerve of *Periplaneta* range from 10μ for the large motor axons to 0·3μ or less for the fine sensory axons.[138] The giant fibres in the nerve cord may be 20μ thick. All the axons contain mitochondria oriented in the long axis. But the sensory axons of *Rhodnius* are so fine (0·5μ or less) that there is no room for mitochondria in the lumen; they may run considerable distances without any contained mitochondria; but at intervals they show fusiform swellings (1·1μ) each of which marks the site of a mitochondrion within the axoplasm[188] (Fig. 111, A). In regenerating axons the mitochondria can be seen passing from the cell body of the sense cell into the axon[188] (Fig. 111, B). The axoplasm of the larger axons commonly has a rope-like structure; it is made up of a bundle of neurofibrils of about 0·5μ diameter with mitochondria between them. The individual fibrils can be traced to the dictyosomes, or Golgi bodies, in the large motor ganglion cells.[18] Under the electron microscope the axon fibrils show a system of very much finer longitudinal, neurofilaments about 100 Å thick.[188] [See p. 202.]

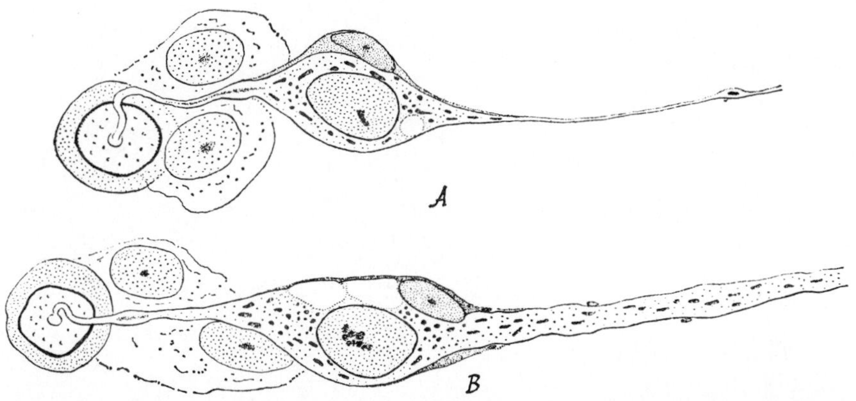

FIG. 111.—*A*, campaniform sensillum in *Rhodnius* showing tormogen, trichogen, neurilemma cell and sense cell with mitochondria. The axon shows a fusiform dilatation containing a mitochondrion. *B*, campaniform organ 4 days after section of the axon. New axon regenerating: cytoplasm and mitochondria streaming out from the sense cell (*after* WIGGLESWORTH)

The axons are enclosed and insulated by the investing Schwann cells. The xon sinks into the Schwann cell invaginating the plasma membrane to form a mesaxon', or suspensory fold, and this old may be wrapped a few times round he axon which is then said to be 'tunicated' Fig. 112).[139] Over a short length of certain f the motor axons as they run through the anglia this wrapping process is continued o produce a typical 'myelinated' axon with thick myelin sheath. That is seen in *Rhodnius*[188] (Fig. 113, A) and *Periplaneta*.[190] Some of the Schwann cell nuclei are deep n the nerve, between the axons; others are uperficial, and these give rise to a connective tissue sheath containing collagen ibres.[138]

In the ganglia the Schwann cells, or lial cells, which probably arise in the ourse of development by differentiation rom ganglionic cells,[157] become specialized or different purposes (Fig. 114).[158, 188] One type are termed 'perineurium cells'.[171] These cells contain abundant mitochondria nd serve as a store for glycogen.[188, 190] They surround the entire ganglion and ay down a tough connective tissue sheath, he 'neural lamella'.[171] This consists of neutral mucopolysaccharide containing ollagen fibrils arranged in layers with lifferring orientations.[115, 117, 162, 188] The lial cells deeper in the ganglion have varied orms; they were formerly described as

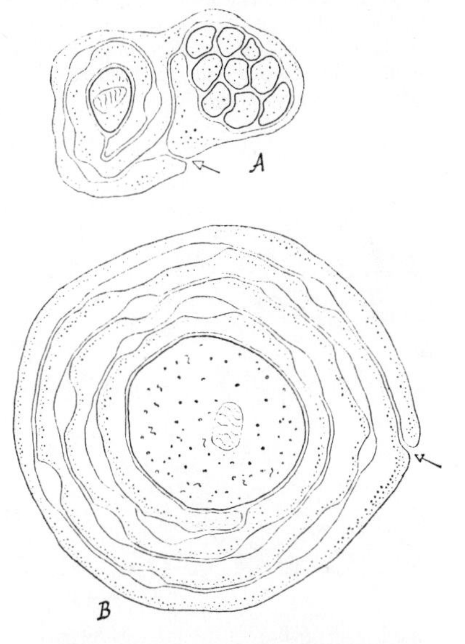

FIG. 112.—Diagrams showing the investment of the axons by the Schwann cells (*after* SMITH and TREHERNE)

The arrows show the points at which the plasma membrane has been invaginated by the axons to form the 'mesaxon'. In *A* the mesaxon divides to enclose a bundle of small axons to the right and a single small axon with a mitochondrion to the left. In *B* the mesaxon is loosely wrapped around a larger axon containing a mitochondrion to form a 'tunicated' nerve. The fusiform dilatations between the two layers of the mesaxon represent 'extracellular space'.

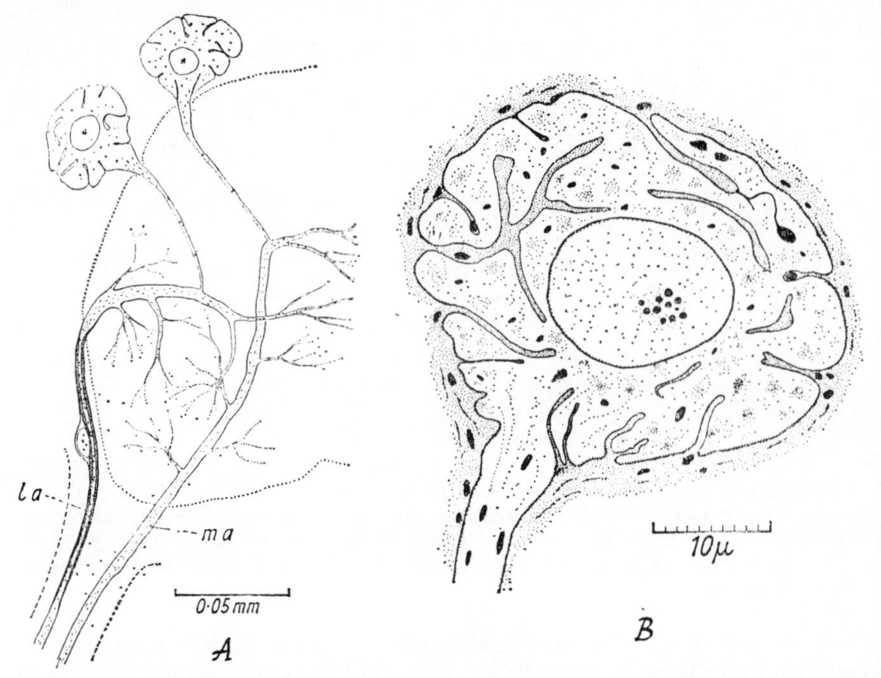

FIG. 113.—Motor neurones from the thoracic ganglion of *Rhodnius* (*after* WIGGLESWORTH)

A, semi-schematic scale drawing of a lateral motor axon with a narrow myelinated segment (*la*) and medial motor axon (*ma*), both containing mitochondria. The dotted lines mark the boundaries of the nerve and of the neuropile. *B*, detail of motor ganglion cell, surrounded by glial cytoplasm containing mitochondria, showing multiple deep invaginations of the plasma membrane (trophospongium).

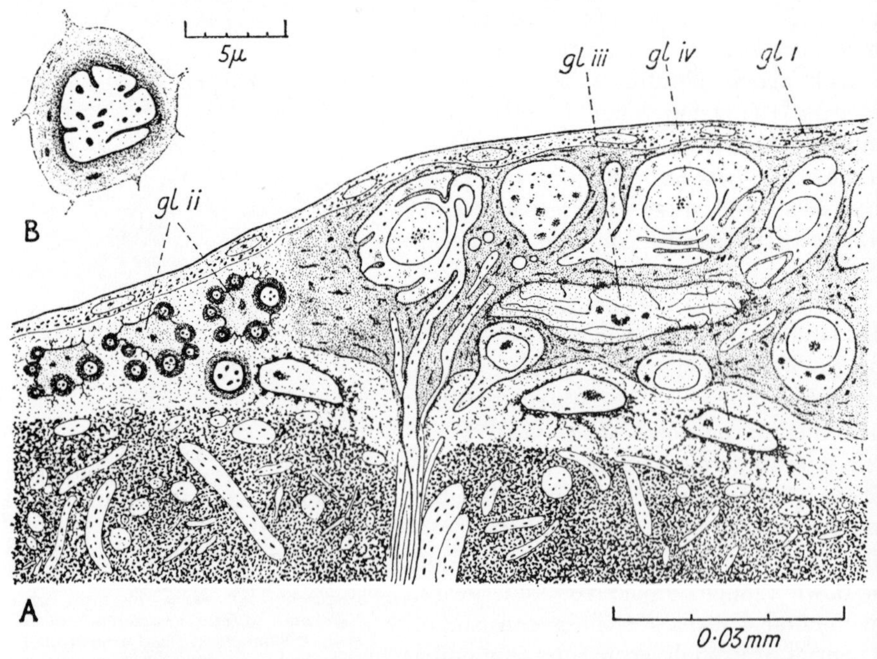

FIG. 114.—*A*, section of thoracic ganglion of *Rhodnius* to show types of glial cells. *B*, transverse section of thick-walled (myelinated) motor axon (*after* WIGGLESWORTH)

gl i, perineurium cells with abundant mitochondria below the neural lamella: *gl ii*, glial cells producing the myelin sheath for the lateral motor axons (*cf*. Fig. 113, *la*): *gl iii*, giant glial nucleus with invaginated membranes; its cytoplasm invests large ganglion cells and axons: *gl iv*, glial cells giving off processes which invest the innumerable axons in the neuropile.

sending out multiple fine branched filaments; but their processes are better described as flattened expansions which are everywhere wrapped around the cells and axons. Extensions from the glial cells, often containing mitochondria, are deeply invaginated into the large ganglion cells (Fig. 113, B); these invaginations are called the 'trophospongium'.[188, 189] The acid mucopolysaccharide hyaluronic acid is present in the extracellular spaces between the glial cells of the cockroach ganglia.[116] Many glial cells are grouped around the central neuropile and send their processes in between the finest nerve endings in the neuropile (Fig. 114, gl iv).[188] [See p. 202.]

In the development of the olfactory organ and chordotonal organs in the labial palp of *Pieris* (Fig. 186) a strand of connective tissue is first formed. Glial cells then migrate along this strand to form a pathway by which the ingrowing nerve fibres reach the nervous system.[148]

The sensory nerves and association fibres do not make contact with the cell bodies of the motor axons, as in vertebrates; all the interneuronal connexions are made within the neuropile. The synaptic junctions between one neurone and another presumably consist of closely apposed membranes with numerous synaptic vesicles (p. 187) within the lumen of the axons.[173, 188] [See p. 203.]

Nutrition and ionic regulation in the ganglia—The ganglia are almost the only solid organs in the insect body; their respiratory needs are provided by the tracheae and tracheoles which penetrate within the ganglia; but there is no circulation of body fluid inside the ganglion so that nutrition and excretion must be provided by diffusion through the sheath of the ganglion and transmission from cell to cell inside.[190] The fibrous 'neural lamella' is quite permeable; the underlying 'perineurium cells' are filled with mitochondria and with stored glycogen: they are probably concerned in the active transport and storage of nutrients. The glial cells receive these nutrients and pass them on to the ganglion cells by way of the invaginations of the 'trophospongium' (Fig. 113, B). In *Periplaneta* the ganglion cells, particularly the axon cone, contain much glycogen. There is a continuous secretion from the cell body of the neurone down the axon.[190] Glucose or trehalose in the circulating blood of *Periplaneta* is rapidly taken up by the ganglia; most of it enters amino acid metabolism and is incorporated chiefly as glutamic acid and glutamine; smaller amounts as glycogen, trehalose and glucose.[178]

Much interest attaches to the uptake and regulation of ions by the ganglia and nerves, because in many insects the level of blood potassium is so high, sometimes exceeding 70 mM per litre (p. 428), that the nerves would be unable to function if they were freely exposed to it.[100, 140] In the peripheral nerves of *Locusta* it may be that the cellular nerve sheath is actively maintaining a suitable ionic balance around the axons, for the efficiency of this dynamic regulation is impaired if the tracheal supply is reduced.[140] In the central nervous system of *Periplaneta*, ions pass rapidly through the neural lamella and the perineurium cells; the regulation around the axons is controlled by the deeper lying glial cells: sodium ions are actively extruded from the nerve cord and potassium ions taken up. The difference in composition between the haemolymph and the extra-cellular space inside the nerve cord is partly due to a Donnan equilibrium in which the acid mucopolysaccharide may be providing the free anion groups.[173, 179, 181] [See p. 203.]

Conduction of impulses in the nervous system—The sensory and motor nerves with the association neurones provide the anatomical basis for behaviour. The disturbance, or impulse, which is propagated along them consists in a change in electrical potential, due to a momentary depolarization of the axon surface, passing like a wave throughout the neurone. These waves succeed one another at a rate which varies with the intensity of the stimulus that is being transmitted; normal contractions in the body muscles are due to trains of motor impulses which rise and decline in frequency as contraction begins and ends.[2, 54] When a group of neurones in a ganglion is discharging impulses simultaneously in this way, they often seem to influence one another so that the potential changes become synchronized in all the axons: the ventral nerve cord of *Dytiscus* isolated from the body shows spontaneous changes in potential of this kind which occur in periodic outbursts from the thoracic and abdominal ganglia, the periods corresponding with the characteristic frequency of the respiratory movements.[2] The same thing is seen in sensory neurones: if the eye of *Dytiscus* is brightly illuminated, all the neurones discharge synchronously and show rhythmic potential oscillations at the rate of 20 to 40 per second, the active neurones tending to stimulate one another until their discharge is synchronized.[3]

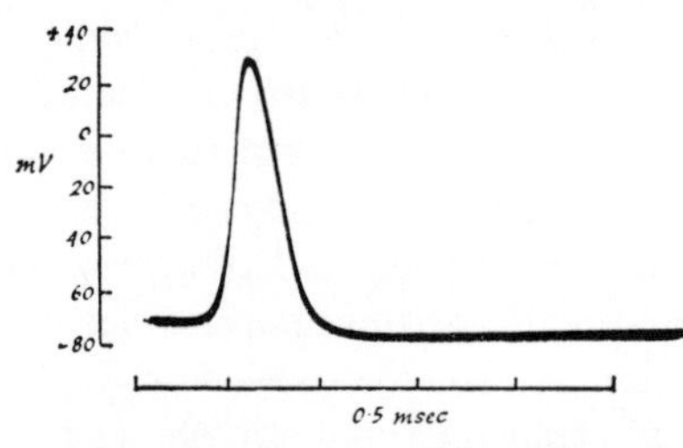

FIG. 115.—Record of the action potential in the giant axon of *Periplaneta* (*after* NARAHASHI)

The giant axon of *Periplaneta* has a resting 'membrane potential' of about 77 mV. On stimulation by cathodal current the membrane is depolarized and an 'action potential' appears with an average value of 99 mV (including an overshoot of about 30 mV (Fig. 115)). These values are of the same order as in other animals. This so-called 'spike potential' can be described in terms of membrane conductance to sodium and potassium ions and the concentration gradients of these ions across the membrane: a high level of potassium and a low level of sodium exists within the resting axon; during the action potential sodium ions flow into the axoplasm while potassium ions flow out; the ionic concentrations are restored to the normal resting level, after activity, by metabolic processes. Calcium ions in the external medium are necessary for the normal maintenance of excitability.[153, 194]

The 'spike potential' developed in this way travels along the nerve by the depolarization of each succeeding stretch. The velocity of conduction increases with the diameter of the fibre: a crural nerve axon of *Locusta* with a diameter of 5–6μ conducts impulses at a velocity of 1·6 metres per second; a giant axon of *Periplaneta*, diameter 45μ, conducts at 6·6–7·2 metres per second.[153] (Myelinated axons in the frog conduct at 25 metres per second.) The biological significance of the giant axons is to provide for rapid evasive movement (p. 188). In the legs of *Schistocerca* and *Locusta* the 'slow' nerve fibres and the 'fast' nerve fibres (p. 137) are contained in separate nerve trunks. When the 'slow' fibres are stimulated, smooth tonic contraction of the muscles starts at 10 impulses per second, rapid contraction at 200 per second. When the 'fast' fibres are stimulated they give a powerful twitch in response to a single impulse and a smooth tetanus at about 30 impulses per second.[141] Although the cell bodies of the motor neurones are not on the direct pathway of conduction, they do show action potentials of the same character as the axons.[137]

The ionic theory of the nerve impulse, as outlined above, faces the difficulty hat the extracellular space in which the axons lie is so restricted as not to allow oom for an adequate ion reservoir outside the nerve. In the axons of *Periplaneta*, oth in the nerve cord and in the peripheral nerves, there are conspicuous dila- ations in the mesaxons of the investing glial cells, visible with the electron iicroscope, which are in fact extracellular space (Fig. 112).[173] [See p. 203.]

Synaptic conduction—The neurones are not continuous with one another. 'he branched terminations of the axon of one neurone come into intimate ssociation with the collateral or terminal arborizations of another neurone to orm a 'synapse'. There is increasing evidence in other groups of animals that ne electrical disturbance does not itself cross the synapse, but causes the libera- ion there of some chemical substance (adrenaline, acetylcholine) which sets off a resh disturbance in the succeeding neurone. The central nervous system of in- ects is particularly rich in acetylcholine;[18] the nerve cord in *Periplaneta* con- ains some 15 times the average concentration in the mammalian central nervous ystem.[46, 84] Similarly, cholinesterase, the specific enzyme which hydrolyses and iactivates acetylcholine, exists in very high concentration in the nervous tissue f *Apis* and *Periplaneta*[60] and *Melanoplus*.[45] In the latter insect cholinesterase rst appears in development when the neuroblasts are differentiated;[76] the same true in the egg of Lepidoptera,[195] and in *Oncopeltus* and *Musca* choline acety- ise first becomes detectable at the time of appearance of the neuroblasts.[150]

In general, in its reactions to drugs, the insect nerve cord shows certain simi- rities to the autonomic nervous system of vertebrates; as in this system, so in ie central nervous system of the insect, acetylcholine (or some similar mediator) nd cholinesterase seem to play an essential part in the transmission of im- ulses.[65] Stimulation of the nerve cord in *Periplaneta* causes an increase in acety- holine, and this increase is still greater if the cholinesterase is inhibited with serine.[123]

Acetylcholine, cholinesterase, and choline acetylase are all present in the in- ect;[172] the greatest activity of cholinesterase occurs in the most active insects ich as *Musca*, *Apis*, and *Periplaneta*.[151] But although it is generally believed that cetylcholine is the transmitter substance in the neuropile,[193] the matter is still ncertain.[122, 123] The ganglia of *Periplaneta* are relatively insensitive to acetyl- holine; but the synaptic sites may be well protected and accessible only to trans- iitter produced in the presynaptic region.[163] In *Rhodnius* cholinesterase is most bundant in the neuropile, especially where synaptic sites are most numerous; ut smaller activity occurs in the glial cytoplasm between the ganglion cells and the Schwann cells along the axons; it is absent from the muscle end-plates). 150).[187] Among the confusion of fine neurones in the neuropile, no constant ructure has yet been recognized as the effective synapse.[180] [See p. 203.]

Transmission across a given synapse can be influenced by many physiological ictors. And since neurones in the central nervous system appear to be connected ith one another in almost every direction by way of synapses and intermediary eurones, it is evident that a given impulse may flow along many possible paths.

Some indication of the properties of the synapse in insects is given by obser- ations on the cockroach. Each nerve after entering the last abdominal ganglion ivides into two roots, one breaking up into fibrils and ending in synapses in the anglion itself, the other passing straight through to the ganglia in front 'ig. 116).[12] If the nerve from the cercus is stimulated electrically or acoustically

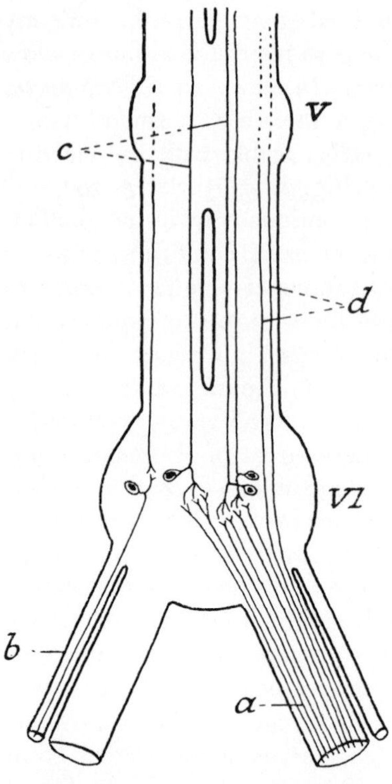

FIG. 116.—Central connexions of cercal nerves in the cockroach (*after* PUMPHREY and RAWDON-SMITH)

a, cercal sensory nerve; *b*, cercal motor nerve; *c*, giant fibres; *d*, through fibres.

(p. 269) and the electrical response in th ganglionic chain recorded, these two type of connexion produce a twofold respons The fibres which pass through withou synapsing produce a small response confine to the same side as the cercus, which per sists as long as the cercal nerve is respondin The synapsing fibres produce a respons of great magnitude which runs withou interruption to the thorax. This is probabl due to stimuli from many fibres of th cercal nerve being collected by way c synapses into a few 'giant fibres' runnin to the thoracic ganglia, where most c the fibres end in association with motc neurones.[63] If the cercal nerve is subjecte to maximal stimulation above a critic frequency, the synapses become 'fatigue and the giant fibres can no longer be excite by stimulation of the cercal nerve; althoug they will still respond if stimulated directl If submaximal electrical stimuli at a lo frequency (25 per second) are applie to the cercal nerve, the postganglioni response in the giant fibres graduall declines until the majority of the stimu fail to get through, although the respons in the cercal nerve is unchanged. If th strength or frequency of the stimulation is then increased it will get throug Thus the synapse shows a state of 'adaptation' (due perhaps to a prolongatio of the relatively refractory state of the synaptic terminations) in which trans mission ceases and is renewed only if the stimuli are strengthened c changed.[56] [See p. 204.]

The giant fibre system is an adaptation by which it is possible to get a ver rapid stereotyped response: the interval between stimulation of the cerci and ex tension of the hind legs in the escape reaction of *Periplaneta* is about 50 milli seconds—about the same as the eye-blink time in man.[164] The giant cockroac *Macropanesthia*, which is extremely slow in its reactions, lacks giant fibres.[15 A similar giant fibre system serving the same purpose is present in other insect such a system is brought into action in the larva of the dragon-fly *Anax* by tacti stimulation of the paraprocts.[129]

Application of weak polarizing currents to the thoracic ganglia of the cock roach will cause predominantly flexion or extension according to the directio of flow of the current. Perhaps this may point to the existence of two types c transmitter substance (acetylcholine and another) involved in central synapti transmission in insects.[55] The sensory nerves of Vertebrates produce a 'stimula tory substance' and an enzyme by which this substance is decomposed. Suc poisons as strychnine or picrotoxin which cause tetanic convulsions produce thi effect by inhibiting the enzyme. A similar system exists in the sensory nervou

'stem of insects; for example, in the eye of the bee. But the enzyme, and pre-
ımably the stimulatory substance, differ from those of vertebrates: strychnine
ill not inhibit the breakdown of the stimulatory substance in the insect and
ıerefore does not act as an insect poison.[97]

Tonus and inhibition—An important function of the muscles, as we have
en (p. 155), is to sustain a constant tonic contraction. In thin-skinned larvae
ıe tonus of the turgor muscles beneath the skin (p. 157) keeps the blood at a
eady pressure, and maintains the shape and attitude of the soft body wall. In
e turgor muscles this state is dependent on the integrity of the nerve supply
om the ganglion of each segment: section of the nerves from a single ganglion
the larva of *Lucanus cervus*[67] or in caterpillars[30, 37] results in immediate loss
tonus and passive distension sharply limited to the one segment (Fig. 117); in
ombyx mori larvae a crystal of cocaine applied to an exposed ganglion causes
mplete flaccidity in the muscles supplied.[51] On the other hand, in *Carausius*
ettigonia, &c., the flexor and extensor muscles of the limbs remain in a state
continuous contraction even after separation from the body.[23] It has been
ggested that in them the peripheral nerve net may be responsible for this
rsistence of tonus; but as yet there is no experimental proof of this.[66]

When a given muscle (such as the flexor of the limb) contracts, its antagonist
ıe extensor) must relax. This active relaxation or diminution
tonus is termed 'inhibition'. In vertebrates it is effected
ntrally by an action upon the neurones which are maintain-
g tonus. In Crustacea it is brought about peripherally; every
eletal muscle has a double innervation, a thick nerve con-
ying motor impulses, a thin nerve conveying impulses which
eak the functional connexion between the stimulatory pro-
ss and the muscles, and hence bring about inhibition.[16] In
sects, inhibition seems to be of this peripheral type (see p. 151)
though in *Periplaneta* muscles no evidence of an inhibitory
rve fibre can be found).[54] In *Libellula*,[44] *Tettigonia*, and
econema[23] weak electrical stimulation of the nerve of the
mur causes extension of the legs with inhibition of the
xor, while strong stimulation causes flexion, the extensors
ing inhibited. In *Carausius* the same result is obtained if the
oracic ganglia are stimulated.[23]

FIG. 117.—Larva of *Cossus* with nerves to left side of the seventh segment cut (*after* BARTH)

Reflex conduction and nerve centres—The simplest
pe of conduction in the central nervous system will con-
t in the transmission of impulses by a sensory neurone
om a receptor or sense organ to the ganglion, through an
sociation neurone to a motor neurone, and thence to a muscle
other effector organ. Stimulation of the sense organ will thus produce
ntraction in the muscle. This path of nervous conduction is termed a 'reflex
' and the response a simple reflex (Fig. 118).

The reflex arc is a physiological abstraction; for even in the simplest response
e course of conduction must be infinitely more complex, involving as it does
hibition of opposing muscles and compensatory movements elsewhere in the
dy.[164] Moreover, the course of reflex conduction is not fixed; as the synapses
come 'fatigued', or 'adapted', the transmission may be blocked or follow some
her reflex arc with a lower threshold; and if the stimuli are excessively strong

they may overflow into many paths and produce a discharge of impulses from large group of motor neurones.

When a reflex response takes place through a single ganglion, this is said contain the centre for that reflex. But the occurrence or non-occurrence of th response can be influenced by more remote regions of the nervous system which may have an inhibitory influence and prevent a response which wou otherwise occur, or a stimulatory influence and thus lower the threshold stimulation necessary to produce the reaction. Where these regions are regarde as having a fixed position they are spoken of as 'stimulatory' or 'inhibito centres'. For example, the respiratory movements of the gills on the first s abdominal segments of *Cloëon* nymphs in response to oxygen want (p. 377) ar said to be controlled by motor centres in their own segments; while these centr are themselves controlled by a stimulatory centre in the 6th abdominal segme and an inhibitory centre in the 2nd thoracic segment.[5] Where movements a made up of an orderly succession of reflexes they are generally considered to under the control of 'co-ordinating centres'.

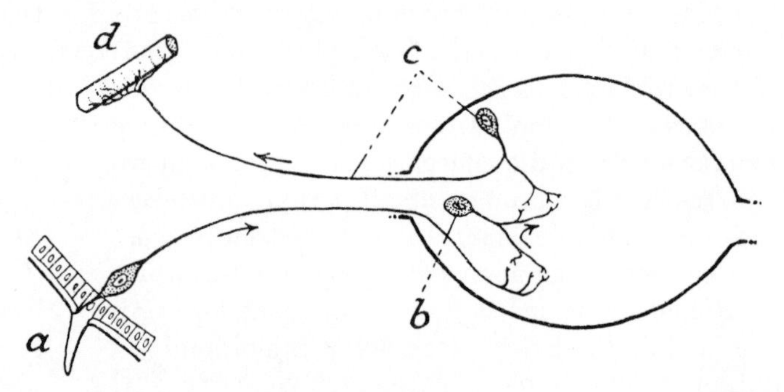

FIG. 118.—Diagram of a simple reflex arc; the arrows show the course of conduction

a, sense organ supplied by a bipolar sense cell which forms the afferent nerve; *b*, association neuro *c*, efferent neurone with unipolar cell body in the ganglion and axon filament forming the efferent ner *d*, muscle.

Centres in the abdominal and thoracic ganglia—The following are sor examples of reflexes obtainable from isolated ganglia, that is, when the ner cord on either side of the ganglion in question has been severed. A single leg *Periplaneta* isolated with its ganglion will make a reflex stepping movement if t tarsus is stimulated by traction;[77] in *Apis* the reflex for insertion and wit drawal of the sting is located in the last abdominal ganglion;[9] the discharge faecal pellets by caterpillars,[37] and oviposition by the silkworm moth in respon to contact of the ovipositor with the surface,[42] are similarly controlled by the la abdominal ganglion; if the skin of a caterpillar is gently stroked with the point a paint-brush there is a local loss of tonus lasting 10 to 30 seconds—a refl inhibition with its centre in the ganglion of the segment in question;[30] and t grasping of any object that they touch which is shown by the prolegs of cate pillars is also a reflex obtainable through a single ganglion.[37]

But most of the reflexes shown by insects require the co-operation of seve ganglia. In *Lymantria* larvae the movements of turning over when placed on t back, of feeling in all directions when the thoracic legs lack support, or avoidance of objects brought into contact with the body, are all reflexes involvi more than one ganglion;[37] and in *Bombyx mori*, although the last abdominal ganglion by itself will control reflex oviposition, the other abdominal ganglia and the thoracic ganglia ordinarily carry out co-ordinated movements (curving of the abdomen, &c.) during the reaction.[42] The same is true of the complicated movements shown by many insects in which a part of the body is cleaned by one or more of the appendages; movements which are carried out readily by the decapitated insect.[75] For example, on touching the gills or abdominal tergites of *Cloëon* nymphs after decapitation, a wiping reflex can be elicited from the 2nd and 3rd pair of limbs.[5]

The posture of an insect is maintained by means of reflex responses to stimuli from the proprioceptive organs (p. 263). In the cockroach, *Periplaneta*, the weight of the body sets up strains in the cuticle of the leg; these strains stimulate the campaniform organs, or 'stress receptors', and cause a reflex discharge of motor impulses to the depressors of the leg. A sudden reduction in the resistance to the depressors causes a momentary arrest of the motor impulses. It is these proprioceptive organs in the cuticle which are responsible for the reflex maintenance of muscle tonus (pp. 155, 189). The quick nerve fibre is called into action by the same types of reflex stimuli that excite the slow fibre, but it requires a greater intensity of stimulus; both fibres appear to have the same central connexions but different thresholds of stimulation.[54]

Characters of reflex responses—Reflex responses are always purposive; they serve some evident aim in the life of the insect. They are also in many cases so deeply seated in the nervous system that they are carried out even when this purpose cannot be achieved. For example, bugs[36, 75] or cockroaches[31] deprived of their antennae will go through the motions of cleaning them; even decapitated insects may perform these movements.[9] And reflexes may be carried out in the normal way even though they are not performing the function for which they were undertaken. If the antenna of a cockroach is stimulated and it is then offered a bristle or the antenna of another insect, it will clean this and neglect its own appendage.[31]

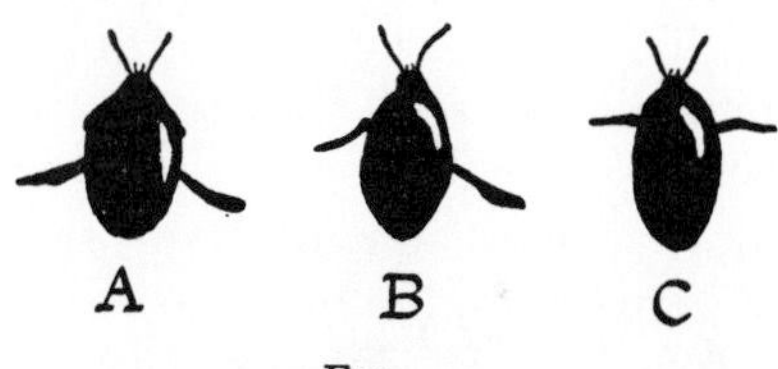

FIG. 119

A, normal swimming movement of *Dytiscus*; both hind legs move simultaneously, the middle legs do the same to a smaller extent. *B*, after amputation of left hind leg; the left middle leg is used to a much greater extent and the remaining hind leg makes compensatory movements. *C*, after amputation of both hind legs; extensive simultaneous movements of the middle pair (diagrammatic) (*after* BETHE).

But reflex responses may be highly plastic. The cockroach normally holds its antenna with one fore leg during cleaning; if the fore legs are removed it may use the middle leg of the opposite side.[31]

Dytiscus normally swims by means of simultaneous movement of both hind legs and, to a lesser extent, the middle legs (Fig. 119). If one hind leg is removed it can be shown, by cinematographic methods, that compensatory movements of a most complex nature ensue. There is increased activity of the middle leg on the operated side, and the movements of the legs on the other side are so modified as to balance exactly the turning moment of this middle leg. As a result, the beetle still maintains a straight course.[32] These changes in co-ordination occur equally in decapitated insects.[10]

Similar plasticity is seen in the responses of the insect in nature. Among Cerambycids one individual may clean the antennae with the legs, another with the mouth parts, and if cleaning cannot be effected with one set of appendages another set is employed.[75] Dragon-fly nymphs deprived of the 'mask', or labium, will seize food in the mandibles.[1] Ants may follow many different procedures in cleaning a given part of the body.[35] And caddis larvae *Neuronia*, which normally build their cases out of fragments of leaves, adopt quite a different procedure when given pine needles and sand or after they have lost the 1st or 2nd pair of legs.[22]

These observations suggest that there are not fixed reflex centres for the various movements but, as was to be expected from the nature of synaptic conduction, the impulses can follow different paths depending on circumstances. Many factors determine what paths shall be followed. Once a given reflex has been set going, conduction in that channel is favoured and the same response may persist a long time;[90] this is termed 'facilitation'. If the end effect of one reflex is prevented mechanically the response may change: when antennal cleaning is prevented in the cockroach the response may be converted into leg cleaning.[31] If stimuli are strong they may flow into other reflex arcs. Thus strong stimuli to the antenna of the cockroach may result in *both* fore legs being employed in cleaning. Or in the earwig, if the leg is gently stimulated with a hair it is cleaned; if it is touched with a needle it is drawn away; if it is pinched with forceps the abdomen is instantly curled to the spot and the pincers brought to bear.[90] And in the dragon-fly, pinching the last abdominal segments results in a generalized reflex, the impulses spreading into many reflex arcs: the abdomen is curved ventrally, the insect sets itself free from its resting place, and the wings begin to flutter.[85]

The nervous system will not always respond to a given stimulus with a given reflex. As we shall see later (p. 310), an important function of the sense organs is to increase the 'nervous tone' of the central nervous system; that is, to bring it into a responsive state. It was shown by Sherrington that in the nervous system of vertebrates, the repetition of a reflex response leads not only to a rise in the threshold level of stimulation necessary to provoke that response, but it lowers the threshold for the antagonistic reflex. This applies also to insects, and not only for so-called simple reflexes but for quite complex acts of behaviour (p. 295). Reflexes often occur in a definite sequence or chain, each reflex in turn apparently bringing the nervous system into such a state that it will respond to the next. Thus in cleaning the antennae, bugs, flies and other insects will first assume a characteristic cleaning attitude, then clean the legs which are to be used for cleaning, clean the antennae, and finally clean the legs again.[75]

These set 'patterns of behaviour', or 'instincts', which are sometimes described as a chain of reflexes, will be considered later (pp. 332, 336).

Co-ordination of walking—The study of co-ordination in walking provides a good example of the nervous regulation of movements. Most insects will both walk and fly after decapitation;[16] the nervous arcs through which co-ordination is effected must therefore all lie in the thoracic ganglia.[78] In *Carausius*, after section through one of the longitudinal commissures between thoracic ganglia, walking movements remain normal; clearly the nervous paths must cross in such a way that when interrupted on one side they can still pass on the opposite side from one segment to the next; whereas walking is abolished if the cord is cut right through between the meso- and metathoracic ganglia.[15] The same is true of *Mantis*.[62]

The normal method of walking on six legs has already been described (p. 156). If the middle pair of legs is removed in *Carausius* a new method and a new sequence of leg movements is adopted: the left fore leg is followed by the right hind leg, the right fore leg moves next and then the left hind leg (Fig. 120);[15] and similar changes in co-ordination appear in *Geotrupes* and many other insects when various legs are amputated.[10, 77] Such changes are so manifold that it is impossible to assume the existence of pre-formed co-ordination centres ready to control each new type of movement; each co-ordination must be brought about anew through the interplay of many factors.[16]

One factor in this co-ordination seems to be the tendency for a reflex stimulus to flow into a muscle which is stretched; in *Libellula* after decapitation, if the leg is flexed stimulation of the tarsus causes extension, if the leg is extended stimulation of the tarsus causes flexion (Fig. 121). But if the stretched muscle cannot respond the stimulus flows into another group.[44] In *Carausius* failure of contact between the amputated limb and the ground seems to be important, for if the middle pair of limbs is cut short, but the stumps are allowed to come

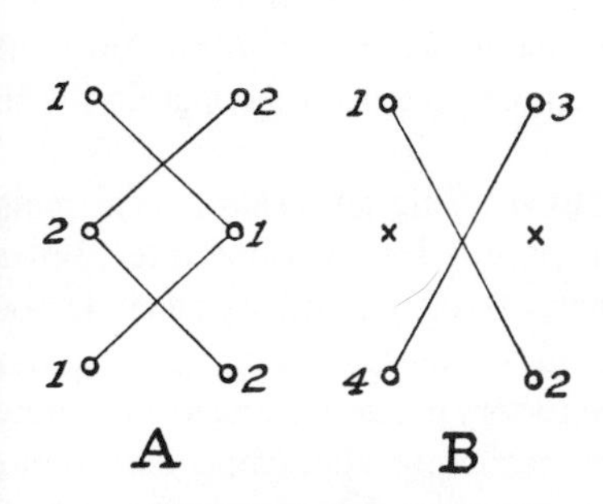

FIG. 120.—Walking movements in *Carausius* (*after* VON BUDDENBROCK)

The numbers indicate the approximate sequence in which the legs are moved (*cf.* p. 102). A, normal insect; *B*, after amputation of the middle pair of legs.

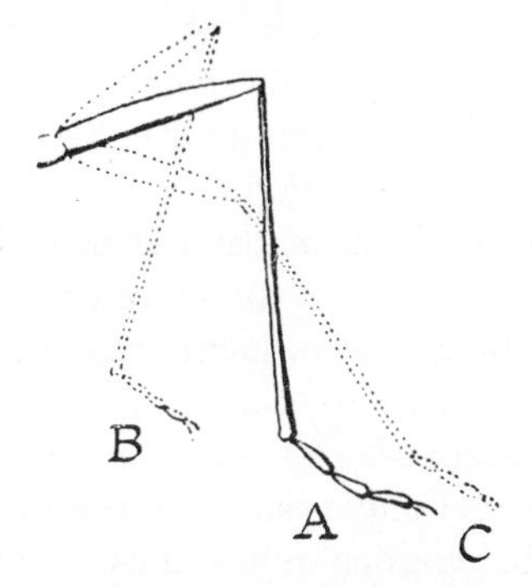

FIG. 121.—Hind leg of *Libellula* nymph isolated with its ganglion

If the tarsus is stimulated when in the normal position, *A*, it is flexed to *B*. If then stimulated it is extended to *C* (*after* MATULA).

in contact with a little platform attached to the ventral surface of the insect, the four-legged insect still shows the normal walking rhythm.[15]

We have seen that in *Periplaneta* a stepping movement can be induced in a single leg isolated with its ganglion when the tarsus is stimulated by traction. This introduces another factor in co-ordination.[85] Unlike *Carausius*, *Periplaneta* still shows co-ordinated walking movements after the cord has been cut right through in the thorax. This seems to result from the traction by the anterior appendages evoking reflex walking movements in the posterior appendages, the excitatory action of the anterior ganglia being replaced by excitation from the periphery.[77] The same thing is seen in *Tettigonia* with the middle legs cut short, and here the movement of the hind leg clearly does not begin until *after* the opposite fore leg, as though stimulated by traction from it.[79] Thus two factors at least are involved in evoking displacement of a given leg in walking: impulses spreading from the ganglia supplying other legs, and impulses from mechanical stimulation of its own sensory endings. Stimulation of one set of muscles causes a reflex inhibition of the antagonists; but if the stimulus is strong the inhibition is often followed by a burst of excitation. Perhaps this is the physiological bas of the rhythmical movements of the legs in walking.[54] The modifications in le movements which follow the amputation of limbs or the cutting of commissure can be explained by changes in the sensory inflow to the central nervous systen It may well be that such afferent stimuli (feed-back loops) are of major import ance in the co-ordination of normal movement.[32] [See p. 204.]

Section of the various commissures between the thoracic ganglia in *Per planeta* proves conclusively that the nervous impulses co-ordinating the activit of one ganglion with another can follow different paths and that peripheral stimu from the proprioceptors in the limbs influence the activity of the respecti ganglia by making the motor centres more or less excitable.[77] By the use o cinematographic methods it can be shown that after such operations there ar small changes in the locomotory rhythms, indicating that there are delays in th passage of impulses along these new paths.[32]

In the crawling of caterpillars[37] and the peristaltic movements of the bod in the larva of *Lucanus cervus*[67] co-ordination between the anterior and posteric halves of the insect ceases if the nerve cord is cut through; it is not affected if onl one commissure is cut. Segments paralysed through section of their nerves d not take part in the movement but the co-ordinated wave of contraction is nc otherwise interfered with. Hence segmental reflexes are not responsible for th wave of stimulation.[67]

Functions of the suboesophageal ganglion—The suboesophageal gang lion contains the motor centres for the mouth parts which it innervates; but i addition it may influence markedly the motor activity of the entire insect. It doe not contain essential co-ordinating centres for movement, since normal walkin may occur after its removal.[9] But it exerts an excitatory influence on the locomoto co-ordinating system in the thoracic ganglia, perhaps facilitating conductio in the various reflex arcs.[16] Thus *Mantis* is very sluggish after decapitation but if the brain is removed and the suboesophageal ganglion left intact it i exceedingly restless; it starts walking on the slightest stimulus and may continu walking for a long time—the suboesophageal centres being perhaps re-excite by impulses from the moving legs.[62] The suboesophageal ganglion has simila properties in *Carausius*;[38] and it is important in maintaining crawling in cater pillars.[37, 51] But it is not essential for movement in any of these insects, for co ordinated crawling can be induced after its removal in caterpillars by exertin tension on the anterior segments of the thorax,[30] and *Carausius*, &c., can b caused to walk by stimulating the thoracic centres by heat.[38] Male Mantid copulate more readily when the suboesophageal ganglion is removed; apparentl this ganglion inhibits to some extent the reflexes involved.[61] The centre respon sible for the copulatory movements in the male mantis is located in the las abdominal ganglion. The efferent nerve activity for many of these movement continues after all the afferent connections of the ganglion have been cut. Indee the activity may increase, because of the removal of inhibition by the suboesopha geal ganglion. The inference is that the whole pattern of behaviour can be gen erated in the isolated ganglion without incoming stimuli to set it off.[165]

Supraoesophageal ganglion—The brain is mainly an association centre It controls the reflex responses by the rest of the body in accordance with th stimuli received from the great sense organs of the head. The brain is therefor responsible for orientation and for all the more complex forms of behaviour t be discussed in Chapter VIII. It may exert these functions by stimulating or by inhibiting reflexes. The excitability of the motor centres in the thorax of the cockroach, in which the giant motor fibres from the cerci end (p. 188), is controlled by descending impulses from the ganglia of the head. The brain is necessary to bring the thoracic centres to a level of excitation which permits an adequate motor discharge. Thus a brief blast of air will take a normal cockroach several feet; after decapitation it makes a few steps only. Co-ordination of the leg movements is not impaired, it is only the ability of the motor centre to respond which is lost. In the same way the evasive response becomes adapted and reduced when the stimulus is applied repeatedly.[63]

This depressive effect of decapitation is sometimes short-lived. After a time the reflex response of *Periplaneta* to draughts is increased and the insect seems generally hypersensitive.[54] In *Mantis* the reflex locomotor activity of the suboesophageal ganglion is subdued when the brain is intact;[62] but without the brain *Mantis* is unable to walk backwards.[62] After removal of the brain in *Hydrophilus*, *Apis*,[9] *Lucanus* larva,[67] &c., reflex responses become exaggerated; responses may be elicited by stimuli which have no effect in the normal insect, and movements (such as cleaning movements) once started may be continued for hours without interruption.[9] Libellulids cannot walk after decapitation because they develop such an active reflex grasping of the surface; a reflex which is normally inhibited by the brain.[85] Females of *Tipula* (Dipt.), after living for some days without egg-laying owing to unsuitable conditions, can be brought by decapitation to immediate oviposition; the inhibitory effect of the brain being removed.[13] In decapitated females of *Bombyx mori* the oviposition response can be evoked repeatedly on pressing the ovipositor: it may be induced before mating, and it persists after all the eggs have been laid. But when the brain is present, oviposition cannot be induced in this mechanical fashion; it is inhibited until after mating. Conversely, it is only in the insect with the brain intact that the normal continued oviposition of a complete batch of eggs takes place.[42] The same type of regulation controls the stinging reflex of the bee.[16]

Selective stimulation of different points in the proto- and deutocerebrum of crickets and grasshoppers can evoke movements of head, antennae, &c., and different types of locomotion. It is generally a co-ordinated pattern of rhythmic movement that appears; each stimulus is clearly acting upon some higher co-ordinating centre. Associated with these movements is the inhibition of other activity.[142] It appears that the mushroom bodies are concerned in the co-ordination of movements in response to stimuli received by the sense organs of the head. Local injuries to the mushroom bodies show that in Orthoptera[142] and Hymenoptera[183] they contain inhibitory systems for general activity. If both mushroom bodies are extirpated, running and leaping movements are increased as the result of removal of inhibition. If the central body is extirpated the removal of activation causes depression of running movements.[142]

Removal of mushroom bodies on both sides in *Gomphocerus* (Acrididae) abolished their song and rendered them unresponsive to the sounds of the female. Excitation of the mushroom bodies may sometimes induce prolonged chirping.[142] Electrical stimulation of certain regions of the mushroom bodies and their afferent tracts evokes normal singing movements in crickets. It appears that in the central body the singing instructions from the mushroom bodies are analysed and transmitted to the effector neurones of the mesothoracic ganglion.[142] In

Orthoptera, the central body and especially the protocerebral bridge are concerned in the control of respiratory movements.[142, 143]

Ants can still learn their way back to the nest after destruction of the mushroom bodies;[183] but their destruction in bees eliminates the capacity for training to colours and scents and to orientation in a labyrinth.[182]

The brain has an effect also on muscle tonus. Removal of the head in *Periplaneta* causes hyperactivity of all the slow nerve fibres which are responsible for muscle tonus (p. 188), but particularly of those to the depressors of the legs.[54] Tonus can be maintained by the segmental ganglia;[30] but the general degree of tonus on the two sides of the body is influenced by the brain, each half of which affects particularly the muscles on its own side. Thus if half the brain is removed, tonus is diminished on the side of the injury and the body becomes flexed towards the sound side (Fig. 122).[4, 9, 44, 62] When such an insect walks it moves in a circle towards the side with the brain intact. After 'pîqure' of the brain on one side in *Dytiscus*, circus movements may persist for months.[12] In *Carabus* and in ants they have been induced temporarily by applying potassium cyanide to one half of the brain.[83]

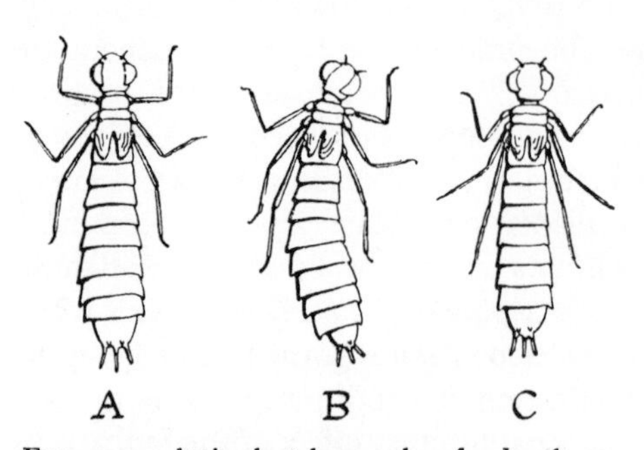

FIG. 122.—Attitude taken up by *Aeschna* larva

A, normal; *B*, after removal of left half of brain; *C*, after removal of entire brain (*after* MATULA).

The mechanism of this circus movement after injury to the brain has been the subject of much controversy. According to one view it is a direct result of the increased tone and contractile power of the muscles on the intact side; these changes affect chiefly the flexor muscles, so that when the insect moves it is carried towards that side.[9, 37] But in *Dytiscus*,[12] *Carabus*, &c.,[6] and *Cloëon* nymphs[4] the changed activity of the limbs which brings about the circus movement is not limited to one side but involves the whole organism. In *Dytiscus*,[21] *Aeschna* nymphs,[7] *Tettigonia*, *Carausius* and many other insects[7] the movement still persists after various limbs have been removed. As in the insect with the brain intact, the reflex paths through which the leg movements are co-ordinated are modified according to the circumstances, but the direction of walking remains the same. From this it has been inferred that the true cause of the circus movement lies in the central nervous system, not in the tonus of the effectors;[7] each half of the brain directing movements particularly towards its own side.[21] In water beetles and dragon-fly larvae which circle in a diffuse light after unilateral injury to the brain, this asymmetrical movement is arrested as soon as they are offered visual marks; they then orient themselves more or less normally.[32]

VISCERAL NERVOUS SYSTEM

The visceral or sympathetic nervous system of insects is generally described as made up of three parts.

(i) **The stomatogastric system** (Fig. 123) arises during embryonic development by an ingrowth from the dorsal wall of the stomodaeum which later acquires connexion with the brain.[48] It consists firstly of a median 'frontal ganglion', joined by bilateral connectives to the anterior surface of the brain, which sends back a median recurrent nerve passing between the brain and the oesophagus. The recurrent nerve may end in a paired or single 'stomachic ganglion' towards the hind end of the oesophagus and on its course there is sometimes a median 'hypocerebral ganglion' behind the brain. Behind the brain, also, is a pair of 'corpora cardiaca', likewise nervous in origin, connected by nerves with the protocerebrum. These structures are variously arranged in different insects;[11, 17, 20, 28] Fig. 123 shows a typical arrangement in the Orthoptera. Often the corpora cardiaca are united in the mid-line to form a single structure. [See p. 204.]

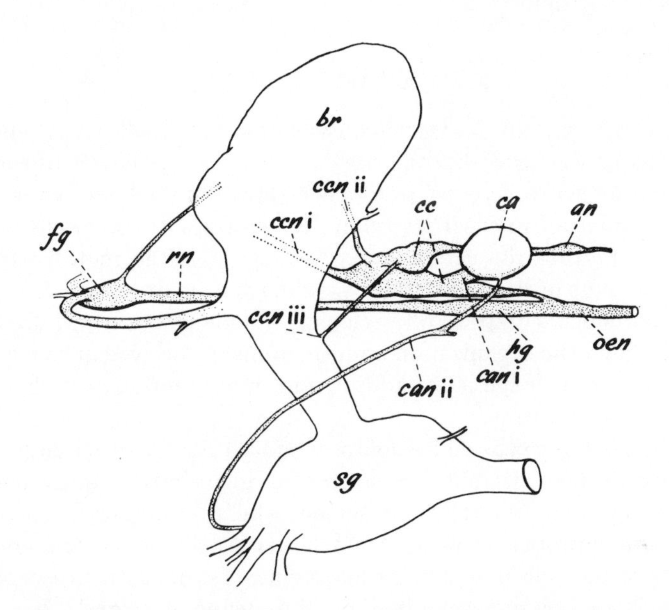

FIG. 123.—The brain and suboesophageal ganglion, the stomatogastric nervous system (shaded) and the corpus allatum in the cockroach (after WILLEY, simplified).

an, aorta nerve; *br*, brain, *ca*, corpus allatum; *can* i and ii, the two corpus allatum nerves; *ccn* i, ii, and iii, the three corpus cardiacum nerves; *fg*, frontal ganglion; *hg*, hypocerebral ganglion; *oen*, oesophageal nerve; *rn*, recurrent nerve; *sg*, suboesophageal ganglion.

The corpora cardiaca receive their nerves from the protocerebrum and innervate the dorsal vessel, thus constituting a 'cardio-aortic system', while the frontal ganglion, hypocerebral ganglion and the gastric nerves, which are of tritocerebral origin, form a 'stomatogastric system' in the strict sense.[17] These systems contain both motor and sensory neurones. The motor fibres end in Doyère's hillocks (p. 150); the sensory fibres in bipolar or multipolar nerve cells on the surface of the muscles (p. 263). They doubtless control the movements of the heart and gut. In *Dytiscus* the frontal ganglion is said to contain the centre for deglutition: swallowing is abolished if it is eliminated; whereas if the frontal ganglion is preserved, swallowing is not affected by destruction of the brain or suboesophageal ganglia.[43] The system has been described in great detail in *Periplaneta.*[192]

(ii) **The unpaired ventral nerves** which arise from the suboesophageal and segmental ganglia and give rise to a pair of transverse nerves supplying the spiracles of their segment are generally included in the sympathetic system. They are typically developed in *Aeschna*[94] and in caterpillars. But they are often absent (as in *Dytiscus*) and the spiracles are then supplied from the lateral abdominal nerves.

(iii) **The caudal sympathetic system** which innervates the sexual organs and the posterior segments of the gut is composed of nerves arising from the composite terminal ganglion of the abdomen.

The mid-gut, the heart, and the Malpighian tubes show continued movements after separation from the central nervous system. The control of these movements is attributed to peripheral ganglion cells; and it has been suggested that it is this autonomous peripheral system that should be called the sympathetic nervous system.[58]

ENDOCRINE SYSTEM

The endocrine system of insects consists of (i) modified nerve cells ('neurosecretory cells'), (ii) 'neuro-haemal organs' which are believed to serve as the sites at which the neurosecretory products escape into the blood-stream and (iii) independent organs of internal secretion derived from nests of cells which have been budded off from the ectoderm in the region of the mouth parts. These various components of the endocrine system have been or will be discussed elsewhere, in accordance with their physiological functions; but it will be convenient to summarize here the organization and functions of the system as a whole. It is a field in which knowledge is expanding and interpretations are changing very rapidly at the present time.

Neurosecretory cells—The neurosecretory cells are often large and lobulated nerve cells with granular cytoplasm or inclusions staining deeply with paraldehyde fuchsin or with chromehaematoxylin after permanganate oxidation. But these staining properties vary;[145] and all nerve cells are continuously passing the products of the cell body along their axons; distinctive neurosecretory cells are therefore becoming increasingly difficult to define. A second criterion, which seems to be constant, is the presence in their cytoplasm of numerous electron dense granules of 1000–3000 Å diameter (p. 66).[152, 169]

Cells with these characters are present in the protocerebrum in medial and lateral groups, which secrete the hormone that activates the thoracic gland during moulting in the young stages (p. 67)[114] and the hormone which accelerates protein synthesis, and thus ensures ripening of the eggs, in the adult female (p. 727).[110] A third group in the brain is located in the deuto- and tritocerebrum of Phasmids; they secrete a humoral factor controlling colour change (p. 621).[161]

The dense 'elementary granules' are formed in association with mitochondria in the Golgi zone of these cells.[154, 175] They pass along the axons from the protocerebral neurosecretory cells to the corpus cardiacum and corpus allatum.[107] In *Periplaneta* there is a group of neurosecretory cells in the suboesophageal ganglion which secretes the hormone controlling the diurnal rhythm of locomotor activity (p. 342, &c.). Judging from observations during moulting, during reproductive activity, and during diapause it would seem that the periods when the neurosecretory cells are laden with secretion are the times of inactivity when the hormone is not being liberated.[191]

In the honey-bee, secretory activity of the neurosecretory cells seems to be ncreased not only during sexual maturity in both sexes but during the foraging ctivity of the workers.[89, 131] In some insects they are concerned in the regulation f water excretion (p. 555). In the higher Diptera their secretion controls the ardening and darkening of the cuticle after emergence (p. 47).

The axons from the medial and lateral groups of neurosecretory cells in the rotocerebrum cross over and leave the brain by two nerves, the medial and ateral nerves, to the corpus cardiacum. The secretory product can be traced long the axons to the corpus cardiacum (where it is believed to be liberated from he swollen nerve endings) and to the corpus allatum (p. 76). But in Aphids copi-us amounts of secretory material can be traced along axons to all parts of the entral nervous system; some end on the pericardial cells, others on muscles, uggesting that transport occurs directly to the organs and tissues.[146] The same videspread distribution is seen in *Corethra* larvae.[132]

Neurosecretory cells occur also throughout the thoracic and abdominal gang-ia of insects. In *Blaberus* three types of cells have been recognized; the secretory roducts from all of them can be seen in axons in the connectives, moving for-vards and backwards from the sites of secretion.[134] The neurosecretory cells in he abdominal ganglia in *Rhodnius* produce a diuretic hormone (p. 555); otherwise heir functions are unknown. [See p. 204.]

Corpus cardiacum—The corpora cardiaca of *Leucophaea* studied with the lectron microscope show the presence of three elements:[169] (i) the distended, ften bulbous endings of the axons from the neurosecretory cells of the proto-erebrum arriving by way of the two corpus cardiacum nerves. These endings are illed with neurosecretory granules; (ii) intrinsic cells, apparently of nervous ature with axon-like processes; they contain rather pale granules; (iii) inter-titial glial cells covering the nervous elements. They are clearly organs which re-eive and presumably liberate the neurosecretory product from the brain, and roduce a neurosecretory product of their own. It has been suggested (in *Peri-laneta*) that the neurosecretory material from the brain is not all liberated into he blood from the corpus cardiacum but some acts on the intrinsic cells and auses them to discharge their secretion down the axons to the target organs such s the pericardial cells.[147]

A number of physiological activities can be stimulated by extracts from the orpus cardiacum containing perhaps the secretion of its intrinsic neuroglandular ells. In *Periplaneta* and *Calliphora* the active substance (soluble in water and lcohol, and resistant to boiling) will bring about contraction of the chromato-hores of Crustacea.[14, 27, 82] The extract from *Corethra* larva will induce expan-ion of the melanophores spread over the air sacs.[96, 98, 102] A saline extract from he corpus cardiacum of *Periplaneta* (1 pair of glands in 10 ml.) will bring about cceleration of the heart-beat and increase in amplitude.[121] This factor, which ertainly comes from the intrinsic cells of the corpus cardiacum, seems to be a eptide or protein; it does not act directly on the heart but on the pericardial ells, causing them to release an active principle which may be an indolalkylamine imilar to serotonin. The same type of action is exerted on the hind-gut of the ockroach.[125] The extract from the corpus cardiacum of *Periplaneta* injected into he same species causes a rise in blood trehalose with a fall in glycogen in fat body nd central nervous system,[174] and a rise in blood glucose.[120] [See p. 205.]

Neurohumoral factors—Pharmacologically active substances of many inds can be extracted from the brain and corpora cardiaca of insects; but some of their effects have been obtained with extracts at concentrations which are quite unphysiological. *Acetylcholine* has been repeatedly demonstrated (p. 187). Concentrated extracts of corpora cardiaca of *Periplaneta* (14–40 pairs of glands in 0·5 ml.) have *adrenaline*-like properties in mammals.[118] *Serotonin* (5-hydroxytryptamine) has been extracted from the central nervous system.[124, 136]

Four neurohormones have been separated (by chromatographic methods) from the nervous system of *Periplaneta*: C_1 stimulates the heart-beat and causes darkening of the integument of *Carausius*. C_2 causes increased frequency of heart-beat but has no effect on colour change. D_1 increases frequency and amplitude of the heart-beat. D_2 acts like C_2.[135] Concentrated extracts of the corpus cardiacum of *Periplaneta* cause decreased frequency of spontaneous nerve impulses in the isolated nerve cord of the cockroach.[155] It is uncertain what part these substances play in normal physiology. [See p. 205.]

Corpus allatum—The corpus allatum is a glandular organ closely associated with the corpus cardiacum. The corpora allata arise by budding of epidermal

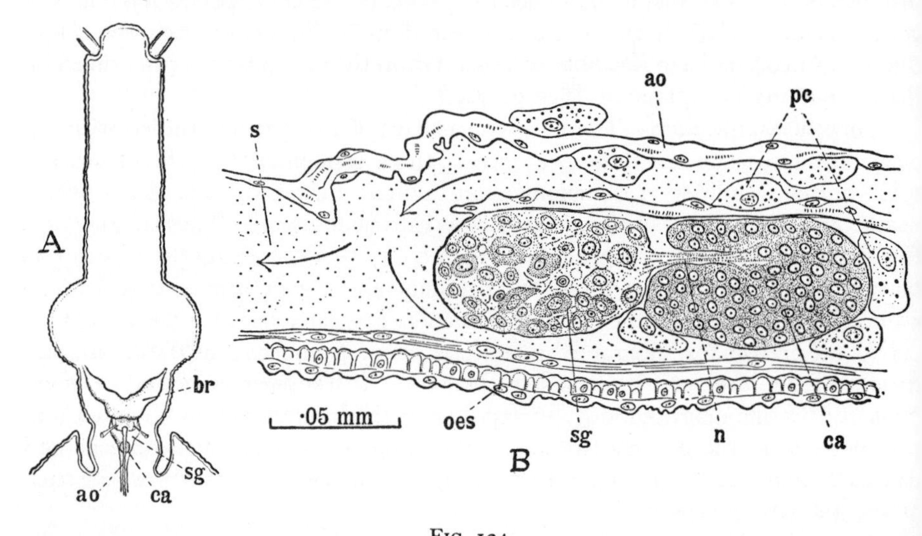

FIG. 124

A, head of *Rhodnius* from above to show position of corpus allatum; *B*, longitudinal section of corpus allatum and related structures (*after* WIGGLESWORTH). *ao*, aorta; *br*, brain; *ca*, corpus allatum; *n*, nerve to corpus allatum; *oes*, oesophagus; *pc*, pericardial cells; *sg*, corpus cardiacum.

cells between the mandibular and maxillary segments. Later these cell nests become separated from the body wall and form compact deeply staining bodies; in Phasmids their ectodermal origin is still apparent in their cyst-like form around a central vacuole. In some insects, e.g. many Hemiptera, they fuse to form a single median structure (Fig. 124).[48] They are absent in Collembola but occur in almost all other insects.[17] In some Lepidoptera the cells form loose clusters like bunches of grapes.[20] The corpora allata are innervated by the nerve which has traversed the corpus cardiacum: the whole 'retrocerebral gland system' has a common nerve supply.[17]

The corpus allatum secretes the juvenile hormone, neotenin (p. 78), which maintains larval characters in the young insect (p. 76), ensures the deposition of yolk in the developing egg (p. 724), induces the type of behaviour appropriate to the type of growth or reproductive activity that is to follow (p. 79), and has some general effects on metabolism (p. 726).[109, 122] Thus in the *Schistocerca* male the corpora allata control the colour change associated with sexual maturation;[149] in this and other locusts they are involved in phase change (p. 105), and in termites (p. 101) and Aphids (p. 102) they are concerned in polymorphism.

The active corpus allatum (e.g. in *Rhodnius*[91] and *Leucophaea*[170]) consists of closely packed cells with much homogeneous cytoplasm; the inactive gland is usually smaller, the cells are shrunken with vacuoles between them. Both light microscope and electron microscope commonly show neurosecretory granules in the nerve axons entering the corpus allatum.[168, 191] [See p. 206.]

There is a striking similarity between this system—consisting of a neurosecretory part (corpus cardiacum) and a glandular part of ectodermal origin

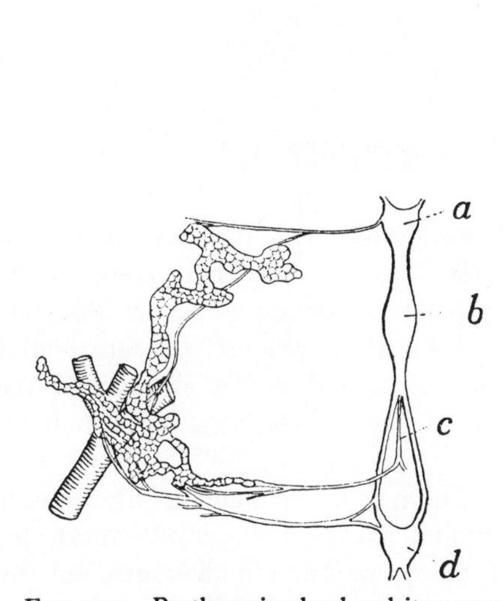

FIG. 125.—Prothoracic gland and its nerve supply in *Saturnia* (*after* LEE)

a, suboesophageal ganglion; *b*, prothoracic ganglion; *c*, median nerve; *d*, mesothoracic ganglion

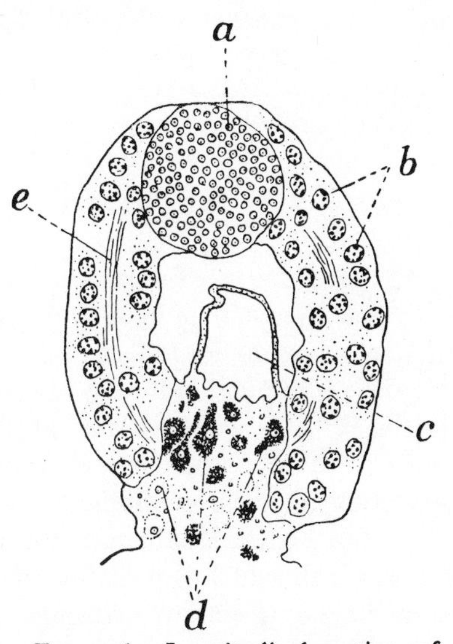

FIG. 126.—Longitudinal section of ring gland in young pupa of *Eristalis* (*after* CAZAL)

a, corpus allatum; *b*, large lateral cells; *c*, aorta; *d*, chromophil and chromophobe cells of corpus cardiacum; *e*, nerve to corpus allatum.

(corpus allatum) innervated by neurosecretory cells in the brain—and the association in vertebrates of the neuro-hypophysis and the glandular hypophysis, supplied by neurosecretory cells in the hypothalamus.[28, 17]

Thoracic glands—Thoracic, or prothoracic, glands of *Bombyx* arise in the embryo from an ingrowth of ectodermal cells at the base of the second maxilla.[177] In some insects, Thysanura, Ephemeroptera, Odonata,[133] Isoptera, Dermaptera, Acrididae, Phasmida,[191] they remain as compact structures in the lower part of the head and are then called 'ventral glands'.[50] In Blattidae, Hemiptera, Coleoptera, Lepidoptera, Hymenoptera, the cells are carried backwards into the thorax, perhaps by the growth of the salivary glands,[184] and here form loose chains of cells or lace-like meshworks. In the larvae of Orthorrhaphous Diptera, they are sometimes termed 'peritracheal glands.[104]

In the larvae of Cyclorrhaphous Diptera these various 'retrocerebral glands' are fused together to form Weismann's ring or the 'ring gland' which encircles the aorta just behind the brain. The ring gland is a composite structure (Fig. 126). A median dorsal mass of small cells is clearly the corpus allatum. The ventral mass of chromophil and chromophobe cells is recognizable as the corpus cardiacum probably combined with the hypocerebral ganglion; the large cells forming the side limbs represent the thoracic glands.[105]

The thoracic glands are closely associated with the tracheal system. They are richly innervated from suboesophageal, prothoracic, and mesothoracic ganglia in Lepidoptera (Fig. 125)[39] and in Blattidae and Acrididae; but a nerve supply is absent in *Rhodnius* and many other Hemiptera[184, 185] and in Coleoptera.

After activation by the hormone from the neurosecretory cells in the brain the thoracic glands secrete the moulting hormone, ecdyson; they go through a cycle of activity in each moulting stage and break down when the insect undergoes metamorphosis to the adult (p. 206). [See p. 206.]

ADDENDA TO CHAPTER V

p. 179. **Fusion of sensory axons**—Examination with the electron microscope has shown that in *Bombyx* and in *Rhodnius* the individual axons in the antennal nerve do not in fact fuse. The average fibre diameter is 0·3μ in *Bombyx*, 0·12μ in *Rhodnius* and they are packed in bundles without individual glial sheaths.[255] The same is true of the beetle *Trypodendron*. In *Periplaneta* the existence of individual axons from each tactile spine has been confirmed by electrophysiological methods.[208]

p. 180. **Multiplication of neurones**—There is little change in the number of nerve cells during growth. In the terminal ganglion of *Acheta* there are about 2,000 association neurones and 100 motor neurons in all stages; but the glial cells increase in number from about 1,000 to 17,000.[216, 225] In Lepidoptera (*Danaus*) new ganglion cells are formed by the division of neuroblasts, or ganglion mother cells, throughout larval life; glial cell production occurs at the time of moulting;[241] and the same is true in *Gryllus*.[246]

p. 182. **Fine structure**—The neurofilaments, or 'neurotubules', in *Periplaneta* closely resemble the 'microtubules' found in many non-nervous tissues; they appear to contain both a carbohydrate and protein component.[232]

p. 182. **Regeneration of axons**—Axons commonly degenerate when they are separated from the nerve cell body; but in the connectives of *Laplatacris*[236] and *Schistocerca* the distal portions of severed axons remain functional with no signs of degeneration for a week after section.[205] When a motor axon is cut in *Periplaneta*, visible changes which precede regrowth of the axon can be seen in the Golgi bodies and neurofibrils of the ganglion cell,[189] and in the cytoplasmic ribonucleic acid.[213] Such changes serve to associate specific nerve fibres with their nerve cell bodies. Not only peripheral but also central nervous structures can regenerate. Connexions between brain and optic lobes recover their function during regeneration; as do the severed olfactory and optic commissures of *Periplaneta*.[128]

p. 185. **Neural lamella**—In Lepidoptera (e.g. *Galleria*[197]) the entire neural lamella around the nerve cord is broken down by 24 hours after pupation; by about 72 hours the new neural lamella of the adult is forming and is complete by 120 hours. The perineurium cells responsible resemble the fibroblasts of vertebrates in their fine structure.[196]

p. 185. **Synapses**—The synapses of insects lie within the neuropile. In *Sarcophaga* and in *Tenebrio*[210] localized foci on adjacent axons with thickenings of the plasma membrane and a cleft about 200 Å wide between them are regarded as synapses. At these points 'synaptic vesicles' of two sorts occur (electron dense and electron lucent, ranging from 300–700 Å in diameter); these are believed to be the immediate source of chemical transmitter substance. In the corpora pedunculata of *Formica* the presynaptic fibres terminate in an 'end knob' with dozens of small postsynaptic end-feet attached to its surface, with the usual vesicles in the presynaptic region.[254]

p. 185. **Glial cells**—The importance of the system of glial cells has been increasingly recognized in recent years. Besides their nutritional function and the regulation of ionic distribution[232] they may well be concerned in gas exchange via the tracheoles, and the energy metabolism of the neurone.[230] The perineurial cells, as well as the cell bodies of the neurones, contain numerous lysosomes which contain stores of phosphatase and other hydrolytic enzymes.[231]

p. 187. **Propagation of action potentials**—Extended studies on *Periplaneta* and *Carausius* have substantiated the important role of the glial sheaths in maintaining an adequate concentration of sodium in immediate contact with the axons to permit the propagation of the action potential in accordance with the conventional theory—despite the low sodium and high potassium content of the haemolymph.[261] Furthermore there are 'tight junctions' in the perineurial intercellular membranes which restrict the inward movement of potassium to the central nervous system.[260] The extracellular fluid surrounding the axon is represented by the 200 Å space between the axon surface and the innermost glial membrane (see Fig. 112*b*). The diffusion pathway linking this space to the overlying extracellular system is the spiral mesaxon. The frequent dilatations of the concentric channels of the mesaxon are believed to serve as reservoirs which increase the extracellular fluid in the neighbourhood of the axon and thus facilitate the dispersal of potassium ions from the axon membrane. In addition the glial cells probably play an active part in regulating the ionic content of the bathing fluid. The whole system makes possible the usual process of sodium-potassium exchange necessary for the action potential, even when the general body fluid would be quite unsuitable in composition for this process.[173, 259] The fact that motor nerve terminals of insects with abnormal ionic concentrations are capped by glial cells, whereas those with normal Na/K ratios can occur as naked structures,[245] affords further evidence of the regulatory function of the glial cells.[259] In *Carausius*[258] and in Lepidoptera[266] a part of the action potential may result from an inward movement of magnesium.

p. 187. **Synaptic transmission**—Noradrenaline and dopamine occur in the brain of *Periplaneta* in concentrations equal to or greater than those found in the brains of vertebrates. Both are active on electrophysiological preparations of the insect nervous system. It seems likely that they play some role as neurotransmitters in insects. The highest concentration of these catecholamines in the brain of *Periplaneta* is in the corpora pedunculata.[219] By the use of histofluorescence and microspectrofluorometric methods it has been possible to demonstrate the presence of dopamine in certain cell bodies in the thoracic ganglia of Trichoptera and in the varicosities of axons in the ganglia.[203]

The distribution of cholinesterase activity in the central nervous system of *Periplaneta* has been defined in detail at the electron microscope level: the distribution agrees in general with that described in *Rhodnius* (p. 187);[187] it occurs particularly in the glial sheaths at the periphery of ganglia and connectives. This system is effective in drastically reducing the acetylcholine in the extracellular fluid.[253, 262] There is increasing evidence that acetylcholine is the excitatory transmitter in the insect central nervous system and that γ-aminobutyric acid (GABA) is the inhibitory transmitter. It may be that the effect of acetylcholine is to increase membrane permeability to sodium ions while the effect of GABA is to increase membrane permeability to chloride ions.[229, 264]

p. 188. **Giant axons**—In *Periplaneta* it has been found that the abdominal giant axons extend continuously to the suboesophageal ganglion, but they taper to become reduced from 60μ to 7μ in diameter as they leave the thorax. They appear to conduct impulses in both directions.[214, 247] At the posterior extremity, in the terminal ganglion, the giant axons break up into innumerable dendrites which make contact with the incoming sensory axons from the cerci.[237] It is doubtful whether the giant fibres in *Periplaneta* directly stimulate the motor neurones to the legs during the escape reaction, or whether they merely activate a general alarm system.[214]

p. 194. **Rhythmic movements in walking**—During rhythmic movements in the legs of *Periplaneta*, bursts of activity in the levator muscles of the femur are matched by reciprocal activity in the slow motor axons to the depressor muscles; and this reciprocal pattern persists even when all sensory input from the legs is eliminated.[248]

p. 197. **Frontal ganglion**—The role of the frontal ganglion is uncertain. It has long been believed to control swallowing. Its removal prevents sexual maturation,[211, 212] but this is probably a nervous effect via the neurosecretory cells and corpus cardiacum and not a direct endocrine effect.[251] Another suggestion is that the ganglion controls by nerve action the neurosecretory cells in the brain that secrete the diuretic factor in *Periplaneta*.[249]

p. 199. **Neurosecretory cells**—Many different types of neurosecretory cells have been described: for example in *Rhodnius*;[198] in *Blaps* 13 types are recognized.[218] Many of these differences probably represent phases in activity. The granules and vesicles passing down the axons and gaining discharge through the axon membrane by exocytosis, or reverse pinocytosis, often have a superficial resemblance to presynaptic vesicles.[242, 252] The axons of neurosecretory cells can conduct nerve impulses like other axons, and electrical stimulation leads to discharge of the secretion.[224] It is generally assumed that they are modified motor neurones, but in *Carausius* and in *Phormia* there are peripheral multipolar neurones, probably sensory neurones, with typical neurosecretory granules in their peripheral dendrites.[217] The product of the neurosecretory cells is polypeptide in character with disulphide (S–S) groups converted by permanganate to sulphydryl (SH) groups which stain with paraldehyde fuchsin, alcian blue or chrome-haematoxylin.[223] In the Pyrrhocorid bug *Iphita*[240] and in *Ranatra*[215] the neurosecretory axons from the pars intercerebralis end not in the corpus cardiacum but in the wall of the aorta where their secretion is liberated into the haemolymph. In adult *Calliphora* storage in the aorta occurs in addition to that n the corpus cardiacum.[257] The neurosecretion from the fused ganglia of *Rhodnius* which contains a diuretic hormone (p. 555) escapes from bulbous endings of the axons which lie just below the nerve sheaths a short distance after the abdominal nerves leave the ganglion (Fig. 126*a*).[233]

The transverse branches of the 'unpaired ventral nerves' in many insects have long been known to have small fusiform swellings. These are called 'perisympathetic organs' and seem to be the neurohaemal organs for neurosecretory cells in the segmental ganglia. They are seen in Phasmids, Blattids etc.[250] and in *Vespula* and other insects.[250] There is no conclusive evidence of their function. Neurosecretory cells in the pars intercerebralis in *Schistocerca*,[226] *Locusta*[207, 222] and *Gryllus*[221] are the source of a diuretic hormone. But in some conditions in *Periplaneta* etc. they exert an 'anti-diuretic' effect.[265] In *Locusta*

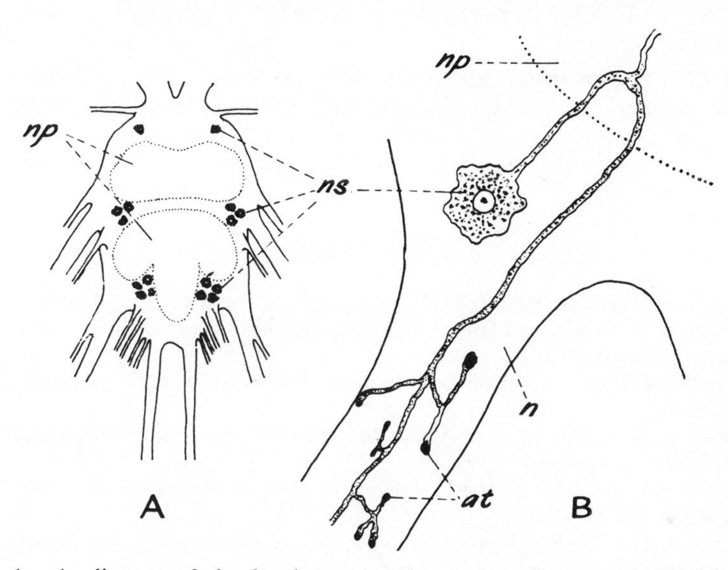

Fig. 126*a*—A, diagram of the fused meso- and metathoracic and abdominal ganglia of *Rhodnius* showing the three paired groups of neurosecretory cells. B, diagram of a neurosecretory cell of the posterior group showing the course of the axon and the swollen endings filled with neurosecretory material which it gives off (after Maddrell).

at, axon terminations; *n*, abdominal nerve; *np*, neuropile; *ns*, neurosecretory cells.

the two factors are present.[238] The darkening and tanning factor has been described (p. 47).

p. 199. **Corpus cardiacum**—Since the corpus cardiacum is a major neurohaemal organ many of the active factors which it liberates are the same as those extractable from neurosecretory cells in the brain. They seem mostly to be small peptides, heat stable and dialysable, but destroyed by chymotrypsin.[206, 239] Besides the hyperglycaemic hormone (p. 199),[243] there is an 'adipokinetic hormone' which increases the level of diglyceride in the haemolymph during flight in *Schistocerca*.[235] The anti-diuretic factor is extractable from both corpora cardiaca and the 'perisympathetic' organs of the ventral cord of *Periplaneta*, Phasmids, &c., and acts upon the Malpighian tubules and rectum.[202, 228]

p. 200. **Neurohumoral factors**—γ-Aminobutyric acid (GABA) (p. 204) can be synthesized in brain homogenates of *Drosophila* and is normally concentrated in the brain.[209]

Cyclic AMP—As in vertebrates so in insects adenosine 3′,5′-monophosphate (cyclic AMP) appears to play an essential role in the action of hormones. According to current theory when a hormone arrives at a cell it reacts with a specific receptor to stimulate the synthesis of cyclic AMP by adenyl cyclase. Cyclic AMP then serves as an 'intracellular second messenger' which is responsible for influencing the effector system. Examples among insects are the action of 5-hydroxytryptamine on salivary secretion in *Calliphora* (p. 528)[201] and the action of the diuretic hormone on the Malpighian tubules of *Rhodnius* and *Carausius* (p. 582).[234]

p. 201. **Corpus allatum**—The most striking feature of the fine structure of the corpus allatum is the abundant network of smooth endoplasmic reticulum recalling the steroid secreting cells of vertebrates: in *Schistocerca*,[244] *Locusta*[22] and Blattids.[252] In adult female *Calliphora* the active gland gives rise to numerous lipid droplets.[256]

p. 202. **Prothoracic gland**—In fine structure the prothoracic gland resembles the steroid secreting cells of vertebrates such as the interstitial cells of the mammalian testis.[252]

REFERENCES

1 ABBOTT, C. E. *Ent. News*, **52** (1941), 47–50 ('plastic behaviour': Odonata).

2 ADRIAN, E. D. *J. Physiol.*, **72** (1931), 132–51 (changes in electrical potential in nervous system: *Dytiscus*, Col.).

3 —— *J. Physiol.*, **91** (1937), 66–89 (electrical discharges in optic ganglion: *Dytiscus*, Col.).

4 ALVERDES, F. *Biol. Zbl.*, **45** (1925), 353–64 (effect of injuries to brain on movements *Cloëon* larva, Ephem.).

5 —— *Z. vergl. Physiol.*, **3** (1926), 558–94 (centres in ventral nerve cord: larvae of *Corethra* and *Cloëon*).

6 BALDI, E. *J. Exp. Zool.*, **36** (1922), 211–88 (circus movements after injury to brain Coleoptera).

7 BALDUS, K. *Z. wiss. Zool.*, **121** (1924), 557–620; *Z. vergl. Physiol.*, **6** (1927), 99–14 (structure and function of brain; circus movement, &c.).

8 BERLESE, A. *Gli Insetti*, Vol. **1**, Milan, 1909.

9 BETHE, A. *Arch. ges. Physiol.*, **68** (1897), 449–545 (brain function).

10 BETHE, A., and WOITAS, E. *Arch. ges. Physiol.*, **224** (1930), 821–35 (plasticity of nervous system: Coleoptera).

11 BICKLEY, W. E. *Ann. Ent. Soc. Amer.*, **35** (1942), 343–54 (stomodoeal nervous system

12 BINET, A. *J. Anat. Physiol.*, **30** (1894), 449–580 (structure and function of gangli of nerve cord).

13 BODENHEIMER, F. *Z. wiss. Zool.*, **121** (1924), 393–441 (biology of *Tipula*, Dipt.).

14 BROWN, F. A., and MEGLITSCH, A. *Biol. Bull.* **79** (1940), 409–18 (action of corpor cardiaca on chromatophores of Crustacea).

15 V. BUDDENBROCK, W. *Biol. Zbl.*, **41** (1921), 41–8 (walking movements: *Carausius*).

16 —— *Grundriss der vergleichenden Physiologie*, Vol. **1**, Borntraeger, Berlin, 1937.

17 CAZAL, P., *et al. Bull. Biol. Fr. Belg.*, **80** (1946), 477–82; 483–6; *Arch. Ziol. exp. gén* **84** (1946), 303–34; *Ibid.*, Notes et Rev., **85** (1947), 55–82; *Bull. Biol. Fr. Belg* Suppl. **32** (1948), 1–227 (retrocerebral endocrine glands of insects).

18 CORTEGGIANI, E., and SERFATY, A. *C.R. Soc. Biol.*, **131** (1939), 1124–6 (acetylcholine i insects).

19 DAY, M. F. *Ann. Ent. Soc. Amer.*, **36** (1943), 1–10 (ring gland of Diptera).

20 EHNBOM, K. *Opusc. Entom. Suppl.*, **8** (1948), 1–162 (sympathetic nervous system, &c.

21 FAIVRE, E. *Ann., Sci. Nat., Zool.* (sér. 4), **8** (1857), 245–74 (brain of *Dytiscus*).

[22] FANNHAUSER, G., and REIK, L. E. *Physiol. Zool.*, **8** (1935), 337–59 (case building by larva of *Neuronia*, Trichopt.).

[23] FRIEDRICH, H. *Z. vergl. Physiol.*, **18** (1933), 536–61 (physiology of peripheral nervous system: *Dixippus*, Orth.).

[24] HANSTRÖM, B. *Vergleichende Anatomie des Nervensystems der Wirbellosen Tiere*, Berlin, 1928.

[25] HANSTRÖM, B. *Hormones in Invertebrates*, Oxford, 1939.

[26] —— *Lunds Univ. Årsskr. N.F.* (2), **34** (1938), No. 16, 1–17 (neurosecretory cells. *Rhodnius*).

[27] —— *Lunds Univ. Årsskrift, N.F.* (2), **36** (1940), No. 12, 1–20 (chromatophorotropic substance in insect head).

[28] —— *Lunds Univ. Årsskrift, N.F.* (2), **37** (1941), No. 4, 1–19; *Biol. Generalis*, **15** (1942), 485–531 (corpora allata and corpora cardiaca).

[29] HOCHREUTHER, R. *Z. wiss. Zool.*, **103** (1912), 1–114 (sensory neurones of insects).

[30] V. HOLST, E. *Z. vergl. Physiol.*, **21** (1934), 395–414 (paths of motor and tonic stimulation: larvae of Lepidoptera).

[31] HOFFMANN, R. W. *Z. vergl. Physiol.*, **18** (1933), 740–95 (reflexes in *Blatta*, Orth.).

[32] HUGHES, G. M. *J. Exp. Biol.*, **29** (1952), 267–84; **34** (1957), 306–33; **35** (1958), 567–83 (co-ordination of locomotion in insects).

[33] JAWLOWSKI, H. *Z. Morph. Oekol. Tiere*, **32** (1936), 67–91 (brain in beetles).

[34] —— *Ann. Univ. Mariae Curie, Sklodowska*, **3**(c) (1948), 1–37 (structure of insect brain).

[35] JUNG, K. *Zool. Jahrb., Syst.*, **69** (1937), 373–416 (cleaning procedures among ants).

[36] KEMPER, H. *Z. Desinfektion*, **21** (1929), 1–13 (sensory physiology: *Cimex*, Hem.).

[37] KOPEC, S. *Zool. Anz.*, **49** (1912), 353–60; *Zool. Jahrb., Physiol.*, **36** (1918), 453–502 (physiology of central nervous system in Lepidoptera larvae and adults).

[38] LANG, J. *Biol. Zbl.*, **52** (1932), 582–4 (reflex walking after decapitation: *Dixippus*, *Tenebrio*).

[39] LEE, H.T.-Y. *Ann. Ent. Soc. Amer.*, **41** (1948), 200–5 (prothoracic glands: Lepidoptera).

[40] LEGHISSA, S. *Arch. Zool. Ital.*, **30** (1942), 289–310 (synapses: *Dixippus*).

[41] de LERMA, B. *Arch. Zool. Ital.*, **17** (1932), 412–33; *Rend. Accad. Lincei Rome Cl. Sci. fis. Nat. Mat.*, **17** (1933), 1105–8; *Arch. Zool. Ital.*, **24** (1937), 339–69 (corpora allata and pharyngeal bodies as endocrine glands).

[42] MCCRACKEN, I. *J. Comp. Neurol. Psychol.*, **17** (1907), 262–85 (oviposition reflexes: *Bombyx mori*).

[43] MARCHAL, P. *Richet's 'Dictionaire de Physiologie'*, **9** (1910), 273–386 (physiology of insects).

[44] MATULA, J. *Arch. ges. Physiol.*, **138** (1911), 388–456 (physiology of central nervous system: Odonata larva).

[45] MEANS, O. W. *J. Cell. Comp. Physiol.*, **20** (1942), 319–24 (cholinesterase: *Melanoplus*, Orth.).

[46] MIKALONIS, S. J., and BROWN, R. H. *J. Cell. Comp. Physiol.*, **18** (1941), 401–3 (acetylcholine in insects).

[47] NEWTON, H. C. F. *Quart. J. Micr. Sci.*, **74** (1931), 647–68 (development of sensory neurones: *Apis*).

[48] NABERT, A. *Z. wiss. Zool.*, **104** (1913), 181–358 (corpora allata).

[49] PFLUGFELDER, O. *Z. wiss. Zool.*, **149** (1937), 477–512 (corpora allata and cardiaca: *Dixippus*).

[50] —— *Z. wiss. Zool.*, **152** (1939), 384–408; *Biol. Zbl.*, **66** (1947), 211–35; *Verh. d. Dtsch. Zool*, 1949, 169–73 (pericardial and ventral glands in insects).

[51] POLIMANTI, O. *Arch. Ital. Biol.*, **47** (1907), 341–72 (physiology of central nervous system: larva of *Bombyx mori*).

[52] POULSON, D. F. *Trans. Conn. Acad. Arts Sci.*, **36** (1945), 449–87 (nature of ring gland in Diptera).

[53] PRINGLE, J. W. S. *J. Exp. Biol.*, **15** (1938), 101–13 (sensory neurones in the cockroach).

[54] —— *J. Exp. Biol.*, **17** (1940), 8–17 (reflexes in cockroach).

[55] PRINGLE, J. W. S., and HUGHES, G. M. *Nature*, **162** (1948), 558 (effect of central stimulation on reflexes: *Periplaneta*).

[56] PUMPHREY, R. J., and RAWDON-SMITH, A. F. *Proc. Roy. Soc.*, B, **122** (1937), 106–18 (transmission of nervous impulses in abdominal ganglia: cockroach).

[57] v. RATH, O. *Z. wiss. Zool.*, **46** (1888), 413–34 (sensory neurones of insects).

[58] REHM, E. *Z. Morph. Oekol. Tiere*, **36** (1939), 89–122 (innervation of internal organs: *Apis*).

[59] RICHARDS, A. G. *Biol. Bull.*, **83** (1942), 300; *J. N.Y. Ent. Soc.*, **51** (1943), 55–69; **52** (1944), 285–310 (lipid nerve sheaths in insects).

[60] RICHARDS, A. G., and CUTKOMP, L. K. *J. Cell. Comp. Physiol.*, **26** (1945), 57–61 (cholinesterase in insect nerves).

[61] ROEDER, K. D. *Biol. Bull.*, **69** (1935), 203–20 (sexual behaviour: *Mantis*, Orth.).

[62] —— J. *Exp. Zool.*, **76** (1937), 353–74 (control of locomotor activity: *Mantis*, Orth.).

[63] —— *J. Exp. Zool.*, **108** (1948), 243–61 (giant fibre system: *Periplaneta*).

[64] —— *J. Cell. Comp. Physiol.*, **32** (1948), 327–38 (effect of potassium and calcium on nervous system: *Periplaneta*).

[65] ROEDER, K. D., *et al. J. Cell. Comp. Physiol.*, **14** (1939), 1–9; *J. Neurophysiol.*, **10** (1947), 1–10; *Bull. Johns Hopkins Hospital*, **83** (1948), 587–99 (acetylcholine in nerve cord: *Periplaneta*).

[66] TONNER, F. *Zool. Anz.*, **113** (1936), 125–36 (cutaneous nervous system of Arthropods).

[67] SASSE, E. *Z. allg. Physiol.*, **13** (1912), 69–104 (physiology of nervous system: larva of *Lucanus*, Col.).

[68] SCHARRER, B. *Naturwiss.*, **25** (1937), 131–8 (secretory activity of nerve cells: review).

[69] —— *J. Comp. Neurol.*, **70** (1939), 77–88 (neuroglia: *Periplaneta*).

[70] SCHARRER, B., *et al. J. Comp. Neurol.*, **74** (1941), 93–108; *Biol. Bull.*, **87** (1944), 242–51; *Physiol. Rev.*, **25** (1945), 171–81; (neurosecretion: *Periplaneta*, &c.).

[71] —— *Biol. Bull.*, **95** (1948), 186–98 (prothoracic glands: *Leucophaea*, Orth.).

[72] SNODGRASS, R. E. *Smithsonian Misc. Coll.*, **77**, No. 8 (1926), 1–80 (sensory nervous system of insects: review).

[73] —— *The Principles of Insect Morphology*, New York, 1934.

[74] STRALLA, D. D. *Bull. Soc. Zool. France*, **73** (1948), 31–6; *C.R. Acad. Sci.*, **227** (1948), 1277–8 (ventral glands: Odonata).

[75] SZYMANSKI, J. S. *Arch. ges. Physiol.*, **170** (1918), 1–244 (cleaning reflexes, &c., in insects).

[76] TAHMISIAN, T. N. *J. Exp. Zool.*, **92** (1943), 199–213 (cholinesterase in egg: *Melanoplus*, Orth.).

[77] TEN CATE, J. *Arch. Néerl. Physiol.*, **12** (1928), 327–35; **25** (1941), 401–9 (physiology of thoracic ganglia: *Periplaneta*, Orth.).

[78] —— *Ergeb. Physiol.*, **33** (1931), 137–336 (physiology of central nervous system: review).

[79] —— *Arch. Néerl. Physiol.*, **21** (1936), 562–6 (nervous control of locomotion: *Locusta*, Orth.).

[80] THOMSEN, E. *Naturwiss.*, **29** (1941), 605–6; *Vidensk. Medd. Dansk Naturhist. Foren.*, **106** (1942), 319–405 (ring gland, &c.: *Calliphora*, Dipt.).

[81] —— *Nature*, **161** (1948), 439 (neurosecretory cells: *Calliphora* adult).

[82] THOMSEN, M. *Kgl. Danske Vidensk. Selskab. Biol. Medd.*, **19** (1943), No. 4, 1–30 (chromatophorotropic action of corpus cardiacum).

[83] TIRELLI, M. *Boll. Ist. Zool. Univ. Roma*, **5** (1927); **6** (1926) (function of brain: Coleoptera, Formicidae).

[84] TOBIAS, J. M., *et al. J. Cell. Comp. Physiol.*, **28** (1946), 159–82 (acetylcholine: *Periplaneta*, &c.).

[85] v. UEXKÜLL, J. *Z. Biol.*, **50** (1908), 168–202 (tonus and reflexes: Odonata).

[86] VOGEL, R. *Z. wiss. Zool.*, **120** (1923), 281–324 (sensory neurones in antennae of Hymenoptera).

[87] VOGT, M. *Arch. Entw. Mech.*, **142** (1942), 129–82; *Biol. Zbl.*, **63** (1943), 56–71 (hormones from ring gland: *Drosophila*, *Calliphora*, Dipt.).

[88] WEBER, H. *Lehrbuch der Entomologie*, Jena, 1933.

[89] WEYER, F. *Zool. Anz.*, **112** (1935), 137–41 (gland-like nerve cells in brain: *Apis*).

[90] WEYRAUCH, W. K. *Z. verg. Physiol.*, **10** (1929), 665–87; *Zool. Jahrb., Physiol.*, **47** (1930), 1–28 (cleaning reflexes, &c.: *Forficula*).

[91] WIGGLESWORTH, V. B. *Quart. J. Micr. Sci.*, **77** (1934), 191–222 (corpus allatum and pericardial cells: *Rhodnius*).

[92] —— *Unpublished observations.*

[93] WILLIAMS, C. M. *Biol. Bull.*, **94** (1948), 60–5; **97** (1949), 111–4 (prothoracic glands: Lepidoptera).

[94] ZAWARZIN, A. *Z. wiss. Zool.*, **122** (1924), 97–115; 323–424 (structure of nervous ganglia: nymph of *Aeschna*, Odonata).

SUPPLEMENTARY REFERENCES (A)

[95] ARVY, L., and GABE, M. *Bull. Soc. Zool. France*, **75** (1950), 267–85 (retro-cerebral endocrine organs in Ephemeroptera).

[96] DUPONT-RAABE, M. *Arch. Zool. exp. gén.* (Notes et Rev.) **86** (1949), 32–9 (control of chromatophores in *Corethra* larva).

[97] FLOREY, E. *Z. vergl. Physiol.*, **33** (1951), 327–77 (stimulatory substance in sensory nerves of insects).

[98] HADORN, E., and FRIZZI, G. *Rev. Suisse Zool.*, **56** (1949), 306–16 (control of melanophores in *Corethra* larva).

[99] HINTON, H. E. *Proc. S. London Ent. Nat. Hist. Soc.*, 1951, 124–60 (endocrine glands in Lepidoptera: review).

[100] HOYLE, G. *Nature*, **169** (1952), 281–2 (high blood potassium and nerve conduction in insects).

[101] KAISER, P. *Arch. Entw. Mech.*, **144** (1949), 99–131 (histology of corpora allata and prothoracic glands in Lepidoptera).

[102] KOPENEC, A. *Z. vergl. Physiol.*, **31** (1949), 490–505 (control of melanophores in *Corethra* larva).

[103] de LERMA, B. *Arch. Zool. Ital.*, **32** (1947); *Ann. Ist. Mus. Zool. Univ., Napoli*, **3** (1951), 1–23 (endocrine organs in Thysanura).

[104] POSSOMPÈS, B. *Bull. Soc. Zool. France*, **71** (1946), 99–109; **73** (1948), 202–6; 228–35 ('peritracheal gland', etc. in *Chironomus, Tabanus, Calliphora*).

[105] —— *C.R. Acad. Sci.*, **228** (1949), 1527–9; **230** (1950), 409–11; **231** (1950), 594–6 (role of brain, 'peritracheal gland', etc. in metamorphosis of Diptera).

[106] REHM, M. *Z. Naturforsch.*, **56** (1950), 167–9; *Arch. Entw. Mech.*, **145** (1951), 205–48 (secretory cycles in neurosecretory cells, &c. in *Ephestia*).

[107] SCHARRER, B. *Anat. Rec.* (Suppl.) **111** (1951), 554–5; **112** (1952), 386–7 (storage of neurosecretory material in corpus cardiacum).

[108] SELLIER, R. *Arch. Zool. exp. gén.*, **88** (1951), 61–72 (prothoracic gland in *Gryllus*).

[109] THOMSEN, E. *J. Exp. Biol.*, **26** (1949), 137–49 (corpus allatum and oxygen consumption in *Calliphora*).

[110] —— *J. Exp. Biol.*, **29** (1952), 137–72 (neurosecretory cells and corpus cardiacum in adult *Calliphora*).

[111] THOMSEN, M. *Kongelig. Danske Vidensk. Selsk. Biol. Skrift.*, **6** (1951), No. 5, 32 pp. (Weismann's ring, &c. in larvae of Diptera).

[112] VOGT, M. *Z. Zelforsch.*, **34** (1948), 160–4 (fat body in *Drosophila* after extirpation of adult ring gland).

[113] WIGGLESWORTH, V. B. *Nature*, **168** (1951), 558; *J. Exp. Biol.*, **29** (1952), 561–70 (thoracic gland and moulting in *Rhodnius*).

[114] WILLIAMS, C. M. *Growth Symposium*, **12** (1948), 61–74 (neurosecretory cells and growth in Lepidoptera).

SUPPLEMENTARY REFERENCES (B)

[115] ASHHURST, D. E. *Quart. J. Micr. Sci.*, **100** (1959), 401–12 (connective tissue sheath of ganglia in locust).

[116] ASHHURST, D. E., and PATEL, N. G. *Ann. Ent. Soc. Amer.*, **56** (1963), 182–4 (hyaluronic acid in *Periplaneta* ganglia).

[117] BACCETTI, B. *Redia*, **40** (1955), 197–212; **46** (1961), 1–7 (fine structure of neural lamella).

[118] BARTON-BROWNE, L., DODSON, L. F., *et al. Gen. Comp. Endocrin.*, **1** (1961), 232 (adrenergic properties of corpus cardiacum of *Periplaneta*).

[119] BODENSTEIN, D. *J. Exp. Biol.*, **139** (1957), 89–116 (nerve regeneration in *Periplaneta*)

[120] BOWERS, W. S., and FRIEDMAN, S. *Nature*, **198** (1963), 685 (corpus cardiacum and glycogen mobilization).

[121] CAMERON, M. L. *Nature*, **172** (1953), 349 (corpus cardiacum and heart acceleration).

[122] CHADWICK, L. E. *Handb. d. exp. Pharmakol. erg. W.*, **15** (1963), 742–98 (acetylcholine review).

[123] COLHOUN, E. H. *J. Ins. Physiol.*, **2** (1958), 108–27 (acetylcholine in *Periplaneta*); *Adv Ins. Physiol.*, **1** (1963), 1–46 (acetylcholine: review).

[124] —— *Experientia*, **19** (1963), 9 (synthesis of 5-hydroxytryptamine in *Periplaneta*).

[125] DAVEY, K. G. *Gen. Comp. Endocrin.*, **1** (1961), 24–9; *Quart. J. Micr. Sci.*, **103** (1962) 349–58; *J. Ins. Physiol.* **8** (1962), 205–8; 579–8; **9** (1962), 375–81 (corpus cardiacum and acceleration of heart beat).

[126] DAY, M. F. *Aust. J. Sci. Res.*, B, **3** (1950), 61–75 (histology of the giant Blattid *Macropanesthia*).

[127] DETHIER, V. G., LARSEN, J. R., and ADAMS, J. R. *Ist Int. Symp. on Olfaction and Taste Stockholm, 1962* (Zottermann, Y., Ed.), Pergamon Press, Oxford, 1963, 105–1c (fine structure of olfactory organs of *Phormia*).

[128] DRESCHER, W. *Z. Morph. Oekol. Tiere*, **48** (1960), 576–649 (regeneration in the brain)

[129] FIELDEN, A. *J. Exp. Biol.*, **37** (1960), 832–44 (giant axons, &c. in *Anax*).

[130] —— *J. Exp. Biol.*, **40** (1963), 553–61 (motor and sensory roots in *Anax*).

[131] FORMIGONI, A. *Ann. Sci. Nat., Zool.*, **18** (1956), 283–91 (neurosecretion in *Apis*).

[132] FÜLLER, H. B. *Zool. Jahrb., Physiol.*, **69** (1960), 223–50 (neurosecretory cells in ganglia of *Periplaneta* and *Corethra*).

[133] GABE, M. *Experientia*, **9** (1953), 352–6 (insect endocrine glands: review).

[134] GELDIAY, S. *Biol. Bull.*, **117** (1959) (neurosecretory cells in *Blaberus*).

[135] GERSCH, M., *et al. Ber. u.d. Hundertjahrfeir d. dtsch. ent. Ges.*, 1956, 146–69; *Naturwiss.* **20** (1957), 525–32; *J. Ins. Physiol.*, **2** (1958), 281–97; *Z. Naturforsch.*, **15b** (1960) 319–22 (neurohormonal factors).

[136] GERSCH, M., *et al. Z. Naturforsch.*, **16** (1961), 351–2 (serotonin in nervous system of *Periplaneta*).

[137] HAGIWARA, S., and WATANABE, A. *J. Cell. Comp. Physiol.*, **47** (1956), 415–28 (synaptic transmission in a cicada).

[138] HESS, A. *J. Morph.*, **103** (1958), 479–502; *Quart. J. Micr. Sci.*, **99** (1958), 334–4c (central and peripheral nerve-fibres in *Periplaneta*).

[139] —— *J. Biophys. Biochem. Cytol.*, **4** (1958), 731–42 (fine structure of ganglia in *Periplaneta*).

[140] HOYLE, G. *Nature*, **169** (1952), 281–2; *J. Exp. Biol.*, **30** (1953), 121–35 (potassium ions and conduction in insect nerves).

[141] —— *Nature*, **172** (1953), 165 ('slow' and 'fast' nerve fibres in locusts).

[142] HUBER, F. *Naturwiss.*, **42** (1955), 566–7; *Z. Tierpsychol.*, **12** (1955), 12–48; *Verh. dtsch zool. Ges.*, 1959, 248–69; *Z. vergl. Physiol.*, **44** (1960), 60–132 (role of higher centres in the control of movements in crickets, &c.).

[143] —— *Fortschr. Zool.*, **15** (1962), 165–213 (physiology of invertebrate nervous system review).

[144] JOHANSSON, A. S. *Trans. Amer. Ent. Soc.*, **83** (1957), 119–83 (nervous system of *Oncopeltus*).

[145] —— *Nyt. Mag. Zool.*, **7** (1958), 5–132 (histological types of neurosecretory cells).

[146] JOHNSON, B. *J. Ins. Physiol.*, **9** (1963), 727–39 (distribution of neurosecretory material in Aphids).

[147] JOHNSON, B., and BOWERS, B. *Science*, **141** (1963), 264–6 (axonal transport of neurosecretion from corpus cardiacum of *Periplaneta*.

[148] KIM, CHANG-WHAN. *Bull. Dept. Biol. Korea Univ.*, **3** (1961), 1–8 (development of sense organs and nerves of the labial palp in *Pieris*).

[149] LOHER, W. *Proc. Roy. Soc.*, B, **153** (1960), 380–97 (corpus allatum and maturation in male *Schistocerca*).

[150] MEHROTRA, K. N. *J. Ins. Physiol.*, **5** (1960), 129–42 (cholinergic system in insect eggs).

[151] METCALF, R. L., MARCH, R. B., and MAXON, M. G. *Ann. Ent. Soc. Amer.*, **48** (1955), 222–8 (cholinesterase in insects).

[152] MEYER, G. F., and PFLUGFELDER, O. *Z. Zellforsch.*, **48** (1958), 556–64 (fine structure of corpora cardiaca in *Carausius*).

[153] NARAHASHI, T. in *Electrical activity of single cells* (T. Katsuki, Ed.) Tokyo, 1960, 119–31; *Adv. Ins. Physiol.*, **1** (1963), 175–256 (properties of insect axons).

[154] NISHIITSUTSUJI-UWO, J. *Z. Zellforsch.*, **54** (1961), 613–30 (fine structure of neurosecretory cells in Lepidoptera).

[155] OSBAS, S., and HODGSON, E. S. *Proc. Nat. Acad. Sci.*, **44** (1958), 825–30 (action of corpus cardiacum extracts).

[156] PANOV, A. A. *Akad. Nauk. SSSR*, **38** (1959), 775–7 (sensory axons and development of ganglia).

[157] —— *Doklady Akad. Nauk. SSSR*, **132** (1960), 689–92 (origin of glial cells).

[158] PIPA, R. L. *J. Comp. Neurol.*, **116** (1961), 15–26 (neuroglia of *Periplaneta*).

[159] POWER, M. E. *J. Exp. Zool.*, **103** (1946), 429–61 (sensory nerves and development of central ganglia in *Drosophila*).

[160] —— *J. Morph.*, **91** (1952), 389–412 (growth of the neuropile, &c. in *Drosophila*).

[161] RAABE, M. *C.R. Acad. Sci.*, **257** (1963), 1171–3; 1552–5; 1804–6 (tritocerebral neurosecretion and colour change in Phasmids).

[162] RICHARDS, A. G., and SCHNEIDER, D. *Z. Naturforsch.*, **13b** (1958), 680–7 (structure of neural lamella).

[163] ROEDER, K. D. *Ann. Rev. Ent.*, **3** (1958), 1–18 (insect nervous system: review).

[164] —— *Smithsonian Misc. Coll.*, **137** (1959), 287–306; *Nerve cells and insect behaviour*, Harvard Univ. Press, 1963 (speed of nervous responses, &c.).

[165] ROEDER, K. D., TOZIAN, L., and WEIANT, E. A. *J. Ins. Physiol.*, **4** (1960), 45–62 ('endogenous nerve activity' and inhibition in mantis, &c.).

[166] RÖNSCH, G. *Z. Morph. Oekol. Tiere*, **43** (1954), 1–62 (differentiation of sense cells in epidermis of Trichoptera).

[167] ROSSBACH, W. *Z. Morph. Oekol. Tiere*, **50** (1962), 616–50 (size of brain in male and female *Deporaus*, &c.).

[168] SCHARRER, B. *Biol. Bull.*, **121** (1961), 370; *Mem. Soc. Endocrin.*, **12** (1962), 89–97; *Z. Zellforsch.*, **62** (1964), 125–48 (fine structure of corpus allatum, &c. in *Leucophaea*).

[169] —— *Z. Zellforsch.*, **60** (1963), 761–96 (fine structure of corpus cardiacum of *Leucophaea*).

[170] SCHARRER, B., and V. HARNACK, M. *Biol. Bull.*, **115** (1958), 508, 520; **121** (1961), 193–8 (histology of corpus allatum in *Leucophaea*).

[171] SCHNEIDER, K. C. *Lehrbuch der vergleichenden Histologie der Tiere*, 1902, 988 pp. Fischer, Jena.

[172] SMALLMAN, B. N. *J. Physiol.*, **132** (1956), 343–57 (acetylcholine, &c. in *Calliphora*).

[173] SMITH, D. S., and TREHERNE, J. E. *Adv. Ins. Physiol.*, **1** (1963), 401–84 (structure and function in central nervous system: review).

[174] STEELE, J. E. *Nature*, **192** (1961), 680–1; *Gen. Comp. Endocrin.*, **3** (1963), 46–52 (hyperglycaemic factor in corpus cardiacum).

[175] STIENNON, J. A., and DROCHMANS, P. *Gen. Comp. Endocrin.*, **1** (1961), 286–94 (fine structure of neurosecretory cells in Phasmidae).

[176] STÜRCKOW, B. *Z. Zellforsch.*, **57** (1962), 627–47 (fusion of axons in labellar nerve of *Calliphora.*, &c.).

[177] TOYAMA, K. *Bull. Coll. Agric. Tokyo*, **5** (1902), 73–117 (development of prothoracic gland in *Bombyx* embryo).

[178] TREHERNE, J. E. *J. Exp. Biol.*, **37** (1960), 513–33 (uptake and metabolism of sugars by ganglia of *Periplaneta*).

[179] —— *J. Exp. Biol.*, **38** (1961), 315–32; 629–36; 729–46; **39** (1962), 631–41 (ionic regulation in central nervous system of *Periplaneta*).

[180] TRUJILLO-CENÓZ, O. *Z. Zellforsch.*, **49** (1959), 432–46 (fine structure of ganglia in *Pholus*, Lepidopt.).

[181] TWAROG, B. M., and ROEDER, K. D. *Biol. Bull.*, **111** (1956), 278–86; *Ann. Ent. Soc. Amer.*, **50** (1957), 231–7 (properties of the sheath in the nerve cord of *Periplaneta*).

[182] VOSKRESENSKAJA, A. K. *Dokl. Acad. Nauk. SSSR*, **112** (1957), 964–7 (role of mushroom bodies in conditioned reflexes in *Apis*).

[183] VOWLES, D. M. *Brit. J. Anim. Behaviour*, **2** (1954), 116; *Quart. J. Micr. Sci.*, **96** (1954), 239–55; *Anim. Behaviour*, **6** (1958), 115–16; in *Current problems in animal behaviour* (W. H. Thorpe and O. L. Zangwill, Edd.), Cambridge Univ. Press, 1961, 5–29 (role of mushroom bodies, &c. in Hymenoptera).

[184] WELLS, M. J. *Quart. J. Micr. Sci.*, **95** (1954), 231–44 (thoracic glands in Hemiptera).

[185] WIGGLESWORTH, V. B. *J. Exp. Biol.*, **29** (1952), 561–70 (thoracic gland and moulting in *Rhodnius*).

[186] —— *Quart. J. Micr. Sci.*, **94** (1953), 93–112 (origin of sensory neurones in *Rhodnius*).

[187] —— *Quart. J. Micr. Sci.*, **99** (1958), 447–50 (esterase in nervous system of *Rhodnius*).

[188] —— *Quart. J. Micr. Sci.*, **100** (1959), 285–98 (histology of peripheral nervous system in *Rhodnius*); 299–313 (ditto of central ganglia).

[189] —— *Quart. J. Micr. Sci.*, **101** (1960), 381–8 (axon structure and Golgi bodies in *Periplaneta*).

[190] —— *J. Exp. Biol.*, **37** (1960), 500–12 (nutrition of ganglia in *Periplaneta*).

[191] —— *Adv. Ins. Physiol.*, **2** (1964), 243–332 (hormones in insect growth and reproduction: review).

[192] WILLEY, R. B. *J. Morph.*, **108** (1961), 219–47 (stomodeal nervous system and associated glands in *Periplaneta*).

[193] YAMASAKI, T., and NARAHASHI, T. *Nature*, **182** (1958), 1805–6; *J. Ins. Physiol.*, **4** (1960), 1–13 (synaptic transmission in *Periplaneta*).

[194] —— *J. Ins. Physiol.*, **3** (1959), 146–58; 230–42 (properties of giant axons in *Periplaneta*).

[195] YUSHIMA, T. *J. Econ. Ent.*, **50** (1957), 441–3 (synthesis of acetylcholine in embryos of Lepidoptera).

SUPPLEMENTARY REFERENCES (C)

[196] ASHHURST, D. E. *Quart. J. micro Sci.*, **105** (1964), 391–403; **106** (1965), 61–73; *Ann. Rev. Ent.*, **13** (1968), 45–74 (formation of neural lamella).

[197] ASHHURST, D. E. and RICHARDS, A. G. *J. Morph.*, **114** (1964), 225–54 (connective tissue of central nervous system in *Galleria*).

[198] BAEHR, J. C. *C.R. Acad. Sc. Paris*, **267** (1968), 2364–7; **268** (1969), 151–4 (neurosecretory cells in brain of *Rhodnius*).

[199] BASSURMANOVA, O. K. and PANOV, A. A. *Gen. comp. Endocr.*, **9** (1962), 245–62 (neurosecretion in *Bombyx*: fine structure and staining).

[200] BAUDRY, N. *C.R. Acad. Sc. Paris*, **267** (1968), 2356–9; **268** (1969), 147–50 (neurosecretory cells in ventral nerve chain of *Rhodnius*).

[201] BERRIDGE, M. J. *J. exp. Biol.*, **53** (1970), 171–86 (role of cyclic AMP in secretion by salivary glands of *Calliphora*).

[202] DE BESSÉ, N. and CAZAL, M. *C.R. Acad. Sc. Paris*, **266** (1968), 615–18 (antidiuretic substance in corpora cardiaca and perisympathetic organs of *Periplaneta* &c.).

[203] BJÖRKLUND, A., FLACK, B. and KLEMM, N. *J. Insect Physiol.*, **16** (1970), 1147–54 (catecholamines in ganglia of Trichoptera).

[204] BLOCH, B., THOMSEN, E. and THOMSEN, M. *Z. Zellforsch. mikrosk. Anat.*, **70** (1966), 185–208 (fine structure of neurosecretory cells in *Calliphora*).

[205] BOULTON, P. S. *et al. Z. Zellforsch. mikrosk. Anat.*, **101** (1969), 98–134 (degeneration in central nervous system of *Schistocerca*).

[206] BROWN, B. E. *Gen. comp. Endocr.*, **5** (1965), 387–401 (active substances from corpus cardiacum of *Periplaneta*).

[207] CAZAL, M. and GIRARDIE, D. *J. Insect Physiol.*, **14** (1968), 655–68 (humoral control of water balance in *Locusta*).

[208] CHAPMAN, K. M. and NICHOLS, T. R. *J. Insect Physiol.*, **15** (1969), 2103–15 (sensory axons from tactile spines of *Periplaneta*).

[209] CHEN, P. S. and WIDNER, B. *Experientia*, **24** (1968), 516–17 (synthesis of GABA in brain of *Drosophila*).

[210] CHIARODO, A. J., KISSEL, J. H. and MACKELL, T. E. *J. Insect Physiol.*, **16** (1970), 361–71 (synapses in neuropile of *Tenebrio* and *Sarcophaga*).

[211] CLARKE, K. U. and GILLOT, C. *J. exp. Biol.*, **46** (1967), 13–34 (metabolic effects of removal of frontal ganglion in *Locusta*).

[212] CLARKE, K. U. and LANGLEY, P. A. *Gen. comp. Endocr.*, **2** (1962), 625–6; *J. Insect Physiol.*, **9** (1963), 411–21 (role of the frontal ganglion and growth in *Locusta*).

[213] COHEN, M. J. and JACKLET, J. W. *Science*, **148** (1965), 1237–9; *Phil. Trans. Roy. Soc. Lond.*, B, **232** (1967), 561–72 (RNA in motor neurones during regeneration of axons in *Periplaneta*).

[214] DAGAN, D. and PARNAS, I. *J. exp. Biol.*, **52** (1970), 313–24 (giant axon function in *Periplaneta*).

[215] DOGRA, G. S. *J. Morph.*, **121** (1967), 223–40 (neurosecretory axons in *Ranatra* ending in the aorta).

[216] EDWARDS, J. S. *Adv. Insect Physiol.*, **6** (1969), 97–137 (postembryonic development and regeneration of insect nervous system: review).

[217] FINLAYSON, L. H. and OSBORNE, M. P. *J. Insect Physiol.*, **14** (1968), 1793–1801 (peripheral neurosecretory cells in *Carausius* and *Phormia*).

[218] FLETCHER, B. S. *J. Insect Physiol.*, **15** (1969), 119–34 (neurosecretory cell types in *Blaps*).

[219] FRONTALI, N. *et al. Acta physiol. scand.*, **66** (1966), 243–4; *J. Insect Physiol.*, **14** (1968), 881–6; *Brain Res.*, **14** (1969), 540–2; *Z. Zellforsch.*, **103** (1970), 341–50 (catecholamines in neurones of brain in *Periplaneta*).

[220] GABE, M. *Neurosecretion*, 1966, 872 pp. Pergamon, Oxford.

[221] GIRARDIE, A. *C.R. Acad. Sci. Paris*, **262** (1966), 1361–4 (diuretic hormone from neurosecretory cells in *Gryllus*).

[222] GIRARDIE, A. *C.R. Acad. Sci. Paris* **271** (1970), 504–7; *Bull. Soc. zool. Fr.*, **95** (1970), 783–802 (diuretic hormone in *Locusta*, &c.).

[223] GIRARDIE, A. and GIRARDIE, J. *Z. Zellforsch.*, **78** (1967), 54–75; *J. Insect Physiol.*, **16** (1970), 1745–56 (histology and histochemistry of neurosecretory cells in *Locusta*).

[224] GOSBEE, J. L., MILLIGAN, J. V. and SMALLMAN, B. N. *J. Insect Physiol.*, **14** (1968), 1785–92 (nature of secretory neurones in brain of *Periplaneta*).

[225] GYMER, A. and EDWARDS, J. S. *J. Morph.*, **123** (1967), 191–7 (postembryonic growth of ganglion in *Acheta*).

[226] HIGHNAM, K. C., HILL, L. and GINGELL, D. J. *J. Zool.*, **147** (1965), 201–15 (neurosecretion and water balance in *Schistocerca*).

[227] JOLY, L., PORTE, A. and GIRARDIE, A. *C.R. Acad. Sci., Paris*, **265** (1967), 1633–5 (fine structure of active and inactive corpus allatum in *Locusta*).

[228] JOLY, P. and CAZAL, M. *Bull. Soc. zool. Fr.*, **94** (1969), 181–94 (functions of corpus cardiacum: review).

[229] KERKUT, G. A., PITMAN, R. M. and WALKER, R. J. *Nature*, **222** (1969), 1075–6 (sensitivity of insect neurones to acetylcholine and GABA).

[230] LANDOLT, A. M. *Z. Zellforsch.*, **66** (1965), 701–36 (relations between glia and neurones in *Formica*).

[231] LANE, N. J. *J. Cell. Biol.*, **37** (1968), 89–104; *Z. Zellforsch.*, **86** (1968), 293–312 (fine structure of perineurium and neuroglia, especially regarding phosphatase, in *Melanoplus*).

[232] LANE, N. J. and TREHERNE, J. E. *Nature*, **223** (1969), 861–2; *J. Cell Sci.*, **7** (1970), 217–31 (fine structure of nervous and non-nervous tissues in ganglion of *Periplaneta*).

[233] MADDRELL, S. H. P. *J. exp. Biol.*, **45** (1966), 499–508; *Symp. Roy. ent. Soc. Lond.*, **5** (1970), 101–16 (diuretic hormone in *Rhodnius* and its neurohaemal release.

[234] MADDRELL, S. H. P., PILCHER, D. E. M. and GARDINER, B. O. C. *Nature*, **222** (1969), 784–785; *J. exp. Biol.*, **54** (1971), 779–804 (pharmacology of Malpighian tubules of *Rhodnius* and *Carausius*).

[235] MAYER, R. J. and CANDY, D. J. *J. Insect Physiol.*, **15** (1969), 611–20 ('adipokinetic hormone' from corpus cardiacum of *Schistocerca*).

[236] MELAMED, J. and TRUJILLO-CENÓZ, O. *Z. Zellforsch.*, **59** (1963), 851–6 (changes in nerve fibres after transection).

[237] MILBURN, N. S. and BENTLEY, D. R. *J. Insect Physiol.*, **17** (1971), 607–23 (dendritic connexions between giant axons and cerci in *Periplaneta*).

[238] MORDUE, W. *J. Endocr.*, **46** (1970), 119–20 (presence of diuretic and antidiuretic hormones in locusts).

[239] NATALIZI, G. M. *et al. J. Insect Physiol.*, **12** (1966), 1279–87; **16** (1970), 1827–36 (physiologically active peptides from corpora cardiaca of *Periplaneta*).

[240] NAYAR, K. K. *Z. Zellforsch.*, **44** (1956), 697–705 (neurosecretory pathways in *Iphita*).

[241] NORLANDER, R. H. and EDWARDS, J. S. *Wilhelm Roux Arch. Entw. Mech. Org.*, **162** (1969), 197–217 (cellular changes in brain of *Danaus* at metamorphosis).

[242] NORMANN, T. C. *Z. Zellforsch.*, **67** (1965), 461–501; *Exp. Cell Res.*, **55** (1969), 285–7 (discharge of neurosecretion through wall of axon in *Calliphora*).

[243] NORMANN, T. C. and DUVE, H. *Gen. comp. Endocr.*, **12** (1969), 449–59 (hyperglycaemic hormone in *Calliphora*).

[244] ODHIAMBO, T. R. *J. Insect Physiol.*, **12** (1966), 995–1002 (fine structure of corpus allatum in *Schistocerca*).

[245] OSBORNE, M. P. *Symp. Roy. ent. Soc. Lond.*, **5** (1970), 77–100 (dendrite terminations in stretch receptors).

[246] PANOV, A. A. *Dokl. Akad. Nauk. SSSR* Moscow, **143** (1962), 471–4 (cell division in neuroblasts of *Gryllus*).

[247] PARNAS, I. and DAGAN, D. *Adv. Insect Physiol.*, **8** (1971), 96–144 (giant axons: review).

[248] PEARSON, K. G. and ILES, J. F. *J. exp. Biol.*, **52** (1970), 139–65 (control of reciprocal contractions during walking in *Periplaneta*).

[249] PENZLIN, H. *J. Insect Physiol.*, **17** (1971), 559–73 (frontal ganglion and indirect control of diuretic hormone in *Periplaneta*).

[250] RAABE, M. *et al. Bull. Soc. zool. Fr.*, **90** (1965), 631–54; *C.R. Acad. Sci., Paris*, **261** (1965), 4040–3; **262** (1966), 303–6; **263** (1966), 2002–5; **264** (1967), 77–80; **271** (1970), 1210–13 ('perisympathetic organs' and neurosecretion in ventral ganglia of insects).

[251] ROUSSEL, J. P. *Bull. Soc. zool. Fr.*, **91** (1966), 379–91 (role of frontal ganglion in insects).

[252] SCHARRER, B. *Z. Zellforsch. mikr. Anat.*, **62** (1964), 125–48; **64** (1964), 301–26; **69** (1966), 1–21; **89** (1968), 1–16; **95** (1969), 177–86; *Arch. Anat. micr.* **54** (1965), 331–42 (fine structure of insect endocrine glands).

[253] SMITH, D. S. and TREHERNE, J. E. *J. Cell Biol.*, **26** (1965), 445–65 (localization of cholinesterase in nervous system of *Periplaneta*).

[254] STEIGER, U. *Z. Zellforsch. mikr. Anat.*, **81** (1967), 511–36 (axon endings in neuropile of corpora pedunculata in *Formica*).

[255] STEINBRECHT, R. A. *J. Cell Sci.*, **4** (1969), 39–53 (absence of axon fusion in antennal nerves of *Rhodnius* and *Bombyx*).

[256] THOMSEN, E. and THOMSEN, M. *Z. Zellforsch. mikr. Anat.*, **110** (1970), 40–60 (fine structure of corpus allatum of female *Calliphora*).

[257] THOMSEN, M. *Z. Zellforsch. mikr. Anat.*, **94** (1969), 205–19 (neurosecretory system in adult *Calliphora*).

[258] TREHERNE, J. E. *J. exp. Biol.*, **42** (1965), 7–27 (ion exchange in central nervous system of *Carausius*).

[259] TREHERNE, J. E. *Symp. Roy. ent. Soc. Lond.*, **5** (1970), 153–64 (ultrastructure and function in insect nervous system).

[260] TREHERNE, J. E. *et al. J. exp. Biol.*, **53** (1970), 109–36 (movements of potassium in nerve cord of *Periplaneta*).

[261] TREHERNE, J. E. and MADDRELL, S. H. P. *J. exp. Biol.*, **46** (1967), 413–21; **47** (1967), 235–47 (ionic regulation and membrane potentials in nerve cord of *Carausius*).

[262] TREHERNE, J. E. and SMITH, D. S. *J. exp. Biol.*, **43** (1965), 13–21; 441–54 (entry and metabolism of acetylcholine in nervous system of *Periplaneta*).

[263] USHERWOOD, P. N. R. *Am. zool.*, **7** (1967), 553–82 (neuromuscular mechanism in insects).

[264] USHERWOOD, P. N. R. and GRUNDFEST, H. *Science*, **143** (1964), 817–18; *J. Neurophysiol.*, **28** (1965), 497–518 (peripheral inhibition in muscles of grasshoppers).

[265] WALL, B. J. *et al. Gen. comp. Endocr.*, **4** (1964), 452–6; *J. Insect Physiol.*, **13** (1967), 565–78 (antidiuretic effect of hormones in *Periplaneta*).

[266] WEEVERS, R. de G. *J. exp. Biol.*, **44** (1966), 163–75 (magnesium and action current in Lepidoptera).

Chapter VI

Sense Organs: Vision

EHAVIOUR IS determined, or at least influenced at every step, by the timuli to which the sense organs are subjected; and the functions of the sense rgans and their powers of discrimination are discovered largely by observations n behaviour. It is not possible therefore to separate these two subjects comletely. But before discussing the general principles in the control of behaviour e must consider the sensory apparatus.

THE COMPOUND EYE

Structure of compound eye—The compound eyes are the chief visual rgans of insects. They are made up of a number of transparent facets in the uticle of the head, each with an longated light-sensitive structure eneath it (Fig. 127). The facet with s underlying receptor is termed an ommatidium'; the whole collection f receptors in the eye forms the retina'. Each ommatidium consists of distal dioptric part, the 'cornea' vith the 'crystalline cone' beneath t, and a proximal receptive part, the retinula'. The retinula is made up usually of eight elongated sensory ells (visual or 'retinal cells') conaining pigment, each continuous vith a post-retinal axon fibre. But n some Lepidoptera there may be s many as ten or eleven retinula ells, and in Diptera the eighth elenent is abortive.[199] They are grouped ound an optic rod or 'rhabdom' ecreted by them collectively. The letailed structure of the rhabdom vill be considered later (p. 220). The vhole ommatidium is surrounded by curtain of pigmented cells; the primary iris cells' covering the crystalline cone, and the 'secondary iris cells' investing both the primary iris cells and the retinula (Fig. 133). The proximal exremities of the ommatidia rest on a fenestrated basement membrane through

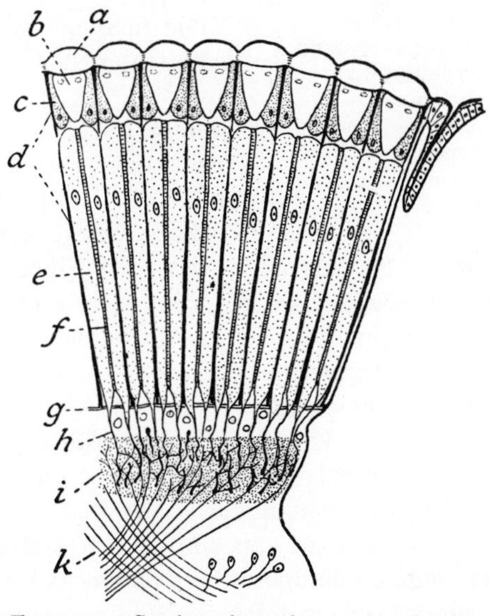

FIG. 127.—Section through margin of compound eye; semi-schematic (*after* SNODGRASS)

a, corneal lens; *b*, crystalline cone; *c*, matrix cells of cornea; *d*, pigment cells; *e*, retinula cells; *f*, rhabdom; *g*, fenestrated basement membrane; *h*, nerves from retinula cells; *i*, lamina ganglionaris; *k*, outer chiasma.

which the nerve fibres and tracheae run; the nerve fibres pass to the periopticon or outermost tract of the optic lobe of the brain.

Optical mechanism of the compound eye—As was originally suggested by Johannes Müller (1829) in his so-called 'mosaic theory', each ommatidium receives the impression of a luminous area corresponding to its projection on the visual field; and it is the juxtaposition of all these little luminous areas, varying in the intensity and quality of the light composing them, which gives rise to the total erect image perceived by the insect. The mechanism can be approximately imitated by holding a bundle of tubes with opaque walls in front of a ground glass screen; an erect image is thrown on the screen, its definition depending on the number of tubes per unit area.[3] In the compound eye the luminous points are formed by the aid of the dioptric apparatus: only those rays which fall normally on the cornea or those which can be brought into this line by refraction will reach the corresponding rhabdom; those rays which enter too obliquely fall on the pigment and are absorbed. Thus each ommatidium makes use of the rays from only a very small part of the visual field. The erect image formed by the apposition of these points of light has been observed in a number of insects and was photographed by Exner in *Lampyris*.

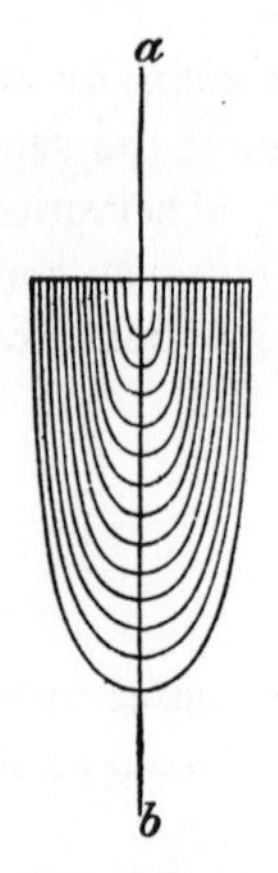

FIG. 128.—Dioptric apparatus of ommatidium of compound eye, made up of superimposed lamellae; schematic (*from* EXNER)

The field of vision of a single ommatidium is, in fact, much greater than has been assumed in the past (in *Locusta* about 20°,[141] in *Calliphora* and *Apis* 20–30°[176]), there is, therefore, a wide overlap in the fields of adjacent ommatidia. Even a light source with an aperture 1/4° illuminates a number of adjacent rhabdoms. But the intensity of illumination falls away progressively and becomes zero in ommatidia with a displacement of 10°.[176] Since each ommatidium is stimulated most effectively by light entering in its central line of vision the proved overlap between ommatidia does not invalidate the mosaic theory. [See p. 245.]

The dioptric apparatus—As was shown by Exner,[35] the cornea and crystalline cone have rather special optical properties. They have a laminated structure and can be pictured as a system made up of a series of cones with their apices turned inwards and superimposed on one another (Fig. 128). Examination with the microrefractometer shows that in this system the refractive index is at a maximum at the axis and decreases progressively towards the periphery. From the optical standpoint the system can be represented by a series of superimposed cylinders whose refringence increases towards the axis, a system called a 'lens-cylinder'. Let *abcd* (Fig. 129) be a cylinder whose refractive index is maximal at *xy* and diminishes towards the periphery. Let *xm* be a ray falling obliquely on the base *ac*. Within the cylinder it encounters the surfaces which separate the layers of unequal refringence, and at each of them (e.g. $a'b'$) it is refracted so that its direction makes a smaller and smaller angle with the axis of the cylinder. Finally it is totally reflected and then follows an inverse course, entering more and more refringent layers, and is brought gradually back to the axis at *y*. An entire spherical wave *mn* emanating from *x* (Fig. 130) will emerge from the cylinder with a concave form m_5n_5 and converge on *y*. In considering the passage of light hrough such lens cylinders the length of the cylinder must be taken into ccount; and there are two special cases that must be considered in connexion vith vision by the compound eye.

(i) When the focus lies at the posterior or retinal base of the cylinder; that

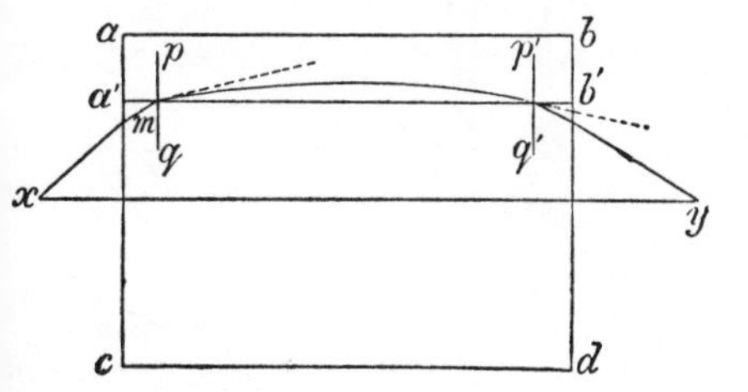

FIG. 129.—Course followed by a ray of light n passing through a lens cylinder (*from* EXNER)

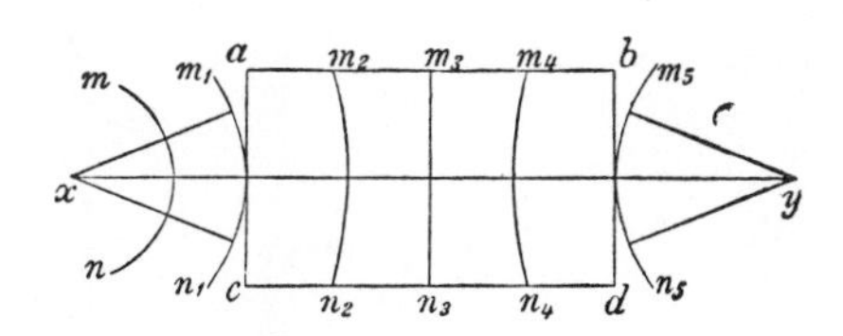

FIG. 130.—Passage of a spherical wave through a lens cylinder (*from* EXNER)

s, when the length of the cylinder is equal to the focal distance (Fig. 131). An nverted image will then be formed at the posterior surface; the principal rays coming from the luminous points of the object emerge from the cylinder parallel.

(ii) When the length of the cylinder is twice the focal distance (Fig. 132). In this case, the inverted image of an object placed at infinity is formed midway along the length of the cylinder at *yz*. The rays continuing from *yz* in the second half of the cylinder follow a course symmetrical with that followed in the

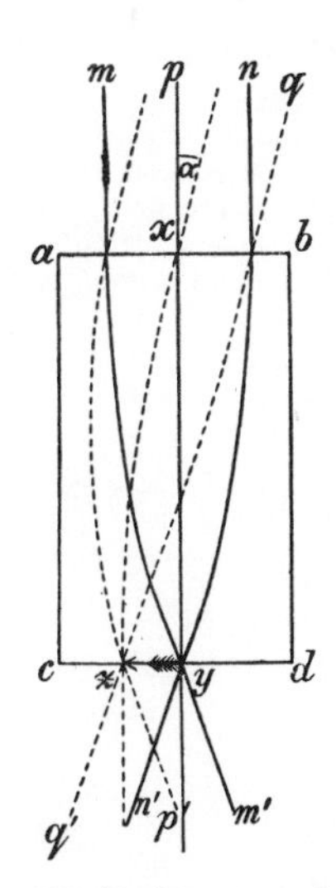

FIG. 131.—Path of rays passing through a lens cylinder of length equal to the focal length (*from* EXNER)

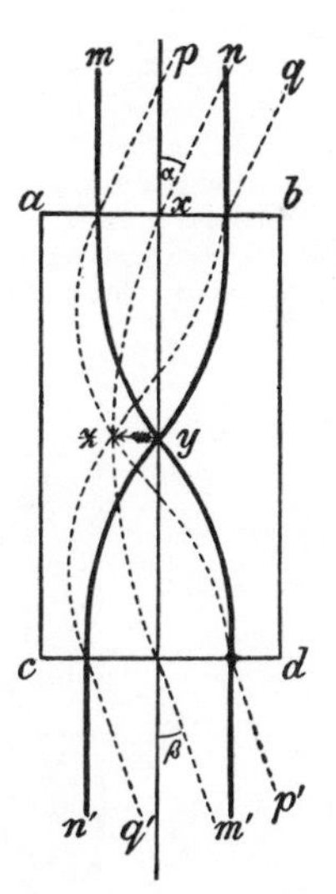

FIG. 132.—Path of rays passing through a lens cylinder twice as long as the focal length (*from* EXNER)

first half and emerge at an angle equal to that at which they entered, and diverted towards the same side of the axis of the cylinder.

Exner showed that there are two types of vision in insects corresponding to these two cases. [See p. 245.]

Formation of images by apposition—In most diurnal insects, Hymen-optera, Diptera, Odonata, many Coleoptera and day-flying Lepidoptera, the

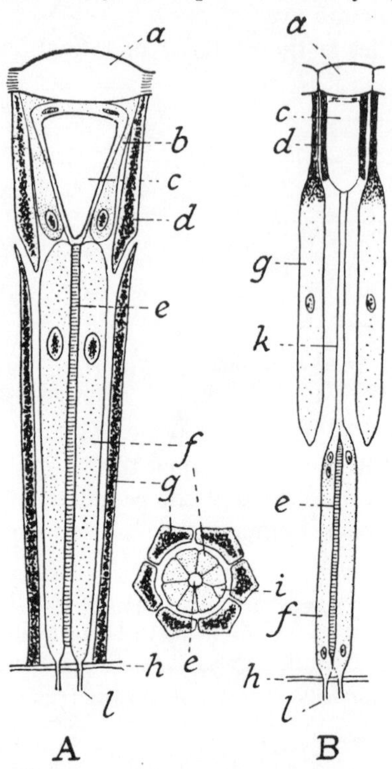

FIG. 133.—Diagrams of the two chief types of ommatidium

A, from eye forming apposition image (*after* SNODGRASS). *B*, from eye of Noctuid forming superposition image (*after* WEBER). *B* is shown in the dark-adapted condition with the pigment in the secondary iris cells almost entirely withdrawn into the outer ends of the cells. *a*, corneal lens; *b*, matrix cell of cornea; *c*, crystalline cone; *d*, iris pigment cells or primary iris cells; *e*, rhabdom; *f*, retinal or sense cells; *g*, retinal pigment cells or secondary iris cells; *h*, fenestrated basement membrane; *i*, eccentric retinal cell; *k*, translucent filament connecting crystalline cone with rhabdom; *l*, nerve fibres.

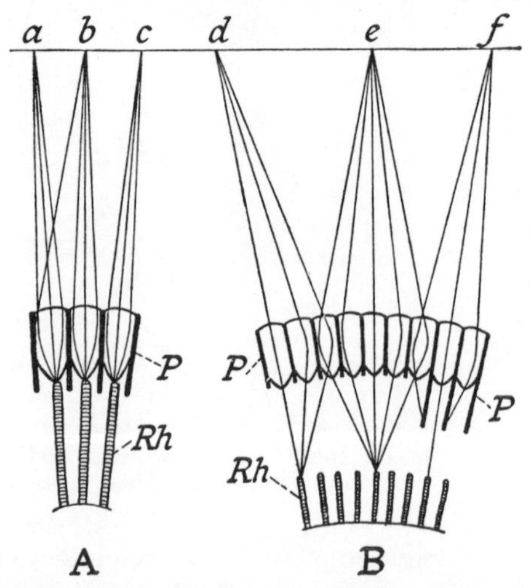

FIG. 134.—Diagram showing image formation by the compound eye (*after* KÜHN)

A, apposition eye; *B*, superposition eye. *a–f*, luminous points with the course of the rays emitted by them; *P*, pigment; *Rh*, rhabdom. At the right side the superposition eye is shown with pigment in the light-adapted position; all rays except those entering the central facet are intercepted.

cones are surrounded by pigment up to their posterior extremities; they allow the light to emerge only at the central point; and the retinulae are short and placed immediately behind the cones (Fig. 133, A). A reversed image of a small part of the visual field is formed where the retinula comes in contact with the apex of the cone and has been observed in various insects (Fig. 134, A). The arrangement therefore corresponds with case (i) above. But this image has no physiological significance, it merely impresses the retina as a simple luminous point, the apposition of all such points as perceived by the different ommatidia forming the erect image perceived by the compound eye as a whole.

Formation of images by superposition—In many nocturnal insects, Lampyridae and other beetles, Noctuidae and other Lepidoptera, the ommatidia are greatly elongated. The retinulae do not lie immediately behind and in contact with the cones but are separated from them by a long interval occupied by a non-refractile transparent medium; and the pigment in the iris cells may be concentrated in front between the crystalline cones (Fig. 133, B).

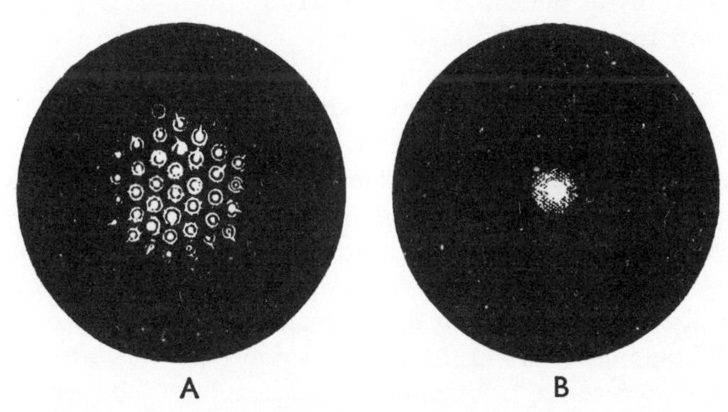

FIG. 135.—Superposition image in *Lampyris* (*after* EXNER)

A, a small point of light as seen through the detached surface of the eye when the microscope is focussed just below the lenses; *B*, as seen when the microscope is focussed at the level of the rhabdoms (*cf.* Fig. 134, B).

The optical system is equivalent to a lens-cylinder twice as long as its focal length. Hence a given rhabdom receives rays not only through its own facet but through neighbouring facets, the rays being refracted by the cones towards the same side from which they have come (Fig. 134, B). For example, if the corneae and adherent cones* of the eye of *Lampyris* are separated from underlying cells and cleaned, and then a minute flame is placed below this group of lenses and the image formed by it is examined with the microscope, it appears as a luminous point (Fig. 135, B). But by focussing the microscope up and down it is seen that the light forming this image does not arise from a single pencil of rays, but from a group of such bundles, each coming through an adjacent facet. So that as the focus is lowered the image point splits up into a number of luminous points (Fig. 135, A). As many as 30 neighbouring ommatidia may unite to concentrate the light in this way upon a single rhabdom. As already described, the image projected upon each rhabdom will be erect. But, again, this is of no physiological significance: the rhabdom receives only a visual stimulus which is presumably the mean of the components of this image. Since each of these elemental images

* The cone in Lampyrid eyes is not a true crystalline cone (eucone type) but an invagination of the cornea (exocone type).[34]

is formed by the superposition of light from a number of adjacent facets, the compound image as received by the entire retina is termed a 'superposition image'. It was this which was photographed by Exner in *Lampyris* (p. 245); it is seen also in *Cantharis*, *Telephorus*, *Hydrophilus*, Cetoniids among Coleoptera, and in Noctuidae and other Lepidoptera.[35] [See p. 245.]

Other types of compound eye—There are doubtless many intermediate stages between these two types of image formation. In butterflies, for example, which form retinal images of the apposition type, the image formed at the apex of each cone is erect; so that in some eyes of this type the lens and cone may evidently act together as a lens-cylinder of twice its own focal length.[34] Exner's interpretations apply particularly to 'eucone' and 'exocone' eyes with well-developed cones. But in certain insects the cones are purely cellular and non-refringent ('acone' eyes of Tipulids, *Forficula*, Hemiptera, various Coleoptera) or the cones may be represented merely by a mass of liquid secreted by the crystalline cells ('pseudocone' eyes of Muscids). In these cases the lens system is composed mainly of the cornea. [See p. 245.]

Diffraction images—Although Exner's observations on the erect and inverted images projected upon the single rhabdoms in the different types of compound eye have been confirmed, some doubt has been cast upon his interpretation of the 'lens cylinder'.[176] Although these images are formed as described, and presumably contribute to the total mosaic image of the retina, each facet also acts as an element in a transparent diffraction grating. Diffraction images are formed deep in the eye, which owe their origin to the interference of oblique rays from many facets.[190] In the excised eye of *Locusta*, besides the first (apposition) image, second and third images formed by diffraction can be seen in the deeper layers of the retina. These images are of the superposition type: they are formed by the diffraction of light received from groups of adjacent facets. It seems that the rays which contribute to them are not excluded by the investing pigment of the apposition eye.[143] Indeed, there is evidence that in *Calliphora* the wave-lengths of the red end of the spectrum can penetrate the pigment sheaths of the ommatidia to provide more diffuse vision but increased sensitivity.[194] It is doubtful whether the diffraction system is concerned in the formation of a general image of the field of vision, but it is exceedingly sensitive to very small movements in the surroundings (p. 245). An alternative view amounts to the abandonment of the mosaic theory and the attribution of all resolution to the individual ommatidia.[20] This scheme will require an immense amount of integration and interpretation in the central nervous system. [See p. 245.]

The retina—The rod-like rhabdom is composed of highly refractile material light entering it will therefore be totally reflected from the walls, and thus it forms an ideal structure to receive and conduct light without loss,[35] even in an eye like that of *Simulium* in which the rhabdoms are curved. This property will also reduce the effect of light reaching the rhabdom obliquely.[193] In the region of the basement membrane there are, in addition, tracheal branches which serve as a 'tapetum' reflecting the light back so that the rays retraverse the rhabdom, and the nervous elements are doubly stimulated. The tracheal tapetum is particularly well developed in Noctuids and other nocturnal insects; in these the tracheae divide up at the base of the retina into innumerable fine branches running a parallel course between the retinulae.[20]

Although the rhabdom is described above as though it were composed of homogeneous material it is in fact a complex structure. It is formed by the juxtaposition of the 'rhabdomeres' of the retinula cells. The rhabdomere represents the microvilli, or 'honey-comb border', which cover the free surface of the retinula cell. These tubular units are 400–1200 Å in diameter, with their long axes at right angles or oblique to the axis of the ommatidium, their distal ends closed, their basal ends open to the retinula cell.[153] In Odonata, Orthoptera, Lepidoptera, Hymenoptera, &c., the rhabdomeres are closely applied against one another and the rhabdom appears as a solid structure. But in Hemiptera,[180] and particularly in the Diptera, the rhabdomeres are widely separated throughout most of their length.[153] In *Drosophila*, for example, there are seven retinula cells with rhabdomeres $1{\cdot}2\mu$ in diameter and about 60μ in length.[204] As seen in horizontal section the rhabdomeres in adjacent ommatidia form an extraordinarily constant pattern over wide areas.[147, 199] In all these insects the orientation of the microvilli also shows a constant pattern: in opposite members of a pair of rhab-

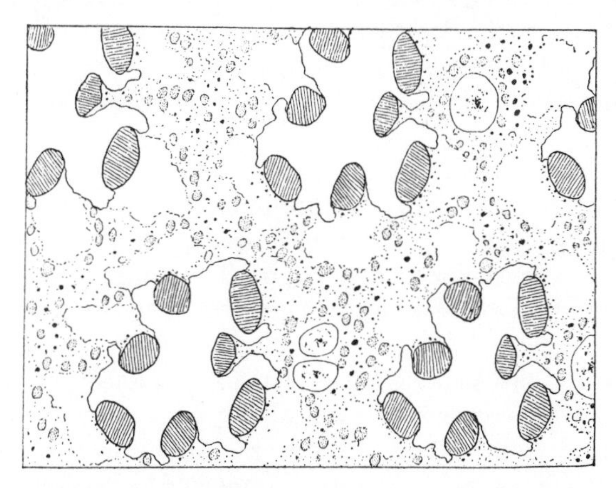

FIG. 136.—Low power electron microscope section through the retina of *Drosophila* (*after* WADDINGTON and PERRY)

The section shows the regular arrangement of the seven retinulae in adjacent ommatidia. The striation of the rhabdomeres indicates the orientation of the microvilli.

domeres the microvilli always lie parallel. This is well seen in *Drosophila* (Fig. 136), or in *Apis*, where the eight retinula cells give rise to a rhabdom of four quadrants with microvilli at right angles in adjacent members.[165, 167] [See p. 246.]

According to current belief the rhabdomere is the site of the chemical reaction, the products of which depolarize the membrane of the retinula cell and initiate impulse formation; but according to an earlier view the function of the striated rhabdom is to scatter the vertical rays of light into the sensory cells of the retinula where photoreception takes place.[189] That the reaction occurs in the rhabdom is made probable by the fact that in the developing pupa of *Bombyx* the first appearance of electrical responses to illumination coincides with the development of the rhabdom.[152]

The most widespread photoreceptor substance in the animal kingdom is rhodopsin, formed by the union of 'retinene' with protein. Retinene is a carotinoid pigment, the aldehyde of vitamin A alcohol. Retinene, partly bound to protein, has been isolated from the head of *Apis*[161] and other insects: *Musca*,

Locusta, *Uropetala* (Odonata), *Cicindela*, and *Nyctemera* (Lep.).[138] It has been suggested that in *Periplaneta* and *Blaberus*, rhodopsin, or its equivalent, is arranged as a monolayer over the surface of the tubular microvilli.[205] During the exposure of the honey-bee to light the retinene in the head region decreases and vitamin A appears: the ratio of vitamin A to retinene was 4 : 1 in light-adapted bees and about 1 : 4 in dark-adapted controls.[161] In houseflies *Musca* reared on diets free of vitamin A neither retinene nor vitamin A alcohol are detectable; but after rearing for several generations on this diet vision seemed to be unaffected.[14]

The whole eye in *Musca*,[198] including the crystalline cone in *Apis*,[159] contains abundant glycogen. In *Calliphora* there is an increased rate of turnover in energy-producing metabolism when the eye is illuminated;[177] oxygen consumption rises immediately in the retina of *Calliphora*, *Apis*, &c., on exposure to light and the increase persists as long as the light continues.[132] [See p. 246.]

Pigment movements in the compound eye—The iris cells contain black pigment which absorbs, and pale or coloured granules which reflect the light. On account of this reflection by the coloured granules the structure is sometimes termed the 'iris tapetum'. Unlike a true tapetum, however, its function is not to increase the illumination of the retina, but to prevent the entry of the oblique rays.

If the eye of a butterfly is examined with an eye mirror at any part of its surface, it shows a central dark spot surrounded by six smaller spots joined to the first by radial dark lines, and sometimes twelve still smaller peripheral spots. This appearance is termed the 'pseudopupil'. The central spot results from the light being absorbed where it falls on the rhabdom or on the black pigment in the retina, while it is reflected from the iris tapetum. In many butterflies, in the middle of the central black spot there is a small luminous spot, due to the light which falls on the rhabdom not being absorbed but reflected from the tracheal tapetum behind.[30] In certain woodland species of butterflies this central luminous spot disappears when the eye is brightly illuminated. This change is due to the movement of pigment in the cells around the basement membrane (Fig. 137). It takes place very rapidly and may be complete within 6 seconds. Perhaps it serves to keep the illumination of the eye approximately constant with rapidly changing light intensities.[28] Pigment movements may occur also in the apposition eyes of Hemiptera (*Notonecta* and *Corixa*).[9]

In Tipulidae the iris pigment migrates radically during dark adaptation so as to enlarge the aperture of each ommatidium 100–200-fold; and at the same time the rhabdomeres (which in *Tipula*, *Culex*, &c., are embedded in the cytoplasm of the retinula cells—except at their distal ends) move distally to the crystalline cone.[196] A similar movement of the rhabdom is very evident in the eye of *Notonecta*[179] where it may extend in between the cells of the crystalline cone; in this case the aperture of the ommatidium increases only about 16-fold during dark adaptation. In *Drosophila* there is evidence that the retinal eye pigments not only absorb, and protect the eye from light entering obliquely, but they also exert a selective action on the wave-lengths intercepted, allowing those to pass which are most effective in stimulating the photoreceptors, notably 366 mμ in the near ultraviolet.[154] [See p. 246.]

But migrations of pigment are much more striking and important in the superposition eyes of nocturnal forms.[88] Most of these may be active also in the daytime, and their eyes possess an arrangement comparable with the mammalian iris by which they can be adapted to different light intensities. The retinal pigment is little affected, but the pigment in the primary and secondary iris cells expands and contracts with the intensity of illumination. The cells themselves do not change in shape or position; the granules of pigment merely become clumped or dispersed. Thus in dim light the pigment in the iris cells is withdrawn upwards and the eye can function as a superposition eye in the manner described above, use being made of all the available light. Whereas in bright light the iris pigment expands downwards beyond the cones, surrounding each retinal element as a black curtain whose extent is proportional to the brightness of the light. In this state the lateral rays are intercepted, each ommatidium receives light only from its own facet, and the eye therefore forms an apposition image (Fig. 134, B). In the living insect examined with an eye mirror in the dark-adapted state, a luminous red reflection from the tapetum is visible from

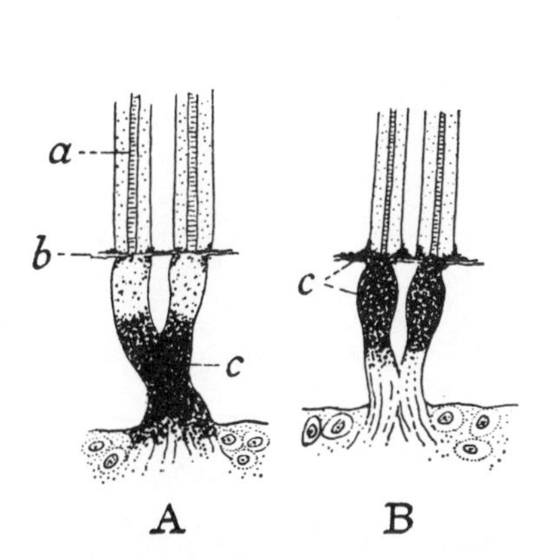

FIG. 137.—Proximal zone of retina in *Vanessa urticae*, showing base of two retinulae

A, in dark-adapted state; *B*, in light-adapted state (*after* DEMOLL). *a*, rhabdom; *b*, fenestrated basement membrane; *c*, pigment.

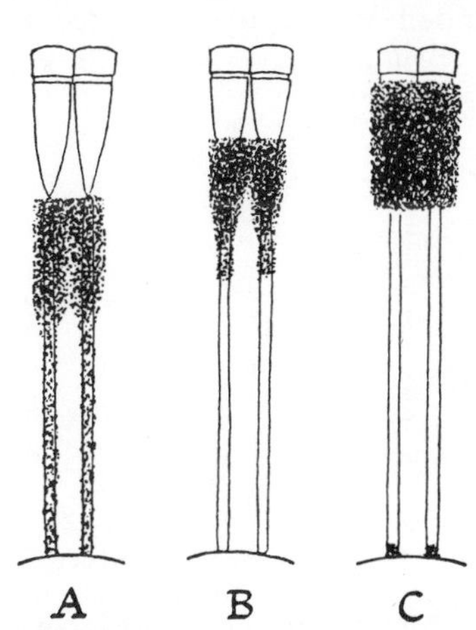

FIG. 138.—Pigment movements in eye of *Carpocapsa*. Semischematic figures of ommatidia, based on photographs (*after* COLLINS)

A, eye completely light-adapted; *B*, intermediate stage; *C*, completely dark-adapted.

a group of ommatidia; in the light-adapted state it is visible only from the central ommatidium of the group examined.[30, 58]

In nocturnal insects the maximum response of the light-adapted eye is about 20 per cent. of that of the dark-adapted eye; in diurnal insects about 60 per cent. (Fig. 139).[136] Under certain conditions the eye in *Cerapteryx graminis* (Noctuidae) is about 1,000 times more sensitive with the pigment in the dark position than in the light position.[136] Two changes seem to be responsible: pigment migration and biochemical events—perhaps the re-synthesis of the photochemical substance. In diurnal insects with insignificant pigment changes, the biochemical events seem alone responsible for the increase in sensitivity during dark adaptation.[136] In the honey-bee, pigment migration is certainly not concerned in dark and light adaptation: light adaptation is complete in a few seconds; dark adaptation is mostly effected in one minute, though it needs 10–15 minutes for completion.[166] [See p. 246.]

In Mantids the anterior parts of the eye, which are used for binocular fixation, are of the apposition type, the lateral parts are of the superposition type.

During the day-time the pigment in the iris cells is expanded; and since this pigment varies in quantity in different parts of the eyes these appear striped or spotted. At night the pigment retracts in all parts and the banding disappears[42] (*cf. Schistocerca*, p. 237). In some Lepidoptera the pigment movements are a direct response to changes in illumination; in nocturnal Lepidoptera the change may be brought about by ultra-violet as well as visible light and takes from 3–17 minutes;[85] in *Cydia* (*Carpocapsa*) *pomonella* (Fig. 138) the movement begins ½ to 1 hour before sunrise or sunset, independently of the length of day, and the change from light to dark adaptation requires about 1 hour for its completion.[23] But apart from this effect of stimulation, many Geometrids, Noctuids, &c., show a diurnal rhythm of pigment migration which persists in complete darkness and does not seem to be due to diurnal changes in temperature, humidity or any other known factor.[58] During the day insects have their pigment expanded in the light-adapted position; but on shaking them the pigment immediately retracts to the dark position. This suggests that the expansion of

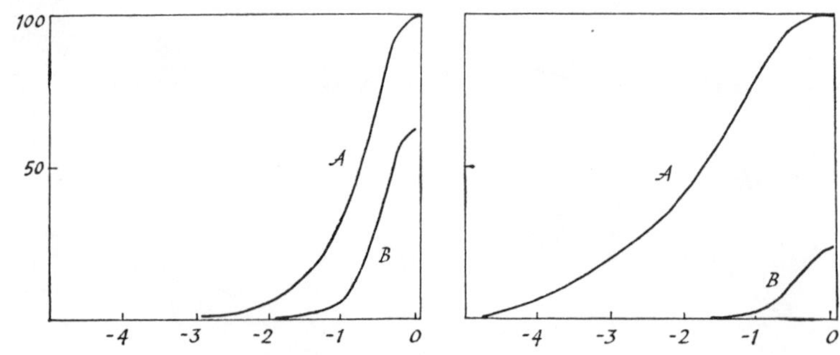

FIG. 139.—Sensitivity curves in the eyes of Lepidoptera in dark-adapted condition (*A*) and light-adapted condition (*B*). To the left; average curves from 4 species of day-flying butterflies. To the right; average curves from 3 nocturnal species (*after* BERNARD and OTTOSON)

Ordinate: amplitude of electrical response, percentage values. Abscissa: log intensity of illumination.

pigment may be simply a phenomenon of 'sleep', and that the rhythm of pigment movement is secondary to a rhythm of general activity.[29]

As regards the controlling mechanism, some authors suppose that the dark-adapted position is maintained by a continuous 'tonus' going out from the brain, and that the movement in response to light is determined by a nervous reflex from the retina. Thus the responses in day-flying butterflies do not occur during the sleep-like state of the insect that has been at rest a long time;[28] the response is abolished by narcosis; and the excised eye always takes up the light-adapted position irrespective of the illumination.[26, 29] Others believe that a photochemical reaction in the pigment cell itself initiates the movement.[23, 88] In *Ephestia*, by illuminating only a few of the ommatidia, movement of individual cells can be induced; there is thus no evidence of control by a hormonal mechanism,[26] and the same conclusion has been reached in *Carausius*.[150]

Retinal responses—The retina shows an electrical response to illumination such that the cornea becomes negative in respect to the back of the eye. This usually takes place in two waves, an 'A' wave which reaches its maximum 0·10–0·12 seconds after the onset of illumination, followed by a 'B' wave which reaches and maintains its maximum in less than 1 second (Fig. 140). On the cessation of stimulation the potential decays asymtotically until it reaches the original level. The actual form of this curve is extremely varied. It is to be regarded as the algebraic summation of two or perhaps three components of potential change, some in the retina, some in the ganglion, whose latent periods, magnitude and time courses differ with the intensity of the stimulating light. The response varies with light of different wave-lengths, with temperature, with the state of dark adaptation in the eye, and with the species of insect. There are also diurnal variations which seem to be independent of the pigment migrations.[24, 68, 104, 117] The negative component of the potential change comes from the retina, the positive component from the optic ganglion. The relative prominence of these two components varies with the anatomy of the eye. When both are well developed, as in *Calliphora*, the eye is found to have a very high 'flicker threshold' or 'fusion frequency' (p. 238).[120]

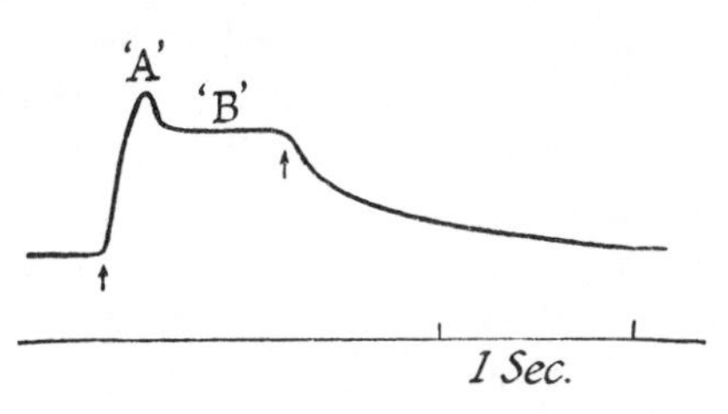

FIG. 140.—Typical form of electrical response to illumination in eye of *Melanoplus femur-rubrum*

'A' and 'B', the two components of the potential wave. The arrows mark beginning and end of illumination (*after* HARTLINE).

There is evidence in *Apis* and *Lucilia* that while the electroretinogram consists of two monophasic components of opposite sign, both of these do in fact originate at the receptor level;[187] indeed, single retinula cells of the honey-bee

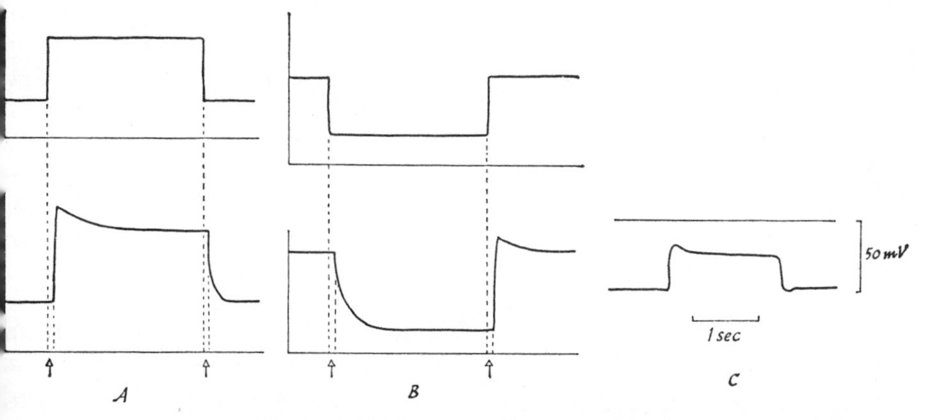

FIG. 141.—The form of the electroretinogram in *Hylobius* (diagrammatic): *A*, in response to increased intensity of illumination. *B*, in response to reduced illumination (*after* KIRSCHFELD). *C*, shows the electroretinogram recorded from a single visual cell in *Calliphora* (*after* BURKHARDT and AUTRUM)

Ordinates: of upper curve in *A* and *B*, relative light intensity: lower curve in *A* and *B*, potential change, negativity of cornea upwards. Abscissa: time. The arrows indicate the latent period.

drone show the complete response: a spike potential and negative after potential (increased at high illumination) followed by a sustained slow potential.[139a, 186] In *Hylobius abietis* the electroretinogram is of the slow monophasic type when the eye is exposed to a bright stimulus (Fig. 141, A); but when exposed to a dark stimulus there is a reversed response with a more or less prolonged fall in potential (Fig. 141, B).[175] [See p. 246.]

A constant response may be obtained with different intensities of illumination provided that the duration of the stimulus is varied in such a way that the product of intensity × duration is constant. This is in accordance with the Bunsen-Roscoe law of photochemistry, according to which the amount of energy (i.e. intensity × duration of exposure) must be constant in order to produce a given amount of photochemical change.[50, 68] The same relation holds in the visual responses of Muscid larvae. The stimulating effect of light on these larvae depends on its brightness (p. 317). If flashes of light following one another at short intervals are employed the effect depends only on the quantity of light received per unit of time. Intermittent light, alternately light and dark, has the same effect as a continuous light of the same average intensity. The eye behaves in this respect exactly as a photographic plate.[90] Applied to the human eye this is known as Talbot's law.

When the eye of *Dytiscus* is stimulated by a bright light, the optic nerve shows rhythmic potential oscillations at the rate of 20–40 per second, showing that all the neurones are discharging simultaneously. But under ordinary illumination this synchronous discharge disappears; the various elements discharge at different rates. It is such differences in the rate of discharge of the separate elements which makes possible the perception of a visual pattern.[2]

Adaptation in the compound eye—The eye is capable of adapting its sensitivity to light over a wide range. Mutants of *Ephestia* with reduced eye pigments show increased sensitivity.[72] Adaptation is due in part to the movements of pigment just described. But it occurs equally in insects in which pigment movements are not very obvious. In these it is probably due chiefly to physico-chemical changes in the receptor mechanism of the retina; partly, perhaps, to central adaptation, that is, a failure of the nervous system to continue to respond to stimuli of a given intensity. Thus in *Eristalis tenax* the sensitivity is judged by reflex turning movements in a beam of light. If this fly after being adapted to a light of 53 metre-candles is placed in complete darkness, the sensitivity increases in the course of one hour to a maximum about 21 times greater than at the outset.[32] *Belostoma* (Hem.) shows a similar increase in sensitivity; when exposed suddenly to light of a given intensity it reacts more rapidly the further adaptation to darkness has proceeded.[101] In the bee a reflex movement of the antennae in the presence of moving stripes is used as a measure of the state of photic adaptation. The sensitivity increases rapidly during the first few minutes of darkness and then more slowly until it reaches a maximum in 25–30 minutes, the total increase in sensitivity being about a thousand-fold (Fig. 142).[115] [See p. 246.]

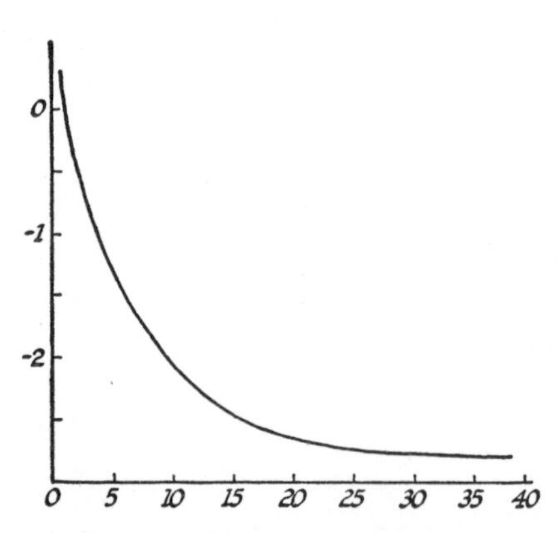

FIG. 142.—Photic adaptation in the honey-bee; relation between threshold intensities and time of dark adaptation (*after* WOLF and ZERRAHN-WOLF)

Ordinate: logarithm of intensity in millilamberts. Abscissa: time in the dark in minutes.

Luminosity discrimination—The ability of the insect to discriminate degrees of luminosity in the different parts of the visual field can be judged only from the differences in intensity necessary to produce certain reflex responses. The true threshold of sensation may be considerably lower. In the bee the intensity discrimination as judged by a response to moving stripes ('optomotor reaction', p. 323) is estimated to be about 20 times worse than that of man; with very strong illumination the least detectable difference is not less than 23 per cent.[113] In *Drosophila* it is even poorer; at low illuminations the stronger light must be 100 times brighter than the weaker; under the optimal conditions obtainable at higher illuminations a difference of 2·5 times can be detected.[51]

When the insect is confronted with very narrow stripes, which subtend an angle much less than that covered by a single ommatidium, these will presumably cause a reduction in the intensity of stimulus received by the ommatidium proportional to the amount of the ommatidial field that is darkened. Making this assumption it has been calculated that *Eristalis* will respond when 4·8 per cent. is darkened, *Pieris* 6 per cent., *Apis* 30 per cent., *Coccinella* 26 per cent. But these, again, are thresholds of reflex reactions, not necessarily thresholds of sensation.[17]

The appreciation of luminosity may be influenced by the brightness of the adjacent part of the visual field, that is, by *simultaneous contrast.* For example, *Macroglossa* seeks dark crevices in which to retreat at night and in the autumn; if a series of dark discs of suitable size are set up in a room it makes for the darker ones; and if two of equal darkness are set up side by side, one on a dull grey ground and one on white, the insects make for the latter with some regularity.[73]

Extent of visible spectrum—Insects are characterized by their sensitivity to the shorter wave-lengths in the spectrum. It was shown many years ago by Lubbock[79] that for ants the visible spectrum extends far into the ultra-violet. Ants which seek the dark will shelter below a flask filled with carbon disulphide, which cuts out the ultra-violet, and avoid a zone covered by a sheet of violet glass which appears quite dark to the human eye .They will choose a light green or yellow zone and avoid a zone covered with deep violet glass. But if a layer of carbon disulphide or quinine sulphate solution is placed over the violet glass (making no perceptible difference to the human eye), they will all collect under the violet shade. If an ants' nest is illuminated with the colours of the spectrum, the ants carry their pupae and deposit them at the infra-red end, just at the limit of our visible spectrum. Whereas at the opposite end they leave vacant a considerable zone beyond our visible range.[79] By covering the eyes Forel[37] showed that they are responsible for this sensitivity to ultra-violet.

Similar conclusions have been reached with many other insects. Moths and other light-seeking insects will fly to glass allowing ultra-violet to pass through, even when it appears opaque to man, in preference to relatively bright blue windows.[81] The honey-bee will respond to light with a wave-length of 297mμ, at the lower limit of the solar spectrum;[10] and *Drosophila* will give an undoubted response even to 257mμ.[83] The bee *Trigona* can distinguish patterns made up of white which reflects ultra-violet and Chinese white which does not, although to the human eye they appear homogeneous.[82]

Ozone in the upper atmosphere filters out almost all the far ultra-violet and little shorter than 300mμ reaches the earth;[164] but bees can orient themselves in relation to the sun even when this is so heavily clouded that its position cannot be detected by the human eye. They depend probably on their sensitivity to ultra-violet.[158] It was shown by Lutz that many flowers have patterns of ultra-violet reflection visible only to insects.[81] Indeed, many flowers such as *Oenothera* or *Potentilla*, which appear uniformly yellow to the human eye have 'honey-guides' around the entrance to the nectaries which are conspicuous to the bee because they fail to reflect ultra-violet (Fig. 143)[81, 149] As v. Frisch has pointed out, the colours of flowers have evolved as an adaptation to the colour sense of insects; they were not designed for the human eye. Some Pierids, &c., notably *Gonepteryx*, show similar patterns of ultra-violet reflection, invisible to man but perhaps important in butterflies,[182] which likewise can perceive ultra-violet.[146]

The tracheal tapetum and other parts of the retina fluoresce when the eye is exposed to ultra-violet; they reflect the light at a wave-length falling within the human range. This affords proof that the ultra-violet can penetrate to the retina (in *Sarcophaga* and *Apis* the cornea is quite transparent to a wave-length of 253mμ);[83] it also suggests the possibility that in some cases fluorescence may be responsible for the apparent perception of the shortest wave-lengths. It is, however, generally accepted that in most insects the retina is sensitive to the ultra-violet itself.[10, 85]

At the other end of the spectrum there are great differences between different

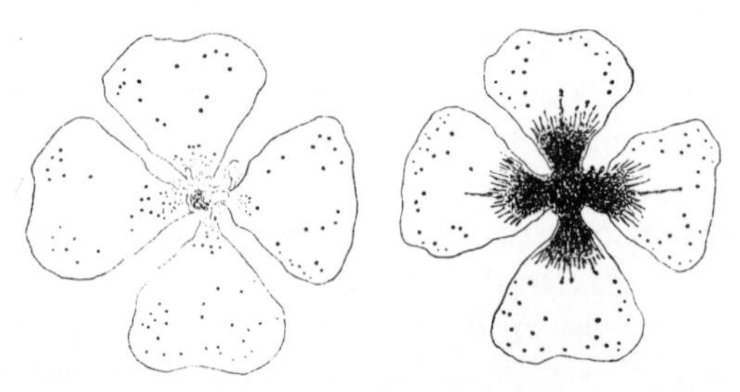

FIG. 143.—Flowers of *Oenothera* photographed in yellow light (on the left) and ultra-violet light (on the right) (*after* DAUMER)

insects. Most insects seem to be insensitive to the deeper shades of red. The honey-bee, for instance, will not respond to light of a wave-length greater than 650mμ (on the borders of red and orange).[41, 75] Wasps (*Vespa*) after being trained to visit a black surface can be diverted equally by black or red.[100] On the other hand, some butterflies (*Pieris brassicae*, *Vanessa urticae*) have an undoubted perception for red, and will visit deep red flowers or red paper models of flowers and may show an actual preference for red.[61] Larvae of *Vanessa* respond to red light,[103] and larvae of the Elaterid *Agriotes* are very sensitive to light of all wave-lengths, including the red.[36] The fire-fly *Photinus pyralis* will respond to flashes of light from some point between 520–560mμ in the green up to 690mμ, at least, in the deep red;[15] and the visible spectrum in *Calliphora* extends as far as 730mμ[131]

Apparent luminosity—The apparent luminosity of different parts of the spectrum has been inferred by comparing the efficiency of different wave-lengths in producing those responses which appear to depend on light intensity. Yellow light has the greatest stimulating value of the visible part of the spectrum in causing the ant *Formica* to remove its pupae.[1] Green light is the most effective part of the ordinary dispersion spectrum in causing colour change in *Carausius* (p. 621); though if light of equal energy is compared the ultra-violet is said to be the most effective and the most luminous region.[93] The most efficient part of the visible spectrum in attracting the bee is the yellow green at about 553mμ or 530mμ, rather close to the region of greatest subjective luminosity for man. The luminosity curve of different colours is similar in larvae and adults of *Dytiscus* and again agrees fairly well with that of man in the light-adapted state.[195] For *Drosophila* the corresponding peak is at 487 mμ;[10] *Calliphora* 504 mμ; while *Tenebrio* is most readily aroused by light of 535 mμ.[10] [See p. 247.]

Insects which are attracted to light seem to perceive ultra-violet better than the other parts of the spectrum;[81] and in the case of the bee[10] and *Drosophila*[10] a second peak of stimulative efficiency, four or five times as high as the peak in the visible part of the spectrum, has been described as existing at about 365mμ in the ultra-violet. But although ultra-violet light is so effective in evoking certain kinds of behaviour this does not necessarily mean that the compound eyes are several times more sensitive to the shorter wave-lengths.[164] In *Musca* no second peak of sensitivity can be detected in the visible range of the spectrum. There is a very high peak at 365mμ; on the short wave side of this peak the sensitivity falls away, but at 302mμ it is still apparently greater than that of either yellow or green.[21] But these conclusions were based on experiments in which light of unequal energy was used over different parts of the spectrum and the relative effect of equal energies determined by calculation. The relative efficiency of wave-lengths of equal energy may, however, be very different for different absolute amounts of energy. Thus, at lower intensities the peak of responsiveness of many insects is in the ultra-violet (365mμ), at higher intensities in the blue-green (470–528mμ);[109] so that the validity of conclusions reached by this method is open to doubt.[97] When light of equal energy is used at all parts of the spectrum there is said to be no trace of the high maximum at 365mμ.[97]

The optomotor reaction, in which the insect responds to moving stripes alternately dark and light, by turning the head or body, is due to the apparent luminosity of the stripes and not to their colour (but see p. 233); it therefore affords a means of testing the apparent luminosity of different wave-lengths. The conclusions reached by this method agree with those recorded above; the luminosity of various parts of the spectrum is very different in different insects. Some, such as *Apis*, *Mantis*, *Coccinella*, agree with the human eye in its dark-adapted state; others, such as Pierids (which, as we have seen, appear to be more sensitive to red), agree with the light-adapted human eye.[63, 99]

Spectral sensitivity—The voltage developed in the electroretinogram can also be used as a measure of the sensitivity of the retina to different wave-lengths. The results obtained in *Melanoplus* in this way agree with those deduced from its behaviour: there is a peak in the blue-green at about 500mμ, a great decrease on the longer wave-length side and a lesser decrease on the shorter wave-length side. The curve is very similar to the absorption curve of visual purple as it occurs in vertebrates.[67] This method has been applied with progressive refinement in recent years—to different species of insects, to different parts of the eye in single species, and to single ommatidia. The results show considerable diversity but certain constant features.[163] A single eye commonly shows several peaks of sensitivity. (i) There is almost always a peak in the near ultra-violet at about 340–350mμ: in *Periplaneta*,[202] in *Calliphora*,[129, 201, 202] in *Apis* compound eye (Fig. 144),[133, 162] and ocellus,[168] and in *Carabus*, *Phalera*, and *Macroglossa*.[169] (ii) In the visible range there are usually one or more peaks which vary in position:

in the *Apis* drone at 447mμ (blue-green) and 530mμ (green-greenish yellow) (Fig. 144);[133] in *Periplaneta* at 500mμ (green);[201] in *Calliphora* at about 507mμ (green) and 630mμ (red);[202] in *Carabus*, *Phalera*, and *Macroglossa* at 430mμ (blue), 500mμ (green), and 620mμ (red).[169]

In *Periplaneta* the high sensitivity to ultra-violet is limited to the dorsal half of the eye, whereas the green-sensitive receptor is found in all parts[201] (cf. *Notonecta*[94] and *Aeschna*,[183] p. 234). Electrical recording from single visual cells in *Calliphora* has shown that most cells are of the so-called 'green type' with two peaks of sensitivity at 350 mμ and 490 mμ. Two scarce types occur, one with maximum sensitivity at 470 mμ ('blue type') and one at 520 mμ ('yellow-green type') (see p. 234).[129, 139]

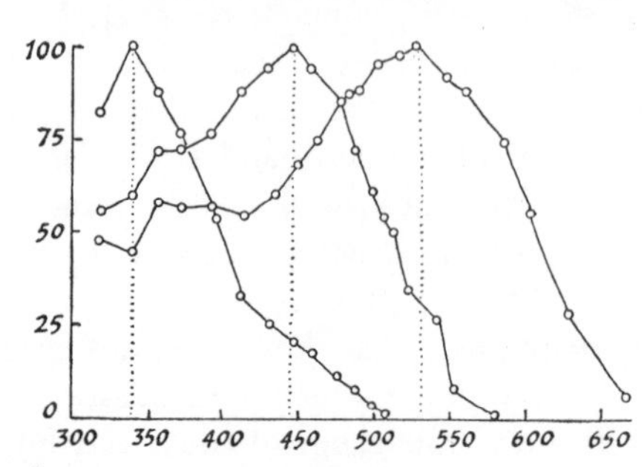

FIG. 144.—Spectral sensitivity of single visual cells in *Apis* drones (*after* AUTRUM and v. ZWEHL)

Ordinate: action potential as percentage of maximum. Abscissa: wave-length in mμ.

These results suggest that there is considerable diversity in the properties of the photoreceptor substance in the insect retina. The peak in the blue around 450mμ agrees with the maximum intensity of radiation from the blue sky[133] and with the absorption curve of the photosensitive pigment containing retinene that has been extracted from the heads of bees;[161, 162] but there must be other visual pigments to account for the other peaks of sensitivity. It is possible to produce photosensitive derivatives of retinene which could serve as ultra-violet receptors.[163]

Wave-length discrimination—Flower-visiting insects often behave as though they have a perception of colour. *Bombylius* visiting blue *Muscari* will fly rapidly from flower to flower; they are equally attracted to flowers enclosed in glass tubes or to pieces of blue paper (Fig. 145).[73] *Argynnis* feeding in the field on flowers of blue bugle will turn aside to visit violet or purple flowers made of paper.[61] Butterflies (Pieridae) during mating recognize one another in the first instance by the wing colouring.[33] It was established by Lubbock that bees can distinguish one colour from another; they can learn to associate the finding of honey with blue or orange papers and they will continue to visit the colour to which they have been accustomed although no honey be present.[79] Forel obtained the same results with paper flowers.[38] But these experiments did not exclude the possibility that the colours might be distinguished not by differences in quality but only by their relative brightness. This doubt was removed by v. Frisch by exposing the colour to which the

FIG. 145.—Course of flight of *Bombylius fuliginosus* (*after* KNOLL)

The chequer board is made up of grey shades ranging from 1 (very pale) to 14 (very dark), a pale blue square (B^1) and a dark blue square (B^2). The insect flies from blue *Muscari* to the blue squares, but disregards the grey shades.

bees had been trained on a chequer board made up of a complete range of grey shades: they recognized their blue or yellow colour among all these greys, even when covered with a glass plate so that smell was excluded; they could *not* be trained to come to any particular shade of grey.[41] The further possibility remained that the coloured papers used might be distinguishable through their reflecting more ultra-violet than any of the greys employed. This was excluded by experiments in which bees were trained to come to bands of spectral light.[75]

Experiments made to discover into how many colour qualities the spectrum is divided have shown that there are wide differences between insect and man and between different insect species. Bees tested by training to bands of spectral light appear to distinguish four regions in the spectrum: 650–500mμ (red, yellow, green); 500–480mμ (blue-green); 480–400mμ (blue and violet); 400–310mμ (ultra-violet). In their perceptions at the red end of the spectrum they approximate to man with red-green colour-blindness.[75] They can distinguish certain red flowers, such as *Papaver rhoeas*, but that is only because these reflect ultra-violet;

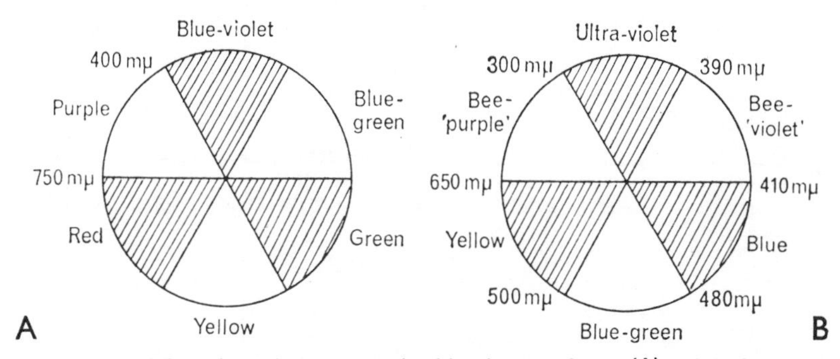

FIG. 146.—The colour circle as perceived by the eye of man (*A*) and the honey-bee (*B*) (*after* DAUMER)

The shaded sectors represent the three primary colours: and the sectors opposite to these are the complementary colours.

bees trained to visit such flowers are equally attracted to a black glass emitting ultra-violet rays invisible to man.[78] The ultra-violet region is probably perceived as a true colour; and from this it follows that reflected daylight probably appears colourless to the bee only when its composition approximates to sunlight; if deficient in ultra-violet, 'white' light will appear coloured.[55] And although the colours of papers remain unchanged to our eyes when covered with glass, they may be totally changed for the bee, grey papers appearing coloured. In fact ultra-violet is just one normal component in a system of three principal colours (Fig. 146); white paper or white flowers which do not reflect much ultra-violet will appear bluish-green to the bee.[55] Bees experience also the phenomenon of simultaneous contrast: a grey area surrounded by yellow appears blue and is visited by bees trained to this colour; surrounded by blue it appears yellow.[75]

But as can be seen from Fig. 146 yellow and blue are not complementary colours for the bee. In the bee, any colour can be matched by the appropriate mixture of three monochromatic stimuli: ultra-violet (360mμ), blue-violet (440mμ), and yellow (588mμ). A mixture of yellow and ultra-violet gives a new colour for the bee which may be called 'bee purple' since it unites the ends of the spectrum to form the complete colour circle (Fig. 146).[148] Among flowers which appear blue or violet to man there is so much variety in ultra-violet reflection that bees see them in not less than four different colour tones. Red flowers appear either 'bee ultra-violet' or 'bee black' depending on the degree of ultra-violet reflection.[149]

Butterflies offered paper flowers of various gay colours obviously prefer these before grey paper flowers. If the number of visits to different colours are counted, *Vanessa urticae* shows two maxima, one in the blue-purple group and one in the yellow-red group; green and greenish-yellow are unattractive. By training for blue at the expense of yellow and red, the blue peak of the curve can be relatively increased and the red-yellow peak decreased or vice versa. *Pieris brassicae* in search of food shows a spontaneous choice of blue and purple papers and, to a less extent, red and yellow; it disregards green, blue-green and grey.[61] But the egg-laying female of the same species shows a peculiar 'drumming' reaction of the fore legs with which it tests the leaves of the food plant; and for this reaction

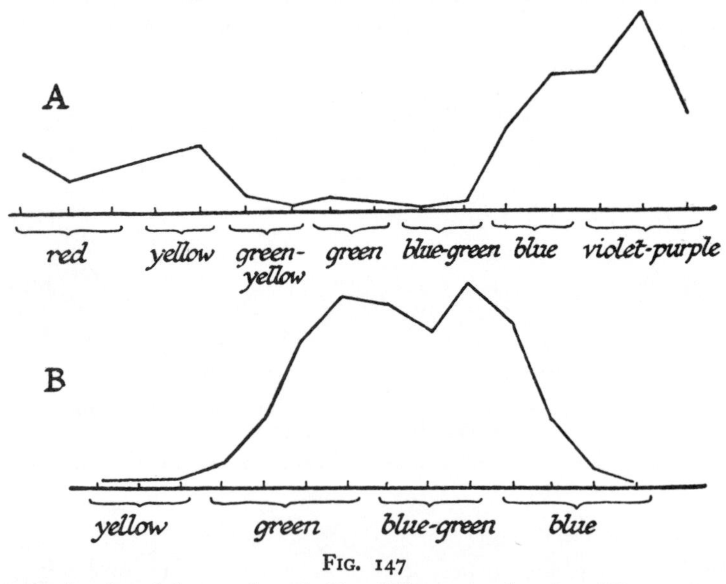

FIG. 147

A, chart showing the relative number of visits of *Gonepteryx rhamni* to different coloured papers during the feeding phase; B, chart showing the number of 'drumming' reactions shown by *Pieris brassicae* females on coloured papers during the egg-laying phase (*after* ILSE).

it chooses emerald green to greenish blue papers, yellow and pure blue are neglected (Fig. 147); though it will sometimes show the 'drumming' reaction at the margin of a purple paper—evidently in response to the green contrast colour.[64] Thus, unlike the bee, *Pieris* shows a definite discrimination between yellow and green; in this and in the discrimination of red its colour vision approximates more closely to that of man. For *Pieris* blue-green and ultra-violet are *not* complementary; indeed, there is no evidence that they can perceive ultra-violet at all,[61] although some butterflies are able to do so.[146]

Celerio can discriminate the blue-violet-purple group of colours of the flowers on which it feeds, from the yellow-green group. It feeds in the dusk and it is able to recognize these colour differences in light so dim that the human eye can no longer see any colour at all.[73] *Macroglossa* shows the same preference for blue flowers during feeding; but during oviposition it is attracted to yellowish-green flowers and plants.[73] Training experiments show that *Aeschna* nymphs can distinguish yellow from violet and grey.[74] By fatiguing the eye specifically to a particular wave-length, it is possible to show that *Drosophila* can detect an interval of 25mμ in the ultra-violet, 20mμ in the blue and 50mμ in the blue-green.[48] The drone-fly *Eristalis* shows a natural preference for yellow in visiting artificial flowers, but can be trained to visit blue.[65] The Aphid *Myzus persicae* can distinguish two colour zones: a longer wave zone, 660–500mμ (red-yellow-green) which excites the insect to probe, and a short wave zone, extending to 400mμ (blue-violet-purple). The occurrence of 'successive contrast' shows that the colours of these two zones are complementary.[125] Some flying Aphids are attracted to the yellow colour of foliage and alight for preference on young leaves; they will settle in large numbers on yellow objects and in yellow traps.[125]

The colour vision of *Drosophila* demonstrated by selective adaptation to different wave-lengths of monochromatic light is present, as in the vertebrate eye, in a receptive mechanism that requires high illumination. The receptive mechanism operative in dim light has no capacity for colour differentiation (as in the rods of the vertebrate eye). Associated with this phenomenon there is a shift in sensitivity towards the shorter wave-lengths as the intensity of the monochromatic light stimulus is decreased—the so-called 'Purkinje shift'.[155] Similarly, in the eyes of *Carabus*, *Phalera*, and *Macroglossa* the fourth peak of spectral sensitivity (at 430 mμ) (p. 230) appears in the blue when the intensity of illumination falls, and at the same time sensitivity to the shorter wave-lengths increases relative to the longer wave-lengths (green and red) so that the overall peak of sensitivity appears to move towards the shorter wave-lengths as in the Purkinje shift.[169] But whether the duplicity theory of vertebrate vision should be extended to insects is considered doubtful.[164] [See p. 247.]

Another method of testing for colour perception is by the use of the optomotor reaction (p. 323). Stripes of a certain blue colour x are alternated with grey stripes of different shades until a grey shade y is found which is equal in luminosity to x and causes no response when the stripes are moved in front of the insect. A yellow colour z is then found which will give no response when alternated with the grey shade y. If now the blue x is set beside the yellow z a response is given: evidently there is a perceptible difference in quality between these two. By this method it has been shown that a colour vision exists in the beetles *Chrysomela*, *Agelastica* and *Geotrupes*, in the fly *Fannia*, and in many Noctuids and Geometrids. The conclusions in the case of the bee agree with those obtained by other means. The plant-feeding Chrysomelids clearly distinguish green tints from one another as well as yellow and orange from blue-violet and green.[98] The optomotor reaction has been used to demonstrate the colour blindness of other beetles and the existence of colour discrimination in various Hemiptera.[185]

The nocturnal *Carausius* and the bug *Troilus* on the other hand appear to possess no colour vision;[98] a conclusion which has been confirmed in the case of *Carausius* by other means.[60] *Trialeurodes* flies only to yellow or green-yellow surfaces, but this seems to be merely because they provide the medium light intensity that is preferred.[108] By means of the optomotor reaction it has been shown that in *Notonecta* it is only the posterior dorsal ommatidia which are fully sensitive to colour. This is the area containing brown pigment, corresponding with the extent of the embryonic eye. The ommatidia of the ventral apex and front margin of the eye are colour blind, and there are transitional zones between.[94] In the adult *Aeschna* also the upper brownish region of the eye shows maximum sensitivity to violet (420mμ) and no capacity for colour vision; the lower yellowish-green region has two receptors: (*a*) sensitive to red and orange (maximum at 610mμ), (*b*) responds from ultra-violet to the far red (maximum at 515mμ).[183]

Whether there is a separate mechanism in the eye for perceiving colour is not known. In many insects one of the retinal cells in each ommatidium differs visibly from the others (Fig. 133, *i*), and the axons from certain of the retinal cells differ in their central connexions. It is possible that these differences correspond with differences in function; but at present this is pure hypothesis.[49, 53] It has been suggested that in *Calliphora*, of the seven visual cells in the ommatidium five belong to the common green receptor group, one to the less frequent blue receptor group and one to the yellow receptor group (see p. 230).[129] According to this conception the ommatidium would be the functional unit for image perception, the individual visual cells would discriminate both for colour and for the plane of polarization (p. 240). This complex information would be analysed by the higher centres.[139]

Perception of form—The retinal image, as we have seen, consists of a mosaic of points of light of varying luminosity and colour; a mosaic which will be coarse or fine depending on the number of facets per unit area. If the insect responds to changes in this visual pattern it may be regarded as having some degree of form perception.

In many insects such perception seems to be of a very elementary kind. If the fly *Eristalis* is allowed to crawl along between two walls, one plain and the other with vertical stripes, it always inclines towards the striped side, and the more so the more broken up the pattern is—unless the stripes become so fine that they can no longer be resolved by the eye.[16] *Coccinella* placed on a white ground with black stripes will follow the stripes and turn corners when they are angulated. This response seems to be best developed in phytophagous insects and is perhaps associated with their habit of following stalks and branches in their search for leaves.[106]

Bees can be trained to associate certain black figures on a white ground with the presence of food. The contrast of such figures against the background, the extent to which the figure is divided up, and the arrangement of the contours, are all properties which seem to be independently perceived.[54] Bees are attracted to any mark which contrasts with the colour of a background to which they have been trained; this response finds its natural use in the attraction of bees by the 'honey-guides' on flowers.[11] We have seen that such honey-guides may be dependent solely on lack of ultra-violet reflection;[149] they are often made up of converging lines which will produce the maximum degree of flicker, and they are often coincident with scent guides.[181]

But such figures as triangles, squares, circles, and ellipses, which are well within the capacity of the retina to differentiate and which appear very different to our eyes, are apparently not distinguished by the bee. On the other hand, it can readily distinguish from these any figures which are markedly broken up into black and white areas—rows of stripes, chequers, flower-like patterns, &c. (Fig. 148). It shows a natural preference for such divided figures, and this preference cannot be overcome by training: the bee will learn to associate the presence of food with a given figure only if this has a more sub-divided form than the control figure;[118] a simple disc is visited only after all other more divided figures have been removed.[54] There is in fact a direct proportionality between the richness of contour in the figures and the number of times they are visited by bees.[118] In butterflies, also, the choice of colour patterns is favoured if they have divided contours.[62]

From these observations it would seem that the choice of forms by these insects is dependent merely on the frequency of change of retinal stimulation; and that they cannot recognize figures by any properties of configuration.[114, 118] In other words, that the perception of form is little more than a perception of different degrees of flicker.[114] Thus, in nature, bees settle more rapidly on flowers if these are being shaken by the wind, or on models of flowers which are artificially kept in motion; they are attracted to moving stripes, and the more rapidly these are moved the greater is the attraction.[114]

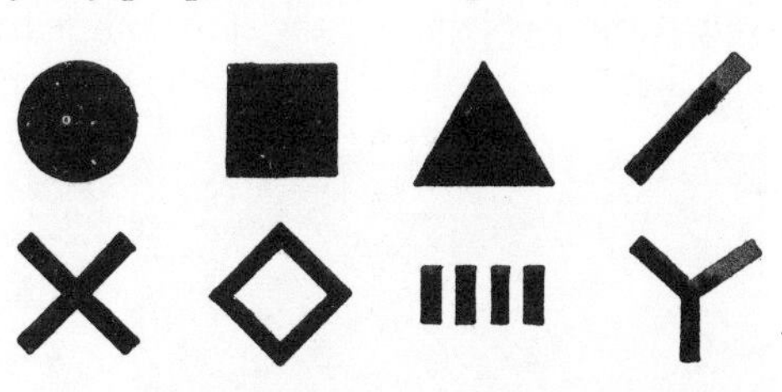

FIG. 148.—Perception of form by the bee

Figures of the upper row are indistinguishable from one another. The same applies to the lower row. But figures of the lower row are readily distinguished from those of the upper (*after* HERTZ).

But in addition to this type of response it is certain that some insects have a more integrated perception of the form or arrangement of objects in the outside world. Whereas *Aeschna* nymphs will snap repeatedly at any moving object[6] and hunting wasps will fly at nails hammered in a wall, mistaking them for flies,[38] *Aeschna* adults, which will turn towards a paper pellet thrown in the air, attracted by its movements, will instantly turn away from it when its image is perceived.[6] Bees and ants utilize visual landmarks in finding their way back to the nest (p. 333). The wasp *Philanthus* locates the entrance to its nest by the arrangement of visual marks around it, and by moving these it can be led astray at will.[105] Bees, which can recognize their hive by coloured marks at the entrance,[41] can appreciate the difference if a given colour is on the right or the left of the entrance hole.[40] In fact, the flying bee is able to differentiate the same three spatial co-ordinates in its visual field as man; it can distinguish left and right, before and behind, above and below, and so form a general appreciation of its surroundings.[111]

Visual acuity—For very small eyes the compound eye of insects is probably a more efficient structure than the single-lens type of eye of vertebrates; but as the size increases the acuity of vision increases more rapidly in the single-lens eye and this becomes more efficient.[134] Fig. 149 shows the well-known photograph obtained by Exner (1891) of the retinal image of *Lampyris splendidula*. It represents a window with a letter R on one pane, at a distance of 2·25 metres, and a church tower beyond. The degree of resolution which this represents has been estimated to correspond with a visual acuity $\frac{1}{60}$ to $\frac{1}{80}$ that of the normal human eye.[84]

The resolving power of the compound eye must depend in the first place upon the number of retinal elements, that is, the number of facets. In *Dytiscus* there are 9,000 facets in each eye; in *Melolontha* 5,100; in *Necrophorus* 3,500; in the winged male of *Lampyris* 2,500, in the wingless female 300; in the workers of the ant *Solenopsis*, which live almost always underground, 6–9, in the male of the same species which pursues the female high into the air, 400;[30] in *Musca* there are 4,000, in *Odonata* 10,000–28,000.[66]

But more important in determining the resolving power will be the angular extent of the visual field that is covered by each element, that is, the ommatidial angle. The ommatidial angle in *Apis* is 1°, in *Forficula* 8°; *Forficula* will obtain only a single point of light from an object which *Apis* will resolve into 64.[18] As the ommatidial angle becomes smaller, although the resolving power will increase, less light will enter each facet and hence the image will be much less luminous. But if the eye is larger the same ommatidial angle will entail a larger facet size and hence a brighter image. Or, conversely, for the same brightness of image, the larger species can have a greater number of facets, with a corresponding increase in resolving power. For example, in the three Lamellicorns *Polyphylla fullo*, *Melolontha vulgaris* and *Phyllopertha horticola* the total angle covered by the eye is about the same, and the size of the facets is about the same, but the number of facets is 12,150, 5,475 and 3,700 respectively; the large *Polyphylla* has three times the facet number of the small *Phyllopertha*.[18] The facet size does in fact remain much more nearly constant than the size of the insect or its eye. The diameter of the facet in *Culex pipiens* is 16μ, in *Melolontha vulgaris* 20μ, in *Periplaneta americana* 32μ, in *Anax formosus* and *Libellula depressa* 40μ; their surface area thus ranging between about 250μ^2 and 1600μ^2.[18] In some insects the facet size is greater in certain parts of the eye: in males of *Tabanus* the facets are larger over the upper and anterior parts; in *Simulium* and *Bibio* the two areas of different sized facets are distinctly separated; and in *Gyrinus* and *Cloëon* they are quite remote.[66] The wide overlap that exists in the visual field of neighbouring ommatidia (p. 216) and the nervous coupling of ommatida (p. 238) will be important factors in visual acuity.

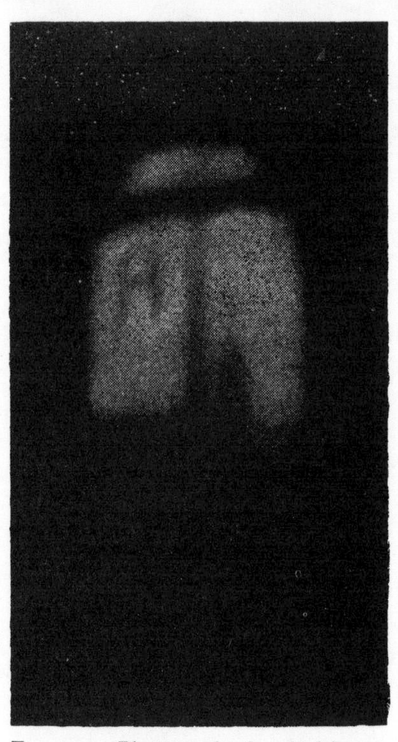

FIG. 149.—Photograph of retinal image in *Lampyris splendidula* (*after* EXNER)

It shows a window with a letter R on one pane and a church beyond.

In nocturnal insects, as we have seen, the luminosity of the image is increased by superposition of stimuli from adjacent facets (p. 219). Presumably this must involve a large sacrifice in the sharpness of the image. In Mantids the lateral parts of the eye form superposition images while the anterior parts, used for binocular fixation, form apposition images.[42] Mutants of *Drosophila* which lack eye pigments fail to respond to a striped pattern—presumably because the spread of light within the eye reduces the visual acuity. Other mutants with reduced numbers of facets may likewise lose their responsiveness.[70] White-eyed mutants of *Musca* and of the Seaweed fly *Coelopa*, in which the isolating eye pigments are lacking, likewise show a very great reduction in visual acuity.[173] In the Desert

Locust *Schistocerca* the compound eye of the solitary phase (p. 224) has broad pale stripes in which the retinal pigment is absent. These eyes are suited to subdued light, the perception of sharp images being impaired. In the gregarious phase on the other hand, the entire eye is pigmented and is thus adapted to the formation of sharp images in bright sunlight in accordance with the changed habits of the insect.[95] In superposition eyes the curvature of the eye surface is usually regular and spherical, as it must be if it is to function correctly. But in apposition eyes the curvature varies in different parts and then the flattest regions, where the angle between adjacent ommatidia is smallest, will be regions of most acute vision.[91] In the bee the ommatidial angle is twice[91] or three times[8] as great in the transverse plane of the eye as it is in the longitudinal plane. This will have the same effect as astigmatism in man. And there is some experimental evidence that the visual acuity of the bee is in fact greater in the vertical axis than in the horizontal.[8, 52]

Where a single object or point of light forms an image on the retina it can be more or less clearly seen even though the image does not cover a single rhabdom completely. In *Formica rufa* the ommatidial angle is 3·5°; but it will respond to a suspended sphere subtending an angle of only 2·5°.[57] (In man the least detectable angle is about 1′.) The same applies to the perception of a narrow stripe; this will slightly reduce the illumination of a succession of ommatidia across the eye, and if there is sufficient contrast between the stripes and the background, so that the stimulus upon adjacent ommatidia exceeds their contrast threshold, it will be perceived.[54] Thus newly hatched *Carausius* nymphs can perceive stripes rather narrower than the ommatidial angle of 7½°;[69] though in *Musca*, with an ommatidial angle of 3°, the narrowest stripe which appears to be perceived subtends an angle of 5°.[43] In *Drosophila* in strong illumination the visual acuity agrees with the value predicted from the visual angle of the ommatidia.[44] At a given intensity of illumination a certain minimum duration of stimulus is necessary to produce an effect on the eye. This must be a factor of some importance in the visual perceptions of flying insects. It has even been argued that, for this reason, the visual acuity of insects in flight will be improved by enlargement of the ommatidial angle (p. 239).[5]

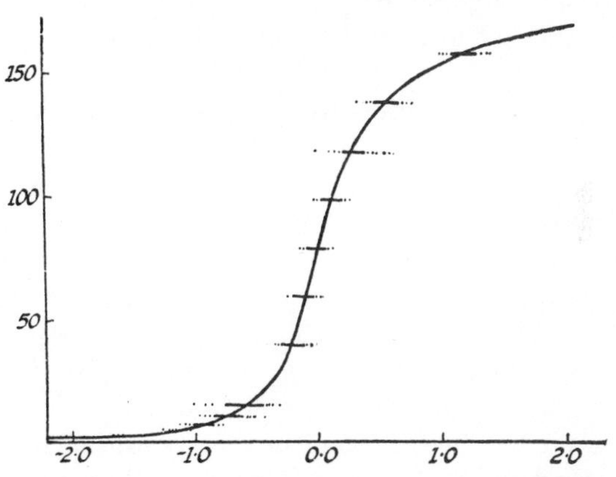

FIG. 150.—Relation between visual acuity and intensity of illumination in the honey-bee (*after* HECHT and WOLF)

Ordinate: units of visual acuity. Abscissa: logarithm of intensity in millilamberts.

The bee[52] and *Drosophila*[51] make characteristic reflex responses when a striped pattern is moved in front of them. Such movement can be perceived only if the pattern is resolved and if the differential illumination of the two components is detectable. It therefore provides a means of testing both intensity discrimination (p. 227) and visual acuity. By this method it has been shown that the visual acuity varies with the logarithm of the intensity along the same type of sigmoid curve as in man (Fig. 150). At low intensities the resolving power is very poor; under optimal conditions it is judged to be something like $\frac{1}{100}$ that of man in the case of the bee, and $\frac{1}{1000}$ in the case of *Drosophila*.[51, 52] The falling off of both acuity and intensity discrimination at low illuminations is attributed by some to the different ommatidia having different thresholds of response: at low illumination only the most sensitive are functional.[51, 52] By others it is attributed to a nervous coupling of groups of ommatidia to form a new unit when the illumination is low.[19]

That leads to another factor in visual acuity: the extent to which the individual ommatidia are normally coupled to a single nerve. From histological studies it has been concluded that in *Calliphora* the coupling of ommatidia is much less pronounced in the anterior region of the eye. Apart from considerations of ommatidial angle, this will therefore constitute a region of most acute vision comparable with the fovea centralis of vertebrates.[8] Possibly there is less coupling in insects with apposition eyes, for in these forms the optic ganglia show more massive development.[91]

Fine resolution in the compound eye—Some insects, as we have seen, can resolve movements or displacements smaller than the ommatidial angle. A formal explanation is based on the assumption of some overlap in the light stimulus received by neighbouring ommatidia.[170] In fact, there is a very large overlap (p. 216): in *Locusta* adjacent ommatidia are inclined to one another at an angle of about 1–2·4° in different planes; but a single ommatidium receives light from a far wider area, a cone of about 20°.[141] As judged by the electrical responses in the eyes of *Locusta* and *Calliphora*, evenly spaced stripes with an angular separation of only 0·3° were resolved by these insects. This capacity could not be explained by the resolution found in the apposition image, but it was explained by the second and third images formed by diffraction in the deeper layers (p. 220), the single retinula cell being regarded as the receptor unit. This system is particularly sensitive to the perception of radial patterns and to movement.[143] It seems doubtful whether it is concerned with the formation of a general image of the field of vision (p. 220). Under suitable conditions it is possible to observe the successive illumination of the individual rhabdomeres in eye slices of *Calliphora*.[176]

Movement perception and temporal resolution—An eye of low resolving power like the compound eye of insects will serve to detect changes in the visual field, that is, movements of objects, rather than their form.[35] *Aeschna* nymphs cannot be induced to snap at an object unless it is moving; the males of many Lepidoptera are not attracted to the females unless these are in motion. Insects also experience the phenomenon of 'induced movement': *Aeschna* nymphs will snap at stationary objects if a striped screen is moved behind them; or sometimes they will snap at such objects as soon as the screen ceases to move.[43] On the other hand, flies do not respond to stripes which are made to appear and disappear in succession round a screen, which gives the semblance of movement (stroboscopic movement) to the human eye.[54]

The optomotor response (p. 323) is the result of a perception of movement; and this response can be used to determine the minimum dark interval between two light impressions which is necessary if the two are to be separately perceived. It is estimated that in the case of *Aeschna* nymphs this minimum interval is 0·0165 second; at higher rates the images fuse. In other words, the 'flicker threshold' is about 60 stimuli per second.[96] In *Anax* nymphs a similar value is obtained;[25] in the bee a value of 55 flickers per second;[114] the corresponding figure in man being 45–53. In *Anax* the value varies with the intensity of illumination and the temperature; it is regarded as a sort of discrimination of intensity between the effect of the flashes of light and the after-effect of these flashes during the intervals of darkness (Fig. 151).[25] This must be a factor of importance for the flying insect. Using electrical methods of recording it has been claimed that the critical interval below which the fusion of successive stimuli occurs in the insect eye is 10–20 milliseconds as opposed to 50 milliseconds in the human eye.[5]

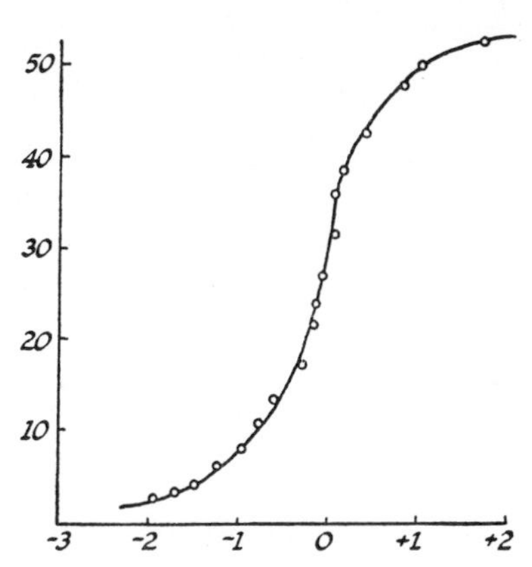

FIG. 151.—Relation between flicker frequency and illumination in the honey-bee (*after* WOLF)

Ordinate: flicker frequency per second. Abscissa: logarithm of intensity in millilamberts. Illuminations required at different flicker frequencies correspond closely to the intensities for threshold response in visual acuity tests (*cf.* Fig. 150).

There seem to be two main types of insect eye characterized by the form of the electroretinogram (p. 225). (i) In those with 'slow' eyes there is a simple negative illumination potential, the usual low flicker fusion frequency (40 per second in *Carausius*), high sensitivity, and a slow rate of adaptation: found in nocturnal insects such as *Carausius*, *Periplaneta*, *Tachycines*, Odonata larva, *Dytiscus*. (ii) In those with 'fast' eyes there is a complex illumination potential, extremely high fusion frequency (up to 300 per second in *Apis*),[121] low sensitivity and almost instantaneous dark adaptation: found in swiftly flying diurnal species such as *Aeschna*, *Vespa*, *Apis*, *Calliphora*.[5, 191] The properties of the ocelli run parallel with those of the compound eyes, and the different rates of flicker frequency seem to be dependent on the properties of the retinula cells.[191] [See p. 247.]

Given this rapid adaptation and high fusion frequency the flying insect may be able to resolve the visual field of a process of 'scanning'. The arrangement of objects in space is converted into a sequence of events in time. If the insect has a sufficiently high 'temporal resolving power' it may resolve a finer pattern in flight than at rest.[5, 121] The locust *Schistocerca* when viewing an object will sway the body from side to side in a movement called 'peering'. This may serve the same purpose: to analyse the field of vision by scanning.[200] [See p. 247.]

Distance perception—Since the eyes of insects are not independently movable, and therefore have no fixation plane upon which they can be converged, stereoscopic vision such as occurs in man cannot exist. But, the eyes being fixed, the distance of any point in the visual field will be determined by its position on the points of intersection of the individual ommatidial axes.[35] The visual field in *Notonecta* amounts to about 246° in the horizontal plane, 94° of which is covered by both eyes; in the vertical plane the field is 360°, of which 120° is binocular dorsally and 80° ventrally.[80] The simultaneous and equal illumination of corresponding points of the retina is probably the chief factor in the perception of distance (Fig. 152).[30] In *Aeschna* nymphs this mechanism is much more efficient when the point of intersection is on the median axis; and the insect, in fact, always turns so that this axis is directed towards any moving object which attracts attention. In the medial region of the eye the visual angle of the ommatidia becomes progressively smaller (1° 12′); this will aid definition as the object is approached. In *Aeschna*, snapping first takes place when the object is within range of the labium;[7] in adults of *Cicindela*, which can observe their prey at 12–15 cm. distance, no attempt is made to seize it until it comes within reach of the mandibles at the point of intersection of the innermost visual axes.[39]

By this mechanism the judgement of distance is absolute and not relative to the surroundings: *Aeschna* is equally accurate in its judgements in a dimly lit tank with plain black walls. The localization is purely visual, for it is effective through glass.[7] In *Agrion* the tactile perceptions of the antennae are normally important, also, in the judgement of distance, but 14 days after the antennae have been removed, the nymphs can seize their prey with the same precision as the normal insects.[4]

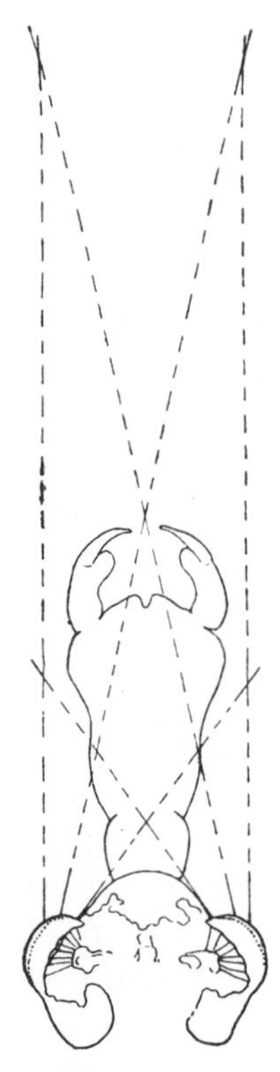

FIG. 152.—Distance perception. Head of *Aeschna* nymph with extended labium

The broken lines are the visual axes of selected ommatidia arrived at by observation of the pseudopupil. The points of intersection of these axes determine the position and distance of objects (*after* BALDUS).

Appreciation of distance is seen, also, in the ant: the visual angle which a moving object must subtend in order to produce a response in *Formica rufa* is smaller if the object is close. This difference disappears when one eye is blackened.[57] *Aeschna* nymphs and *Cicindela* adults likewise lose most of their power of judging distance after unilateral blinding. But a somewhat indefinite monocular localization still persists, dependent perhaps on the greater definition by the ommatidia at the inner margin of the eye.[7, 39] The peering or scanning procedure of *Schistocerca*, described above, probably has the function also of determining the relative distance of objects.[200] [See p. 247.]

Perception of the plane of polarized light—The unaided human eye has very little ability to perceive the direction of light polarization, but the compound eye contains a more efficient analyser.[127] Bees and ants utilize the pattern of polarization which exists in the blue sky for purposes of orientation and navigation.[123] The location and nature of the analyser have not been fully determined. But it would appear that the mechanism exists within a single ommatidium: polarized light produces a greater stimulating effect in a single ommatidium than non-polarized light of the same intensity.[122] There is evidence that in the eyes of Diptera, as in the nicol prism, the rays in one plane are reflected within the ommatidium.[124] In orientation by the polarization plane the bee can use ultra-violet light alone (300–400mμ) or visible light up to a wavelength of 500mμ (green-yellow).[156] This same ability to recognize the plane of polarization and to utilize the pattern of polarization for orientation is shown by larvae of sawflies and Lepidoptera which have only lateral ocelli.[128]

It is claimed by some authors[135, 172, 174] that polarized light is not perceived directly but only when reflected from a dark background. But in orienting themselves by the aid of the pattern of polarized light in the blue sky, bees appear to use the pattern itself and not a reflection pattern around them.[157] In the eye of *Notonecta* the height of the 'on-effect' in the electroretinogram depends on the plane of vibration of the polarized light, the extreme values being separated by an angle of 90°.[180]

One difficulty in interpreting the perception of polarized light is to know how one of the rays is eliminated before it emerges from the birefringent element. Hence the suggestion that it might be absorbed; as happens in dichroic media such as polaroid.[127] The precise orientation of the microvilli of the rhabdomeres whose paired members are usually diametrically opposite one another, with parallel microvilli (p. 221), suggested that these might provide the basis for the analysis of polarized light, and that each rhabdomere may be made up of molecular layers of dichroic visual pigments oriented in a regular fashion in individual rhabdomeres.[153, 204] The rhabdom in Diptera and Hymenoptera is demonstrably anisotropic; it may be that anisotropic absorbing substances are in the form of oriented micelles in the rhabdom.[197, 203] Records of action potentials from single visual cells of *Calliphora* show that the magnitude of the response depends on light intensity and on the plane of polarization; there is a fall of about 50 per cent in going from the plane of maximum to minimum efficiency.[140, 188]

From the electric responses of the eye of *Calliphora*, with the ommatidia illuminated with polarized light from the side (after cutting the eye vertically), it has been concluded that the axis of symmetry of the visual pigments is perpendicular to the optical axis of the ommatidium, and parallel to the axis of the microvilli of the rhabdomeres. The analyser for polarized light is clearly in the rhabdomeres.[160] [See p. 247.]

SIMPLE EYES

Stemmata—The eyes of larval and pupal forms are termed stemmata or lateral ocelli. These are very variable in structure. In larvae of Lepidoptera, Trichoptera, *Sialis*, *Myrmeleon*, &c., they form a group, each member of which has a structure something like a single ommatidium of a compound eye consisting of a cornea and crystalline lens and 7 retinal cells[31] (Fig. 153). The eyes of Collembola are of the same kind; there are about eight on either side, each with the structure of an ommatidium of eucone type. On the other hand, in larvae of Tenthredinidae, many Coleoptera, &c., each eye, of which there is usually only one on either side, consists of a single transparent lens-like thickening of the cuticle (the cornea) with underlying epidermis, and below this a number of retinulae each composed of two or three innervated visual cells grouped round a rhabdom (Fig. 154). The eyes are of this type in *Pediculus*.[112] The visual cells may themselves contain pigment; or there may be pigment cells of variable distribution. In some larvae the eyes are rudimentary; for example, in *Ceratopogon* they consist only of a pair of visual cells and two overlying pigment cells.[66]

There is no doubt that in all cases the stemmata are organs for the perception of light; and in some insects, in spite of their very simple structure, they also subserve colour vision and a rudimentary perception of form. An appreciation of colour is seen in caterpillars of *Vanessa* and *Pieris*, which are attracted by the green colour of leaves or pieces of paper, irrespective of the colour of the surroundings, but more so on a white than a black background. This attraction to green disappears at pupation, when brown and black are preferred.[46]

An appreciation of form is shown by *Lymantria* larvae which will walk towards vertical silhouettes, such as cylinders of brown paper against a light background.[77] When such pillars are of equal width, the newly hatched larvae of this species prefer the taller; pillars of 9 and 10 cm. can be differentiated with certainty at a distance of 30 cm., those of 10 cm. being chosen. It can be shown

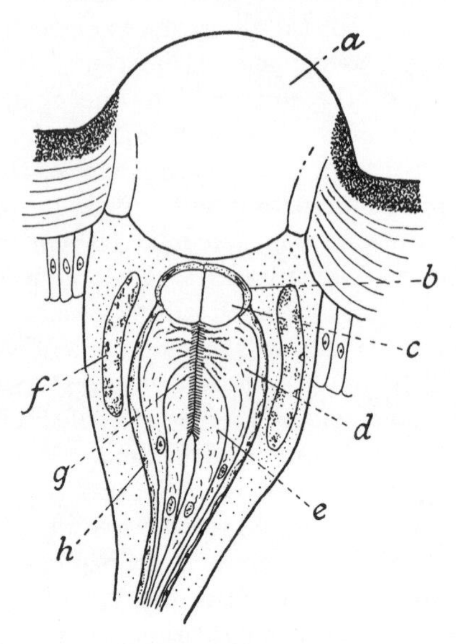

FIG. 153.—Simple eye, type i. Stemma of larva of *Gastropacha rubi* (*after* DEMOLL)

a, cornea; *b*, cells of crystalline body; *c*, crystalline body; *d*, distal sense cell; *e*, proximal sense cell; *f*, giant or mantle cell; *g*, rhabdom; *h*, pavement epithelium.

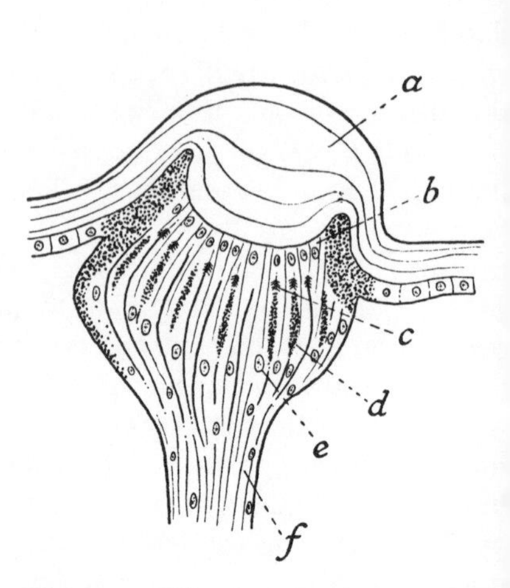

FIG. 154.—Simple eye, type ii. Ocellus of *Aphrophora spumaria* (*from* IMMS *after* LINK)

a, lens; *b*, matrix cells of lens; *c*, rhabdom; *d*, pigment; *e*, nuclei of retinula cells; *f*, nerve.

that objects of the same outline which subtend the same angle appear of equal size to the larvae. When different colours are compared blue appears the most attractive.[59]

The dioptric apparatus in the stemmata of these Lepidopterous larvae admits a large amount of light (f=0·93) and forms a clear inverted image, objects at all distances being focussed on the retinal elements. Since the rhabdom is not constructed for the reception of images it is the light-concentrating function of the ocellus which is important. Each is a constituent of the group of 6 on either side of the head; altogether they form 12 points of light. They are so arranged on the head that there is little or no overlapping of their visual fields—each spot represents the mean intensity of light from a different area. The mosaic is exceedingly coarse, but it is compensated to some extent by the larva

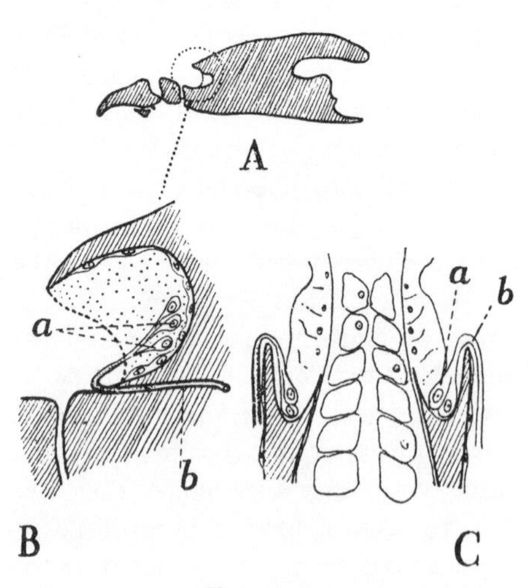

FIG. 155

A, mouth hooks and cephalo-pharyngeal skeleton of *Musca* larva; the small cavity in the ringed area marks the position of the light-sensitive cells. *B*, detail showing the light-sensitive cells *a*, and their nerve, *b*. *C*, horizontal section of anterior end of pharyngeal skeleton, showing pockets with the light-sensitive cells (*after* BOLWIG).

moving its head from side to side as it advances; it is thus able to achieve a considerable degree of form perception. If all the ocelli are covered except one, the normal phototactic responses remain, but form perception is seriously impaired.[31]

In larvae of *Cicindela* there are six stemmata on each side, two large, two small and two vestigial. In the large type there are up to 6,350 visual cells, and the single lens is quite efficient. Therefore, although a clear perception of images is not possible, they can certainly form visual patterns of some detail. In the capture of prey by these larvae the axis of direction of each stemma is very important. The prey is perceived (though only if it is moving) at a distance of 3–6 cm.; the larva snaps at it only when it comes within reach of the mandibles; in the perception of distance all the stemmata clearly function together as a unit.[39]

In Muscid larvae there are light-perceiving organs somewhere in the region of the head which have proved difficult to localize;[27, 110] it has even been suggested that it is the imaginal discs of the adult compound eyes which are the sensitive organs.[92] Careful experiments have shown that no point on the cephalic lobes is sensitive to light, and removal of the whole of these lobes does not affect the light reactions. In order to eliminate the reaction to light it is necessary to remove the anterior end of the main section of the pharyngeal skeleton. The light-sensitive cells lie, in fact, in a pocket on either side of the condyle-spine of this skeleton (Fig. 155), so placed that they are shaded by the body and by the pharyngeal skeleton from behind, but exposed to light from in front. These cells are supplied by a nerve running on the outer surface of the pharyngeal skeleton, and if this nerve is cut the light response is eliminated. The advancing larva moves its anterior end so as to reduce the illumination of these cells to a minimum; it therefore retreats from the light, keeping the cells in the shadow of its own body. The cells are most sensitive to green light, about 520 mμ; they do not react to red light and only to very strong blue or violet.[12] [See p. 247.]

Ocelli, or dorsal ocelli, are the simple eyes of adult insects which occur, usually on the frontal region of the head, alongside the compound eyes. There is a correlation between the presence of ocelli and the development of wings.[71] Their structure varies, but in general it agrees with that of the second type of stemma described above, in which there is a group of visual cells below a common lens. Where the ocelli are best developed, the lens and retina are well differentiated; but in some cases the ocellus may be nothing more than an unpigmented spot in the cuticle with a few irregularly arranged rhabdoms beneath.

There can be no doubt that the ocelli are light-sensitive organs. The ocelli of Agrionids and Libellulids have a tapetum, and when brought from the dark into a bright light they show a rapid movement of pigment which spreads in a few seconds like a brown curtain across the white fundus.[56, 76] Although the relation does not always hold,[116] in many insects the ocelli seem designed to admit much more light than the compound eyes. The ocelli of *Eristalis* are estimated to have an aperture of *f* 1·8, in females of *Formica f* 1·5; whereas the facets of the compound eyes have an aperture of *f* 2·5–4·5 and will therefore have only $\frac{1}{16}$ of the luminosity of the ocelli. On the other hand, the superposition eyes of Sphingids and Noctuids are more luminous than ocelli, and in these forms ocelli are absent.[57] But the ocelli seem very ill designed for the perception of images; for not only is the angular separation of the rhabdoms very great (it varies from 3°–10° in various insects),[57] but the image is usually focussed far behind the retina:[116] in *Eristalis* the retina lies 0·11 mm. from the lens; the focal plane lies at 0·29 mm.[57]

From these considerations of structure it has been concluded that the ocelli are adapted to the immediate perception of small changes in light intensity.[56, 57] But although they must be stimulated by light the insect often shows no outward response to such stimulation. Ants with their ocelli alone uncovered behave as though blind;[57] bees[86] and *Drosophila*[13] after blackening the compound eyes no longer show any reactions to light. In the cicada *Cryptotympana*, on the other hand, the ocelli as well as the compound eyes play an obvious part in light perception, and asymmetrical covering of the ocelli leads to circus flight.[22] In some insects, although the ocelli by themselves are incapable of evoking reflex movements in response to light, they have the effect of accentuating the responses of the insect to light stimuli received through the compound eyes. In both *Drosophila*[13] and *Apis*[86] the insect with ocelli uncovered responds more rapidly to changes in light intensity. The ocelli are therefore regarded as 'stimulatory organs' which increase photokinesis (p. 310). Results in *Calliphora* and *Locusta* support the view that covering the ocelli causes a decrease in the sensitivity to light perceived in the anterior part of the compound eye.[145] Honey-bees with the ocelli blackened begin to forage later in the morning and cease earlier in the evening than normal workers.[178]

In *Locusta* there is a spike potential in the ocellar nerve when the light is switched off[142] and a continuous discharge during darkness.[171] The development of a continuous impulse discharge on darkening the ocelli is seen also in *Periplaneta*, Odonata[191] and *Calliphora*.[130, 184] In successive flashes of light the fusion frequency in *Calliphora* is about 250 per second, so that the rate of temporal resolution is as high as in the compound eyes.[184] In *Periplaneta*, with 'slow' compound eyes, the fusion frequency in the ocelli is also slow: 45–60 per second.[192] The ocelli in these insects can therefore be described as 'darkness receptors'. If their afferent impulses have an inhibitory effect on the insect as a whole that would account for the 'stimulatory action' of illumination of the ocelli. In *Periplaneta* the characteristic persistent daily rhythm of activity depends on stimulation of the ocelli (p. 341). [See p. 247.]

The ocelli of *Periplaneta* appear to have only a single photoreceptor type with maximal sensitivity about 500mμ; they lack the ultra-violet sensitivity of the compound eye.[168] *Apis* on the other hand does have a second type of receptor with a peak sensitivity about 335–340mμ.[168]

Dermal light sense—In some insects the general body surface seems to be sensitive to light. Blind cavernicolous beetles of the genus *Anophthalmus* respond to the light of a candle.[84] Blinded *Blattella*[47] and *Periplaneta*[14] still settle for preference in the dark after complete blinding. Caterpillars will go towards a source of light after the eyes are blackened.[76] And the bug *Neides* in a state of partial hypnosis turns over when the eyes are blackened much more readily in the light than in the dark, as though activated by a dermal light sense.[45] Many caterpillars which feed in exposed situations and are coloured like their surroundings, place themselves so that the light always falls upon them from a particular angle. In this response the light receptors appear to lie in the general body surface.[102] In the case of *Chironomus* larvae the responses to light after blackening the eyes are due to light entering the translucent head capsule and reaching the eyes from below.[87] But this cannot be a general explanation of the results, for *Tenebrio* larvae avoid the light even after decapitation.[107] A dermal light sense is present in all the tergites of the larvae of *Acilius* and *Dytiscus* but is particularly well developed in the region of the spiracles at the end of the abdomen.[126] The antennae of *Aphis fabae* appear to be the site of a dermal light sense which is a source of kinetic stimulation.[137] In all these cases it is important to be sure that it is the light itself which is the effective stimulus, and not the heat into which it degenerates after absorption. [See p. 248.]

ADDENDA TO CHAPTER VI

p. 216. **Optical mechanism of compound eye**—The responses of the retinula cell appear to be dominated by rays which are within 2° of the axis of each cell.[240] In accordance with Exner's theory of internal reflection of the rhabdom and of the path leading to it, the visual field of a single ommatidium in the locust is about twice as wide in dark-adapted as in light-adapted eyes; this change being correlated with changes in the degree of packing of mitochondria around the rhabdom.[240] Acuity is governed by the angle of acceptance, which is a property of the optical pathway in the ommatidium.[218, 229]

p. 217. **Dioptric apparatus**—Exner's conception of the lens cylinder, originally derived from the superposition eye of the Lampyrid beetle applies equally to the eucone superposition eye of *Ephestia*.[227] The corneal facets in many Lepidoptera, Neuroptera, mosquitos, &c., are raised in the form of multiple minute nipples in perfect hexagonal array with a centre to centre spacing of about 200 mμ. This nipple array functions as an 'impedance transformer' equalizing by gradual transition the refractive index of air and cornea and thus decreasing reflection and increasing the transmission of visible light.[209] The nipples are formed by the evagination of the minute plaques of outer epicuticle which, as in all insect cuticles, are first laid down at the tips of the microvilli of the epidermal cells.[215]

p. 220. **'Superposition' eye**—By electrical recording from single receptors during illumination of single facets it has been shown that in the 'apposition' eye of the drone honey-bee[236] and locust[240] optical interaction is extremely small; in the superposition eye of the crayfish more than half the light captured enters through neighbouring facets, even when the screening pigments are in the fully light-adapted position.[236] The path of light passing through the 'pseudocone' of *Photuris* agrees with the description of Exner in *Lampyris*, but no functional superposition image is formed at the level of the receptors.[220] Indeed, Exner had himself called attention to the fact that the superposition image as obtained in his experiments fell *beyond* the level of the rhabdoms.[235] In compound eyes with a 'clear zone' between the cones and the receptor layer in the dark-adapted state, such as exists in *Ephestia*, elongated crystalline tracts from the cones provide an individual optical pathway for each ommatidium, as in the light-adapted state. Light entering through numerous neighbouring facets is scattered, and refracted as described by Exner towards the side from which it has been received; and not being intercepted by screening pigment, it provides an additional stimulus to the deeper parts of the retina. But the image produced at the level of the receptors by this 'superposition route' must be highly diffuse.[221]

p. 220. **Composition of the cone**—In *Pieris* the cones consist of a mucopolysaccharide protein complex like the cornea of vertebrates; fluid when first secreted but becoming insoluble by polymerization.[225]

p. 221. **Image reception in retina and optic ganglia**—The neuroanatomical work of Cajal and Sánchez (1915) had suggested that visual excitation is projected upon progressively deeper layers of the optic lobes. Detailed re-investigation along these lines have refined this conclusion. In apposition eyes, with fused rhabdoms, the axon bundles from each ommatidium can be followed along a regular pathway first to the lamina ganglionaris and then via the chiasma to the medulla where they are projected exactly in order but reversed about the vertical plane.[222] In the higher flies (*Musca* &c.) the retinula cells are widely separated (Fig. 136). The six peripheral members receive light from six different directions, or six different points in the field of vision. The geometry of the eye is such that each of these six points lies in the line of vision of one retinula cell in each of six neighbouring ommatidia. The axons from the cells with the same line of vision sort themselves out and run together to the same point in the lamina, so that the *Musca* compound eye is a kind of 'neural superposition eye'.[210, 226] The two central retinula cells (not a single one as in Fig. 136) receive light from a seventh direction and they have long axons which pass through the lamina to the medulla.[239] This superposition of light stimuli, from single points in the field of vision, upon a number of widely separated retinula cells, and their recombination by the neural route at a single point in the optic ganglia, demands an almost incredibly precise control of both the geometry of the eye and the reassortment of the axons. The resulting resolution is very much finer than that expected from the mosaic theory of Müller and Exner.[234]

p. 222. *Drosophila* reared on a sterile diet without carotene shows a reduction of visual sensitivity to one thousandth of the normal; but the photosensitivity of the circa-dian rhythm (p. 343) is not affected;[246] and *Manduca* reared for several generations without vitamin A and its precursors suffers extreme degeneration in the retinal epithelium and associated nervous tissues.[213]

p. 222. **Pigment movements**—In the eyes of *Dytiscus* and of *Lethocerus* (Hemiptera) the retinula cells retract during light adaptation; and in consequence the crystalline thread from cone to rhabdom increases in length and decreases in diameter. Thus the light passing down to the rhabdom will be reduced and the radial migration of pigment to the crystalline thread will probably absorb most of the light which passes down.[241]

p. 223. **Dark-adaptation**—In *Schistocerca* sensitivity of the compound eye increases 5–10 times after 15 min. dark adaptation; the change consists of neural (both peripheral and central) as well as photochemical factors.[214] In the fully dark-adapted eye of the honey-bee the threshold of stimulation is 0·26 Lux; whereas immediately after complete light adaptation it was 2 Lux.[207]

p. 225. **Retinal responses**—On exposure to light the distal ends of the visual cells of *Calliphora* are the most strongly depolarized and thus become potential 'sinks' in respect to the proximal parts. Simultaneous illumination thus causes the entire distal part of the retina to become negative to the proximal part and so to give rise to the electroretinogram.[245]

p. 226. **Adaptation in the compound eye**—The speed of dark adaptation is very sensitive to temperature. The honey-bee adapted to 2,000 Lux for a short period reaches its maximum sensitivity (dark adaptation) in 5 sec. at 40° C.; in 1 min. at 15° C.[233]

p. 229. **Spectral sensitivity**—The apparent specific sensitivity of the eye of *Calliphora* at the red end of the spectrum may simply be due to the fact that neither pteridine nor ommochrome screening pigments will prevent the spread of light of these wave lengths.[216, 228] On the other hand there is strong evidence that in the butterflies *Heliconius* and *Papilio* there is a red receptor with a peak at 600–645 mμ as well as other receptors maximally sensitive to blue and blue-green (483–578 mμ).[237] In *Notonecta* the spectral maxima, relative frequency, and the shape of the sensitivity curves are strikingly similar to those in the honey-bee, whereas *Calliphora* and *Locusta* have quite different colour receptor systems.[211] In *Ascalaphus* the ultra-violet sensitive pigment is confined to the frontal eye. It is converted by ultra-violet light to meta-rhodopsin but is quickly reconverted by exposure to light in the 400–600 mμ range, so that the ultra-violet sensitivity is always nearly maximal.[219] There is some evidence that in addition to retinal pigments based on rhodopsin (retinene) with sensitivity in the visible range and in the ultra-violet, there may be unrelated fluorescent materials in the retinal cells, some emitting radiations acting upon rhodopsin, some non-radiating but with chromophores specifically sensitive to ultra-violet.[223]

p. 233. **Wave-length discrimination**—In the honey-bee, with its trichromatic receptor system, the position of the sensitivity maxima for each receptor type is independent of the state of adaptation: there is no Purkinje shift. On the other hand, the highest sensitivity lies in the green (545 mμ) for dark-adapted eyes and in the blue (449 mμ) for light-adapted eyes; so that although the relative brightness changes, bees can see colour in every state of adaptation.[238]

p. 239. **'Fast' and 'slow' eyes**—Comparative studies on the form of the electroretinogram in a wide range of insects from eight orders have shown that it is not possible to classify 'slow' eyes and 'fast' eyes on the slope of the electrophysiological response.[244]

p. 239. **'Scanning' by compound eye**—Lowne (1892) had described a muscle within the compound eye of *Calliphora*. The occurrence of a response in this muscle to moving striped patterns suggests that the muscle causes scanning movements of the compound eye (such as occur in salticid spiders) by displacing the photoreceptors.[212]

p. 240. **Distance perception**—The praying mantis has a 'fovea' in each retina where the radius of curvature is smaller and the ommatidia more concentrated. It clearly estimates the catching distance of the prey by a process of triangulation with the fovea as the site of image formation.[230]

p. 241. **Perception of polarized light**—Direct microspectrophotometric measurements in isolated crayfish rhabdoms support the belief that the dichroism arises from the fine structural geometry of the absorbing dipoles of the chromophores.[242]

p. 243. **Illumination and eye structure**—Eyes of mosquito larvae grown in complete darkness show defects in fine structure which are quickly rectified (within 12 hours or so) on exposure to light.[243]

p. 244. **Function of ocelli**—It has been suggested that rather than serving as 'stimulatory organs' ocelli may act as moderators of the intensity of neural signals from the compound eyes, thus serving as extra-ocular adjusters of visual sensitivity.[224] Electrophysiological records show that interaction occurs between stimuli received by ocelli and compound eyes.[217] In conformity with the view that the ocelli serve to regulate brain activity, it has been found that the response to antennal stimulation is greatly facilitated by photic stimulation of the ocelli.[232] In the Sarcophagid fly, *Boettcherisca*, recordings of potential discharges from areas in the brain support the view that the level of illumination of the ocelli regulates brain activity.[232]

p. 245. **Response of ganglia to illumination**—The abdominal ganglia of *Periplaneta*, after adaptation to darkness, give rise to action potentials in the nerve cord when exposed to brief light stimuli.[208]

REFERENCES

1 ABBOTT, C. E. *Ann. Ent. Soc. Amer.*, **20** (1927), 117–22 (effect of monochromatic light on *Formica*).

2 ADRIAN, E. D. *J. Physiol.*, **91** (1937), 66–89 (electrical response of optic ganglion: *Dytiscus*).

3 ALTENBERG, E. *Brit. J. Exp. Biol.*, **4** (1926), 38–45 (model of compound eye).

4 ALVERDES, F. *Biol. Zbl.*, **43** (1924), 577–605 (sensory perceptions of Ephemerid and Odonata larvae).

5 AUTRUM, H. *Nachr. Akad. Wiss. Göttingen, Math.-Phys. Kl.*, 1948, 8–12; *Experientia*, **5** (1949), 271–7; *Naturwiss.*, **39** (1952), 290; *Klim. Wockenschr.*, **31** (1953), 241; *Exp. Cell Res.*, Suppl. **5** (1958), 426–39 (visual resolving power in flying insect).

6 BALDUS, K. *Z. wiss. Zool.*, **121** (1924), 557–620 (function of brain: larvae and adults of Odonata).

7 —— *Z. vergl. Physiol.*, **3** (1926), 475–505 (distance localization: *Aeschna* larva).

8 BAUMGÄRTNER, H. *Z. vergl. Physiol.*, **7** (1928), 56–143 (visual acuity: honey-bee).

9 BEDAU, K. *Z. wiss. Zool.*, **97** (1911) (compound eyes of *Notonecta*, &c., Hem.).

10 BERTHOLF, L. M. *J. Agric. Res.*, **42** (1931), 379–419; **43** (1931), 703–13 (extent of visible spectrum: honey-bee); *Z. vergl. Physiol.*, **18** (1932), 32–64 (ditto: *Drosophila*).

11 BOLWIG, N. *Entom. Medd.*, **20** (1938), 80–101 (response of bee to 'honey-guides').

12 —— *Vidensk. Medd. Dansk Naturhist. Foren.*, **109** (1946), 81–207 (sense organs of *Musca* larva).

13 BOZLER, E. *Z. vergl. Physiol.*, **3** (1925), 145–82 (function of ocelli: *Drosophila*, Dipt.)

14 BRECHER, G. *Z. vergl. Physiol.*, **10** (1929), 495–526 (dermal light sense: *Periplaneta* Orth.).

15 BUCK, J. B. *Physiol. Zool.*, **10** (1937), 45–58 (perception of red light by firefly, *Photinus* Col.).

16 v. BUDDENBROCK, W. *Naturwiss.*, **23** (1935), 98–100 (sense of form: *Eristalis*, Dipt.)

17 —— *Biol. Rev.*, **10** (1935), 283–316; *Naturwiss.*, **23** (1935), 154–7 (physiology of compound eye).

18 —— *Grundriss der vergleichenden Physiologie*, Berlin, 1937.

19 v. BUDDENBROCK, W., and SCHULZ, E. *Zool. Jahrb., Physiol.*, **52** (1933), 513–36 (ommatidium as functional unit).

20 BUGNION, E., and POPOFF, N. *Arch. Anat. Microsc.*, **16** (1914), 261–304 (eyes of nocturnal insects).

21 CAMERON, J. W. MacB. *Canad. J. Res.*, D, **16** (1938), 307–42 (reactions to light of different wave-lengths: *Musca*, Dipt.).

22 CHEN, S. H., and YOUNG, B. *Sinensia*, **14** (1943), 55–60 (vision and flight: *Cryptotympana* Hem.).

23 COLLINS, D. L. *J. Exp. Zool.*, **69** (1934), 165–97 (iris pigment migration: *Carpocapsa* Lep.).

24 CRESCITELLI, F., and JAHN, T. L. *J. Cell. Comp. Physiol.*, **13** (1939), 105–12; **19** (1942) 47–66 (electrical response of insect eye).

25 CROZIER, W. J., *et al. J. Gen. Physiol.*, **20** (1937), 363–410 (response to flickering light larva of *Anax*, Odonata).

26 DAY, M. F. *Biol. Bull.*, **80** (1941), 275–91 (pigment movements in eye: *Ephestia* Lep.).

[27] DEBAISIEUX, P. *Ann. Soc. Sci. Bruxelles*, **59** (1939), 2–23 (sense organs in head of *Lucilia* larva, Dipt.).

[28] DEMOLL, R. *Arch. ges. Physiol.*, **129** (1909), 461–75 (pigment migration in eyes of day-flying Lepidoptera).

[29] —— *Zool. Jahrb., Physiol.*, **30** (1911), 169–80 (movements of iris pigment in compound eye).

[30] —— *Die Sinnesorgane der Arthropoden ihr Bau und ihre Funktion*, Braunschweig, 1917.

[31] DETHIER, V. G. *J. Cell. Comp. Physiol.*, **19** (1942), 301–13; **22** (1943), 116–26 (lateral ocelli of caterpillars).

[32] DOLLEY, W. L. *Physiol. Zool.*, **2** (1929), 483–90 (dark adaptation: *Eristalis*, Dipt.).

[33] EGGERS, F. *Zool. Jahrb.; Abt. Syst.*, **71** (1938), 277–90 (vision in mating: Pieridae, Lep.).

[34] ELTRINGHAM, H. *The Senses of Insects*, Methuen's Biological Monographs, London, 1933.

[35] EXNER, S. *Die Physiologie der facettierten Augen von Krebsen und Insekten*, Leipzig, 1891.

[36] FALCONER, D. S. *J. Exp. Biol.*, **21** (1945), 33–8 (sensitivity to light: *Agriotes* larva, Col.).

[37] FOREL, A. *Rev. Sci.* (*Paris*), **38** (1886), 660–1 (vision of ultra-violet by ants).

[38] —— *Das Sinnesleben der Insekten*, Munich, 1910.

[39] FRIEDERICHS, H. F. *Z. Morph. Oekol. Tiere*, **21** (1931), 1–172 (vision in *Cicindelidae*, Col. larva and adult).

[40] FRIEDLAENDER, M. *Z. vergl. Physiol.*, **15** (1931), 193–260 (visual training of honey-bee).

[41] v. FRISCH, K. *Zool. Jahrb., Physiol.*, **35** (1914), 1–182 (sense of form and colour in honey-bee).

[42] FRIZA, F. *Z. vergl. Physiol.*, **8** (1928), 289–336 (colours and pigment movements in compound eyes).

[43] GAFFRON, M. *Z. vergl. Physiol.*, **20** (1934), 299–337 (perception of movement: Odonata larvae and Muscidae).

[44] v. GAVEL, L. *Z. vergl. Physiol.*, **27** (1939), 80–135 (visual acuity: *Drosophila*).

[45] GODGLÜCK, U. *Biol. Zbl.*, **55** (1935), 187–97 (dermal light sense: *Neides*, Hem.).

[46] GÖTZ, B. *Z. vergl. Physiol.*, **23** (1936), 429–503 (perception of colour and form in larvae of Lepidoptera).

[47] GRABER, V. *Sitz. Ber. Akad. Wiss. Wien, Abt. 1*, **87** (1883), 201–36 (dermal light sense: *Blattella*, Orth.).

[48] HAMILTON, W. F. *Proc. Nat. Acad. Sci. U.S.A.*, **8** (1922), 350 (colour vision: *Drosophila*).

[49] HANSTRÖM, B. *Z. vergl. Physiol.*, **6** (1927), 566–97 (functional differentiation of colour-perceiving elements in retina).

[50] HARTLINE, H. K. *Amer. J. Physiol.*, **83** (1928), 466–83 (electric response to illumination in compound eye).

[51] HECHT, S., and WALD, G. *J. Gen. Physiol.*, **17** (1934), 517–47 (visual acuity and intensity discrimination: *Drosophila*, Dipt.).

[52] HECHT, S., and WOLF, E. *J. Gen. Physiol.*, **12** (1929), 727–60 (visual acuity: honey-bee).

[53] HERTWECK, H. *Z. wiss. Zool.*, **139** (1931), 559–663 (anatomy of sense organs: *Drosophila*, Dipt.).

[54] HERTZ, M. *Z. vergl. Physiol.*, **8** (1929), 693–748; **11** (1929), 107–45; **14** (1931), 629–74; **21** (1934), 579–603; *Naturwiss.*, **23** (1935), 618–24 (perception of form by honey-bee).

[55] —— *Z. vergl. Physiol.*, **25** (1938), 239–50; *J. Exp. Biol.*, **16** (1938), 1–8 (colour vision in bee: technique).

[56] HESS, C. *Arch. ges. Physiol.*, **181** (1920), 1–16 (function of ocelli: Odonata).

[57] HOMANN, H. *Z. vergl. Physiol.*, **1** (1924), 541–78 (function of ocelli: *Eristalis*, Dipt., *Formica*, Hym.).

[58] HORSTMANN, E. *Biol. Zbl.*, **55** (1935), 93–7 (diurnal periodicity of pigment movements in eyes of Noctuidae).

[59] HUNDERTMARK, A. *Z. vergl. Physiol.*, **24** (1936), 42–57; 563–82 (perception of form and colour by larvae of *Lymantria*, Lep.).

[60] —— *Biol. Zbl.*, **57** (1937), 228–33 (luminosity discrimination: *Dixippus*, Orth.).

[61] ILSE, D. *Z. vergl. Physiol.*, **8** (1928), 658–92 (colour sense in butterflies).
[62] —— *Z. vergl. Physiol.*, **17** (1932), 537–56 (perception of form in butterflies).
[63] —— *Biol. Zbl.*, **52** (1932), 660–7 (apparent luminosity of colours: Lepidoptera).
[64] —— *Nature*, **140** (1937), 544 (responses to colours in egg-laying butterflies).
[65] —— *Nature*, **163** (1949), 255–6 (colour vision: *Eristalis*, Dipt.).
[66] IMMS, A. D. *Textbook of Entomology*, Methuen, London, 1935.
[67] JAHN, T. L. *J. N.Y. Ent. Soc.*, **54** (1946), 1–8 (wave-length sensitivity: *Melanoplus* Orth.).
[68] JAHN, T. L., *et al. J. Cell. Comp. Physiol.*, **12** (1938), 35–55; **13** (1938), 113–9; *Biol Bull.*, **78** (1940), 42–52; *Proc. Soc. Exp. Biol. Med.*, **48** (1942), 656–60; *J. Gen Physiol.*, **26** (1942), 75–88; *Physiol. Zool.*, **16** (1943), 101–9 (electrical responses of compound eye).
[69] KALMUS, H. *Z. vergl. Physiol.*, **24** (1937), 644–55 (visual acuity: *Dixippus*, Orth.).
[70] —— *J. Genetics*, **45** (1943), 206–13 (optomotor responses of *Drosophila* mutants).
[71] —— *Proc. Roy. Ent. Soc. Lond.*, A, **20** (1945), 846 (correlation between wings and ocelli).
[72] KLINGSBEIL, K. H. *Biol. Zbl.*, **58** (1938), 631–46 (reactions to light in *Ephestia* mutants).
[73] KNOLL, F. *Z. vergl. Physiol.*, **2** (1921), 329–80; *Abhandl. zool.-bot. Ges. Wien*, **12** (1926), 1–616 (insects and flowers; colour vision).
[74] KOEHLER, O. *Verhandl. deutsch. Zool. Ges.*, **29** (1924), 83–91 (colour vision: nymphs of *Aeschna*, Odonata).
[75] KÜHN, A. *Z. vergl. Physiol.*, **5** (1927), 762–800 (colour sense of honey-bee).
[76] LAMMERT, A. *Z. vergl. Physiol.*, **3** (1925), 225–78 (pigment movements in ocelli).
[77] de LÉPINEY, J. *Bull. Soc. Zool. Fr.*, **53** (1928), 479–90 (vision in larvae of *Lymantria* Lep.).
[78] LOTMAR, R. *Z. vergl. Physiol.*, **19** (1933), 671–723 (perception of ultra-violet light honey-bee).
[79] LUBBOCK, SIR J. *Ants, Bees and Wasps*, London, 1885.
[80] LÜDTKE, H. *Z. vergl. Physiol.*, **22** (1935), 67–118 (visual responses of *Notonecta*, Hem.).
[81] LUTZ, F. E. *Ann. N.Y. Acad. Sci.*, **29** (1924), 181–283 (perception of ultra-violet by flower-visiting insects).
[82] —— *Amer. Mus. Novit.*, No. **641** (1933), 1–26 (perception of ultra-violet patterns by *Trigona*, Hym.).
[83] LUTZ, F. E., and GRISEWOOD, E. N. *Amer. Mus. Novit.*, No. **706** (1934), 1–14 (extent of visible spectrum for *Drosophila*, Dipt.).
[84] MARCHAL, P. *Richet's 'Dictionaire de Physiologie'*, **9** (1910), 273–386 (physiology of insects).
[85] MERKER, E. *Zool. Jahrb., Physiol.*, **46** (1929), 297–374; *Biol. Rev.*, **9** (1934), 49–7[illegible] (visibility of ultra-violet light).
[86] MÜLLER, E. *Z. vergl. Physiol.*, **14** (1931), 348–84 (function of ocelli: bees and ants).
[87] OEHRING, W. *Zool. Jahrb., Physiol.*, **53** (1934), 342–66 (dermal light sense: *Chironomus*, Dipt. larva).
[88] PARKER, G. H. *Ergeb. Biol.*, **9** (1932), 239–91 (movements of retinal pigment: review).
[89] PARRY, D. A. *J. Exp. Biol.*, **24** (1947), 211–9 (function of ocelli: *Locusta*, Orth.).
[90] PATTEN, B. M. *J. Exp. Zool.*, **17** (1914), 213–80; **20** (1916), 585–98; *Amer. J. Physiol.*, **38** (1915), 313–38 (photic responses of blow-fly larva).
[91] del PORTILLO, J. *Z. vergl. Physiol.*, **23** (1936), 100–45 (ommatidial angle and visual acuity).
[92] POUCHET, G. *Rev. Mag. Zool.* (sér. 2), **23** (1872), 110; 312 (the perception of light by blow-fly larvae).
[93] PRIEBATSCH, I. *Z. vergl. Physiol.*, **19** (1933), 453–88 (light and colour change in *Dixippus*, Orth.).
[94] ROKOHL, R. *Z. vergl. Physiol.*, **29** (1942), 638–76 (sensitivity of regions of the retina *Notonecta*, Hem.).
[95] ROONWAL, M. L. *Proc. Roy. Soc.*, B, **134** (1947), 245–72 (eyes of locust phases: *Schistocerca*, Orth.).
[96] SÄLZLE, K. *Z. vergl. Physiol.*, **18** (1932), 347–68 (perception of movement: *Odonata* larvae).

[7] SANDER, W. *Z. vergl. Physiol.*, **20** (1933), 267–86 (response to light of different wave-length in honey-bee).

[8] SCHLEGTENDAL, A. *Z. verg. Physiol.*, **20** (1934), 545–81 (colour sense demonstrated by optomotor response).

[9] SCHLIEPER, C. *Z. vergl. Physiol.*, **6** (1927), 453–72; **8** (1928), 281–8 (colour sense of insects studied by the optomotor reaction).

[0] SCHREMMER, F. *Z. vergl. Physiol.* **28** (1941), 457–66 (red-blindness: *Vespa*, Hym.).

[1] STEHR, W. C. *J. Exp. Zool.*, **59** (1931), 297–335 (activating influence of light on aquatic Hemiptera).

[2] SÜFFERT, F. *Z. Morph. Oekol. Tiere*, **26** (1932), 147–316 (dermal light sense: caterpillars).

[3] SÜFFERT, F., and GÖTZ, B. *Naturwiss.*, **24** (1936), 815 (colour vision: *Vanessa* larva, Lep.).

[4] TAYLOR, I. R., *et al. Physiol. Zool.*, **16** (1943), 213–22; **17** (1944), 193–9 (electrical responses of retina of insects).

[5] TINBERGEN, N. *Z. vergl. Physiol.*, **16** (1932), 305–34; **21** (1935), 699–716; *Biol. Zbl.*, **58** (1938), 425–35 (visual orientation of *Philanthus*, Hym.).

[6] TISCHLER, W. *Zool. Jahrb., Physiol.*, **57** (1936), 157–202 (sense of form).

[7] TUCOLESCO, J. *Bull. Biol. Fr. Belg.*, **67** (1933), 480–514 (dermal light sense: larva of *Tenebrio*, Col.).

[8] WEBER, H. *Z. Morph. Oekol. Tiere*, **23** (1931), 575–753 (biology of *Trialeurodes*, Hem.).

[9] WEISS, H. B., *et al. J. N.Y. Ent. Soc.*, **49** (1941), 1–20; 149–59; **50** (1942), 1–35; **51** (1943), 117–31; **52** (1944), 267–71; **54** (1946), 17–30; *J. Econ. Ent.*, **36** (1943), 1–17 (colour perception in insects).

[0] WELSH, J. H. *Science*, **85** (1937), 430–1 (sensory papillae of larvae of *Lucilia*, Dipt.).

[1] WIECHERT, E. *Z. vergl. Physiol.*, **25** (1938), 455–93 (perception of space by bee).

[2] WIGGLESWORTH, V. B. *Parasitology*, **33** (1941), 67–109 (senses of *Pediculus*, Anopl.).

[3] WOLF, E. *J. Gen. Physiol.*, **16** (1933), 407–22 (intensity discrimination: honey-bee).

[4] WOLF, E., *et al. Z. vergl. Physiol.*, **20** (1933), 151–61; *J. Gen. Physiol.*, **17** (1933), 7–19; **20** (1937), 511–18; *Naturwiss.*, **23** (1935), 369–71 (responses of honey-bee to moving objects or flicker).

[5] WOLF, E., and ZERRAHN-WOLF, B. *J. Gen. Physiol.*, **19** (1935), 229–38 (dark adaptation of eye in honey-bee).

[6] WOLSKY, A. *Z. vergl. Physiol.*, **12** (1930), 783–7; **14** (1931), 385–9 (function of ocelli).

[7] WULFF, V. J., and JAHN, T. L. *J. N.Y. Ent. Soc.*, **55** (1947), 65–83 (electroretinogram: *Cynomyia*, Dipt.).

[8] ZERRAHN, G. *Z. vergl. Physiol.*, **20** (1933), 117–50 (perception of form by honey-bee).

SUPPLEMENTARY REFERENCES (A)

[9] AUTRUM, H. *Z. vergl. Physiol.*, **32** (1950), 176–227 (potential changes in eye of *Calliphora* and *Dixippus*).

[0] AUTRUM, H., and GALLOWITZ, U. *Z. vergl. Physiol.*, **33** (1951), 407–35 (potential changes in eyes of insects).

[1] AUTRUM, H., and STOECKER, M. *Z. Naturforsch.*, **5b** (1950), 38–43 (fusion frequency in eye of bee).

[2] AUTRUM, H., and STUMPF, H. *Z. Naturforsch.*, **5b** (1950), 116–22 (perception of polarized light in compound eye).

[3] V. FRISCH, K. *Experientia*, **6** (1950), 210–21 (perception of polarized light by insects).

[4] MENZER, G., and STOCKHAMMER, K. *Naturwissenschaften*, **38** (1951), 190–1 (analysis of polarized light by insect eye).

[5] MOERICKE, V. *Z. Tierpsychol.*, **7** (1950), 265–74; *Z. Naturforsch.*, **7b** (1952), 304–9; *Mitt. Biol. ZentAnst.*, 1952, 90–7 (colour perception by Aphids).

[6] SCHÖNE, H. *Z. vergl. Physiol.*, **33** (1951), 63–98 (dermal light sense in *Acilius* and *Dytiscus* larvae).

[7] WATERMAN, T. H. *Trans. N.Y. Acad. Sci.*, **14** (1951), 11–14 (polarized light and orientation by insects).

[128] WELLINGTON, W. G., *et al. Canad. J. Zool.*, **29** (1951), 339–51 (polarized light an orientation in larvae of Lepidoptera and Hymenoptera).

SUPPLEMENTARY REFERENCES (B)

[129] AUTRUM, H., and BURKHARDT, D. *Nature*, **190** (1961), 639 (spectral sensitivity of sing visual cells in *Calliphora*).

[130] AUTRUM, H., and METSCHL, N. *Z. Naturforsch.*, **16b** (1961), 385–8; *Z. vergl. Physiol.*, **4** (1963), 256–73 (electrical responses of ocelli in *Calliphora* to light and darkness

[131] AUTRUM, H., and STUMPF, H. *Z. vergl. Physiol.*, **35** (1953), 71–104 (extent of the visib spectrum in *Calliphora*).

[132] AUTRUM, H., and TSCHARNTKE, H. *Z. vergl. Physiol.*, **45** (1962), 696–710 (oxygen cor sumption of the insect retina on stimulation).

[133] AUTRUM, H., and V. ZWEHL, V. *Z. vergl. Physiol.*, **46** (1962), 8–12; *Naturwiss.*, **50** (1963 698 (spectral sensitivity of single visual cells in *Apis* drone).

[134] BARLOW, H. B. *J. Exp. Biol.*, **29** (1952), 675–84 (comparison between insect and vert brate eye).

[135] BAYLOR, E. R., and SMITH, F. E. *Amer. Nat.*, **87** (1953), 97–102 (perception of the plar of polarized light in insects).

[136] BERNHARD, C. G., and OTTOSON, D. *J. Gen. Physiol.*, **44** (1960), 195–203; 205–15; **4** (1964), 465–78 (dark adaptation in the eyes of Lepidoptera).

[137] BOOTH, C. O. *Nature*, **197** (1963), 265–6 (photokinetic function of antennae in Aphids

[138] BRIGGS, M. H. *Nature*, **192** (1961), 874–5 (retinene in insects).

[139] BURKHARDT, D., *et al. Verh. dtsch. zool. Ges.*, 1961, 182–5; *Symp. Soc. Exp. Biol.*, **1** (1962), 87–109 (spectral sensitivity of single visual cells in insects).

[139a] BURKHARDT, D., and AUTRUM, H. *Z. Naturforsch.*, **15b** (1960), 612–16 (electric responses of single visual cells in *Calliphora*).

[140] BURKHARDT, D., and WENDLER, L. *Z. vergl. Physiol.*, **43** (1960), 687–92 (perception the plane of polarized light by single visual cells in *Calliphora*).

[141] BURTT, E. T., and CATTON, W. T. *J. Physiol.*, **125** (1954), 566–80 (effective ommatidi angle in *Locusta*).

[142] —— *J. Physiol.*, **133** (1956), 68–88 (electrical responses to stimulation of compoun eyes and ocelli in *Locusta*).

[143] —— *Proc. Roy. Soc.*, B, **157** (1962), 53–82; *Symp. Soc. Exp. Biol.*, **16** (1962), 73–8 (diffraction theory of insect vision).

[144] COHEN, C. F., and BARKER, R. J. *J. Cell. Comp. Physiol.*, **62** (1963), 43–7 (vitamin A an spectral response in *Musca*).

[145] CORNWELL, P. B. *J. Exp. Biol.*, **32** (1955), 217–37 (function of ocelli in *Calliphora* an *Locusta*).

[146] CRANE, J. *Zoologica*, **39** (1954), 85–115; **40** (1955), 167–96 (ultra-violet sensitivity an reflectance in butterflies).

[147] DANNEEL, R., and ZEUTZSCHEL, B. *Z. Naturforsch.*, **12b** (1957), 581–3 (fine structure retinula in *Drosophila*).

[148] DAUMER, K. *Z. vergl. Physiol.*, **38** (1956), 413–78 (colour sense in the honey-bee).

[149] —— *Z. vergl. Physiol.*, **41** (1958), 49–110 (ultra-violet reflection from flowers).

[150] DUPONT-RAABE, M. *C.R. Acad. Sci.*, **230** (1950), 873–4 (pigment migration in eyes *Carausius*).

[151] DE VRIES, H., and KUIPER, J. W. *Ann. N.Y. Acad. Sci.*, **74** (1958), 196–203 (optics of th insect eye).

[152] EGUCHI, E., NAKA, K., and KUWABARA, M. *J. Gen. Physiol.*, **46** (1962), 143–57 (fir appearance of electrical responses in the developing eye of *Bombyx*).

[153] FERNÁNDES-MORAN, H. *Nature*, **177** (1956), 742–3; *Exp. Cell. Res.*, **5** (1958), 586–64 (fine structure of the compound eye).

[154] FINGERMAN, M. *J. Exp. Zool.*, **120** (1952), 131–64 (role of eye pigments in *Drosophila*

[155] FINGERMAN, M., and BROWN, F. A. *Physiol. Zool.*, **26** (1953), 59–67 (colour vision an 'Purkinje shift' in *Drosophila*).

6 V. FRISCH, K. *S. B. Bayer Akad. Wiss. Math. Nat. Kl.* 1953, 197–9 (wave-length and polarized light in *Apis*).

7 V. FRISCH, K., LINDAUER, M., and DAUMER, K. *Experientia*, **16** (1960), 289 (direct perception of polarized light by the honey-bee).

8 V. FRISCH, K., LINDAUER, M., and SCHNEIDER, F. *Naturwiss. Rdsch.*, **5** (1960), 169–72 (perception of the sun through cloud in the honey-bee).

9 FYG, W. *Schweiz. ent. Ges.*, **33** (1960), 185–94 (glycogen in crystalline cone of *Apis*).

0 GIULIO, L. *Z. vergl. Physiol.*, **46** (1963), 491–5 (dichroitic properties of the retina in *Calliphora*).

1 GOLDSMITH, T. H., *et al. Ann. N.Y. Acad. Sci.*, **74** (1958), 223–9; *Proc. Nat. Acad. Sci.*, **44** (1958), 123–6; *J. Gen. Physiol.*, **46** (1962), 357A–367A; **47** (1964), 433–41 (retinene and vitamin A in compound eye of honey-bee).

2 GOLDSMITH, T. H. *J. Gen. Physiol.*, **43** (1960), 775–99 (spectral sensitivity in compound eye of *Apis*).

3 —— In *Sensory Communication* (W. A. Rosenblith, Ed.), Wiley, New York, 1961, 357–75 (wave-length discrimination and types of retinene).

4 —— In *Light and life* (W. D. McElroy and B. Glass, Edd.) Johns Hopkins Press, 1961, 771–94 (colour vision in insects: review).

5 —— *J. Cell. Biol.*, **14** (1962), 489–94 (fine structure of retinulae in *Apis*).

6 —— *Comp. Biochem. Physiol.*, **10** (1963), 227–37 (light and dark adaptation in *Apis*).

7 GOLDSMITH, T. H., and PHILPOTT, D. E. *J. Biophys. Biochem. Cytol.*, **3** (1957), 429–40 (fine structure of retina in *Apis*).

8 GOLDSMITH, T. H., and RUCK, P. R. *J. Gen. Physiol.*, **41** (1958), 1171–85 (spectral sensitivity of ocelli in cockroaches and honey-bee).

9 HASSELMANN, E. M. *Zool. Jahrb., Physiol.*, **69** (1962), 537–76 (spectral sensitivity in compound eyes of Coleoptera and Lepidoptera).

0 HASSENSTEIN, B., and REICHARDT, W. *Z. Naturforsch.*, **11b** (1956), 513–24 (perception of small movements by the compound eye).

1 HOYLE, G. *J. Exp. Biol.*, **32** (1955), 397–407 (response of the ocellar nerve in *Locusta* to darkening).

2 KALMUS, H. *Nature*, **182** (1958), 1526–7 (perception of polarized light by insects).

3 —— *Vision Res.*, **1** (1961), 192–7 (optomotor response in white-eyed mutants of *Musca*, &c.).

4 KENNEDY, D., and BAYLOR, E. R. *Nature*, **191** (1961), 34–7 (analysis of polarized light by the eye of *Apis*).

5 KIRSCHFELD, K. *Z. vergl. Physiol.*, **44** (1961), 371–413 (electric response of retina in *Hylobius* to bright and dark stimulus).

6 KUIPER, J. W. *Symp. Soc. Exp. Biol.*, **16** (1962), 58–71 (optics of the compound eye).

7 LANGER, H. *J. Ins. Physiol.*, **4** (1960), 283–303 (energy metabolism in retina of *Calliphora* on stimulation).

8 LINDAUER, M., and SCHRICKER, B. *Biol. Zbl.*, **82** (1963), 721–5 (role of ocelli in twilight vision in *Apis*).

9 LÜDTKE, H. *Z. vergl. Physiol.*, **35** (1953), 129–52 (dark adaptation in retina of *Notonecta*).

0 —— *Z. vergl. Physiol.*, **40** (1957), 329–44 (fine structure of retina in *Notonecta* and analysis of polarized light).

1 MANNING, M. *Behaviour*, **9** (1956), 114–39 (acrion of honey-guides).

2 MAZOKHIN-PORSHNIAKOV, G. A. *Biofizika*, **2** (1956), 352–62; **4** (1959), 46–57 (ultra-violet reflection from wings of butterflies and role of ultra-violet in insect vision).

3 —— *Biophysica*, **4** (1959), 427–36 (colour vision in Odonata).

4 METSCHL, N. *Z. vergl. Physiol.*, **47** (1963), 230–55 (electrical responses in ocelli of *Calliphora*).

5 MOLLER-RACKE, I. *Zool. Jahrb., Physiol.*, **63** (1952), 237–74 (optomotor reaction and colour vision in insects).

6 NAKA, K., and EGUCHI, E. *J. Gen. Physiol.*, **45** (1962), 663–80 (electroretinogram from single retinula cells).

7 NAKA, K., and KUWABARA, M. *J. Ins. Physiol.*, **3** (1959), 41–9; (analysis of electroretinogram).

[188] NAKA, K., and KUWABARA, M., *Nature*, **184** (1959), 455–6 (analysis of polarized ligh
by single visual cell in *Calliphora*).

[189] NOWIKOFF, M. *Biol. Zbl.*, **51** (1931), 325–9 (model of the compound eye).

[190] ROGERS, G. L. *Proc. Roy. Soc.*, B, **157** (1962), 83–98 (diffraction theory of insect vision

[191] RUCK, P. *J. Ins. Physiol.*, **1** (1957), 109–23; **2** (1958), 189–98; *J. Gen. Physiol.*, **44** (1961
605–57 (electrophysiology of insect ocellus).

[192] —— *J. Ins. Physiol.*, **2** (1958), 261–74 (ocellar nerve responses in insects with 'fas
and 'slow' eyes).

[193] —— *Ann. Rev. Ent.*, **9** (1964), 83–102 (retinal structure and photoreception in insect
review).

[194] SCHNEIDER, G. *Verh. dtsch. zool. Ges.*, (1954), 346–51; *Z. vergl. Physiol.*, **39** (1956
1–29 (spectral sensitivity in *Calliphora* and penetration of longer wave-lengths

[195] SCHÖNE, H. *Z. vergl. Physiol.*, **35** (1953), 27–55 (luminosity curve in *Dytiscus*, &c.).

[196] SOTAVALTA, O., TUURALA, O., and OURA, A. *Ann. Acad. Sci. Fennicae*, **62** (1962), 6–1
(pigment migration, &c., in eyes of Tipulidae).

[197] STOCKHAMMER, K. *Z. vergl. Physiol.*, **38** (1956), 30–83; *Ergebn. Biol.*, **21** (1959), 23–5
(analysis of polarized light by the compound eye).

[198] SUZUKI, K. *J. Fac. Sci. Hokkaido Univ.*, **15** (1962), 137–47 (glycogen, &c. in the com
pound eye).

[199] WADDINGTON, C. H., and PERRY, M. M. *Proc. Roy. Soc.*, B, **153** (1960), 155–78 (fin
structure of developing eye in *Drosophila*).

[200] WALLACE, G. K. *J. Exp. Biol.*, **36** (1959), 512–25 (visual scanning in *Schistocerca*).

[201] WALTHER, J. B. *Biol. Zbl.*, **77** (1958), 63–104; *J. Ins. Physiol.*, **2** (1958), 142–51 (spectr
sensitivity in eyes of *Periplaneta*).

[202] WALTHER, J. B., and DODT, E. *Experientia*, **13** (1957), 333; *Z. Naturforsch.*, **14b** (1959
273–8 (ultra-violet sensitivity in *Calliphora* and *Periplaneta*).

[203] WATERMAN, T. H. *Trans. N.Y. Acad. Sci.*, **14** (1951), 11–14; *Progress in Photobiol*
1961, 214–6 (polarized light orientation).

[204] WOLKEN, J. J., CAPENOS, J., and TURANO, A. *J. Biophys. Biochem. Cytol.*, **3** (1957), 441–
(photoreceptor structures in *Drosophila*).

[205] WOLKEN, J. J., and GUPTA, P. D. *J. Biophys. Biochem. Cytol.*, **9** (1961), 720–4 (phot
receptor structures in cockroaches).

[206] YAGI, N., and KOYAMA, N. *The compound eye of Lepidoptera*, Shinkyo-Press, Toky
1963, 319 pp.

SUPPLEMENTARY REFERENCES (C)

[207] AUTRUM, H. and SEIBT, U. *Naturwissenschaften*, **20** (1965), 566 (dark adaptation in e
of honey-bee).

[208] BALL, H. J. *J. Insect Physiol.*, **11** (1965), 1311–15 (photosensitivity of terminal ganglic
of *Periplaneta*).

[209] BERNHARD, C. G. *et al. Acta physiol. scand.*, **63** (1965), 9–79; *Z. vergl. Physiol.*, **67** (1970
1–25 (corneal nipple array and its function).

[210] BRAITENBERG, V. *Z. vergl. Physiol.*, **52** (1966), 212–14; *Exp. Brain Res.*, **3** (1967
271–98 (projection of axons from retinula cells to lamina in eye of *Musca*).

[211] BRUCKMOSER, P. *Z. vergl. Physiol.*, **59** (1968), 187–204 (spectral sensitivity of sing
visual cells in *Notonecta*).

[212] BURTT, E. T. and PATTERSON, J. A. *Nature*, **228** (1970), 183–4 (internal muscle in eye
Calliphora).

[213] CARLSON, S. D. *et al. Science*, **58** (1967), 628–30 (vitamin A deficiency and eye degener
tion in *Manduca*).

[214] COSENS, D. J. *J. Insect Physiol.*, **12** (1966), 871–90 (dark adaptation in eye of *Schist
cerca*).

[215] GEMNE, G. *Phil Trans. Roy. Soc.*, B, **262** (1971), 343–63 (fine structure of cornea
nocturnal Lepidoptera).

[216] GOLDSMITH, T. H. *J. Gen. Physiol.*, **49** (1965), 265–87 (probable absence of red recept
in *Calliphora*).

[217] GOODMAN, L. J. *Adv. Insect Physiol.*, **7** (1970), 97–195 (structure and function of dorsal ocelli in insects: review).

[218] GÖTZ, K. G. *Kybernetik*, **2** (1964), 77–92, 215–22 (optomotor responses and light reception in *Drosophila*).

[219] HAMDORF, K., SCHWEMER, J. and GOGALA, M. *Nature*, **213** (1971), 458–9 (visual pigment sensitive to ultra-violet in *Ascalaphus*).

[220] HORRIDGE, G. A. *Proc. Roy. Soc. Lond.*, B **171** (1969), 445–63 (the path of light rays entering the 'pseudocone' eye of *Photuris*).

[221] HORRIDGE, G. A. *Proc. Roy. Soc. Lond.*, B, **179** (1971), 87–124 (retina of *Ephestia*: image reception in the dark-adapted eye).

[222] HORRIDGE, G. A. and MEINERTSHAGEN, I. A. *Z. vergl. Physiol.*, **66** (1970), 369–78 (projection of axon pathways from retina to medulla).

[223] KAY, R. E. *J. Insect Physiol.*, **15** (1969), 2021–38 (ultra-violet sensitive pigments in eye of *Manduca*).

[224] KERFOOT, W. B. *Nature*, **215** (1967), 305–7 (ocelli as moderators of compound eye sensitivity).

[225] KIM, CHANG-WHAN, *Korean J. Zool.*, **7** (1964), 90–4 (mucopolysaccharide substance of crystalline cone of *Pieris*).

[226] KIRSCHFELD, K. *Exp. Brain Res.*, **3** (1967), 248–70 (projection of visual field on the separated rhabdomeres of *Musca* eye).

[227] KUNZE, P. and HAUSEN, K. *Nature*, **231** (1971), 392–3 (graded refractive index in crystalline cone of *Ephestia* eye).

[228] LANGER, H. and HOFFMANN, C. *J. Insect Physiol.*, **12** (1966), 357–87 (screening effect of pteridines and ommochromes in eye of *Calliphora*).

[229] McCANN, G. D. and MACGINITIE, G. F. *Proc. Roy. Soc. Lond.*, B, **163** (1965), 369–401 (change in width of ommatidial field with dark- and light-adaptation in *Musca*).

[230] MALDONADO, H. *et al.* *Z vergl. Physiol.*, **67** (1970), 58–101 (foveal vision in praying mantis).

[231] MAZOKHIN-PORSHNYAKOV, G. A. *Insect Vision* (tranl. T. H. Goldsmith) Plenum Press: New York. 1969, 306 pp.

[232] MIMURA, K. *et al.* *Z. vergl. Physiol.*, **62** (1969), 382–94; **68** (1970), 301–10 (regulation of insect brain excitability by the ocelli).

[233] SEIBT, U. *Z. vergl. Physiol.*, **57** (1967), 77–102 (temperature and rate of dark-adaptation in *Apis*).

[234] SEITZ, G. *Z. vergl. Physiol.*, **59** (1968), 205–31 (optical path of light entering the apposition eye of *Calliphora*).

[235] SEITZ, G. *Z. vergl. Physiol.*, **62** (1969), 61–74 (dioptric apparatus of *Lampyris*).

[236] SHAW, S. R. *Science*, **165** (1969), 88–90 (optics of compound eye in honey-bee and locust).

[237] SWIHART, S. L. *J. Insect Physiol.*, **13** (1967), 1679–88; **16** (1970), 1623–36 (red receptors in eyes of butterflies).

[238] THOMAS, I. and AUTRUM, H. *Z. vergl. Physiol.*, **51** (1965), 204–18 (Purkinje phenomenon in eyes of insects).

[239] TRUJILLO-CENÓZ, O. and MELAMED, J. *J. Ultrastruct. Res.*, **16** (1966), 395–8; **21** (1968), 313–34 (axon tracts from retinula cells of Cyclorrhapha).

[240] TUNSTALL, J. and HORRIDGE, G. A. *Z. vergl. Physiol.*, **55** (1967), 167–82 (optics of locust retina).

[241] WALCOTT, B. *Nature*, **223** (1969), 971–2 (movement of retinula cells in light adaptation in *Dytiscus* and *Lethocerus*).

[242] WATERMAN, T. H., FERNANDEZ, H. R. and GOLDSMITH, T. H. *J. gen. Physiol.*, **54** (1969), 415–32 (dichroism of the rhabdoms in the compound eye).

[243] WHITE, R. H. *et al.* *J. exp. Zool.*, **164** (1967), 461–78; **166** (1967), 405–26; **169** (1968), 261–78 (effect of prolonged darkness on the larval eye in mosquito).

[244] YINON, U. *J. Insect Physiol.*, **16** (1970), 221–5 (electroretinograms of 'slow' and 'fast' eyes).

[245] ZETTLER, F. *Z. vergl. Physiol*, **56** (1967), 129–41 (analysis of electroretinogram in *Calliphora*).

[246] ZIMMERMANN, W. F. and GOLDSMITH, T. H. *Science*, **171** (1971), 1167–9 (circadian rhythm in carotenoid-depleted *Drosophila*).

Chapter VII

Sense Organs: Mechanical and Chemical Senses

MECHANICAL SENSES

THE SIMPLEST mechanical sense is that of touch. This may be subserved by (i) receptors of contact projecting from the surface of the body, (ii) sense organs stimulated by the strains set up in the cuticle by pressure, or (iii) organs sensitive to changes of tension within the body. These three types of receptors merge rather indefinitely into one another: in the last analysis they are all stimulated by pressure changes in the end organ of the nerve. Further, any of them may serve to detect either stimuli from the outside world, or stimuli resulting from the position or movements of the insect's own body; that is, they may act as exteroceptors or as proprioceptors. Proprioceptive stimuli may result from the action of gravity; so that the same receptors may subserve a static sense. Finally, the perceptions of these same sense organs may be so refined that they respond to the pressure changes and movements in the air, which we perceive as sounds; that is, they serve as organs of hearing.

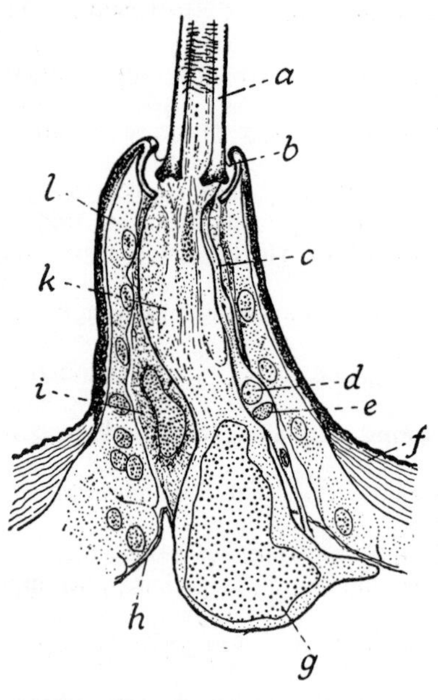

FIG. 156.—Section through base of tactile bristle in larva of *Vanessa urticae* (*after* HSÜ)

a, base of hair; *b*, articular membrane; *c*, scolopoid body; *d*, sense cell; *e*, neurilemma cell; *f*, cuticle; *g*, trichogen cell; *h*, basement membrane; *i*, tormogen cell; *k*, vacuoles in trichogen cell; *l*, epidermis.

Tactile spines and hairs—Nearly all the spines and hairs of the body surface are sensory end organs. Each individual organ was termed by Haeckel a 'sensillum'; as we shall see later, these sensilla may assume many forms and serve many different senses. The simplest are the widely spread articulated setae (Fig. 156). As a rule, each of these is supplied by a single bipolar sense cell, invested by neurilemma, situated beneath the epidermis and sending its distal process to end in the margin of the socket, its proximal process to the central ganglion. The distal process (the 'terminal filament') is often deeply staining, and at moulting it is shed with the cuticle; it is sometimes stated to be a chitinized product of the sense cell.[123] This distal process extends in *Psychoda* from the cuticle that is being moulted to the new cuticle (whether larval, pupal or adult) and thus enables the tactile hairs to function after the old cuticle has separated.[200] In *Rhodnius* they remain functional in this way until the moment of ecdysis.[288]

Sensory hairs of this type occur all over the body. They appear to be stimulated by movement of the spine in its socket. Thus if the electrical changes in the crural nerve of the cockroach are recorded while the hairs on the tibia are deflected with a fine glass needle, two types of response can be obtained. (i) In the case of the finer hairs sudden deflection causes a short burst of impulses which is over in a fraction of a second (that resembles the response of the touch receptors in vertebrates). (ii) In the case of the stout spines adaptation is much slower; the response to the same angular deflection varies according to its direction, being maximal towards the proximal end of the segment; and the initial frequency of discharge depends on the speed of deflection. (In these characters it resembles more closely the response from the proprioceptors of vertebrates.)[105]

These two types are sometimes called respectively 'velocity sensitive' and 'pressure sensitive'. On stimulation a 'receptor' or 'generator' potential builds up in the sense organ, presumably as the result of a graded depolarization of part of the neurone. When this reaches a certain level it will initiate a propagated impulse in the nerve axon; and the greater the degree of depolarization the greater

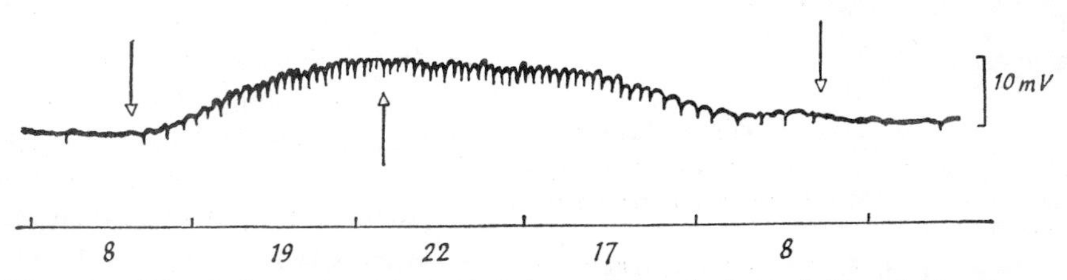

FIG. 157.—Electrical response of tactile hair on the anal plate of *Phormia* female to mechanical stimulation (*after* WOLBARSHT)

The left-hand arrow indicates the beginning of deformation; the middle arrow, the maximum; and the right-hand arrow, the return to the resting position. Time intervals of 0·2 second are shown on the abscissa, where the figures indicate approximate numbers of impulses per 0·2 second of stimulation.

the frequency of nervous discharge.[291] Tactile receptors are commonly of the 'phasic-tonic' variety: they show a high rate of discharge on initial stimulation, followed by a continuous discharge at a lower rate if the pressure persists (Fig. 157).[179] Often, as in the honey-bee,[277] the response depends on the direction, degree, and velocity of bending.

These articulated hairs are clearly tactile receptors. They are very numerous at the tip of the tibia and on the tarsal segments, where they evidently perceive contact stimuli. They occur all over the cerci of crickets and other Orthoptera in which they perceive, among other stimuli, earth-borne vibrations when they rest on the ground, and air currents when they are held aloft.[107] The antennae of some insects are richly supplied with such hairs and are then obviously tactile organs: this is particularly so in cave-dwelling insects such as crickets, in the cockroach[22] or in the earwig. When *Forficula* moves along it palpates in every direction with its antennae; if one is removed it palpates all sides with the remaining antenna.[149] The antennae of *Rhodnius* have a limited number of such spines, which cause instant withdrawal of the antennae if they are touched; they stand out in all directions in such a way as to prevent any object from coming in contact with the more delicate sensilla with which the antennae are richly supplied (Fig. 158).[151] In *Gerris* they are numerous on the trochanter and femur where these rest on the water; here they enable the insect to maintain its normal position on the water surface and serve to perceive vibrations in the surface film set up by other insects on which they prey.[142] Lepidoptera have a group of long innervated hairs (Eltringham's organ) on the dorsum of the head, which are perhaps stimulated by air currents during flight.[45] (A current of air directed against groups of trichoid sensilla on the head is necessary to maintain flight in locusts.)[146a] In *Schistocerca* these receptor hairs show a steady continuous discharge of impulses for a long period if they are held deflected.[217] Along the margin of the wing in Lepidoptera there is a fringe of sensory hairs of the same type, and some of the wing scales particularly on the marginal veins and the lower surface of the wing are innervated: these have been regarded as responding to air movements during flight;[51, 13] hairs on the leading edge of the wings of flies, &c., may fire at the rate of 600 or more impulses per second;[190] they probably respond, also, to contact stimuli during the folding of the wings.

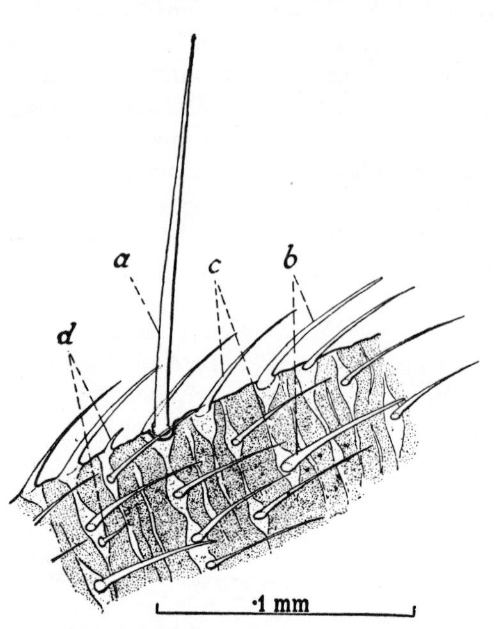

FIG. 158.—Surface view of part of an antennal segment in *Rhodnius* (*after* WIGGLESWORTH and GILLETT)

a, tactile spine; *b*, thin-walled sensilla, probably olfactory (*cf.* Fig. 184); *c*, fine thick-walled sensilla, perhaps temperature receptors (*cf.* Fig. 191); *d*, short sensilla showing the same structure as *c*.

Where very numerous and delicate hair-like sensilla occur on the antennae it has been suggested that they serve to detect the changes in air currents when the insect in flight approaches some solid object, and so enable it to avoid collisions in the dark.[104] The importance of air movements is seen also in flies. As is well known, flies are instantly disturbed by a sudden movement of the hand in their vicinity; they are not so disturbed if they are protected by a glass container.[56] The antennal aristae of Calliphorid flies are reputed to be sensitive to slight changes in atmospheric pressure; but the sense organs concerned have not been determined.[147] [See p. 294.]

Campaniform sensilla—In the campaniform sensilla no outgrowth from the cuticle remains. The terminal filament from the sense cell ends in a 'sense rod' or 'scolopale' which is inserted into a dome-shaped area of relatively thin cuticle (Fig. 159). In certain mutants of *Drosophila* all intermediate stages between normal campaniform organs and fully formed bristles can be found on a single wing: the dome of the organ is clearly the homologue of a bristle.[81] This dome is usually overhung by the more rigid cuticle around. Sometimes it is circular and symmetrical; sometimes it is elliptical; and sometimes, notably in the so-called 'Hicks' papillae' on the halteres of Diptera, in which the dome is elliptical, the surface is thickened in the longitudinal axis so as to form a stiffening rod that has been compared with the crest of a helmet (Fig. 160).[33, 99] They are similar in the legs of *Periplaneta* (Fig. 161).[101]

These organs occur in all parts of the body. In *Dytiscus* they occur particularly on the basal joints of the extremities and head appendages.[73] They are regularly distributed on the wings of Lepidoptera and other insects.[185] In the altered form described above they are present in great numbers on the halteres of Diptera.[99] They are very numerous on the gills of *Caenis* nymphs (Ephemeroptera).[40]

Almost all authors are agreed in regarding the campaniform sensilla as being sensitive to mechanical stimuli. They have been thought to respond to water pressure in aquatic insects, as in the antennae of *Dytiscus*[73] or the gills of the Ephemerid *Caenis* (in which they may be stimulated perhaps by vibrations in

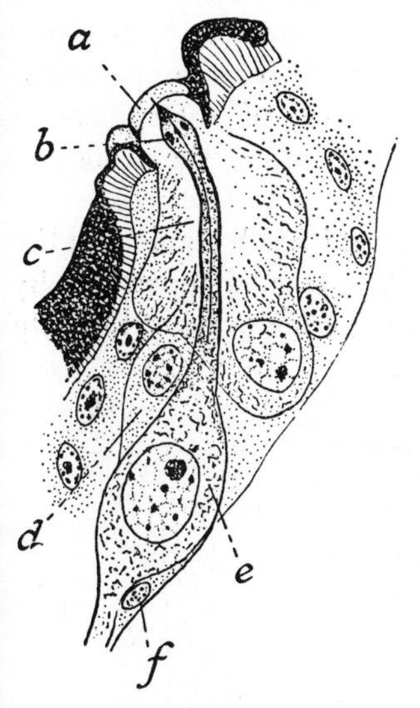

FIG. 159.—Campaniform sensillum on the cercus of *Acheta domestica* (*after* HSÜ)

a, dome-shaped plaque; *b*, scolopale; *c*, vacuole in membrane-forming cell; *d*, accessory cell; *e*, sense cell; *f*, neurilemma cell.

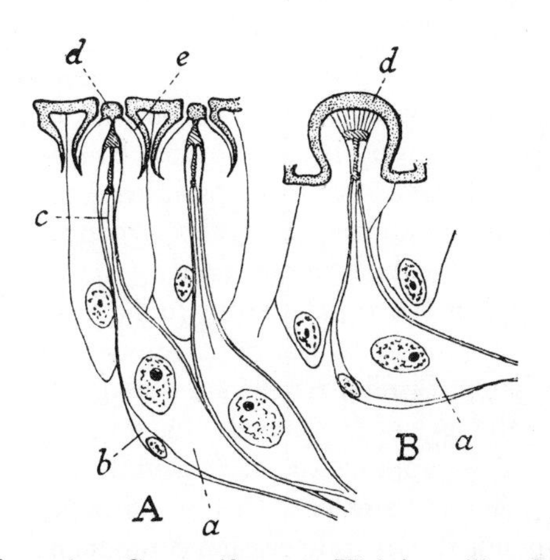

FIG. 160.—Campaniform or Hicks' papillae from halteres of Diptera (*from* DEMOLL *after* PFLUGSTAEDT)

A, from *Calliphora*, cut transversely to the long axis of the papilla; *B*, from *Sarcophaga*, cut in the long axis of the papilla. *a*, sense cell; *b*, neurilemma; *c*, terminal filament running to scolopale; *d*, longitudinal thickening of the dome; *e*, thin membrane forming side of dome.

the water or by the impact of suspended particles).[40] It has often been suggested that the organs on the wings and elsewhere may be sensitive to changes in air pressure during flight, and thus serve to correlate such changes with the wing vibrations—as in *Dytiscus*,[73] in the bee,[96] and in Lepidoptera.[51, 119] They are much reduced in number on the wings of *Calandra granaria* which has lost the power of flight, as compared with *C. oryzae* which still flies;[113] but they persist in female Lepidoptera with vestigial wings and in Pupipara where the wings are reduced in both sexes, and in ants they are confined to the region proximal to the preformed line of cleavage.[15]

Alternatively they have been thought to respond to bending of the cuticle. Where they are radially symmetrical they will react to flexion equally in all directions; where they are elliptical, as in Hicks' papillae, they will be affected only by bending in their long axis, i.e. in the direction of the stiffening rod. The side walls of the dome are so thin and delicate that they will not influence the movement of this rod to which the scolopale is attached; whereas compression in the long axis will exert tension on the sensory ending (Fig. 160).[33] This conception is supported by the fact that in the halteres a single group of papillae all have their long axes disposed in the same direction, but different groups are set at an angle of approximately 90° with each other.[33] And on the legs of *Periplaneta* the organs occur in groups all elongated in the same direction (Fig. 161).[101] In the cerci of *Gryllus*, &c., it has been suggested that they may be stimulated by the bending of the cuticle set up by pressure upon adjacent spines.[123] The campaniform organ on the antennae of caterpillars is stimulated by pressure upon the adjoining cuticle.[35]

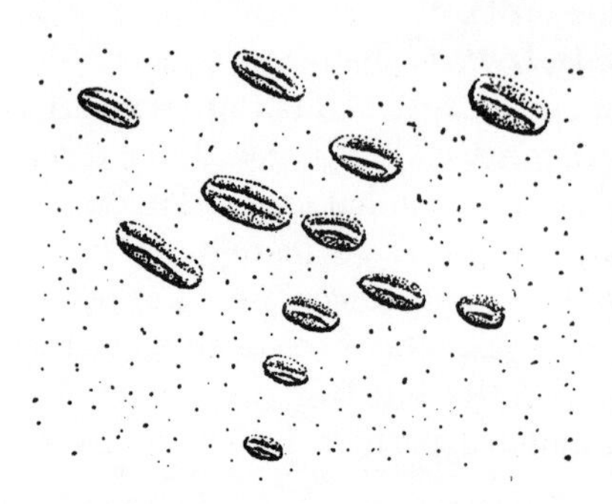

FIG. 161.—Group of campaniform sensilla on hind leg of *Periplaneta*, showing the longitudinal thickening of the dome (*after a photomicrograph by* PRINGLE)

In *Periplaneta* there is a group of campaniform sensilla in each segment of the palp, each group being supplied by a large sensory fibre, formed apparently by fusion of the axons of the peripheral sense cells. When the segments of the palp are forcibly moved a regular synchronous electrical discharge is set up in this nerve; but the organs seem to be very little stimulated by active movements of the joints. Similar results are obtainable with the legs of *Periplaneta*. Each group of sensilla appears to act as a unit, and the evidence suggests that it probably responds to the compression component of the shear force. For example, in the tarsi the sensilla occur on the dorsal aspect of the segments; they will therefore be compressed when the foot is applied to any surface, and are perhaps responsible in part for the perception of contact stimuli.[101] In *Pediculus* the campaniform organs, at the distal end of the relatively fixed trochanter, are well placed to detect strains in the limb caused by any resistance to its movements.[150] [See p. 295.]

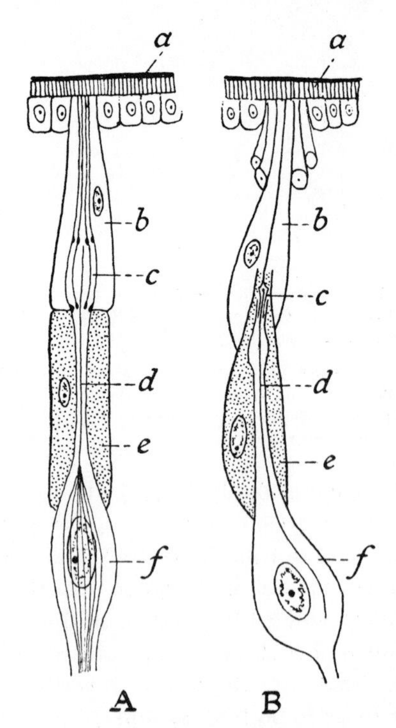

FIG. 162.—Scolopidia, schematic

A, from chordotonal organ (*from* DEBAUCHE *after* SNODGRASS); *B*, from Johnston's organ (*from* DEBAUCHE *after* EGGERS). *a*, cuticle; *b*, distal or cap cell; *c*, scolopale or sense rod; *d*, terminal filament; *e*, middle or sheath cell; *f*, sense cell.

Chordotonal sensilla or scolopidia—These sensilla are generally believed to be derived from campaniform sensilla through their component parts becoming elongated and deeply sunk within the body.[18, 32, 33] In the usual type, the sensory rod and terminal filament, in which the distal process of the sense cell ends, lie in the axis of an elastic strand stretched between two points in the body wall (Fig. 162). One end at least is nearly always inserted into some pliable region of the cuticle, such as the intersegmental membranes.[42] They are exceedingly widespread in all parts of the body: in the legs, antennae (Fig. 163, *b*, *e*), palps, wing bases, &c., besides the general body cavity.[31, 42] They can be seen most readily in larvae of Nematocera, in which each segment except the first and last contains two pairs, a large and small. The respiratory siphon of *Culex* larvae contains three pairs.[33] The larva of

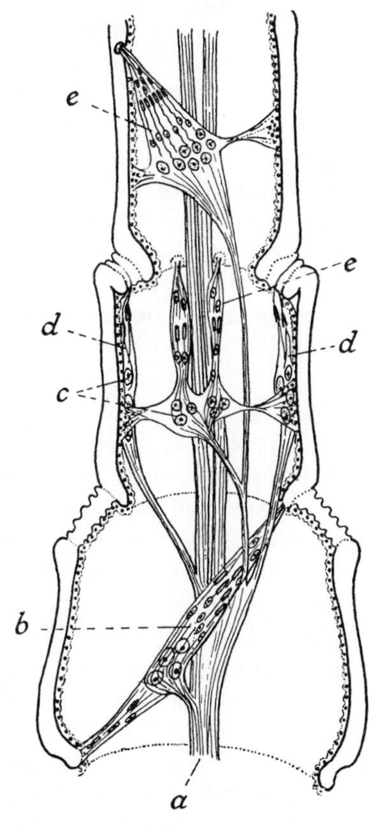

FIG. 163.—Arrangement of sense organs in the three basal segments of the antenna of *Meconema varium* (*after* DEBAUCHE)

a, antennal nerve; **b**, chordotonal organ of scapus; **c**, sense cells; **d**, Johnston's organ; **e**, chordotonal organ.

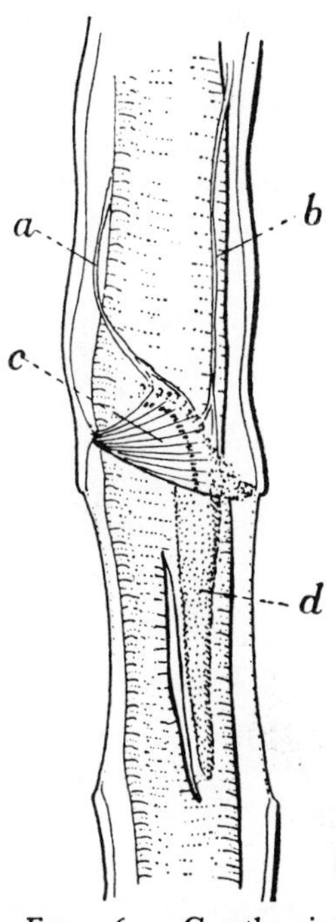

FIG. 164.—Greatly simplified figure showing position of subgenual organ in a grasshopper (*after* DEBAISIEUX)

a, nerve to subgenual organ; **b**, dorsal femoral nerve; **c**, subgenual organ; **d**, crista acoustica.

Drosophila has 90 chordotonal organs, attached at both ends to the epidermis.[70] They occur in the base of the wing in Lepidoptera[136] and at the base of the wing and halteres in Diptera.[70, 99]

In all cases there seems no doubt that the integumental scolopidia* are stimulated by changes in tension. Such changes may be induced by external pressure, or by the passive movements of the segments;[33] by the pressure or tension of the contracting muscles;[70] by general changes of internal pressure in the blood or perhaps in the tracheal system;[68] or, in some special cases, by

* A special 'subintegumental type' will be mentioned later (p. 272).

sound vibrations (p. 271). They are not to be regarded as structures of constant function, but as nervous organs, peculiar to insects, which may be adapted to diverse physiological purposes.[32]

There are groups of scolopidia present in the legs of most insects, always associated with the articulations and perhaps stimulated by movements of the joints. But there is a special group in the proximal region of the tibia, not associated with any joint, forming the so-called 'subgenual organ' which is specially adapted to perceive vibrations (Fig. 164).[31] This organ is most highly developed in Orthoptera, moderately in Hymenoptera and Lepidoptera, very poorly in Hemiptera, and is completely lacking in Coleoptera and Diptera. The sensitivity to vibrations, as estimated by the electrical response in the nerves to the leg, is in the same order: in *Locusta* the smallest amplitude of vibration that will stimulate the subgenual organ is 0·36 Å (at 2,000 cycles per second) a value approximating to atomic magnitudes;[7] it will even perceive airborne sounds if these are of low pitch and of sufficient intensity to set the substrate on the insect in vibration.[7] Hemiptera, Coleoptera, and Diptera detect coarse vibrations by means of tibio-tarsal chordotonal organs and hair sensilla in the tarsal joints.[9] [See p. 295.]

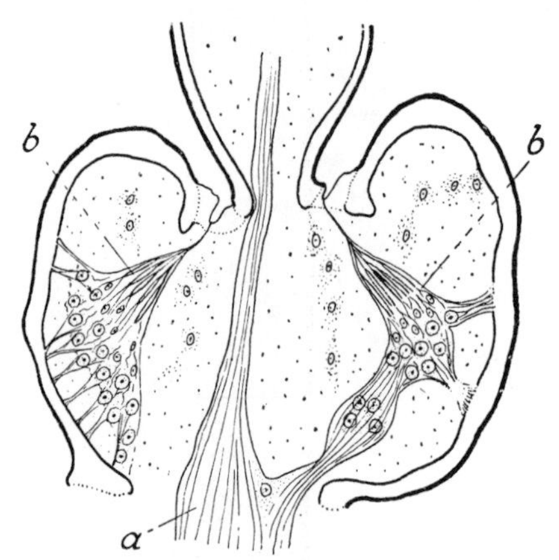

FIG. 165.—Longitudinal section of pedicellus of antenna in *Chironomus viridis* (*after* DEBAUCHE)

a, antennal nerve; *b*, Johnston's organ.

Johnston's organ—The second antennal segment or pedicellus always contains a group of sense organs very similar in structure to the chordotonal sensilla (Fig. 162, B; 163, *d*). These are radially arranged, with their lower insertion on the wall of the pedicellus and their upper insertion in the third intersegmental membrane of the antenna. They are referred to as Johnston's organ.[42] The antenna, as we have seen, contains also a number of campaniform sensilla and independent chordotonal organs.[32] But in those insects in which Johnston's organ becomes more highly developed these begin to recede; and in Culicids and Chironomids (Fig. 165), in which it shows its highest degree of complexity, completely filling the great globular second segment, chordotonal organs and campaniform sensilla are entirely absent.[32]

Judging from its structure, Johnston's organ must be stimulated by movements of the flagellum of the antenna on its base. It is therefore described in the bee, for example, as a 'statical organ' to register the flagellar movements.[84, 137] The honey-bee regulates the speed of flight in response to stimuli perceived by Johnston's organ, which responds to the strength of the air-stream. Increasing the air-stream in a bee in fixed flight results in a reduction in the amplitude of the wing-beat.[220] Such movements may be active movements induced by the muscles, or passive displacements due to contact or to air currents.[42, 142, 151] In male mosquitos it is the auditory organ[115a] (p. 270). It is very well developed in the whirligig *Gyrinus*, and since the antenna is carried in the surface film as the beetle moves rapidly about, it has been suggested that the Johnston's organ enables it to perceive the curvature and finer vibrations of the water surface and so avoid collision. Beetles deprived of the antennae collide repeatedly with the sides of their aquarium.[44] Johnston's organ in *Notonecta* and *Corixa* is used to maintain the normal swimming position by means of the tension exerted by the air bubble trapped between the antennae and the lower surface of the head.[256] *Muscina* suspended in stationary flight is caused to raise the legs and fold them against the sides of the thorax in response to a stream of air over the antennae, probably as the result of stimulation of Johnston's organ. The same stimulus influences the path travelled by the wing tip and so controls the equilibrium of the flying insect (p. 166).[74] If the flagellum of the antenna in an Aphid is cut off beyond the second segment, flight becomes quite erratic; normal flight is restored if an artificial antenna is re-attached.[227] [See p. 295.]

The antennae of *Calliphora* are highly sensitive and extraordinarily fast-reacting sense organs for air movements. The organs concerned include the arista and Johnston's organ. The antenna as a whole combines the properties of an organ of hearing and an organ for detecting currents of air. At a frequency of 150–250 cycles per second (that is, in the region of the wing-beat frequency) the antennae are a little less sensitive than the human ear. They are responsible for controlling the speed of flight and may even serve to detect the acceleration that takes place during each wing-beat.[180, 181]

Proprioceptive organs—It is clear from the foregoing account that many different sense organs responding to mechanical stimuli may serve as proprioceptors—sense organs stimulated by the movements of the insect's own body or by the strains set up in it by the tension of its muscles.

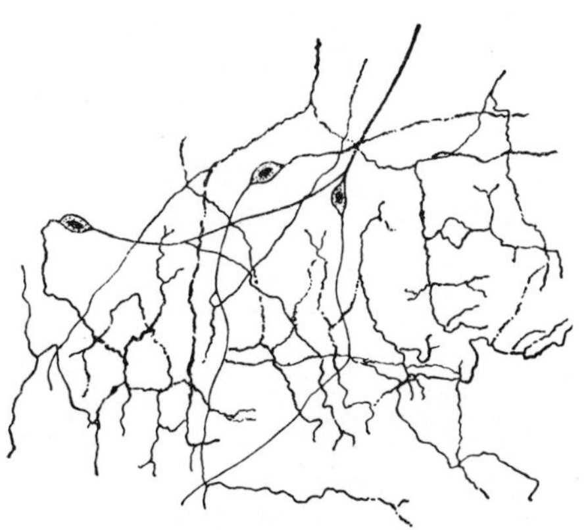

FIG. 166.—Subepithelial nerve plexus with bipolar sensory cells in the larva of *Melolontha* (*after* ZAWARZIN)

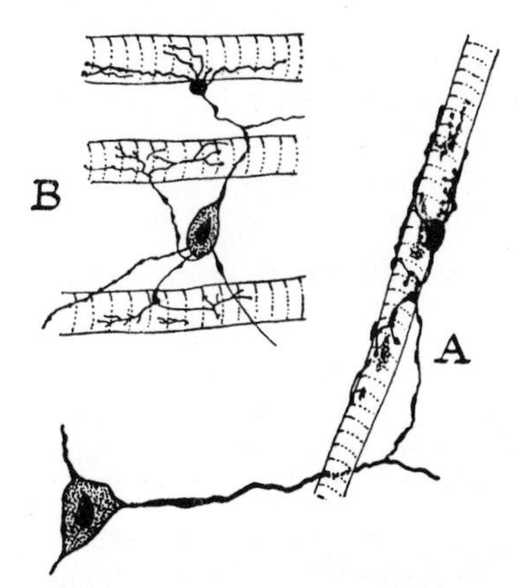

FIG. 167.—Two types of sensory nerve endings in the muscle fibres of the oesophagus of *Oryctes* larva

A, muff-like investment of the fibre; *B*, fine diffuse nerve endings (*after* ORLOV).

(i) The peripheral nerve plexus (p. 179) contains bipolar and multipolar sense cells which give out fine branches ending freely under the general body surface in soft-skinned larvae, as in caterpillars[75] and in *Melolontha* (Fig. 166),[155] or in the intersegmental membranes of forms with a rigid cuticle,[33] for example, in the articulations of the limbs in *Aeschna* nymphs;[155] these diffuse endings are particularly evident in termites.[258] These endings are believed to be stimulated by tension in the body wall or movements of the joints.[33] In larvae of *Phormia* there are abundant bipolar and multipolar neurones with ramifying processes below the basement membrane. Although commonly referred to as a nerve net, or plexus, there is in fact no fusion of the terminal processes. The terminations are naked and without any Schwann cell investment; these naked regions are probably the receptor sites.[247, 248]

(ii) A similar nerve net is present in the peripheral part of the visceral nervous system in *Melolontha* and *Oryctes* larvae, where this supplies the muscles of the mid-gut and oesophagus. It contains two types of sensory ending: (*a*) a network of varicose branches enveloping the individual muscle fibres in a sheath or muff, very similar to the sensory muscle endings of vertebrates, and (*b*) diffuse branching endings between the muscle fibres (Fig. 167).[98] Similar endings occur in the muscles of the stomach, hind gut, &c., but whether they exist in the skeletal muscles is not known.

(iii) In some insects erect tactile bristles serve to detect the degree of flexion at the joints.[33] Such groups of hairs were termed by Doflein 'postural hairs' (Stellungshaare);[192] they had been described by Lowne (1892) on the prothorax of *Calliphora*.[253] When *Periplaneta* falls through the air the thoracic segments move against one another and flight begins; this response is attributed to stimulation of the tactile hairs in this region.[39] Tactile hairs doubtless serve as proprioceptors during the cleaning of appendages, &c.[105] In Mantids there are two cushions covered with sensory hairs just behind the head, which serve to inform the insect of the position of the head and so enable it to strike accurately at a fly to one side or the other of the body.[243] In bees such hair patches serve as gravity receptors (p. 266).[237]

(iv) Many of the chordotonal organs must serve to perceive the movements and internal strains in the body;[31, 33, 108] for example, in the base of the wings and halteres;[136, 145] in Johnston's organ, which must often provide the chief proprioceptor of the antennae;[42] in larvae of Diptera, where the organs are pressed upon by the contracting muscles.[70] It is conceivable that some of the chordotonal organs act as 'manometers' detecting pressure in the tracheal system or the blood.[68] The stretch receptors in the wing hinge of *Locusta* (p. 164) are of phasic-tonic type—but the tonic component is of no significance in flight control. The phasic component probably makes it possible for the insect to estimate the difference in phase between the wing strokes of fore and hind wings.[250]

(v) The slightest movement of the limb in *Periplaneta* causes an outburst of impulses in the nerve in the femur; this proves the existence of tension receptors in the limb.[11] The campaniform sensilla are probably the chief organs concerned in this response.[101] Thus it would seem as though the insect responds chiefly to strains in the cuticle of the limb as a whole, whereas the vertebrate responds to tension changes in the individual muscles.[101] The campaniform sensilla in the wings are also regarded as proprioceptive organs.[15, 102]

(vi) 'Stretch receptors' consisting of connective tissue strands with associated neurones, or sensory neurones in close association with abdominal muscles, exist in many insects. In the larvae of Lepidoptera the structure resembles the vertebrate muscle spindle. A modified muscle fibre has a middle innervated region from which the large nerve cells send dendrites into a tube of connective tissue which lies along the muscle (Fig. 168, A).[202, 248] In caterpillars one of these organs occurs on each side of every segment from the mesothoracic to the ninth abdominal.[201] In the dorsal region of most abdominal segments in *Ephemera*, *Isoperla*, *Periplaneta*, *Forficula*, *Dytiscus*, &c., there is a pair of connective tissue strands, vertical and longitudinal, each with a single sensory neurone (Fig. 168, B).[248] In *Phormia* larvae, besides chordotonal organs, there are neurones with branching dendrites running over skeletal muscles, and innervated strands of connective tissue.[246] In the cockroach *Blaberus* the axon, cell body, and dendrites associated with the connective tissue strand are invested by Schwann cells; but the extreme ending of the axon is naked and embedded in the connective tissue matrix. It is probably stimulated by compression.[247] Receptors of this type, subjected to phasic stimulation, are incapable of responding accurately if stretched more frequently than five times a second.[238] They are probably involved in rhythmical movements of the abdomen and thorax, as in the respiration of *Dytiscus* and *Locusta*[225] and the flight movements of *Schistocerca* (p. 164).[211] Exposed to steady stretching, the sensory neurone discharges impulses at a frequency that is linearly proportional to the tension applied.[201] Dragonflies and

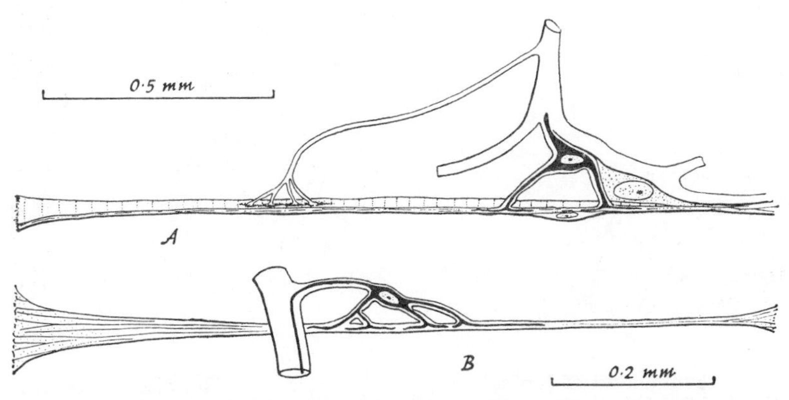

FIG. 168.—Stretch receptors (*after* OSBORNE and FINLAYSON). A, *Phryganea* larva (Trichoptera). B, *Forficula*. In *Phryganea* it consists of an innervated strand of connective tissue applied to an innervated muscle; in *Forficula* an innervated band of connective tissue. Sensory neurone in black.

other insects continue feeding indefinitely if the abdomen is removed. Stretching in the abdomen of *Rhodnius* serves both to terminate ingestion of the blood meal[239] and to initiate secretion of the moulting hormone.[287] [See p. 295.]

Organs of equilibrium—The stimulation of these various proprioceptive organs by the weight of the body acting upon the limbs must be an important factor in the equilibration of the insect in respect to gravity.[11] The sense of gravity is probably the sense of position and the sense of contact distributed over the whole body. Thus *Notonecta*, *Naucoris* and *Macrocorixa*, blinded and with weights attached to the body, are able to maintain their direction up and down in the water, and they can still do so after removal of the antennae and after section of the nerves to the abdomen.[97]

The beetles *Bledius* and *Dyschirius* which make vertical burrows in the sand can judge the direction of gravity with an error of 1–3°.[178] Johnston's organ in the antennae has been regarded as the chief gravity sense organ in ants;[280] but the sensory hair-cushions at the neck, petiolus and abdomen, as well as at the joints between the thorax and the limbs, appear to be more important.[178, 241] Patches of hairs between head and thorax and between thorax and abdomen serve as gravitational sense organs in the honey-bee: if the centre of gravity of the bee is displaced by attaching a weight to the top of the head, the bee confuses 'up' with 'down'.[237] *Dytiscus* tends always to climb upwards on an inclined slope (negative geotaxis). This response is not merely a mechanical effect of the weight of the abdomen, for it persists after the centre of gravity has been entirely changed by attaching a weight to the fore part of the body.[156] *Carausius morosus* keeps the body free from the ground by feed-back control from hair-plates which measure the angle between coxa and femur. These same axons compensate for an increase in load on the insect.[284] [See p. 295.]

In addition to this general sense, a few aquatic insects possess special static sense organs. The larva of *Limnophila fuscipennis* has a pair of sacs in the last abdominal segment, each with a sensory hair at its blind end and granules, or

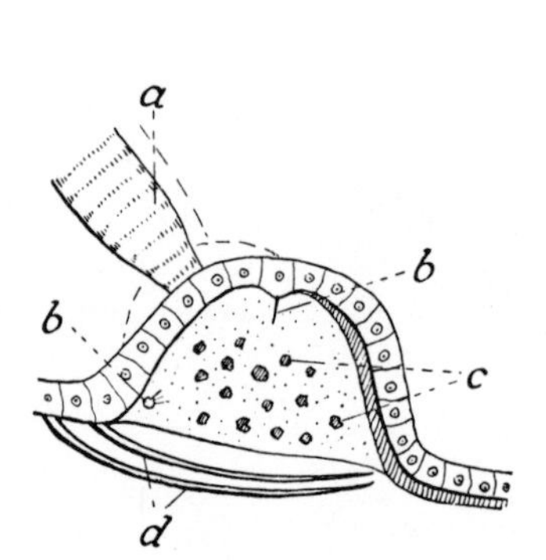

FIG. 169.—Static organ of Limnobiid larva *Ephelia*. Schematic section (*from* WEBER *after* WOLFF)

a, muscle; *b*, sensory hairs; *c*, statoliths; *d*, bristles closing the organ below.

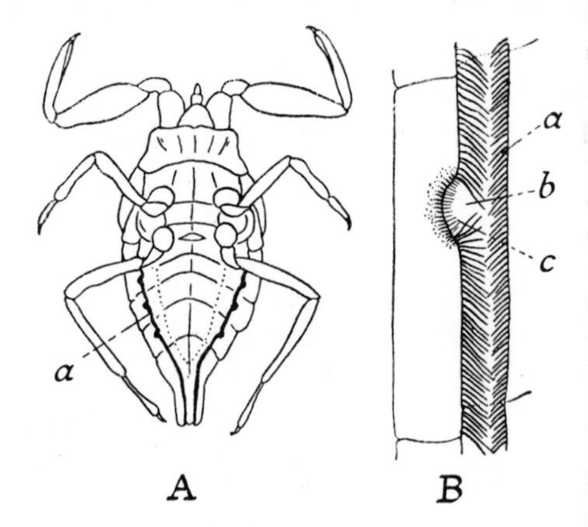

FIG. 170

A, nymph of *Nepa*, ventral view to show respiratory furrow, *a*. *B*, detail of part of respiratory furrow in nymph of *Ranatra* to show sensory pit (*after* BAUNACKE). *a*, respiratory furrow with roof of bristles; *b*, point of contact between air and water; *c*, sensory hairs.

statoconia, in its lumen (Fig. 169). As this sac is rhythmically pumped up with water and relaxed, the statoconia fall upon the sensory hair. The normal larva is positively geotactic; this response is abolished if the organ is removed.[128] In the larva of *Ptychoptera contaminata* there is a typical vesicular statocyst with movable statoliths in the 10th and 11th segments of the abdomen.[143] And it has been suggested that the peculiar Graber's organ of Tabanid larvae is a static organ; but experiment does not appear to support this idea.[26]

Static organs of a totally different character occur in some aquatic Hemiptera. They consist of localized collections of tactile hairs placed at interfaces where the air continuous with that in the tracheal system comes in contact with the water. Any movement at this interface due to changes in pressure will be detected by the sensory hairs; and since the pressure of the water will be greatest in the most dependent parts, they will serve to orientate the insect in respect to gravity. Organs of this type are best developed in *Nepa*. In the nymphs the air is held in grooves, roofed over by long hairs, along each side of the ventral surface of the abdomen. On four of the segments there is a small interval where the long hairs are replaced by short sensory hairs, and it is here that the air comes in contact with the water (Fig. 170). In the adult a new sensory apparatus based on the same physical principle is developed in connexion with three of the abdominal spiracles (Fig. 171). If a normal *Nepa* is blinded (for vision certainly assists in the orientation) and allowed to crawl on a small submerged seesaw towards the water

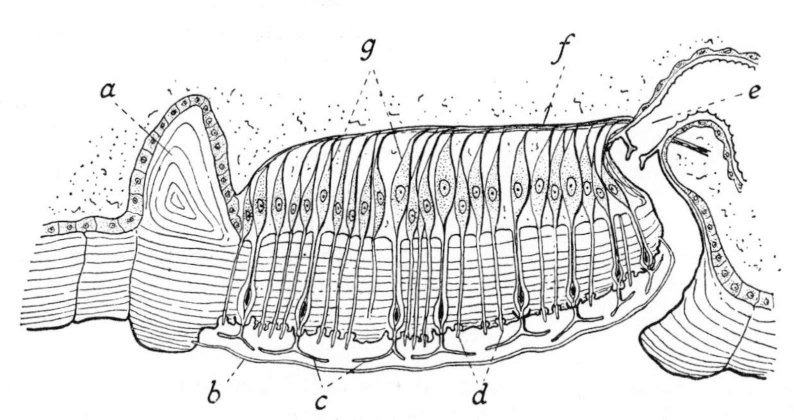

FIG. 171.—Transverse section through abdominal spiracle of *Nepa* adult showing the adjacent sense organ, apparently adapted to detect the hydrostatic pressure (*after* HAMILTON)

a, chitinous ridge around the organ; b, membrane covering the organ; c, flattened scale-like sense organs probably responding to pressure; d, peg-like sensilla; e, spiracle; f, nerve; g, bipolar sense cells. (The membrane b should be shown extending over the spiracle, so as to put the organs into communication with one another by way of the tracheal system.)[30]

surface, it turns round at once when the inclination is reversed; whereas the insect with the organs put out of action walks up or down indiscriminately.[14] These organs in *Nepa* are adapted to detect relative differences in pressure at the different spiracles; they function only as a differential manometer and fail to respond to changes in the absolute pressure.[130] In addition, *Nepa* possesses a general sense of position resident in the appendages and persisting after section of the nerves to the abdomen.[97] In *Lethocerus* (Belostomatidae) the static organ consists of a pair of deeply sunk abdominal spiracles with overlying tactile hairs.[142]

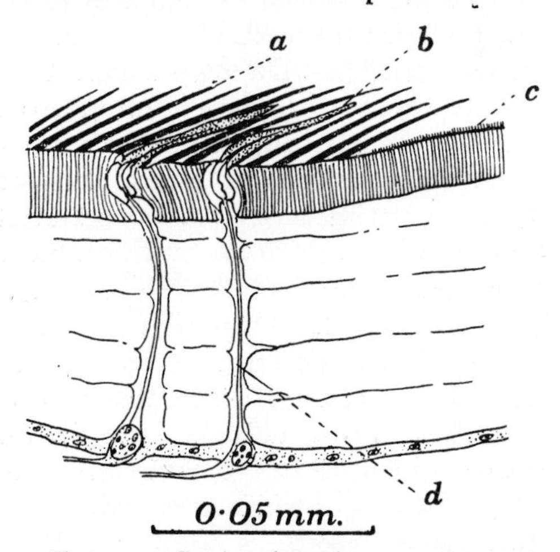

FIG. 172.—Section through organ of pressure sense in *Aphelocheirus* (*after* THORPE and CRISP)

a, hair pile; b, sensory hair; c, plastron hair pile (see Fig. 202); d, sensory nerve.

The Naucorid *Aphelocheirus*, unlike most other aquatic insects, is heavier than water; it carries no air bubble such as will normally provide a ready means of perceiving pressure changes (p. 398), and therefore needs special pressure receptors. It uses three means of orientation: the ordinary tactile sensilla, the light-sensitive organs (p. 326), and pressure receptors developed only in the adult insect These consist of a pair of oval depressions on the sternum of the seconc abdominal segment, closely associatec with the spiracles and with a pil of long recumbent hydrofuge hairs about 60,000 per sq. mm., holding a film of air (Fig. 172). Tactil sensilla are interspersed among the hydrofuge hairs and are moved by changes in pressure at the air-water interface. These organs are very sensitive both to uniform and to differential pressure changes and can thus direct the swimming movements upwards or downwards, and maintair the insect on an even keel.[130]

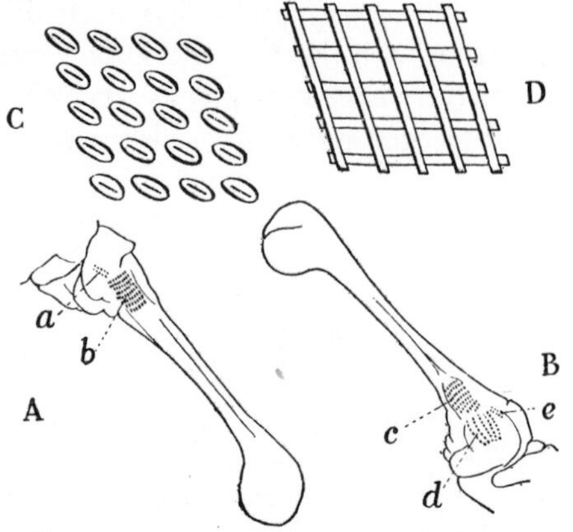

FIG. 173.—*A*, ventral view of haltere of *Lucilia*; *B*, dorsal view; *C*, diagrammatic surface view of basal plate sensilla; *D*, trellis with similar mechanical properties (*after* PRINGLE)

a, ventral Hicks' papillae; *b*, ventral scapal plate; *c*, dorsal scapal plate; *d*, basal plate; *e*, dorsal Hicks' papillae

The numerous proprioceptive organs on the wings (chordotonal anc campaniform organs and tactile hairs' probably assist in the maintenance o equilibrium during flight. These organs occur chiefly at the base of the radius and subcosta, but there are a few also in distal parts of the wing. A well-marked correlation exists between the number of organs and the capacity for flight.[91, 21] In Diptera this function seems to be centred particularly in the halteres, which are well provided with such receptors (Fig. 173). Some Diptera, such as Tabanids, are able to fly without the halteres;[23] but others, such as *Calliphora*, show complete loss of equilibrium in the air after their removal; they are able to fly only if equilibrium is restored by attaching a strand of cotton to the tip of the abdomen[50] (p. 167). It is suggested that the mechanism by which the halteres serve as 'balancers' during flight is as follows. During the normal vertical vibration the obliquely placed rows of elongated campaniform organs at the base are not subjected to any shearing strain; but any lateral movement of the haltere will stimulate them. The vibrating haltere with its terminal knob acts as an alternating gyroscope; consequently, if the insect is rotated out of the plane of their vibration lateral shearing forces will be set up in the cuticle of the base and these will be perceived by the campaniform sensilla.[50, 94a, 102] The campaniform organs of the 'scapal plates' and the 'Hicks' papillae' are probably excited by the normal movements of the haltere as shown by electrical records they are stimulated only when the oscillation has reached its maximum. Perhaps they provide a means of controlling the amplitude of oscillation. It is the highly sensitive campaniform sensilla of the 'basal plate that are stimulated by the strains set up in the cuticle by the gyroscopic forces (Fig. 173). Organs which, in the wing, are probably stimulated by aerodynamic forces, have become specialized in the haltere for the perception of inertia forces.[102]

Among other methods of maintaining equilibrium (p. 167) the dragon-fly *Anax* has a 'dynamic organ' in the neck. This consists of a sensory cushion with abundant sensory hairs. Because of the large moment of inertia of the head any rotation of the trunk will produce a slight twisting of the neck and this causes the hard parts of the head to press upon the sensory hairs. The flying insect reacts by changing the angle of the wings and thus rotating the trunk so that it again becomes symmetrical with the head.[164] Hair-plates on the cervical sclerites of locusts and other Saltatoria serve to register movements of the head relative to the body and thus play an essential part in the dorsal light reaction (p. 325), which locusts (like *Anax*) exhibit during flight.[212]

HEARING

The waves of air pressure which we recognize as sounds may be perceived by insects by two quite different mechanisms. The movements of the air which are secondary to the pressure changes may stimulate the more delicate tactile hairs, or special auditory organs (tympanal organs) may respond to the pressure waves themselves.[6] It is held by some that even the tympanal organs are displacement rather than pressure receptors;[106] displacement receptors respond to the direction as well as the magnitude of the stimulus.[218]

An important element in hearing is the 'time constant' of the sense organ concerned; that is, the time taken for return to the original state after the stimulus ceases. For the human ear this is about 20 milliseconds, so that sounds that follow one another at intervals of less than about $\frac{1}{10}$ second will not be distinguished. The 'time constant' in insects may be as low as 2 milliseconds or less, and they can distinguish sounds separated by intervals as small as $\frac{1}{100}$ second. The insect ear is therefore well adapted to recognize the intervals between very rapid pulses of sound.[218]

Acoustic function of hair sensilla—The perception of vibratory air movements is very widespread. Ants react to sounds so long as their antennae are intact.[6] Caterpillars respond to sounds by cessation of movement and some contraction of the body; in *Vanessa antiopa* partial removal of the body hairs diminishes this response;[92] in *Datana* larvae the response fails when the hairs are loaded with water droplets or dry flour or when the body surface is anaesthetized with procaine.[1] Even relatively hairless forms such as *Danais plexippus* are sensitive; perhaps very small hairs are concerned.[92] Hearing of this type has been demonstrated also in *Blattella germanica*, Ephemeroptera, *Crioceris* (Col.), &c.[10] And in insects provided with tympanal organs (*Tettigonia*,[62] *Liogryllus*[111] and *Thamnotrizon*[112] (Orth.), *Agrotis* and *Catocala* (Lep.)[43]) it is probably responsible for the slight sensitivity to sounds which persists after these organs have been removed. Indeed, many insects devoid of tympanal organs show movements ('phonokinesis') in response to sounds.[184] In the moth *Ctenucha* destruction of the tympanal membranes on both sides of the abdomen abolished responses above 15,000 cycles per second, but did not affect those at lower frequency. The capacity for hearing these lower frequencies agrees with that of the butterfly *Cercyonis* and is probably served by tactile hairs and perhaps chordotonal organs in the body wall.[205a] In grasshoppers, the tympanal organs, long hair sensilla on the anal cerci, and hair sensilla on thorax and abdomen all play a part in acoustic reception.[216]

These vibrating hairs will respond to a wide range of tones. Caterpillars react at least from 32 vibrations per second to 1,024 vibrations.[92] Their properties have been studied in most detail in the anal cerci of *Acheta* and *Periplaneta*, by amplifying the electrical responses in the sensory nerve and studying them by means of the cathode ray oscillograph. The nervous response to sounds observed in the cercal nerve is probably derived from the hairs on the cercus, since it is abolished if vaseline is applied to the ventral surface from which they arise, whereas vaseline applied to the upper surface bearing numerous campaniform sensilla has no effect. There seems to be no lower limit to the frequency which is effective; they will respond to low frequencies quite inaudible to the human ear. With pure tone sound stimuli from 50 up to 400 cycles per second the frequency of response is synchronous with the stimulus (Fig. 174, B, C). Occasional synchrony, due perhaps to the individual fibres responding to every other stimulus, may occur even up to 800 per second. Above this the response is asynchronous (Fig. 174, D); but it persists up to about 3,000 cycles per second or rather more.[107] It is possible, however, that the acoustic function of these organs is merely incidental; in the cricket the main function of the cercus is

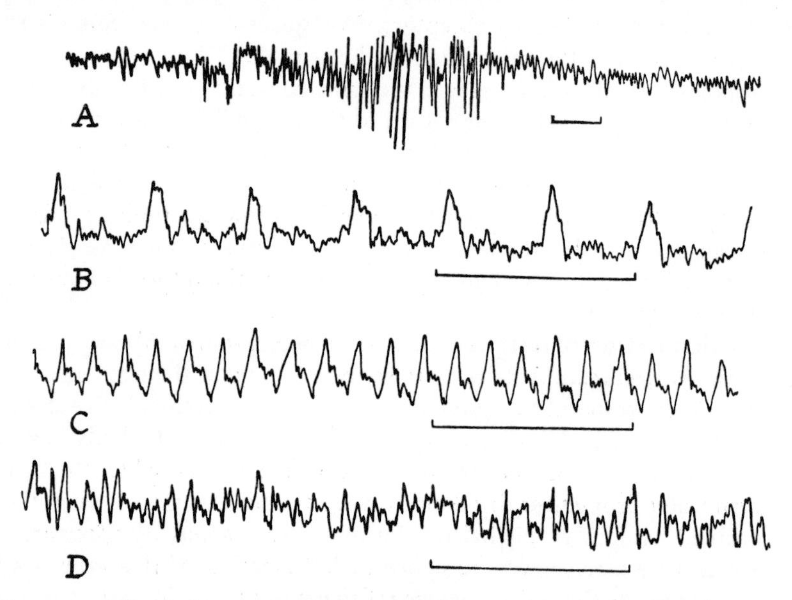

FIG. 174.—Electrical responses from cercal nerve of *Acheta* (*from* PUMPHREY and RAWDON-SMITH)

A, response to a light puff of air; *B*, synchronous response to pure tone stimuli of frequency 100 cycles per second; *C*, synchronous response to pure tone stimuli of 300 cycles per second; *D*, irregular or asynchronous response to pure tone stimuli of 900 cycles per second. The straight line indicates 20 milliseconds in each case.

perhaps the detection of earth-borne vibrations; in the cockroach, in which the cercus is held erect, it may be a wind gauge.[107]

When the antennae are very light it is possible that the entire flagellum may vibrate in response to air-borne sounds, and then Johnston's organ may be an auditory receptor. This has been suggested in the case of Culicids[115a] and Chironomids[42] and in *Reduvius personatus*.[142] [See p. 295.]

Hearing in mosquitos—It was shown by Mayer (1874) that male mosquitos respond to such tones as will set the bushy antennae in motion. Hearing in male *Aëdes* and other mosquitos is certainly dependent on the vibration of the antennal shaft which stimulates Johnston's organ (Fig. 175) and not on the movements of the antennal hairs themselves; but the organ does not react to sounds if the hairs are procumbent, as in the newly emerged mosquito. The sound emitted by the female evokes a mating response in the flying male (p. 709) and this response can be elicited by tuning forks producing sound frequencies between 100–800 cycles per second.[115a, 171] Males of *Aedes aegypti* are strongly attracted to pure sounds, particularly in the range 500–550 cycles per second. These were as attractive as the flight tone of the female which extended over 449–603 c.p.s.[289] In *Culex* Johnston's organ becomes fully effective as an organ of hearing only during flight; in the female *Aedes* on the ground it seems to perceive gravity and air currents.[173] Electric recordings from Johnston's organ in mosquitos show an alternating potential of the same frequency as the acoustic stimulus (up to 600 c.p.s. in some species) with the greatest sensitivity at the fundamental of the female flight tone of about 380 c.p.s. The location of the sound may depend on the direction of movement of the shaft; that is, on the relative extent to which it vibrates in the long axis or rocks from side to side (Fig. 175), this ratio being termed the 'clatter factor' (Klirrfaktor).[230, 261, 278]

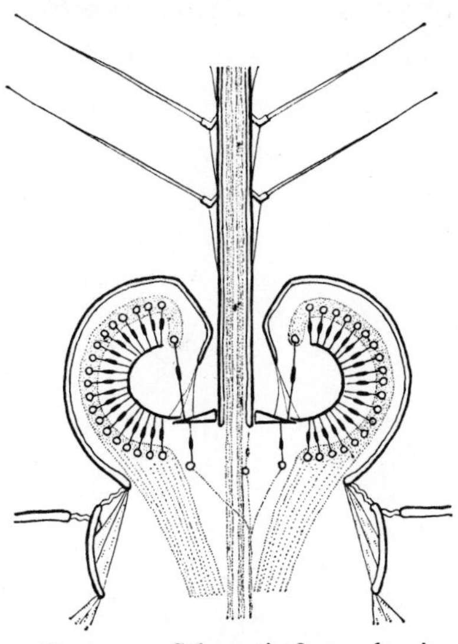

FIG. 175.—Schematic figure showing the arrangement of scolopidia in Johnston's organ in the male mosquito. The nerves are shown dotted (*after* RISLER)

Tympanal organs—The chordotonal organs were so named by Graber[62] in the belief that they are always sensitive to sounds. That view is not supported at the present time. But all the evidence goes to show that where the chordotonal organs are associated with a cuticular drum or tympanum they do respond to

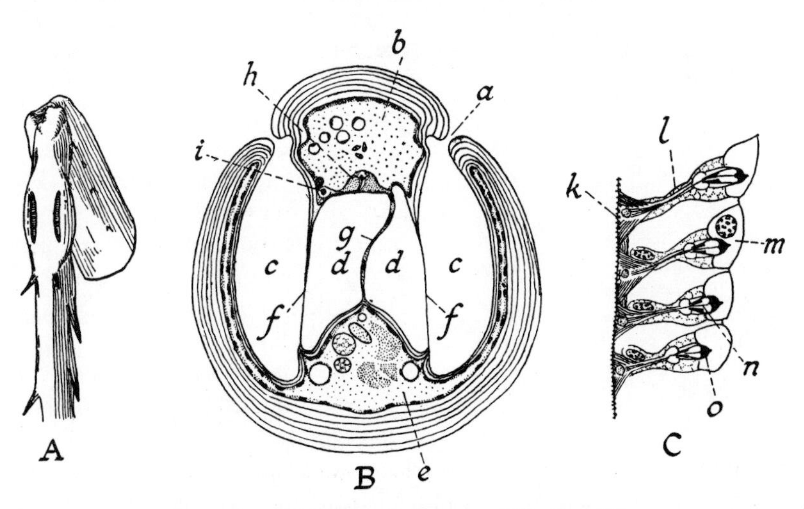

FIG. 176.—Tibial tympanal organ of Tettigoniids

A, fore leg of *Tettigonia* showing the slit-like openings of the tympanal sacs (*after* WEBER). *B*, transverse section of anterior tibia of *Decticus* at the level of the tympanal organ (*after* SCHWABE). *a*, cleft in anterior surface of tibia; *b*, blood channel; *c*, tympanal cavities; *d*, tracheae; *e*, blood channel with muscles, nerves, &c.; *f*, tympanum; *g*, rigid wall between the enlarged tracheae; *h*, crista acoustica with one chordotonal organ; *i*, sense cell. *C*, longitudinal section through a part of the crista showing four chordotonal organs (*after* SCHWABE). *k*, trachea; *l*, sheath cell; *m*, cap cell; *n*, axial filament; *o*, apical body.

sounds. The scolopidia in these organs may be of the type already described, with a filamentous insertion at both ends (integumental scolopidia) as in Acrididae, Cicadidae, Lepidoptera, &c., or one of these attachments may be replaced by a large cap cell ending freely in the body cavity ('subintegumental scolopidia'), as in Gryllidae and Tettigoniidae.

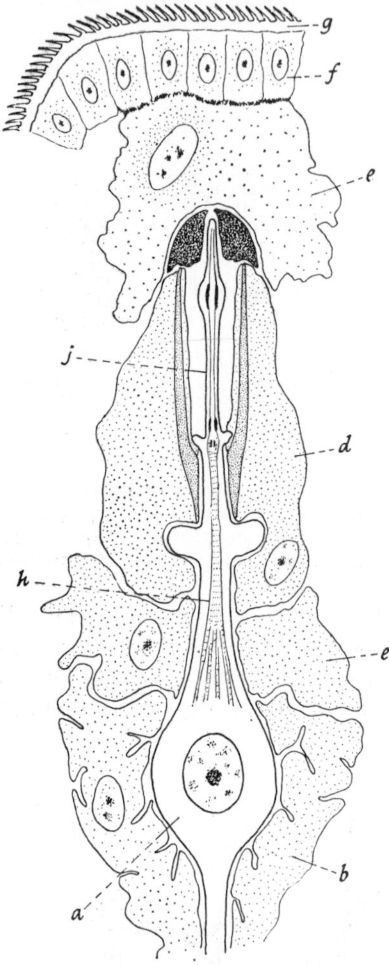

Most tympanic organs are fundamentally alike; a thin cuticular drum is exposed on one side to the external air, and a group of scolopidia is either directly connected to its inner surface or lies on a secondary membrane so situated as to move with it. In Acrididae a pair of such organs occurs in the first abdominal segment; a large tracheal air sac covers most of the inner surface of the drum, and a group of integumental scolopidia is attached to the drum near one margin. In Tettigoniidae a very complex organ is present near the base of each anterior tibia (Fig. 176). This part of the tibia is dilated, and has two fine slits directed forwards. These lead on each side into an invagination or pocket. Two tracheae separated by a rigid wall fill the space between the two pockets. Where each invagination is applied to the trachea its cuticle is very thin and forms the drum. No scolopidia are inserted into the drums themselves, but a row of subintegumental scolopidia is attached to the anterior surface of one of the tracheae. These constitute the 'crista', the individual scolopidia of which become progressively shorter from above downwards (Fig. 176, C).[3, 122] In Gryllidae the structure is similar but the drums are not protected by an infolding of the cuticle.[33] In the Drepanidae and Cymatophoridae (Lep.) the tympanal organs are peculiar in that the true tympanum is formed from two layers of tracheal epithelium enclosing the scoloparia between them.[59]

FIG. 177.—Auditory scolopale of *Locusta* (*after* GRAY)

a, sense cell; *b*, Schwann cell; *c*, fibrous sheath cell; *d*, scolopale cell; *e*, attachment cell; *f*, epidermal cells; *g*, cuticle of tympanum; *h*, root of cilium inside the dendrite; *j*, cilium with dilatation above.

The abdominal tympanal organ of *Locusta* has about 70 bipolar sensory neurones, around each of which are wrapped three satellite cells (the Schwann cell around the nerve cell, a fibrous sheath cell, and a scolopale cell investing the scolopale) and an attachment cell which links the structure to the epidermis of the tympanum (Fig. 177). The fine structure of these sensilla as seen with the electron microscope is exceedingly complex. Two points may be singled out for mention: (*a*) the apex of the dendrite as it runs upwards in the scolopale shows the characteristic structure of a cilium, with an outer ring of nine fibrils; (*b*) the root of the cilium within the dendrite becomes a solid rod with transverse striations like those of collagen, and this rod breaks up into some forty striated rootlets which run into the cell body.[214]

Rudimentary tympanic organs occur in the mesothorax of *Plea*, *Corixa*, &c. (Hem.), and in the mesothorax and metathorax of *Nepa* and *Naucoris*.[154, 235a] The chordotonal organs at the base of the wing in some butterflies, notably Satyridae, have their ventral insertion attached to a thin area of cuticle which constitutes a drum.[136] There are well-developed tympanal organs in the metathorax of many Lepidoptera (Notodontids, Lymantriids, Noctuids, &c.) (Fig. 178) and in the first abdominal segment of Geometrids and some other Lepidoptera.[43] A highly complex tympanal organ occurs in the second abdominal segment of *Cicada*; here the drum lies between the outer air and an abdominal air sac; the tension in the drum is regulated by a muscle; and about 1,500 slender scolopidia are present. In most of the tympanal organs it is obvious how the movements of the drum will stimulate the scolopidia; in *Cicada* it is uncertain

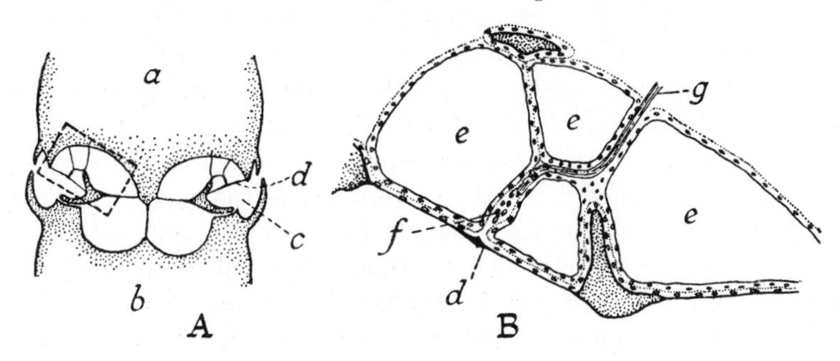

FIG. 178

A, horizontal section of thoracic tympanal organs of *Agrotis*; *B*, detail of organ on the left side (*after* EGGERS). *a*, thorax; *b*, abdomen; *c*, tympanal pit; *d*, tympanum; *e*, tracheal air sacs; *f*, chordotonal organs; *g*, nerve.

whether the vibrating drum causes the organs to swing laterally, or to be extended and relaxed in their long axis, or whether vibrations are set up in the fluid parts of the organ and the scolopidia respond by resonance. Their length varies in the ratio of 2 : 3; it is therefore possible, if they act as resonators, that they might distinguish tones over about one octave.[138]

There is much evidence to prove that the tympanal organs are sensitive to sounds. Various Noctuid moths (*Catocala*, *Agrotis*, *Amphipyra*) will vibrate their wings or fly away in response to the high-pitched notes of a Galton whistle, under conditions where all but air conduction is excluded.[132] They do not respond if the tympanal organ at the base of the abdomen is pierced with a needle.[43] The same behaviour is shown by *Pyrausta* and *Larentia*, but never by Rhopalocera, Sphingids, Hepialids or other forms without drums.[43, 77] Whether the normal function of these organs in Lepidoptera is the perception of sounds has been doubted; perhaps they respond to other air vibrations; for example, vibrations set up by the wings. (But see p. 276.) They are poorly developed or absent in wingless females, and they are reduced in females of those Noctuids with large ovaries, in which wing reduction is only just beginning.[60] [See p. 296.]

Tettigoniids of the genus *Thamnotrizon* normally chirp alternately with one another when their tympanal organs are intact; but when these are removed they chirp irregularly. Here again the sound is air-borne; for a pair of individuals placed in megaphones directed towards each other alternate properly; they chirp irregularly if the megaphones are directed away from each other.[112] Females of *Gryllus* are attracted to the male by his chirping, from a distance of more than 30 feet, and males without their stridulating organs do not attract. Chirps transmitted by telephone will cause the female to go to the receiver; but this response is no longer shown if both tympanal organs are removed; whereas removal of the antennae is without effect.[111] Similar evidence exists in the case of *Stenobothrus* and other Orthoptera.[10] The tympanal organs of *Tettigonia* give an immediate initial 'phasic' response in the nerve, for which only a single cycle of the flight tone is necessary. This is followed by a 'tonic' response which requires about 25–30 cycles to set it off and which continues so long as the sound lasts.[172]

Worker bees respond by crouching down on the comb when the queen 'pipes'; and they will respond in the same way when the sound is transmitted through a microphone and loudspeaker. No tympanum is present but it is possible that the whole integument serves as a resonating drum for the chordotonal organs in the legs.[158] The plate organs, or placoid sensilla of the honey-bee which have often been regarded as olfactory organs (p. 289) have certain features which recall the tympanal organ of locusts, including the cilia-like endings of the dendrites. It has been suggested that they may respond to pressure and thus serve to detect sound waves or other vibrations.[270, 272] Many Homoptera-Auchenorrhyncha respond to the songs of their own species although they have no tympanal organ. Here the vibrations are probably conducted mainly by the substratum.[165]

Auditory range of tympanal organs—The range of sensitivity of the tympanal organs has been investigated by observing the electrical response in the sensory nerves or central nervous system during stimulation by sounds of known frequency (Fig. 179). In an Acridiid, in which the response in the nervous system is found to disappear after destruction of the abdominal tympanal organ, the range extends from about 300 to 20,000 cycles per second. At low frequencies the tympanal organ is very insensitive as compared with the human ear: near the lower limit it requires for threshold stimulation a sound pressure more than 30,000 times that required by the human ear. But with increasing frequency the sensitivity increases, and at 10,000 cycles it is close to that of man. Sensitivity is greatest at this frequency, as contrasted with 1,500 in man.[148] In the katydids *Ambycorypha* and *Pterophylla* a response was obtained over a range from 800 cycles up to 45,000 cycles, which is well above the human limit;[148] in *Gryllus* it extended from 300 to 8,000 cycles.[148] In *Locusta* the sensitivity appeared to be still increasing above 10,000 cycles per second,[107] and in this zone the sensitivity seems to have reached the limits set by the physical nature of the air.[8] In *Locusta* and *Decticus* the upper limit is above 90,000 cycles per second, more than two octaves higher than man; frequencies of this order occur in the stridulation of grasshoppers.[7] [See p. 296.]

Observations on the reflex responses of intact insects support these conclusions. The katydid *Thamnotrizon* will respond to frequencies from 435 to 25,000 per second,[112] *Acridium* from about 1,740 to 20,000.[5] In *Locusta* the type of nervous response is asynchronous at all frequencies; it is probable, therefore, that these organs cannot distinguish tones. But being paired and, for sounds of low absolute intensities, very sensitive to intensity fluctuations, they are well suited to enable the insect to localize a source of sound.[107] In man, the perception of sound direction is dependent on the time difference in the stimulation of the two ears; differences in intensity play a secondary role. *Tettigonia* scans the surroundings by swinging the legs. At certain points of this move-

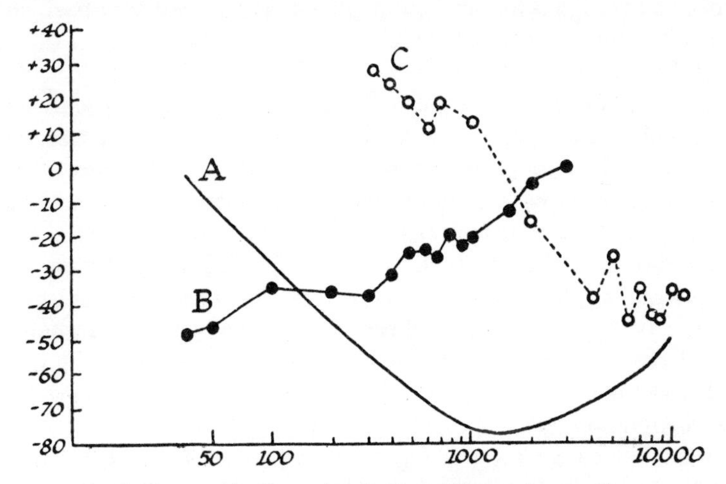

FIG. 179.—Auditory threshold at different frequencies

A, normal human threshold; *B*, threshold of electrical response in cercal nerve of *Gryllus*; *C*, threshold of electrical response in tympanal nerve of *Locusta*. Ordinate: sound pressures expressed as decibels above or below a zero approximating to 10 dynes per sq. cm. Abscissa: frequency in cycles per second on a logarithmic scale (*after* PUMPHREY and RAWSON-SMITH).

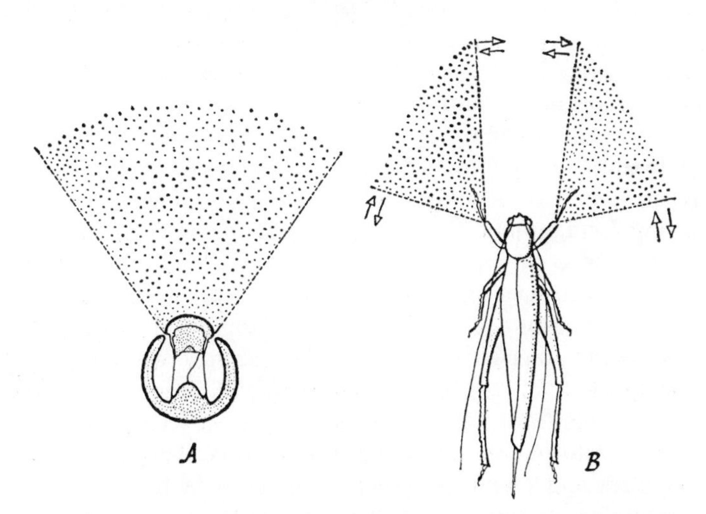

FIG. 180.—Sound location in *Tettigonia cantans* (*after* AUTRUM)

Directly in front of the fore leg, as seen in section *A*, there is a sector of reduced sound, about 70° wide. *B*, the locust scans the surroundings by swinging the fore legs as it walks.

ment there will be a rapid change in sensitivity, which makes accurate localization possible, even with a single leg (Fig. 180).[7] In crickets the important factors in sound localization seem to be differences in intensity and in time of stimulation on the two sides; and these differences are exaggerated by ipse-lateral stimulation of the nerve cord by sound received in one tympanal organ being combined with contra-lateral inhibition.[274]

The fact that females of *Liogryllus* are attracted to chirps transmitted through the telephone, in which the high frequencies will be lost, suggests that it is the rhythm of the sound rather than the frequency as such which is the essential stimulus for orientation.[148] Indeed, the sound is so distorted by the telephone as to be quite unrecognisable to the human ear. It has therefore been suggested that the quality which the female perceives is not the frequency of the different components in the sound (the insect, indeed, has no peripheral mechanism permitting harmonic analysis as in the cochlea) but the frequency with which the sound as a whole undergoes modulation in amplitude. This character would remain no matter how much the sound was distorted.[106, 107] According to this suggestion the high frequencies emitted by the stridulating insect act as a carrier for a low frequency modulation. It has been shown experimentally in *Locusta* that when a pure tone of 8,000 cycles per second is modulated at any frequency up to 300 per second, the nervous impulse as detected with an oscillograph consists of bursts of activity at the modulated frequency; and this response is unaffected by large changes in the frequency of the carrier note. The insect thus stands in marked contrast to man who is almost insensitive to changes in the frequency of modulation.[107]

The frequency range for the tympanal organs of *Locusta* is from 600 to 45,000 cycles per second, the most sensitive range being from 5,000 to 9,000 cycles per second. The frequency of discharge in the nerves is characteristic for each neurone; it is increased up to a point by the intensity of the sound, but it is not related with sound frequency: the organ has no well-developed ability for analysing sound frequencies.[273] But Orthoptera can discriminate pitch to a very small degree;[224] the Tettigoniid *Gampsocleis* has two classes of neurones in the tympanic nerve with peak sensitivity at about 10,000 cycles per second and 6,000 to 7,000 cycles per second respectively.[228] [See p. 296.]

Similarly, the response in the auditory nerve of the male cicada to the song of another cicada shows a synchronous volley of nerve impulses corresponding to the pulse frequency: it is responding to the 'modulation envelope'.[255]

Night-flying Lepidoptera respond to sounds in the high frequency range, especially from 40,000 to 80,000 cycles per second, by a flight response or by 'death-feigning'. This reaction may serve for perception and escape from bats, which emit orientation sounds in this frequency range.[169] The tympanic nerve of *Prodenia eridania* and other Noctuid moths contains only two acoustic cells which differ in their threshold of response. One or both will respond, depending on the intensity of the sound, from 3,000 to 100,000 cycles per second, with a maximum sensitivity at 13,000 to 16,000. They appear to be specialized for the reception of brief sound pulses and readily respond to the cries of bats hunting by echolocation; these can be detected at a distance of 100 ft. or more. It is possible that these organs are used also to detect the flight sounds of other insects or even the reflections of an insect's own flight sounds from neighbouring objects.[263, 279]

Sound production—Sounds may be produced by insects in many different ways.[76, 103]

(i) By tapping, as in *Anobium*, which strikes its head against the woodwork where it burrows, or the soldiers of some termites, such as *Leucotermes*, which hammer the ground in unison at about ten times per second, producing a faint drumming sound.[218]

(ii) By friction of the so-called stridulatory organs. In Acrididae the upper surface of the costal margin of the hind wing is rubbed against the lower surface of the fore wing, or the inner aspect of each hind femur bearing a series of pegs is rubbed against the outer surface of the corresponding fore wing; in Tettigoniidae and Gryllidae two modified areas of the fore wings are rubbed together;[232] ants stridulate by rubbing a hard piece on the petiole up and down on a striated plaque at the anterior constriction of the abdomen;[109] some Coleoptera, Lepidoptera and Hemiptera have various other means of stridulating.[10] *Corixa* uses the same stridulating organ to produce two quite different sounds: an intermittent chirping and a continuous brushing sound. These sounds are produced during the mating period; they can be distinguished by the opposite sex and give rise to

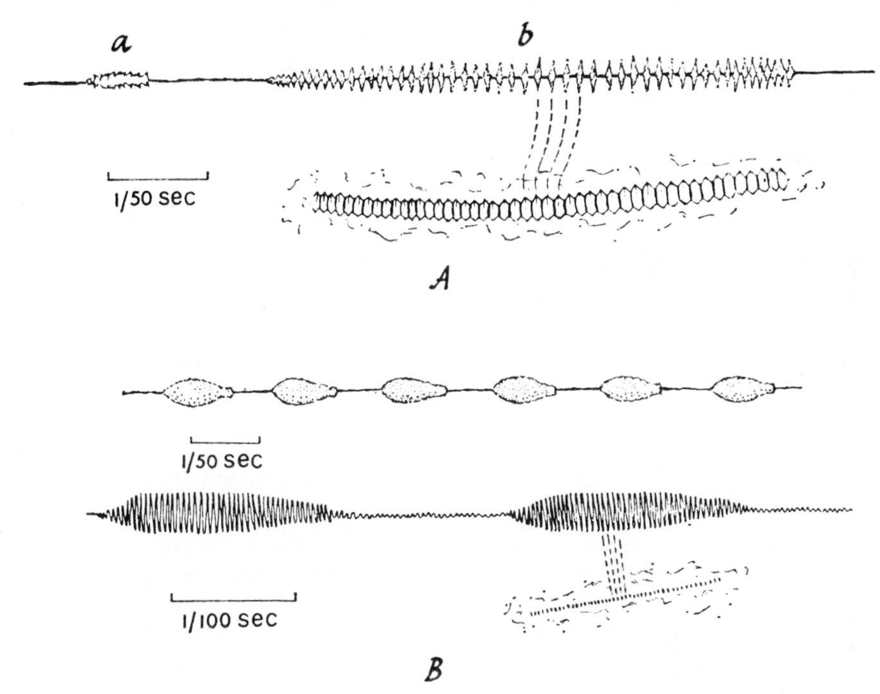

FIG. 181.—A, the file on the tegmina of *Ephippiger bitterensis* and oscillogram of the song. a represents the sound caused by opening the tegmina; b the sound caused by closing them. B, the file on the tegmina of *Oecanthus pellucens* and a slow and fast oscillogram of the song. Explanation in the text (*after* BUSNEL)

different responses; but all reactions are eliminated if the tympanal organs are destroyed.[168]

The spectral analysis of the sounds produced by Orthoptera shows that in *Gryllotalpa* the song consists of a single pure note between 3,500 to 4,500 cycles per second. But in most species there are multiple bands of varying width in both the audible frequencies and the ultra-sonic range. *Decticus*, for example, emits ultra-sounds of very high frequency up to 100,000 cycles per second.[157] The stridulation noise of *Locusta migratoria* is concentrated within the frequency range 5,000 to 16,000 cycles per second (Fig. 182, B).[194]

Most commonly, in grasshoppers, the impact of each tooth produces a succession of vibrations in the wing which are very rapidly damped before the next tooth strikes and a new chain of vibrations begins (as in *Ephippiger*, Fig. 181, A).[186] As a consequence of this rapid damping a quick succession of trills, or pulses of sound, are produced, each of which records the impact of a new tooth. The whole pattern of sound is decided by the anatomy of the insect (the nature of the vibrating wing-membrane and the number and spacing of the teeth) and by the frequency and rapidity of movement of the legs or the other structures that carry the teeth.[216]

In the songs of some crickets, the Tree Cricket *Oecanthus* for example (Fig. 181, B) each trill is produced by some ten or twenty oscillations of the wing membrane; but here the structure is so arranged that the time between the impact of one tooth and the next corresponds with the resonant frequency of the wing. The result is that one sound wave is given out as each tooth strikes the wing, and a pure note of 2,000 to 3,000 cycles per second is produced.[251] [See p. 296.]

(iii) Vibration of the wings in flight (in *Bombus*, *Apis*, *Melolontha*, &c.) produces a note whose pitch is determined by the rate of vibration (p. 165).

(iv) Males of Cicadidae have a pair of drums at the base of the abdomen; these are convex outwards and a powerful muscle is attached to the inner concave surface; when this drum is repeatedly drawn inwards and released it produces sound by the same mechanism as a rounded tin lid pressed inwards by the finger.[142] The normal song in cicadas consists of a train of damped oscillations at a fundamental frequency of about 4,500 per second, determined by the natural period of vibration of the tymbals; the frequency of repetition of the pulse being about 390 per second (Fig. 182, A). At low frequencies of stimulation the muscle responds with a series of contractions to each new impulse. At frequencies above 60 per second it goes into continuous activity at a frequency unrelated to the frequency of stimulation (see p. 154).[255] [See p. 296.]

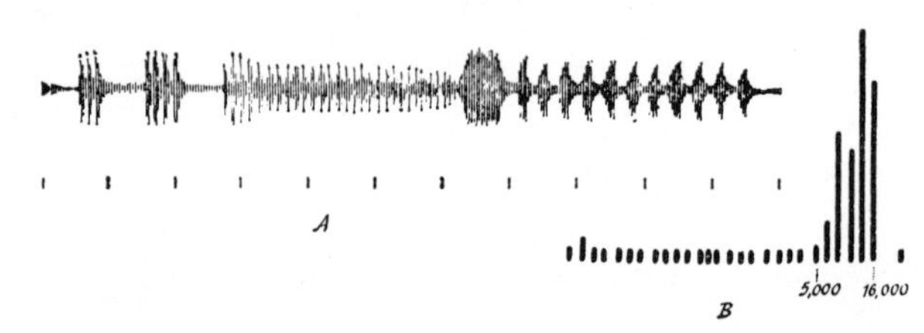

FIG. 182.—*A*, oscillogram of the song of the cicada *Terpnosia ridens*. Time marker indicates 1-second intervals (*after* PRINGLE). *B*, oscillogram of stridulation sound of *Locusta migratoria* analysed by means of an acoustic spectrometer. Sound components confined to frequencies 5,000 to 16,000 cycles per sec. and with maximum at 12,500 (Band 2 at 50 cycles per sec. is an artefact) (*after* EVANS)

The Arctiid moth *Melese laodania* has metathoracic tymbal organs which produce ultrasonic signals in the same range of frequencies as those used by bats; they may possibly be used to confuse hunting bats.[175]

(v) High-pitched sounds may be produced by mechanisms which are not in all cases understood. In *Acherontia*, air appears to be forced out from the oesophageal diverticulum and sets a fold in the mouth cavity in vibration.[185] In the 'piping' of the queen bee the sound has commonly been attributed to air driven out of the trachae in pulses;[143, 158] but piping continues with all the spiracles blocked save one; it results from vibration of the thorax by the flight muscles without the wings being spread.[269] The sounds produced by *Calliphora* probably arise in the same way or perhaps by friction of the wing bases against the schlerites.[76] [See p. 296.]

Many of these sounds are adventitious and appear to have no physiological significance. Interest attaches chiefly to the stridulatory organs. In Orthoptera these are associated with the presence of auditory tympanal organs and, as we have seen, they are of undoubted importance in bringing the sexes together, and indeed in many other phases of sexual behaviour.[186, 198, 218] In Cicadidae the function of the song is to assemble the local population of a particular species (males and females) into a small group.[255] But in beetles, Hemiptera, &c., it is very doubtful if the sounds are heard by other insects. Although in some, large Cerambycid beetles for example, the sounds may well have a protective function. In ants it is doubtful if any sound is produced at all.[6] Stridulation in some insects seems to be a nervous reflex, of unknown significance. [See p. 296.]

CHEMICAL SENSES

Nerve endings of many kinds are doubtless sensitive to irritant chemical substances. From this has been evolved the common chemical sense of primitive invertebrates. In insects and vertebrates this sense has become further differentiated into the two senses of taste and smell.[63] There is no absolutely satisfactory distinction between these two. For in insects both are subserved by primary sense cells (p. 178); their distribution varies from one species to another, and they may be located in the same part of the body; although the sensitivity is usually far greater in the case of smell, as estimated by the lowest molecular concentration that can be detected, yet there are very great differences between species; for terrestrial organisms taste is a sense of solutes and smell a sense of vapours, but this does not apply to aquatic forms. In the amphibious beetle *Laccophilus* the same sensilla basiconica on the tips of the antennae are the principal chemoreceptors for both gaseous and liquid stimuli.[221] Here we shall regard as taste the perception of the qualities of sweet, sour, salt, and bitter, which are called tastes in man; other chemical perceptions will be included in the sense of smell.[38, 55] Whether these four tastes recognized by man are really 'primary modalities', or merely 'points of familiarity in a continuous spectrum based on stimulative effect',[53] only further work can show.

General chemical sense—Irritant substances such as ammonia or chlorine fumes may be perceived in many parts of the body. Probably they stimulate sensory endings which normally respond to other stimuli. The legs of termites are sensitive to such irritants;[66] the cerci of *Gryllus*, which show no true sense of smell, will react to essential oils,[123] as will also the basal segments of the antennae and many other parts of the body in *Tenebrio*.[133] It appears to be this general sensitivity which has given rise to the mistaken belief that the widespread campaniform sensilla are organs of smell.[54, 101]

Taste—Organs of taste are associated with the **mouth** in many insects. Collembola are repelled by bitter, alkaline, acid or salt tastes in the food;[127] bees reject honey treated with quinine or salt;[152] caterpillars make vigorous spitting movements of the mouth in response to salt, acid or bitter substances.[35, 41] In the bee sensory pits at the base of the tongue are perhaps the receptors;[152] in caterpillars the response is abolished if the epipharynx and hypopharynx are removed.[35] The oral lobes of the proboscis in *Calliphora* bear gustatory hairs along their margins.[65, 92] In *Periplaneta*, the contact chemoreceptors are on the tips of the maxillary and labial palps, and either the inner surface of the mouth parts or the wall of the buccal cavity.[53] Hemiptera have a perforated area of cuticle in the pharynx with flask-like cells below; this is often regarded as an organ of taste, although no nerve supply to it has been demonstrated.[24]

After a course of training, in which the beetles are either rewarded with sweetened meat or punished with quinine-treated meat, *Dytiscus* can learn to associate a particular taste with the presence of food. After several months, for example, if accustomed to being fed after a salt taste, *Dytiscus* will disregard cotton wool containing sugar but respond to wool containing salt. In this way it can be shown that *Dytiscus* distinguishes the four taste qualities. Many of the receptors concerned are borne by the **palps** but others (inside the mouth) remain after the palps have been removed.[13, 118] Similar results have been obtained with *Hydrophilus*; in this beetle sweet, salt and bitter tastes are perceived only with the apices of the maxillary palps, sour tastes with the tip of the labial palps, which are rich in rod-like sensilla.[114] The dung beetle *Geotrupes*,[141] the bark beetle *Rhagium* and the cricket *Liogryllus*[80] also detect sweet tastes by means of the palps. (See further, p. 283.) [See p. 296.]

In a few insects such as ants,[120] bees, and wasps, the **antennae** have been shown to bear organs of taste. The honey-bee and the wasps *Vespa* and *Polistes* can distinguish plain and sweetened water by means of the antennae, and this response to sugar is prevented if acid or quinine is added. The receptors are confined to the eight distal segments and are perhaps the pore-plates (*cf.* p. 288).[80, 92] They occur also in Lepidoptera.[206]

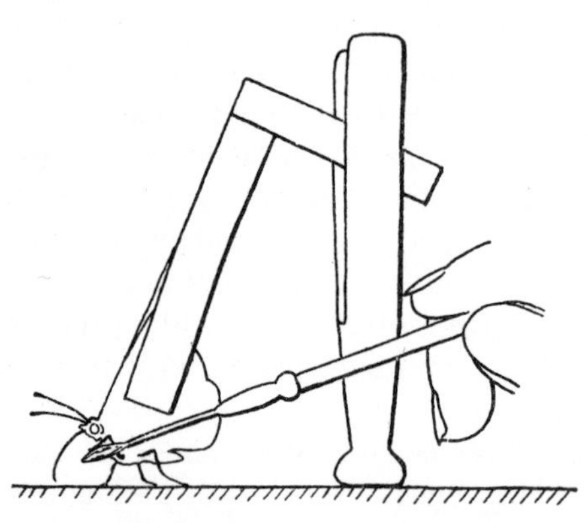

FIG. 183.—Demonstration of tarsal organs of taste in *Vanessa* (*after* MINNICH)

On touching the tarsus with a paint-brush moistened with sugar water the proboscis is immediately extended.

A more frequent site is the **tarsus** and distal end of the tibia. In the butterflies *Pyrameis*,[4, 92, 146] *Danais*,[4] *Pieris*,[92] *Plusia* and many other Noctuids,[121] the flies *Drosophila*,[12] *Calliphora*,[30, 92] *Musca*,[34] and in the honey-bee *Apis*,[55] immediately the tarsi come in contact with sweetened water the proboscis is extended (Fig. 183); this response is given by insects which pay no attention to water alone. The tarsi of *Pyrameis* are also very sensitive to quinine; $\frac{1}{32000}$M quinine will prevent the response to sugar; but 4 per cent. sodium chloride is taken as readily as distilled water.[146] The tarsi of *Calliphora* are much less sensitive to acid and bitter repellent tastes than the tongue; they seem quite insensitive to salt.[65] In the case of *Pyrameis* the sensilla concerned consist of thin-walled tubular hairs in sockets, each associated with a large trichogen or glandular cell and a group of sense cells with a bundle of distal processes (*cf.* p. 288).[47] Besides their sensitivity to touch and vibration the modified forelegs of *Nepa* bear numerous thin-walled sensilla which are chemoreceptors used in recognizing prey. On the corresponding legs in the Corixids *Sigara* and *Corixa* there is likewise a thin-walled organ which can recognize sugar and salt solutions and aids the insect in seeking food.[2] They occur also in Trichoptera.[206] (See further, p. 279.)

Contact chemoreceptors, with about the same sensitivity as those elsewhere on the body, apparently occur in the **ovipositor** of parasitic Hymenoptera; but the sensilla concerned have not been identified.[37] Chemoreceptors on the ovipositor are important also in *Phormia* (p. 289).

The nature of the taste stimulus is obscure. The unacceptability of salts is correlated with the mobilities of their kations, the hydrogen ion being the most repellent. Whether this is an effect on penetration, or on adsorption at some lipoprotein interface is uncertain.[53] Studies of the chemical structure of test substances and tarsal stimulation in *Phormia* fail to prove whether adsorption or penetration is the decisive factor,[38] but there does seem to be some relation between stimulating efficiency and lipid solubility.[38] It has been suggested that the sugar molecule, for example, combines with a specific receptor site by weak forces, such as van der Waals' forces, to form a complex which depolarizes the membrane.[189] The salt-combining sites of the labellar hairs in *Phormia* are anionic and strongly acidic; it seems that the monovalent cations of salts are chiefly responsible for the stimulation.[197]

Oral taste in the bee—The sense of taste has been studied in most detail in the mouth parts of the bee. Experiments in which the effects of training are combined with the repellent action of certain tastes have proved that sweet, bitter, acid and salt are recognized as separate qualities. As in the case of man, the sourness of solutions of hydrochloric acid is proportional to the hydrogen ion concentration, but weakly dissociated acids such as acetic are relatively more sour at a given *p*H. The bee is more sensitive to sodium chloride than man, but quinine is less repellent.[55] It may be that very dilute solutions of salts taste sweet;[53] honey-bees prefer dilute sodium chloride or ammonium chloride to distilled water.[25]

Most substances which appear sweet to us are tasteless to the bee. For example, pentoses (arabinose, xylose, rhamnose), sugar alcohols (mannitol, dulcitol, sorbitol, erythritol) and many true hexose sugars (mono-, di- and trisaccharides) are all tasteless. Out of 34 sugars and related substances tested, 30 appear sweet to man, only 9 to the bee; all these nine being present in the natural food of the bee and capable of being metabolized by it (p. 516).[55] All artificial sweet substances, such as saccharin, are tasteless in dilute solution and repellent at higher concentrations; whereas acetylsaccharose, which appears very bitter to man, has no repellent effect on the bee (it has been proposed as a suitable substance to add to sugar so that this might be made unfit for human use and supplied free of duty to bee-keepers).[140]

The threshold of taste for cane sugar, as judged by the lowest concentration that is imbibed, varies from 1M to M/8 according to the richness and abundance of nectar in the field. The sensitivity falls with increasing age and the threshold is raised if a sweet mixture has been given shortly before the test; whereas in bees starved for some hours it will fall to M/16. The sweeter the mixture the more of it does the bee imbibe. For example, in one set of experiments bees took up 42 ml. of a M/2 sucrose solution, 49 ml. of 1M and 56 ml. of 2M. And the sweeter the mixture the more do the bees excite the other members of the hive on their return, by 'round dances' and extrusions of their scent glands.[55]

Labellar hairs and pharyngeal receptors in Diptera—The sweetest sugar for the oral lobes in *Calliphora* is sucrose, with a threshold concentration of M/100–M/400; glucose appears less sweet; lactose and maltose, with a threshold concentration of 128/100–64/100M less sweet still.[92] The threshold falls somewhat during fasting. These flies taste as sweet many more sugars than the bee, but they metabolize about an equal number. Some sugars, such as the pentoses arabinose and xylose, are not utilized although they taste sweet; others, such as galactose (which is not utilized by the bee), are metabolized by *Calliphora* although tasteless (p. 516). In the blow-fly *Phormia* there is no good correlation between the acceptability of sugars and their metabolic value. For example, the pentose fucose has a high stimulatory value but is not utilized; whereas mannose has a relatively low stimulatory value.[162] *Phormia* will ingest increasing volumes of sugar solution as the concentration increases up to an optimum point, after which there is a decrease; but the total weight of sugar ingested continues to increase with rising concentration.[191] The threshold concentration for glucose ingestion falls with increasing duration of flight; but the injection of glucose into the blood has no effect on threshold. This seems to depend chiefly on the quantity of sugar in the gut.[222]

There are large numbers of minute 'inter-pseudo-tracheal' sensilla of basiconic type over the surface of the lobes of the labellum,[235] but the sensilla studied in most detail are the thin-walled hairs arising from sockets around the margins. These hairs have a double lumen; they have four sense cells at the base in *Phormia*,[196] and in *Calliphora* four distal nerve filaments can be traced along the small lumen to the tip of the hair.[252] In *Phormia* there may be more than one sort of hair: the thick-walled canal sometimes contains five nerve fibres.[234] It is only the extreme tip that is sensitive to stimulation by chemicals. These organs consist of four receptors built into a single hair:[188] (*a*) mechanical and responsive to bending, (*b*) stimulated by certain sugars, (*c*) responds to monovalent inorganic cations, (*d*) responds only to water.[196, 290, 292] The lowest threshold value obtained with sucrose for a single hair was 1×10^{-5}M.[188] Repeated stimulation of one hair in *Phormia* can lead to its complete adaptation; it becomes apparently insensitive without neighbouring hairs being affected.[188]

Electric recordings from single chemoreceptive hairs of *Phormia*, *Sarcophaga*, *Musca*, and *Drosophila* show that there exists a peripheral mechanism for discrimination between sugars and non-sugars. Two types of response are given: (*a*) small voltage spike potentials follow stimulation of the sugar receptor and lead to immediate extension of the proboscis. (*b*) Spike potentials of higher voltage follow stimulation by salts, acids, alcohols, &c.[222, 223] Receptors of this type with multiple sensitivities seem to be widely spread: receptors sensitive to chemical, tactile, and temperature stimuli occur in various Lepidoptera and Diptera.[222]

By electric recording from single sensilla in the larva of *Bombyx*, on the labella of *Calliphora*, and on the tarsi of *Vanessa*, it has been shown that chemical stimuli depolarize the receptor membrane near the tip. These changes in membrane potential control the rate of discharge of impulses initiated near the base of the sensillum, probably at the body of the sense cell. Such impulses are carried to the tip of the sensillum and to the central nervous system.[244] [See p. 297.]

Oral taste in other insects—Among ants there is much variation in the sense of taste. *Lasius*, *Formica* and the different species of *Myrmica* differ in the number of sugars that are apparently sweet, and in the relative sweetness of those sugars to which they are sensitive.[120]

Lymantria larvae, after being offered water until it is refused, will accept sucrose at a minimum concentration varying from M/2–M/10 and drink it vigorously. The M/10 solution may be accepted after water but refused after the M/2 solution.[35] Various caterpillars will ingest sodium chloride up to a maximum concentration of M/8, hydrochloric acid up to M/320 and quinine up to M/10,240; but these thresholds are raised if sucrose also is present.[41] Larvae of *Anosia plexippus*, which feed on milkweed, will commence eating other leaves if these are enclosed between slices of milkweed leaves, but almost at once they are rejected.[35] Presumably this indicates a perception of flavour not included in the four taste qualities. Similarly, larvae of *Pieris brassicae* and *P. rapae* attack any plants smeared with mustard oil, and even starch or paper so treated;[134] and the same is true of *Plutella maculipennis*.[275] The larvae of the saw-fly *Priophorus padi* feed on various Rosaceae, all containing the glucoside amygdalin; the beetle *Gastroidea viridula* feeding on *Rumex* appears to be attracted by oxalic acid and oxalates;[134] butterflies of the genus *Papilio* have been able to transfer from Rutaceae to Umbelliferae as food plants because there are identical attractive chemicals in the essential oils of the two families.[35]

Stahl (1888) developed the thesis that the secondary plant substances (glucosides, tannins, alkaloids, essential oils, &c.) served originally for protection against phytophagous animals; but later they became tolerated or even attractive for the few animals that feed on the species in question.[204] The essential feeding stimulants of widespread distribution in plants are sapid nutrients. In addition certain species require a specific 'piquant' stimulus probably of no metabolic importance but peculiar to certain genera or families of plants; perhaps the 'piquant' substances serve to lower the threshold of response to the sapid nutrients.[276] In Aphids this has been called the 'dual discrimination theory'.[229] A glucoside has been extracted from the tomato and related Solanaceae which is a principal gustatory stimulant both for the Sphingid *Protoparce* and for *Leptinotarsa. Protoparce* will attack filter paper, &c., treated with this substance; but only if sugars or other nutrients are present;[293] after removal of the maxillae *Protoparce* can be successfully reared on non-solanaceous plants.[281] Likewise, *Bombyx mori* larvae will feed, for a time at least, on plants such as cherry and cabbage, not normally touched, after extirpation of the maxillae.[226] Although these caterpillars are attracted from a distance by the β-γ-hexenal and α-β-hexenal in leaves their feeding is determined by more specific flavours.[215] β-sitosterol and other sterols seem also to exert an effect.[245] [See p. 297.]

The Cabbage Aphid *Brevicoryne brassicae* can be induced to feed on *Vicia faba* if the mustard oil glucoside sinigrin is introduced into the leaves; this specific stimulus is received via the stylets after they have penetrated the leaf surface.[285] In *Rhodnius*, which ingests saline or serum through a membrane much more readily if a little haemoglobin is added to it, the maxillary and mandibular stylets both carry sense cells with dendrites running to the tip.[254]

There are sensilla inside the pharynx of *Musca* which send impulses to the proventricular ganglion, which then controls the passage of the ingested fluid: 5 per cent. sugar solution or above goes to the crop; 1 per cent. or below goes to the mid-gut.[286] But this may be a response to osmotic pressure (p. 477).

Tarsal and antennal organs of taste—Of all the gustatory organs those on the tarsus appear to be the most sensitive to sugars. They are tested by allowing the insect to drink distilled water until contact of the tarsus with water no longer causes the proboscis to be extended; in such insects sweetened water above a threshold concentration will still evoke a response. The level of this threshold depends on the state of nutrition. In *Pyrameis* fed regularly with sucrose the threshold for this sugar is around M/10–M/100, which agrees with the threshold of the human tongue (about M/50). But if sugar is withheld the threshold falls; after prolonged sucrose inanition it may be M/3,200, M/6,400 or even M/12,800 (200 times the sensitivity of the human tongue).[92] *Danais* has been reported to give occasional responses to M/120,400.[4] This high sensitivity perhaps indicates that these tarsal organs serve to *call attention* to the presence of sweet substances rather than to test them as food.[55] In all cases sucrose is the sweetest sugar, as it is for the mouth parts; but sugars which appear sweet to the mouth parts may fail to cause any response by the tarsi; thus lactose, to which the oral lobes of *Calliphora* are very sensitive, has no effect on the tarsi even in 1M concentration.[92] The threshold for sucrose for the tarsal organs of the bee is given as 1M, and for the antennal organs M/12.[87]

The structure of the tarsal chemoreceptors of *Phormia* agrees closely with that of the labellar hairs.[213, 236] They adapt within a few seconds to stimulation by sucrose; but recovery from adaptation is also very rapid; in large part it is probably taking place within the central nervous system.[187] The threshold of the tarsal receptors in *Phormia* is raised by feeding the insect with sugars, even sugars like mannose and fructose that are tasteless, or like lactose and fucose that are non-nutritious. They seem to act on some region of the gut other than the crop.[195] The electric responses in these organs are the same as in the labellar hairs (p. 282).

Adults of *Anopheles*, *Aëdes*, and *Culex* can distinguish between distilled water and sodium chloride solution by tarsal chemoreceptors;[283] there are also receptors on the labella that respond to sugars,[199, 209] and taste organs in the cibarium.[249] Likewise, in *Oncopeltus*, *Nezara*, and other Heteroptera there are external contact chemoreceptors on the tips of the antennae and on the fore- and mid-tarsi; but in addition chemoreception has been proved to take place also in the pharynx after passage of food through the stylet canal.[242]

Smell—The sense of smell in insects is located chiefly in the **antennae**. This was clearly established by Lefebvre (1838), who showed that the antennae of the bee are extended in any direction towards a needle dipped in ether; while in the absence of the antennae no response to odours is given; observations which have been confirmed on other insects (Ichneumonids, Sphegids, Muscids, beetles, &c.) by many authors.[49, 85] *Periplaneta* will locate cheese from a distance by means of the antennae, with which it reaches out in all directions as the source of smell is moved.[58] In termites the antennae alone are the organs for distinguishing true odours.[66] The louse *Haematopinus* extends its antennae towards a source of smell and with these intact can distinguish between the finger and a warmed rod.[144] The males of Saturniid moths, such as *Callosamia*, locate the female by smell, and this sense is confined to the highly branched antennae;[89] in the antenna of a male *Telea* there are some 70,000 sensilla with about 150,000 sense cells.[177] Females of *Habrobracon* discover their host *Ephestia* by smell, and the males find the females by the same means; if the antennae are amputated or covered with varnish these reactions fail.[94] Muscid flies find flowers or dung and the males locate the females by smell; accordingly the males and the dung-feeding species have antennae more plentifully supplied with sensory pits and sensilla.[83] The tsetse-fly *Glossina* can hunt and locate its host by sight alone, but the antennae are generally necessary to call forth the probing response after settling on the host.[95] *Drosophila* mutants 'antennaless' do not respond to olfactory stimuli although their other reactions are normal.[16] *Scarabaeus* searching for dung holds up the antennae with the clubs spread open.[72] In *Tenebrio* the organs of smell lie especially in the last four antennal segments.[133]

But in many insects the **palps** also bear olfactory organs. At the apex of the palps of butterflies there is an olfactory pit well supplied with sensilla; and it is estimated that the sense of smell in *Pieris* is not reduced by much more than half if the antennae are amputated.[92] *Periplaneta*[58] and *Gryllus*[123] can locate food at close quarters by olfactory organs on the palps. *Geotrupes* has organs of smell on the maxillary and labial palps; but these are much less important than the antennae in finding dung at a distance.[141] The same applies to *Necrophorus* and *Silpha*[36] and to *Dytiscus*, except that here apparently the labial palps are not concerned: beetles trained to associate a given scent (coumarin, musk, &c.) with the presence of food can still respond so long as one maxillary palp remains intact.[118] While in *Hydrophilus*, where the antennae are specialized for purposes of respiration, the sense of smell appears to be confined to the palps; if these are removed, although the antennae are intact, the insect can no longer be trained to associate odours with the presence of food.[114]

In the blow-fly *Cynomyia* olfactory receptors are confined to the **antennae** and the **labella**. This is proved by training the fly to associate contact of the tarsi with sugar and the odour of coumarin. On exposure to the odour alone they will then extend the proboscis. If both antennae and labella are removed they cannot learn, and the response is eliminated in conditioned flies; but either alone will mediate the response. The antennae function as directional chemoreceptors. It may be that the labellar organs function as non-directional receptors which, in conjunction with the contact chemoreceptors also on the labella, can serve in the reception of combined olfactory and gustatory stimulation during feeding—like taste and flavour in man.[53]

In adult female *Phormia* the anal leaflets of the **ovipositor** carry olfactory pegs which are used in the selection of sites for egg-laying.[282]

Sense of smell in the honey-bee—The olfactory perceptions of the honey-bee have been studied in great detail by training foraging bees to associate particular scents, mostly flower scents and essential oils, with the presence of sugar.[54] By this means it has been shown that the bee can distinguish, for example, essence of orange from 43 other ethereal oils. There seems to be much in common between this sense in the bee and the human sense of smell. Thus the threshold concentration of these scents is of the same order as in man, though the bee seems better able to detect a given scent among a mixture of others. Curves of sensitivity to homologous series of compounds follow the same course in bee and man, which suggests that these must be a physico-chemical similarity between the olfactory organs.[268] The olfactory acuity in the honey-bee is approximately the same as in man; but man is five times more sensitive to oil of rosemary, and the bee more sensitive to the odour of wax and of the bee scent organ.[203] The honey-bee is particularly sensitive in distinguishing mixtures of odours.[257]

Some 32 odorous substances of no biological significance appear scented to the bee; whereas odourless flowers such as *Ribes rubrum*, *Vaccinium myrtillus*, &c., appear unscented; and substances such as nitrobenzol and oil of bitter almonds, or amylacetate and methyl heptenone which smell alike to man are also confused by the bee. Bees trained to particular scents lose the effect of training if the antennae are removed, whereas such treatment does not affect bees trained to colours. It is of interest, also, to note that while bees will learn to associate many ethereal scents with the presence of sugar, they cannot be trained in this way with foul-smelling substances such as skatol or asafoetida.[54]

Sense of smell in other insects—The sense of smell is of some importance in caterpillars. Larvae of *Danais plexippus* separated by wire gauze from leaves of *Asclepias*, make searching movements as they crawl across these leaves, or come to rest over them, whereas they move straight ahead without stopping when passing over other leaves; and they refuse to feed on *Asclepias* leaves treated with scent or methyl alcohol.[35] The gregarious larvae of *Pieris brassicae* utilize the species odour, which they can detect at a distance of 5 cm., for collecting into colonies, and they will follow trails of this scent leading to such a colony.[153] Ovipositing Lepidoptera are presumably guided by smell in selecting their food plant: the Oleander hawk, *Daphnis nerii*, always chooses plants of the family Apocynaceae, *Nerium oleander* in the Mediterranean region, *Vinca major* in Northern Europe, *Trachelospermum jasminoides* in India; *Acherontia* chooses Solonaceae—potato or nightshade.[143] But the scents in question are not perceptible to man. Similarly, *Rhyssa* can recognize the larva of *Sirex* through several centimetres of solid wood (perhaps in this case the perception of vibrations is responsible). And males of Saturniid and Lasiocampid moths will assemble in large numbers around newly emerged females, attracted, apparently from great distances, by scents not detectable by man. The olfactory sense of the male *Bombyx mori* seems to be specialized to perceive only the female odour: the male is quite unresponsive to various essential oils and is not deterred when the female is coated with such oils.[28] Among ants, smell is enormously important both inside and outside the nest in a hundred different ways. Ants, like bees, distinguish by smell the members of their own community from others (ants of different genera which normally fight will feed amicably together after removal of the antennae);[49] they follow trails of scent to and from the nest; the sense of smell amounts in them almost to a sense of form and topography —if the trail followed by ants to and from a group of Aphids is reversed 180°, the ants are at once thrown into confusion.[49] The Hymenopterous egg parasite *Trichogramma* can detect (perhaps by the antennae, perhaps by the tarsal receptors) the odour imparted to eggs by another female which has walked over them, and in consequence avoids them for oviposition. If a host egg already contains a *Trichogramma* egg, this can also be detected, again perhaps by chemical sense, after inserting the ovipositor.[117] [See p. 297.]

Chemoreceptive sensilla—All the sensilla of insects are covered by a continuous cuticular sheath;[116] but in certain of them the cuticle is excessively delicate, and it is natural to regard these as the chemoreceptors responsible for taste and smell.* The cuticular parts of these thin-walled sensilla may be

* The campaniform sensilla have occasionally been regarded as olfactory organs but their distribution does not agree with that of the sense of smell,[54] and in the cockroach they show no electrical response to olfactory stimuli.[101]

:longated in the form of slender hairs (sensilla trichodea olfactoria) as in the ıntennae of the wasp,[137] of *Rhodnius* (Fig. 184),[151] and of Saturniidae[104] or the .arsi of *Pyrameis*,[47] or the labella ınd tarsi of *Tabanus*,[53] *Calliphora*,[131] ınd *Musca*.[67] Or the hairs may be ;hortened to cones or pegs (sensilla ;tyloconica and basiconica) as in he antennae of *Periplaneta*,[58] *Pediculus* (Fig. 192),[150] *Tenebrio*,[133] *Geotrupes*,[141] *Necrophorus* and *Silpha*,[36] *Dytiscus*, Hymenoptera Fig. 187),[137] and the apex of the ›alps of *Hydrophilus*.[114] The large ,ensilla basiconica are probably the :hemoreceptors on the antennae of :aterpillars.[35] The chemoreceptors ısed in the orientation and biting eactions of wireworms (*Agriotes*) ›ppear to be peg-organs on the labial and maxillary palps and galeae and a cup-

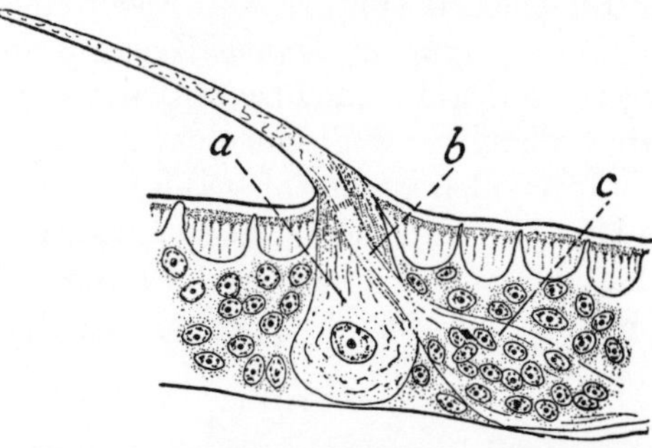

FIG. 184.—Section through the thin-walled type of sensillum (probably olfactory) on antenna of *Rhodnius* (*cf.* FIG. 158, *b*) (*after* WIGGLESWORTH and GILLETT)

a, trichogen or glandular cell; *b*, bundle of terminal filaments from the sense cells; *c*, fusiform mass of sense cells numbering about 15.

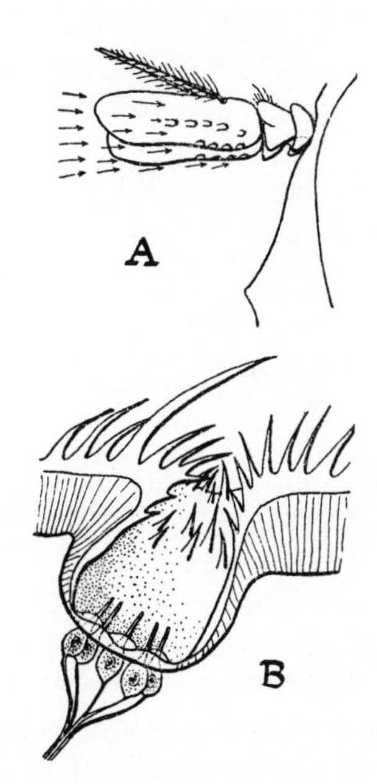

FIG. 185

A, antennae of Muscid fly erected during ight so as to expose the olfactory pits to the ir stream; *B*, detail of olfactory pit showing sen- ›ry rods each supplied by a single sense cell *ıfter* LIEBERMANN).

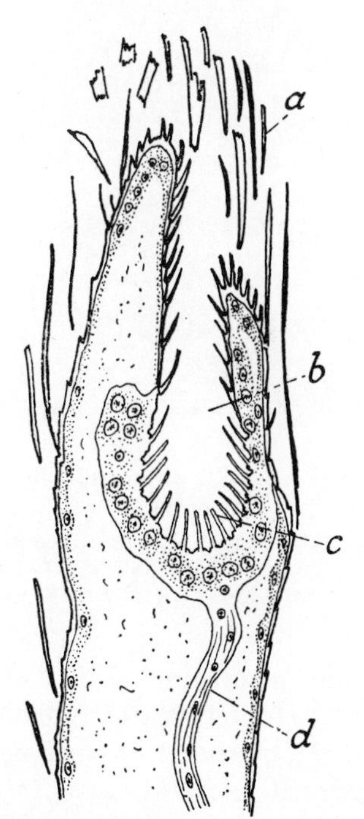

FIG. 186.—Longitudinal section of tip of labial palp of *Pieris rapae* showing pit containing olfactory rods (*after* HSÜ)

a, covering scales; *b*, pit; *c*, sensory rods; *d*, nerve.

haped organ on the distal segment of each antenna.[29] Sometimes the thin- ʋalled cones may arise from the floor of a single small pit in the cuticle ;ensilla coeloconica) as in many dung beetles,[141] *Dytiscus*,[73] Lepidoptera,[19,104,119]

&c. Sometimes a number of cones or rods arise together from the floor of a deep pit, as in the antennae of Muscids (Fig. 185)[83] or at the apex of the palps in Rhopalocera (Fig. 186).[110] Fine thin-walled pegs and cones over the surface of the antennae of *Drosophila* are probably the chemoreceptors. These occur also in the leg-like antennae of 'aristopedia' flies (p. 111); but the pit organs, which are probably long-distance chemoreceptors, are absent from these antennae.[16]

In Hymenoptera, thin-walled hairs of this type may become applied to the surface of the antenna. Sometimes such hairs are stiffened by a longitudinal thickening. And it has been possible to trace the evolution of these structures until the hair has disappeared and they consist of a flattened plate of fairly

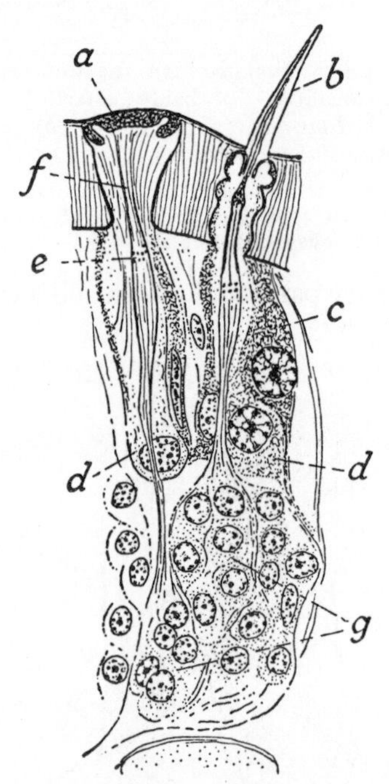

FIG. 187.—Sensilla on antenna of *Apis* (*after* HSÜ)

a, placoid sensillum; b, thin-walled trichoid sensillum; c, distal sheath cell; d, basal sheath cell; e, sensory rodlets; f, terminal filament; g, sense cells.

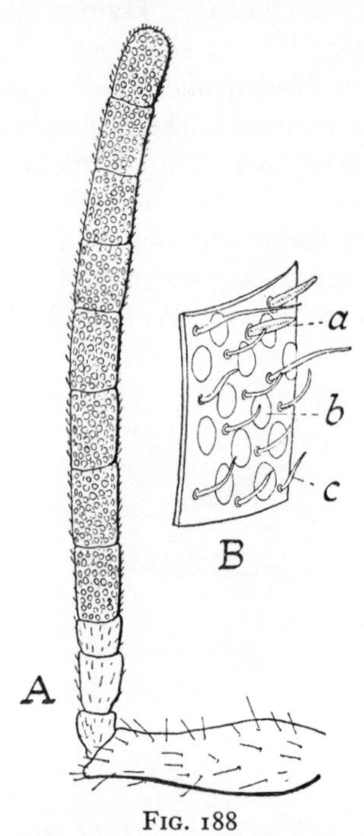

FIG. 188

A, antenna of honey-bee worker, medial surface, to show distribution of pore-plates (*after* V FRISCH); B, detail of a part of one segment (*after* SNODGRASS). a, sensory peg; b, pore-plate (*cf* Fig. 187, *a*); c, thin-walled sensory hairs (*cf* Fig. 187, *b*).

thick cuticle surrounded by a very delicate membrane ('pore-plates' or sensilla placodea) (Fig. 187, *a*).[139] Such plates may be rounded, as in *Apis*,[137] *Melolontha*,[141] *Dytiscus*,[73] or so elongated that each will occupy the entire length of an antennal segment, as in wasps and many Ichneumonids.[137] The male honey-bee has about 30,000 of these organs on each antenna, the worker about 6,000.[137]

These thin-walled sensilla generally have no socket. The tormogen cell is often absent. But the trichogen cell is usually very large and secretes a clear fluid which fills the cavity of the hair or cone. The sense cells are generally multiple; they may form a cluster of twenty or thirty, ensheathed in a nucleated coat continuous with the neurilemma of the nerve.[110] The proximal processes of these cells form the afferent nerve; the distal processes are united to form a bundle (the terminal filament), often with a group of little rod-like structures on its course, which penetrates the cavity of the sensillum to its tip in the case of the cones and hairs[125] or ends on the delicate marginal membrane in the case of the pore-plates.[137] There are sensory organs of varying form on the cephalic lobes of fly maggots, which are almost certainly olfactory.[20]

The evidence that the thin-walled sensilla are in fact the chemoreceptors is for the most part indirect. In Lepidoptera the sensilla styloconica and basiconica are most numerous on the antennae of the male.[19] Among Muscids the rods in the so-called olfactory pits average 820 in dung-feeding species, 494 in flower-haunting forms which orientate themselves in part by sight, and they are more numerous in the males than the females.[83] During flight these flies erect the antennae so as to expose these pits most effectively to the air (Fig. 185): in *Drosophila* the sensitivity to odours is reduced if the antennae are fastened down—they will respond to a given scent at 5 cm., when normal insects with free antennae under the same conditions will respond at 22–35 cm.[48] The elongated pore-plates in *Habrobracon* are much more conspicuous in the male.[94] In *Dytiscus* the distribution of the organs of smell on the antennae and palps has been determined by eliminating the various segments; this distribution agrees closely with that of the 'goblet-shaped organs' which approximate to the pore-plates of Hymenoptera.[118] In the honey-bee the pore-plates are confined to the eight terminal segments of the flagellum (Fig. 188); a bee with only one of these segments remaining can still recognize the scent to which it has been trained and can be trained again to a new scent; whereas this power is lost if this one segment is removed. Bees in which all the sense organs except a group of pore-plates on one antennal segment have been extirpated still show slight olfactory responses.[53, 54] Examination of the fine structure of the plate organs of the honey-bee has cast doubt on their olfactory function; and it has been pointed out that the thin-walled hairs (Fig. 187, B) have the same distribution as the pore-plates and in their fine structure they closely resemble the thin-walled pegs of the grasshopper which are known to be olfactory.[272] (But see p. 291.) The louse *Pediculus* no longer responds to odours if the little group of thin-walled sensilla at the tip of the antenna (Fig. 192) is covered over.[150] [See p. 297.]

Fine structure of olfactory sensilla—The fine structure of the various thin-walled hairs and pegs on the antennae of insects, as studied with the electron microscope, leaves little doubt that these are olfactory organs. The details vary in different examples but in all that have been examined the thin cuticle has a sieve-like structure: it is pierced by a great number of minute pores. The distal tips of the dendrites, after branching repeatedly within the lumen of the hair or peg, end in these pores and are freely exposed to the air. That has been observed in *Melanoplus* (Fig. 189),[271] *Apis mellifera*,[272] *Aedes aegypti*,[272] *Lygaeus kalmii*.[272] In the large thin-walled pegs of *Lygaeus* the dendrites from a large group of neurones enter a set of cuticular sheaths invaginated from the wall of the peg at its base; each dendrite acquires the structure of a cilium on entering the sheath; it then loses the ciliary structure, emerges into the lumen of the peg and branches as described above to end in pores with external openings of about 100 Å diameter.[272] [See p. 297.]

In the antennal sense organ of the *Calliphora* larva the apical dome is a chemoreceptor. It is provided with 21 bipolar sense cells, the distal processes of which break up into groups which divide repeatedly until their finest extremities pierce the epicuticle by way of some 36,000 canals with a diameter of about 200 Å. Thus the endings of the sensory processes are in direct contact with the environment; in sum they occupy about 5 per cent. of the total surface of the dome.[259]

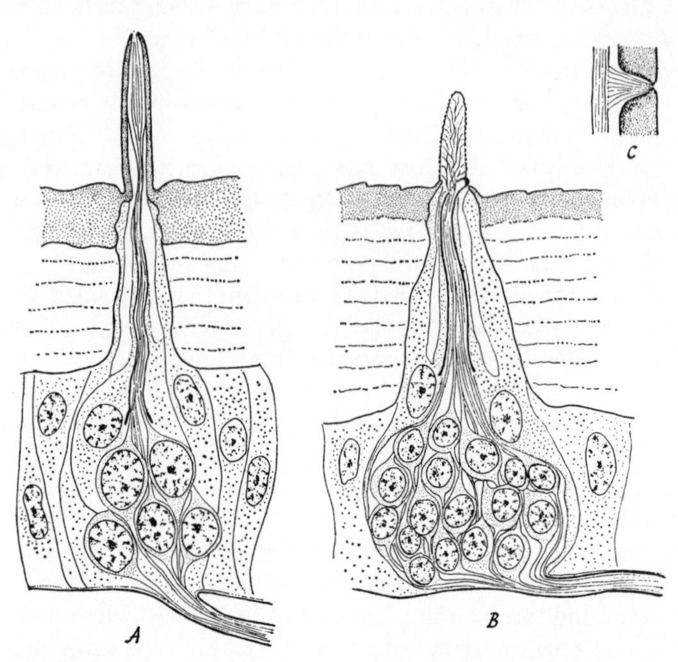

FIG. 189.—A, section of thick-walled, basiconic sensory peg of *Melanoplus*. The dendrites of the sense cells enter a funnel-like cuticular sheath and end in an apical pore. B, section of short, thin-walled, basiconic sensillum of *Melanoplus*. The numerous dendrites in the cuticular sheath end at the base of the sensillum but they give off fine branches which end in pores over the surface of the organ (*after* SLIFER, PRESTAGE, and BEAMS). C, detail of the finest branches coming off a dendrite and converging upon a pore in the thin cuticle of an olfactory sensillum in *Oncopeltus* (*after* SLIFER and SEKHON)

Electrical responses in olfactory receptors—When an odorous air current impinges on the antenna of *Bombyx*, the distal part of the flagellum becomes electrically negative to the basal part ('electroantennogram').[266] Isolated antennae of male *Bombyx* show some spontaneous electrical activity in the chemoreceptors when at rest; but on stimulation with a concentrated extract of the female sex attractant they show a striking increase in the frequency of discharge, which is reversibly suppressed by ether or chloroform. No such effects are given by the antenna of the female.[265] Electrical records obtained from single basiconic sensilla of *Necrophorus*, *Tenebrio*, and *Sphinx pinastri* exposed to different stimulants have shown that qualitative and quantitative differences can be produced in single olfactory units. These differences might be sufficient to enable the central nervous system to distinguish quite a large number of different odours.[267]

The olfactory receptors in the antennal clubs of *Necrophorus* when exposed to the odour of putrid meat give rise to electrical responses which can be recorded with microelectrodes. There is a receptor potential set up in the sense organs and a discharge of nerve impulses. Fig. 190 shows a typical example. The receptor potential shows a rapid maximum, falls to perhaps 50 per cent. and persists. The discharge of nerve impulses may be of two types: (*a*) 'Phasic-tonic', with a rapid initial rate of discharge on stimulation, falling quickly to a persistent lower rate of discharge; (*b*) 'phasic', with a brief rapid discharge and then no further impulses. The concentrated odour of putrid flesh causes persistent responses for at least half an hour.[176]

Electrophysiological evidence suggests that the pore-plates in the antenna of

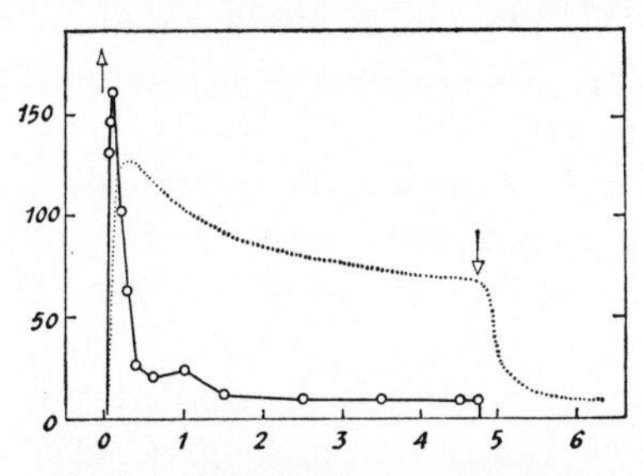

FIG. 190.—Phasic-tonic response to the odour of putrid flesh in a single sensillum on the antenna of *Necrophorus* (*after* BOECKH)

The arrows mark the onset and ending of stimulation. Continuous line: rate of potential discharge in impulses per second (left-hand ordinate). Dotted line: generator potential in relative units (right-hand ordinate). Abscissa: time in seconds.

the honey-bee are indeed olfactory organs. A single pore-plate will respond to odours in a phasic-tonic fashion. It will not respond to light, sound, humidity or carbon dioxide. No response to odour could be obtained from the other types of sensillum and responses are invariably obtained from regions on the antennae in worker bees where plate-organs alone are present.[233] The overlying cuticle of the plate-organs on the antennae of *Megoura* and other Aphids is very thin and is perforated by small openings into which the dendrites of the sense cells penetrate.[272] [See p. 297.]

TEMPERATURE AND HUMIDITY

Temperature sense—All insects are sensitive to high temperatures and avoid places which are unduly hot. In grasshoppers this sensitivity is distributed over the entire body, but is most marked in the proximal half of the antennae and on the pulvilli and tarsi of the hind and fore legs.[57] In the arolium and pulvilli of *Periplaneta* there is a temperature-sensitive region which becomes more active when the temperature is lowered below 13° C.; indeed it actively increases with the fall in temperature: it appears to be a 'cold receptor'. Over the critical range below 13° C. it is sensitive to a drop in temperature of 1° C.[231] *Liogryllus* perceives thermal stimuli chiefly with

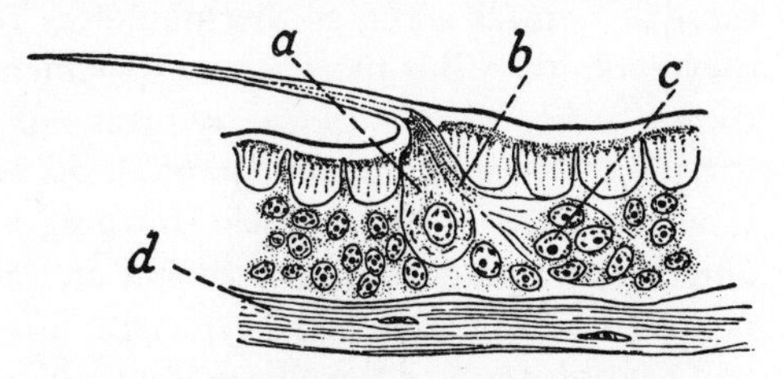

FIG. 191.—Sensillum on antenna of *Rhodnius* (*cf.* Fig. 158, *c*); perhaps temperature receptor (*after* WIGGLESWORTH and GILLETT)

a, trichogen or glandular cell; *b*, bundle of terminal filaments from sense cells; *c*, fusiform group of sense cells numbering about 5; *d*, nerve.

the antennae but also with cerci, wings, abdomen, &c.[69] The avoidance of high temperature by the bugs *Pyrrhocoris* and *Lygaeus* is greatly upset if the last segment alone of the antenna is removed.[69] The antennae of Phasmids are sensitive to excessively high temperatures;[27] and Collembola perceive warmth chiefly through the antennae.[127] [See p. 298.]

Bees can be trained to come to warm places; they can remember a temperature difference as small as 2° C. The thermotactile receptors are mainly, but not exclusively, in the antennae. No evidence of orientation of bees to radiant heat could be obtained (see p. 326).[219]

In some insects which suck the blood of mammals the perception of warmth by the antennae is an important factor in the location of the host. The mosquito *Anopheles* will attempt to probe warmed glass;[86] the lice *Pediculus*[52, 88, 150] and *Haematopinus*[144] respond to warmth as well as to smell, and will follow a warm glass rod in all directions; the sense is distributed over the body but its chief site is the antenna. The blood-sucking bugs *Cimex*,[115, 124] *Triatoma*,[64] and *Rhodnius*,[151] and the flies *Stomoxys* and *Lyperosia*[78] are attracted to their host by warmth as well as smell and will probe warmed cotton wool or a tube of warm water. Female *Anopheles* will attempt to probe a warm host even when the proboscis is removed and all the sense organs on the head are lacking. It appears that the hind legs are utilized either for the direct perception of warmth or possibly of convection currents.[166]

In most of these insects it is doubtful whether there are any special sensory endings which are stimulated by high temperature; but in an insect such as *Rhodnius*, in which the antennae are exceedingly sensitive to very slight differences in the temperature of the air (*Cimex* orientates itself to a warm tube 1° C. above the surroundings from a distance of 1 cm.[124]), it is certain that specialized receptors must exist. The antennae of *Rhodnius*, besides the long tactile hairs and the thin-walled trichoid sensilla which are presumably olfactory, bear a vast number of very fine but relatively thick-walled trichoid sensilla, each provided with a group of about 6 sense cells; these are probably the temperature receptors (Figs. 158, *c*; 191).[151] Similar fine hair sensilla appear to be responsible also in *Pyrrhocoris*.[210]

Perception of radiant heat—It has been suggested that some of the sensory pits or tubular cavities on the antennae of insects may serve as dielectric wave guides and resonators for infra-red rays of particular wave-length bands;[159] or that these structures serve as thermometers responding to the expansion of the air they contain.[163] But there is no experimental evidence for these hypotheses. The thermoreceptors of *Locusta*, by means of which it orientates itself in relation to the radiant heat of the sun (p. 671), do not have this structure. They consist of localized areas of thin cuticle, forming a crescent around the insertion of the antenna and patches on the thorax and abdomen, beneath which are richly innervated cells of an unusual type with interdigitating processes.[170] Although only half the thickness of the adjacent cuticle, the 'fenestrae' of grasshoppers transmit less infra-red radiation; they may act like an imperfect black body producing a heated surface for the thermosensitive receptors.[193] There is, however, some doubt as to whether these structures really are heat sensitive.[240] If Pentatomids or Coreids at a low temperature are illuminated from above they stop movement and direct their backs to the light, apparently with the object of obtaining radiant heat.[161]

Humidity sense—Some insects certainly choose their resting places according to the humidity (p. 312) but little is known of the sensory mechanism concerned. *Sminthurus* and other Collembola are very strongly 'hydrotactic'; they seek out moist places, perceiving not merely the humidity of the ground but of the air.[127] The earwig avoids humid resting places, the chief site of the moisture receptors being apparently the ventral surface of the abdomen.[149] The beetle *Ptinus tectus* shows reactions to humidity which vary with its state of desiccation.[17] The honey-bee,[71] and the blood-sucking Muscids *Stomoxys* and *Lyperosia*,[78] are able to perceive water from a distance, perhaps by means of special receptors in the antennae. Blow-flies (*Phormia*) certainly detect moisture with the antennae; if these are removed they still react normally to stimuli of sight and taste but not to odours or aqueous vapour.[126] *Drosophila* of the mutant form 'antennaless' show a negative reaction to moist air, whereas normal flies show a preference for about 90 per cent. relative humidity. Perhaps the thin-walled organs on the antennae normally produce a positive response, while organs elsewhere produce an antagonistic negative response to high humidity.[16] The mosquito *Culex fatigans* avoids high humidities above 95 per cent. relative humidity. Over this part of the humidity scale it is sensitive to differences of 1 per cent. R.H.; whereas between 30 and 85 per cent. R.H., humidity differences of 40 per cent. cause no response.[129]

The adult *Tenebrio* likewise avoids a high humidity, the intensity of the reaction increasing rapidly as the humidity approaches saturation. This response appears to be to relative humidity rather than saturation deficiency—perhaps because it is the hygroscopic substance in the wall of the sensillum which is being affected.* This reaction is abolished by removal of the antennae. Five types of sensillum are present, of which the 'pit peg organs' and perhaps others are probably the hygroreceptors.[100] In adult beetles of many kinds the intensity of the response to dry air or to moist air after operations to the antennae is correlated with the distribution of the thin-walled sensilla of chemoreceptive type.[167] In *Tenebrio*, *Dermestes*, *Blattella*, and *Aëdes*, the humidity receptors appear to be thin-walled (trichoid or basiconic) sensilla;[264] and in *Melanoplus* the coeloconic sensilla seem to be the receptors: the humidity reaction is eliminated if they are removed.[260]

In *Pediculus*, by covering different parts of the antenna, it has been possible to prove that the humidity receptors are four 'tuft organs' each bearing four tiny hairs (Fig. 192). The louse with these organs exposed usually avoids high humidities; but it tends to become adapted to a given humidity and may then avoid a change in either direction.[150] The larva of *Musca* during the feeding stage, if offered two conditions of humidity, always avoids the drier, the sensitivity of the response increasing as the humidity approaches saturation. This response is to relative humidity and not to saturation deficiency, suggesting that the receptors act hygroscopically. The probable receptors bear some slight resemblance to those of *Pediculus* and consist of three pairs of sense organs on the ventral surface of the thorax, each organ bearing three minute hairs.[160] The same applies to the tufted organs of the *Drosophila* larva.[174]

Wireworms, larvae of *Agriotes*, avoid low humidities. Even when offered a choice between 100 per cent. R.H. and 99·5 per cent. R.H. a significant response

* The uptake of moisture by hygroscopic substances is proportional to the relative humidity of the air.

is obtained. As in other insects, sensitivity is much greater in moist air: the reaction is very strong between 80 and 90 per cent.; there is no reaction between 60 and 70 per cent. or lower. The sensilla concerned are distributed over antennae, maxillary and labial palps. No specific humidity receptor has been recognized. Perhaps it is the evaporation from the various thin-walled receptors

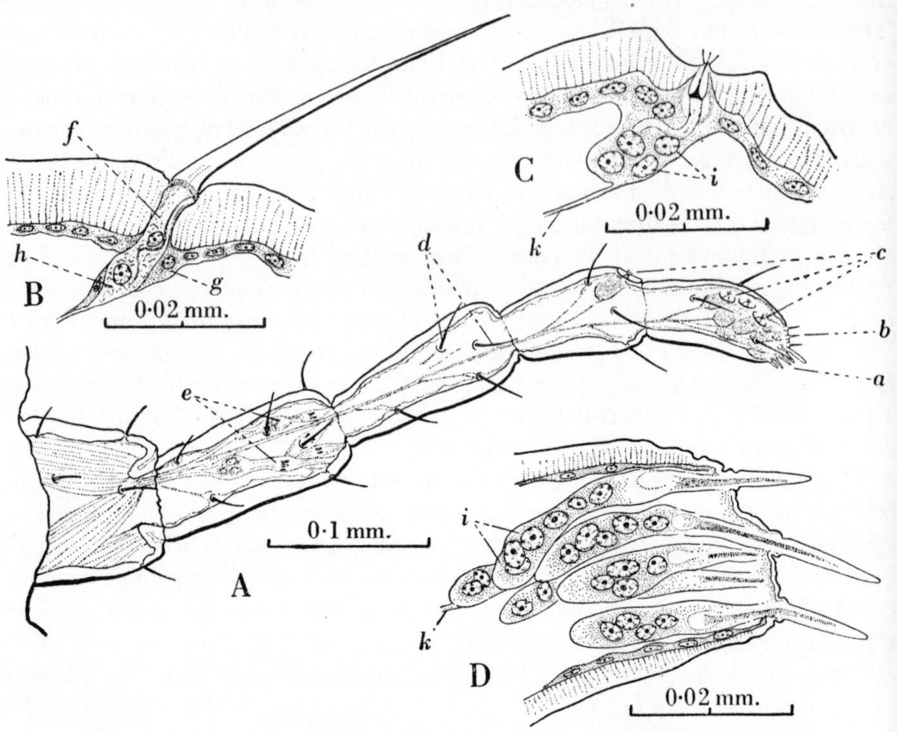

FIG. 192.—Sense organs on the antenna of *Pediculus* (*after* WIGGLESWORTH)

A, general view of antenna. *a*, peg organs with rounded tips; *b*, peg organs with sharp tips; *c*, tuft organs; *d*, tactile hairs; *e*, scolopidial organs. *B*, detail of tactile hair. *f*, trichogen cell; *g*, tormogen cell; *h*, sense cell. *C*, detail of tuft organ (humidity receptor). *D*, detail of peg organs (olfactory receptor). *i*, sense cells, *k*, nerve.

on the head appendages which initiates the humidity response. An 'evaporimeter' type of receptor of this kind could produce an avoidance only of dry air, whereas the 'hygrometer' type, which seems to occur in *Tenebrio*, *Culex*, and *Pediculus*, might react in both directions.[82] Almost all species of ants when placed in a humidity gradient choose a saturated atmosphere.[61] The tsetse-fly *Glossina* has humidity receptors in the spiracular filters on the thorax, which regulate the level of activity of the fly.[182] [See p. 298.]

ADDENDA TO CHAPTER VII

p. 258. **Tactile hairs**—The tactile hairs on the head of *Locusta* are closely invested by the socket. Deformation of the hair stretches the membrane joining the hair to the cuticle; the elastic force from a given degree of deflection being the same in all directions.[340] The sensitivity of the hairs on the cerci of *Peri-*

planeta is greatly influenced by the direction in which the hair is bent. The ultimate receptor appears to be an assembly of microtubules within the base of the cuticular hair.[324] Reflex stimulation from hair sensilla in the tarsi[333] and the chordotonal organs in the femur[343] of *Schistocerca* are important in both postural and walking activity. Hair sensilla on the surface of the compound eyes are concerned in the regulation of flight in honey-bees – along with Johnston's organ in the antennae (p. 262).[323] *Notonecta,* like *Gerris,* locates its prey by the detection of surface waves in the water, perceived by trichoid sensilla on the legs.[318]

p. 260. **Campaniform organs**—The fine structure of campaniform organs has been described from the halteres of *Drosophila.*[302] In hair sensilla, campaniform organs, and chordotonal organs the effective stimulus seems to be the compression of the numerous microtubules which are bound together to form a 'tubular body'.[339, 342] These microtubules may be involved in the transduction of the stress stimulus into nerve impulse.

Tactile spines on the legs of *Periplaneta* have a single campaniform sensillum in the thick cuticular wall where it joins the soft cuticle of the socket. It is probably stimulated by pressure of the cap of the sensillum against the lip of the socket.[301]

p. 262. **Chordotonal organs**—In Plecoptera the subgenual organs are responsible for detecting the vibrations of the substratum set up by the drumming which leads to the meeting of the sexes.[334]

p. 263. **Johnston's organ**—Further evidence has been obtained for the role of Johnston's organ in the detection of waves and menisci in the water surface, notably in *Gyrinus.*[332]

p. 265. **Stretch receptors**—In *Periplaneta* there are multipolar stretch receptors near the joints of the legs which probably play an important part in the leg reflexes.[307] In *Blaberus* there are mechanoreceptors of unusual type in the ventral mid-line of most of the abdominal segments. These respond in a phasic manner to many mechanical stimuli. They may well act as ventilation receptors.[312] Stretch receptors consisting of multiterminal neurones with dendrites desheathed at their tips are present in the cibarial pump of *Calliphora.* They monitor and perhaps control, the rate of cibarial pumping.[328] The muscle receptors in caterpillars convey the same detailed information as the muscle spindles of vertebrates. A major function of these stretch receptors is to mediate a negative feedback reflex which tends to stabilize the position of the body irrespective of the load.[344]

p. 266. **Sense of gravity**—Honey-bees can be trained to keep a fixed angle of 60° to gravity on their way to a food source; a perfect menotactic orientation (p. 332) of this kind is possible only if the hair cushions at the neck, and either the petiole or the coxal joint, are fully functional.[316]

p. 270. **Antennae and palps as auditory organs**—Leaf-hoppers, *Oncopsis,* have complex courtship songs which resemble those of cicadas (p. 278) although they lack hearing organs with tympanal membranes. But the slender antennal flagellum is inserted on a radial membrane in the pedicel. This may well be a sound receptor in which the flagellar movements are detected by Johnson's organ.[309] In *Celerio* and other Choerocampine Sphingid moths the bulbous labial palp serves as an auditory organ responsive in the 15 kHz range (like Noctuid tympanic organs responding to bat cries, p. 276). Apparently the entire palp vibrates in response to the sound.[331]

p. 273. **Sound location**—There is evidence that Noctuid moths can locate a sound source in three dimensional space by analysis of the changing intensity of the sound in the two ears during the wing-beat cycle.[326] The turning response evoked by such stimuli has been analysed in detail in the laboratory.[330]

p. 274. **Auditory range**—In *Locusta* an auditory centre has been located at the border of proto- and deutocerebrum. Its sensitivity to ultra-sound clearly has biological significance, for *Locusta migratoria* practises 'mandible snapping' which produces wave frequencies of 10,000–30,000 cycles per second with a duration sometimes less than 0·5 msec.[294] *Chrysopa* can perceive ultra-sonic frequencies up to at least 100 kilocycles per second modulated at a pulse rate of 150 per second. The receptor site is a pair of small bilateral swellings in a vein of the forewings.[320]

p. 276. **Frequency discrimination**—Locusts can discriminate sound frequency to some degree:[346] in *Locusta* some central units in the auditory centre are sensitive only to the 4–8 kcs range, others only to the 12–30 kilocycles per second range.[295] In *Schistocerca* there are different areas of the tympanum with different resonances which are picked up by the attached receptor cells; they can make an approximate estimate of frequency with very few neurones.[319] The Tettigoniid *Homorocoryphus* is most sensitive to 30 kilocycles per second whereas the dominant song frequency is 15 kilocycles per second. The ear may provide a warning system.[315] In some butterflies the inflated sac-like structures at the base of the wings are responsive to low frequency sounds.[341]

p. 278. **Sound production**—In *Homorocoryphus* experimental sound production by isolated tegmina has shown that the area of the right tegmen responsible for the radiation of sound is the mirror frame, the vein enclosing the mirror membrane. When the tooth impact approaches 12–15 kilocycles per second the mirror frame is thrown into resonant vibration and the characteristic pure tone of 15 kilocycles per second is emitted.[297] Mole crickets, *Gryllotalpa* spp., construct horn-shaped burrows which direct and concentrate the sound.[298] Besides the myogenic oscillation of the tymbal muscles, two other ways of obtaining a high frequency of clicks are used by cicadas: (i) while activated neurogenically with a one-to-one relation between nerve impulse and muscle contraction, the tymbal muscles on the two sides may contract alternately; (ii) the tymbals may be so constructed that they produce a *series* of clicks as they go from the resting to the fully distorted state.[296]

p. 279. In flies and bees the sounds emitted contain a mixture of vibrations of wing-beat frequency with high frequency sounds probably due to skeletal vibration.[305] *Heliothis* in flight emits ultra-sonic bursts at 50 kilocycles per second every 25 msec. These emissions are synchronized with the wing beat but are not invariably produced. It has been suggested that these pulses may be used for the detection of objects by echolocation or possibly for assembling of the sexes.[311]

p. 279. The stridulation sound produced by the leaf-cutting ant *Atta* is conducted through the ground and then acts as a distress alarm attracting other workers in the colony, which dig where the intensity of vibration is highest. The receptors appear to be campaniform sensilla at the joint between the trochanter and femur.[317]

p. 280. **Taste**—Sense organs of the clypeo-labrum and hypopharynx as well as the palps are responsible for feeding responses in *Schistocerca*.[308]

p. 282. **Origin of receptor potentials**—The mode of action of uncharged molecules (such as glucose) in producing electrical changes in the receptor membrane is not fully understood; but it appears likely that the occupation of receptor sites by the sugar leads to a diffusion of cations which is responsible for the receptor membrane current.[321] Likewise in the 'water' receptor of *Phormia* the receptor potential is believed to be a streaming potential governed by cations.[327]

p. 283. **Receptors in piercing and sucking insects**—In Aphids, there are sensory axons in each mandibular stylet.[325] In mosquitos cibarial sensilla control the acceptance of blood, sugar, &c.[335]

p. 286. **Molecular basis of chemoreception**—No single molecular property such as chain length, double bonds, or functional groups, is alone responsible for excitation. A combination of such properties seems to be involved.[299]

p. 289. **Odour receptors in Lepidoptera**—In *Danaus* the short thin-walled sensilla, widely distributed on the antennae of the female, alone can receive the male sex pheromone. About 1,000–1,500 of these sensilla (about 5 per cent. of the total) must be exposed for successful courtship. The same sensilla will serve to detect the odour of honey.[322]

p. 289. **Fine structure of olfactory sensilla**—The dendrites of the sense cells do not appear to enter the olfactory pores. These contain extracellular material, formed by the trichogen cell long before the dendrites invade the sense hair.[304] There is some evidence that in male *Antheraea* and *Hyalophora* the olfactory pores may be occupied by a specific receptor protein for the odorous substance, which then diffuses into the sensillum to stimulate the dendritic endings.[329] At the moulting of thick-walled chemoreceptors the distal parts of the dendrites are shed along with the old cuticle—as they are in tactile sensilla (p. 256).[338]

p. 291. **Specialized sensory neurones**—A sense organ may either increase or decrease the rate of impulse discharge in response to a stimulus. Excitation is associated with a negative (depolarization) receptor potential; inhibition is correlated with a positive (hyperpolarization) receptor potential. In olfactory receptors, of three neighbouring cells with their dendrites ending at the open pores of the same olfactory hair, one may be excited (+), one inhibited (−) and one not affected (0) by the same substance. An individual cell exposed to different chemicals shows a reproducible +, − or 0 pattern. In many insects there are specialized cells which respond only to single substances, for example, sex attractants ('Bombycol' in silkworm, 'Queen substance' in honey-bee) or food odours (e.g. carrion in burying beetles, &c.). A Bombycol receptor needs about 100 molecular strikes per second for a significant increase in impulses, but in the intact male moth a behaviour response (p. 286) can be elicited if 40 out of 40,000 specialized receptor cells receive one Bombycol strike per second.[300] In male *Antheraea* only a limited number of the sense cells of the trichoid sensilla respond to the female sex attractant and none of the basiconic sensilla, all of which are of the generalist type.[336] Only one of the sense cells of the lateral styloconic sensilla of the *Pieris* larva responds to mustard oil glucosides; a second cell is stimulated by sucrose and a third by amino acids.[337] Specialized neurones exist in the olfactory lobe of *Periplaneta* which respond specifically to the sex pheromone.[345] In *Apis* two specialized types of receptor cells could be recognized in the pore plates of the antennae; one responds solely to the queen substance (9-oxodecenoic acid), the other to the scent of the Nassonoff gland.[310]

p. 292. **Cold receptor**—The highly sensitive 'cold receptor' of *Periplaneta* consists of a ring-shaped structure, about 8μ in diameter, with a short hair in the centre.[314]

p. 294. **Humidity and temperature receptors**—In addition to olfactory receptors the antennae of caterpillars (*Manduca*, *Pieris*, &c.) carry temperature receptors which increase their rate of discharge as temperature falls, and humidity receptors which increase their rate of firing in moist air. They can thus get information about the turgidity of a leaf and thus its quality as food.[303] In the antennae of *Apis* some of the pit pegs are carbon dioxide receptors, some are hygroreceptors, and some are cold receptors.[313]

REFERENCES

1 ABBOTT, C. E. *Psyche*, **34** (1927), 129–33 (reaction of caterpillars to sounds).

2 ABRAHAM, G. R. *Z. vergl. Physiol.*, **30** (1944), 321–42 (chemoreceptors on fore legs: *Nepa*, &c., Hem.).

3 v. ADELUNG, N. *Z. wiss. Zool.*, **54** (1892), 316–49 (tympanal organs in tibiae of Tettigoniidae).

4 ANDERSON, A. L. *J. Exp. Zool.*, **63** (1932), 235–59 (sensitivity of tarsi to sugars in Lepidoptera).

5 AUGER, D., and FESSARD, A. *C.R. Soc. Biol.*, **99** (1928), 400–1 (sensitivity of tympanal organ in *Acridium*, Orth.).

6 AUTRUM, H. *Z. vergl. Physiol.*, **23** (1936), 332–73 (production and perception of sounds: ants, &c.).

7 —— *Z. vergl. Physiol.*, **28** (1940), 326–52; 580–637 (sense of hearing and vibration: Tettigoniidae, &c.).

8 —— *Biol. Zbl.*, **63** (1943), 209–36 (physical limits to sensory stimuli).

9 AUTRUM, H., and SCHNEIDER, W. *Z. vergl. Physiol.*, **31** (1948), 77–88 (perception of vibrations by insects).

10 BAIER, L. J. *Zool. Jahrb.*, *Physiol.*, **47** (1930), 151–248 (hearing and stridulation in insects).

11 BARNES, T. C. *Ann. Ent. Soc. Amer.*, **24** (1931), 824–6 (proprioceptors in insects).

12 BARROWS, W. M. *J. Exp. Zool.*, **4** (1907), 515–37 (reactions of *Drosophila*, Dipt. to odours, &c.).

13 BAUER, L. *Z. vergl. Physiol.*, **26** (1938), 107–20 (sense of taste: Dytiscidae).

14 BAUNACKE, W. *Zool. Jahrb.*, *Anat.*, **34** (1912), 179–346 (static organs in Nepidae, Hem.).

15 BAUS, A. *Z. Morph. Oekol. Tiere*, **32** (1936), 1–46 (campaniform organs in insects with reduced wings).

16 BEGG, M., and HOGBEN, L. *Proc. Roy. Soc.* B, **133** (1946), 1–19 (chemoreceptors in *Drosophila*).

17 BENTLEY, E. W. *J. Exp. Biol.*, **20** (1944), 152–8 (humidity reactions: *Ptinus*, Col.).

18 BERLESE, A. *Gli Insetti*, Vol. 1, Milan, 1909.

19 BOHM, L. K. *Arb. Zool. Inst. Wien*, **19** (1911), 219–46 (antennal sense organs of Lepidoptera).

20 BOLWIG, N. *Vidensk. Medd. Dansk Naturhist. Foren.*, **109** (1946), 81–207 (sense organs: *Musca* larva).

21 BRAUNS, A. *Zool. Jahrb.*; *Abt. Physiol.*, **59** (1939), 245–390 (halteres).

22 BRECHER, G. *Z. vergl. Physiol.*, **10** (1929), 495–526 (sense organs and orientation in *Periplaneta*, Orth.).

23 v. BUDDENBROCK, W. *Grundriss der vergleichenden Physiologie*, Borntraeger, Berlin, 1937.

24 BUGNION, E., and POPOFF, N. *Arch. Zool.*, **7** (1911), 643–74 (possible organ of taste in Hemiptera).

[25] BUTLER, C. G. *J. Exp. Biol.*, **17** (1940), 253–61 (taste in *Apis*).

[26] CAMERON, A. E. *Trans. R. Soc. Edin.*, **58** (1934), 211–50 (Graber's organ: Tabanidae larvae).

[27] CAPPE DE BAILLON, P. *Bull. Biol. Fr. Belg.*, **70** (1936), 1–35 (antennal organ of unknown function in Phasmidae).

[28] CHEN, S. H., and YOUNG, B. *Sinensia*, **14** (1943), 45–8 (mating behaviour: *Bombyx*).

[29] CROMBIE, A. C., and DARRAH, J. H. *J. Exp. Biol.*, **24** (1947), 95–109 (chemoreceptors: *Agriotes* larva, Col.).

[30] CROW, S. *Physiol. Zool.*, **5** (1932), 16–35 (sensitivity of tarsi to sucrose in Calliphoridae).

[31] DEBAISIEUX, P. *Ann. Soc. Sci. Bruxelles*, **54** (1934), 86–91; 338–45; *La Cellule*, **44** (1935), 273–314; **47** (1938), 79–202 (scolopidial organs in insects).

[32] DEBAUCHE, H. *La Cellule*, **44** (1935), 45–83 (chordotonal organs in antenna: *Hydropsyche*, Trichopt.).

[33] DEMOLL, R. *Die Sinnesorgane der Arthropoden ihr Bau und ihre Funktion*, Braunschweig, 1917.

[34] DEONIER, C. C., and RICHARDSON, C. H. *Ann. Ent. Soc. Amer.*, **28** (1935), 467–74 (tarsal chemoreceptors in *Musca*, Dipt.).

[35] DETHIER, V. G. *Biol. Bull.*, **72** (1937), 7–23; **76** (1939), 325–9; **80** (1941), 403–14; *Amer. Nat.*, **75** (1941), 61–73 (taste and smell in caterpillars).

[36] —— *J. N.Y. Ent. Soc.*, **55** (1947), 285–93 (chemoreceptors in carrion beetles).

[37] —— *J. Exp. Zool.*, **105** (1947), 199–207 (oviposition of *Nemeritis*; chemoreceptors).

[38] DETHIER, V. G., and CHADWICK, L. E. *J. Gen. Physiol.*, **30** (1947), 247–53; *Physiol. Rev.*, **28** (1948), 220–54 (chemoreceptors in insects).

[39] DIAKONOFF, A. *Arch. néerl. Physiol.*, **21** (1936), 104–29 (tactile hairs as proprioceptors: *Periplaneta*, Orth.).

[40] EASTHAM, L. E. S. *Trans. R. Ent. Soc. Lond.*, **85** (1936), 401–14 (campaniform organs on gills of *Caenis* nymphs, Ephem.).

[41] EGER, H. *Biol. Zbl.*, **57** (1937), 293–308 (sense of taste in caterpillars).

[42] EGGERS, F. *Zool. Anz.*, **57** (1923), 224–40; *Z. Morph. Oekol. Tiere*, **2** (1924), 259–349 (chordotonal and Johnston's organs).

[43] —— *Z. vergl. Physiol.*, **2** (1925), 297–314; *Zool. Bausteine*, **2** (1928) (hearing in Lepidoptera).

[44] —— *Zool. Anz.*, **68** (1926), 184–92 (function of Johnston's organ in *Gyrinus*, Col.).

[45] EHNBOM, K. *Opusc. Entom. Suppl.*, **8** (1948), 1–162 ("Eltringham's organ": Lep.).

[46] ELTRINGHAM, H. *The Senses of Insects*, Methuen, London, 1933.

[47] —— *Trans. Ent. Soc. Lond.*, **81** (1933), 33–6 (tarsal sense organs of Lepidoptera).

[48] FLÜGGE, C. *Z. vergl. Physiol.*, **20** (1934), 463–500 (orientation by smell in *Drosophila*, Dipt.).

[49] FOREL, A. *Das Sinnesleben der Insekten*, Munich, 1910.

[50] FRAENKEL, G., and PRINGLE, J. W. S. *Nature*, **141** (1938), 919–20 (function of halteres).

[51] FREILING, H. H. *Z. wiss. Zool.*, **92** (1909), 210–90 (scent-producing organs and sense organs on the wings in Lepidoptera).

[52] FRICKLINGER, H. W. *Z. angew. Entom.*, **3** (1916), 263–81 (sense of smell and temperature in *Pediculus*, Anoplura).

[53] FRINGS, H., *et al.* Taste and smell in insects. *J. Exp. Zool.*, **88** (1941), 65–93 (*Cynomyia*, Dipt.); *Ibid.*, **97** (1944), 123–34 (*Apis*, Hym.); *Ibid.*, **102** (1946), 23–50 (*Periplaneta*); *Ibid.*, **103** (1946), 61–79 (*Tabanus*, Dipt.); *Biol. Bull.*, **88** (1945), 37–43 (*Platysamia Lep.*); *J. Comp. Physiol. Psychol.*, **41** (1948), 25–34 (general).

[54] v. FRISCH, K. *Zool. Jahrb., Physiol.*, **37** (1920), 1–238; **38** (1921), 449–514; *Handbuch norm. path. Physiol.*, **11** (1926), *Receptions-organe* i, 223–31 (sense of smell in honey-bee).

[55] —— *Z. vergl. Physiol.*, **21** (1934), 1–156 (sense of taste in honey-bee).

[56] GAFFRON, M. *Z. vergl. Physiol.*, **20** (1934), 299–337 (perception of movements: Muscidae, Odonata larvae).

[57] GEIST, R. M. *Ann. Ent. Soc. Amer.*, **21** (1928), 614–18 (heat-sensitive areas in grasshoppers).

[58] GLASER, R. W. *Psyche*, **34** (1927), 209–15 (sense of smell in antennae of cockroach, *Periplaneta*).

[59] GOHRBANDT, I. *Z. wiss. Zool.*, **149** (1937), 537–600 (tympanal organs: Drepanidae and Cymatophoridae, Lep.).

[60] —— *Z. wiss. Zool.*, **151** (1938), 1–21 (tympanal organs in Noctuidae with reduced wings).

[61] GÖSSWALD, K. *Z. wiss. Zool.*, **154** (1941), 247–72 (humidity responses in ants).

[62] GRABER, V. *Arch. Mikr. Anat.*, **20** (1882), 506–640; **21** (1882), 65–145 (chordotonal organs and hearing).

[63] HANSTRÖM, B. *Z. vergl. Physiol.*, **4** (1926), 528–44 (senses of taste and smell).

[64] HASE, A. *Z. Parasitenk.*, **4** (1932), 585–652 (stimuli to feeding in *Triatoma*, &c., Hem.).

[65] HASLINGER, F. *Z. vergl. Physiol.*, **22** (1935), 614–40 (sense of taste and utilization of sugars in *Calliphora*, Dipt.).

[66] HARTWELL, R. A. *Ann. Ent. Soc. Amer.*, **17** (1924), 131–62 (sense of smell in termites).

[67] HAYES, W. P., and LIU, Y. S. *Ann. Ent. Soc. Amer.*, **40** (1947), 401–6 (tarsal chemoreceptors: *Musca*, Dipt.).

[68] HEITMANN, H. *Zool. Jahrb., Anat.*, **59** (1934), 135–200 (tympanal organs in Lepidoptera).

[69] HERTER, K. *Biol. Zbl.*, **43** (1923), 282–5; *Z. vergl. Physiol.*, **1** (1924), 221–88; *Handb. norm. path. Physiol.*, **11** (1926), *Receptions-organe* i, 173–9 (sense of temperature in *Acheta* (Orth.), *Formica* (Hym.), *Pyrrhocoris* (Hem.), &c.).

[70] HERTWECK, H. *Z. wiss. Zool.*, 139 (1931), 559–663 (sense organs in *Drosophila* larva and adult).

[71] HERTZ, M. *Z. vergl. Physiol.*, **21** (1934), 463–7 (detection of water at a distance by the honey-bee).

[72] HEYMONS, R. *Biol. Zbl.*, **47** (1927), 164–87 (antennae as organs of smell in *Scarabaeus*, Col.).

[73] HOCHREUTHER, R. *Z. wiss. Zool.*, **103** (1912), 1–114 (sensilla in *Dytiscus*, Col.).

[74] HOLLICK, F. S. J. *Phil. Trans. Roy. Soc.*, B, **230** (1940), 357–90 (antennae as receptors during flight: *Muscina*, Dipt.).

[75] HOLMGREN, E. *Anat. Anz.*, **12** (1896), 449–57 (cutaneous nervous system of insects).

[76] IMMS, A. D. *Textbook of Entomology*, Methuen, London, 1935.

[77] v. KENNEL, J., and EGGERS, F. *Zool. Jahrb., Anat.*, **57** (1933), 1–104 (abdominal tympanal organs in Lepidoptera).

[78] KRIJGSMAN, B. J., *et al. Z. vergl. Physiol.*, **11** (1930), 702–29 (stimuli to feeding in *Stomoxys*, Dipt.); *Ibid.*, **13** (1930), 61–73 (ditto in *Lyperosia*, Dipt.).

[79] KRÖNING, F. *Naturwiss.*, **18** (1930), 380–7 (hearing in insects: review).

[80] KUNZE, G. *Zool. Jahrb., Physiol.*, **52** (1933), 465–512 (sense of taste: antennae of honey-bee, &c., palps of *Liogryllus*, &c.).

[81] LEES, A. D. *Nature*, **150** (1942), 375 (campaniform organs: *Drosophila*).

[82] —— *J. Exp. Biol.*, **20** (1943), 43–60 (reactions to temperature and humidity: *Agriotes* larva, Col.).

[83] LIEBERMANN, A. *Z. Morph. Oekol. Tiere*, **5** (1926), 1–97 (antennal organs of smell in Muscidae).

[84] MCINDOO, N. E. *J. Comp. Neurol.*, **34** (1922), 173–99 (sense of hearing in honey-bee).

[85] MARCHAL, P. *Richet's 'Dictionaire de Physiologie'*, **9** (1910), 273–386.

[86] MARCHAND, W. *Psyche*, **25** (1918), 130–5; *Ent. News*, **31** (1920), 159–65 (temperature and feeding in *Anopheles*, Dipt.).

[87] MARSHALL, J. *J. Exp. Biol.*, **12** (1935), 17–26 (chemoreceptors on antenna and tarsus in honey-bee).

[88] MARTINI, E. *Z. angew. Entom.*, **4** (1917), 34–70 (reactions of *Pediculus* to warmth).

[89] MAYER, A. G. *Psyche*, **9** (1900), 14–20 (sense of smell in mating of *Callosamia*, Lep.).

[90] MAYER, A. M. *Amer. Nat.*, 8 (1847), 577 (hearing in mosquitos).

[91] MELIN, D. *Uppsala Univ. Årssk.*, 1941, No. 4, 1–247 (campaniform organs on wings and halteres).

[92] MINNICH, D. E. *J. Exp. Zool.*, **33** (1921), 173–203; **35** (1922), 57–82; **36** (1922), 445–57 (contact chemoreceptors in tarsi of *Pyrameis*, Lep.); *Z. vergl. Physiol.*, **11** (1929), 1–55 (in tarsi of *Calliphora*, Dipt.); *J. Exp. Zool.*, **60** (1931), 121–39 (proboscis of *Calliphora*); **61** (1932), 375–93 (tarsi and antennae of *Apis*, Hym.).

[93] —— *J. Exp. Zool.*, **42** (1925), 443–69; **72** (1936), 439–53 (responses of caterpillars to sounds).

[94] MURR, L. *Z. vergl. Physiol.*, **11** (1930), 210–70 (sense of smell in *Habrobracon*, Hym.).

[94a] NAGEOTTE, J. *C.R. Acad. Sci.*, **215** (1942), 509–10; *Arch. Zool. exp. gén.*, **83** (1944), Notes et Rev., 99–111 (function of halteres).

[95] NASH, T. A. M. *Bull. Ent. Res.*, **21** (1930), 201–56 (biology of *Glossina*, Dipt.).

[96] NEWTON, H. C. F. *Quart. J. Micr. Sci.*, **74** (1931), 647–68 (campaniform organs in honey-bee).

[97] OEVERMANN, H. *Z. wiss. Zool.*, **147** (1936), 595–628 (static perceptions in aquatic bugs).

[98] ORLOV, J. *Z. wiss. Zool.*, **122** (1924), 425–502 (innervation of gut: Lamellicorn larvae, Col.).

[99] PFLUGSTAEDT, H. *Z. wiss. Zool.*, **100** (1912), 1–59 (halteres of Diptera).

[100] PIELOU, D. P., and GUNN, D. L. *J. Exp. Biol.*, **17** (1940), 286–306 (humidity receptors: *Tenebrio*, Col.).

[101] PRINGLE, J. W. S. *J. Exp. Biol.*, **15** (1938), 101–31 (function of campaniform sensilla in cockroach).

[102] —— *Phil. Trans. Roy. Soc.*, B, **233** (1948), 347–84 (sense organs in halteres).

[103] PROCHNOW, O. *Schröder's Handbuch d. Entomologie*, **1** (1910), 61–75 (sound production).

[104] PRÜFFER, J. *Zool. Jahrb., Anat.*, **51** (1929), 1–46; *Zool. Anz.*, **115** (1936), 157–9 (sense organs on antennae of *Saturnia*, Lep.).

[105] PUMPHREY, R. J. *J. Physiol.*, **86** (1936) (*Proc. Physiol. Soc.*, Mar. 14, 1936) (electrical responses in nerves from tactile hairs of cockroach).

[106] —— *Biol. Rev.*, **15** (1940), 107–32 (insect hearing: review).

[107] PUMPHREY, R. J., and RAWDON-SMITH, A. F. *Nature*, **137** (1936), 990; *J. Physiol.*, **87** (1936) (*Proc. Physiol. Soc.*, Mar. 14, 1936); *Proc. Roy. Soc.*, B, **121** (1936), 18–27; *Nature*, **143** (1939), 806 (electrical responses in nerves from organs of hearing: *Periplaneta*, *Gryllus*, *Locusta*).

[108] RÁDL, E. *Biol. Zbl.*, **25** (1905) (chordotonal organs and hearing).

[109] RAIGNIER, A. *'Brotéria' série de Ciências Naturais*, Lisbon, **2** (1933), 51–82 (stridulation and hearing in ants).

[110] V. RATH, O. *Z. wiss. Zool.*, **46** (1888), 413–34; **61** (1896), 499–539 (cutaneous sense organs of insects).

[111] REGEN, J. *Zool. Anz.*, **40** (1912), 305–16; *Arch. ges. Physiol.*, **155** (1913), 193–200; *Sitzgsber. Akad. Wiss. Mat. Nat. Kl. Abt. 1, Wien*, **132** (1923), 81–88 (hearing in *Gryllus*, Orth.).

[112] —— *Sitzgsber. Akad. Wiss. Mat. Nat. Kl. Abt. 1, Wien*, **123** (1914) (hearing in *Thamnotrizon*, Orth.).

[113] REUTER, E. *Zool. Jahrb., Anat.*, **62** (1937), 449–506 (sense organs on wings and elytra: *Calandra*, Col.).

[114] RITTER, E. *Z. vergl. Physiol.*, **23** (1936), 543–70 (sense of taste and smell in *Hydrous*, Col.).

[115] RIVNAY, E. *Parasitology*, **24** (1932), 121–36 (responses of *Cimex*, Hem., to warmth, &c.).

[115a] ROTH, L. M. *Amer. Mid. Nat.*, **40** (1948), 265–352 (hearing in *Aëdes*, Culicidae).

[116] RULAND, F. *Z. wiss. Zool.*, **46** (1888), 602–28 (antennal sense organs).

[117] SALT, G. *Proc. Roy. Soc.*, B, **122** (1937), 57–75 (senses used by *Trichogramma*, Hym., in distinguishing parasitized and unparasitized hosts).

[118] SCHALLER, A. *Z. vergl. Physiol.*, **4** (1926), 370–464 (sense of taste and smell in *Dytiscus*, Col.).

[119] SCHENK, O. *Zool. Jahrb., Anat.*, **17** (1903), 573–618 (sensilla on antennae: Lep. and Hym.).

[120] SCHMIDT, A. *Z. vergl. Physiol.*, **25** (1938), 351–78 (sense of taste in ants).

[121] SCHREMMER, F. *Zool. Jahrb.; Abt. Syst.*, **74** (1941), 375–434 (tarsal chemoreceptors: *Plusia*, Lep.).

[122] SCHWABE, J. *Zoologica*, **20** (1906), No. 6 (tympanal sense organs in Orthoptera: histology).

[123] SIHLER, H. *Zool. Jahrb., Anat.*, **45** (1924), 519–76 (sense organs on cerci: Orth.).

[124] SIOLI, H. *Zool. Jahrb., Physiol.*, **58** (1937), 284–96 (perception of warmth by bed-bug *Cimex*).

[125] SNODGRASS, R. E. *Smithsonian Misc. Coll.*, **77**, No. 8 (1926), 1–80 (sense organs of insects: review).

[126] STEINER, G. *Z. vergl. Physiol.*, **30** (1942), 1–38 (humidity perception: *Phormia*, Dipt.).

[127] STREBEL, O. *Z. Morph. Oekol. Tiere*, **25** (1932), 31–153 (biology of Collembola).

[128] V. STUDNITZ, G. *Zool. Jahrb., Physiol.*, **50** (1932), 313–478 (static organs in larvae of *Limnobiidae*, Dipt.).

[129] THOMSON, R. C. M. *Bull. Ent. Res.*, **29** (1938), 125–40 (response of *Culex fatigans* to temperature and humidity).

[130] THORPE, W. H., and CRISP, D. J. *J. Exp. Biol.*, **24** (1947), 310–28 (pressure receptors: *Nepa*, *Aphelocheirus*, Hem.).

[131] TINBERGEN, L. *Arch. néerl. Zool.*, **4** (1939), 81–92 (contact chemoreceptors: *Calliphora*, Dipt.).

[132] TURNER, C. H., *et al.* *Biol. Bull.*, **27** (1914), 275–93; 325–32 (hearing in *Catocala* and Saturniidae, Lep.).

[133] VALENTINE, J. M. *J. Exp. Zool.*, **58** (1931), 165–220 (sense of smell in *Tenebrio* adult, Col.).

[134] VERSCHAEFFELT, E. *Proc. K. Akad. Wetensch. Amsterdam Ent. Sci.*, **13** (1911), 536–42 (selection of food in phytophagous insects).

[135] VOGEL, R. *Z. wiss. Zool.*, **98** (1911), 68–134 (sense organs on wings of Lepidoptera).

[136] —— *Z. wiss. Zool.*, **100** (1912), 210–44 (chordotonal organs in wings of Lepidoptera).

[137] —— *Zool. Anz.*, **53** (1921), 20–8; *Z. wiss. Zool.*, **120** (1923), 281–324 (sense organs in antennae of wasps and bees).

[138] —— *Z. ges. Anat.*, *Abt. 1*, **67** (1923), 190–231 (tympanal organ in *Cicada*, Hem.).

[139] WACKER, F. *Z. Morph. Oekol. Tiere*, **4** (1925), 739–812 (antennal sense organs in Hymenoptera).

[140] WAHL, O. *Z. vergl. Physiol.*, **24** (1936), 116–42 (acetyl-saccharose as repellent in bee sugar).

[141] WARNKE, G. *Z. vergl. Physiol.*, **14** (1931), 121–99; *Zool. Anz.*, **108** (1934), 217–24 (sense of smell in *Geotrupes*, &c., Col.).

[142] WEBER, H. *Biologie der Hemipteren*, Berlin, 1930.

[143] —— *Lehrbuch der Entomologie*, Jena, 1933.

[144] WEBER, HEINZ. *Z. vergl. Physiol.*, **9** (1929), 564–612 (sensory physiology: *Haematopinus*, Anoplura).

[145] WEINLAND, E. *Z. wiss. Zool.*, **51** (1891) (halteres of Diptera).

[146] WEIS, J. *Z. vergl. Physiol.*, **12** (1930), 206–48 (taste receptors in tarsi of *Pyrameis*, Lep.).

[146a] WEIS-FOGH, T. *Nature*, **164** (1949), 873–4 (aerodynamic sense organ: *Schistocerca*, Orth.)

[147] WELLINGTON, W. G. *Canad. J. Res.*, D, **24** (1946), 105–17 (reactions to atmospheric pressure: Muscidae).

[148] WEVER, E. G., *et al.* *J. Comp. Cell. Physiol.*, 4 (1933), 79–93; *J. Comp. Psychol.*, **20** (1935), 17–20 (hearing in Orthoptera studied by electric recording from tympanal nerve).

[149] WEYRAUCH, W. K. *Z. vergl. Physiol.*, **10** (1929), 665–87 (sensory physiology of the earwig, *Forficula*).

[150] WIGGLESWORTH, V. B. *Parasitology*, **33** (1941), 67–109 (sense organs of *Pediculus*).

[151] WIGGLESWORTH, V. B., and GILLETT, J. D. *J. Exp. Biol.*, **11** (1934), 120–39; 408–10 (function of antennae: *Rhodnius*, Hem.).

[152] WILL, F. *Z. wiss. Zool.*, **42** (1885), 674–707 (sense of taste in bees and wasps).

[153] WOJTUSIAK, R. J. *Biol. Abstr.*, **6** (1932), 24933 (sense of smell in orientation of larvae of *Pieris*, Lep.).

[154] WOTZEL, F. *Z. wiss. Zool.*, **143** (1933), 241–62 (tympanal organs in aquatic Hemiptera).

[155] ZAWARZIN, A. *Z. wiss. Zool.*, **100** (1912), 245–89; 447–58 (sensory nervous system: larvae of *Aeschna*, Odon., and *Melolontha*, Col.).

[156] ZEISER, T. *Zool. Jahrb., Physiol.*, **53** (1934), 501–20 (responses to gravity: *Dytiscus*, Col.).

SUPPLEMENTARY REFERENCES (A)

[157] BUSNEL, R. G., and CHAVASSE, P. *Nuovo Cimento* (suppl.), **7** (ser. 9) (1950), 470–86 (physical analysis of sounds produced by Orthoptera).

[158] GOODING, S. *J. Roy. Soc. Arts*, **99** (1951), 597–612 (hearing in honey-bee).

[159] GRANT, G. R. M. *Proc. Roy. Soc. Queensland*, **60** (1950), 89–98 (sensory pits as possible resonators to infra-red rays).

[160] HAFEZ, M. *Parasitology*, **40** (1950), 215–36 (humidity sense in *Musca* larva).

[161] HASSENSTEIN, B. *Verhandl. Deutsch. Zool.* (1948), 371–6 (perception of radiant heat: Hemiptera).

[162] HASSETT, C. C., *et al. Biol. Bull.*, **99** (1950), 446–53 (nutritive value and taste thresholds of carbohydrates in *Phormia*, Dipt.).

[163] MARCUS, H. *Folia Univ. Cochabamba*, No. 2 (1948), 13–22 (temperature perception in Hymenoptera).

[164] MITTELSTAEDT, H. *Z. vergl. Physiol.*, **32** (1950), 422–63 (static sense organ in *Anax*, Odonata).

[165] OSSIANNILSSON, F. *Opusc. entom.* (Suppl.), **10** (1949), 1–146 (sound production and hearing in Homoptera).

[166] ROTH, L. M. *Ann. Ent. Soc. Amer.*, **44** (1951), 59–74 (temperature perception in mosquitos).

[167] ROTH, L. M., and WILLIS, E. R. *J. Exp. Zool.*, **117** (1951), 451–87 (hygroreceptors in Coleoptera).

[168] SCHALLER, F. *Z. vergl. Physiol.*, **33** (1951), 476–86 (sound production and hearing in *Corixa*, Hem.).

[169] SCHALLER, F., and TIMM, C. *Z. vergl. Physiol.*, **32** (1950), 468–81 (hearing in nocturnal Lepidoptera).

[170] SLIFER, E. H. *Proc. Roy. Soc.*, B., **138** (1951), 414–37 (thermoreceptors and radiant heat perception in *Locusta*).

SUPPLEMENTARY REFERENCES (B)

[171] ARNAL, A. *Bol. Soc. Esp. Hist. Nat.*, **56** (1958), 5–20 (stimulation of Johnston's organ in mosquitos).

[172] AUTRUM, H. *Acustica*, **10** (1960), 340–8 (phasic-tonic response of tympanal organ in *Tettigonia*).

[173] BÄSSLER, U. *Z. Naturforsch.*, **16b** (1961), 264–7 (functions of Johnston's organ in *Culex*).

[174] BENZ, G. *Experientia*, **12** (1956), 297 (humidity receptors in *Drosophila* larva).

[175] BLEST, A. D., COLLETT, T. S., and PYE, J. D. *Proc. Roy. Soc.*, B, **158** (1963), 196–207 (ultrasonic signals produced by *Melese*, Arctiidae).

[176] BOECKH, J. *Z. vergl. Physiol.*, **46** (1962), 212–48 (electrical responses from single olfactory sensilla in *Necrophorus*).

[177] BOECKH, J., KAISSLING, K. E., and SCHNEIDER, D. *Zool. Jahrb., Anat.*, **78** (1960), 560–84 (sensilla on antennae of Saturniidae).

[178] BÜCKMANN, D. *Z. vergl. Physiol.*, **36** (1954), 488–507; *Naturwiss.*, **42** (1955), 78–9 (gravity sense in *Bledius* and *Dyschirius*, Col.); *Naturwiss.*, **49** (1962), 28–33 (gravity sense: review).

[179] BURKHARDT, D. *Ergebn. Biol.*, **22** (1960), 226–67 (properties of sense organs).

[180] —— *J. Ins. Physiol.*, **4** (1960), 138–45 (antennae of *Calliphora* as mechanical sense organs).

[181] BURKHARDT, D., and SCHNEIDER, G. *Z. Naturforsch.*, **12b** (1957), 139–43 (antennae of *Calliphora* as indicators of flight velocity).

[182] BURSELL, E. *J. Exp. Biol.*, **34** (1957), 42–51 (humidity receptors in thoracic spiracles of *Glossina*).

[183] BUSNEL, M. V. In *L'acoustique des Orthoptères. Ann. épiphyt.*, suppl. 1955, 175–202 (sound production in *Oecanthus*).

[184] BUSNEL, M. C., DUMORTIER, B., and BUSNEL, R. G. *Bull. Soc. Zool. France*, **84** (1959), 351–70 ('phonokinesis' in insects without tympanal organs).

[185] BUSNEL, R. G., and DUMORTIER, B. *Bull. Soc. Ent. France*, **64** (1959), 44–50 (sound production in *Acherontia*).

[186] BUSNEL, R. G., DUMORTIER, B., and BUSNEL, M. C. *Bull. Biol. Fr. Belg.*, **90** (1956), 219–86 (acoustic behaviour of *Ephippiger*).

[187] DETHIER, V. G. *Biol. Bull.*, **103** (1952), 178–89 (adaptation in tarsal chemoreceptors of *Phormia*).

[188] —— *Quart. Rev. Biol.*, **30** (1955), 348–71 (physiology and histology of contact chemoreceptors of *Phormia*).

[189] —— In *Molecular structure and functional activity of nerve cells*. A.I.B.S. Washington, D.C., 1956, 1–30; *Symp. Soc. Exp. Biol.*, **16** (1962), 180–96 (chemoreceptor mechanisms in insects).

[190] —— *The physiology of insect senses*. Methuen, London, 1963, 255 pp.

[191] DETHIER, V. G., EVANS, D. R., and RHOADES, M. V. *Biol. Bull.*, **3** (1956), 204–22 (factors controlling ingestion of sugars in *Phormia*).

[192] DOFLEIN, H. *Festschr. f. R. Hertwig.*, **3** (1910), 217–92 ('Stellungshaare' in Crustacea).

[193] DUNHAM, J. *Physiol. Zool.*, **25** (1962), 297–303 (infra-red transmission in cuticle of grasshopper).

[194] EVANS, E. J. *Proc. Roy. Ent. Soc. Lond.*, A, **27** (1952), 39–42 (sound spectrum of locust stridulation).

[195] EVANS, D. R., and DETHIER, V. G. *J. Ins. Physiol.*, **1** (1957), 3–17 (regulation of taste threshold for sugars in *Phormia*).

[196] EVANS, D. R., and MELLON, D. *J. Gen. Physiol.*, **45** (1962), 487–500 (water receptor in taste sensillum of *Phormia*).

[197] EVANS, D. R., and MELLON, D. *J. Gen. Physiol.*, **45** (1962), 651–61 (stimulation of taste receptor in *Phormia* by salts).

[198] FABER, A. *Laut- und Gebärdensprache bei Insekten Orthoptera* (*Geradflüger*). Part 1. Stuttgart, 1953, 198 pp.

[199] FEIR, D., LENGY, J. I., and OWEN, W. B. *J. Ins. Physiol.*, **6** (1961), 13–20 (chemoreceptors on labella of *Culiseta*).

[200] FEUERBORN, H. J. *Zool. Anz.*, **70** (1927), 167–84 (formation of new tactile bristles in *Psychoda*).

[201] FINLAYSON, L. H., and LOWENSTEIN, O. *Nature*, **176** (1955), 1031; *Proc. Roy. Soc.*, B, **148** (1958), 433–49 (stretch receptors and their function in various insects).

[202] FINLAYSON, L. H., and MOWAT, D. J. *Quart J. Micr. Sci.*, **104** (1963), 243–51 (histology of stretch receptors of Saturniid moths at different stages).

[203] FISCHER, W. *Z. vergl. Physiol.*, **39** (1957), 634–59 (olfactory acuity in the honey-bee).

[204] FRAENKEL, G. *Biochemistry of Insects*, Pergamon Press, London, 1959, 1–14 (chemistry and host specificity of phytophagous insects).

[205] FRINGS, H., and FRINGS, M. *Amer. Midl. Nat.*, **41** (1949), 602–48 (contact chemical receptors in insects: historical review).

[205a] —— *Ann. Ent. Soc. Amer.*, **49** (1956), 611–17 (reaction to sounds by *Cercyonis*, Lep.).

[206] —— *Biol. Bull.*, **110** (1956), 291–9; *J. N.Y. Ent. Soc.* **67** (1959), 97–105 (site of chemoreceptors in Lepidoptera); *Biol. Bull.*, **111** (1956), 92–100 (ditto in Trichoptera).

[207] —— *Ann. Rev. Ent.*, **3** (1958), 87–106 (uses of sounds by insects: review).

[208] —— *Sound production and sound perception by insects—a bibliography*, Pennsylvania State Univ. Press, 1960.

[209] FRINGS, H., and HAMRUM, C. L. *J. N.Y. Ent. Soc.*, **58** (1950), 133–42 (contact chemoreceptors in *Aëdes*).

[210] GEBHARDT, H. *Zool. Jahrb., Physiol.*, **63** (1953), 558–92 (thermoreceptors in *Pyrrhocoris*, &c.).

[211] GETTRUP, E. *J. Exp. Biol.*, **40** (1963), 323–33 (stimulation of thoracic stretch receptor in *Schistocerca* during flight).

[212] GOODMAN, L. J. *Nature*, **183** (1959), 1106–7 (hair plates on cervical sclerites of locusts).

[213] GRABOWSKI, C. T., and DETHIER, V. G. *J. Morph.*, **94** (1954), 1–17 (structure of tarsal chemoreceptors in *Phormia*).

[214] GRAY, E. G. *Phil. Trans. Roy. Soc.*, B, **243** (1960), 75–94 (fine structure of auditory sensilla in *Locusta*).

[215] HAMAMURA, Y. *Nature*, **183** (1959), 1746–7 (food selection by *Bombyx* larva).

[216] HASKELL, P. T. *J. Exp. Biol.*, **33** (1956), 756–76 (hearing in grasshoppers).

[217] HASKELL, P. T. *Nature*, **183** (1959), 1107 (physiology of prothoracic hair receptors in *Schistocerca*).

[218] —— *Insect sounds*, Witherby: London, 1961, 189 pp.

[219] HERAN, H. *Z. vergl. Physiol.*, **34** (1952), 179–206 (thermoreception in honey-bee).

[220] —— *Z. vergl. Physiol.*, **42** (1959), 103–63 (perception and regulation of speed of flight in the honey-bee).

[221] HODGSON, E. S. *Biol. Bull.*, **105** (1953), 115–27 (chemoreception in the amphibious beetle *Laccophilus*).

[222] —— *J. Ins. Physiol.*, **1** (1957), 240–7; *Biol. Bull.*, **115** (1958), 114–25 (responses of chemoreceptors in *Phormia*); *Ann. Rev. Ent.*, **3** (1958), 19–36 (chemoreception in insects: review).

[223] HODGSON, E. S., and ROEDER, K. D. *J. Cell. Comp. Physiol.*, **48** (1956), 51–76 (properties of labellar chemoreceptors of Diptera).

[224] HORRIDGE, G. A. *Nature*, **185** (1960), 623–4; *Proc. Roy. Soc.*, B, **155** (1961), 219–31 (pitch discrimination in locusts).

[225] HUGHES, G. M. *Nature*, **170** (1952), 531–2 (stretch receptors in rhythmical movements in *Dytiscus* and *Locusta*).

[226] ITO, T., HORIE, Y., and FRAENKEL, G. *J. Sericult. Sci. Japan*, **28** (1959), 108–13 (effect of removing maxillae on food selection by *Bombyx* larva).

[227] JOHNSON, B. *Aust. J. Sci.*, **18** (1956), 199–200 (function of antennae of Aphids during flight).

[228] KATSUKI, Y., and SUGA, N. *Proc. Jap. Acad.*, **34** (1958), 633–8; *J. Exp. Biol.*, **37** (1960), 279–90 (sound frequency discrimination in Tettigoniidae).

[229] KENNEDY, J. S., and BOOTH, C. O. *Ann. Appl. Biol.*, **38** (1951), 25–64 (dual discrimination theory of host-plant selection in Aphids).

[230] KEPPLER, E. *Z. Naturforsch.*, **13b**, 280–6 (hearing in mosquitos).

[231] KERKUT, G. A., and TAYLOR, B. J. R. *J. Exp. Biol.*, **34** (1957), 486–93 (temperature receptor in tarsus of *Periplaneta*).

[232] KEVAN, D. K. MCE. *Ann. Epiphyt.*, Suppl. Coll. 1955, 103–41 (sound production in Orthoptera).

[233] LACHER, V., and SCHNEIDER, D. *Z. vergl. Physiol.*, **47** (1963), 274–8 (pore-plates in *Apis* as olfactory receptors).

[234] LARSEN, J. R. *J. Ins. Physiol.*, **8** (1962), 683–91 (fine structure of chemosensory hairs of *Phormia*).

[235] —— *Science*, **139** (1963), 347 (interpseudotracheal papillae of *Phormia*).

[235a] LARSÉN, O. *Lunds Univ. Årssk.*, N.F. Avd. 2. **53** (1957), 1–67 (tympanal organs in *Corixa*, &c.).

[236] LEWIS, C. T. *Nature*, **173** (1954), 130–1; *Bull. Ent. Res.*, **45** (1954), 711–22 (structure of chemosensory hairs on tarsi of *Phormia*, &c.).

[237] LINDAUER, M., and NEDEL, J. O. *Z. vergl. Physiol.*, **42** (1959), 334–64 (hair patches as gravity sense organs in *Apis*).

[238] LOWENSTEIN, O., and FINLAYSON, L. H. *Comp. Biochem. Physiol.*, **1** (1960), 56–61 (responses of stretch receptors).

[239] MADDRELL, S. H. P. *Nature*, **198** (1963), 210 (control of ingestion in *Rhodnius*).

[240] MAKINGS, P. *J. Exp. Biol.*, **40** (1964), (in the press) (Slifer's patches and thermal sense in Acrididae).

[241] MARKL, H. *Z. vergl. Physiol.*, **45** (1962), 475–569 (hair patches as gravity receptors in ants, &c.); *Naturwiss.*, **17** (1963), 559–65; *Nature*, **198** (1963), 173–5 (gravity receptors: review).

[242] MILES, P. W. *J. Ins. Physiol.*, **2** (1958), 338–47 (chemoreception in pharynx of Heteroptera).

[243] MITTELSTAEDT, H. *Verh. dtsch. zool. Ges.*, 1952, 102–6; In *Recent Advances in Invertebrate Physiology*, Univ. Oregon Publ., 1957, 51–71 (hair cushions as proprioceptors in Mantids).

[244] MORITA, H., *et al. J. Cell. Comp. Physiol.*, **54** (1959), 177–87, 189–202; *J. Exp. Biol.*, **38** (1961), 851–61 (receptor potentials in chemoreceptors of *Vanessa*, *Calliphora*, and *Bombyx* larva).

[245] NAYAR, J. K., and FRAENKEL, G. *J. Ins. Physiol.*, **8** (1962), 505–25 (sterols and host-plant selection in *Bombyx*).

[246] OSBORNE, M. P. *Quart. J. Micr. Sci.*, **104** (1963), 227–41 (stretch receptors in abdomen of *Phormia* larva).

[247] —— *J. Ins. Physiol.*, **9** (1963), 237–45 (fine structure of stretch receptor nerve endings in *Periplaneta*); *Nature*, **201** (1963), 526–7 (ditto in blow-fly larva).

[248] OSBORNE, M. P., and FINLAYSON, L. H. *Quart. J. Micr. Sci.*, **103** (1962), 227–42 (stretch receptors in seven orders of insects).

[249] OWEN, W. B. *J. Ins. Physiol.*, **9** (1963), 73–87 (contact chemoreceptors in mosquitos).

[250] PABST, H., and SCHWARZKOPFF, J. *Z. vergl. Physiol.*, **45** (1962), 396–404 (stretch receptors in wing hinge of *Locusta*).

[251] PASQUINELLY, F., and BUSNEL, M. C. *Ann. épiphyt.*, Suppl.: *L'acoustique des Orthoptères*, 1955, 145–53 (sound production in *Oecanthus*).

[252] PETERS, W. *Verh. XI Int. Kongr. Ent. Wien 1960*, 407–9 (structure of chemoreceptor hairs in labellum of *Calliphora*).

[253] —— *Z. Morph. Oekol. Tiere*, **51** (1962), 211–26 (proprioceptive hair patches in *Calliphora*).

[254] PINET, J. M. *C.R. Acad. Sci.*, **257** (1963), 3666–8 (sensory innervation of stylets in *Rhodnius*).

[255] PRINGLE, J. W. S. *Nature*, **172** (1953), 248–9; *J. Exp. Biol.*, **31** (1954), 525–60 (physiological analysis of cicada songs).

[256] RABE, W. *Z. vergl. Physiol.*, **35** (1953), 300–25 (role of Johnston's organ in *Notonecta* and *Corixa*).

[257] RIBBANDS, C. R. *Proc. Roy. Soc.*, B, **143** (1955), 367–79 (scent perception in the honey-bee).

[258] RICHARD, G. *Ann. Sci. Nat., Zool.* (sér. 11), **12** (1950), 65–82; **13** (1951), 397–411; *Bull. Soc. Zool. Fr.*, **77** (1952), 99–106 (sensory nerve endings in termites, &c).

[259] RICHTER, S. *Z. Morph. Oekol. Tiere*, **52** (1962), 171–96 (fine structure of chemoreceptive antennae of *Calliphora* larva).

[260] RIEGERT, P. W. *Canad. Ent.*, **92** (1960), 561–70 (humidity receptors in *Melanoplus*).

[261] RISLER, H. *Zool. Jahrb., Anat.*, **73** (1953), 165–86; **74** (1955), 478–90 (Johnston's organ and hearing in male mosquitos).

[262] ROEDER, K. D. *Anim. Behav.*, **10** (1962), 300–4 (response of flying moths, &c., to ultrasonic pulses).

[263] ROEDER, K. D., and TREAT, A. E. *J. Exp. Zool.*, **134** (1957), 127–58; *Anat. Rec.*, **134** (1959), 630 (ultrasonic reception by tympanic organ of Noctuidae).

[264] ROTH, L. M., and WILLIS, E. R. *J. Exp. Zool.*, **117** (1951), 451–87 (hygroreceptors in Coleoptera); *J. Morph.*, **91** (1952), 1–14 (ditto in *Aëdes* and *Blattella*).

[265] SCHNEIDER, D. *Experientia*, **13** (1957), 89; *Z. vergl. Physiol.*, **40** (1957), 8–41 (electrical responses from chemoreceptors of *Bombyx* antenna).

[266] SCHNEIDER, D. *Ann. Rev. Ent.* **9** (1964), 103–22 (insect antennae: review).

[267] SCHNEIDER, D., and BOECKH, J. *Z. vergl. Physiol.*, **45** (1962), 405–12 (responses from single chemoreceptors on antennae of *Tenebrio*, *Necrophorus*, and *Sphinx*).

[268] SCHWARZ, R. *Z. vergl. Physiol.*, **37** (1955), 180–210 (olfactory acuity in the honey-bee).

[269] SIMPSON, J. *Z. vergl. Physiol.*, **48** (1964), 277–82 (mechanism of piping in the queen bee).

[270] SLIFER, E. H. *Int. Rev. Cytol.*, **11** (1961), 125–59 (fine structure of insect sense organs: review).

[271] SLIFER, E. H., PRESTAGE, J. J., and BEAMS, H. W. *J. Morph.*, **101** (1957), 359–81; **105** (1959), 145–92 (fine structure of chemoreceptors in *Melanoplus*).

[272] SLIFER, E. H., and SEKHON, S. S., *et al. Exp. Cell. Res.*, **19** (1960), 410–4; *J. Morph.*, **109** (1961), 351–81 (fine structure of insect sense organs: plate organs, &c., of *Apis* antenna); *J. Morph.*, **111** (1962), 49–56 (ditto: on antenna of *Aëdes*); **112** (1963), 165–93 (ditto on antenna of *Lygaeus*); *Quart. J. Micr. Sci.*, **105** (1964), 21–9 (ditto: on antenna of *Megoura*).

[273] SUGA, N. *Jap. J. Physiol.*, **10** (1960), 533–46 (peripheral analysis of sound in *Locusta*).

[274] SUGA, N., *et al. J. Exp. Biol.*, **38** (1961), 545–58; *J. Ins. Physiol.*, **9** (1963), 867–73 (central mechanism of hearing and sound localization in Orthoptera).

[275] THORSTEINSON, A. J. *Canad. J. Zool.*, **31** (1953), 52–72 (chemotaxis and host-plant selection in *Plutella*).

[276] —— *Ann. Rev. Ent.*, **5** (1960), 193–218 (host-plant selection: review).

277 THURM, U. *Z. vergl. Physiol.*, **46** (1963), 351–82; **48** (1964), 131–56 (properties of tactile hairs in the honey-bee).

278 TISCHNER, H., *et al. Acustica*, **3** (1953), 335–43; *Verh. dtsch. zool. Ges. Tübingen, 1954*, 1955, 444–60 (hearing in mosquitos).

279 TREAT, A. E. *Ann. Ent. Soc. Amer.*, **48** (1955), 272–84 (response of Lepidoptera to sounds).

280 VOWLES, D. M. *J. Exp. Biol.*, **31** (1954), 356–75 (orientation of ants to gravity).

281 WALDBAUER, G. P. *Ent. Exp. Appl.*, **5** (1962), 147–58 (removal of maxillae and feeding of *Protoparce* larvae).

282 WALLIS, D. I. *J. Exp. Biol.*, **39** (1962), 603–15; *J. Ins. Physiol.*, **8** (1962), 453–67 (olfactory and tactile sensilla on ovipositor of *Phormia*).

283 WALLIS, R. C. *Amer. J. Hyg.*, **60** (1954), 135–68 (tarsal chemoreceptors of mosquitos sensitive to sodium chloride solutions).

284 WENDLER, G. *Naturwiss.*, **48** (1961), 676–7; *Z. vergl. Physiol.*, **48** (1964), 198–250 (proproceptive hairs and position of limbs in *Carausius*).

285 WENSLER, R. J. D. *Nature*, **195** (1962), 830–1 (chemical factors and host selection in *Brevicoryne*).

286 WIESMANN, R. *Mitt. schweiz. ent. Ges.*, **36** (1964), 249–73 (sense organs controlling the disposal of the meal in *Musca*).

287 WIGGLESWORTH, V. B. *Quart. J. Micr. Sci.*, **77** (1934), 191–222 (stretching of abdomen as stimulus to moulting in *Rhodnius*).

288 WIGGLESWORTH, V. B. *Quart. J. Micr. Sci.*, **94** (1953), 93–112 (origin of sensory neurones in *Rhodnius*).

289 WISHART, G., *et al. Canad. Ent.* **91** (1959), 181–91; **94** (1962), 614–26 (orientation of male *Aëdes* to sound).

290 WOHLBARSHT, M. L. *Science*, **125** (1957), 1258 (water as specific stimulus to labellar hairs of *Phormia*).

291 —— *J. Gen. Physiol.*, **44** (1960), 105–22 (electrical characteristics of tactile hair receptors in insects).

292 WOHLBARSHT, M. L., and DETHIER, V. G. *J. Gen. Physiol.*, **42** (1958), 393–412 (responses of chemoreceptors of *Phormia* to chemical and mechanical stimuli).

293 YAMAMOTO, R. T., and FRAENKEL, G. *Ann. Ent. Soc. Amer.*, **53** (1960), 499–503 (specific gustatory stimulant for *Protoparce* in *Solanaciae*).

SUPPLEMENTARY REFERENCES (C)

294 ADAM, L. J. *Z. vergl. Physiol.*, **63** (1969), 227–89 (neurophysiology of hearing in *Locusta*).

295 ADAM, L. J. and SCHWARTZKOPFF, J. *Z. vergl. Physiol.*, **54** (1967), 246–55 (separate representation of sound frequencies in brain of *Locusta*).

296 AIDLEY, D. J. *J. exp. Biol.*, **51** (1969), 325–37 (sound production in cicadas).

297 BAILEY, W. J. and BROUGHTON, W. B. *J. exp. Biol.*, **52** (1970), 495–517 (mechanics of stridulation in *Homorocoryphus*, Tettigoniidae).

298 BENNET-CLARK, H. C. *J. exp. Biol.*, **52** (1970), 619–52 (projection of sound by *Gryllotalpa* spp.).

299 BOECKH, J. *Z. vergl. Physiol.*, **55** (1967), 378–406 (molecular basis of olfactory stimulus in *Locusta*).

300 BOECKH, J., KAISSLING, K. E. and SCHNEIDER, D. *Cold Spring Harb. Symp. Quant. Biol.*, **30** (1965), 263–80 (sensitivity of insect olfactory receptors).

301 CHAPMAN, K. M. *J. exp. Biol.*, **42** (1965), 191–203 (campaniform sensilla of tactile spines of *Periplaneta*).

302 CHEVALIER, R. L. *J. Morph.*, **128** (1969), 443–64 (fine structure of campaniform organs in halteres of *Drosophila*).

303 DETHIER, V. G. and SCHOONHOVEN, L. M. *J. Insect Physiol.*, **14** (1968), 1049–54 (evaluation of evaporation by cold and humidity receptors in caterpillars).

304 ERNST, K. E. *Z. Zellforsch, microsk. Anat.*, **94** (1969), 72–102 (fine structure of antenna of *Necrophorus*, Col.).

305 ESCH, H. and WILSON, D. *Z. vergl. Physiol.*, **54** (1967), 256–67 (sounds produced by bees and flies).

[306] FINLAYSON, L. H. *Symp. zool. Soc. Lond.*, **23** (1968), 217–49 (abdominal stretch receptors in insects: review).

[307] GUTHRIE, D. M. *J. Insect Physiol.*, **13** (1967), 1637–44 (multipolar stretch receptors in legs of *Periplaneta*).

[308] HASKELL, P. T. *et al. Ent. exp. & appl.*, **12** (1969), 432–40; 591–610 (taste receptors in locusts).

[309] HOWSE, P. E. and CLARIDGE, M. F. *J. Insect Physiol.*, **16** (1970), 1665–75 (Johnston's organ and song reception in leaf-hopper, *Oncopsis*).

[310] KAISSLING, K. E. and RENNER, M. *Z. vergl. Physiol.*, **59** (1968), 357–61 (specialized chemoreceptors in the pore plates of *Apis*).

[311] KAY, R. E. *J. Insect Physiol.*, **15** (1969), 989–1001 (acoustic signalling by *Heliothis* in flight).

[312] KEHLER, J. G. *et al. J. Insect Physiol.*, **16** (1970), 483–97 (ventral phasic mechanoreceptors in *Blaberus* abdomen).

[313] LACHER, V. *Z. vergl. Physiol.*, **48** (1964), 587–623 (varied sense receptors on antennae of *Apis*).

[314] LOFTUS, R. *Z. vergl. Physiol.*, **59** (1968), 413–55 (cold receptor on antenna of *Periplaneta*).

[315] McKAY, J. M. *J. exp. Biol.*, **51** (1969), 787–802 (frequency sensitivity in ear of *Homorocoryphus*, Tettigoniidae).

[316] MARKL, H. *Z. vergl. Physiol.*, **53** (1966), 328–52 (perception of the angle of orientation to gravity by honey-bee).

[317] MARKL, H. *Z. vergl. Physiol.*, **57** (1967), 299–330; **69** (1970), 6–37 (perception of stridulation by *Atta*).

[318] MARKL, H. and WIESE, K. *Z. vergl. Physiol.*, **62** (1969), 413–20 (perception of surface waves in water by *Notonecta*).

[319] MICHELSON, A. *Nature*, **220** (1968), 585–6; *Z. vergl. Physiol.*, **71** (1971), 49–128 (frequency discrimination by the ear of *Schistocerca*).

[320] MILLER, L. E. and MACLEOD, E. G. *Science*, **154** (1966), 891–3 (ultrasonic perception by *Chrysopa*).

[321] MORITA, H. *et al. Mem. Fac. Sci. Kyushu Univ. Ser. E. (Biol.)* **4** (1966), 123–35; *J. gen. Physiol.*, **52** (1968), 559–83 (nature of sensory stimulation by sugars in the flesh fly *Boettcherisca*).

[322] MYERS, J. H. *et al. J. Insect Physiol.*, **15** (1969), 2117–30; **16** (1970), 573–8 (odour perception in *Danaus*).

[323] NEESE, V. *Z. vergl. Physiol.*, **52** (1966), 149–54 (eye hairs in *Apis* and regulation of speed of flight).

[324] NICKLAUS, R. *et al. Z. vergl. Physiol.*, **50** (1965), 331–62; **56** (1967), 412–15 (sensory apparatus in tactile hairs of *Periplaneta*).

[325] PARRISH, W. B. *Ann. ent. Soc. Am.*, **60** (1967), 273–6 (innervation of stylets of Aphids).

[326] PAYNE, R. S., ROEDER, K. D. and WALLMAN, J. *J. exp. Biol.*, **44** (1966), 17–31 (directional sensitivity in ears of Noctuid moths).

[327] REES, C. J. C. *Proc. Roy. Soc. Lond.*, B, **174** (1970), 469–90 (importance of ions in 'water' receptor of *Phormia*).

[328] RICE, M. J. *J. Insect Physiol.*, **16** (1970), 277–89 (cibarial stretch receptors in *Calliphora*).

[329] RIDDIFORD, L. *J. Insect Physiol.*, **16** (1970), 636–60 (antennal proteins as possible receptors for odours in *Antheraea*).

[330] ROEDER, K. D. *J. Insect Physiol.*, **13** (1967), 873–8; **15** (1969) 825–38 (turning responses to sound in flying Noctuid moths).

[331] ROEDER, K. D. and TREAT, A. E. *J. Insect Physiol.*, **16** (1970), 1068–86 (auditory organ in labial palp of some Sphingidae).

[332] RUDOLPH, P. *Z. vergl. Physiol.*, **56** (1967), 341–75 (Johnston's organ used to perceive wave motion in water by *Gyrinus*).

[333] RUNION, H. I. and USHERWOOD, P. N. R. *J. exp. Biol.*, **49** (1968), 421–36 (tarsal receptors and leg reflexes in *Schistocerca*).

[334] RUPPRECHT, R. *Z. vergl. Physiol.*, **59** (1968), 38–71 (subgenual organ as vibration receptor in Plecoptera).

[335] SALAMA, H. S. *J. Insect Physiol.*, **12** (1966), 1051–60 (cibarial chemoreceptors in mosquitos).

[36] SCHNEIDER, D., LACHER, V. and KAISSLING, K. E. *Z. vergl. Physiol.*, **48** (1964), 632–62 (specialization of chemoreceptors in antenna of male *Antheraea*).

[37] SCHOONHOVEN, L. M. *Koninkl. Nederl. Akad. Wetensch.*, **70** (1967), 556–68; *Nature*, **221** (1969), 1268; *Ent. exp. and appl.*, **12** (1969), 555–64 (specialization of chemoreceptors on mouth parts of caterpillars).

[38] SLIFER, E. H. *Ann. Rev. Entom.*, **15** (1970), 121–42 (structure of insect chemoreceptors: review).

[39] SMITH, D. S. *Tissue & Cell*, **1** (1969), 443–84 (fine structure of haltere sensilla in *Calliphora*).

[40] SMOLA, U. *Z. vergl. Physiol.*, **67** (1970), 382–402 (mechanics of wind-sensitive hairs in *Locusta*).

[41] SWIHART, S. L. *J. Insect Physiol.*, **13** (1967), 469–76 (hearing in butterflies).

[42] THURM, U. *Science*, **145** (1964), 1063–5; *Cold Spring Harbor Symp. Quant. Biol.*, **30** (1965), 75–94 ('tubular body' and mechanics of mechanoreceptors in insects).

[43] USHERWOOD, P. N. R., RUNION, H. I. and CAMPBELL, J. I. *J. exp. Biol.*, **48** (1968), 305–23 (chordotonal organs and leg reflexes in *Schistocerca*).

[44] WEEVERS, R. de G. *J. exp. Biol.*, **44** (1966), 177–208; **45** (1966), 229–49 (physiology of stretch receptors in caterpillars).

[45] YAMADA, M., ISHII, S. and KUWAHARA, Y. *Nature*, **227** (1970), 855 (specialized sex pheromone receptor neurones in olfactory lobe of cockroach).

[46] YAMAGISAWA, K., HASHIMOTO, T. and KATSUKI, Y. *J. Insect Physiol.*, **13** (1967), 635–643 (frequency discrimination in locusts).

Chapter VIII

Behaviour

THE PHYSIOLOGICAL study of behaviour consists in the analysis of th movements of the whole organism into a series of reflexes or observed correlations between stimulus and response. Theoretically, in a reflex, the response i a function of the stimulus. But, as we saw in Chapter V, the relation betwee the two may be affected by a great variety of third variables—the number o times the stimulus has been given, the changes of adaptation or fatigue in th synapses, or the preceding activities of the nervous system, as in the reflex chain In the behaviour of intact organisms such third variables lead to much apparen inconsistency in the relation between stimulus and response. In practice, therefore, the analysis of behaviour along these lines does not go very far, and i many descriptions of behaviour psychological conceptions, particularly conation the innate seeking of some end or goal, are introduced in order to supplement th physiological mechanisms.

KINESIS AND RELATED PHENOMENA

Stimulatory organs—In addition to receiving specific stimuli, the majorit of sense organs, when they are stimulated, have the effect of increasing the refle excitability of the nervous system. Their action seems to be necessary in orde to open the nervous paths, allow reflexes to take place, and so enable the nerv centres to carry out their normal function. Sense organs having this effect ar spoken of as 'stimulatory organs';[173] and the increased activity due to their actio is termed 'kinesis'.

The **compound eyes**, for example, have an important stimulatory function In *Periplaneta*, exposure to light greatly reduces the threshold of response to othe stimuli.[15] The rate of walking in *Popillia japonica* is accelerated as the ligh intensity is increased.[109] The moth *Larentia truncata* responds to sounds by flying away when it is on a pale background, but it cannot be shifted by sound when on a dark background.[83] And many day-flying butterflies (such as *Erebi* species) are entirely incapable of flight in the absence of bright sunlight. If th eyes of *Macroglossa* are darkened it settles at once, folds the wings and lays bac the antennae; if it is flying in a room lit with electric light, it immediately fal to the ground if the light is switched off.[61, 87]

The **ocelli** of adult insects which, as we have seen (p. 243), evoke by themselves no visual responses, are perhaps specific stimulatory organs which increas the reflex reactions to stimuli received by the compound eyes, and thus enabl the insect to respond to changes in light intensity at a level of general intensity at which they are normally incapable of movement (Fig. 193). In *Drosophil* suddenly exposed to the light, movements begin in an average of 4 seconds; if the ocelli are blackened the latent period is prolonged to 5 seconds; and after cutting off the light, activity diminishes more rapidly in the insect with ocelli intact.[10] Similarly, bees wake up very quickly in the light with functional ocelli, very slowly if the ocelli are covered. (See p. 216.) If they are exposed to two lights of different intensities bees normally go to the brighter; but if the median ocellus and one lateral ocellus are occluded, they always go to the side with the sound ocellus, even if the light on this side is considerably weaker, as though the optic centre of this side were more strongly stimulated.[110] In the Cicada *Cryptotympana* both eyes and ocelli are necessary for flight.[26] [See p. 345.]

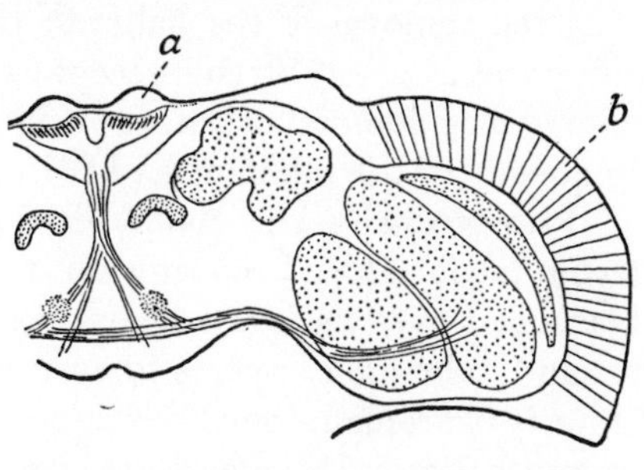

FIG. 193.—Transverse section of head of *Drosophila* showing the close association of the nerve tracts of the compound eyes and the ocelli (*after* BOZLER)

a, ocelli; **b**, compound eyes.

An example of the **antennae** serving as 'stimulatory organs' is afforded by the blood-sucking bug *Rhodnius*. If both antennae are removed, *Rhodnius* settles into a state of sleep, or 'akinesis', from which it is aroused with difficulty, not only by such stimuli as air currents (which are normally perceived by the tactile spines and perhaps the Johnston's organ of the antenna), but by vibrations or violent stimuli to the abdomen. The antennae clearly have a kinetic function and normally keep the nervous system in a state of 'tone' in which it will respond readily to stimuli of all kinds. Air movements probably provide the chief stimuli concerned in this function.[169] In *Muscina* a flow of air on the antennae is normally necessary to maintain the wing vibrations of flight; these are difficult to elicit if a capsule is placed over the antennae.[63] A similar kinetic effect is seen in Orthoptera and Odonata.

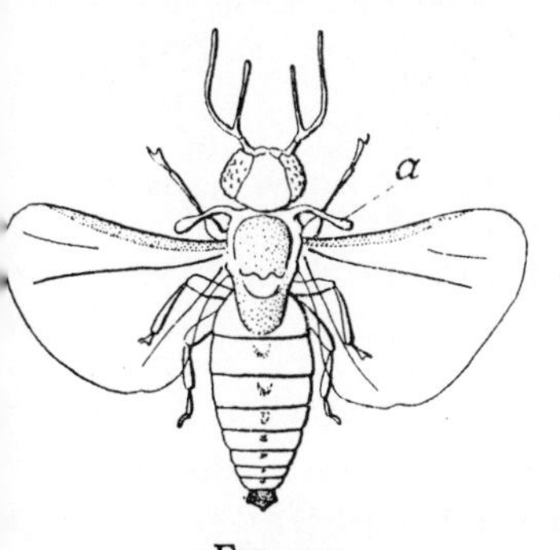

FIG. 194

Male of *Elenchinus* (Strepsiptera) showing the fore wings (a) reduced to halteres (*after* ULRICH).

In addition to their function as organs of equilibrium (p. 268) the **halteres** of Diptera have been regarded as important stimulatory organs. It is supposed that the chordotonal and campaniform sensilla which they carry are stimulated by their vibratory movements, and that these stimuli facilitate the conduction of the reflexes concerned in flight.[14, 20, 21, 106] Suspended flies continue flying for a much shorter time after each stimulus if the halteres are removed;[38] in *Tipula*, both wings and legs are weakened by removal of the halteres, and *Sarcophaga* deprived of halteres can scarcely flutter the wings.[20] In Strepsiptera the fore wings have become similarly transformed into club-shaped halteres richly supplied with campaniform sensilla (Fig. 194), and they appear to have the same function: if they are amputated the insect loses the power of flight and finally the hind wings just stay at rest.[154] The vibrations of the tarsi of the middle pair of legs in the dung beetle *Sisyphus* before and probably during flight has been regarded as serving a stimulatory function;[94] as have also the vibrations of the wings of many insects, and the vigorous pumping movements made by *Melolontha* before flight[21] (*cf.* p. 670). Many Diptera, such as *Tabanus*[21] and *Calliphora* (if it is suitably balanced and allowed a day or so to recover from the operation)[40] can still fly without reduction in the frequency of wing beat after the removal of the halteres; their stimulatory function, therefore, is not always necessary.[40] Stretch receptors at the base of the wing in *Schistocerca*, containing a scolopidial organ, provide a feed-back mechanism which modifies the frequency of the inherent flight rhythm in *Schistocerca*.[215]

Phototonus—The Asilid fly *Proctacanthus* becomes immobile or makes feeble incoordinated movements if the eyes are blackened. In the dark or with blackened eyes its legs collapse; in the light it raises its body well above the ground. The light appears to exert a general effect on muscle tonus. If one eye alone is blackened, tonus is exaggerated on the uncovered side; and if the upper or lower parts of both eyes are covered, different groups of muscles are affected so that the insect takes up characteristic attitudes (Fig. 195).[49] The same response is shown by the aquatic bug *Ranatra*[64] and the Syrphid *Eristalis*.[104] It is some-

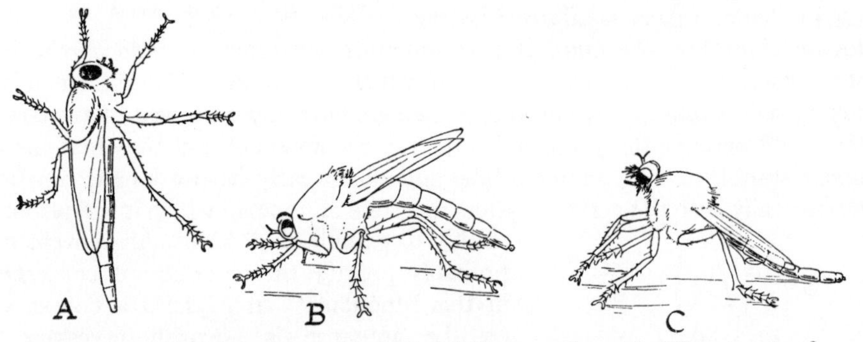

FIG. 195.—Attitudes taken up by the Asilid *Proctacanthus* with different parts of the eye blackened (*after* GARREY)

A, right eye blackened; fly viewed from above; *B*, upper halves of eyes blackened; *C*, lower halves of eyes blackened.

times termed 'phototonus', being regarded as a stimulatory effect of different regions of the eye upon different groups of muscles.[49, 173] But, as we shall see, this type of behaviour is susceptible of other explanations (p. 318). Locusts take up elaborate postures at right angles to the sun's rays, or parallel to these rays, depending upon the body temperature (p. 671). Here there is a complex interaction of thermal and visual stimuli.[160]

Hygrokinesis and chemokinesis—In the responses of insects to odours and to moisture there is a large kinetic element. This will be discussed more fully when we deal with mechanisms of orientation (p. 315), but a few examples may be noted here. In an insect like *Forficula*, which will not settle in contact with moist surfaces, the effect of moisture appears to be stimulatory or kinetic;[16] but, if desiccated, earwigs choose moist air, and during the winter they burrow into moist soil and remain there.[252] The same applies to *Locusta migratoria* which, when offered two alternative humidities, always shows a preference for the drier at all parts of the humidity range, although dry air is not necessarily optimal for development or breeding. For when tested in an actograph, designed to give an automatic record of spontaneous movements, *Locusta* always shows most activity in moist air;[80] and *Schistocerca* flies more consistently when the humidity is high.[268] The insect will therefore come to rest more frequently in dry surroundings. The reactions of the adult *Tenebrio* are the same.[56] In the wire-worm *Agriotes* the reaction is reversed: the larva is sluggish in saturated air, but activity and rate of movement increase as soon as the humidity falls below 100 per cent. R.H.; and there is a similar response to the quantity of water in the soil: intense burrowing in dry soil, immobilization in moist.[93] Tsetse-flies, *Glossina*, have humidity receptors in the spiracular filters of the thorax: at high humidity these cause a central inhibition of locomotion; in dry air, or if the sense organs are destroyed, the level of activity is increased.[198]

Stereokinesis or thigmotaxis—In all the stimulatory effects so far considered, the stimuli received in the sense organs have served to increase the excitability of the nervous system. But in the case of the mechanical senses the reverse effect is often seen: movements and reflex sensitivity are inhibited by stimulation. When the earwig comes to rest it brings as many of its tactile receptors as possible into contact with some object; it may then require very strong stimuli to induce it to move. Although light ordinarily has a kinetic effect, the inhibitory influence of these contact stimuli may successfully compete with this, so that the insect remains at rest even when the eyes are brightly illuminated (Fig. 196).[165] This inhibitory effect of contact is termed 'stereokinesis' or 'thigmotaxis'. It is seen, also, in *Cimex* which, like the earwig, may settle in the light if its contact receptors are sufficiently stimulated;[123] and in the moth *Amphipyra*, which normally creeps into dark crevices, but will come to rest equally between glass plates.[96] Many insects, e.g. *Haematopinus*,[163] remain constantly in motion on smooth glass, but soon come to rest ('sleep' or 'akinesis') on a rough surface.

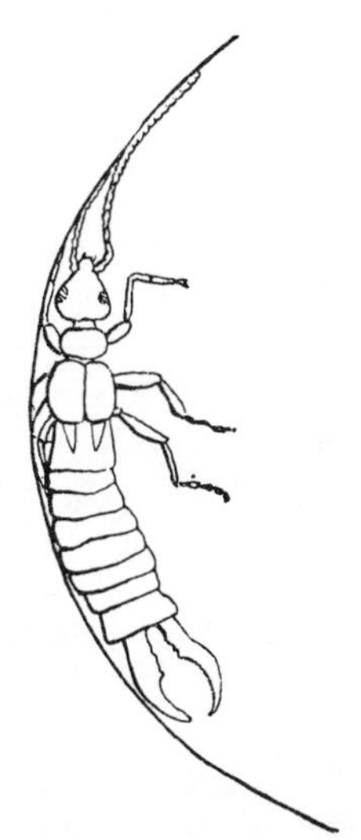

FIG. 196.—Example of thigmotaxis. Position taken up by *Forficula* in a circular glass container (*after* WEYRAUCH)

The inhibitory effect of contact is seen very strikingly in flying insects. Loss of contact between the tarsi and the ground, in Odonata[153] and in Muscid flies and other insects,[38] immediately induces flight. A suspended fly with vibrating wings instantly comes to rest if a little ball of cotton-wool is placed between the feet;[38] in *Musca*, contact with a single claw of one leg is sufficient to inhibit flight, and this inhibition disappears after amputation of the tarsi.[31] If any part of the body of a suspended insect is touched, it will extend the legs to this point, secure contact, and so inhibit the flight movements.[62] Similarly, the swimming movements of *Aeschna* nymphs are reflexly inhibited if a paper ball is placed between the feet[150] and the turning over of beetles or caterpillars which have fallen on their backs can be prevented if they are given some object to hold.[21] [See p. 345.]

Reflex immobilization, hypnosis, &c.—Akinesis induced by contact stimuli is the normal state of rest or sleep in insects. But in response to various mechanical stimuli the nervous system may be thrown into a state of abnormal inhibition. This state is sometimes compared with the local fatigue or inhibition which occurs in a reflex when the stimulus has been repeated a number of times; but instead of being localized it spreads throughout the nervous system.[166] It is termed 'hypnosis' or 'reflex immobilization'; but it is not always clearly distinguishable from stereokinesis or immobilization through passive contact stimuli.[61] It is sometimes regarded as a reflex with a precise peripheral localization, the receptors being constant in a given species;[118] but it can often be induced by a great variety of stimuli—repeated pressure, vibration, sudden seizure. In the earwig it is most readily induced by holding the pincers in forceps and drawing the insect along on the surface of cloth or paper;[166] in *Ranatra* it can be brought on by touching, rolling between the fingers, &c.[65] In many insects, beetles, bugs, caterpillars, it may be induced by sudden loss of contact between the feet and the ground, as when the plant on which the insect is walking is suddenly jarred; it is then commonly termed 'thanatosis' or 'death feigning'.[9, 50] In *Triatoma* the state may supervene naturally during feeding.[59] In *Carausius* it shows a diurnal rhythm and comes on during daylight; in Gerridae it may occur naturally or can be induced by pressure on the leg;[62] in these insects it appears to be a protective adaptation to life among plants.[137] The rolling or 'spiral reflex' of some caterpillars, and the 'warning attidues' assumed by other caterpillars and saw-fly larvae, seem to be special cases of the same response.[97, 140] The reflex centres for clinging, and turning over, in caterpillars are located in the suboesophageal ganglion or in the thorax; but they are subject to inhibition by a centre in the protocerebrum responsible for inducing reflex immobilization (thanatosis). If this centre is eliminated these reflexes at once become evident.[218]

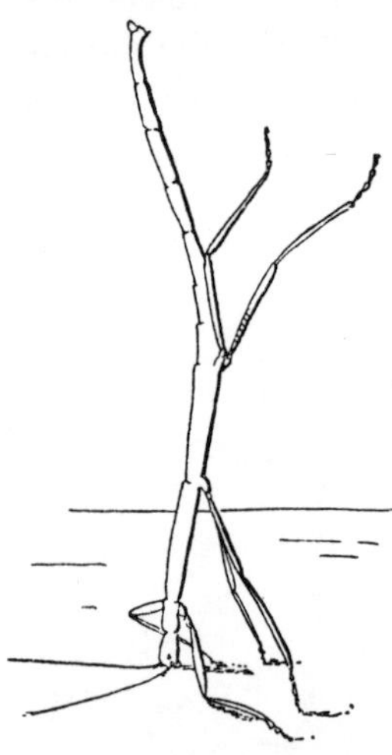

FIG. 197.—Catalepsy in *Carausius*. The appendages have been forced into appropriate positions and the insect balanced on its head (*after* SCHMIDT)

During this condition of immobilization, all reflexes for locomotion and the correction of posture are inhibited, sensation is reduced, and the muscles are in a state of tonic contraction.[61] In *Ranatra*[65] and *Carausius*[129] the condition shows a striking resemblance to the state of 'catalepsy' in man. There is the same insensibility to stimuli, the same absence of fatigue, and the muscles show the same characteristic '*flexibilitas cerea*'—they are plastic and yielding so that the limbs retain any bizarre position into which they are forced (Fig. 197).

If *Carausius* is cut through between the meso- and metathorax while it is immobilized, the posterior fragment quickly recovers its activity; it becomes excessively responsive to reflex stimulation, and cannot be immobilized again; while the anterior fragment remains in its cataleptic state.[129] Similarly, if *Ranatra* is divided between the 1st and 2nd thoracic ganglia, both separated fragments will show the response, but it is of much shorter duration in the posterior fragment; insects without the head can be immobilized only for brief periods.[65] Thus the chief nervous centre concerned seems to lie in the brain.[137] If the response is induced repeatedly by appropriate stimulation it persists a shorter and shorter time until finally the nervous system becomes too excitable and the insect is quite refractory.[65, 88]

ORIENTATION

A great part of behaviour consists in orientation—the direction of the movements of the organism in space in response to external stimuli. For the purposes of physiological description such responses are regarded as being made up of a series of 'forced movements' or 'taxes', that is, reactions which invariably follow a given stimulus when the organism is in a given physiological state. Such 'forced movements' are themselves made up of a succession of reflex responses. But in dealing with whole organisms, in which the central mechanism of the response is quite obscure, it is often convenient to classify the mode of orientation without attempting to press the analysis to the level of these component reflexes.

Mechanisms of orientation—Mechanisms of orientation have been classified by Loeb[96] and Kühn.[37, 90] The classification set out here is that of Fraenkel and Gunn.[39] These classifications are convenient, but it must be realized that several of the mechanisms may be acting at one and the same time, and that there are orientated movements which do not fall naturally into the scheme.

(i) **Kineses.** In the first place the insect may reach its destination without being truly orientated. The direction of its movement is not precisely related with the direction of the source of stimulus. Sense organs which will discriminate different intensities of stimulation are all that is required. Reactions of this type are termed 'kineses' or undirected reactions.

(*a*) The simplest form is **orthokinesis** where the average speed of movement or the frequency of activity depends on the intensity of stimulation. While the stimulus acts the insect moves; when it ceases to act the insect comes to rest, and here the animals tend to aggregate. We have already considered examples of orthokinesis in response to light, humidity, &c., earlier in this chapter.

(*b*) A more complex form of undirected response is **klinokinesis** in which the insect moves in a straight line in a favourable environment but as soon as it enters a mildly unfavourable environment it begins to make turns, the frequency of turning increasing with the strength of the stimulus. If the stimulus is very strong the insect turns aside instantly, giving an 'avoiding reaction'. After a time it becomes 'adapted' to the adverse stimulus; it then goes straight, and continues to do so as long as the stimulus remains the same or diminishes. If the stimulus increases, it once more begins to turn. These two factors, of random turning and adaptation, lead the insect to a favourable environment (Fig. 198). Klinokinesis is a mechanism well suited for orientation to diffuse types of stimuli of temperature, smell, humidity, or texture[170] where steep gradients are wanting.

(ii) **Taxes.** These are directed reactions for which a more complex system of sense organs is required in order to make the discrimination of direction possible. They are divided into:

(*a*) **Klinotaxis** in which the insect compares the intensity of stimulation in its vicinity by alternate movements to the left and right. These may be movements of the whole body or of some mobile receptor organ such as the antennae. The characteristic feature in klinotaxis is that stimulation is compared successively on the two sides.

(*b*) **Tropotaxis** is similar, but the intensity of stimulation on the two sides is compared simultaneously by means of symmetrically placed receptor organs.

These terms merely describe the mechanisms by which the sense organs are used to bring about the orientation of the animal. In the past, klinokinesis and klinotaxis have commonly been included together under the terms 'avoiding reaction', 'shock reaction', 'trial and error' or 'phobotaxis'. In many responses the whole series of reactions, orthokinesis, klinokinesis, klinotaxis and tropotaxis, come into action in turn as the animal approaches the source of stimulus and enters a steeper and steeper gradient of stimulation. Even at the same instant more than one mechanism may be operating.

The tropotactic mechanism resembles that advocated by Loeb[96] under the name of the 'tropism theory'.* Loeb explained 'tropisms' or 'forced movements' in terms of the 'tonus hypothesis', according to which the muscular tone on the two sides of the body is proportional to the intensity of stimulus received in bilaterally symmetrical sense organs, and, this being the case, the animal when it moves turns towards the side on which the muscular action is strongest. It continues to turn until the sense organs on the two sides are equally stimulated.

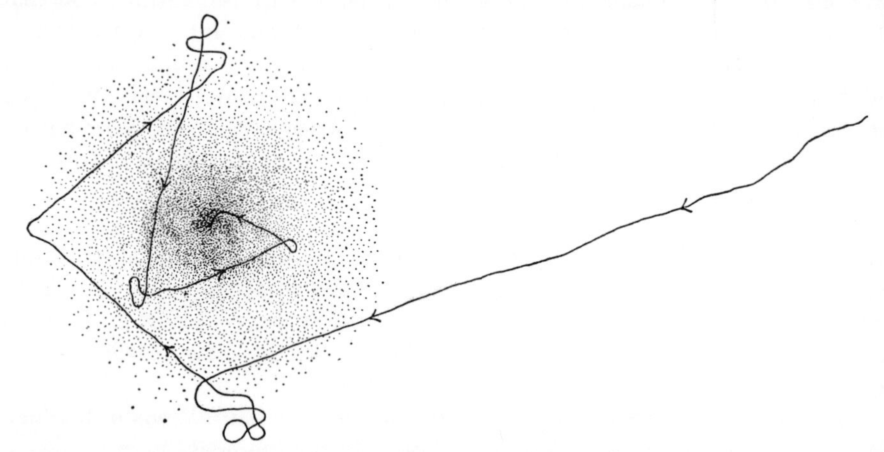

FIG. 198.—Hypothetical track of an insect approaching the centre of some favourable diffuse stimulus. It goes straight while the stimulus remains constant or is increasing, it turns at random when the stimulus is decreasing (*after* WIGGLESWORTH)

Muscle tone is then equal also and the insect moves in a straight line towards or away from the source of stimulus. Probably few authors would to-day accept this as a satisfactory description.

(*c*) **Telotaxis.** This includes all those responses in which the organism 'fixates' one source of stimulus with its sense organs and advances towards it, or orientates itself in respect to it, in such a way that a certain region of the receptor apparatus is always acted upon by the chosen stimulus, other sources of stimulus being disregarded. The mechanism is defined as a series of reflexes, evoked and ended by the localization of the stimulus on the sensory surface.

Theoretically, these types of orientation may be effected through any of the senses. Thus the response to light may be 'orthokinetic', 'klinotactic', 'tropotactic', &c.; response to smell may be a 'chemoklinotaxis' or 'chemotropotaxis' and so forth.

Orientation by light—The kinetic effects of light have already been discussed; in addition to these, orientated movements to sources of light play a large part in the behaviour of insects. In the simplest cases these movements

* The word 'tropism' has been used in so many different senses that it is best avoided.[103]

are responses to differences in light intensity perceived by the light sensitive organs. Where the eyes are more complex, the localization of the source of stimulus on the retina becomes important. And this leads on to examples in which the insect is orientated by the visual pattern.

Some insects are photopositive; they move towards the source of light; others are photonegative. Some, such as Muscid larvae, will always show the same esponse. In others it appears only when they are in a given physiological state. Young caterpillars of *Porthesia* (*Euproctis*) are strongly positive to light before hey have fed; this response normally leads them upwards to the leaves of their ood plant; but it is lost almost completely after feeding.[96] And at the time of the nuptial flight male and female ants are strongly positive, but this ceases as soon as they have shed their wings.[96] In the tsetse-fly *Glossina* and to a less extent *Stomoxys*, the response is affected by temperature. These flies are usually positive, and go to the light half of a cage; but if the temperature is raised towards 40° C. the response is reversed and they make for the dark half, even though the emperature there be so high that they soon drop dead.[71] As in *Glossina* the adult *Eristalis* reacts positively to light within the normal temperature range of 10–30°C. Both above and below this range it becomes highly negative. The upper temperature at which this change occurs becomes lower if the illumination is increased.[177] (This same increase in illumination also lowers the resistance to high emperature (p. 671).) The flight of certain Carabid beetles towards the west is due to the fact that they fly particularly on calm warm evenings and start flying owards the sun.[95] In many insects the response to light occurs only if they are alarmed' by some external stimulus: flies in a room are usually indifferent to ight, but make for the window if disturbed; mosquito larvae show an immediate negative phototaxis and positive geotaxis if their vessel is jarred.[35, 66] In such cases the response is sometimes regarded as an 'escape reaction' enabling the insect to flee straight to the light or to the dark in circumstances of danger.[42]

Phototaxis in insects with simple eyes—The larvae of Muscid flies show a well-marked negative phototaxis. The light receptors have already been described (p. 243). they are incapable of any kind of image formation; the response s dependent solely on the perception of general light intensity. Exposed to a single light, the larvae move directly away from it. Exposed to two lights they move away along the resultant line dividing the angle between the two sources. If the two lights are of unequal intensity this line deviates away from the stronger; and although the degree of deviation is not exactly in accordance with the Weber-Fechner law (which lays down hat the effective difference between two stimuli s determined by the percentage difference between them regardless of their absolute intensities), yet all the results obtained in such experiments can be explained by supposing that here is an effect on the musculature of the two sides of the body which is proportional to the strength of stimulus received by their sense organs.[116] This insect therefore has been regarded as affording the ideal example of 'tropotaxis'; the insect behaves as though it were being forced

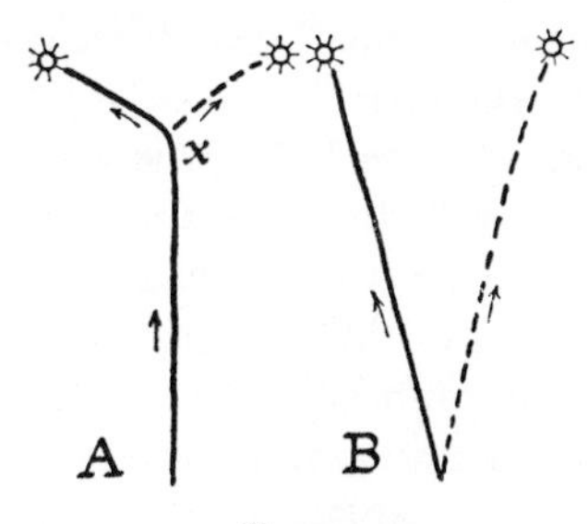

FIG. 199

A, *B*, two types of course followed by a positively phototactic caterpillar in the presence of two sources of light. *x*, deciding point ('Entscheidungspunkt'), (*after* LUDWIG).

by the effect on muscle tone to turn into such a position that the symmetrical organs receive the same illumination.[96] But if the behaviour of these maggots is observed more closely it is seen that they swing the head alternately to the left and right and then choose the side on which the photoreceptors are most shaded. The larvae, in fact, are orientating themselves by 'klinotaxis'. If they are in a diffuse light which is made suddenly brighter each time the larva swings its head to the left, then it will crawl in a circle to the right.[39] This, however, does not exclude the possibility that the maggot can on occasion use a tropotactic mechanism.

In the louse *Pediculus*, tropotaxis is certainly the most important element in the avoidance of light; it makes circus movements to the covered side if one eye is blackened.[170] Young caterpillars of *Euproctis*, *Orgyia*, *Lymantria*, &c., before their first meal are positively phototactic; they will crawl straight towards a single source of light. If they are offered two sources of light they generally approach a variable distance along the resultant between them and then, usually when they lie at an angle of about 115° to the direction of the light, they begin to make searching movements and go straight to one light or the other (Fig. 199).[13, 9] This response is sometimes described as a 'tropotaxis' while they advance along the resultant, and a 'telotaxis' when they begin to move towards one only of the two lights. But the apparent neglect of the one source of light in the later stages may result from its falling outside the field of vision during the searching movements. The same response occurs, and the larva may go to either light, if the eyes are covered on one side.[13, 92] Some individual caterpillars when offered two lights will go straight to one of them. This again appears like a telotaxis or orientation towards a chosen source of stimulus. But it has been proved to depend on a natural bias of these individuals to incline in one direction as they advance.[51, 10] Larvae of *Ephestia* are negatively phototactic; but they show this response to light stimuli only when they are in a definite state of stimulation. They then move directly away from a single light source and along the resultant if two lights are present.[11] Larvae of *Vanessa*, which are positively phototactic, blinded on both sides, still find their way to a source of light—perhaps by a cutaneous light sense (p. 244); a single stemma suffices for the complete series of orientation reactions.[113]

Phototaxis in insects with compound eyes—In insects with compound eyes capable of perceiving accurately the direction of a source of light, the mechanism of orientation becomes more complex. In a number of insects, such as the flies *Proctacanthus*[49] and *Eristalis*,[104] in *Ranatra*,[64] and nymphs of *Agrion* and *Cloëon*,[1] if the eyes on one side are blackened, the muscular tone on the uncovered side is relatively increased, so that the insect at rest inclines towards that side. This has already been quoted as an example of the stimulatory effect of light, or 'phototonus' (p. 312). Sometimes this effect becomes apparent only when muscular movements are made; the muscles on the uncovered side are then more active, and the insect moves in a circle in that direction.[1, 49] According to this explanation the orientation of these insects to light would be an example of 'tropotaxis'; the insect would be pictured as being orientated through the 'kinetic' effect of the sense organs being unequally distributed on the two sides of the body, a view which would unite 'kinetic' and 'orientating' responses into a single mechanism.[96] As the exposed eye becomes light-adapted, the exaggeration of tonus on that side is diminished; and in the insect at a given state of light-adaptation, the effect is dependent on the intensity of illumination: the greater the illumination the smaller the circle in which the moving insect turns.[49]

After a time, however, in such an insect with one eye blackened, the circus

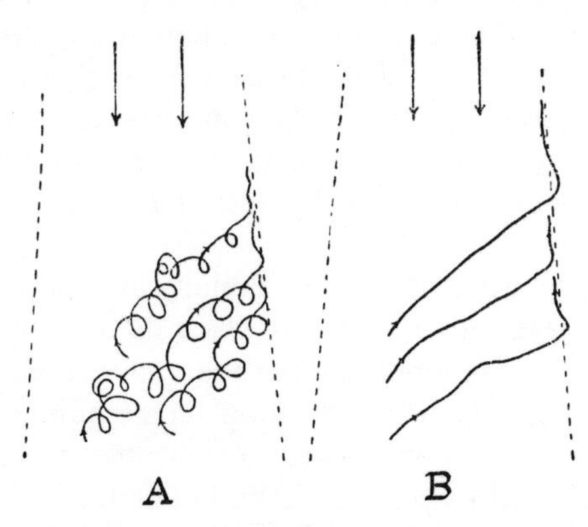

FIG. 200.—Tracks followed by *Dineutes assimilis* (Col.) with left eye blackened, in a divergent beam

Arrows show direction of light; broken lines show margin of beam. *A*, with eye dark-adapted; circus movements to the margin of beam, then straight to the source. *B*, after being in the beam some minutes (eye light-adapted); elimination of circus movements (*after* CLARK).

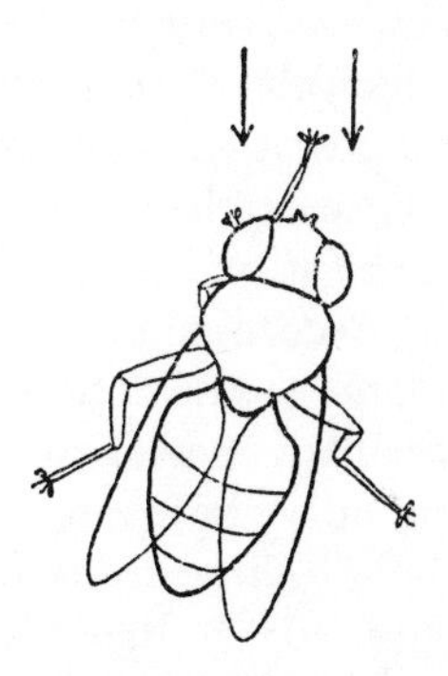

FIG. 201.—The fly *Eristalis* advancing towards a source of light after removal of the fore and hind leg on the right side (*after* MAST)

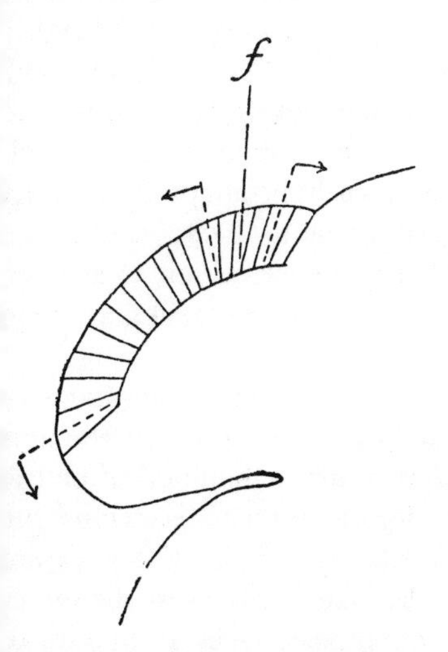

FIG. 202.—Diagram of the telotactic turning response in the compound eye (*from* LÜDTKE *after* KÜHN)

Arrows show direction of turning induced by illumination of different regions of eye. *f*, line of fixation.

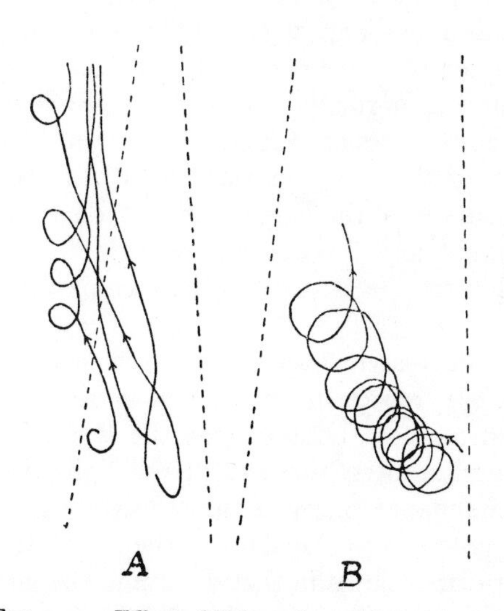

FIG. 203.—Effect of light reflected from the background on circus movements in *Notonecta* (*after* LÜDTKE)

A, tracks followed by four insects in dark-adapted condition in a beam of light, with a black background; *B*, track followed by similar insects in the presence of a white background.

movements cease; the insect will go straight to the light, or turn in either direction (Fig. 200). Orientation cannot, therefore, be wholly dependent on the relative intensity of light received in the two eyes. Moreover, insects with one or more legs removed from one side can still walk straight to the light (Fig. 201). If the lower half of the eyes in the cicada *Cryptotympana* are coated, it will fly straight upwards and vanishes into the air; if the upper half is coated it hits the ground.[26] Orientation, therefore, is not necessarily dependent on balanced action of the locomotor appendages. These observations have been made on *Vanessa antiopa* (Lep.),[32] *Eristalis* and *Proctacanthus* (Dipt.),[104] *Notonecta* (Hem.),[27, 99] *Dineutes* (Col.)[27] and *Apis* (Hym.).[155] They are best explained by supposing that illumination of the eye brings about a series of co-ordinated reflexes in the legs of both sides of the body specifically related to the localization of the stimulus in the retina.[104, 105]

According to this view, the eye consists of two functionally distinct regions: a large lateral zone which causes turning movements towards its own side, and a narrow medial zone which causes movements towards the opposite side. The boundary between these two constitutes the line of 'fixation', corresponding with the *fovea centralis* in the vertebrate eye; it evokes no turning movements (Fig. 202).[99] In the unilaterally blinded insect, in the dark-adapted condition, the sensitivity of the lateral region is so great (in *Eristalis tenax* the central part of the eye is often more than 50 times as sensitive as the anterior part,[33] in *Notonecta* the eye 78° behind the median axis may be 11 times more sensitive than 12° in front)[99] that the insect exposed to a beam of light turns repeatedly in circles towards the uncovered side. When the eye is in this sensitive state the light reflected from the background also contributes to the turning effect; turning is less marked if the background is black, with a white background it is continuous (Fig. 203).[27] Or if one eye is dark-adapted and the other light-adapted the same turning to the dark-adapted side is seen.[27] The beetle *Dineutes* which is normally positively phototactic may become negatively phototactic when the eyes are fully dark-adapted; this is particularly evident if only a small part of the eye is exposed.[121] But as the eye becomes light-adapted, and the sensitivity of the lateral region decreases, the insect is able to 'fixate' the source of light and go straight to it.[27, 104, 105] It is this response which is termed 'telotaxis'; it is the limiting condition between two sets of reflex turning movements.[99]

On this view the tilted positions taken up by flies after blackening parts of the eye, represent attempts to turn towards the light and bring it into the axis of fixation without moving the feet. These positions are maintained after the light has been turned off; the light itself is therefore not necessary for the maintenance of the tonus effect.[104] In the *Libellula* nymph, if the eyes are covered and the head is then bent over to one side and fixed with wax, the insect performs spirals in that direction: the circus motion appears to be secondary to the position of the head.[150] The axis of fixation in these insects is not morphologically established; it varies with the physiological state of adaptation, &c.;[99] in *Notonecta* the various ommatidia change their function in successive nymphal stages as the eye grows;[99] and insects with the legs removed from one side move obliquely towards the light, fixating it with a new part of the retina. It is to explain adjustments of this type in the mechanism of orientation that psychological conceptions such as the 'attainment of the end in view' are sometimes invoked.[105]

'Light-compass' orientation—Insects may sometimes orientate themselves obliquely or at right angles to a source of light in such a way that one side is nore strongly illuminated than the other. This behaviour was first observed in nts, which have been found in nature to maintain a straight course by moving t a fixed angle to the sun's rays (p. 333); if ants are detained on one of their xcursions for an hour or two and then liberated, they follow a route which nakes an angle with their outgoing route exactly equal to the angle that the un has moved round (Fig. 204).[17] (But see p. 340.) This behaviour was thereore termed 'light-compass' orientation.[126] It has since been observed in many ther insects under experimental conditions.[242]

Larvae of *Vanessa* or *Pieris* exposed to the sun will walk in a straight line; f the ground on which they are walking is rotated through 90°, they will turn hrough the same angle; if they are covered with a dark box, they will turn in ends and spirals, but on exposure to the sun again they move in a straight line n the old direction (Fig. 205).[51] If *Geotrupes*, *Coccinella*, *Forficula*, *Carausius*, cc., are walking in the presence of a single source of light from a given point,

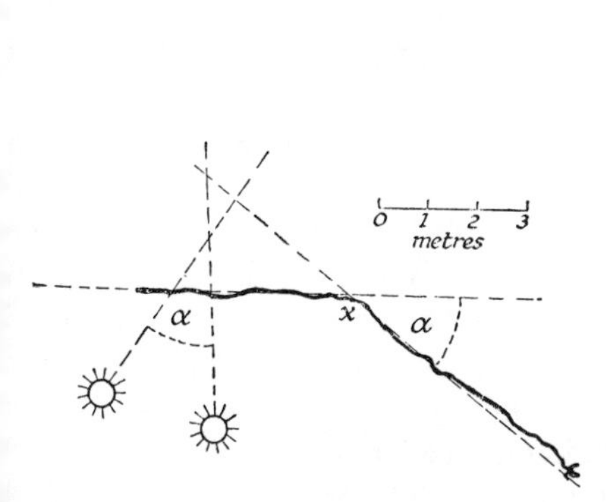

IG. 204.—Course followed by an ant, *Lasius niger*

On reaching *x* at 2.39 p.m. it was placed in a box ntil 5.09 p.m. It was then released and followed course making the same angle α with the original ourse as the sun had moved through in this time fter BRUN).

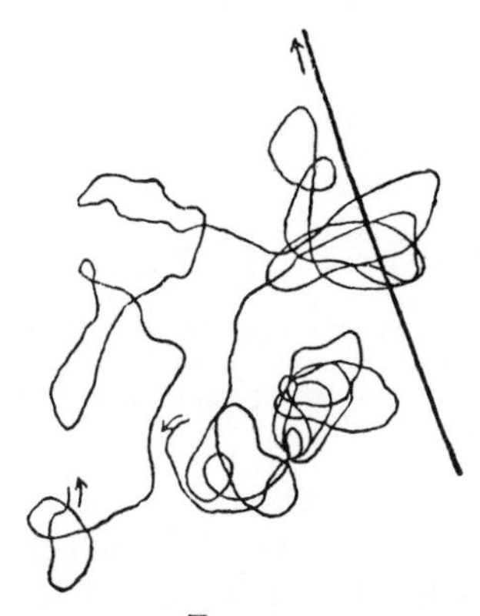

FIG. 205

The straight line shows the course followed by a Chrysomelid beetle exposed to a bright light. The convoluted line shows the course followed in the dark (*after* v. BUDDENBROCK).

nd this light is switched off as another is switched on, they will turn through the mallest angle possible so as to receive the new source of illumination from the ame quarter.[22] It has, however, been pointed out that in many circumstances the pparent orientation by sun-compass is in fact orientation with reference to the un as a *heat* source, and this may be combined with orientation to the polarizaion pattern in the sky.[270]

This oblique type of orientation to light has sometimes been regarded, for xample in caterpillars,[92, 100] as an 'asymmetrical tropotaxis'; that is, the nsect maintains equal activity on the two sides of the body only when there s a constant degree of inequality in the stimulation of the two eyes. In insects vith primitive eyes capable of perceiving only differences in luminosity this ould be the only mechanism by which orientation could be maintained.[100] But ve have seen (p. 241) that even caterpillars are capable of some degree of pereption of form; and most authors consider that the response is, rather, an asymmetrical telotaxis', dependent on the insect keeping constant the incidence f the light rays on a given region of the retina. For example, if *Geotrupes* s moving obliquely in respect to one light and a second is switched on, it pays no attention to the second and remains orientated by the first.[68] The response probably depends therefore on the maintenance of a constant pattern of stimulation in the retina.

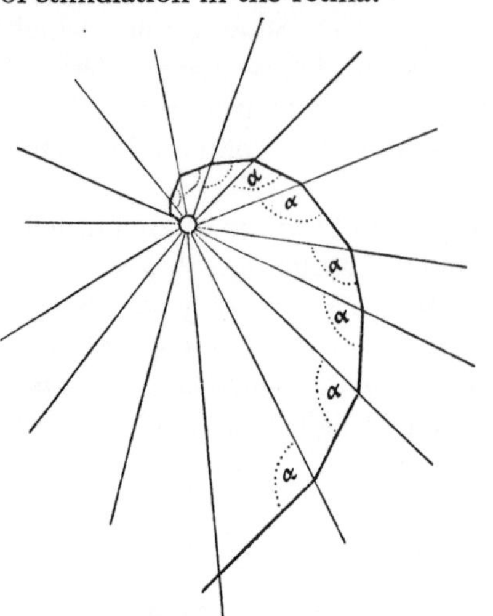

FIG. 206.—Menotaxis (*after* V. BUDDENBROCK)

An insect, moving so that its course makes a constant angle with the beam of light from a near source, approaches the light along a logarithmic spiral.

If a bright light is employed, the insect may be caused to turn as soon as the beam of light passes from one ommatidium to the next. Good agreement exists between the ommatidial angle and the smallest angle through which the light must be moved in order to cause a change in direction. This affords further evidence that the ommatidium is the visual unit. At low illumination, in *Geotrupes* and some other beetles, the light must be moved through a larger angle; it appears as though in dim light seven adjacent ommatidia become linked to form a new visual unit.[22]

In this type of orientation, if the source of light is distant, like the sun or moon, the insect will continue in a straight line. But if the light is close, its incidence on the retina will soon change if the insect goes straight. The angle of incidence of the light can be kept constant only if the insect continually turns towards the source. It will thus move along a logarithmic spiral ending at the light itself (Fig. 206).[19] The flight of moths to a flame is explained on these lines; the insect utilizes a near source of light for a mechanism of orientation which is adapted only to distant sources.[21] Caterpillars will sometimes crawl to a light in exactly this type of spiral.[100]

Menotaxis—Insects may orientate themselves obliquely not merely to a bright point of light, as in the light-compass response, but to more or less complex patterns. The basis of all such responses appears to be the same: the maintenance of a constant visual pattern. They are commonly grouped together under the term 'menotaxis'.

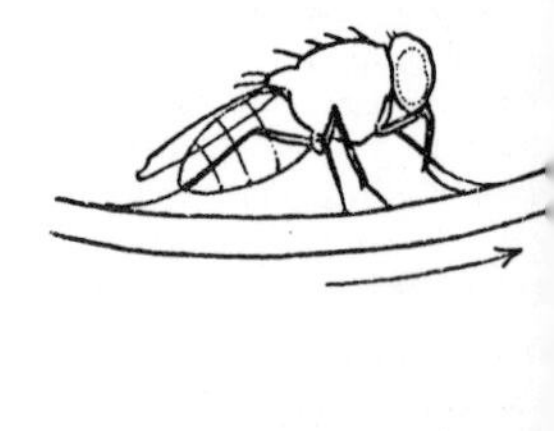

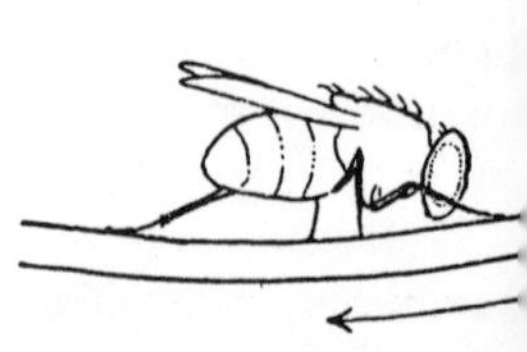

FIG. 207.—Attitudes taken up by the fly *Pollenia* when a striped pattern is moved below it in the direction of the arrows (*after* GAFFRON)

As the result of such photic orientation, insects placed on a turn-table move round when this rotates; they will often remain, for example, facing a window or other conspicuous object.[119] They will show the same response, also, if they themselves are stationary and a striped screen is moved past them ('optomotor response').[128] If stripes are moved underneath a resting fly (*Pollenia*) it will take up an attitude, recalling that of *Proctacanthus* with partially covered eyes, reaching out in the direction of movement (Fig. 207).[48] *Drosophila* gives the same responses.[78] Aquatic insects such as *Aeschna*,[125] *Dytiscus*[174] or *Notonecta*[132] will swim around a glass container if vertical stripes are rotated around it, keeping themselves in line with a given stripe. And various insects, Orthoptera, Coleoptera, and Hymenoptera, if they are fixed, and stripes are moved in front of them, will turn the head and extend the antennae towards the oncoming pattern.[171] If *Calandra* walks between vertical stripes which move at different speeds, it follows an oblique straight course, showing that the action of the two sets of stripes are simply summated.[197]

The orientation of flying insects in a wind is brought about by the same mechanism.* Flying mosquitos will turn towards vertical stripes perceived by the lateral ommatidia of the eyes; this response is discussed under the heading 'skototaxis'. But if the stripes are below and are perceived by the ventral ommatidia, the reaction is quite different. If the stripes are moving the mosquito flies over them in the direction of movement, gaining on the stripes. That is equivalent to flying against the wind over a stationary background. And, in fact, if the air is moving and the stripes are still the mosquito will fly upwind. In *Aedes aegypti* the maximum speed of flight is 3·3 miles per hour (150 cm. per second). If the wind speed exceeds this they are carried backwards and the stripes will move from behind forwards over the retina. This the insect will not tolerate; as soon as the wind speed exceeds this value the mosquito alights. This special optomotor response to images perceived by the ventral parts of the eye explains the upwind orientation of insects near the ground, and the refusal of flying insects to take off if the wind exceeds a certain velocity. When far above the ground this mechanism will not operate and the insect will be carried along by the wind.[81] In still air bees fly to and from the hive at about 29 km. per hour, when they are offered rich sugar solutions. When the solution is more dilute the speed of flight in both directions may be reduced to 4–7 km. per hour. In an opposing wind they increase this rate of flight, and in a following wind they reduce it so that their movement over the ground tends to remain unchanged. They compensate for a cross-wind by directing their body axis into the wind.[212]

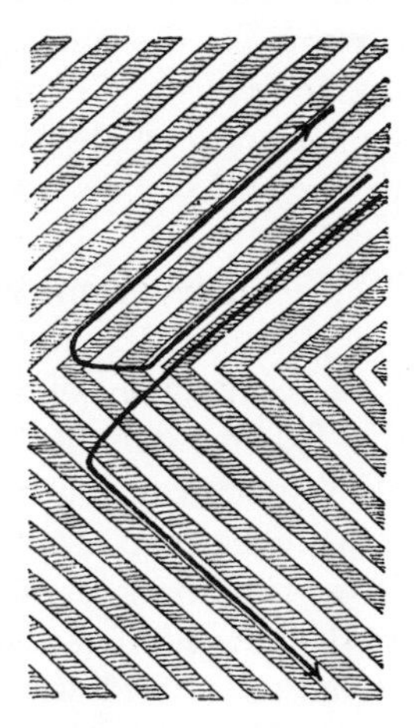

FIG. 208.—Tracks followed by *Coccinella* on a pattern made up of black and white stripes (*after* TISCHLER)

The tendency for insects to walk along black and white stripes, or along the border of a wide black stripe, even when set obliquely in a vertical surface, which is seen in *Lymantria* larvae,[69] *Carausius*,[74] *Coccinella*, &c.,[149] is perhaps in part menotaxis of the same kind (Fig. 208). The circus movement due to covering one eye in *Coccinella* can be eliminated in the presence of such a pattern.[149]

'Gregariousness' in locusts affords another example of the important part played by these 'compensatory reactions' in insect behaviour. When locust populations increase to a high density the larvae ('hoppers') come together in bands where they continually touch one another and quickly learn to form coherent groups. It is in these groups that gregarious behaviour develops.[206] Gregarious

* *Stomoxys* and *Lyperosia*[89] will sit at rest facing a current of air. This is presumably a tactile response.

behaviour depends on visual reactions: each locust moves in such a way that it tends to oppose or nullify the movement of visual images across the retina. When one locust advances, its neighbour advances with it. Hence the swarming insects keep moving along together. This has been termed 'gregarious inertia'. Associated with it is a 'light compass reaction' to the incidence of the sun's rays. If the locust hoppers are shaded with a blanket and the light of the sun is reflected on them from the other side by a mirror, they may be caused to turn about and march off in the opposite direction. Or they will march round and round a central source of light. The combination of these two reactions ensures the continued migration of whole bands in one direction. Hoppers basking in the sun (p. 671) probably 'fix' the angle of incidence of the light and preserve it when they move off.[82]

This **'optomotor response'** appears to be a reaction to changes in the general visual pattern, not to a single point, and it is effected through the lateral and peripheral regions of the eye. It is quite distinct from, and is, indeed, inhibited by the 'fixation response', which is a reaction to single points falling on the region of most acute vision.[151] If an insect is suspended in mid-air and allowed to hold in its feet a ring, it makes walking movements with its legs and the ring rotates. If the ring forks its motions reflect the orientation responses of the insect. This method has been used to study the optomotor response in various Coleoptera and Hemiptera. The lower threshold of movement for the optomotor response in the weevil *Chlorophanus* is 1·3° per minute or less, that is, about one quarter of the speed of the large hand of a clock.[180] The perception of movement in the eye of the beetle results from the multiplication between stimulus intensities in adjacent ommatidia.[180] [See p. 345.]

Normal flies (*Eristalis*) are not greatly influenced by a stationary striped pattern when they move near it: there is a central mechanism which largely compensates for the spontaneous forward movements of the insect. But if the head is rotated on the neck through 180° the fly shows circus movements: the effect of its own motions are no longer compensated. Flies treated in this way turn *against* the movement of a vertically striped cylinder, whereas normal flies follow this movement. But eventually the flies become adapted to the reversed position of the head and are then able to move in a straight line towards a source of light.[184] The eye does not distinguish between its own movements and those of its surroundings, but the animal with its central nervous system intact can integrate all the impressions it receives and perceive the difference very well.[182]

By the same mechanism *Notonecta* can maintain itself at a fixed point in a flowing stream: it turns and faces the current and swims so as to keep constant the visual impressions it receives. If landmarks on the bank are moved it will move with them; if the eyes are blackened, in the dark, or in a vessel with plain white sides, the reaction fails and the insect is carried down-stream. In this visual response, movement across the eye from behind forwards offers the greater stimulus; were this not so, active movements of the insect forwards would be inhibited.[132] *Dytiscus* with one eye covered makes circus movements towards the uncovered side; these movements can be arrested or even reversed by the orientation to revolving stripes.[174] Even Gyrinid beetles swimming in all directions in the water surface have been shown to maintain their position in flowing water by means of the surrounding pattern; although in this case the pattern of retinal stimulation is continually changing. Swarms of Gyrinids are completely disorientated in the dark or if a sudden change in landmarks is made.[16] The orientation against a current of water ('rheotaxis') is thus in fact a visual response. [See p. 345.]

Transference of menotactic responses—Bees which have oriented their movements in relation to the sun can transpose this orientation into a relation to gravity: movement directly towards the sun becomes converted into movement directly upwards on a verticle plane; 30° right of the sun becomes 30° right of the vertical (p. 265). Other insects besides bees are able to make this same transference and to convert the photomenotactic course they have been following on a horizontal surface into a geomenotactic course on a vertical surface. *Geotrupes* walking towards the light, when transferred to the vertical plane in diffuse light, walks *downwards*. If walking 30° left of the light source, it goes 30° left of the vertical in the downward direction. The angle is retained as in the bee, but the relation to the light source is inverted. On the other hand, *Coccinella* and *Melasoma* behave like the bee and identify 'towards the light' with the 'upwards' direction.[192] These same interchanges of stimuli in menotactic responses occur in ants; they can interchange a source of light and a polarized light pattern, or light and gravity; but they cannot interchange gravity and polarized light.[267] Trichoptera can effect the transference between light and gravity, but the angles are transposed on a proportional basis: the geotaxis angle is related to the phototaxis angle as 2 to 3.[229] A similar though smaller 'error' in transposition is seen also in the honey-bee.[214]

Skototaxis—Various insects exposed in a diffusely lit arena with a black screen in it, will orientate themselves towards the black screen; and among a series of grey screens they will make for the darkest. This response is termed 'skototaxis'. It has been observed in *Lepisma*,[108] *Forficula*,[86] caterpillars,[51] in the ant *Lasius*,[167] in mosquitos (*Aedes*),[81, 120] and in the louse *Pediculus*.[170] If *Forficula* is offered a dark screen and a single light, two modes of orientation may be made to compete, with the result that the insect may either go directly away from the light (negative phototaxis) or towards the screen (skototaxis) (Fig. 209).[86]

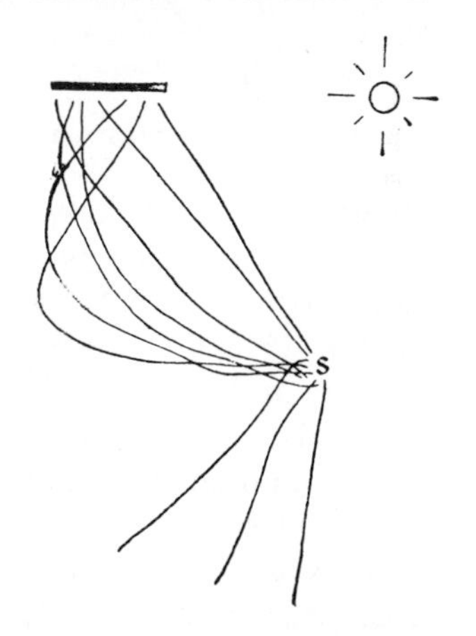

FIG. 209.—Tracks followed by *Forficula* in presence of a single source of light and a small black screen. Some insects show skototaxis; others show negative phototaxis (*after* KLEIN)

There are two elements in this response: a phototactic reaction towards a region of reduced luminosity and a telotactic reaction towards the form or movement of the dark object. It is questionable whether a special term is needed for its description.[39] In some cases, such as the mosquito[81, 120] or the louse,[170] the phototropotactic element predominates; in others it can be shown that form is important in the orientation. The small Carabid *Notiophilus* will ignore a black screen measuring 15×15 cm. near to it, and make for a screen measuring 4×4 cm. some distance away.[167] We have already seen that caterpillars prefer dark vertical forms of particular proportions (p. 241).

Dorsal light response—Certain aquatic insects, such as Ephemerid nymphs and Dytiscid larvae, maintain their position in the water by orientation to illumination from above. This is termed the 'dorsal light reflex'.[21] A nymph of *Cloëon* illuminated from below is caused to turn over on its back. This response persists if one eye is blackened, which suggests that it is a menotactic response. It is inhibited if the feet are in contact with some object (thigmotaxis).[2] The same reaction is well developed in the Naucorid *Aphelocheirus*.[147] The dorsal light response operates also in the dragon-fly *Anax* during flight (p. 167). The compound eyes are mainly responsible, but there is some disturbance in the response if the ocelli are covered. The reaction is unaffected if one eye is blackened: it is not a tropotactic response.[185] Larvae of *Acilius* and *Dytiscus* possess a dermal light sense in the tergites which enables them to retain the dorsal light response after total blinding.[188]

Orientation by temperature—The avoidance of high temperatures which is shown by all insects is clearly an extreme example of 'klinokinesis': as soon as the insect comes in contact with an unfavourable temperature it makes a reflex 'avoiding reaction' and turns away. The orientation of blood-sucking insects by

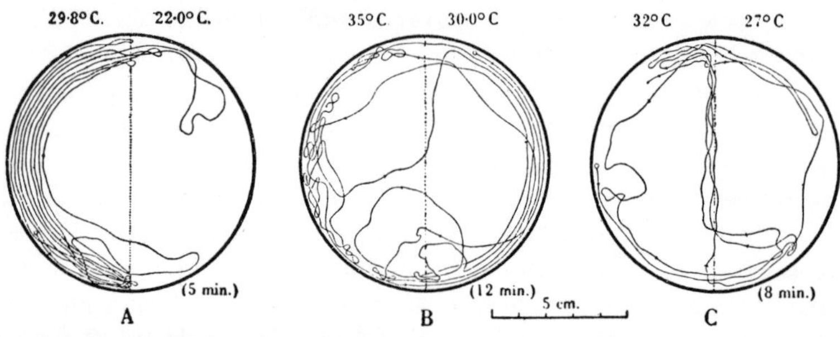

FIG. 210.—Tracks followed by louse *Pediculus* in an arena with the two halves at different temperatures. *A* shows 'avoiding reactions' to the low temperature. *B* shows 'klinokinesis'—an increased rate of random turning at the higher temperature. *C* shows klinokinesis in both halves of the arena and accurate orientation by 'klinotaxis' in the optimal temperature zone along the boundary (*after* WIGGLESWORTH)

the warmth of their host is a different problem. The importance of temperature for this purpose has been recognized in many such insects (p. 292) but the mechanism of orientation has been studied only in the blood-sucking bugs *Rhodnius*[169] and *Cimex*.[135] It has been proved in *Rhodnius* that it is the air temperature and not the radiant heat which is the source of stimulus: this bug is attracted equally by a test-tube full of warm water and by a similar tube covered with aluminium foil—although, as may be shown with a suitable thermopile connected to a galvanometer, the radiant heat emitted by the metal-covered tube is less than one-tenth that from the other. Placed at a distance of 4 or 5 cm. from a test-tube at 37° C., *Rhodnius* first cleans the antennae, then moves them in each direction in turn, and finally extends them towards the warm tube, advances to it and probes it with the labium. It makes exactly the same response and locates the tube in the same way if one antenna is removed. The stationary insect is thus able to judge with its antennae or with a single antenna in which direction the source of stimulus lies (Fig. 211).[169] This response is an example of 'klinotaxis': using a single antenna or both, the insect tests the air in each direction and then turns towards the preferred side. In the steep gradient of stimulation close to the source there may, however, be an element of 'tropotaxis', for the insect may suddenly incline towards the side with the antenna (Fig. 211, A).

The louse *Pediculus* lives in an environment between the clothes and the skin where the temperature is about 30° C., and it shows a very decided preference for this temperature (p. 676). The mechanism of orientation is mainly by 'klinokinesis'. If it is enclosed in an arena with a difference in temperature in the two halves, it follows a straight course at 30° C., it makes frequent turns in the regions above or below this temperature (Fig. 210, A, B). When one-half of the arena is above the preferred temperature and the other below, the louse may show an accurate orientation along the boundary (Fig. 210, C). It does this by 'klinotaxis', by swinging the head and antennae alternately to right and left as

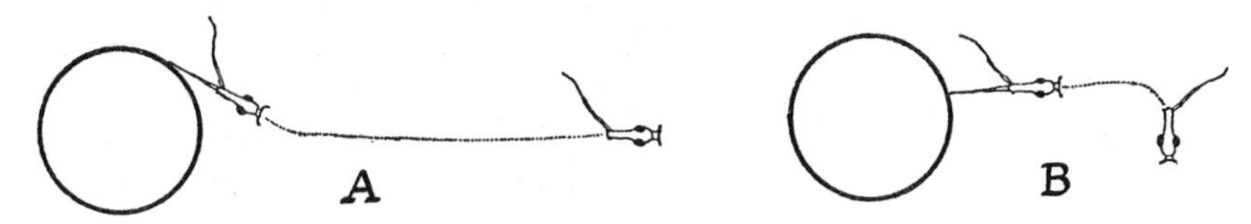

FIG. 211.—Course followed by *Rhodnius* with a single antenna in the presence of a test-tube of warm water

In *A*, the insect inclined abruptly towards the intact antenna when close to the source of stimulus (*after* WIGGLESWORTH and GILLETT).

it moves along. It shows no response to radiant heat within the range of temperatures normally encountered.[170] The reactions of *Musca* larvae to combined stimuli, during the feeding stages, show that their behaviour is dominated by temperature and humidity and is little influenced by light and smell. Both klinokinetic and klinotactic responses occur.[179]

Orientation by smell has much in common with orientation by temperature; indeed in blood-sucking insects the two factors are concerned at the same time.[201] In both cases the insect perceives with its antennae the qualities of the air in its immediate vicinity, and by means of these perceptions it is able to reach the source of stimulus. The insect is at first excited by the attractive odour whether from food or from the opposite sex; thescent has a stimulatory or kinetic effect. In this state it may move hither and thither and reach the source of stimulus by klinokinesis, assisted in the last stage of the approach, where the gradient of stimulation is steep, by klinotaxis or tropotaxis. That the antennae of the honey-bee can be used tropotactically is proved by crossing the antennae and fixing them in this position; a bee trained to the scent of aniseed and treated in this way will always go to the wrong side when exposed to this scent in a Y-shaped osmometer.[244] Of course the antennae may well be used klinotactically also.

In *Pediculus* the reactions to smell are the same as those to temperature. It shows no evidence of being 'attracted' by a favourable stimulus, it merely shows an increase in random turning movements upon entering a zone of adverse stimulation. And a zone may be 'adverse' because some repellent odour is present or because a favourable stimulus (to which the insect has become accustomed by recent experience) is absent (Fig. 213, B). Klinotaxis may again be involved when the gradient is steep; there is no evidence of tropotaxis.[170]

Drosophila deprived of the wings, in the presence of a strong source of odour such as fermenting fruit, with a concentration gradient in still air, can steer in a straight line towards it from a distance of 15–20 cm.;[4, 34] *Habrobracon* which locates its host, *Ephestia* larvae, by smell, advances to it in a straight line over the last 5 cm.,[111] and the same applies to the beetle *Geotrupes* in the presence of dung.[162] The males of Arctiid or Lasiocampid moths fly great distances along a comparatively straight course as they approach the female upwind; they wheel around at once when they pass to windward of the source of scent.[84] The excited insect will, in fact, continue in a straight line if the gradient of stimulation is rising, but turns aside if the gradient falls. The orientation of male Lepidoptera to the female is a flight against the wind evoked by a scent-laden air-stream.[263] The mechanism in all these cases is klino-kinesis. [See p. 345.]

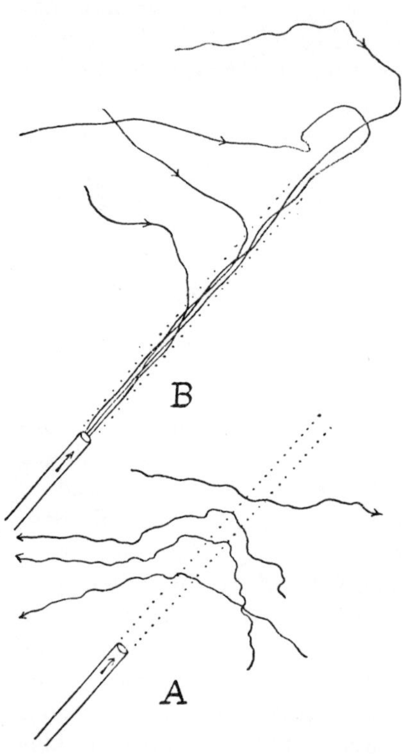

FIG. 212.—Tracks followed by *Drosophila* deprived of their wings when exposed to a current of air

A, odourless stream of air; *B*, air carrying odour of pears (*after* FLÜGGE).

If one antenna be removed in *Drosophila*[4] or *Calliphora*[58] in the presence of food, or in males of *Bombyx mori*[79] or *Tenebrio*[157] in the presence of the female, the insect turns more frequently towards the intact side; and *Geotrupes* so treated inclines towards that side when it gets close to the dung,[162] as does *Rhodnius* as it approaches a source of warmth and smell.[169] In *Geotrupes* the asymmetrical response of insects deprived of one antenna disappears after 4–8 days.[200] While in the presence of a repellent odour *Drosophila* with one antenna turns chiefly to the operated side.[34] The orientation is therefore sometimes said to be in part tropotactic. But whether such differential stimulation of the antennae is important in the normal insect is uncertain; for the difference on the two sides can never approach that caused by complete removal of one antenna; the gradient of smell must usually be greatly upset by air currents; and in insects with mobile antennae capable of testing the air on all sides, such as *Habrobracon*[111] and *Rhodnius*,[169] the insect can orientate itself to the source of smell almost equally well with one antenna. From a distance of 5 cm. *Rhodnius* can determine the direction of a source of warmth and smell (a warm tube of water covered with fresh mouse skin) before moving, by reaching out with its antennae; it will then advance straight to the source, sometimes inclining towards the intact antenna when close to it. It is more probable that any directed element in the orientation to smell is based on successive comparisons of stimulation on the two sides, that is klinotaxis. [See p. 345.]

It is possible in some cases that the current of air bearing an attractive odour may assist in orientation. *Geotrupes* generally approaches dung upwind, detecting it from a distance of 2–4 metres.[162] *Drosophila* deprived of the wings turns sharply into such a current of air and moves rapidly and accurately against it.

It is merely disturbed without being orientated by an odourless stream (Fig. 212)[34]—at least in *D. melanogaster*;[77] and *Drosophila* in flight likewise turns into the wind on encountering an attractive scent, using its eyes to establish the direction of the air-stream relative to the ground.[233] The male *Bombyx mori* produces an air current with its wings, drawing the air towards it. By testing the air from all directions it locates the source of the female scent.[262] [See p. 345.]

Another method of orientation by odours consists in following trails of scent. This is extensively used by ants which, as they go along, repeatedly touch the surface with the abdomen and so leave a delicate trail of formic acid. If artificial routes are made by painting the earth or plant stems with dilute formic acid, ants can easily be led astray along them.[21] Similarly, the ovipositing female of *Habrobracon* may find the larva of *Ephestia* by following its trail, bringing the antennae closely in contact with the ground.[111]

Orientation by Humidity—The part played by orthokinesis in the responses of insects to moisture has already been discussed (p. 312). In addition, klinokinetic and klinotactic reactions occur as with temperature and smell. The

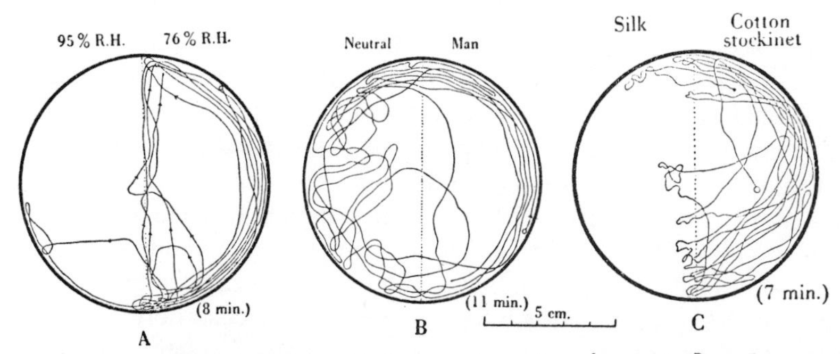

FIG. 213.—Orientation of the louse *Pediculus* to humidity (*A*), smell (*B*) and contact (*C*)

In each case there is klinokinesis, an increased rate of random turning on the unfavourable side, that is, the humid side in *A*, the side which lacked the smell of human skin secretions in *B*, the smooth side in *C*. There is some indication of 'klinotaxis' along the boundary in *A* (*after* WIGGLESWORTH).

adult *Tenebrio* makes more frequent turning movements in the moist half of an arena, and along the boundary it will explore the two sides in turn with its antennae and then move towards the drier side. As in *Rhodnius* responding to warmth and smell,[169] the scolopidial organs (Johnston's organ) in the second antennal segment probably serve as proprioceptors to indicate the position of the antennae at the moment of greatest stimulation. There is no evidence of hygrotropotaxis in *Tenebrio*: amputation of one antenna causes no tendency to circus movement.[56] Exactly the same types of response to humidity are shown by *Pediculus* (*cf.* p. 293) (Fig. 213, A).[170] *Tribolium* normally shows a preference for the drier of two humidities, the response becoming more intense as the moisture conditions approach saturation. But if the beetle is starved, and particularly if it is also desiccated, it shows within three days an intense preference for the higher humidities. This response is reversed within an hour if the insect is given water.[187] Loss of water rather than a change in the proportion of water seems to be the important factor in the reversal.[187]

In Carabid and other beetles the preference for dry or humid air is correlated, with the moisture conditions in the natural habitat of the species. Many species, however, which first show a preference for dry air, develop a reversed response after desiccation.[181, 186]

The locomotor activities of the spruce budworm *Choristoneura* vary greatly with the evaporating power of the air. Indeed results with this insect suggest that many of the apparent temperature preferences recorded in the literature may really be preferences for evaporation rates.[190]

Orientation by Contact—Orthokinesis induced by loss of contact stimuli (stereokinesis or 'thigmotaxis') has been described above (p. 313). Contact stimuli may also produce a klinokinesis. This is well seen in the louse *Pediculus* which follows a convoluted course on a smooth surface but goes straight as soon as it regains a comparatively rough surface (Fig. 213, C).[170] Similar responses have been obtained with the Pigeon Louse *Columbicula* (Mallophaga).[255]

Orientation by sound has been discussed already (p. 271). Males of *Aëdes aegypti* which are attracted to the flight tone of the female, direct their movements so that the maximum stimulation is received equally in the two antennae.[271]

CO-ORDINATED BEHAVIOUR

In the experimental study of orientation, conditions are usually so arranged that the insect receives only one kind of stimulus at one time; the response then appears as a kinesis or taxis or some other form of 'forced movement'. But under normal conditions the insect is exposed to a great variety of stimuli which compete with one another in the production of reactions. According to some authors (the 'Gestalt' school) the organism responds to changes in the whole complex or pattern of stimuli to which it is exposed, and hence it is impossible to obtain a true description of behaviour by considering responses to single stimuli. Females of *Rhizopertha* infesting grain will oviposit normally only when the general pattern of excitation has the appropriate character, involving both olfactory and tactile sensations.[28] In finding the entrance to its nest *Philanthus* makes use of the general configuration of the surrounding landmarks as well as the properties of the individual elements.[8, 148] But the physiological study of behaviour is necessarily analytic, and in this section we shall endeavour to indicate some of the ways in which normal behaviour can be resolved into a succession of responses of the sort already considered in the present chapter.

Relation of simple responses to biology—The newly hatched larvae of *Euproctis* or *Lymantria* show two dominating reactions: a negative geotaxis and a positive phototaxis. These responses ordinarily serve to lead them up to the leaves of their food plant. Of the two reactions the phototaxis is the stronger and if, under experimental conditions, they are illuminated from below, they will migrate downwards and die of starvation.[92, 96] The newly hatched larva of the weevil *Sitona lineata* shows no directed responses to light, but it is positively geotactic and very strongly thigmotactic; these responses cause it to burrow in the soil.[3] Larvae of the Elaterid *Agriotes*, which also burrow in the soil, show a photokinesis and a negative phototaxis in addition to geotaxis and thigmotaxis.[138] If the mosquito is activated by the scent of its host and is induced to fly upwind by the visual reaction already described (p. 323), it will be provided with an effective host-finding mechanism so long as it remains near the ground.[81] *Forficula* shows photokinesis, thigmotaxis, hygrokinesis, and negative geotaxis; responses which will explain the normal hiding places of earwigs, and are utilized by gardeners when they trap earwigs in flower-pots containing dry straw inverted on canes.[165] Many other examples of the same kind could be quoted.

Succession and inhibition of responses—How varied are the perceptions which bring about even a simple reaction is shown by the feeding behaviour of blood-sucking insects.[89] In *Stomoxys* and *Lyperosia*, for example, movement to the host is determined by smell from the skin, warmth, and moisture; extension of the proboscis by the same stimuli at close quarters; probing, by smell and warmth; sucking, by taste.[89] Reactions of this kind consist of series of responses each of which prepares the nervous system to react to the next. A simple example of this is seen during the feeding response of *Rhodnius*. Normally, if the antenna of *Rhodnius* is touched by some cold object it is quickly withdrawn and the insect retreats. But if it is touched by such an object during the feeding reaction, while the insect is advancing towards a source of stimulus, the proboscis is at once extended—this being the ensuing response in the reflex chain.[169] [See p. 346.]

In the feeding of butterflies, the flower scent sometimes merely excites the insect, which then uses the coloured object as a visual mark: smell from a distance, vision, and smell or taste at close quarters are all concerned in different proportions. In some butterflies the colour stimulus alone may initiate the whole reflex chain of the feeding reaction.[70] A greenish-yellow colour will attract the ovipositing female of *Macroglossa*, but only if the odour of *Galium* is added to the visual stimulation is egg-laying started: eggs are immediately deposited on a test-tube filled with a solution of chlorophyll and covered with *Galium* extract.[87]

The landing responses of flying insects are complex and may take place at very high speeds. In *Lucilia* the legs are lowered from their retracted flight position as the result of decreasing light intensity moving rapidly across the retina as the object is approached. Contact with the feet then leads to instant inhibition of flight and simultaneous seizure of the surface by the claws.[261] [See p. 346.]

The multiplicity of perceptions may have another important effect; responses to stimuli received through one set of sense organs may be inhibited by the perceptions of another group. An example of this may again be taken from the behaviour of *Rhodnius*. In the normal bug it is very difficult to elicit the probing reaction by visual stimuli, because objects moved in front of the eye set up air currents perceived by the antennae and the insect moves away. But if the antennae are removed, it is possible to obtain a very striking response; the bug will run in all directions after a moving pencil and probe it immediately the movement ceases.[169] The louse *Pediculus* will likewise probe repeatedly on any warm smooth surface if the antennae are removed.[67, 170]

In the spinning of the cocoon the larva of the giant silk moth *Hyalophora cecropia* has two mutually exclusive patterns of movement which may be termed 'stretch-bend' and 'swing-swing' and which are separated by the caterpillar turning around in the cocoon. The total pattern of behaviour is stored in the corpora pedunculata, but the successive stages are produced by specific stimuli, some internal, resulting from the amount of silk that has been discharged from the silk gland; some external.[266] By applying a rigid band of adhesive tape around the full-grown larva of the silkworm *Bombyx* the degree of movement of the head can be restricted, with corresponding changes in the size and form of the cocoons produced.[209]

Sometimes, during response to the perceptions of one sense the perceptions of another sense seem to be centrally inhibited. An example of this is afforded by the hunting wasp *Philanthus*, which stocks its nest solely with honey-bees. It hunts these bees by sight, and is often attracted momentarily to bumble-bees or other insects. It may perceive a bumble-bee at 30 cm. distance, but on approaching within 5–15 cm. it is recognized by its smell and left; whereas a honey-bee is seized. Smell is therefore exceedingly important in obtaining prey; but it is brought into play only when the presumed victim has been seen; during the process of searching *Philanthus* is completely insensitive to bee odour.[148]

Succession of responses is a characteristic also of the building behaviour of solitary bees and wasps. Each act in this process can be performed only in its proper sequence. The succession of duties discharged by workers of the honey-bee, at first in the hive and later foraging, is perhaps another example of the same phenomenon in its most complicated form; though these duties can be altered greatly, abbreviated or extended beyond their time, according to the needs of the community.[46, 124, 239] [See p. 346.]

Reproductive cycles and 'successive induction' of responses—Many insects, very shortly after emergence as adults, undertake a migratory or dispersal flight[231] during which their reproductive activities are inhibited.[234] In Aphids the act of flying leads to a changed response to vegetation: they settle down, feed and reproduce, changing from an active locomotory phase of behaviour to a sedentary vegetative phase.[230, 234] The change from the solitary to the gregarious phase in locusts is similar, but extends over several generations.[234] The Scolytid beetle *Trypodendron* behaves somewhat like the flying Aphid; but the latent responsiveness to the chemical and contact stimuli of the host tree seems to be released by the act of swallowing air during the dispersive flight.[217]

A well-known phenomenon in the action of the nervous system was named by Sherrington 'rebound' or the 'successive induction' of antagonistic reflexes: the operation of a reflex lowers the threshold of stimulation for its antagonist. This is an important element in the 'reflex chain'; it is to be seen not only in so-called 'simple reflexes' but in complex acts of behaviour like walking and probing in Aphids: the diverse types of stimulus which elicit probing may often do so indirectly by first inhibiting locomotion and liberating the probing response.[227] Similar antagonistic responses are flight and settling in Aphids; these become progressively more ready to settle as flight continues.[235] Another factor which operates in nature in this case is an attraction to the short-wave light of the sky during the active migratory phase and a strengthening of the positive response to the long-wave light from the ground during the alighting phase.[236] Whether this antagonistic induction in behaviour is physiologically similar to the rebound in simpler reflex action is not known.[235]

Hunger and thirst—Starvation in *Phormia* will increase the sensitivity of the taste response to sugars, the intake of food and the locomotor activity; each of these changes can be defined as 'hunger' and each is experimentally dissociable from the others.[208] Hunger for sugar seems to be independent of the storage of reserves, the blood sugar level, &c., and to be determined by nerve impulses coming from food in the fore-gut and passing via the recurrent nerve to the brain, where they inhibit the effect of sensory stimuli from the oral receptors. If the recurrent nerve is cut this inhibition no longer occurs, and feeding on sugar continues until the fly dies.[203] The readiness of *Phormia* to drink water is quickly abolished by injection of water into the haemocoele, and it is induced by bleeding. It is also affected by cutting the recurrent nerve just in front of the hypocerebral ganglion; flies treated in this way become bloated with water.[204] *Lucilia cuprina* is at once induced to drink on injection of sodium chloride solutions exceeding 0·3M sugar solutions, and sodium salts other than chloride are without effect.[191] [See p. 346.]

Route finding by Hymenoptera—The methods used by Hymenoptera in finding their way to and from the nest furnish one of the best examples of the co-ordination of multiple impressions. Ants which forage far from the nest utilize many factors: the direction of the sun's rays[98] ('light-compass' orientation), the visual perception of landmarks, the odour of trails, a topochemical sense[36] (by which is meant the perception of scents to either side of a given track, perceptions which will be reversed on the return journey), tactile sense, proprioceptive senses and perhaps fatigue.[126] Fig. 214 illustrates a simple experiment with the ant *Messor barbarus*. It went north from the nest to *a*; it was then displaced to *b*, and orientating itself by the direction of the light it returned south to *c*. Here it encountered a known track which it followed to the nest, the direction of the nest being determined probably by distant landmarks and by topochemical perceptions.[126] In *Lasius*, continuous odour trails are built up by the gradual accumulation of spots of odour along them. This odour is probably deposited by the abdomen; it is peculiar to the colony, and workers of another colony pay no attention to it.[102] *Acanthomyops* (*Lasius*) *niger* orientates itself largely by visual means, making use of the positions of near and distant objects and of the plane of polarization of the blue sky (p. 240).[175, 189] But it also pays attention to the olfactory background, and if blinded it can be guided by odour trails alone.[175] *A.* (*Lasius*) *fuliginosus*, on the other hand, uses the odorous streaks deposited by other workers almost exclusively carrying its antennae close to the ground. Although the individual streaks taper in the direction of movement of the ant which laid them the ants cannot recognize the direction or 'polarization' of the track as a whole.[175, 176] [See p. 346.]

Ants can utilize the polarization pattern of the blue sky. They can remember a direction in relation to a light souce for at least 5 days, and with the help of landmarks they can remember at least four localities and the corresponding direction of the run. With the help of their sense of time ants can calculate and allow for movements of the sun in their direction finding.[228]

Bees likewise make use of many factors in returning to their hive. In a new locality or on an open sheet of water without landmarks the direction of the sun's rays is the chief factor. But they can also appreciate the *distance* from the collecting ground to the hive (Fig. 215). Later they make use of landmarks and are able to remember and utilize the angles through which they have to turn at particular points. In open country they can be made to follow a series of artificial landmarks. In the immediate vicinity of the hive, orientation marks are used; if the hive is moved, conspicuous landmarks will accelerate the accumulation of a crowd of bees at its original site. But scent is also important, for if the bees are accustomed, for example, to a scent of aniseed in the hive, they will locate it in its new position considerably sooner.[172] When good landmarks are present these are preferred; but in a bare landscape the 'sky compass' is used;[211] and the pattern of polarized light in the blue sky can be utilized in place of the sun itself.[178]

It has been shown repeatedly that the return flight of the bee from a source of food must be learned separately from the outward flight. The same can be

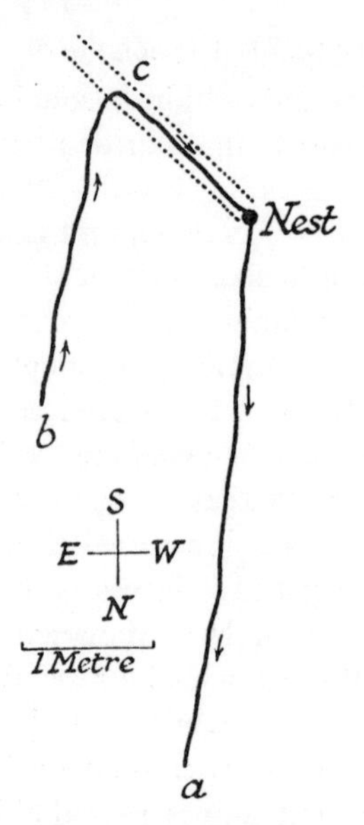

Fig. 214.—Finding of nest by worker of *Messor barbarus* (*after* Santschi)

For explanation see text.

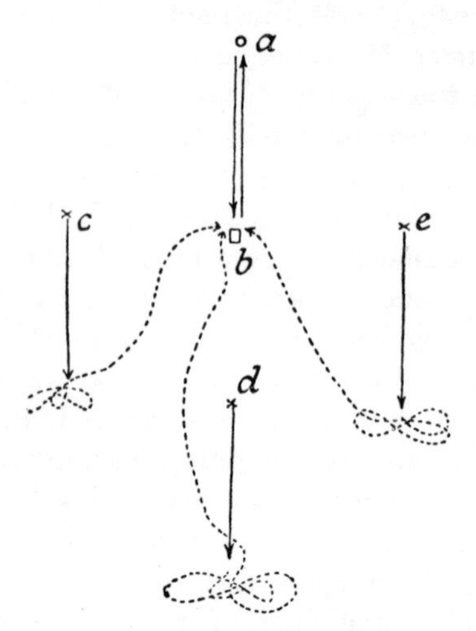

Fig. 215.—Return of bees to the hive (*after* Wolf)

Bees feeding at **a** fly straight to and from the hive at *b*. If removed from **a** to *c*, *d*, or *e*, they always fly in the same direction and for the same distance before returning to the hive.

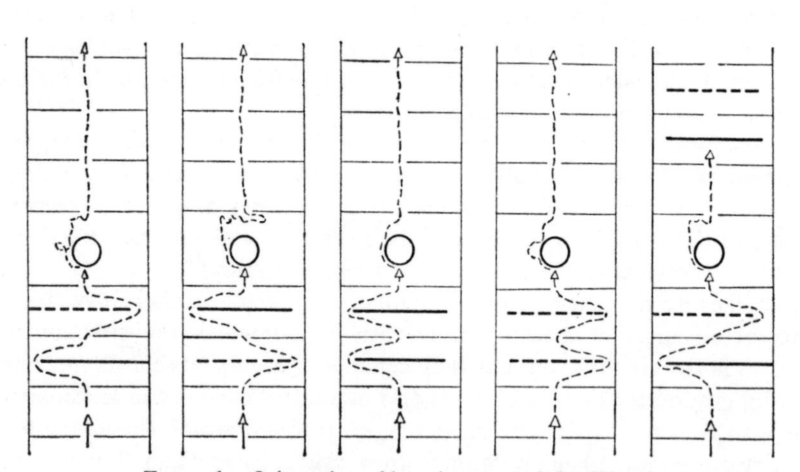

Fig. 216.—Orientation of bees in a maze (*after* Weiss)

A dish of syrup is in the centre. Bees entering the maze learn to turn left before a blue screen (continuous thick line) and right before a yellow screen (broken thick line) irrespective of the order of the screens. When (in the experiment to the extreme right) they meet screens of blue or yellow on their way out, they are confused and do not know which way to turn.

demonstrated with bees or wasps in a labyrinth. They would soon learn to turn left to get round a blue obstruction and right to get round a yellow obstruction (Fig. 216); whereas for the exit they could run straight forward. But if opposed by a blue or a yellow obstruction on the exit journey they had to learn the problem all over again.[269]

Communication in social Hymenoptera—In the honey-bee community food transmission is the most primitive and perhaps the most important method of communication. It serves to keep the members of the hive informed of many details about the state and the needs of the society.[259] Specific secretions from the surface of the body are used by insects for many purposes: protection, sexual excitation, sexual attraction, as stimuli to hormone production or possibly as hormone-like substances themselves. These diverse materials are mentioned in many places in this book (p. 609). Despite their variety in nature and function they have been given the unifying name of 'pheromones'.[232, 245] A few examples may be selected from the many that are concerned in communication among social Hymenoptera.

Bumble-bees, *Bombus*, mark the region of the nest with the secretion of the mandibular glands[219] which is scented like hydroxycitronellal.[237] The queen bee *Apis mellifera* secretes in its mandibular glands the 'queen substance' (9-oxodecenoic acid) which serves as a sex attractant to the drone during mating (p. 710) and keeps the workers in the hive informed of the continued presence of the queen. The secretion is obtained by licking the body of the queen and passed throughout the community by 'food exchange'; so long as it is present the building of queen cells by the workers is inhibited. In conjunction with another scent from the queen, the queen substance suppresses the maturation of the ovaries of the workers;[199] and yet another secretion keeps the workers in the swarm in close contact with the queen. The scent gland (Nassonow's gland) is used by worker bees as a guide for other bees; this scent seems to be the same in all colonies, but the hive odour adhering to the surface of the body is used by bees for distinguishing the members of other colonies.[257] Similar substances, both for trail formation and for communication within the nest, are produced by ants.[238, 264] Some of the active substances may conceivably act as hormones ('ectohormones') in other members of the community; but for the most part they seem to act as signals calling forth the appropriate act of behaviour. [See p. 346.]

Bees which have located a source of nectar or pollen communicate the fact to other foraging bees in the hive by performing certain characteristic movements, or dances. If the source is near the hive they perform a 'round dance', if the distance exceeds 100 metres they perform a 'tail-wagging dance' and as the distance increases still further the rhythm of the dance changes. The intensity of dancing varies with the richness of the collecting ground and the concentration of the nectar. At the same time the direction of the source may be indicated: on a horizontal surface the direction of the straight run made in the course of the dance marks the direction of the source; on a vertical surface the deviation from the perpendicular during the straight run marks the direction of flight in relation to the sun.

Bees performing their dances on a horizontal surface become disorientated in the dark: but they are correctly orientated as soon as a piece of blue sky becomes visible; they are able to recognize the position of the sun from the pattern of polarization in the light from the blue sky (p. 240): the direction of dancing can be changed at will by interposing a polarizing film between them and the light.[178]

The distance to the collecting ground is indicated chiefly by the number of flicks of the abdomen during the straight run; there are about 13 complete flicks per second and this number increases with increasing distance of the goal at the rate of about 1 per 75 metres over the range 100–700 metres. The dance may be regarded as a highly ritualized 'intention movement' leading to communication which is mainly kinaesthetic: the bees fly to the appropriate goal because, by following the dancer, they have automatically carried out the intention movements for the flight.[210, 221, 225] During the straight run of the tail-wagging dance the bee gives out a series of short pulses of sound; by indicating the duration of the tail-wagging period they may help in communicating information about distance.[207] At the same time the flower scent clinging to the dancing bee, and the scent in the nectar which it administers to other bees, enables it to lead its fellows to the appropriate flowers. The scent glands of the bee itself are also important in exciting and guiding other bees.[47, 213] The stingless bees, Meliponini, perform the alerting 'round dance' but give no indication of the direction of the collecting ground; for this purpose they lay trails of scent marks.[243]

After inbibing sugar the blow-fly *Phormia* goes through a convoluted movement on the ground which strikingly resembles a dance. When exposed to a beam of light this movement becomes oriented in the direction of the beam and on a vertical surface it becomes directed by gravity. As in the bee, the number of turns per unit time increases with more concentrated sugar solutions. Methylene blue taken up with the sugar is disgorged and spreads rapidly through neighbouring flies. These observations suggest that the communication dances of bees may represent a highly evolved form of a primitive innate search pattern.[202] Similarly, after each flight, the Brazilian Saturniid moth *Automeris* performs rhythmic rocking movements, and the total number of these movements is proportional to the length of the flight completed.[194]

Inborn associations in behaviour—The greater part of behaviour in insects is inborn, being made up of responses which are performed with the same precision on the first occasion in the life of the individual as later (p. 191). Insects frequently make use of 'token' stimuli of many kinds to guide them in their reactions; stimuli, that is, which have no necessary connexion with what they seek but are commonly associated with it in nature. For blood-sucking insects, warmth is an important factor in locating the host;[89] females of Hymenopterous parasites are often attracted to the food of the insects in which they seek to lay their eggs: *Nemeritis* to oatmeal,[141] *Alysia* and *Mormoniella* to decaying meat.[72, 91] *Pimpla ruficollis*, a parasite of the pine shoot moth *Rhyacionia buoliana* is repelled by pine oil early in adult life, but as soon as the ovaries ripen it is attracted by the oil.[142] Oil of turpentine will make wood more attractive to *Hylotrupes* for oviposition although it actually reduces its nutritive value.[5] The cherry fly *Rhagoletis cerasi* discovers cherries solely by visual stimuli; it will accept for oviposition, balls of wax which are smooth and of appropriate size.[168] Wireworms, *Agriotes*, show a high degree of sensitivity to substances, such as asparagine, likely to be given off by damaged roots in the soil. A klinokinetic response to such substances tends to keep the larva in the vicinity of the plant; it proceeds to bite and feed as soon as it encounters an appropriate taste.[145]

Memory and conditioned responses—To some extent behaviour can be nfluenced by experience. The simplest example of this is termed 'motor earning'. If *Ephestia* larvae are kept walking round a circular dish for ten ninutes or so, they will continue to move in an rc when set free, even in a beam of light in which they normally go straight; and if they are laced in a track with a series of T-shaped urnings and forced always to turn the same way at the bends (Fig. 217), they will continue uch turns when given a free choice.[12]

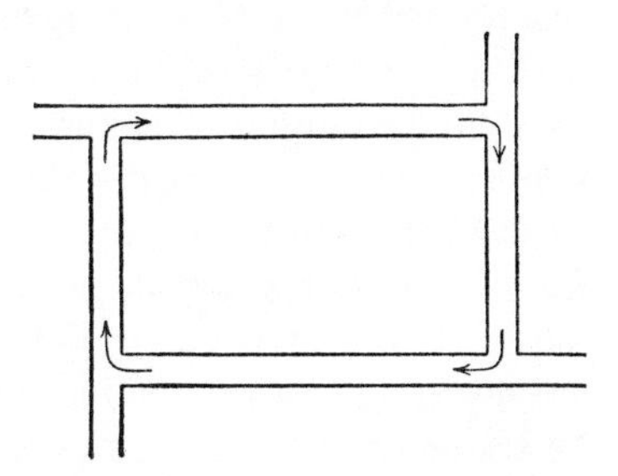

FIG. 217.—Simple maze consisting of successive T turnings in which larva of *Ephestia* is obliged always to turn to the right (*after* BRANDT)

We have seen that the efficiency of the klinokinetic' mechanism of orientation depends n the phenomenon of 'adaptation' (Fig. 198) p. 316). The insect responds by an increased ate of turning as soon as it leaves a more avourable zone of stimulation. This is a rimitive sort of 'memory'. After a time the memory fades', the insect becomes 'adapted' and it follows a straight course nce more. This kind of learning is termed *habituation.*[144]

Ants of a given nest become habituated to the odour of other members of he colony. They will fight and kill ants of the same species from other colonies; ut if pupae from two nests are placed together in the formicary they will hatch ut and live amicably together. The odour seems to be derived from the queen nd all workers from the eggs of a given queen are habituated to the same dour.[143]

In some special cases habituation may be highly persistent and remain hrough metamorphosis. The Ichneumonid *Nemeritis canescens* during oviposiion shows a strong preference for the odour of *Ephestia* which it normally arasitizes; but this preference can be reduced if it has been reared in the larvae f the wax moth *Meliphora grisella*[141] or if the adults are exposed to the odour f *Meliphora* for some time.[143] But a similar conditioning of *Mormoniella* to oviosit in *Musca* or in *Calliphora* seems to depend on the contact of the emerging dult with the puparia of these hosts, and not on the host in which the *Mormonilla* had actually been reared.[249] *Drosophila* are normally repelled by essence of eppermint; but if the larvae have been reared in a medium containing peppermint the resulting adults are strongly attracted to it for oviposition: the fly has ecome habituated to the repellent constituent (menthol) and the attractive contituents (esters) can now exert their effect.[143] Similar results have been obtained n *Calliphora.*[29] The natural preference of *Drosophila guttifera* for oviposition in fungus-containing medium may be weakened if the larva has been reared in he standard laboratory medium.[30] And similar responses are seen in some leafating caterpillars and sawflies in which the oviposition site of the female is nfluenced by the nature of the food during larval life.[156] [See p. 346.]

But learning can also result from an association of sensory experiences. This is ermed *association learning* and is sometimes divided into two classes: 'conditionng', where the stimulus and the reward accompany one another, and 'trial nd error learning', where the stimulus and response precede the reward.[144, 265]

Nearly all blow-flies, *Cynomyia*, can learn to associate the odour of coumarin with the presence of sugar (p. 285) but there are great individual variations in the rate of learning.[43] Gyrinid beetles (*Dineutes*) can be trained to associate surface waves, set up in the water by a tuning fork, with the presence of food.[134] *Tenebrio* can be trained to avoid either rough or smooth surfaces.[159] The cockroach normally avoids the light and seeks the dark; but by setting up a chamber with light and dark ends, and giving the insects electric shocks at the dark end, they can be taught to remain in the light section, and return there from the dark section, without repeating the shocks. Different individuals require from 18 to 118 shocks before they acquire this change in response. They retain the change for 4–55 minutes, but acquire it more rapidly a second time.[139, 152] If such an experiment is arranged in the form of a T-shaped maze in which electric shocks are given on one side and a dark shelter provided on the other, *Periplaneta* can be trained to turn always to the right or the left. This result is not affected by blinding; but removal of the antennae causes immediate loss of the training; and after removal of the left antenna an insect trained to go to the left almost always turns to the right. The cockroach clearly depends on the antennae for the appreciation of right and left.[15] Likewise in *Blattella*.[25] The cockroach *Periplaneta* can learn to associate the position of the leg with an electric shock, and can transfer the effect to another leg, even after decapitation.[226] [See p. 346.]

This type of association, where the insect is 'punished' after responding to a given stimulus, is sometimes termed 'negative' training, as opposed to 'positive' training in which it is 'rewarded' after a particular response. Both methods have been used in *Dytiscus*, which may learn to associate a given taste or scent with bitter meat to follow (negative training) or with sweetened meat to follow (positive training). Positive training is generally found to give results more rapidly.[57, 127] By these means ants can master a maze of ten blind alleys.[130]

The length of time required to produce such responses and the time they are retained varies with the species and with the senses employed. *Dytiscus* can retain the effects of teaching for some days.[127] Newly emerged *Plusia gamma* use the sense of smell alone in finding flowers, flying always upwind; they cannot find scentless flowers. But after their first visit to flowers, as the result of the visual experience, they can locate flowers which have no scent.[131] Butterflies require several days' training before they will associate particular colours with the presence of food, and after an interval of one day they retain only traces of the acquired response[70]; whereas bees may be successfully trained to come to a given colour within 2 hours and retain this training for 4 days.[44] And they are still more retentive of odours: a scent used for training for only a few hours may be remembered after several weeks.[45] Bees can remember a collecting site after an interruption of a week of rainy days;[242] they can be trained to visit two different sites (or even four different sites) at different times of day.[240] It has been claimed that honey-bees in the spring will visit the site of a drinking fountain that they used the previous autumn, even though the fountain has been removed.[24] It is obvious that memory, both visual and olfactory, plays a large part in the finding of the way to and from the nest among Hymenoptera (p. 333). Male bumble-bees, *Bombus*, fly regularly round a number of fixed landmarks, in the same sequence for weeks on end.[41] The ant *Formica* can remember the route back to the nest for 5 days.[228]

It has been proved experimentally that in associating a given colour with a source of food, the bee uses only those perceptions which *precede* its discovery—those stimuli which it receives during the 'approach flight' occupying perhaps 3 seconds. It is unaffected, for example, by any changes made in the colour of the background while it is feeding, or by the colour present during the longer 'orientation flight', occupying perhaps 10 seconds, which precedes the return to the hive. This is precisely what happens in the so-called conditioned reflexes of mammals. But if the food is removed to a new site during feeding, the bee usually returns to the place where feeding was finished, that is, the site 'memorized' during the orientation flight.[114]

Learning and instinct—The type of learning which is seen in the homing of bees, learning without obvious reward, is termed *latent learning*. It usually involves responses to a complex combination of features. It is seen also in dragonflies hawking for prey over the same terrain for days at a stretch, or in Gyrinids occupying the same patch of water all the season or even from one year to the next.[144] Whether insects are ever capable of *insight learning*, involving the production of new adaptive responses in circumstances not previously encountered —what is commonly called 'intelligent' behaviour—is a subject on which authors are not agreed.[144, 265] The acts of co-operative behaviour among termites exhibit a degree of purposiveness and plasticity which makes it impossible to describe them adequately in physiological terms;[53] but even in such complex sequences as the swarming, mating and nesting of termites, some progress can be made by analysing the consecutive acts into fixed reactions to known stimuli.[53] The apparent co-operation among termites in the building of the nest seems to result from the product of their labour evoking similar activities by themselves and other members of the community. This process has been termed 'stigmergy'; the impression of co-operation is illusory.[217a] Many of these complex acts are termed '*instincts*', that is, inborn patterns of movement, or automatisms, brought into operation by some appropriate condition. 'Instincts' are normally manifested in response to a specific 'releaser stimulus'; but in the absence of such releaser, the threshold of stimulation falls and the whole elaborate response may be set off by some quite abnormal circumstance.[146]

These inborn patterns of behaviour are genetically determined, or inherited; they are built into the central nervous system in the same way as morphological patterns are built into the growing body. By far the greater part of insect behaviour consists in producing the pattern of behaviour suited to the circumstances of the moment. It can be argued that in an animal as short-lived as most insects, ability to learn has little survival value when compared with a set of complex 'built-in' reactions.[261] Where the number of neurones is limited, behaviour will be dominated by innate or instinctive patterns; this economy in nerve units may be an important factor in the operation of the insect nervous system.[261] There is, of course, no hard and fast line between 'inborn patterns of behaviour' and 'reflex responses': many of the acts described under 'characters of reflex responses' (p. 191) could equally well be described as inborn patterns of behaviour.

Appreciation of time—Bees can appreciate time and can be trained to come for food at any hour of the day (Fig. 218).[6] If the concentration of sugar which is available differs at different times of the day, they are able to remember the time at which it is greatest.[161] Under natural conditions this memory for time is connected with the fact that the flowers visited offer their pollen or nectar at particular hours.[85] Wasps, also, have some appreciation of time[158] and it is claimed that ants can remember periods ranging from 3 hours up to 5 days.[52] But in the case of foraging ants it is claimed by some that the rhythmic activity is dependent directly upon climatic factors and not upon any sense of time.[122]

The nature of such memory for time is uncertain. In the bee it seems to be bound up with the 24-hour rhythm; if bees are fed at a 48-hour period, they come for food also at the 24-hour intervals.[6, 46] Yet it appears to be independent of periodic changes in the outer world; bees can still recognize the hour of day if the illumination is kept constant artifically.[6] It is possible that they are perceiving some periodic factor in the environment as yet unrecognized; but the evidence seems to point to their depending on metabolic changes in their own bodies. For if bees trained to come for food at a given hour are kept cold, their time of arrival is delayed; whereas narcosis is without effect.[52, 73] Bees firmly trained to forage at a particular hour spend the rest of the day in a remote part

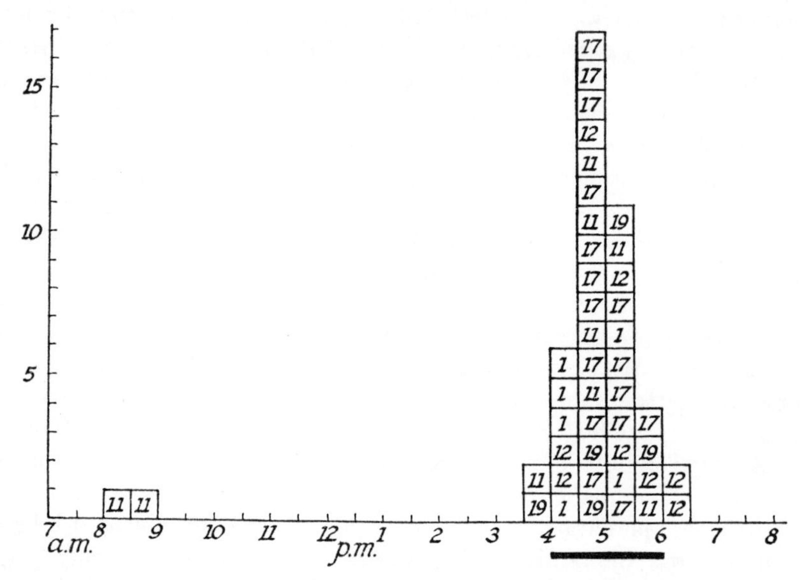

FIG. 218.—Appreciation of time by the honey-bee (*after* BELING)

Bees were given syrup at an artificial feeding place for 3 weeks between 4 and 6 p.m. On the day of experiment the syrup was left out all day. Ordinate: number of bees visiting the feeding place during half-hour periods. Abscissa: time of day. The numbers in the squares indicate individual marked bees.

of the hive. Here they are not disturbed by the dancing of bees which have discovered a rich source of nectar or pollen. These discoverer bees excite, attract and lead out only those foraging bees whose habits have not become fixed in this way.[46] Bees transported overnight from Paris to New York, or from New York to California, at first adhere to their original feeding time; they become adapted to the new sun-time in the course of 3 or 4 days.[258] [See p. 346.]

Astronomical navigation—The time sense of the honey-bee, combined with the capacity for steering by the sun or by the pattern of polarized light in the blue sky, makes possible a simple form of astronomical navigation.[258] If bees trained to steer their way to and from a feeding source by the sun are imprisoned for a few hours and then released, they fly off at once in the correct direction of the hive. They are able to calculate the position of the sun according to the time of day, and orient themselves accordingly.[246] Indeed, so efficient is this sun navigation that bees trained in the afternoon to a particular source will set off in the right direction next morning although the apparent position of the sun is reversed.[241] In the Northern Hemisphere the sun appears to move round in a clockwise direction; but in the Southern Hemisphere the apparent direction of movement is reversed. Honey-bees south of the equator, trained to forage at a collecting ground south of the hive, will fly north if the hive is moved overnight to a position north of the equator. But a stock of bees moved in this way becomes adapted in a few weeks to the changed apparent movement of the sun.[241] There is some doubt about the ability of bees to navigate in the tropics during the time the sun is in the zenith.[242, 247]

An appreciation of time and a capacity to perceive the plane of plane-polarized light (p. 240) are used by various insects for purposes of navigation. The Gerrid *Velia currens* will orient differently to the same light source at different times of day: it has an internal clock compensating for the sun's movements.[193]

A similar astronomical navigation based on the sun or on the pattern of polarized light in the blue sky is seen in certain riparial Carabids which when disturbed always run in a direction away from the sea—independently of the slope of the shore.[251]

Diurnal rhythms—Many of the activities of insects show a diurnal periodicity.[164] This is seen in the movements of the iris pigment in Noctuids (p. 223), and in the general colour change in *Carausius*, which is light by day and dark by night (p. 621). *Carausius*, also, is active at night and in a state of catalepsy by day. This rhythm may persist for some days in permanent darkness or even with reversed illumination,[137] and it shows a daily rhythm in defaecation and oviposition.[75] The fire-fly *Luciola sinensis* kept for several days in the dark will still light up at the normal time of 7 p.m.;[136] and *Photinus pyralis* shows periods of flashing which recur at intervals of 24 hours and persist for at least 4 days in the uniform conditions of the darkroom.[18] *Tettigonia* and *Decticus* show a diurnal rhythm of singing in the latter part of the day; the rhythm can be reversed by reversed illumination; singing begins 12 hours after the last period of darkness.[112] *Gryllus* shows a daily rhythm in general activity which will persist at least two weeks in continuous darkness at constant temperature and humidity. Reversed illumination will cause reversal of activity which, again, will continue under constant conditions.[101, 115] The same is seen in *Periplaneta*, in which the behaviour, such as the reaction to light, is changed in the different phases of the activity rhythm;[55, 107] and in *Ptinus tectus* (Fig. 219).[7] If the light is kept constant and the temperature fluctuates, the greatest activity occurs when the temperature is falling, and after transfer to a constant temperature, this period still occurs at the same hour for several days. Perhaps humidity may likewise influence the rhythm. *Leptinotarsa* has a daily rhythm of feeding activity which is acquired in the larval stage in response to diurnal changes in light; it persists through the pupal stage and determines the hours of diurnal activity in the adult.[54] All such rhythms appear to indicate a memory for time; but they are usually regarded as being dependent on the rate of metabolism in the cells,[23] which results perhaps in a rhythm of hormone secretion.[75] The diurnal activity rhythm of *Carausius* persists in continuous darkness, but is arrested by continuous light. It is eliminated by extirpation of the brain and by cutting the connectives with the suboesophageal ganglion, but is not affected by removal of the corpus allatum or corpus cardiacum.[205]

The existence of a hormonal cycle controlling a diurnal rhythm was first demonstrated in *Periplaneta*. Here, as in other insects, the diurnal rhythm of activity eventually disappears if the insect is kept in continuous light; but a rhythm can be imposed on such an insect by parabiosis with a second cockroach with an established rhythm.[222] The source of the hormone concerned is a group of neurosecretory cells in the suboesophageal ganglion; and these cells, which seem to constitute the main 'clock' controlling the activity rhythm, are regulated by means of the ocelli, which discharge nerve impulses continuously in darkness (p. 244). An insect without a rhythm acquires an imposed rhythm if this group of cells, taken from another cockroach with a normal rhythm, is implanted into it.[222] The rhythm of 'activity' includes not only increased locomotion, but increased growth in the replacement cells (p. 488) of the mid-gut. If the cockroach is ex-

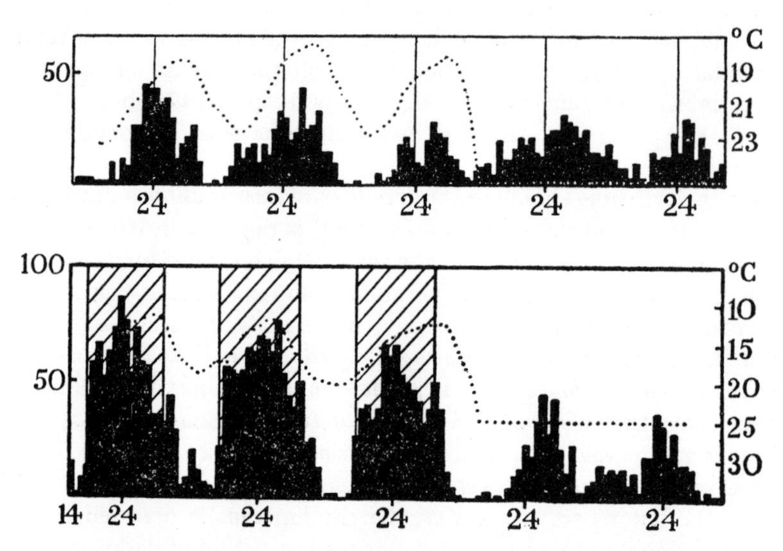

FIG. 219.—Activity rhythm in *Ptinus tectus* (*after* BENTLEY, GUNN and EWER)

Ordinates: temperature, on right; percentage of insects active, on left. Abscissa: time, in 24-hour periods.

A shows an activity rhythm in constant light with fluctuating temperature, continuing at constant temperature. *B* shows an activity rhythm in alternating light and darkness (shaded period) coupled with fluctuating temperature, persisting in constant light and temperature.

posed experimentally to the hormone twice a day, growth in the mid-gut becomes disorganized and transplantable tumours develop.[222]

For their full activity the neurosecretory cells require a second neurosecretory substance, which comes via the nerve from the corpus cardiacum; if this nerve is cut the rhythm fades, even in alternating light and darkness.[222] They are subject also to integration with 'physiological clocks' elsewhere in the body. For it would appear that a basic 24-hour rhythm is present in the cells of all animals: the effect of the environment is merely to position the phases of these rhythms within the 24-hour cycle.[223] The period of these inherent rhythms is not exactly 24 hours in length; they are therefore sometimes called 'circa-dian' (or 'about a day') rhythms. When the rhythm is allowed to persist under constant environmental conditions (the 'free running' state) the activity therefore begins a few minutes earlier or a few minutes later each day (Fig. 220). These small departures from 24 hours are constant within a given individual.[224, 253] In *Leucophaea* this departure from 24 hours is well seen; here, as in most insects, the period of the rhythm is compensated for temperature changes between 20–30° C.[260] Local cooling to near 0° C. stops the functioning of the clock completely.[223] *Acheta domestica*, which had lost the rhythm of activity under constant conditions, could re-establish it in alternating conditions if either the compound eyes or the ocelli were covered, but not if both were covered.[248] [See p. 346.]

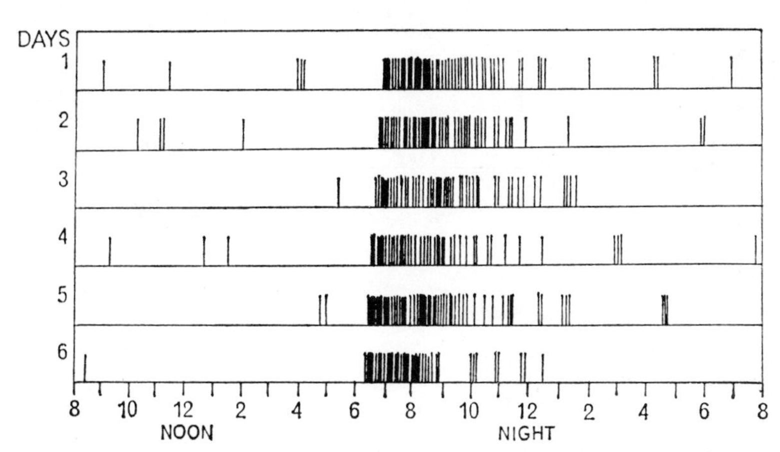

FIG. 220.—Record of activity of a single *Periplaneta* kept in continuous darkness in an unchanging environment. Activity began between 6 and 7 p.m. But measurement of the 24 hours was not exact; activity began about 7 minutes earlier each day (*after* HARKER)

Eclosion rhythms—A somewhat different type of rhythm is seen in the emergence of insects from the pupa, or in moulting, or hatching from the egg, as in *Carausius*.[75] Most species emerge at a definite time of day, and will continue to do so even when those factors (temperature, light, barometric pressure, relative humidity, electrical potential and electrical conductivity of the air) which show a daily periodicity are kept constant. *Ephestia* kept in a constant temperature room for three generations continue to emerge chiefly in the evening. But this inborn rhythm can be overcome by external stimuli. When there is a diurnal cycle of temperature, emergence in *Ephestia* occurs when the temperature begins to fall (Fig. 221); and if larvae or pupae are exposed to an artificial rhythm of this kind, emergence can be induced at an abnormal time of day. Such a rhythm may be produced if the period of the cycle is 16 hours or 20 hours, but not if it exceeds 24 hours; evidently there is some close relation with the normal daily period.[133] In *Drosophila*, which emerges chiefly in the morning, the periodicity disappears after rearing for 15 generations in uniform dimly lit conditions; but a single exposure to daylight is sufficient to initiate a diurnal periodicity in emergence.[23, 76] An emergence rhythm can be established by exposure of any of the larval stages to brief illumination;[195] indeed, a flash as brief as $\frac{1}{2000}$ second will have this effect.[254] Here again the metabolic rate seems to lie at the base of this apparent memory. The length of period can be prolonged to 30 hours by exposure to 10° C. and prolonged to 30–35 hours if the oxygen tension is reduced to 5 per cent. of the normal.[23, 76] The emergence period of *Drosophila* in constant darkness is about 24·5 hours; it can be entrained to a new timing by either light or temperature, but light has the stronger effect.[196] [See p. 347.]

On the other hand, the emergence of *Chironomus* adults from the pupa during

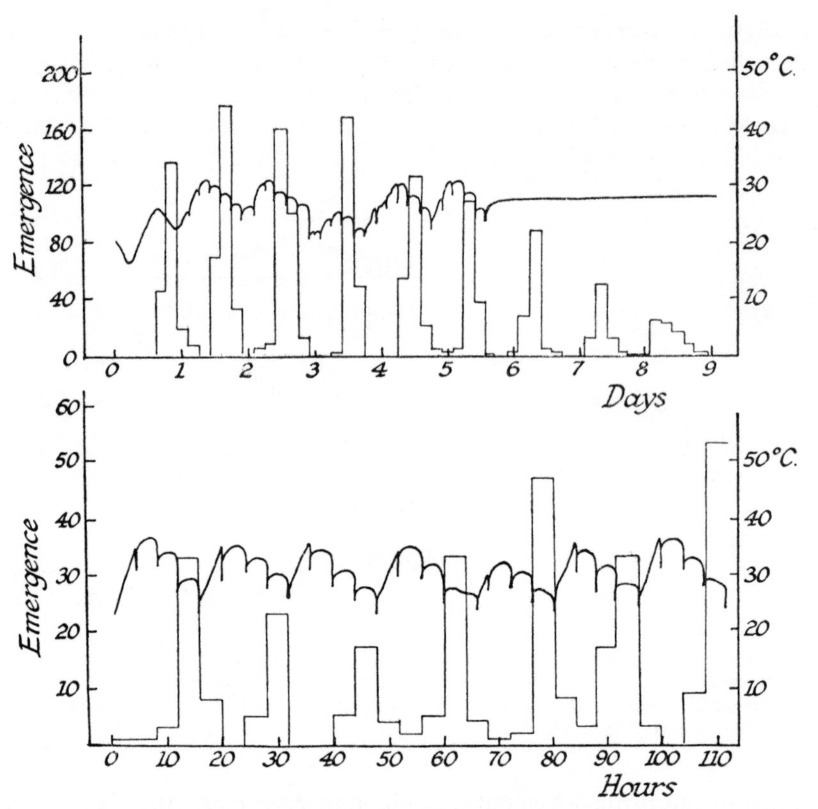

FIG. 221.—Rhythm of emergence in pupae of *Ephestia kühniella* (*after* SCOTT)

Ordinates: to the left, number of insects emerging during a given period (indicated by the block in the figures); to the right, temperature in ° C. (indicated by the continuous line). Abscissae: upper chart, days; lower chart, hours.

Upper chart shows 24-hour temperature rhythm. Maximum emergence occurs when the temperature begins to fall and this emergence rhythm persists with constant temperature. Lower chart shows 16-hour rhythm of temperature and emergence.

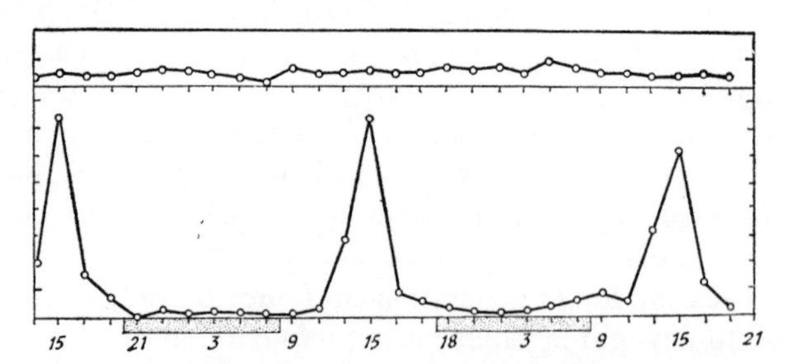

FIG. 222.—Emergence of *Pseudomittia* (Chironomidae) from the pupa (*after* REMMERT)

Upper chart; pupae exposed to continuous light: low rate of emergence throughout the 24 hours. Lower chart; pupae exposed to 12 hours light and 12 hours dark: the emergence rate rises to a peak in the middle of the light period. Ordinate: relative numbers emerging. Abscissa: hour of day.

the daytime seems to be a direct effect of light and not an inborn rhythm.[117] The synchronization is independent of temperature within ecological limits.[250] In the Chironomid *Pseudomittia* there is normally a peak of adult emergence 6–8 hours after the beginning of illumination. In continuous darkness *no* imagines emerge; in continuous light, emergence is evenly spread throughout the day; wavelengths from 470–641 mμ are effective (Fig. 222).[256] [See p. 347.]

It is likely that rhythms of oviposition of the same kind are widespread in insects. In *Aëdes aegypti* and *A. africanus* there is a regular cycle of oviposition with a peak in the afternoon. Light is the chief controlling factor; but the rhythm soon breaks down under constant conditions of illumination.[220]

ADDENDA TO CHAPTER VIII

p. 311. **Ocelli as stimulatory organs**—*Schistocerca* or *Anacridium* flying on a roundabout require illumination for flight. Blackening of the ocelli reduces the flight speed; blackening of the compound eyes does not;[279] and the speed of climbing in *Mantis* and *Anacridium* is diminished if the ocelli are covered, but not if the compound eyes are covered.[279]

p. 313. **Thigmotaxis**—Tethered *Schistocerca* begin opening their wings about 45 msec. after wind stimulation of the facial setae begins; or at least 30 msec. after release of tarsal contact with the ground. But an untethered locust in a takeoff jump commences wing opening within 15 msec. of the first detectable motion towards jumping. It is likely that initial wing opening in normal flight has a central origin and is not evoked by stimulation of tarsi or facial setae.[274]

p. 324. **Optokinetic memory**—*Locusta*, like other arthropods, shows an optokinetic memory: if a contrasting visual field is moved in the dark and then re-illuminated the locust will turn its head as in an optomotor response; it can retain the original position to within 0·1° for many seconds in the dark.[285]

p. 325. Optomotor reactions control the stability of flying locusts, *Schistocerca*: the head is oriented so that the horizon is horizontal with the brighter light above, and then the body is aligned with the head by differential wing movements.[282] Crickets appear to possess an inhibitory nervous mechanism by means of which movements of an object are distinguished from movements of the eye itself.[295]

p. 328. **Klinokinesis in the flying mosquito**—In host-finding, the flying mosquito *Aëdes* makes use of the usual klinokinetic response on leaving a favourable air stream or on entering a repellent vapour.[277]

p. 328. **Klinotaxis**—If one tarsus of a hungry fly *Phormia* comes into contact with a sugar solution, or a more concentrated sugar solution, it will turn in that direction; and adult *Carabus* behaves in the same manner if its antennae make contact with sugar solutions.[296]

p. 329. **Anemotaxis**—The normal upwind movement of locusts in response to the odour of grass is likewise due to 'anemotaxis': the grass odour activates the insect and leads to positive anemotaxis; a repellent odour activates and leads to negative anemotaxis.[288] Flying males of *Plodia* responding to the sex pheromone (p. 710) of the female respond by anemotaxis: they depend on visual orientation and on extinguishing the light they are at once dispersed and carried away by the wind.[276]

p. 331. **Successive responses**—*Stomoxys* is induced to probe by a rapid increase in relative humidity.[281]

p. 331. Another component in the stimulus for landing during flight is the expansion of a pattern on the compound eye of the fly: *Musca*, in suspended flight, exhibits the 'landing reaction' if a rotating spiral is held in front of it.[30-]

p. 332. Building behaviour in termites appears to be a direct consequence of low-level alarm stimuli; its immediate function is defence. Other workers are recruited by trail laying and transmission of the alarm; primary construction ceases when building has eliminated the original causal stimulus.[299]

p. 333. **Inhibition of feeding**—Inhibition of the ingestion of sugar in *Phormia* results from the over-stimulation of stretch receptors in the fore-gut and even more in the body wall.[278] Likewise in the female mosquito *Aëdes*, if the nerve cord is cut in front of the second abdominal ganglion the stretch receptors are put out of action, the ingestion of blood will increase fourfold and the body wall is ruptured.[283]

p. 333. **Route finding by Hymenoptera**—The scent trail of *A.* (*L.*) *fuliginosus* is species specific; if the antennae are crossed experimentally the workers turn towards the side where the trail is weaker or absent.[284] Close experimental analysis of the topochemical sense in *Apis* shows that they use their antennae as contact chemoreceptors and can recognize a defined sequence of scents and can resolve two different scents side by side. These learning abilities are closely tied with kinaesthetic orientation.[291]

p. 335. **Communication in social Hymenoptera**—When honey-bees sting an object they release pheromones (2-heptanone from the mandibular glands and iso-amyl acetate from the base of the sting) which induce neighbouring bees to join in the attack.[280] The pheromone responsible for stabilizing swarm clusters in the honey-bee is 9-hydroxydecenoic acid. Crawling worker honey-bees deposit a 'footprint substance' which is attractive to other workers and wasps, *Vespula*, likewise produce a different footprint pheromone.[273]

p. 337. **Persistence of habituation through moulting and metamorphosis**—Polyphagous caterpillars, such as *Manduca sexta* or *Heliothis zea* show a clear preference for the plant previously eaten; such a preference will persist over several larval moults while feeding on other food.[287] Both conditioned avoidance responses and maze learning can be retained over metamorphosis in *Tenebrio*; this implies that information is stored in parts of the central nervous system which persist through pupal and imaginal moults.[298]

p. 338. **Learning in headless cockroach**—This type of learning in *Periplaneta* is associated with a reduction in cholinesterase activity in the relevant part of the nervous system, a greater turnover of RNA, enhanced protein synthesis and a slower production of γ-amino butyric acid (GABA).[28-]

p. 340. **Appreciation of time**—The endogenous 'clock' of the honey-bee fails during deep and prolonged CO_2 narcosis (2–5 hours in 20–100 per cent CO_2).[292]

p. 343. **Circadian rhythm of activity**—Recent work has failed to confirm in detail the control of the circadian rhythm of activity in the cockroach as described (p. 342).[272, 290, 297] It has been suggested that the neurosecretory cells in the pars intercerebralis may be the site of the centre controlling rhythmi ctivity in *Periplaneta* &c.,[294] or that an electric pace-maker in the brain may o-ordinate neuroendocrine rhythms in the nerve cord ganglia.[272] The compound eyes alone are effective for entraining light signals.[294] The circadian ycle in *Acheta* is correlated with a cycle of RNA and protein synthesis in the eurosecretory cells of the pars intercerebralis and the suboesophageal ganglion, nd these changes are associated with the diurnal changes in locomotor activity.[275]

p. 343. **Eclosion rhythm**—The entrainment of the eclosion cycle in *)rosophila* by temperature cycles combined with its invariance at different onstant temperatures implies some mechanism for temperature compensation which is not fully explained.[301]

p. 345. The intertidal midge *Clunio* has a semilunar (15 day) emergence hythm to synchronize with the cycle of spring and neap tides, combined with a liurnal rhythm.[293]

REFERENCES

[1] ALVERDES, F. *Biol. Zbl.*, **43** (1924), 577–605 (function of eyes in behaviour: larvae of *Cloëon* (Ephem.) and *Agrion* (Odon.)).

[2] —— *Z. vergl. Physiol.*, **5** (1927), 598–606 ('dorsal light reflex': *Cloëon* nymphs, Ephem.).

[3] ANDERSEN, K. T. *Z. vergl. Physiol.*, **15** (1931), 749–83 (sense physiology and biology: *Sitona* larvae, Col.).

[4] BARROWS, W. M. *J. Exp. Zool.*, **4** (1907), 515–37 (orientation by smell: *Drosophila*, Dipt.).

[5] BECKER, G. *Z. vergl. Physiol.*, **30** (1944), 253–99 (sensory physiology: *Hylotrupes*, Col.).

[6] BELING, I. *Z. vergl. Physiol.*, **9** (1929), 259–338 (appreciation of time in honey-bee).

[7] BENTLEY, E. W., *et al. J. Exp. Biol.*, **18** (1941), 182–95 (diurnal rhythm: *Ptinus*, Col.).

[8] van BEUSEKOM, G. *Behaviour*, **1** (1948), 195–225 (visual orientation: *Philanthus*, Hym.).

[9] BLEICH, O. E. *Z. Morph. Oekol. Tiere*, **10** (1928), 1–61 (hypnosis: Coleoptera).

[10] BOZLER, E. *Z. vergl. Physiol.*, **3** (1925), 145–82 (ocelli as stimulatory organs).

[11] BRANDT, H. *Z. vergl. Physiol.*, **20** (1934), 646–73 (orientation to light: larva and imago of *Ephestia*).

[12] —— *Z. vergl. Physiol.*, **21** (1934), 545–51 ('motor learning': *Ephestia* larva, Lep.).

[13] —— *Z. vergl. Physiol.*, **24** (1937), 188–97 (reactions of caterpillars to light: *Lymantria*).

[14] BRAUNS, A. *Zool. Jahrb., Abt. Physiol.*, **59** (1939), 245–390 (halteres).

[15] BRECHER, G. *Z. vergl. Physiol.*, **10** (1929), 495–526 (orientation in the cockroach: *Periplaneta*).

[16] BROWN, C. R., and HATCH, M. H. *J. Comp. Psychol.*, **9** (1929), 159–89 (visual orientation: Gyrinidae).

[17] BRÜN, R. *Die Raumorientierung der Ameisen und das Orientierungs-problem im allgemeinen*, Jena, 1914.

[18] BUCK, J. B. *Physiol. Zool.*, **10** (1937), 45–58 (diurnal rhythm in flashing of *Photinus*, Col.).

[19] v. BUDDENBROCK, W. *S. B. Heidelb. Akad. Wiss., Math. Nat. Kl.*, 1917 ('light-compass' orientation in caterpillars, &c.).

[20] —— *Arch. ges. Physiol.*, **175** (1919), 125–64 (function of halteres).

[21] —— *Grundriss der vergleichenden Physiologie*, Borntraeger, Berlin, 1937.

[22] v. BUDDENBROCK, W., *et al. Z. vergl. Physiol.*, **15** (1931), 597–612; *Zool. Jahrb., Physiol.*, **52** (1933), 513–36 (light-compass orientation).

[23] BÜNNING, E. *Ber. deutsch. bot. Ges.*, **53** (1935), 594–623 (diurnal rhythms in insects).

[24] BUTLER, C. G. *J. Exp. Biol.*, **17** (1940) 253–61 ('memory' in *Apis*).

[25] CHAUVIN, R. *Bull. Biol. Fr. Belg.*, **81** (1947), 93–128 (behaviour of *Blattella* in maze).

[26] CHEN, S. H., and YOUNG, B. *Sinensia*, **14** (1943), 55–60 (vision and flight: *Crypto tympana*, Hem.).
[27] CLARK, L. B. *J. Exp. Zool.*, **51** (1928), 37–50 (reactions to light: *Notonecta*, Hem.) *Ibid.*, **58** (1931), 31–41; **66** (1933), 311–33 (ditto: *Dineutes*, Col.).
[28] CROMBIE, A. C. *J. Exp. Biol.*, **18** (1941), 62–79 (oviposition behaviour: *Rhizoperth* Col.).
[29] —— *J. Exp. Biol.*, **20** (1944), 159–66 (olfactory habituation: *Calliphora*, Dipt.).
[30] CUSHING, J. E. *Proc. Nat. Acad. Sci.*, **27** (1941), 496–99 (olfactory habituation *Drosophila*).
[31] DIAKONOFF, A. *Arch. néerl. Physiol.*, **21** (1936), 104–29 (flight reflexes: *Periplaneta* Orth.).
[32] DOLLEY, W. L. *J. Exp. Zool.*, **20** (1916), 357–420 (reactions to light: *Vanessa*, Lep.
[33] DOLLEY, W. L., and WIERDA, J. L. *J. Exp. Zool.*, **53** (1929), 129–39 (sensitivity to ligh of different parts of eye: *Eristalis*, Dipt.).
[34] FLÜGGE, C. *Z. vergl. Physiol.*, **20** (1934), 463–500 (orientation by smell: *Drosophila* Dipt.).
[35] FOLGER, H. T. *Physiol. Zool.*, **19** (1946), 190–202 (reactions of *Culex* larvae).
[36] FOREL, A. *The Senses of Insects*, London, Methuen, 1908.
[37] FRAENKEL, G. *Biol. Rev.*, **6** (1931), 36–87 (mechanisms of orientation).
[38] —— *Z. vergl. Physiol.*, **16** (1932), 371–417 (nervous co-ordination during flight).
[39] FRAENKEL, G., and GUNN, D. L. *The Orientation of Animals*, Oxford, 1940.
[40] FRAENKEL, G., and PRINGLE, J. W. S. *Nature*, **141** (1938), 919–20 (function of halteres
[41] FRANK, A. *Z. vergl. Physiol.*, **28** (1941), 467–84 (flight paths: *Bombus*).
[42] FRANZ, V. *Zool. Jahrb., Physiol.*, **33** (1913), 259–86 (role of phototaxis in nature).
[43] FRINGS, H. *J. Exp. Zool.*, **88** (1941), 65–93 (learning in *Cynomyia*, Dipt.).
[44] V. FRISCH, K. *Zool. Jahrb., Physiol.*, **35** (1914), 1–182 (visual training in honey-bee
[45] —— *Zool. Jahrb., Physiol.*, **37** (1920), 1–238; **38** (1921), 449–514; *Handb. norm path. Physiol.*, **11**, *Receptionsorgane*, i, 223–31 (training of honey-bee to odours
[46] —— *Z. Tierpsychol.*, **1** (1937), 9–21 (psychology of honey-bee).
[47] —— *Österreich. zool. Z.*, **1** (1946), 1–48; *Experientia*, **2** (1946); *Naturwiss.*, **35** (1948 12–23; 38–43 (dance language of honey-bee).
[48] GAFFRON, M. *Z. vergl. Physiol.*, **20** (1934), 299–337 (responses to moving object Odonata larvae, Muscid adults).
[49] GARREY, W. E. *J. Gen. Physiol.*, **1** (1918), 101–25 (light and muscle tonus: *Proctacanthu* Dipt.).
[50] GODGLÜCK, U. *Biol. Zbl.*, **55** (1935), 187–97 (catalepsy: *Neides*, Hem.).
[51] GÖTZ, B. *Z. vergl. Physiol.*, **23** (1936), 429–503 (visual responses of caterpillars *Vanessa*).
[52] GRABENSBERGER, W. *Z. vergl. Physiol.*, **20** (1933), 1–54 (appreciation of time in ant and termites); *Ibid.*, 338–42 (the same in bees and wasps).
[53] GRASSÉ, P. *J. Psychol. norm. path.*, 1939, 370–96; *Biol. Bull. Fr. Belg.*, **76** (1942), 347–8 (co-operative behaviour in termites).
[54] GRISON, P. *C.R. Acad. Sci.*, **217** (1943), 621–2; *Bull. Soc. Zool. France*, **68** (1943 100–7 (diurnal rhythms: *Leptinotarsa*, Col.).
[55] GUNN, D. L. *J. Exp. Biol.*, **17** (1940), 267–77 (diurnal rhythms: *Blatta*, Orth.).
[56] GUNN, D. L., and PIELOU, D. P. *J. Exp. Biol.*, **17** (1940), 307–16 (humidity reactio *Tenebrio*, Col.).
[57] HANNES, F. *Zool. Jahrb., Physiol.*, **47** (1930), 89–150 (types of 'learning' in hone bee, &c.).
[58] HARTUNG, E. *Z. vergl. Physiol.*, **22** (1935), 119–44 (orientation by smell: *Calliphor* Dipt.).
[59] HASE, A. *Z. Parasitenk.*, **5** (1933), 708–23 (catalepsy during blood-sucking: *Triatom* Hem.).
[60] HOFFMANN, R. W. *Handb. norm. path. Physiol.*, **17** (iii) (1926), 644–58 (diurn periodicity, &c., in insects).
[61] —— *Handb. norm. path. Physiol.* **17** (iii) (1926), 690–714 (reflex immobilizatio review).
[62] —— *Z. vergl. Physiol.*, **23** (1936), 504–42 (effect of contact stimuli on insects).

HOLLICK, F. S. J. *Phil. Trans. Roy. Soc.*, B, **230** (1940), 357–90 (flight in *Muscina*, Dipt.).

HOLMES, S. J. *J. Comp. Neurol. Psychol.*, **15** (1905), 305–49 (reactions to light: *Ranatra*).

—— *J. Comp. Neurol.*, **16** (1906), 200–16; *Biol. Bull.*, **12** (1907), 158–64 ('death-feigning': *Ranatra*, Hem.).

—— *J. Animal Behaviour*, **1** (1911), 29–32 (reactions to light: larvae and adults of mosquitos).

HOMP, R. *Z. vergl. Physiol.*, **26** (1938), 1–34 (orientation to warmth: *Pediculus*, Anopl.).

HONJO, I. *Zool. Jahrb., Physiol.*, **57** (1937), 375–416 ('light-compass' orientation).

HUNDERTMARK, A. *Z. vergl. Physiol.*, **24** (1937), 563–82 (perception of form in caterpillars: *Lymantria*).

ILSE, D. *Z. vergl. Physiol.*, **8** (1928), 658–92 (colour sense in butterflies).

JACK, R. W., and WILLIAMS, W. L. *Bull. Ent. Res.*, **28** (1937), 499–503 (effect of temperature on light reactions: *Glossina*, Dipt.).

JACOBI, E. F. *Arch. néerl. Zool.*, **3** (1939), 197–282 (behaviour of *Mormoniella*, Hym.).

KALMUS, H. *Z. vergl. Physiol.*, **20** (1934), 405–19 (appreciation of time in honey-bee).

—— *Z. vergl. Physiol.*, **24** (1937), 644–55 (response to visual forms: *Dixippus*, Orth.).

—— *Z. vergl. Physiol.*, **25** (1938), 494–508 (diurnal periodicity: *Dixippus*, Orth.).

—— *Z. vergl. Physiol.*, **26** (1938), 362–5; *Nature*, **145** (1940), 72 (diurnal rhythm: *Drosophila*).

—— *Nature*, **150** (1942), 405 (anemotaxis: *Drosophila*).

—— *Physiol. Compar. Oecol.*, **1** (1948), 127–47 (optomotor responses: *Drosophila*).

KELLOGG, V. L. *Biol. Bull.*, **12** (1907), 152–4 (reflexes and orientation: *Bombyx mori*).

KENNEDY, J. S. *J. Exp. Biol.*, **14** (1937), 187–97 (reactions to humidity: *Locusta*).

—— *Proc. Zool. Soc. Lond.*, A, **109** (1939), 221–42 (visual responses of mosquitos: *Aëdes*).

—— *Trans. Roy. Ent. Soc. Lond.*, **89** (1939), 385–542; **95** (1945), 247–62 (behaviour of gregarious locusts, &c.: *Schistocerca*, Orth.).

V. KENNEL, J., and EGGERS, F. *Zool. Jahrb., Anat.*, **57** (1933), 1–104 (tympanal organs in Lepidoptera).

KETTLEWELL, H. B. D. *Entomologist*, **79** (1946), 8–14 ('assembling' in Lepidoptera).

KLEBER, E. *Z. vergl. Physiol.*, **22** (1935), 221–62 (appreciation of time in honey-bee).

KLEIN, K. *Z. wiss. Zool.*, **145** (1934), 1–38 (skototaxis: *Forficula*).

KNOLL, F. *Abhandl. Zool.-bot. Ges. Wien*, **12** (1921), 1–616 (insects and flowers: sense organs and behaviour).

KOZHANTSHIKOV, J. *Bull. Plant Protection*, **3**, 233–40 (thanatosis in *Lochmaea capreae L.*).

KRIJGSMAN, B. J., *et al. Z. vergl. Physiol.*, **11** (1930), 702–29 (orientation of blood-sucking insects to their host: *Stomoxys*, Dipt.); *Ibid.*, **13** (1930), 61–73 (ditto: *Lyperosia*, Dipt.); *Z. Parasitenk.*, **9** (1937), 549–58 (ditto: review).

KÜHN, A. *Die Orientierung der Tiere im Raum*, Fischer, Jena, 1919; *Handb. norm. path. Physiol.*, **12** (i) (1929), *Receptionsorgane*, ii, 17–35 (mechanisms of orientation).

LAING, J. *J. Anim. Ecol.*, **6** (1937), 298–317 (attraction of parasites to food of host).

LAMMERT, A. *Z. vergl. Physiol.*, **3** (1925), 225–78 (responses of caterpillars to light and gravity).

LEES, A. D. *J. Exp. Biol.*, **20** (1943), 43–60 (reactions to temperature and humidity *Agriotes*, Col.).

V. LENGERKEN, H. *Biol. Zbl.*, **54** (1934), 646–50 (legs as stimulatory organs: *Sisyphus* Col.).

LINDROTH, C. H. *Ent. Tidskr.*, **69** (1948), 132–4 (orientation of beetles by the sun).

LOEB, J. *Forced Movements, Tropisms and Animal conduct*, Philadelphia, 1918.

LÖHNER, L. *Z. allg. Physiol.*, **16** (1914) ('death-feigning' reflex).

LUBBOCK, J. *Ants, bees, and wasps*, London, Kegan Paul, 1884.

LÜDTKE, H. *Z. vergl. Physiol.*, **22** (1935), 67–118; **26** (1938), 162–99; *Biol. Zbl.*, **62** (1942), 220–6 (visual responses: *Notonecta*, Hem.).

LUDWIG, W. *Z. wiss. Zool.*, **144** (1933), 469–95; **146** (1934), 193–235 (responses of caterpillars to light).

LUTZ, F. E. *Amer. Mus. Nov.*, **550** (1932), 1–24 (diurnal rhythms: Orth.).

MACGREGOR, E. G. *Behaviour*, **1** (1948), 267–96 (odour and orientation in ants).

[103] MAST, S. O. *Arch. Entw. Mech.*, **41** (1915), 251–63 (the term 'tropism').
[104] —— *J. Exp. Zool.*, **38** (1923), 109–205; *Amer. J. Physiol.*, **68** (1924), 262–7 (orientation to light: *Eristalis* and *Proctacanthus*, Dipt.).
[105] —— *Biol. Rev.*, **13** (1938), 186–224 (orientation to light: review).
[106] MELIN, D. *Uppsala Univ. Årskr.*, 1941, No. 4, 1–247 (insect flight).
[107] MELLANBY, K. *J. Exp. Biol.*, **17** (1940), 278–85 (daily rhythm: *Blatta*).
[108] MEYER, A. E. *Z. wiss. Zool.*, **142** (1932), 254–312 (skototaxis: *Lepisma*, Thysan.).
[109] MOORE, A. R., and COLE, W. H. *J. Gen. Physiol.*, **3** (1921) (response of *Popillia* to ligh Weber-Fechner law).
[110] MÜLLER, E. *Z. vergl. Physiol.*, **14** (1931), 348–84 (function of ocelli in ants an bees).
[111] MURR, L. *Z. vergl. Physiol.*, **11** (1930), 210–70 (orientation by smell: *Habrobracon* Hym.).
[112] NIELSEN, E. T. *Ent. Medd.*, **20** (1938), 121–64 (daily rhythm: Tettigoniidae).
[113] OEHMIG, A. *Z. vergl. Physiol.*, **27** (1939), 492–524 (orientation to light: *Vanessa* larv. Lep.).
[114] OPFINGER, E. *Z. vergl. Physiol.*, **15** (1931), 431–87 (orientation of bee at feeding place
[115] PARK, O., and KELLER, J. G. *Ecology*, **13** (1932), 335 (diurnal activity rhythm in fore insects).
[116] PATTEN, B. M. *J. Exp. Zool.*, **17** (1914), 213–80; **20** (1916), 585–98; *Amer. J. Physiol* **38** (1915), 313–38 (reactions of blowfly larvae to light).
[117] PHILLIPP, P. *Zool. Anz.*, **122** (1938), 237–45 (emergence rhythm: *Chironomus*).
[118] RABAUD, E. *Bull. Biol. Fr. Belg.*, **53** (1919), 1–149 (reflex immobilization).
[119] RÁDL, E. *Biol. Zbl.*, **22** (1902), 728–32 (visual responses of insects on a turntable).
[120] RAO, T. R. *J. Exp. Biol.*, **24** (1947), 64–78 (visual responses of *Culex*, Dipt.).
[121] RAYMONT, J. E. G. *Biol. Bull.*, **77** (1939), 354–63 (light responses: *Dineutes*, Col.).
[122] REICHLE, F. *Z. vergl. Physiol.*, **30** (1943), 227–51 (activity rhythms in ants).
[123] RIVNAY, E. *Parasitology*, **24** (1932), 121–36 (behaviour in *Cimex*, Hem.).
[124] RÖSCH, G. A. *Z. vergl. Physiol.*, **2** (1925), 571–631; **6** (1927), 264–98 (succession duties in the beehive).
[125] SÄLZLE, K. *Z. vergl. Physiol.*, **18** (1932), 347–68 (perception of moving object Odonata larvae).
[126] SANTSCHI, F. *Rev. Suisse Zool.*, **19** (1911), 303–38; **21** (1913), 347–426 (orientatic among ants).
[127] SCHALLER, A. *Z. vergl. Physiol.*, **4** (1926), 370–464 (training and response to taste an smell: *Dytiscus*, Col.).
[128] SCHLIEPER, C. *Z. vergl. Physiol.*, **6** (1927), 453–72 (optomotor response).
[129] SCHMIDT, P. *Biol. Zbl.*, **33** (1913), 193–207 (catalepsy: *Dixippus*).
[130] SCHNIERLA, T. C. *Comp. Psychol. Monogr.*, **6** (1929), No. 4 (learning and orientation i ants).
[131] SCHREMMER, F. *Zool. Jahrb.; Abt. Syst.*, **74** (1941), 375–434 (flower visiting behaviou *Plusia*, Lep.).
[132] SCHULZ, W. *Z. vergl. Physiol.*, **14** (1931), 392–404 (visual orientation: *Notonecta* Hem.).
[133] SCOTT, W. N. *Trans. Roy. Ent. Soc. Lond.*, **85** (1936), 303–30 (factors governing hou of emergence of *Ephestia* pupae, Lep.).
[134] SHIMA, T. *Annot. Zool. Japon.*, **19** (1940), 294–300 (training experiments: *Dineute* Col.).
[135] SIOLI, H. *Zool. Jahrb., Physiol.*, **58** (1937), 284–96 (orientation to warmth: *Cime* Hem.).
[136] V. STEIN-BELING, I. *Biol. Rev.*, **10** (1934), 18–41 (appreciation of time: review).
[137] STEINIGER, F. *Z. Morph. Oekol. Tiere*, **26** (1933), 591–708; **28** (1933), 1–51; (cata lepsy: Phasmidae, Gerridae, &c.); *Ergeb. Biol.*, **13** (1936), 348–451 (hypnosi review).
[138] SUBKLEW, W. *Z. vergl. Physiol.*, **21** (1934), 157–66 (sense physiology and biolog *Agriotes* larva, Col.).
[139] SZYMANSKI, J. S. *Arch. ges. Physiol.*, **144** (1912), 132–4 (learning in cockroach).
[140] —— *Arch. ges. Physiol.*, **170** (1918), 1–244 (reflex responses in insects).

141 THORPE, W. H., and JONES, F. G. W. *Proc. Roy. Soc.*, B, **124** (1937), 56–81 (host during larval life affecting selection of host by adult: *Nemeritis*, Ichneumonidae).

142 THORPE, W. H., and CAUDLE, H. B. *Parasitology*, **30** (1938), 523–8 (olfactory responses of insect parasites to food plant of host).

143 THORPE, W. H. *Proc. Roy. Soc.*, B, **126** (1938), 370–97; **127** (1939), 424–33 (pre-imaginal olfactory conditioning).

144 —— *Brit. J. Psychol.*, **33** (1943), 220–34; **34** (1943), 21–31; **34** (1944) 66–67 (learning in insects).

145 THORPE, W. H., *et al.* *J. Exp. Biol.*, **23** (1947), 234–66 (behaviour of wireworms in response to chemical stimuli: *Agriotes*, Col.).

146 THORPE, W. H. *Bull. Anim. Behaviour*, No. 7 (1948), 1–12 (instinctive behaviour).

147 THORPE, W. H., and CRISP, D. J. *J. Exp. Biol.*, **24** (1947), 310–28 (orientation during swimming: *Nepa*, *Aphelocheirus*, &c., Hem.).

148 TINBERGEN, N., *et al.* *Z. vergl. Physiol.*, **16** (1932), 305–34; **21** (1935), 699–716; **25** (1938), 292–334; *Biol. Zbl.*, **58** (1938), 425–35 (orientation in *Philanthus*, Hym.).

149 TISCHLER, W. *Zool. Jahrb., Physiol.*, **57** (1936), 157–202 (response of insects to visual forms).

150 TONNER, F. *Z. vergl. Physiol.*, **22** (1935), 517–23 (swimming reflexes in larvae of *Aeschna*, Odonata).

151 —— *Z. vergl. Physiol.*, **25** (1938), 427–54 (reflexes and perception of movement: *Aeschna* larva).

152 TURNER, C. H. *Biol. Bull.*, **23** (1912), 371–86 (learning in cockroach).

153 v. UEXKÜLL, J. *Z. Biol.*, **50** (1908), 168–202 (reflex behaviour: Odonata).

154 ULRICH, W. *Z. Morph. Oekol. Tiere*, **17** (1930), 552–624 (halteres in Strepsiptera).

155 URBAN, F. *Z. wiss. Zool.*, **140** (1932), 291–355 (responses to light: honey-bee).

156 UVAROV, B. P. *Vᵉ Congrès Internat. d'Entom. Paris, 1932*, 1933 (conditioned reflexes in insects).

157 VALENTINE, J. M. *J. Exp. Zool.*, **58** (1931), 165–220 (orientation by smell: *Tenebrio*, Col.).

158 VERLAINE, L. *Bull. Ann. Soc. Ent. Belg.*, **69** (1929), 115–25 (appreciation of time in wasps).

159 DU VERNAY, W. von B. *Z. vergl. Physiol.*, **30** (1942), 84–116 (learning in *Tenebrio*, Col.).

160 VOLKONSKY, M. *Arch. Inst. Pasteur d'Algérie*, **17** (1939), 194–220 (photo-akinesis in locusts).

161 WAHL, O. *Z. vergl. Physiol.*, **18** (1933), 709–17 (appreciation of time in honey-bee).

162 WARNKE, G. *Z. vergl. Physiol.*, **14** (1931), 121–99 (orientation by smell: *Geotrupes*. Col.).

163 WEBER, HEINZ. *Z. vergl. Physiol.*, **9** (1929), 564–612 (sensory physiology: *Haematopinus*, Anoplura).

164 WELSH, J. H. *Quart. Rev. Biol.*, **13** (1938), 123–39 (diurnal rhythms).

165 WEYRAUCH, W. K. *Z. vergl. Physiol.*, **10** (1929), 665–87 (sense physiology and ecology: *Forficula*).

166 —— *Z. Morph. Oekol. Tiere*, **15** (1929), 109–55 (hypnosis: *Forficula*).

167 —— *Rev. Suisse Zool.*, **43** (1936), 455–65; *Zool. Anz.*, **113** (1936), 115–25 ('skototaxis': *Lasius* (Formicidae), *Notiophilus* (Carabidae), &c.).

168 WIESMANN, R. *Landwirtschaft. Jahrb. d. Schweiz*, 1937, 1080–1109 (sensory physiology of *Rhagoletis*, Dipt.).

169 WIGGLESWORTH, V. B., and GILLETT, J. D. *J. Exp. Biol.*, **11** (1934), 120–39; 408–9 (orientation by warmth and smell: *Rhodnius*, Hem.).

170 WIGGLESWORTH, V. B. *Parasitology*, **33** (1941), 67–109 (orientation in *Pediculus*).

171 de WILDE, J. *Arch. néerl. Physiol.*, **25** (1941), 277–86 (antennal responses to visual stimuli).

172 WOLF, E. *Z. vergl. Physiol.*, **3** (1926), 615–91; **6** (1927), 221–54; **14** (1931), 746–62 (orientation of bees returning to the hive).

173 WOLSKY, A. *Biol. Rev.*, **8** (1933), 370–417 (stimulatory organs: review).

174 ZEISER, T. *Zool. Jahrb., Physiol.*, **53** (1934), 501–20 (responses to light and gravity: *Dytiscus*, Col.).

SUPPLEMENTARY REFERENCES (A)

[175] CARTHY, J. D. *Behaviour*, **3** (1951), 275–18 (orientation in ants).
[176] CHAUVIN, R. *L'Année psychol.*, **46** (1945), 148–55 (odour trails in ants).
[177] DOLLEY, W. L., *et al. Biol. Bull.*, **92** (1947), 178–86; **100** (1951), 84–9; 90–4 (temperature and responses to light: *Eristalis*, Dipt.).
[178] v. FRISCH, K. *Experientia*, **5** (1949), 142–8; **6** (1950), 210–21; *Naturwissenschaften*, **38** (1951), 105–12 (polarization of light and orientation in honey-bee).
[179] HAFEZ, M., *et al. Parasitology*, **40** (1950), 215–36; *J. Exp. Zool.*, **124** (1953), 199–226; *Bull. Soc. Entom. Egypte*, **42** (1958), 123–61 (orientation of *Musca* larvae to temperature, &c.).
[180] HASSENSTEIN, B. *Naturwissenschaften*, **37** (1950), 45–6; *Verhandl. Deutsch. Zool.* (1948), 371–6; *Z. vergl. Physiol.*, **33** (1951), 301–26; *Z. Naturforsch.*, **14b** (1959), 659–74 (optomotor responses in Coleoptera, &c.).
[181] van HEERDT, P. F. *Koninkl. Nederl. Akad. Wetensch.*, **53** (1950), No. 3, 1–16 (temperature and humidity preferences in Coleoptera).
[182] v. HOLST, E., and MITTELSTAEDT, H. *Naturwissenschaften*, **37** (1950), 464–76 (reciprocal adaptation between central nervous system and sense organs: *Eristalis*, Dipt.).
[183] KENNEDY, J. S. *Phil. Trans. Roy. Soc.*, B, **235** (1951), 163–290 (locust migration).
[184] MITTELSTAEDT, H. *Naturwissenschaften*, **36** (1949), 90 (optomotor reaction in *Eristalis* with eyes inverted).
[185] —— *Z. vergl. Physiol.*, **32** (1950), 422–63 (dorsal light response during flight: *Anax*, Odonata).
[186] PERTTUNEN, V. *Ann. entom. Fennici*, **17** (1951), 72–84 (humidity preferences in Carabidae).
[187] ROTH, L. M., and WILLIS, E. R. *J. Exp. Zool.*, **115** (1950), 561–87; **117** (1951), 451–87 (humidity reactions in *Tribolium* and other Coleoptera).
[188] SCHÖNE, H. *Z. vergl. Physiol.*, **33** (1951), 63–98 (dermal light sense and dorsal light response in *Dytiscus* larvae, &c.).
[189] VOWLES, D. M. *Nature*, **165** (1950), 282–3 (sensitivity of ants to polarized light).
[190] WELLINGTON, W. G. *Sci. Agric.*, **29** (1949), 201–15 (response to temperature and moisture: *Choristoneura* larva, Lep.).

SUPPLEMENTARY REFERENCES (B)

[191] BARTON-BROWNE, L. B. *Ann. Rev. Ent.*, **9** (1964), 63–82 (water regulation in insects: review).
[192] BIRUKOW, G., *et al. Naturwiss.*, **40** (1953), 60–2, 611–12; *Z. vergl. Physiol.*, **36** (1954), 176–211 (transference of reactions between light and gravity in *Melolontha*, &c.).
[193] —— *Z. Tierpychol.*, **13** (1957), 463–84; **14** (1957), 184–203 (appreciation of time and perception of plane-polarized light in *Velia*).
[194] BLEST, A. D. *Nature*, **181** (1958), 1077–8; *Behaviour*, **16** (1960), 188–253 (settling behaviour of Lepidoptera in relation to distance flown).
[195] BRETT, W. J. *Ann. Ent. Soc. Amer.*, **48** (1955), 119–31 (persistent emergence rhythm in *Drosophila*).
[196] BRUCE, V. G., and PITTENDRIGH, C. S. *Amer. Nat.*, **91** (1957), 179–95 (endogenous rhythms in insects, review).
[197] v. BUDDENBROCK, W., and MOLLER-RACKE, I. *Zool. Jahrb., Physiol.*, **65** (1954), 219–36 (optomotor responses in *Calandra*).
[198] BURSELL, E. *J. Exp. Biol.*, **34** (1957), 42–51 (humidity perception and activity in *Glossina*).
[199] BUTLER, C. G. *J. Ins. Physiol.*, **7** (1961), 258–64 ('queen substance', &c., and communication in the honey-bee).
[200] DANZER, A. *Z. vergl. Physiol.*, **39** (1956), 76–83 (orientation of *Geotrupes* to odours).
[201] DETHIER, V. G. *Exp. Path.*, **6** (1957), 68–122 (sensory physiology of blood-sucking insects: review).
[202] —— *Science*, **125** (1957), 331–6 (dance-like behaviour in *Phormia*).

[203] DETHIER, V. G., and BODENSTEIN, D. *Z. Tierpsychol.*, **15** (1958), 129–40 (hunger in the blow-fly, *Phormia*).

[204] DETHIER, V. G., and EVANS, D. R. *Biol. Bull.*, **121** (1961), 108–16 (physiological control of water ingestion in *Phormia*).

[205] EIDMANN, H. *Z. vergl. Physiol.*, **38** (1956), 370–90 (diurnal rhythms in *Carausius*).

[206] ELLIS, P. E. *Symp. Gen. Biol. Ital.*, **10** (1961), 226–34 (development of gregarious behaviour in locusts).

[207] ESCH, H. *Z. vergl. Physiol.*, **45** (1961), 1–11 (sound production during the tail-wagging dance of the honey-bee).

[208] EVANS, D. R., and BROWNE, L. B. *Amer. Midl. Nat.*, **64** (1960), 282–300 (physiology of hunger in *Phormia*).

[209] FRAISSE, R., and ALLEGRET, P. *Rev. Zool. Agric. Appl.* 1957, 1–6 (induced changes in the form of the cocoon in *Bombyx*).

[210] V. FRISCH, K., and JANDER, R. *Z. vergl. Physiol.*, **40** (1957), 239–63 (the tail-wagging dance of the honey-bee).

[211] V. FRISCH, K., and LINDAUER, M. *Naturwiss.*, **41** (1954), 245–53 (landmarks and the sky-compass in orientation of honey-bee).

[212] —— *Naturwiss.*, **42** (1955), 377–85 (speed of flight and orientation of bees in a wind).

[213] —— *Ann. Rev. Ent.*, **1** (1956), 45–59 ('language' and orientation in the honey-bee: review).

[214] —— *Naturwiss.*, **48** (1961), 585–94 (systematic 'errors' in the indication of direction by the honey-bee).

[215] GETTRUP, E. *Nature*, **193** (1962), 498–9 (thoracic proprioceptors in the flight system of locusts).

[216] GOODMAN, L. J. *J. Exp. Biol.*, **37** (1960), 854–78 (landing response in *Lucilla*).

[217] GRAHAM, K. *Nature*, **184** (1959), 283–4; **191** (1961), 519 (egg-laying responses in *Trypodendron*, Scolytidae).

[217a] GRASSÉ, P. P. *Insectes Sociaux*, **6** (1959), 41–84 (co-operative building among termites).

[218] GROSS, F. J. *Bonn. zool. Beitr.*, **10** (1959), 160–71 (reflexes in caterpillars).

[219] HAAS, A. *Naturwiss.*, **39** (1952), 484 (pheromone in mandibular glands of *Bombus*).

[220] HADDOW, A. J., and GILLETT, J. D. *Ann. Trop. Med. Parasit.*, **51** (1957), 159–74 (oviposition-cycle in *Aëdes* spp.).

[221] HALDANE, J. B. S., and SPURWAY, H. *Insectes Sociaux*, **1** (1954), 247–81 (communication in the honey-bee).

[222] HARKER, J. E. *Nature*, **173** (1954), 689; **175** (1955), 733; *J. Exp. Biol.*, **33** (1956), 224–34; **37** (1960), 154–63; *Cold Spring Harbor Symp. Quant. Biol.*, **25** (1960), 279–87 (hormonal control of diurnal rhythm in *Periplaneta*).

[223] —— *Biol. Rev.*, **33** (1958), 1–52; *Ann. Rev. Ent.*, **6** (1961), 131–45 (diurnal rhythms: review).

[224] —— *The physiology of diurnal rhythms*, Cambridge Univ. Press, 1964, 114 pp.

[225] HERAN, H. *Z. vergl. Physiol.*, **38** (1956), 168–218 (perception of distance in the honey-bee).

[226] HORRÍDGE, G. A. *Nature*, **193** (1962), 698–9; *Proc. Roy. Soc.*, B, **157** (1962), 33–52 (learning of leg position in headless insects).

[227] IBBOTSON, A., and KENNEDY, J. S. *J. Exp. Biol.*, **36** (1959), 377–90 (interaction between walking and probing in *Aphis fabae*).

[228] JANDER, R. *Z. vergl. Physiol.*, **40** (1957), 162–238 (visual orientation in *Formica*).

[229] —— *Z. vergl. Physiol.*, **43** (1960), 680–6 (transference of the menotaxis angle between light and gravity in *Trichoptera*).

[230] JOHNSON, B. *Anim. Behav.*, **6** (1958), 2–26 (factors affecting locomotion and settling in Aphids).

[231] JOHNSON, C. G. *Nature*, **186** (1960), 348–50 (dispersal flights of newly emerged insects).

[232] KARLSON, P., and BUTENANDT, A. *Ann. Rev. Ent.*, **4** (1959), 39–58 (pheromones in insects: review).

[233] KELLOG, F. E., FRIZEL, D. E., and WRIGHT, R. H. *Canad. Ent.*, **94** (1962), 884–8 (orientation to odours in flying *Drosophila*).

[234] KENNEDY, J. S. *Nature*, **189** (1961), 785–91 (physiology and insect migration).

[235] KENNEDY, J. S., *et al. Proc. X. Int. Congr. Ent. 1956*, **2** (1958), 397–404; *J. Exp. Biol.*, **40** (1963), 351–69 (co-ordination of successive activities in Aphids).

[236] KENNEDY, J. S., BOOTH, C. O., and KERSHAW, W. J. S. *Ann. Appl. Biol.*, **49** (1961), 1–21 (visual responses in migration and alighting in Aphids).
[237] KULLENBERG, B. *Zool. Bidr. Uppsala*, **31** (1956), 254–354 (sex pheromones in Hymenoptera).
[238] LANGE, R. *Naturwiss.*, **45** (1958), 196 (pheromones and communication in ants).
[239] LINDAUER, M. *Z. vergl. Physiol.*, **34** (1952), 299–345 (division of labour in the hive).
[240] —— *Cold Spring Harbor Symp. Quant. Biol.*, **25** (1960), 371–7 (solar orientation of bees at different times of day).
[241] —— *Communication among social bees*, Harvard Univ. Press, 1961, 143 pp.
[242] —— *Fortschritte d. Zool.*, **16** (1963), 58–140; *Ergeb. Biol.*, **26** (1963), 158–81 (orientation in insects: review).
[243] LINDAUER, M., and KERR, W. E. *Bee World*, **41** (1960), 29–41; 65–71 (communication in stingless bees, Meliponini).
[244] LINDAUER, M., and MARTIN, H. *Naturwiss.*, **50** (1963), 509–14 (orientation of bees to odours).
[245] KARLSON, P., and LÜSCHER, M. *Nature*, **183** (1959), 55–6 ('pheromones').
[246] MEDER, E. *Z. vergl. Physiol.*, **40** (1958), 610–41 (role of the sense of time in solar navigation by the honey-bee).
[247] NEW, D. A. T., and NEW, J. K. *J. Ins. Physiol.*, **6** (1961), 196–208; *J. Exp. Biol.*, **39** (1962), 271–91 (navigation by the honey-bee with the sun at the zenith).
[248] NOWOSIELSKI, J. W., and PATTON, R. L. *J. Ins. Physiol.*, **9** (1963), 401–10 (diurnal rhythms in *Acheta*).
[249] OHGUSHI, R. *Physiol. et Ecol.*, **9** (1960), 11–31 (host selection in *Mormoniella*).
[250] PALMEN, E. *Ann. Zool. Soc. 'Vanamo'*, **17** (1955), No. 3, 30 pp. (emergence rhythm in Chironomidae).
[251] PAPI, F. *Att. Soc. Toscana Sci. Nat. (Mem. Ser. B.)*, **62** (1955), 83–97 (astronomical navigation in Carabids).
[252] PERTTUNEN, V. *Nature*, **170** (1952), 209–10 (humidity response of *Forficula* at different seasons).
[253] PITTENDRIGH, C. S. *Harvey Lectures*, **56** (1962), 93–125 (diurnal rhythms: review).
[254] PITTENDRIGH, C. S., and BRUCE, V. G. in *Photoperiodism and related phenomena in plants and animals*, (Withrow, Ed.) Washington: A.A.A.S., 1959, 475–505 (nature of diurnal rhythms).
[255] RAKSHPAL, R. *Parasitology*, **49** (1959), 232–41 (orientation in *Columbicola*, Mallophaga).
[256] REMMERT, H. *Z. vergl. Physiol.*, **37** (1955), 338–54 (emergence rhythm in *Pseudomittia*, Chironomidae).
[257] RENNER, M. *Naturwiss.*, **21** (1955), 589 (action of scent glands of honey-bee).
[258] —— *Naturwiss.*, **19** (1955), 540–1; *Z. vergl. Physiol.*, **40** (1957), 85–118; *Ergebn. Biol.* **20** (1958), 127–58; **42** (1959), 449–83 (sense of time and solar orientation in the honey-bee).
[259] RIBBANDS, C. R. *et al. Proc. Roy. Soc.*, B., **140** (1952), 32–59 (food transmission and communication in the honey-bee).
[260] ROBERTS, S. K. de F., *et al. J. Cell. Comp. Physiol.*, **55** (1960), 99–110; **59** (1962), 175–86 (diurnal rhythms in *Leucophaea*).
[261] ROEDER, K. D. *Smithsonian Misc. Coll.*, **137** (1959), 287–306 (operation of the central nervous system in the insect).
[262] SCHNEIDER, D. *Z. vergl. Physiol.*, **40** (1957), 8–41 (chemoreception and mechanoreception in antennae of *Bombyx*).
[263] SCHWINK, I. *Z. vergl. Physiol.*, **37** (1954), 19–59 (orientation to odours in Lepidoptera).
[264] STUMPER, R. *C.R. Acad. Sci.*, **242** (1956), 2487–9 (scent trails of ants).
[265] THORPE, W. H. *Learning and instinct in animals*. Methuen: London, 1962.
[266] VAN der KLOOT, W. G., and WILLIAMS, C. M. *Behaviour*, **5** (1953), 141–74; **6** (1954), 233–255 (nervous control of cocoon spinning in *Cecropia*).
[267] VOWLES, D. M. *J. Exp. Biol.*, **31** (1954), 341–85 (orientation of ants to light, gravity and polarized light).
[268] WALOFF, Z. *Bull. Ent. Res.*, **43** (1953), 575–80 (humidity and flight in locusts).
[269] WEISS, K. *Z. Tierpsychol.*, **10** (1953), 29–44; *Z. vergl. Physiol.*, **36** (1954), 2–20; 531–42 (learning ability of bees and wasps in a labyrinth).

[270] WELLINGTON, W. G., *et al. Canad. Ent.*, **86** (1954), 529–42; *Ann. Ent. Soc. Amer.*, **48** (1955), 67–76 (polarized light and temperature in the solar orientation of insects).

[271] WISHART, G., van SICKLE, G. R., and RIORDAN, D. F. *Canad. Ent.*, **94** (1962), 614–26 (orientation of male *Aëdes* to sound).

SUPPLEMENTARY REFERENCES (C)

[272] BRADY, J. *Nature*, **223** (1969), 781–4 (control of insect circadian rhythms).

[273] BUTLER, C. G. *et al. Anim. Behav.*, **17** (1969), 142–7 (footprint pheromone in wasp and honey-bee).

[274] CAMHI, J. M. *J. exp. Biol.*, **50** (1969), 335–73 (sensory stimulation and flight initiation in *Schistocerca*).

[275] CYMBOROWSKI, B., DUTKOWSKI, A. *et al.*, *J. Insect Physiol.*, **16** (1970), 341–8; **17** (1971), 99–108; 1763–72 (circadian activity rhythm in *Acheta*).

[276] DAHM, K. H. *et al. Life Sciences Pt. II*, **10** (1971), 531–9 (response of flying *Plodia* males to sex attractant).

[277] DAYKIN, P. N., KELLOG, F. E. and WRIGHT, R. H. *Canad. Entom.*, **97** (1965), 240–63, (host-finding and repulsion in *Aëdes*).

[278] DETHIER, V. G. and GELPERIN, A. *J. exp. Biol.*, **47** (1967), 191–200 (hyperphagia and stretch receptors in *Phormia*).

[279] ERGENE, S. B. *Z. vergl. Physiol.*, **48** (1964), 467–80; **51** (1965), 96–102 (ocelli as stimulatory organs in locusts and mantids).

[280] FREE, J. B. and SIMPSON, J. *Z. vergl. Physiol.*, **61** (1968), 361–5 (alerting pheromones in honey-bee).

[281] GATEHOUSE, A. G. *J. Insect Physiol.*, **16** (1970), 61–74 (probing response of *Stomoxys*).

[282] GOODMAN, L. J. *J. exp. Biol.*, **42** (1965), 385–407 (optomotor reactions in flight stability of *Schistocerca*).

[283] GWADZ, R. W. *J. Insect Physiol.*, **15** (1969), 2039–44 (control of blood meal size in *Aëdes*).

[284] HANGARTNER, W. *Z. vergl. Physiol.*, **57** (1968), 103–36 (use of scent trails in the ant *Lasius*).

[285] HORRIDGE, G. A. *J. exp. Biol.*, **44** (1966), 255–61 (optokinetic memory in *Locusta*).

[286] HOYLE, G. *Adv. Insect Physiol.*, **7** (1970), 349–444 (cellular mechanisms underlying behaviour).

[287] JERMY, T., HANSON, F. E. and DETHIER, V. G. *Ent. exp. & appl.*, **11** (1968), 211–30 (induction of specific food preference in caterpillars).

[288] KENNEDY, J. S. and MOORHOUSE, J. E. *Ent. exp. & appl.*, **12** (1969), 487–503 ('anemotaxis' of locusts to odours).

[289] KERKUT, G. A. *et al. Nature*, **227** (1970), 722–3 (chemical change during learning in headless cockroach).

[290] LEUTHOLD, R. *J. Insect Physiol.*, **12** (1966), 1303–31 (activity cycles of female *Leucophaea*).

[291] MARTIN, H. *Z. vergl. Physiol.*, **50** (1965), 254–92 (topochemical sense in honey-bee).

[292] MEDUGORAC, I. and LINDAUER, M. *Z. vergl. Physiol.*, **55** (1967), 450–74 (effect of narcotics on sense of time in honey-bee).

[293] NEUMANN, D. *Z. vergl. Physiol.*, **53** (1966), 1–61; **60** (1968), 63–78 (semilunar and diurnal emergence rhythm in intertidal midge *Clunio*).

[294] NISHIITSUTSUGI-UWO, J. *et al. Biol. Bull. mar. biol. Lab., Woods Hole*, **133** (1967), 679–96; *Z. vergl. Physiol.*, **58** (1968), 1–46 (control of circadian activity rhythm of cockroach).

[295] PALKA, J. *J. exp. Biol.*, **50** (1969), 723–32 (discrimination between movements of eye and object in crickets).

[296] PFLUMM, W. *Z. vergl. Physiol.*, **68** (1970), 49–59 (chemotropotaxis in *Phormia* and *Carabus*).

[297] ROBERTS, S. K. de F. *Science*, **148** (1965), 958–9; *J. cell. Physiol.*, **67** (1966), 473–86 (circadian activity rhythm in cockroaches).

[298] SOMBERG, J. C., HAPP, G. M. and SCHNEIDER, A. M. *Nature*, **228** (1970), 87–8 (retention of conditioned response through metamorphosis in *Tenebrio*).

[299] STUART, A. M. *Science*, **156** (1967), 1123–5 (building behaviour in termites).
[300] de TALENS, A. F. P. and FERRETTI, C. T. *J. exp. Biol.*, **52** (1970), 233–56 (landing reaction in *Musca*).
[301] ZIMMERMAN, W. F., PITTENDRIGH, C. S. and PAVLIDIS, T. *J. Insect Physiol.*, **14** (1968), 669–84 (eclosion rhythm in *Drosophila*).

Chapter IX

Respiration

IN THIS chapter we shall consider how oxygen is conveyed from the atmosphere to the tissues, and how the carbon dioxide resulting from oxidation in the tissues is eliminated. The original respiratory organ of all metazoa is the skin; but we have seen that the skin of insects, in association with their terrestrial existence, has become impermeable to water and thereby ill-adapted for respiration. The great majority of insects, as was shown by Malpighi (1669) in the silkworm, breathe by means of tracheal tubes, which usually open at the surface of the body through a number of spiracles, and convey air directly to the tissues. Respiration solely by the skin, without the intervention of an air-filled system of this type, has been acquired, probably secondarily,[154] by a few of the primitive Poduridae from moist surroundings, and in the young stages of some aquatic and parasitic forms (p. 385).

THE TRACHEAL SYSTEM

The tracheae are invaginations of the cuticle, which branch everywhere among the tissues. In the primitive *Campodea* each spiracle gives rise to a tracheal tree which is quite independent, and shows no anastomoses with the tracheae of adjacent segments; but in all other insects the tracheae anastomose freely to form longitudinal and transverse trunks (Fig. 223). The number and arrangement of the functional spiracles varies enormously in the different groups of insects.[80, 90, 191] With the exception of some Collembola living in wet places, and in the larvae of some insects (the bee, many Diptera), the spiracles are almost always provided with closing mechanisms of varied design—first observed by Burmeister (1832).[6, 75, 102, 110, 191]

Structure of tracheae—The histological structure of the tracheae is essentially the same as that of the body surface from which they are derived; they consist of a matrix of discrete epithelial cells, relatively thick near the spiracles, flattened along the deeper branches, and a thin but complex cuticular lining[252] (Fig. 224). The cuticle is thrown into folds, which typically run a spiral course round the tubes for a short distance before a new fold begins. Often the margins of the folds fuse to form a thread, the so-called spiral filament or taenidium;[52] or a rather dense deposit of chitin and protein may be laid down between the folds.[252] Near the spiracles the folds are often irregular and sometimes hardly more than a collection of wrinkles in the crumpled membrane.[165] Branched projections may run laterally from the folds so that the cuticle shows quadrangular areas.[52, 120] Sometimes the spiral thread gives off hair-like processes into the lumen of the trachea.[52, 165]

As already noted (p. 30) the cuticle lining the tracheae has the same composition as that covering the surface of the insect and the ducts of the dermal

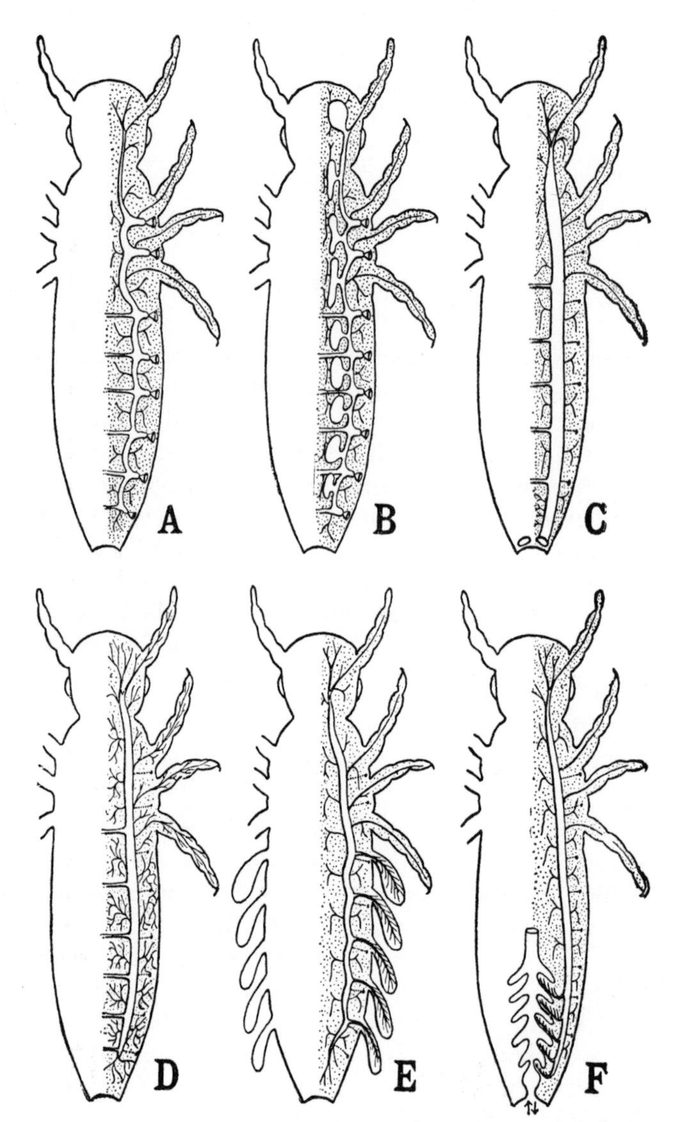

FIG. 223.—Some types of respiratory system in insects, schematic

A, simple anastomosing tracheae, with sphincters in the spiracles; *B*, mechanically ventilated air sacs developed; *C*, metapneustic respiration, terminal spiracles alone functional; *D*, tracheal system entirely closed, cutaneous respiration; *E*, the same with abdominal tracheal gills; *F*, the same with rectal tracheal gills.

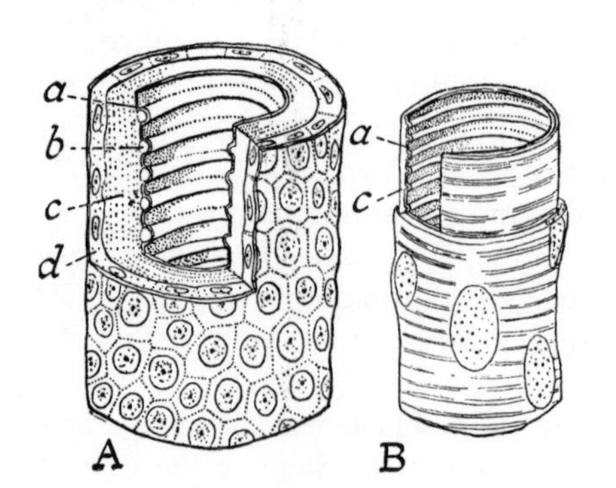

FIG. 224.—Structure of tracheae (*after* WEBER)

A, tracheal branch close to spiracle; *B*, small tracheal branch more highly magnified. *a*, epicuticle thrown into spiral folds; *b*, exocuticle; *c*, endocuticle; *d*, epithelial matrix.

glands. Near the spiracle the lining of cuticulin has a thick layer containing chitin beneath it. In the smaller branches and the air sacs chitin is absent. In the finest terminations the lining membrane is excessively delicate, and is freely permeable to water.[203] Its composition is not known.

Types of tracheae—In form the tracheae are extremely varied.[25] Typically they are circular in cross-section and prevented from collapsing by their spiral folds. Often they are elliptical, as in the main tracheae of *Dytiscus* larvae,[144] or mosquito larvae,[7] and the 'spiral thread' then tends to atrophy so that the tubes collapse when the air pressure within them is reduced. Saccular dilatations may occur along their course, as in *Melolontha*; and such dilatations may be enlarged by the fusion of the matrix of adjacent branches to form great air sacs,[51] as in *Orthoptera*, *Dytiscus*, *Apis* (Fig. 225), *Musca*, &c., and become bound to the surrounding tissues, usually muscles, by numberless small branches which they give off. Some Scarabaeid beetles even have small air sacs along the tracheae of the elytra.[34] These air spaces are generally flattened and often collapsed; but if they are so placed that they cannot collapse, as in the head or parts of the thorax, they form permanent air chambers. Nerves or muscles may run across such chambers and they are then invested with the tracheal coverings, but in the reverse order, with the cuticle outside ('inverted tracheae').[86]

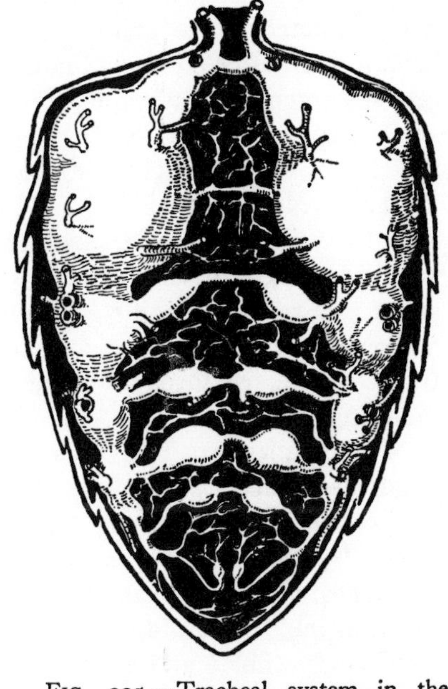

Fig. 225.—Tracheal system in the abdomen of the honey-bee worker, showing the air sacs. Dorsal tracheae and air sacs have been removed (*after* Snodgrass)

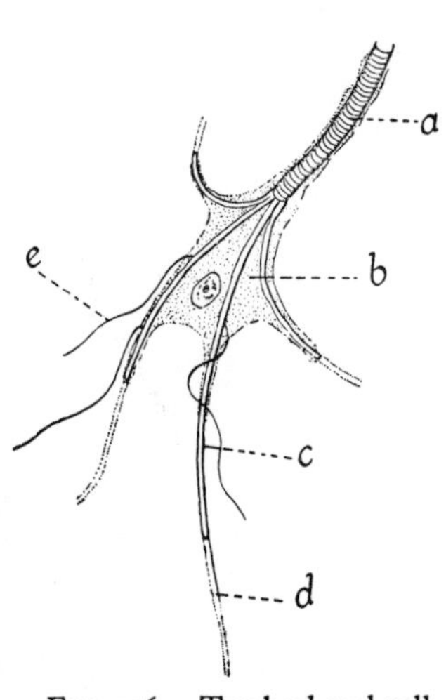

Fig. 226.—Tracheal end cell in head of mosquito larva *Aëdes aegypti* (*after* Wigglesworth)

a, trachea; **b**, tracheal end cell; **c**, tracheole containing air; **d**, terminal part of tracheole containing fluid; **e**, fine branch of tracheole.

Tracheoles—The mode of ending of the tracheae shows an equal diversity.[100] Typically, when the trachea has been reduced by repeated branching to a diameter of 2–5μ it enters a large stellate cell, the 'tracheal end cell' or 'transition cell',[82] and there breaks up abruptly into a number of tracheal capillaries or tracheoles, which are less than 1μ in thickness and are characterized by the absence of a spiral fold as seen with the light microscope (Fig. 226). When examined with the electron microscope, however, a taenidium about 0·025μ broad is visible in the tracheoles (of *Culex*, *Apis*, &c.) right up to the point where they end blindly at a diameter of 0·2μ.[152]

In the tracheoles, as in many of the smallest tracheae, the folds are more often annular than spiral, the form depending presumably on whether or not a longitudinal stress component is added to the radial stress which is probably responsible for the appearance of the taenidium.[221, 252] The tracheal cells (Fig. 226) are probably homologous with the epithelial cells that lay down the tracheal cuticle; they have become spread out into a web-like form through being situated at the point of furcation, and they are very large because they have to look after so many branches.[118, 202] When the trachea passes gradually into the tracheole, as is the case at many tracheal endings, tracheal end cells are not conspicuous;[19, 42, 67] nor are they where a single tracheole arises from the side of a large tracheal stem.

Tracheoles in the epidermis of *Rhodnius* are simple or branched tubes; they arise from the endings of small tracheae 1·5–2μ in diameter. They are usually 200–350μ in length, each formed within a single cell the nucleus of which lies at about 75μ from the point of origin. They taper gradually from a diameter of 0·6–0·9μ to end blindly at a diameter of 0·2μ or less.[273] Once laid down the tracheoles persist throughout the life of the insect; the cuticle is not shed at moulting (p. 364).

The palmate tracheal cells may unite with one another by means of the finger-like processes they give off, and so form a fenestrated membrane, rich in tracheae, in which the organs are invested. That is so around the ovaries and testes.[150, 153, 162] It is this membrane, variously modified, which forms the 'peritoneal membranes' and some of the 'connective tissues' of insects, as around the intestine.[51] It is said that the membranes so formed may be differentiated to form muscle fibres and elastic fibres,[150, 153] but this probably results from the incorporation of haemocytes and other mesodermal cells.

In many tissues, membranes of this kind cannot be detected. In the salivary glands and gut of some insects the tracheoles ramify to form a rich network around and between the epithelial cells without penetrating the cytoplasm.[28, 186, 212] In other insects or other tissues the tracheoles enter the cells—as in the photogenic cells in the luminous organs of beetles,[118, 202] the fat body sometimes,[176] in the cells of the *organe rouge* of *Gasterophilus* larvae,[146] the salivary glands of Hemiptera,[58] in the rectal glands of Muscids, and the anal papillae of mosquito larvae.[205] There is great variation in different organs.[204] When studied with the electron microscope it is clear that the formative cells, or 'tracheoblasts', may deeply indent the walls of other cells to give rise to 'intracellular' tracheolation; but whether the tracheole is interstitial or 'intracellular' it is always separated from the tracheolated cell by the cytoplasm and membranes of the tracheoblast.[239, 264]

There are many differences, also, in the mode of ending in the muscles. In the ordinary muscles of the limb and body wall the tracheoles lie on the surface, or penetrate a short distance into the superficial layers of the fibre—as in the leg muscles of *Hydrophilus*[5, 93] or the muscles of the bee, excluding the flight muscles.[83] On the other hand, in the flight muscles there is often a network of intracellular tracheoles running and anastomosing chiefly longitudinally;[5, 93, 129] and in wing muscles of the fibrillar type (p. 147) the tracheoles ramify in the substance of the sarcoplasm, closely investing each fibril[5] so that each sarcomere is ringed almost completely by a tracheole or by portions of adjoining tracheoles.[132]

It is uncertain whether the tracheoles commonly end by anastomosing with one another.[204] There seems little doubt that they may do so within the muscle fibres;[129] but they certainly end separately as a rule in the fenestrated membranes on the surface of organs;[150, 153] and even in the gill plates of *Aeschna* nymphs, where they appear to form closed loops, careful observation has shown that each capillary ends blindly.[99] It may be that despite appearances to the contrary, the tracheoles almost always end blindly.[271]

Movements of fluid in the tracheal endings—After death the tracheoles quickly fill with fluid, which creeps along them from the tissues. This has led to an old-standing controversy as to whether, during life, the endings contain air or liquid.[204] The question cannot be decided by dissection of the tissues, because they are so easily affected by the fluid in which they are immersed. But examination of living insects under the microscope by transmitted light has shown that there is much variation in different tissues and different insects. In the abdomen

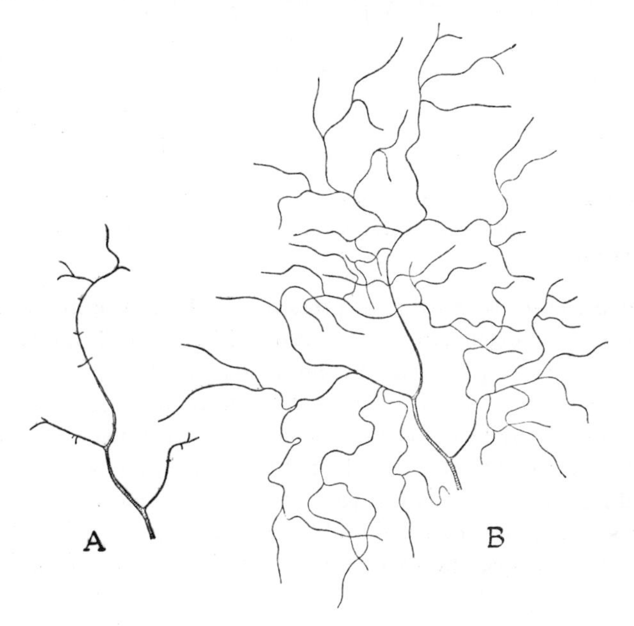

FIG. 227.—Tracheal ending on surface of gut of living mealworm larva

A, at rest; *B*, after asphyxiation by flooding with water (*after* WIGGLESWORTH). The tracheoles are visible as black threads when they contain air, invisible when they contain fluid.

of the flea, for example, the column of air in the tracheoles ends abruptly while they are still quite large (perhaps 0·2–0·3μ) and beyond that they contain fluid; whereas among the muscles of the legs they contain air as far as they can be resolved with the microscope. In the young mealworm larva (*Tenebrio*) the air extends further among the muscles of the body wall than on the surface of the gut (Fig. 227, A);[203] while in the Anoplura all the tracheoles seem to contain air as far as they can be traced. The same is true in larvae of *Sciara*,[92] and in the tracheoles of the epidermis in *Rhodnius*.[273]

Since liquid tends to rise up the tracheoles from the tissues their walls must be wetted by water. There must therefore be a capillary force drawing liquid upwards from the endings of the tracheoles. How great this force may be is not known; but since the wall of the tracheole is wetted by oil as well as by water it is probably not very great.[31] This force must be opposed by an equal force holding the fluid in the tissues. It is probable that imbibition by the colloid substance of the tracheole wall and the cytoplasmic layer around it is responsible for holding the liquid back, and that it is the differences in these properties in different organs and insects which are responsible for the differences in the extent to which water rises up the tracheoles.[13, 207]

The water-binding power around the endings is influenced by changes in the osmotic pressure of the blood in which they are bathed. During muscular contraction, especially if the oxygen supply is deficient, the osmotic pressure

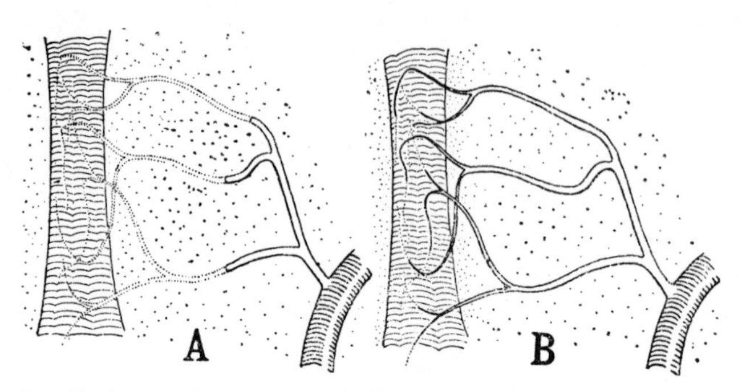

FIG. 228.—Tracheoles running to a muscle fibre: semi-schematic (*after* WIGGLESWORTH)

A, muscle at rest; terminal parts of tracheoles (shown dotted) contain fluid. B, muscle fatigued; air extends far into tracheoles.

of the blood increases (the osmotic pressure in the mosquito larva, expressed by the equivalent concentration of sodium chloride, will increase from 0·85 per cent. to 1·0 or 1·1 per cent. during asphyxiation[207]), and this increase is associated with the absorption of more fluid from the endings so that the air extends into finer branches of the tracheoles, to retreat again when the muscles rest and the metabolites responsible for the increased osmotic pressure are oxidized and removed[203, 272] (Figs. 227, 228). But if the cell membranes around the tracheole endings are rendered unduly permeable, by an excess of potassium ions or by other means, the extraction of fluid from the tracheoles does not occur.[31]

DEVELOPMENT OF THE TRACHEAL SYSTEM

During embryonic development the tracheae arise as more or less solid ingrowths of cells from the ectoderm, enclosing between them only a potential lumen.[195] The ingrowths from adjacent segments divide and unite to produce the longitudinal trunks. Stems from these grow deeply into the tissues dividing as they go, and a lumen appears, at first in the outer trunks, then extending inwards and gradually increasing in size. The cuticular intima is then laid down and becomes thrown into spiral folds. These folds may overlie many separate cells without showing any break in their continuity; they probably arise spontaneously by the action of some simple physical force, perhaps the expansion or contraction of the previously uniform membrane, rather than by the directing influence of the epithelial cells themselves.[128, 170] The spiral folding certainly appears before any air has entered the tracheae;[195, 273] and the periodicity of the folds agrees approximately with what would be expected if the cylindrical tube were undergoing axial compression as the result of expansion of its walls.[252]

Development of tracheoles—At the terminations of the tracheae, certain cells separate from the general tracheal epithelium and grow out as finger-like processes towards the tissues. It is in these cells that the tracheoles are laid down as fine intracellular canals.[141, 177, 195] They may be seen penetrating the fully developed muscle fibres, and there giving rise to the intracellular tracheoles;[4, 141] and it seems probable that the intracellular tracheoles of other tissues arise in the same way, by the ingrowth of the tracheal cells.[183] The intracellular formation of tracheoles within the tracheal cells or 'tracheoblasts' can be observed in the living larva of *Sciara* with the aid of the phase contrast microscope. In the earliest stages the main tracheal trunks in this insect also arise intracellularly and are then indistinguishable from tracheoles.[92]

New tracheae and tracheoles arise in *Rhodnius* by the outgrowth of columns

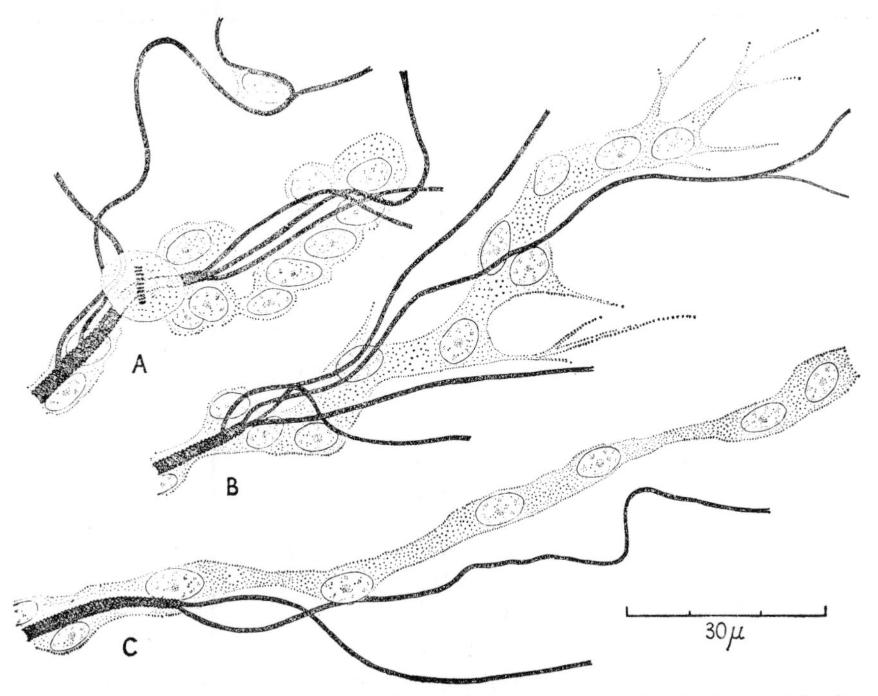

FIG. 229.—Early stages in the growth of new tracheae and tracheoles during moulting in *Rhodnius* (*after* WIGGLESWORTH)

A, tracheal epithelial cells multiplying to form an elongated cluster. *B*, the cluster of cells extending outwards. *C*, tracheal cells fully extended to form a column in which the new trachea will form. Existing tracheae and tracheoles injected with cobalt sulphide.

of cells from the sides and endings of existing tracheae at the time of moulting, the tracheoles being formed by the development of elongated canals within the filamentous outgrowths of the terminal cells (Fig. 229).[273]

Appearance of air in the tracheal system—The tracheal system is completely formed before any air appears in it. About the time of hatching from the egg the fluid, which until then fills the lumen of the tubes, is absorbed into the tissues.[42, 88, 195] Sometimes this does not happen until the spiracles are exposed to the air, either within the shell, as in the flea or mealworm,[161] or after hatching, as in the Hemiptera or Anoplura.[161] But often, both in aquatic insects such as Chironomids[88] or Odonata[178] in which the tracheal system is closed, and in Muscids,[195] *Drosophila*,[87] and Lepidoptera[161] in which the system is later open to the air, the tracheae fill with gas while the insect is still bathed in fluid. In the one case the air enters from outside, in the other it is liberated from solution in the tissue fluids.

In the larva of the mosquito *Aëdes aegypti* the fluid is absorbed from the tracheal system by the secretory activity of the cells bounding it (Fig. 230). This larva is able to defer absorption for several days after hatching, if it is not permitted to expose its spiracles at the water surface; in fact, the nervous system seems to be responsible for initiating the absorptive activity of the tracheal walls.[208] These observations make it probable that the filling of the tracheal system in other insects also is under the control of the cells. In *Sciara* larvae the appearance of gas in the tracheal system is completely inhibited in the absence of oxygen; but oxygen concentrations of 0·3–0·5 per cent. merely immobilize the larvae without preventing gas filling.[217] It looks as though powerful physical

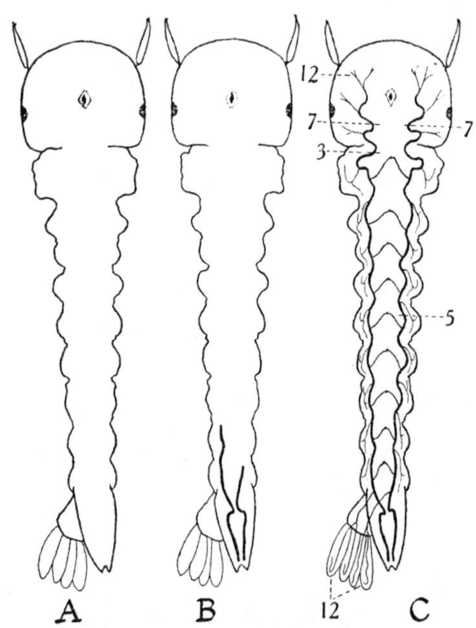

FIG. 230.—Entry of air into the tracheal system of the mosquito larva *Aëdes aegypti* after hatching from the egg (*after* WIGGLESWORTH)

A, before entry of air; *B*, system partially filled; *C*, system completely filled. The numbers in *C* show the time in minutes after exposure of the spiracles to the air; they mark the level to which the columns of air extended at these times.

forces (osmotic pressure or imbibition perhaps) would be needed; forces sufficient to lead to the rupture of the continuous column of liquid in the closed tracheal system.[88, 161] But if, as seems likely, the tracheal walls become less hydrophil at the time of filling, as though coated with wax (p. 606), the liberation of gas will be brought about by quite small negative pressures.[272]

Moulting of the tracheal system—At each moult a similar process takes place. The space between the old tracheal cuticle and the new becomes filled with fluid continuous with the moulting fluid of the body surface. This fluid plays the same role as that beneath the surface cuticle: the taenidia as well as the chitinous endocuticle are digested and dissolved; and when the process is complete, only the very thin and fragile epicuticle remains to be shed.[252, 273]

When moulting occurs the old linings of the tracheae are drawn out through the spiracles and the fluid is actively absorbed by the cells forming the new tracheal wall.[208, 211] In the early larval stages of *Sciara* the intima is moulted to the finest extremities, but in the pupal moult only the tracheal lining is shed: the existing tracheole persists and becomes continuous with the new trachea.[92] In *Rhodnius*, once the tracheoles have been laid down they persist throughout the life of the insect; the only part of the tracheole cuticle that is shed at moulting is that which runs between the old tracheal wall and the new (Fig. 231, A, B). The existing tracheole is joined to the new tracheal cuticle by a ring of cement (Fig. 231, C). The trachea thus grows like a plant, the 'nodes' of which represent the terminal tracheoles of successive instars (Fig. 231, D).[273, 274]

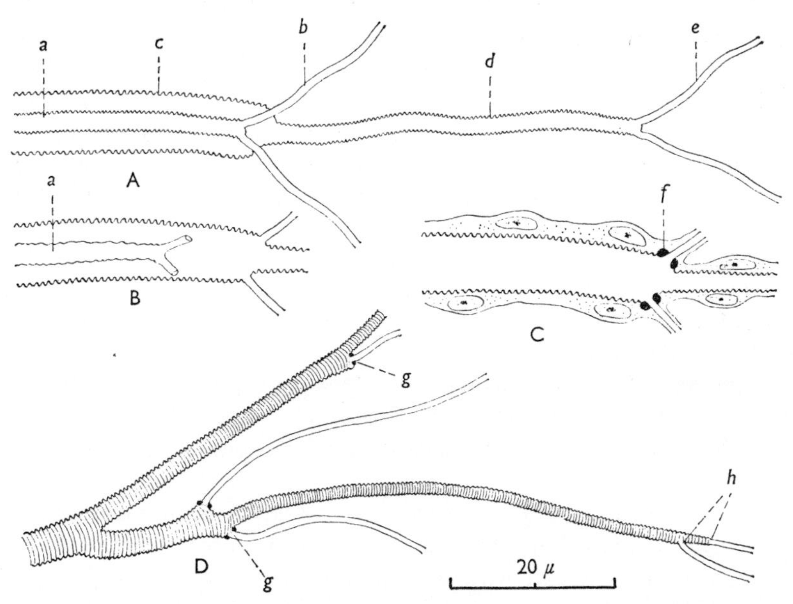

FIG. 231.—The moulting of the lining cuticle of tracheae and tracheoles (*after* WIGGLESWORTH)

A, diagram of tracheal ending, shortly before moulting. *a*, old tracheal cuticle: *b*, old terminal tracheoles: *c*, new tracheal cuticle: *d*, newly formed terminal trachea: *e*, new terminal tracheoles. *B*, the same at the moment of moulting. The old tracheal cuticle (*a*) with short segments of tracheole attached is being withdrawn. *C*, the same showing tracheal epithelium and the rings of cement (*f*) securing the old tracheoles to the new tracheal cuticle. *D*, actual tracheae and tracheoles in *Rhodnius*. *g*, shows tracheole attachments where moulting has occurred (ring of cement present): *h*, shows attachments of recently formed tracheoles (no rings of cement).

In most insects the tracheal linings are drawn out through the functional spiracles, the tubes breaking where they anastomose one with another. In Culicid larvae there are preformed points of weakness along the tracheal trunks where rupture occurs; each segment is drawn out through the corresponding pair of lateral spiracles.[208] Immediately afterwards the tubes to these spiracles collapse to form the 'stigmatic cords' of Palmen; only the terminal spiracles remain functional in respiration. Where a complex atrial chamber is developed this acquires a secondary connexion with the exterior. The primary spiracular opening now serves only for the withdrawal of the linings and then collapses to leave a scar. This scar may be central as in Tipulids, or lie to the inner side of the functional spiracle as in Cyclorrhapha.[90] The same mechanism has been developed independently in many beetles, the scars of the ecdysial tubes being sometimes easily mistaken for the true spiracles.[80]

THE TRANSPORT OF OXYGEN TO THE TRACHEAL ENDINGS

Site of oxygen uptake—It seems probable that in most insects, under ordinary conditions, comparatively little respiration takes place through the skin.[64, 249] In *Phormia regina* larvae with both pairs of spiracles tied off, the partial pressure of oxygen in the tracheal system is zero even when the larva is exposed to pure oxygen. In the normal larva in air probably less than 2·5 per cent. is taken up through the skin.[234] The cockroach, *Blatta*, will remain alive many hours with the spiracles blocked with paraffin.[148] But it will survive similarly in the complete absence of oxygen (p. 630). On anatomical grounds it is probable that most of the oxygen taken up enters the tissues through the walls of the tracheoles; for these are always most abundant in such organs as the wing muscles, ovaries, &c., with high oxygen requirements. And if reduced indigo (indigo white) is injected into the body of a living insect, it is oxidized to indigo blue almost solely around the terminal network of tracheoles.[121, 150] On the other hand, the tracheal walls are permeable to gases,[41, 65, 203] and it is impossible to say how much of the oxygen consumed passes through them; but the volume of tissue around the tracheae which could be supplied by diffusion through the tissue fluids is certainly insignificant in comparison with that supplied by the tracheoles.[270] [See p. 400.]

Mechanism of oxygen absorption—It is generally supposed that passage of oxygen from the tracheoles into the tissues takes place by physical diffusion. But several authors have suggested that the tracheal epithelium, and particularly the tracheal end-cells, play a more active part.[6, 215] In many insects (Orthoptera, Odonata) the cellular matrix of the tracheae is filled with granules of pigment; these show reversible colour changes on treatment with reducing agents[99] and have been believed (though without experimental evidence) to play a part in furthering oxidations or in storing up oxygen.[153] Similarly, the tracheal end-cells, which are very active in oxidizing injected indigo white, and in reducing osmic acid when the living insect has been exposed to this vapour, have also been credited with important functions in respiration.[118, 150]

Diffusion theory—It is certain that the greater part of the uptake of oxygen occurs in the tracheoles; and the main problem in insect respiration is the supply of oxygen to these endings. It was assumed long ago by Treviranus and by Thomas Graham (1833) that many insects must be dependent on diffusion; but this hypothesis was not generally accepted until Krogh,[108] taking into consideration the average diameter and length of the tracheae, the oxygen consumption of the insect and the diffusion constant of oxygen, showed by calculation that diffusion alone would supply the tissues with those quantities of oxygen actually consumed, and yet maintain at the commencement of the tracheoles a partial pressure of oxygen not more than 2 or 3 per cent. below that in the atmosphere. These calculations were made on large caterpillars which show no respiratory movements, it being assumed that diffusion takes place from the spiracles; in the case of *Dytiscus* larvae, diffusion was considered to take place from the periphery of the great flattened tracheae which are ventilated mechanically. It seems to be fairly generally true, in *Tenebrio*,[108] *Aphelocheirus*,[175] *Rhodnius*,[252] that the sum of the area of cross-section of the branches is equal to the area of cross-section of the main tracheae, so that diffusion may be pictured as taking place in a series of uniform cylinders; but it is doubtful whether this always applies to the tracheoles.[232]

In these calculations it was assumed also that the spiracles were open all the time. But the spiracles are almost always provided with sphincters (Fig. 232), the chief function of which is to protect the insect from loss of water. For, as in other terrestrial animals, those conditions which favour the supply of oxygen, favour also the loss of water; therefore, during rest, the spiracles of insects are kept closed most of the time, being opened only just enough to supply the insect with sufficient oxygen.[76] Hence the average tension of oxygen at the tracheal endings is probably lower than that calculated by Krogh.

In insect flight muscle the tracheole supply is so dense that between 0·1–0·001 per cent. of any cut area is occupied by air tubes. Gaseous diffusion will account for the gaseous exchanges observed in this system.[270] In *Drosophila* diffusion will provide for supply from the spiracles; in larger insects mechanical ventilation of the thoracic tracheae and air sacs is needed. But the muscle fibres cannot exceed about 20μ in diameter unless the tracheoles indent the surface and become 'internal'.[270]

'Diffusion control'—The median nerves which come off the ventral ganglia and then bifurcate to supply the nerves to the spiracles[237, 262] were called by Newport[134] 'respiratory nerves'. Each spiracle (in *Periplaneta* and *Blaberus*) receives a second nerve supply from one of the ordinary lateral nerves, but section of the ventral nerve to one side puts that spiracle out of action.[237]

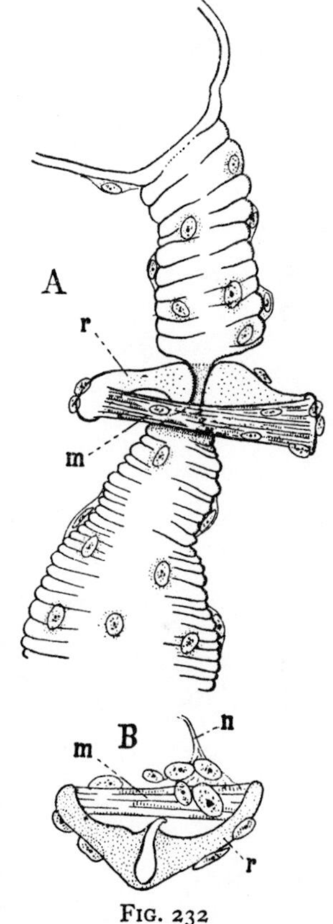

FIG. 232

A, closing mechanism on trachea of flea (*cf*. Fig. 233); *B*, transverse section through trachea at level of closing mechanism (*after* WIGGLESWORTH). *m*, muscle; *n*, nerve; *r*, flexible rod compressing trachea.

The control of respiration by the opening and closing of the spiracles, which is regulated by the nerves, is termed 'diffusion control' (p. 392).[76] Its importance in regulating respiration is best seen in a small insect like the flea, in which there is no mechanical ventilation of the tracheal system.[206] In the flea (*Xenopsylla*) at rest the whole of respiration takes place through two pairs of spiracles, the 1st* and 8th abdominal (Fig. 233). The sphincters in these spiracles often show a rhythmical opening and closing, with periods of five or ten seconds (Fig. 234). If the flea struggles, the thoracic spiracles open, and remain open for some seconds after the muscular movements cease. During the height of digestion, and while the eggs are ripening, the 1st and 8th abdominal spiracles keep open all the time, and the remaining spiracles show a rhythmical opening and closing, or may perhaps be held permanently open also. The rhythm of all the spiracles is quickened if the temperature is

* In the original paper the author followed Lass[111] in regarding this spiracle as the 3rd thoracic. It is more likely that the flea, like other insects, has only two pairs of thoracic spiracles.[163]

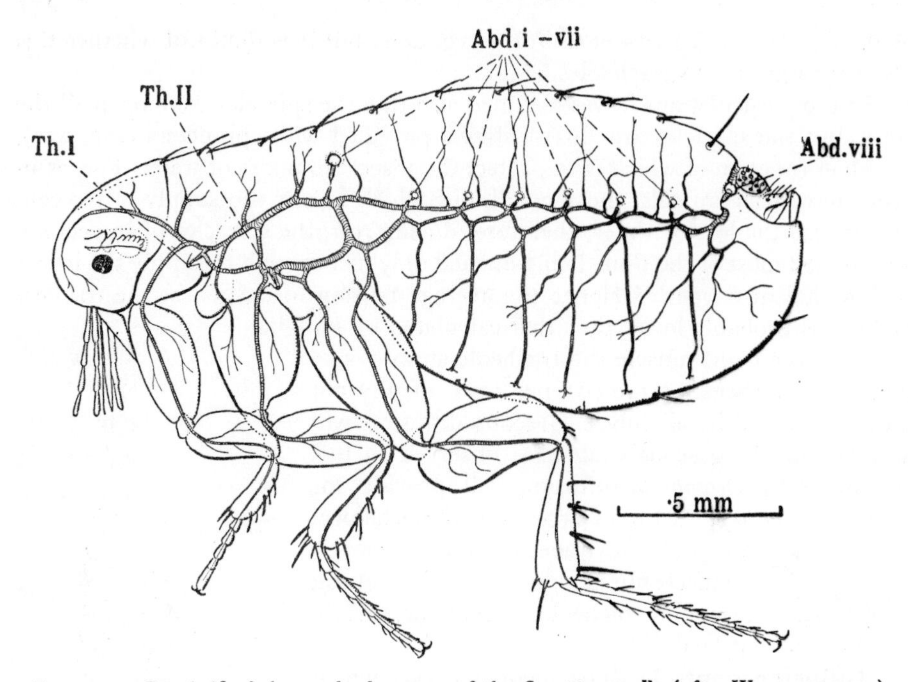

FIG. 233.—One-half of the tracheal system of the flea *Xenopsylla* (*after* WIGGLESWORTH) *Th.I*, *Th.II*, thoracic spiracles; *Abd. i–viii*, abdominal spiracles.

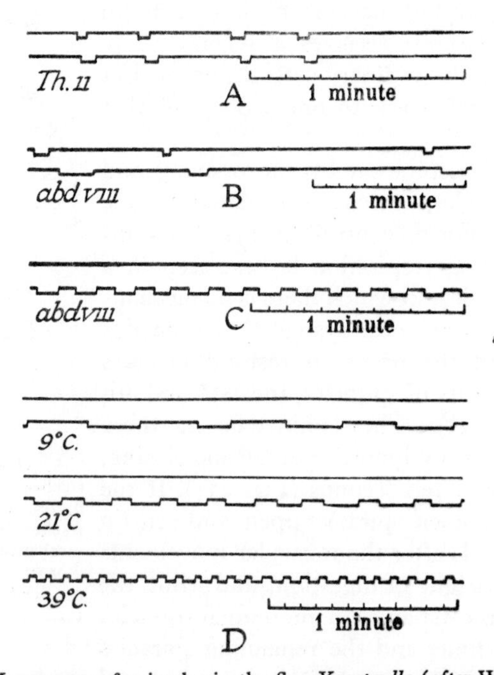

FIG. 234.—Movements of spiracles in the flea *Xenopsylla* (*after* WIGGLESWORTH)

The upper line of each tracing indicates periods of struggling; the lower line shows opening (downwards) and closing of the spiracles. A–C, at 22° C. *A*, second thoracic spiracle opening soon after struggling has begun; *B*, eighth abdominal spiracle in fasting flea opening after struggling has ceased; *C*, eighth abdominal spiracle in flea after feeding, showing rhythmical opening and closing in the absence of muscular movement; *D*, effect of temperature on the rhythmical opening and closing of the eighth abdominal spiracle.

raised.[206] During the period of closure of the spiracles, as the gas within the tracheal system is used up, many of the larger tracheae in the flea collapse and are flattened, while the smaller tracheae are compressed in the long axis.[78] The same thing is seen in larvae of mosquitos if they remain long under water.[7]

Spiracles and loss of water—The importance of the spiracular closing mechanisms in conserving water is well seen if the evaporation of water from the living insect is measured quantitatively.[125] In one experiment, a batch of adult *Xenopsylla* were caused to double their rate of loss of water when the spiracles were kept open by exposure to 5 per cent. carbon dioxide (Fig. 235, B). One recently fed mealworm larva increased its rate of loss from 5·1 hundredths of a milligram per hour in air, to 12·5 in 5 per cent. CO_2 in air; and after starving for four months, as the result of which its rate of metabolism was reduced so that the spiracles were normally opened less frequently, it increased its water loss from 1·5 hundredths of a milligram per hour in air, to 11·1 in 5 per cent. CO_2 in air (Fig. 235, A).[125] The bug *Rhodnius*, which is normally very resistant to desiccation, dies in about 3 days in a dry atmosphere if it is caused to keep its

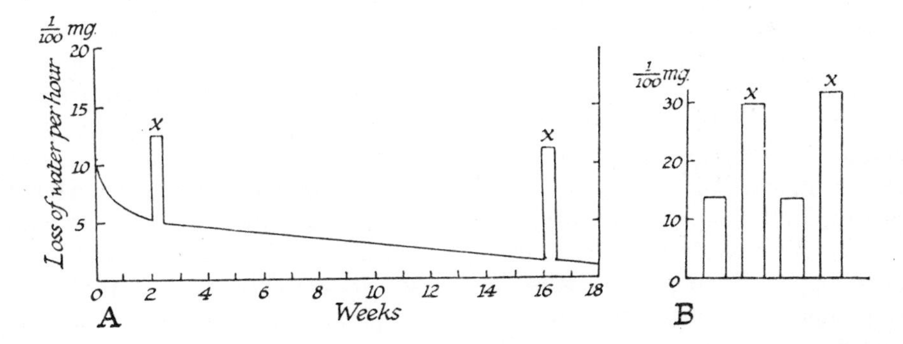

FIG. 235.—The effect of opening the spiracles on the rate of loss of water (*after* MELLANBY)

A, rate of loss in 1/100 mg. per hour from a single mealworm starved over a period of 4 months. At the point *X* the insect was caused to open the spiracles. *B*, rate of loss from a number of adult fleas (*Xenopsylla*) in 1/100 mg. per hour per mg. of insect. Columns marked *X* obtained when the spiracles were kept open with 5% carbon dioxide.

spiracles open.[209] In newly emerged males of *Anopheles maculipennis* the loss of weight amounted to 1·9 per cent. in 30 minutes in the normal insect; the loss increased to 23·1 per cent. if the spiracles were kept open with 2–3 per cent. CO_2.[268]

If the spiracles of a given group of beetles are compared, the xerophilous forms accustomed to a dry environment are found to have spiracles better adapted to prevent loss of water than the hygrophilous forms; in the former the openings are often small and deeply sunk; and in xerophilous Buprestids there is a basketwork of outgrowths across the openings, an arrangement which is believed to impede the diffusion of the aggregated molecules of water more than the carbon dioxide and oxygen.[15, 75] In many beetles from very dry environments (Tenebrionidae, *Niptus*, *Gibbium*) the subelytral space into which the spiracles open is considerably more humid than the atmosphere; they lose water much more rapidly if the subelytral space is opened.[250] In the sheep ked *Melophagus* and in the Anoplura the arrangement of atrial chambers and spines in the spiracles appears to be an adaptation to trap dust and prevent this from entering the system.[192, 193]

The seive plate of the silkworm spiracle reduces the total cross-sectional area to $\frac{1}{10}$ and the total area of perforation is the smallest in the tracheal system. But by virtue of the pore-diffusion effect of Brown and Escombe this reduction will not lower the rate of gaseous diffusion through it.[232]

'Ventilation control'—In addition to this regulation of diffusion, many insects effect a varying degree of mechanical ventilation of the larger tracheal trunks, or 'ventilation control'. It has often been suggested[7, 110, 149] that the muscular movements of the insect, and the contractions of the heart, intestine, Malpighian tubes and so forth, by pressing upon the tracheae, will assist their ventilation, and will do so more or less in proportion to the needs of the animal. But in the large caterpillar, *Cossus*, even the most violent contractions of the body cause very little ventilation—at most 2–9 c.mm. in a total tracheal volume of 50 c.mm.;[108] and in *Schistocerca* the pumping of air and of haemolymph brought about by the shortening of the muscles is of little importance for the exchange of gases, though of major importance for the supply of fuel for combustion.[270]

In many insects, however, particularly in active adult forms, specialized movements of the body wall occur, which serve the purpose of ventilating the tracheal system. These may consist of peristaltic waves over the abdomen as in *Tipula*, dorso-ventral flattening of the abdomen in many forms: grasshoppers, beetles, &c., telescoping movements of abdominal segments in Hymenoptera and Diptera (Figs. 236, 237).[6, 143, 149] In *Dytiscus* and *Hydrophilus*,[26] in *Carausius*[29] and doubtless other insects the thorax also may take part in the movements of expansion and contraction. As a rule expiration is the active movement and takes place more brusquely than inspiration, which is often effected by the elasticity of the body wall; but in certain cases, as in the larva of *Aeschna*[187] and some grasshoppers,[113] inspiratory muscles also are present (Fig. 237). If there is a respiratory pause it is always at the end of inspiration. In *Schistocerca*, in addition to dorso-ventral and longitudinal movements in the abdomen, there are ventilating mechanisms in the neck and prothorax which account for 14 per cent. of the air pumped.[253] [See p. 400.]

The function of the respiratory movements is partially to renew the air in the tracheae by alternate compression and expansion of the tracheal system, the changes in pressure being transmitted from the body wall by the haemolymph. At no stage is there any material compression of the gas in the tracheal system.[48, 108] In *Schistocerca*, *Bombus*, *Tabanus*, *Passalus* and many other insects the commencement of expiration before the spiracles open leads to a phase of 'compression' in the respiratory cycle; but the pressure rises by only 0·6–1·0 mm. of mercury in normal respiration and by a maximum of 10 mm. of mercury in extreme dyspnoea induced by activity or very low concentrations of oxygen.[226] These pressure changes will cause a compression ranging from less than 1 per thousand to a maximum of less than 1 per cent. In *Hydrophilus* during violent respiration the pressure in the tracheal system rises by 8–15 mm. of mercury. A suggested physiological explanation is that the spiracular openings must be kept small during expiration in order that the air sacs near the spiracles may not collapse and prevent the emptying of the system.[216]

In virtue of their spiral folding the tubular tracheae resist compression from side to side, but they are very extensible; they can be stretched to double their length without injury[156] or can be shortened 20–30 per cent. by a pressure of 5·6 cm. of mercury.[48] And many of the larger tracheae, such as the longitudinal trunks in *Dytiscus* larvae.[144] or the tracheae between the flight muscles of Muscid flies, are oval in cross section and cannot resist collapse. If the ventral surface of the adult water-beetle, *Cybister*, is examined with a lens while breathing at the surface, the tracheae can be seen through the transparent integument alternately flattened and dilated.[24] In the larva of *Eristalis* the enlarged tracheal trunks show both changes during expiration; they not only flatten but become shorter, the taenidia closing up like the coils of a spring.[50] The main tracheae of *Phormia*

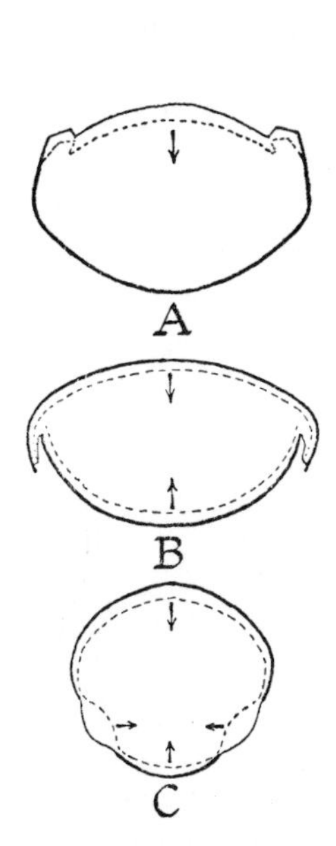

FIG. 236.—Types of expiratory movements in abdomen of insects (*from* SNODGRASS *after* PLATEAU)

A, movement almost confined to the terga (Heteroptera and Coleoptera); *B*, terga and sterna both take part in the movement (Odonata, Acridiidae, Aculeate Hymenoptera, Diptera); *C*, terga and sterna separated by ample membranous areas which are drawn inwards as the terga and sterna approach (Tettigoniidae, Neuroptera, Trichoptera, Lepidoptera).

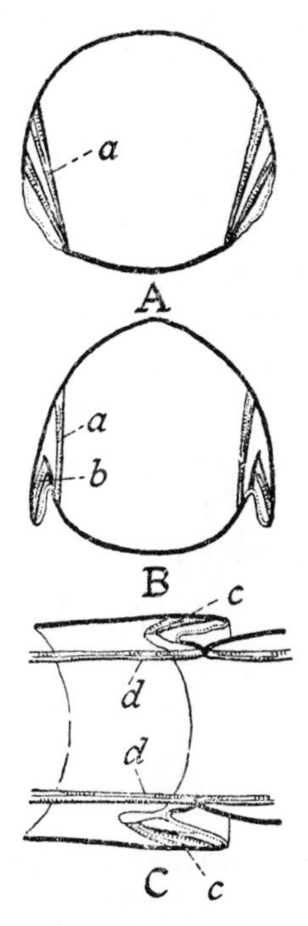

FIG. 237.—Diagrams of respiratory muscles in abdomen of insects (*after* SNODGRASS)

A, segment with compressor muscles only; *B*, with compressor and dilator muscles; *C*, longitudinal section showing protractor and contractor muscles which bring about the telescoping movements. *a*, compressor muscle; *b*, dilator; *c*, protractor; *d*, contractor.

larvae flatten by means of a longitudinal shearing movement in which the taenidial coils lie upon one another in sequence.[232]

A most unusual pumping mechanism for the intake of air during intra-uterine life has been described in the posterior spiracles, the 'polypneustic lobes', of the larva of *Glossina*.[235]

Functions of air sacs—The ventilation of the tracheal system is further helped by the special modifications of the tracheae present in those insects which show the most pronounced respiratory movements. The innumerable thin-walled saccular dilatations which occur all along the larger tracheae in the cockchafer, *Melolontha*, are collapsed and empty during expiration, whereas the tubular tracheae themselves appear unchanged.[48] In the larva of *Eristalis*, the tracheal sacs formed as dilatations of the main trunks are rhythmically collapsed and expanded by the integumental muscles.[2] In other insects, especially those with well-developed powers of flight, similar dilatations form large air sacs which extend into all parts of the body. Even where these lie in rigid parts such as the head or legs, the changes in pressure produced by the respiratory movements of the abdomen will be transmitted to them by the blood.[37, 70, 114] In locusts and dragon-flies each flight muscle receives a direct supply ending in a large air sac, which is automatically ventilated by thoracic pumping due to the movements of the nota and pleura during wing movements.[269] Even when the air sacs are of no great importance in ventilation they will serve to reduce the mechanical damping of the wing movements by the haemolymph.[269]

There can be little doubt, therefore, that the chief function of the air sacs is to increase the volume of the 'tidal air' which is renewed at each respiration. But this is certainly not their only function. As pointed out by John Hunter[134] they also serve to lower the specific gravity of the insect (the massive head and mandibles of such beetles as *Lucanus cervus* are largely filled with tracheal air sacs) and to that extent they will assist in flight. But their supposed buoyant action as

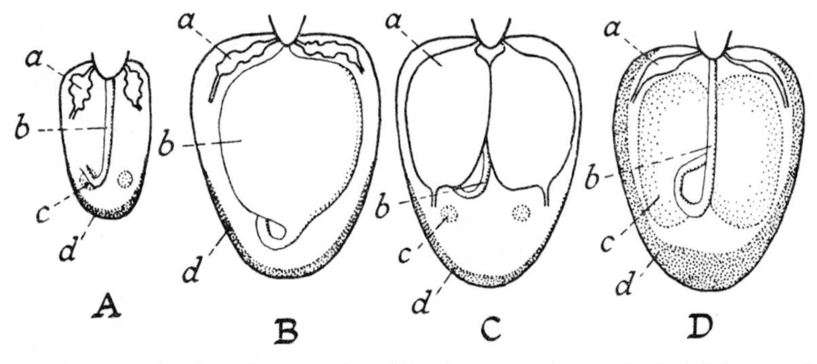

FIG. 238.—Diagram showing the part played by the great air sacs in the abdomen of *Lucilia* during the growth changes in the adult fly (*after* EVANS)

A, fly just emerged. *B*, after 5 minutes; gut filled with air, abdomen distended. *C*, after 10 hours; gut collapsed, air sacs distended. *D*, fully fed for 6 days; ovaries and fat body fill abdomen, air sacs collapsed again. *a*, air sacs; *b*, gut; *c*, ovaries; *d*, fat body.

the result of the air within them being heated by muscular contractions is certainly not significant. In the abdomen of Muscid flies (*Calliphora*, &c.) there is a pair of vast air sacs, the main function of which seems to be simply to provide a space into which the abdominal organs can grow without influencing the outward form of the abdomen. They are fully distended in the immature female fly; as the ovaries ripen and the fat body enlarges they become collapsed without interfering with respiration (Fig. 238).[56, 63] A similar compensatory role of the tracheal system is seen in *Locusta*.[238]

It is often said that the air sacs serve as reservoirs for oxygen during flight. But in most insects studied, *Vespa*, *Schistocerca*, Sphingids, respiratory movements of great amplitude take place all the time the insect flies.[62] It is true that in certain beetles, *Cetonia*, *Melolontha*, the respiratory movements are arrested at least during short flights;[62] the explanation of this is not known; but the spiracles are said to remain open[30] and it is possible that the tracheae are being ventilated by the direct action of the wing muscles.[149] The automatic ventilation of the intermuscular air sacs during flight is an essential element in the supply of oxygen to the flight muscles of *Schistocerca*.[269]

A further function that has been suggested is that of reducing the volume of the circulating blood: in *Drosophila*, *Musca*, *Calliphora*, &c., there are elaborate ramifications of the air sacs which occupy almost every free space in the thorax. This results in a great reduction of the blood volume, so that the thin layers of circulating blood carrying reserves of sugars and other fuels are brought into immediate contact with the muscles.[275]

Efficiency of tracheal ventilation—Some indication of the efficiency of the mechanical ventilation of the tracheal system is given by measurements of the respiration volume or tidal air in relation to the total capacity of the system. In the larva of *Dytiscus* the total capacity of the tracheal system is some 107 c.mm. (6–10 per cent. of the total volume of the animal); the vital capacity is 64 c.mm. In other words, during strong expiration the tracheal system is emptied of nearly two-thirds of its total capacity.[108] In the adult cockchafer, *Melolontha*, with a total capacity of 630 c.mm., or about 39 per cent. of the body volume, the vital capacity is about 210 c.mm. or one-third of the total capacity.[48] The tidal air, the volume actually inspired or expired at each respiration, varies within wide limits in both these insects. It has been calculated that in the grasshopper each deep respiration will effect a 20 per cent. renewal of the air in the hind leg.[106] For the sake of comparison with pulmonary respiration, it may be recalled that the vital capacity, the extreme degree of ventilation possible, in man is about two-thirds of the total capacity; in normal quiet respiration the tidal air is very much less than this.

Streams of air through the tracheal system—In some insects the mechanical ventilation of the tracheal system doubtless takes place through all the spiracles indiscriminately. But in many forms the opening and closing of the spiracles in different parts of the body is so timed in relation to the pumping movements that inspiration occurs predominantly in one region, expiration in another; with the result that there is a more or less directed flow of air through the tracheal system. If the 4 anterior spiracles in *Schistocerca* are separated by a diaphragm from the 6 posterior pairs, there is a transference of air from the anterior half to the posterior varying from 5–20 c.mm. per second (Fig. 239).[62] In another grasshopper, *Chortophaga*, the stream also enters chiefly by the thorax and leaves by the abdomen;[113] at 28° C. an average of about 2·5 c.mm. and a maximum of 12·5 c.mm. are passed through per second.[124] But only part of the movement is a through movement; for if the carbon dioxide given off from the two parts is compared, 20 per cent.

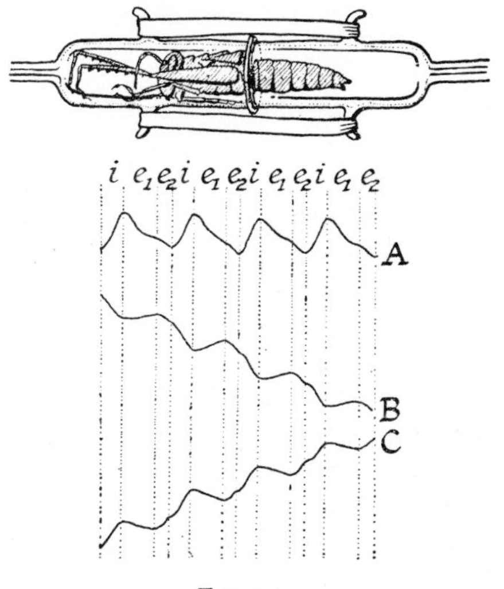

FIG. 239

Above, apparatus for demonstrating the directed stream of air through the tracheal system; grasshopper with thorax and abdomen in two halves of gas chamber separated by rubber partition. Below, volume changes in the apparatus during normal respiration; schematic. *A*, respiratory movements; *B*, fall in volume in the thoracic half of gas chamber; *C*, increase in volume in abdominal half. *i*, inspiratory phase; thoracic spiracles opened, abdominal spiracles closed; e_1, first part of expiratory phase, all spiracles closed; e_2, second part of expiratory phase, abdominal spiracles open; air streams out into posterior part of chamber (*after* FRAENKEL).

comes from the thorax and 80 per cent. from the abdomen;[124] and the rhythm of spiracle movements and the direction of flow may change from time to time.[122] When the respiration of *Chortophaga* is strongly stimulated by 15 per cent. carbon dioxide, the flow may be reversed or the passage of air may cease.[124] In the large cockroach, *Nyctobora*, the normal flow is in the same direction and in similar quantities; when increased by carbon dioxide the flow is maintained against a pressure of 20 cm. of water.[94]

During flight in *Schistocerca* oxygen consumption is increased 24 times, but the volume of air pumped by the abdomen increases only 4–5 times. The auxiliary mechanisms of ventilation which occur in the insect on the ground do not appear in the insect during flight; but the dorso-ventral abdominal pumping continues at increased frequency and amplitude.[253] The increased oxygen supply is effected by direct ventilation of the tracheae with elliptical cross-section lying between the flight muscles.[269] In the adult locust many of the cross links in the tracheal system have become occluded, with the result that during flight there are two largely independent ventilating systems. (i) A two-way system, which ventilates the flight muscles through the open spiracles 2 and 3, and is pumped by the flight movements. (ii) A one-way system, which ventilates primarily the central ner-

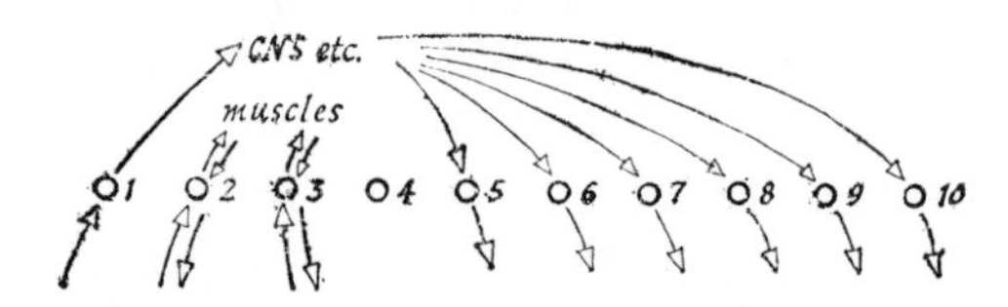

FIG. 240.—Diagram showing the course of the ventilating air-streams in the locust *Schistocerca* during flight (*after* MILLER)

vous system and is pumped by the abdomen in through spiracle 1 and out through spiracles 5–10 (Fig. 240).[253] [See p. 400.]

In the honey-bee, in normal air, there is a strong tidal flow in and out of the 1st thoracic spiracles, effected by the pumping movements of the abdomen, but only a very weak and variable directed air-stream. The flying bee probably changes to the other condition, that can be produced on exposure to carbon dioxide, in which air is inhaled at the 1st thoracic spiracles and largely exhaled via the propodial spiracles, providing a rapid flow of fresh air to the flight muscles.[227] In dragon-flies there is probably a one-way air-stream: inward through the thoracic spiracles to the flight muscles and then posteriorly along the abdomen.[255]

In the sheep ked *Melophagus* there is a flow of air directed forwards, expiration being effected mainly by the thoracic spiracles. In this insect the direction of flow is assisted by a unique structure in the form of a cuticular fold in the anterior region of the dorsal longitudinal trachea, so arranged that it serves as a valve restricting the passage of air from thorax to abdomen. It is of interest that in this insect the abdominal spiracles are the more effectively protected against the entry of dusts.[192]

In many aquatic insects, although on occasion any of the spiracles can be used either for inspiration or expiration, the structure and balance of the body is such that in the normal position of respiration at the water surface some spiracles are more important in inspiration than others.[17, 84] In *Hydrophilus*, which comes head first to the surface, inspiration is believed to occur chiefly by the mesothoracic spiracles and expiration by the abdominal spiracles; in *Dytiscus*, which rises tail first, the terminal spiracles of the abdomen are chiefly used in inspiration, and so on.[24]

But in all these insects, whether the respiratory movements cause a simple tidal flow in and out of the spiracles, or whether they bring about a directed stream of air through the system, they can ventilate only the larger tracheal trunks. The finer branches must always be dependent on diffusion.

THE ELIMINATION OF CARBON DIOXIDE

We have seen that diffusion through the smaller tracheal branches, combined in many insects with mechanical ventilation of the larger trunks, will explain the supply of oxygen to the tissues through the tracheal system. Since the volume of carbon dioxide produced in metabolism is generally somewhat less than the oxygen consumed, and its rate of diffusion through air is not much slower, there is no difficulty in accepting the elimination of carbon dioxide by the same route, particularly as the partial pressure of carbon dioxide in the surrounding atmosphere is almost zero. But carbon dioxide diffuses through animal tissues very much more readily than oxygen; it will diffuse through the connective tissues of vertebrates nearly thirty-five times as rapidly.[107] Although exact measurements have not been made, it is probable that the same applies to the cuticular coverings of insects. It is therefore almost certain that both the body surface and the linings of the tracheae play a much larger part in the elimination of carbon dioxide than they do in the uptake of oxygen.

This belief is supported by several observations. In the soft-skinned larvae of beetles and other insects, carbon dioxide is easily demonstrated escaping from the general body surface; in the heavily sclerotized adult Coleoptera, this applies only to the intersegmental membranes.[171] In the endoparasitic Agromyzid larva, *Cryptochaetum*, whereas the uptake of oxygen occurs most actively in the respiratory tails, the elimination of carbon dioxide takes place more or less equally over the general body surface.[172] From measurements of the output of carbon dioxide from insects or parts of insects in which the spiracles were occluded, it has been estimated that in the stick insect *Carausius*, about 25 per cent. of the total escapes through the skin; in the larvae of *Dytiscus* and *Eristalis* somewhat less.[29] The absorption of oxygen through the skin in these insects is certainly far less than that.

It has often been observed that the greater part of the carbon dioxide is given off by the posterior half of the body. In some cases this is due to the backward circulation of air through the tracheal system—as in the locust *Chortophaga*, in which 80 per cent. comes from the abdomen.[124] Where it happens in insects without respiratory movements (in the large caterpillar, *Dicranura*, nearly 75 per cent. comes from the posterior half) it may be due to the carbon dioxide carried backwards by the circulating blood being eliminated by the rich tracheal network connected with the last two pairs of spiracles;[214] or it may be due simply to the fact that in quiet respiration certain spiracles alone are functional—in the flea the 8th abdominal spiracles.[206]

THE RESPIRATION OF AQUATIC INSECTS

The structural adaptations of aquatic insects for respiration are immensely varied,[45, 126, 144] but here we shall deal only with examples which serve to illustrate the problems involved and the principles upon which their solution has been based.

Cutaneous respiration—The most complete adaptation to life in water is shown by certain larvae in which the tracheal system, though developed in the usual way by ectodermal invaginations, has become entirely cut off or 'closed' from the exterior by the obliteration of the stigmatic branches.[136] During the early stages of some larvae of this type, *Chironomus*,[139] *Simulium*,[185] *Acentropus* (Lep.),[135] the system may remain filled with fluid; and then respiration must be accomplished by a simple diffusion of gases through the skin. In a few insects, for example the aquatic Hymenopteron *Polynema*,[79] the tracheal system may remain entirely devoid of air even in the adult; but as a rule, after the first instar, if not sooner, the system contains gas.

So-called 'blood gills'—In some larvae certain regions of the integument are often exceedingly thin and project from the body surface (*Chironomus*), or are evaginated from the rectum (*Simulium*), as delicate blood-filled sacs, devoid of tracheae or provided with only a few small branches. These are often termed 'blood gills', through which it is supposed that the gaseous exchanges chiefly occur.[127, 185] But papillae of this type exist equally in larvae, such as *Eristalis* and other Syrphids, Culicids, &c., which breathe atmospheric air at the water surface; and what experimental evidence exists is against their having any respiratory function. Various protozoa (*Bodo*, *Polytoma*, &c.) which form aggregations at a particular concentration of oxygen in the water[55] have been used as indicators for this purpose. When the larva is immersed in a culture of such organisms they first congregate at those points where oxygen is being most actively absorbed or consumed. Then, as the oxygen concentration falls below the optimum, they form a band or shell which stands out farther from such points than elsewhere on the body surface (Fig. 241). Experiments on these lines have shown that the anal and ventral papillae of *Chironomus*,[60] the anal papillae of mosquito larvae[205] and the rectal blood gills of Trichoptera, *Simulium*, *Corethra*, &c.,[173] are not the site of particularly active oxygen uptake, though it has been suggested that the ventral gills of *Chironomus* may be of value during recovery from oxygen want.[74]

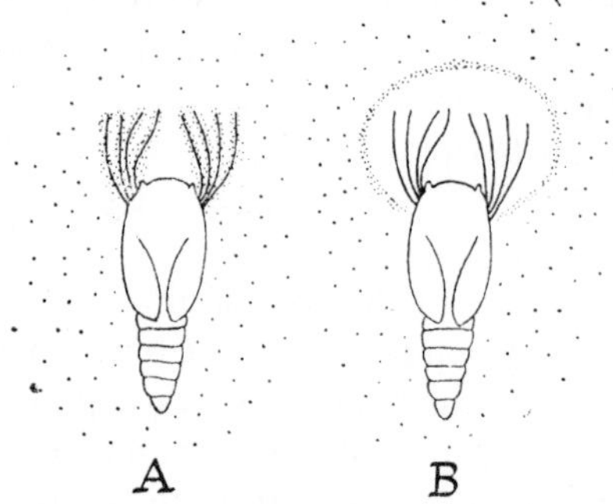

FIG. 241.—Respiratory function of the cuticular gills in *Simulium* pupae demonstrated by means of the flagellate *Bodo* (*after* Fox)

A, soon after immersion in the culture; flagellates collecting on the surface of the gills. B, later; flagellates concentrated in a band some distance from the gills.

In some Culicine mosquito larvae such as *Aëdes argenteopunctatus* the anal papillae are elongated and very richly supplied with tracheoles; the respiratory siphon is reduced and the papillae are held in the current of water created by the mouth brushes. It seems highly probable that in such species the papillae do function as gills.[220] The rectal papillae of *Eristalis* are not extruded more frequently during asphyxiation;[2] although it has been claimed that they are an important site of elimination of carbon dioxide.[194] We shall consider the function of all these organs in a later chapter (p. 514).

Tracheal gills—On the other hand there are many modifications of the closed tracheal system undoubtedly directed to obtaining oxygen from the water. The skin is often supplied with a rich network of fine capillaries, as in *Simulium*,[126] *Chironomus*,[215] *Acentropus*,[135] or the 1st instar of *Corixa*.[72] This network may be most abundant where the cuticle is thin, as in Trichoptera[117] or Plecoptera,[159] or in the larva of the Ceratopogonid *Atrichopogon*, which has a series of richly tracheated cushions along the back.[32] Then, in various regions of the body wall, or within the rectum, there may be evaginations well supplied with tracheae which form true 'tracheal gills';[45, 136] as in Ephemeroptera,[181] Trichoptera,[117] Plecoptera,[112, 116] and Odonata.[151, 178]

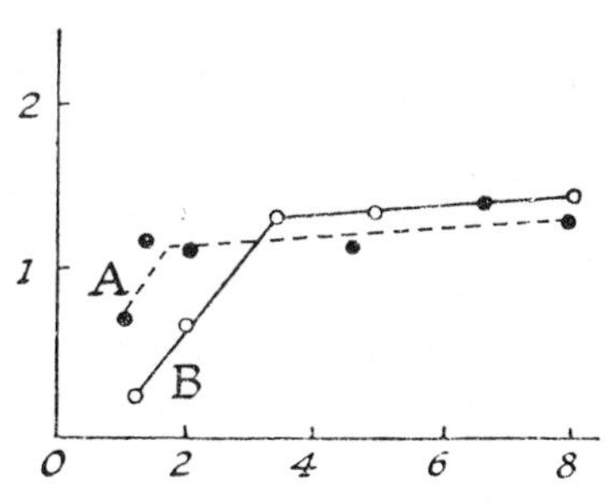

FIG. 242.—Oxygen uptake in nymphs of *Cloëon dipterum* in still water

A, normal insects; *B*, insects with gills removed. Ordinate: oxygen consumption (c.c./gm./hr.). Abscissa: oxygen concentration (c.c./litre) (*after* WINGFIELD).

But cutaneous respiration is still important in most aquatic larvae even when efficient tracheal gills are present.[174] The same applies of course to other animals; a large part of the normal respiration of frogs takes place through the skin and at low temperatures they can live without the lungs. It is this fact which has led to much controversy as to the importance of the tracheal gills in respiration. The sixty tracheal gills of the caddis larva (*Macronema*) can be removed and cause almost no reduction in the oxygen consumption.[131] The lateral gill plates of Ephemerid nymphs can be removed without causing the death of the insect; cutaneous respiration by itself is sufficient in winter, during rest and in highly oxygenated water; but the gills become indispensable in summer, during activity and in water poor in oxygen.[39, 130] The gill plates of these nymphs are moved more vigorously in water that is poor in oxygen[9] and when different species are compared there is found to be an inverse correlation between gill area and the oxygen content of the water in which they live.[49] In normal nymphs of *Cloëon* at 10° C. the rate of oxygen uptake does not begin to fall off steeply until the oxygen in the water has fallen to 1·5 c.c. per litre, whereas in gill-less nymphs under the same conditions this occurs at about 3 c.c. per litre (Fig. 242). At these low concentrations the gills beat actively; their function in this insect seems to be mainly to drive the water over the true respiratory surfaces; for if the water is artificially stirred, removal of the gills has no effect on the oxygen uptake.[210] In *Baetis*, a nymph found in swift streams with a high oxygen content, the gills are non-mobile and play no part in respiration: oxygen uptake is the same in normal and gill-less nymphs between 8·0 and 5·0 c.c. per litre at 10° C.; below 4·0 c.c. per litre both groups become moribund. In *Ephemera*, which burrows in the mud of stagnant ponds, the gills are filamentous and mobile; even at a concentration of 7·0 c.c. per litre the oxygen uptake of the gill-less nymphs is about one-quarter of the normal; the gills probably serve as true respiratory organs as well as paddles.[210]

Similarly, it has long been known[70, 204] that Agrionid nymphs can survive without the caudal lamellae; yet it has been estimated from measurements of the gaseous diffusion into the tracheal system of normal and gill-less nymphs, that in *Agrion pulchellum* 32–45 per cent. of the oxygen uptake normally enters by the lamellae.[98] In *Enellagma* (Odonata Zygoptera) the caudal gills are clearly more efficient in oxygen absorption than the general body surface: in closed containers the normal nymphs could reduce the oxygen concentration to about 2·4 per cent. of saturation; those without the caudal lamellae, only to about 14·5 per cent.[140]

The closed tracheal system—The mechanism by which the exchange of gases in the closed tracheal system is effected has been studied only in the nymphs of Anisopterous dragon-flies. These have an elaborate system of tracheal gills, the branchial basket, consisting of six double rows of imbricated lamellae, situated in the rectum; they breathe by passing water in and out of this gill

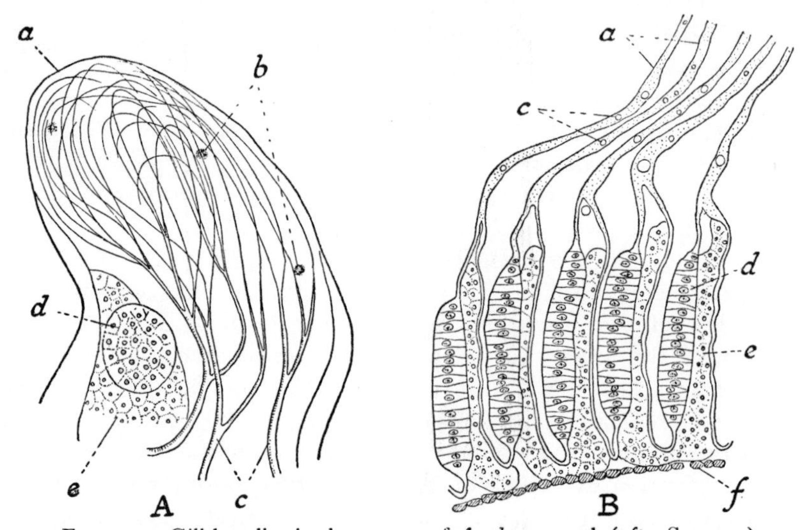

FIG. 243.—Gill lamellae in the rectum of *Aeschna* nymph (*after* SADONES)

A, surface view of lamella; *B*, longitudinal section through series of lamellae. *a*, gill plate; *b*, tubercles; *c*, tracheae and tracheoles; *d*, epithelial pad at base of gill plate; *e*, fat body; *f*, circular muscles.

chamber.[156] Sometimes the water is 'swallowed' in small quantities, and mixed by peristaltic movements;[179] but in an actively respiring mature nymph of *Aeschna*, each pumping movement of the abdominal wall will effect about 83 per cent. renewal of the contents of the hind-gut, the ventilation volume per minute amounting to about 1 c.c.[187]

The gaseous exchange in these dragon-fly nymphs was attributed by Dutrochet (1837) to simple diffusion; and that view is now pretty generally accepted. A closed tracheal system can only function on this plan provided it has non-compressible walls; and in fact it has been shown that in *Aeschna* a pressure increase of half an atmosphere causes at most a compression of 4–5 c.mm.;[108] although the spiracles are open, their structure resists the entry of water: even at a pressure of 10 metres of water none passes in.[99] The gills themselves are well suited to maintain a diffusion equilibrium between the water and the tracheal gases, the membrane being very thin and the tracheoles lying immediately below it (Fig. 243). And it has been shown quantitatively that oxygen diffuses with equal rapidity in the two directions: the membrane behaves like a dead membrane, complying with the laws of diffusion.[99] Moreover, in the living nymph the partial pressure of oxygen in the system is always less than the tension of oxygen in the water: 10 per cent. of an atmosphere in the water, 5–6 per cent. in the tracheae; 35 per cent. in the water, 20–30 per cent. in the tracheae.[187] We have seen that if the water pressure is increased, none enters the spiracles; but these do not resist the outward passage of air. If the larva is placed in alcohol and water, in which the solubility of air is reduced, bubbles of gas immediately escape from one or more of the spiracles. This is because the pressure of gas in the tracheal system now exceeds the pressure of the dissolved gases; it affords confirmatory evidence that the exchanges are normally effected by diffusion.[99]

Spiracular gills—There is another type of gill in aquatic insects in which there can be no question of any secretory activity. These are termed 'spiracular gills', 'tube gills' or 'cuticular gills'; they occur in the larva of *Teichomyza* (Dipt.),[184] and in the pupae of *Simulium*,[169] *Deuterophlebia* and probably the Blepharoceridae.[147] They arise as filamentous outgrowths of the ectoderm and are, therefore, covered with a layer of cuticle. The lumen of the filament loses all connexion with the body cavity, and the epithelium more or less degenerates; but the cuticular wall becomes much thickened and is excavated to form air spaces which establish a connexion with the tracheal system. Thus the air in the tracheae is separated from the water only by a delicate wall of cuticle about $0{\cdot}1\mu$ thick. The pupae of *Simulium* can develop after removal from water; this type of gill is probably an adaptation to life in mountain streams which are liable to dry up.[147] Tubular gills of the same type are found in the pupae of certain aquatic beetles of the genus *Psephenoides* which pupate beneath the surface of the water in fast-flowing rivers and torrential streams.[81]

Spiracular gills have been evolved on a number of occasions in Coleoptera and Diptera.[244] Their advantage in waters liable to dry up lies in the enormous surface area available for diffusion of oxygen inwards from the surrounding water when the structure is immersed, whereas the area available for evaporation when the structure is dry is limited to the narrow respiratory neck connecting the tracheal system with the gill. They are also adapted in various ways to resist high pressures. A remarkable feature about the pupal spiracle in *Simulium* is that the closing mechanism is operated by a muscle inside the developing adult, after the adult cuticle is formed (the so-called 'pharate adult').[244] [See p. 401.]

Surface breathing; hydrofuge structures—So far we have considered only aquatic insects which breathe dissolved oxygen, but the great majority of aquatic forms are dependent on the atmospheric air. These show many structural adaptations which facilitate the connexion of certain spiracles with the atmosphere, such as the terminal respiratory tube of *Nepa*, the modified antennae of *Hydrophilus* and its allies which come head first to the surface, the balance of the body in Dytiscids which rise tail first, the respiratory siphons of *Eristalis*, Culicids, &c., bearing the spiracles at the apex. These we cannot discuss here in detail. But a problem common to all these insects is the breaking of the 'surface film' of the water so that the spiracles may be exposed to the air above. This problem has been solved in all cases by the same device; by the provision of regions of the body wall whose physical properties are such that they have a greater affinity for air than for water, which are in fact 'hydrofuge'—the 'angle of contact' between the water and the region in question being relatively large.[23] The cohesion of the water is then greater than the adhesion to the body; when the insect reaches the surface film, the water therefore falls away and leaves the cuticle dry. It seems that the elytra of *Dytiscus* or *Gyrinus* owe their hydrofuge properties to a monolayer of grease or wax molecules held to the cuticle by their polar ends so that the hydrofuge paraffinic head is in contact with the water.[229] In order to furnish this property in the larvae of Diptera (*Eristalis*,[50, 186] Tachinids,[137] Trypetids,[142] Culicids, &c.[89, 90]) there are always small glands at the spiracular openings (perispiracular glands) which produce an oily secretion; and in *Dytiscus* similar glands probably occur all over the body surface.[18] In *Drosophila* larvae there are three unicellular glands associated with each posterior spiracle; their oily secretion extends into the felt chamber and the tracheal trunks and possibly over the general surface of the cuticle.[261] If soap solution is added to water until the surface tension is reduced from the normal 70 dynes per cm. to about 27–36 dynes, Culicid larvae are unable to hold to the surface, but sink and drown.[155]

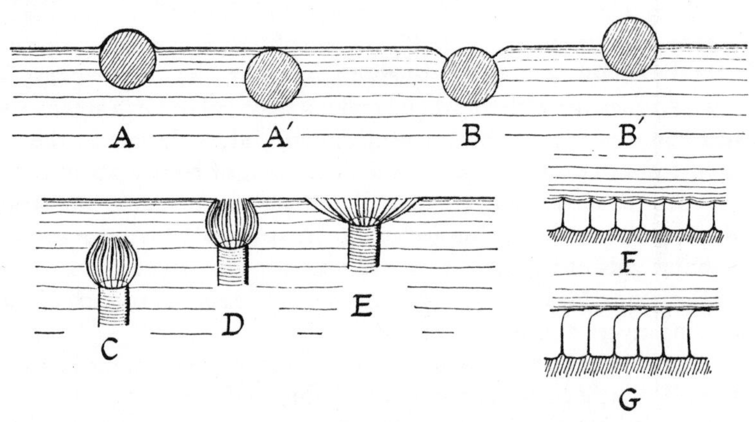

FIG. 244

A, A′, hydrophile hair in the water surface seen in cross-section; *A′* is the position of equilibrium *B, B′*, hydrophobe hair in the water surface; *B′* is the position of equilibrium. *C*, spiracle surrounded by crown of hydrophobe hairs (as in *Stratiomys* larva); this is the position of equilibrium below the surface. *D, E*, the same at the water surface; *E*, shows the position of equilibrium. *F*, short hydrophobe hairs standing erect on the cuticle (as on ventral surface of *Notonecta*). *G*, hydrophobe hairs curved over at the tip (as in *Elmis* and *Haemonia*).

Another device consists in the provision of a fringe or crown of long hairs (around the spiracles of *Notonecta* [24, 84] or along the antennae of *Hydrophilus*) which are held in the surface film as though they were hydrophile on one side and hydrofuge on the other. On reaching the water surface a fringe of this kind is spread outwards by surface tension, and the air space within put into communication with the atmosphere. The physical explanation of these 'semi-hydrofuge' hairs is given in Fig. 244, which shows cross-sections of hairs of different types placed in the water surface. In Fig. 244, A, the hair is hydrophile with an angle of contact approaching zero. In such a case the free energy of the surface is at a minimum, that is, the surface area is smallest when the hair is completely wetted (Fig. 244, A′). In Fig. 244, B, the hair is hydrophobe, the angle of contact being 90°. Here the surface energy is least when the hair is held in the water surface with half in air and half in water; hence that is the position of equilibrium. When below the surface, the boundary between air and water will be smallest when the crown of hairs is drawn together (Fig. 244, C); on exposure to the air, the interface will be smallest when the fringe is widely spread (Fig. 244, E).

All such hydrofuge structures are subject to hysteresis: the surface may be hydrofuge on immersion but appears to be hydrophile when withdrawn from the water.[228] This effect results from reversal in the orientation of polar molecules which may be arranged with their hydrophilic groups either attached to the cuticle or exposed to the water.[229]

These hydrofuge structures around the spiracles also have the function of preventing water from entering the tracheal system. The protective action of the spiracles of the larva of *Gasterophilus* is destroyed by bile in a dilution of 1 per cent., and the tracheae fill with fluid. But while water cannot get in, oil can do so readily, and spreads throughout the tracheal system. It is this property which lies at the back of the observation made by Aristotle, and explained by Malpighi, that oil is always fatal when applied to the surface of insects; and it is this property which forms the basis of the classical method of controlling mosquito larvae by spraying oil on the surface of the water.[204]

Air stores of aquatic insects—Besides retaining a film of air immediately around the spiracles, many aquatic insects are able to carry bubbles or films of air on other parts of the body. In the larva of *Hydrocampa* (Lep.),[144] and on the ventral surface of *Notonecta* and many other insects, there is a fine pile of erect hydrofuge hairs, which hold between them a layer of air, like velvet when immersed in water (Fig. 244, F). In the aquatic Coleoptera *Haemonia* and *Elmis*, there are long 'semi-hydrofuge' hairs all over the surface of the body which are bent over at the tips; hence, as explained above, the outer surface of the free ends appears hydrophile, and so provides a hydrophile surface enclosing beneath it a thin film of air (Fig. 244, G).[24] Dytiscids carry considerable quantities of air beneath the elytra, or as a bubble attached to the posterior extremity. We must now consider the function of such air stores.

In the first place these stores of air have an important **hydrostatic function** in enabling the insect to rise to the water surface, and to do so in the correct position for renewing its supply of oxygen. By filling the rectal ampulla to a varying extent with water *Dytiscus* can control the size of the air-filled subelytral space and so regulate their buoyancy.[243] The air stores on the ventral aspect of the abdomen in *Notonecta* are said to be used for respiration, whereas those between head and prothorax, and between the elytra and the back are mainly hydrostatic.[260] If weights of different magnitude are attached to *Notonecta*, the diminution in buoyancy which this causes is compensated by an increase in the volume of air carried in the air store.[20] Indeed some writers have regarded the air stores of *Dytiscus*,[24, 26, 199] *Notonecta*, *Corixa*, *Nepa*, *Hydrophilus*[24] as containing only expired air which is normally of no further value in respiration.

That, however, is an overstatement of the case; the air in the stores can certainly be breathed in again through the spiracles,[24] and to this extent they constitute an oxygen reserve. Thus in *Dytiscus*, the respiratory pumping movements continue after submergence,[8] and the oxygen content of the elytral air falls rapidly from 19·5 per cent. at the moment of diving, to 1 per cent. or less in three or four minutes.[53, 104] If the capacity of the subelytral space is increased by removing the hind wings, the average duration of submergence is prolonged.[77] In the large aquatic bug *Hydrocyrius* (Belostomatidae) the first two pairs of spiracles are not in contact with the subelytral air store; but spiracle 3, the most permeable of all the spiracles, opens directly into the store which thus supplies the main tracheae to the thoracic muscles. All the abdominal spiracles likewise communicate with the air store.[254]

But the air stores have a third function, which consists in separating by diffusion the oxygen dissolved in the water: they serve as **'physical gills'**. This function has been suggested by many authors for a great many different insects;[204] but the most instructive paper from a theoretical and quantitative standpoint is by Ege,[53] who shows that the capacity of such a film of air to function as a 'gill' in this way depends on the fact that the invasion coefficient of oxygen between water and air is more than three times as great as that of nitrogen. Consequently, if the tension of oxygen in the water is higher than the partial pressure of oxygen in the air film, there will be a greater tendency for equilibrium to be restored by the diffusion of oxygen into the bubble, than by the diffusion of nitrogen out of it. The so-called 'invasion coefficient' between water and air is not a physical constant: it is dependent on the rate of diffusion of gas through the layer of still water in immediate contact with the interface; and the thickness of this diffusion boundary will vary widely in different circumstances.[223] The uptake of oxygen by such a system will be greatly influenced by the current over the surface of the gas store. The air stores of *Naucoris* and *Notonecta*, insects which set up currents of water with the hindlegs, are far more efficient as physical gills than that of *Hydrophilus* in which ventilating movements are absent.[222]

During summer, when the insects are actively swimming, they cannot obtain enough oxygen by diffusion from the water in this way; but the mechanism is of more or less value in all aquatic insects, which will benefit to the same extent as though they were provided with about 13 times as much air as they actually carry. *Corixa* can obtain enough oxygen in this way even at summer temperatures so long as it does not swim actively; and this applies even to the much larger *Dytiscus* during the winter.[53] The proportion of the total oxygen uptake in *Naucoris* obtained from the physical gill at different seasons was as follows: December (10° C.) 96 per cent.; April (11° C.) 59 per cent.; May (15° C.) 54 per cent.; June (20° C.) 30·5 per cent.; August (20° C.) 28 per cent.[276] To quote a single experiment showing the value of the process: *Notonecta* lived 7 hours in water saturated with atmospheric air, 35 minutes in water saturated with oxygen, and 5 minutes in water saturated with nitrogen. Thus, in the second case, although the insect had five times as much oxygen in its store as in the first case, it survived only one-tenth of the time, because, in the absence of nitrogen, invasion of the air store by oxygen could not take place.[53] Of course the process cannot continue indefinitely because the nitrogen in the air store is gradually dissolved; and if the insect is compelled to remain submerged, all the gas in its store eventually disappears into solution. In fact, with the smaller insects, in which the invasion process plays a more significant part, it is really of more importance for the insect to renew its store of nitrogen than to obtain a fresh supply of oxygen.

In Corixids the gas from the tracheal system is expired through the anterior thoracic spiracles, moved backwards over the body surface by the limbs, and inspired through the anterior abdominal spiracles. By these means the thoracic tracheal system is automatically ventilated during swimming.[258] The larva of *Helodes* carries an air bubble applied to the anal papillae and in direct communication with the last abdominal spiracles. Occlusion of the papillae does not affect respiration; but removal of the air bubble halves the time of survival of submerged larvae.[266]

Plastron respiration—In a few cases, for example in the beetles *Haemonia*, *Elmis*,[24] *Coxelmis*[43] and some others[175] (see p. 381), the air film is so held by the hydrofuge hairs that it cannot be replaced by water. Conditions then approximate to the closed tracheal system, and the insect becomes independent of the atmospheric air. *Elmis*, for instance, will live for months below the surface, if the water is well aerated.

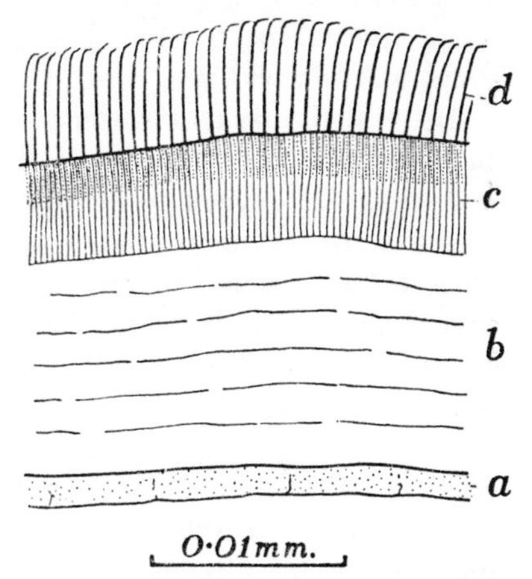

FIG. 245.—Section through abdominal sternum of *Aphelocheirus* showing the plastron hairs bent over at the tip (*after* THORPE and CRISP)

a, epidermis; *b*, endocuticle; *c*, exocuticle; *d*, plastron hairs

This very thin firmly held layer of gas is termed a 'plastron'. In it the volume of gas is negligible, but the interface is held in position by surface forces in such a way that this volume remains constant. Functionally it resembles a tracheal gill more nearly than an air store. The arrangement finds its highest development in the adult of the Naucorid *Aphelocheirus*, a flattened bug with reduced wings which is entirely subaquatic.[175] Most of the body surface is covered with a very fine plastron held in position by an epicuticular hair pile with some 2,000,000 hairs per sq. mm., each bent over at right angles at the tip (Fig. 245). The spiracles are protected by a screen of long cuticular filaments beset with fine hairs; they open into a chamber from which radiating channels run outwards in the form of a rosette. These channels communicate with the exterior, that is, with the plastron, by minute openings along their course, the whole structure being lined by the very fine plastron hairs. The air film forming this plastron is unaffected no matter how long the insect is submerged in gas-free water; under a pressure of four or five atmospheres the hair pile collapses but is not wetted. On the other hand the air is displaced at once on contact with alcohol or with wetting agents. The structure, dimensions and packing of the plastron hairs in *Aphelocheirus* are such as to give what must be nearly the most favourable compromise between the conflicting requirements of high resistance to wetting and to mechanical collapse on the one hand, and of a large water-gas interface for respiratory exchange on the other.[175] The related Hemipteron *Cheirochela* combines plastron respiration with specialized areas of cutaneous respiration.[214a] The plastron of *Elmis*, carried by relatively coarse hairs, is much less efficient than that of *Aphelocheirus*. It cannot withstand a pressure difference as great as one atmosphere. In the aquatic Curculionid *Phytobius* the hairs are borne on overlapping flattened scales giving a highly efficient plastron which, as in *Aphelocheirus*, needs no replacement.[224]

The small wasp *Agriotypus* which is parasitic upon various species of Trichoptera utilizes the same principle in an unusual way. The larva constructs an air-containing silk filament extending from the submerged caddis case below the stones. This filament serves as a physical gill; it enables the wasp to live for some eight months as an imago in the case under water.[201]

The aquatic beetle *Potamodytes*, a stream-living species in West Africa, swims an inch or two below the surface and carries an elongated respiratory bubble much larger than itself, which serves as a permanent physical gill. This steady state is attributed to the increased velocity of water flow immediately around the bubble causing a decrease in pressure, and therefore a temporary supersaturation with air, which permits diffusion of gases into the bubble.[265]

Aquatic plants as source of oxygen—There is yet another way in which aquatic insects have become independent of the water surface. This is by obtaining their oxygen from aquatic plants. The free bubbles of gas given off by plants are frequently taken up by the hydrofuge surfaces of insects;[24, 211] and with the approach of winter many aquatic insects congregate in those pools which are most rich in vegetation.[199] Other insects obtain their oxygen from the intercellular air spaces of water plants, either by biting into the air-containing tissues, as in *Elmis* (Col.)[24] and the pupating larva of *Hydrocampa* (Lep.),[54] or by inserting a specially modified respiratory siphon into the air-containing cells. Different modifications for piercing plants and obtaining oxygen in this way have been evolved independently by larvae in two families of Coleoptera and three families of Diptera.[180] The best-known examples are the larva of the beetle *Donacia*, first described by v. Siebold in 1859,[46, 180] and the larva of the mosquito *Mansonia*. The oxygen content of the gas in the rhizomes and roots from which *Donacia* gets its air, seldom exceeds 10 per cent., and in winter it may fall below 1 per cent.;[53] though values as high as 45 per cent. have occasionally been found in *Potamogeton*.[54] [See p. 401.]

Cutaneous respiration—In all these aquatic insects which breathe by open spiracles, as in those which breathe by gills, the skin is more or less important in respiration; and the part it plays is naturally greater if the insect is obliged to remain submerged. Some species of mosquito larvae can develop completely in well aerated water without access to the surface or even with the tracheal system filled with a non-toxic oil.[205] When the Tipulid larva *Pedicia* has its spiracular process in connexion with the air, the uptake of oxygen by the skin is very slight, but it at once becomes greatly increased if the larva is denied access to the air, the exchange being greater in the region of the 'gills'; though this larva is said to be unable to survive permanently the occlusion of the spiracles.[174]

Hydrostatic function of tracheal system—In addition to its respiratory cunction the closed tracheal system has a hydrostatic function;[136] and in a few fases, notably the Nematocerous larvae *Mochlonyx* and *Corethra*, the latter function is the chief one. In *Corethra* the tracheal system is practically reduced

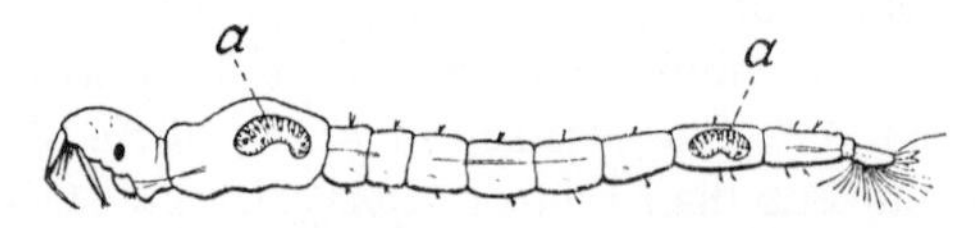

FIG. 246.—Larva of *Corethra*, showing the tracheal system reduced to two pairs of kidney-shaped air sacs (a) covered with pigment cells

to two pairs of bean-shaped sacs with a number of contractile pigment cells spread over the surface (Fig. 246). The gases in these air bladders come into diffusion equilibrium with any gas mixture in the surrounding water.[105] But they can vary in size (shrinking to 90 per cent. and expanding to 122 per cent. of their normal volume[41]) so that the buoyancy of the larva accords with the density and pressure of its environment, and it is enabled to float at any desired level. The substance of the wall of the sac seems to differ from that of most tracheae in being able to imbibe water; it is described as consisting of a colloid material ('trachein') which contracts on drying and swells up instantly on moistening with the insect's blood or with solutions of sodium chloride.[36, 65, 95] The living insect is able to control its buoyancy apparently by inducing an active swelling or shrinkage of these cuticular walls. The chemical mechanism of this change is not known, nor is the source of the sensory impulses by which it is evoked.[41, 65]

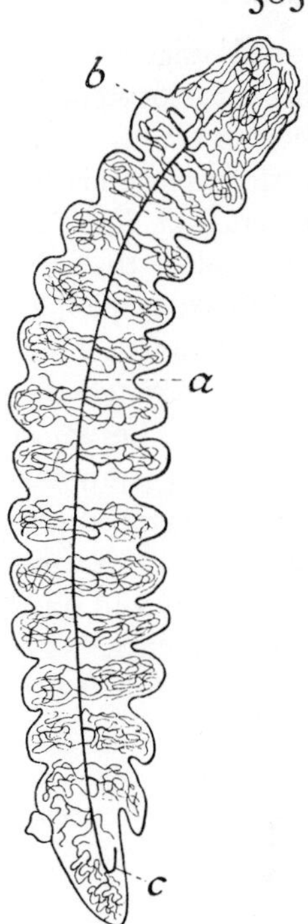

FIG. 247.—First-stage larva of *Macrocentrus gifuensis* (Ichneumonidae), lateral view, to show the rich tracheal supply beneath the skin (*after* PARKER)

a, lateral tracheal trunk; *b*, anterior commissure; *c*, posterior commissure.

RESPIRATION OF ENDOPARASITIC INSECTS

The respiration of endoparasitic insect larvae shows some striking parallels with the respiration of aquatic forms. Thus, in the 1st instar of many hymenopterous parasites,[160] and in the Agromyzid (Dipt.) parasite *Cryptochaetum*,[172] the tracheae are filled with fluid and not functional; the exchange of gases takes place directly between the tissue fluids of parasite and host. When the tracheae fill with gas, usually in the 2nd instar, they supply a rich network of fine branches to the skin (Fig. 247), and it is not until the larva is about to quit its host that the spiracles become open and functional.[160] The survival of parasitic larvae (e.g. *Nemeritis* in caterpillars of *Ephestia*) may be limited by the supply of oxygen; if parasites are too numerous some of them are killed by asphyxiation.[240]

'Blood gills' and 'tracheal gills'—Many Braconid, Ichneumonid, and Chalcid larvae possess a 'tail' of varied form, usually best developed in the early

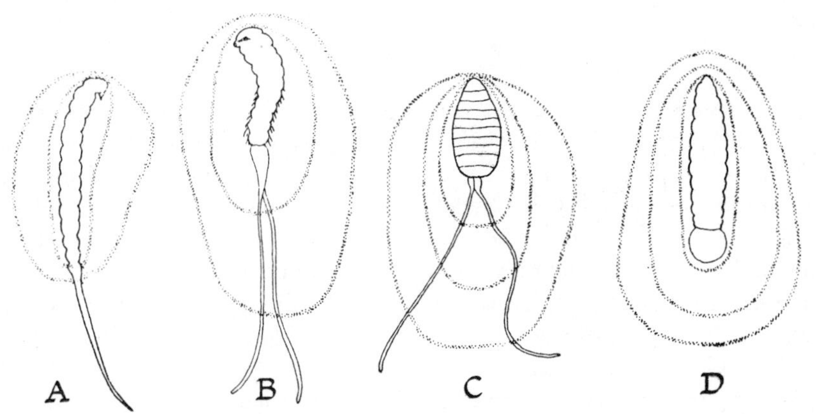

FIG. 248.—Parasitic larvae immersed in cultures of *Polytoma uvella*, showing successive positions at which bands of aggregation are formed (*after* THORPE)

A, 1st instar larva of *Omorgus mutabilis* (Hym. Ichneumonidae); *B*, 3rd instar larva of *Cryptochaetum iceryae* (Dipt. Agromyzidae); *C*, last instar larva of the same; *D*, late 2nd instar larva of *Apanteles popularis* (Hym. Braconidae).

larval stages, which recalls the 'blood gills' of aquatic larvae. This structure has had locomotor, absorptive and excretory functions assigned to it;[71, 197] but so far as its respiratory function is concerned, it can be said that it is of little importance. Using the aggregation the flagellate *Polytoma*, and the luminescence of *Bacillus phosphorescens*, as indicators of oxygen uptake, Thorpe[172] concluded that the 'tail' of 1st instar Ichneumonids is of no importance in respiration; the flagellates pay no attention to it at all (Fig. 248, A). The caudal vesicle of Braconids (an evagination of the wall of the hind gut which is prolapsed through the anus, and is filled with blood that is constantly taken up by the hindmost chamber of the heart (Fig. 249)) is the site of rather more active oxygen uptake than the rest of the body surface (Fig. 248, D); but even when large and supplied with a good blood circulation, as in *Apanteles* and *Microgaster*, it cannot be responsible for more than about one-third of the total uptake,[172] and much of this may perhaps be associated with its own oxygen requirements.[97] The 'tails' of the Agromyzid *Cryptochaetum*, which are well supplied with fine tracheae, are also rather more important in oxygen absorption, especially towards their base, than other parts of the cuticle (Fig. 248, B, C); carbon dioxide, as in aquatic forms, being given off more or less equally over the general body surface.[172] The main function of these organs is perhaps to be sought elsewhere (p. 514).

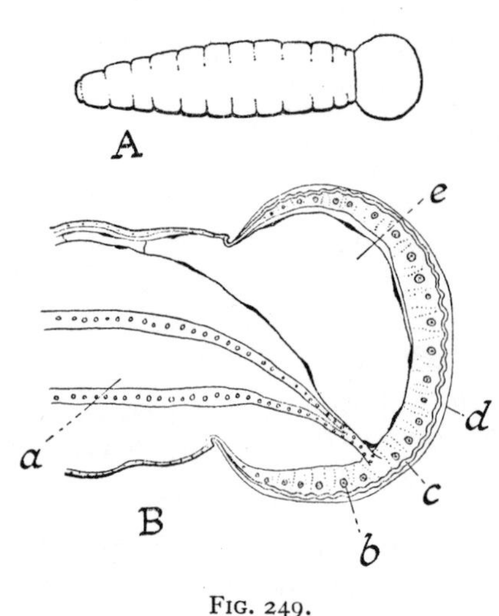

FIG. 249.

A, larva of *Apanteles* showing vesicle at hind end; B, longitudinal section of vesicle in *Apanteles glomeratus* (*after* THORPE). a, mesenteron, closed behind; b, proctodoeum covering the vesicle; c, cuticle; d, outer membrane formed from persistent amnion; e, posterior chamber of heart occupying the vesicle.

In two species of *Cryptochaetum*, however, which are parasites of the giant scale *Aspidoproctus*, the terminal filaments are enormously developed. In the first stage larva, which obtains its nutriment through the cuticle, they are no more than three times the length of the body; but in the third stage larva they are ten times the length of the body and are packed with a great mass of fine tracheal branches which extend at least two-thirds of the way to the tip. These filaments are commonly entangled among the tracheae of the host; they are obviously of importance in respiration, and by the same sort of calculation as that applied by Krogh to the diffusion of oxygen to the tracheal endings (p. 366), it can be shown that the filaments have reached the maximum effective dimensions.[172]

Respiration of atmospheric air—We have already seen that many aquatic larvae are 'metapneustic',[136] breathing only through the hindmost pair of spiracles. This condition is paralleled among the parasitic Hymenoptera by *Blastothrix*, *Encyrtus* and other Chalcids during their 1st instar. In these the pedicel of the egg protrudes externally through the body wall of the host and functions as a kind of respiratory tube into which the larva inserts its posterior spiracles in order to breathe the atmospheric air (Fig. 251, A).[172] The remains of the egg may serve this purpose throughout the first three larval stages.[245]

In the eggs of *Oöencyrtus* and related forms the part implanted into the host bears an elongated thickened plate excavated to form air-containing cells, the 'aeroscopic plate' (Fig. 250). The outer bulb of the egg collapses and occludes the lumen of the stalk; apparently it is the special band on the chorion and not the hollow stalk which is used to conduct air to the posterior spiracles.[119] The respiration of many Tachinid larvae is similar; but here the respiratory tube is formed by the epidermis of the host spreading inwards over the surface of the larva and so investing it in a cuticular sheath (Fig. 251, B).[138]

Finally, just as certain aquatic insects obtain their oxygen from the intercellular gases of aquatic plants, so there are parasitic larvae which tap the tracheal system of their host. This is very common among Tachinids (*Gymnosoma*, &c.[109]). Where the trachea is ruptured by the larva, its epithelium again spreads inwards to form a cuticular siphon or sheath around the parasite (Fig. 251, C).[14, 35, 109, 138] In the last larval stage of the Chalcid *Encyrtus infelix*, the tracheae of its Coccid host become perforated by some unexplained mechanism in the neighbourhood of the anterior and posterior spiracles of the larva, and a sheath is formed which completely encloses it.[172] Whatever the

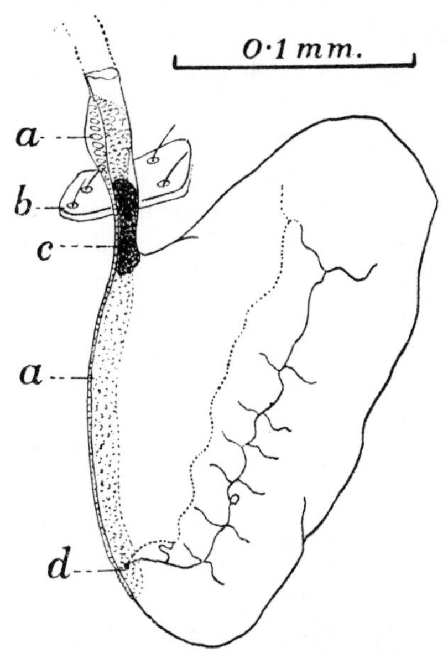

FIG. 250—Recently hatched larva of the Encyrtid *Anagyrus* with posterior spiracles applied to the 'aeroscopic plate' of the egg-shell (*after* MAPLE)

a, air cells of the 'aeroscopic plate'; b, integument of host; c, neck of egg-shell blocked by plug of material; d, spiracles of parasite.

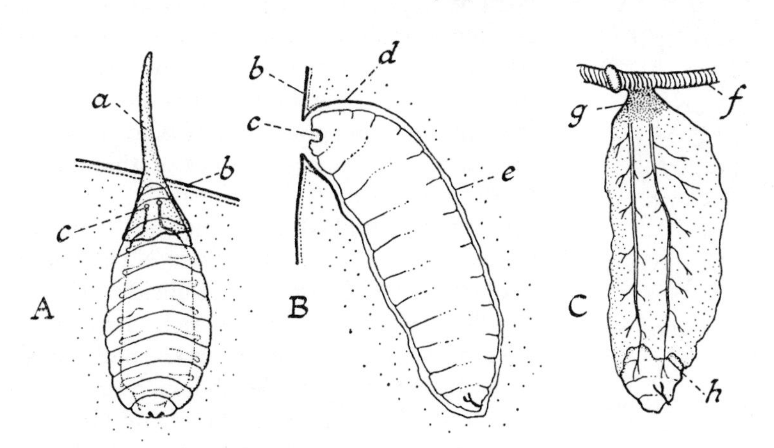

FIG. 251.—Breathing of atmospheric air by endoparasitic larvae

A, 1st instar larva of *Blastothrix* with spiracles still in chorion of egg whose pedicel protrudes through the integument of the host (*after* IMMS); B, larva of *Thrixion* enclosed in integumental sheath and breathing through respiratory funnel at the site of the entrance hole into the host; C, Tachinid larva partially enclosed in tracheal sheath and attached to tracheal trunk by means of secondary respiratory funnel (*from* IMMS *after* PANTEL). a, pedicel of egg; b, integument of host; c, spiracles; d, primary respiratory funnel; e, integumental sheath; f, trachea of host with spiracle; g, secondary respiratory funnel; h, margin of tracheal sheath.

mechanism of formation of this sheath, the final result is analogous to the sheaths formed by the outgrowth of epidermis or tracheal wall around Tachinids.

THE RESPIRATORY FUNCTION OF THE BLOOD

As we have already seen, the walls of the tracheae are permeable to gases,[41, 105, 203] particularly to carbon dioxide. There must therefore be some respiratory exchange between the haemolymph and the air throughout the tracheal system. And at the tracheal endings, except where these are intracellular or very closely applied to the surface of the cells, the blood must always play the part of intermediary between the lumen of the trachea and the cytoplasm of the cells. How important this part may be will depend on the anatomical arrangements. Thus in the ovary of some insects there is a considerable blood sinus between the membrane bearing the tracheal capillaries and the ovarioles; whereas in other species the membrane lies close against the ovaries. In yet other cases, for example the pupa of *Sphinx ligustri*, there are no tracheae going to the ovary; the blood, aerated elsewhere, plays the sole part in the supply of oxygen.[153] This must apply also, of course, to those insects in which the tracheae are wanting or filled with fluid.

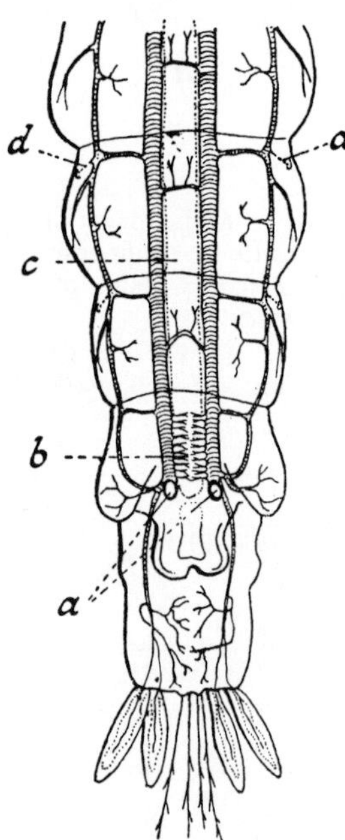

Fig. 252.—Hind end of mosquito larva, *Anopheles*, showing the rich tracheal supply to the posterior extremity of the heart (*after* Imms)

a, spiracles; *b*, tracheae to posterior chamber of heart; *c*, position of heart; *d*, stigmatic cords to the functionless spiracles.

The so-called 'blood gills' of aquatic and parasitic insects seem to be of small importance in respiration (p. 376); but the blood which circulates through them must carry away some oxygen in solution; and the increased rate of heart beat which occurs in water poor in oxygen in the larvae of Trichoptera,[117] and in *Acentropus* (Lep.),[135] has been ascribed to this cause.

'Tracheal lungs'—In some insects the anatomy of the tracheal system suggests that it is adapted to aerate the blood; that it serves as a lung. In the larva of *Hypoderma* (Dipt.) the tracheae end abruptly and break up into bunches of long, slender capillaries which seem to lie free in the body cavity and not to invest the cells;[190] and the same is seen in the tracheae from the anterior spiracles of Muscid pupae.[196] Where the aorta of the honey-bee passes through the stalk of the abdomen, it is excessively convoluted and very richly supplied with tracheae.[66] In the metapneustic larvae of Diptera-Nematocera (in *Ctenophora*,[182] *Psychoda*,[47] Culicidae (Fig. 252),[85] Tipulidae,[27, 69, 200] &c.) a rich network of fine tracheae arises from the main tracheal trunks at the posterior extremity of the abdomen and invests the hindmost chamber of the heart, or forms a meshwork around the posterior opening. And in the pupa of *Bombyx mori* the posterior bulb of the heart is richly tracheated.[68] But in none of these cases is it known whether a significant fraction of the respiratory exchange takes place by this route.

In certain insects which have lost the power of flight, such as *Nepa* and

Gryllus, the dorsal longitudinal muscles of the thorax atrophy, while their abundant tracheal supply persists. This has given rise to the impression that these vestigial muscles function as 'tracheal lungs'.[73] But they are not adapted structurally to the rapid aeration of the blood, and this does not, in fact, flow through them.[101] On the other hand, the blood sometimes circulates along the course of the tracheal system. In *Sphinx convolvuli* and *Vespa crabro* it circulates around the air sacs; and wherever these sacs give off tracheoles, the circulating blood follows their course between the muscle bundles (p. 372). It circulates also along the 'inverted tracheae' (p. 359) which traverse the rigid air chambers in the thorax of these insects.[25] And it has been suggested that the wings of Lepidoptera serve as respiratory organs, the blood which circulates through them being aerated from the tracheal system.[145]

Haemoglobin in insect blood—It is evident from the foregoing discussion that the blood of insects is often called upon to transport oxygen; the question arises whether it contains an oxygen carrier analogous to the haemoglobins or haemocyanins of other animals. In the larva of the honey-bee the oxygen capacity of the blood is no greater than can be accounted for by physical solution;[16] the blood of *Dytiscus* and *Hydrophilus* normally contains only minute quantities of oxygen;[12] but it is possible that it may be otherwise in other insects.

Appreciable amounts of copper are present in the blood and tissues,[133] but there is no evidence that this is in the form of haemocyanin. On the other hand, haemoglobin occurs free, that is, not contained in corpuscles, in the blood of certain Chironomid larvae; and particularly in those species which live in mud at the bottom of pools, in an environment very poor in oxygen.[126] It has been suggested that this haemoglobin serves as a store for oxygen during times when dissolved oxygen is scarce;[127] but the quantity of oxygen that could be stored in this way by a larva of *Chironomus* would last the insect only about 12 minutes.[115] At all ordinary tensions of oxygen the haemoglobin in the blood is fully saturated, and therefore plays no part as a carrier; but at tensions of 7 mm. and under, that is, at less than 1 per cent. of an atmosphere pressure of oxygen, the haemoglobin becomes partially reduced. There is clear evidence, therefore, that this pigment does function as a carrier, enabling the blood to bind chemically sufficient oxygen for the needs of the animal when the tension of oxygen becomes so low that the necessary amount cannot be supplied by physical solution.[115]

The amount of haemoglobin varies in different species. In most *Chironomus* species of the *genuinus* and *connectans* group it is almost 25–30 per cent. of the concentration in normal human blood. In the *plumosus* group it is only about 14·5 per cent., and in those Tanypinae which have haemoglobin it is lower still —10–14 per cent. In those types which appear to have possessed haemoglobin longest (*Chironomus*) the tracheal system has become much reduced, while groups in which haemoglobin seems to have been more recently acquired (Tanypinae) still have a highly developed tracheal system.[158] In larvae from a given environment, a decrease in the amount of haemoglobin is correlated with an increase in the length and diameter of the tracheal trunks.[166] *Chironomus thummi* reared on a diet with an adequate iron content have a haemoglobin content in the blood that may reach 25 per cent. of the value for human blood. If the diet is deficient in iron the haemoglobin falls to a very low level and the resulting pale larvae are less tolerant of very low oxygen tensions.[257] Larvae of *Chironomus* and of another Chironomid *Anatopynia* form more haemoglobin in poorly aerated water.[241]

The oxygen tension at which the oxygen consumption begins to fall off, varies greatly in different species of Chironomid larvae. If the insect is kept moist and exposed to the air, this happens in *Prodiamesa praecox* at an oxygen pressure of 6–7 per cent. of an atmosphere; in *Chironomus plumosus* at 1–2 per cent.; in *Eutanytarsus inermipes* at a still lower level.[74] These differences are dependent in part upon the thickness of the body wall; but they are also related to the presence or absence of haemoglobin, the species with haemoglobin being the more resistant.[139] In these the oxygen uptake begins to fail when the haemoglobin ceases to be saturated. It is possible, though this has not been proved, that the haemoglobin of each species has a different dissociation curve, and is therefore functional over a different range of oxygen tension.[74] In *Chironomus gregarius*, haemoglobin first appears during the 2nd instar, and the resistance of the larva to oxygen lack increases as the formation of haemoglobin proceeds.[139] In *Chironomus thummi*, exposure to 20 per cent. carbon monoxide (which renders the haemoglobin functionless by combining firmly with it) first affects the oxygen uptake markedly if less than 11 per cent. of oxygen is present; whereas in larvae recovering from asphyxia, the effect is visible when 21 per cent. of oxygen is present.[74] This demonstrates again that it is only under conditions of oxygen want that the haemoglobin is active as a carrier.

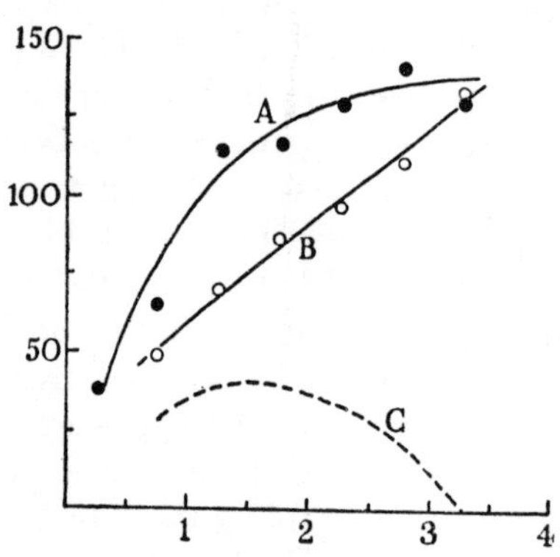

FIG. 253.—Rates of oxygen consumption in *Chironomus* at 17° C. at varying concentrations of dissolved oxygen

A, normal animals; *B*, animals with haemoglobin blocked with carbon monoxide; *C*, represents the difference between curves *A* and *B* (*after* EWER).
Ordinate:
oxygen consumption in mm.^{3}/gm./hour.
Abscissa:
oxygen concentration in c.c./litre.

In *Chironomus* larvae of the *plumosus* group, at 17° C., the oxygen consumption begins to fall off when the oxygen concentration drops below 3 c.c. per litre. If the larvae are carefully treated with carbon monoxide in such a way that the haemoglobin is put out of action but the respiratory enzymes (p. 624) are unlikely to be affected, the oxygen uptake is the same as the normal down to 3 c.c. of oxygen per litre; below this it falls more steeply than in the normal insects (Fig. 253).[57] In *Tanytarsus brunnipes* the haemoglobin has such a high affinity for oxygen that the tissues suffer oxygen lack at oxygen pressures which will not cause dissociation. The haemoglobin is deoxygenated at an external concentration of oxygen equal to 5 per cent. of air saturation; it can function as a carrier over the range 5–25 per cent. air saturation, but in the normal life of *Tanytarsus* it probably plays no part as an oxygen carrier.[188] When *Chironomus* larvae are returned to oxygenated water after a prolonged period of asphyxia, there is a brief rise in the oxygen uptake, and during this period haemoglobin is functional in the carriage of oxygen.[188] The resistance to oxygen lack is much greater in species of Chironomid larvae from stagnant water than from streams (p. 629); but this difference is not correlated with the presence or absence of haemoglobin.[189] In *Chironomus plumosus* under natural conditions haemoglobin has been proved to be of functional importance. It functions in oxygen transport at very low oxygen tensions; it thus enables the larva to maintain active 'filter feeding' when very little oxygen is present; and it increases the rate of recovery from oxygen lack, thus making recovery possible under adverse conditions.[225] It has the same importance in leaf-mining Chironomid larvae.[225]

In *Gasterophilus* there are modified fat body cells (tracheal cells) filled with haemoglobin and richly supplied with tracheae (p. 447). This respiratory pigment enables the larva to make better use of an intermittent contact with air bubbles by taking up a much larger amount of oxygen than is necessary for its immediate requirements.[90] The haemoglobin in the full-grown larva will fix about 2·7 c.mm. of oxygen, which will last the insect 0·5–4 minutes depending on the respiratory activity.[91] In the Notonectid *Anisops* the haemoglobin contained in the modified fat body cells (p. 447) serves as an oxygen store which enables the insect to prolong its stay under water; after exposure to 8 per cent. carbon monoxide in air these bugs perform dives of less than half the normal duration. Large spiracles ensure the rapid oxygenation of this haemoglobin.[256]

Properties of insect haemoglobin—The haemoglobin of *Chironomus plumosus* has a molecular weight of about 31,400, half that of vertebrate haemoglobin, and presumably contains only two haemin groups.[167] There are spectroscopic differences between the haemoglobins of different species, even within a single genus of Chironomids.[61] Its dissociation curve varies with the concentration of haemoglobin, the affinity for oxygen increasing with dilution.[59] We have already noted its very high affinity for oxygen: 50 per cent. saturation occurs at an oxygen tension of 0·6 mm. Hg. as opposed to 27 mm. Hg. in human blood. In human blood this affinity is reduced in the presence of carbon dioxide; but *Chironomus* haemoglobin has not this property.[61] The affinity for carbon monoxide is even greater than in mammals.[61] [See p. 401.]

Carriage of carbon dioxide—The enzyme carbonic anhydrase which accelerates the hydration and dehydration of carbon dioxide ($H_2CO_3 \rightleftarrows H_2O + CO_2$) is present in *Chironomus* blood, though it is much less active than in mammalian blood.[22] It has also been claimed that insect blood contains a factor which inhibits the hydration reaction of carbon dioxide.[103] If the existence of such a factor is confirmed it must have the effect of reducing the capacity of the blood to carry carbon dioxide. The fresh blood of the caterpillar *Prodenia* contains 10.03 volumes of CO_2 per cent.; of this about 78 per cent. is combined as bicarbonate, 22 per cent. as carbonic acid.[10] In the blood of *Gasterophilus* larvae carbonic anhydrase is absent and there is no trace of any factor inhibiting the hydration reaction of carbon dioxide either in this insect or in *Locusta*. In *Gasterophilus* 30–50 per cent. of the carbon dioxide is in solution, the rest as bicarbonate.[219] [See p. 401.]

THE REGULATION OF RESPIRATORY MOVEMENTS

We must now consider the nature of the stimuli which control the respiratory movements of insects, and the nervous mechanisms by which they are coordinated. The simplest case is that of an insect which shows no pumping movements and regulates its respiration solely by opening and closing the spiracles. The control of these movements was first studied in the cockroach,[76] but the most detailed investigation has been carried out in the flea.[206] This insect is so small that diffusion takes place very rapidly, and the respiratory responses are correspondingly quick.

Chemical control of the spiracles—The spiracular movements of the normal flea have already been described (p. 367); we shall now consider the regulation of their rhythmical opening and closing. Each act of opening or closing is the immediate result of chemical stimuli; there is no rhythmical discharge from

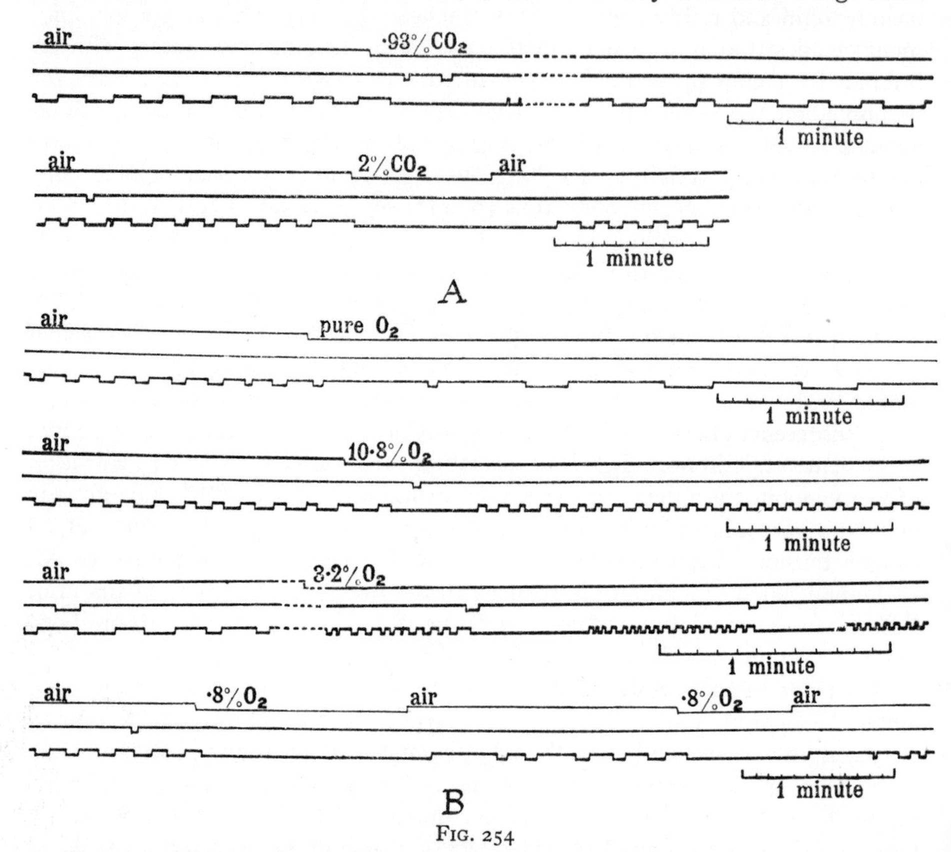

FIG. 254

A, effect of different concentrations of carbon dioxide on the opening and closing of the eighth abdominal spiracle of *Xenopsylla*; *B*, effect of different concentrations of oxygen on the same spiracle (*after* WIGGLESWORTH). In each tracing the upper line shows the change in gas mixture, the second line shows periods during which the flea struggled, the lower line shows opening (downwards) and closing of the spiracle. Note the prolonged opening caused by struggling in 3·2 per cent. oxygen.

the nervous system causing opening and closing at a given rate. Further, both the exhaustion of the supply of oxygen in the tracheal system during the period of closure, and the accumulation of carbon dioxide in the tissues, contribute to the regulation. In a given flea the length of time the spiracles remain closed is more or less proportional to the oxygen content of the air breathed, and below 1 per cent. of oxygen they are open almost all the time. But the duration of the open period is determined largely by the amount of carbon dioxide that has accumulated during the period of closure. Hence the length of the open period is more or less proportional to the length of the closed period which has preceded it; in 1·8 per cent. of oxygen the spiracles close again almost instantly, in pure oxygen they remain open for perhaps 15 or 20 seconds (Fig. 254, B). This is because during the time the spiracles are closed, the carbon dioxide is present mainly in solution in the tissue fluids, and some time is necessary for it to diffuse out when the spiracles open. Thus carbon dioxide has the same effect in slowing up the responses as it has in regulating the respiration of mammals. The addition of say 1 per cent. of carbon dioxide to the outside air prolongs the open period some 50 per cent., by delaying this process of diffusion; at a concentration of 2 per cent. the spiracles are kept permanently open (Fig. 254, A).[206] For this reason the susceptibility of insects to fumigants is increased by a deficiency of oxygen or an excess of carbon dioxide.[38]

In *Musca*, likewise, the effect of carbon dioxide on spiracular opening is influenced by the partial pressure of oxygen: over the range 2·3–54·5 per cent. O_2, the effective concentration of CO_2, increased from 3·6 per cent. to 14·7 per cent. As the carbon dioxide concentration increases the metathoracic spiracle of *Musca* shows a graded opening.[236] In dragonflies the readiness of the spiracles to open in response to carbon dioxide depends on the state of hydration of the insect: in partially desiccated insects they do not open in less than about 5 per cent. CO_2 in air, and after flight they close at once; in well hydrated insects they open in 1–2 per cent. CO_2 and may remain open for as long as 2 minutes after flight.[255]

The effective stimulus both in oxygen want and carbon dioxide excess is probably the acidity of the tissue fluids. For lactic acid, along with other acid metabolites, is known to be produced in insects in the absence of oxygen (p. 630); the effect of oxygen want on the spiracles can be simulated by introducing lactic acid into the blood of the insect; and the spiracles of the flea remain widely open even when all the CO_2 liberated is absorbed, along with the oxygen in the air, by enclosing the insect in a bubble in alkaline pyrogallol.[206] [See p. 401.]

Injection of lactic and citric acids at pH 4·2 does not cause spiracular opening in *Musca* and *Sarcophaga*, whereas carbonic acid at pH 6·25–6·5 does;[236] but this is probably because these acids fail to penetrate to the centres; undissociated organic acid molecules will enter and stimulate the central nervous system in *Blaberus*.[236]

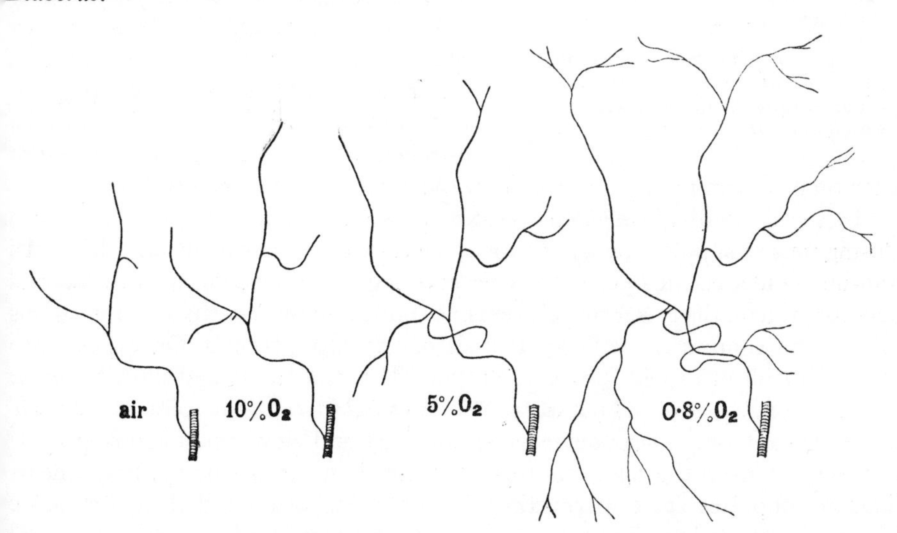

FIG. 255.—Extent of gas in one group of tracheoles in abdomen of flea at rest, exposed to different concentrations of oxygen (*after* WIGGLESWORTH)

This same accumulation of metabolites through incomplete oxidation causes an increase in the osmotic pressure of the blood, which in turn, in some insects, results in a removal of fluid from the tracheal endings (p. 361); and this effect, by increasing the respiratory surface, will provide a further adaptation to conditions of oxygen want. The extent of the removal of fluid from the tracheoles in the insect at rest is more or less constant for each concentration of oxygen (Fig. 255); on exposure to pure oxygen, the fluid rises higher than in air.[206] The same removal of fluid from the tracheoles will occur if aerobic metabolism is blocked with carbon monoxide. But whether the metabolites themselves are responsible for absorbing the water is uncertain. It seems more likely that it results from swelling of the cell proteins[207, 272]—due perhaps to a reduction in the redox potential.[31]

Nervous control of the spiracles—The spiracular sphincters are controlled by segmental nerves coming from the nerve cord, and the stimuli affecting their movements presumably act upon respiratory centres in the nervous system. In most insects, with a tracheal system of very small capacity as compared with the body volume, oxygen want will provide a more sensitive index of increased oxygen consumption than the production of carbon dioxide. For if, in Fig. 256, the tissues at **A** become active, oxygen will flow to this point from the entire tracheal system; the centre at **B** will suffer from oxygen want (with consequent local formation of acid metabolites) and bring about opening of the spiracle at **C**, before carbon dioxide in sufficient quantity has had time to reach **B** either by the circulating blood or by the tracheal system.[206]

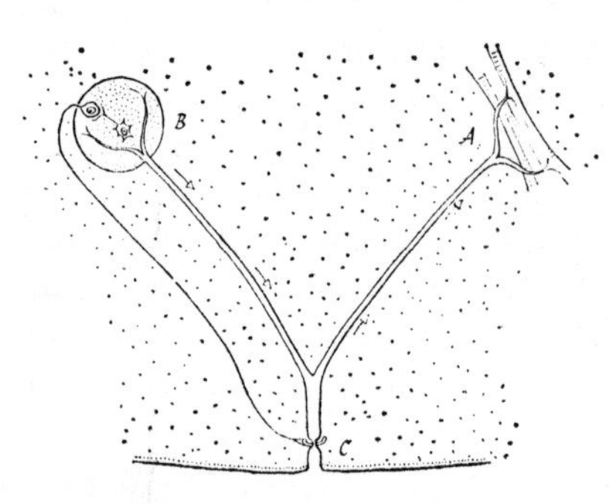

FIG. 256.—Diagram to illustrate the principles of nervous control of the spiracles (*after* WIGGLESWORTH) Explanation in the text.

In the flea the main centres controlling the abdominal spiracles lie in the abdominal ganglia, but the sensitivity of these centres to carbon dioxide or to oxygen want is increased if the ganglia of the thorax are intact. The nervous system, so far as it controls respiration, is highly integrated; this is apparent from the fact that with weak stimuli the opening is confined to particular spiracles (p. 367). But slow reactions of the spiracles to nitrogen or to carbon dioxide may persist after they are completely severed from the central nervous system.[76, 206]

In the pupa of *Hyalophora cecropia* the spiracular apparatus consists of a closing muscle supplied by a network of nerves without any ganglion cells, and a non-innervated elastic opener.[230] After severing the connection with the central nervous system, the closer muscle remains contracted for months and during this time shows continuous action potentials of 12–15 per second. On exposure to carbon dioxide or to acid Ringer's solution (pH 5) the denervated spiracle opens; in oxygen or Ringer at pH 9 it closes.[231, 267] In *Schistocerca* the spiracular muscle of the second thoracic segment receives both 'fast' and 'slow' motor axons (p. 150). In the intact insect the valves are regulated chiefly by the fast axon. From time to time the impulses cease, permitting the muscle to relax, and then the elastic forces open the valves. There is no evidence of a peripheral sensory-motor nervous system situated at the spiracle itself. In the presence of high concentrations of carbon dioxide the muscle will relax and the spiracle open no matter what impulses are coming from the central nervous system. This seems to be a direct effect on neuromuscular junctions in the muscle.[246] In dragon-flies this peripheral control of spiracular movements seems to play an important part even in the normal insect.[255] [See p. 401.]

The prolonged opening of the spiracles which follows muscular exertion in the flea, seems always to be the result of chemical stimulation, probably again the combined effect of carbon dioxide production and oxygen want, and is never due to a nervous stimulus coincident with the onset of muscular action. The rapid rhythm of spiracular movements which occurs when the temperature is raised can be explained by the combined effect of temperature on the intensity of metabolism and on the rate of diffusion.[206]

Cyclic discharge of carbon dioxide—In the larva of *Papilio machaon* there is a more or less continuous discharge of carbon dioxide, but in the pupa the elimination of CO_2 becomes periodic. At first there are about three periods of discharge per hour, but within a few days the frequency of these 'CO_2 outbursts' has fallen to one per day, and very little CO_2 escapes betweenwhiles. The frequency of discharge increases again towards emergence.[259] In *Gryllus* and *Periplaneta* and many other insects there is a continuous fluctuating discharge of CO_2; in *Triatoma* and *Carabus* at rest there may be a CO_2 outburst every 30 minutes; the very infrequent outbursts are characteristic of the pupae of Lepidoptera, particularly of large species.[259]

The phenomenon has been studied particularly in the large diapausing pupae of *Agapema*[233] and *Hyalophora cecropia*.[263] It is dependent, of course, upon retention of carbon dioxide by firm closure of the spiracles: if the spiracles are intubated and kept open artificially the cyclic discharge ceases.[233] The phenomenon occurs only when the rate of metabolism is low: in the *Hyalophora* pupa in diapause the cycles of discharge of CO_2 occur once in 3 days at 10° C., every 7 hours at 25° C.; and when development begins again and oxygen uptake rises twelve-fold, the cyclic discharge comes to an end; it is eliminated also by low O_2 concentration (6 per cent.) or high CO_2 (10 per cent.).[263] Exposure to nitrogen after a normal CO_2 outburst in the *Agapema* pupa causes a further discharge equal to double that of the original outburst.[233] The uptake of oxygen by these insects is more or less continuous, but only a very small amount of CO_2 escapes between the 'outbursts' (Fig. 257).[263] During the interburst period the spiracular valves are held closed but they open to a very minute extent and 'flutter' from time to time; such 'fluttering' is absent in pure oxygen.[263]

This cyclic release of CO_2 represents an exaggeration of the respiratory phenomena described above (p. 392) in the flea.[206] Spiracular opening is induced by oxygen deficiency (presumably due to acid metabolites at the sensitive respiratory centres) plus carbonic acid accumulation (Fig. 254). Control by oxygen deficiency alone (as in very low oxygen mixtures) causes a rapid fluttering of the spiracles (Fig. 254, B, 3·2 per cent. O_2). At high oxygen levels the accumulation of CO_2 plays a dominant part in the final stimulation of the spiracular opening: once the spiracles are opened they then remain open for a long period while the accumulated CO_2 diffuses out from the tissues (Fig. 254, B, pure O_2). We also saw that during the time the spiracles of the flea are closed a negative pressure develops in the tracheal system and the tubes show a partial collapse (p. 369).[78]

In the large diapausing pupae of Lepidoptera these effects are exaggerated

(i) by the large fluid content of the blood and tissues which retain much CO_2; (ii) by a very high tolerance for CO_2 (15 per cent. CO_2 is needed to cause the spiracles to open); (iii) by a very low rate of metabolism which extends enormously the time over which CO_2 accumulates to threshold level. During the inter-burst period the oxygen in the tracheal system falls to threshold level; a negative pressure develops; the spiracles open a minute degree and air rushes in.[233] The oxygen lack in the centre is relieved very rapidly and the spiracles close quickly

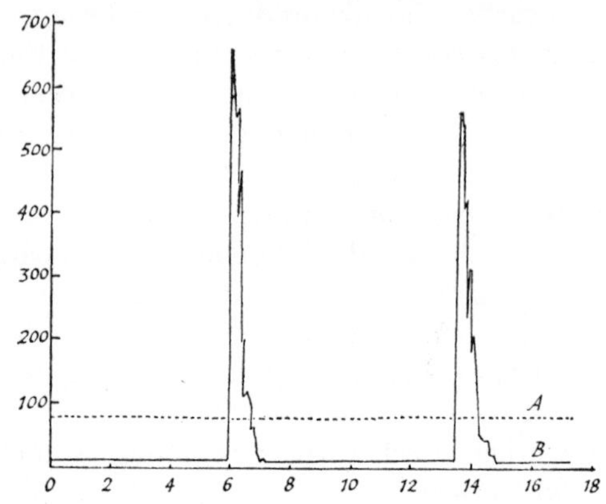

FIG. 257.—Chart showing the course of oxygen uptake (A) and carbon dioxide output (B) in diapausing pupa of *Hyalophora* at 25° C. (*after* SCHNEIDERMAN and WILLIAMS)

Ordinate: rate of oxygen uptake and carbon dioxide output in mm.^{3}/pupa/hour. Abscissa: time in hours.

again before time has been allowed for CO_2 to come out of solution and escape. Exposure to nitrogen during a CO_2 outburst increases the production of acid metabolites, prolongs the open period, and allows far more CO_2 to escape than would otherwise be necessary. [See p. 401.]

The puzzling feature about this phenomenon, as about the effect of pure oxygen on the spiracles of the flea, is the nature of the 'triggering' of the CO_2 discharge: once this has been initiated the spiracles remain open for a long period although the CO_2 must have fallen well below the level at which 'triggering' occurred.

Chemical control of pumping movements—When we turn to insects in which pumping movements are added to regulation by diffusion, the phenomena become more complicated (Fig. 258). In the cockroach[76] at low concentrations of carbon dioxide, regulation of the spiracles resembles that in the flea: 2 per cent. CO_2 in air causes all the spiracles to remain permanently open. But on raising the CO_2 to 10 per cent., slow pumping movements of 8–10 per minute begin, even in the insect at rest; and then the spiracles show rhythmic movements—the second thoracic and the last seven abdominal spiracles closing for a period at the beginning of expiration, the first thoracic and first abdominal remaining open all the time. At 15 per cent. CO_2, the pumping movements increase to 90–120 per minute and the spiracle movements continue. At 20–30 per cent. CO_2, respiration is at the rate of 150–180 per minute and the spiracles no longer close. Besides varying in rate, the respirations vary also in depth of movement.[62]

These results show how complex is the regulation of respiration in such an insect. At concentrations of carbon dioxide great enough to cause pumping movements, the spiracles no longer remain permanently open, but a rhythmic closure of certain of them supervenes; a process which, as we have seen (p. 373), drives a stream of air through the tracheal system. There is a true nervous co-ordination between ventilating and spiracular movements; for both persist unchanged in insects with all the abdominal contents removed with the exception of the ganglia and muscles.[62]

In the control of the pumping movements in the cockroach, as in the control of the spiracles, both oxygen want and carbon dioxide excess can act as stimulus. At 5 per cent. of oxygen in nitrogen, pumping movements and the spiracular rhythm begin; and at a given partial pressure of carbon dioxide, dyspnoea is greater, the less the partial pressure of oxygen: in 15 per cent. CO_2 in *air* dyspnoea begins in 8 seconds and shows a maximum frequency of 120 per minute; whereas in 15 per cent. CO_2 in *oxygen* dyspnoea begins in 60 seconds and reaches a frequency of only 45 per minute.[76]

Stimulation of respiratory movements by both high carbon dioxide and low oxygen concentrations, has also been observed in other insects; grasshoppers

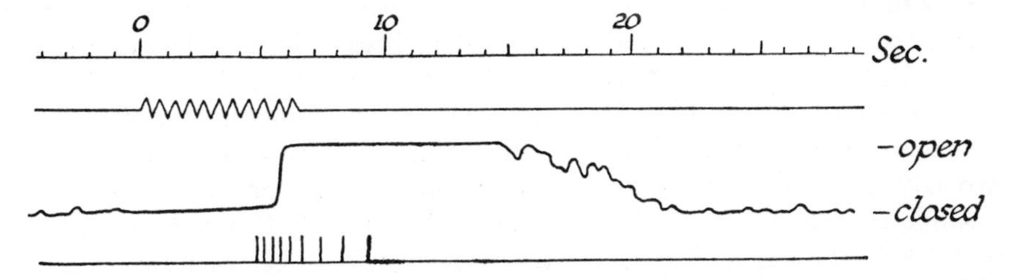

FIG. 258.—Effect of struggling on respiratory movements in *Periplaneta* (*after* HAZELHOFF)

Upper line: time in seconds. Second line: struggling movements. Third line: degree of opening of first thoracic spiracle. Lower line: ventilating movements.

show a violent hyperpnoea in oxygen containing 10 per cent. of carbon dioxide;[106] *Carausius morosus* shows respiratory movements when the oxygen content of the air falls below 3–4 per cent., besides responding to high carbon dioxide concentrations;[29, 164] in *Schistocerca* pure oxygen slows the respiratory rhythm and there is an enormous increase in frequency on going back to air;[33] in the sheep ked *Melophagus*, 5 per cent. CO_2 causes an increase in both the rate and amplitude of the respiratory movements of the abdomen; pure oxygen reduces both and causes the prothoracic spiracle to remain closed.[192] The ventilation of the gill chamber of *Aeschna* larvae is brought about by water poor in oxygen or water containing minute quantities of free carbonic acid, changing the *p*H from 7 to 6·75.[164]

Control of respiratory activities of aquatic insects—On the other hand, the rhythmical vibrations of the gill lamellae of Ephemerid nymphs,[9] the body movements of the aquatic larva of *Nymphula* (Lepidoptera),[198] and the movements by which *Phryganea* larvae (Trichoptera) ventilate their tubes,[40] seem to be induced solely by the low oxygen content of the water; and the same is true of *Limnephilus*.[242] When the aquatic bug *Corixa* is submerged, it will often direct a current of water over its ventral air store (p. 381) by movements of its legs. These movements are said to be due always to the accumulation of carbon dioxide and not to lack of oxygen. Whereas the stimulus which causes this insect to rise to the surface to replenish its supply of air, is said to be the diminishing size of the air bubble it carries;[21] though it will also rise in response to a deficiency of oxygen and an associated phototaxis.[258] In *Aphelocheirus* there is a small blind air sac associated with the second abdominal spiracle, which is supplied with a large sense cell and perhaps serves to detect any shrinkage in the volume of gas in the tracheal system.[251]

In *Notonecta*, likewise, it is the diminution of buoyancy due to the contraction of the air store which is the stimulus driving it to the surface; an artificial increase in air pressure above the water causes *Notonecta* to rise immediately.[20] On the other hand, *Naucoris* is said to be made to rise by lack of oxygen, and if kept in water exposed to pure oxygen it will die, because this stimulus never occurs until all its air store has been consumed and water has entered the tracheal system.[213] It is lack of oxygen, also, and not carbon dioxide excess, which causes the surface breathing ('Notatmung') of *Aeschna* larvae, which come to the surface to breathe if the oxygen content of the tracheae falls to 3·2 per cent.,[187] and the ascent of the mosquito larva,[96] and the larva[108] and adult[77] of *Dytiscus* to the water surface. The integument of *Eristalis* larvae is highly impermeable to oxygen and relatively impermeable to carbon dioxide. The air in the larger air sacs, which form 9 per cent. of the body volume, can be renewed to the extent of 50 per cent. by a deep breath. Both carbon dioxide accumulation and oxygen want serve as stimuli for ventilation. The anal filaments have no respiratory function. If the larvae are kept submerged for a time they become more buoyant and float up to the surface. The gas responsible for this is carbon dioxide which may reach a concentration of 20–40 per cent. in the tracheal air while the oxygen falls to 3·5 per cent.[218] Presumably this carbon dioxide is produced by the action of acid metabolites on the calcium carbonate accumulated in the Malpighian tubes (p. 560).

Nervous control of pumping movements—These chemical stimuli exert their action upon respiratory centres in the nervous system, the distribution of which recalls that of the centres controlling the spiracles in the flea. It was shown by Marshall Hall (1842), and has been repeatedly confirmed since,[204] that the isolated abdominal segments of *Libellula* and other insects can still perform respiratory movements; and the isolated ganglia of *Dytiscus* and *Aeschna* show rhythmical changes in electrical potential which doubtless indicate the activity of respiratory centres.[1] The abdominal centres are sometimes termed 'primary centres'; they are always rather insensitive to stimuli. In the larva of *Aeschna* the isolated abdomen is quite indifferent to changes in the oxygen tension of the medium;[164, 187] in *Carausius* the abdominal centres will respond to oxygen want with pumping movements, but they are far less sensitive than the intact insect. The 'secondary centres', which are responsible for the more sensitive reactions to chemical stimuli, are situated in the thorax, usually in the prothoracic ganglion (in *Aeschna*[187] and in *Carausius*[164]). Decapitation, involving removal of the brain and suboesophageal ganglion, causes only a temporary and variable disturbance of the respiratory movements.[204]

Probably there is only a quantitative difference between the 'primary' and 'secondary' respiratory centres. If the thorax of the locust *Schistocerca* is separated from the abdomen, the respiratory rhythm of the thorax is unchanged, while that of the abdomen is slowed. The 'secondary' centres in the thorax are really the 'primary' centres for the thoracic respiratory movements; and being more sensitive to chemical stimuli they set the general pace of the respiratory rhythm.[62] In *Schistocerca* the metathoracic ganglion probably contains a pacemaker controlling the frequency and amplitude of all forms of ventilation; and each ganglion of the head and thorax contains carbon dioxide receptors which modify the action of this pacemaker; oxygen lack is normally less important as a stimulus.[253] The gas in the thoracic air sacs during flight is commonly about 5 per cent. CO_2 and 15 per cent. O_2.[269] The isolated nervous system shows a rhythm of electric discharge which is modified and possibly initiated by carbon dioxide.[253]

In *Dytiscus* and *Locusta* there are proprioceptive mechanoreceptors, perhaps chordotonal organs or other stretch receptors, in which impulses are set up during inspiration and expiration; they probably serve to co-ordinate and regulate the rhythmical movements.[248] In *Gryllus*, as in other insects, each abdominal ganglion can maintain the respiratory movements of its own segment. The suboesophageal ganglion seems to contain the pacemaker for the whole system. Localized stimulation of different parts of the protocerebrum evoke specific respiratory effects: respiratory co-ordination as a whole is carried out by the entire central nervous system.[247] On the other hand, the movements of the gills of *Cloëon* nymphs are said to be controlled by a series of stimulatory and inhibitory centres with definite localization.[3] [See p. 401.]

But besides being influenced by chemical stimuli the rhythmic discharges of the respiratory centres are affected by nervous stimuli from elsewhere in the system. Acceleration of respiration may be brought about by any outside stimulus (mechanical, sudden illumination, &c.) to which the insect may be exposed;[11, 62, 164] and respiration may be accelerated, or the movements increased in amplitude, before or at the very outset of muscular effort. Thus an analysis of the respiratory movements during and after flight in a great number of insects, has shown that both reflex nervous stimuli from other motor centres, and chemical stimuli resulting from increased metabolism, contribute to the augmented ventilation.[62] [See p. 401.]

Control of the tracheal supply—The pattern of tracheal supply is a part

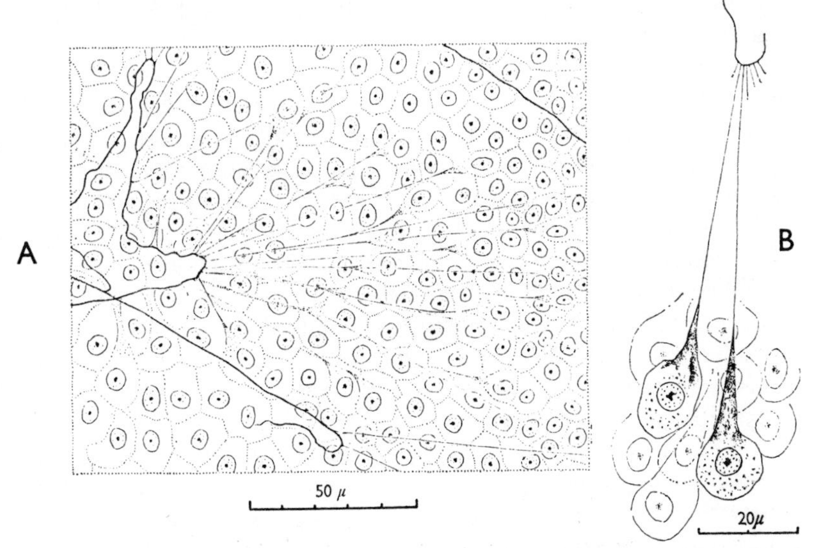

FIG. 259.—*A*, Epidermal cells in *Rhodnius*, in an area deprived of its tracheal supply, showing contractile filaments attached to tracheole loops. *B*, detail of the filaments (*after* WIGGLESWORTH)

of the normal predetermined pattern of growth; but this pattern can be changed by the oxygen requirements of the tissues.[273] Insects reared in an atmosphere containing only 7 or 10 per cent. of oxygen show a greatly increased formation of new tracheae and tracheoles during moulting.[252, 273]

New formation of tracheoles cannot take place between moults; but there are striking changes in the *distribution* of the existing tracheoles to meet local needs; for example, if a tracheal branch is cut. In the epidermis of *Rhodnius* these movements are brought about by the epidermal cells themselves; they give rise to contractile filamentous processes (up to 100μ long) which attach themselves to the tracheoles and pull these towards the oxygen-deficient cells (Fig. 259).[274] This reaction provides the mechanism for the even distribution of the tracheoles in the epidermis.

ADDENDA TO CHAPTER IX

p. 366. **Site of oxygen uptake**—It is calculated that in *Schistocerca* the rate of diffusion of oxygen into the flight muscles via the tracheoles is up to 10^5 times faster than it would be if all diffusion were in the liquid phase.[270] In *Aeschna* during flight, with an oxygen consumption of 1·8 ml O_2/g. muscle/min., the oxygen in the primary tubes would be adequate for 1–2 wing strokes only; it must be renewed at every wing beat. In *Aeschna* there are no indenting tracheoles and the diameter of the fibres (11·8μ) is close to the theoretical maximum for diffusion from the tracheoles of the oxygen actually consumed. At higher metabolic rates (e.g. in Diptera) the size of fibres must be reduced or the tracheoles must indent the cells. In general, diffusion is adequate to meet all needs provided the primary tubes are adequately ventilated.[270]

p. 370. **Ventilation of tracheae**—In certain non-flying Prionine beetles pumping movements occur in the head, prothorax and metathorax in phase with abdominal ventilation.[291]

p. 374. **Air stream in tracheal system**—Spiracle 1 in *Schistocerca* provides the central nervous system with a mechanism for sampling the pterothoracic gases. Thus abdominal ventilation can be varied in flight to meet the demands while the thorax continues to pump in and out a fixed volume.[290] Simultaneous measurements of the unidirectional flow of air in *Schistocerca*, and the total abdominal pumping, have shown that during rest, flight and recovery the flow caused by abdominal pumping remains small and almost constant at 30 l. air/kg./hour in through the thoracic spiracles and out through the abdominal. During flight the total pumping amounts to about 180 l./kg./hour—of which 70 litres ventilate the thorax and 80 litres the other parts of the body. During average horizontal flight thoracic ventilation goes up to 320 l. air/kg./hour of which 250 litres are moved by the thoracic pump. But, if need be, the capacity of this pump can be increased to at least 950 l./kg./hour. This rate of ventilation permits sustained flight by the locust without risk of desiccation. In contrast with *Schistocerca*, large dragonflies (*Aeschna*) depend almost exclusively on thoracic pumping during flight; while large wasps (*Vespa*) depend on abdominal pumping.[293]

In the Cerambycid beetle *Petrognatha* during flight a stream of air enters by spiracle 2 and leaves by spiracle 3. It may well be that this serves as an air-cooling device for the flight motor.[277]

p. 379. **Cuticular gills**—In most if not all cuticular gills the air spaces are not separated from the water by a membrane but are in direct contact with it through numerous perforations. The structure affords therefore an example of plastron respiration.[284] Plastron-bearing spiracular gills occur also in the respiratory horns of some pupae that occupy the intertidal zone.[283]

p. 384. **Oxygen from aquatic plants**—Even adult *Donacia* remaining in the gas-filled cocoon during the winter obtains sufficient oxygen by diffusion from the gas spaces in the roots.[285]

p. 391. **Insect haemoglobin**—Up to nine different haemoglobins have been detected in single species of *Chironomus*.[292] The so-called 'myoglobin fold' in the polypeptide chain of mammalian haemoglobin serves to hold the haem group so that its iron atom can combine reversibly with molecular oxygen. This same pattern of folding has been found in the haemoglobins of *Chironomus*.[286]

p. 391. **Carbonic anhydrase**—The presence of carbonic anhydrase has been confirmed in *Acheta*; it is widely distributed but does not occur in all tissues. Its presence in different insects appears to be correlated with the discontinuous elimination of CO_2.[279]

p. 393. **Control of spiracles**—In adult dragonflies the closure of the spiracles is more rigorous in the desiccated insect, and the threshold of spiracular response to carbon dioxide and oxygen lack are raised.[289]

p. 395. **Nervous control of spiracles**—In *Glossina* and in *Galleria* there is a multipolar neurone situated close to the spiracle, which sends one of its processes to the spiracular muscle. It remains to be established whether this neurone is concerned both in sensory perception and in initiation of muscle activity.[281]

p. 396. **Cyclic discharge of CO_2**—The periodic discharge of carbon dioxide helps in the conservation of water; in *Schistocerca* it occurs in the newly emerged adult; it does not occur after feeding starts; but prior to death there is a return to the cyclic process.[282] In *Hyalophora* the system functions so efficiently that water loss is reduced to a value which is approximately the same as that produced in metabolism.[287] Direct measurement of gas pressure within the tracheal system during the respiratory cycles in the diapausing pupa of *Hyalophora* show that during 95 per cent. of the entire cycle the intratracheal pressure is below atmospheric. During periods of spiracular 'flutter' the pressure ranges from 0·025–0·9 mm. Hg; during the constriction period preceding the CO_2 'outburst' the pressure falls to between -3 and -5 mm Hg.[278]

p. 399. **Nervous control of pumping movements**—The location of the pacemaker for tracheal ventilation in the metathorax and the influence of mechanoreceptors in the abdomen in modulating its activity have been confirmed in *Periplaneta*. Internuncial fibres running from the metathoracic to the terminal ganglion produce synchronous contraction of the abdominal expiratory muscles.[280]

Recent study has tended to confirm the existence of a rhythmic system in the ganglia of larvae of *Aeschna* and *Anax*, which controls ventilation of the rectum and whose activity is modified by endogenous stimuli; the pacemaker appears to be located in the eighth ganglion.[288]

REFERENCES

[1] ADRIAN, E. D. *J. Physiol.*, **72** (1931), 132–51 (rhythmical potential changes in isolated nervous system: *Dytiscus*, Col.).
[2] ALTERBERG, G. *Biol. Zbl.*, **54** (1934), 1–20 (biology of *Eristalis* larvae, Dipt.).
[3] ALVERDES, F. *Z. vergl. Physiol.*, **3** (1926), 558–94 (co-ordinating centres: *Cloëon*, Ephem.).
[4] ANGLAS, J. *C. R. Soc. Biol.*, **1** (1904), 175–6 (tracheae during metamorphosis).
[5] ATHANASIU, I., and DRAGOIU, I. *C. R. Soc. Biol.*, **75** (1913), 578–82; *Arch. Anat. Microsc.*, **16** (1914), 345–61 (tracheal endings in muscles).
[6] BABÁK, E. *Winterstein's Handb. d. vergl. Physiol.*, **1**, Part 2, 362 (mechanics and control of tracheal respiration).
[7] —— *Int. Rev. Ges. Hydrobiol. u. Hydrog.*, **5** (1912), 81–90 (respiration in *Culex*, Dipt.).
[8] —— *Arch. ges. Physiol.*, **147** (1912), 349–74 (respiration in *Dytiscus*, Col.).
[9] BABÁK, E., and FOUSTKA, O. *Arch. ges Physiol.*, **119** (1907), 530–48 (respiratory stimulus in Odonata nymphs).
[10] BABERS, F. H. *J. Agric. Res.*, **57** (1938), 697–706 (respiratory function of blood: *Prodenia*, Lep.).
[11] BARLOW, W. F. *Phil. Trans. Roy. Soc.*, **145** (1855), 139–48 (respiratory movements).
[12] BARRATT, J. O. W., and ARNOLD, G. *Quart. J. Micr. Sci.*, **56** (1910), 149–65 (blood of *Dytiscus* and *Hydrophilus*, Col.).
[13] BEADLE, L. C. *J. Exp. Biol.*, **16** (1939), 346–62 (anal papillae: *Aedes detritus*, Dipt.).
[14] BEARD, R. L. *Ann. Ent. Soc. Amer.*, **35** (1942), 68–72 (formation of tracheal funnel in parasitized *Anasa*, Hem.).
[15] BERGOLD, G. *Z. Morph. Oekol. Tiere*, **29** (1935), 511–36 (form of spiracles in beetles from different environments).
[16] BISHOP, G. H., *et al. J. Biol. Chem.*, **58** (1923), 543–65; **66** (1926), 77–88 (blood of honey-bee larva).
[17] BLUNCK, H. *Z. wiss. Zool.*, **117** (1917), 1–129; *Z. wiss. Zool.*, **121** (1924), 171–391 (biology and physiology of *Dytiscus*, Col.).
[18] —— *Zool. Anz.*, **55** (1922), 45–66 (biology of *Cybister*, Col.).
[19] BONGARDT, J. *Z. wiss. Zool.*, **75** (1903), 1 (tracheal endings in luminous organs: Lampyridae).
[20] BOTH, M. P. *Z. vergl. Physiol.*, **21** (1934), 167–75 (regulation of buoyancy in *Notonecta*, Hem.).
[21] BOTJES, J. O. *Z. vergl. Physiol.*, **17** (1932), 557–74 (regulation of respiration: *Corixa*, Hem.).
[22] BRINKMAN, R., *et al. J. Physiol.*, **75** (1932), 3 P. (carbonic anhydrase in *Chironomus*, Dipt.).
[23] BROCHER, F. *Ann. Biol. Lac.*, **4** (1909), 89–138 (surface forces and aquatic insects),
[24] —— *Ann. Biol. Lac.*, **4** (1909), 9–32 (respiration in adult aquatic insects: *Notonecta*, Hem.); *Ibid.*, **4** (1910), 383–98 (*Dytiscus*, Col.); *Ibid.*, **5** (1912), 5–26 (*Haemonia*, Col.); 136–79 (*Elmis*, Col.); 218–19 (*Cybister*, Col.); 220–56 (*Hydrophilus*, Col.); *Ibid.*, **6** (1913), 120–46 (larva of *Dytiscus*, Col.); *Zool. Jahrb., Physiol.*, **33** (1913), 225–34 (*Notonecta*); *Ann. Biol. Lac.*, **7** (1914), 1–39 (*Dytiscus*); *Arch. Zool.*, **55** (1916), 483–514 (*Nepa*); *Ibid.*, **56** (1916), 1–24 (*Dytiscus*); *Ann. Biol. Lac.*, **9** (1918), 31–50 (résumé).
[25] —— *Arch. Zool.*, **60** (1920), 1–45 (types of tracheae: *Sphinx*, Lep.); *Ann. Soc. Ent. Fr.*, **89** (1920), 209–32 (ditto: *Vespa*, Hym.).
[26] —— *Arch. Zool.*, **74** (1931), 25–32 (respiratory movements: *Hydrophilus*, *Dytiscus*, Col).
[27] BROWN, J. M. *Trans. Linn. Soc. (Zool.)*, **11** (1910), 125–35 (respiration in larva of *Tipula*, Dipt.).
[28] BRUNTZ, L. *C.R. Acad. Sci.*, **146** (1908), 871–3 (tracheal endings in labial kidneys: *Machilis*, Thysan.).
[29] v. BUDDENBROCK, W., and v. ROHR, G. *Z. allg. Physiol.*, **20** (1922), 111–60 (respiration in *Dixippus*, Orth.).

[30] BUISSON, M. du. *Bull. Acad. Roy. Belg. Cl. Sci.*, sér. 5, **10** (1924), 373–91; 635–56; **12** (1926), 127–38 (mechanism of tracheal ventilation).

[31] BULT, T. *Thesis, Assen*, 1939, 1–143 (movement of fluid in tracheoles).

[32] BURTT, E. T. *Proc. Roy. Ent. Soc. Lond.*, A, **11** (1936), 61–5 (respiration in larva of *Atrichopogon*, Ceratopogonidae, Dipt.).

[33] CHAUVIN, R. *Ann. Ent. Soc. Fr.*, **110** (1941), 133–272 (physiology of *Schistocerca*, Orth.).

[34] CHEN, S. H. *Sinensia*, **16** (1945), 31–5 (air sacs in elytra: Scarabaeidae).

[35] CHOLODKOWSKY, N. *Zool. Anz.*, **7** (1884), 316–19 (respiration of Tachinid larva).

[36] CHRISTENSEN, P. J. H. *Dansk. Naturhist. For.*, **86** (1928), 21–48 (hydrostatic mechanism of *Corethra* larva).

[37] CONTEJEAN, C. *C.R. Acad. Sci.*, **111** (1890), 361–3 (respiration in grasshopper).

[38] COTTON, R. T. *J. Econ. Entom.*, **25** (1932), 1088–1103 (respiration and susceptibility to fumigants).

[39] CUÉNOT, L. *L'Adaptation*, pp. 193–9, Gaston Doin, Paris, 1925.

[40] van DAM, L. *Zool. Anz.*, **118** (1937), 122–8 (respiratory movements: *Phryganea* larva, Trichopt.).

[41] DAMANT, G. C. C. *J. Physiol.*, **59** (1924), 345–56 (swim bladders of *Corethra*, Dipt.).

[42] DAVIES, W. M. *Quart. J. Micr. Sci.*, **71** (1927), 15–30 (tracheal system of *Sminthurus*, Collemb.).

[43] DAVIS, C. *Proc. Linn. Soc., N.S.W.*, **67** (1942), 1–8 (aquatic respiration: *Coxelmis*, Col.).

[44] DAY, M. F. *J. Coun. Sci. Ind. Res. (Australia)*, **11** (1938), 317–27 (respiratory conditions in nest of *Eutermes*).

[45] DEEGENER, P. *Schröder's Handb. d. Entomologie*, **1** (1913), 317–82 (respiration).

[46] DEIBEL, J. *Zool. Jahrb., Anat.*, **31** (1911), 107–60 (respiration of *Donacia* larvae, Col).

[47] DELL, J. A. *Trans. Ent. Soc. Lond.*, 1905, 293–311 (larva of *Psychoda*, Dipt.).

[48] DEMOLL, R. *Z. Biol.*, **86** (1927), 45–66; **87** (1927), 8–22; *Zool. Jahrb., Physiol.*, **45** (1928), 513–34 (function of air sacs: evolution of carbon dioxide, &c.).

[49] DODDS, G. S., and HISAW, F. L. *Ecology*, **5** (1924), 262–71 (size of gills in relation to environment).

[50] DOLLEY, W. L., and FARRIS, E. J. *J. N.Y. Ent. Soc.*, **37** (1929), 127–33 (glands of respiratory siphon: *Eristalis* larva, Dipt.).

[51] DREHER, K. *Z. Morph. Oekol. Tiere*, **31** (1936), 608–72 (respiratory system of honey-bee).

[52] DUJARDIN, F. *C. R. Acad. Sci.*, **28** (1849), 674–77 (structure of tracheae).

[53] EGE, R. *Z. allg. Physiol.*, **17** (1915), 18–124 (function of air stores of aquatic insects).

[54] —— *Vidensk. Meddel. Dansk. Naturh. Foren. Copenhagen*, **66** (1915), 183–96 (respiration in larva and pupa of Donaciae, Col.); *Physiological Papers dedicated to August Krogh, London*, 1926, 25–39 (respiration in *Hydrocampa* larva and pupa, Lep.).

[55] ENGELMANN, T. W. *Arch. néerl. Sci.*, **28** (1895), 358–71 (bacteria and protozoa as indicators of oxygen tension).

[56] EVANS, A. C. *Bull. Ent. Res.*, **26** (1935), 115–22 (function of abdominal air sacs: *Lucilia*, Dipt.).

[57] EWER, R. F. *J. Exp. Biol.*, **18** (1942), 197–205 (function of haemoglobin in *Chironomus*).

[58] FAURÉ-FREMIET, E. *Ann. Sci. Nat., Zool.*, **12** (1910), 217–40 (labial glands of aquatic Hemiptera).

[59] FLORKIN, M., *et al. Acta Biol. Belg.*, **2** (1941), 305–6 (dissociation curve of *Chironomus* haemoglobin).

[60] FOX, H. M. *J. Gen. Physiol.*, **3** (1920), 565–74 (protozoa as indicators of site of oxygen uptake).

[61] —— *Nature*, **156** (1945), 475; **162** (1948), 20; *J. Exp. Biol.*, **21** (1945), 161–5 (properties of haemoglobin in *Chironomus*, &c.).

[62] FRAENKEL, G. *Z. vergl. Physiol.*, **16** (1932), 394–460 (nervous control of respiration).

[63] —— *Proc. Zool. Soc. Lond.*, 1935, 893–904 (function of abdominal air sacs: *Lucilia*, Dipt.).

[64] FRAENKEL, G., and HERFORD, G. V. B. *J. Exp. Biol.*, **15** (1938), 266–80 (respiration through the skin).

[65] v. FRANKENBERG, G. *Zool. Jahrb., Physiol.*, **35** (1915), 505–92; *Zool. Anz.*, **76** (1928), 237–40 (swim bladders of *Corethra*, Dipt.).

[66] FREUDENSTEIN, K. *Z. wiss. Zool.*, **132** (1928), 404–75 (heart and circulation: honey-bee).

[67] GEIPEL, E. *Z. wiss. Zool.*, **112** (1915), 239–90 (tracheal endings in luminous organs: Coleoptera).

[68] GEROULD, J. H. *Acta Zool.*, **19** (1938), 297–352 (circulation and respiration: *Bombyx*).

[69] GERBIG, F. *Zool. Jahrb., Syst.*, **35** (1913), 127–84 (respiration in Tipulid larvae).

[70] GRABER, V. *Die Insekten*, Munich, 1877.

[71] GRANDORI, R. *Redia*, **7** (1911), 363–428 (biology of *Apanteles*, Braconidae).

[72] HAGEMANN, J. *Zool. Jahrb., Anat.*, **30** (1910), 373–426 (respiration in *Corixa*, Hem.).

[73] HAMILTON, M. A. *Proc. Zool. Soc. Lond.*, 1931, 1063–1136 (tracheal lungs, &c.: *Nepa*, Hem.).

[74] HARNISCH, O. *Z. vergl. Physiol.*, **11** (1930), 285–309; **13** (1930), 280–99; *Verhandl. deutsch. Zool. Ges.*, 1933, 209–17; *Z. vergl. Physiol.*, **24** (1937), 198–209 (anal papillae, haemoglobin, &c. in respiration of Chironomid larvae, Dipt.).

[75] HASSAN, A. A. G. *Trans. Roy. Ent. Soc. Lond.*, **94** (1943), 103–53 (spiracular mechanisms).

[76] HAZELHOFF, E. H. *Regeling der Ademhaling bij Insekten en Spinnen*, Utrecht, 1926; *Z. vergl. Physiol.*, **5** (1927), 179–90 (control of respiration by the spiracles).

[77] HEBERDEY, R. F. *Z. Morph. Oekol. Tiere*, **33** (1938), 667–734 (respiration of Dytiscidae).

[78] HERFORD, G. M. *J. Exp. Biol.*, **15** (1938), 327–38 (rhythmical collapse of tracheae in the flea).

[79] HEYMONS, R. *Deutsch. Ent. Z.*, **52** (1908), 137–50 (respiration, &c.: aquatic Hymenoptera).

[80] HINTON, H. E. *Trans. Roy. Ent. Soc. Lond.*, **98** (1947), 449–73 (number of spiracles and their moulting process).

[81] —— *Proc. Roy. Ent. Soc. Lond.*, **22** (1947), 52–60 (tubular gills: Psephenidae, Col.).

[82] HOLMGREN, E. *Anat. Anz.*, **11** (1895), 340–6 (tracheal endings in spinning glands of caterpillars).

[83] —— *Arch. Mikr. Anat.*, **71** (1907), 165–247 ('trophospongium' in muscle fibres).

[84] HOPPE, J. *Zool. Jahrb., Physiol.*, **31** (1911), 189–244 (respiration in *Notonecta*).

[85] IMMS, A. D. *J. Hyg.*, **7** (1907), 291–318 (respiration in *Anopheles* larva, Dipt.).

[86] JANET, C. *C.R. Acad. Sci.*, **152** (1911), 110–2 ('inverted tracheae' in head of bee).

[87] KALISS, N. *Genetics*, **24** (1939), 244–70 (filling of tracheal system with air: *Drosophila*).

[88] KEILIN, D. *Proc. Camb. Phil. Soc.* (Biol. Sci.), **1** (1924), 63–70 (appearance of air in tracheae).

[89] KEILIN, D., *et al. Parasitology*, **27** (1935), 257–62 (perispiracular glands: mosquito larvae).

[90] KEILIN, D. *Parasitology*, **36** (1944), 1–66 (respiratory system in larvae and pupae of Diptera).

[91] KEILIN, D., and WANG, Y. L. *Biochem. J.*, **40** (1946), 855–66 (haemoglobin in *Gasterophilus*).

[92] KEISTER, M. L. *J. Morph.*, **83** (1948), 373–423 (development of tracheae and tracheoles: *Sciara*, Dipt.).

[93] KIELICH, J. *Zool. Jahrb., Anat.*, **40** (1918), 515–36 (tracheoles in muscles).

[94] KITCHEL, R. L., and HOSKINS, W. M. *J. Econ. Entom.*, **28** (1935), 924–33 (stream of air through tracheal system in cockroach).

[95] KOCH, A. *Mitt. Zool. Inst. Westfäl. Wilhelms-Univ. Münster*, **1** (1918), 11–13 (tracheal system of *Mochlonyx* larva, Dipt.).

[96] —— *Zool. Jahrb., Physiol.*, **37** (1920), 361–492 (respiration of *Culex* larvae, Dipt.).

[97] KOCH, H. *Ann. Soc. Sci. Bruxelles*, **54** (1934), 346–61 (function of cells of anal papillae, &c.).

[98] —— *Natuurvet. Tijdsch.*, **16** (1934), 75–80 (relative importance of gill lamellae in Agrionid nymphs).

[99] —— *Recherches sur la physiologie du système trachéen clos.*, Brussels, 1936, 98 pp.

[100] KOEPPEN, A. *Zool. Anz.*, **52** (1921), 132–9 (tracheal endings: *Dytiscus*, Col.).

[101] KRAMER, A. S. *Zool. Anz.*, **117** (1937), 181–91 ('tracheal lungs': *Gryllus*, Orth., and *Nepa*, Hem.).

[102] KRANCHER, O. *Z. wiss. Zool.*, **35** (1881), 505–74 (structure of spiracles).
[103] KREPS, E. M., and CHENIKAEVA, E. J. *C. R. Acad. Sci. U.R.S.S.*, **34** (1942), 142–5; *Bull. Acad. Sci. U.R.S.S.*, 1942, 310–21 (carbon dioxide in insect blood).
[104] KREUGER, E. *Lunds. Univ. Årsskr.*, **10** (1914), No. 13, 20 pp. (function of elytral air in *Dytiscus*, Col.).
[105] KROGH, A. *Skand. Arch. Physiol.*, **25** (1911), 183–203 (hydrostatic mechanism of *Corethra* larva).
[106] —— *Skand. Arch. Physiol.*, **29** (1913), 28–36 (composition of air in tracheal system of grasshoppers).
[107] —— *J. Physiol.*, **52** (1919), 391–408 (diffusion of gases through tissues).
[108] —— *Arch. ges. Physiol.*, **179** (1920), 95–120 (diffusion theory of tracheal respiration).
[109] KUNCKEL d'HERCULAIS, J. *Ann. Soc. Ent. Fr.*, **9** (1879), 349–57 (respiration of larva of *Gymnosoma*, Tachinidae).
[110] LANDOIS, H., and THELEN, W. *Z. wiss. Zool.*, **17** (1867), 185–214 (closing mechanisms in spiracles).
[111] LASS, M. *Z. wiss. Zool.*, **79** (1905), 73–131 (anatomy of the flea).
[112] LAUTERBORN, R. *Zool. Anz.*, **26** (1903), 637–42 (tracheal gills in Plecoptera larvae).
[113] LEE, M. O. *Amer. J. Physiol.*, **68** (1924), 135; *J. Exp. Zool.*, **41** (1925), 125–54; **49** (1927), 319–20 (respiration in Orthoptera).
[114] —— *Science*, **69** (1929), 334–5 (function of air sacs).
[115] LEITCH, I. *J. Physiol.*, **50** (1916), 370–9 (haemoglobin in *Chironomus* larva, Dipt.).
[116] LESTAGE, J. A. *Ann. Biol. Lac.*, **10** (1920), 231–60 (biology of Plecoptera).
[117] LÜBBEN, H. *Zool. Jahrb., Anat.*, **24** (1907), 71–128 (respiration, &c.: Trichoptera).
[118] LUND, E. J. *J. Exp. Zool.*, **11** (1911), 415–68 (tracheal endings in photogenic organs: Lampyridae).
[119] MAPLE, J. D. *Ann. Ent. Soc. Amer.*, **30** (1937), 123–54 (egg-shell structure and respiration of Encyrtid larvae, Hym.).
[120] MARCU, O. *Zool. Anz.*, **85** (1929), 239–32; **89** (1930), 186–9; **93** (1931), 61–3 (structure of tracheae).
[121] MARTIN, J. *Bull. Soc. Philomat. Paris*, **4** (1892), 122–4; *C. R. Soc. Philomat. Paris*, **6** (1893), 3 (entry of oxygen through tracheal endings).
[122] MCARTHUR, J. M. *J. Exp. Zool.*, **53** (1929), 117–28 (functions of different spiracles: Orthoptera).
[123] MCCUTCHEON, F. H. *Ann. Ent. Soc. Amer.*, **33** (1940), 35–55 (respiratory mechanism: *Dissosteira*, Orth.).
[124] MCGOVRAN, E. R. *Ann. Ent. Soc. Amer.*, **24** (1931), 751–61; *J. Econ. Entom.*, **25** (1932), 271–6 (stream of air through tracheal system in grasshoppers).
[125] MELLANBY, K. *Proc. Roy. Soc.*, B, **116** (1934), 139–49 (site of water loss from insects).
[126] MIALL, L. C. *The Natural History of Aquatic Insects*, Macmillan, London, 1903.
[127] MIALL, L. C., and HAMMOND, A. R. *Life History of the harlequin fly (Chironomus)*, Oxford, 1900.
[128] MINOT, C. S. *Arch. Physiol. norm. path.*, **3** (1876), 1–10 (formation of spinal thread: *Hydrophilus*).
[129] MONTALENTI, G. *Boll. Ist. Zool. Roma*, **4** (1926), 133–50 (tracheal endings in muscle).
[130] MORGAN, A. H., and GRIERSON, M. C. *Physiol. Zool.*, **5** (1932), 230–45 (gills in nymphs of *Hexagenia*, Ephem.).
[131] MORGAN, A. H., and O'NEIL, H. D. *Physiol. Zool.*, **4** (1931), 361–79 (tracheal gills in *Macronema* larva, Trichopt.).
[132] MORISON, G. D. *Quart. J. Micr. Sci.*, **71** (1927), 395–463 (tracheal endings in muscle: *Apis*).
[133] MUTTKOWSKI, R. A. *Ann. Ent. Soc. Amer.*, **14** (1921), 150–6; *Trans. Amer. Microsc. Soc.*, **40** (1921), 144–57 (copper in insect blood).
[134] NEWPORT, G. *Phil. Trans. Roy. Soc.*, 1836, 529–66; *Trans. Linn. Soc.*, **20** (1851), 419–23 (function of air sacs, &c.).
[135] NIGMANN, M. *Zool. Jahrb., Syst.*, **26** (1908), 489–560 (biology of *Acentropus*, Lep.).
[136] PALMÉN, J. A. *Zur Morphologie des Tracheensystems*, Helsingfors, 1877.
[137] PANTEL, J. *Bull. Soc. Ent. Fr.*, 1901, 57–61 (perispiracular glands, &c.: Tachinid larvae).
[138] —— *La Cellule*, **26** (1910), 25–212 (endoparasitic larvae of Diptera).

[139] PAUSE, J. *Zool. Jahrb., Physiol.*, **36** (1918), 339–452 (respiration, &c. in larvae of *Chironomus*, Dipt.).

[140] PENNAK, R. W., and MCCOLL, C. M. *J. Cell. Comp. Physiol.*, **23** (1944), 1–10 (respiration of Agrionid nymphs).

[141] PÉREZ, C. *Arch. Zool.*, sér. 5, **4** (1910), 1–274 (tracheae, &c. during metamorphosis: *Calliphora*, Dipt.).

[142] PHILLIPS, M. E. *Ann. Ent. Soc. Amer.*, **32** (1939), 325–8 (peristigmatic glands: Trypetidae).

[143] PLATEAU, F. *Mém. Acad. Roy. Belg.*, **45** (1884), 1–219 (respiratory movements).

[144] PORTIER, P. *Arch. Zool.*, sér. 5, **8** (1911), 89–379 (respiration of aquatic insects).

[145] PORTIER, P., *et al. C. R. Soc. Biol.*, **105** (1930), 760–64; **122** (1936), 1292–93 (wings and respiration in Lepidoptera).

[146] PRENANT, A. *Arch. Anat. Microsc.*, **3** (1900), 293–336 (tracheal cells of *Gasterophilus* larva, Dipt.).

[147] PULIKOVSKY, N. *Z. Morph. Oekol. Tiere*, **7** (1927), 384–443 (respiratory adaptations in *Simulium* pupae, Dipt.).

[148] RAMSAY, J. A. *J. Exp. Biol.*, **12** (1935), 373–83 (evaporation of water from the cockroach).

[149] RATHKE, H. *Ann. Mag. Nat. Hist.*, ser. 3, **9** (1862), 81–106 (physiology of respiration in insects).

[150] REMY, P. *Contribution à l'étude de l'appareil respiratoire et de la respiration chez quelques invertébrés*, Vagner, Nancy, 220 pp.

[151] RICH, S. G. *J. Morph.*, **31** (1918), 317–49 (gill chamber of Odonata nymphs).

[152] RICHARDS, A. G., and ANDERSON, T. F. *J. N.Y. Ent. Soc.*, **50** (1942), 147–67; 245–7 (structure of tracheoles).

[153] RIEDE, E. *Zool. Jahrb., Physiol.*, **32** (1912), 231–310 (oxygen supply to ovary of insects).

[154] RIPPER, W. *Z. wiss. Zool.*, **138** (1931), 303–69 (homology of tracheal system).

[155] RUSSELL, P. F., and RAO, T. R. *Amer. J. Trop. Med.*, **21** (1941), 767–77 (surface tension and respiration of mosquito larvae).

[156] SADONES, J. *La Cellule*, **11** (1896), 273–324 (respiration in Odonata larvae).

[157] SAKURAI, M. *C. R. Acad. Sci.*, **187** (1928), 614–15 (tracheal glands).

[158] SCHEER, D. *Arch. Hydrobiol.*, **27** (1934), 359–96 (haemoglobin in Chironomids).

[159] SCHOENEMUND, E. *Arch. Hydrobiol. Planktonk.*, **15** (1925), 339–69 (respiration of Plecoptera larvae).

[160] SEURAT, L. G. *C.R. Acad. Sci.*, **127** (1898), 636–8; *Ann. Sci. Nat., Zool.*, **10** (1899), 1–159 (respiration of parasitic larvae, Hymenoptera).

[161] SIKES, E. K., and WIGGLESWORTH, V. B. *Quart. J. Micr. Sci.*, **74** (1931), 165–92 (first appearance of air in tracheal system).

[162] de SINÉTY, R. *La Cellule*, **19** (1901), 119–278 (anatomy of Phasmidae).

[163] SNODGRASS, R. E. *Principles of Insect Morphology*, New York, 1935.

[164] STAHN, I. *Zool. Jahrb., Physiol.*, **46** (1928), 1–86 (regulation of respiration: *Dixippus*, *Aeschna* larva).

[165] STOKES, A. C. *Science*, **21** (1893), 44–6 (structure of tracheae).

[166] STUART, T. A. *Trans. Roy. Soc. Edin.*, **60** (1941), 475–502 (haemoglobin and tracheal system in Chironomids).

[167] SVEDBERG, T., *et al. J. Amer. Chem. Soc.*, **56** (1934), 1700–6 (molecular weight of Chironomid haemoglobin).

[168] SZABÓ-PATAY, J. *Ann. Mus. Nat. Hungary*, **21** (1924), 33–55 (respiration in *Aphelocheirus*, Hem.).

[169] TAYLOR, T. H. *Trans. Ent. Soc. Lond.*, 1902, 701–16 (tracheal system in *Simulium*, Dipt.).

[170] THOMPSON, W. R. *Trans. Ent. Soc. Lond.*, 1929, 195–244 (form of tracheae: Muscid larvae).

[171] THORPE, W. H. *Science*, **68** (1928), 433–4 (elimination of carbon dioxide).

[172] —— *Proc. Zool. Soc. Lond.*, 1930, 929–71 (respiration of parasitic larvae: *Cryptochaetum iceryae*, Dipt.); *Proc. Roy. Soc.*, B, **109** (1932), 450–71 (ditto: Ichneumonids and Braconids); *Quart. J. Micr. Sci.*, **77** (1934), 273–304 (ditto: *Cryptochaetum grandicorne*); *Parasitology*, **28** (1936), 517–40 (ditto: *Encyrtus infelix*,

Chalcid, Hym.); *Ibid.*, **33** (1941), 149–68 (ditto: *Cryptochaetum* spp. in *Aspidoproctus*).

[173] THORPE, W. H. *Nature*, **131** (1933), 549 (blood gills of aquatic insects).

[174] —— *Ve Congr. Internat. Entom. Paris, 1932* (1933), 345–51 (respiration of aquatic and parasitic larvae).

[175] THORPE, W. H., and CRISP, D. J. *J. Exp. Biol.*, **24** (1947), 227–328; *Discussions of Faraday Soc.*, **3** (1948), 210–20 (plastron respiration: *Aphelocheirus*, Hem.) *J. Exp. Biol.*, **26** (1949) 219–60 (ditto: Coleoptera).

[176] THULIN, I. *Anat. Anz.*, **33** (1908), 193–205 (tracheal endings in fat body, &c.).

[177] TIEGS, O. W. *Trans. R. Soc. S. Australia*, **46** (1922), 319–527 (metamorphosis in *Nasonia*, Hym.).

[178] TILLYARD, R. J. *The Biology of Dragonflies*, Cambridge, 1917.

[179] TONNER, F. *Z. wiss. Zool.*, **147** (1936), 433–54 (respiratory and swimming movements: Libellulid nymphs).

[180] VARLEY, G. C. *Proc. Roy. Ent. Soc. Lond.*, A, **12** (1937), 55–60; **14** (1939), 115–23 (aquatic larvae obtaining oxygen from plants).

[181] VAYSSIÈRE, A. *Ann. Sci. Nat., Zool.*, **13** (1882), 1–137 (anatomy of Ephemerid nymphs).

[182] VIALLANES, H. *C. R. Acad. Sci.*, **90** (1880), 1180–2 (respiration and circulation: larva of *Ctenophora*, Dipt.).

[183] VIEWEGER, T. *Arch. Biol.*, **27** (1912), 1–33 (tracheal cells: *Hypocrita*, Lep.).

[184] VOGLER, C. H. *Illustr. Z. Entom.*, **5** (1900), 17–20 (gills in *Teichomyza* larva, Dipt.).

[185] WAGNER, W. *Zool. Jahrb., Physiol.*, **42** (1926), 441–86 (respiratory system of *Simulium* larvae, Dipt.).

[186] WAHL, B. *Arb. Zool. Inst. Wien*, **12** (1899), 1–54 (respiratory system: *Eristalis*, Dipt.).

[187] WALLENGREN, H. *Lunds. Univ. Årsskr.*, N.F. Afd., 2, **9** (1913), No. 16, 30 pp.; **10** (1914), No. 4, 24 pp.; *Ibid.*, No. 8, 28 pp.; **11** (1915), No. 11, 12 pp. (physiology of respiration in larva of *Aeschna*, Odonata).

[188] WALSHE, B. M. *J. Exp. Biol.*, **24** (1947), 329–51 (haemoglobin in Chironomids).

[189] —— *J. Exp. Biol.*, **25** (1948), 35–44 (oxygen requirements and habitat: Chironomidae).

[190] WALTER, E. *Zool. Jahrb., Syst.*, **45** (1922), 587–608 (respiration in larvae of *Hypoderma*; and *Gasterophilus*, Dipt.).

[191] WEBER, H. *Lehrbuch der Entomologie*, Jena, 1933.

[192] WEBB, J. E. *Proc. Zool. Soc. Lond.*, **115** (1945), 218–50 (respiratory mechanism: *Melophagus*, Dipt.).

[193] —— *Proc. Zool. Soc. Lond.*, **116** (1947), 49–119 (spiracle structure: *Anoplura*).

[194] WEISE, H. *Z. wiss. Zool.*, **151** (1938), 467–514 (respiration: *Eristalis* larva, Dipt.).

[195] WEISMANN, A. *Z. wiss. Zool.*, **13** (1863), 159–220 (development of tracheal system: *Calliphora*, Dipt.).

[196] —— *Z. wiss. Zool.*, **14** (1865), 187–336 (pupal respiration: *Calliphora*, Dipt.).

[197] WEISSENBERG, R. *S. B. Ges. naturf. Freunde Berlin*, 1908, 1–18 (biology of *Apanteles*, Braconidae).

[198] WELCH, P. S., and SEHON, G. L. *Ann. Ent. Soc. Amer.*, **21** (1928), 243–58 (respiration in larva of *Nymphula*, Lep.).

[199] WESENBERG-LUND, C. *Int. Rev. Biolog. Suppl.*, 1912 (biology of *Dytiscus*).

[200] WETTINGER, O. *Z. wiss. Zool.*, **129** (1927), 453–82 (circulatory system in Tipulid larvae).

[201] WHITEHEAD, H. *The Naturalist*, 1945, 123–6 (respiration of *Agriotypus*, Hym.).

[202] V. WIELOWIEJSKI, H. R. *Z. wiss. Zool.*, **37** (1882), 354; *Zool. Anz.*, **12** (1889), 594–600 (tracheal endings in luminous organ: Lampyridae).

[203] WIGGLESWORTH, V. B. *Proc. Roy. Soc.*, B, **106** (1930), 229–50 (movements of fluid in tracheoles in aquatic insects); *Ibid.*, **109** (1931), 354–9 (the same in terrestrial insects).

[204] —— *Biol. Rev.*, **6** (1931), 181–220 (respiration: review).

[205] —— *J. Exp. Biol.*, **10** (1933), 1–37 ('anal gills' of mosquito larva).

[206] —— *Proc. Roy. Soc.*, B, **118** (1935), 397–419 (respiration in the flea, *Xenopsylla*).

[207] —— *J. Exp. Biol.*, **15** (1938), 235–47 (osmotic pressure and fluid in tracheoles: mosquito larva).

[208] WIGGLESWORTH, V. B. *J. Exp. Biol.*, **15** (1938), 248–54 (filling of tracheal system with air: *Aëdes*, Dipt.).

[209] WIGGLESWORTH, V. B., and GILLETT, J. D. *Proc. Roy. Ent. Soc. Lond.*, A, **11** (1936), 104–7 (loss of water at moulting: *Rhodnius*, Hem.).

[210] WINGFIELD, C. A. *Nature*, **140** (1937), 27; *J. Exp. Biol.*, **16** (1939), 363–73 (function of gills in *Cloëon* nymph, Ephem.).

[211] WINTERSTEIN, H. *Winterstein's Handb. d. vergl. Physiol.*, **1**, part 2 (1911) (physical chemistry of respiration).

[212] v. WISTINGHAUSEN, C. *Z. wiss. Zool.*, **49** (1890), 564–82 (tracheal endings in spinning glands of caterpillars).

[213] WREDE, F., and KRAMER, H. *Arch. ges. Physiol.*, **212** (1926), 15–25 (respiration in *Naucoris*, Hem.).

[214] WREDE, F., and TREECK, A. *Arch. ges. Physiol.*, **211** (1926) 228–43 (tracheal respiration in caterpillars).

[214a] YOUNG, B. *Sinensia*, **15** (1944), 141–4 (plastron respiration: *Cheirochela*).

[215] ZAVŘEL, J. *Bull. Intern. Acad. Sci. Prague*, **22** (1920), 120–9 (respiration in larvae of Chironomidae).

SUPPLEMENTARY REFERENCES (A)

[216] HRBÁČEK, J. *Věstnik. Čsl. Zool. společnosti*, **13** (1949), 136–76 (compression phase and ventilation of tracheal system: *Hydrous*, Col.).

[217] KEISTER, M. L., and BUCK, J. B. *Biol. Bull.*, **97** (1949), 323–30; *J. Exp. Biol.*, **32** (1955), 681–91 (tracheal filling in *Sciara*).

[218] KROGH, A. *Ent. Medd.*, **23** (1943), 49–65 (respiration in *Eristalis* larva).

[219] LEVENBOOK, L., *et al. J. Exp. Biol.*, **27** (1950), 158–85 (carriage of carbon dioxide in blood of *Gastrophilus* larva).

[220] LEWIS, D. J. *Proc. Roy. ent. Soc. Lond.* A, **24** (1949), 60–6 (tracheal gills in mosquito larvae).

[221] RICHARDS, A. G., and KORDA, F. H. *Ann. ent. Soc. Amer.*, **43** (1950), 49–71 (structure of tracheoles).

[222] de RUITER, L., WOLVEKAMP, H. P., *et al. Acta physiol. pharm. neerl.*, **2** (1952), 180–213 ('physical gills' in aquatic insects).

[223] THORPE, W. H. *Biol. Rev.*, **25** (1950), 344–90 (plastron respiration: review).

[224] THORPE, W. H., and CRISP, D. J. *J. Exp. Biol.*, **26** (1949), 219–60 (plastron respiration in Coleoptera).

[225] WALSHE, B. M. *J. Exp. Biol.*, **27** (1950), 73–95; **28** (1951), 57–61 (function of haemoglobin in Chironomid larvae).

[226] WATTS, D. T. *Ann. ent. Soc. Amer.*, **44** (1951), 527–38 (intratracheal pressure in insect respiration).

SUPPLEMENTARY REFERENCES (B)

[227] BAILEY, L. *J. Exp. Biol.*, **31** (1954), (air currents in the tracheal system of *Apis*).

[228] BAUDOIN, R. *Bull. Biol. Fr. Belg.*, **89** (1955), 15–164 (surface properties of insect cuticle).

[229] BEAMENT, J. W. L. *Nature*, **186** (1960), 408–9; *Biol. Rev.* **36** (1961), 281–320 (wetting properties of insect cuticle).

[230] BECKEL, W. E. *Proc. Xth Int. Congr. Ent. 1956*, **2** (1958), 87–115 (spiracular closing mechanism in *Hyalophora*).

[231] BECKEL, W. E., and SCHNEIDERMAN, H. A. *Science*, **126** (1957), 352–3 (regulation of spiracles in *Hyalophora*).

[232] BUCK, J. *Ann. Rev. Ent.*, **7** (1962), 27–56 (insect respiration: review).

[233] BUCK, J., *et al. Biol. Bull.*, **109** (1955), 144–63; **114** (1958), 118–40; *J. Ins. Physiol.*, **1** (1958), 327–40; **2** (1958), 52–60 (cyclical release of CO_2 in large pupae of Lepidoptera).

[234] BUCK, J., and KEISTER, M. L. *Biol. Bull.*, **105** (1953), 402–11 (cutaneous respiration in *Phormia* larva).

[235] BURSELL, E. *Proc. Roy. Soc.*, B, **144** (1955), 275–86 (respiration of *Glossina* larva).

[236] CASE, J. F. *Physiol. Zool.*, **29** (1956), 163–71; *J. Cell. Comp. Physiol.*, **49** (1957), 103–14; *Biol. Bull.*, **121** (1961), 285 (effects of acid metabolites on respiratory centres in insects).

[237] —— *J. Ins. Physiol.*, **1** (1957), 85–94 (median nerves and spiracular control in cockroaches).

[238] CLARKE, K. U. *Proc. Roy. Ent. Soc. Lond.*, A, **32** (1957), 67–79 (air sacs and body-volume in *Locusta*).

[239] EDWARDS, G. A., RUSKA, H., and de HARVEN, E. *Arch. Biol.*, **69** (1958), 351–69 (fine structure of tracheae and tracheoles).

[240] FISHER, R. C. *J. Exp. Biol.*, **40** (1963), 531–40 (suppression of insect parasites by asphyxiation).

[241] FOX, H. M. *Nature*, **174** (1954), 355 (oxygen and haem in insects, &c.).

[242] FOX, H. M., and SIDNEY, J. *J. Exp. Biol.*, **30** (1953), 235–7 (oxygen lack and respiratory movements in *Limnephilus* Trichopt.).

[243] HEUMANN, L. *Z. vergl. Physiol.*, **31** (1948), 58–76 (regulation of buoyancy in *Dytiscus*).

[244] HINTON, H. E. *Science Progress*, No. **180** (1957), 692–700; *Proc. Xth Int. Cong. Ent., 1956*, **1** (1958), 543–8; *Proc. Zool. Soc. Lond.*, **138** (1962), 111–22 (spiracular gills in insects).

[245] —— *Science Progress*, **50** (1962), 97–112 (respiratory systems of insect eggs).

[246] HOYLE, G. *J. Ins. Physiol.*, **3** (1959), 378–94; **4** (1960), 63–79; **7** (1961), 305–14 (control of spiracular muscles in *Schistocerca*).

[247] HUBER, F. *Z. vergl. Physiol.*, **43** (1960), 359–91 (nervous control of respiration in *Gryllus*).

[248] HUGHES, G. M. *Nature*, **170** (1952), 531–2 (abdominal stretch receptors and respiration in *Dytiscus* and *Locusta*).

[249] ITO, T. *Biol. Bull.*, **105** (1953), 308–15 (permeability of cuticle of *Bombyx* larva to oxygen and carbon dioxide).

[250] KÜHNELT, W. *Stzb. Osterr. Akad. Wiss. Mat.—Naturw. Kl. Abt. l.* **164** (1955), 49–64 (water relations of Tenebrionidae, &c.).

[251] LARSEN, O. *Acta Univ. Lund.*, **51** (1955), 1–59 (tracheal mechanoreceptor in *Aphelocheirus*).

[252] LOCKE, M. *Quart. J. Micr. Sci.*, **98** (1957), 487–92; **99** (1958), 29–46, 373–91 (structure, formation and growth of the tracheal system in *Rhodnius*).

[253] MILLER, P. L. *J. Exp. Biol.*, **37** (1960), 224–78 (control of ventilation and spiracle movements in *Schistocerca*).

[254] —— P. L. *J. Ins. Physiol.*, **6** (1961), 243–71 (respiratory system of *Hydrocyrius*, Belostomatidae).

[255] —— *Nature*, **191** (1961), 622; *J. Exp. Biol.*, **39** (1962), 513–35 (spiracle control in Odonata).

[256] —— *Nature*, **201** (1964), 1052 (function of haemoglobin in *Anisops*).

[257] NEUMANN, D. *Z. Naturforsch.*, **16b** (1961), 820–4 (effect of iron in the diet on haemoglobin in *Chironomus* larvae).

[258] POPHAM, E. J. *Proc. Zool. Soc. Lond.*, **135** (1960), 209–42 (respiration in Corixidae).

[259] PUNT, A., *et al. Acta. Brev. Neerl. Physiol.*, **16** (1948), 30; *Physiol. Comp. Oecol.*, **2** (1949), 59–74; *Biol. Bull.* **112** (1957), 108–19 (cyclical discharge of carbon dioxide in hibernating pupae, &c.).

[260] QADRI, M. A. H. *Pakistan J. Sci.*, **3** (1951), 119–23 (use of air stores in Notonectidae).

[261] RIZKI, M. T. M. *J. Morph.*, **98** (1956), 497–511 (spiracular glands in *Drosophila*).

[262] SCHMITT, J. B. *Ann. Ent. Soc. Amer.*, **47** (1957), 677–82 (nerve supply to spiracles in Orthoptera).

[263] SCHNEIDERMANN, H. A., *et al. Anat. Rec.*, **117** (1953), 540; *Biol. Bull.*, **109** (1955), 123–143; *Nature*, **182** (1958), 491–3; *Biol. Bull.*, **119** (1960), 394–528 (discontinuous respiration in insects).

[264] SHAFIQ, S. A. *Quart. J. Micr. Sci.*, **104** (1963), 135–40 (fine structure of tracheoles).

[265] STRIDE, G. O. *Nature*, **171** (1953), 885–6; *Ann. Ent. Soc. Amer.*, **48** (1955), 344–51 (respiratory bubble of *Potamodytes*, Coleopt.).

[266] TREHERNE, J. E. *Trans. IXth Int. Congr. Ent.*, **1** (1952), 311–14 (respiration of *Helodes* larva).

[267] VAN DER KLOOT, W. *Comp. Biochem. Physiol.*, **9** (1963), 317–33 (nervous control of spiracular muscles in *Hyalophora* pupa).

[268] VINOGRADSKAYA, O. N. *Ent. Obozr.*, **33** (1955), 157–60 (role of spiracles in water loss in *Anopheles*).

[269] WEIS-FOGH, T. *Proc. XIVth Int. Congr. Zool.* (*Copenhagen*) *1953*, 1956, 283–5; *J. Exp. Biol.*, **41** (1964), 207–28 (ventilation of the thoracic tracheal system in flying insects).

[270] —— *J. Exp. Biol.*, **41** (1964), 229–56 (diffusion and gaseous exchanges in insect wing muscles).

[271] WIGGLESWORTH, V. B. *Quart. J. Micr. Sci.*, **91** (1950), 217–24 (injection methods for the tracheal system).

[272] —— *Quart. J. Micr. Sci.*, **94** (1953), 507–22 (surface forces in the tracheal system).

[273] —— *Quart. J. Micr. Sci.*, **95** (1954), 115–37 (growth and regeneration in the tracheal system).

[274] —— *J. Exp. Biol.*, **36** (1959), 632–40 (role of the epidermal cells in the distribution and movement of tracheoles.

[275] —— *Nature*, **198** (1963), 106 (functions of air sacs).

[276] WOLVEKAMP, H. P. *Experientia*, **11** (1955), 294–301 (physical gills of aquatic insects).

SUPPLEMENTARY REFERENCES (C)

[277] AMOS, W. B. and MILLER, P. L. *Entomologist*, **98** (1965), 88–94 (air current in thorax of Cerambycid beetle in flight).

[278] BROCKWAY, A. P. and SCHNEIDERMAN, H. A. *J. Insect Physiol.*, **13** (1967), 1413–51 (intratracheal pressure in pupa of *Hyalophora*).

[279] EDWARDS, L. J. and PATTON, R. L. *J. Insect Physiol.*, **13** (1967), 1333–41 (carbonic anhydrase in *Acheta*).

[280] FARLEY, R. D. *et al. J. Insect Physiol.*, **13** (1967), 1713–28; **14** (1968), 591–601 (sensory effects on respiratory pacemaker in *Periplaneta*).

[281] FINLAYSON, L. H. *J. Insect Physiol.*, **12** (1966), 1451–4 (sensory innervation of spiracle muscle in *Glossina* and *Galleria*).

[282] HAMILTON, A. G. *Proc. Roy. Soc. Lond*, B, **160** (1964), 373–95 (periodic carbon dioxide discharge in *Schistocerca*).

[283] HINTON, H. E. *Nature*, **209** (1966), 220–1 (plastron respiration in insects of intertidal zone).

[284] HINTON, H. E. *Adv. Insect Physiol.*, **5** (1968), 65–162 (spiracular gills: review).

[285] HOULIHAN, D. F. *J. Insect Physiol.*, **15** (1969), 1517–36; **16** (1970), 1607–22 (respiratory physiology of *Donacia*).

[286] HUBER, R., FORMANEK, H. and EPP, O. *Naturwissenschaften*, **75** (1968), (haemoglobin structure in *Chironomus*).

[287] KANWISHER, J. W. *Biol. Bull. mar. biol. Lab., Woods Hole*, **130** (1966), 96–105 (tracheal gas dynamics in pupa of *Hyalophora*).

[288] MILL, P. J. *et al. J. exp. Biol.*, **44** (1966), 297–316; **52** (1970), 167–75 (nervous control of ventilation in larvae of *Aeschna* and *Anax*).

[289] MILLER, P. L. *J. exp. Biol.*, **41** (1964), 331–57 (control of spiracles in adult dragonflies).

[290] MILLER, P. L. In *The physiology of the insect central nervous system*, J. E. Treherne and J. W. L. Beament (Eds.), Academic Press, London & New York, 1965, 141–55; *Adv. Insect Physiol.*, **3** (1966), 279–354 (regulation of insect respiration: review).

[291] MILLER, P. L. *J. Insect Physiol.*, **17** (1971), 395–405 (thoracie ventilation in non-flying beetles).

[292] THOMPSON, P. E. and ENGLISH, D. S. *Science*, **152** (1966), 75–6 (multiplicity of haemoglobins in *Chironomus* spp.).

[293] WEIS-FOGH, T. *J. exp. Biol.*, **47** (1967), 561–87 (tracheal ventilation in locusts and other insects during flight).

Chapter X

The Circulatory System and Associated Tissues

THE BODY cavity of insects is of twofold origin; being formed in the embryo by the fusion of the lumen of the coelom sacs with the epineural sinus or haemocoele. It lacks the epithelial lining of a true coelom, and it contains circulating blood. The organs and tissues are thus exposed freely to a stream of fluid which percolates among them. Most of the organs in question, the epidermis and tracheal system, the alimentary canal and Malpighian tubes, the nervous system, muscles, glands and gonads, are dealt with elsewhere; but there are a number of tissues, the haemocytes or blood cells, the pericardial cells and other 'nephrocytes', the oenocytes, fat body and mycetocytes, which are in relation only with the blood. These must perform their functions solely through exchanges with the circulating haemolymph; and although they doubtless play widely different parts in metabolism, these tissues may be conveniently considered here.

CIRCULATORY SYSTEM

The Dorsal Vessel—The circulation of the blood is maintained by a system of muscular pumps and fibro-muscular septa (Fig. 260). The most important and in many insects the only organ responsible for its transport is the dorsal vessel, which collects blood from the abdominal cavity and discharges it in the head. The dorsal vessel is divided rather indefinitely into a posterior region, the heart, which has a pair of valved openings or ostia in each body segment, and an anterior region, the aorta, a closed tube. In some insects, such as the cockroach, the heart may extend throughout the abdomen into the hind part of the thorax, and possess as many as 13 pairs of ostia; in other insects, such as the nymphs of Agrionid dragon-flies, it is reduced to a single chamber with a single pair of ostia; and between these extremes all intermediate stages exist (in *Aeschna* nymphs, a single chamber with 2 pairs of ostia;[257] in different species of Ephemeroptera 1–4 pairs of functional ostia;[150] in the honey-bee 5 pairs;[65] in the larva of *Corethra* 7 pairs,[128] &c.) (Fig. 261).

In its simplest form, as seen in the larvae of Diptera, the wall of the dorsal vessel is made up of a single layer of cells, their substance differentiated into circular fibrillae, indistinctly striated, enclosed between two homogeneous membranes probably derived from the sarcolemma.[167] Or the muscles may consist of interlacing striated fibrillae embedded in sarcoplasm, as in *Aeschna*;[257] or longitudinal and circular fibrillae, as in *Nepa*;[80] and outside the outer membrane there is generally an adventitia of connective tissue carrying the tracheoles (Fig. 262).[196] The heart and aorta commonly have the same histological structure; both are contractile. [See p. 456.]

The heart is tied to adjacent structures by radiating filaments. It may be

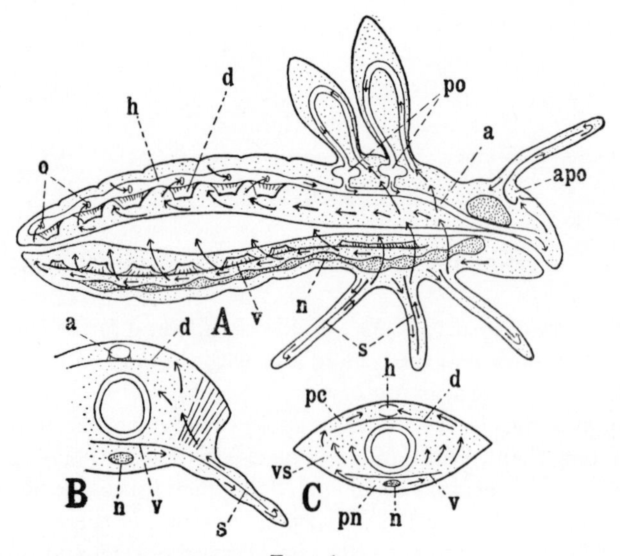

FIG. 260

A, insect with fully developed circulatory system, schematic; *B*, transverse section of thorax of the same; *C*, transverse section of abdomen. Arrows indicate course of circulation (*based largely on* BROCHER). **a**, aorta; *apo*, accessory pulsatile organ of antenna; *d*, dorsal diaphragm with aliform muscles; *h*, heart; *n*, nerve cord; *o*, ostia; *pc*, pericardial sinus; *pn*, perineural sinus; *po*, meso- and metathoracic pulsatile organs; *s*, septa dividing appendages; *v*, ventral diaphragm; *vs*, visceral sinus.

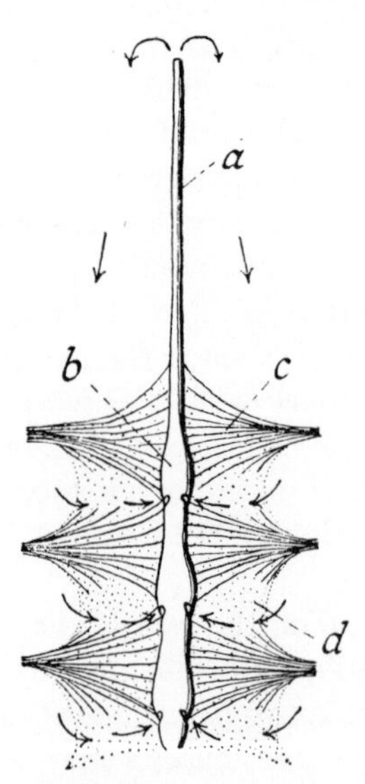

FIG. 261.—Diagram of aorta with three chambers of the heart, dorsal view (*after* SNODGRASS)

a, aorta; *b*, heart; *c*, alary muscles; *d*, dorsal diaphragm.

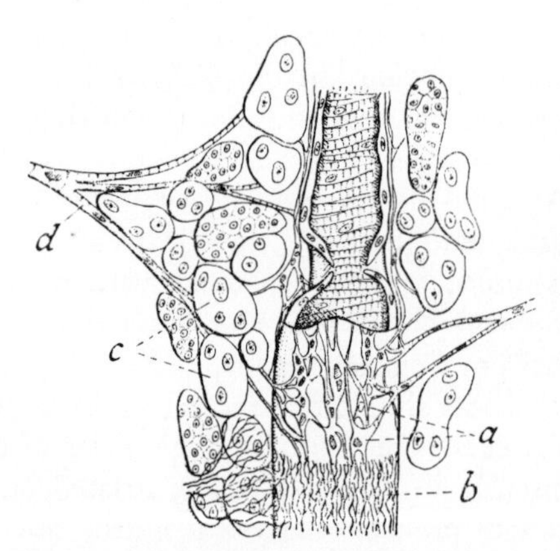

FIG. 262.—Part of heart and adjacent tissues in *Calandra*, dissected (*after* MURRAY and TIEGS)

Cavity of heart exposed above to show contractile tissue and a pair of valves guarding two ostia. *a*, adventitia (pericardium); *b*, elastic tissue (removed elsewhere); *c*, pericardial cells; *d*, alary muscles.

bound directly to the dorsal body wall, as in *Aleurodes* (Fig. 263, A),[228] or attached to it by threads, as in *Lepisma*,[25] *Corethra*,[128] *Nepa* (Fig. 263, C),[80] and many other insects. It may be connected to the lateral body wall in each segment by fan-shaped strings, as in *Corethra*,[128] or by the pericardial septum (p. 415), when this is inserted directly into the sides of the heart, as in *Nepa*,[80] *Aleurodes*,[228] &c. More often this membrane lies below the heart, which is then attached to it by elastic fibres running downwards and outwards, as in Ephemeroptera,[181] *Dytiscus* (Fig. 263, C),[22] &c.

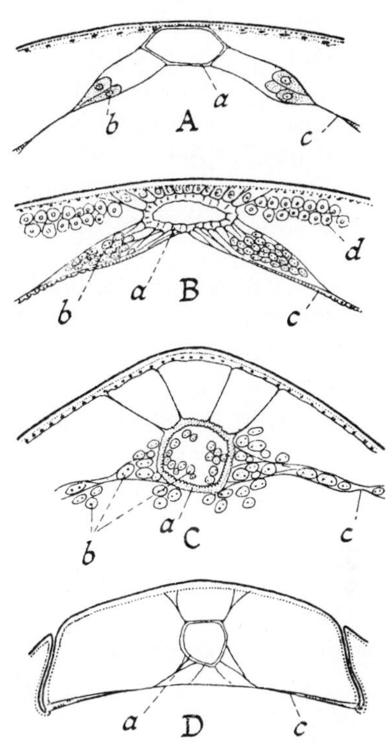

FIG. 263.—Examples of different methods of support for the heart, showing the effect of the pericardial septum in causing dilatation

A, *Aleurodes* (*after* WEBER); *B*, *Apis* (*after* SNODGRASS); *C*, *Nepa* (*after* HAMILTON); *D*, *Dytiscus* (*from* WEBER *after* OBERLE). *a*, heart; *b*, pericardial cells; *c*, pericardial septum; *d*, fat body cells.

Ostial valves—The ostia are slit-like openings in the sides of the heart, with their margins prolonged inwards to form valves. These valves serve primarily to prevent the reflux of blood into the body cavity when the heart contracts. Their action is well seen in the larva of *Corethra*.[219] Here each ostial lip has a unicellular thickening which runs into a thread attached to the inner wall of the heart. When the heart dilates the valves are widely separated and the blood enters (Fig. 264, A); when dilatation is complete (diastole) the valves are closed and stand out at right angles to the wall (Fig. 264, B); during contraction or systole they become evaginated as far as their attached threads will permit, and are forced together so that no blood can escape (Fig. 264, C).

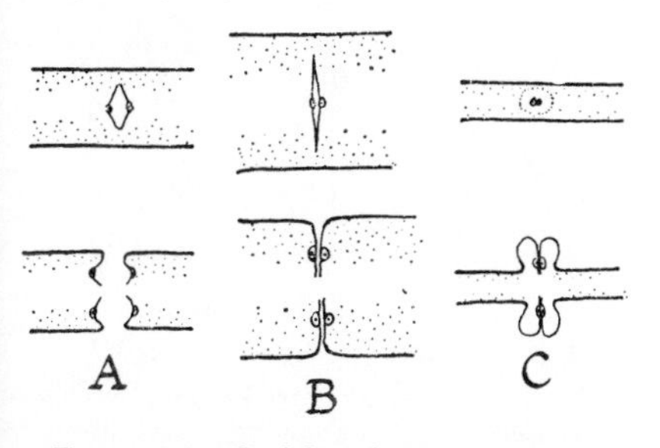

FIG. 264.—Ostial valves in larva of *Corethra* at different phases of the heart beat

Above, lateral view; below, dorsal view. *A*, beginning of diastole; *B*, complete diastole; *C*, complete systole (*after* TZONIS).

Heart chambers—In most insects the heart is a continuous tube (Fig. 265, A); rarely is it divided into chambers. Chambers are sometimes formed by the prolongation inwards of the ostial valves. In nymphs of *Cloëon* the anterior lips of these valves are so elongated that they meet across the lumen, and serve to some extent to prevent reflux of blood within the heart (Fig. 265, B); but they are not entirely competent, for some blood passes back again during diastole.[75, 181] Occasionally, valves may be developed from folds of the inner wall of the heart, independent of the ostia. In *Chironomus dorsalis* a pair of such valves separates the heart from the aorta (Fig. 265, C);[181] in Tipulids there is a pair in each segment just anterior to the ostia (Fig. 265, E).[230] In some species of *Chironomus* and *Tanypus*[181] a pair of muscular pads projects inwards between successive ostia; the heart contracts most strongly at this level, causing an apparent division into chambers (Fig. 265, D). Another type of intra-cardiac valve is seen in the larvae of *Sarcophaga carnaria*, the Tachinid *Compsilura*, &c., and consists of two cushions of large vacuolated cells projecting far into the lumen (Fig. 265, F).[168] In the pupa of *Bombyx mori* the ostia have posterior valve flaps only. These do not restrict backward flow; the entering blood may go forwards or backwards.[70] In the larva of the flea *Nosopsyllus* both internal valves and alary muscles are completely absent.[197]

Segmental blood vessels arising from the sides of the dorsal vessel have been described in many insects.[145] In the adult *Periplaneta* there are five pairs in the abdomen and two in the thorax, coming off somewhat irregularly from the sides of the heart. The cardiac end of each is enlarged and filled with a sponge-like mass of cells. Indian ink taken in at the posterior end of the dorsal vessel is forced into these lateral vessels and discharged into the abdominal cavity.[145] In *Carausius* there are minute openings below the heart at the anterior end of the abdomen from which canaliculi come off and ramify among a mass of cells of unknown nature.[162] These cells are richly innervated; perhaps they constitute yet another endocrine organ (p. 176). [See p. 456.]

Extremities of the dorsal vessel—In most insects the heart is closed behind; but posterior ostia occur in *Aleurodes*,[228] *Corethra*, *Culex*, *Ctenophora*,

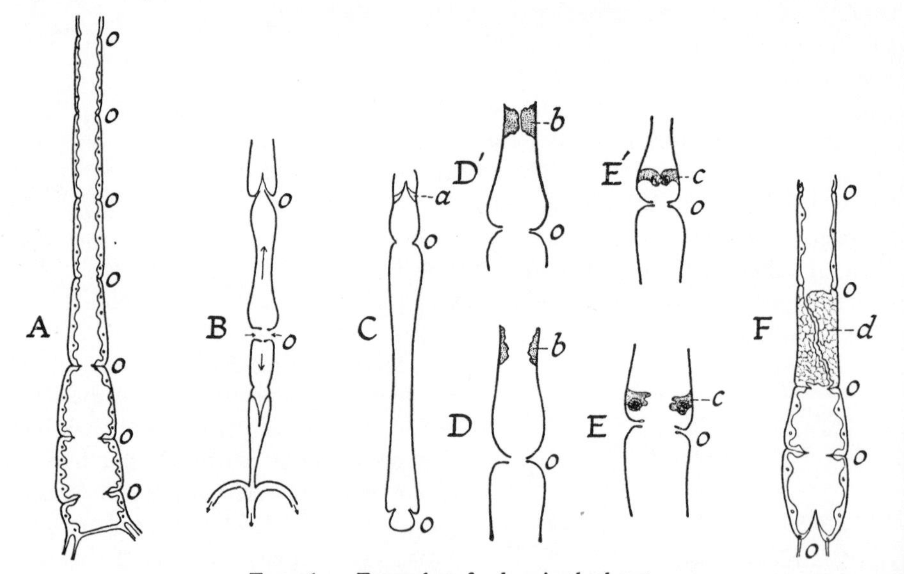

FIG. 265.—Examples of valves in the heart

A, larva of *Thrixion* (Tachinidae) a heart with no valves between chambers (*after* PANTEL). *B*, *Cloëon dipterum*; the ostial valves modified to provide interventricular valves; in the hindmost chamber they are directed backwards, the blood flowing into the three vessels to the caudal filaments. *C*, Chironomid larva with separate interventricular valves (*a*) between heart and aorta. *D*, Chironomid larva with muscular pads (*b*) which separate the chambers during systole (*D′*) (*after* POPOVICI-BAZNOSANU). *E*, larva of *Tipula* with separate valve flaps (*c*) in front of ostia. *E′*, the same in systole (*after* WETTINGER). *F*, larva of *Ceromasia* (Tachinidae) showing 'preventricular cushion' (*d*) and terminal posterior ostium *after* PANTEL). *o*, ostia.

&c., and a median posterior opening with well developed valves is present in the Tachinid *Ceromasia* (Fig. 265, F),[168] and in the larva of *Tipula maxima*.[23]

In Ephemerid nymphs the blood flows backwards in the hindmost chamber, through a pair of backwardly directed valves, into the vessels to the caudal filaments (Fig. 265, B).[181]

In front, the aorta ends by discharging into the body cavity, often by way of a sinus which accompanies the oesophagus beneath the brain; in the Reduviid *Rhodnius*, it discharges downwards just behind the brain into a sinus which runs forwards through the oesophageal ring and backwards around the corpus allatum (Fig. 266).[237] In the cockroach,[172] Phasmids,[198] &c., the aorta continues as an inverted gutter to fuse with a transverse band of muscle in the head and discharges ventrally in front of the brain (Fig. 269). In the honey-bee, some blood escapes from the aorta before it enters the oesophageal ring, and bathes the back of the brain.[65]

The pericardial septum—The direction of the blood stream outside the dorsal vessel is controlled in part by fibro-muscular septa. A dorsal diaphragm

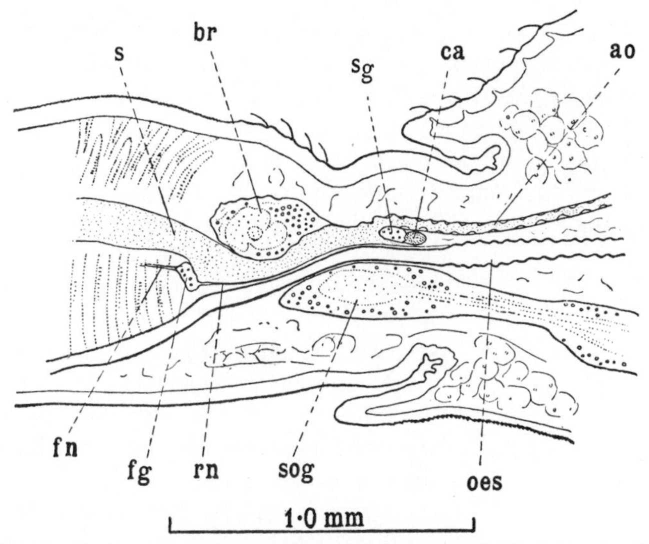

FIG. 266.—Longitudinal section of part of head and thorax of *Rhodnius* showing the ending of the aorta (*after* WIGGLESWORTH)

ao, aorta; *br*, brain; *ca*, corpus allatum; *fg*, frontal ganglion; *fn*, frontal nerve; *oes*, oesophagus; *s*, blood sinus into which aorta discharges; *sg*, corpus cardiacum; *sog*, suboesophageal ganglion.

or pericardial septum lies immediately below the heart and is attached laterally to the terga. It often contains muscle fibres; typically these form the alary muscles arranged in a fan-shaped manner in each segment, widely spread beneath the heart, converging towards the body wall. The ostia of the heart generally open into that part of the body cavity, the pericardial sinus, which is cut off by this septum; and in order to reach this space the blood must either pass through openings in the membrane, which is usually fenestrated, as in the cockroach,[22] or, if the membrane is imperforate, as in grasshoppers,[141] it must flow round the posterior border. In some Orthoptera, *Pachytylus*, *Locusta*, in addition to the ordinary lateral ostia of the heart, there are five pairs which open ventrally and collect blood directly from the perivisceral cavity.[115] In some insects (many Dipterous larvae, Hemiptera, Anoplura, &c.) the pericardial septum is attached directly to the heart; sometimes it is reduced to a few fibrous strands, as in *Corethra* larvae.[128] [See p. 456.]

Circulatory mechanism—The blood is aspirated through the ostia by the dilatation of the heart. In the honey-bee, and doubtless other insects, this aspirating movement is greatest in the wide posterior segment with muscle-thickened walls[65] and here it sets up transverse currents towards the ostia. Dilatation is brought about by the elasticity of the fibres radiating from the heart. The pericardial septum is convex upwards; hence when the alary muscles contract they will enlarge the pericardial sinus,[75] displace the blood towards

FIG. 267.—Course of blood circulation in the silkworm (*after* YOKOYAMA)
The direction of flow is marked by arrows the length of which indicates approximately the relative quantity of blood passing through the ostia.

the heart, and by drawing upon the ventral wall of the heart, aid its dilatation.[70, 167, 198] Or, if they merely maintain a constant tension in the septum, they will provide an elastic pull upon the fibres to the heart, and will have the same effect (Ephemeroptera,[181] *Dytiscus*,[22] *Apis*[65]).

In the silkworm larva the blood flows backwards through the body cavity, it enters the heart to a slight extent only by the lateral ostia of the abdominal segments i–vi, but almost wholly by the ostia of segments vii and viii (Fig. 267). It leaves the heart not only by the anterior extremity but by the ostia of the thoracic segments II and III and sometimes in the abdominal segments i and ii. There is no contrivance for preventing the blood from leaving the ostia.[254]

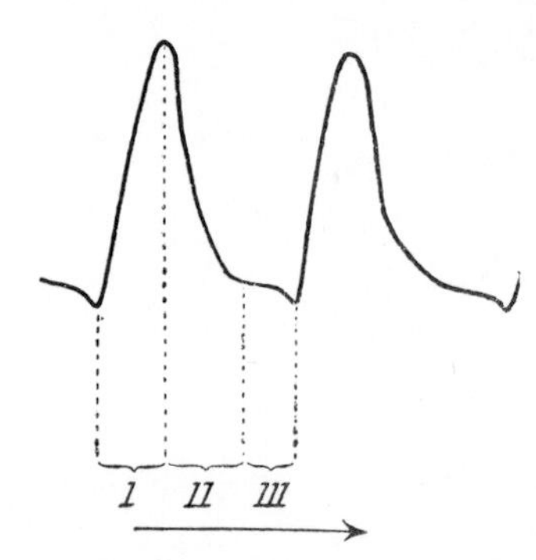

FIG. 268.—Diagram showing type of mechanical record obtainable from the isolated heart of *Periplaneta americana* (*after* YEAGER)
i, phase of systole; *ii*, phase of diastole; *iii*, phase of diastasis, with 'presystolic notch' before the ensuing systole.

When the heart is filled with blood, a steady wave of contraction passes from behind forwards; there is no alternation of diastole and systole in successive chambers. The blood is carried to the head and there discharged. This causes a relative increase in pressure at the anterior end of the insect; so that at the same time as blood passes forwards in the heart, it is moved backwards towards the abdomen in the general body cavity. The pressure gradients which maintain this circulation are produced by the muscular work of the heart; such gradients can exist whether the average pressure in the body cavity is greater or less than the atmospheric pressure. The circulation can be maintained in the narcotized insect, in which the general pressure is below that of the atmosphere (in *Dytiscus*[22] and in the honey-bee *Apis*[65]); but this is no reason for regarding the heart primarily as a suction pump.[150]

Three phases in the cycle of the heart's activity at a given point in its course are described: (i) contraction or systole, (ii) relaxation or diastole and (iii) rest in the relaxed condition or diastasis.[247] The diastolic relaxation of the heart is attributed to the elasticity of its walls. But at the end of the diastolic rest, it may show a sudden further dilatation before the ensuing systole (the 'presystolic notch'). Perhaps this is caused by a contraction of the alary muscles preceding the contraction of the heart wall,[119] more probably by distension with blood driven forwards from behind (Fig. 268).[247] Very similar records have been obtained by photoelectric methods using reflected light in the intact insect.[381]

The heart of *Periplaneta* can be induced to give a steady tetanic contraction in response to electrical or mechanical stimuli. It can thus be caused to contract again during systole; and after such a contraction it does not show the compensatory pause that always occurs in the vertebrate heart if it is induced to make an extra systole.[247] The wave of electrical potential which accompanies the contraction, the 'electrocardiogram', is normally of a very complex type; but by treatment with cold physiological saline it can be transformed into the type characteristic of vertebrates, a spike slightly preceding contraction and a slow wave.[40, 100] [See p. 456.]

In the heart muscle of the Saturniid moth *Telea* the sodium ion is not the major current carrier as it is generally supposed to be; nor is the resting potential determined primarily by the intra-extracellular concentrations of the potassium ion.[350]

The ventral diaphragm—In Orthoptera, Odonata, Ephemeroptera, Hymenoptera, and particularly in Lepidoptera, there is another well developed fibro-muscular septum stretched across the abdomen just above the nerve cord, thus enclosing below it a perineural sinus. This membrane is capable of undulatory movements, which direct the flow of blood beneath it backwards and laterally,[22, 65, 70] and ensure the irrigation of the nerve cord. In Libellulids this membrane is convex downwards; when it contracts the perineural sinus is enlarged;[75] in nymphs of *Cloëon* (Ephem.) it is convex upwards; its contraction will drive blood out of the ventral sinus.[181] Like the dorsal diaphragm, this membrane is often incomplete (as in the cockroach[22]) so that blood can pass through it into the perivisceral cavity; in many insects it is absent altogether.

In *Schistocerca* the activity of the ventral diaphragm is myogenic and propagated largely by tension; inhibitory fibres from the ganglia slow the rate of beat, and stretch receptors consisting of sense cells with dendritic endings in the muscle fibres (Fig. 168) are probably responsible for this inhibition.[322] Although absent from all insects which lack an abdominal nerve cord (Hemiptera, some Coleoptera, and higher Diptera) and although never extending beyond the posterior extremity of the abdominal nerve cord, there are groups which lack a ventral diaphragm but do possess an abdominal nerve cord (cockroaches, some Orthoptera, Isoptera).[362] In *Anopheles* it contracts at a rate of 3–15 per minute;[335] in *Melanoplus*, about 30 per minute, *Bombyx*, 80 per minute; *Arctias luna* up to 120 per minute.[362]

The effect of respiratory movements on the circulation—The changes in blood pressure brought about by the respiratory pumping movements (p. 370) can seldom be equal throughout the body; they must, therefore, influence the circulation of the blood. In *Aeschna* nymphs there are two vertical transverse diaphragms in the thorax, provided with sphincters through which the oesophagus passes. At each inspiratory movement of the abdomen the blood flows backwards; during expiration it is prevented from returning by contraction of these sphincters; so that respiration actively assists the circulation. In this insect, and in Agrionid nymphs, the pulses of blood in the femur coincide with the respiratory movements in the rectum.[22] In Ephemerid nymphs the pumping of water in and out of the rectum causes a backward and forward displacement of the blood in the abdomen, but does not bring about any transport. In the legs, on the other hand, these rhythmic changes in pressure do seem to assist the circulation.[150]

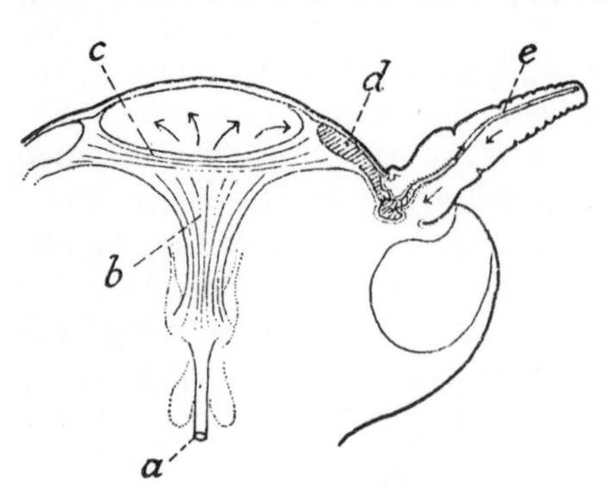

FIG. 269.—Circulation of blood in head of *Blatta* (*after* PAWLOWA)

a, aorta; *b*, longitudinal muscle in form of an inverted trough; *c*, transverse muscle; *d*, ampulla; *e*, blood vessel running from ampulla along antenna.

Circulation in appendages; accessory hearts—Special arrangements exist for irrigating the appendages. Rarely the blood is conveyed by closed vessels from the aorta: in May-fly nymphs, the blood entering the hindmost chamber of the heart divides into a forward and a backward current, and the hind end of the heart splits into three arteries which carry blood along the caudal filaments.[181] In *Vespa*, the aorta ends in the head in a transverse vessel which gives off branches to the eyes.[22] In the *Bombyx mori* adult the aorta ends in a frontal sac* connected by 'arteries' to the antennae, eyes, and maxillae;[70] the antennal vessel has a pump at its base.[368]

But as a rule the appendages are supplied with blood from the sinus cavities, and it is driven through them by independent pumps. In *Periplaneta*, *Locusta*, *Stenobothrus*, &c. there are two little ampullae beneath the clypeus, each giving off a vessel which runs to the tip of the antenna (Fig. 269). Each ampulla has a valvular opening communicating with the blood space in front of the brain into which the aorta discharges. A transverse muscle with striated fibres connects the inner walls of these two sacs and acts as a dilator; the contraction of the sacs, which drives the blood up the antenna, is due perhaps to their own elasticity, perhaps to intrinsic muscles in their walls. All along these antennal vessels are small perforations through which the blood escapes.[22, 172] Similar vessels and ampullae supply the antennae of the honey-bee[65] and doubtless many other insects. In Culicidae and most other families of Nematocera there is a myogenic pulsating organ which drives the blood through a blood vessel to the antenna; it is dilated by muscles and compressed by elastic contraction.[299]

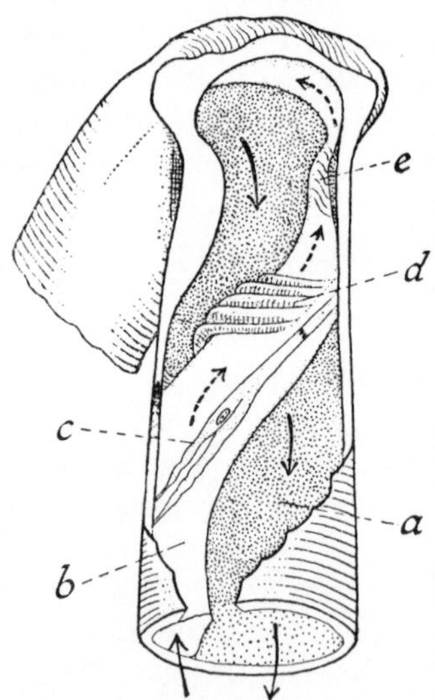

FIG. 270.—Pulsatile organ in fore leg of *Notonecta* (*after* DEBAISIEUX, simplified)

a, ventral blood sinus; *b*, dorsal blood sinus; *c*, scoloparium; *d*, compressor muscle; *e*, valvular membrane. Arrows indicate direction of blood flow.

Pulsating organs are particularly numerous in Hemiptera: they were discovered by Behn in 1834 in the legs of *Notonecta*, and have been seen in many other bugs;[47] they occur in the wings of *Nepa*;[80] there are four in each wing in

* On the wall of this sinus there is a pair of small oval clusters of deeply staining cells, perhaps another endocrine organ[70] (p. 198).

Musca, lying on the course of the efferent veins;[213] five have been observed in the basal part of the wing in *Drosophila*,[358] and no doubt many more remain to be discovered. Their mode of action is not well understood, but Fig. 270 represents

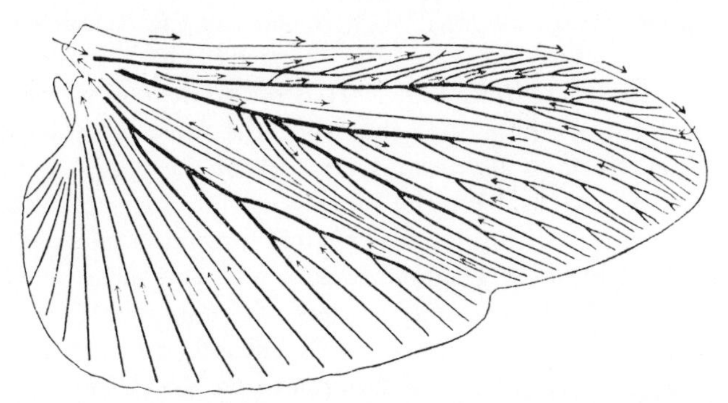

FIG. 271.—Hind wing of *Periplaneta americana* with the course of circulation of the blood indicated by arrows (*after* YEAGER and HENDRICKSON)

the probable arrangement in the limbs of *Notonecta*. A dorsal vessel running up the leg crosses the tibia obliquely just below the knee, and at this point a muscle curves round its upper border. This muscle contracts spasmodically, driving the blood along the vessel, past a valvular membrane, to the body cavity, and aspirating blood from this cavity into the ventral sinus of the limb.[47] In the legs and gills of Ephemerid nymphs the movements of the appendages themselves seem to aid the flow of blood.[150]

Circulation in the wings of insects was observed by Henry Baker in the grasshopper as early as 1744,[244] and is probably universal. The blood flows between the tracheae and the walls of the 'veins'; it enters the wing by the costa as a rule, and returns to the body by the posterior margin, following a fairly constant path along the larger channels in the wing, a very variable course in the minor passage-ways (Fig. 271).[36, 209] This circulation is necessary for the normal sclerotization of the wing and for maintaining the wings in a healthy condition; parts deprived of circulating haemolymph become dry and brittle and often crack away.[36, 209]

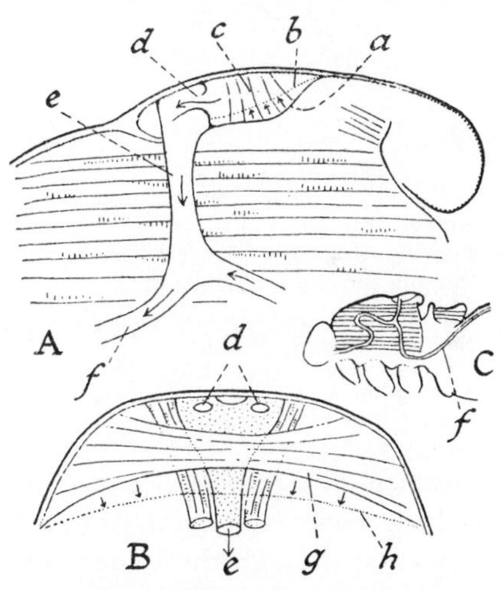

FIG. 272.—Mesothoracic pulsatile organ in *Sphinx convolvuli* (*after* BROCHER)

A, side view; *B*, posterior view; *C*, course of aorta through the thorax as seen in side view. *a*, pulsatile membrane contracted; *b*, dotted line showing position of relaxed membrane; *c*, elastic fibres connecting membrane to dorsal wall; *d*, valved orifice; *e*, connecting vessel running to aorta; *f*, aorta; *g*, pulsatile membrane relaxed; *h*, dotted line showing position of the membrane when contracted.

As we have seen, this circulation may be furthered by pumps in the wing itself, but it is effected chiefly by pulsating organs in the thorax. Such organs are present in most insects. They have been described, for example, in the meso- and metathorax of

Odonata nymphs,[22] *Dytiscus*[22] and other beetles,[22] in the mesothorax of Sphingidae,[22] *Vespa*,[22] *Apis*,[65] *Tabanus*,[22] *Musca*,[213] and in the cicada *Cryptotympana*.[255] In *Tipula* and *Scatophaga* it is claimed that the organs in the wings are not actively pulsatile but respond passively to pressure changes set up by the pulsatile organs in the thorax.[403] The organs in the thorax seem to consist of a muscular plate, which encloses a blood space beneath the dorsal wall of the thorax, often in the scutellum (Fig. 272). This space communicates with the veins of the posterior wing margin, and when the muscles relax, it dilates and blood is aspirated through the venous network of the wings; blood enters the wings, by the anterior veins, from the lateral intermuscular spaces of the thorax. When the organ contracts, the blood is expelled through a valved opening below; this may lead directly or by a tubular vessel into the aorta (in Odonata,[22, 231] *Dytiscus*,[22] *Coccinella*,[22] the mesotergal organ of Sphingidae[22]), or it may simply discharge into the body cavity (in *Musca*,[213] *Tabanus*, and the metatergal organ of Sphingidae[22]). In *Bombyx mori* the mesotergal organ discharges directly into the dorsal loop of the aorta, the two limbs of which are so close together that they are readily mistaken for a single vessel.[70] In Odonata some blood also enters the mesothoracic organ through ostia from the haemocoele.[231]

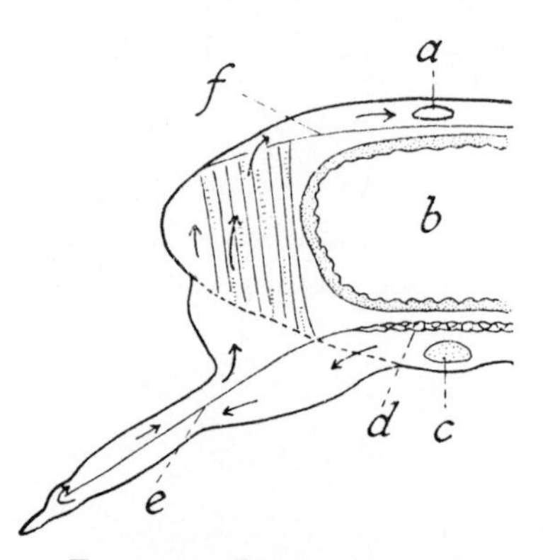

FIG. 273.—Diagrammatic cross-section through thorax of *Blatta*, showing the course of circulation through the leg (*after* BROCHER)

a, heart; **b**, gut; **c**, ventral nerve cord; **d**, ventral diaphragm roofing over the perineural sinus; **e**, septum dividing the leg; **f**, dorsal diaphragm forming floor of pericardial sinus.

From time to time the direction of flow in the wings may be reversed (in *Coccinella*,[22] in *Periplaneta*[244]) quite independently of the reversal in the dorsal vessel (p. 422). The mechanism of this change is not known. In May-fly nymphs the wings are irrigated by an evagination of the dorsal wall of the aorta which is itself closed and takes up no blood; but it is surrounded by a blood space communicating with the wing veins, and when it contracts blood is sucked from the wings and pressed forward towards the head.[150] In the pupal wing of *Ephestia* there is a to and fro movement along all the longitudinal veins, mixing the blood in the wings and the body cavity, alternating with periods of arrest.[258] In addition, there is a true circulation through the lacunae of the wing, the blood entering as a rule by the radial stem and returning to the body by the axillary stem; but occasionally the direction is reversed (Fig. 61, p. 86).[204]

The pulsating organs of the thorax may have far reaching effects upon the movements of the blood; in Sphingidae they are said to be more important than the heart itself.[22] In most insects the legs are divided by a membrane into two compartments, and in many forms in which pulsating organs have not been demonstrated in the legs, the blood flows down the posterior and up the anterior compartment. This movement is probably due to the fact that the posterior compartment is in connexion with the perineural sinus in which the pressure is relatively high, while the anterior compartment communicates with the lateral intermuscular spaces of the thorax, in which the aspiration of the pulsating tergal organs maintains a lower pressure (in Sphingidae,[22] and in the honey-bee[65]). In the cockroach perhaps the difference in pressure between the perineural sinus and the perivisceral sinus is sufficient to maintain this flow (Fig. 273).[22]

The pulse rate in the dorsal vessel varies from many causes. In the larvae of *Sphinx ligustri* it falls during the course of development from a rate of 82 per minute in the 1st instar, 89 in the 2nd, 63 in the 3rd, 45 in the 4th, to 39 per minute shortly before maturity. During the resting period before each moult it will fall to 30. In the pupa it falls from 20 to 10 per minute, and during the winter rest ceases almost completely.[158] Similar results have been obtained with *Bombyx mori.*[142] The rate of beat is increased as the temperature is raised.[41, 66, 205] It is increased when the insect becomes active: in the imago of *Sphinx* the rate is 40–50 per minute during rest, 110–140 during activity.[158] And it is increased if the metabolic rate is high from any other cause: the respiratory metabolism of *Baetis rhodani* (Ephem.) in rapid streams is three or four times that of *Cloëon dipterum* in ponds, and the ratio of their pulse rate is 3:1.[64]

The accessory pulsating organs, though their movements are quite independent of the heart,[47] respond in the same way. The metathoracic organ in *Bombyx mori* may beat at 150 per minute when the rate in the dorsal vessel is only 40 per minute.[70] The leg pumps in *Notonecta* are accelerated by high temperature;[42] the mesotergal organ in *Macroglossa* beats 10–12 per minute with frequent pauses during rest, 75 per minute when the insect is disturbed, and at a rate too fast to count when it is about to fly.[22] Circulation in the wing of *Blattella* is much more rapid during flight.[36]

The rate of propagation of the beat also varies very much. In the mature larva of *Lucanus cervus* at 18° C. the pulse rate is 12–20 per minute (average 14) and the rate of propagation 19·5–44·5 mm. per second (average 27·2).[126] In the larva of *Corethra* at 17–18° C., with a pulse rate of 15–16 per minute, the waves do not move at more than 1 mm. per second, so that three systolic waves may be passing along the dorsal vessel at the same time,[219] and simultaneous pulsations take place in the backward flow of blood in the body cavity.

Automatism of the heart—The property of rhythmical contraction certainly resides in the heart muscle: there are no nerve cells or ganglia associated with the heart in *Aeschna*,[257] and *Anax*[136] nymphs, or in the adult of *Belostoma* (Hem.);[136] and yet the isolated heart, or even fragments of the heart (*Dytiscus*), continue to beat rhythmically.[56, 141] And the propagation of the contraction wave seems also to be effected wholly by the muscles; waves are still conducted after section of the heart unless it be cut through entirely.[54]

It has been claimed that the elastic tension of the alary muscles, or even the pull of their contractions, is necessary to ensure rhythmical pulsations (in *Chironomus* larvae and Agrionid nymphs).[54] This stretching of the wall certainly modifies the contraction; for the heart of *Dytiscus*, which beats at 30–70 per minute when intact, is slowed to 15 per minute when all the alary muscles have been cut;[56] but it is not arrested;[75, 181] and in *Apis* the heart beat continues after section of the entire pericardial septum.[181] In nymphs of *Anax* it is said that the suspensory ligaments running to the dorsal cuticle must be intact if the heart muscle is to function.[136] In the larva of *Cossus* the contractions of the alary muscles alternate regularly with those of the heart. The heart muscle is certainly the seat of the automatism; the alary muscles are activated by its contraction and the tension which they exert modifies the heart beat. The same applies to *Hydrophilus*.[242]

Likewise in *Corethra* it seems clear that the heart beat is myogenic.[304] It may be that the type of automatism in the heart varies in different insects: *Anopheles* appears to have a non-innervated myogenic heart, *Anax* an innervated myogenic heart, *Periplaneta* a neurogenic heart.[335]

Reversal of the heart beat—All segments of the heart may show automatism.[54] But normally it is best developed at the hind end. This sets the pace; and waves pass forwards from the hindmost chamber. It is, however, not uncommon for the beat to be reversed. This was observed by Malpighi in the silkworm, in which it begins about 48 hours before pupation, and continues at intervals until the death of the adult.[69] Reversal of this kind has been described in the pupae and imagines of many Diptera, Coleoptera, Lepidoptera and Hymenoptera.[69] It is an interesting phenomenon, firstly because it emphasizes the unity of the insect heart, the point already stressed that the heart acts as a single tube and not as a series of chambers,[69] and secondly because of the light it throws on the cause of the normal dominance of the hind end. In normal adults of *Corethra* the beat consists of a forward pulsation followed, after a pause of some seconds, by a backward pulsation.[340] There seem to be pacemakers at the two ends of the heart.[383]

It has been suggested that the chief cause of reversal may be mechanical: the blockage of the ostia by the disintegrating fat body, in *Galerucella*;[183] or the obstruction of flow by the histolyzing tissues in the body cavity increasing the tension in the head and starting an antiperistalsis, in *Saturnia*.[216] But the heart, even an isolated segment of the heart, upon the excised dorsum of the pupa continues to exhibit a periodic reversal.[69] It seems more likely that there is a real disturbance of automatism.

In the silkworm embryo the heart beat begins 2 days before the entrance of air in the tracheae; but it is slow and shows no definite direction. When air appears, the beat becomes regular, and its forward direction becomes constant.[251] In the larva of the silkworm, in which reversal does not normally occur, it can be evoked by warming the anterior region of the heart, or by occluding the last two pairs of spiracles.[251] Thus it may be that the normal dominance of the posterior extremity in the larva is due to its rich tracheal supply; in the pupa the respiratory supply to the vessel is very feeble behind; in the imago it is dense in both regions.[251] The posterior extremity of the heart in Dipterous larvae is invested by a rich basketwork of tracheae (p. 388); perhaps the function of this is to ensure that this region shall be the pace-maker of the heart.

Certainly the heart is very sensitive to asphyxiation. If the last seven pairs of spiracles in the silkworm larva are occluded, the heart generally stops.[251] Nymphs of *Cloëon* show immediate stoppage of the heart in water saturated with carbon dioxide;[62] and the pulsations in the embryo of *Melanoplus* are slowed 50 per cent. by 4 per cent. of carbon dioxide.[225]

Nervous and humoral control of the heart—The heart receives a double innervation; (i) from the paired cardiac ganglia of the stomatogastric system, (ii) from the segmental ganglia of the ventral chain;[121, 257] and in the cockroach a third supply of supposedly sensory fibres comes off the sensory branches to the dorsal body wall and reaches the heart from above.[3] All these fibres unite to form a pair of lateral nerves, which run along the sides of the heart and give off a plexus of branches ending in its muscular wall and in the alary muscles (Fig. 274). Nerve cells and ganglia are usually completely absent from the heart itself;[121, 136, 257] but in the cockroach nerve cells scattered all along the lateral nerves have been described,[3] and in *Carausius* there are two sorts of nerve cells surrounded by a characteristic basketwork of fine nerve fibres.[162]

In discussing the automatism of the heart we saw that this is probably myogenic; but the rate and amplitude of beat are under nervous control. Electrical stimulation of the brain in the grasshopper,[141] and in the larva of *Lucanus*,[126] usually arrests the heart. Faradic stimulation of the nerve cord in the neck of the cockroach after decapitation, causes immediate acceleration; the impulses reaching the heart through both the ventral nerve cord and the lateral nerves.[205] And, as we have seen, any disturbance of the intact insect will increase the rate of beat. Acetylcholine causes acceleration and irregularity in the heart beat in *Melanoplus*, an effect antagonized by atropine; nicotine causes a great increase in amplitude.[79] Nothing is known of the control of the accessory hearts.

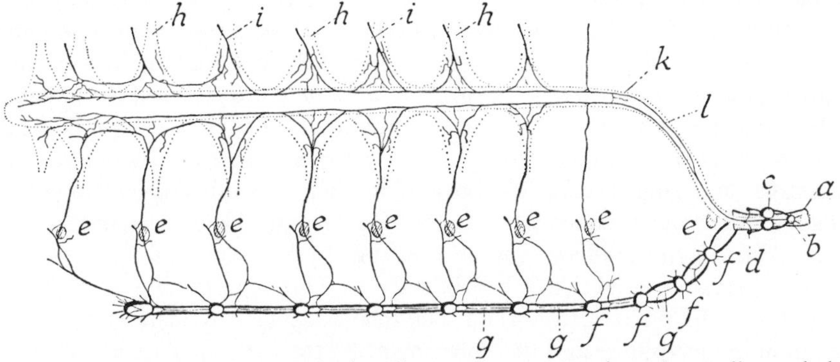

FIG. 274.—Innervation of the heart in the silkworm larva as seen in an insect dissected along the spiracular line of the right side (*after* KUWANA)

a, gut; *b*, frontal ganglion; *c*, supraoesophageal ganglion; *d*, pharyngeal ganglion of sympathetic system; *e*, spiracles; *f*, ganglia of ventral nerve cord; *g*, unpaired ventral nerve; *h*, alary muscles of heart; *i*, nerve from ventral nerve cord; *k*, dorsal vessel; *l*, lateral nerve of the heart.

From the action of drugs on the heart of *Periplaneta* it has been concluded that the rate of beat is determined by a neurogenic pace-maker with cholinergic properties; that is, it is stimulated by acetylcholine and by substances which inhibit the enzymatic breakdown of acetylcholine. The neurones from the pace-maker appear to have adrenergic properties, their terminations being stimulated by adrenaline.[264] [See p. 456.]

An extremely dilute aqueous extract of the corpus cardiacum causes an immediate acceleration of the heart in *Periplaneta* with an increase in amplitude.[294] This active factor appears to be a peptide. It does not act directly upon the heart, but causes the pericardial cells to secrete a cardiac stimulator. This is believed to be an indolalkyl-amine related to serotonin. The material seems to be produced as required by the pericardial cells, perhaps by the action of a decarboxylase on an amino acid—as serotonin is produced by the action of a specific decarboxylase on 5-hydroxytryptophane. If the cockroach is fed with sugar there is an immediate acceleration of the heart; this results from stimulation of the taste receptors on the inner surface of the labrum leading to liberation of the active factor from the corpus cardiacum.[303] In *Anopheles* larvae, also, the rate of heart beat is dependent on feeding rather than activity.[335]

This appears to be a normal physiological process; but there are a great number of substances extractable for the brain, &c., which have an effect on the heart at high concentrations, the significance of which in normal physiology remains to be proved. These include acetylcholine itself (p. 200), the neurohormones C & D (p. 200), serotonin,[301] adrenalin-like substances,[287] and a series of perhaps six factors extracted with ethanol from the corpora cardiaca and allata of *Periplaneta*.[360]

THE HAEMOLYMPH

The blood or haemolymph is a clear fluid, sometimes colourless, often tinged with green or yellow pigment. In the cockroach it forms only some 5 or 6 per cent. of the total body weight;[246] but in soft bodied larvae it comprises a quarter or a third: 25–30 per cent. in the larva of the honey-bee;[15] 20–40 per cent. in various caterpillars.[187, 343] It is the medium through which all the chemical exchanges between the organs are effected, hormones conveyed, food carried from the gut, and waste products to the excretory organs. It plays a certain part in respiration, which has already been discussed (p. 388), and a most important part in transmitting pressure from one region of the body to another—as in the ventilation of the tracheal system (p. 370), hatching, ecdysis, and the expansion of the wings (pp. 19, 45). An important function of the haemolymph is to serve as a reserve of water which can be drawn upon as needed without the tissues themselves suffering desiccation. So long as there is sufficient haemolymph to circulate, metabolism is not affected by large changes in water content.[147]

In *Schistocerca* the blood volume rises during the latter half of an instar and attains its maximum level just before ecdysis. This high volume is maintained for about 24 hours after ecdysis and then falls sharply.[342] We have already noted the fall in blood volume and its replacement by the expanded air sacs that takes place in Diptera after emergence; this probably aids the transport of nutrients to the flight muscles (p. 372).[395]

Composition of the blood:* protein—In the larva of the honey-bee the total protein in the blood, 6·6 per cent., approximates to that in human plasma; and its two chief components, 'albumen' (3·46 per cent.) and 'globulin' (3·10 per cent.) are in about the same proportions.[15] But in most insects the quantity of protein is less than this: 5 per cent. in *Bombus*, 3–4 per cent. in *Hydrophilus* and other Coleoptera, 2·6–3·3 per cent. in *Aeschna* nymphs, 1·3–2·6 per cent. in *Limnephilus* larvae (Trichopt.), 1·0 per cent. in *Carausius*.[60] In the silkworm it rises from just over 1 per cent. in the 4th instar, to nearly 6 per cent. at the time of spinning the cocoon; it then falls to 2·6 per cent. in the ten-day-old pupa, and below 2 per cent. in the adult at emergence.[60, 400] By 'salting out' by means of increasing concentrations of phosphate, it has been possible to separate two albumen fractions and one globulin fraction in the haemolymph of *Hydrophilus*.[61] The distribution of haemoglobin has already been discussed (p. 389). As seen in *Bombyx mori*, above, there is a several-fold increase in protein concentration during larval life, becoming rapid in the last instar. There is little change at spinning and in the early pupa, but a sharp fall during development of the adult. In silk production the haemolymph proteins are little used; amino acids are the efficient precursors of the silk.[398] [See p. 456.]

By the use of electrophoresis and antigen-antibody precipitation, a great num-

* For reviews of the composition of insect blood see references.[139, 215, 398]

ber of protein fractions can be demonstrated in insect blood. The relative intensity of the different protein bands changes strikingly as growth proceeds, as during the last instar of *Bombyx*.[361] At least 7 blood proteins can be separated in the *Drosophila* larva;[297] at least 4 in *Culex* larva;[297] a total of 19 in *Drosophila* in the course of development from egg to adult.[309] There are well-marked quantitative and perhaps minor qualitative differences in the blood proteins of the two sexes in various insects.[375] The cell-free haemolymph of *Hyalophora cecropia* larvae contains at least 9 antigens, each with its own characteristic pattern of concentration change during development. One fraction is characteristic of the adult female; it first appears in the prepupal stage, and is present in the male at only one thousandth the concentration in the female.[382] [See p. 456.]

Many of the protein fractions are conjugated proteins: five or six fractions separated in *Periplaneta* contain neutral lipid, sterols, phospholipid or carbohydrate. Here again there is a general increase during the premoult stage and a gradual reversion after moulting.[373] Almost half the carbohydrate of the haemolymph in *Periplaneta* is bound to protein: this includes glucosamine, galactosamine, mannose, galactose, and smaller amounts of glucose.[398]

Enzymes—Many enzymes (protease, amylase, sucrase, lipase) have been reported in the haemolymph of insects.[215] They become particularly active at metamorphosis when they are liberated by the disintegrating tissues.[139]

After the blood is shed it rapidly darkens, absorbing much oxygen in the process. This results from the interaction of tyrosine and the enzyme tyrosinase. In the blowfly *Sarcophaga* these appear simultaneously in the blood of the full-grown larva, but they do not interact until oxidation of the tyrosine is necessary for the hardening and darkening of the cuticle (p. 34). Virtually all the proteins isolated by electrophoresis in the haemolymph of the growing larvae of *Hyalophora cecropia*, and *Samia cynthia* show enzymic activity: esterases of several kinds, phosphatases, carbohydrases, sulphatases, tryosinase, chymotrypsin, malic dehydrogenase; these become strikingly reduced in the diapausing pupa.[341] In the silkworm chitinase and chitobiase (p. 504) appear in the haemolymph, reaching a maximum 2 to 3 days after the pupal moult.[332] [See p. 456.]

Amino acids—The ratio of non-protein nitrogen to total nitrogen is about 1 : 2, in the larva of the honey-bee[15] and in the mature larva of *Celerio* (Lep.);[86] and of the non-protein fraction some 50–85 per cent. is in the form of amino acids. The amino acid nitrogen is 1·34 gm. per litre in the *Dytiscus* adult, 3·27 in the pupa of *Attacus*, 3·22 in the pupa of *Sphinx*, 2·34 in the larva of *Cossus*,[55] 1·7 in the larva of *Celerio*[86]—values 50 or 100 times greater than the normal for mammalian plasma. Vertebrates have about 0·05 g. per cent. free amino acids in the blood; in insects the amount ranges from 0·29–2·43 g. per cent.[398] The blood of the young larva of *Attacus*, that is, until it is near the third moult, contains only traces of amino acids. After the fourth moult it becomes laden with proteins, peptones, amino acids, and reducing substances. On emergence of the adult the amino acids show an âbrupt fall.[52] Cauterization of the spinneret in the silkworm, leading to lysis of the silk glands, causes a great increase in the amino acids of the haemolymph, notably glycine, alanine, serine, and tyrosine.[277]

In the blood of *Melolontha* and *Oryctes* larvae there is some 300–400 mg. of non-protein amino nitrogen per cent., consisting of free amino-acids, di- and tri-peptides, and the amides glutamine and asparagine.[220] Somewhat lower values are recorded in the young adult *Bombyx mori.*[61] There are considerable differences in the relative concentration of individual amino acids in different species of insects.[61, 215, 306] In nymphs of *Aeschna*, glycine, alanine, valine, and leucine are in greatest concentration, proline and tyrosine in moderate amounts, serine and lysine in small quantities only; about two fifths of the total N is non-protein-N in this insect.[186] Analysis of insect blood by means of paper partition chromatography has revealed the presence of certain amino acids (e.g. β-alanine, α-amino-*n*-butyric acid) that are not known as constituents of protein. In *Musca*, &c., taurine is relatively abundant.[269]

There are many striking differences in the pattern of amino acid concentration in different insects and in the same insect at different stages of its life cycle.[314] The larva of *B. mori* has a high concentration of the basic amino acids, lysine, and histidine;[316, 400] and in *Rhodnius* histidine accounts for about 35 per cent. of the free amino acids in the haemolymph.[326] In the silkworm the tyrosine content rises during each intermoult period and falls immediately after moulting; it is also taken up by the silk gland in the 5th-stage larva.[308] But there seems to be no very exact regulation of the amino acid patterns: the haemolymph of the Aphid *Megoura* contains the same amino acids as honey dew and in similar concentrations;[310] and if *Rhodnius* is fed on horse serum with added alanine, alanine appears in the haemolymph within a few minutes, and in the urine in half an hour, but the high level in the haemolymph persists for weeks.[327] [See p. 456.]

Lipids are present in the blood in the form of 'lipomicrons', or minute fat particles. In the cockroach these are most plentiful 14–19 hours after a meal, especially if this contains much fat.[78] In the larva of the honey-bee at the time of metamorphosis, when the fat body cells rupture and liberate their inclusions in the blood, the fat content rises enormously.[15, 154] But a large part of the lipid (neutral fat, sterols, phospholipid, &c.) is conjugated with protein (lipoprotein);[373] and in the pupa of *Hyalophora* lipid from the fat body seems to be transported in the haemolymph largely as diglyceride linked to protein.[298] [See p. 457.]

Carbohydrates—A substantial amount of the carbohydrate in insect blood is conjugated with protein (glycoprotein); in *Periplaneta* this represents about half the total carbohydrate of the haemolymph (p. 425). In *Schistocerca* a small amount of polysaccharide other than glycogen is present.[331] The blood of many insects contains little reducing sugar, but on acid hydrolysis this shows a striking rise.[122] The major blood sugar in insects is in fact α-trehalose, a non-reducing glucose-glucose disaccharide, formerly regarded as characteristic of certain lower plants.[399] In Lepidoptera trehalose makes up over 90 per cent. of the blood sugar (0·2–1·5 g. per cent.); in sawfly larvae about 80 per cent. *Galleria* fed on honey had 1·5 per cent. trehalose in the blood and only 0·02 per cent. glucose.[399] In the larva of the bee *Anthophora* trehalose reaches a concentration of 6·55 g. per cent.[307] and the blood of *Megoura*, in which the total sugar amounts to 5–8 g. per cent., trehalose makes up two thirds, and glucose one third.[310] The enzyme trehalase is present in the haemolymph and fat body of *Schistocerca.*[331] In the blood of *B. mori* trehalose is kept at a more or less constant level (0·25–0·63 per cent.)[307] by breakdown and synthesis of fat body glycogen.[364]

The very high content of trehalose is not universal in insects: in the adult honey-bee free reducing sugars amount to 1–4 per cent. in the haemolymph (p. 595),[398] in the Tachinid *Agria affinis* the total blood sugar amounts to 0·1–0·17 per cent. of which 80 per cent. is glucose, 5–6 per cent. fructose, 1–2 per cent.

trehalose; pentose (arabinose and ribose) 1–2 per cent., mannose and unidentified 3–4 per cent.[285]

The haemolymph may also contain large amounts of glycerol. At the pupal moult of *Hyalophora* the glycerol content is about 0·015M but it accumulates during diapause to reach 0·1–0·3M. It disappears again during adult development. In certain overwintering insects it can become the most abundant solute in the plasma (p. 681). It is perhaps a produce of glycolysis under the conditions of cytochrome deficiency in diapause.[401]

The non-fermentable reducing substances, which are plentiful in insect blood, are largely amino acids, particularly tyrosine.[331] They show a great increase before ecdysis.[49] In the body fluid of *Philosamia* the chief reducing substance is said to be ascorbic acid.[325]

Organic acids, esters etc.—Insects carry dissolved in their plasma substantial amounts of metabolites which other animals retain in their cells. Free amino acids are in higher concentration in the blood plasma than in the tissues (p. 425). Succinate is present in the blood of *Gasterophilus* larvae at a concentration of 230 mg. per cent.[131] In *Bombyx* larva malate accounts for 43 per cent. of the total anions in the blood.[312] Fumarate, citrate, lactate, pyruvate, α-ketoglutarate, &c., can be readily demonstrated.[398] Organic phosphates include α-glycerophosphate, phosphoethanolamine, phosphocholine in *Hyalophora*,[402] sorbitol-6-phosphate, glucose-6-phosphate, uridine diphosphate n-acetylglucosamine in *Bombyx*.[402]

Uric acid and other nitrogenous wastes—Ammonia is probably absent from the circulating blood, but in *Hydrophilus* it is rapidly formed (by some unknown mechanism) after the blood is shed.[61] Uric acid ranges from 5·3 mg. per cent. in the larva of *Apis*,[15] 10·7–14·5 in *Hydrophilus*,[60] to 20 in the larva of *Celerio*.[86] In the silkworm it is about 10 mg. per cent. while the larva feeds, less than 5 mg. at other times;[123] but it may go up to more than 16 mg. per cent. in the mature larva when accumulated waste products are being discharged into the blood.[102] But other nitrogenous waste products may play a larger part than uric acid. In *Popillia japonica* uric acid nitrogen decreases during the first week of starvation from 3·4 to 2·0 mg. per cent., and then remains constant; allantoin nitrogen remains constant at values between 33–39 mg. per cent.; and urea nitrogen increases from 27·3 to 74·1 mg. per cent.[348]

Pigments—Haemoglobin imparts a bright red colour to some Chironomid larvae (p. 389), and the haemolymph of *Rhodnius* contains traces of kathaemoglobin derived from the blood meal;[241] apart from this the blood is usually colourless, green or pale amber. In many phytophagous insects the blood is green, due to a mixture of blue and yellow chromoproteins, the blue containing biliverdin, the yellow derived from carotene or xanthophyll (p. 614). Fluorescence studies on the blood of Orthoptera have revealed the presence of isoxanthopterine (p. 612) and riboflavine (p. 612). The two substances may be connected since xanthinoxidase which is concerned in pterine metabolism is a compound of flavoprotein nature.[265]

In the larvae, pupae, and imagines of Lepidoptera the green colour is more common in the female; the blood of the male is often colourless or pale yellow.[72] The significance of this difference is unknown; nor whether it is due to sexual differences in pigment absorption or in metabolism;[203] chemical differences can certainly be demonstrated in the blood proteins of male and female *Pieris*.[203]

Salts—The ionic composition of insect haemolymph shows some striking contrasts with the blood of mammals, and within the insects themselves there is remarkable diversity. There seem to be two extreme patterns of ionic balance (Table 1). (*a*) A primitive pattern, well seen in Odonata (*Libellula* larva, Table 1) with high sodium, low potassium and magnesium. (*b*) A specialized condition, characteristic of Lepidoptera (*Bombyx* larva, Table 1), Chrysomelidae (*Leptinotarsa* larva, Table 1) *Carausius*, *Pteronidea*, with low sodium, high potassium and magnesium. Most insects fall into one or other of these groups.[305]

In most exopterygote insects sodium and chloride account for the greater

TABLE 1

Ionic concentrations in haemolymph (milliequivalents per litre)

	Na	K	Ca	Mg
Libellula larva	178·3	3·8	18·4	12·0
Leptinotarsa larva . .	2·0	54·9	43·4	146·9
Bombyx mori larva . .	14·6	46·1	24·5	101·0
Rat	90·0	4·0	4·1	1·8

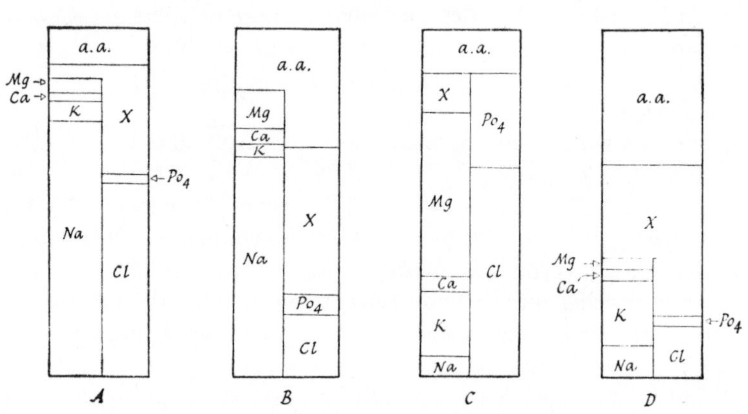

FIG. 275.—Charts summarizing the ionic composition of the haemolymph in different groups of insects (*after* SUTCLIFFE)

A, exopterygotes (Ephemeroptera, Odonata, Plecoptera, Dictyoptera, Heteroptera). *B*, endopterygotes (Megaloptera, Neuroptera, Mecoptera, Trichoptera, Diptera). *C*, Phasmida. *D*, Lepidoptera, Hymenoptera. **a.a.**, amino acids; **x**, unknown; to the left the chief cations, sodium, potassium, calcium, magnesium; to the right the chief anions, chloride and phosphate.

part of the total osmolar concentration; potassium, calcium, magnesium, inorganic phosphates, and free amino acids account for minor proportions only (Fig. 275, A). In endopterygote insects chloride is reduced and organic acids, including free amino acids, are increased (Fig. 275, B).[380] Apart from these generalizations there are various special types: for example, in Phasmida the sodium level is very low, potassium and magnesium are very high (Fig. 275, C) and in Lepidoptera and Hymenoptera inorganic ions contribute only a minor proportion of the total osmolar concentration; free amino acids and other organic components contribute the greater part (Fig. 275, D).[380] The concentration of inorganic phosphates varies from 5 to 20 times that in mammals.

At one time it was considered that a high potassium level was characteristic of phytophagous insects, a high sodium of carnivorous species. There is much variation from one species to another; the following are typical examples: Na/K in *Dytiscus*=13·3, *Cimex*=15·5; *Melolontha* adult=0·42, *Bombyx mori* larva=0·40 (man=29·2).[20] Although this association with the nature of the food seems to be fairly constant there are many phytophagous species with a high Na/K ratio. In *Tenebrio* the ratio remains the same whether reared on cabbage and flour or on meat powder.[20] In *Periplaneta* the Na/K ratio is 6·2; it is depressed by feeding on leaves but not nearly down to unity.[217] In *Bombyx mori* the sodium is almost all eliminated before pupation; the pupal blood contains only vanishingly small amounts.[217] *Periplaneta* shows great tolerance to wide changes in the ionic composition of the haemolymph.[282] The heart of *Chortophaga* can tolerate great variations, but for normal beating the K/Ca ratio must be between 1 : 1 and 3 : 1, and the Na/K ratio must be at least 3 : 1 and can be increased to 34 : 1.[286] The nervous system is very resistant to variations in the sodium potassium ratio.[189] [See p. 457.]

Calcium is in rather higher concentration in the blood of insects than in man; 0·014 gm. per cent. in the bee larva,[15] 0·033 gm. per cent. in the pupa of *Sphinx pinastri*,[21] 0·028 in *Saturnia pavonia*.[51] And magnesium is far higher; 0·19–0·22 gm. per cent. in the bee larva[15] (about eight times the value in human blood), 0·056 gm. per cent. in the pupa of *Sphinx*,[21] 0·044 in *Celerio*.[86] There is evidence that some of the metals in insect haemolymph are present in part, not as free ions but as organic complexes. In the haemolymph of *Telea polyphemus* 15–20 per cent. of the Ca and Mg are bound to macromolecules, whereas Na and K are all in the ionized state.[296] The amino acids in insect blood make possible the holding of calcium and magnesium in solution without their precipitation as phosphates.[398]

There is an excess of fixed base in the blood. In the pupa of *Sphinx pinastri* there are 0·096 basic equivalents, against 0·046 acid equivalents; giving an excess of 0·050 gm. equivalents per litre. This excess is combined to a small extent as bicarbonate, but chiefly with proteins and organic acids.[21] Correspondingly the chloride content is low. In the imago of *Dytiscus*, chlorides, expressed as sodium chloride, have a concentration of 0·37 per cent., in the larva of *Cossus* 0·04 per cent.,[182] in the larva of *Pieris brassicae* 0·16 per cent.,[21] in the bee larva 0·19 per cent.,[15] in the larva of *Culex* 0·28 per cent.[239]—as opposed to about 0·6 per cent. in mammalian plasma. *Tenebrio* larvae and adults, with their high osmotic pressure in the blood, equivalent to 1·85–2·3 per cent. NaCl, have an exceptionally high chloride content equivalent to 0·85 per cent. NaCl in the larva, and 0·77–1·02 per cent. in the adult.[352]

Hydrogen ion concentration—The blood is usually very slightly acid. The pH in the bee larva, at the carbon dioxide tension of 35–60 mm. of mercury normal in the blood, is 6·77–6·93 (average 6·83);[15] in *Melanoplus*, *Chortophaga*, and other grasshoppers, 6·4–7·0 (usually about 6·7);[17] in the silkworm larva, 6·7–6·8 and in the imago 6·6–6·7;[48] 6·2–7·2 among beetles.[60, 108] These values are typical of those obtained in other insects,[6, 139, 215] save that in larvae of Diptera the values are slightly alkaline, *Drosophila* 7·1, *Sciara* 7·2, *Chironomus* 7·48,[16] and the blood may become very slightly alkaline in the last instar larva of *Pieris* (7·17) and *Heliothis* (7·23).[38]

Most metabolic products are acid; it is not surprising therefore that the buffer capacity of the blood is greater on the acid side of neutrality. The buffering power is probably due to bicarbonates, inorganic phosphates, proteins, amino acids, and to a small extent, urates.[6] In insect blood the buffer capacity is always lowest in the region of the normal *p*H of the blood, so that when buffer capacity is plotted against *p*H a **U**-shaped curve is obtained. This is exactly the reverse to that obtained in vertebrates. In *Gasterophilus* larvae bicarbonate is much the most important dialysable buffer; proteins account for about 62 per cent. of the total.[267]

Osmotic pressure—The total molecular concentration in insect blood is rather high. The values have usually been expressed in terms of depression of freezing point (Δ). If they are converted into concentrations of sodium chloride of the same osmotic pressure, we get instead of 0·9 per cent., the usual value in mammals, in the bee larva 1·5 per cent.,[15] in the adult *Dytiscus* 1·25 per cent. and *Hydrophilus* 1·05,[10] in various species of water beetles 0·8–1·9,[7] in the larva of *Saturnia carpini* 0·9, in the pupa of *Sphinx ligustri* 1·15,[182] in the various Ephemerid nymphs 0·69–1·0[63] and in the larvae of mosquitos 0·75–0·89,[239] *Pieris* larva 1·38, *Tenebrio* larva 2·12.[170a] Expressed in atmospheres, the osmotic pressure ranges from a lower limit of 6·8 found in *Carausius* to a maximum of 15. The average value is about 10·5 atmospheres. It tends to be higher in insects from dry environments (14·1 in the larva of *Tenebrio* for example) but this is not a constant rule.[193]

In the larva of *Drosophila melanogaster* the total concentration of the haemolymph averages 378 mM/l (equivalent to a 1·1 per cent. solution of NaCl). Of this total, chloride makes up 38 mM/l (10 per cent.); glucose 2·5 mM/l (0·7 per cent.); amino acids 112·5 mM/l (30 per cent.); 59 per cent. of the total concentration is therefore made up of other substances.[405] There is a similar deficiency in amino acids in *Libellula*.[379] Of the other organic acids concerned, citric acid may be important, for it is in far higher concentration than in other animals.[344] In *Petrobius maritimus* the osmotic pressure reaches a level of 232 mM/l NaCl; this is mainly accounted for by sodium chloride, which is unusual among insects. The Na/K ratio is very high.[347]

The osmotic pressure of the colloids in the blood, which in human plasma has a value of 30–40 cm. of water, has the low value of 5 cm. in *Mantis*, the only insect studied.[151]

Osmotic and ionic regulation in insects[372]—If the insect struggles violently, in the absence of oxygen, the osmotic pressure of the blood is increased by metabolites set free from the muscles; and as we have seen, this change may have important effects upon respiration in the tracheoles (p. 394). In the mosquito larva *Culex* the osmotic pressure of the blood may rise during asphyxiation from 0·85 per cent. 'sodium chloride' to 1·1 per cent.[239] On readmission of air the normal level is soon restored.

Mosquito larvae, *Culex pipiens* and *Aëdes aegypti*, which breed normally in fresh water, when they are reared in sea water hypertonic to the normal blood, show an increase in osmotic pressure so that the blood becomes isotonic with the external medium (Fig. 276, A). Larvae with their osmotic pressure raised in this way to the equivalent of 1·5 per cent. of sodium chloride may survive. Conversely, if the larvae are starved, the osmotic pressure falls from the normal 0·85 to about 0·7. If they are starved in chloride-free distilled water, the chloride in the blood falls from the normal level of 0·3 per cent. (expressed as NaCl) to 0·05 per cent.; yet the total osmotic pressure is scarcely any lower (0·65–0·7) (Fig. 277). Clearly the loss of chloride has been compensated by the liberation of some other substance into the blood, perhaps amino-acids.[239] Water beetles, *Dytiscus*, immersed for 24 hours in 3 per cent. sodium chloride, show an increase in osmotic pressure in the blood from 1·2 per cent. NaCl to 2·0 per cent.[7]

The aquatic larva of *Sialis* has no mechanism for the active uptake of ions (p. 515) but it loses chloride extremely slowly into distilled water. As in mosquito larvae (p. 432),[239] if the chloride in the blood is reduced from the normal value of 0·3 per cent. NaCl to 0·1 per cent., by starvation or by repeated bleeding, the non-protein nitrogen is increased so as to maintain a total osmotic pressure equivalent to 0·80–1·0 per cent. NaCl. This compensatory increase appears to be made at the expense of the blood protein which can be reduced from the normal 6–7 per cent. almost to zero without ill effect.[260] Similar compensatory changes occur in Odonata larvae.[273] [See p. 457.]

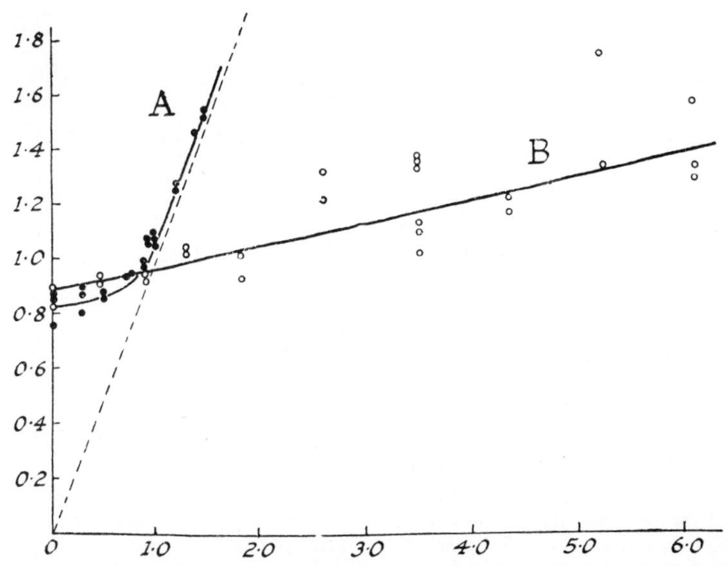

FIG. 276.—Relation between osmotic pressure of the medium and of the haemolymph in mosquito larvae

Ordinate: values in haemolymph. Abscissa: values in medium. Both expressed as equivalent concentrations of NaCl in gm. per 100 c.c. The broken line shows where the points would fall if blood and medium had the same composition. *A*, dots, larvae of *Aëdes aegypti*, which breeds normally in rain water (*after* WIGGLESWORTH); *B*, circles, larvae of *Aëdes detritus*, which breeds normally in brackish water (*after* BEADLE).

The nature of this compensatory change doubtless varies in different insects. In *Drosophila* larvae placed in 2·9 per cent. NaCl the rise in blood chloride is compensated by an equivalent fall in amino acids.[405] In *Tenebrio* adults the osmotic pressure runs more or less parallel with the sugar content; but sugar and amino acid levels may vary so as to compensate for one another.[352] In *Bombyx* larvae the different amino acids vary widely in amount, but the osmotic pressure remains fairly constant, largely by regulation in the histidine and methionine levels.[333] In the haemolymph of *Libellula* there are not enough amino acids in the haemolymph to make good the deficiency in inorganic ions for maintaining the osmotic pressure, either in the normal insect or in the insect which has lost chloride; non-amino organic acids must be responsible.[379]

Insects adapted to life in salt water doubtless have greater powers of regulating the osmotic pressure of the blood. Thus the aquatic bug *Sigara lugubris* is a brackish water insect occurring in pools with a salt content of 0·5–1·8 per cent. Throughout this normal range the osmotic pressure of the blood remains constant at −0·7° C. (=1·2 per cent. NaCl); the osmotic pressure does not increase until that range is exceeded. Whereas the related forms *S. distincta* and *S. fossarum* are homoiosmotic only in the fresh water which they normally inhabit.[37] And the mosquito larva *Aëdes detritus*, which breeds in salt or brackish water, is able to maintain the normal osmotic pressure in the blood, although living in a hypertonic medium, in a way that *A. aegypti* cannot. Even when in water of a salinity equivalent to 6 per cent. NaCl it is able to prevent the osmotic pressure of its blood from rising above 1·3–1·4 per cent. (Fig. 276, B).[11] The difference lies in the properties of the gut wall which is able to counteract the penetration of salts much more effectively than in *A. aegypti*. This is

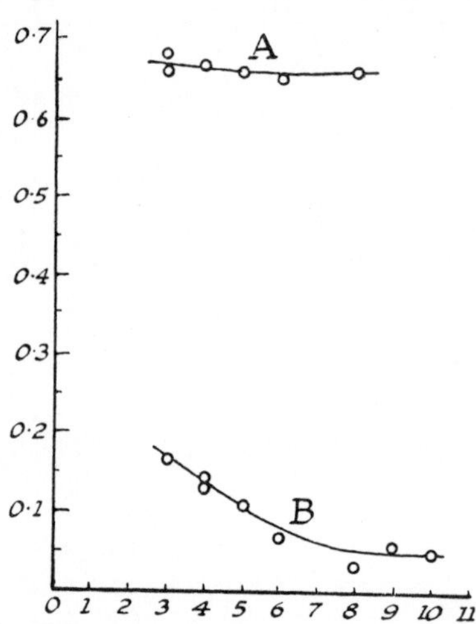

FIG. 277.—Effect of starvation in distilled water on osmotic pressure (A) and chloride content (B) in the haemolymph of the mosquito larva (*Aëdes*) (*after* WIGGLESWORTH)

Ordinate: values expressed as equivalent concentrations of NaCl in gm. per 100 c.c. Abscissa: days after beginning of starvation.

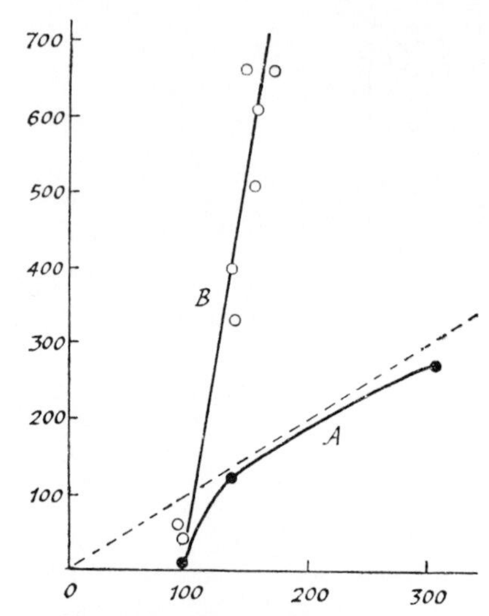

FIG. 278.—The relationship between osmotic pressure in the rectal fluid and the haemolymph in *Aëdes aegypti* (A) and *A. detritus* (B) (*after* RAMSAY)

Ordinate: osmotic pressure of rectal fluid as mM/l NaCl. Abscissa: osmotic pressure of haemolymph in the same units. The broken line indicates the iso-osmotic relation.

probably combined with active excretion of excess salt by the Malpighian tubes.[11] (See also p. 515.)

In *Aëdes aegypti* larvae the Malpighian tubes can contribute to the work of salt retention by excreting a fluid containing less sodium than the haemolymph, but never a fluid containing more sodium than the haemolymph. In the regulation of total osmotic pressure, however (as distinct from possible differences in chemical composition), the rectum is the most important organ in mosquito larvae. In *Aëdes aegypti* the rectal epithelium actively absorbs salts, leaving a hypotonic fluid. In *A. detritus* there are two segments in the rectum showing histological differences: the posterior region resembles the rectum in *A. aegypti* and probably reabsorbs salts when the larva is in a dilute medium; the anterior region reabsorbs water and so concentrates the contents when the larva is in a strongly saline medium (Fig. 278).[270] Similar mechanisms for osmoregulation are described in the larva of *Corethra*.[271] [See p. 457.]

There are somewhat similar differences between osmoregulation in the fresh water caddis larvae *Limnephilus stigma* and *Anabolia nervosa* on the one hand, and the salt water caddis *Limnephilus affinis* on the other, as there are between *Aëdes aegypti* and *A. detritus*. *L. affinis* restricts its drinking, and the gut cells resist damage by water of high salt content (410 mM NaCl or 75 per cent. sea water). It produces a rectal fluid hyperosmotic to the external medium by excreting chloride; but by increasing the non-electrolyte fraction in the haemolymph it allows the blood to become roughly iso-osmotic with the medium. The fresh water caddis larvae, on the other hand, drink large amounts of water; their gut wall is damaged as soon as the salt concentration rises above 60 mM/l NaCl; the salt content of the haemolymph increases and the blood becomes hypertonic to the external medium; rectal fluid is not produced, and the larvae soon die.[378]

Although the larva of *Sialis* has an impermeable body wall it cannot live in saline solutions more concentrated than the blood, because it lacks the power of preventing entry of salts through the wall of the gut.[371] *Ephydra riparia* can produce excretory fluid considerably more concentrated than the external medium. It can be kept in tap water, or in sea water evaporated to half its original volume. It has no functional anal papillae and can regulate its internal medium even more effectively than *Aëdes detritus*.[377] Larvae of *Ephydra cinerea* living in the Great Salt Lake in Utah (U.S.A.) has an osmotic pressure in the haemolymph of about 20·2 atm. Regulation is aided by a highly impermeable cuticle.[354]

'Reflex bleeding'—Certain insects when handled discharge small drops of fluid from various points of their body. Sometimes this is the product of special groups of dermal glands, as in the Mexican bean beetle (*Epilachna*) which ejects a bitter yellow fluid from the femoro-tibial articulations.[144] But in many cases the fluid is actually blood, as was originally supposed by Leydig. It appears around the mouth, the limb joints, and the base of the elytra, in such beetles as *Coccinella*, *Timarcha*, and Meloids, and the Orthoptera *Eugaster* and *Ephippiger*. Apparently the blood is strongly compressed by contraction of the abdomen, and escapes by rupture of the skin at points of least resistance.[44] Some grasshopper nymphs can eject fluid as much as two inches.[139] In the Acridiid *Dictyophorus* the blood is mixed with air to form a brownish foam as it escapes; apparently both tracheae and body wall are perforated.[76] But in many insects, preformed pores or ostioles exist (perhaps these mark the earlier site of glands); in the Tenthredinid larvae *Cimbex* and *Trichiosoma* these pores are closed by valves when the blood pressure is relaxed.[93] Some moths produce a nauseous froth, largely blood and air, although the noxious properties may be due to glandular secretion.[33] In all these forms the blood contains cantharidin, or other caustic or repellent substance; and there is evidence that the habit is of some biological value as a protection from predators.[44] On the back of *Pseudococcus*, and at the apex of the abdominal tubes or siphunculi of Aphids, are ostioles, opened by a special muscle, which allow wax-laden blood cells to escape.[226]

HAEMOCYTES

The haemocyte count—The loose cells in the body cavity of insects are called haemocytes or blood cells. Most of these are normally resting on the surfaces of various organs, but some circulate freely in the blood. In the living larva of *Corethra* it is very evident that most of the haemocytes are sedentary;

chiefly they lie along the outer surface of the heart, only a few are being carried through the dorsal vessel and along the body cavity by the blood stream. Cells travelling along the vessels in the wings of the cockroach often adhere to the walls and cease to move.[244] In the silkworm,[69] in *Aleurodes*,[228] and in *Rhodnius*, no haemocytes enter the heart through the ostia so that only the cell-free haemolymph can be said truly to circulate. Haemocytes are carried into the lumen of the heart in *Periplaneta*, *Blattella*, *Tenebrio*, *Galleria*, *Pieris*, and *Apis*; but not in *Cimex*, *Phormia*, or *Drosophila*,[334] In *Blaberus* the wing veins gradually become occluded by accumulating proleucocytes and plasmatocytes, and in old insects the circulation continues only in the main veins; the blood cells can be seen to force their way into the narrow spaces.[280] In *Periplaneta* there is a reduction of the numbers of cells free in the blood at the time of moulting and a rapid recovery after the skin is shed.[357] In *Rhodnius* at this time large numbers of blood cells insinuate themselves between the muscle fibres.[391] There are always haemocytes associated with the membranes around the ovaries.

The numbers of circulating cells vary enormously from time to time. In the cricket *Gryllus assimilis*, figures ranging from 15,000 to 275,000 per cubic millimetre have been obtained (average 70,000), in *Periplaneta* 15,000–60,000 (average 30,000).[210] The highest values are obtained, however, after injury and haemorrhage, in insects attacked by parasites, and particularly during ecdysis, when there is a real increase in the number of haemocytes present. They are most numerous during the phases of active growth in *Ephestia*,[279] in *Leptinotarsa*[5], and *Locusta*.[387]

Haemopoietic organs—There are all intermediate stages between freely circulating blood cells, cells loosely adherent to the tissues or forming transient aggregations, and collections of cells which form distinct organs more or less constant in position. These permanent aggregations are regarded as blood-forming, or haemopoietic organs.[393] Organs of this type occur behind the prothoracic spiracles of young caterpillars; mitoses take place in them and haemocytes are liberated.[281] In *Musca* the organ is applied to the dorso-lateral epidermis near the posterior spiracles.[281] In saw-fly larvae there are cell masses in the outer layer of the fat body in every body segment; the transition from proleucocytes to haemocytes takes place in these organs which disappear at the beginning of the pupal stage.[345] [See p. 457.]

Types of haemocytes—The haemocytes arise in the embryo from undifferentiated mesodermic tissue.[232] As occasion demands, both the independent cells in the blood,[12, 25] and the aggregations of cells along the heart,[43, 141] continue to multiply by mitosis throughout the life of the insect.

When the cells attach themselves to other tissues they tend to become pear-shaped, or fusiform, or to spread themselves out in a stellate manner applied as closely as possible to the surface.[236] They may thus assume an infinite variety of appearances (Fig. 279). But when floating freely in the blood they become round or oval and can then be grouped into fairly definite classes.[337, 393]

(i) *Proleucocytes*, small cells with deeply staining cytoplasm rich in ribonucleic acid and a nucleus which almost fills the cell (Fig. 279, *b*). These have been recognized in all groups of insects.[190] In Lepidoptera they are abundant only in the young larva;[94] and in all insects they are often seen undergoing mitosis. They are therefore regarded as young growing forms.

(ii) *Plasmatocytes* derived by way of transitional forms from the proleucocytes. They now have an increased amount of cytoplasm. They are exceedingly polymorphic: rounded, spindle-shaped or pear-shaped when free in the blood, flattened or drawn out into elongated form when attached to other organs (Fig. 279, *a*). A very large number of these forms have been described in Lepidoptera in *Prodenia* larva, 10 classes subdivided into 32 types and many more variants have been named.[249] They often contain inclusions of neutral mucopolysaccharides, in *Rhodnius*,[391] *Xylotrupes*,[356] and *Bombyx*.[355] Examined in the fresh state or in osmium-fixed material they show abundant filiform prolongations.[390, 391]

(iii) *Granular cells* and *Spherule cells* seem to be derivatives of the plasmatocytes with more conspicuous inclusions. In *Bombyx* they are forms with still larger inclusions of mucopolysaccharide.[355] Cells of this class are evident in *Ephestia*,[279] *Pieris*,[94] and other Lepidoptera,[46, 97] *Sarcophaga*,[49, 336] *Drosophila*,[363] *Tenebrio*,[336] and other Coleoptera.[89, 183] In *Sialis* there appears to be only one

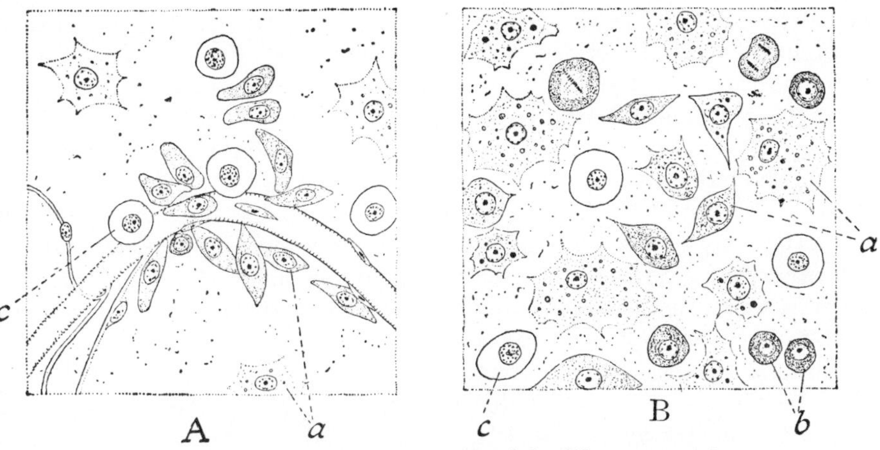

FIG. 279.—Haemocytes in *Rhodnius* (*after* WIGGLESWORTH)

A, group of haemocytes around trachea where this is passing through the basement membrane of the epidermis; a few are spread out in stellate form on the basement membrane. *B*, haemocytes below the basement membrane at the height of moulting; many of the phagocytes contain basophil droplets from disintegrated cells; proleucocytes dividing mitotically. *a*, phagocytes (plasmatocytes); *b*, proleucocytes; *c*, oenocytoids.

type of blood cell which passes through a series of stages; only the first stage is capable of division; they become phagocytic and finally granular haemocytes or lipid-filled morula cells.[369] [See p. 458.]

(iv) *Oenocytoids*. In preparations fixed with acid fixatives and stained with haematoxylin and eosin, these have a characteristic appearance: they are round or oval cells, 8–12μ in diameter, with darkly staining nucleus and clear, uniform, weakly acidophil cytoplasm. They are present in Coleoptera,[183] Hemiptera (Fig. 279, *c*),[89, 236] Lepidoptera[148] (they form 3 per cent. of the circulating cells in *Pieris* larvae[94]), but are wanting in Orthoptera[43, 89, 153] and perhaps in Hymenoptera.[149] These cells seem to be quite distinct from types i, ii, and iii and possibly do not come from the proleucocytes; they have been thought to be derived from the true oenocytes (p. 449),[177] but in *Rhodnius* that is certainly not so.[236] Although in fixed preparations the cytoplasm of the oenocytoids appears homogeneous, in the fresh state, and particularly under phase contrast, the cytoplasm shows spherical inclusions; examined in this manner they have sometimes been described as 'granular cells'.[336]

(v) *Other types* of blood cells that occur in some insects are (*a*) *Adipocytes* ranging from 12–35μ, filled with droplets of fat, observed in *Pyrrhocoris apterus.*[89] They occur in *Rhodnius* where they are indistinguishable from fat body cells that are set free in the blood.[391] (*b*) *Wax cells.* Similar cells laden with wax in the blood of Coccids and Aphids,[89] e.g. in *Pulvinaria.*[179] (*c*) *Crystal cells* described in the blood of *Drosophila*[363] (p. 397). (*d*) *Unstable hyaline corpuscles* (*coagulocytes*), not easily recognizable histologically, concerned in the coagulation of the haemolymph (p. 438). (*e*) *Other specialized types*: for example, in *Pulvinaria* there are spherical cells stuffed with eosinophil granules, large cells which take up ammonia carmine, giant cells, and cells with phenolic inclusions.[179]

Functions of haemocytes: (i) phagocytosis—The most obvious activity of the haemocytes is phagocytosis, or the ingestion of small solid particles. They will take up injected particles of Indian ink or carmine,[43] or dead bacilli,[148] or the dead bodies of one another,[89, 236] or the histolysing tissues in moulting and metamorphosis.[173, 183] In the Thysanura all the blood cells appear to be phagocytic;[25] but usually the proleucocytes are not;[43, 89, 236] the oenocytoids when present never are.[236] The most active phagocytes are the plasmatocytes.[94, 149]

Sedentary blood cells tend to accumulate along the dorsal vessel. In Gryllidae, Acrididae, *Forficula,* they form dense aggregations, 'phagocytic organs', through which the blood must work its way as it passes to the heart.[43] These organs are devoid of tracheae; they differ from accidental accumulations of haemocytes only in the constancy of their position and form. They have the same properties as the free cells in the blood, multiplying by mitosis and ingesting foreign particles. In Blattids, Mantids and Tettigoniids, indeed in most insects, compact phagocytic organs are wanting; there are only simple groups of phagocytes around the pericardial cells and in the interstices of the pericardial septum.[43] In Thysanura the pericardial septum seems itself to be composed of a syncytium of phagocytic cells.[25, 176]

The transition from free haemocytes to phagocytic tissue is well seen in Chironomid larvae. In *Tanypus*, haemocytes alone are present; in *Chironomus plumosus* and *lobiferus*, circulating cells are entirely wanting, being replaced by a reticular phagocytic tissue at the hind end of the abdomen, through which the blood must pass to reach the heart; in *Tanytarsus roseiventris* free cells and tissue are both present.[124] There seems to be no very sharp distinction between 'phagocytic organs' and 'haemopoietic organs'.

Under normal circumstances a chief function of the haemocytes appears to be the removal of the larger particles of solid matter set free into the body cavity. Hence they increase enormously in numbers during moulting (Fig. 279),[177, 236] and metamorphosis;[173, 183] and at these times they tend to leave their fixed stations along the heart and become scattered.[155] [See p. 458.]

We have seen that the histolysis of tissues may be achieved completely without the intervention of the blood cells; but that, in many insects, once degeneration has set in, they play a part in removing the dying cells and tissue residues[27] (p. 74). The activity of the haemocytes in removing debris from the haemolymph at the time of moulting is well seen in the bed-bug *Cimex*. Sometimes, after a meal of blood, the stomach of this insect may rupture and red corpuscles be set free into the body cavity. Here they may persist unaltered for several weeks;[235] but if the insect moults they disappear completely.[107] In the later stages of moulting[236] and metamorphosis[183] the excess haemocytes themselves decay and are ingested by their fellows. The oenocytoids, which play no part in phagocytosis, show the same proliferation during moulting and metamorphosis and suffer the same fate when the process is complete.[183, 236] Much of the normal phagocytic activity may be imperceptible. The cytoplasmic processes on the cells in the *Drosophila* larva are so fine that they are visible only by phase contrast; these delicate filaments are in movement and minute particles in the haemolymph adhere to them.[363] [See p. 458.]

The haemocytes also collect at the site of injuries, forming a plug which helps to seal the wound, proliferating, and removing the dead and discarded cells.[188, 238]

ii. Resistance to micro-organisms—The natural immunity of insects to bacteria is chiefly a phagocytic immunity,[35] but it varies enormously in different insect species and in respect of different micro-organisms.[96, 206] The numerous bacteria in the gut of *Calliphora* larvae are liberated into the body cavity of the pupa during histolysis. But these are actively destroyed by the phagocytes; so that at the time of emergence, or within a few hours afterwards, the adult flies are bacteriologically sterile.[8] Non-pathogenic organisms such as tubercle bacilli, or feebly pathogenic forms such as staphylococci, are rapidly taken up by phagocytes of *Galleria* and other caterpillars;[148] whereas the pathogenic coccobacilli, &c. are not.[94] In *Blatta* infected with bacteria the percentage of mitotically dividing blood cells may rise from a normal maximum of 0·5 per cent. up to 3·1 per cent., the total number of haemocytes increasing greatly.[208]

iii. Resistance to metazoan parasites—In the larvae of Lepidoptera, the eggs and young larvae of Hymenopterous parasites become encysted by blood cells; such parasites induce mitosis and capsule formation[18] (though around the larger larvae of parasites, blood cells may be absent altogether[43]). Trematodes in *Dytiscus*,[92] and in the flea,[34] become invested in a thick mantle of haemocytes.

Entomophagous parasites show considerable specificity as regards the hosts in which they will develop. The caterpillars of *Loxostege sticticalis* are non-susceptible to *Eulimneria* spp. (Hym.). This type of immunity is due to the very active encapsulation by the blood and tissue cells.[125] But the immunity of *Pseudococcus* to Hymenopterous parasites is not necessarily accompanied by phagocytic encapsulation; this occurs in some resistant hosts but not in others.[13] When the egg of *Diplazon fissorius* (Ichneumonidae) is laid in its normal Syrphid host *Epistrophe bifasciata* its serosa appears to produce toxins which prevent the accumulating haemocytes from forming a capsule. In abnormal hosts the egg is encapsulated and fails to hatch: in *E. balneata* a tough brown capsule without nuclei is formed; in *Syrphus ribesii* a thick gelatinous nucleated capsule.[272]

The defence reactions of insects to entomophagous and other metazoan parasites are more effective than has been generally supposed.[366] The haemocytes are their sole agent and most parasites entering an unusual host are overcome by this means, but when in their usual host some parasites elicit no definite reaction; or if they do they are able to resist it. Eggs or young larvae of *Nemeritis canescens* which develop readily in *Ephestia*, their normal host, are usually encapsulated by haemocytes, which give rise to melanin formation around them, when they are introduced into a wide variety of other insects. It seems to be the surface properties of the parasite which determine whether or not it shall be encapsulated.[365] The cause of death seems to be asphyxiation.[313, 365, 393] [See p. 458.]

In the saw-fly *Pristiphora* the developing egg of the Ichneumonid *Mesoleius* is invested by phagocytic haemocytes containing inclusions of mucopolysaccharide, and the encapsulated larvae die, probably from asphyxiation;[292] but in certain geographical races no such encapsulation occurs.[353] Likewise in *Drosophila* it is possible to select for resistance to the parasite *Pseudeucoila* and thus to increase the percentage of individuals in which protective capsules of blood cells are formed around the egg of the parasite.[323]

iv. Blood coagulation—The blood of various beetles, when collected in glass capillaries, usually clots within 5–10 minutes;[108] and clotting is particularly rapid in the vesicant species and others subject to 'reflex bleeding'.[141] On the other hand, the blood of the honey-bee larva,[15] of *Rhodnius* and some other insects, seems never to coagulate. And in many forms, such as the cockroach, the apparent coagulum is merely a clump of haemocytes; there is no gelation of the plasma.[245] The blood cells become round, develop thread-like pseudopodia, and agglutinate to form a plug which is the essential factor in the formation of the clot.[156] But in other insects, such as *Gryllus*, there is a true coagulation of the plasma with the formation of fibrin.[245] Blood clotting seems to concern particularly the lipoproteins of the haemolymph: these react to form new lipoproteins, one of which is the relatively insoluble coagulum network.[373]

In the haemolymph of almost all insects there are certain 'hyaline haemocytes', indistinguishable in ordinary stained preparations but easily recognized with the phase contrast microscope. In insects in which blood clotting does not occur these cells show no special change. But when the blood coagulates they produce a fine precipitate in the plasma round them, or they extrude thread-like pseudopodial expansions which fuse to form a meshwork in which the other cells are entangled. Varying degrees of coagulation in the plasma may follow these changes. The whole process resembles that produced by the 'explosive' corpuscles described by W. B. Hardy many years ago in Crustacea.[263]

In most Hemiptera the hyaline haemocytes show no change in the shed blood, and the plasmatic reaction is absent. Various substances, including agents that bind calcium ions, which act as anticoagulants in vertebrates, will also inhibit visible changes in the hyaline haemocytes and thus prevent clotting.[263] The percentage of 'coagulocytes' is much increased in the newly moulted adult of *Periplaneta*, and this is associated with a thirteen-fold increase in the speed of clotting.[388] [See p. 458.]

v. Connective tissue formation—All the organs and tissues of insects are separated from the body cavity by connective tissue membranes, or basement membranes, which are characterized by staining with the periodic acid—Schiff reaction (PAS).[283, 359, 391] There is no doubt that many of these membranes, such as those surrounding the epithelium of the gut, the Malpighian tubules, the neural lamella around the nerves and central nervous system (p. 183) are produced by the cells which they cover.[391] Some authors go further and consider that all these investing membranes are solely the product of the invested cells.[38] But tubercle bacilli in the body cavity of insects are walled off by aggregates of blood cells to form small 'tubercles'[43] in which the central parts break down while fresh cells congregate at the periphery.[58, 148] Blood cells collect also round foreign bodies to form a capsule. Usually the cellular character of these capsules persists, but sometimes the nuclei disappear and a capsule of connective tissue is produced. This led to the suggestion that in the normal insect the haemocytes might be concerned in connective tissue formation.[125, 127] [See p. 458.]

The plasmatocytes ('amoebocytes') of *Rhodnius* contain inclusions of neutral mucopolysaccharide with the same staining properties as the connective tissues and basement membranes (Fig. 280).[391] These cells become exceedingly numerous during the stages of growth and moulting when these membranes are formed; they apply themselves to the basement membranes, the sarcolemma of the developing muscles, &c., and can be seen to be discharging their mucopolysaccharide inclusions. There is no doubt that this material is added to the membranes lining the body cavity. What proportion of these membranes is actually contributed by the blood cells is not known.[391] These sheaths and membranes are far more fragile than the sheaths of fibrous tissue formed in vertebrates; but not only the neural lamella (p. 183) but the basement membranes of Malpighian tubules and fat body (in *Aiolopus*) do contain collagen-like fibrils.[284] [See p. 458.]

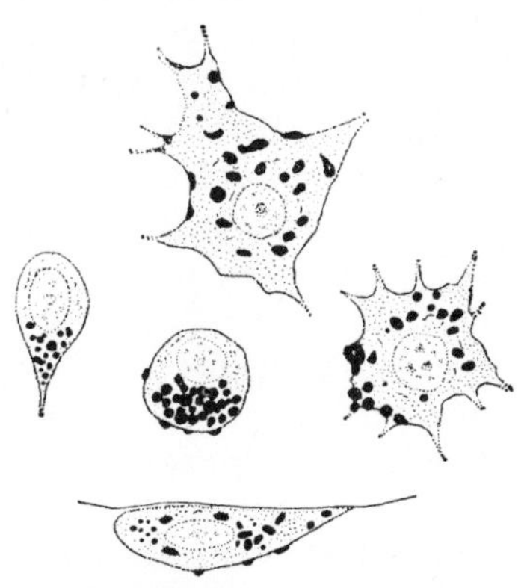

FIG. 280.—Haemocytes (plasmatocytes) in *Rhodnius* showing mucopolysaccharide inclusions after PAS staining (*after* (WIGGLESWORTH)

vi. Haemocytes and intermediary metabolism—It seems likely that some of the most important functions of the haemocytes may be in intermediary metabolism. It has often been suggested that they may be concerned in the production of certain of the blood proteins.[302] They may be concerned in the transfer of nutrients;[12, 302] blood cells in *Prodenia* may contain considerable amounts of glycogen;[250] and the cells in *Ephestia* have been described as carrying fat, &c., into the developing wing and there breaking down and giving up their reserves to the epidermis;[258] large vacuolated haemocytes appear to play a large part in transporting food material during metamorphosis in *Apis*.[134]

In *Rhodnius* the haemocytes seem to play an essential role in the initiation of hormone secretion in the thoracic gland (p. 67): if the plasmatocytes are blocked by the injection of particulate matter, growth is delayed. But the nature of this action is uncertain.[390, 393]

There seems little doubt that the haemocytes are concerned in the metabolism of tyrosine which leads to the production of the polyphenolic substances concerned in the melanization and sclerotization of the cuticle (p. 47).[49, 336] There is evidence that the 'crystal cells' in the larva of *Drosophila melanogaster* are concerned in melanin production at metamorphosis; ^{14}C-tyrosine injected into the larva becomes concentrated in the inclusions of the cells; and if the blood cells are incubated in a solution of 'dopa' the cytoplasm of the crystal cells is blackened, but the crystals remain colourless. It appears that these cells carry both enzyme and substrate for melanin formation.[363a] [See p. 459.]

Immunity—Acquired immunity, resulting from infection with pathogenic organisms, is a general immunity in which the essential response seems to be an increased sensitivity towards the organism on the part of the body cells; the nervous system being apparently connected in some way with the change.[148] Phagocytosis, giant cell formation, and encapsulation, are more active. And various antibodies are produced: agglutinins for the pathogenic *Coccobacillus acridiorum* appear in the blood of immune *Melanoplus*;[73] bacteriolysins, apparently non-specific, can be evoked by many organisms or even, it is said, by injections of foreign proteins;[259] specific antitoxins can be formed.[35] Passive immunity is said to be conferred by the transference of blood from one insect to another, both cells and plasma by themselves being effective; but the chief factor in acquired immunity, also, seems to be an increase in the rate of phagocytosis.[96] These changes may perhaps result from alterations in the electric charge on haemocytes or micro-organisms, alterations due to changes in plasma constitution, such as are known to be important in mammalian blood.[147]

Caterpillars do not respond to foreign proteins with the formation of circulating antibodies; this is a factor which greatly facilitates the interspecific transplantation of organs and tissues in insects.[289] After injection of various bacteria, caterpillars develop a largely non-specific, heat stable, anti-bacterial principle of a totally different nature from the antibodies of vertebrates.[291] Larvae of *Galleria* immunized against the bacterium *Pseudomonas* do not develop antibodies of globulin type; but their blood does acquire specific bactericidal properties which persist for about 3 days. The active factor is dialysable, heat stable, unaffected by trypsin, weakly acidic and of relatively small molecular weight. Besides giving protection against *Pseudomonas* it prevents melanization by inhibiting tyrosinase.[376] *Oncopeltus* develops lytic substances in the haemolymph after a single injection of *Pseudomonas*, and these substances afford increased resistance to infection.[321] [See p. 459.]

Giant cell formation—Giant cells may appear around the sites of chronic inflammation (around foreign bodies or slowly healing wounds), originating in two ways. (i) By the fusion of haemocytes or other cells into multinucleate masses;[99] (ii) by the hypertrophy of single cells. Giant cells of the second type, ranging up to 150μ in diameter, have been termed 'teratocytes' and regarded as hypertrophied macronucleocytes.[94, 386] But it is certain that the cells of many other tissues, epidermis,[236] fat body, Malpighian tubes, gonads, can suffer the same change in the proximity of internal parasites. In *Pieris* attacked by *Apanteles* and Aphids invaded by *Aphidius* the giant cells are derived from the 'amniotic membrane' which surrounds the embryo.[166, 329] [See p. 459.]

PERICARDIAL CELLS AND SO-CALLED 'NEPHROCYTES'[338]

Anatomy and Histology—The pericardial cells which were recognized by Leydig in 1866 are of mesodermic origin like the haemocytes. But they are incapable of migrating in the blood stream, and occupy fairly constant positions. In *Panorpa* the pericardial cells are said to increase greatly in number during the larval stages; the new cells may come from blood cells which settle between the large pericardial cells and enlarge simultaneously with them.[27]

In *Machilis* they resemble fat cells, and lie along the borders of the connective tissue around the pericardial sinus; in *Lepisma* they are suspended on the fibres which attach the heart to the dorsal body wall.[25] They are usually arranged upon the surface of the heart; often they are scattered also over the pericardial septum and the alary muscles; in Hemiptera-Heteroptera they are very abundant and lie inside the heart as well (Fig. 263, *c*). They may be large, few and isolated, as in the larvae of mosquitos; or small, numerous and closely packed.[1] The fully formed cells usually contain more than one nucleus;[92a] in *Galleria* larvae they are large syncytia containing as many as six nuclei of variable size.[148] They often contain vacuoles and granular inclusions, which vary in size from time to time; sometimes they are filled with yellow, brown, red or green pigments. In Lepidoptera, the cells show peripheral striations.[92a]

It has been pointed out already (p. 199) that the pericardial cells of *Periplaneta* liberate a cardiac stimulator, perhaps an indolalkylamine, when exposed to the secretion for the corpus cardiacum. Under the action of the corpus cardiacum hormone the pericardial cells enlarge and form droplets of secretion; and at the same time become more strongly argentaffin. After prolonged exposure these activities cease; apparently the supply of the necessary metabolites becomes exhausted.[303]

Function of pericardial cells—The most obvious property of the pericardial cells is their ability to absorb colloidal particles from the blood. Haemoglobin, chlorophyll, egg-white, trypan blue, ammonia carmine, injected into the blood, appear as inclusions in the pericardial cells.[92a, 129] They chiefly take up particles with a radius of 16–20Å.[133] When they take up litmus these cells are always red or acid;[114] and when ammonia carmine is absorbed it is believed that this acidity causes the liberation of the insoluble free carmine[92a] with the result that the dye persists indefinitely in the cells. Indeed when the pericardial cells become completely saturated with the dye, they break down and are destroyed by phagocytes (in Muscid larvae[114] and *Galleria* larvae[148]).

Injection of ammonia carmine reveals that there are other cells with similar properties in the insect body. Cells scattered through the fat body in *Lepisma*, Odonata nymphs[24] and *Pediculus*;[159] in groups around the mouth appendages or at the bases of the limbs in *Gryllus*, *Mantis*, *Periplaneta*,[26] or on either side of the oesophagus in *Pediculus*;[159] and forming a garland of cells that hangs between the salivary glands of Muscid larvae.[104]

All these clearly belong to the same system as the pericardial cells. The whole are often classed as 'nephrocytes' in the belief that the accumulation of injected dyes within them proves that their function is the segregation and storage of waste products.[114] But with few exceptions (p. 581) the contents of these cells do not increase with age;[43] and if they play a part in excretion it is likely to be rather in the intermediary metabolism of waste substances.[43, 92a]

The current view on this system is that it is analogous with the reticulo-endothelial system of vertebrates;[180] that as the haemocytes are concerned in the removal of gross fragments from the blood, so these cells are responsible for the 'micro-phagocytosis' of colloidal particles—the absorption of a given particle being determined partly by its charge (electronegative colloids such as acid dyes being most readily seized), partly by the size of particle.[129] But other undetermined physico-chemical factors can influence absorption; for the uptake of dyes is somewhat capricious in different species,[92a] and it may be influenced by whether the dye is injected or given by the mouth.[77] At the time of metamorphosis (in Coleoptera, Lepidoptera, Trichoptera) albuminoid inclusions appear within the cells[92a, 164]—as was to be expected if they are freeing the blood of foreign colloids. Whether they liberate the products of their activity in simpler form, or whether, as has been suggested,[92a] they produce the specific proteins of the blood, is quite unknown.

In *Rhodnius* the pericardial cells are filled with blue-green deposits of biliverdin, which is derived from traces of haemoglobin absorbed into the blood without being digested. Transition from haemoglobin, to haematin, to a verdohaem pigment, to biliverdin and free iron can be observed in the cells.[24] These are the same changes which happen to haemoglobin in the Kupfer cells of the reticulo-endothelial system in the liver. [See p. 459.]

THE FAT BODY

Anatomy and Histology—The fat body is a more definite organ than the other tissues of the body cavity (Fig. 281). It is derived from the mesoderm of the walls of the coelomic cavities; but its original segmental arrangement is generally lost when the embryonic coelom breaks down; so that the fat body is arranged as a loose meshwork of lobes, invested in delicate connective tissue membranes (p. 438), and joined by connective tissue strands, so as to expose the maximum of surface to the blood. In spite of the looseness of its texture, the fat body is arranged in a constant manner in each species. Often there is a peripheral layer beneath the skin, and a central layer as a sheath round the gut, but in many insects the peripheral part alone is present. There may also be cylindrical cords in the thorax, compact masses in the head, and so on.[97, 118, 234] Though present in all insects, the fat body is most conspicuous in the larvae of holometabolous forms; in the full-grown larva of the honey-bee it makes up 65 per cent. of the total body weight.[14]

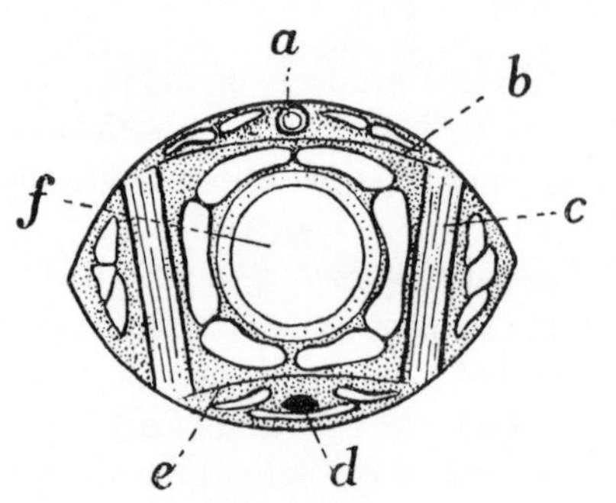

FIG. 281.—Diagram showing typical arrangement of the fat body as seen in a cross section of the abdomen (*after* PARDI)

a, heart; *b*, pericardial septum; *c*, muscle; *d*, ventral nerve cord; *e*, ventral diaphragm *f*, gut. Fat body shown in white.

In the embryo and newly hatched insect the cells are in loose strings; they are round with a homogeneous chromophil cytoplasm, free from vacuoles or inclusions.[118, 177, 183, 195] When the larva starts to feed the cells become vacuolated and increase in size (Fig. 285). At moulting, they multiply by mitosis and gradually become so large and distended that the cell boundaries are no longer visible. But they do not actually form a syncytium; for in the larva of *Dytiscus*, after prolonged starving, the cell limits reappear; vacuoles and granules disappear completely, and the cells revert to their embryonic form;[118] and when the larva of the honey-bee becomes full-grown, the cell boundaries can again be seen.[195]

Relation to the blood cells—In the very young insect the fat body cells are scarcely distinguishable from the haemocytes; and the relations between these two tissues have often been considered to be very close.[127] What appear to be ordinary haemocytes enter the histolysing flight muscles of queen ants and become converted into fat cells; and the imaginal fat body of Muscids arises from 'mesodermal leucocytes' in the same way.[101] In the Chrysomelid *Haemonia*, blood cells enter the elytra and later become vacuolated and adhere together to form fat body.[116, 117] In aquatic Hemiptera, the fat body is said to increase in size throughout life by the inclusion in its substance of free adipocytes, that is, haemocytes charged with fat (p. 436).[177] And in *Aleurodes*, since all the fat cells are freely floating in the body cavity, it is impossible to draw any line of distinction between them and the haemocytes.[228] In Lepidoptera,[97] and Coleoptera[155, 183] the fat cells are said to be capable of phagocytosis and to ingest the debris of histolysed tissues; but most of the formed elements appearing in the fat body during metamorphosis are undoubtedly taken up from the blood in solution.[174]

Storage in the fat body during growth—The most obvious function of the fat body is the storage of reserve materials. 'Fat body' is a misnomer, for it is equally important as a store for proteins and glycogen. Fat droplets and protein granules usually appear soon after feeding has begun. If the larva is starved, the cells become completely emptied (*Dytiscus*,[118] *Tenebrio*[112]), to fill again when feeding is resumed. In the mosquito larva *Aëdes* the character of the reserves may be strikingly influenced by the nature of the food. If the starved larva is given carbohydrate alone the cells are stuffed with glycogen and there are relatively few fat droplets; if given fat alone, droplets of fat are numerous but no glycogen appears.[240] Towards the end of larval life the cells are filled with drops of fat, spheres of protein, and glycogen in granules or dense peripheral deposits, (Fig. 283),[19, 170, 258] and among these are occasional watery vacuoles containing no reserves.[240] In the fat body of the silkworm the glycogen is estimated to range from 2–17 per cent.;[252] in the mature larva of the bee it forms more than 33 per cent. of the dry weight of the whole insect.[200] The same accumulations appear in the young stages of hemimetabolous forms.[177]

These reserves bear a special relation to moulting (in *Galerucella*,[183] for example). The changes at moulting have been studied in detail in *Rhodnius*:[240] the fat appears to be a general energy reserve; the glycogen is formed and mobilized only during the time when the chitin of the new cuticle is being laid down; the changes in protein run more or less parallel with those in glycogen and are related to the deposition of protein in the cuticle.

The fat body during metamorphosis—(Figs. 282, 285)—At the approach of metamorphosis, the fat cells become intensely active. They are now stuffed with reserves. The nuclear surface is commonly increased by amitosis,[157] or by

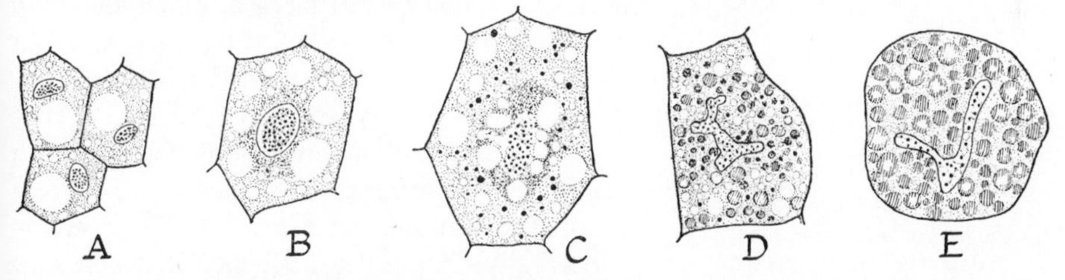

FIG. 282.—Changes in fat body cells during growth and metamorphosis in the honey-bee (*after* BISHOP)

A, shortly after hatching; cells with one large fat-vacuole. *B*, in the growing larva; cells with peripheral ring of fat-vacuoles, central densely staining cytoplasmic zone, and densely granular nucleus. *C*, in larva just becoming quiescent; central ring of fat-vacuoles indenting the nuclear vesicle, nuclear membrane absent, and nuclear granules dispersed into the cytoplasm. *D*, in early pupa; nuclear membrane reformed, fat-vacuoles disappearing, nuclear granules growing to albuminoid globules. *E*, in the middle of pupal development; cell about to disintegrate, albuminoid globules matured, their centres acidophile, fat almost absent.

the elongation or branching of the nucleus.[14, 173] The nuclear membrane may become indefinite, and small basophil granules appear around it. These are variously regarded as chromatoid granules discharged from the nucleus,[14] comparable with the chromidia of the oöcyte,[117] as transformed mitochondria,[163] or as products of the nucleoli extruded through the nuclear membrane. They do not give the Feulgen reaction for nucleoprotein.[165] Gradually they spread outwards from the nucleus, increase in size, and stain less deeply with basic dyes[14] or become frankly eosinophil.[195] As these 'albuminoid spheres' enlarge, reaching

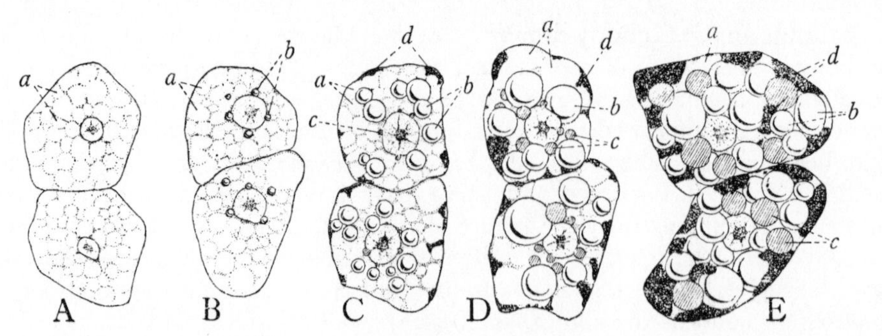

FIG. 283.—Build up of reserves in fat body cells of the mosquito larvae *Aëdes* fed on casein (*after* WIGGLESWORTH)

A, cells in starved larva; *B*, in larva after feeding for 24 hr.; *C*, after 2 days; *D*, after 3 days; *E*, after 7 days. *a*, watery vacuoles; *b*, fat droplets; *c*, protein droplets; *d*, glycogen stained with iodine.

a diameter of 12–15μ in the pupa of the bee,[195] the fat droplets diminish, and the spheres begin to stain with osmic acid.[14] This whole process, as observed in the bee,[14, 195] *Dytiscus*,[118] Muscids,[173] Tenthredinids[194] and Lepidoptera,[90, 97] gives the impression that the free fat in the cells is being incorporated with the protein elements to produce a raw material comparable with the yolk spheres of the egg.[117, 174] In *Rhodnius*[240] and in the larvae of *Aëdes*,[240] on the other hand, although the protein droplets darken to some extent in osmic acid, there is no evidence that fat is being incorporated in them (Fig. 283).[170]

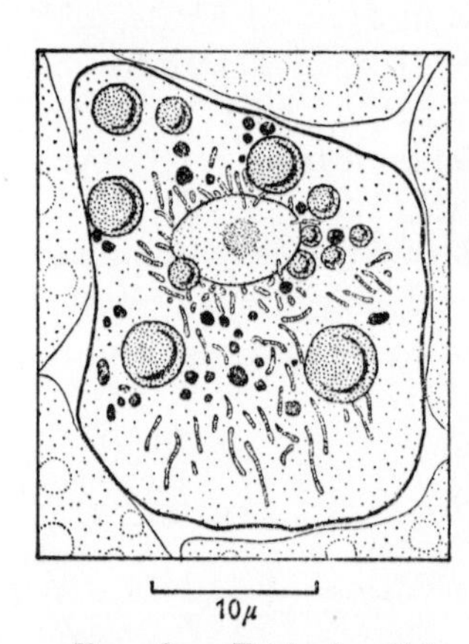

FIG. 284.—Fat body cell in *Rhodnius* one day after feeding the starved insect, stained with osmium and ethyl gallate, showing fat droplets, rapidly increasing mitochondria, and osmiophil inclusions (lysosomes) probably derived from degenerated mitochondria (*after* WIGGLESWORTH)

When the fully starved larva of *Rhodnius*, with the fat body cells almost devoid of reserves, is given a meal of blood, the first changes that become apparent within 24 hours are enlargement of the nucleolus with accumulation of ribonucleic acid, the deposition of RNA around the nucleus, and a great increase in the numbers of filamentous mitochondria radiating out from the nucleus. Glycogen and droplets of fat have begun to appear within 24 hours (Fig. 284).[394] During the final stages of larval development in *Galleria* and other Lepidoptera the mitochondria in the fat body undergo an apparent autolysis to form amorphous inclusions which then enlarge and accumulate the protein reserves to be carried over into the pupa.[386a] In the larva of *Drosophila*, below the investing connective tissue sheath, the cells (as seen with the electron microscope) show deep and branching invaginations of the cell membrane, apparently the site of pinocytosis. Fat is accumulated and stored in the lacunae of the granular endoplasmic reticulum; protein inclusions arise (as in *Galleria*, &c., see above) from breakdown products of the mitochondria (lysosomes) and appear in sections as mitochondrial residues mixed with glycogen, multiple laminated thin membranes, and endoplasmic reticulum with ribosomes. The nucleus and nuclear membrane are closely involved in these changes.[318] [See p. 459.]

The appearance of 'pseudonuclei' of uric acid within the albuminoid spheres of Muscids (Fig. 285) and the deposition of uric acid in the fat body of other insects is discussed elsewhere (p. 579); likewise the occurrence of biliverdin and other pigments (p. 610). These pigment deposits play a major part in the coloration of mosquito larvae and of *Simulium* larvae.[311, 330]

During metamorphosis and moulting, these reserve materials are liberated into the blood; the cells for the most part survive.[12] But in some Hymenoptera, and in the higher Diptera, there is an almost complete dissolution of the fat body in the pupa or young imago; it is then formed anew, from a few surviving cells in the bee,[195] or from embryonic cells in the Muscids[173] and *Drosophila*.[240]

In the honey-bee, shortly before pupation, the fat cells, now filled with

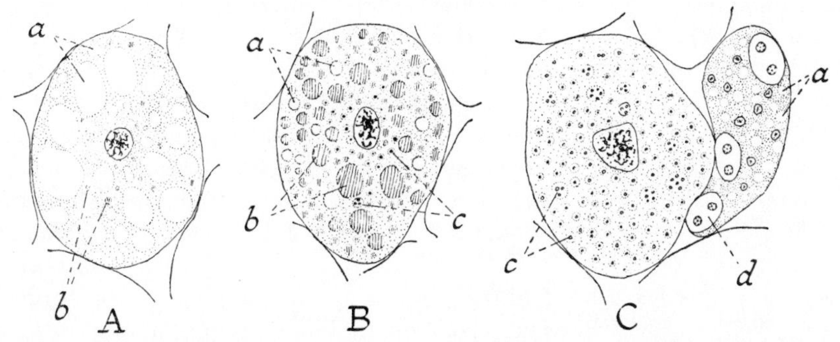

FIG. 285.—Fat body cells of *Calliphora* at different stages of growth (*after* PÉREZ)

A, growing larva; much fat, little albumen. *B*, early in pupal development; little fat, many albuminoid spheres of all sizes, some with pseudo-nuclei of uric acid. *C*, shortly before emergence from pupa. On left, remains of larval fat body, with no fat and all albuminoid spheres full of pseudo-nuclei. On right, imaginal fat body with oenocytes. **a**, fat globules; **b**, albuminoid spheres; **c**, pseudo-nuclei; **d**, oenocytes.

eosinophil inclusions, are set free into the blood, which becomes a thick creamy liquid. In the young pupa these fragile cells begin to disintegrate spontaneously. Fat droplets and albuminoid spheres float freely in the blood. As time goes on few intact cells remain; the albuminoid spheres, and to a less extent the oil drops, are removed; and by the time of emergence almost all the cells have disappeared. In the adult, the detritus is all dissolved, and to some extent ingested by the phagocytes, so that the blood becomes quite clear again.[200]

The fat body in the imago—Metamorphosis so far as it concerns the fat body is often still uncompleted when the adult emerges. The body cavity of newly hatched Muscid flies is full of little floating spheres which are the remains of the larval fat body. These are laden with protein granules, which disappear during egg development; they are then attacked by haemocytes.[59] In *Drosophila* they have disappeared within three or four days after emergence.[240] In the young adult of *Culex*, the abdomen still contains the residue of larval muscles; in a few days after emergence these disappear, and the fat body (or the ovaries) grows at their expense.[192] Thus in the adult insect the fat body provides reserve material for egg production; it is often more prominent in the female.[118] It also furnishes the reserves for hibernation; in September and October overwintering females of *Culex* with a wet weight of 3 mg. contain 0·91 mg. of fat; in March and April with a wet weight of 2 mg. they have a fat content of 0·13 mg.[32] In *Drosophila* the massive reserves of glycogen in the fat body are the chief source of energy for flight (p. 601).[240] As a rule there are no conspicuous reserves of protein in the fat body of the adult insect, but in the honey-bee, as the result of pollen uptake in the autumn, protein stores in the form of droplets are laid down. These are not utilized during the winter, but serve to build up the salivary glands for feeding the brood in the spring.[134] [See p. 459.]

The fat body in intermediary metabolism—The fat body has often been compared with the liver of vertebrates: a general centre of intermediary metabolism which serves also for the storage of reserves. The isolated fat body of *Periplaneta* shows oxidative and dehydrogenase activities towards a wide range of substrates.[404] The complete citric acid cycle operates in the fat body.[339] The fat body of *Schistocerca* can incorporate a wide variety of precursors into fat, protein, and glycogen;[300] and it acts like vertebrate liver in transaminating amino acids and making them available for metabolism in other tissues.[300, 339] The fat body of *Schistocerca* is the chief site of the synthesis of trehalose from glucose; glucose-6-phosphate and uridine diphosphate glucose are probably concerned as intermediates.[295] Lipase and alkaline phosphatase occur in the fat body of *Schistocerca*.[319] In *Rhodnius* and other insects, esterase (lipase) appears in the form of localized 'caps' applied to the surface of every fat droplet. These may be the sites of reversible hydrolysis of triglycerides controlling the movement of fat to and from the droplet (Fig. 286).[392] Cell-free preparations of the fat body of *Locusta*, provided with co-enzymes, &c., will incorporate acetate into fatty acids to yield mainly palmitic acid.[385] [See p. 460.]

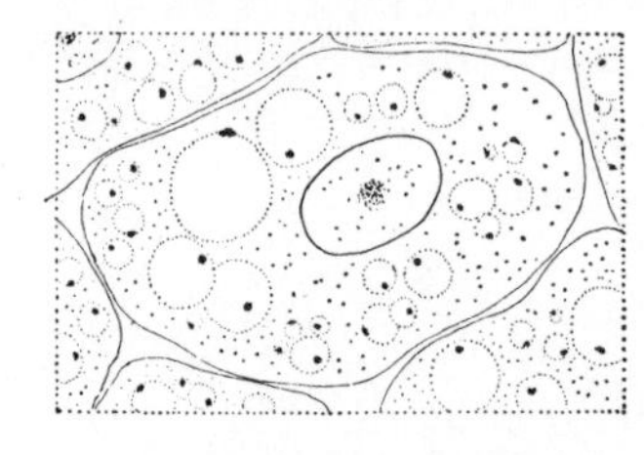

FIG. 286.—Fat body cell of *Rhodnius* showing the localization of esterase (lipase) in the form of 'caps', one on each droplet of fat. 5-bromoindoxyl acetate method (*after* WIGGLESWORTH)

We saw that the synthesis of blood proteins in *Bombyx* becomes very active during the fifth larval stage (p. 424). The chief site of this synthesis is the fat body, which takes up amino acids and incorporates them chiefly into two globulins and an albumen, and discharges them into the haemolymph. This synthesis of protein can be demonstrated in the isolated fat body; it is inhibited by cyanide, 2–4, dinitrophenol, and other agents which interfere with oxidative metabolism.[372a] In *Schistocerca* the fat body incorporates isotopically labelled leucine into protein which is then liberated into the haemolymph and taken up by the developing oöcytes (p. 704). The rate of incorporation of ^{14}C-leucine into the fat body of mature larvae of *Hyalophora cecropia* is very high; it falls to about 0·3 per cent. of the larval rate in the diapausing pupa but it can be caused to rise again to a high level by injury to the integument. There is also a renewal of incorporation when adult development begins.[376a] [See p. 460.]

In *Drosophila*, where kynurenine plays an important role as the precursor of many of the eye pigments, the essential conversion of tryptophane to kynurenine appears to take place in the fat body in which fluorescent granules of kynurenine accumulate towards the time of pupation.[363a] The fat body contains enzymes active in purine metabolism: in *Prodenia* and *Tenebrio* it will synthesize uric acid from many purine substrates, and contains adenase, guanase, and xanthine oxidase.[278] In *Leucophaea* it contains uricase (converting uric acid to allantoine), guanase and xanthine oxidase.[346]

So-called 'tracheal cells'—During the 1st instar in the parasitic larva of *Gasterophilus*, when the cells in the anterior part of the fat body begin to increase in size and accumulate fat, those in the hind part become specialized to form the '*organe rouge*'.[50] The cells of this organ are pyriform, 350–400μ in length, and richly supplied with tracheae and intracellular tracheoles. They are rich in glycogen; they have a strong succinic-cytochrome system; and their cytoplasm is filled with haemoglobin. This pigment has a molecular weight of about 34,000, that is, about half the value in vertebrates, and it is synthesized by the larva. It has a high affinity for oxygen, a low affinity for carbon monoxide (of the same order as cytochrome oxidase and polyphenoloxidase) and is thus to some extent intermediate in properties between the two groups of haematin-protein compounds—the oxygen carriers (p. 389) and certain oxidizing catalysts (p. 624).[105] Its function is uncertain.[106]

Somewhat similar very large 'tracheal cells', sometimes as much as 120 μ in length, abundantly supplied with tracheae and containing haemoglobin, occur in the aquatic bugs *Buenoa*[98] and *Anisops*.[178] But they are absent in the closely related genera *Plea*, *Notonecta*, *Corixa*, &c. For this reason it has been thought that haemoglobin arises in these cells as an accidental product of metabolism without functional significance.[178] It is not impossible that it is derived from haemoglobin taken in with the food.[241] For its role in respiration see p. 391.

THE OENOCYTES

Anatomy—Unlike the other tissues of the haemocoele, which are of mesodermal origin, the oenocytes appear early in embryonic development by the enlargement of metameric groups of ectodermal cells. With the exception of the Thysanura *Lepisma*, *Campodea*,[232] oenocytes, which were discovered originally by Fabre, are universal among insects. In many forms their segmental arrangement is retained: in Ephemeroptera, Odonata, Plecoptera, and Termites, they occur as clumps in the pleural epidermis near the spiracles;[232] in Lepidoptera they are in close relation with the spiracular tracheae;[207] in the Tachinid larva *Thrixion* they form groups of 3–6 large cells, strictly metameric, between the muscles and the body wall.[167] But often they are dispersed at random. They may remain in close relation with the epidermis: they lie scattered between the epidermis and the basement membrane in *Tomocerus* (Collembola),[243] *Blatta*,[232] Phasmids,[198] *Melolontha*,[234] Hemiptera-Heteroptera.[177, 234, 236] But often they tend to lose this primitive connexion: in the larvae of the beetles *Galerucella*,[1] *Melasoma*, *Haemonia*,[117] and *Calandra*,[155] though they are massed particularly in the neighbourhood of the spiracles, they extend inwards and merge with the fat body; in *Dytiscus* larva they form long bands between the tracheal trunks;[118] in *Geotrupes* (Col.),[234] Aphids,[232] Hymenoptera (Fig. 287),[113, 173, 195] adult Muscids,[173] &c. they are scattered throughout the fat body. In *Tenebrio* there are sub-epidermal oenocytes below the abdominal sternites and segmental groups on the main tracheae.[240]

Histology—In the fresh state the oenocytes are often of a wine-yellow

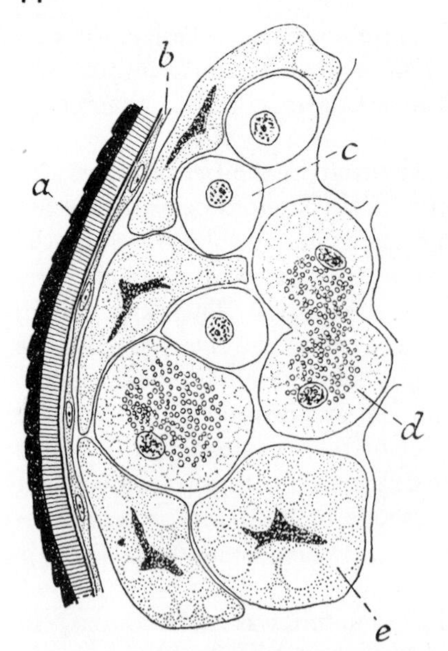

FIG. 287.—Section through dorsal region of the abdomen in the ant *Tapinoma* to show the underlying tissues (*after* BERLESE)

a, cuticle; *b*, epidermis; *c*, oenocytes; *d*, urate cells with uric acid concretions; *e*, ordinary fat body cells.

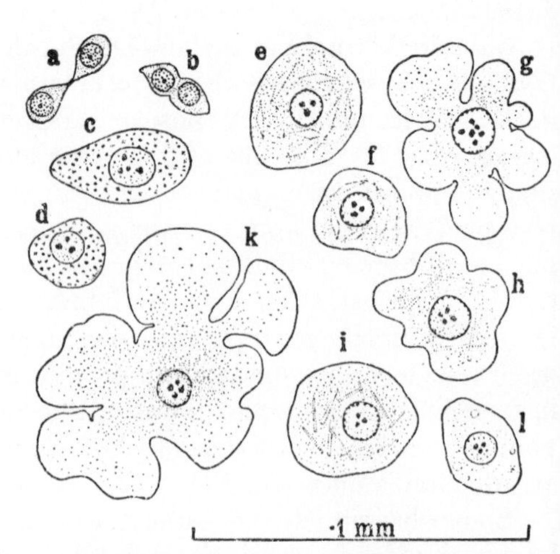

FIG. 288.—Oenocytes in *Rhodnius* (*after* WIGGLESWORTH)

a, *b*, young forms; *c*, *d*, resting forms with granular inclusions; *e*, *f*, forms with spindle-shaped clefts; *g*, *k*, forms at height of secretory activity; *h*, *i*, forms with needle-like crystals in cytoplasm.

colour to which they owe their name;[234] in *Schistocerca* they contain a resistant yellowish pigment, perhaps a lipochrome; but in many insects they are colourless. Usually they are large cells few in number: in some Cynipid larvae they are 150μ across, one-fifth of the whole length of the insect;[191] in the pupa of the queen bee 176μ;[113] quite commonly they are 60–100μ. In adult insects they tend to be more numerous and small (15–25μ in mosquitos,[95] 8–30μ in *Melasoma* (Col.)[117]). They adhere closely to one another, or to the fat body (in Muscid flies they are completely buried in the fat cells[173]) but they are not usually bound by tracheae or connective tissue. They are often lobulated, and certainly seem to migrate about the body; but they have not been seen to make amoeboid movements in the living state.[91] In stained preparations the oenocytes are characterized by their dense, eosinophil and usually homogeneous cytoplasm; but sometimes this is granular or vacuolated, and in fixed material it may show spindle-shaped clefts (which may be artefacts[167, 236]), or radiating canals;[118, 207] and sometimes it contains elongated apparently crystalline bodies best seen in fresh preparations;[91, 236] and granules of pigment of many kinds may occur[91] (Fig. 288). In the young adult of *Tribolium* they are filled with rod-like inclusions which disappear at the time of sexual maturity.[109]

Generations of oenocytes—In the aquatic Hemiptera, fresh oenocytes continue to arise from small subepidermal cells throughout post-embryonic life;[177] in *Rhodnius* (Hem.) a new generation arises in this way during every moult except the last.[236] In *Rhodnius* they show no further multiplication once they are developed, but it has been claimed that in *Gerris* they produce the oenocytoids (p. 435) by a process of budding.[177] In holometabolous insects there is only one generation throughout larval life (larval oenocytes); a second generation arises at pupation (imaginal oenocytes).

Larval oenocytes—In *Ephestia* (Lep.), when newly hatched, there are one or two cells round each trachea; these increase by amitosis to about 10 cells;[207] in *Bombyx mori*, on the other hand, no such increase occurs.[221] In Cynipids the larval oenocytes multiply by amitosis;[191] while in *Hyponomeuta* (Lep.) they are said to bud off small amoebocytes.[97] In *Galerucella* the larval oenocytes contain two or three nuclei, suggesting amitosis;[183] but they show no increase in *Dytiscus*[118] or *Polistes* (Hym.);[173] and in Nematocerous larvae (*Chironomus*, *Culex*, &c.) the same 5 cells at each side of the abdominal segments persist until pupation.[95, 234, 240]

Imaginal oenocytes—With the possible exception of the ant *Formica*[173] and the beetle *Galerucella*,[183] in which they are said to arise by a process of budding from the larval cells, the oenocytes of the imago always develop, like the oenocytes in the embryo, from segmental groups of cells in the epidermis of the abdomen, usually around the spiracles. Metameric thickenings or imaginal discs of this kind are described in the ant *Lasius*,[103] *Apis*,[113] *Polistes*,[173] the Chalcid *Torymus*,[229] and Cynipids[191] among Hymenoptera; in *Hyponomeuta*,[97] the silkworm,[222] and *Ephestia*[207] among Lepidoptera; and in *Dytiscus*[118] and *Calandra*[155] among Coleoptera, in *Sialis*[161] and *Drosophila*.[110] At first they are enclosed beneath the basement membrane, but later this is broken through and they are scattered throughout the fat body. Apart from their smaller size, the imaginal oenocytes resemble those of the larva; but in Chironomidae they have been regarded as distinctive organs, the 'synoenocytes'.[256]

Changes in oenocytes during growth—In the larva of *Thrixion* (Tachinidae) the oenocytes increase progressively in size: 15μ in the 1st instar, 45μ in the 2nd, 85μ in the 3rd.[167] But in some insects they undergo a cycle within each instar. In the silkworm and in *Ephestia* they increase enormously in size before moulting; the vesicular nucleus contracts and becomes branched and distorted,

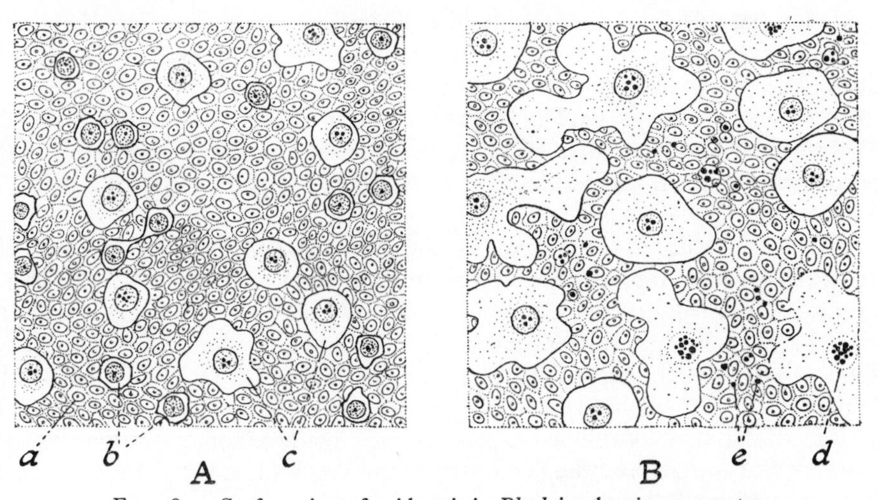

FIG. 289.—Surface view of epidermis in *Rhodnius* showing oenocytes

A, in nymph before feeding; *B*, in nymph at height of moulting process, shortly before new cuticle is laid down. *a*, epidermis; *b*, new generation of oenocytes; *c*, preceding generation of oenocytes; *d*, oenocyte undergoing cytolysis; *e*, chromatin droplets derived from broken down oenocytes and other cells.

and the cytoplasm fills with vacuoles; as these vacuoles disappear the nucleus regains its normal form.[207, 222] A similar cycle coincident with the moulting cycle occurs in *Galerucella* (Col.),[1] *Apis*,[349] and probably in aquatic Hemiptera, in which the oenocytes are small after each moult.[177] This cycle is particularly evident in *Rhodnius* (Fig. 289)[236, 240] and in *Tenebrio*:[240] the cells swell and throw out lobes like pseudopodia while the epidermis is growing; they reach a maximum just before the new cuticle is laid down; then they shrink, many of the older generation undergoing chromatolysis and disappearing, to be replaced by the new generation that has arisen (Fig. 289).[236] The same cycle occurs in the embryo of Lepidoptera before hatching,[207] and in *Rhodnius* in the egg before the embryonic cuticle is shed.[146]

During pupal development the oenocytes show similar changes. Larval and imaginal oenocytes hypertrophy enormously; then they diminish, and many of the larval cells disintegrate, only a few surviving in the adult. A cycle of this kind has been observed in Trichoptera,[68] Lepidoptera,[97, 207] Hymenoptera,[103, 113, 173, 191] Coleoptera,[117, 183] (in *Dytiscus* the surviving cells diminish from 68μ to 25μ in the course of pupal development[118]) Diptera Nematocera,[95] and Cyclorrhapha.[173]

Oenocytes in the adult—Oenocytes always persist in the adult. In *Dytiscus* they increase in size again during the weeks following emergence.[118] In aquatic Hemiptera they show signs of secretory changes in the female at the times of oviposition;[177] and this is very obvious in the blood-sucking bug *Rhodnius*, in which they swell up enormously in the female after a meal, when eggs are being developed, but show comparatively little increase in the male.[236] The same enlargement during egg development is seen in *Apis*,[349] *Formica*,[367] and *Calliphora*.[384]

Function of oenocytes—Most authors are agreed that the oenocytes are organs of intermediary metabolism which discharge their secretion into the blood;[91, 117] but there is little to indicate what the nature of this secretion may be. Glycogen[91, 169, 177] as well as protein and, occasionally, fat[240] have been demonstrated in the cells, and their needle-like crystals have been thought to be wax.[91] The oenocytes may hypertrophy in larvae invaded by parasites,[18] and in Aphids during the lysis of their symbionts, for which they have indeed been held responsible.[218] They may accumulate pigments in old insects[113] and absorb certain dyes injected into the blood, and have therefore been thought to remove waste products or toxins from the blood, or generally to regulate its physico-chemical composition.[177]

But it is evident that the oenocytes bear some special relation with growth, and perhaps with reproduction. Possibly they secrete enzymes which further cytolysis[4] or dissolve stored substances in the fat cells.[214] It was suggested at one time that they produce the hormones which initiate moulting.[111] Certainly in the silkworm they are the first cells to show activity during the moulting cycle, while they show no such changes in a strain of silkworm which fails to moult;[253] and in Chironomidae their activity runs parallel with the development of the imaginal discs.[256] But in *Rhodnius* the oenocytes do not reach the height of their activity until long after the moulting cycle has begun (p. 40).

In *Rhodnius* the secretory changes in the oenocytes reach their peak just before the new epicuticle is laid down. At this time they stain deep grey with osmic acid and diffusely with fat stains. As they shrink and discharge their contents the same staining reactions are shown by the epidermal cells and then by the newly formed cuticulin layer. In the *Tenebrio* pupa oenocytes are present only on the sternites where the cuticulin layer is very much thicker. The cells at the height of their development contain a large amount of bound lipid revealed only by destructive oxidation, as in the epicuticle. For all these reasons it seems likely that the oenocytes may be regarded as epidermal cells that have become specialized for the production of the lipoproteins which form the cuticulin layer.[240] Indeed this is the one suggested function of the oenocytes for which there is definite evidence: in *Calliphora*,[397] and in the *Rhodnius* epidermis stained with osmium and ethyl gallate,[396] the products of the oenocytes can be seen passing into the epidermal cells and being transferred to their outer surface to form the epicuticle. It is worth noting that in the blood of *Carcinus maenas* there are circulating 'lipoprotein cells' which have staining properties similar to those of the oenocytes and which go through a similar cycle in relation to the secretion of the epicuticle.[370]

There is some evidence that the numerous subepidermal oenocytes in the cockroach are the source of the free wax which is so plentiful on the cuticle of this insect.[115a] In the young honey-bee worker there is a correlation between the size of the oenocytes and the height of the columnar epithelium of the wax gland[115a, 290, 315] which again suggests a role in lipoprotein or wax metabolism.

The imaginal oenocytes seem to be related to development in the gonads. In *Tribolium* the rod-like inclusions disappear in both sexes at the time of sexual maturity.[109] It has been suggested that they may be concerned in hormone production in adult Chironomids,[256] or in the production of materials that go to form the eggs or their shells.[117, 236] In *Rhodnius* the egg-shell is rich in lipoproteins (p. 1) and the activity of the oenocytes is much greater in the female.[236] The greater enlargement of the oenocytes in the queen bee,[349] female *Calliphora*,[384] and the female *Formica*,[367] would conform with this hypothesis; but there is no direct evidence, as there is in the case of the formation of the epicuticle, and they may be contributing something else to the development of the ovaries. [See p. 460.]

LIGHT-PRODUCING ORGANS

Luminescence occurs in many insects; notably Collembola, the larvae of some Nematocera, and the larvae and imagines of certain beetles, mostly Malacodermata and Elateridae. In origin it was probably an accidental accompaniment of some specific process in metabolism (for many substances slowly oxidized at low temperature in the dark are luminous) and only secondarily and in some instances has the light itself become of biological importance.[140, 82] In many insects, such as caterpillars of Lepidoptera,[175] occasional examples of luminosity are the result of an abnormal infection with luminous bacteria;[45] but insects which show the phenomenon constantly, generate the light in the course of their own metabolism.[141]

Nature of photogenic organs—A general distribution of luminescence throughout the body, except the legs and antennae, is seen in the Collembolan *Anurida* sp. which gives out a continuous pale greenish glow[9] and in *Onychiurus* (*Lipura*) *armatus*[85] and *Achorutes muscorum*,[202] also Collembola. In these insects it seems to be the fat body that is the source of the light; and that is certainly the case in the Mycetophilid *Ceroplatus testaceus*, in which the peripheral layer and the anterior part of the visceral layer of the fat body in the larva gives out a very faint persistent light. The light in *Ceroplatus* is still present in the pupa but disappears soon after the adult emerges; that is, at the time when the cells of the larval fat body dissolve.[201]

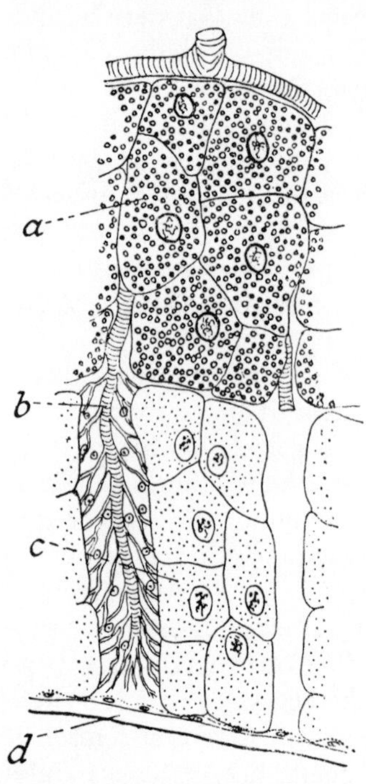

FIG. 290.—Section of a small part of the photogenic organ of *Luciola italica* (*after* BERLESE)

a, opaque reflector layer; b, tracheae, c, photogenic cells; d, translucent cuticle.

The photogenic organs of beetles show all degrees of specialization.[31] In *Phengodes* there are loose independent giant cells, apparently without tracheae, perhaps identical with the oenocytes. In *Phryxothrix* there are small rounded masses of cells with a specific tracheal supply. In Lampyrids the organs lie on the ventral aspect of the posterior abdominal segments; in the Elaterids *Pyrophorus*, &c. there is a pair at the posterior angles of the prothorax on the dorsal surface, and a third which lies mid ventral on the first abdominal segment. The histological details vary in different forms, but the organ consists in general of a deep layer of 'reflector cells', their cytoplasm stuffed with uratic granules, a more superficial mass of large photogenic cells richly supplied with nerves and tracheae and tracheoles and a translucent unpigmented window in the overlying cuticle (Fig. 290).[45, 88] In the most complex types there are conspicuous tracheal end cells (p. 359) which probably play an important part in the control of light production.[31] These organs are probably derived, in the course of development, from the fat body;[45] for in *Lampyris*, the luminous organ of the adult beetle is certainly produced by modification of fat body cells already existing in the larva.[224] In this same insect, and in *Pyrophorus*, the female often shows a diffuse luminosity throughout the abdomen which is caused by the generation of light in the yolk of the unfertilized eggs.[53]

An unusual type of photogenic organ exists in the larva of the Mycetophilid *Bolitophila* (*Arachnocampa*) *luminosa*, in which the distal ends of the four long coiled Malpighian tubes are converted into thick rod-shaped light-producing structures, and rest upon a mass of modified connective tissue (derived perhaps from tracheal cells) filled with reflecting granules.[233, 317] Another predaceous Mycetophilid larva has small luminous black bodies in the head and tail; it lives below a sticky web and its nocturnal light is believed to lure insects to this trap.[67]

It was suggested by Dubois[53] that the production of light in many animals is brought about by luminous bacteria which maintain a symbiotic existence within the cells.[28] In the case of insects, the photogenic cells are in fact filled with uniform granules which bear a resemblance to micro-organisms: in the male of *Photinus* the granules are round like cocci, in the female they are rod-shaped like bacilli.[45] But if the luminous organ is excised from the larva of *Photuris*, although the usual glowing of the pupa is eliminated, yet perfectly normal photogenic organs develop in the adult. Unless, therefore, it be supposed that the bacteria go through a developmental stage in which they are not luminous, it must be concluded that micro-organisms are not responsible for the light production.[84]

Chemistry and physics of light production—It is generally agreed that the production of light accompanies an oxidative process in which a substrate, called by Dubois 'luciferin', is oxidized in the presence of an enzyme 'luciferase'. Luciferin can be oxidized by other agents, such as potassium ferricyanide, but no light is then produced. It is therefore believed that it is the enzyme and not the substrate which gives out the light. Luciferin and luciferase have been isolated in crystalline form. The luminescent system of *Photinus* requires luciferin, luciferase, adenosine triphosphate (ATP), magnesium ions and oxygen. Briefly, the reduced form of luciferin (LH_2) is activated by ATP with liberation of pyrophosphate. This step is anaerobic and reversible and quite independent of light production. The activated luciferin, probably adenyl-luciferin, i.e. LH_2-adenosine monophosphate (AMP), is very labile. It reacts with oxygen in the presence of luciferase to form the excited adenyloxyluciferin molecule (L*AMP). Decay of this excited state is marked by the release of a quantum of light energy and the fall of adenyloxyluciferin to a lower energy level (L.AMP). All these changes are catalysed by the purified enzyme luciferase. Although ATP is essential for light production in the fire-fly it does not supply the energy for light emission. The free energy change in the formation and hydrolysis of adenyl luciferin is relatively small. The energy for light emission is apparently supplied directly by the oxidative process.[320, 351]

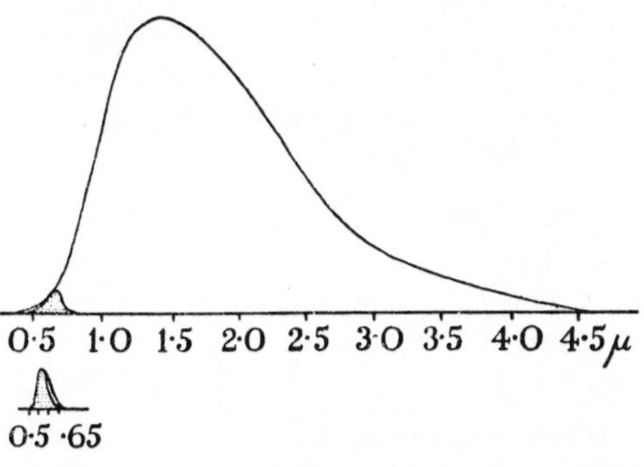

FIG. 291.—Luminous efficiency of a carbon filament lamp (upper curves) and a fire-fly (lower curves).

The shaded area of each curve represents the amount of visible light energy (the luminous flux). The total area represents the total radiant energy (the radiant flux). The ratio of luminous to radiant flux is the luminous efficiency: 0·5 per cent. for the carbon filament lamp, 95 per cent. for the fire-fly (*after* IVES)

The quantity of heat set free in this reaction is exceedingly small. In the case of *Pyrophorus* it is judged to be less than $\frac{1}{80,000}$ of that produced by a candle flame of equivalent brightness:[140] at least 98 per cent. of the radiant energy emitted appears in the form of light (Fig. 291).[53] In colour the light varies according to the species from greenish blue to reddish golden; the dorsal thoracic organs in *Pyrophorus* give out a greenish light, the ventral organ on the abdomen is orange in colour; the larva of *Phryxothrix* (Phengodidae) emits a red luminescence from the head, a greenish yellow along the sides of the body segments.[83] Analysed spectroscopically it is found to be entirely free from ultra-violet, and to form a continuous spectrum over a very narrow range of wave lengths (520–650mμ in *Photinus*, 486–720mμ in *Pyrophorus*, 518–656mμ in *Lampyris*) occupying just about that zone to which the human eye at least is most sensitive.[140] In *Pyrophorus* the light from the thoracic organs has been estimated as 505–650mμ and from the abdominal organ 540–645mμ. Perhaps these differences indicate the existence of several types of luciferin and luciferase.[30] It is estimated that some 37–38 *Pyrophorus*, the most luminous of insects, with all their organs functioning, can illuminate a room with about the same intensity as one candle.[53]

Control of light production—Even in the primitive forms, such as Collembola, the production of light is subject to some degree of regulation by the insect. It may be increased by mechanical stimuli (in *Onychiurus armatus*[85]) or take place only as a result of stimulation (in *Achorutes muscorum*[202]). Since oxygen is needed for its production this effect of stimulation is perhaps simply the result of increased oxygenation of the tissues following increased activity. But in the luminescent beetles, the production of light is fully under the control of the insect: in *Lampyris* and *Pyrophorus* the steady glow is maintained only during certain hours of the day; in *Luciola* and *Photinus* the light is produced in rhythmic flashes with a characteristic frequency and duration.[29] Here the control lies in the central nervous system. The rhythmical flashing of *Luciola* ceases after decapitation, but can be induced again if the severed nerve cord is stimulated;[223] and illumination of the head alone, with the remainder of the body in darkness, causes an immediate reflex extinction of the light.[71] In the fire-fly it is possible to excite small areas of the lantern by stimulating selected branches of the segmental nerves.[324]

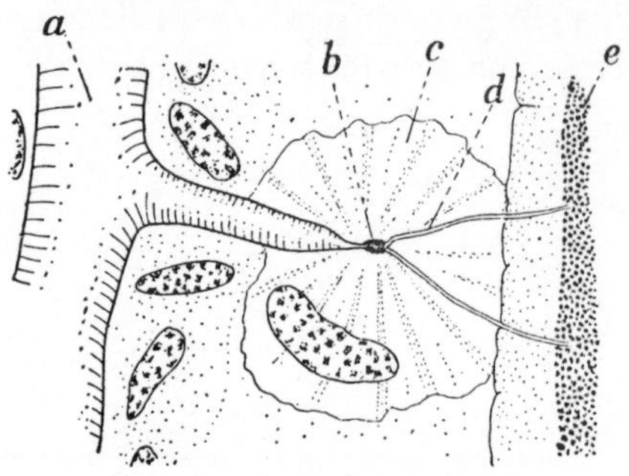

FIG. 292.—Tracheal supply to the light organ of *Photinus* (*after* DAHLGREN)

a, trachea; *b*, supposed sphincter at the tracheal ending; *c*, tracheal end cell with radial fibrillae; *d*, tracheoles; *e*, luminous granules.

If the light emitted by *Pyrophorus* is analysed with a photo-electric cell it nearly always shows rhythmic changes in intensity at a rate of about 300 per minute at the outset, falling to 150 per minute, and amounting to 5–6 per cent. of the maximal intensity. The records of such changes recall those of an incomplete tetanus in a contracting muscle, and are regarded as due to rhythmic discharges from the nerve centres to the organs.[81] But the mechanism by which the nervous system brings about control is uncertain. Dubois claimed that in Pyrophorids the luminous organs are enclosed by special muscles, which control the entry of blood. When this enters the organ the luciferase in solution in the blood reaches the luciferin in the granules of the photogenic cells and light is produced; when the blood flow is arrested the light becomes dim or is extinguished; and while the light is shining the muscles are said to make rhythmic movements.[53] But in Lampyridae the emission of light may be general throughout the luminous organ or confined to a few discrete areas.[135] There is clearly some form of local control. [See p. 460.]

This control reaches its highest development in those forms in which the histology of the light organs is most complex and in which the tracheal end cells are most evident.[31] It is believed by some that the nerve impulses act by releasing the photogenic substances and allowing them to react within the photogenic cells. Others hold that the limiting factor is the supply of oxygen: a very high tension of oxygen will cause the flashing type of organ to emit a steady glow. The control of oxygen admission has been attributed to the tracheal end cells in which elaborate contractile structures have been described (Fig. 292).[45, 71, 199] The tracheal ending may be kept closed between flashes by tonic contraction of an inner ring; for the production of a flash the ring may be relaxed and radial elements contract to open the sphincter. According to this view the flash results from nerve impulses to the tracheal end cells admitting a sudden burst of oxygen to the photogenic cells.[2] In *Photinus* the injection of adrenaline produces an intense steady glow which is brilliant in oxygen, extinguished in nitrogen. This is attributed to contraction of the radial fibres.[39, 57] Another suggestion is that the nerves cause changes in osmotic pressure in the photogenic cells which remove fluid from the tracheoles and admit oxygen.[137, 138, 227] But such removal of fluid is probably a secondary effect of activity, improving the oxygen supply; the flash produced by exposure to pure oxygen occurs more rapidly after keeping in nitrogen than in air.[2]

Studies with the electron microscope have shown that the radiating structure of the end-cell cytoplasm as seen with the light microscope is due not to contractile fibres but to elongated mitochondria and villus-like folding of the end-cell membrane. No nerves run to the tracheal end-cells.[288] The photogenic cells as well as the tracheal end-cells are exceedingly rich in mitochondria with associated respiratory enzymes, an abundant supply of ATP and of alkaline phosphatase.[328] In *Photinus* the nerves, containing 'synaptic vesicles' and 'neurosecretary droplets', end, not at the surface of the photocytes, but between the tracheal end-cell and the cell body of the tracheolar cell.[374] [See p. 460.]

When produced by electrical stimulation the flashes of light of adult *Photinus* have latencies of 25–250 milliseconds in different species, and durations of 100–1,000 milliseconds. They can be repeated many hundred times with little fatigue, and the response shows summation, 'staircase effect', tetanus, &c., as in conventional neuro-effector systems, and a striking neuro-effector facilitation is evident. From the electro-physiological standpoint the flash responses to stimulation resemble ordinary neuro-effector excitation systems so closely that direct neural control can scarcely be doubted.[293] The ultimate control of luminescence seems to be by chemical reactions not directly involving oxygen, so that the light organ can be well oxygenated at all times. The nerve impulse probably brings about release of ATP which in turn stimulates the light-yielding reaction and generates the flash.[320, 351, 374]

Function of light production—There is little doubt that in the most primitive forms the light is an adventitious side-product of the chemical activities within the cells and has no biological significance. In the Mycetophilid larvae mentioned it may serve to lure prey. In the glow-worms and fire-flies it is certainly a mating signal. The large-eyed male of *Lampyris* is attracted by the brightly glowing female. In *Photinus pyralis*, which is one of the flashing types of fire-fly, the response of the male is of two kinds. If there is a flash about two seconds after his flash, he turns in that direction and proceeds towards the lure. But if the flashes are close to him and occur at about the time when he would be expected to flash again, he responds by flashing synchronously with them.[29] This appears to be the basis of the spectacular pulsating swarms of male fire-flies in tropical Asia.

ADDENDA TO CHAPTER X

p. 411. **Dorsal vessel**—In the grasshopper *Aiolopus*, the heart consists of a cylinder of circular striated muscle fibres between two sheaths of connective tissue. These sheaths contain fibres of collagenous and elastic type embedded in a polysaccharide matrix.[410]

p. 414. **Blood vessels over the brain**—In adult *Sialis* there is a network of fine blood vessels over the surface of the brain, opening at intervals into the haemocoele;[463] and analogous vessels occur in the head of *Bombyx*[368] and *Blaberus*.[280]

p. 415. **Pericardial septum**—In *Locusta* one contraction cycle of the dorsal diaphragm occupies about 25 heart beats, i.e. about 20 seconds.[413] The alary muscles of *Hyalophora* share certain structural features with the 'slow' striated somatic muscles.[461]

p. 417. **Blood pressure in Locusta**—Changes in pressure in the circulatory system have been measured in *Locusta*. Diastolic pressure in the aorta has a mean value of +3·2 cm. H_2O; systolic pressure +8·6 cm. H_2O; the frequency of heart beat at 23° C. being 79·5 beats/min. Body movements produce maximum values of +13·0 and minimum of −3·9 cm. H_2O in the dorsal vessel. In the resting animal the pressure in the body cavity of the abdomen varies from −2·4 to −3·5 cm. H_2O; positive values occur only during muscular movements. The lowest pressure (−15·2 cm. H_2O) occurs in the dorsal ampullae under the metascutellum; respiratory movements of about 20 per min. produce a fluctuation of about 4 cm. H_2O in the general internal pressure.[413]

p. 423. **Innervation of heart**—While it is agreed that the heart muscle of *Periplaneta* will contract rhythmically after removal of all nerve connexions it is richly supplied with nerves. These are of two kinds: (i) motor axons arising from spontaneously active cardiac ganglion cells in the lateral cardiac cords, which provide the 'pacemaker' of the heart beat; (ii) neurosecretory axons, containing granules of varying type, some of which have their cell bodies in the lateral cardiac cords whereas others arise from the ventral ganglia and reach the lateral cardiac nerves by way of the segmental nerves. The precise relation between these two systems in controlling the heart beat is uncertain. Neither system seems to exert an inhibitory action. Acetylcholine and atropine act on the cardiac nervous system; most other drugs affect the mycardium.[449, 450, 466]

p. 424. **Blood protein**—Haemolymph proteins in the caterpillar of *Diatraea* are synthesized not only by the fat body (p. 446) but by the cells of the mid-gut; probably not by the haemocytes.[424] In *Rhodnius*, the increase in haemolymph protein after injury suggests that some is formed in the haemocytes.[426]

p. 425. In *Phormia*, at least 19 protein bands have been demonstrated in the haemolymph. Total protein concentration rises some 24-fold during larval life, reaching a maximum of about 20 per cent. shortly before pupation. It then falls during pupal development to about 3·5 per cent. in the newly emerged adult.[421] Likewise in *Calliphora stygia*: the 25 protein bands in the third-stage larva are reduced to 14 in the prepupa when the cuticle of the puparium is forming.[442]

p. 425. **Enzymes**—The spontaneous activation of prophenoloxidase in the haemolymph of *Calliphora* larvae is prevented by an inhibitor, perhaps derived from the salivary glands.[469]

p. 426. **Amino acids**—In the late third instar larva of *Phormia* about

60 per cent. of the total free amino acid is in the haemolymph, 40 per cent. in the tissues. In addition, there are many low molecular weight peptides, mostly composed of two amino acids only.[444] Tyrosine and phenylalanine accumulate in the haemolymph of higher Diptera (*Drosophila*, *Sarcophaga*, *Musca*) in conjugated form: tyrosine-phosphate, alanyl-tyrosine, glutamyl-phenylalanine. These substances disappear during hardening and darkening of the puparium; they serve as reservoirs for phenols to be utilized for sclerotization and melanin formation.[416]

p. 426. **Lipids in haemolymph**—In *Galleria* larvae the fat body releases free fatty acids to the haemolymph; these are partly re-synthesized to triglycerides and partly transported as fatty acids bound to protein.[481] In the adult *Manduca*, ingested glucose is converted to lipid and released from the fat body as diglyceride.[415] The plasma lipid of *Calliphora* rises in concentration up to the end of the third stage; there is then a marked drop at the onset of puparium formation,[442] when lipid is incorporated in the cuticle.[477] A variety of lipoproteins have been isolated from the pupal haemolymph of *Hyalophora* and *Philosamia*, which carry not only diglyceride but free cholesterol, a series of phospholipids, and carotenoids. They may be concerned also in transporting lipoidal hormones such as ecdysone and juvenile hormone.[423] Certain of these lipoproteins are probably included among the substances transferred to the oöcytes to contribute to the yolk protein.[468]

p. 429. **Blood potassium**—A large part of the potassium in haemolymph is contained within the haemocytes and therefore does not contribute to the ionic concentration in the circulating blood.[418] Although the blood of *Periplaneta* has a high K^+ level which is subject to regulation, ionic content does vary with the diet and it may even vary in different parts of the body.[456]

p. 431. **Osmotic regulation**—The use of amino acids to compensate for changes in chloride concentration in the haemolymph is seen also in the terrestrial insect *Dysdercus*: Na^+ and Cl^- concentrations increase in the middle of the last larval instar and the amino acid level falls.[414] *Periplaneta* can regulate haemolymph concentration during dehydration by the removal of NaCl and other solutes. The fate of the excess NaCl is unknown; it is not excreted; and on rehydration the haemolymph volume increases with almost no decrease in osmolarity.[471]

p. 432. In *Aëdes detritus* a part of the Malpighian secretion is carried forwards by retroperistalsis and reabsorbed in the mid-gut. Stretch receptors in the body wall responding to changes in body volume control this retroperistalsis.[467] In the salt marsh hemipteran *Halosalda* the hind-gut contains cells of two types, as in the larva of *Aëdes detritus*, and probably concerned in ionic regulation.[434]

p. 434. **Haemopoietic organs**—Haemopoietic organs in *Bombyx* are associated with the imaginal wing discs. They are enclosed by a perforated acellular sheath through which the new haemocytes escape into the blood.[408] The so-called 'phagocytic organs' along the dorsal vessel of Orthoptera (p. 436) are clearly haemopoietic organs; they are made up of reticular cells of mesodermic origin, with phagocytic activity, which differentiate into free blood cells of several types.[439] The so-called 'lymph glands' of *Drosophila* also are haemopoietic organs releasing several types of haemocytes and perhaps secreting into the haemolymph.[412]

p. 435. **Types of haemocytes**—In *Locusta* two types of 'granular' cells are recognized with the electron microscope, one with dense structureless contents, the other with large globules, up to 1·5μ diameter, which contain bundles of tubes 100–150 Å in diameter.[438] In *Rhodnius* the oenocytoids (under the electron microscope) have dense cytoplasm and very dense spherical granules with structureless contents; the plasmatocytes have large dense granules and others less dense with hexagonally arranged tubules.[443, 480] It is this tubular material of the plasmatocyte granules which is discharged and incorporated in the substance of the basement membrane during a sharply defined period during moulting.[480]

Likewise in *Hyalophora* and other Saturniids there are two dominant types of haemocyte: plasmatocytes which spread over wound sites and foreign bodies and 'granulocytes' which (like the oenocytoids) are completely passive *in vitro* under all conditions.[472]

p. 436. **Phagocytosis**—The phagocytic blood cells in *Ephestia* are continually ingesting small droplets of haemolymph by 'pinocytosis';[436] in *Rhodnius* pinocytosis by the plasmatocytes leads to granule formation.[480]

p. 437. During the histolysis of larval tissues at metamorphosis in *Calliphora*[428] and *Galleria*[429] the haemocytes contain diverse cytolysome inclusions with strong acid phosphatase activity.

p. 437. **Encapsulation of parasites**—Nematodes invading larvae of *Chironomus* and *Aëdes* are encapsulated by a deposit of dense material from the haemolymph without the participation of blood cells; this encapsulation can even take place *in vitro*.[435] The egg of *Nemeritis* taken from the distal part of the ovarioles is encapsulated by the blood cells of *Ephestia*; in the calyx an outer coating is applied to the chorion and this enables the egg to escape investment by the haemocytes. Likewise the young larva acquires a resistant covering to the cuticle shortly before hatching (or artificially by exposure to fragments of the lower part of the female genital tract). The nature of the coating is not known.[459]

p. 438. **Blood coagulation**—In the pupa of *Tenebrio* two main types of blood cell are recognized: (i) highly polymorphous 'phagocytic cells', and (ii) oval or rounded explosive corpuscles which react at once on contact with foreign bodies or the outer atmosphere to induce clotting.[447] In *Locusta* the 'coagulocytes' represent a further differentiation of the plasmatocytes. Coagulation is initiated without any change in the granules of these cells, the function of which is not known[441] (see 'connective tissue formation' p. 439).

p. 438. **Connective tissue formation**—The capsule formed within three days around foreign bodies in *Ephestia* is a compact tissue 50–60 cells thick. The innermost cells form attenuated sheets about 0·3μ thick; the integrity of the cells is maintained but they adhere closely together with a conspicuous electron-dense material between.[436]

p. 439. Further evidence of the contribution of the plasmatocytes to connective tissue formation has been published.[409] In the emerging adult of *Sarcophaga* the spherule cells seem to accumulate material that is laid down as connective tissue strands.[475] The most convincing evidence is that all stages in the discharge of the 'tubular' inclusions in the plasmatocytes of *Rhodnius* and the incorporation of their substance into the basement membranes can be observed in electron microscope sections at the appropriate moment in the moulting process. These inclusions contain bound lipids.[480] On the other hand autoradiographic studies on haemocytes in *Galleria* gave no evidence of blood cells being involved in the formation of the neural lamella (p. 183).[464]

p. 439. **Enzymes in blood cells**—In *Antheraea* pupae, phenoloxidase occurs in intact corpuscles, as well as in the blood plasma, as an inert precursor, or prophenoloxidase.[431] In *Locusta* the 'coagulocytes' are involved.[440] In *Sarcophaga* the haemocytes also contain dopa-decarboxylase converting 'dopa' into dopamine (p. 47). Bursicon (p. 47) accelerates this synthesis, perhaps by breaking the barrier between tyrosine and the enzymes.[474]

p. 440. **Anti-bacterial factor**—The factor in the blood of *Galleria* active against a wide range of bacteria is a lysozyme. This is claimed to be the fundamental factor in humoral immunity in insects. Lysozyme is produced also in the mid-gut and regulates the bacterial flora.[452]

p. 440. **Teratocytes**—The teratocytes may draw upon the nutrients in the body of the host and weaken it so as to prevent an effective haemocytic reaction to the young parasite.[460] The 'pseudogerms', or growing fragments of the trophamnion (p. 729), may exercise a similar delibitating effect before being eaten by the parasite.[460]

p. 442. **Pericardial cells**—Studies with the electron microscope support the view that the pericardial cells are equivalent to the reticulo-endothelial system.[478] In *Myzus persicae*,[417] *Drosophila*,[407, 451] *Locusta*[437] and *Hyalophora*[462] the cells show exceedingly active pinocytosis producing 'coated vesicles' which appear to fuse and lead to the formation of abundant 'residual bodies', or cytolysomes. They show selective uptake of haemolymph protein and are perhaps a major site of protein turnover—both foreign proteins absorbed from the gut and waste proteins liberated in metabolism.[338, 478]

p. 445. **Cytological changes in the fat body**—In the advanced stages of starvation most of the mitochondria in *Rhodnius* fat body degenerate to form cytolysomes. The multiplication and division of mitochondria which follows feeding is inhibited by reduced oxygen supply following tracheal section; and so is the formation of watery vacuoles and the synthesis of RNA.[476] The disc-like caps which are present on every fat droplet (p. 446) are also the site of a very active NAD-diaphorase and other dehydrogenases; these enzyme sites have been named 'catalysomes'.[476] In *Calpodes* larva, proteins (including foreign proteins such as plant peroxidase) are taken up by pinocytosis and enclosed within paired membranes from the Golgi complex to form 'multi-vesicular bodies'. At pupation haemolymph proteins are taken up and sequestered in the same way, and the bodies enlarge rapidly to form storage protein granules. There is thus a close relation between the storage granules and the cytolysomes produced by autolysis of mitochondria within the cell. The protein storage granules arise mainly by uptake of protein rather than synthesis within the fat body. In *Calpodes* larvae much of this protein comes from the larval endocuticle which is being resorbed at this time.[446]

p. 446. **Fat body in the adult**—The accumulation and storage of fat is markedly affected by the juvenile hormone. To a large extent the accumulation which follows allatectomy is a feed-back effect from the ovary which ceases to take up lipid for yolk formation,[479] as in *Musca*;[406] but it may be that in some insects the hormone also has a direct effect on the fat body. In *Locusta* and other Orthoptera there is a sexual dimorphism in the fat body: in the female it is more rich in RNA, amino acids and phenolic compounds; in the male it has a higher content of glycogen and fat. The female characters are doubtless related with synthesis of vitellogenins (p. 704).[448]

p. 446. **Lipid and carbohydrate metabolism**—The release of diglyceride from the fat body (p. 426) is an endergonic process requiring ATP; it is inhibited by metabolic inhibitors (nitrophenol, azide &c.) whereas release of free fatty acids is accelerated by these chemicals.[422] Lipase activity can be demonstrated in the fat body.[432] In *Hyalophora* synthesis of trehalose is inhibited by excess trehalose in the medium;[454] it is much increased in *Leucophaea* in the presence of the hyperglycaemic peptide hormone from the corpus cardiacum (p. 199) which increases glycogen breakdown and diverts the hexose residues to trehalose synthesis.[473]

p. 446. **Protein metabolism**—That the fat body is an important site of protein synthesis is supported by observations on *Rhodnius*,[426] on *Calliphora*, where the protein 'calliphorin' which accounts for more than 50 per cent. of the soluble protein in the late larva is synthesized, along with other proteins, in the fat body,[453, 458] and on adult *Tenebrio* where at least five proteins are common to the fat body and the oöcytes.[455] On the other hand we have seen (p. 424) that proteins are taken up from the haemolymph and stored in the fat body. That is seen at the time of pupation in *Pieris*[425] and *Diatraea*;[424] in *Calpodes* and *Galleria*[427] and in *Philosamia*.[470]

p. 451. **Oenocytes and lipid metabolism**—Electron microscope studies of oenocytes in *Dacus*,[430] *Calpodes*[445] and the pupa of *Culex*[433] show that the cytoplasm is always densely packed with smooth tubular endoplasmic reticulum. In *Culex* lipid-containing vesicles are continually produced during cuticle formation. These observations support the view that the oenocytes are the site of synthesis of precursors for the production of wax or other lipids contributed to the integument or to the eggs.[477] In young workers of *Apis* isotopically labelled acetate is concentrated in the oenocytes and the secreted wax is strongly labelled.[457]

p. 454. **Inhibition of light production**—In *Luciola* it is claimed that there are two mechanisms of inhibition of flashing by illumination: a nervous effect acting at the level of the central nervous system; and a humoral effect due to a factor released in the male gonads acting on the photogenic organ itself.[411, 419]

p. 455. **Humoral control of luminescence**—Injected adrenaline or noradrenaline will induce glowing in the light organ of *Photinus*.[465] Extracts of the normal lantern of *Photuris* have adrenaline-like properties, and reserpine (a drug which drains vertebrate nerve endings of adrenergic transmitters) abolishes the luminescence following nervous stimulation.[420] It may be that the effect of adrenaline on light production is to stimulate ATP production.[420]

REFERENCES

[1] ALBRO, H. T. *J. Morph.*, **50** (1930), 527–52 (oenocytes in *Galerucella*, Col.).

[2] ALEXANDER, R. S. *J. Cell. Comp. Physiol.*, **22** (1943), 51–71 (control of fire-fly luminescence).

[3] ALEXANDROWICZ, J. S. *J. Comp. Neurol.*, **41** (1926), 291–309 (innervation of heart in cockroach).

[4] ANGLAS, J. *Bull. Sci. Fr. Belg.*, **34** (1900), 1–111 (histology during metamorphosis: *Vespa* and *Apis*).

[5] ARVY, L. *et al. Bull. Soc. Zool. France*, **69** (1944), 230–5; **70** (1945), 144–8; *Bull. Biol. Fr. Belg.*, **82** (1948), 37–60 (haemocytes: *Forficula*, *Leptinotarsa*).

[6] BABERS, F. H. *J. Agric. Res.*, **57** (1938), 697–706; **63** (1941), 183–90 (*p*H, buffer capacity of blood, &c.: *Prodenia* larva, Lep.).

[7] BACKMAN, E. L. *Arch. ges. Physiol.*, **149** (1912), 93–114 (osmotic pressure of blood: aquatic Coleoptera).

[8] BALZAM, M. *Ann. Inst. Pasteur*, **58** (1937), 181–211 (fate of the bacterial flora of *Calliphora*, Dipt. during metamorphosis).

[9] BARBER, H. S. *Proc. Ent. Soc. Wash.*, **15** (1913), 46–50 (luminous Collembola).

[10] BARRATT, J. O. W., and ARNOLD, G. *Quart. J. Micr. Sci.*, **56** (1910), 149–66 (composition of blood: *Dytiscus*, *Hydrophilus*, Col.).

[11] BEADLE, L. C. *J. Exp. Biol.*, **16** (1939), 346–62 (osmoregulation: *Aëdes detritus*).

[12] BERLESE, A. *Zool. Anz.*, **23** (1900), 441–9; **24** (1901), 515–21 (role of haemocytes during metamorphosis).

[13] BESS, H. A. *Ann. Ent. Soc. Amer.*, **32** (1939), 189–226 (immunity to parasites: *Pseudococcus*, Hem.).

[14] BISHOP, G. H. *J. Morph.*, **36** (1922), 567–60; **37** (1923), 533–53 (changes in fat body of honey-bee during growth and histolysis).

[15] BISHOP, G. H., *et al. J. Biol. Chem.*, **58** (1923), 543–65; **66** (1926), 77–88 (blood of honey-bee larva).

[16] BOCHE, R. D., and BUCK, J. B. *Physiol. Zool.*, **15** (1942), 293–303 (*p*H of insect blood).

[17] BODINE, J. H. *Biol. Bull.*, **51** (1926), 363–9 (*p*H in blood of grasshoppers).

[18] BOESE, G. *Z. Parasitenk.*, **8** (1936), 243–84 (response of blood cells and oenocytes in caterpillars towards Hymenopterous parasites).

[19] BOGOJAWLENSKY, K. S. *Z. Zellforsch. mikr. Anat.*, **22** (1935), 206–12 (fat body: *Bombyx mori*).

[20] BONÉ, G. J. *Ann. Soc. Roy. Zool. Belg.*, **75** (1944), 123–32; *Nature*, **160** (1947), 679–80 (sodium-potassium ratio in insect blood).

[21] BRECHER, L. *Z. vergl. Physiol.*, **2** (1925), 691–713; *Biochem. Z.*, **211** (1929), 40–64 (chemistry of blood: larvae and pupae of *Pieris* and *Vanessa*, Lep.).

[22] BROCHER, F. *Arch. Zool.*, **55** (1916), 347–73; **56** (1917), 347–58 (circulation of blood: *Dytiscus*, Col.); *Ibid.*, **56** (1917), 445–90 (the same: Odonata larvae); *Ibid.*, **60** (1920), 1–45 (the same: *Sphinx*, Lep.); *Ann. Ent. Soc. Fr.*, **89** (1920), 209–32 (the same: *Vespa*, Hym.); *Ibid.*, **91** (1922), 156–64 (the same: *Periplaneta*, Orth.); *Rev. suisse Zool.*, **36** (1929), 593–607 (circulation in wings: *Coccinella*, Col.); *Arch. Zool.*, **74** (1931), 25–32 (circulation of blood: résumé).

[23] BROWN, J. M. *Trans. Linn. Soc.* (*Zool.*), **11** (1910), 125–35 (anatomy of heart: *Tipula* larva, Dipt.).

[24] BRUNTZ, L. *Arch. Biol.*, **20** (1904), 217–422 (storage excretion in insects, &c.).

[25] —— *Arch. Zool.*, **8** (1908), 471–88 (phagocytosis and excretion: Thysanura).

[26] —— *Arch. Zool.*, (sér. 5), **2** (1909), xvii–xix (pericardial cells in Orthoptera).

[27] de BRUYNE, C. *Arch. Biol.*, **15** (1898), 181–300 (role of phagocytes in metamorphosis).

[28] BUCHNER, P. *Tier und Pflanze in Symbiose*, Berlin, 1930.

[29] BUCK, J. B. *Physiol. Zool.*, **10** (1937), 45–58; *Quart. Rev. Biol.*, **13** (1938), 301–14 (control of flashing in the fire-fly *Photinus*, Col.).

[30] —— *Proc. Rochester Acad. Sci.*, **8** (1941), 14–21 (spectrometric data on fire-fly light).

[31] —— *Ann. N.Y. Acad. Sci.*, **49** (1948), 397–482 (anatomy and physiology of light organs of fire-flies: review).

[32] BUXTON, P. A. *Parasitology*, **27** (1935), 263–5 (fat content of *Culex*, Dipt. during hibernation).

[33] CARPENTER, G. D. H., and ELTRINGHAM, H. *Proc. Zool. Soc. Lond.*, A, **108** (1938), 243–52 (reflex bleeding: *Lep.*).

[34] CHEN, H. T. *Z. Parasitenk.*, **6** (1934), 603–37 (reactions of blood cells in the flea, *Ctenocephalides* to *Dipylidium*).

[35] CHORINE, V. *Bull. Biol. Fr. Belg.*, **65** (1931), 291–393 (immunity in insects: *Galleria* Lep. &c.).

[36] CLARE, S., and TAUBER, O. E. *Ann. Ent. Soc. Amer.*, **35** (1942), 57–67; *Iowa State Coll. J. Sci.*, **16** (1942), 349–56 (circulation in wings of *Blattella*).

[37] CLAUS, A. *Zool. Jahrb., Physiol.*, **58** (1937), 365–432 (osmo-regulation in *Sigara* spp., Hem.).

[38] CRAIG, R., and CLARK, J. R. *J. Econ. Ent.*, **31** (1938), 51–4 (*p*H and buffer value in blood of Lepidoptera larvae).

[39] CREIGHTON, W. S. *Science*, **63** (1926), 600–1 (adrenaline and fire-fly luminescence).

[40] CRESCITELLI, F., and JAHN, T. L. *J. Cell. Comp. Physiol.*, **11** (1938), 359–76 (electrocardiogram: *Melanoplus*, Orth.).

[41] CROZIER, W. J., and FEDERIGHI, H. *J. Gen. Physiol.*, **7** (1925), 565–70 (effect of temperature on heart beat in *Bombyx mori*).

[42] CROZIER, W. J., and STIER, T. J. B. *J. Gen. Physiol.*, **10** (1927), 479–500 (temperature and rate of beat in 'accessory hearts' of *Notonecta*, Hem.).

[43] CUÉNOT, L. *Arch. Biol.*, **14** (1895), 293–341 (haemocytes, phagocytic organs: Orthoptera).

[44] —— *Arch. Zool.*, **4** (1896), 655–80 (reflex bleeding).

[45] DAHLGREN, U. *J. Franklin Inst.*, **183** (1917), 79–94; 211–20; 323–48; 593–624 (light production: review).

[46] DAKHNOFF, A. *C. R. Soc. Biol.*, **128** (1938), 520–3 (regeneration of blood: *Galleria* Lep.).

[47] DEBAISIEUX, P. *Ann. Soc. Sci. Bruxelles*, B, **56** (1936), 77–87 (pulsatile organs in tibia: *Notonecta*, Hem.).

[48] DEMJANOWSKI, S., *et al. Biochem. Z.*, **247** (1932), 386–405 (*p*H in blood of *Bombyx mori*).

[49] DENNELL, R. *Proc. Roy. Soc.*, B, **134** (1947), 79–110; 136 (1949), 94–109 (oenocytoids as source of tyrosinase: *Sarcophaga*, Dipt.),

[50] DINULESCU, G. *Ann. Sci. Nat., Zool.*, **15** (1932), 1–183 (physiology of *Gastrophilus*, Dipt.).

[51] DRILHON, A. *C. R. Soc. Biol.*, **115** (1934), 1194–5; **118** (1935), 131–2 (chemistry of haemolymph: Lepidoptera).

[52] DRILHON, A., and FLORENCE, G. *Bull. Soc. Chim. Biol.*, **28** (1946), 160–7 (amino acids in blood: *Attacus*, Lep.).

[53] DUBOIS, R. *Bull. Soc. Zool. Fr.*, **11** (1886), 1–275 (light production in Elateridae).

[54] DUBUISSON, M. *Arch. Biol.*, **39** (1929), 247–70; **40** (1930), 83–97; *Arch internat. Physiol.*, **32** (1930), 416–22 (automatism of heart: *Chironomus*, Dipt., *Agrion*, Odonata, *Hydrophilus*, Col.).

[55] DUVAL, M., and PORTIER, P., *et al. C. R. Soc. Biol.*, **99** (1928), 1831–2 (phosphorus in blood of insects); *C.R. Acad. Sci.*, **186** (1928), 652–3 (amino acids in blood of insects).

[56] DUWEZ, Y. *C. R. Soc. Biol.*, **122** (1936), 84–7 (automatism of heart: *Dytiscus*, Col.).

[57] EMERSON, G. A., and EMERSON, M. J. *Proc. Soc. Exp. Biol. Med.*, **48** (1941), 700–3 (effect of adrenaline on fire-fly luminescence).

[58] ERMIN, R. *Z. Zellforsch. mikr. Anat.*, A, **29** (1939), 613–69 (haemocytes: *Periplaneta*).

[59] EVANS, A. C. *Bull. Ent. Res.*, **26** (1935), 115–22 (fat body, &c. in *Lucilia* adult, Dipt.).

[60] FLORKIN, M. *Mém. Cour. Acad. Roy. Belg.*, **16** (1937), 69 pp. (chemistry of haemolymph).

[61] FLORKIN *et al.* Chemistry of insect blood; *Ass. Française avance. Sci.*, **63** (1939); *Arch. intern. Physiol.*, **50** (1940), 197–202 (ammonia); *Bull. Acad. roy. Belg. Cl. Sci.*, (sér 5), **28** (1942), 373–6 (amino-acids); *Acta. biol. Belg.*, 1943, 1 (proteins); *Bull. Soc. roy. Sci. Liège*, 1943, 301–4 (ionic composition).

[62] FOX, H. M. *Proc. Roy. Soc.*, B, **112** (1933), 479–95 (effect of carbon dioxide on heart beat: *Cloëon* nymphs, Ephem.).

[63] FOX, H. M., and BALDES, E. J. *J. Exp. Biol.*, **12** (1935), 174–8 (osmotic pressure of blood: Ephemeroptera nymphs).

[64] FOX, H. M., and SIMMONDS, B. G. *J. Exp. Biol.*, **10** (1933), 67–74 (metabolic rates for Ephemerid nymphs from different habitats).

[65] FREUDENSTEIN, K. *Z. wiss. Zool.*, **132** (1928), 404–75 (heart and circulation: honey-bee).

[66] FRIES, E. F. B. *J. Gen. Physiol.*, **10** (1929), 227–37 (temperature and rate of heart beat in cockroach).

[67] FULTON, B. B. *Ann. Ent. Soc. Amer.*, **34** (1941), 289–302 (luminous Mycetophilid).

[68] GEE, W. P. *Biol. Bull.*, **21** (1911), 222–34 (oenocytes of *Platyphylax*, Trichopt.).

[69] GEROULD, J. H. *J. Morph.*, **48** (1929), 385–430; *Biol. Bull.*, **56** (1929), 215–25; *Science*

71 (1930), 264–5; **73** (1931), 323–5; *Biol. Bull.*, **64** (1933), 424–31 (reversal of heart beat: silkworm, &c.).

[70] GEROULD, J. H. *Acta Zool.*, **19** (1938), 297–352 (heart of *Bombyx mori*).

[71] GERRETSEN, F. C. *Biol. Zentralbl.*, **42** (1922), 1–9 (control of light production: *Luciola*, Col.).

[72] GEYER, K. *Z. wiss. Zool.*, **105** (1913), 249–499 (sexual differences in insect blood).

[73] GLASER, R. W. *Psyche*, **25** (1918), 39–46 (immunity: caterpillars and grasshoppers).

[74] —— *J. Gen. Physiol.*, **7** (1925), 599–602 (*p*H in blood of insects).

[75] GRABER, V. *Arch. mikr. Anat.*, **9** (1873), 129–96; **12** (1876), 575–82 (mechanism of circulation).

[76] GRASSÉ, P. P. *C.R. Acad. Sci.*, **204** (1936), 65–7 (reflex bleeding: *Dictyophorus*, Orth.).

[77] GRASSÉ, P. P., and LESPERON, L. *C.R. Acad. Sci.*, **201** (1935), 618–20 (storage of dyes in the silkworm).

[78] HABER, V. R. *Bull. Brooklyn Ent. Soc.*, **21** (1926), 61–100 (blood of cockroach, *Blattella*).

[79] HAMILTON, H. J. *J. Cell. Comp. Physiol.*, **13** (1939), 91–103 (action of drugs on heart: *Melanoplus*, Orth.).

[80] HAMILTON, M. A. *Proc. Zool. Soc. Lond.*, 1931, 1063–1136 (anatomy of *Nepa*, Hem.).

[81] HARVEY, E. N. *J. Gen. Physiol.*, **15** (1931), 139–45 (photocell analysis of light from *Pyrophorus*, Col.).

[82] —— *Ergeb. Enzymforsch.*, **4** (1935), 333 (light production: review); *Living Light*, Princeton, 1940.

[83] —— *J. Cell. Comp. Physiol.*, **23** (1944), 31–8; **26** (1945), 185–7 (luminescence: *Phryxothrix*).

[84] HARVEY, E. N., and HALL, R. T. *Science*, **69** (1929), 253–4 (effect of removal of luminous organs in larva of fire-fly).

[85] HEIDT, K. *Biol. Zentralbl.*, **56** (1936), 100–9 (light production in Collembola).

[86] HELLER, J., *et al. Biochem. Z.*, **219** (1930), 473–89; **255** (1932), 205–21 (blood in metamorphosis: *Deilephila*, Lep.).

[87] HEMMINGSEN, A. M. *Skand. Arc. Physiol.*, **45** (1924), 204–10 (sugar in blood: Lepidoptera larvae).

[88] HENNEGUY, L. F. *Les Insectes*, Paris, 1904.

[89] HOLLANDE, A. C. *Arch. Zool.*, (sér. 5), **2** (1909). 271–94; *Ibid.*, **6** (1911), 283–323 (blood cells).

[90] —— *Arch. Zool.*, **5** (1914), 559–78 (albuminoid and urate deposits in fat body of *Vanessa* larvae, Lep.).

[91] —— *Arch. Anat. Microsc.*, **16** (1914), 1–64 (oenocytes: review and original).

[92] —— *Arch. Zool.*, **59** (1920), 543–63 (encapsulation of parasites by blood cells).

[92a] —— *Arch. Anat. Microsc.*, **18** (1922), 85–307 (pericardial cells: review and original).

[93] —— *Arch. Anat. Microsc.*, **22** (1927), 374–412 ('reflex bleeding').

[94] HOLLANDE, A. C., *et al. C. R. Soc. Biol.*, **99** (1928), 120–2; *Arch. Zool.*, **69** (1929), 1–11; **70** (1930), 231–80 (blood cells and phagocytosis of bacteria, &c.: Lepidoptera larvae).

[95] HOSSELET, C. *C.R. Acad. Sci.*, **180** (1925), 399–401 (cytology of oenocytes: *Culex*, Dipt.).

[96] HUFF, C. G. *Physiol. Rev.*, **20** (1940), 68–88 (immunity in insects).

[97] HUFNAGEL, A. *Arch. Zool.*, **57** (1918), 47–202 (blood cells in metamorphosis: *Hyponomeuta*, Lep.).

[98] HUNGERFORD, H. B. *Canad. Entom.*, **54** (1922), 262–3 (haemoglobin in cells of *Buenoa*, Hem.).

[99] IWASAKI, Y. *Arch. Anat. Microsc.*, **23** (1927), 319–46 (encapsulation of foreign bodies in caterpillars).

[100] JAHN, T. L., *et al. J. Cell. Comp. Physiol.*, **10** (1937), 439–60 (electrocardiogram: *Melanoplus*).

[101] JANET, C. *Bull. Soc. Ent. Fr.*, 1907, 350–1 (origin of imaginal fat body in Muscids).

[102] JUCCI, C., and DEIANA, G. *Bol. Soc. Ital. Biol. Sperim.*, **5** (1930), 167–70 (uric acid in blood of silkworm).

[103] KARAWAIEW, W. *Z. wiss. Zool.*, **64** (1898), 385–478 (post-embryonic development in *Lasius*, Formicid.; oenocytes, &c.).

[104] KEILIN, D. *Parasitology*, **9** (1917), 325–450 (nephrocytes in larvae of Anthomyidae, Dipt.).

[105] KEILIN, D., and WANG, Y. L. *Biochem. J.*, **40** (1946), 855–66 (haemoglobin of *Gastrophilus*, Dipt.).

[106] v. KEMNITZ, G. A. *Z. Biol.*, **67** (1916), 129–244 (metabolism in *Gastrophilus* larva, Dipt.).

[107] KEMPER, H. *Z. Parasitenk.*, **5** (1932), 112–37 (fate of mammalian blood cells in body cavity of the bed-bug *Cimex*, Hem.).

[108] KOCIAN, V., and ŠPAČEK, M. *Zool. Jahrb., Physiol.*, **54** (1934), 180–90 (*p*H in blood of Coleoptera).

[109] KOCH, A. *Z. Morph. Oekol. Tiere*, **37** (1940), 38–62 (oenocytes: *Tribolium*, Col.).

[110] KOCH, J. *Rev. Suisse Zool.*, **52** (1945), 415–20 (oenocytes: *Drosophila*).

[111] KOLLER, G. *Biol. Rev.*, **4** (1929), 269–306 (internal secretions in insects: review).

[112] KOLLMANN, M. *Bull. Soc. Zool. Fr.*, **34** (1909), 149–55 (albuminoid reserves in fat body: *Tenebrio*, Col.).

[113] KOSCHEVNIKOV, G. A. *Zool. Anz.*, **23** (1900), 337–53 (fat body and oenocytes: honeybee).

[114] KOWALEVSKY, A. *Biol. Centralbl.*, **9** (1889), 33–47; *Congr. internat. Zool. Moscow*, **1** (1892), 187–234 (pericardial cells and storage excretion).

[115] —— *Arch. Zool.*, (sér. 3), **2** (1894), 485–90 (ventral ostia in heart of Orthoptera).

[115a] KRAMER, S., and WIGGLESWORTH, V. B. *Quart. J. Micr. Sci.*, **91** (1950), 63–72 (function of oenocytes: *Periplaneta*).

[116] KREMER, J. *Zool. Jahrb., Anat.*, **40** (1917), 105–54; **41** (1920), 175–272 (fat body and oenocytes in Coccinellidae, Col.).

[117] —— *Z. microsk. anat. Forsch.*, **2** (1925), 536–81; **4** (1926), 290–345 (oenocytes and fat body in Coleoptera).

[118] KREUSCHER, A. *Z. wiss. Zool.*, **119** (1922), 247–84 (fat body and oenocytes in *Dytiscus*, Col.).

[119] KREY, J. *Zool. Jahrb., Physiol.*, **58** (1937), 201–25 (physiology of heart: Trichoptera larvae).

[120] KUHL, W. *Zool. Jahrb., Anat.*, **46** (1924), 75–198 (heart, &c. in *Dytiscus*, Col.).

[121] KUWANA, Z. *Bull. Imp. Sericult. Exp. Sta.* (Tokyo), **8** (1932), 116–20 (innervation of heart: *Bombyx mori*).

[122] —— *Japanese J. Zool.*, **7** (1937), 273–303 (reducing power in blood of *Bombyx mori*).

[123] —— *Japanese J. Zool.*, **7** (1937), 305–9 (uric acid in blood, &c., in *Bombyx mori*).

[124] LANGE, H. H. *Z. Zellforsch. mikr. Anat.*, **16** (1932), 753–805 (phagocytic tissue in Chironomid larvae).

[125] LARTSCHENKO, K. *Z. Parasitenk.*, **5** (1933), 679–707 (encapsulation of parasites in larvae of *Loxostege* and *Pieris*, Lep.).

[126] LASCH, W. *Z. allg. Physiol.*, **14** (1913), 312–19 (control of heart beat: *Lucanus* larva, Col.).

[127] LAZARENKO, T. *Z. mikr. anat. Forsch.*, **3** (1925), 409–99 (connective tissue and encapsulation of foreign bodies: *Oryctes* larva, Col.).

[128] LEBRUN, H. *La Cellule*, **37** (1926), 183–200 (heart, &c., in *Corethra* larva, Dipt.).

[129] LESPERON, L. *Arch. Zool.*, **79** (1937), 1–156 (excretion, pericardial cells, &c.: *Bombyx mori*).

[130] LEVENBOOK, L. *Biochem. J.*, **47** (1950), 336–46 (composition of blood in *Gastrophilus* larva).

[131] LEVENBOOK, L., and WANG, Y. L. *Nature*, **162** (1948), 731–2 (succinic acid in blood of *Gastrophilus*).

[132] LIEBMAN, E. *Growth*, **10** (1946), 291–330 (haemocytes).

[133] LISON, L. *Mém. Acad. Roy. Belg. Cl. Sci.* (8), **19** (1942), 1–106 (athrocytosis of dyes in insects).

[134] LOTMAR, R. *Landw. Jb.*, **53** (1939), 34–70; *Schweiz. Bienen-Zeitung*, Beiheft. **1** (1945), 443–506 (protein in fat body; and role of blood cells in transport of reserves: *Apis*).

[135] LUND, E. J. *J. Exp. Zool.*, **11** (1911), 415–68 (photogenic organs: Lampyridae).

[136] MALOEUF, N. S. R. *Ann. Ent. Soc. Amer.*, **28** (1935), 332–7 (automatism in heart: *Anax*, Odonata; *Belostoma*, Hem.).

[137] MALOEF, N. S. *Science Progress*, 1937, 228–45 (light production: review).
[138] —— *Ann. Ent. Soc. Amer.*, **31** (1938), 374–80 (flashing of *Photuris*, Col.).
[139] —— *Quart. Rev. Biol.*, **14** (1939), 149–91 (insect blood: review).
[140] MANGOLD, E. *Handb. norm. path. Physiol.*, **8** (2), 1072–92 (light production: review).
[141] MARCHAL, P. *Richet's 'Dictionaire de Physiologie'*, **9** (1910), 273–386 (insect physiology).
[142] MASERA, E. *Riv. Biol.*, **15** (1933), 225–34 (rate of heart beat: *Bombyx mori*).
[143] MAY, R. M. *Bull. Soc. Chim. Biol.*, **17** (1935), 1045–53 (reducing substance and chloride in blood of Orthoptera).
[144] MCINDOO, N. E. *Smithson. Misc. Coll.*, **82** (1931), No. 18, 1–70 ('reflex bleeding' in Coleoptera).
[145] —— *J. Morph.*, **65** (1939) 323–51 (segmented blood vessels: *Periplaneta*).
[146] MELLANBY, H. *Quart. J. Micr. Sci.*, **79** (1936), 1–42 (later embryology: *Rhodnius*, Hem.).
[147] MELLANBY, K. *Biol. Rev.*, **14** (1939), 243–60 (functions of insect blood).
[148] METALNIKOV, S. *Arch. Zool.*, (sér. 4), **8** (1908), 489–588 (blood cells and phagocytosis: *Galleria*, Lep.); *C. R. Soc. Biol.*, **83** (1920), 214–15 (phagocytosis of tubercle bacilli; *Galleria*); *V^e congr. Internat. Entom. Paris, 1932*, **2** (1933), 209–20 (immunity in insects).
[149] METALNIKOV, S., and TOUMANOV, C. *C. R. Soc. Biol.*, **103** (1930), 965–7 (blood cells and phagocytosis: larva of honey-bee).
[150] MEYER, E. *Z. Morph. Oekol. Tiere*, **22** (1931), 1–52 (circulation: Ephemeroptera).
[151] MEYER, P. *J. Physiol. Path. Gen.*, **34** (1936), 448–53 (colloid osmotic pressure of blood: *Mantis*, Orth.).
[152] MEYER, P. F. *Z. vergl. Physiol.*, **11** (1930), 173–209 (plant pigments in blood of caterpillars).
[153] MILLARA, P. *Bull. Biol. Fr. Belg.*, **81** (1947), 129–53 (haemocytes).
[154] MUNSON, S. C., and YEAGER, J. F. *Ann. Ent. Soc. Amer.*, **37** (1944), 396–400 (fat in blood cells: *Prodenia*, Lep.).
[155] MURRAY, F. V., and TIEGS, O. W. *Quart. J. Micr. Sci.*, **77** (1935), 405–95 (histology of *Calandra*, Col. during metamorphosis).
[156] MUTTKOWSKI, R. A. *Bull. Brooklyn Ent. Soc.*, **18** (1923), 127–36; **19** (1924), 4–19; 128–44 (blood cells and coagulation).
[157] NAKAHARA W. *J. Morph.*, **30** (1917), 483–525 (amitosis in fat body cells).
[158] NEWPORT, G. *Phil. Trans. Roy. Soc.*, **127** (1837), (temperature and circulation in insects).
[159] NUTTALL, G. H. F., and KEILIN, D. *Parasitology*, **13** (1921), 184–92 (nephrocytes in *Pediculus*).
[160] OCHSÉ, W. *Rev. Suisse Zool.*, **53** (1946), 534–7 (connective tissue).
[161] —— *Rev. Suisse Zool.*, **53** (1946), 39–71 (oenocytes: *Sialis*, Megal.).
[162] OPOCZYŃSKA-SEMBRATOWA, Z. *Bull. Int. Acad. Polon. Sci. Lett. Cl. Sci. Nat.*, B, 1936, 411–36 (anatomy and nerves to heart: *Dixippus*, Orth.).
[163] PAILLOT, A., and NOEL, R. *C.R. Acad. Sci.*, **182** (1926), 1044–46 (origin of albuminoid inclusions in fat cells).
[164] —— *Bull. Histol. Appl. Physiol., Path.*, **5** (1928), 1–20; 56–78; 105–28 (histophysiology of pericardial cells, fat body, and haemocytes: Lepidoptera larvae).
[165] —— *C.R. Acad. Sci.*, **205** (1937), 1095–6 (fat body: *Bombyx*).
[166] —— *C. R. Soc. Biol.*, **127** (1938), 1504–6 (giant cells in parasitized insects).
[167] PANTEL, J. *La Cellule*, **15** (1898), 7–290 (anatomy of *Thrixion*, Tachinid, Dipt.).
[168] —— *La Cellule*, **29** (1914), 381–9 (posterior region of dorsal vessel in Muscid larvae).
[169] PARDI, L. *Monit. Zool. Ital.*, **49** (1938), 108–15; **50** (1939), 88–93 (oenocytes: *Melasoma* Col., *Polistes* Hym., &c.).
[170] —— *Redia*, **25** (1939), 87–288 (fat body of insects).
[170a] PATTON, R. L., and CRAIG, R. *J. Exp. Zool.*, **81** (1938), 437–57 (osmotic pressure of blood: *Tenebrio*).
[171] PAUSE, J. *Zool. Jahrb., Physiol.*, **36** (1918), 339–452 (circulation, &c.: larva of *Chironomus*, Dipt.).
[172] PAWLOWA, M. *Zool. Anz.*, **18** (1895), 7–13 (accessory hearts in head of Orthoptera).
[173] PÉREZ, C. *Bull. Sci. Fr. Belg.*, **37** (1902), 195–427 (histological changes in metamorphosis: *Formica*, Hym.); *Arch. Zool.*, **4** (1910), 1–274 (ditto: *Calliphora*, Dipt.); *Mém. Cl. Sci. Acad. Roy. Belg.*, **3** (1911), 1–103 (ditto: *Polistes*, Hym.).

[174] PEREZ, C. *Arch. Zool.*, **59** (1920), 5–10 (fat body inclusions during metamorphosis).
[175] PFEIFFER, H., and STAMMER, H. J. *Z. Morph. Oekol. Tiere*, **20** (1930), 136–71 (pathogenic luminescence in insects).
[176] PHILIPTSCHENKO, J. *Z. wiss. Zool.*, **88** (1907), 99–116 (pericardial septum and phagocytic organs: *Ctenolepisma*, Thysanura).
[177] POISSON, R. *Bull. Biol. Fr. Belg.*, **58** (1924), 49–202 (haemocytes, &c.: aquatic Hemiptera).
[178] POISSON, R. *Arch. Zool.*, **65** (1926), 181–208 ('tracheal cells' in *Anisops*; Hem.).
[179] POISSON, R., and PESSON, P. *Arch. Zool. exp. gén.*, **81** (1939), Notes et Rev. 23–32 (blood cells in *Pulvinaria*, Hem.).
[180] POLL, M. *Rec. Inst. Zool. Torley-Rousseau*, **5** (1934), 73–126 (storage excretion, &c. in *Tenebrio*, Col.).
[181] POPOVICI-BAZNOSANU, A. *Jena Z. Naturwiss.*, **40** (1905), 667–96; *Arch. Zool.* (sér. 4), **5** (1906), notes et revue, 66–78; *Zool. Anz.*, **35** (1910), 628–30 (heart and mechanism of circulation: Chironomid larvae, Ephemeroptera, &c.).
[182] PORTIER, P., and DUVAL, M. *C. R. Soc. Biol.*, **97** (1927), 1605–6 (osmotic pressure and chloride content in blood of insects).
[183] POYARKOFF, E. *Arch. Anat. Microsc.*, **12** (1910), 333–474 (histology during moulting and metamorphosis: *Galerucella*, Col.).
[184] PRENANT, A. *Arch. Anat. Microsc.*, **3** (1900), 292–336 ('tracheal cells' of *Gastrophilus* larva, Dipt.).
[185] RADU, V. *Arch. Anat. Microsc.*, **28** (1932), 107–20 ('tracheal cells' of *Gastrophilus*, Dipt.).
[186] RAPER, R., and SHAW, J. *Nature*, **162** (1948), 999 (amino-acids in blood: *Aeschna*, Odon.).
[187] RICHARDSON, C. H., *et al. Ann. Ent. Soc. Amer.*, **24** (1931), 503–7 (blood volume: Lepidoptera larva).
[188] RIES, E. *Z. Morph. Oekol. Tiere*, **25** (1932), 184–234 (haemocytes during wound healing and removal of dead tissue).
[189] ROEDER, K. D. *J. Cell. Comp. Physiol.*, **32** (1948), 327–38 (effect of potassium and calcium on nervous system: *Periplaneta*).
[190] ROOSEBOOM, M. *Arch. Néerl. Zool.*, **2** (1937), 432–559 (haemocytes: review).
[191] RÖSSIG, H. *Zool. Jahrb., Syst.*, **20** (1904), 19–20 (oenocytes, &c., Cynipid larvae).
[192] ROUBAUD, E. *C.R. Acad. Sci.*, **194** (1932), 389–91 (development of fat body in imago of *Culex*, Dipt).
[193] ROUSCHAL, W. *Z. wiss. Zool.*, **153** (1940), 196–218 (osmotic pressure of insect blood).
[194] SCHMIEDER, R. G. *J. Morph.*, **45** (1928), 121–86 (fat body during growth and metamorphosis: Tenthredinidae).
[195] SCHNELLE, H. *Zool. Anz.*, **57** (1923), 172–9 (fat body of honey-bee).
[196] SEMICHON, L. *Bull. Sci. Fr. Belg.*, **40** (1906), 281–439; *Bull. Soc. Ent. Fr.*, **38** (1933), 319–20 (inclusions in fat body).
[197] SHARIF, M. *Phil. Trans. Roy. Soc.*, B, **227** (1937), 465–538 (heart of flea larva, *Nosopsyllus*).
[198] de SINÉTY, R. *La Cellule*, **19** (1901), 119–278 (anatomy of Phasmids).
[199] SNELL, P. A. *Science*, **73** (1931), 372–3 (control of flashing in Lampyrid fire-flies).
[200] SNODGRASS, R. E. *Anatomy and physiology of the honey-bee*, New York, 1925.
[201] STAMMER, H. J. *Z. Morph. Oekol. Tiere*, **26** (1932), 135–46 (light production in *Ceroplatus*, Mycetophilidae, Dipt.).
[202] STAMMER, H. J. *Biol. Zentralbl.*, **55** (1935), 178–82 (light production in *Achorutes*, Collemb.).
[203] STECHE, O. *Verh. deutsch. zool. Ges.*, **23** (1913), 272–81 (sexual differences in insect blood).
[204] STEHR, G. *Rev. Suisse Zool.*, **54** (1947), 573–608 (circulation in pupal wings: *Ephestia*, Lep.).
[205] STEINER, G. *Z. vergl. Physiol.*, **16** (1932), 290–304 (control of heart beat: *Periplaneta*, Orth.).
[206] STEINHAUS, E. A. *Insect Microbiology*, Ithaca, N.Y., 1946; Steinhaus, E. A., Ed. *Insect Pathology, an Advanced Treatise*, Academic Press, 1963.
[207] STENDELL, W. *Z. wiss. Zool.*, **102** (1912), 136–68 (oenocytes in *Ephestia*, Lep.).

[208] TAUBER, O. E., *et al. Ann. Ent. Soc. Amer.*, **33** (1940), 113–19; *Proc. Soc. Exp. Biol. Med.*, **51** (1942), 45–7 (bacterial immunity: *Blatta*).
[209] TAUBER, O. E., and CLARE, S. *Trans. Amer. Micr. Soc.*, **61** (1942), 290–6 (circulation in wings: *Blattella*).
[210] TAUBER, O. E., and YEAGER, J. F. *Ann. Ent. Soc. Amer.*, **28** (1935), 229–40; **29** (1936), 112–8; *Iowa State Coll. J. Sci.*, **9** (1934), 13–24 (total blood cell count in various insects).
[211] TAYLOR, A. *Ann. Ent. Soc. Amer.*, **28** (1935), 135–45 (changes in blood cell types: *Periplaneta*, Orth.).
[212] TAYLOR, I. R., *et al. Physiol. Zool.*, **7** (1934). 593–9 (*p*H changes in *Galleria*, Lep. during metamorphosis).
[213] THOMSEN, E. *Z. Morph. Oekol. Tiere*, **34** (1938), 416–38 (circulation in wings of Muscidae).
[214] TIEGS, O. W. *Trans. R. Soc. S. Australia*, **46** (1922), 319–527 (histology during metamorphosis: *Nasonia* (*Mormoniella*), Hym.).
[215] TIMON-DAVID, J. *Ann. Fac. Sci. Marseille*, **13** (1945), 237–308; *Année Biol.*, **21** (1945) 134–54 (internal medium of insects).
[216] TIRELLI, M. *Arch. Zool. Ital.*, **22** (1936), 279–307 (reversal of heart beat during meta morphosis: *Saturnia*, Lep.).
[217] TOBIAS, J. M. *J. Cell. Comp. Physiol.*, **31** (1948), 125–42; 143–48 (sodium–potassium ratio: *Periplaneta*, *Bombyx mori*).
[218] TÓTH, L. *Z. Morph. Oekol. Tiere*, **33** (1937), 412–37 (role of oenocytes in nutrition: *Pemphigus*, Aphid.).
[219] TZONIS, K. *Zool. Anz.*, **116** (1936), 81–90 (heart beat in larva of *Corethra*, Dipt.).
[220] USSING, H. H. *Acta. Physiol. Scand.*, **11** (1946), 61–84 (amino-acids and amides in blood: *Melolontha*, &c., Col.).
[221] VERSON, E. *Zool. Anz.*, **38** (1911), 295–301 (oenocytes, pericardial cells, &c.).
[222] VERSON, E., and BISSON, E. *Bull. Soc. Ent. Ital.*, **23** (1891), 3–20; *Zool. Anz.*, **23** (1900), 657–61 (oenocytes in silkworm).
[223] VERWORN, M. *Centralbl. Physiol.*, **6** (1892), 69–74 (control of light production: *Luciola*, Col.).
[224] VOGEL, R. *Zool. Anz.*, **41** (1913), 325–32 (development of photogenic organs: *Lampyris*, Col.).
[225] WALKER, J. F. *J. Cell. Comp. Physiol.*, **6** (1935), 317–34 (effect of carbon dioxide on beat of lateral body walls of grasshopper embryo: *Melanoplus*).
[226] WEBER, H. *Biologie der Hemipteren*, Springer, Berlin, 1930.
[227] —— *Lehrbuch der Entomologie*, Jena, 1933.
[228] —— *Zoologica*, **33** (1935), No. 6, 71 pp. (structure of imago in Aleurodids).
[229] WEISSENBERG, R. *Zool. Jahrb., Anat.*, **23** (1907), 231–68 (oenocytes in *Torymus*, Chalcid.).
[230] WETTINGER, O. *Z. wiss. Zool.*, **129** (1927), 453–82 (circulatory system in Tipulid larvae).
[231] WHEDON, A. D. *J. Morph.*, **63** (1938), 229–49 (aortic diverticula: Odonata).
[232] WHEELER, W. M. *Psyche*, **6** (1892), 216–20; 233–6; 253–8 (blood cells, fat body, oenocytes, &c.).
[233] WHEELER, W. M., and WILLIAMS, F. Q. *Psyche*, **22** (1915), 36–43 (luminous organ of *Bolitophila*, Tipulid).
[234] WIELOWIEJSKI, H. R. *Z. wiss. Zool.*, **43** (1886), 512–36 (fat body, pericardial cells and oenocytes).
[235] WIGGLESWORTH, V. B. *Nature*, **127** (1931), 307–8 (fate of mammalian blood cells in body cavity of the bed-bug *Cimex*, Hem.).
[236] —— *Quart. J. Micr. Sci.*, **76** (1933), 269–318 (haemocytes, oenocytes, &c.: *Rhodnius*, Hem.).
[237] —— *Quart. J. Micr. Sci.*, **77** (1934), 191–222 (anterior ending of dorsal vessel: *Rhodnius*, Hem.).
[238] —— *J. Exp. Biol.*, **14** (1937), 364–81 (epidermal cells and haemocytes during wound healing: *Rhodnius*, Hem.).
[239] —— *J. Exp. Biol.*, **15** (1938), 235–47 (chloride and osmotic pressure in blood of mosquito larvae).
[240] —— Fat body and oenocytes. *J. Exp. Biol.*, **19** (1942), 56–77 (larva of *Aëdes*, Dipt);

Proc. Roy. Soc., B, **134** (1947), 163–181 (*Rhodnius*, Hem.); *Quart. J. Micr. Sci.*, **89** (1948), 197–217 (*Tenebrio*, Col.); *J. Exp. Biol.*, **26** (1949), 150–63 (adult *Drosophila, Dipt.*).

[241] WIGGLESWORTH, V. B. *Proc. Roy. Soc.*, B, **131** (1943), 313–39 (biliverdin in pericardial cells: *Rhodnius* Hem.).

[242] de WILDE, J. *Arch. Néerl. Physiol.*, **28** (1948), 530–42 (function of alary muscles).

[243] WILLERS, W. *Z. wiss. Zool.*, **116** (1916), 43–74 (oenocytes, &c., during moulting).

[244] YEAGER, J. F., and HENDRICKSON, G. O. *Proc. Soc. Exp. Biol. Med.*, **30** (1933), 858–60; *Ann. Ent. Soc. Amer.*, **27** (1934), 257–72 (circulation in wings of cockroach).

[245] YEAGER, J. F., and KNIGHT, H. H. *Ann. Ent. Soc. Amer.*, **26** (1933), 591–602 (blood coagulation in various insects).

[246] YEAGER, J. F., and TAUBER, O. E. *Ann. Ent. Soc. Amer.*, **25** (1932), 315–27 (total blood volume in the cockroach).

[247] YEAGER, J. F. *J. Agric. Res.*, **56** (1938), 267–76; *Ann. Ent. Soc. Amer.*, **32** (1939), 44–8 (mechanical records of cardiac activity: *Periplaneta*, Orth.).

[248] —— *J. Agric. Res.*, **59** (1939), 121–38 (electrical stimulation of the heart: *Periplaneta*).

[249] —— *J. Agric. Res.*, **71** (1945), 1–40 (haemocytes: *Prodenia*, Lep.).

[250] YEAGER, J. F., and MUNSON, S. C. *J. Agric. Res.*, **63** (1941), 257–94 (glycogen in haemocytes: *Prodenia*, Lep.).

[251] YOKOYAMA, T. *Proc. Imp. Acad.* (*Tokyo*), **5** (1929), 483–6; *Bull. Imp. Sericult. Exp. Sta.* (Tokyo), **8** (1932), 100–2; (normal and reversed heart beat in *Bombyx mori*).

[252] —— *Bull. Imp. Sericult. Exp. Sta.* (Tokyo), **8** (1934), 539–50 (glycogen in fat body, &c.: *Bombyx mori*).

[253] —— *Proc. Roy. Ent. Soc. Lond.* A, **11** (1936), 35–44 (histology of a non-moulting strain of silkworm).

[254] —— *Proc. Imp. Acad. Tokyo*, **15** (1939), 94–7 (circulation in silkworm).

[255] YOUNG, B. *Sinensia*, **16** (1945), 37–43 (circulation in wings: *Cryptotympana*, Hem.).

[256] ZAVŘEL, J. *Publ. Fac. Sci. Univ. Masaryk*, No. 213, 1935, p. 18, No. 257 (1938), 1–23 (oenocytes, &c., in *Syndiamesa*, Chironomid.).

[257] ZAWARZIN, A. *Z. wiss. Zool.*, **97** (1910), 481–510 (histology and nerve supply of heart in *Aeschna* larva, Odonata).

[258] ZELLER, H. *Z. Morph. Oekol. Tiere*, **34** (1938), 663–738 (blood cells and fat body in wing: *Ephestia*, Lep.).

[259] ZERNOFF, V. *Ann. Inst. Pasteur*, **46** (1931), 565–71 (bacteriolysins and immunity in insects).

SUPPLEMENTARY REFERENCES (A)

[260] BEADLE, L. C., and SHAW, J. *J. Exp. Biol.*, **27** (1950), 96–109 (osmoregulation in larva of *Sialis*).

[261] BEARD, R. L. *Physiol. Zool.*, **23** (1950), 47–57 (coagulation of insect blood).

[262] FINLAYSON, L. H., and HAMER, D. *Nature*, **163** (1949), 843–4 (amino-acids in blood of *Calliphora*).

[263] GRÉGOIRE, C., and FLORKIN, M. *Physiol. compar. oekol.*, **2** (1950), 126–39; *Blood*, **6** (1951), 1173–98; *Biol. Bull.*, **104** (1953), 372–93; *Arch. Biol.*, **66** (1955), 103–48 ('hyaline haemocytes' and coagulation of insect blood).

[264] KRIJGSMAN, B. J., *et al. Nature*, **165** (1950), 936–7; *Bull. Ent. Res.*, **41** (1950), 141–51; **42** (1951), 143–55; *Biol. Rev.*, **27** (1952), 320–46 (nervous control of heart beat in insects).

[265] de LERMA, B. *Boll. Zool.*, **16** (1949), No. 4; *Boll. Soc. Ital. Biol. Sperim.*, **26** (1950), No. 4 (riboflavine and pterines in blood of Orthoptera).

[266] LEVENBOOK, L. *Biochem. J.*, **47** (1950), 336–46 (composition of blood of *Gastrophilus* larva).

[267] —— *J. Exp. Biol.*, **27** (1950), 184–91 (buffer capacity of blood of *Gastrophilus* larva).

[268] MCELROY, W. D., *et al. Proc. Nat. Acad. Sci.*, **33** (1947), 342–5; *J. cell. comp. Physiol.*, **32** (1948), 421–5 (adenosine triphosphate and light production in *Photinus*).

[269] PRATT, J. J. *Ann. ent. Soc. Amer.*, **43** (1950), 573–80 (free amino-acids in insect blood).

[270] RAMSAY, J. A. *J. Exp. Biol.*, **27** (1950), 145–57; **28** (1951), 62–73 (osmotic regulation in mosquito larvae).

[271] SCHALLER, F. *Z. vergl. Physiol.*, **31** (1949), 684–95 (osmotic regulation in *Corethra* larva).

[272] SCHNEIDER, F. *Vierteljahrsschr. d. Naturforsch. Gesell. Zurich*, **95** (1950), 22–44; *Z. angew. Entom.*, **33** (1951), 150–62 (resistance of Syrphid larvae to entomophagous parasites).

[273] SCHOFFENIELS, E. *Arch. intern. Physiol.*, **58** (1950), 1–4 (osmotic regulation in Odonata larvae).

[274] SCHWINCK, I. *Arch. Entw. Mech.*, **145** (1951), 62–108 (pericardial cells etc. in *Panorpa*).

[275] STREHLER, B. L., and MCELROY, W. D. *J. cell. comp. Physiol.*, **34** (1949), 457–66 (firefly luciferin).

[276] TOUMANOFF, C. *Rev. Canad. Biol.*, **8** (1949), 343–67 (natural immunity in insects review).

SUPPLEMENTARY REFERENCES (B)

[277] AMANIEU, M., *et al. Arch. Int. Physiol. Biochem.*, **64** (1956), 518; **69** (1961), 628–37 (effect of silk retention on amino acids in the haemolymph of *Bombyx*).

[278] ANDERSON, A. D., and PATTON, R. L. *J. Exp. Zool.*, **128** (1955), 443–51 (uric acid synthesis in fat body).

[279] ARNOLD, J. W. *Canad. J. Zool.*, **30** (1952), 355–64 (haemocytes in *Ephestia*).

[280] —— *Ann. Ent. Soc. Amer.*, **52** (1959), 229–36; *Canad. J. Zool.*, **39** (1961), 755–66 (haemocytes in wing veins of *Blaberus*).

[281] ARVY, L. *C.R. Acad. Sci.*, **235** (1952), 1539–41 (haemopoietic organs: Lepidoptera); *Bull. Soc. Zool. Fr.*, **78** (1953), 158–71; *Proc. Roy. Ent. Soc. Lond.*, A, **29** (1954), 39–41 (ditto: Diptera).

[282] van ASPEREN, K., and van ESCH, I. *Arch. Neerl. Zool.*, **11** (1958), 342–60 (ionic composition of haemolymph in *Periplaneta*).

[283] BACCETTI, B. *Redia*, **40** (1955), 197–212, 269–79; **41** (1956), 75–104, 259–76; **46** (1961), 1–7; *Exp. Cell. Res.*, **13** (1957), 158–60; *Atti. Accad. Torino*, **95** (1961), 343–50 (connective tissue in insects).

[284] —— *Monit. Zool. Ital.*, **71** (1963), 361–7 (collagen in basement membranes of *Aiolopus*).

[285] BARLOW, J. S., and HOUSE, H. L. *J. Ins. Physiol.*, **5** (1960), 181–9 (carbohydrates in haemolymph of *Agria* (Dipt.)).

[286] BARSA, M. C. *J. Gen. Physiol.*, **38** (1954), 79–92 (effect of ions on heart of *Chortophaga* and *Samia*).

[287] BARTON-BROWNE, L., DODSON, L. F., *et al. Gen. Comp. Endocrin.*, **1** (1961), 232 (adrenergic properties of corpus cardiacum of *Periplaneta*).

[288] BEAMS, H. W., and ANDERSON, E. *Biol. Bull.*, **109** (1955), 375–93 (fine structure of light organ of *Photinus*).

[289] BERNHEIMER, A. W., CASPARI, E., and KAISER, A. D. *J. Exp. Zool.*, **119** (1952), 23–35 (failure of anti-body formation in caterpillars).

[290] BOEHM, B. *Naturwiss.*, **48** (1961), 675–6 (fat body, wax glands, and oenocytes in *Apis*).

[291] BRIGGS, J. D. *J. Exp. Zool.*, **138** (1958), 155–88 (humoral immunity in caterpillars).

[292] BRONSKILL, J. F. *Canad. J. Zool.*, **38** (1960), 769–75 (encapsulation of eggs of *Mesoleius* (Ichneumonidae) by the saw-fly *Pristiphora*).

[293] BUCK, J., and CASE, J. F. *Biol. Bull.*, **121** (1961), 234–56 (nervous control of flashing in *Photinus*).

[294] CAMERON, M. L. *Nature*, **172** (1953), 349 (heart accelerating factor in corpus cardiacum of *Periplaneta*).

[295] CANDY, D. J., and KILBY, B. A. *Nature*, **183** (1959), 1594–5; *Biochem. J.*, **74** (1960), 19p; **78** (1961), 531–6 (biosynthesis of trehalose in *Schistocerca*).

[296] CARRINGTON, C. B., and TENNEY, S. M. *J. Ins. Physiol.*, **3** (1959), 402–13 (binding of ions in haemolymph of *Telea*).

[297] CHEN, P. S. *J. Ins. Physiol.*, **2** (1958), 38–51, 128–36; **3** (1959), 335–44; **9** (1963), 453–62; *Rev. Suisse Zool.*, **66** (1959), 280–9 (separation of blood proteins in *Culex* and *Drosophila*).

[298] CHINO, H., and GILBERT, L. I. *Science*, **143** (1964), 359–36 (lipid transport in *Hyalophora*).

[299] CLEMENTS, A. N. *Quart. J. Micr. Sci.*, **97** (1956), 429–33 (antennal pulsating organ in mosquitos, &c.).

[300] —— *J. Exp. Biol.*, **36** (1959), 665–75 (metabolism of locust fat body).

[301] COLHOUN, E. H. *Experientia*, **19** (1963), 9 (synthesis of 5-hydroxytryptamine (serotonin) in *Periplaneta*).

[302] CUÉNOT, L. *Arch. Zool. Exp. Gén.*, **9** (1891), 593–670 (functions of haemocytes).

[303] DAVEY, K. G. *Nature*, **192** (1961), 284; *Gen. Comp. Endocrin.*, **1** (1961), 24–9; *Quart. J. Micr. Sci.*, **103** (1962), 349–58; *J. Ins. Physiol.*, **8** (1962), 205–8, 579–83; **9** (1962), 375–81; *J. Exp. Biol.*, **40** (1963), 343–50 (mode of action of corpus cardiacum in accelerating the heart beat in *Periplaneta*).

[304] DAVIS, C. C. *J. Cell. Comp. Physiol.*, **47** (1956), 449–68 (automation of heart-beat in *Corethra*).

[305] DUCHÂTEAU, G., FLORKIN, M., and LECLERCQ, J. *Arch. Int. Physiol.*, **61** (1953), 518–49 (fixed bases in insect haemolymph).

[306] DUCHÂTEAU, G., and FLORKIN, M. *Arch. Int. Physiol. Biochim.*, **66** (1958), 573–91 (amino acids in insect haemolymph).

[307] DUCHÂTEAU, G., and FLORKIN, M., *et al. Arch. Int. Physiol. Biochim.*, **66** (1958), 434; **67** (1959), 306–14; (trehalose in insect haemolymph).

[308] DUCHÂTEAU-BOSSON, G., JEUNIAUX, C., and FLORKIN, M. *Arch. Int. Physiol. Biochim.*, **70** (1962), 287–91 (tyrosine in haemolymph of *Bombyx*).

[309] DUKE, E. J., and PANTELOURIS, E. M. *Comp. Biochem. Physiol.*, **10** (1963), 351–5 (haemolymph proteins in *Drosophila*).

[310] EHRHARDT, P. *Z. vergl. Physiol.*, **46** (1962), 169–211 (amino acids in haemolymph and honey-dew of the Aphid *Megoura*).

[311] EPURE, E. *Arch. Zool. Exp. Gén.*, **79** (1937), 17–23 (pigmented fat body cells in *Simulium* larvae).

[312] FAULKNER, P. *Biochem., J.*, **64** (1956), 430–6 (malate in haemolymph of *Bombyx*).

[313] FISHER, R. C. *J. Exp. Biol.*, **40** (1963), 531–40 (asphyxiation of Hymenopterous parasites by haemocytes).

[314] FLORKIN, M. *IVth Int. Congr. Biochem.*, **12** (1960), 63–77 (free amino acids in insect haemolymph).

[315] FREUDENSTEIN, H. *Zool. Jahrb., Physiol.*, **69** (1960), 95–124; *Biol., Zbl.*, **80** (1961), 479–92 (oenocytes and wax glands in the honey-bee).

[316] FUKUDA, T., *et al. J. Biochem.*, Tokyo, **42** (1955), 341–6 (free amino acids in blood of *Bombyx*).

[317] GATENBY, J. B. *Trans. Roy. Soc. N.Z.*, **87** (1959), 291–314 (light production in *Bolitophila*).

[318] V. GAUDECKER, B. *Z. Zellforsch.*, **61** (1963), 56–95 (formation of fat body reserves in *Drosophila* larva: electron microscopy).

[319] GEORGE, J. C., and EAPEN, J. *Nature*, **183** (1959), 268; *J. Cell. Comp. Physiol.*, **54** (1959), 293–5 (lipase and alkaline phosphatase in fat body of *Schistocerca*).

[320] GILMOUR, D. *The biochemistry of insects*, Academic Press, New York, and London, 1961. 343 pp.

[321] GINGRICH, R. E. *J. Ins. Physiol.*, **10** (1964), 179–94 (acquired immunity in *Oncopeltus*).

[322] GUTHRIE, D. M. *Nature*, **196** (1962), 1010–2 (control of the ventral diaphragm).

[323] HADORN, E., and WALKER, I. *Rev. Suisse Zool.*, **67** (1960), 217–25 (encapsulation of parasites by *Drosophila*; selection experiments).

[324] HANSON, F. E. *J. Ins. Physiol.*, **8** (1962), 105–11 (electrical stimulation of light organ in the fire-fly).

[325] HARADA, M. *Mem. Coll. Sci., Kyoto*, **27** (1960), 17–23 (reducing activity in haemolymph of *Philosamia*).

[326] HARINGTON, J. S. *Nature*, **178** (1956), 268 (histidine in haemolymph of *Rhodnius*).

[327] —— *J. Ent. Soc. S. Africa*, **24** (1961), 218–19 (fate of alanine fed to *Rhodnius*).

[328] HARVEY, E. N. *Bioluminescence*, Academic Press, New York, 1952.

[329] HINTON, H. E. *Science Progress*, No. 168 (1954), 684–96 (giant cells in haemolymph, &c.).

[330] HINTON, H. E. *Quart. J. Micr. Sci.*, **100** (1959), 65–71 (pigmented fat body cells (chromatocytes) In *Simulium*).

[331] HOWDEN, G. F., and KILBY, B. A. *J. Ins. Physiol.*, **4** (1960), 258–69; **6** (1961), 85–95 (reducing power and trehalose in insect haemolymph).

[332] JEUNIAUX, C. *Arch. Int. Physiol. Biochim.*, **69** (1961), 750 (chitinase in haemolymph of *Bombyx*).

[333] JEUNIAUX, C., *et al. Arch. Int. Physiol. Biochim.*, **69** (1961), 617–27 (amino acids and osmoregulation in *Bombyx*).

[334] JONES, J. C. *Ann. Ent. Soc. Amer.*, **46** (1953), 336–72 (circulation of haemocytes in insects).

[335] —— *J. Morph.*, **94** (1954), 71–124; *J. Exp. Zool.*, **131** (1956), 223–4 (heart beat in *Anopheles*).

[336] —— *Ann. Ent. Soc. Amer.*, **47** (1954), 307–15 (haemocytes as seen with phase contrast microscopy: *Tenebrio*); *J. Morph.*, **99** (1956), 233–58 (ditto: *Sarcophaga*); *Anat. Rec.*, **128** (1957), 571 (ditto: *Periplaneta*).

[337] —— *Amer. Zoologist*, **2** (1962), 209–46 (types of haemocytes: review).

[338] KESSEL, R. G. *J. Morph.*, **109** (1961), 289–321; **110** (1962), 79–103 (pericardial cells in *Melanoplus* and historical review).

[339] KILBY, B. A. *Adv. Ins. Physiol.*, **1** (1963), 111–174 (biochemistry of the insect fat body).

[340] LAGERSPETZ, K., and PERTTUNEN, V. *J. Ins. Physiol.*, **8** (1962), 621–5 (heart-beat in *Corethra*).

[341] LAUFER, H. *Ann. N.Y. Acad. Sci.*, **89** (1960), 490–516 (enzymic activity of blood proteins in insects).

[342] LEE, R. M. *J. Ins. Physiol.*, **6** (1961), 36–51 (blood volume in *Schistocerca*).

[343] LEVENBOOK, L. *J. Cell. Comp. Physiol.*, **52** (1958), 329–39 (blood volume in *Prodenia* larva).

[344] LEVENBOOK, L., and HOLLIS, V. W. *J. Ins. Physiol.*, **6** (1961), 52–61 (citrate in haemolymph of *Prodenia*).

[345] L'HELIAS, C. *Bull. Soc. Zool. France*, **78** (1953), 76–83 (haemopoietic organ in Tenthredinidae).

[346] LISA, J. D., and LUDWIG, D. *Ann. Ent. Soc. Amer.*, **52** (1959), 548–51 (purine metabolism in fat body of *Leucophaea*).

[347] LOCKWOOD, A. P. M., and CROGHAN, P. C. *Nature*, **184** (1959), 370–1 (osmotic pressure in *Petrobius*).

[348] LUDWIG, D., and CULLEN, W. P. *Physiol. Zool.*, **29** (1956), 153–7 (nitrogen metabolism during starvation in *Popillia*).

[349] LUKOSCHUS, F. *Z. Morph. Oekol. Tiere*, **45** (1956), 157–97 (oenocytes during development in *Apis*).

[350] MCCANN, F. V. *J. Gen. Physiol.*, **46** (1963), 803–21 (electrophysiology of the heart in *Telea*).

[351] MCELROY, W. D., *et al. Arch. Biochem. Biophys.*, **46** (1953), 399; **64** (1956), 257; **72** (1957), 358; *Biochim. Biophys. Acta*, **20** (1956), 170; **27** (1958), 519; *J. Biol. Chem.*, **233** (1958), 1528; in *The Physiology of Insects*, Rockstein, M., Ed. Academic Press, 1964, 463–508 (biochemistry of light production).

[352] MARCUZZI, G. *R. C. Accad. Lincei*, **18** (1955), 654–62; **20** (1956), 492–500 (osmotic pressure in *Tenebrio*).

[353] MULDREW, J. A. *Canad. J. Zool.*, **31** (1953), 313–32 (encapsulation of *Mesoleius* in the sawfly *Pristiphora*).

[354] NEMENZ, H. *J. Ins. Physiol.*, **4** (1960), 38–44 (osmoregulation in *Ephydra*).

[355] NITTONO, Y. *Bull. Sericult. Exp. Sta.* Tokyo, **16** (1960), 261–6 (blood cells in *Bombyx*).

[356] OHUYE, T., and HORIKAWA, M. *Mem. Ehime Univ. Sect. ii* (*Sci.*). *Biol.*, **2** (1956), 283–91 (haemocytes in *Xylotrupes*).

[357] PATTON, R. L., and FLINT, R. A. *Ann. Ent. Soc. Amer.*, **52** (1959), 240–2 (blood cell count in *Periplaneta* during growth).

[358] PERTTUNEN, V. *Ann. Ent. Fernnicae*, **21** (1955), 78–88 (pulsatile organs in wings of *Drosophila*).

[359] PIPA, R. L., and COOK, E. F. *J. Morph.*, **103** (1958), 353–85 (connective tissue in *Anoplura*).

[360] RALPH, C. L. *J. Ins. Physiol.*, **8** (1962), 431–9 (factors from the nervous system of *Periplaneta* acting on the heart beat).

[361] REALI, G. *Boll. Zool. Agr. Bach.*, **21** (1955), 185–8 (haemolymph proteins in *Bombyx*).

[362] RICHARDS, A. G. *J. Morph.*, **113** (1963), 17–18 (ventral diaphragm of insects).

[363] RIZKI, M. T. M. *J. Exp. Zool.*, **123** (1953), 397–411; *J. Morph.*, **100** (1957), 437–58; *J. Biophys. Biochem. Cytol.*, **5** (1959), 235–9 (blood cells in *Drosophila*).

[363a] RIZKI, M. T. M., and RIZKI, R. M. *J. Biophys. Biochem. Cytol.*, **5** (1959), 235–9 (role of 'crystal cells' in *Drosophila* larva).

[364] SAITO, S. *J. Ins. Physiol.*, **9** (1963), 509–19 (trehalose in blood of *Bombyx*).

[365] SALT, G. *Proc. Roy. Soc.*, B, **144** (1955), 380–98; **146** (1956), 93–108; **147** (1957), 167–84; **151** (1960), 446–67 (encapsulation of *Nemeritis* larvae in various insects).

[366] —— *Parasitology*, **53** (1963), 527–642 (defence reactions of insects to metazoan parasites: review).

[367] SCHMIDT, G. H. *Z. Zellforsch.*, **55** (1961), 707–23 (function of oenocytes in *Formica*).

[368] SCHNEIDER, D., and KAISSLING, K. E. *Zool. Jahrb., Anat.*, **77** (1959), 111–32 (pulsatile organ in antenna of *Bombyx*).

[369] SELMAN, B. J. *J. Ins. Physiol.*, **8** (1962), 209–14 (haemocytes in *Sialis*).

[370] SEWELL, M. T. *Quart. J. Micr. Sci.*, **96** (1955), 73–83 (lipoprotein cells in blood of *Carcinus*).

[371] SHAW, J. *J. Exp. Biol.*, **32** (1955), 355–82 (ionic regulation in *Sialis*).

[372] SHAW, J., and STOBBART, R. H. *Adv. Ins. Physiol.*, **1** (1963), 315–99 (osmotic and ionic regulation in insects: review).

[372a] SHIGEMATSU, H. *Nature*, **182** (1958), 881–2; *Bull. Sericult. Exp. Sta.*, **16** (1960), 165–170 (protein synthesis in fat body of *Bombyx*).

[373] SIAKOTOS, A. N. *J. Gen. Physiol.*, **43** (1960), 999–1030 (conjugated plasma proteins in *Periplaneta*).

[374] SMITH, D. S. *J. Cell. Biol.*, **16** (1963), 323–59 (fine structure and innervation of light organ in *Photuris*).

[375] STEPHEN, W. P., and STEINHAUER, A. L. *Physiol. Zool.*, **30** (1957), 114–20 (sexual differences in blood proteins).

[376] STEPHENS, J. M. *Canad. J. Microbiol.*, **5** (1959), 203–28; **8** (1962), 492–9, 598–602, 720–5 (acquired bacterial immunity in *Galleria*).

[376a] STEVENSON, E., and WYATT, G. R. *Arch. Biochem. Biophys.*, **99** (1962), 65–71 (incorporation of leucine into proteins of *Hyalophora*).

[377] SUTCLIFFE, D. W. *Nature*, **187** (1960), 331–2 (osmotic regulation in *Ephydra riparia*).

[378] —— *J. Exp. Biol.*, **38** (1961), 501–19, 521–30; **39** (1962), 141–60 (salt and water balance in Trichoptera larvae).

[379] —— *J. Exp. Biol.*, **39** (1962), 325–43 (amino acids, &c., and osmotic pressure of the haemolymph in *Libellula* larvae).

[380] —— *Comp. Biochem. Physiol.*, **9** (1963), 121–35 (comparative survey of the ionic composition of the blood in insects).

[381] TACHIBANA, K., and NAGASHIMA, C. *Jap. J. Appl. Ent. Zool.*, **1** (1957), 155–63 (photoelectric recordings of heart beat in *Periplaneta*).

[382] TELFER, W. H., and WILLIAMS, C. M. *J. Gen. Physiol.*, **36** (1953), 389–413 (sexual differences in antigens in the blood of *Hyalophora*).

[383] TENNEY, S. M. *Physiol. Comp. Oecol.*, **3** (1953), 286–306 (pace-makers in heart of Lepidoptera).

[384] THOMSEN, E. *Bertil Hanström*, Wingstrand, K.G., Ed. Zool. Inst., Lund, Sweden, 1956, 298–306 (oenocytes in adult *Calliphora*).

[385] TIETZ, A. *J. Lipid Res.*, **2** (1961), 182–7 (fat synthesis by cell-free preparations of *Locusta* fat body).

[386] TUZET, O., and MANIER, J. F. *Bull. Biol. Fr. Belg.*, **91** (1957), 264–70 (origin of teratocytes in *Pieris* larvae parasitized by *Apanteles*).

[386a] WALKER, P. A. *Thesis, Univ. Birmingham* 1963 (changes in fat body of Lepidoptera during metamorphosis).

[387] WEBLEY, D. P. *Proc. Roy. Ent. Soc. Lond.*, (A) **26** (1951), 25–37 (blood-cell counts in *Locusta*).

[388] WHEELER, R. E. *J. Ins. Physiol.*, **9** (1963), 223–35 (haemocyte count in *Periplaneta* during the moulting cycle).

[389] WHITTEN, J. M. *Quart. J. Micr. Sci.*, **103** (1962), 359–67 (breakdown and formation of connective tissue in pupae of Diptera).
[390] WIGGLESWORTH, V. B. *J. Exp. Biol.*, **32** (1955), 649–63 (haemocytes and moulting in *Rhodnius*).
[391] —— *Quart. J. Micr. Sci.*, **97** (1956), 89–98 (haemocytes and connective tissue formation in *Rhodnius*).
[392] —— *Quart. J. Micr. Sci.*, **99** (1958), 441–50 (esterase in the fat body cells, &c.).
[393] —— *Ann. Rev. Ent.*, **4** (1959), 1–16 (insect blood cells: review).
[394] —— *J. Exp. Biol.*, **40** (1963), 231–45 (early changes in the fat body cells of *Rhodnius* during growth).
[395] —— *Nature*, **198** (1963), 106 (air sacs and blood volume).
[396] —— unpublished observations.
[397] WOLFE, L. S. *Quart. J. Micr. Sci.*, **95** (1954), 49–66 (oenocytes and epicuticle formation in *Calliphora*).
[398] WYATT, G. R. *Ann. Rev. Ent.*, **6** (1961), 75–102 (biochemistry of insect haemolymph: review).
[399] WYATT, G. R., and KALF, G. F. *J. Gen. Physiol.*, **40** (1957), 833–47 (trehalose, &c., in insect blood).
[400] WYATT, G. R., LOUGHHEED, T. C., and WYATT, S. S. *J. Gen. Physiol.*, **39** (1956), 853–68 (amino acids, proteins, organic phosphates, &c., in blood of *Bombyx*).
[401] WYATT, G. R., and MEYER, W. L. *J. Gen. Physiol.*, **42** (1959), 1005–11 (glycerol in insect haemolymph).
[402] WYATT, G. R., *et al. Fed. Proc.*, **17** (1958), *J. Ins. Physiol.*, **9** (1963), 137–52 (organic phosphates and glycerol in insect haemolymph).
[403] YOUNG, B. *Sinensia*, **17** (1947), 37–42 (circulation in wings of Diptera).
[404] YOUNG, R. G., *et al. Ann. Ent. Soc. Amer.*, **52** (1959), 567–73 (oxidative and dehydrogenase activities in fat body of *Periplaneta*).
[405] ZWICKY, K. *Z. vergl. Physiol.*, **36** (1954), 367–90 (osmoregulation in *Drosophila* larva).

SUPPLEMENTARY REFERENCES (C)

[406] ADAMS, T. S. and NELSON, D. R. *J. Insect Physiol.*, **15** (1969), 1729–47 (corpus allatum and ovaries affecting fat body storage in *Musca*).
[407] AGGARWAL, S. K. and KING, R. C. *Protoplasma*, **63** (1967), 344–51 (fine structure of pericardial cells of *Drosophila* larva).
[408] AKAI, H. and SATO, S. *J. Insect Physiol.*, **17** (1971), 1665–76 (fine structure of haemopoietic organs in *Bombyx*).
[409] ASHHURST, D. E. *Ann. Rev. Entom.*, **13** (1968), 45–74 (connective tissues of insects: review).
[410] BACCETTI, B. and BIGLIARDI, E. *Z. Zellforsch.*, **99** (1969), 13–24 (fine structure of heart-wall in *Aiolopus*, Orthopt.).
[411] BAGNOLI, P. *et al. Arch. ital. Biol.*, **108** (1970), 181–206 (peripheral inhibition of flashing in *Luciola*).
[412] BAIRATI, A. *Z. Zellforsch.*, **61** (1964), 769–802 (haemopoietic organs in *Drosophila*).
[413] BAYER, R. *Z. vergl. Physiol.*, **58** (1968), 76–135 (blood pressure in *Locusta*).
[414] BERRIDGE, M. J. *J. exp. Biol.*, **43** (1965), 523–53 (ionic regulation in *Dysdercus*, Hemipt.).
[415] BHAKTHAN, N. M. G. and GILBERT, L. I. *Comp. Biochem. Physiol.*, **33** (1970), 705–6 (lipid transport in *Manduca*, Sphingidae).
[416] BODNARYK, R. P. *J. Insect Physiol.*, **16** (1970), 919–29; **17** (1971), 1201–10; *Gen. comp. Endocr.*, **16** (1971), 363–8 (incorporation of tyrosine and phenylalanine in conjugated amino acids in *Musca*, *Sarcophaga*, &c.).
[417] BOWERS, B. *Protoplasma*, **60** (1964), 352–67 (fine structure of pericardial cells in *Myzus*).
[418] BRADY, J. *J. exp. Biol.*, **47** (1967), 313–26 (distribution of ions in cells and plasma of *Periplaneta* blood).
[419] BRUNELLI, M. *et al. Archs. ital. Biol.*, **106** (1968), 85–112 (inhibition of flashing in fireflies).

[420] CARLSON, A. D. *J. exp. Biol.*, **48** (1968), 381–7; **48** (1968), 195–9; *Adv. Insect Physiol.*, **6** (1969), 51–96 (neural and humoral control of luminescence).

[421] CHEN, P. S. and LEVENBOOK, L. *J. Insect Physiol.*, **12** (1966), 1595–1627 (haemolymph proteins in *Phormia*).

[422] CHINO, H. and GILBERT, L. I. *Biochim. Biophys. Acta*, **98** (1965), 94–110 (lipid release and transport from the fat body).

[423] CHINO, H., MURAKAMI, S. and HARASHIMA, K. *Biochim. Biophys. Acta*, **176** (1969), 1–26 (lipoproteins carrying diverse lipids in *Philosamia*).

[424] CHIPPENDALE, G. M. *J. Insect Physiol.*, **16** (1970), 1909–20 (biosynthesis of haemolymph proteins in *Diatraea*, Lep.).

[425] CHIPPENDALE, G. M. and KILBY, B. A. *J. Insect Physiol.*, **15** (1969), 905–26 (uptake of proteins from haemolymph by fat body in *Pieris*).

[426] COLES, C. G. *J. Insect Physiol.*, **11** (1965), 1317–23 (changes in haemolymph protein in *Rhodnius*).

[427] COLLINS, J. V. *et al. J. Insect Physiol.*, **15** (1969), 341–52; **16** (1970), 1967–1708 (protein uptake from haemolymph by fat body in *Calpodes* and *Galleria*).

[428] CROSSLEY, A. C. S. *J. exp. Zool.*, **157** (1964), 375–98 (origin and physiology of haemocytes in *Calliphora*).

[429] EDWARDS, J. S., and FIORE, C. *Ent. exp. & appl.*, **9** (1966), 419–27 (hydrolases in *Galleria* haemolymph at metamorphosis).

[430] EVANS, J. J. T. *Z. Zellforsch.*, **81** (1967), 49–61 (fine structure of oenocytes in *Dacus*).

[431] EVANS, J. J. T. *J. Insect Physiol.*, **14** (1968), 107–19 (phenoloxidase in haemolymph of *Antheraea*).

[432] GILBERT, L. I., CHINO, H., and DOMROESE, K. A. *J. Insect Physiol.*, **11** (1965), 1057–70 (lipase in insect tissues).

[433] GNATZY, W. *Z. Naturforsch.*, **24** (1969), 1209–11 (oenocytes and cuticle formation in pupa of *Culex*).

[434] GOODCHILD, A. J. P. *Proc. R. ent. Soc. Lond.*, A, **44** (1969), 62–70 (rectal glands in *Halosalda*, Hemipt.).

[435] GÖTZ, P. *Verhandl. dtsch. zool. Ges.*, **33** (1969), 610–17 (encapsulation of parasites in *Chironomus* larvae).

[436] GRIMSTONE, A. V., ROTHERAM, S., and SALT, G. *J. Cell Sci.*, **2** (1967), 281–92 (capsule formation by blood cells in *Ephestia*).

[437] HOFFMANN, J. A. *C.R. Acad. Sci. Paris*, **262** (1966), 1469–71 (fine structure of pericardial cells in *Locusta*).

[438] HOFFMANN, J. A. *C.R. Acad. Sci. Paris*, **263** (1966), 521–4; *J. Microsopie*, **5** (1966), 269–72; *Arch. Zool. exp. gén.*, **108** (1967), 251–91 (types of haemocytes in *Locusta*).

[439] HOFFMANN, J. A. *C.R. Acad. Sci. Paris*, **266** (1968), 1882–3; *J. Insect Physiol.*, **15** (1969), 1375–84; *Z. Zellforsch.*, **106** (1970), 451–72 (haemopoietic organs in Orthoptera).

[440] HOFFMANN, J. A., PORTE, A., and JOLY, P. *C.R. Acad. Sci. Paris*, **270** (1970), 629–31 (phenoloxidase in haemolymph of *Locusta*).

[441] HOFFMANN, J. A. and STOECKEL, M. E. *C.R. Soc. Biol.*, **12** (1968), 2257 (coagulocytes in *Locusta*).

[442] KINNEAR, J. F. *et al. Aust. J. biol. Sci.*, **21** (1968), 1033–45 (protein and lipid in haemolymph of late larva of *Calliphora*).

[443] LAI-FOOK, J. *J. Morph.*, **130** (1970), 297–314 (haemocytes in *Rhodnius*).

[444] LEVENBOOK, L. *Comp. Biochem. Physiol.*, **18** (1966), 341–51 (small molecular weight peptides in haemolymph of *Phormia*).

[445] LOCKE, M. *Tissue & Cell*, **1** (1969), 103–54 (fine structure of oenocytes during growth in *Calpodes*).

[446] LOCKE, M. and COLLINS, J. V. *J. Cell Biol.*, **26** (1965), 857–84; **36** (1968), 453–83; *Nature*, **210** (1966), 552–3; *Science*, **155** (1967), 467–9 (uptake and sequestration of protein in fat body of *Calpodes*).

[447] MARSCHALL, K. J. *Z. Morph. Oekol. Tiere*, **58** (1966), 182–246 (blood cells in *Tenebrio*).

[448] MARTOJA, R. and LAUVERJAT, S. *Bull. Soc. zool. Fr.*, **89** (1964), 339–81 (sexual differences in fat body of *Locusta*).

[449] MILLER, T. *et al. J. Insect Physiol.*, **14** (1968), 383–94; 1099–1104, 1265–75; *Experientia* (Suppl. **15**), 1969, 206–18 (innervation of heart in *Periplaneta*).

[450] MILLER, T. and USHERWOOD, P. N. R. *J. exp. Biol.*, **54** (1971), 329–48 (regulation of heart beat in *Periplaneta*).

[451] MILLS, R. P. and KING, R. C. *Quart. J. micro. Sci.*, **106** (1965), 261–8 (pericardial cells of *Drosophila*).

[452] MOHRIG, W. and MESSNER, B. *Biol. Zbl.*, **87** (1968), 439–70; 705–18 (lysozyme as antibacterial factor in *Galleria*).

[453] MUNN, E. A. *et al. J. Insect Physiol.*, **15** (1969), 1601–5; 1935–50 (soluble protein 'calliphorin' in *Calliphora*).

[454] MURPHY, T. A. and WYATT, G. R. *J. biol. Chem.*, **240** (1965), 1500–8 (glycogen and trehalose synthesis in fat body of *Hyalophora*).

[455] PEMRICK, S. M. and BUTZ, A. *J. Insect Physiol.*, **16** (1970), 1443–53 (proteins common to fat body and oöcytes in *Tenebrio*).

[456] PICHON, Y. *J. exp. Biol.*, **53** (1970), 195–209 (ionic content of haemolymph in *Periplaneta*).

[457] PIEK, T. *J. Insect Physiol.*, **10** (1964), 563–72 (lipid synthesis in oenocytes).

[458] PRICE, G. M. and BOSMAN, T. *J. Insect Physiol.*, **12** (1966), 741–5 (proteins of fat body and haemolymph in *Calliphora*).

[459] SALT, G. *Proc. R. Soc. Lond.* B, **162** (1965), 303–18; **165** (1966), 155–78; **176** (1970), 105–14 (resistance of insect parasitoids to encapsulation).

[460] SALT, G. *Biol. Rev.*, **43** (1968), 200–32; *The Cellular Defence Reactions of Insects*, Cambridge University Press, London, 1970, 118 pp. (cellular defence reactions: review).

[461] SANGER, J. W. and MCCANN, F. V. *J. Insect Physiol.*, **14** (1968), 1539–44 (fine structure of alary muscles in *Hyalophora*).

[462] SANGER, J. W. and MCCANN, F. V. *J. Insect Physiol.*, **14** (1968), 1839–45 (fine structure of pericardial cells in *Hyalophora*).

[463] SELMAN, B. J. *Proc. Zool. Soc. Lond.*, **144** (1965), 487–535 (circulatory system in *Sialus*).

[464] SHRIVASTAVA, S. C. and RICHARDS, A. G. *Biol. Bull. mar. biol. Lab., Woods Hole*, **128** (1965), 337–45 (haemocytes and neural lamella in *Galleria*).

[465] SMALLEY, K. N. *Comp. Biochem. Physiol.*, **16** (1965), 467–77 (adrenergic transmission in light organ of *Photinus*).

[466] SMITH, N. A. *Experientia* (Suppl. **15**), 1969, 200–5 (corpus cardiacum and neural rhythmicity of *Periplaneta* heart).

[467] STOBBART, R. H. *J. exp. Biol.*, **54** (1971), 67–82 (factors controlling body volume in *Aëdes* larvae).

[468] THOMAS, K. K. and GILBERT, L. I. *Archs. Biochem. Biophys.*, **127** (1968), 512–21; *Physiol. Chem. Physics* **1** (1969), 293–311 (haemolymph lipoproteins in *Hyalophora*).

[469] THOMSON, J. A. and SIN, Y. T. *J. Insect Physiol.*, **16** (1970), 2063–74 (activation of prophenoloxidase in haemolymph of *Calliphora*).

[470] WALKER, P. A. *J. Insect Physiol.*, **11** (1965), 1625–31; **12** (1966), 1009–28 (fine structure of fat body in *Blaberus* and *Philosamia*).

[471] WALL, B. J. *J. Insect Physiol.*, **16** (1970), 1027–42 (osmoregulation in *Periplaneta*).

[472] WALTERS, D. R. *J. exp. Zool.*, **174** (1970), 441–50 (types of haemocytes in *Hyalophora*).

[473] WIENS, A. W. and GILBERT, L. I. *Science*, **150** (1965), 614–16; *J. Insect Physiol.*, **13** (1967), 779–94 (corpus cardiacum and carbohydrate mobilization in *Leucophaea*).

[474] WHITEHEAD, D. L. *FEBS Letters*, **7** (1970), 263–6 (dopadecarboxylase in haemocytes of *Sarcophaga*).

[475] WHITTEN, J. *J. Insect Physiol.*, **10** (1964), 447–69; **15** (1969), 763–78 (haemocytes and connective tissue formation in *Sarcophaga* and *Drosophila*).

[476] WIGGLESWORTH, V. B. *J. Cell Sci.*, **2** (1968), 243–56 (cytological changes in fat body of *Rhodnius*).

[477] WIGGLESWORTH, V. B. *Tissue & Cell*, **2** (1970), 155–79 (oenocytes and structural lipids in cuticle).

[478] WIGGLESWORTH, V. B. *J. Reticuloendothelial Soc.*, **7** (1970), 208–16 (pericardial cells: review).

[479] WIGGLESWORTH, V. B. *Insect hormones*, Oliver and Boyd, Edinburgh, 1970, 159 pp.

[480] WIGGLESWORTH, V. B. unpublished observations.

[481] WLODAWER, P. and LAGWINSKA, E. *J. Insect Physiol.*, **13** (1967), 319–31 (uptake and release of lipids by isolated fat body of *Galleria*).

Chapter XI

Digestion and Nutrition

ORGANIC MATERIAL of almost every kind in nature may serve as food for insects.[44] The mouth parts in the different groups are adapted for chewing solid substances such as foliage or wood, the bodies of other creatures, or organic refuse of any kind; for collecting exposed fluids such as the nectar of flowers, honey dew, or exudates from decaying matter; or for piercing the integument of animals or the tissues of plants and sucking out of their juices.[291, 292, 339]

The digestive system is correspondingly modified. In some insects it must cope with an abundant dilute fluid; in others, almost every trace of water must be conserved. In some it has to digest only a few simple sugars; in others even the resistant skeletal proteins, such as keratin or collagen, or carbohydrates such as cellulose must be broken down.

The alimentary canal is a tube of epithelium running a straight or convoluted course from the mouth to the anus. In the head it is connected to the body wall

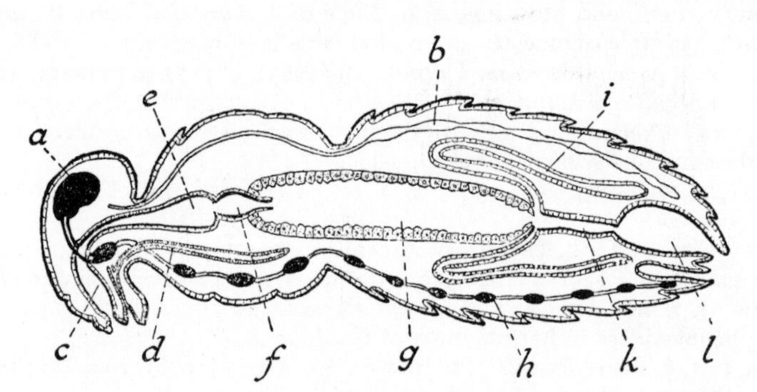

FIG. 293.—Diagram of alimentary system of an insect (*after* SNODGRASS)

a, brain; *b*, heart; *c*, pharynx; *d*, salivary gland; *e*, crop; *f*, proventriculus; *g*, mid-gut; *h*, ventral nerve cord; *i*, Malpighian tubes; *k*, small intestine (hind-gut); *l*, rectum.

by muscles; elsewhere its coils are supported as a rule only by tracheal branches. It consists of three primary divisions (Fig. 293): a 'fore-gut' derived from the stomodoeum, lined with cuticle continuous with that covering the surface of the body, a 'hind-gut' derived from the proctodoeum, again with a cuticular lining, and an endodermal 'mid-gut' uniting these two. All these parts have a muscular coat innervated from the visceral nervous system; the fore-gut and the greater part of the mid-gut from the frontal and pharyngeal ganglia, the hind-gut and the posterior region of the mid-gut from the hindmost ganglion of the ventral nerve cord.[175] All segments of the gut show peristaltic and churning movements which serve to mix the contents and carry them along. The digestive enzymes which hydrolyse the food into products capable of absorption, are secreted partly by the 'salivary glands' (dermal glands associated with the mouth appendages), but chiefly by the mid-intestine. All segments of the gut may take part in absorption to a varying extent.

THE FORE-GUT

Anatomy and histology—The 'buccal cavity', forming the commencement of the fore-gut, is followed by the 'pharynx' with an elaborate musculature concerned with the ingestion and deglutition of the food.[291, 339] Then follows the 'oesophagus', which may be a simple narrow tube leading to the mid-gut, as in Collembola, Hemiptera, &c., but which is often modified to form a 'crop'. Typically the crop is a symmetrical dilatation of the hind part of the oesophagus, as in the cockroach, in caterpillars, in predaceous Coleoptera, in the bee, &c.; sometimes it forms a lateral dilatation, as in *Gryllotalpa*, certain termites, the larva of the ant lion, and in Curculionidae; sometimes it is a diverticulum separated from the oesophagus by a long and narrow tube, as in adult Diptera, the larvae of Cyclorrhapha, and in most adult Lepidoptera (Fig. 294). The fore-gut is separated from the mid-gut by the 'cardiac sphincter', and at this point it may be modified to form a muscular 'proventriculus'. The epithelium of the fore-gut is generally continued beyond the cardiac sphincter as a fold, the 'oesophageal invagination', which projects into the mid-gut. [See p. 526.]

Histologically, the fore-gut is made up of a cuticular coat, smooth, spiculated, or bearing hairs or teeth as the case may be, laid down by a rather thin epidermis resting on a basement membrane. Outside this there is a relatively thick muscular coat with longitudinal muscles inside and circular muscles outside; the whole enclosed in a connective tissue sheath bearing nerves and tracheae.

Function of the crop—The crop may simply form a temporary reservoir for food. In caterpillars the food is merely stored and passed on from time to time as the mid-gut is evacuated. In the tsetse-fly *Glossina*, great quantities of blood are taken into the crop and transferred to the mid-gut as required.[181] The same is true of *Musca*, *Calliphora* and other flies that feed on exposed fluids of all kinds.[129] In *Haematopota* the blood goes first into the stomach, but when this is full it is diverted to the crop;[57] and in mosquitos, whereas blood is always passed into the stomach, and appears only in traces in the diverticula of the oesophagus, sweet juices or exudates from fruit go first into the crop, and are then conveyed at intervals to the mid-gut.[27, 33, 194] Chemoreceptors on the tip of the labium, when stimulated by sugars, seem to induce swallowing to the diverticula, whereas contact of blood with the labrum causes swallowing to take place with opening of the cardiac sphincter;[461] and a similar control operates in *Culicoides*.[492] In *Musca* as we have seen (p. 283) the direction in which sugary solutions are passed is controlled by receptors in the pharynx.[558]

In *Periplaneta* the rate of crop emptying is little affected by the viscosity of the contents; the chief controlling factor is osmotic pressure. Emptying is delayed by high osmolar concentration, a sense organ in the pharynx serving as an osmoreceptor.[420] [See p. 526.]

But in some insects the food in the crop is mixed with the digestive juices. In Orthoptera the secretion of the salivary glands is swallowed with the food, the secretion from the mid-gut is passed forward, and the crop is probably the chief region where digestion takes place.[242, 348] It is the main seat of digestion also in Carabids and other beetles; [242, 273] and probably in the ant lions and their allies. The food of blow-fly maggots contains the enzymes present in their own excreta (p. 506) and these must continue to act while the food is in the crop. In

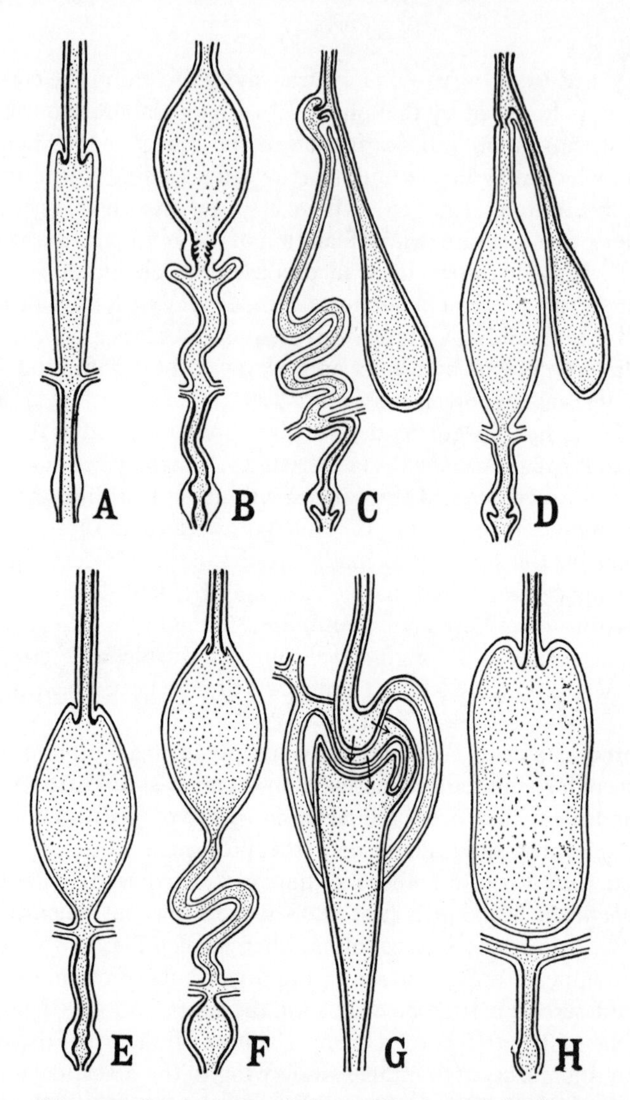

FIG. 294.—Diagrams showing some of the modifications of the alimentary system

Fore-gut and hind-gut indicated by a heavy line internally. *A*, in primitive insects and many larvae; *B*, in Orthoptera, Odonata, Hymenoptera, many Coleoptera; *C*, in higher Diptera; *D*, in Diptera Nematocera, Lepidoptera; *E*, in Siphonaptera, Siphunculata; *F*, in many Hemiptera Heteroptera; *G*, in Coccidae; *H*, in larvae of Hymenoptera Apocrita and in larvae of *Myrmeleon* and other Neuroptera.

the bee, the nectar of flowers is mixed with enzymes from the salivary glands and is converted into honey in the crop or 'honey stomach'.[269]

On the other hand, the crop is of little importance in absorption. Aqueous solutions can be exposed in the crop dissected from the mosquito for several days without drying up.[33] The crop of the cockroach used as an osmometer is practically impermeable to water,[93] and no sugar will pass through it.[1] Olive oil, however, will pass through, and in the living cockroach fat may be absorbed in it; for if the crop is ligated and the insect fed with fat, this will appear only in the cells in front of the ligature,[1] or in cells overlying localized fat.[268] Some lipase activity exists in the crop of *Periplaneta* as the result of transfer of secretions forward from the mid-gut. The presence of the resulting fatty acids facilitates a small amount of absorption through the cuticle,[427] such as occurs also through the cuticle of the body wall.[559]

Diverticula from the oesophagus may have some accessory functions. In the larva of the saw-fly *Lophyrus*, a pair of small pouches serve as receptacles for the useless resin from the surface of the pine needles on which the larvae feed.[266] The infrabuccal chamber ('cibarium') of *Camponotus* serves to filter off and withhold solid particles so that the liquid in the communal crop supply, available to the society, is kept particle-free.[429] In mosquitos the diverticula seem to take up any air swallowed with the food.[194] In many insects the crop is the chief receptacle for the air that is swallowed and distends the body at ecdysis (p. 45); in those Lepidoptera which take no food in the adult state it retains this as its sole function.[301]

Function of the proventriculus—The proventriculus shows all degrees of development, from a simple sphincter lined with soft cuticle, to a powerful muscular organ armed with spines and teeth. It occludes the passage from the fore-gut to the mid-gut. In many beetles it probably transmits the contents of the crop in small quantities at a time, and perhaps acts as a sieve.[315] In *Calandra* it serves to retain minute particles in the crop while allowing the digestive fluid from the mid-gut to pass forwards.[79] In ants the proventriculus, though subject to much variation in different groups, seems to be concerned mainly with actively controlling the passage of fluid to the mid-gut.[428] But in some insects it undoubtedly serves to crush and triturate the food.[88] In the Cerambycid larva *Xystrocera globosa* no proventriculus is present, and the particles of wood in the mid-gut are relatively large; whereas in *Macrotoma palmata*, which has a well developed proventriculus, the contents of the gut are finely pulverized.[199] In *Ips* it serves at once for grinding and for straining the woody plant tissue.[92] Among the water-beetles all stages of evolution between a crushing organ and a filter are found.[12] The proventriculus of the larva of *Galleria* is covered with small teeth which serve to grind the wax particles, &c., in the food.[403]

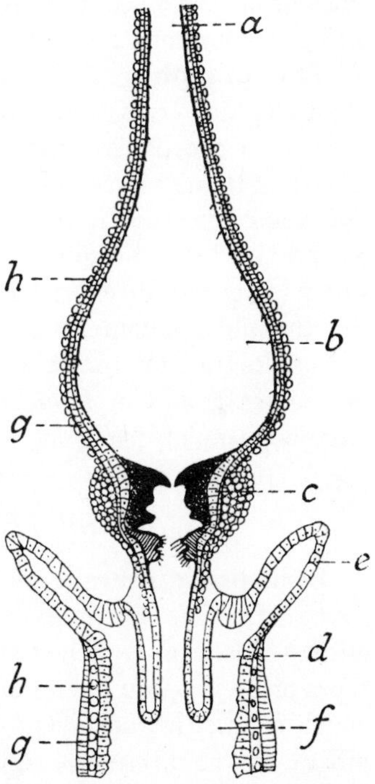

FIG. 295.—Fore-gut of cockroach *Blatta orientalis* (*after* SNODGRASS)

a, oesophagus; *b*, crop; *c*, proventriculus with large teeth in front and hairy cushions behind; *d*, oesophageal invagination; *e*, caeca of mid-gut; *f*, mid-gut; *g*, longitudinal muscles; *h*, circular muscles.

This gizzard-like function has been most clearly demonstrated in the cockroach. Here the proventriculus bears in front six powerful radial teeth invested by a strong compressor muscle; behind these are two corresponding rows of hairy cushions (Fig. 295). The hind part is just a sphincter, the fore part is certainly a crushing apparatus. For the solid particles are always smaller in the mid-gut than in the crop; and the old lining of the proventriculus which is cast off at moulting, is crushed and broken up and passed in small fragments in the faeces, as soon as the new proventriculus has hardened. After crushing, the products are first returned to the crop, exposed to the digestive juices, and then admitted to the mid-intestine.[94]

In the flea, the proventriculus is lined by long curved backwardly directed spines. No crop is present, and during digestion the proventriculus contracts rhythmically, driving the spines backwards into the mid-gut and probably breaking up the corpuscles in the ingested blood.[101]

In the bee the proventriculus is called the 'honey stopper'. Here, and in other Hymenoptera, it forms an elongated plug with an elaborate internal musculature, which projects far into the crop, its lumen occluded by four converging lips with short curved spines. In spite of its strong muscles, this organ seems incapable of crushing grains of pollen.[342] But it will seize these grains and transfer them into the stomach without removing fluid with them. If the bee is fed with syrup containing a suspension of pollen grains, the crop remains turgid with fluid long after all the grains have been removed.[346] The extraction of pollen grains and other particles from the contents of the crop is very efficient: all particles from 3μ or less, up to 50μ or more are completely filtered off in 20–25 minutes.[366, 536]

The oesophageal invagination, which follows the proventriculus or cardiac sphincter, is often referred to as the 'oesophageal valve', in the belief that it forms a mechanism for preventing regurgitation from the mid-gut to the fore-gut. It may have this function in the bee,[5, 346] but in most insects such reflux is prevented by the cardiac sphincter,[287] which, indeed, in many forms extends into the invagination.[350] In most insects the invagination is quite unsuited by its structure to act as a valve; having too rigid walls, or being so disposed that the mid-gut contents cannot exert pressure on its outer face.[349, 350] We shall discuss its function in connexion with the peritrophic membrane (p. 481). In the queen bee[445] and in *Drosophila*[560] the oesophageal invagination is the site of large deposits of glycogen.

THE PERITROPHIC MEMBRANE

Peritrophic membrane: type i—The epithelial cells of the mid-gut are devoid of cuticle, but in most insects they are protected from the contents of the gut by a delicate detached sheath, the 'peritrophic membrane'. There are two types of peritrophic membrane, formed in two different ways. In the first type, which occurs in Phasmids,[61] Acridiids,[76] Ephemeroptera,[78] Odonata,[334] caterpillars,[314, 331] *Tenebrio*, *Hydrophilus* and other beetles,[78, 331] and *Apis*,[232] *Vespa*[130] and their larvae,[254] the membrane is made up of concentric lamellae, independent or loosely attached to one another. It is produced by the separation of thin sheets from the surface of the cells throughout the length of the mid-gut. The mid-gut cells generally bear a striated border (p. 485), and in these insects each new sheet appears as a limiting membrane at the surface of this border (in the bee,[327] in *Galleria* larva[314]) which is raised and detached by the pouring out of secretion below it (Fig. 296). When the newly formed lamellae in the bee are stained, they often show polygonal areas corresponding to the cell surfaces by which they were laid down.[78] As in the formation of the body cuticle, this process is sometimes regarded as a secretion of substance which condenses to a membrane[331] sometimes as the transformation of a part of the cell surface; but there is no real distinction between these alternatives. Sometimes the lamellae carry with them a part of the striation from the cells (in *Aeschna*,[334] in *Apis*[232, 346]), but they cannot be regarded merely as detached striated borders, because the lamellae contain chitin while the striated border does not.[78]

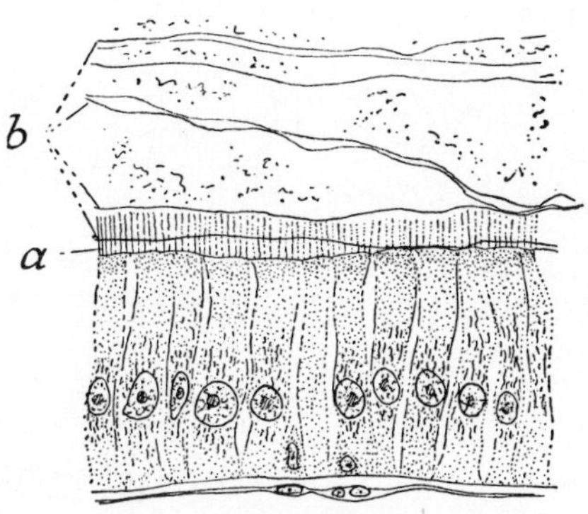

FIG. 296.—Section through mid-gut of nymph of *Cloëon* sp., showing origin of peritrophic membrane of type i (*after* v. DEHN)

a, striated border; *b*, successive laminae of peritrophic membrane shed off from the mid-gut cells.

In the larvae of the wasp and bee, at least half a dozen membranes of this kind are set free each day;[254] and since in these larvae the mid-gut forms a closed sac, the contents of the gut come to be invested in a great number of superimposed envelopes. In the larva of *Aeschna*, even in the absence of feeding, about two such peritrophic sacs may be discharged each day.[10]

Peritrophic membrane: type ii—In the second type, which occurs in the larvae and adults of Diptera,[350] in the earwig[73, 350] and perhaps in termites,[209, 350] the peritrophic membrane consists of a single uniform layer. It is secreted in viscous form by a group of cells at the anterior limit of the mid-gut (Fig. 297), passes through an annular cleft between the oesophageal invagination and the mid-gut, and in so doing solidifies to form a homogeneous tube.

An annular mould of this type was first recognized in larvae of *Ptychoptera*[117] and *Chironomus*;[331] it has since been described in the larvae of mosquitos (Fig. 298, A) and many other Nematocera,[350] in the larva of *Eristalis*[10] and in adults of *Glossina* (Fig. 298, B),[349] *Calliphora*,[129] *Drosophila*,[526] and other flies. In these insects the mid-gut forms the outer wall of the mould, its constrictor muscles being generally exaggerated at this level and so disposed as to provide the pressure necessary for moulding; this is well seen in the earwig (Fig. 298, C),[350] in *Calliphora* adult[129] and in the larva of the Syrphid *Syritta*.[171] The inner wall of the mould is provided by the oesophageal invagination, which may be composed of large tense cells forming a solid plug (as in the larvae of *Calliphora*,[236] and *Syritta*[171] and the adult of *Glossina*[349]), or the cells upon the outer face of the invagination may lay down a ring of thick and rigid cuticle (as in *Forficula*,[350] *Anopheles*,[350] *Rhyphus*[350] and several other Nematocera[271, 304]). Longitudinal muscles may be inserted into this ring and serve to draw it forwards;[350] and between the walls of the invagination are often blood sinuses, which can be distended and so force it back again.[10] On the outer wall of the invagination there are often rows of spines (in *Mycetophila*,[149, 195] *Simulium*,[304] and *Tabanus*[294] larvae) and then the backward and forward movements will serve to draw out the peritrophic membrane. In other cases, such as the Muscoid flies, the membrane is probably drawn backwards by the peristaltic movements of the gut.[129] In the unfed larva of *Eristalis* the membrane is continuously secreted at the rate of about 6 mm. an hour.[10]

The peritrophic membrane of this second type is always of uniform circumference throughout its length. Where there are diverticula from the mid-gut, as

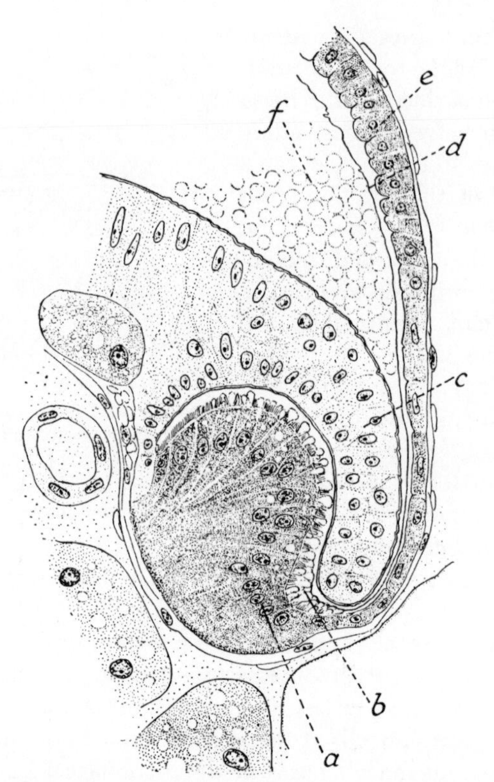

FIG. 297.—Formation of peritrophic membrane (type ii) in *Glossina* (*after* WIGGLESWORTH) (*cf.* Fig. 298 *B*)

a, cells secreting substance of membrane; *b*, vesicles of secretion; *c*, invagination of fore-gut, forming press; *d*, peritrophic membrane; *e*, mid-gut epithelium; *f*, freshly ingested blood in lumen of gut.

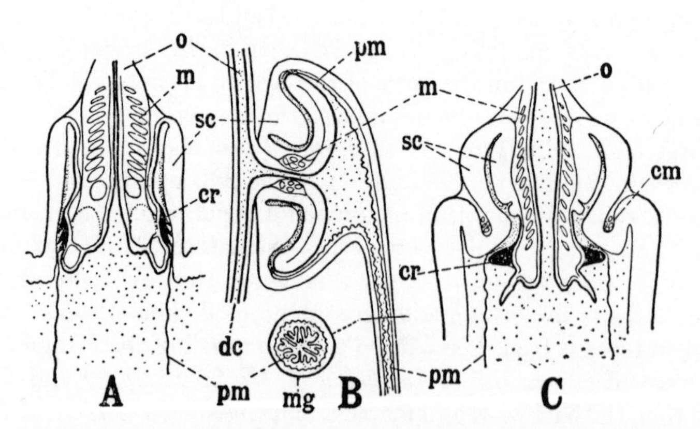

FIG. 298.—Annular moulds producing peritrophic membrane (type ii)

A, larva of mosquito, *Anopheles*; *B*, tsetse fly, *Glossina*; *C*, earwig, *Forficula* (*after* WIGGLESWORTH), *cr*, cuticular ring forming inner wall of press; *cm*, circular muscle compressing outer wall against this ring; *dc*, duct of crop; *m*, sphincter muscle; *mg*, mid-gut in cross-section; *o*, oesophagus; *pm*, peritrophic membrane; *sc*, cells secreting the substance of the membrane.

in Muscid or Culicid larvae, it bridges their openings without entering them. Where the gut is wider than the mould in which the tube is formed, the membrane may form loose coils within the lumen, as in *Tabanus*[294] and *Lucilia*[144] larvae (Figs. 299 A, 305); while when the gut is narrower than the mould, as in many parts of the intestine in *Glossina* adults,[349] the membrane is thrown into longitudinal folds (Fig. 299, B).

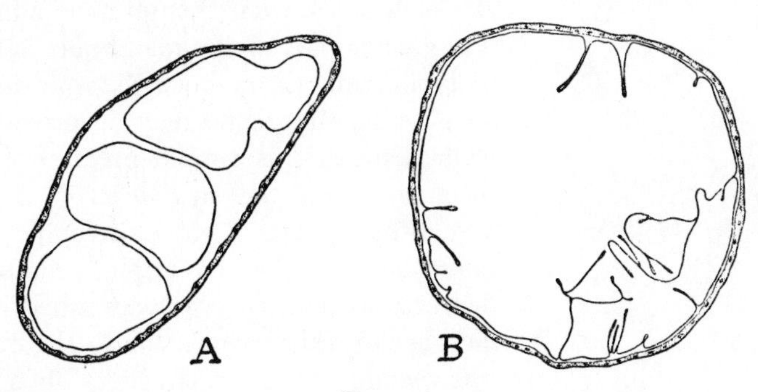

FIG. 299

A, transverse section of dilated segment of mid-gut in larva of *Musca*. Peritrophic membrane is thrown into coils and is then cut through in several places (*cf*. Fig. 305). *B*, transverse section of mid-gut of *Glossina* adult. The lumen is smaller than the peritrophic membrane, which is thrown into longitudinal folds (*after* WIGGLESWORTH).

The oesophageal invagination in most insects is a thin-walled structure. In caterpillars it is often cleft along one side as in *Vanessa*,[138] or split up into two or three lobes as in *Galleria*,[204] *Chimabache* (Fig. 300), *Ephestia*, &c.[350] When food passes through, these lobes are pressed against the wall of the mid-gut, and

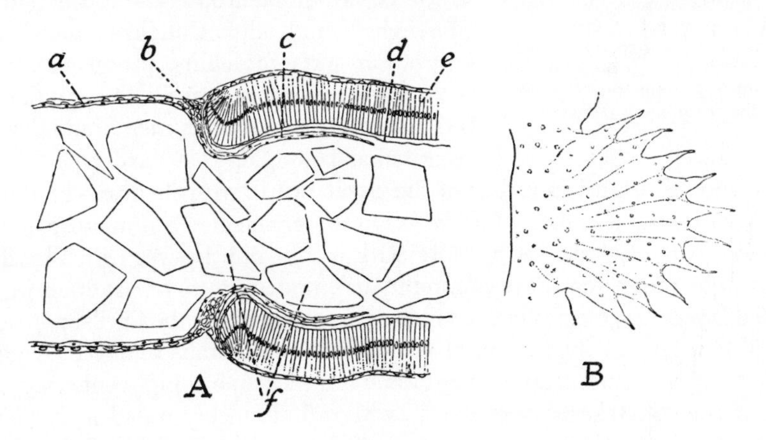

FIG. 300

A, longitudinal section of junction of fore-gut and mid-gut in the caterpillar of *Chimabache*; *B*, one of the three leaflets of the oesophageal invagination seen in surface view (*after* WIGGLESWORTH). *a*, fore-gut; *b*, circular muscle; *c*, oesophageal invagination; *d*, peritrophic membrane; *e*, mid-gut; *f*, fragments of foliage in lumen of gut.

thus ensure that the contents enter within the peritrophic membrane. That is perhaps the primary function of the invagination. But where the peritrophic membrane is covered by the invagination, it is compressed against the surface of the gut; in this region it is more compact and less obviously laminated; and at the anterior limit of the mid-gut it generally remains attached to the cells.

It may thereby give a false impression of taking origin solely from these cells (as in *Aeschna*,[334] *Chrysopa*,[201] the larva of the bee,[214] *Xylocopa*[36] and in caterpillars[38]).

In other insects, such as saw-fly larvae,[350] the adults of Neuroptera[201] and Hymenoptera,[350] the thin-walled invagination can be distended with blood, and in this way also may exert pressure upon the wall of the mid-gut (Fig. 301). In such cases, even though it be admitted that the greater part of the membrane is formed by delamination throughout the mid-intestine, it is possible that some may be provided by the cells beneath this simple press. It has in fact been suggested that in *Polistes*,[237] *Bombus*, *Apis* and other insects[350] the peritrophic membrane may have a double origin, combining the features of the two types described above, and perhaps representing a stage in the evolution of the second type from the first. Thus in Blattids and other Orthoptera and in *Forficula* the proventricular membrane is added to by delaminations from the mid-gut further back.[10] In the gut of *Acridium* finger-like processes of the peritrophic membrane can be drawn out from the gastric caeca.[21] A similar dual origin is seen in *Calotermes*,[10] *Bombyx mori*,[10] *Tineola*,[190] and the larva of *Apis*,[10, 174] and *Myrmica*.[557] In such insects as the larvae of Neuroptera and ants, in adult *Simulium*[381] and adult Culicidae, in which the entire mid-gut epithelium produces a viscous secretion enveloping entirely the gut contents, the resulting peritrophic membrane has been regarded as belonging to a third type.*[542]

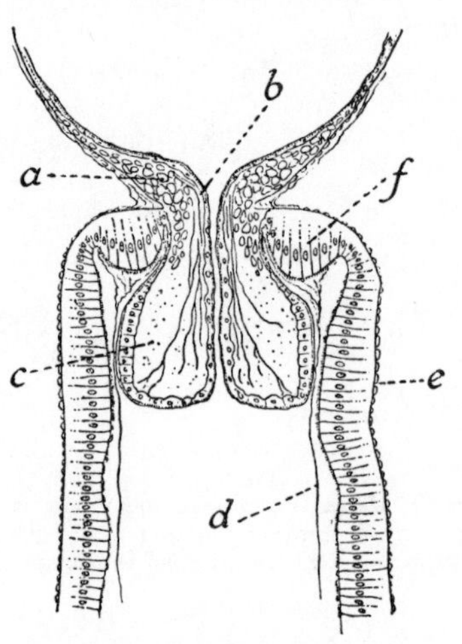

FIG. 301.—Oesophageal invagination of larva of saw-fly (Tenthredinidae) (*after* WIGGLESWORTH)

a, circular sphincter muscle between fore-gut and mid-gut; **b**, longitudinal muscle running into oesophageal invagination; **c**, blood sinus distending the invagination; **d**, peritrophic membrane; **e**, circular muscle of mid-gut; **f**, ring of cells with peritrophic membrane adherent to them.

Composition and function of the peritrophic membrane—Examination of the peritrophic membrane of *Carausius* under the electron microscope shows that it consists of a fibrous network with a thin film between.[375] The fibrillar strands of less than 100 Å are commonly oriented at 60° to one another in superimposed layers, to give perfect hexagonal symmetry;[494] or the fibrils may be quite irregularly arranged. The amorphous film between them is removed by proteinase.[479, 553, 564] The peritrophic membrane has the same components as the inner layers of the cuticle: a basis of chitin, with protein incorporated in it.[128, 349, 350] This applies both to the laminae separated from the surface of the mid-gut, and to the membranes secreted by the cardiac cells.[78] The chitin content of the membrane in different insects ranges from 3·7–12·9 per cent.; the protein content from 21 per cent. in *Aeschna* to 47 per cent. in *Carausius*. The remainder of the dry weight is probably made up of other mucopolysaccharides.[422] The

* The grouping of peritrophic membranes into types (i) and (ii) was originally introduced (in the first edition of this book) in an attempt to resolve the controversy between rival supporters of the two modes of formation. It is clear from the description given above that in fact types (i) and (ii) merge into one another; to introduce a type (iii) is perhaps unlikely to clarify the issue.

glycogen in the mid-gut cells of many insects may contribute to the formation of the membrane,[226, 228] [See p. 526.]

The peritrophic membrane is generally regarded as protecting the mid-gut cells from abrasion by hard fragments in the food, replacing in this respect the mucous involucrum which lubricates the gut of vertebrates.[20, 209, 350] Indeed, from the biochemical standpoint, mucoprotein and chito-protein are closely allied. This belief is supported by the fact that the membrane is absent in Hemiptera and in adult Lepidoptera, which feed only upon fluids, and in many of the blood-sucking insects such as Tabanids, fleas, and lice, while in Culicids[159] and *Phlebotomus*[85] it is exceedingly tenuous. It is absent in many carnivorous Coleoptera (Carabids and Dytiscids) in which the mid-gut cells break down completely during secretion (p. 488). It is commonly said to be absent in *Gryllotalpa*, in which the mid-gut is very short and is protected by four delicate chitinous lamellae, attached to the hind margin of the oesophageal invagination, which reach as far back as the commencement of the hind-gut.[73] But it is curiously missing also in the ant,[78] and in the larva of *Anthrenus* (Dermestidae),[208] in which its protective function might seem to be required. In fact, peritrophic membranes are being found in many insects in which they were formerly believed to be absent. An ill organized membrane occurs in *Gryllotalpa*;[551] one is present in certain *Corixids* feeding on detritus and other solid matter;[388] and in *Anthrenus* and other Dermestids;[551] a striking type (i) membrane is found in adult *Pieris*, a type (ii) membrane in other families of Lepidoptera.[551] [See p. 526.]

The permeability of the membrane varies somewhat from one insect to another; it has been tested with different dyes in *Carausius*,[61] termites,[209] *Apis*, *Vanessa* larva, and *Calliphora*.[78] As a rule only dyes with large colloidal particles such as congo red or Berlin blue are arrested by it. In mosquito larvae colloidal gold particles 2–4 mμ in diameter diffuse through; particles of 20 mμ are retained.[534] To that extent it serves as an ultra-filter. But it offers no hindrance to digestion; for digestive enzymes and the products of their action readily pass through it.[78, 349]

THE MID-GUT

Histology—The mid-gut is made up of cubical or columnar cells, usually much thicker than the epidermis of the fore-gut, a basement membrane, an inner circular muscle coat, and an outer longitudinal. As a rule, the lattice work of muscle fibres covers but a part of the epithelial cells; in many places these are separated from the circulating blood only by the basement membrane and a delicate connective tissue sheath.

Internally the cells are generally bounded by a striated border, though this may be absent. As in the Malpighian tubes (p. 566) two types of border probably occur;[217] one made up of rod-like elements which appear in the light microscope as though held together to form a rigid structure (the 'honey-comb border'), the other composed of independent hair-like filaments which, though not motile, can be moved passively about (the 'brush border'). The former is more frequent; the latter is best seen in caterpillars and in saw-flies: in *Cimbex* (Tenthredinidae) the hair-like filaments are 45μ in length, longer than the cells which carry them.[113] In the cardiac end of the mid-gut of *Phlebotomus* (Dipt.) the independence of the rods is demonstrated by the fact that intestinal flagellates multiply among them;[4] in *Ptychoptera* larvae they are often inclined at an angle to one another.[117]

In the caterpillar *Malacosoma* the striated border lining the cells is seen with the electron microscope to be made up of a multitude of independent microvilli. The striated appearance at the base of the cells results from the numerous deep perpendicular invaginations of the plasma membrane.[398] In representatives of adult Orthoptera, Coleoptera, and Diptera the mid-gut epithelium shows a very rich endoplasmic reticulum (ergastoplasm) and Golgi complex in the perinuclear cytoplasm, well marked nuclear pores, and abundant invaginated plasma membranes in the basal zone.[393] [See p. 526.]

Goblet cells—Among the cylindrical epithelial cells there may occur a smaller number of 'goblet cells', in which the cytoplasm is reduced, and the cell surface with its striated border is invaginated to form a deep cavity;[138] or, alternatively, that a cavity with striated border is formed within the developing cell.[550] These cells occur in Ephemeroptera, Plecoptera, &c., but they are most conspicuous in the larvae of Lepidoptera, becoming more numerous towards the hind end of the gut (Fig. 302).[282] No satisfactory function has been ascribed to them. By some authors they are regarded merely as a resting or senescent stage of the cylindrical cells;[77, 282] in *Pyrausta* larvae transitional forms from one type to the other are described.[45] Others consider them as quite distinct, arising independently from the embryonic cells; in *Galleria*[313] and *Vanessa*[138] larvae no intermediate forms can be detected. Both types are present in the newly hatched larva,[138] or even during the last few days of embryonic life;[313] but this does not exclude the possibility of their being exhausted cells, for even at this stage digestion of the yolk and of the egg-shell has begun.[282] In *Tineola*, in which the mid-gut epithelium is completely renewed at each moult, the two cell types can be traced back to the earliest stage and are always distinct.[19]

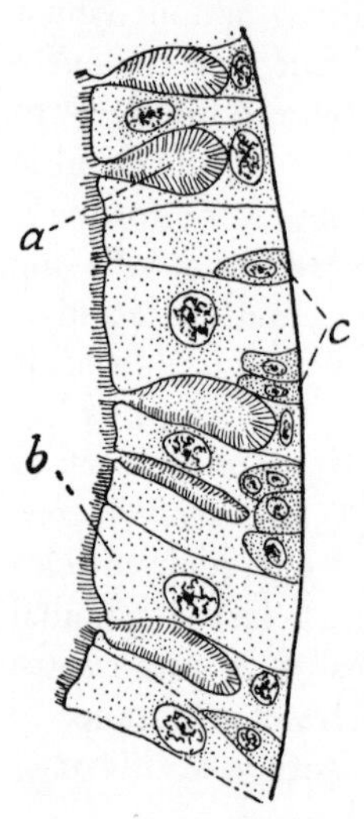

FIG. 302.—Epithelium of mid-gut in larva of *Vanessa urticae* (*after* HENSON)
a, goblet cell; **b**, columnar cell; **c**, interstitial or regenerative cells.

Secretory changes—There is no doubt that secretion in the mid-gut can take place without the cells showing any alteration, either in the living state (in *Chironomus*[33] and mosquito larvae[359]), or in fixed preparations (in *Vespa*,[130] in caterpillars and other insects[138, 216]). But many kinds of visible changes may occur. The nucleus is sometimes described as giving out deeply staining granules or chromidia from its inner pole, which migrate towards the lumen and give rise to vacuoles (in the saw-fly *Nematus*,[150] in *Acridium*,[148] in *Musca* larva[45]). In the beetle larva *Rhagium*[95] and in the cockroach[131] the nucleus shows no change, but the mitochondria in the basal region of the cells are believed to be associated with the separation of secretory material, which then, under the influence of the Golgi apparatus, appears in the form of granules in the region of the nucleus. The granules become vacuoles, which may remain separate or fuse into a single cavity. Similar cytological changes occur in the cardiac cells of *Drosophila*; the granules which eventually form the substance of the peritrophic membrane are secreted by the Golgi element.[285] The contents of the separate vacuoles may escape as little vesicles between the rods of the striated border; or the apex of the cell may become so distended that the striated border disappears, and a single large globule is set free (in *Ptychoptera*,[117] *Pyrausta*,[45] *Chrysopa*,[201] *Galerucella*[247]). Or the vacuolated tips of the cells may be nipped off and break down in the lumen (in *Tabanus*,[70] *Glossina*,[349] *Calliphora*,[129] *Aphis*,[336] and other Hemiptera[244]). Or little spheres of cytoplasm may be extruded through the striated border (in *Celerio*,[77] *Gastropacha rubi*,[137] and other Lepidoptera[301, 190]). In *Stomoxys*, *Musca*, &c., many of the vacuoles in the wall of the mid-gut contain fat. It may be that some of the vacuoles seen in the mid-gut of other insects are of this nature.[382] The part played by the mid-gut epithelium in the storage of fat, glycogen, and protein is described elsewhere (p. 596).

These changes have been observed sometimes in fresh material;[117, 137] but to what extent they are a part of normal physiology has been a subject of much controversy. It is certain that vesicular secretion of this kind can be produced artificially. In the living larva of *Chironomus* no vesicles occur; but they appear at once if the larva is dissected or compressed.[330] In the bee[232, 239] and in other insects[300] they can be seen forming during fixation. In the mid-gut caeca of mosquito larvae they are absent during life,[359] and in sections they appear only after certain fixatives.[33] But it is possible that these technical procedures cause the cell contents to protrude in a manner resembling the normal secretory changes.[300] In the larva of *Vanessa*[138] the 'secretion globules' are thought to result from disintegration of the cells, and not to represent a secretory process; they occur only in very large columnar cells, and are not seen earlier than the end of the 2nd instar. These epithelial changes may appear indiscriminately throughout the gut. But sometimes, as in *Celerio*[77] and in *Aphis*,[336] waves of secretory activity seem to pass from before backwards when a meal is taken; vacuolation, distension, and rupture of the cells apparently succeeding one another in an orderly manner. In Coccinellid beetles secretion tends to be monophasic and synchronous throughout the mid-gut in carnivorous forms, polyphasic and asynchronous in herbivorous species.[248] In *Corethra* the cells of the uniform mid-gut go through alternate phases of vesicular secretion and absorption.[535] Vesicular changes in the gut of *Dysdercus*, formerly regarded as evidence of enzyme secretion, are the result of starvation.[475]

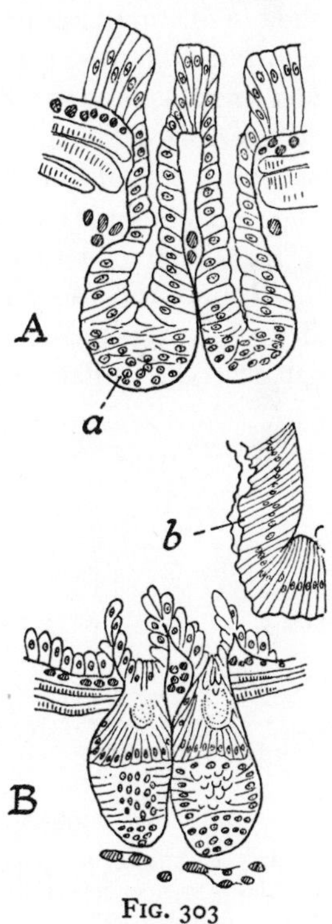

FIG. 303

A, mid-gut epithelium in *Hydrophilus* during resting stage; showing two regenerative crypts (*a*). *B*, the same with the old epithelium (*b*) cast off, and new epithelium being regenerated (*after* RENGEL).

In the striated border of Orthoptera, Coleoptera, and Diptera, as studied with the electron microscope, vesicular extrusion can be seen at the apical ends of the microvilli, and pinocytosis in the border zone between the microvilli.[393] In the mid-gut cells of adult *Aëdes* after a meal of blood, the whorls of granular endoplasmic reticulum condensed around the nucleus unfold to form a complex system ramifying throughout the cytoplasm and reverting to whorls when blood digestion is complete. These changes are presumably related with secretion of protease.[401]

In the types of secretion so far described, the nucleus remains intact and the cells can recover; this is termed 'merocrine secretion'. But in some insects a variable number of the cells break down completely; this is called 'holocrine secretion'. There is no sharp distinction between the two types; in the bee, both occur together;[342] in Orthoptera, secretion is merocrine during continuous small meals, holocrine when a meal follows a period of fasting.[282] In *Tabanus*[70] and *Dytiscus*,[29] vacuoles alone may be discharged; or the inner border of the cell may give way and a mass of granular cytoplasm, often carrying the nucleus with it, be set free into the lumen. In *Dytiscus* the cells go through a definite secretory cycle which is repeated an uncertain number of times before degeneration and replacement occur.[91] In the flea,[101] and in *Galerucella* (Col.),[247] while some cells show little change, others break away entire and disintegrate in the cavity of the gut. In *Hydrophilus* (Fig. 303),[113] *Tenebrio*,[24] and other beetles,[216] the entire lining of the mid-gut is shed off and replaced every forty-eight hours.

Control of enzyme secretion—Protease activity in the mid-gut of *Aëdes aegypti* is increased 26 fold in 18 hours after a meal of blood, but there is only a brief twofold rise after a meal of syrup.[434] Invertase likewise increases fourfold after blood feeding, but shows only a slight change after feeding on sucrose.[434] A protein meal is clearly necessary for enzyme formation. The question arises whether a hormone is involved. There is evidence in *Aëdes* that some fraction in the blood stimulates protease secretion.[537] Protease activity fails to develop in the mid-gut of adult *Tenebrio* if the insect is decapitated one day before emergence, and there is some evidence that injection of haemolymph from normal adults will relieve this deficiency.[417] There is evidence that the hormone from the neurosecretory cells in the brain of *Calliphora* is necessary for synthesis of protease in the mid-gut (p. 198); but it is necessary to exclude the possibility that this is a 'feed-back' effect from the ovaries (p. 727). [See p. 527.]

Regeneration—Among the active epithelial cells there are small basal, embryonic or replacement cells (Fig. 304). These may be scattered singly along the gut as in caterpillars and in Diptera; or they may be collected at intervals in small

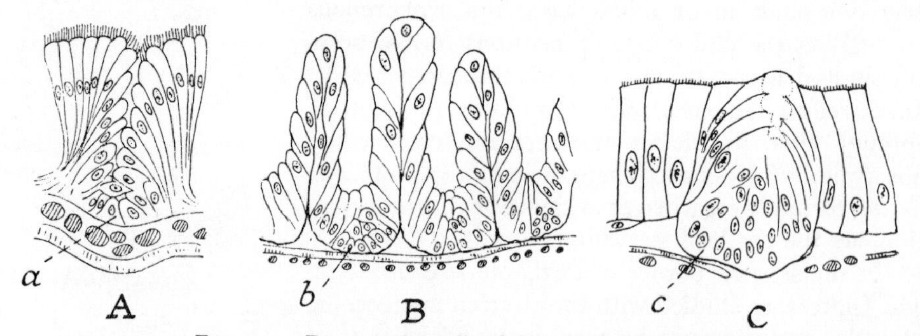

FIG. 304.—Regenerative cells of the mid-gut epithelium

A, *Stenopelmatus* (Orthopt.) (*after* DAVIS); *B*, *Apis* (*after* SNODGRASS); *C*, *Dytiscus* (*after* RUNGIUS). *a*, nidus of regenerative cells; *b*, cells in the folds of the gut; they are probably both secretory and regenerative; *c*, regenerative crypt.

groups or nidi, as in Orthoptera, Odonata, Plecoptera, &c.; or they may form large crypts projecting through the muscular coat and standing out like tags or villi all over the outer surface of the gut, as in Carabids and many other Coleoptera.[282] In *Dytiscus* these three arrangements may appear at successive stages in development.[29] As the epithelial cells degenerate, these basal cells grow to replace them.[113, 148, 248] This may be a gradual process, as in *Galleria*,[313] *Vanessa*,[138] or *Popillia*,[192] where all through larval life they are actively dividing to provide for renewal of the epithelium; or it may occur rhythmically, as in those Coleoptera showing holocrine secretion, and sometimes (perhaps in association with moulting) in the cockroach.[131, 344]

In some insects the epithelium of the mid-gut appears to be renewed completely at each moult, as in Thysanura and Collembola, *Galleria*,[313, 190] and *Psychoda*,[134] termites,[344] *Blattella*,[344] *Dermestes* and *Anthrenus*;[208] in others complete replacement takes place only at pupation, the larval epithelium being generally shed into the lumen, where it breaks down to form an amorphous mass known as the 'yellow body', as in *Celerio*,[77] *Vanessa*,[138] *Popillia*,[192] *Ptinus*, &c.[197] The imaginal epithelium may be formed from the replacement cells of the mid-gut, or from groups of interstitial cells at the points of union with the fore-gut and hind-gut. In some insects there is a true pupal epithelium in the gut between those of the larva and the adult; in others the larval epithelium goes over to the imago; and a third type is intermediate, showing a partial development and partial shedding of a pupal epithelium.[190, 51]

Absorption—There is no phagocytosis of food particles by the mid-gut cells of insects; all the products of digestion are absorbed in solution. In most insects there is no doubt that secretion and absorption are carried out by the same cells. Sometimes the cells are thought to go through alternate phases of secretion and absorption, as in *Celerio*[77] and the cockroach[131]; but there is no clear proof of this. In *Dytiscus* the young epithelial cell is believed to be concerned in absorption and the storage of digestive enzymes; on feeding it goes through a cycle of secretion and extrusion and regeneration, and then functions again as an absorbing cell and begins to store its secretion anew. After a number of these cycles it undergoes degeneration.[91] The distribution of absorbing cells has been tested by feeding insects with iron salts and with dyes; but these methods have given contradictory results on the same insect, notably the cockroach, in the hands of different authors.[73, 131, 300] The histological stages in absorption have been described (in *Chrysopa*,[201] *Musca* larva,[45] &c.) but the conclusions rest on a more or less arbitrary seriation of the appearances seen in sections. Fat is absorbed in the mid-gut;[73, 300] in *Galleria* it is taken up by the columnar cells, never by the goblet cells;[313] and there is evidence that it is first hydrolysed; for if *Tenebrio* is fed with fat dyed red, the fat appearing in the cells is colourless.[24] In *Tineola* iron saccharate is taken up only by the goblet cells; the columnar cells remain colourless.[190] In Aphids absorption occurs particularly in the hind end of the mid-gut.[41]

Vacuolated mid-gut cells are generally thought to be engaged in secretion; but it is probable that the vacuoles may sometimes represent absorbed material. Thus in *Notonecta* the dilated anterior part of the mid-gut is regarded as the chief seat of secretion on account of the great vacuolation of the cells;[32] but in other bugs (*Cimex* and *Rhodnius*[354]), it is known that this region is concerned only in absorption of the fluid from the meal; and the vacuoles in the mid-gut of *Stomoxys* are filled with droplets of fat.[191a]

In *Periplaneta* and *Schistocerca* the absorption of glucose is largely confined to the mid-gut caeca; the rate of absorption is related linearly with the rate of crop emptying which seems to be the limiting process; and the rate of crop emptying is controlled by the osmotic pressure of the solution: more concentrated solutions are retained longer in the crop.[545] The concentration of glucose in the haemolymph is very low (0·024 per cent.); any glucose absorbed is quickly converted to trehalose, notably in the fat body;[519] this process maintains a steep concentration gradient for glucose across the mid-gut wall and thus facilitates the absorption of glucose by diffusion.[545] Fructose and mannose are absorbed in the same way, but they are converted into trehalose less readily and therefore diffuse inwards more slowly.[545] There is no significant absorption of tripalmitin and its break-down products in the crop of *Periplaneta*; here again absorption is confined to the caeca and anterior part of the ventriculus, and the rate of crop emptying is again the limiting factor in absorption.[545] The absorption of amino acids from the mid-gut of *Schistocerca* seems likewise to be dependent on a diffusion gradient across the mid-gut wall, created perhaps by the relatively rapid movement of water into the haemolymph.[545] [See p. 527.]

Aphids bring about differential absorption of the mixed amino acids in the phloem sap on which they feed.[405] They convert dietary sugars in part to oligo-saccharides (the trisaccharide melizitose and others (p. 498)); perhaps this serves to restrain the absorption of unwanted carbohydrates.[473]

Cell inclusions—Globules of fat are often present in the cells of the mid-gut. In many cases this is probably storage fat; for in the cockroach it appears after feeding with sugar,[300] and in the mid-gut of blow-fly larvae it is present after feeding on meat.[144] Crystals, apparently composed of protein, occur within the nuclei in larvae of Lamellicorn beetles[206] and *Tenebrio*,[24, 112] and in other insects,[238, 245] but their significance is not known. Granules of many kinds are often present in the cells, and in the mid-gut of the bee and some other insects these inclusions are composed of lime.[166]

Functional subdivisions of the mid-gut—In most insects the structure of the mid-gut is more or less uniform throughout; though there may be one or more groups of diverticula or caeca which serve to increase its secretory and absorptive area, as in Orthoptera, Lamellicornia, and many Dipterous larvae. In some, however, there is a clear division into regions. We have already discussed the large deeply staining columnar cells in the cardia of Diptera, which secrete the peritrophic membrane (p. 481). In most Nematocerous larvae, such as *Ptychoptera*,[117] *Anopheles*,[111] *Culex*,[267] this zone is followed by a group of caeca in which both secretion and absorption are believed to take place. The remainder of the mid-gut is a long straight tube composed of two segments: an anterior in which the cells are clear and free from granules, regarded on histological grounds as absorbing cells, and a posterior in which the cells, regarded as secreting cells, are opaque white when fresh and filled with deeply staining granules. In *Culex* larvae the absorption of iron saccharate is limited to the posterior half of the mid-gut.[33] In the larva of *Aëdes* after feeding on carbohydrates and certain amino acids there is a massive deposition of glycogen in the epithelium of the *posterior half* of the mid-gut, and a small amount appears in the cells of the gastric caeca. After feeding on olive oil, droplets of fat are limited to the cells of the *anterior half* of the mid-gut and to some cells in the caeca. There seems to be a sharp distinction between the absorptive (or storage) zones for fat and carbohydrate.[356] A similar accumulation of glycogen, limited to the posterior half of the mid-gut, is seen in larvae of *Simulium*[511] and of *Bombyx mori*[567] and may be widespread.

In *Lucilia* larvae the long coiled gut consists of three segments: the anterior and posterior, which are similar histologically and consist of vacuolated cells, secrete digestive enzymes; the short middle zone is strongly acid (p. 502) but secretes no enzymes (Fig. 305). In these larvae the food is introduced in a fluid state from the crop into the mid-gut. Here it is mixed with digestive enzymes; and water and simple products are absorbed. By the time it reaches the posterior segment it has been concentrated to a pasty consistency; it dissolves in the alkaline fluid of that zone and is digested and absorbed.[144] The middle segment of the mid-gut of *Drosophila*[520] and the second zone of the middle mid-gut of *Lucilia*[555] is a mosaic of two types of cell: (*a*) cells with a striated border of unusual type, their cytoplasm packed with lipid spheres and glycogen and rich in acid phosphatase; (*b*) cells with an inconspicuous striated border, rich in esterases, cytochrome oxidase and dehydrogenase, which specifically accumulate iron and copper,[550] but the significance of this differentiation is unknown. The cuprophilic cells have an unexpected origin; they are derived from residual pole cells (p. 10) left behind at the posterior end of the egg after the formation of the gonads.[520] In all, five zones in the middle mid-gut of the *Lucilia* larva have been distinguished histochemically.[520]

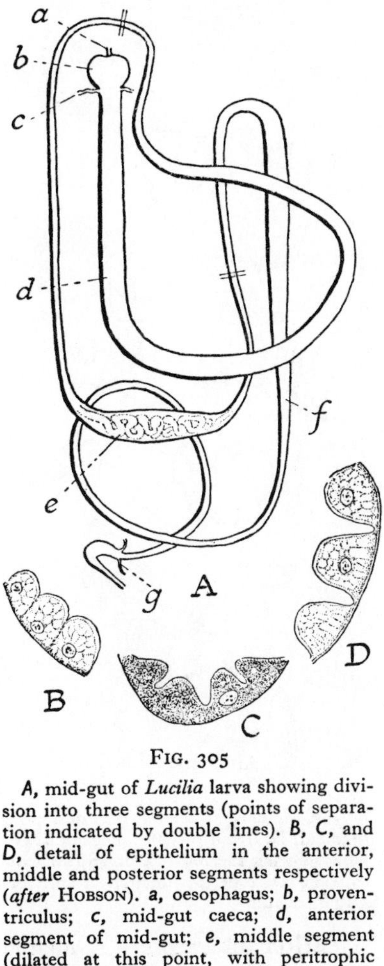

FIG. 305

A, mid-gut of *Lucilia* larva showing division into three segments (points of separation indicated by double lines). *B*, *C*, and *D*, detail of epithelium in the anterior, middle and posterior segments respectively (*after* HOBSON). *a*, oesophagus; *b*, proventriculus; *c*, mid-gut caeca; *d*, anterior segment of mid-gut; *e*, middle segment (dilated at this point, with peritrophic membrane thrown into coils); *f*, posterior segment; *g*, Malpighian tubules.

In the adult tsetse-fly *Glossina*, there is a still more obvious division of labour in the mid-gut. The gut is long and coiled (Fig. 306). In the anterior half the cells are small and pale staining, and in this segment the blood is thickened to a friable consistency by the absorption of water; but no enzymes are present and no digestion takes place. This region gives way abruptly to a middle segment of large deeply staining cells which secrete digestive enzymes; the blood is blackened when it comes in contact with this epithelium. This segment gradually changes to a narrow posterior segment which is probably absorptive.[349] A similar though less striking subdivision occurs in *Calliphora* and other Muscid flies.

In *Belostoma*, *Ranatra*,[187] *Notonecta*[32] and other Hemiptera-Heteroptera, the mid-gut consists of a dilated stomach separated by a sphincter from a long intestine (Fig. 307). The stomach is often regarded as the chief secretory part because of the vacuolation of its cells,[32, 69] but in the blood-sucking forms *Cimex* and *Rhodnius* it serves merely as a crop; the blood may be retained in it for several weeks, the excess fluid is absorbed, but the blood is not blackened or digested until it is transferred to the intestine.[354, 357] The same three divisions are seen in the mid-gut of Capsid bugs: a sac-like region in which the ingested food is concentrated, a tubular region which is the chief site of digestion, and a third region specialized for absorption.[449]

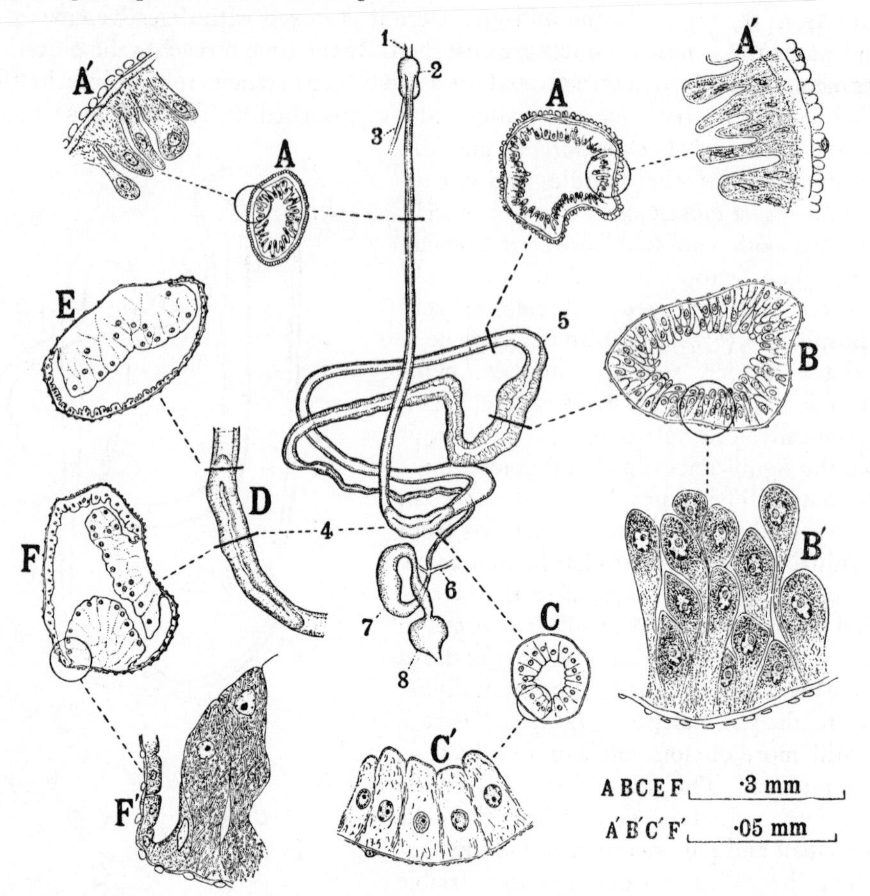

FIG. 306.—Alimentary canal of tsetse-fly *Glossina*, showing the structure of the different parts of the mid-gut (*after* WIGGLESWORTH)

A,A,′ transverse section of anterior segment; *B,B′*, ditto of middle segment; *C,C′*, ditto of posterior segment; *D*, detail of zone of giant cells containing symbionts; *E,F,F′*, transverse sections through this zone showing the rod-like organisms within the cells. *1*, oesophagus; *2*, proventriculus; *3*, duct of crop; *4*, giant-cell zone; *5*, junction of anterior segment with middle segment; *6*, Malpighian tubes; *7*, hind-gut; *8*, rectum.

Filter chamber—A characteristic arrangement of the mid-gut exists in Homoptera, in association it is supposed with their habit of feeding on the copious watery juices of plants. The terminal region of the mid-gut comes into intimate relation with the lower end of the oesophagus or the commencement of the mid-gut.[337] In Membracids the end of the mid-gut simply forms a loop which penetrates the outer wall and applies itself to the epithelial layer of the fore-gut. In Psyllids the hind-gut and oesophagus are spirally wound round one another. In Aleurodids a still more intimate relation exists.[340] In Cercopids the first part of the mid-gut forms a dilated pouch, and the lower end of the mid-gut and a part of the Malpighian tubes lie coiled between its epithelium and connective tissue coat (Fig. 308).[182] In the Cicadoidea there is extensive and intimate contact between the anterior and posterior ends of the mid-gut sac and the proximal parts of the Malpighian tubules. The two sets of walls are very thin, with flattened cells, and the rectum begins where the mid-gut emerges from the peritoneal covering of this 'filter complex'.[450] In the Fulgorid *Phalix* a filter chamber is absent but the mid-gut is tubular and coiled within a membranous sheath made up of a double cell layer.[450]

In Coccids the arrangement is more complex, for the associated coils of mid-gut are enveloped in the hind-gut. This is well seen in *Pseudococcus*, where the initial and terminal parts of the mid-gut form two loops invested in a connective tissue sheath; these lie in an invagination of the surface of the rectum; the intermediate segment of the mid-gut, with a very different histological structure,

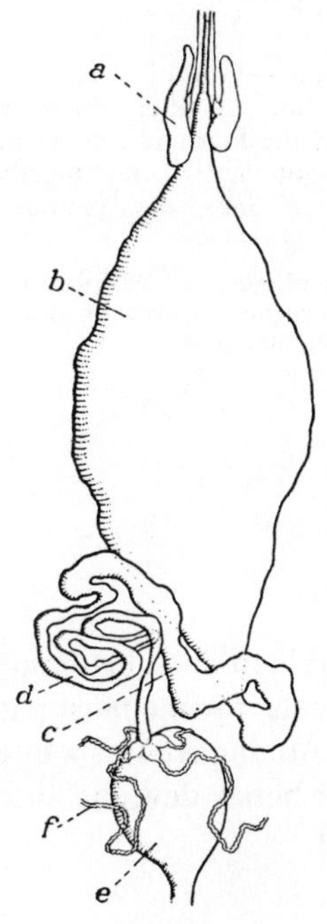

FIG. 307.—Alimentary system of *Rhodnius* (*after* WIGGLESWORTH)

a, salivary glands; *b*, stomach (mid-gut); *c*, anterior segment of intestine (mid-gut); *d*. posterior segment of intestine (mid-gut); *e*, rectum (hind-gut); *f*, Malpighian tubes.

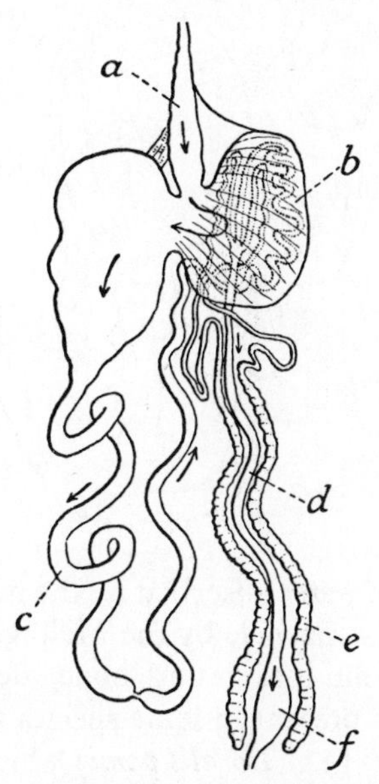

FIG. 308.—Gut of *Thomaspis saccharina* (Cercopidae) (*after* KERSHAW)

a, oesophagus; *b*, filter chamber consisting of coiled endings of mid-gut and Malpighian tubes invaginated into a diverticulum of the oesophagus; *c*, mid-gut; *d*, hind-gut; *e*, Malpighian tubes (two only of the four are shown); *f*, rectum.

being free in the body cavity.[238] In *Lecanium* the first and last parts of the mid-gut form a closely interwoven coil like a glomerulus within the invagination of the rectal wall (Fig. 309).[238]

It is believed that these arrangements have a common purpose: they enable the excess fluid in the food to pass directly from the first part to the last part of the mid-intestine without general dilution of the haemolymph. The main digestive segment of the gut is thereby short-circuited, and receives only the valuable constituents of the plant juices.[20, 182, 450, 500] This transference of fluid cannot of course be brought about by simple physical 'filtration'; the different epithelia must play an active part, as in any other secretory process.[238] Where the related coils are invaginated into the rectum, as in *Pseudococcus* and *Lecanium*, it is possible that a further passage of fluid may take place through the walls of this invagination into the lumen of the rectum;[20] but since the mid-gut is continuous with the hind-gut this is not a necesy*.[238] Perhaps the invagination merely serves to isolate the 'filter chamber' from the general body cavity. It is doubtful whether the term 'filter chamber' is appropriate for this arrangement.[238] In Cercopids there is a rapid elimination of fluids from the gut—but even here most

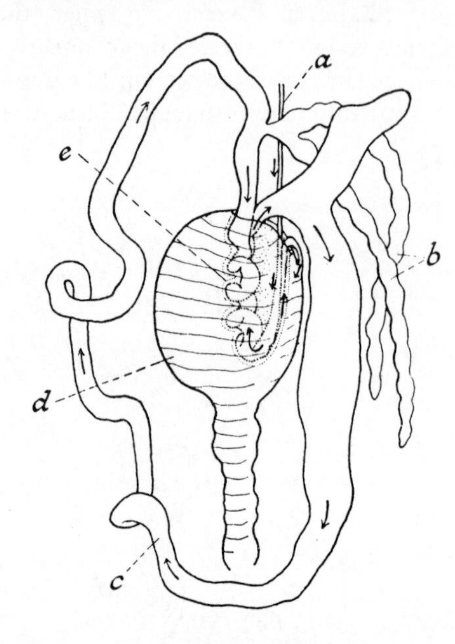

FIG. 309.—Gut of *Lecanium* (Coccidae); showing filter chamber consisting of the first and last segments of the mid-gut closely coiled together and invaginated into the rectum (*after* PESSON)

a, oesophagus; *b*, Malpighian tubes; *c*, intermediate segment of mid-gut; *d*, rectum; *e*, filter chamber.

of the water absorbed by the first part of the mid-gut loop must be removed from the haemocoele by the Malpighian tubes. The Coccids, for the most part, retain their fluid, excreting 'honey dew' rich in sugars.[238] Among Aphids a filter chamber is present in some species which do not produce honey dew and absent from others (e.g. *Doralis pomi*) which do.[41] [See p. 527.]

THE HIND-GUT AND 'RECTAL GLANDS'

Histology—The hind-gut is made up of the same layers as the fore-gut, but the epithelial cells are usually larger and they often show a conspicuous vertical striation of the inner border. The cuticle is thinner and, unlike that of the fore-gut, it is readily permeable to water.[1, 93, 128]

Towards the hind end the epithelial cells tend to increase in size and to be arranged in six longitudinal folds, as in Thysanura. This wider part is termed the 'rectum', in distinction from the narrower 'ileum' which precedes it. Smaller cells may lie between the longitudinal folds, as in *Tenebrio* (Fig. 310, A); or the large columnar cells may be collected into six cushions radially disposed and occupying a relatively small part of the total surface area, as in Dermaptera, Orthoptera,[37] Carabidae (Fig. 310, B).[40] These discrete epithelial pads are termed 'rectal glands'; but all intermediate stages exist between them and a uniform epithelium.[20] Perhaps they are an adaptation permitting distension of

* Only in the Diaspinae (*Aspidiotus*, &c.) is the gut discontinuous, the anterior and rectal parts being connected merely by a ligament.[238]

the rectum.[62] Rectal glands become most obvious when the rectum is a capacious sac; in the adults of Siphonaptera and Diptera[39, 129] they form large conical papillae (Fig. 310, D); in Lepidoptera they consist of numerous projections each made up of two or three large cuneiform cells.[38] Sometimes, in the course of development, other cells from the body cavity apply themselves to the outer surface of the epithelial pads, and produce a two-layered structure with a cavity between, as in the bee (Fig. 310, C)[98, 327, 191] and in *Chrysopa*.[201] They always have a rich tracheal supply, indicating a high rate of metabolism.[62]

When stained by routine methods the cytoplasm of the rectal gland cells appears pale and empty; but when stained by methods to show up mitochondria and plasma membranes,[562, 563] or examined in thin sections with the electron

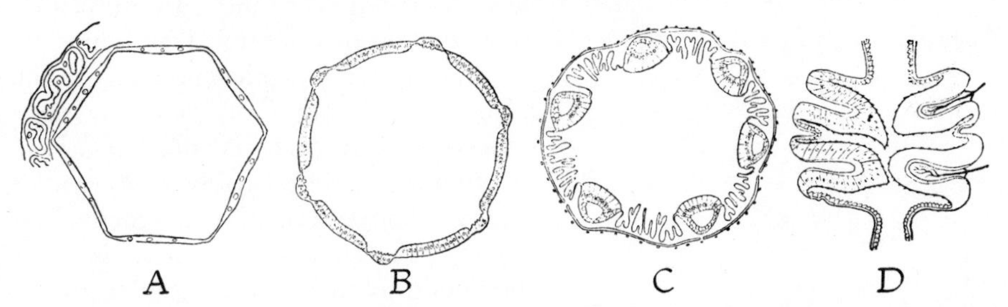

FIG. 310.—Sections through the rectum of some insects to show the types of rectal glands (*modified after various authors*)

A, *Tenebrio*, showing a part of the investing sheath carrying the Malpighian tubes; *B*, *Pterostichus* (Carabidae); *C*, *Apis*; *D*, *Calliphora* (longitudinal section).

microscope,[392, 505] the cells in termites, *Gryllus*, *Periplaneta*, *Calliphora*, &c., are characterized by extensive invaginations of the apical plasma membranes extending far into the body of the cell, and by an enormous concentration of mitochondria between these membranes. The ileum in adult Orthoptera likewise shows a complex system of invaginated plasma membranes in the apical region of the cells and many mitochondria in the basal region. This agrees with the belief that this region is absorbing water and perhaps other substances from the excreta.[393]

The cuticle in the rectum is sometimes thinner than in the ileum, as in the Syrphid larva, *Syritta*;[171] and that over the rectal glands may be particularly thin: in *Carabus* it is only 1·5μ thick, being 3μ elsewhere in the rectum.[40] As seen with the electron microscope, the epicuticle over the rectal glands of *Oryctes* larvae differs from that elsewhere in having a porous structure.[479]

Function of the hind-gut and rectal glands—The mid-gut is separated from the hind-gut by a 'pyloric sphincter'; when this is closed the hind-gut receives only the contents of the Malpighian tubes (p. 561). Normally the sphincter opens from time to time and admits a portion of the mid-gut contents. When these enter they are quite fluid, and the most obvious function of the hind-gut in many insects is the absorption of water. In *Tenebrio* and other beetles,[112] Thysanura, Dermaptera, Orthoptera, Neuroptera, &c.,[352] the contents become progressively drier as they pass along the hind-gut. This desiccation is particularly evident in the rectum, where the material is in contact with the high epithelium or with the rectal glands; and here it is converted into a more or less dry pellet before it is discharged through the anus. The rectal glands probably play an important part in this process of absorbing water (Fig. 311).[352] In the larvae of Lepidoptera the hind-gut consists of a variable number of chambers separated by sphincters.[38] In *Vanessa* there are three such chambers, the ileum, the colon, and the rectum (Fig. 324),[138] and the gradual drying of each bolus of moist residue into a faecal pellet as it passes through them is very obvious.[329]

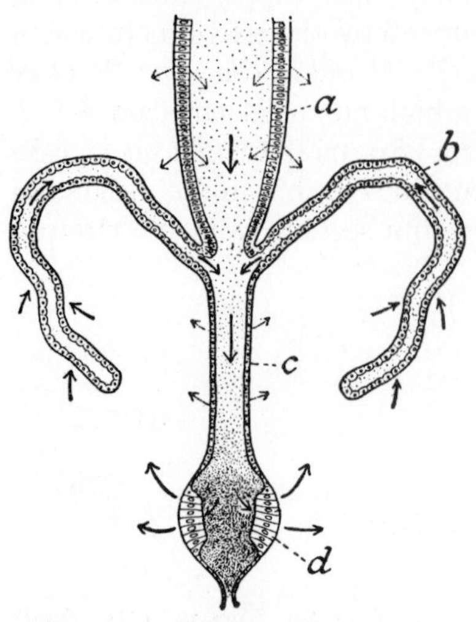

FIG. 311.—Diagram of water-circulation in alimentary and excretory systems of an insect (*after* WIGGLESWORTH)

a, mid-gut; b, Malpighian tubes; c, hind-gut; d, rectal glands.

In those insects, such as the adults of Diptera, Siphonaptera, Lepidoptera and Hymenoptera, in which the rectal contents are always fluid, the absorption of water is less apparent; but as we shall see in discussing excretion (p. 553) there is some evidence that the rectal glands are engaged in reabsorption here also. In *Lucilia* the accumulation of indicator dyes around the bases of the rectal papillae probably results from the absorption of water by these organs.[335] But this is not their sole function; they are probably concerned in recovering other valuable constituents from the excreta. In *Tenebrio* there is a re-absorption of sodium in the rectum[231] and in *Chironomus* and *Limnephilus* (Trichoptera) chloride is taken up.[34] Chloride is completely removed from the rectal fluid of *Sialis* larvae.[538] In *Schistocerca*, along with the active uptake of water there is an absorption of sodium, potassium, and chloride ions against steep concentration gradients. In this locust the cuticular intima lining the rectum acts as a molecular sieve restricting the diffusion of large molecules and ions: it is 10–100 times more permeable to water than to KCl or NaCl; and the rate of diffusion of disaccharides such as trehalose is only 0·3 per cent. of that of NaCl. It is estimated that the water-filled pores have a radius of about 7 Å; unlike the cuticular intima of the fore-gut it is specialized to allow absorption of small physiologically important molecules such as water, various salts, and amino acids, but excludes larger organic molecules.[517] In the hind-gut of *Lucilia* larvae there is a striking functional differentiation among the epithelial cells; certain of these cells appear to take up ammonia from the haemolymph and transfer it to the hind-gut as bicarbonate.[552] [See p. 527.]

In wood-feeding termites and Lamellicorn beetles[343] the hind-gut is very large, and one segment is enormously dilated; as we shall see later (p. 511) this appears to be the chief site of digestion and perhaps of absorption. Whether other substances besides water and inorganic ions are absorbed in the hind-gut of most insects is not known. Experiments with fats, dyes, and iron salts have generally proved negative;[300] though the hind-gut of some beetles is said to be permeable to glucose and to some dyes.[128] In such blood-sucking insects as mosquitos, *Glossina*, *Cimex*, &c., absorption is to all appearances complete in the mid-gut, and nothing enters the hind-gut but a little haematin. We shall refer again later to the part played by the hind-gut in the uptake of salts (pp. 514, 573).

The capacious rectal ampulla of the *Dytiscus* larva has quite a different function. It serves to hold the large quantities of water which are swallowed by this insect at the time of moulting; just as the crop or mid-gut serve to receive the air swallowed by terrestrial insects (p. 45). And throughout larval life the contents of this ampulla appear to vary so as to compensate for the varying state of nutrition of the insect,[265] serving in this respect the same function as the abdominal air sacs of Muscid flies (p. 372).

The peritrophic membrane in the hind-gut—In those insects which possess a peritrophic membrane, this continues into the hind-gut and often forms a sheath in which the faecal pellets are invested. The cuticle of the hind-gut frequently bears small backwardly directed spicules, which probably assist in drawing back the membrane and its contents when peristaltic waves pass along it. In Muscid flies a complex muscular organ, or 'rectal valve', divides the hind-gut into two segments; it consists of a fold of cuticle bearing sharp spines, and is preceded and followed by circular muscles which act as sphincters (Fig. 312). During each movement of this organ short lengths of peritrophic membrane are seized and dragged towards the rectum.[129] The rectal papillae of these flies, which are also covered with spicules, have likewise been thought to grip the peritrophic membrane and draw it to the anus[20] or, by their churning movements, to tear the membrane into fragments before it is discharged.[96] In mosquito larvae, the membrane may occasionally extrude from the anus as an unbroken tube almost as long as the larva itself.

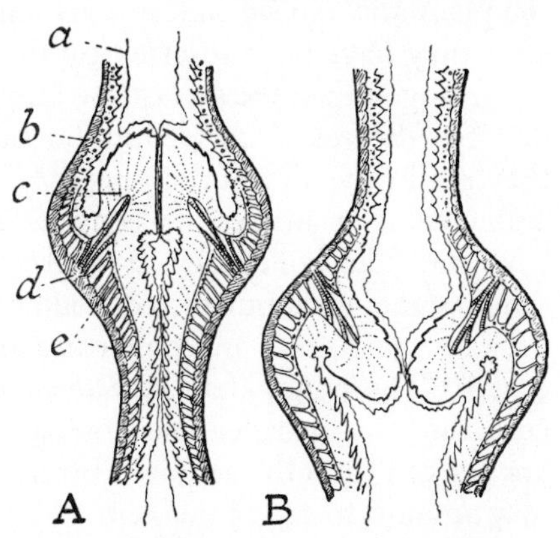

FIG. 312.—Rectal valve of *Calliphora* (*after* GRAHAM-SMITH). Two phases of its action are shown, to illustrate its function of drawing backwards the peritrophic membrane

a, peritrophic membrane; *b*, muscular layer of hind-gut; *c*, thick 'opalescent' layer of the rectal valve with fibrillae; *d*, point of attachment of elongated cells (resembling muscles) inserted into the valve; *e*, epithelial cells.

The faeces—The excrement is often in the form of dry pellets, which are passed at more or less regular intervals by insects that feed continuously. If the cockroach is fed with soft food, such as a sweetened banana paste, this begins to enter the mid-gut within half an hour, and the passage through the entire alimentary canal requires from 9–33 hours.[290] The average time required for the passage in the silkworm is from 2–3 hours,[284] in the Noctuid larva *Prodenia* $3\frac{1}{4}$ hours,[71] and in the clothes moth *Tineola* 2–4 days.[319] The rate of passage has been studied in a number of insects by the use of X-rays.[219]

In other insects, such as Diptera, Hymenoptera and Lepidoptera, the excrement is always fluid, and it may be retained for long periods in the rectum: in the young honey-bee the rectum is not evacuated until foraging begins, that is, for about 3 weeks after emergence; nor during the whole winter period. It is at such times that the reabsorption of water is probably important.[352]

In plant-sucking Hemiptera the excrement is a copious clear fluid. In Aphids and Coccids it contains much unabsorbed organic matter, particularly carbohydrates, and furnishes the 'honey-dew' that collects on the leaves of plants in dry weather. Manna is a similar product from the Coccid *Trabutina mannipara* feeding on tamarisk in Sinai; it contains 55 per cent. of cane sugar,

25 per cent. of invert sugar, and 19·3 per cent. of dextrin.[337] The significance of this wasteful method of feeding is generally supposed to lie in a deficiency of some component in the plant juice. It was suggested that the component might be protein,[50] but although proteins and peptides are absent from the sieve tube juice on which the Aphid feeds, there are abundant amino acids and amides.[421, 499] In any case, the flow of sap into the Aphid is effected largely by tugor pressure; to obtain it costs the Aphid almost no expenditure of energy.[472, 473] It is possible that there is some other necessary substance, inorganic salt, vitamin, &c., which can only be obtained in the required amount by sucking excessive quantities of juice. But when *Trialeurodes* feeds on the same plants as Aphids, it has neither carbohydrate nor protein in its excrement.[338] The disposal of droplets of honey-dew by dense colonies of Aphids in confined quarters, as in a gall, presents a physical problem which is solved by coating the excretory droplets with wax as they are formed.[406]

Aphids feed on the phloem sieve-tube juice which exists under pressure in the plant and can be collected by cutting through the embedded stylet-bundles after they have been inserted by the insect.[472] The honey-dew always contains less nitrogen than the sieve-tube juice.[488] The amino acids excreted in the honey-dew by *Tuberolachnus* feeding on *Salix* are in much lower concentration than they are in the sieve-tube juice.[499] The phloem juice of *Vicia faba* contains 19 amino acids or amides, no peptide or protein and has a total N content of 0·24 per cent. The honey-dew from *Megoura* feeding on this juice contains all the same amino acids but in reduced concentration, with a total N content of 0·11 per cent., and there are some extra amino acids including tryptophane and histidine.[426] The honey-dew of *Tuberolachnus* also contains sucrose, fructose, glucose, the trisaccharide melezitose, and various other oligosaccharides all of which are derived from the action of invertase on the sucrose of the plant sap,[499] which may amount to 10–15 per cent.[421] This is a usual phenomenon in the feeding of Aphids and Coccids.[394, 426, 431, 452]

The excrement finds many uses in the economy of insects. The sugary excreta of Aphids and Coccids is much sought after as food by ants and other insects. The food residue is often utilized in constructing the cocoon: swallowed earth mixed with the Malpighian secretion hardens to form a cement for this purpose in *Cetonia* (Col.) larvae;[343] the Javanese Tenebrionid *Platydema* spins its cocoon of threads made up of chitin fragments glued together, derived from the fungus on which it feeds.[278]

SECRETIONS OF THE ALIMENTARY CANAL

Salivary glands and their secretions—Of the dermal glands associated with the mouth appendages, the labial pair most commonly serve as salivary glands; mixing their secretion with the food as this is taken in. They are of varied form,[162] and in some insects, notably in Hemiptera,[13, 49, 103] they are made up of several lobes with quite different structure and with different functions. In *Rhodnius* the salivary glands are cherry red in colour; they contain a pigment resembling 'haemalbumen' derived from traces of haemoglobin absorbed from the gut.[357]

The saliva is generally a clear watery approximately neutral fluid (*p*H 6·9 in the cockroach).[348] It is used in the first place for moistening the food: when the cockroach feeds, its mouth parts are wet with the secretion; while Lepidoptera commonly extrude a drop of saliva in order to dissolve dried sugars, &c.[301] In forms with piercing and sucking mouth parts, such as the Hemiptera or the blood-sucking Diptera, in which the saliva is driven down the hypopharyngeal or other duct, and poured out at the tip of the proboscis, it probably serves to keep the mouth parts moist and clean in the intervals between feeding.[181] But in many insects it contains active constituents. [See p. 528.]

The labial glands are frequently modified for the production of silk (p. 604) for the formation of 'stylet sheaths' (p. 500), and for other purposes. In Psocids two pairs are present; one pair of silk glands, the other serving to moisten the food and perhaps to produce digestive enzymes.[341] [See p. 528.]

Enzymes in salivary glands—Amylase and invertase are the enzymes most frequent in the salivary glands. Amylase is exceedingly potent in the saliva of the cockroach; it occurs also in *Calliphora*,[349] Aphids,[75] Jassids,[139] &c. Invertase is present in Lepidoptera,[301] and in the cockroach *Blattella* (though not in *Periplaneta*).[348] In the larva of *Corethra*, the greater part of digestion takes place in the dilated pharynx, the residue of the prey being ejected; in this case the proteolytic digestive fluid appears to come, not from the salivary glands but from the mid-gut secretion.[447] In the Hemiptera-Heteroptera the enzymes present in the salivary glands vary with the nature of the food: protease and lipase in *Notonecta*, *Naucoris*, and *Nepa*; protease alone in *Gerris*, *Nabis*, and *Anthocoris*; invertase and amylase in *Corixa* and *Lygus*; invertase alone in *Pentatoma*, *Gastrodes*, *Dysdercus*, and *Pyrrhocoris*; amylase alone in *Metacanthes* and *Corizus*;[13] in *Oncopeltus* the watery saliva containing the enzymes: amylase, protease, invertase, and lipase comes from the posterior lobe of the salivary glands;[407, 496] in *Rhodnius*, *Triatoma*, and *Cimex* no enzymes can be detected.[13]

The saliva of the assassin bug *Platymeris* rapidly abolishes the excitability of nerves and muscles of its prey. A general lysis of the tissues follows quickly; lipid layers in the cells of nervous tissue are rapidly dispersed and only cuticular and collagenous structures are unaffected. Hyaluronidase is present, and at least three proteolytic fractions, and phospholipase which is perhaps responsible for the toxicity. This saliva can be projected by spitting, up to a distance of 30 cm., through the powerful action of the salivary pump.[425]

Aphids which reach the phloem sieve tubes by passing their stylets *between* the cells rather than through them possess pectinase in the saliva.[490] The Pea and Bean Aphid *Megoura* has weak α-glucosidase activity and very strong pectin polygalacturonase, doubtless associated with this mode of feeding.[426] This enzyme, active in the pH range 4·4–5·6, is present in Myridae but not in all plant-sucking bugs.[483]

It has been suggested that the toxic substances (proteases and auxins) in the saliva of some plant sucking bugs, such as *Plesiocoris rugicollis*, and many other Capsids (Myridae),[449] may originate in the host plant and be taken up by the insect and transferred to the salivary glands.[507] But some species certainly produce their own protease in the salivary glands. Commonly, protease and amylase occur in species such as the leafhopper *Empoasca devastans* which feed on the mesophyll; and are absent from the salivary secretion of phloem-feeding Jassids such as *E. flavescens*.[507, 532] The lack of injury to wheat grains as the result of feeding by *Lygus rugulipennis* is attributed to the absence of protease in the saliva of this species.[507] In *Dolycoris baccarum* the presence of albumen in the diet stimulates secretion of salivary protease, which may otherwise be absent.[508] Other examples of the digestive functions of the salivary glands will be considered under the heading of extra-intestinal digestion (p. 505). [See p. 528.]

Factors in the saliva are sometimes responsible for gall formation. The gall-forming agent in *Hormaphis* feeding on *Hamamelis* is Feulgen-positive and is considered to be a virus which brings out latent growth potentialities in the plant.[486] The Vine Phylloxera (*Viteus vitifolii*) secretes a viscous saliva which induces primary polyploid changes in the meristematic tissue of the host-plant and thus leads to gall formation.[390] The larva of the Cecidomyid *Mikiola fagi* produces substances similar to the growth hormones of plants and moulds the leaf by inducing cell divisions and cell lengthening.[465] [See p. 528.]

Stylet sheaths—In Aphids[75] and Aleurodids[338] the track of the stylets as they pass between or through the cells, becomes walled off by a resistant sheath which is produced, at least in part, from the salivary secretion. This sheath seems to have the function of filtering off particles in the sap which might lead to obstruction of the sucking canal,[306] and of preventing the sap from exuding to the exterior around the stylets.[499] A similar gelatinous sheath is formed by the salivary secretion in *Cicadulina* (Jassidae).[302] *Oncopeltus* and *Dysdercus* produce two types of saliva, the viscous sheath material which quickly coagulates to form a lining to the stylet path, and the watery saliva containing the digestive enzymes.[496] The sheath material represents a mixture of secretions from anterior and lateral lobes; under the action of an oxidase from the accessory gland it forms a polymerized, perhaps 'self-tanned' product, a condensed lipoprotein similar to that which forms the basis of the 'cuticulin' of the epicuticle (p. 35).[496] In the Pentatomid *Eumecopus* the stylet sheath is stabilized by hydrogen bonds and disulphide bonds. The precursor material comes from the anterior lobe of the salivary glands; it is rich in sulphydryl groups which are prevented from forming disulphide bonds by the reducing conditions set up by the tyrosine and DOPA present. On ejection the lipoprotein gels rapidly even under water.[497] *Aphis craccivora* discharges an exceedingly viscous secretion which solidifies almost immediately to form the stylet sheaths, and it then pours out watery saliva which presumably contains the digestive enzymes.[495] [See p. 529.]

Drosophila and *Phormia* larvae cement the puparium to the substrate by means of a copious proteinaceous secretion from the salivary glands,[438] apparently a mixture of mucoproteins.[516] [See p. 529.]

Saliva of blood-sucking insects—In the blood-sucking insects *Chrysops* (Tabanidae)[349] and the tsetse-fly *Glossina*,[349] the saliva contains no digestive enzymes; but in *Glossina*, at least, a very active anticoagulin is present.[363] If the glands are removed from the living tsetse-fly, feeding and digestion take place normally for some time; but eventually the proboscis and the crop become blocked with clots of blood.[181] Anticoagulin is present in the salivary glands of some other blood-sucking insects, but not in all: it occurs in *Anopheles maculipennis* but not in *A. bifurcatus*, *Culex* or *Aëdes*;[205, 363] it occurs in *Philaematomyia* (though not in the non-blood-sucking species of *Musca*), yet it is absent in *Stomoxys*;[67] it is present in the blood-sucking bugs *Rhodnius*, *Triatoma*, and *Cimex*;[13] and it is present in the salivary glands of *Gasterophilus*[83] although these larvae seldom seem to ingest blood.[261] When the flexible stylet bundle of *Aëdes* enters a capillary there is an enormous acceleration in the blood flow; saliva appears to be injected throughout the act of biting.[127] [See p. 529.]

Salivary glands of the honey-bee—The honey-bee has a particularly complex system of salivary glands (Fig. 313).[539] (i) A pair of *mandibular glands* with an acid secretion (*p*H 4·6–4·8) which are very active in the queen, less so in workers, and more or less vestigial in drones;[168] perhaps their secretion serves to soften the cocoon at emergence;[86] and it serves as a solvent during the collection and working of propolis.[510] In the queen bee they are the source of the 'queen substance' and other pheromones (p. 335). (ii) The *pharyngeal glands*, which, in the early days of life of the adult bee, produce the 'royal jelly' used for feeding the young brood and the larvae of queens. The bee requires pollen for these glands to become fully active, which happens at 3–6 days after emergence. Their secretion is acid (*p*H 4·5–5·0)[168] and proteolytic.[158] Later in life, when the bee begins foraging at about 3 weeks after emergence, the pharyngeal glands begin to secrete the enzymes amylase and invertase, becoming most active in this respect in bees about one month old.[167] They are probably responsible for the invertase and amylase in the honey;[168] even honey made from pure cane-sugar syrup contains these enzymes;[269] and they are said to secrete an oxidizing enzyme which brings about the formation of acid (perhaps gluconic acid) from glucose. The resulting acidity (*p*H 3·6–4·6) is sometimes regarded as serving to preserve the honey.[115] The pharyngeal glands of forager bees can redevelop to enable them to feed the brood again if the state of the colony should require it.[443] (iii) The *labial glands*, each with two divisions: (*a*) the 'posterior cephalic glands' which produce a clear oil used for working the wax,[158, 509] and the saliva used to moisten materials for chewing;[539] (*b*) the 'thoracic glands' which are active throughout life, and produce a watery secretion, with a *p*H of 6·3–7, probably used in building the comb.[142] They are said to produce invertase in the foraging bee.[158]

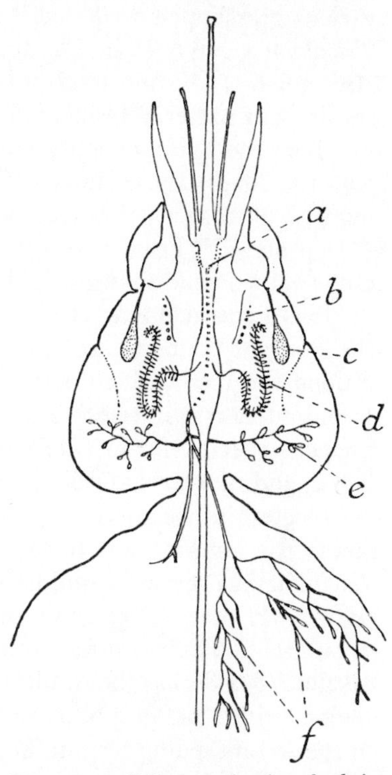

FIG. 313.—Salivary glands of *Apis*, schematic (*after* HESELHAUS)

a, opening of labial glands; b, circumoral glands; c, mandibular glands; d, pharyngeal glands; e, post cephalic part of labial glands; f, thoracic part of labial glands.

Royal jelly—*Apis* larvae destined to become queens receive a much higher proportion of 'royal jelly' in their diet and receive it throughout their growth. It is this difference in diet which determines the difference in caste (p. 99); but in spite of extensive analyses of royal jelly the nature of the determining factor is uncertain. Royal jelly contains 40 per cent. of dry matter comprising lipoproteins, neutral glycerides, free fatty acids (largely hydroxy fatty acids all with ten carbon atoms),[556] sugar, amino acids, adenine, adenosine, guanosine, adenosine monophosphate, &c.; all the B-vitamins[466, 523] and other components. It has a very high content of acetylcholine (1 mg./g.).[459] The food given to the queen larvae is characterized by a much higher content of pantothenic acid, folic acid, and biopterine.[523]

Hydrogen ion concentration of gut contents—The digestive juices of most insects are weakly acid or weakly alkaline.[172, 299, 311] Thus the *p*H of the mid-gut contents in the cockroach averages 6·2,[348] in *Carausius* 7·0–7·5,[61] in various grasshoppers 5·8–6·9,[31] in Odonata 6·8–7·2,[283] in *Popillia* 7·4–7·5,[310] in *Tribolium* 5·2–6·0,[478] in *Chironomus* larvae 7·2–7·8,[72] in *Glossina* 6·5–6·6,[349] in *Apis* 5·6–6·3,[5, 154] and in *Apis* larva 6·8.[5] More acid conditions may sometimes result from bacterial fermentation, notably in the crop of Orthoptera[242, 299, 300] in which the *p*H commonly ranges from 5·0–5·9.[31] This effect naturally varies with the diet: in the cockroach *Blattella* the *p*H in the crop falls to 4·4 after feeding with glucose which is readily broken down to lactic acid, 4·8 after lactose, 6·3 after protein.[348] Perhaps this is the cause also of the rather acid contents of the crop (*p*H 5·2) and mid-gut (*p*H 5·2) in termites.[252]

It is a characteristic of larvae of Trichoptera[283] and Lepidoptera[160, 309] that the mid-gut contents are always strongly alkaline (*p*H 9·0–9·4). Similar conditions are met with in some herbivorous Coleoptera;[283] but in the case of Lepidoptera it is interesting to note that this alkalinity is not confined to phytophagous species, for in *Galleria* feeding on honeycomb the *p*H of the mid-gut is 8·4, and in *Tineola* feeding on hair, &c., it is 9·9.[90]

In a few insects there are localized differences within the mid-gut. In *Tenebrio* larvae the anterior two thirds turn litmus red, the posterior third blue.[24] In *Lucilia* larvae the short middle segment of the mid-gut (p. 491) is strongly acid, with a *p*H of 3·0–3·5, while the segments which precede and follow this are about *p*H 7·5, except towards the hind end, where the contents become strongly alkaline (8·0–8·5) as the result of ammonia production (p. 556).[144, 335] A similar region, with a *p*H of 3·0, occurs near the junction of the mid-gut and hind-gut in the wood-feeding termite *Zootermopsis*.[252] In the *Lucilia* adult also there is a very short middle segment with strongly acid contents (*p*H 3·5–3·9), an anterior segment with *p*H of 4·8–5·3, and an alkaline posterior segment (7·8–8·0) containing ammonia; the *p*H of the hind-gut being 4·4–5·3.[335] The nature of the acid in these cases is not known with certainty, but it is probably phosphoric acid;[144] in the normal bee the concentration of phosphate in the mid-gut contents (0·046 M) is nearly five times that in the blood (0·01 M), whereas the excreta contain only 0·005 M.[154] In *Carausius* the buffering capacity of the gut contents is greatly increased in the starving insect by a secretion of unknown nature.[299] The reaction of the hind-gut will be considered in connexion with the urine (p. 555).

Digestive enzymes: adaptation to the diet—Broadly speaking the digestive enzymes in insects are adapted to the diet on which they feed. Omnivorous insects like the cockroach secrete protease, lipase, amylase, invertase and maltase, hydrolysing respectively natural proteins, fats, starch, cane sugar and maltose.[308, 348] The same series of enzymes appear in the mealworm[24] and in the phytophagous Orthoptera[61, 273] and Coleoptera[40] (though amylase is occasionally absent, as in *Popillia*[310]) in the larvae of Lepidoptera[284, 301, 274] (though here again amylase is said sometimes to be wanting, as in *Laspeyresia*[305] and in some races of *Bombyx mori*[284]), and in the honey-bee.[99, 232] These enzymes are secreted, as we have seen, chiefly by the mid-gut and its caeca, though invertase and amylase are occasionally contributed by the salivary glands (p. 499).

Insects which live on food rich in some particular substance generally produce the appropriate enzymes in particular abundance; and in those which live on a highly restricted diet, the enzymes present are correspondingly limited. Thus amylase is exceptionally active in the saliva of the cockroach.[273] In the bee, invertase first appears in the pharyngeal glands (p. 501) when the young bee starts foraging; it is already present in the mid-gut of the newly emerged bee, but later it increases greatly in amount.[167] Protease and lipase predominate in most carnivorous and necrophorous beetles;[40, 211] in the predatory Carabids lipase is weak, carbohydrases are quite absent except a very feeble amylase, but proteases are extremely active.[273] Among Orthoptera, protease and lipase predominate in the predatory *Decticus*, in which lichenase acting on stored cellulose is absent; whereas carbohydrases, including lichenase, are very active in Acrididae.[273] We shall discuss in a later section (p. 506) the distribution of enzymes which attack the skeletal substances of plants and animals.

These differences according to the composition of the diet are sometimes very evident when the developmental stages of an insect are compared. We have seen the wide range of enzymes present in caterpillars. In adult Lepidoptera, on the other hand, (*Celerio*, *Macroglossa*, &c.,[301] *Laspeyresia*,[309] *Ephestia*[222]) only invertase remains (in saliva and mid-gut); and in insects such as *Dicranura*, *Lymantria*, &c., which have vestigial mouth parts, and which take no food, even invertase is wanting.[301] Larvae of the blow-fly *Lucilia* are adapted primarily to feed on meat; the mid-gut secretes protease and lipase; carbohydrases are absent save a very feeble amylase in the salivary glands.[144] Whereas in the adult blow-fly *Calliphora*, protease is relatively feeble, while amylase is active in the saliva and mid-gut, and invertase and maltase in the mid-gut.[349]

Calliphora, feeding largely on sweet substances, forms a marked contrast with *Glossina*, feeding exclusively on blood. In *Glossina*, protease is very active in the mid-gut, carbohydrases are absent except a very feeble amylase in the same region.[349] Whereas *Chrysops* (Tabanidae), which feeds on both blood and nectar, occupies an intermediate position: an active protease and invertase and a weak amylase are present in the mid-gut.[349] We have seen (p. 499) that the presence or absence of protease in the salivary secretion of Aphids, Jassids, &c., is dependent upon whether they feed on the mesophyll or on the sieve-tube juice.

The enzymes increase in quantity in the mid-gut in the early stages of digestion; they are clearly produced under the stimulus of feeding. Later they decrease and they diminish towards the end of the intestine, being practically absent from the contents of the hind-gut; their fate is not known.[273] In *Dytiscus* the tryptic protease of the mid-gut is confined to the active epithelial cells between the crypts; during the secretory cycle in these cells (p. 488) their tryptase content may fall to one third or one twelfth of its original value.[91] In adult mosquitos the presence of blood in the mid-gut is necessary to stimulate proteolytic activity; almost no protease is present in the unfed insect or after feeding on sugar.[369] Attempts have been made to alter the relative activity of the enzymes in the cockroach by feeding it on diets with different proportions of protein and carbohydrate, but so far with negative results.[273]

Properties of the enzymes—In general, the digestive enzymes of insects resemble those of other animals: they are activated and inhibited by the same reagents, they are similarly affected by changes in *p*H and so forth.[348] The amylase in the saliva of the cockroach, for example, is inactivated if it is freed from chloride by dialysis, like the corresponding enzyme in human saliva;[348]

the lipase has properties similar to that of vertebrates, &c.[273] But they are usually adapted to work best at the hydrogen ion concentration normal in the insect: the amylase of the cockroach, with its acid crop, has an optimum *p*H of 5·9;[348] that of the silkworm, with its very alkaline mid-gut, has an optimum *p*H of 9·5.[183, 284] All the digestive enzymes in the gut of *Tribolium* show maximum activity at about *p*H 5·5.[478] [See p. 529.]

A wide range of carbohydrases are commonly present in insects. The gut of the cockroach *Blaberus* contains amylase but no cellulase, together with an α-glucosidase (sucrose and maltose), a β-glucosidase (cellobiose), a α-fructofuranoside (raffinose and sucrose), an α-galactosidase (melibiose), and a β-galactosidase (lactose).[397] Similar results are given in *Schistocerca*.[430] The classification of all carbohydrases as either α- or β-glucosidase, as proposed by Weidenhagen, is an oversimplification.[430, 546] In the Cuban burrowing cockroach *Byrsotria* the same carbohydrases are present as in *Blaberus*, plus cellulase and chitinase.[433] β-glucosidase is very evident in insects: as cellobiase in *Periplaneta*,[503] *Lepisma*,[482] and *B. mori*.[464a] [See p. 529.]

In connexion with the composition of 'honey-dew' (p. 498) we saw that the hexose residue split from the sucrose in the diet is not transferred exclusively to water, as in chemical hydrolysis, but will combine with other sugars present to produce various oligosaccharides: gluco-sucrose, fructomaltose, &c.[553] This 'transglucosidase' activity of invertase is found not only in the Aphids and Coccids as described, but in the Lygaeid *Oxycarenus*,[533] in the excreta of *Phormia*,[452] and in honey.[491]

The protease, like that of vertebrates, is made up of several components: a *proteinase* acting upon natural proteins, and a group of *peptidases* by which the products of protein digestion are further hydrolysed. The proteinase is always of the tryptic type; enzymes acting like pepsin in a strongly acid medium do not occur in insects. The proteinase of the cockroach,[348] tsetse-fly,[349] and blow-fly larva,[144] when acting upon gelatin, has a *p*H optimum of about 7·5; but it remains active well on the acid side of neutrality. In *Stenobothrus* and *Tettigonia* (Orth.) it has an optimum at 6·2;[273] in the silkworm, with its alkaline juice, at *p*H 9·5.[283] It is activated by 'enterokinase' extracted from the small intestine of the pig; but no activator of this type can be demonstrated in any part of the gut of insects. It is inhibited by 'zookinase' or reduced glutathione, behaving in these respects like the trypsin of vertebrates.[273] In higher Diptera, besides the trypsin-like enzyme with a *p*H optimum around 8·5, there is a second enzyme with an optimum in the acid range: in *Calliphora* adult female at about *p*H 3;[442] in *Stomoxys* larva it is described as a pepsin-like enzyme with an optimum at *p*H 2·4;[481] in adult *Musca* it is described as having characters intermediate between pepsin and cathepsin.[453] Three proteolytic enzymes have been separated by electrophoresis in *Musca*[514] and *Stomoxys*;[515] but the third one seems to be an intracellular enzyme. [See p. 529.]

The proteinase liberates peptones, polypeptides, and some free amino acids. The first two provide the substrate for the peptidases: (i) carboxypolypeptidase, which attacks the peptide chain from the –COOH end, and depends for its action upon the presence in the chain of tyrosin or certain other specific amino acids, (ii) aminopolypeptidase, which attacks the chain from the NH_2– end, and will attack chains made up of any natural amino acids, and (iii) dipeptidase, which hydrolyses all dipeptides. These enzymes have been demonstrated in

Carabids,[273] Orthoptera,[273] *Dytiscus*,[91] *Tenebrio*,[391] *Bombyx mori*[183] and other caterpillars.[90] But whereas the proteinase is active in the contents of the gut, the peptidases occur much more abundantly within the epithelium; this suggests that the final hydrolysis of proteins may take place within the cells. But in *Locusta* and *Dysdercus* a dipeptidase is present also in the lumen of the gut.[474]

Digestive enzymes in the Malpighian tubes—It is generally accepted at the present time that the Malpighian tubes, which open at the junction of the mid-gut with the hind-gut, are excretory organs (p. 561). In many insects their secretion never mixes with the mid-gut contents; and by gross tests digestive enzymes are absent from them.[84] It was observed in the beetles *Gnaptor* and *Necrophorus*,[128] that although extracts from the Malpighian tubes could not themselves digest protein, they could facilitate protein breakdown by the mid-gut secretion. The explanation of this appears to lie in the fact that while the Malpighian tubes do not secrete proteinase, they contain varying quantities of peptidase, particularly dipeptidase (in Carabids and Orthoptera).[273] These enzymes, as we have seen, are mainly intracellular; it is therefore uncertain whether those in the Malpighian tubes are concerned in intestinal digestion, or merely in intermediary or cellular metabolism.

If the mosquito larva *Aëdes* is fed solely upon olive oil, starch or amino acids, and if the gut contents happen to enter the Malpighian tubes, these develop small deposits of fat (after oleic acid) or glycogen (after carbohydrate or glycine).[356] The Malpighian tubes certainly retain some of the properties of the mid-gut.

Extra-intestinal digestion—Since the saliva is usually ejected during feeding, a certain amount of digestion must commonly take place outside the body; notably in the plant-sucking Homoptera (p. 499). But in some insects solid foods of all kinds are to a large extent predigested in this way and absorbed in fluid form. The larvae of the Chalcid *Mormoniella* live as ectoparasites between the wall of the puparium and the pupa of Muscid flies, perforating the pupal integument, digesting and absorbing the tissues.[68] The larva of the wasp *Pseudagenia*, which feeds on spiders, emits a digestive fluid which causes complete solution not only of the protein contents but of the chitin.[251] *Cryptochaetum* (Agromyzidae),[316] and other internal parasites, contain no solid matter in the gut in their young larval stages; some extra-intestinal digestion of the tissues probably occurs. When the larvae of ants, or of the hornet *Belonogaster*, &c., are fed, they discharge a strongly proteolytic secretion. This fluid, which has a somewhat sweet taste, is greedily eaten by the nurse; and it has been suggested that such exchange of nourishment may form the basis of the social habit.[345] Other examples are the larva of *Syrphus pyrastri*, which preys on Aphids,[179] the larva of *Miastor metraloas*, which pours out a secretion, probably from the salivary glands, which causes solution of some of the constituents of wood,[293] and the predaceous Heteroptera. In the case of the predaceous Pentatomid *Troilus* even the cuticle is said to be dissolved;[337] and see *Platymeris* (p. 499).

In many of these cases there is some doubt whether the digestive fluid comes from the salivary glands or from the gut. But some of the most striking examples of extra-intestinal digestion occur in beetles, in which salivary glands are wanting. In the larva of *Dytiscus* the contents of the gut are regurgitated to the oesophagus and through the perforated mandibles. As this larva feeds on some transparent insect, a black fluid can be seen to come from the tips of the jaws and spread among the organs. At once the larva is paralysed and dies; the secretion appears to act as a nerve poison, making the prey quiescent. Very quickly the tissues melt away into a liquid with floating granules, which can be seen flowing back into the mandibular hooks.[246] A caddis larva 12 mm. long may be completely emptied in 10 minutes.[29] A similar process occurs in *Hydrophilus*, *Myrmeleon*, *Chrysopa*, &c.;[179] and a large part of the digestion in Carabids,[161, 169] Cicindelids,[351] and the larvae of *Lampyris* feeding upon slugs,[333] takes place outside the body by ejection of the stomach juices from the mouth. In the larvae of *Lucilia* and *Calliphora*, the proteolytic enzymes persist in the excreta and are responsible for some liquefaction of the meat, even in the absence of bacteria.[144, 361]

The softening of the cocoon by means of fluids ejected at the time of emergence forms a special type of extra-intestinal digestion. The newly emerged silkworm moth extrudes a liquid containing a very active protease capable of attacking the sericin layer of the silk fibres;[151] the origin of this fluid is uncertain. Imagines of *Dicranura* and other Lepidoptera emit from the mouth a strongly alkaline fluid, containing some 1·4 per cent. of potassium hydroxide, which appears to come from the gut.[178]

DIGESTION OF SOME SKELETAL AND OTHER SUBSTANCES OF PLANTS AND ANIMALS

In the section on digestive enzymes, we have considered the mechanisms of digestion for the ordinary constituents of the food. Here we shall discuss the adaptations to deal with special substances.

Digestion of leaves and pollen—Great quantities of the foliage of plants are devoured by insects, but for the most part the cell walls and skeletal parts pass through the gut apparently unchanged. The fragments of leaves ingested by caterpillars often appear in the faeces with the cell contents still intact, except at the margins where the cells have been actually cut open.[3, 242] But sometimes the contents of the cells can be completely digested without mechanical or chemical breakdown of the walls; in *Gastropacha rubi* quite thick pieces can be emptied in this way;[26] and we have referred already to the phenomenon in plant-sucking forms (p. 499). The same change is seen in pollen in the stomach of the bee: some pollen grains pass through unchanged; the contents of many are completely dissolved; yet none are crushed. Here digestion may take place perhaps through the micropylar membranes.[346] Starch grains often escape digestion unless broken open; but the pollen starch is easily digested by the bee; perhaps it lacks a protective sheath.[189] Some indication of the efficiency with which leaves may be utilized for growth is given by figures obtained in the Noctuid caterpillar *Prodenia*: of 10·8 gm. of food eaten by 60 larvae, 5·5 gm. appeared in the excreta, 3·6 gm. was added to the weight.[71] In the larvae of *Phalera bucephala* 60 per cent. of the protein, 80 per cent. of the soluble sugars, 60 per cent. of the fat, and 30 per cent. of the ash was utilized; polysaccharides were excreted unchanged.[97]

About 41–46 per cent. of the chlorophyll and 29–34 per cent. of the carotenoids of mulberry leaves are said to be digested and absorbed by the silkworm. Carotin and xanthophylls appear in the blood; chlorophyll is said to be broken down, with opening of the porphyrin ring.[568]

Enzymes capable of attacking cellulose have not been found in phytophagous insects,[273] but hemicellulase seems to be present in the gut contents of *Forficula*, and various Acridids,[26] and in the fluid from the crop of *Carausius*;[19] and lichenase, producing glucose from lichenin or stored cellulose, occurs in *Stenobothrus*, *Tettigonia*, &c.[273] On the other hand, the Silver-fish *Ctenolepisma* secretes a cellulase and a cellobiase into the mid-gut; bacteria-free silver-fish fed with cellulose uniformly labelled with ^{14}C will respire $^{14}CO_2$; cellulose is clearly digested and metabolized;[482] and cellulase is present also in the gut of the Wheat-stem Saw-fly *Cephus cinctus*.[460]

The chlorophyll of plants, for the most part, passes unchanged through the gut of phytophagous insects but is partially broken down in the silkworm. Phytol and methoxyl are removed and perhaps utilized by the organism, and a crystalline derivative named 'phyllobombycin' can be isolated from the faeces. The breakdown does not extend to the liberation of porphyrins as it does in ruminants.[106] Indeed, phyllobombycin may be an artificial product, for it is said to be absent from fresh faeces.[281] In *Leptinotarsa* the breakdown of chlorophyll extends only to the removal of magnesium with the formation of phaeophorbide-*a*.[52, 249] Complicated changes are described in the chlorophyll in the gut of *Vanessa* larvae, leading it is claimed to the production of a red pigment;[185] but these changes have not been studied chemically. [See p. 429.]

Digestion of wood—The chief constituents of wood are cellulose (comprising 40 to 62 per cent. of the dry weight) and lignin (18 to 38 per cent.). Hemicelluloses, a mixture of polysaccharides, both hexosans and pentosans, come next. Pentosans (6 to 23 per cent.) yielding xylose and arabinose on hydrolysis, occur chiefly in the cell walls; hexosans (2 to 14 per cent.) yielding glucose, chiefly as reserve material. Starch ranges from 0 to 5·9 per cent.; sugar, expressed as glucose, from 0 to 6·2 per cent.; protein from 1·1 to 2·3 per cent.[199]

Of these constituents, the lignin seems never to be digested by insects. Termites feeding on wood containing cellulose 54·6 per cent., pentosans 18·0 per cent., lignin 27·4 per cent., produced faecal material containing cellulose 18·0 per cent., pentosans 8·5 per cent., lignin 75·5 per cent.[252] Here an extensive digestion of cellulose has taken place. As we shall see later (p. 512) many termites are dependent for this activity upon the protozoal fauna in the gut; these forms produce no cellulase in their own secretions. The same is true of the wood-feeding cockroach *Cryptocercus* (p. 512);[324] and the larvae of Lamellicorn beetles apparently depend on the cellulose-digesting bacteria in the food;[343, 347] though it is uncertain to what extent cellulose is in fact broken down by them (p. 511).[258] On the other hand a true cellulase, producing glucose from filter paper, secreted by the mid-gut of the insect itself, occurs in the larvae of some beetles: *Cerambyx cerdo*,[258] *Hylotrupes bajulus*,[102] and *Macrotoma palmata*[199] and other Cerambycidae,[80, 423] and in *Xestobium rufovillosum* (Anobiidae).[258] The cellulase in the gastric juice of *Stromatium* larvae (Col.), with a *p*H optimum of 5·5, will attack cellulose even when bound with lignin as lignocellulose.[199] Cellulose and pentosans are present in the frass of *Xestobium* in much smaller amounts than in the original wood;[58] the cellulose/lignin ratio varied from 2·38 to 2·81 in sound oak wood on which the larvae were feeding; it was between 0·86 and 1·24 in the frass; so that assuming no lignin was attacked, one-third of the total weight ingested had been assimilated, and 80 per cent. of this was cellulose.[221] The wood adjacent to the borings (with a ratio of 1·13) may be almost as much affected as the frass, suggesting that the enzyme is still active in the excreta.

But many wood-boring insects cannot digest cellulose. In the case of *Lyctus*,[58] *Xystrocera* (Col.),[199] and *Cossus* (Lep.)[258] (but see p. 512) there is no difference between the cellulose content of the food and excrement. Hemicellulases are probably more widespread: the Cerambycid larva *Phymatodes variabilis* contains a 'xylanase' hydrolysing xylosan to xylose, and thus reducing the pentosan content from 23·54 per cent. in the beech wood on which it feeds, to 18·48 per cent. in the excreta.[280] The larva of *Cossus* secretes a 'lichenase' acting on reserve cellulose;[199] but in *Lyctus* neither cellulase nor hemicellulase occurs.[229]

Three types of wood feeding are thus recognized in beetles. (i) Larvae able to utilize only the cell contents and perhaps part of the polysaccharides intermediate between starch and the hemicelluloses: Lyctidae and Bostrychidae. (ii) Larvae using the cell contents and the carbohydrates of the cell wall as far as the hemicelluloses: Scolytidae. (iii) Larvae able to use all the carbohydrates of the cell wall including cellulose: Anobiidae and most Cerambycidae.[229]

Where cellulose is not attacked the insects become increasingly dependent on the starch and sugar in the wood. *Cossus* larvae must make great burrows and ingest huge quantities of wood in order to obtain sufficient carbohydrates: there were 2·27 per cent. of reducing sugars in poplar wood on which these larvae fed, none in the excreta.[258] Assimilable carbohydrate is probably a limiting factor in the rate of growth, for when fed on beetroot they will complete their development in one year instead of three.[258] Egg-laying females of *Lyctus* 'taste' and select wood in which starch is present in the cells; wood slowly seasoned, and so free from starch, is immune against attack.[221] If the wood is extracted with water at 60° C., so that sugars are removed but starch remains, *Lyctus* larvae can live but fail to grow; if the starch also is extracted, by using boiling water, the larvae die; but growth takes place if sugar is restored.[229] The Cerambycid larva *Macrotoma palmata* which, as we have seen, contains an active cellulase, can invade the heartwood of trees with a starch and sugar content of only 0·47 to 0·7 per cent.; whereas when the Cerambycid *Xystrocera globosa*, from which cellulase is absent, attacks these same trees, it is found only in the sapwood with 10 per cent. of starch and sugar.[199] *Anobium* larvae can maintain themselves on cellulose alone, but the addition of sugar to the wood accelerates growth.[17] With *Hylotrupes*, on the other hand, the supply of protein is the limiting factor for growth. The addition of 2 per cent. of peptone to wood will increase the rate of growth 10–15 times.[17] [See p. 530.]

Digestion of collagen, silk and keratin—These animal proteins are very resistant to most digestive enzymes. Collagen, the main constituent of fibrous tissues, is completely unaffected by proteinases of the tryptic type. In the blow-fly larva *Lucilia*, however, there is a distinct enzyme, a 'collagenase', separable from the tryptase, which can digest collagen and elastin in alkaline solution (optimum *p*H 8·5). This enzyme is present in the excreta of bacteria-free larvae, and is doubtless of importance in the extra-intestinal digestion of the fibrous septa of the muscles.[144] Collagenase is secreted also by *Hypoderma* larvae[487] which migrate through the tissues of their mammalian host by enzymic dissolution of collagen fibres and ground substance.[566]

Silk is digested by the various clothes-moth larvae, and by the larvae of *Anthrenus*, but nothing is known of the enzymes concerned.[2] Indeed it is doubtful whether Dermestid or Tineid larvae can digest the water-insoluble fraction of the silk fibre.[550]

Keratin, the chief constituent of hair and feathers, forms the basis of the diet in the clothes moths, in *Anthrenus* and other Dermestids, and in Mallophaga. Its digestion in *Tineola* is evidenced by the fact that the excreta contain 4·0–4·6 per cent. of sulphur, whereas the wool on which they feed contains 2·2–2·6 per cent.;[319] yet extracts from the gut of clothes-moth larvae are almost without effect on keratin.[276] That is because the keratinase system is inhibited by oxygen. Keratin is believed to be made up of long peptide chains, folded so as to allow elasticity; and these are bound together by S—S linkages, the two halves of the cystine molecule being shared by two adjacent chains. The peptide linkages cannot be attacked by hydrolytic enzymes until these S—S bonds have been broken by reduction and the chains set free. In the mid-gut secretion of *Tineola*, *Anthrenus*, and Mallophaga, there is a strong reducing agent capable of opening these S—S bonds; the protein is then digested by the proteinase. Most proteinases of the tryptic type are markedly inhibited by –SH compounds; but the proteinase of these insects is peculiar in being insensitive in this respect. It is these two adaptations, the low oxidation-reduction potential of the medium, and the low sensitivity to –SH compounds, which enable these insects to digest keratin.[186] The mid-gut in *Tineola* and in Dermestid larvae is very poorly supplied with tracheae.[550] It is interesting to note that *Galleria* larvae secrete a proteinase which is less affected by –SH compounds than is the trypsin of vertebrates; if a suitable reducing agent is added to the extract, it also is capable of digesting keratin.[90]

Those fractions of the wool which contain most cystine are most resistant to digestion, so that the undigested residue in the excreta shows an increased cystine content.[550] *Tineola* protease can digest some wool in the absence of reducing agents, whereas trypsin cannot; but for normal digestion in the insect reducing agents are necessary.[521] Urea is present in considerable quantities in the gut of *Tineola* and *Attagenus*, and up to 3 per cent. appears in the excreta. It probably favours wool digestion.[550] Raw wool fragments in the gut of *Tineola* are highly birefringent. The onset of digestion is associated with a great reduction in this birefringence. This change occurs abruptly part way along the mid-gut, at the point where the number of goblet cells is abruptly reduced. This change may mark the beginning of the strongly reducing zone. The xanthine oxidase present may possibly contribute to the very low redox potential in the mid-gut;[368] but cystine reductase (linked to reduced triphosphopyridine nucleotide) is believed to be the chief agent concerned in reducing the disulphide bonds of keratin.[521]

Digestion of wax—The diet of *Galleria* larvae, which feed on honey-comb, consists largely of wax. The honey-comb contains about 60 per cent. of wax, the excreta only about 28 per cent.;[204] part of the wax is certainly utilized. This fraction is variously estimated as 34–43 per cent.[89] and 50 per cent.[82] It is believed to comprise all the alcohol components of the wax, a part of the fatty acids and esters of high molecular weight, but none of the paraffins.[89] The lipase extracted from the gut of the larvae will act on tributyrin, olive oil, &c., but has no action on 'myricin', i.e. the esters of higher alcohols (p. 606).[89, 170] Possibly the breakdown of some of the components is begun by the bacteria present in the gut.[82] A bacterium which utilizes the fatty acids and some of the esters in beeswax, but not the hydrocarbons and higher alcohols, has been isolated from the gut contents of *Galleria* larvae.[370]

A large part of the digestion of the wax does appear, however, to be effected by the secretions of the larva itself. Lipase, lecithinase, and cholesterol esterase are present in extracts of the larva,[553] and larval extracts can hydrolyse beeswax.[448] Axenic *Galleria* larvae can digest some but not all of the wax constituents;[553] considerable desaturation of the long chain alcohols is effected: the iodine value of the wax is 15; that of the lipids in the gut lumen is 50.[565] The formation of phospholipids constitutes an important step in the utilization of wax by *Galleria* larvae.[504] About half of all the lipids in the gut tissue are phospholipids; it seems that the digestible components penetrate the cells in phospholipid form.[565]

THE ROLE OF LOWER ORGANISMS IN DIGESTION *

Bacteria—The gut of many insects contains a bactericidal principle of unknown nature, which greatly restricts the bacterial flora.[87] Thus in the gut of the blow-fly larva *Lucilia*, the common saprophytic organisms are wanting (perhaps they are killed off in part in the acid segment of mid-gut (p. 502) and the flora is limited to a few kinds of aerobic bacilli. These are non-proteolytic, and are therefore of no importance in digestion. Digestion takes place equally well in larvae reared from sterilized eggs in the absence of bacteria;[145] though in the early stages of feeding on muscle, the alkalinity produced by bacteria is of some importance in liquefying the tissue.[145] The lactase, maltase and invertase which occur along with the natural enzymes of *Hypoderma* larvae are probably the products of bacteria in the gut.[286]

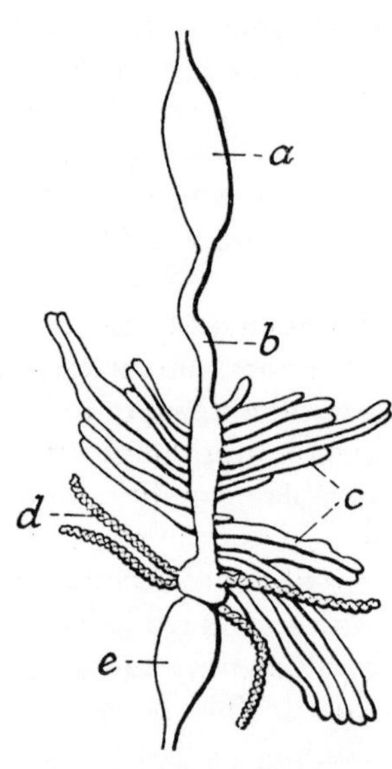

FIG. 313*a*.—Gut of *Aphanus* (Lygaeidae) showing bacterial crypts (*after* KUSKOP)

a, dilated segment of mid-gut; *b*, narrow segment of mid-gut; *c*, bacterial crypts; *d*, Malpighian tubes; *e*, rectum.

Yeasts capable of breaking down starch and sucrose, are present in the gut of Jassids, and are ejected with the saliva during feeding.[139] In those Hemiptera-Heteroptera which feed on plant-juices (Pentatomids, Coreids, Lygaeids, &c.) there are numerous caeca in the mid-gut (Fig. 313*a*) which are uniformly inhabited by bacteria of varied form, each characteristic of its own host species;[123, 173] but there is no evidence that these play any part in digestion. The same applies to the bacteria in the gut of the blood-sucking Reduviidae; for these form their colonies in the dilated segment of the mid-gut where digestion does not occur (p. 491);[354] and it applies to the bacteria in the mid-gut caeca of Trypetidae.[295]

Bacteria may be of importance in the breakdown of some of the constituents of wax by larvae of *Galleria*,[82] yet these larvae can be reared under sterile conditions.[361] Bacilli and cocco-bacilli capable of oxidizing paraffins occur in

* The part played by micro-organisms in nutrition is considered in a later section (p. 421); but effects on digestion and nutrition are sometimes difficult to separate.

the mineral oils of California. It is doubtless the presence of these organisms which makes possible the colonization of the petroleum wells by larvae of the Ephydrid, *Psilopa petrolei*; the organisms in question are abundant in the gut of this larva.[317]

Bacterial fermentation chambers—In the digestion of cellulose by many mammals, bacteria play an important part; but in most phytophagous insects, as we have seen, the food is passed through the gut so rapidly (p. 497) that no appreciable fermentation can take place. Cellulose-fermenting bacteria are, however, always present in rotting wood and vegetation, and must form an important element in the diet of insects feeding on such materials. They occur in the gut of *Rhagium* (Cerambycidae) and *Tipula*.[243] In some insects they may

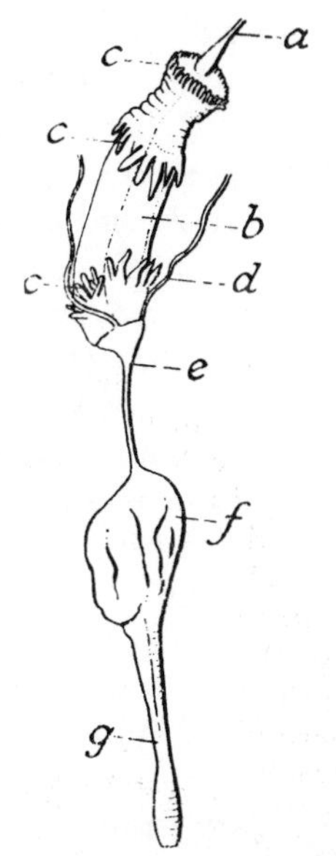

FIG. 314.—Gut of larva of *Oryctes nasicornis* showing the bacterial fermentation chamber in the hind-gut (*after* MINGAZZINI)

a, oesophagus; *b*, mid-gut; *c*, rings of mid-gut caeca; *d*, Malpighian tubes; *e*, hind-gut; *f*, fermentation chamber; *g*, rectum.

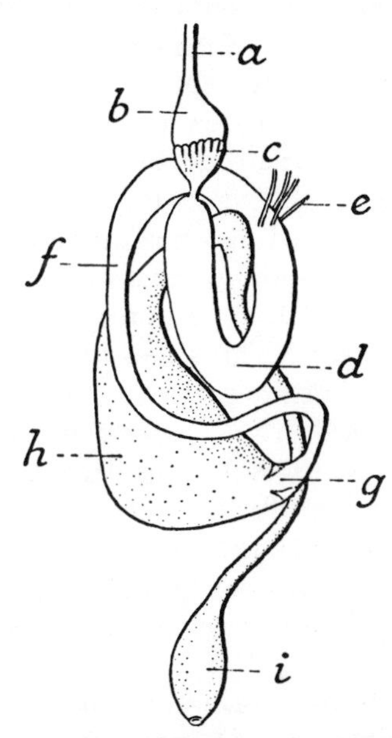

FIG. 315.—Gut of *Eutermes* showing rectal pouch (*from* WEBER *after* HOLMGREN)

a, oesophagus; *b*, crop; *c*, proventriculus; *d*, mid-gut; *e*, Malpighian tubes; *f*, hind-gut; *g*, rectal valve; *h*, rectal pouch; *i*, terminal segment of rectum.

continue their action in the gut. In the hind-gut of Lamellicorn larvae there is a dilated sac, the 'fermentation chamber', which is probably important for this purpose (Fig. 314). On the walls of this chamber are areas of cuticle bearing branched spines; between these areas the cuticle is thin and pierced by fine canals; this chamber is regarded as being the chief site of digestion and absorption.[206] The woody particles may be held up for weeks within this sac, to be fermented by the bacteria. The enzymes secreted in the mid-gut accumulate here also, and ultimately digest the dead bodies of the bacteria, and of the flagellates which feed upon them.[347] Larvae of *Cetonia*, feeding on the decaying pine needles of ant heaps, are said to grow only at those temperatures at which the bacterial mixture present can ferment cellulose.[343] On the other hand, in the larvae of *Dorcus* and *Osmoderma*, although the fermentation chamber is full of bacteria, no evidence of cellulose breakdown could be obtained.[258] Certain

Tipulids, also, have fermentation chambers in the hind-gut;[48] whereas the wood-feeding cockroach, *Panesthia*, depends for the decomposition of its cellulose on bacteria in the crop.[64] Although *Cossus* larvae appear to have little power of cellulose digestion (p. 508), they do harbour micro-organisms in the crop which can break down cellulose: they survive far longer when fed on clean filter paper than when they are starved.[541]

Protozoa—In wood-feeding termites, although bacteria are present,[18] intestinal protozoa are the chief agents in the breakdown of cellulose (though it has been claimed that these protozoa themselves contain the bacteria which are ultimately responsible.[241] The hind-gut is enlarged to form a pouch which is often teeming with many kinds of flagellate protozoa (Fig. 315).* These take up and digest particles of wood, different species perhaps playing a different part in the process. Such termites can live for long periods on a diet of pure cellulose; but if they are deprived of their protozoa by starvation, by exposure to a high tension of oxygen (3·5 atmospheres of oxygen) or by a high temperature (24 hours at 36° C.) this ability is lost, and they die in 3–4 weeks if given a diet of wood. But they can still live on fungus-digested cellulose; and if their protozoal fauna is restored they can live on wood again.[63] It has been calculated that, of the material normally absorbed from the intestine by *Zootermopsis*, about two-thirds has been rendered assimilable by the digestive activity of the protozoa.[156] This seems to be a true symbiosis: the termite needs the metabolic services of the protozoa, and provides the anaerobic environment which these require.[156] The wood-feeding cockroach *Cryptocercus*, also contains abundant cellulose-digesting flagellates; and cellulase can be extracted from the hind-gut where the flagellates are confined; as it can be from the flagellates of termites. But cellulase is absent from insects that have been freed of protozoa by high temperature or oxygen pressure.[324] In the hind-gut of *Cryptocercus* the flagellates maintain a maximally anaerobic environment by the secretion of reduced glutathione.[525]

In what form the products of cellulose digestion are assimilated by the termite is not fully known. Most of the glucose formed by the action of the protozoa is rapidly fermented; it is unlikely that glucose is absorbed to any great extent. The chief carbon compounds utilized are probably the fermentation products of glucose, chiefly perhaps acetic acid, the formation of which has been demonstrated. From the amount of hydrogen produced by the termite it is possible to estimate the quantity of fermentation products formed; and these are sufficient to account for the oxygen consumption observed.[156]

Fungi—Insects sometimes make use of fungi in the conversion of woody materials to an assimilable form. These organisms may grow outside the body, and themselves provide the immediate source of nourishment. Certain ants and termites maintain 'fungus gardens' cultured upon fragments of leaves or the comminuted vegetable matter in the excrement. The egg-laying females of bark beetles (Scolytidae) regularly infect their burrows with moulds ('ambrosia fungi') which grow upon the dead wood;[118] but though these moulds are frequently eaten they do not appear necessary for the normal development of the larvae.[132] The saw-fly *Sirex* also transmits a fungus which infects the burrow, but instead of forming a dense outgrowth on the wall like the ambrosia fungus, it penetrates

* These are by no means universal; the most serious wood-eating termite in Australia, *Eutermes exitiosus*, has no protozoa, but only bacteria.[147]

the surface layers of the wood.[47] The larvae cannot digest cellulose but thrive upon the mould hyphae and the invaded wood.[60, 110, 230] Larvae of *Sciara* which burrow into wood infected with dry rot, feed not on the wood but on the fungus.[15] The termite *Zootermopsis* is unable to grow on sound wood containing 0·03–0·06 per cent. of nitrogen; protein is the limiting factor and if this is added there is a great increase in growth. Certain fungi which grow into wood from the soil bring in nitrogen and then normal growth is possible. No evidence of nitrogen fixation can be observed in colonies in the laboratory (see p. 524), but cannibalism may be important.[156] The growth of *Anobium*[17] and *Hylotrupes*[17] in wood is favoured by fungal attack; the same beneficial effect of fungal decay is seen in *Xestobium*, where it is ascribed mainly to the mechanical weakening of the timber.[58] [See p. 530.]

NUTRITION* [326, 462, 484, 489]

Water requirements—The amount of water needed in the food depends upon the rate at which it is lost by the body; and this depends, on the one hand, upon the properties of the cuticle (p. 36), respiratory system (p. 369) and excretory system (p. 553), and on the other, upon the drying power of the air (p. 664). Insects such as the honey-bee or Muscid flies, which produce liquid excrement, must drink frequently if they are to survive; whereas insects such as the mealworm, which extract almost every trace of water from their excrement, can live on very dry materials. Such insects can cover much of their water requirements by the water produced in metabolism from the oxidation of the dry food (p. 667). But since foodstuffs are mostly hygroscopic, completely dry foods can exist only in a completely dry atmosphere; and it becomes difficult to separate these two factors.[56] Even among insects living in dry stored products striking differences can be found: *Tribolium*, *Silvanus*, &c. grow fairly well in food with 6 per cent. water content; *Ephestia kuehniella* will grow even at 1 per cent.; but *Lasioderma*, *Sitodrepa* and *Ptinus* require at least 10 per cent. of water.[108] Insects in deserts are able to obtain water by eating fragments of dead vegetation which have absorbed water from the atmosphere during the night, when the temperature falls and the relative humidity is high.[56, 157] And there is some evidence that mealworms may eat excessively, and pass large amounts of undigested food, in order to benefit from the small proportion of water it contains.[277] Although *Tenebrio* larvae given dry food can survive and produce one generation a year, if allowed to drink water they will undergo six generations in the year and will lay down much more fat.[493]

Salt requirements—It is possible that salts may constitute a limiting factor in the growth of insects upon some diets. In the larva of *Tribolium* (Col.), phosphorus composes about 0·19 per cent. of the wet weight at all stages; growth is delayed if the phosphorus content of the flour is below 0·1 per cent.[215] Phosphorus may also be a limiting factor in the growth of *Lucilia* larvae on mammalian blood.[146] *Drosophila* has been reared on a diet containing K_2HPO_4 and $MgSO_4$ as the sole salts; NaCl and $CaCl_2$ being present only in traces as impurities. The potassium and phosphate were essential, and no flies were raised if sodium was substituted for potassium; a fly occasionally developed with potassium phosphate alone.[188] It has been suggested that the widespread association of many insects with yeasts, such as is very evident in *Drosophila*, may have

* Nutrition in relation to fecundity is discussed on pp. 720–3.

originated in the extent to which yeasts ensure the efficient utilization of the phosphorus of the environment.[476] *Drosophila* in normal media contained 5·8 mg. Na and 1·3 mg. Ca per gram dry weight; when reared in media containing minimal quantities of these ions, the Na was reduced to less than 5 per cent. and the Ca to 1 per cent. of the normal values; yet the motility and sensitivity of the insects was not affected. The normal molecular ratio of Na/Ca of 7·9 could be altered in this way to more than 300 or less than 1·6 without any influence on activity.[263] Adult Lepidoptera frequently drink water where this is contaminated with excrement or sweat, and will suck up sweat from the skin.[223] It is possible that this habit may be connected with salt requirements.

Potassium is an essential constituent in the diet of *Tenebrio*. Of the trace elements tested, only zinc (which is present in the molecular structure of a number of important enzymes) is needed, at a level of 6 parts per million, for the optimum growth of *Tenebrio*.[437] [See p. 530.]

Uptake of salts by acquatic insects—The thin-walled anal papillae of Culicid and other Nematocerous larvae are, as we have seen, of little importance in respiration (p. 376). They are permeable to water and to salts; and water is in fact continuously taken up by them into the blood and continuously eliminated by the Malpighian tubes (p. 553).[353] But their most important function appears to be the uptake of chloride ions.[165] If *Culex* larvae have the chloride content of their blood, expressed as NaCl reduced to 0·05 per cent. by keeping them in distilled water, and they are then transferred to tap water containing less than 0·006 per cent. of NaCl, they can absorb the chloride and raise the concentration in the blood to the normal 0·3 per cent.[355] This is an active process requiring the expenditure of energy; *Chironomus* larvae cannot absorb or retain chloride under anaerobic conditions.[140] If the larvae are reared in water with a very low chloride content, the anal papillae become greatly hypertrophied (Fig. 316). The anal papillae of *Chironomus* larvae are reduced in size with increasing chloride content of the water;[457] they are greatly enlarged in *C. thummi* from Alpine lakes with very low chloride content. The same hypertrophy is provoked by lack of alkalis and general ion deficiency; but calcium lack causes no hypertrophy.[543] In mosquito larvae the cells contain numerous mitochondria which are applied in pairs to invaginated canaliculi of the cell membrane to form 'mitochondrial pumps' presumably concerned in the active transport of ions.[416a] [See p. 530.]

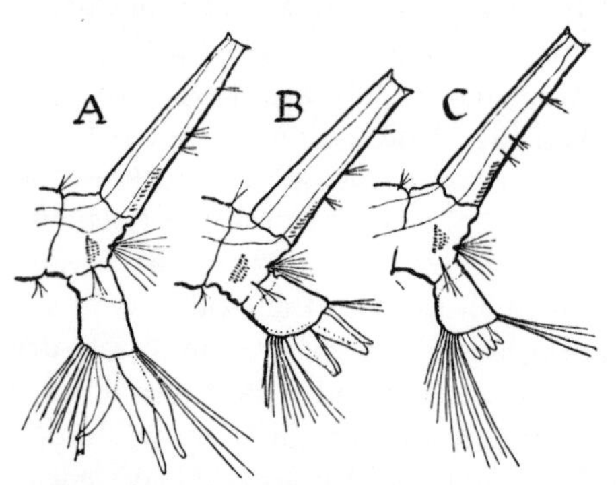

FIG. 316.—Posterior extremity of larvae of *Culex pipiens* showing variation in size of anal papillae (*after* WIGGLESWORTH)

A, reared in distilled water; *B*, in tap water with a chloride content equivalent to 0·006% NaCl; *C*, in salt water equivalent to 0·65% NaCl.

The mosquito larva *Aëdes aegypti*, which lives in small collections of clear rain water, is much more efficient at collecting chloride in this way than *Culex pipiens*, which commonly develops in contaminated water. The chloride concentration in the water can be a limiting factor in growth, *Aëdes aegypti* can grow in water with a smaller chloride content than *Culex pipiens* can.[355] The relative efficiency of this mechanism may perhaps be the factor which limits the breeding of certain species of *Anopheles* to slightly brackish waters. Even closely related forms of *Anopheles maculipennis* differ in the mineral requirements in the waters in which they breed.[14] In the larva of *Aëdes detritus*, on the other hand, a species adapted to live in water with a high salt content, the anal papillae are much reduced and apparently impermeable to salts. Chloride can be taken up only from water containing 0·04 per cent. NaCl; it is absorbed not by the papillae but by the gut wall.[16] [See p. 530.]

It is probable that many other thin-walled outgrowths of the body wall in insects will prove to have this same function; such organs as the anal gills in larvae of *Eristalis* and other Syrphids[296] or in *Pantophthalmus*,[318] the numerous ventral papillae of Blepharocerid larvae, the gills of Trichoptera, &c. In the larva of *Helodes* (Col.) there is an active absorption of chloride ions from dilute media, both gut and anal papillae playing a part in this process.[544] In most Dipterous larvae the anal papillae are represented by patches of enlarged epidermal cells, with thin cuticle over their surface, in the neighbourhood of the anus.[379] In the larva of *Drosophila* there is evidence that these cells, like the anal papillae of mosquito larvae (p. 514) are concerned in the uptake of chloride.[373] Under experimental conditions they are very active in accumulating iodine.[389] We discuss elsewhere the possibility that the rectal glands may be concerned in reabsorbing chloride from the excreta (p. 495); there is the further possibility that in such aquatic insects as the larvae of *Aeschna*,[104] Ephemeroptera, *Corixa*,[244] &c., in which water is constantly passed in and out of the rectal chamber, the enlarged cells corresponding to the rectal glands may be absorbing chloride or other ions from the environment. The larva of *Sialis* has no means of absorbing chloride from the surrounding water; this is correlated with a remarkable capacity for conserving chloride.[538] [See p. 530.]

Nutrition through the body surface—In Nematocerous larvae it is not known whether substances other than chloride, sodium, and potassium ions are absorbed through the cuticle;[180] but some parasitic insects obtain all their nourishment in this way. In larvae of Strepsiptera,[213] the early stages of many Tachinid larvae (e.g. *Thrixion*[227]), the first stage of some Hymenopterous parasites, and in *Cryptochaetum* (Agromyzidae)[316] there is no buccal opening. It may be that the 'tails' of many parasitic larvae, or the terminal ampulla of Braconids (p. 386) are specially concerned in the uptake of salts or other food substances. In *Cryptochaetum* the permeability of the tails for water and salt resembles that of the anal papillae of mosquito larvae.[316]

Organic substances are required for growth and reproduction on the one hand and for energy production on the other. It is characteristic of many insects that, if given a diet adequate for the second purpose, they may survive for a long time in an apparently normal state, but without making any growth. But during long periods (3 months) of protein starvation the endogenous output of nitrogen is about 84 mg. N/Kg./24 hours in *Blatta*, and 159 mg. in *Tenebrio*. These figures are of the same order as in vertebrates. Given a complete protein, such as casein, they require only about 30 per cent. more N intake than this to satisfy the minimum protein requirements.[480] Cockroaches can maintain their body weight for many months on a nitrogen-free diet;[364] termites can exist for considerable periods on pure cellulose; but for growth they must have a source of organic nitrogen, sulphur, phosphorus, and salts.[66] *Tribolium*, *Lasioderma*, and *Ptinus* do not need carbohydrate in their diet; they can grow normally on meat meal or yeast. *Ephestia*, *Silvanus*, and *Sitodrepa*, which must have carbohydrate, fail to grow on these materials.[108]

Two factors determine the requirements for either purpose:

(i) The organic substances must either be directly assimilable by the tissues (like glucose or the natural amino acids), or they must be susceptible of being hydrolysed to assimilable components by the digestive system. Cellulose or keratin are of value in nutrition only to those few insects which can break them down; albumen is of no benefit to adult Lepidoptera because they lack proteolytic enzymes.

(ii) The insect must be able to synthesize from the raw materials provided, all the complex organic substances in its body. Where it is incapable of carrying out such syntheses from the simple components of proteins, fats and carbohydrates, the complex compounds themselves, or special substances which can serve as precursors for them, must be included in the diet. These are called accessory substances.

Available carbohydrates—The extent to which organic materials can be digested by insects has been dealt with in preceding sections. But one aspect of this question which may be considered here is the availability of various sugars

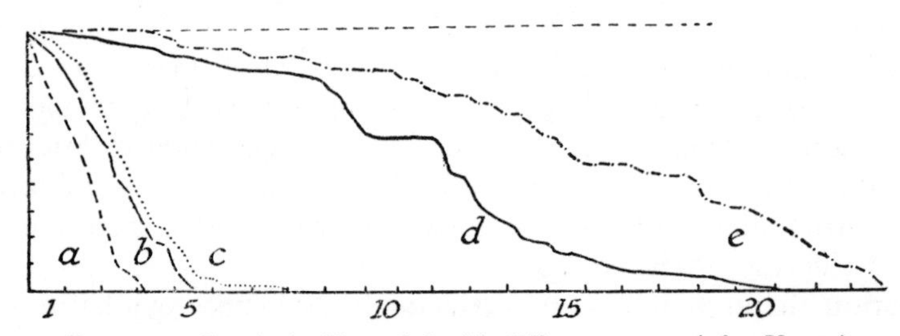

FIG. 317.—Survival of bees fed with different sugars (*after* VOGEL)

Ordinate: percentage of bees surviving. Abscissa: days from commencement of experiment. *a*, fasting; *b–d*, received one part of 2M sucrose; in addition, *b* received seven parts 1M lactose; *c*, seven parts water; *d*, seven parts 1M sucrose; *e*, abundant M/2 sucrose.

as sources of energy. This has been tested by comparing the length of time insects can survive when given water alone or a mixture of sugar and water. The larva of the honey-bee can survive from 4 to 7 times as long when given sucrose, fructose, maltose, melizitose, dextrose or honey, than when given pure water. It can survive about 3 times as long on trehalose and dextrin, less than twice as long on lactose and galactose and no longer on glycogen or starch.[22] In the adult honey-bee, lactose is again found to be of no value in prolonging life;[240] the same is true of a number of other sugars and sugar alcohols which are tasteless to the bee (Fig. 317). Of the sugars and sugar alcohols commonly met with in nature, as nectar (sucrose, glucose and fructose) or honey-dew (sucrose, glucose, fructose, melizitose, trehalose, mannitol, dulcitol, raffinose) all are utilized with the exception of dulcitol. It is worth noting that the tasteless pentose sugars xylose and arabinose are almost equal in value to cane sugar.[332] [See p. 530.]

The adult fly *Calliphora* dies in 2–3 days at 26° C. if given water alone; it will live 1–2 months on water and cane sugar. Almost all other sugars are as effective as cane sugar with the exception of lactose and the pentose xylose, which enable the fly to live 1–2 weeks, and cellobiose, which is of no value. These results are explicable on the assumption that enzymes hydrolysing α-glucosides and α-galactosides are present, together with a weak β-galactosidase (acting on lactose) but no β-glucosidase to act on cellobiose.[107] Similar results have been obtained in *Drosophila*.[135]

In general it has been found that the adult and larval bee, the adult *Calliphora* and *Drosophila*, are almost identical in their ability to utilize sugars. There are, however, some minor differences: mannose is used by flies, not by the bee; dextrin, starch and glycerol are not used by the bee but are used by *Calliphora* and *Drosophila*; inositol is used by *Drosophila* alone, arabinose only by the bee. Mannose is actually poisonous for the bee and for *Vespa*, but not for the other insects tested. This probably results from the 'competitive inhibition' of enzyme action; mannose is sufficiently similar to glucose to block the enzyme surface in the bee, but not similar enough to be broken down.[298] Galactose exerts a similar inhibitory action in *Tenebrio*.[436]

The utilization of sugars in the mosquito larva *Aëdes* has been studied by observing the formation of glycogen in the gut wall and fat body of starved insects fed upon the pure substances.[356] And the ability of *Drosophila* to use different carbohydrates for flight has been tested by giving them to the exhausted flies.[358] The results are in more or less close agreement with those obtained in the experiments on survival. Other insects studied in this context are *Aëdes*, *Sarcophaga*, and *Musca*.[446]

Amino acids—The amino acids resulting from protein digestion are among the most important raw materials for growth. Aphids move away from feeding sites when the nitrogen content of the sieve-tube juice is low.[488] They feed for preference on the phloem sap of growing or of senescent leaves, or on galled leaves or virus infected leaves, in all of which the translocation of amino acids and amides is most active.[471] The utilization of dietary ammonia for amino acid synthesis, and the exchange of amino groups among amino acids, take place in *Calliphora*, &c., much as in mammals.[462]

In mammals, certain of the amino acids are essential in the sense that they cannot be synthesized in the body but must be included in the diet. *Drosophila* larvae have been reared with ammonium salts[188, 328] and the cockroach with the simple amino acid glycine[364] as the sole source of nitrogen; but there is no reason to doubt that the numerous bacteria and yeasts present in the medium and the gut were responsible in these cases for synthesizing more complex substances. Both tryptophane and methionine are essential for rapid growth in *Blattella* but this cockroach can grow slowly even in the absence of these amino acids; methionine can apparently be synthesized by the intestinal micro-organisms, tryptophane perhaps by the intracellular symbionts (p. 522).[384] The amino acids essential for growth in the rat are also essential for growth and pupation in *Tribolium*[380] and in *Drosophila*.[377] When *Drosophila* larvae are reared under sterile conditions, tryptophane and cystine are essential, and probably also lysine, arginine, and histidine.[177] In these same larvae casein is inadequate as a protein unless cystine is added to make good its deficiency in that amino acid.[152] The same applies to *Dermestes*.[116]

In *Aëdes aegypti* the amino acids indispensable for growth are: glycine, leucine, iso-leucine, histidine, arginine, lysine, tryptophane, threonine, phenylalanine, and methionine; the status of valine is uncertain.[126] Glycine can often be dispensed with, but the remaining 10 amino acids on this list are the essential requirements in the diet of most insects: *Tribolium*,[439] *Hylemyia* (Dipt.),[444] *Apis mellifera* adult worker,[454] *Agria affinis*,[463] and many others.[462] The same amino acids are essential for egg production in *Aëdes aegypti*[424] and *Anthonomus grandis*.[547] Approximately the same conclusions have been arrived at in *Phormia* by the indirect method of feeding the insect with ^{14}C-glucose: glycine, alanine, serine, glutamic acid, aspartic acid, and proline were found to be radioactive; the essential amino acids were not.[468] [See p. 530.]

Nucleic acids—All insects appear to be capable of growing without nucleic acids in the diet; but an additional supply of ribonucleic or desoxyribonucleic acid may often accelerate the rate of growth, because this can be limited by the rate of synthesis of individual nucleotides.[414]

Fat soluble accessory substances; sterols, &c.—It is characteristic of insects that they do not require the ordinary fat soluble vitamins that are needed by vertebrates. Vitamin A is not required for growth by *Drosophila*[11, 176] or by *Lucilia*.[146] Nor is calciferol (vitamin D) required by these insects or by *Tribolium*. [See p. 530.]

Sterols—Insects almost always require a source of sterol in the diet, whereas vertebrates are able to synthesize their cholesterol from acetic acid. *Lucilia* needs a fat soluble substance present in wool-wax and muscle oil which is certainly a sterol and is probably cholesterol itself;[146] and a substance of similar distribution is required by *Ephestia*[256] and *Blattella*.[203] The need for some sterol is general among insects but there are qualitative and quantitative differences in different species. In *Drosophila* the necessary unsaponifiable fraction of the fat can be replaced by a number of sterols: cholesterol, sitosterol, stigmasterol, ergosterol, &c., but calciferol (vitamin D) is ineffective.[152] The same applies to *Aëdes aegypti*, which seems to need lecithin in addition.[202] Similarly, the flour and grain infesting insects can grow equally well on the sitosterol of higher plants or the ergosterol of micro-organisms; although the total requirements of *Sitodrepa* and *Ephestia* are greater than those of *Tribolium*, *Lasioderma*, *Silvanus*, and *Ptinus*.[109] On the other hand, *Dermestes*, which grows only in food of animal origin, cannot use the sterols of plants and yeasts and must have cholesterol, of which it requires about 1 mg. to 3 gm. of food.[116] The beetle *Callosobruchus chinensis* will develop in flour made from the bean *Phaseolus angularis* but not in that from *P. vulgaris*. The latter is deficient in sterols: it becomes adequate for normal growth if 1 per cent. of stigmasterol, β-sitosterol, or cholesterol is added to it.[378] [See p. 530.]

There is much variation in detail in the sterol requirements of insects. The Apterygote *Ctenolepisma* is unusual in being able to synthesize labelled cholesterol from ^{14}C-acetate.[416] Cholesterol, sitosterol, and ergosterol are all satisfactory as a source of sterol for *Phormia*; although this subsists naturally on carrion.[411] *Musca* utilizes β-sitosterol directly as a precursor of 7-dehydro-β-sitosterol without conversion to cholesterol;[467] on the other hand, *Blattella* converts β-sitosterol to cholesterol.[527] Sterol seems particularly necessary in *Musca* for pupation and seems also important in resistance to bacterial infection.[146, 485] (The steroid nature of the moulting hormone has already been noted, p. 71.) Various substances related to cholesterol will inhibit its utilization by *Blattella*; they seem to act by competitive inhibition of the cholesterol esterase which esterifies cholesterol during absorption.[506] Labelled acetate is incorporated by *Dermestes* into squalene but not into cholesterol; it may be that biosynthesis of cholesterol is interrupted at this stage;[404] but cholesterol synthesis is probably blocked at various stages.[462] (The incorporation of labelled acetate into farnesol, which is the immediate precursor of squalene, has already been noted, p. 80.)

In general, it may be said that a series of phytophagous species distributed among Coleoptera, Hemiptera, Hymenoptera, Lepidoptera, and Orthoptera can convert the C_{28-29} sterols (β-sitosterol, ergosterol, &c.) of their food to cholesterol; whereas obligatory carnivora like *Dermestes* and *Attagenus* must have a supply of cholesterol itself.[485] [See p. 530.]

Lipids—A special requirement which is recorded for *Ephestia kuehniella* and some other insects is the unsaturated fatty acid linoleic acid which is normally present in wheat germ oil. If too little is present in the diet the moths emerge with naked wings, most of the scales being left behind in the pupa. (A second non-saponifiable fraction in the wheat germ oil is important: α-tocopherol (vitamin E); but this does not seem to exert any specific effect on growth; by reason of its reducing properties it simply serves as an antioxidant, preserving the unsaturated fatty acids.) Beetles, the related moth *Plodia*, even other species of *Ephestia* seem to require no fat soluble factor apart from sterol;[109] but there is evidence that linoleic acid is essential for normal development and reproduction in *Loxostege* (Pyralidae) and that it cannot be synthesized in the amounts required.[234]

Highly unsaturated fatty acids are needed also by the pink bollworm *Platyedra*; linolenic acid is more effective than linoleic, particularly in promoting emergence of the adult moths.[548] Unsaturated fatty acids are required by locusts, and also inositol which may be important as a component of the phospholipids concerned in the mobilization and transport of fats.[419] The unsaturated lipid in wheat-germ oil is necessary for reproduction in the male cricket *Acheta*;[524] here the associated vitamin E is said to provide a necessary factor for spermatogenesis. [See p. 530.]

Locusts which have been suitably fed transmit sufficient carotene to the eggs to mask any deficiencies of carotene in the growing insect. But if eggs deficient in carotene are obtained the resulting larvae show inferior growth in the absence of carotene in the diet, and their colour is highly abnormal (p. 614).[418] This need of β-carotene for growth is unusual among insects.

Ascorbic acid—Ascorbic acid (vitamin C) is not required by *Drosophila*[11] or Tribolium;[307] and cockroaches (*Blattella*), reared aseptically for 15 years on a diet free from vitamin C, were found to contain the same amount of this substance as newly captured insects.[362] Homogenates of *Periplaneta*, and particularly homogenates of the fat body (including the mycetocytes, p. 522) can synthesize ascorbic acid from mannose, glucose, fructose, and other sugars.[529] But ascorbic acid is needed by *Schistocerca* and *Locusta* for normal growth,[418] and it improves growth in the silkworm in which it perhaps acts as a phagostimulant.[464] Ascorbic acid is indispensable for growth and for normal egg production in the boll-weevil *Anthonomus grandis* and the boll-worm *Heliothis*. If the adult boll-weevil is given ascorbic acid in the food some of the offspring will develop to adults even though their diet lacks the vitamin.[549] [See p. 531.]

Vitamins of the B group, &c.—Most of the 'accessory factors', or vitamins required by insects belong to the B group, that is, the group comprising thiamine (B_1, aneurine), riboflavine (B_2), nicotinic acid, pyridoxine, pantothenic acid, choline, inositol, *p*-amino-benzoic acid, and biotin. The extent to which these nine substances are required by *Tribolium* and other insects on the one hand and by vertebrates on the other, is strikingly alike. For both groups thiamine (B_1), riboflavine (Fig. 318), nicotinic acid, pyridoxine, and pantothenic acid are indispensable; the growth rate is greatly affected if any one is deficient. Choline and biotin have somewhat less effect, but are still important. Inositol and *p*-amino-benzoic acid are of minor or perhaps no importance.[108] A need for inositol has been demonstrated, however, in a variety of insects: *Anthonomus*, *Periplaneta*, *Schistocerca*, and *Gryllus*.[462]

Results essentially in agreement with these have been obtained on other insects from stored products,[108, 114, 224] on *Drosophila*,[11, 133, 176, 312] *Lucilia*,[146] mosquito larvae *Aëdes*, &c.,[325, 305, 126, 202] the Pyralid *Corcyra cephalonica*,[270] and the Dermestid *Attagenus*.[212] *Galleria* is exceptionally sensitive to deficiencies of nicotinic acid.[264] (Mammals are able to synthesize nicotinic acid from tryptophane; the insects studied appear unable to do this; indeed the need for nicotinic acid increases with an increase of protein in the diet.[440]) *Tineola*, on the other hand, is relatively indifferent to the absence of some B vitamins;[108] it is apparently able to synthesize riboflavine.[53] All the water-soluble accessory factors of the B group are stored in the fat body of *Tenebrio* when they are present in adequate amounts in the food.[469] In *Corcyra* pyruvic acid accumulates in the blood if the food is deficient in thiamine, as it does in birds and mammals.[23, 270] *Lasioderma* and *Sitodrepa* are much less sensitive to vitamin deficiencies in the diet than *Tribolium*, *Ptinus* or *Ephestia*; but that is only because they obtain certain vitamins from their symbionts (p. 525).[26] Folic acid 'antagonists' will delay pupation or prevent adult emergence in *Drosophila*.[374] In non-sterile *Blattella* riboflavine is synthesized, perhaps by intestinal micro-organisms; choline, pantothenic acid and nicotinic acid are essential for complete growth, but choline can be replaced by betaine; absence of pyridoxine, thiamine, and riboflavine delays growth but does not prevent some insects reaching maturity; omission of inositol, *p*-amino-benzoic acid, vitamin K or biotin usually has no effect.[384] These results must be considered alongside the nutritional requirements of cockroaches deprived of their symbionts (p. 526). *Triatoma infestans* will not grow to maturity on blood serum alone; but fertile adults are produced if this is supplemented with thiamine, riboflavine, pantothenic acid, nicotinic acid, and ascorbic acid.[193, 218] The requirements of *Musca*[408] and *Phormia*[411] reared under sterile conditions are similar. [See p. 531.]

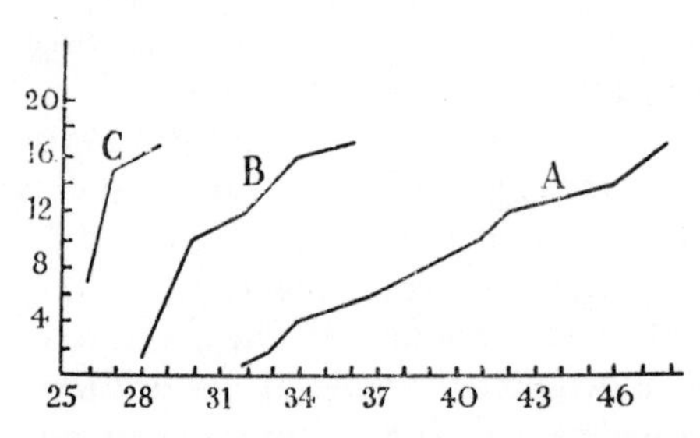

FIG. 318.—Rate of larval growth of *Tribolium*

A, on highly refined 'Patent flour'; *B*, on the same supplemented with 0·25 mg. riboflavine per 2 gm. of food; *C*, on wholemeal flour.

Ordinate: number of pupae. Abscissa: number of days (*after* FRAENKEL and BLEWETT).

Pyridoxine (in the form of pyridoxal phosphate) serves as a co-enzyme in amino acid metabolism; tryptophane is normally converted to kynurenine and this to various further products (p. 616); but if tryptophane is fed to pyridoxine-deficient larva of *Corcyra* they produce yellow faeces containing unmetabolized kynurenine and 3-hydroxykynurenine.[531]

The 'folic acid' fraction from yeast (pteroyl-glutamic acid) is a necessary constituent in the diet of *Aëdes* or *Anopheles* if normal pupation is to occur.[126] It is necessary also to *Ephestia*, *Plodia*, *Tribolium*, and *Tenebrio*. Aminopterin, a structural analogue of folic acid which acts as an 'anti-vitamin', will cause delay in larval development in *Musca* and may prevent adult emergence altogether.[498]

Certain additional accessory factors are required by insects. In addition to the eight B vitamins and folic acid, *Tenebrio* requires a new factor B_T[108], which has proved to be carnitine.[371] In its absence a severe necrosis develops in the epithelial cells of the mid-gut.[371] Carnitine is almost universal in living tissues but nothing is known of its function; like choline it may perhaps be a constituent of phospholipids. Some insects (*Dermestes*, *Phormia*) are able to synthesize carnitine and do not require it in the diet.[371] *Tribolium* is exceedingly sensitive to its absence.[502] *Pyrausta* requires a stable water-soluble and dialysable 'corn leaf factor' which has not been identified.[399] Some forms of polymorphism among insects have been ascribed to an ill-defined substance termed 'vitamin T' present in the fatty tissues of insects and yeasts (p. 98).[124] [See p. 531.]

Vitamin deficiences in the diet may have important delayed effects on growth. Different strains of yeast which vary in vitamin content will influence both the fecundity of female *Drosophila* and the viability of the eggs.[259] It has been claimed that in *Tribolium destructor* the rate of development of the offspring is influenced by the nutritional value of the food on which their parents were fed;[255] but these results have not been confirmed.[415]

Detailed studies of complete nutritional requirements, with the development of wholly synthetic diets, have been made on *Aëdes aegypti*,[540] *Drosophila melanogaster*,[530] *Hylemya antiqua*, and *H. cilicrura*,[444] *Agria* (*Pseudosarcophaga*) *affinis*,[462, 463] and *Schistocerca* and *Locusta*.[418]

Micro-organisms as a source of accessory substances—The eggs of insects can be sterilized, and the resulting larvae reared on sterile media. *Drosophila*,[121] *Calliphora*,[361] *Lucilia*,[146] *Aëdes*,[126] and *Blattella*[362] have been so raised. But this is possible only if they are provided with all the necessary accessory factors. If these are deficient, infection of the food with micro-organisms (in the case of *Drosophila*, particularly the introduction of yeasts) improves the rate of growth.[15, 121] Sterile *Lucilia* larvae will grow on beef muscle; they fail to grow on guinea-pig muscle; but if this is infected with *Bacillus coli* or if a yeast extract is added to it, normal growth takes place (Fig. 319).[146] *Lucilia* larvae cannot grow on sterile blood, but growth is much improved by the presence of bacteria.[146] A sterile blood medium is insufficient for larvae of *Culex* and *Aëdes*; it becomes adequate if infected with bacteria.[262] In these cases there is little doubt that the micro-organisms are synthesizing the necessary vitamins of the 'B' group. How important micro-organisms may be in nutrition is well shown by the observations of Payne,[233] who placed full grown larvae of the beetles *Synchroa* and *Dendroides* in vials with sterilized oak bark. None pupated, although they survived for 6 years. As soon as they were fed on unsterilized oak they pupated within 5 days. As originally suggested by Arthur Bacot the larvae of fleas feed on the faeces of the adults because these serve as a source of bacteria. The larvae of the rat flea *Xenopsylla* requires vitamins of the B group to supplement a diet of pure blood.[387]

We have considered the bacteria and protozoa in the hind-gut of Lamellicorn larvae and termites from the standpoint of digestion (p. 511). These organisms digest cellulose and utilize it for their own growth. The insect probably subsists to a great extent upon their dead bodies; when regurgitated into the mid-gut in

Oryctes nasicornis, they are utilized as a direct source of protein.[528] They may therefore be regarded as synthesizing essential matter for their hosts, using as raw materials not only the cellulose-containing food, but excretory substances poured into the gut by the Malpighian tubes. Given a balanced diet and a source of vitamins, defaunated *Zootermopsis* will grow normally.[65] It has even been

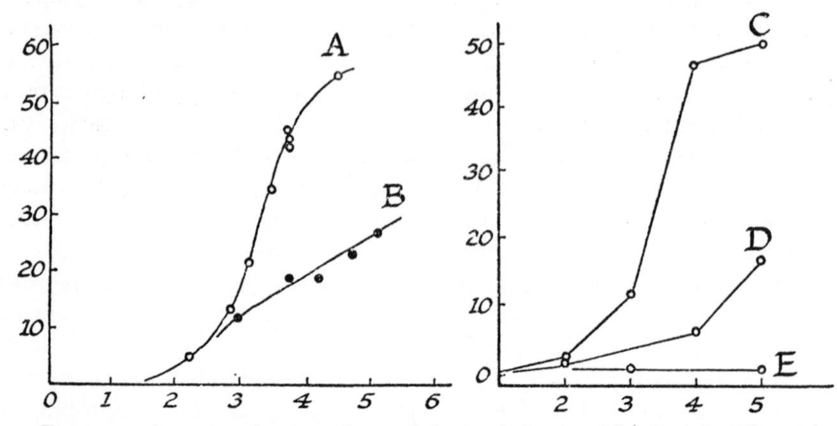

FIG. 319.—Accessory factors and growth in axenic larvae of *Lucilia* (*after* HOBSON)

Ordinates: mean weight per larva in mg. Abscissae: days after hatching. *A*, grown on sterile guinea-pig muscle + aqueous yeast extract; *B*, on sterile guinea-pig muscle + equivalent amounts of yeast extract ash. *C*, grown on blood + marmite (autolysed yeast); *D*, blood + yeast extract; *E*, blood alone.

suggested that the free bacteria in the gut of termites may utilize atmospheric nitrogen,[241, 125] and the same suggestion has been made in regard to the caecal bacteria of the olive fly *Dacus* (see p. 525).[235]

Hereditary micro-organisms—Many insects contain specific micro-organisms, which are present in every individual, and are transmitted from one generation to the next by elaborate mechanisms (p. 741). Such organisms may be confined to the lumen of the gut, as in the mid-gut diverticula of Pentatomids, Lygaeids, &c.,[123, 173] and Trypetids,[295] and in the blood-sucking Reduviids *Rhodnius*, *Triatoma*, &c.[81, 354, 42] They may occur in cells of the gut wall and be constantly set free into the lumen, as in the larvae of some Cerambycids,[136] in pouches of the mid-gut in Anobiids,[43, 136] and in Pupipara.[7, 365] In adult *Glossina* they are confined to special cells or 'mycetocytes' grouped to form a 'mycetome' in the wall of the mid-gut (Fig. 306);[260, 349] and mycetomes in the gut wall occur also in Anoplura[9, 257] and *Camponotus* (Formicidae).[48] Modified Malpighian tubes form mycetomes in some species of *Apion* (Curculionidae)[220] and *Donacia*, &c. (Chrysomelidae).[297] But frequently the cells containing micro-organisms are entirely separate from the gut: mycetocytes are scattered singly throughout the fat body of cockroaches[119] and the primitive termite *Mastotermes*,[164] strings or masses of modified fat cells form mycetomes in Aphids, Coccids, Aleurodids, Cicadids, &c. Mycetomes away from the gut, often connected with the fat body or the gonads, occur in *Cimex*,[46] *Formica fusca*,[184] and the beetles *Oryzaephilus surinamensis*,[163] *Sitodrepa panicea*,[163] *Lyctus*,[229] *Calandra* and *Rhizopertha*[198] among others.

These micro-organisms are of many kinds. In Anobiid and Cerambycid beetles they are evidently yeasts; and perhaps also in Aphids. In *Stegobium* (*Sitodrepa*) the natural yeasts can be eliminated by sterilizing the surface of the eggs. The young larvae can then be reinfected with a foreign yeast *Torulopsis utilis* which successfully invades the epithelial cells of the mid-gut; but it is not transferred to the offspring.[435] In cockroaches[122, 153] they resemble diphtheroid bacilli, or Rickettsiaceae.[120] In *Rhodnius* they are Actinomyces (*Nocardia rhodnii*).[42] In some they appear unlike any free-living micro-organisms. Those bacteria and yeasts which spend a part of their life cycle in the lumen of the gut (in Pentatomids, &c., Anobiids, Cerambycids, *Rhodnius*, &c.) are readily cultured outside the body; but with the possible exception of those from Aphids, the intracellular forms have always failed to grow on artificial media. Some doubt, however, exists in this respect about the intracellular bacteria of Blattidae; most authors[120, 119, 141, 155] have been unable to obtain growth in culture; but it is claimed that by using special methods, this organism can become adapted to grow on artificial media, and then appears as a diphtheroid bacillus named *Corynebacterium periplanetae*.[122, 153] But the fact that cultures from newly hatched insects are unsuccessful, raises some doubt as to whether these micro-organisms are indeed the symbiont.[30] Another isolated organism, which is claimed to be serologically identical with the symbiont, has been named *Rhizobium uricophilus* on account of the activity with which it metabolizes uric acid.[470, 518] The Australian relict termite *Mastotermes darwiniensis* has two sets of symbionts; the intestinal flagellates characteristic of wood-eaters (p. 507) and intracellular bacteria like those of cockroaches.[409] [See p. 531.]

The role of hereditary micro-organisms in nutrition—The constant occurrence of these micro-organisms in a given species, the elaborate mechanisms which ensure their hereditary transmission, and the fact that during development the mycetomes commonly develop in readiness *before* they become invaded, have led to the belief that the organisms are symbionts which contribute something of value in the insect's economy.

This is a subject on which it is impossible to generalize. In origin the organisms have doubtless been parasites; they are sometimes regarded simply as parasites to which the insect has developed an immunity.[198, 225, 258] But it is possible that in the course of evolution they may have become a useful adjunct in metabolism, or even a necessity. Each case requires physiological investigation to decide whether any such benefit to the host exists.[279]

The association of symbionts with the gut suggested that they might be useful in digestion.[48, 253, 260] But there is little support for this view. In the wood-feeding insects there is no relation between the ability to digest cellulose* and the occurrence of hereditary symbionts:[258] some Cerambycid larvae have no yeast-containing mycetocytes in the gut and yet are able to digest cellulose;[199] and cellulose is not attacked by yeasts isolated from the gut of those Anobiids and Cerambycids which possess them.[136] Among blood-sucking insects, symbionts are in close relation with the gut in Anoplura, *Rhodnius* and its allies, *Glossina* and Pupipara. In *Glossina* the mycetome lies in the middle of the anterior segment of the mid-gut, where absorption but no digestion takes place (p. 491) (Fig. 306). Digestion begins abruptly where the type of epithelium changes, long after the blood has passed the mycetome.[349] In *Rhodnius*, the bacteria form large colonies in the blood stored in the dilated stomach; but, as we have seen the blood does not suffer digestion until it enters the intestine.[354]

* The part played by extraneous micro-organisms in the digestion of wood and other substances has already been discussed (p. 510).

If these organisms are of value it seems more likely that they contribute to nutrition or metabolism. It has often been suggested that they serve to fix atmospheric nitrogen, like the bacteria in the root nodules of leguminous plants.[136] Micro-organisms isolated from termites and from Aphids and other Homoptera, incubated with oxalacetic acid and nitrogen gas, will form amino-acids, like the root-nodule bacteria,[322] and the yeast isolated from the mycetome in the gut of *Rhagium* is said to assimilate atmospheric nitrogen.[272] But direct observations on Aphids (*Myzus persicae* and *Doralis fabae*) have afforded no evidence of nitrogen fixation; there is no fixation of the stable isotope of nitrogen (^{15}N). Nor does such fixation seem necessary in these insects; for the nitrogen content of the food is probably adequate to double the weight of an Aphid in 58 hours.[288] *Tuberolachnus* feeding on the phloem sap of *Salix* assimilates only so much nitrogen as is available in the sap; it does not supplement the dietary nitrogen by fixation of nitrogen from the atmosphere.[499] The intracellular symbionts of *Pseudococcus*, regarded as a *Mykobacterium*, can become adapted to grow on artificial media; they will then fix atmospheric nitrogen and produce amino acids, vitamins, enzymes, &c.[432] [See p. 531.]

Symbiotic micro-organisms have also been thought to synthesize protein from nitrogenous waste products;[163] for example, in Aphids which lack Malpighian tubes. In the Aphid *Pemphigus*, the reproductive activity of the host and of the symbionts go hand in hand during the year. At the height of viviparous reproduction in *Aphis sambuci*, when the fundatrix is producing offspring equal to her body weight each day, the symbionts are continually growing and undergoing lysis. They perhaps synthesize protein for their host.[320] [See p. 531.]

Another possibility is that they provide accessory factors, and thus enable their host to live permanently on a restricted or highly specialized diet which is incomplete in some respects.[257] It is notable that symbionts occur in practically all the plant-sucking Hemiptera.[48] Among the blood-sucking insects they occur only in those which take no food but blood at all stages of growth; they are absent in mosquitos, Tabanidae, *Stomoxys*, fleas, the larvae of which enjoy a varied diet.[7, 349, 561] We have seen that sterile mammalian blood is an incomplete diet for insects, but that it becomes adequate if infected with bacteria or if extracts of yeast are added to it. The growth of larvae of *Lucilia* on blood is enormously improved if this is infected with the symbiotic bacteria from *Rhodnius*; these seem to furnish vitamins of the B group.[354] [See p. 531.]

Larvae of *Rhodnius* deprived of their *Nocardia rhodnii* by sterilizing the surface of the egg-shell (p. 742) rarely grow beyond the 4th or 5th instar and if they do become adult they fail to produce eggs. Normal growth and reproduction are resumed if they are reinfected.[42] The same is true of *Triatoma* when re-infected with the natural species of *Nocardia*, or another species of the same genus taken from culture. *Streptococcus liquefaciens*, which is always present in *Triatoma*, is not effective.[402] The *Nocardia* sp. from *Triatoma* produces in culture an excess of folic acid[455] and thiamine.[456] *Rhodnius*, indeed, when fed on mouse blood, is dependent on the symbiotic bacteria for its supply of pyridoxine, calcium pantothenate, nicotinic acid, and thiamine; biotin and folic acid are present at adequate concentration in this diet to meet the insect's requirements.[396] There is some evidence that the symbionts in *Cimex* can provide their host with thiamine, riboflavine, and folic acid when the insect is fed on the blood of mammals deficient in these vitamins.[383] [See p. 531.]

In the louse *Pediculus* the symbionts are necessary both for growth and reproduction. Young larvae deprived of them by centrifuging the eggs always die in about 6 days.[8, 9] In the adult male they degenerate more or less completely; but in the female they are necessary for egg production; and females deprived of them by excision of the mycetome in the last nymphal stage, die in about a week, producing either no eggs or a few non-viable eggs. Partial extirpation produces intermediate effects.[9] *Pediculus* deprived of its symbionts by centrifuging the egg will complete its development if given a single dose of pure B vitamins.[522]

Young larvae of the Pentatomid *Coptosoma* deprived of their bacteria never effect their first moult and die after some weeks;[275] but these defects can be made good by feeding the bugs on very young green seedlings of *Vicia faba*.[501] *Eurydema*, on the other hand, when treated in the same way shows normal development and produces normal offspring which develop crypts without any bacteria.[35] Perhaps the difference lies in the nutritive value of the leaves on which the two species feed. In general the presence of symbionts in the Pentatomids seems to be associated with plant feeding; the Asopinae which feed on insects are devoid of symbionts.[35]

Larvae of the beetle *Stegobium paniceum* have likewise been obtained free from symbionts by sterilizing the outside of the egg-shell (p. 742). Such larvae, reared on 'pea sausage', were still minute at the end of 10 weeks; normal larvae on this diet, and symbiont-free larvae on the same diet plus 25 per cent. of dried yeast, had grown to a large size.[163] *Lasioderma* and *Stegobium* (*Sitodrepa*) deprived of their symbionts do not grow well on flour deficient in vitamin B, but grow normally on this flour if they retain their symbionts. The intracellular micro-organisms probably supply at least five or six members of the B group; without symbionts their vitamin requirements are the same as those of *Tribolium* or *Ptinus*.[28] The yeasts from *Stegobium* and from *Lasioderma* are readily grown in culture; they have quite different morphological characters, but either can readily reinfect each of the two species of beetle. They have been proved to be a rich source of the 9 vitamins of the B group and they are also a source of sterols for their hosts;[512] they probably supply their hosts also with essential amino acids.[513] These yeasts need a small supply of B-vitamins in order to induce growth, after which they produce an excess of vitamins of the same type: thiamine, riboflavine, pyridoxine, folic acid, β-biotin, and carnitine, most of which are liberated into the medium.[451] *Anobium* larvae freed from symbionts by feeding with 4-amino-benzylsulphonamide can no longer mature in timber.[400] *Calandra* freed from symbiotic yeasts by heat treatment of the adult female develop normally in the first generation; but in the next generation the larvae die immediately after leaving the egg (unless fed on a vitamin-rich diet). The essential nutrient provided is believed to be distinct from any of the vitamins needed by *Tribolium*.[534a] [See p. 531.]

On the other hand, *Oryzaephilus surinamensis* which had their symbionts killed by exposure to 36° C. developed normally, and the females gave rise to symbiont-free offspring in which the mycetome developed in the usual way. These grew as rapidly as normal insects and produced an equal number of progeny, even on very poor diets.[163] In *Periplaneta* the presence of the intracellular symbionts will not compensate entirely for an inadequate diet.[30] The species of Cerambycid larvae which have no symbionts develop just as rapidly as those which possess them.[258] Clearly, in these cases the symbionts are not essential to life. But that does not necessarily mean that they serve no useful purpose. It is interesting to note that the persistence of a mycetome without symbionts occurs naturally in certain ants (*Formica rufa*, &c.),[184] and in the weevil *Calandra granaria*;[198] symbionts being still present in *Formica fusca* and *Calandra oryzae*. This again proves that the symbionts are not always essential. The Coccid *Hippeococcus* develops a mycetome like most other Coccids, but no endosymbiotic micro-organisms are present. In the females of this myrmecophilous insect the ovaries and embryos fail to develop until the insect has been carried by ants into their nests—where it is presumably fed on juice rich in vitamins.[413]

In *Periplaneta* the presence of the intracellular symbionts in the fat body has been said not to compensate for an inadequate diet.[30] But if young cockroaches *Blattella* are obtained free of symbionts by feeding the parents with aureomycin over a long period, or by exposing them to high temperatures, they are incapable of growth on a natural diet adequate for normal insects; they can be induced to grow slowly to maturity if given large amounts of yeast in the diet. The reproductive activity of both sexes is impaired by the absence of symbionts. They can be reinfected by the implantation of normally infected fat body.[410] Symbiont-free *Periplaneta* fail to produce ripe eggs.[372, 441] The cockroach can utilize sulphate sulphur for the synthesis of methionine and cystine, but this is no longer possible if the intracellular symbionts are eliminated. *Blattella* retaining their intracellular symbionts but sterile in respect of free-living micro-organisms can synthesize most amino acids; whereas similar cockroaches deprived of their symbionts fail to synthesize tyrosine, phenylalanine, isoleucine, valine, arginine, and possibly threonine from labelled glucose.[458] In *Leucophaea* the synthesis of ascorbic acid is effected by the symbionts in the fat body; it fails to occur in insects deprived of their symbionts.[518] [See p. 532.]

ADDENDA TO CHAPTER XI

p. 477. **Cibarium**—The cibarium of Heteroptera forms a sucking pump which conveys fluid foods to the pharynx. In the Hydrocorisae these pumps have the added function of grinding and filtering particulate matter; such a system is highly developed in *Aphelocheirus*.[634]

p. 477. **Crop emptying**—In *Leucophaea* it appears that the stomatogastric nervous system controls crop emptying in response to the degree of stretch of the crop and consistency of the food.[597]

p. 485. **Peritrophic membrane**—A comparative study of the fine structure in some 75 insect species has confirmed that the peritrophic membrane always contains microfibrils which are composed of chitin. These can be arranged in a felt-like texture, in a hexagonal or in an orthogonal, or grid, texture[635] perhaps produced by the formation of microfibrils between the microvilli of the mid-gut cells.[494]

A peritrophic membrane composed of protein, mucopolysaccharide and chitin is present in the plant-sucking insect *Cicadella*.[601]

p. 486. **Fine structure of mid-gut**—In *Schistocerca*[580] and *Periplaneta*[589] the mid-gut epithelial cells contain numerous autophagic vacuoles, or cytolysomes, derived from mitochondria &c., and these increase in size and number during starvation.

p. 488. **Proteinase**—In *Leucophaea*[598] and *Sarcophaga*[598] proteolytic activity is strictly correlated with the quantity of protein ingested. In *Glossina* the presence of protein (serum) in the meal is necessary to provoke secretion of mid-gut protease, but for a high level of protease activity distension of the crop is needed, and this acts by way of the visceral nerves and the neurosecretory system.[616]

p. 490. **Absorption**—In the cockroach *Eurycotis* cholesterol and its aliphatic esters are absorbed largely in the crop and in the mid-gut caeca.[587]

The active transport of ions may also occur across the mid-gut epithelium. In the larva of *Hyalophora* the mid-gut actively transports potassium and sodium from the haemolymph to the lumen. It is believed that the goblet cells are concerned in the process and that ion accumulation by mitochondria in close association with folded plasma membranes is responsible.[603] It appears that an electrogenic potassium pump is located at the apical membrane of the epithelial cells.[650] This secretory activity may be related with the high Na : K ratio in the haemolymph of *Hyalophora*. In *Periplaneta* in which there is a low Na : K ratio there is a sodium pump which transports water and sodium from the lumen to the haemolymph. The flow of water so induced may assist absorption of other molecules by a solvent drag effect.[633] The theory of solute recycling used to describe the reabsorptive action of the rectal epithelium (p. 496) may be applicable to the mid-gut.[625, 640]

p. 490. **Gastric caeca**—The fine structure of the cells of the gastric caeca of *Aëdes* larvae supports the view that they are concerned mainly in absorption.[609]

p. 494. **Filter chamber**—By the use of dyes and radio-active components in the diet it has been confirmed that in the Jassid *Euscelidius* and the Cercopid *Triecophora* most of the water and sugars pass by way of the filter epithelia to reach the hind-gut by the short route and are rapidly eliminated as honey-dew; whereas amino acids and protein remain in the digestive tract.[629]

p. 496. **Rectal absorption**—In the 5th-stage larva of *Dysdercus* the excretory activity of the Malpighian tubules ceases when feeding stops. Thereafter the rectum contains clear fluid, always hypotonic to the haemolymph, which serves as a water reserve.[576] In the rectal gland cells of *Dacus* there is a complicated system of tubules which is probably concerned in the transfer of electrolytes as well as water.[570] In the cuticle overlying the rectal glands pore canals are always absent, and it is specialized in various other ways[630] presumably related with the absorptive function of the cells.

Water absorption in the rectum of *Schistocerca* can take place without accompanying solute or even when the net movement of solute (KCl, NaCl) is in the opposite direction. It would appear that active transport of water is independent of sodium transport.[607, 645] The pores in the intima have an important filtering role in the exclusion of large organic molecules.[636] There is no evidence that the cuticle has 'rectifying', or valve-like, properties (such as occur in the external cuticle of the insect) which would allow osmosis to occur more readily in the inward than in the outward direction.[637]

In other insects there is evidence that the uptake of water may result from an osmotic flow into intercellular spaces where the osmotic pressure is kept up by the secretion of sodium or other ions in high concentration. The process was first described in epithelia of vertebrates in which water is transported by way of the lateral spaces between the cells. Basal infoldings and secretory canaliculi serve the function of preventing actively transported solute from diffusing away until water can follow in isotonic proportions.[594] According to this theory the 'mitochondrial pumps' (p. 514) in the anal papillae of mosquito larvae are engaged in the secretion of ions into deeply invaginated intercellular spaces. Similar but even more elaborate 'plasmalemma-mitochondrial complexes'[632] occur in the rectal pads of insects;[602] the stacked membranes are the site of adenosine triphosphatase (ATP-ase) and are perhaps actively pumping ions (notably KCl) into the narrow intercellular spaces.[578] In the rectal papillae of *Calliphora* this system links up with a continuous intercellular sinus which discharges into the infundibular sinus within the cone-like papilla and so into the body cavity by way of a valved opening. This is the route which the water osmotically absorbed from the lumen of the rectum is believed to follow.[577] The visible changes in the system which follow changes in water supply support this interpretation.[578]

In the rectum of *Periplaneta* samples of fluid obtained by micropuncture of the intercellular spaces are consistently more concentrated than the fluid of the rectal lumen; it is presumably the osmotic distension of these spaces which brings about the flow of fluid into the haemolymph.[632, 647] Moreover a continued flush of water through these spaces could provide gradients for a passive movement of other molecules against concentration gradients.[577] There is evidence that this process may be regulated by hormones of neuroendocrine origin in *Periplaneta*[647] and perhaps by neurosecretory nerve terminals in the rectal glands of *Calliphora*.[602] In *Periplaneta* there is an active absorption of amino acids in the absence of water movements.[572]

p. 499. **Salivary gland secretions**—The three lobes of the salivary glands of *Acricotopus* (Chironomidae) produce proteins with different amino acid composition. This is correlated with differences in the patterns of puffs and Balbiani rings in the giant chromosomes of each lobe (p. 70).[574] 5-Hydroxytryptamine (5HT) will induce secretion by the salivary glands of *Calliphora*—apparently by the mediation of cyclic AMP (p. 206).[577] Secretion involves the cycling of potassium ions in long narrow channels of the transporting epithelium with the formation of a dilute saliva.[631]

p. 499. Silkmoths produce proteolytic secretions which soften the silk of the cocoon and facilitate emergence (p. 605). In *Antheraea* the moth secretes virtually pure proteinase from the maxillary galeae;[610] this crystalline material is then dissolved by a copious alkaline (pH 8·5) secretion containing $KHCO_3$ isotonic with the haemolymph, produced by the labial glands.[610]

p. 500. On tactile stimulation by the workers, wasp larvae discharge the contents of the labial glands, containing small amounts of amino acids and proteins, 9 per cent. sugar (trehalose and glucose) and carbohydrase and protease enzymes.[622]

p. 500. **Gall formation**—It is possible to simulate very closely the galls produced by phylloxera by injecting into vines solutions of amino acids and indolacetic acid approximating to those naturally present in the saliva.[641] A bug which is normally not cecidogenic can be induced to form galls by injecting it with precursors of indolacetic acid. In gall-producing Hemiptera, tryptophane and phenylalanine in the blood are probably natural precursors of indolacetic acid in the saliva.[624] The further oxidation of quinones, produced by the plant, to non-toxic polymers by the salivary phenolases of Hemiptera may be another factor in gall formation.[624]

p. 500. **Stylet sheaths**—The formation of stylet sheaths is described in *Acyrthosiphon*: the stylets pierce the plant tissues, are then withdrawn slightly and a drop of sheath material discharged and distended by the secretion of a watery fluid; the stylets then pierce the closed end of this sheath and the process is repeated to give a typically beaded structure.[619] The mode of hardening in Aphids appears to be the same as in Heteroptera (p. 500): by means of hydrogen bonding, disulphide bonds and phenolic tanning.[623]

p. 500. In *Chironomus* the salivary gland proteins are in part synthesized by the gland cells, in part synthesized elsewhere and taken up by the gland from the haemolymph.[595]

p. 500. **Saliva of blood-sucking insects**—The salivary glands of *Glossina* contain a plasminogen activator; its anti-coagulant activity is an antithrombin effect. In *Rhodnius* and *Entriatoma* plasminogen activity is restricted to the gut.[604]

The salivary glands of robber flies, Asilidae, secrete a neurotoxic venom which kills locusts and produces effects in mice resembling those of colubrid snake venom.[611]

p. 504. **Properties of enzymes**—It has been confirmed that the lipase of *Periplaneta* and *Blaberus* is secreted in the mid-gut and that, like pancreatic lipase, it preferentially liberates the fatty acids located on the external positions of the triglyceride molecule.[579]

p. 504. The chitinase from the gut of *Periplaneta* tested on the cell walls of *Micrococcus* resembles 'lysozyme' in its effects;[639] indeed the bacteriolytic component in the gut contents of many insects is considered to be lysozyme, which has chitinolytic activity.[628] It is claimed that in *Periplaneta* and in other cockroaches cellulase is secreted by the salivary glands but that in the gut it is produced only by the microflora; this enzyme seems equally active in omnivorous and in wood-eating cockroaches.[649] In *Dermestes* at least two enzymes are involved in the digestion of starch, one for hydrolysis of the α 1–4 bonds and the other for the α 1–6 bonds; and two invertases: α 1–2 glucosidase and β 2–1 fructosidase; there is a specific α-galactosidase which will not hydrolyse α 1–6 galactosidic bonds, and also a specific trehalase.[584] In the mid-gut of *Calliphora*, apart from amylase and trehalase all the enzymic hydrolysis found could be brought about by α-glucosidase and α-galactosidase activity (or by a single enzyme in which these activities are combined). There is no need for β-fructosidase activity; indeed this is not present.[648]

p. 504. In the Carabid beetle *Pterostichus* both trypsin and chymotrypsin with pH optimum at about 8·0 have been demonstrated.[600] In *Aëdes aegypti* both enzymes are present in all stages, but only trypsin is highly active in adult females after a blood meal.[651] Acid phosphatase and ATP-ase activity occur in both goblet cells and cylindrical cells of caterpillars; both cell types may thus be concerned in active fluid transport during the absorption of digested materials.[569]

p. 507. **Protective materials in plants**—Many plants contain accumulations of toxic substances of many kinds. These are commonly regarded as a protective adaptation against plant eating insects. A normal function of the microsomal oxidase enzymes in the mid-gut epithelium is probably to detoxify these 'natural insecticides'. Such oxidases show much higher activity in polyphagous species.[614]

p. 508. **Digestion of wood**—*Anobium* developing in sound dry wood utilized about 7–8 per cent. of the lignin ingested, about 40 per cent. of the cellulose and 42 per cent. of the insoluble protein but virtually all the soluble protein.[571]

p. 513. **Fungus gardens**—The food fungus cultured by *Atta* in its 'fungus gardens' is deficient in protease activity. This is made good by the supply of the faecal material of the ant containing amino acids, nitrogenous waste and some protease; while the fungus contributes cellulose degrading activity.[621]

p. 514. **Trace elements**—Trace amounts of iron, zinc, manganese and copper are required by *Myzus persicae*, *Aphis fabae*, &c., for normal growth.[591]

p. 514. **Salt uptake by Aëdes, &c.** In *Aëdes aegypti* the folding of the plasma membrane below the cuticle of the anal papillae and the number of mitochondria are reduced as the chloride content of the outside medium is increased.[642]

p. 515. In the larvae of *Aëdes aegypti*[644] and the salt-water mosquito *A. campestris*[638] the anal papillae are able to transport actively both sodium and chloride independently. The movements of these ions may be effected by anionic and cationic carriers located in an osmotic barrier in the papillae, in conjunction with Na and Cl pumps located at the inner surface of the barrier.[644] There is evidence that the uptake of ions is controlled by humoral factors from within the larva, perhaps a hormone from the thoracic ganglia or the retrocerebral complex.[646]

p. 515. In *Corixa* starved in de-ionized water the rectal fluid is strongly hypotonic to the haemolymph; it is virtually chloride-free, ammonium bicarbonate being responsible for almost the total osmotic pressure (see p. 556).[643]

p. 516. **Available carbohydrates**—Sorbitol dehydrogenase is particularly active in *Calliphora*, notably in the thoracic muscles, mid-gut and Malpighian tubules.[648]

p. 518. **Amino acids**—For normal growth, larvae of *Bombyx* require either aspartic or glutamic acid in addition to the usual eleven essential amino acids.[608]

p. 518. **Fat-soluble vitamins**—Vitamins A and E accelerate growth and increase the number of emerging adults of *Agria* (*Pseudosarcophaga*) *affinis*. Vitamin E is essential to the female for the production of viable offspring.[606] These larvae take synthetic diets satisfactory for growth in preference to diets that are nutritionally inadequate.[606] 'Retinene' in the visual purple (rhodopsin) is a derivative of vitamin A. If *Aëdes* is reared for a generation on a diet containing neither vitamin A nor its precursor β-carotene, the functions of the eyes are impaired and the fine structure of the retina is defective. Addition of β-carotene to the diet eliminates these defects.[581]

p. 518. **Sterols**—The molecular structure of sterols usable by *Drosophila* has been defined: phytosterols are superior to cholesterol and the general pattern of sterol utilization is like that of other phytophagous insects.[588]

p. 519. *Dermestes* is able to convert the phytosterol campesterol into cholesterol.[612]

p. 519. **Lipids**—Linolenic acid is an essential nutrient for normal wing development in *Trichoplusia* and cannot be replaced by linoleic acid.[586] Although carbohydrate can be replaced entirely by beeswax in the food of *Galleria* larvae, beeswax itself or its derivative myricin seems to be an essential nutrient.[590]

p. 519. **Ascorbic acid**—There is evidence that the 'corn-leaf factor' required by *Pyrausta* (*Ostrinia*) and *Trichoplusia* is ascorbic acid.[585, 586]

p. 520. **Vitamins of B group**—Choline (for inclusion in phospholipids and in acetylcholine) is an essential dietary requirement for growth in *Drosophila*, particularly for reproduction in the adult female.[599] Acetylcholine may take the place of choline in the diet.[605]

p. 521. **Carnitine**—It has been claimed that carnitine activates fatty acids in flight muscle,[575] but that does not seem to be so in the flight muscles of *Phormia*; here acetyl carnitine increases four-fold on the initiation of flight, in parallel with the increase in pyruvate.[583]

p. 523. **Hereditary micro-organisms**—The histochemistry and ultrastructure of the inclusions in the mycetocytes of *Psylla* suggests that they are probably degenerate or aberrant micro-organisms.[582] The bacteroids of *Periplaneta* have a murein basal structure containing glucosamine, indicating that they are indeed procaryotic microbes.[593]

p. 524. **Micro-organisms and symbionts in nutrition**—*Aphis fabae* feeding on *Vicia faba* excretes 30 per cent. of the ingested nitrogen in the adult stage, 50 per cent. in the larval.[573] *Pseudomonas* an obligatory extracellular symbiont of the apple maggot *Rhagoletis* probably provides essential amino acids absent in apple tissue.[627] *Anobium* larvae feeding on sound wood acquire 2½ times the amount of nitrogen provided by the wood consumed, which again raises the question of the fixation of atmosphoric nitrogen.[571]

p. 524. Carbon from labelled glucose injected into *Periplaneta* is rapidly incorporated into essential amino acids—probably by the intracellular symbionts.[617] *Myzus* growing on wholly synthetic diets require only three essential amino acids: methionine, histidine and isoleucine, over a test period of two generations. Phenolic compounds are unnecessary. It seems likely that the intracellular symbionts are concerned;[592] for Aphids (*Myzus*) deprived of their symbionts by antibiotic treatment can no longer grow in the absence of the 'essential' amino acids.[626] The symbionts of *Aphis fabae* can be eliminated by the injection of lysozyme;[596] and in *Neomyzus* by feeding for a couple of generations on an artificial diet lacking trace elements. Without addition of iron, zinc, manganese, copper and calcium the second generation lacks symbionts and is reproductively sterile. Normal *Neomyzus* require a dietary source of only ascorbic acid, nicotinic acid, folic acid and thiamine. Diets lacking riboflavin, pantothenic acid, biotin and choline are adequate for normal growth and reproduction. These vitamins are probably provided by the symbionts.[596] There is evidence, however, that with adequate levels of vitamins in the food sufficient reserves of all vitamins to complete larval growth are passed from adults to their developing embryos.[592]

p. 524. Further studies on *Rhodnius* suggest that *Nocardia* supplies its host with folic acid, pantothenic acid, pyridoxine and thiamine perhaps; and nicotinic acid and riboflavin.[615]

p. 525. *Stegobium* artificially deprived of its symbionts, could be re-infected with a culture of the yeast *Torulopsis utilis*; this yeast was both localized in the symbiote organs and spread to all the mid-gut cells of the larva, but was eliminated in the pupa so that the emerging adults were sterile.[613]

p. 526. *Periplaneta* can be rendered free of symbionts in the fat body by the injection of lysozyme.[620] The fat body of normal *Periplaneta* contains considerably larger amounts of ascorbic, folic and pantothenic acids than a fat body of insects deprived of symbionts.[618]

REFERENCES

1 ABBOTT, R. L. *J. Exp. Zool.*, **44** (1926), 219–54 (digestion and absorption in the crop of the cockroach).

2 ABDERHALDEN, E. *Z. physiol. Chem.*, **142** (1925), 189–90 (digestion of silk: *Anthrenus* larva, Col.).

3 ACQUA, C. *Boll. Labor. Zool. Gen. Agr. Portici*, **11** (1916), 3–44 (utilization of plant cell contents by insects).

4 ADLER, S., and THEODOR, O. *Ann. Trop. Med. Parasit.*, **20** (1926), 109–42 (gut and salivary glands: *Phlebotomus*, Dipt.).

5 ARMBRUSTER, L. *Handb. d. Ernährung u. Stoffwechsels d. Landwirtschaftlichen Nutztiere*, Berlin, **7** (1931), 478–563 (nutrition, digestion and metabolism of the honey-bee).

6 ARNDT, W. *Zool. Anz.*, **93** (1931), 199–207 (digestion of spongin by beetle larvae).

7 ASCHNER, M. *Z. Morph. Oekol. Tiere*, **20** (1931), 368–442 (symbionts in Pupipara).

8 —— *Naturwiss.*, **27** (1932), 501–4; *Parasitology*, **26** (1934), 309–14 (symbionts in *Pediculus*).

9 ASCHNER, M., and RIES, E. *Z. Morph. Oekol. Tiere*, **26** (1933), 529–90 (symbionts in *Pediculus*).

10 AUBERTOT, M. *C. R. Soc. Biol.*, **111** (1932), 743–5 (peritrophic membrane: *Eristalis* larva, Dipt.); *Ibid.*, 746–8 (ditto: *Aeschna* larva, Odonata); *Ibid.*, **111** (1933), 1005–1007; Thesis, Nancy, 1934, 1–357, and *Arch. Zool. Exp. Gén.*, **79** (1938), 49–57 (ditto: larvae of Diptera Nematocera).

11 BACOT, A. W., and HARDEN, A. *Biochem. J.*, **16** (1922), 148–52 (vitamin requirements: *Drosophila*).

12 BALFOUR-BROWN, F. *J. Roy. Micr. Soc.*, **64** (1944), 68–117 (proventriculus of Coleoptera Adephaga).

13 BAPTIST, B. A. *Quart. J. Micr. Sci.*, **82** (1941), 91–139 (salivary glands: Hemiptera).

14 BATES, M. *Amer. J. Trop. Med.*, **19** (1939), 357–84 (adaptation to salt solutions: *Anopheles*, Dipt.).

15 BAUMBERGER, J. P. *J. Exp. Zool.*, **28** (1919), 1–81 (micro-organisms and nutrition: *Drosophila*, *Musca*, *Sciara*, Dipt.).

16 BEADLE, L. C. *J. Exp. Biol.*, **16** (1939), 346–62 (adaptation to salt water: *Aëdes detritus*, Dipt.).

17 BECKER, G. *Naturwiss.*, **26** (1938), 462–3; *Z. vergl. Physiol.*, **29** (1942), 315–88 (nutrition *Hylotrupes* larva, Col.); *Z. Morph. Oekol. Tiere*, **39** (1942), 98–152 (ditto: *Anobium* larva, Col.).

18 BECKWITH, T. D., and ROSE, E. J. *Proc. Soc. Exp. Biol. Med.*, **27** (1929), 4–6 (bacteria and cellulose digestion in termites).

19 BELEHRÀDEK, J. *Arch. Intern. Physiol.*, **17** (1922), 260–5 (cellulase and amylase in *Dixippus*, Orth.).

20 BERLESE, A. *Gli Insetti*, Vol. **1**, Milan, 1909

21 BERRETTA, L. *Boll. Ist. Zool. Palermo*, **2** (1935), 117–32; *Boll. Soc. Sci. Nat. Econ. Palermo*, **19** (1937), 1–5 (peritrophic membrane: Orthoptera).

22 BERTHOLF, L. M. *J. Agric. Res.*, **35** (1927), 429–52 (utilization of carbohydrates by honey-bee larvae).

23 BHAGVAT, K., and SARMA, P. S. *Ind. J. Med. Res.*, **31** (1943), 173–181 (pyruvic acid formation and vitamin B_1; *Corcyra*, Lep.).

24 BIEDERMANN, W. *Arch. ges. Physiol.*, **72** (1898), (digestion in *Tenebrio* larva, Col.).

25 —— *Winterstein's Handb. d. vergl. Physiologie*, **2**, i (1911), (digestion in insects).

26 —— *Arch. ges. Physiol.*, **174** (1919), 392–425 (digestion of plant cell contents).

27 BISHOP, A., and GILCHRIST, B. M. *Parasitology*, **37** (1946), 85–100 (function of crop: *Aëdes*, Dipt.).

[28] BLEWETT, M., and FRAENKEL, G. *Proc. Roy. Soc.*, B, **132** (1944), 212–21 (symbionts as source of vitamins: *Lasioderma* and *Sitodrepa*, Col.).

[29] BLUNCK, H. *Z. wiss. Zool.*, **121** (1924), 171–391 (digestion, nutrition, &c.: *Dytiscus* larva, Col.).

[30] BODE, H. *Arch. Microbiol.*, **7** (1936), 391–403 (symbiotic bacteria of Blattidae).

[31] BODINE, J. H. *Biol. Bull.*, **48** (1925), 79 (*p*H in gut of grasshoppers).

[32] BOGIAWLENSKY, K. *Rev. Zool. Russe*, **5** (1925), 30 (structure of gut: *Notonecta*, Hem.).

[33] de BOISSEZON, P. *C. R. Soc. Biol.*, **103** (1930), 567–8; *Arch. Zool.*, **70** (1930), 281–431 (histophysiology of gut: *Culex* larva, Dipt.).

[34] BONÉ, G., and KOCH, H. J. *Ann. Soc. Roy. Zool. Belg.*, **73** (1942), 73–87 (reabsorption of chlorides in rectum of insects).

[35] BONNEMAISON, L. *Bull. Soc. Ent. France*, 1946, 40–2 (symbionts in nutrition: Pentatomidae, Hem.).

[36] BORDAS, L. *Bull. Soc. Sci. Med. de l'Ouest*, **14** (1905), 1–18 (fore-gut of *Xylocopa*).

[37] —— *C.R. Acad. Sci.*, **126** (1898), 911–12; *Ann. Sci. Nat., Zool.*, (sér. 8), **5** (1898). 1–208 (rectal glands in Orthoptera); *C.R. Acad. Sci.*, **158** (1914), 1930–1 (ditto in Carabidae); *Insecta, Rennes*, **5** (1915), 137–40 (ditto in Lepidoptera).

[38] —— *Ann. Sci. Nat., Zool.*, **14** (1911), 191–273 (gut of Lepidoptera larvae); *Ibid.*, **3** (1920), 175–250 (ditto: adults).

[39] BORRI, C. *Atti della Soc. Toscana Sci. Nat.*, **36** (1925), 226–80 (rectal glands of insects).

[40] BOUNOURE, L. *Aliments, chitine et tube digestif chez les Coleoptères*, Paris, 1919, 294 pp.

[41] BRAMSTEDT, F. *Z. Naturforsch.*, **3b** (1948), 14–24 (digestion in Aphids).

[42] BRECHER, G., and WIGGLESWORTH, V. B. *Parasitology*, **35** (1944), 220–4 (symbiotic actinomyces in nutrition of *Rhodnius*).

[43] BREITSPRECHER, E. *Z. Morph. Oekol. Tiere*, **11** (1928), 495–538 (symbionts in Anobiidae, Col.).

[44] BRUES, C. T. *Insect Dietary*, Cambridge, Mass., 1946.

[45] BUCHMANN, W. W. *Zool. Anz.*, **79** (1928), 223–43 (changes in mid-gut cells during secretion and absorption: *Pyrausta* larva, Lep.); *Z. Desinfektion*, **21** (1929), 1–10; **22** (1930), 1–12 (the same in *Musca* and *Stomoxys* larvae, Dipt.).

[46] BUCHNER, P. *Arch. Protistenk.*, **46** (1923), 225–63 (symbionts in *Cimex*, Hem.).

[47] —— *Holznahrung und Symbiose*, Berlin, 1928.

[48] —— *Tier und Pflanze in Symbiose*, Berlin, 1930.

[49] BUGNION, E., and POPOFF, N. *Arch. Anat. Microsc.*, **10** (1908), 227–68; **11** (1910), 435–56 (salivary glands of Hemiptera).

[50] BÜSGEN, M. *Jena, Z. Naturwiss.*, **25** (1891), 339–428 (formation of honey dew).

[51] BUSHNELL, R. J. *J. Morph.*, **60** (1936), 221–41 (metamorphosis of mid-gut: *Acanthoscelides*, Col.).

[52] BUSNEL, R. G. *Études physiologiques sur le Leptinotarsa decemlineata* Say. Thesis, Paris, 1939.

[53] BUSNEL, R. G., and DRILHON, A. *C.R. Acad. Sci.*, **216** (1943), 213 (riboflavine in *Tineola*, Lep.).

[54] BUTLER, C. G. *Trans. Roy. Ent. Soc. Lond.*, **87** (1938), 291–311 (feeding of *Aleurodes*, Hem.).

[55] BUXTON, P. A. *Proc. Roy. Soc.*, B, **96** (1924), 123–31 (water relations of desert insects).

[56] —— *Biol. Rev.*, **7** (1932), 275–320 (insects and humidity: review).

[57] CAMERON, A. E. *Trans. Roy. Soc. Edin.*, **58** (1934), 211–50 (biology of *Haematopota*, Dipt.).

[58] CAMPBELL, W. G., *et al. Biochem. J.*, **23** (1929), 1290–3; **34** (1940), 1404–14; **35** (1941), 1200–8 (digestion of wood: *Lyctus* and *Xestobium*, Col.).

[59] CARTER, W. *J. Econ. Ent.*, **38** (1945), 33–8 (oral secretions: *Pseudococcus* (Hem.).

[60] CARTWRIGHT, K. ST. G. *Ann. App. Biol.*, **16** (1929), 184–7; **25** (1938), 430–2 (fungus associated with Siricidae).

[61] CASTELNUOVO, G. *Arch. Zool. Ital.*, **20** (1934), 443–66 (digestive enzymes and peritrophic membrane: *Dixippus*, Orth.).

[62] CHUN, C. *Abh. Senckenb. Ges.*, **10** (1876), 27–52 (rectum and rectal glands).
[63] CLEVELAND, L. R. *Biol. Bull.*, **48** (1925), 284–326; 455–68; *Quart. Rev. Biol.*, **1** (1926), 51–60 (flagellates and digestion of cellulose in termites).
[64] CLEVELAND, L. R., *et al. Mem. Amer. Acad. Arts Sci.*, **17** (1935), 185 (protozoa and bacteria and cellulose digestion: *Cryptocercus*, *Panesthia*, Blattidae, Orth.).
[65] COOK, S. F. *Physiol. Zool.*, **16** (1943), 123–8 (utilization of carbohydrates: *Zootermopsis*).
[66] COOK, S. F., and SCOTT, K. G. *J. Cell. Comp. Physiol.*, **4** (1933), 95–110 (food requirements of *Zootermopsis*, Isopt.).
[67] CORNWALL, J. W., and PATTON, W. S. *Ind. J. Med. Res.*, **2** (1914), 569–96 (anticoagulins in salivary glands of blood-sucking insects).
[68] COUSIN, G. *Bull. Biol. Fr. Belg.*, **67** (1933), 371–400 (biology of *Mormoniella*, Chalcid.: extra-intestinal digestion).
[69] CRAGG, F. W. *Ind. J. Med. Res.*, **2** (1915), 706–20 (gut and digestion: *Cimex*, Hem.).
[70] —— *Ind. J. Med. Res.*, **7** (1920), 648–63 (secretion and regeneration in mid-gut: *Tabanus*, Dipt.).
[71] CROWELL, H. H. *Ann. Ent. Soc. Amer.*, **34** (1941), 503–12; **36** (1943), 243–9 (feeding and nutrition: *Prodenia*, Lep.).
[72] CROZIER, W. J. *J. Gen. Physiol.*, **6** (1924), 289–93 (*p*H of gut: *Psychoda*, *Chironomus* larvae, Dipt.).
[73] CUÉNOT, L. *Arch Biol.*, **14** (1895), 293–341 (peritrophic membrane, digestion and absorption in Orthoptera); *Arch. Zool.*, (sér. 3), **6** (1898), lxv–lxix (absorption in cockroach).
[74] DAUBERSCHMIDT, K. *Z. angew. Entom.*, **20** (1933), 204–67 (anatomy of gut: Lepidoptera larvae and adults).
[75] DAVIDSON, J. *Ann. Appl. Biol.*, **10** (1923), 35–54 (penetration of plant tissues by Aphids).
[76] DAVIS, A. C. *Univ. California Publ. Ent.*, **4** (1927), 159–208 (gut and peritrophic membrane: *Stenopelmatus*, Orth.).
[77] DEEGENER, P. *Arch. Naturgesch.*, **75** (1909), 71–110 (changes in mid-gut cells during secretion: *Deilephila*, Lep.).
[78] v. DEHN, M. *Z. Zellforsch. Mikr. Anat.*, **19** (1933), 79–105; **25** (1936), 787–91 (peritrophic membrane in various insects).
[79] DENNELL, R. *Phil. Trans. Roy. Soc.*, B, **231** (1942), 247–9 (foregut, &c.: *Calandra*, Col.).
[80] DESCHAMPS, P. *Bull. Soc. Ent. France*, **49** (1944), 104–8 (cellulase in Cerambycid larvae).
[81] DIAS, E. *Mem. Inst. Osw. Cruz.*, **32** (1937), 165–8 (symbionts in *Triatoma*, Hem.).
[82] DICKMAN, A. *J. Cell. Comp. Physiol.*, **3** (1933), 223–46 (digestion of wax: *Galleria* larvae, Lep.).
[83] DINULESCU, G. *Ann. Sci. Nat., Zool.*, **15** (1932), 1–183 (physiology of *Gastrophilus* larva, Dipt.).
[84] DIRKS, E. *Arch. Naturgesch.*, **88** (1922), 161–220 (Malpighian tubes and digestive enzymes).
[85] DOLMATOVA, A. V. *Med. Parasit.*, **11** (1942), 52–70 (peritrophic membrane: *Phlebotomus* adult, Dipt.).
[86] DREHER, K. *Zool. Anz.*, **113** (1936), 26–8 (function of mandibular glands: *Apis*).
[87] DUNCAN, J. T. *Parasitology*, **18** (1926), 238–52 (bactericidal principle in gut of insects).
[88] DU PORTE, E. M. *Psyche*, **25** (1918), 117–22 (function of proventriculus: *Gryllus*, Orth.).
[89] DUSPIVA, F. *Z. vergl. Physiol.*, **21** (1934), 632–41 (digestion of wax: *Galleria* larvae, Lep.).
[90] —— *H. S. Zeitschr. Physiol. Chem.*, **241** (1936), 168–76; 177–200 (*p*H and protein digestion: *Tineola* and *Galleria* larvae, Lep.).
[91] —— *Protoplasma*, **32** (1938), 211–50 (digestion: *Dytiscus*, Col.).
[92] EATON, C. B. *Ann. Ent. Soc. Amer.*, **35** (1942), 41–9 (proventriculus of *Ips*, Col.).
[93] EIDMANN, H. *Biol. Zentralbl.*, **42** (1922), 429–35 (permeability of cuticle in crop and hind gut of cockroach).
[94] —— *Z. wiss. Zool.*, **122** (1924), 281–307 (structure and function of proventriculus: *Periplaneta*, Orth.).
[95] EKBLOM, T. *Skand. Arch. Physiol.*, **61** (1931), 35–48 (cytology of mid-gut and symbiotic yeasts: *Rhagium*, Col.).

[96] ENGEL, E. O. *Z. wiss. Zool.*, **122** (1924), 503–33 (rectal glands in Diptera).

[97] EVANS, A. C., *et al. Trans. Roy. Ent. Soc. Lond.*, **89** (1939), 13–22; *Proc. Roy. Ent. Soc. Lond.*, A, **14** (1939), 25–30; 57–62 (utilization of food by Lepidopterous larvae and *Tenebrio*).

[98] EVENIUS, C. *Zool. Jahrb., Anat.*, **56** (1933), 349–72 (development of rectal glands: *Vespa*, Hym.).

[99] EVENIUS, J. *Arch. Bienenk.*, **7** (1926), 229–44 (enzymes in mid-gut of honey-bee).

[100] EVENIUS, J., and EVENIUS, C. *Zool Anz.*, **62** (1925), 250–6 (epithelial regeneration in mid-gut: *Apis*).

[101] FAASCH, W. J. *Z. Morph. Oekol. Tiere*, **29** (1935), 559–84 (gut and digestion in Aphaniptera).

[102] FALCK, R. *Cellulosechemie*, **11** (1930), 89 (digestion of cellulose: *Hylotrupes* larva, Col.).

[103] FAURÉ-FREMIET, E. *Ann. Sci. Nat., Zool.*, (sér. 9), **12** (1910), 217–40 (labial glands of aquatic Hemiptera).

[104] FAUSSEK, V. *Z. wiss. Zool.*, **45** (1887), 694–712 (histology of gut: grasshopper, dragon-fly nymphs, &c.).

[105] FINK, D. E. *J. Agric. Res.*, **45** (1932), 471–82 (digestive enzymes in *Leptinotarsa*, Col.).

[106] FISCHER, H., and HENDSCHEL, A. *H. S. Zeitschr. physiol. Chem.*, **198** (1931), 33–42 (breakdown of chlorophyll in *Bombyx mori*).

[107] FRAENKEL, G. *Nature*, **137** (1936), 237; *J. Exp. Biol.*, **17** (1940), 18–29 (utilization of sugars by *Calliphora*, Dipt.).

[107a] FRANKEL, G., and BLEWETT, M. *Bull. Ent. Res.*, **35** (1944), 127–39 (metabolic water).

[108] —— —— *Nature*, **147** (1941), 716–17; **149** (1942), 301; **150** (1942), 177–8; **151** (1943), 703–4; **157** (1946), 697; **161** (1948), 981–3; *Trans. Roy. Ent. Soc. Lond.*, **93** (1943), 457–90; *J. Exp. Biol.*, **20** (1943), 28–34; **22** (1946), 156–61; 162–71; *Biochem. J.*, **37** (1943), 686–92; **41** (1947), 469–75; *Proc. Roy. Ent. Soc. Lond.*, A, **19** (1944), 30–5 (vitamin requirements: *Tribolium* and other insects).

[109] —— —— *Biochem. J.*, **35** (1941), 712–20; **37** (1943), 692–5; *Nature*, **155** (1945), 392–3; *J. Exp. Biol.*, **22** (1946), 172–90 (sterol and linoleic acid requirements; *Dermestes*, Col., *Ephestia*, Lep., &c.).

[110] FRANCKE-GROSMANN, H. *Z. angew. Ent.*, **25** (1939), 647–80; *Verhandl. vii. internat. Kongr. Ent.*, 1938, *Berlin*, **2** (1939), 1120–37; *WandVersamml. dtsch. Ent.*, **8** (1957), 37–43 (relation between fungi and Siricidae).

[111] FREDERICI, E. *Rend. Accad. Lincei*, **31** (1922), 264–8; 394–7 (mid-gut epithelium: *Anopheles* larva, Dipt.).

[112] FRENZEL, J. *Berlin. Entom. Z.*, **26** (1882), 267–316 (digestion in *Tenebrio* larva, Col.).

[113] —— *Arch. microsk. Anat.*, **26** (1886), 229–306 (epithelial regeneration in mid-gut).

[114] FRÖBRICH, G. *Z. vergl. Physiol.*, **27** (1938), 335–83 (vitamin requirements: *Tribolium*, &c.).

[115] GAUHE, A. *Z. vergl. Physiol.*, **28** (1941), 211–53 (oxidation of glucose by pharyngeal glands: *Apis*).

[116] GAY, F. J. *J. Exp. Zool.*, **79** (1938), 93–107 (nutrition: *Dermestes*, Col.).

[117] van GEHUCHTEN, A. *La Cellule*, **6** (1890), 185–290 (histology of the gut: peritrophic membrane, &c.: *Ptychoptera* larva, Dipt.).

[118] GERMER, F. *Z. wiss. Zool.*, **101** (1912), 684–735 (nutrition in 'ambrosia beetles', Lymexylonidae).

[119] GIER, H. T. *Biol. Bull.*, **71** (1936), 433–52 (bacteroids in cockroaches).

[120] —— *Ann. Ent. Soc. Amer.*, **40** (1947), 303–17 (nutrition: *Periplaneta*).

[121] GLASER, R. W. *Amer. J. Trop. Med.*, **4** (1924), 85–107 (role of micro-organisms in nutrition of Diptera).

[122] —— *J. Exp. Med.*, **51** (1930), 59–82 (bacteroids in cockroaches).

[123] GLASGOW, H. *Biol. Bull.*, **26** (1914), 101–70 (bacteria in mid-gut caeca: Heteroptera).

[124] GOETSCH, W. *Oesterr. zool. Zeitschr.*, **1** (1947), 193–274 ('vitamin T').

[125] —— *Oesterr. zool. Zeitschr.*, **1** (1947), 58–96 (fixation of atmospheric nitrogen in termites).

[126] GOLDBERG, L., *et al. Nature*, **154** (1944), 609–10; **160** (1947), 22; 582–5; *J. Exp. Biol.*: **21** (1945), 90–6 ('folic acid', amino-acids, and other nutritional requirements, *Aëdes*, Dipt.).

[127] GORDON, R. M., and LUMSDEN, W. H. R. *Ann. Trop. Med. Parasit.*, **33** (1939), 257–78 (feeding of mosquito *Aëdes*).

[128] V. GORKA, A. *Zool. Jahrb., Physiol.*, **34** (1914), 233–338 (digestive enzymes in gut and Malpighian tubes: *Gnaptor*, *Necrophorus*, Col.).

[129] GRAHAM-SMITH, G. S. *Parasitology*, **26** (1934), 176–248 (gut of *Calliphora*, Dipt. musculature; proventriculus; rectal valve; rectal papillae).

[130] GREEN, T. L. *Proc. Zool. Soc. Lond.*, 1931, 1041–66; 1933, 629–44 (anatomy and histology of mid-gut: *Vespo*, Hym.).

[131] GRESSON R. A. R. *Quart. J. Micr. Sci.*, **77** (1934), 317–34 (cytology of mid-gut: *Periplaneta*, Orth.).

[132] GROSMANN, H. *Z. Parasitenk.*, **3** (1930), 56–102 (moulds and nutrition in Ipidae, Col.)

[133] GUYENOT, E. *C. R. Soc. Biol.*, **61** (1906), 634–5; **74** (1913), 97 *et seq.*; *Bull. Biol. Fr Belg.*, **41** (1907), 353–70; **51** (1917), 1–330 (role of micro-organisms in digestion and nutrition: larvae of *Lucilia*, *Drosophila*, &c.).

[134] HASEMAN, L. *Ann. Ent. Soc. Amer.*, **3** (1910), 277–313 (histology of mid-gut: *Psychoda* larva, Dipt.).

[135] HASSETT, C. C. *Biol. Bull.*, **95** (1948), 114–23 (utilization of sugars, &c.: *Drosophila*)

[136] HEITZ, E. *Z. Morph. Oekol. Tiere*, **7** (1927), 279–305 (symbionts in wood-feeding beetle larvae).

[137] HENSCHEN, F. *Anat. Hefte*, **26** (1904), 575–94 (vesicular secretion in mid-gut).

[138] HENSON, H. *Quart. J. Micr. Sci.*, **73** (1929), 87–105; **74** (1931), 321–60 (mid-gut peritrophic membrane, &c.: *Vanessa* larva, Lep.).

[139] HERFORD, G. V. B. *Ann. Appl. Biol.*, **22** (1935), 301–6 (secretion of amylase and invertase: *Empoasca*, Jassid. Hem.).

[140] HERS, M. J. *Ann. Soc. Roy. Zool. Belg.*, **73** (1942), 173–9 (chloride uptake: *Chironomus*)

[141] HERTIG, M. *Biol. Bull.*, **41** (1921), 181–7 (bacteroids in Blattidae).

[142] HESELHAUS, F. *Zool. Jahrb., Anat.*, **43** (1922), 369–464 (glandular system of honey bee).

[143] HICKERNELL, L. M. *Ann. Ent. Soc. Amer.*, **13** (1920), 223–48 (gut of cicada, *Tibicen* Hem.).

[144] HOBSON, R. P. *J. Exp. Biol.*, **8** (1931), 109–123 (structure of gut, *p*H and enzymes *Lucilia* larva, Dipt.); *Biochem. J.*, **25** (1931), 1458–63 (collagenase in *Lucilia*).

[145] —— *J. Exp. Biol.*, **9** (1932), 128–38; 359–65 (role of bacteria in digestion: *Lucilia* larva, Dipt.).

[146] —— *J. Exp. Biol* , **9** (1932), 366–77; *Biochem. J.*, **27** (1933), 1899–1909; **29** (1935), 1286–96 (micro-organisms and vitamin requirements: *Lucilia* larvae).

[147] HOLDAWAY, F. G. *J. Coun. Sci. Ind. Res. Australia* (1933), 160–5 (biology of *Eutermes*)

[148] HOLLANDE, A. C. *C. R. Acad. Sci.*, **184** (1927), 1476–8 (epithelial regeneration in mid-gut: *Acridium*, Orth.).

[149] HOLMGREN, N. *Z. wiss. Zool.*, **88** (1907), 1–77 (formation of peritrophic membrane: *Mycetophila* larva, Dipt.).

[150] HOLTZ, H. *Anat. Hefte*, **39** (1909), 683–94 (changes in mid-gut cells: *Nematus*, Tenthredinidae).

[151] HONDA, M. *Centralbl. Bakt.*, Abt. 2, **67** (1926), 365–9 (softening of cocoon by intestinal fluid in silkworm).

[152] van T'HOOG, E. G. *Z. Vitaminforsch.*, **4** (1935), 300–23; **5** (1936), 118–25 (vitamin requirements: *Drosophila*).

[153] HOOVER, S. C. *J. Morph.*, **76** (1945), 213–25 (bacteroids in *Cryptocercus*, Orth.).

[154] HOSKINS, W. M., and HARRISON, A. S. *J. Econ. Entom.*, **28** (1934), 924–42 (*p*H and buffering power of stomach contents: *Apis*).

[155] HOVASSE, R. *Arch. Zool.*, **70** (1930), Notes et Rev., 93–6 (bacteroids in *Periplaneta*).

[156] HUNGATE, R. E. *Ecology*, **19** (1938), 1–25; **20** (1939), 230–45; *Ann. Ent. Soc. Amer.*, **34** (1941), 468–89; **36** (1943), 730–9 (protozoa and nutrition in the termite, *Zootermopsis*).

[157] HUSAIN, M.A., and MATHUR, C. B. *Ind. J. Agric. Sci.*, **6** (1936), 263–7 (water uptake: *Schistocerca*, Orth.).

[158] INGLESENT, H. *Biochem., J.*, **34** (1940), 1415–8 (enzymes in salivary glands: *Apis*).

[159] JAGUJINSKAIA, L. W. *Med. Parasit.*, **9** (1940), 601–2 (peritrophic membrane: *Anophelse* adult).

[160] JAMESON, A. P., and ATKINS, W. R. G. *Biochem. J.*, **15** (1921), 209–12 (*p*H and digestion in silkworm).

[161] JORDAN, H. *Biol. Centralbl.*, **30** (1910), 85–96 (extra-intestinal digestion: *Carabus*, Col.).

[162] KNÜPPEL, A. *Arch. Naturgesch.*, **52** (1886), 271–303 (salivary glands of insects).

[163] KOCH, A. *Z. Morph. Oekol. Tiere*, **23** (1931), 389–424; **32** (1936), 137–80 (symbionts and nutrition: *Oryzaephilus surinamensis*, Col.); *Biol. Zentralbl.*, **53** (1933), 199–203 (ditto: *Sitodrepa panicea*, Col.).

[164] —— *Z. Morph. Oekol. Tiere*, **34** (1938), 584–609 (intracellular symbionts: *Mastotermes*).

[165] KOCH, H. J. *J. Exp. Biol.*, **15** (1938), 152–60 (chloride uptake by anal papillae: *Chironomus* and *Culex* larvae).

[166] KOEHLER, A. *Z. angew. Entom.*, **7** (1920), 68–91 (inclusions in mid-gut cells: *Apis*).

[167] KOSMIN, N. P., and KOMAROW, P. M. *Z. vergl. Physiol.*, **17** (1932), 267–78 (invertase in salivary glands and mid-gut of honey-bee).

[168] KRATKY, E. *Z. wiss. Zool.*, **139** (1931), 120–200 (glands in head and thorax of honey-bee).

[169] KRAUSSE, A. *Z. allg. Physiol.*, **17** (1918), 164–7 (extra-intestinal digestion: Carabidae).

[170] KRAUT, H., *et al. Biochem. Z.*, **269** (1934), 205–10 (digestion of wax: *Galleria* larva, Lep.).

[171] KRÜGER, F. *Z. Morph. Oekol. Tiere*, **6** (1926), 149 (gut, peritrophic membrane, &c.: *Syritta* larva, Syrphidae, Dipt.).

[172] KRÜGER, P. *Ergeb. Physiol.*, **35** (1933), 538–72 (digestive enzymes; *p*H in gut of insects: review).

[173] KUSKOP, M. *Arch. Protistenk.*, **47** (1923), 350–84 (bacterial symbiosis: Heteroptera).

[174] KUSMENKO, S. *Zool. Jahrb.*; *Abt. Anat.*, **66** (1940), 463–530 (peritrophic membrane: *Apis* larva).

[175] KUWANA, A. *Annot. Zool. Japon*, **15** (1935), 247–60 (innervation of gut in *Bombyx mori*).

[176] LAFON, M. *C. R. Soc. Biol.*, **124** (1937), 798–802 (vitamin requirements: *Drosophila*).

[177] LAFON, M. *Ann. physiol. physicochim. biol.*, **15** (1939), 215–60 (amino-acid requirements: *Drosophila*).

[178] LATTER, O. H. *Trans. Ent. Soc. Lond.*, 1892, 287–92; 1895, 399–412; 1897, 113–25 (potassium hydroxide and emergence from cocoon; *Dicranura*, Lep.).

[179] v. LENGERKEN, H. *Biol. Zbl.*, **44** (1924), 273–95 (extra-intestinal digestion).

[180] LENZ, F. *Mitt. Georg. Ges. u. Naturhist. Mus. Lübeck*, **31** (1926), 153–69 (nutrition in Chironomid larvae, Dipt.).

[181] LESTER, H. M. O., and LLOYD, L. *Bull. Ent. Res.*, **19** (1928), 39–60 (digestion in *Glossina*, Dipt.).

[182] LICENT, P. E. *La Cellule*, **28** (1912), 7–161 (gut and Malpighian tubes: Cercopidae, Hem.).

[183] LICHTENSTEIN, N. *Enzymologia*, **12** (1947), 156–65 (proteases in silkworm).

[184] LILIENSTERN, M. *Z. Morph. Oekol. Tiere*, **26** (1932), 110–34 (bacterial symbiosis in ants).

[185] v. LINDEN, M. *Ann. Sci. Nat. Zool.* (sér. 8), **20** (1905), (breakdown of chlorophyll in larvae of *Vanessa*, Lep.).

[186] LINDERSTROM-LANG, K., and DUSPIVA, F. *Nature*, 1935, 1039–40; *H.S. Zeitschr. physiol. Chem.*, **237** (1935), 131–58 (digestion of keratin: *Tineola*, Lep.).

[187] LOCY, W. A. *Amer. Nat.*, **18** (1884), 250–5; 353–67 (anatomy and physiology of Nepidae, Hem.).

[188] LOEB, J. *Science*, **41** (1915), 169–70; *J. Biol. Chem.*, **23** (1915), 431–4 (salt requirements: *Drosophila*).

[189] LOTMAR, R. *The Bee World*, **15** (1934), 34–36; *Arch. Bienenk.*, **16** (1935), 195–204 (utilization of starch and dextrin by the honey-bee).

[190] —— *Mitt. Schweiz. Entom. Ges.*, **18** (1941), 233–48; 445–55 (mid-gut of *Tineola* during moulting and metamorphosis).

[191] LOTMAR, R. *Schweiz. Bienen-Zeitung*, Beiheft **1** (1945), 443–506 (rectal glands, &c.: *Apis*).

[192] LUDWIG, D., and ABERCROMBIE, W. F. *J. Morph.*, **59** (1936), 113–22 (histology of mid-gut during growth: *Popillia*, Col.).

[193] LWOFF, M., and NICOLLE, P. *C. R. Soc. Biol.*, **138** (1944), 205–6; **139** (1945), 879–81; *C.R. Acad. Sci.*, **219** (1944), 246–8; *Bull. Soc. Path. exot.*, **36** (1943), 110–24; **37** (1944), 38–51; **39** (1946), 206–21 (vitamin requirements: *Triatoma*, Hem.).

[194] MacGREGOR, V. E. *Trans. R. Soc. Trop. Med. Hyg.*, **23** (1930), 329–31 (function of oesophageal diverticula in mosquitos).

[195] MADWAR, S. *Phil. Trans. Roy. Soc.*, **227** (1937), 1–110 (formation of peritrophic membrane, &c.: Mycetophilid larvae).

[196] MALOEUF, N. S. R. *Riv. Biol.*, **24** (1938), 1–62 (physiology of the alimentary canal: review).

[197] MANSOUR, K. *Quart. J. Micr. Sci.*, **71** (1927), 313–52 (development of mid-gut in larva and adult: *Calandra*, Col.).

[198] —— *Quart. J. Micr. Sci.*, **77** (1934), 243–71 (intracellular micro-organisms: *Calandra*, *Rhizopertha*, &c., Col.).

[199] MANSOUR, K., and MANSOUR-BEK, J. J. *J. Exp. Biol.*, **11** (1934), 243–56 (digestion of wood: *Macrotoma*, *Xystrocera*, Col.); *Biol. Rev.*, **9** (1934), 363–82 (digestion of wood: review); *Enzymologia*, 4 (1937), 1–6 (cellulose, &c.: *Stromatium* larvae, Cerambycidae, Col.).

[200] MARCHAL, P. *Richet's 'Dictionaire de Physiologie'*, **9** (1910), 273–386 (insect physiology).

[201] McDUNNOUGH, J. *Arch. Naturgesch.*, **75** (1909), 313–60 (gut and appendages in larva and adult: *Chrysopa*, Neuropt.).

[202] de MEILLON, B., *et al. J. Exp. Biol.*, **21** (1945), 84–9; *Nature*, **158** (1946), 269 (vitamin requirements, &c.: *Aëdes*, Dipt.).

[203] MELAMPY, R. M., and MAYNARD, L. A. *Physiol. Zool.*, **10** (1937), 36–44 (vitamin requirements in cockroach, *Blattella*).

[204] METALNIKOV, S., *et al. Arch. ges. Physiol.*, **102** (1904), 269–86; *Arch. Zool.*, **8** (1908), 489–588 (digestion, nutrition, &c.: *Galleria* larvae, Lep.).

[205] METCALF, R. L. *J. Nat. Malar. Soc.*, **4** (1945), 271–8 (salivary glands: *Anopheles*).

[206] MINGAZZINI, P. *Mitt. Zool. Stat.*, Naples, **9** (1889), 1–112 (gut of Lamellicorn larvae).

[207] MISRA, A. B. *Proc. Zool. Soc.*, 1931, 293–323; 1359–81 (gut, &c., in the lac insect, *Laccifer*, Coccidae).

[208] MÖBUSZ, A. *Arch. Naturgesch.*, **63** (1897), 89–128 (gut of *Anthrenus* larva, Col.).

[209] MONTALENTI, G. *Boll. Ist. Zool. R. Univ. Roma*, **8** (1930), 3–31 (origin and function of peritrophic membrane).

[210] —— *Arch. Zool. Ital.*, **16** (1931), 859–64 (digestive enzymes and absorption in gut of termites).

[211] MÜLLER, F. P. *Zool. Jahrb.: Abt. Syst.*, **71** (1938), 291–318 (digestive enzymes: Coleoptera).

[212] MOORE, W. *Ann. Ent. Soc. Amer.*, **36** (1943), 483–5 (vitamin requirements. *Attagenus*, Col.).

[213] NASSONOW, N. V. *Ber. naturw.-med. Ver. Innsbruck*, **33** (1910), (nutrition: Strepsiptera).

[214] NELSON, J. A. *J. Agric. Res.*, **28** (1924), 1167–1213 (gut, peritrophic membrane, &c.: honey-bee larva).

[215] NELSON, J. W., and PALMER, L. S. *J. Agric. Res.*, **50** (1935), 849–52 (phosphorus and vitamin D requirements: *Tribolium*, Col.).

[216] NEWCOMER, E. J. *Ann. Ent. Soc. Amer.*, **7** (1914), 311–22 (changes in mid-gut cells during secretion).

[217] NEWELL, G. E., and BAXTER, E. W. *Quart. J. Micr. Sci.*, **79** (1936), 123–50 (nature of striated border in mid-gut: *Locusta*, Orth. and *Chironomus*, Dipt.).

[218] NICOLLE, P., *and* LWOFF, M. *Bull. Soc. Path. exot.*, **35** (1942), 219–32; **36** (1943), 155–67; *C. R. Soc. Biol.*, **138** (1944), 341–3 (vitamin requirements: *Triatoma*, Hem.).

[219] NIELSEN, E. T. *Ent. Medd.*, **23** (1943), 255–72 (X-ray studies on passage of food: Orthoptera).

[220] NOLTE, H. W. *Z. Morph. Oekol. Tiere*, **33** (1937), 165–200 (symbionts in *Apion*, Col.).

[221] NORMAN, A. G. *Biochem. J.*, **30** (1936), 1135–7 (digestion of wood: *Xestobium*, Col.).

[222] NORRIS, M. J. *Proc. Zool. Soc. Lond.*, 1934, 333–60 (nutrition in adult *Ephestia*, Lep.).

[223] NORRIS, M. J. *Trans. Roy. Ent. Soc. Lond.*, **85** (1936), 61–90 (feeding habits of adult Lepidoptera: review).

[224] OFFHAUS, K. *Z. vergl. Physiol.*, **27** (1939), 384–428 (vitamin requirements: *Tribolium* Col.).

[225] PAILLOT, A. *C. R. Acad. Sci.*, **188** (1929), 1118–20; **193** (1931), 300; 676 (bacterial symbiosis and immunity in Aphids).

[226] PAILLOT, A. *Ann. Epiphyt. Phytogén.*, **5** (1939), 339–86 (glycogen in mid-gut cells, &c.: *Bombyx*).

[227] PANTEL, J. *La Cellule*, **15** (1898), 7–290 (anatomy and biology: *Thrixion*, Tachinid.).

[228] PARDI, L. *Redia*, **25** (1939), 87–288 (glycogen in mid-gut cells, &c.).

[229] PARKIN, E. A. *Ann. Appl. Biol.*, **23** (1936), 369–400; *J. Exp. Biol.*, **17** (1940), 364–77 (digestion of wood by *Lyctus* and other beetles).

[230] —— *Nature*, **147** (1941), 329; *Ann. App. Biol.*, **29** (1942), 268–74 (symbiosis in Siricidae).

[231] PATTON, R. L., and CRAIG, R. *J. Exp. Zool.*, **81** (1939), 437–57 (reabsorption inhind-gut: *Tenebrio*, Col.).

[232] PAVLOWSKY, E. N., and SARIN, E. J. *Quart. J. Micr. Sci.*, **66** (1922), 509–56 (histology and digestive enzymes in gut a of honey-bee).

[233] PAYNE, N. N. *Ent. News*, **42** (1931), 13 (micro-organisms in nutrition: Coleoptera).

[234] PEPPER, J. H., and HASTINGS, E. *Montana State Coll., Agric. Exp. Sta. Tech. Bull.*, No. 413 (1943), 1–36 (linoleic acid requirements: *Loxostege*, Lep.).

[235] PETRI, L. *Atti R. Acad. Lincei*, **14** (1905), 399–404; *Centralbl. Bakt.*, (ii) **26** (1910), 357–67 (intestinal bacteria: *Dacus*, Trypetidae).

[236] PEREZ, C. *Arch. Zool.* (sér. 5), **4** (1910), 1–274 (histology during metamorphosis: *Calliphora*, Dipt.).

[237] —— *Mém. Cl. Sci. Acad. Roy. Belg.*, **3** (1911), 1–103 (histology during metamorphosis: *Polistes*, Hym.).

[238] PESSON, P. *Bull. Soc. Zool., Fr.*, **58** (1933), 404–22 (gut of *Pseudococcus*, Hem.); *Ibid.*, **69** (1935), 137–52 (gut of *Lecanium*, Hem.); Thesis, Paris, 1944, 1–266 (gut of Coccidae).

[239] PETERSEN, H. *Arch. ges. Physiol.*, **145** (1912), 121–51 (digestion in honey-bee).

[240] PHILLIPS, E. F. *J. Agric. Res.*, **35** (1927), 385–428 (utilization of carbohydrates in honey-bee).

[241] PIERANTONI, U. *Arch. Zool. Ital.*, **22** (1936), 135–73 (bacteria and protozoa in wood digestion by termites).

[242] PLATEAU, F. *Mém Acad. Roy. Belg.*, **41** (1874), 1–124 (digestion in insects).

[243] POCHON, J. *C.R. Acad. Sci.*, **208** (1939), 1684–6 (cellulose digesting bacteria: *Rhagium*, Col., *Tipula*, Dipt.).

[244] POISSON, R. *Bull. Biol. Fr. Belg.*, **58** (1924), 49–204 (biology and histology of aquatic Hemiptera).

[245] —— *Arch. Zool. exp. gén.*, **80** (1939), Notes et Rev. 77–89 (protein crystals in hind-gut cells: *Nepa*, Hem.).

[246] PORTIER, P. *Arch. Zool.* (sér. 5), **8** (1911), 89–379 (physiology of aquatic insects).

[247] POYARKOFF, E. *Arch. d'Anat. Microsc.*, **12** (1910), 333–474 (secretion in gut of *Galerucella* larva, Col.).

[248] PRADHAN, S. *Quart. J. Micr. Sci.*, **81** (1939), 451–78 (secretion and regeneration: Coccinellidae).

[249] RAFFY, A., and BUSNEL, R. G. *C. R. Soc. Biol.*, **126** (1937), 682–5 (chlorophyll break-down: *Leptinotarsa*, Col.).

[250] RAMME, W. *Zool. Jahrb.*, *Anat.*, **35** (1912), 419–53 (function of proventriculus: Coleoptera and Orthoptera).

[251] —— *S. B. Ges. Naturf. Freunde. Berlin*, 1920, 130–2 (extra-intestinal digestion: *Pseudogenia* larva, Hym.).

[252] RANDALL, M., and DOODY, T. C. *Termites and Termite Control*, Univ. California Press, 1934, 99–104 (*p*H in gut of *Zootermopsis*).

[253] REICHENOW, E. *Arch. Protistenk.*, **45** (1922), 95–116 (intracellular symbionts in blood-sucking insects).

[254] RENGEL, C. *Z. wiss. Zool.*, **75** (1903), 221–32 (mid-gut and peritrophic membrane in larvae of aculeate Hymenoptera).

[255] REYNOLDS, J. M. *Proc. Roy. Soc.*, B, **132** (1945), 438–51 (inheritance of food effects: *Tribolium*, Col.).

[256] RICHARDSON, C. H. *J. Agric. Res.*, **32** (1926), 895–929 (vitamin requirements in *Ephestia*, Lep.).

[257] RIES, E. *Z. Morph. Oekol. Tiere*, **20** (1931), 233–367; *Z. Parasitenk.*, **6** (1933), 339–49 (symbiosis in Anoplura, &c.).

[258] RIPPER, W. *Z. vergl. Physiol.*, **13** (1930), 313–33 (cellulose digestion in wood-feeding insects).

[259] ROBERTSON, F. W., and SANG, J. H. *Proc. Roy. Soc.*, B, **132** (1944), 258–90 (nutrition and fecundity, &c.: *Drosophila*).

[260] ROUBAUD, E. *Ann. Inst.Pasteur*, **33** (1919), 489–536 (nutrition and symbionts: *Glossina*, Dipt.).

[261] ROY, D. N. *Parasitology*, **29** (1937), 150–62 (digestion in *Gastrophilus* larva, Dipt.).

[262] ROZEBOOM, L. E. *Amer. J. Hyg.*, **21** (1935), 167–79 (role of bacteria in nutrition of mosquito larvae).

[263] RUBENSTEIN, D. L., *et al. Biochem, Z.*, **278**, 418–27; **284** (1936), 437–42 (sodium and calcium requirements: *Drosophila*).

[264] RUBENSTEIN, D., and SHEKUN, L. *Nature*, **143** (1939), 1065–6 (nicotinic acid requirements: *Galleria*, Lep.).

[265] RUNGIUS, H. *Z. wiss. Zool.*, **98** (1911), 179–287 (gut of *Dytiscus* larva and adult, Col.).

[266] SAINT-HILAIRE, K. *Z. Morph. Oekol. Tiere*, **21** (1931), 608–16 (diverticula of fore-gut in *Lophyrus* larva, Tenthredinidae).

[267] SAMTLEBEN, B. *Zool. Anz.*, **81** (1929), 97–109 (histology of mid-gut: mosquito larva).

[268] SANFORD, E. W. *J. Exp. Zool.*, **25** (1918), 355–401 (fat absorption in crop of cockroach).

[269] SARIN, E. J. *Biochem. Z.*, **135** (1923), 58–84 (invertase and formation of honey in honey-bee).

[270] SARMA, P. S. *Ind. J. Med. Res.*, **31** (1943), 161–71 (vitamin requirements: *Corcyra*, Lep.).

[271] SAUNDERS, L. G. *Parasitology*, **16** (1924), 164–212 (gut, oesophageal valve, &c.: *Forcipomyia* larva, Ceratopog. Dipt.).

[272] SCHANDERL, H. *Z. Morph. Oekol. Tiere*, **38** (1942), 526–33 (assimilation of atmospheric nitrogen by symbionts of *Rhagium*, Col.).

[273] SCHLOTTKE, E. *Z. vergl. Physiol.*, **24** (1937), 210–47 (distribution of digestive enzymes in gut and Malpighian tubes: Carabidae); *Ibid.*, 422–50 (relation between enzymes and diet in different species of Orthoptera); *Ibid.*, 463–92 (the same: *Periplaneta*, Orth.).

[274] —— *Biol. Gen.*, **18** (1944), 204–39 (digestive enzymes: larvae and pupae of Lepidoptera).

[275] SCHNEIDER, G. *Z. Morph. Oekol. Tiere*, **36** (1940), 595–644 (symbionts in nutrition: Pentatomidae).

[276] SCHULZ, F. N. *Biochem. Z.*, **156** (1925), 124–9 (digestion in the clothes-moth larva *Tinea*, Lep.).

[277] —— *Biochem. Z.*, **227** (1930), 340–53 (water relations of *Tenebrio*, Col.).

[278] SCHULZE, P. *Z. Morph. Oekol. Tiere*, **9** (1927), 333–40 (chitinous threads spun by larva of *Platydema*, Tenebrionidae, Col.).

[279] SCHWARTZ, W. *Arch. Microbiol.*, **6** (1935), 369–460 (symbionts in insects, &c.: review).

[280] SEILLIÈRE, G. *C. R. Soc. Biol.*, **58** (1905), 940–1 (xylanase in mid-gut of *Phymatodes* larva, Col.).

[281] SEYBOLD, A. and EGLE, K. *H.-S. Z. physiol. Chem.*, **257** (1938), 49–53 (digestion of chlorophyll in insects).

[282] SHINODA, O. *Z. Zellforsch. mikr. Anat.*, **5** (1927), 278–92 (comparative histology of mid-gut).

[283] —— *Anniversary volume dedicated to Prof. Masumi Chikashige*, Kyoto, 1930 (*p*H in gut of various insects).

[284] —— *Annot. Zool. Japon*, **13** (1931), 117–25 (starch digestion: *Bombyx mori*).

[285] SIANG-HSU, W. *J. Morph.*, **8** (1947), 161–84 (cytology of mid-gut cells: *Drosophila*).

[286] SIMMONS, S. W. *Ann. Ent. Soc. Amer.*, **32** (1939), 621–7 (digestive enzymes: *Hypoderma*, Dipt.).

[287] de SINÉTY, R. *La Cellule*, **19** (1901), 119–278 (anatomy of Phasmids).
[288] SMITH, J. D. *Nature*, **162** (1948), 930–1 (failure of Aphids to fix atmospheric nitrogen).
[289] SMITH, K. M. *Ann. Appl. Biol.*, **13** (1926), 109–39 (effect of plant-sucking Hemiptera on plant tissue).
[290] SNIPES, B. T., and TAUBER, O. E. *Ann. Ent. Soc. Amer.*, **30** (1937), 277–84 (time required for passage of food through the gut: *Periplaneta*, Orth.).
[291] SNODGRASS, R. E. *Principles of insect Morphology*, New York, 1935.
[292] —— *Smithsonian Misc. Coll.*, **104** (1943), No. 1, 1–51 (mouth parts of biting flies).
[293] SPRINGER, F. *Zool. Jahrb., Syst.*, **40** (1915), 57–118 (biology of *Miastor*, Dipt.).
[294] STAMMER, H. J. *Z. Morph. Oekol. Tiere*, **1** (1924), 121–70 (gut, peritrophic membrane, &c.: larvae of Tabanidae).
[295] —— *Z. Morph. Oekol. Tiere*, **15** (1929), 481–523 (bacterial symbiosis in Trypetidae, Dipt.).
[296] —— *Z. Morph. Oekol. Tiere*, **26** (1933), 437–43 (anatomy of wood-boring Syrphid larva: *Temnostoma*, Dipt.).
[297] —— *Z. Morph. Oekol. Tiere*, **29** (1935), 585–608 (symbionts in Donaciinae Chrysomelidae, Col.).
[298] STAUDENMAYER, T. *Z. vergl. Physiol.*, **26** (1939), 644–68 (mannose poisonous for bee, &c.).
[299] STAUDENMAYER, *et al. Anz. Schädlingsk.*, **16** (1940), 114–19; 125–32; *Z. angew. Ent.*, **26** (1940), 589–607 (*p*H in gut of insects).
[300] STEUDEL, A. *Zool. Jahrb., Physiol.*, **33** (1913), 165–224 (absorption and secretion in insect gut).
[301] STOBER, W. K. *Z. vergl. Physiol.*, **6** (1927), 530–65 (digestion and nutrition: Lepidoptera larvae and adults).
[302] STOREY, H. H. *Proc. Roy. Soc.*, B, **127** (1939), 526–43 (saliva: Jassidae).
[303] STREET, H. R., and PALMER, L. S. *Proc. Soc. Exp. Biol. Med.*, **32** (1935), 1500–1 (vitamin requirements: *Tribolium*, Col.).
[304] STRICKLAND, E. H. *J. Morph.*, **24** (1913), 43–106 (peritrophic membrane: *Simulium* larva, Dipt.).
[305] SUBBAROW, Y., and TRAGER, W. *Biol. Bull.*, **75** (1938), 75–84; *J. Gen. Physiol.*, **23** (1940), 561–8 (vitamin requirements: *Aëdes*, Dipt.).
[306] SUKHOV, K. S. *C.R. Acad. Sci. U.R.S.S.* (N.S.), **42** (1944), 226–8 (salivary sheath: Aphididae).
[307] SWEETMAN, W. D., and PALMER, L. S. *J. Biol. Chem.*, **77** (1928), 33–52 (vitamin requirements: *Tribolium*, Col.).
[308] SWINGLE, H. S. *Ohio J. Sci.*, **25** (1925), 209–18 (enzymes in the cockroach, *Blatta*).
[309] —— *Ann. Ent. Soc. Amer.*, **21** (1928), 469–75 (digestive enzymes: *Laspeyresia* larva and adult, Lep.).
[310] SWINGLE, M. C. *J. Agric. Res.*, **41** (1930), 181–96 (physiology of gut: *Popillia*, Col.).
[311] —— *Ann Ent. Soc. Aner.*, **24** (1931), 489–95 (*p*H in gut of various insects).
[312] TATUM, E. L. *Proc. Nat. Acad. Sci.*, **25** (1939), 490–7; **27** (1941), 193–7 (vitamin requirements: *Drosophila*).
[313] TCHANG YUNG-TAI. *Bull. Soc. Zool. Fr.*, **53** (1928), 56–60; *C. R. Soc. Biol.*, **100** (1929), 809–12; *C. R. Acad. Sci.*, **188** (1929), 93–5; *Bull Biol. Fr. Belg.* Suppl. **12** (1929), 144 pp. (histophysiology of mid-gut epithelium: *Galleria*, Lep.).
[314] TCHANG YUNG-TAI. *Bull. Soc. Zool. Fr.*, **54** (1929), 255–63 (peritrophic membrane in larvae of Lepidoptera).
[315] THIEL, H. *Z. wiss. Zool.*, **146** (1936), 395–432 (function of proventriculus: Coleoptera).
[316] THORPE, W. H. *Proc. Zool. Soc. Lond.*, 1930, 929–71; *Quart. J. Micr. Sci.*, **77** (1934), 273–304 (nutrition, &c.: *Cryptochaetum*, parasitic Agromyzid larva, Dipt.).
[317] THORPE, W. H. *Trans. Ent. Soc. Lond.*, **78** (1930), 331–43; *Nature*, 1932, Sept. 17th (petroleum bacteria and nutrition: *Psilopa petrolei*, Ephydridae, Dipt.).
[318] —— *Trans. Roy. Ent. Soc. Lond.*, **82** (1934), 5–22 (structure and biology: *Pantophthamus*, Dipt.).
[319] TITSCHACK, E. *Z. techn. Biol.*, **10** (1922), 1–166; *Zool. Anz.*, **93** (1931), 4–6 (digestion, &c.: *Tineola* larvae, Lep.).
[320] TÓTH, L. *Z. Morph. Oekol. Tiere*, **27** (1933), 692–731; **33** (1937), 412–37 (symbionts and nutrition in *Pemphigus*, Aphid.).

[321] TÓTH, L. *Ann. Mus. Nat. Hungarici. Zool.*, **33** (1940), 167–71 (protein metabolism: Aphididae).

[322] TÓTH, L., *et al. Z. vergl. Physiol.*, **30** (1942), 67–73; 300–20; *Trav. Inst. Hongrois Rec. Biol.*, **16** (1944), 7–34; *Monographs on Natural Sciences*, 5, Budapest 1946; *Tijdsch. Ent.*, **95** (1952), 43–62 (fixation of atmospheric nitrogen by symbionts of Aphids, termites, and other insects).

[323] TÓTH, L. *Trav. Inst. Hungrois Rec. Biol.*, **14** (1942), 397–440 (gut of Collembola).

[324] TRAGER, W. *Biochem J.*, **26** (1932), 1762–71 (cellulose digestion: termites; *Cryptocercus*, Blattidae).

[325] TRAGER, W., *et al. Amer. J. Hyg.*, **22** (1935), 18–25; 475–93; *J. Exp. Biol.*, **14** (1937), 240–51; *Biol. Bull.*, **75** (1938), 75–84 (accessory growth factors: *Aëdes*, Dipt.).

[326] TRAGER, W. *Physiol. Rev.*, **21** (1941), 1–35; *Biol. Rev.*, **22** (1947), 148–77 (insect nutrition: review).

[327] TRAPMANN, W. *Arch. Bienenk.*, **5** (1923), 204–12 (formation of peritrophic membrane in honey-bee).

[328] UVAROV, B. P. *Trans. Ent. Soc. Lond.*, 1928, 255–343 (nutrition and metabolism: review).

[329] VERSON, E. *Z. wiss. Zool.*, **82** (1905), 523–600 (development of gut in *Bombyx mori*: absorption in hind-gut, &c.).

[330] VIGNON, P. *Arch. Zool.*, **7** (1899), xvii–xxv (vesicular secretion).

[331] —— *Arch. Zool.*, (sér. 3), **9** (1901), 371–715 (secretion; formation of peritrophic membrane: *Chironomus* larva, silkworm larva, &c.).

[332] VOGEL, B. *Z. vergl. Physiol.*, **14** (1931), 273–347 (utilization of sugars by honey-bee).

[333] VOGEL, R. *Z. wiss. Zool.*, **112** (1925), 291–432 (biology of *Lampyris* larva, Col.: extra-intestinal digestion).

[334] VOINOV, D. N. *Bull. Soc. Sci. Bucharest*, **7** (1898), 473–93 (gut and peritrophic membrane in Odonata larvae).

[335] WATERHOUSE, D. F. *Coun. Sci. Ind. Res. Australia*, Pamph. 102 (1940), 7–27 (*p*H in gut of *Lucilia*, larva and adult, Dipt.).

[336] WEBER, H. *Zoologica*, **28** (1928), 1–120 (gut of *Aphis fabae*).

[337] —— *Biologie der Hemipteren*, Springer, Berlin, 1930.

[338] —— *Z. Morph. Oekol. Tiere*, **23** (1931), 575–753 (biology of *Trialeurodes*, Hem.).

[339] WEBER, H. *Lehrbuch der Entomologie*, Jena, 1933.

[340] —— *Zoologica*, **33** (1935), No. 6, 71 pp. (anatomy of Aleurodidae).

[341] —— *Zool. Jahrb.*; *Abt. Anat.*, **64** (1938), 243–86 (labial glands: Copeognatha).

[342] WEIL, E. *Z. Morph. Oekol. Tiere*, **30** (1935), 438–78 (gut of bees and wasps: peritrophic membrane, secretion, rectal glands, &c.).

[343] WERNER, E. *Z. Morph. Oekol. Tiere*, **6** (1926), 150–206 (nutrition and cellulose digestion in larva of *Potosia* (*Cetonia*) Col.).

[344] WEYER, F. *Z. Morph. Oekol. Tiere*, **30** (1935), 648–72; *Verhandl. deutsch., zool. Ges.*, 1936, 157–63 (ditto: *Blattella*, Orth.), (epithelial regeneration in mid-gut of termites during moulting).

[345] WHEELER, W. M. *Proc. Amer. Phil. Soc.*, **57** (1918), 293–343 (salivary secretions of ant larvae and origin of social habit).

[346] WHITCOMB, W., and WILSON, H. F. *Agric. Exp. Sta. Univ. Wisconsin, Res. Bull.*, **92** (1929), 27 pp. (digestion of pollen in honey-bee).

[347] WIEDEMANN, J. F. *Z. Morph. Oekol. Tiere*, **19** (1930), 229–58 (cellulose digestion: Lamellicorn larvae, Col.).

[348] WIGGLESWORTH, V. B. *Biochem. J.*, **21** (1927), 791–811; **22**, 150–61 (*p*H in gut and digestion: cockroach).

[349] —— *Parasitology*, **21** (1929), 288–321 (digestion; peritrophic membrane; symbionts: *Glossina*, Dipt.); *Ibid.*, **23** (1931), 73–6 (digestion: *Chrysops*, Dipt.).

[350] —— *Quart. J. Micr. Sci.*, **73** (1930), 593–616 (formation of peritrophic membrane).

[351] —— *Bull. Ent. Res.*, **20** (1930), 403–6 (extra-intestinal digestion: *Cicindela* larva, Col.).

[352] —— *Quart. J. Micr. Sci.*, **75** (1932), 131–50 (function of rectal glands).

[353] —— *J. Exp. Biol.*, **10** (1933), 1–37 (water uptake by anal papillae: mosquito larvae).

[354] —— *Parasitology*, **28** (1936), 284–9 (symbiotic bacteria in digestion and nutrition: *Rhodnius*, Hem.).

[355] WIGGLESWORTH, V. B. *J. Exp. Biol.*, **15** (1938), 235–47 (chloride uptake by anal mosquito larvae).
[356] —— *J. Exp. Biol.*, **19** (1942), 56–77 (absorption and storage in gut of *Aëdes* larva, Dipt.).
[357] —— *Proc. Roy. Soc.*, B, **131** (1943), 313–39 (breakdown of haemoglobin: *Rhodnius*, &c.).
[358] —— *J. Exp. Biol.*, **26** (1949), 150–63 (utilization of sugars by *Drosophila* during flight).
[359] —— Unpublished observations.
[360] WILSON, S. E. *Ann. Appl. Biol.*, **20** (1933), 661–90 (starch and invasion of wood by *Lyctus*, Col.).
[361] WOLLMANN, E. *Ann. Inst. Pasteur*, **25** (1911), 79–88; *C. R. Soc. Biol.*, **82** (1919), 593–4; 1208–10; *Arch. intern. Physiol.*, **18** (1921), 194–9 *Ann. Inst. Pasteur*, **36** (1922), 784–8 (rearing of Muscidae under sterile conditions).
[362] WOLLMANN, E. *et al. C. R. Soc. Biol.*, **95** (1926), 164–5; **124** (1937), 434–5 (rearing of cockroach *Blattella* under sterile conditions).
[363] YORKE, W., and MACFIE, J. W. S. *Ann. Trop. Med. Parasit.*, **18** (1924), 103–8 (anti-coagulins in salivary secretion: *Anopheles*, *Glossina*, Dipt.).
[364] ZABINSKI, J. *Brit. J. Exp. Biol.*, **6** (1929), 360–85 (nutrition in cockroaches).
[365] ZACCHARIAS, A. *Z. Morph. Oekol. Tiere*, **10** (1928), 676–737 (symbionts in Pupipara).

SUPPLEMENTARY REFERENCES (A)

[366] BAILEY, L. *J. Exp. Biol.*, **29** (1952), 310–26 (function of proventricultes in honey-bee).
[367] DAY, M. F. *Australian J. Sci. Res.* B., **4** (1951), 136–43 (secretion in salivary gland of *Periplaneta*).
[368] DAY, M. F., *et al. Australian J. Sci. Res.* B., **4** (1951), 42–74 (digestion of wool in *Tineola*).
[369] FISK, F. W. *Ann. Ent. Soc. Amer.*, **43** (1950), 555–72 (protein digestion in mosquitos).
[370] FLORKIN, M., *et al. Arch. intern. Physiol.*, **57** (1949), 71–88 (digestion of wax in *Galleria*).
[371] FRAENKEL, G., *et al. Physiol. Zool.*, **23** (1950), 92–108; *Arch. Biochem. Biophys.*, **34** (1951), 457–76; **35** (1952), 241–2; **50** (1954), 486–95; *Biol. Bull.*, **104** (1953), 359–371; *Physiol, Zool.*, **27** (1954), 40–56 (nutrition of *Tenebrio*).
[372] GLASER, R. W. *J. Parasit.*, **32** (1946), 483–9 (intracellular symbionts of cockroach).
[373] GLOOR, H., and CHEN, P. S. *Rev. Suisse Zool.*, **57** (1950), 570–6 (anal organ and chloride uptake in *Drosophila*).
[374] GOLDSMITH, E. D., *et al. Anat. Rec.*, **101** (1948), 93 (folic acid antagonists and growth in *Drosophila*).
[375] HAASER, C. *Nature*, **165** (1950), 397 (structure of peritrophic membrane in *Dixippus*).
[376] HASSETT, C. C., *et al. Biol. Bull.*, **99** (1950), 446–53 (nutritive values of carbohydrates in *Calliphora*).
[377] HINTON, T., *et al. Physiol. Zool.*, **24** (1951), 335–53 (essential amino-acids in nutrition of *Drosophila*).
[378] ISHII, S. *Botyu-Kagaku*, **13** (1949), 30–7; **16** (1951), 83–90 (sterol requirements in *Callosobruchus*, Col.).
[379] KÜHNELT, W. *Österreich. zool. Z.*, **2** (1949), 223–41 (reducing substances in integument of insects).
[380] LEMONDE, A., and BERNARD, R. *Canad. J. Zool.*, **29** (1951), 71–83 (essential amino-acids in nutrition of *Tribolium*).
[381] LEWIS, D. J. *Nature*, **165** (1950), 978; *Bull. Ent. Res.*, **43** (1953), 597–644 (peritrophic membrane in *Simulium*).
[382] LOTMAR, R. *Mitt. Schweiz. entom. Gesell.*, **22** (1949), 97–115 (digestion in *Stomoxys*).
[383] de MEILLON, B., *et al. S. Afr. J. Med. Sci.*, **12** (1947), 111–16; *J. Exp. Biol.*, **24** (1947) 41–63 (nutrition of *Cimex* on vitamin-deficient hosts).
[384] NOLAND, J. L., *et al. Ann. Ent. Soc. Amer.*, **42** (1949), 154–64; **44** (1951), 184–8 (vitamin and protein requirements of *Blattella*).

[385] SCHANDERL, H., *et al. Z. Naturforsch.* **4b** (1950), 50–3 (symbiotic yeasts in gut of Aphids).

[386] SCHLOTTKE, E. *Zool. Jahrb., Abt. Physiol.*, **61** (1945), 88–140 (enzymes in wood-eating beetle larvae).

[387] SHARIF, M. *Parasitology*, **38** (1948), 254–63 (nutritional requirements of flea larvae, *Xenopsylla*).

[388] SUTTON, M. F. *Proc. Zool. Soc. Lond.*, **121** (1951), 465–99 (feeding and peritrophic membrane in Corixidae).

[389] WHEELER, B. M. *J. Exp. Zool.*, **115** (1950), 83–104 (uptake of iodine by anal organ of *Drosophila*).

SUPPLEMENTARY REFERENCES (B)

[390] ANDERS, F. *Biol. Zbl.*, **79** (1960), 679–700 (gall formation by the vine phylloxera *Viteus*).

[391] APPLEBAUM, S. W., *et al. Comp. Biochem. Physiol.*, **11** (1964), 85–103 (proteolytic enzymes in *Tenebrio*).

[392] BACCETTI, B. *Redia*, **45** (1960), 263–78; **47** (1962), 105–18 (fine structure of hind-gut and rectal papillae in a grasshopper).

[393] —— *Redia*, **46** (1962), 157–65 (fine structure of mid-gut epithelium in Orthoptera, Coleoptera, and Diptera).

[394] BACON, J. S. D., and DICKINSON, B. *Biochem. J.*, **66** (1957), 289 (origin of melizitose in honey-dew).

[395] BAILEY, L. *J. Exp. Biol.*, **29** (1952), 310–26 (filtering action of proventriculus in *Apis*).

[396] BAINES, S. *J. Exp. Biol.*, **33** (1956), 533–41 (symbiotic bacteria in the nutrition of *Rhodnius*).

[397] BANKS, W. M. *Science*, **141** (1963), 1191–2 (carbohydrate digestion in *Blaberus*).

[398] BEAMS, H. W., and ANDERSON, E. *J. Morph.*, **100** (1957), 601–19 (fine structure of mid-gut epithelium of *Melanoplus*).

[399] BECK, S. D. *J. Gen. Physiol.*, **36** (1953), 317–25 ('corn leaf factor' in diet of *Pyrausta*).

[400] BEHRENZ, W., and TECHNAU, G. *Z. angew. Ent.*, **44** (1959), 22–8 (nutrition of symbiont-free *Anobium*).

[401] BERTRAM, D. S., and BIRD, R. G. *Trans. Roy. Soc. Trop. Med. Hyg.*, **55** (1961), 404–23 (fine structure of mid-gut epithelium in *Aëdes* adult).

[402] BEWIG, F., and SCHWARTZ, W. *Arch. Mikrobiol.*, **24** (1956), 174–208 (symbionts and nutrition in *Triatoma* and *Rhodnius*).

[403] V. BITTNER, A. *Z. Univ. Greifswald*, **3** (1954), 212–23 (function of proventriculus in *Galleria* larva).

[404] BLOCH, K., *et al. Biochim. Biophys. Acta*, **21** (1956), 176 (failure in biosynthesis of cholesterol in *Dermestes*).

[405] BRAGDON, J. C., and MITTLER, T. E. *Nature*, **198** (1963), 209–10 (differential absorption of amino acids by *Myzus persicae*).

[406] BROADBENT, L. *Proc. Roy. Ent. Soc. Lond.*, (A) **26** (1951), 97–103 (Aphid excretion).

[407] BRONSKILL, J. F., SALKELD, E. H., and FRIEND, W. G. *Canad. J. Zool.*, **36** (1958), 961–8 (enzymes in saliva of *Oncopeltus*).

[408] BROOKES, V. J., and FRAENKEL, G. *Physiol. Zool.*, **31** (1958), 208–23 (nutritional requirements of *Musca*).

[409] BROOKS, M. A. *Symp. Soc. Gen. Microbiol.*, **13** (1963), 200–31 (symbiosis in insects: review).

[410] BROOKS, M. A., and RICHARDS, A. G. *Biol. Bull.*, **109** (1955), 22–39; *J. Exp. Zool.*, **132** (1956), 447–66 (intracellular symbiosis in *Blattella*).

[411] BRUST, M., and FRAENKEL, G. *Physiol. Zool.*, **28** (1955), 186–204 (nutritional requirements in *Phormia*).

[412] BUCHNER, P. *Endosymbiose der Tiere mit pflanzlichen Mikroorganismen*, Birkhäuser: Basel and Stuttgart, 1953, 771 pp.

[413] —— *Z. Morph. Oekol. Tiere*, **45** (1957), 379–410 (endosymbiosis in *Hippeococcus*).

[414] BURNET, B., and SANG, J. H. *J. Ins. Physiol.*, **9** (1963), 553–62 (nucleic acid requirements in *Drosophila*).

[415] CASHMAN, E. F. *82nd Ann. Rep. Ent. Soc. Ontario*, 1951, 74–7 (effect of parental feeding on growth of offspring in *Tribolium*).

[416] CLAYTON, R. B., EDWARDS, A. M., and BLOCH, K. *Nature*, **195** (1962), 1125–6 (biosynthesis of cholesterol in *Ctenolepisma*).

[416a] COPELAND, E. *J. Cell Biol.*, **23** (1964), 253–60 ('mitochondrial pump' in anal papillae of mosquito larvae).

[417] DADD, R. H. *J. Exp. Biol.*, **38** (1961), 259–66 (humoral regulation of mid-gut secretion in *Tenebrio*).

[418] —— *Proc. Roy. Soc.*, B, **153** (1960), 128–43 (nutritional requirements of locusts: ascorbic acid); *J. Ins. Physiol.*, **4** (1960), 319–47 (lipids); **5** (1960), 161–8 (sterols), 301–16 (carbohydrates); **6** (1961), 1–12 ('B' vitamins), 126–45 (fatty acids, chlorophyll, salts, amino acids); *Bull. Ent. Res.*, **52** (1961), 63–81 (carotene).

[419] —— *Adv. Ins. Physiol.*, **1** (1963), 47–109 (locust feeding and nutrition: review).

[420] DAVEY, K. G., and TREHERNE, J. E. *J. Exp. Biol.*, **40** (1963), 763–80 (control of crop emptying in *Periplaneta*).

[421] v. DEHN. M. *Z. vergl. Physiol.*, **45** (1961), 88–108 (composition of sieve-tube juice and honey-dew).

[422] DE METS, R., and JEUNIAUX, C. *Arch. Int. Physiol. Biochim.*, **70** (1962), 93–6 (composition of peritrophic membrane).

[423] DESCHAMPS, P. *Ann. Sci. Nat., Zool.*, **15** (sér, 11) (1953), 449–533 (digestion and nutrition in Cerambycidae).

[424] DIMOND, J. B., LEA, A. O., and DeLONG, D. M. *Proc. Xth Int. Congr. Ent. 1956*, **2** (1958), 135–8 (nutritional requirements of *Aëdes aegypti* for egg production).

[425] EDWARDS, J. S. *XI. Int. Kongr. Ent. Wien 1960*, **3** (1960), 259–63; *J. Exp. Biol.*, **38** (1961), 61–77 (properties of saliva in *Platymeris*, Reduviidae).

[426] EHRHARDT, P. *Z. vergl. Physiol.*, **46** (1962), 169–211 (feeding and metabolism in the Aphid *Megoura*).

[427] EISNER, T. *J. Exp. Zool.*, **130** (1955), 159–81 (digestion and absorption of fats in the crop of *Periplaneta*).

[428] —— *Bull. Mus. Comp. Zool. Harvard*, **116** (1957), 440–90 (proventriculus of ants).

[429] EISNER, T., and HAPP, G. M. *Psyche*, **69** (1962), 108–16 (filtration device in buccal cavity of ants).

[430] EVANS, W. A. L., *et al. Nature*, **177** (1956), 478 (carbohydrases in *Lucilia* larva); *Biochem. J.*, **76** (1960), 50–1 (ditto in *Schistocerca*).

[431] EWART, W. H., and METCALF, R. L. *Ann. Ent. Soc. Amer.*, **49** (1956), 441–7 (sugars and amino acids in honey-dew of Coccids).

[432] FINK, R. *Z. Morph. Oekol. Tiere*, **41** (1952), 78–146 (isolation and culture of symbionts from *Pseudococcus*).

[433] FISK, F. W., and RAO, B. R. *Ann. Ent. Soc. Amer.*, **57** (1964), 40–4 (carbohydrases in the cockroach *Byrsotria*).

[434] FISK, F. W., and SHAMBAUGH, G. F. *Ohio J. Sci.*, **52** (1952), 80–8 (digestion in *Aëdes*: protease); **54** (1954), 237–9 (ditto: invertase).

[435] FOECKLER, F. *Z. Morph. Oekol. Tiere*, **50** (1961), 119–62 (replacement of symbiotic yeasts in *Stegobium*).

[436] FRAENKEL, G. *J. Cell. Comp. Physiol.*, **45** (1955), 393–408 (sugars inhibiting growth in *Tenebrio*).

[437] —— *J. Nutrition*, **65** (1958), 361–95 (zinc and potassium in nutrition of *Tenebrio*).

[438] FRAENKEL, G., *et al. Biol. Bull.*, **105** (1953), 442–9; **106** (1954), 178–84 (salivary glands and puparium cement in *Drosophila*, *Phormia*, &c.).

[439] FRAENKEL, G., and PRINTY, G. E. *Biol. Bull.*, **106** (1954), 149–57 (amino acid requirements of *Tribolium*).

[440] FRAENKEL, G., and STERN, H. R. *Arch. Biochem.*, **30** (1951), 438–44 (nicotinic acid requirements of *Tribolium* and *Tenebrio*).

[441] FRANK, W. *Z. Morph. Oekol. Tiere*, **44** (1956), 329–66 (nutritional influence of intracellular symbionts in *Periplaneta*).

[442] FRASER, A., RING, R. A., and STEWART, R. K. *Nature*, **192** (1961), 999–1000 (proteases in mid-gut of *Calliphora*).

[443] FREE, J. B. *Proc. Roy. Ent. Soc. Lond.*, A, **36** (1961), 5–8 (activity of pharyngeal glands in honey-bee).

[444] FRIEND, W. G., *et al.* *Canad. J. Zool.*, **34** (1956), 152–62; **35** (1957), 235–43; **36** (1958), 931–6 (nutritional requirements in *Hylemyia antiquia* (Dipt.)).

[445] FYG, W. *Z. Bienenforsch.*, **5** (1961), 213–19 (glycogen in proventriculus of *Apis*).

[446] GALUN, R., and FRAENKEL, G. *J. Cell. Comp. Physiol.*, **50** (1957), 1–23 (effects of sugars on nutrition of *Aëdes*, *Musca*, and *Sarcophaga*).

[447] GERSCH, M. *Z. vergl. Physiol.*, **34** (1952), 346–69 (digestion in *Corethra* larva).

[448] GOOD, M. E., *et al.* *Canad. Ent.*, **85** (1953), 252–3 (wax digesting enzymes in *Galleria*).

[449] GOODCHILD, A. J. P. *Proc. Zool. Soc. Lond.*, **122** (1953), 543–72 (digestion in Capsid bugs).

[450] —— *Trans. Roy. Ent. Soc. Lond.*, **115** (1963), 217–37; *Proc. Zool. Soc. Lond.*, **141** (1963), 851–910 (water disposal in sap-sucking Hemiptera).

[451] GRÄBNER, K. E. *Z. Morph. Oekol. Tiere*, **41** (1954), 471–528 (symbionts of Anobiidae, &c., as a source of vitamins).

[452] GRAY, H. E., and FRAENKEL, G. *Science*, **118** (1953), 304–5; *Physiol. Zool.*, **27** (1954), 56–65 (carbohydrates in honey-dew).

[453] GREENBERG, B., and PARETSKY, D. *Ann. Ent. Soc. Amer.*, **48** (1955), 46–50 (proteases in *Musca*).

[454] de GROOT, A. P. *Experientia*, **8** (1962), 192–4 (amino acid requirements of *Apis*).

[455] HALFF, L. A. *Acta Tropica*, **13** (1956), 225–53 (gut symbionts and growth of *Triatoma*).

[456] HARINGTON, J. S. *Parasitology*, **50** (1960), 279–86; *Nature*, **188** (1960), 1027–8 (symbionts of *Rhodnius* and vitamin synthesis).

[457] HAAS, H., and STRENZKE, K. *Biol. Zbl.*, **76** (1957), 513–28 (anal papillae of *Chironomus* in various saline media).

[458] HENRY, S. M., and BLOCK, R. J. *Federation Proc.*, **21** (1962), Abst. 9. (intracellular symbionts of *Blattella* and amino acid synthesis).

[459] HENSCHLER, D., *et al.* *Naturwiss.*, **41** (1954), 142; **47** (1960), 326–7 (acetylcholine content of royal jelly).

[460] HOLMES, N. D. *Canad. Ent.*, **86** (1954), 159–67 (digestive enzymes in *Cephus* larva).

[461] HOSOI, T. *Annot. Zool. Jap.*, **27** (1954), 82–90 (sensory control of the distribution of food in the mosquito).

[462] HOUSE, H. L. *Ann. Rev. Ent.*, **6** (1961), 13–26; *Ann. Rev. Biochem.*, **31** (1962), 653–72 (insect nutrition: review).

[463] HOUSE, H. L., *et al.* *Canad. J. Zool.*, **32** (1954), 331–65; **34** (1956), 182–9; *J. Nutrition*, **72** (1960), 409–14 (nutritional requirements of *Agria affinis*, Sarcophagidae).

[464] ITO, T. *Bull. Sericult. Exp. Sta.*, **17** (1961), 119–36; *Nature* **192** (1961), 951–2 (ascorbic acid in nutrition of *Bombyx*).

[464a] ITO, T., and TANAKA, M. *Biol. Bull.*, **116** (1959), 95–105 (β-glucosidase in mid-gut of *B. mori*).

[465] JENSEN, P. B. *Physiol. Plantarum*, **1** (1948), 95–108; *Dan. Biol. Medd.*, **18** No. 18 (1952), 1–18 (gall formation by *Mikiola*, Cecidomyidae).

[466] JOHANSSON, T. S. K. *Bee World*, **36** (1955), 3–13, 21–32 (composition of royal jelly).

[467] KAPLANIS, J. N., *et al.* *Ann. Ent. Soc. Amer.*, **56** (1963), 198–201 (sterol metabolism in *Musca*).

[468] KASTING, R., and MCGINNIS, A. J. *Nature*, **182** (1958), 1380–1 (essential amino acids in *Phormia*).

[469] KAUDEWITZ, H. *Z. vergl. Physiol.*, **35** (1953), 380–415 (storage of vitamins in the fat body of *Tenebrio*).

[470] KELLER, H. *Z. Naturforsch.*, **5b** (1950), 269–73 (culture of intracellular symbionts of *Periplaneta*).

[471] KENNEDY, J. S., *et al.* *Nature*, **168** (1951), 825; *Ann. Appl. Biol.*, **38** (1951), 25–64 (preferred feeding sites of Aphids).

[472] KENNEDY, J. S., and MITTLER, T. E. *Nature*, **171** (1953), 528 (obtaining phloem sap via the stylets of Aphids).

[473] KENNEDY, J. S., and STROYAN, H. L. G. *Ann. Rev. Ent.*, **4** (1959), 139–60 (biology of Aphids review).

[474] KHAN, M. R. *Comp. Biochem. Physiol.*, **6** (1962), 169–70 (dipeptidase in mid-gut of *Locusta* and *Dysdercus*).

[475] KHAN, M. R., and FORD, J. B. *J. Ins. Physiol.*, **8** (1962), 597–608 (cytology of mid-gut in *Dysdercus*, Hemipt.).

[476] KING, R. C. *Amer. Naturalist*, **88** (1954), 155–8; *Evolution*, **9** (1955), 93–6 (yeasts and phosphorus uptake by *Drosophila*).

[477] KOCH, A. *Forsch. Fortschr.*, **28** (1954), 33–7; *Exp. Parasitol.*, **5** (1956), 481–518 (symbionts as source of vitamins: review).

[478] KRISHNA, S., and SAXENA, K. N. *Physiol. Zool.*, **35** (1962), 67–78 (digestion and absorption in *Tribolium*).

[479] KÜMMEL, G. *Z. Morph. Oekol. Tiere*, **45** (1956), 309–42 (fine structure of the peritrophic membrane and of the cuticular lining of the insect gut).

[480] LAFON, M. *Physiol. Comp. Oecol.*, **2** (1952), 225–40 (nitrogen requirements of *Blatta*).

[481] LAMBREMONT, E. N., *et al. Science*, **129** (1959), 1484–5 (pepsin-like enzyme in *Stomoxys* larva).

[482] LASKER, R., and GIESE, A. C. *J. Exp. Biol.*, **33** (1956), 542–53 (cellulose digestion in *Ctenolepisma*).

[483] LAUREMA, S., and NUORTEVA, P. *Ann. Ent. Fennicae*, **27** (1961), 89–93 (pectin polygalacturonase in salivary glands of Capsidae).

[484] LEVINSON, Z. H. *Riv. Parasit.*, **16** (1955), 1–48 (nutritional requirements of insects: review).

[485] LEVINSON, Z. H., *et al. Biochem. J.*, **58** (1954), 291–7; **65** (1957), 254–60; *J. Ins. Physiol.*, **8** (1962), 191–8 (dietary sterols in insects).

[486] LEWIS, I. F., and WALTON, L. *Trans. Amer. Micr. Soc.*, **77** (1958), 146–200 (gall formation by *Hormaphis*).

[487] LIENERT, E., and THORSELL, W. *Parasitology*, **4** (1955), 117–22 (digestion of connective tissue by *Hypoderma* larva).

[488] LINDEMANN, C. *Z. vergl. Physiol.*, **31** (1948), 112–33 (nutrition of Aphids).

[489] LIPKE, H., and FRAENKEL, G. *Ann. Rev. Ent.*, **1** (1956), 17–44 (insect nutrition: review).

[490] MCALLAN, J. W., and ADAMS, J. B. *Canad. J. Zool.*, **39** (1961), 305–10 (pectinase and probing by Aphids).

[491] MAURIZIO, A. *Bee World*, **38** (1957), 14–17 ('transglucosidase' activity of invertase in *Apis*).

[492] MEGAHED, M. M. *Bull. Soc. Ent. Egypte*, **42** (1958), 339–55 (distribution of ingested food in gut of *Culicoides*).

[493] MELLANBY, K., and FRENCH, R. A. *Ent. Exp. et Appl.*, **1** (1958), 116–24 (effect of water drinking in insect larvae).

[494] MERCER, E. H., and DAY, M. F. *Biol. Bull.*, **103** (1952), 384–94 (fine structure of peritrophic membrane).

[495] MILES, P. W. *Nature*, **183** (1959), 756 (two types of salivary secretion in *Aphis*).

[496] —— *J. Ins. Physiol.*, **3** (1959), 243–55; **4** (1960), 209–19, 271–82; *Nature*, **191** (1961), 911–12 (formation of stylet-sheath in *Oncopeltus* and *Dysdercus*).

[497] ——*J. Ins. Physiol.*, **10** (1964), 147–60 (formation of stylet sheath in *Eumecopus*, Pentatomidae).

[498] MITLIN, N., KONECKY, M. S., and PIQUETT, P. G. *J. Econ. Ent.*, **47** (1954), 932–3 (effect of a folic acid antagonist on *Musca*).

[499] MITTLER, T. E. *J. Exp. Biol.*, **34** (1957), 334–41; **35** (1958), 74–84; 626–38 (feeding and nutrition in *Tuberolachnus*, Aphididae).

[500] MUKHARJI, S. P. *Ind. J. Zoot.*, **2** (1961), 56–121 (alimentary canal of some Homoptera).

[501] MÜLLER, H. J. *Z. Morph. Oekol. Tiere*, **44** (1956), 459–82 (role of symbionts in *Coptosoma*, Heteroptera).

[502] NATON, E. *Z. angew. Ent.*, **48** (1961), 58–74; *Zool. Beitr.*, **8** (1963), 96–123, 174–86, 448–82 (growth of *Tribolium*; role of carnitine).

[503] NEWCOMER, W. S. *J. Cell. Comp. Physiol.*, **43** (1954), 79–86 (β-glucosidase in gut of *Periplaneta*).

[504] NIEMIERKO, W., WLODAWER, P., and PRZELECKA, A. *III Congr. Int. Biochim., Bruxelles*, 1955, Abstr. 12–32 (role of phosphorus compounds in absorption of wax in *Galleria*).

[505] NOIROT, M. C., and NOIROT-TIMOTHÉE, C. *C. R. Acad. Sci.*, **251** (1960), 2779–81 (fine structure of hind-gut epithelium in termites and crickets).

[506] NOLAND, J. L. *Arch. Biochem. Biophys.*, **48** (1954), 370–9; **52** (1954), 323–30 (sterol metabolism in *Blattella*).

[507] NUORTEVA, P. *Ann. Ent. Fennicae.*, **20** (1954), 76–9 (salivary enzymes in *Empoasca flavescens*); 102–24 (ditto in *Lygus rugulipennis*); **21** (1955), 33–7 (ditto in *Plesiocoris rugicollis*); *Ent. Exp. et Appl.*, **1** (1958), 41–9 (ditto, general discussion).

[508] NUORTEVA, P., and LAUREMA, S. *Ann. Ent. Fennicae*, **27** (1961), 93–7 (salivary proteases in *Dolycoris*, Pentatomidae).

[509] OHNESORGE-HUMPERDINCK, I., *Naturwiss.*, **40** (1953), 61 (head glands of *Apis*).

[510] ÖRÖSI-PÁL, Z. *Bee World*, **38** (1957), 70–3 (function of mandibular glands of *Apis*).

[511] PACAUD, A. *C. R. Acad. Sci.*, **228** (1949), 1664–5; *Ann. Sci. Nat. Zool.*, (Sér. 11) **12** (1950), 1–13 (glycogen in mid-gut epithelium of *Simulium*).

[512] PANT, N. C., and FRAENKEL, G. *Science*, **112** (1950), 498–500; *Biol. Bull.*, **107** (1954), 420–32 (function of symbiotic yeasts in *Stegobium* and *Lasioderma*).

[513] PANT, N. C., and KAPOOR, S. *Ind. J. Ent.*, **25** (1963), 10–16; 312–15 (symbionts of *Lasioderma* and supply of amino acids and cholesterol).

[514] PATEL, N. G., and RICHARDS, A. G. *J. Ins. Physiol.*, **4** (1960). 146–53 (proleolytic enzymes in *Musca*).

[515] PATTERSON, R. A., and FISK, F. W. *Ohio J. Sci.*, **58** (1958), 299–310 (protease of *Stomoxys*).

[516] PERKOWSKA, E. *Exp. Cell. Res.*, **32** (1963), 259–71 (salivary secretion of *Drosophila* larva).

[517] PHILLIPS, J. E. *J. Exp. Biol.*, **41** (1964), 15–80 (rectal absorption of salts and water in *Schistocerca* and *Calliphora*).

[518] PIERRE, L. L. *Nature*, **193** (1962), 904–5 (symbionts of *Leucophaea*: synthesis of ascorbic acid); **201** (1964), 54–5 (ditto: uricase activity).

[519] PILLAI, M. K. K., and SAXENA, K. N. *Physiol. Zool.*, **32** (1959), 293–8 (fate of fructose in gut of *Periplaneta*).

[520] POULSON, D. F., and WATERHOUSE, D. F. *Aust. J. Biol. Sci.*, **13** (1960), 541–67 (pole cells and mid-gut differentiation in Diptera).

[521] POWNING, R. F., and IRZYKIEWICZ, H. *Nature*, **184** (1959), 1230–1; *J. Ins. Physiol.*, **8** (1962), 275–84 (reducing system and proteases in keratin digestion in *Tineola*).

[522] PUCHTA, O. *Z. Parasitenk.*, **17** (1955), 1–40; *Z. Morph. Oekol. Tiere*, **44** (1956), 416–41 (morphology and function of symbionts in *Pediculus*).

[523] REMBOLD, H., *et al. Verh. XI Int. Kongr. Ent.*, (*Wien*) Symp. III 1961, Chemie der Insekten, 77–81; *Hoppe-Seyl. Z.* **319** (1960), 200–19 (chemistry of royal jelly).

[524] RITCHOT, C., and MCFARLANE, J. E. *Canad. J. Zool.*, **40** (1962), 371–4 (unsaturated lipid and reproduction in male *Acheta*).

[525] RITTER, H. *Biol. Bull.*, **121** (1961), 330–46 (glutathione and anaerobiosis in gut of *Cryptocercus*).

[526] RIZKI, M. T. M. *J. Exp. Zool.*, **131** (1956), 203–18 (formation of peritrophic membrane in *Drosophila*).

[527] ROBBINS, W. E., *et al. Ann. Ent. Soc. Amer.*, **55** (1962), 102–5 (sterol metabolism in *Blattella*).

[528] RÖSSLER, M. E. *J. Ins. Physiol.*, **6** (1961), 62–80 (nutrition in *Oryctes* and *Melolontha* larvae).

[529] ROUSELL, P. G. *Trans. N.Y. Acad. Sci.*, Ser. 2. **19** (1956), 17–18; *J.N.Y. Ent. Soc.*, **66** (1958), 49–58 (ascorbic acid synthesis in *Periplaneta*).

[530] SANG, J. H. *J. Exp. Biol.*, **33** (1956), 45–70; **38** (1961), 793–809; *Proc. Roc. Soc. Edinb.*, B, **66** (1957), 339–59 (nutritional requirements of *Drosophila*).

[531] SARMA, P. S. *Proc. Soc. Exp. Biol. Med.*, **58** (1945), 140–1 (pyridoxine and tryptophane metabolism in *Corcyra*).

[532] SAXENA, K. N. *Experientia*, **10** (1954), 383 (feeding of leafhoppers, Jassidae).

[533] SAXENA, K. N., and BHATNAGAR, P. *J. Ins. Physiol.*, **7** (1961), 109–26 (effects of invertase in digestion in *Oxycarenus*, Lygaeidae).

[534] SCHILDMACHER, H. *Biol. Zbl.*, **69** (1950), 390–438 (permeability of peritrophic membrane in mosquito larvae).

[534a] SCHNEIDER, H. *Z. Morph. Oekol. Tiere*, **44** (1956), 555–625 (symbionts and nutrition in *Calandra*).

[535] SCHÖNFELD, C. *Zool. Jahrb., Physiol.*, **67** (1958), 337–64 (digestion in *Coretha* larva).

[536] SCHREINER, T. *Z. vergl. Physiol.*, **34** (1952), 278–98 (transport in the gut and function of proventriculus in *Apis*).

[537] SHAMBAUGH, G. F. *Ohio J. Sci.*, **54** (1954), 151–60 (protease stimulation by feeding in *Aëdes*).

[538] SHAW, J. *J. Exp. Biol.*, **32** (1955), 355–82 (ionic regulation and water balance in *Sialis* larva).

[539] SIMPSON, J. *J. Ins. Physiol.*, **4** (1960), 107–21; *J. Apicult. Res.*, **2** (1963), 115–16 (functions of salivary glands in *Apis*).

[540] SINGH, K. R. P., and BROWN, A. W. A. *J. Ins. Physiol.*, **1** (1957), 199–220 (nutritional requirements of *Aëdes aegypti*).

[541] SLABÝ, O. *Vestnik. Čsl. Zool. Společnosti*, **12** (1948), 184–209 (cellulose digestion in Cossidae and Sesiidae).

[542] STOHLER, H. *Acta Tropica*, **14** (1957), 303–52 (peritrophic membrane in adult *Aëdes*).

[543] STRENZKE, K., and NEUMANN, D. *Biol. Zbl.*, **79** (1960), 199–225 (salt content of the medium and anal papillae of Chironomid larvae).

[544] TREHERNE, J. E. *Trans. Roy. Ent. Soc. Lond.*, **105** (1954), 117–30 (chloride uptake by anal papillae of *Helodes*, Col.).

[545] —— *J. Exp. Biol.*, **34** (1957), 478–85; **35** (1958), 297–303, 611–25 (absorption in *Periplaneta* and *Schistocerca*: sugars); **35** (1958), 862–70 (ditto: fats); **36** (1959), 533–45 (ditto: amino acids).

[546] —— in *Viewpoints in Biology* (J. D. Carthy and C. L. Duddington, Edd.) **1** (1962), 201–41 (absorption from the gut in insects: review).

[547] VANDERZANT, E. S. *J. Ins. Physiol.*, **9** (1963), 683–91 (amino acid requirements in *Anthonomus grandis*).

[548] VANDERZANT, E. S., KERUR, D., and REISER, R. *J. Econ. Ent.*, **50** (1957), 606–8 (dietary fatty acids required by *Platyedra*).

[549] VANDERZANT, E. S., POOL, M. C., and RICHARDSON, C. D. *J. Ins. Physiol.*, **8** (1962), 287–97 (ascorbic acid in nutrition of *Anthonomus* and *Heliothis*).

[550] WATERHOUSE, D. F. *Aust. J. Sci. Res.*, B, **5** (1952), 143–88; **6** (1953), 257–75; *Advances in pest control research*, **2** (1958), 207–62 (digestion of keratin (wool) by *Tineola* and other insects).

[551] WATERHOUSE, D. F. *Aust. J. Zool.*, **1** (1953), 299–318; *Aust. J. Biol. Sci.*, **7** (1954), 59–72 (formation of peritrophic membrane in insects).

[552] —— *Aust. J. Biol. Sci.*, **8** (1955), 514–29 (functional differentiation of the hind-gut in *Lucilia* larva).

[553] —— *Ann. Rev. Ent.*, **2** (1957), 1–18 (digestion in insects: review).

[554] WATERHOUSE, D. F., and STAY, B. *Aust. J. Biol. Sci.*, **8** (1955), 253–77 (histochemistry of mid-gut in *Lucilia* larvae).

[555] WATERHOUSE, D. F., and WRIGHT, M. *J. Ins. Physiol.*, **5** (1960), 230–9 (fine structure of mid-gut in *Lucilia* larvae).

[556] WEAVER, N. *Nature*, **188** (1960), 938–9 (fatty acids &c., in royal jelly).

[557] WEIR, J. S. *Quart. J. Micr. Sci.*, **98** (1957), 499–506 (anatomy of mid-gut of *Myrmica*).

[558] WIESMANN, R. *Mitt. Schweiz. Ent. Ges.*, **36** (1964), 249–73 (direction of ingested foods in the alimentary canal of *Musca*).

[559] WIGGLESWORTH, V. B. *Bull. Ent. Res.*, **33** (1942), 205–18 (absorption of lipids through the cuticle of *Rhodnius*).

[560] —— *J. Exp. Biol.*, **26** (1949), 150–63 (glycogen in proventriculus of *Drosophila*).

[561] —— *Tijdschrift Ent.*, **95** (1952), 63–8 (symbiosis in blood-sucking insects: review).

[562] —— *Quart. J. Micr. Sci.*, **100** (1959), 315–20 (osmium and ethyl gallate staining for mitochondria, &c.).

[563] —— Unpublished observations.

[564] WILDBOLZ, T. *Mitt. schweiz. ent. Ges.*, **27** (1954), 194–240 (fine structure of peritrophic membrane in *Melolontha*).

[565] WLODAWER, P., *et al. Acta Biol. Exper.*, **16** (1952), 157–70; **17** (1956), 221–35 (role of phosphatides in wax digestion in *Galleria*).

[566] WOLFE, L. S. *Canad. J. Anim. Sci.*, **39** (1959), 145–57 (digestion of collagen by *Hypoderma* larvae).

[567] YAMAGUCHI, S. *Res. Rep. Fac. Textile and Sericult.*, **6** (1956), 45–60; **7** (1957), 69–72 (distribution of glycogen in the mid-gut of *Bombyx* larva).

[568] YOSHIDA, T. *Bull. Sericult. Exp. Sta.*, **14** (1955), 353–426 (digestion of chlorophyll, &c., in *Bombyx*).

SUPPLEMENTARY REFERENCES (C)

[569] AKAI, H. *J. Insect Physiol.*, **15** (1969), 1623–28 (localization of phosphatases in mid-gut of *Bombyx*).

[570] BACCETTI, B., MAZZI, V. and MASSIMELLO, G. *Redia*, **48** (1963), 265–87 (fine structure of rectal gland cells of *Dacus*).

[571] BAKER, J. M. *Proc. R. Ent. Soc. Lond.*, C, **33** (1969), 31 (digestion of wood by *Anobium*, &c.).

[572] BALSHIN, M. and PHILLIPS, J. E. *Nature*, **233** (1971), 53–5 (rectal absorption of amino acids in *Schistocerca*).

[573] BANKS, C. J. and MACAULAY, E. D. M. *Ann. appl. Biol.*, **55** (1965), 207–18 (ingestion of nitrogen by *Aphis fabae*).

[574] BAUDISCH, W. *Biol. Zbl.*, **82** (1963), 351–61; **86** (1967), 157–62 (diverse proteins from different lobes of salivary glands of *Acricotopus*, Chironomidae).

[575] BEENAKKERS, A. M. T., *et al. Naturwissenschaften*, **50** (1963), 361; *Biochim. biophys. Acta*, **84** (1964), 205–7 (carnitine and fatty acid catabolism in insects).

[576] BERRIDGE, M. J. *J. exp. Biol.*, **43** (1965), 511–21 (rectal fluid as water reserve in *Dysdercus*).

[577] BERRIDGE, M. J. *Symp. Roy. Ent. Soc. Lond.* **5** (1970), 135–51 (structural analysis of intestinal absorption).

[578] BERRIDGE, M. J. and GUPTA, B. L. *J. Cell Sci.*, **2** (1967), 89–112; **3** (1968), 17–32 (fine structure and localization of ATP-ase in rectal papillae of *Calliphora*).

[579] BULLADE, D., PARIS, R. and MOULINS, M. *J. Insect Physiol.*, **16** (1970), 45–53 (properties of intestinal lipase in cockroach).

[580] BOWEN, I. D. *Histochem. J.* **1** (1968), 141–51 (cytochemistry of gut of *Schistocerca*).

[581] BRAMMER, J. D. and WHITE, R. H. *Science*, **163** (1969), 821–3 (vitamin A deficiency and eye structure of *Aëdes*).

[582] CHANG, K. P. and MUSGRAVE, A. J. *Tissue & Cell*, **1** (1969), 597–606 (symbionts of *Psylla* as degenerate micro-organisms).

[583] CHILDRESS, C. C., SACKTOR, B. and TRAYNOR, D. R. *J. biol. Chem.*, **242** (1966), 754–60 (carnitine in flight muscle of *Phormia*).

[584] CHINNERY, J. A. B. *J. Insect Physiol.*, **17** (1971), 47–61 (carbohydrases in mid-gut of *Dermestes*).

[585] CHIPPENDALE, G. M. and BECK, S. D. *Ent. exp. & appl.*, **7** (1964), 241–8 (ascorbic acid as corn leaf factor for *Ostrinia*).

[586] CHIPPENDALE, G. M., BECK, S. D. and STRONG, F. M. *Nature*, **204** (1964), 710–11; *J. Insect Physiol.*, **11** (1965), 211–23 (linolenic acid and other requirements for *Trichoplusia*).

[587] CLAYTON, R. B. *et al. Comp. Biochem. Physiol.*, **11** (1964), 333–50 (absorption of cholesterol in cockroach *Eurycotis*).

[588] COOKE, J. and SANG, J. H. *J. Insect Physiol.*, **16** (1970), 801–12 (sterol utilization in *Drosophila*).

[589] COUCH, E. F. and MILLS, R. R. *J. Insect Physiol.*, **14** (1968), 55–62 (autophagic vacuoles in mid-gut of *Periplaneta*).

[590] DADD, R. H. *J. Insect Physiol.*, **12** (1966), 1479–92 (beeswax as essential nutrient for *Galleria*).

[591] DADD, R. H. and KRIEGER, D. L. *J. Insect Physiol.*, **13** (1967), 763–78; *J. econ. Ent.*, **60** (1967), 1512–14 (trace metal requirements of Aphids).

[592] DADD, R. H., KRIEGER, D. L. and MITTLER, T. E. *J. Insect Physiol.*, **13** (1967), 249–72; **14** (1968), 741–64 (dietary amino acid needs of Aphids).

[593] DANIEL, R. S. and BROOKS, M. A. *Experientia*, **23** (1967), 1–11 (bacteroids of *Periplaneta* as procaryotic microbes).

[594] DIAMOND, J. M. and TORMEY, J. M. *Nature*, **210** (1966), 817–20 (role of long extracellular channels in fluid transport across epithelia).

[595] DOYLE, D. and LAUFER, H. *Exp. cell. Res.*, **57** (1969), 205–10 (source of salivary gland proteins in *Chironomus*).

[596] EHRHARDT, P. *Z. vergl. Physiol.*, **53** (1966), 130–41; **58** (1968), 47–75; **60** (1968), 416–26 (nutritional requirements of Aphids deprived of symbionts).

[597] ENGELMANN, F. *J. Insect Physiol.*, **14** (1968), 1525–31 (feeding and crop emptying in *Leucophaea*).

[598] ENGELMANN, F. *et al. Naturwissenschaften*, **4** (1966), 113–14; *J. Insect Physiol.*, **15** (1969), 217–35; *Nature*, **222** (1969), 798; (stimulus to protease secretion in *Leucophaea* and *Sarcophaga*).

[599] GEER, B. W., OLANDER, R. M. and SHARP, P. L. *J. Insect Physiol.*, **16** (1970), 33–43 (choline requirements of *Drosophila*).

[600] GOODING, R. H. and HUANG, C. T. *J. Insect Physiol.*, **15** (1969), 325–39 (proteases in *Pterostichus*, Carabidae).

[601] GOURANTON, J. and MAILLET, P. L. *C.R. Acad. Sci. Paris*, **261** (1965), 1105–5 (peritrophic membrane in *Cicadella*).

[602] GUPTA, B. L. and BERRIDGE, M. J. *J. Morph.*, **120** (1966), 23–81; *J. Cell. Biol.*, **29** (1966), 376–82 (fine structure of rectal papillae in *Calliphora*).

[603] HARVEY, W. R. *et al. Proc. Nat. Acad. Sci.*, **51** (1964), 757–65; *J. Cell. Biol.*, **31** (1966), 107–34; *J. exp. Biol.*, **50** (1969), 297–306, **54** (1971), 269–74 (active transport of ions in mid-gut of *Hyalophora*).

[604] HAWKINS, R. I. *et al. Nature*, **212** (1966), 738–9; *Brit. J. Haem.*, **12** (1966), 86–91 (factors affecting blood clotting in *Glossina* and *Rhodnius*).

[605] HAYASHIYA, K. *Nature*, **205** (1965), 620–1 (acetylcholine as growth factor in *Bombyx*).

[606] HOUSE, H. L. *J. Insect Physiol.*, **12** (1966), 409–17; *Can. Ent.* **99** (1967), 1310–21 (vitamin E needs, and preference for complete diets in Sarcophagid larvae).

[607] IRVINE, H. B. and PHILLIPS, J. E. *J. Insect Physiol.*, **17** (1971), 381–3 (water absorption in rectum of *Schistocerca*).

[608] ITO, T. and ARAI, N. *J. Insect Physiol.*, **23** (1966), 861–9 (amino acid requirements in *Bombyx*).

[609] JONES, J. C. and ZEVE, V. H. *J. Insect Physiol.*, **14** (1968), 1567–75 (fine structure of gastric caeca in *Aëdes* larva).

[610] KAFATOS, F. C. *et al. Science*, **146** (1964), 538–40; **161** (1968), 470–2; *J. biol. Chem.* **7** (1967), 1477–94; *J. exp. Biol.*, **48** (1968), 435–53 (proteinase acting on cocoon in *Antheraea*).

[611] KAHAN, D. *Israel J. Zool.*, **13** (1964), 47–57 (toxic saliva of Asilidae).

[612] KATZ, M. *et al. J. Insect Physiol.*, **17** (1971), 1295–1303 (sterol requirements of *Dermestes*).

[613] KOCH, A. *Symp. Genet. Biol.*, **12** (1961), 360–78 (*Torulopsis* yeast replacing symbionts in *Sitodrepa*).

[614] KRIEGER, R. I. *et al. Science*, **172** (1971), 579–81 (detoxication enzymes in gut of caterpillars).

[615] LAKE, P. and FRIEND, W. G. *J. Insect Physiol.*, **14** (1968), 543–62 (*Nocardia* as source of vitamins in *Rhodnius*).

[616] LANGLEY, P. A. *J. Insect Physiol.*, **12** (1966), 439–48; **13** (1967), 477–86 (stimulus to protease secretion in *Glossina*).

[617] LIPKE, H. *et al. J. Insect Physiol.*, **11** (1965), 1225–32 (carbohydrate amino acid conversions in *Periplaneta*).

[618] LUDWIG, D. and GALLAGHER, M. R. *J.N.Y. Ent. Soc.*, **74** (1966), 134–9 (symbionts and vitamin synthesis in *Periplaneta*).

[619] McLEAN, D. L. and KINSEY, M. G. *Ann. Ent. Soc. Amer.*, **61** (1968), 730–9; **62** (1969), 287–94 (stylet sheath formation in Aphids).

[620] MALKE, H. *Nature*, **204** (1964), 1223–4 (elimination of symbionts from cockroaches by lysozyme).

[621] MARTIN, M. M. and MARTIN, J. S. *J. Insect Physiol.*, **16** (1970), 109–19 (biochemical value of food fungus for *Atta*).

[622] MASCHWITZ, U. *Z. vergl. Physiol.*, **53** (1966), 228–52 (nutrient value of larval saliva for adult wasps).

[623] MILES, P. W. *J. Insect Physiol.*, **11** (1965), 1261–8; *Aust. J. biol. Sci.*, **22** (1969), 1271–6 (stylet sheath formation in Aphids, &c.).

[624] MILES, P. W. *et al. Nature*, **213** (1967), 802–3; *J. Insect Physiol.*, **14** (1968), 97–106; *Ent. exp. & appl.* **12** (1969), 736–44 (factors in gall production by Hemiptera).

[625] MILLS, R. R. *et al. J. Insect Physiol.*, **16** (1970), 417–27 (mechanisms of water transport in mid-gut of *Periplaneta*).

[626] MITTLER, T. E. *J. Insect Physiol.*, **17** (1971), 1333–47 (symbionts and amino acid supply in *Myzus*).

[627] MIYAZAKI, S. *et al.* *J. Insect Physiol.*, **14** (1968), 513–18 (*Pseudomonas*, symbiont in *Rhagoletis*, as source of amino acids).

[628] MOHRIG, W. and MESSNER, B. *Biol. Zbl.*, **87** (1968), 705–18 (lysozyme as antimicrobial agent in gut of insects).

[629] MUNK, R. *Z. vergl. Physiol.*, **58** (1968), 423–8; **61** (1968), 129–36 (function of filter chamber in a Jassid and Cercopid).

[630] NOIROT, C. *et al. Z. Zellforsch.*, **101** (1969), 477–509 (fine structure of rectal cuticle).

[631] OSCHMAN, J. L. and BERRIDGE, M. J. *Tissue & Cell*, **1** (1970), 281–310 (secretory mechanism in salivary gland of *Calliphora*).

[632] OSCHMAN, J. L. and WALL, B. J. *J. Morph.*, **127** (1969), 475–510 (fine structure of rectal pads of *Periplaneta*).

[633] O'RIORDAN, A. M. *J. exp. Biol.*, **51** (1969), 699–714 (transport of water and ions across mid-gut wall in *Periplaneta*).

[634] PARSONS, M. C. *J. Morph.*, **129** (1969), 17–30 (grinding function of cibarium in *Aphelocheirus*, Naucoridae).

[635] PETERS, W. *Z. Morph. Oekol. Tiere*, **62** (1968), 9–57; **64** (1969), 21–58 (fine structure of peritrophic membrane).

[636] PHILLIPS, J. E. *Am. Zool.*, **5** (1965), 662; *Trans. Roy. Soc. Can.*, **3** (1965), 237–54 (role of cuticle intima in rectal function in locust).

[637] PHILLIPS, J. E. and BEAUMONT, C. *J. exp. Biol.*, **54** (1971), 317–28 (osmotic flow across rectal cuticle of *Schistocerca*).

[638] PHILLIPS, J. E. and MERIDETH, J. *Nature*, **222** (1969), 168–9 (sodium and chloride transport in anal papillae of *Aëdes campestris*).

[639] POWNING, R. F. and IRZYKIEWICZ, H. *J. Insect Physiol.*, **13** (1967), 1293–9 (lysozyme-like action of enzymes in *Periplaneta* &c.).

[640] SAUER, J. R. and MILLS, R. R. *J. Insect Physiol.*, **15** (1969), 1489–98 (movement of ions across mid-gut epithelium of *Periplaneta*).

[641] SCHÄLLER, G. *Zool. Jahrb. Physiol.*, **71** (1965), 385–92; **74** (1968), 54–87 (indolacetic acid and gall formation by phylloxera).

[612] SOHAL, R. S. and COPELAND, E. *J. Insect Physiol.*, **12** (1966), 429–39 (fine structure of anal papillae in *Aëdes* related with salinity).

[643] STADDON, B. W. *J. exp. Biol.*, **41** (1964), 609–19 (water balance in *Corixa*: function of rectum).

[644] STOBBART, R. H. *J. exp. Biol.*, **42** (1965), 29–43; **47** (1967), 35–57; **54** (1971), 19–27 (sodium and chloride uptake by larva of *Aëdes aegypti*).

[645] STOBBART, R. H. *J. Insect Physiol.*, **14** (1968), 269–75 (ion and water transport in rectum *Schistocerca*).

[646] STOBBART, R. H. *J. exp. Biol.*, **54** (1971), 29–66 (control of sodium uptake by larva of *Aëdes*).

[647] WALL, B. J. *et al. J. Insect Physiol.*, **13** (1967), 565–78; *Science*, **167** (1970), 1497–8 (water transport in rectum of *Periplaneta*).

[648] WENZL, H. *Z. vergl. Physiol.*, **62** (1969), 167–82; 411–12; **65** (1969), 340–50 (hydrolysis and metabolism of sugars and sugar alcohols in *Calliphora*).

[649] WHARTON, D. R. A. *et al. J. Insect Physiol.*, **11** (1965), 947–59; 1401–5 (cellulase in cockroach species).

[650] WOOD, J. L. *et al. J. exp. Biol.*, **50** (1969), 169–78 (active transport of potassium in mid-gut of *Hyalophora*).

[651] YANG, Y. J. and DAVIES, D. M. *J. Insect Physiol.*, **17** (1971), 117–31 (trypsin and chymotrypsin in *Aëdes aegypti*).

Chapter XII

Excretion

THE FUNCTION of the excretory system is to maintain a constant internal environment in the body, by the elimination or segregation of unwanted substances present in the blood, and by the retention or reabsorption of constituents needful to the organism. Many organs and tissues contribute to this end, but the most important are the Malpighian tubes, which discharge into the intestine at the junction of the mid-gut with the hind-gut. The urine, the product of the Malpighian tubes, is commonly mixed with the residue of food from the mid-gut; these together make up the excrement or faeces of insects, the formation and disposal of which have already been discussed (p. 495). But it is possible to study the properties of the urine alone, either by observing the contents of the Malpighian tubes, or by taking advantage of the fact that in many insects there are times when the excrement is made up wholly of the Malpighian secretion.

THE URINE

The role of water in excretion—The character of the urine depends in the first place upon the water relations of the insect. In blood-sucking and plant-sucking forms, immediately after a meal, when there is an abundance of water to be eliminated, the urine is a crystal-clear fluid. Mosquitos will begin passing clear drops of this kind within a few minutes of feeding; the tsetse fly, *Glossina*, will excrete in this way 43 per cent. of its meal in the course of one hour;[82] in two or three hours the blood-sucking bug *Rhodnius*, which takes very large meals, may produce its own weight of urine, and get rid of 75 per cent. of the water in the ingested blood.[152] In aquatic insects, also, the urine is generally copious and clear. Mosquito larvae are continuously absorbing water, partly with the food but chiefly by osmosis through the anal papillae (p. 514) and this water is continuously excreted again as urine.[154] [See p. 582.]

When the water available becomes less, the urine is concentrated. If mosquito larvae are kept in salt water, in which the osmotic uptake of water ceases, the Malpighian tubes are often filled with solid matter (Fig. 320).[154] In the blood-sucking insects, within a few hours after feeding, the urine becomes cloudy; and the sediment increases until the urine attains a creamy consistency. Such insects as *Rhodnius* or the bed-bug *Cimex* live for weeks without ingesting any water, and during this period the urine is in the form of a pultacious mass which dries as a yellow powder. During the pupal stage of insects, all excretion takes place without any addition to the water supply. The meconium which accumulates in the rectum, notably in Lepidoptera, and is discharged soon after emergence from the pupa, is composed wholly of urine; and it is of this same type, a heavy, whitish deposit in a yellow fluid.

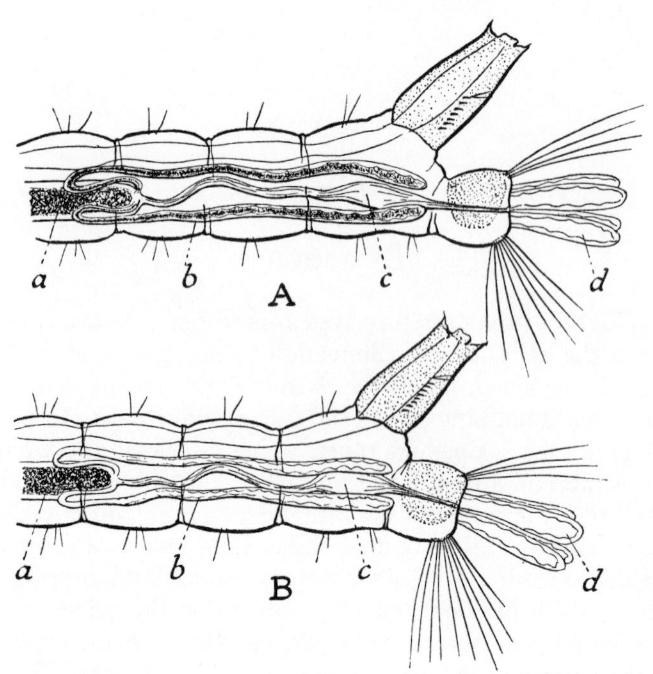

FIG. 320

A, hind end of larva of the mosquito *Aëdes aegypti* reared in salt water, showing solid uric acid in lumen of Malpighian tubes. *B*, the same larva a few minutes after transfer to fresh water; the Malpighian tubes have been flushed out by water entering through the anal papillae. *a*, mid-gut; *b*, Malpighian tubes; *c*, hind-gut; *d*, anal papillae.

We have seen that the excrement of many insects, *Lepisma*, cockroaches, Carabids, caterpillars, &c. is in the form of rather dry pellets from which the water has been extracted by the epithelium of the hind-gut and particularly by the rectal glands (p. 495). In such pellets the urine is present in semi-solid form, sometimes incorporated with the mass, often forming a whitish deposit outside the residue of the peritrophic membrane. And in forms like the mealworm, *Tenebrio*, which live in very dry materials, the urine, like the food residue, is reduced to a bone-dry powder before it is discharged.[34, 153]

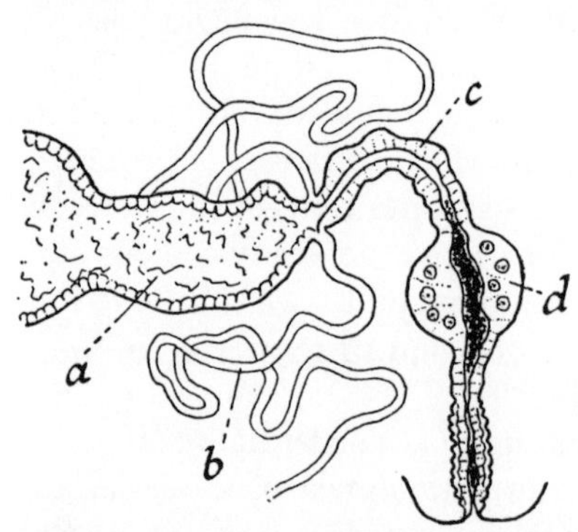

FIG. 321.—A part of the excretory system of *Pediculus*, showing the Malpighian tubes free from solid matter, and solid uric acid separating out in contact with the rectal glands

a, mid-gut; *b*, Malpighian tubes; *c*, hind-gut; *d*, rectal glands.

The urine as it leaves the Malpighian tubes may already contain more or less sediment (in the larvae and pupae of Lepidoptera, pupae of Muscid flies, in mosquitos, *Rhodnius*, &c.); the amount of sediment increases as water is reabsorbed in the hind-gut and rectum. But this reabsorption of water is particularly evident in insects in which the excrement is solid and yet the Malpighian secretion is perfectly clear. In the human louse, *Pediculus*, which will sometimes produce urine free from faecal admixture, no solid matter appears until the fluid enters the hind-gut. A granular deposit then separates out, accumulates particularly in the region of the rectal glands, and is finally discharged as a moist, white pellet (Fig. 321).[153] In cockroaches, Neuroptera, *Tenebrio*, &c., in spite of the dry state of the excrement, the urine as it is discharged into the gut contains no suspended matter.[153] In *Oncopeltus* the residue of the meals accumulates in the mid-gut throughout larval life, and the Malpighian tube secretion is alone discharged; only in the adult does the connection between mid-gut and hind-gut become functional.[194]

Physical characters of the urine—It is obvious that the physical characters of urine which ranges in appearance from a clear fluid to a dry solid, must be extremely variable. The osmotic pressure of the clear urine excreted by *Rhodnius* immediately after feeding is little greater than that of the ingested blood, being equivalent to about 1·0 per cent. of sodium chloride. The fluid accumulating in the rectum at 24 hours after feeding has an osmotic pressure equivalent to 1·7 per cent., and at 48 hours it equals 2·2 per cent. of sodium chloride.[152] The rectal contents can be much concentrated in *Carausius*[207], in *Schistocerca*,[203] and particularly in insects living in dry surroundings (p. 495).

The reaction must also vary extensively with the diet. The contents of the Malpighian tubes are often stated to be alkaline (in *Galleria*,[95] in *Gnaptor* (Col.)[27, 38]) but in the larvae of *Psychoda* (Dipt.) and *Chironomus* the gut contents change in reaction from alkaline (*p*H 7·2–7·8) to acid (*p*H 6·0–6·6) at the point of discharge of the Malpighian tubes;[21] and the meconium of *Celerio* (Lep.) has a *p*H of 5·8–6·3, though it is liable to become alkaline as the result of ammonia production through bacterial decomposition.[43] The clear urine of the newly fed *Rhodnius* is alkaline, *p*H 7·8–8·0, but in the later stages of excretion it gradually becomes more acid until it is about *p*H 6·0.[152]

The stimulus to urine formation—Excretion in insects is primarily a secretory process; it is inhibited by poisons which suppress oxidative metabolism.[201] There is some uncertainty as to what is the immediate stimulus to secretion. Saline solution (Ringer's solution) injected into the body cavity of *Rhodnius* in large amounts does not cause a rapid flow of urine.[225] Potassium ions induce some secretion but this is not continued.[207, 225] There is some evidence that the neurosecretory cells in the brain control water excretion: in *Anisotarsus* (Col.) extirpation of the dorsum of the brain causes water retention;[199] and decapitation of *Rhodnius* immediately after feeding leads to considerable reduction in the elimination of water by diuresis.[199] The isolated system of Malpighian tubules from *Rhodnius*, immersed in haemolymph from an unfed *Rhodnius*, show no signs of secretory activity; but haemolymph from a newly fed *Rhodnius* will induce a rapid flow of urine. The factor responsible is a hormone derived from the central nervous system; there is some activity in extracts from the brain, but most of the activity seems to be concentrated in the groups of neurosecretory cells in the metathoracic ganglia.[189] [See p. 582.]

The chemical composition of the urine depends upon what substances are present in the diet in excess of the needs of the body, and upon the production of waste products in metabolism. In the nutrition of *Rhodnius* and other blood-sucking insects, the food contains an excess of water and salts, which are rapidly eliminated as a clear solution of bicarbonates and chlorides of sodium and potassium; phosphates of magnesium and calcium appear later and in smaller quantities. Those substances which arise as end products in metabolism do not begin to appear until some hours after feeding. Much of the phosphate doubtless arises in this way, as the end product in the katabolism of materials such as lecithin and nuclein which contain phosphorus in organic combination. Sulphate occurs in small quantities as the end product of sulphur metabolism, coming mostly from the cystine and methionine components of the protein molecule. But the most important end product, always present in excess in the proteins of the food, is nitrogen; the elimination of nitrogen is the most important function of the excretory system.

Nitrogenous excretion: ammonia—The simplest form in which nitrogen can be excreted is ammonia. But free ammonia is a toxic substance, and large quantities of water are therefore needed for its elimination; it is seldom important as an excretory substance in terrestrial animals.[25] Ammonia is present, however, in large amounts in the excreta of the meat-eating larvae of *Calliphora* and *Lucilia* during their period of growth,[151, 12] and though a part of this is doubtless the result of bacterial decomposition, much comes from the tissues themselves; for it is formed equally by larvae reared under sterile conditions.[50] This ammonia is produced in the mid-gut, absorbed into the haemolymph, excreted by the Malpighian tubes and concentrated by the removal of water during the passage of the excreta down the hind-gut. Analyses of the ammonia (in mg. ammonia N per cent.) in the different regions are as follows: anterior mid-gut 11; middle segment of mid-gut 27; posterior mid-gut 4; haemolymph 10; Malpighian tubes 77; anterior hind-gut 206; posterior hind-gut 577. Expressed as ammonium carbonate it represents about one quarter of the total solids in the hind-gut.[81] Similar results are given in *Musca*.[211]

Ammonia combined in salts occurs in the excrement of many insects. In the dried excreta (urine and food residue) of the clothes moth, *Tinea pellionella*, 10·2 per cent. of the nitrogen is present as ammonia, 47·3 per cent. as uric acid;[1] in *Tineola bisselliella* 13·2 per cent. as ammonia, 47·2 per cent. as uric acid;[57] in the grasshopper *Melanoplus* 0·6–0·7 mg. of nitrogen is excreted as ammonia to every 10·6–15·6 mg. eliminated as uric acid;[13] in the meconium of the moth *Antheraea pernyi*, which is composed only of urine, about 0·5 mg. of nitrogen is present as ammonia to 8·7 mg. as uric acid.[20] But it is possible that in some of these cases bacterial decomposition may be in part responsible; for the meconium of *Celerio* (Lep.) soon comes to contain much ammonia from this cause, whereas when fresh the ammonia nitrogen averages 0·001 per cent., the uric acid nitrogen being about 1·8 per cent.;[43] and the urine of *Rhodnius* does not contain a trace of ammonia at any stage.[152]

Free amino acids and amides, the unabsorbed residue from the phloem sap, will account for up to 90 per cent. of the total nitrogen in the honey-dew of *Brevicoryne brassicae*. Neither uric acid, urea, allantoine, allantoic acid, creatine, nor creatinine can be detected. The only form of excretory nitrogen to be detected is ammonia, which accounts for less than 0·5 per cent. of the total nitrogen. Ammonia produced in metabolism may be the nitrogenous substrate for the intracellular symbionts.[186]

Ammonia is often the most important medium for nitrogen excretion in aquatic insects. In the larva of *Sialis* 90 per cent. of the waste nitrogen appears as ammonia. But the terrestrial pupa and adult of the same insect go over to uric acid excretion. If excretion is prevented by ligature, there is no accumulation of ammonia in the body; presumably it is stored in the form of amino groups, perhaps in glutamine.[216] Ammonia is the main nitrogenous component in the urine of larvae of *Aeschna* and *Phryganea*, adults of *Acilius* and *Dytiscus*, and adults of *Notonecta*. In *Notonecta* the bulk of the ionic material in the rectal fluid is ammonium bicarbonate. Water output is regulated so that the concentration of ammonia in the rectal fluid is limited to about 120 mM/litre, water being taken up as needed through the cuticle and the gut.[216] [See p. 582.]

Urea, the main nitrogenous waste product of mammals and aquatic vertebrates, is always present in the urine of insects, but in small quantities only; as a trace in the meconium of *Antheraea pernyi*;[20, 80] 0·4 per cent. (uric acid 28 per cent.) in dried excrement of *Tineola*,[57] 0·3–0·4 per cent. (uric acid 3·2–4·7 per cent.) in the dried excrement of *Melanoplus*;[13] in *Rhodnius* urea is plentiful in the watery urine soon after feeding, and present in traces later. Some urea is excreted by *Tenebrio* and *Oncopeltus*.[196] In the mosquitos *Aëdes*, *Anopheles*, and *Culex*, about 85 per cent. of the total nitrogen in the excreta can be accounted for; this is made up approximately as follows: uric acid N, 50 per cent.; urea N, 10 per cent.; ammonia N, 10 per cent.; amino N, 5 per cent.; protein N (probably glycoprotein, 10 per cent.[185]

Of the **amino acids,** leucine has often been claimed as a constituent of insect urine.[127] But the evidence is purely microscopical, soft yellowish spheres present in the lumen of the Malpighian tubes, and remaining after treatment with uric acid solvents, being judged to be crystals of leucine.[136] Chemical analyses have usually shown only a small amount of nitrogen in amino form: 3·1 per cent. (uric acid nitrogen 8·7 per cent.) in the dried meconium of *Antheraea*[20]; 0·35 per cent. (uric acid nitrogen 1·1–1·6 per cent.) in the dried excreta of *Melanoplus*, of which arginine formed a small part;[13] undetermined but small quantities in the urine of *Rhodnius*. Chromatography of the excreta of *Rhodnius* has revealed the presence of leucine, histidine, and histamine, giving particularly intense spots, and smaller amounts of taurine, glycine, valine, phylalanine, and alanine.[182] Histidine is conspicuous also in the excreta of *Bombyx mori*. In *Ephestia* the presence of tryptophane in the excreta is confined to the white-eyed mutant *a* in which kynurenine formation is deficient (p. 91).[176] [See p. 582.]

In *Tineola*, feeding on wool, 55 per cent. of the total water soluble sulphur in the excreta is cystine sulphur, and only 8 per cent. is sulphate sulphur. Cystine is the chief sulphur-containing excretory product in both *Tineola* and the carpet beetle *Attagenus*. When fed on the same diet, *Attagenus* excretes about twice as much cystine as *Tineola*; that is because *Tineola* contains an enzyme which decomposes cysteine with the production of hydrogen sulphide.[204] Clothes-moth excreta also contain 0·28 per cent. of elemental sulphur, probably produced by the reaction between sulphides and disulphide bonds during the process of digestion.[204]

Uric acid is by far the most important nitrogenous constituent in the urine of insects, as it is in birds and reptiles. Uric acid contains less hydrogen than any other nitrogenous compound excreted by animals, and is therefore well adapted for the conservation of water;[1] and being highly insoluble as the free acid or as the ammonium salt, it requires little water for its elimination. These advantages are manifest in free living insects in dry environments; but they are perhaps even more important during the periods of development in the egg or in the pupa, where the insect often has no means of replenishing its water supply.[101]

In the silkworm 85·8 per cent. of the nitrogen is excreted as uric acid;[30] the dried excreta of *Tineola* contain 28 per cent. of uric acid;[57] those of the fasting

TABLE 2

NITROGENOUS CONSTITUENTS OF INSECT EXCRETA IN GM. PER CENT.

	Tineola bisselliella (dried excreta)[57]	*Melanoplus bivittatus* (dried excreta)[13]	*Antheraea pernyi* (dried meconium)[20]
Uric acid	28	3·2–4·7	26·2
Ammonia	3·0	0·07–0·08	0·6
Amino-N	—	0·35	3·1
Urea	0·4	0·3–0·4	trace

mealworm *Tenebrio* over 50 per cent.;[153] the meconium of *Antheraea pernyi* contains 5 per cent. of dry matter, 26·2 per cent. of which is uric acid;[20, 80] the meconium of *Celerio* contains 5·5 per cent. of uric acid in suspension;[43] uric acid, as we have seen, predominates in the excreta of *Melanoplus* (Table 2);[13] in *Rhodnius*, from 64–84 per cent. of the dried urine is composed of uric acid; this insect, which weighs less than 100 mg., excreting from 0·5–0·6 mg. of uric acid daily during the first week or two after feeding.[152]

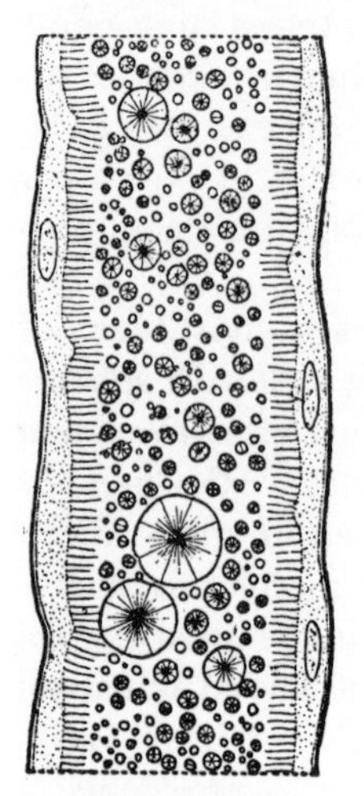

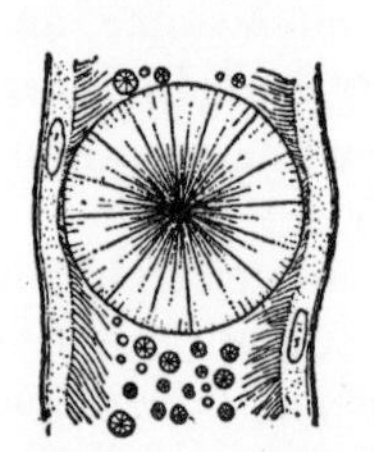

FIG. 322.—Large uric acid spheres in Malpighian tubes of *Rhodnius* starved for several weeks (*after* WIGGLESWORTH)

When the insect has plenty of water available, the uric acid is in solution (p. 553); when water becomes scarce, it separates out into crystalline spheres. These spherical granules, with a radial striation, ranging in diameter from a fraction of 1μ up to 60μ or more, with an average size of 3–4μ, represent the form in which uric acid and various urates crystallize out of impure solutions (Fig. 322). Their precise composition doubtless varies in different insects.[137] Where much ammonia is present they may contain ammonium urate, as in *Tinea pellionella*;[130] some may contain calcium urate, as in the meconium of *Osmia* (Hym.);[90] or urates of sodium and potassium. But it is certain that in many cases they consist of free uric acid: in *Rhodnius* 80–90 per cent. of the uric acid is free, the rest presumably as sodium and potassium acid urate.[152] In the meconium of *Celerio* about half the uric acid is free and half as acid potassium urate.[42] If these spheres are dissolved in very dilute alkali they leave behind a husk or stroma of organic material of unknown composition; if they are immersed in dilute acetic acid or in distilled water, rhombic crystals of pure uric acid separate out.[152]

Other nitrogenous constituents—Uric acid has a dual origin in metabolism: (*a*) as an end-product of purine metabolism; (*b*) as a product of the waste nitrogen derived from protein. Some of the uric acid (*a*) may be replaced by other, less fully oxidized purines (indeed, in some instances these other purines may represent the end product of protein nitrogen metabolism); and some of the uric acid (*b*) may suffer further breakdown, notably to allantoine. [See p. 582.]

Among the **other purines,** guanine, which is the main nitrogenous end-product in some other Arthropods, has not been found in insects. Xanthine and hypoxanthine are excreted by *Melophagus*, the xanthine being in about the same quantity as uric acid.[197] Xanthine and hypoxanthine are present at all stages in

Galleria; in the larva the ratio of uric acid to hypoxanthine is about 10 : 1.[196] Hypoxanthine appears in the excreta of a *Drosophila* mutant.[195] And in the pupa of *Pieris brassicae* xanthine synthesis has largely replaced uric acid synthesis—though less than 1 per cent. is excreted in the urine.[181]

Allantoine and its further breakdown products are derived from uric acid as follows:

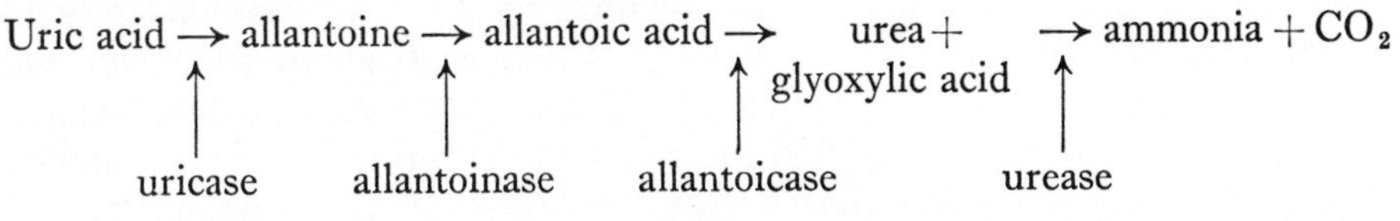

It may be that the excretion of ammonia was the primitive state and uricotelic excretion has been achieved by the successive loss of the uricolytic enzymes.[209] The Collembolan *Xenylla welchi* contains uricase, allantoinase, and allantoicase; it produces both allantoic acid and urea. The same is true of *Gryllus*. *Carausius* degrades at least part of its purines to allantoic acid. In *Bombyx mori* allantoinase is more active so that allantoic acid is more plentiful; in *Aglais urticae* larvae uric acid makes up 2·5 per cent. of the excreta, allantoic acid 2·3 per cent.; but in the adult the corresponding figures are 25·3 per cent. and 0·75 per cent. In many other Lepidoptera allantoic acid is absent from the excreta. Uricase is particularly active in Heteroptera which excrete a large part of their nitrogen as allantoine.[171, 209] The enzymes are located in the fat body and Malpighian tubules.

Allantoine is present in the excreta of blow-fly maggots;[121] uric acid accumulates in the tissues of these larvae, but they excrete only allantoine; the pupae on the other hand produce much uric acid and no allantoine; the adults excrete both.[14] Differences in uric acid content in different genotypes of *Drosophila melanogaster* are related with differences in uricase activity; uric acid is absent in adults of 'white' and 'brown' mutants, in which uricase is greatly increased.[192] In the carnivorous Dytiscids and Carabids, and in various Orthoptera, uricase and allantoinase are present and break down uric acid to allantoine and allantoic acid; this appears to be the end product; no urea or glyoxylic acid are produced.[122] *Tenebrio* larvae secrete some allantoine.[196] [See p. 583.]

Pteridines, which are important pigments in insects (p. 612) are synthesized in a manner closely analogous to uric acid (p. 612); in most insects they are present in small amounts only. In the dried faeces of *Oncopeltus* there are 0·2 per cent. of pteridines.[174] But in *Pieris brassicae* much larger quantities of the pteridines leucopterine and isoxanthopterine are produced; most of this, however, is stored in the wings, fat body, &c. (p. 613); less than 1 per cent. is excreted in the urine.[181]

Creatine and creatinine are absent in the excreta of *Melanoplus*[13] and the meconium of *Antheraea*;[80] but creatine appears in small quantities in the urine of *Rhodnius*.[152]

Leaf hoppers (Jassidae) produce solid excretory material in the form of minute dodecahedra, deposited within the cytoplasm of the Malpighian tubule and later set free into the lumen. These so-called 'brochosomes' appear to consist mainly of protein.[175, 215]

Lime and calcium oxalate—The urine of plant-feeding caterpillars is loaded with granules and crystals of many sorts.[126] Besides uric acid and urate granules, calcium carbonate and calcium oxalate are plentiful.[9] The same is true of Phasmids.[136] Amorphous granules of calcium carbonate are produced abundantly, also, by the larvae of many Diptera: they were observed by Lyonet in *Eristalis*;[72] they occur in Stratiomyids,[146] in *Ptychoptera*,[105] *Calliphora* and *Drosophila*,[26] in leaf-mining Agromyzids[140] (in *Acidia heraclei* lime occurs in the Malpighian tubes in the form of 'calcospherites', concentrically laminated granules 8–140μ in diameter (Fig. 323)[63]) and in larvae of *Cerambyx* (Col.).[93] In *Lucilia* the epithelium of the granule-containing region of the Malpighian tubes seems to be specially adapted to secrete into the lumen magnesium, calcium and other alkaline earth metals, as well as phosphates, bicarbonates and carbonates, all of which together with uric acid contribute to the formation of the granules. The granules appear to serve no purpose other than storage excretion.[164]

The significance of the excretion of oxalate is uncertain; possibly it is a method for eliminating oxalic acid present in the food or produced in metabolism. But the lime is probably to be regarded simply as a mechanism for getting rid of the excess of calcium in the food.[73, 105] If there is an excess of fixed base in the food, as is the case with vegetable matter, this excess is most readily neutralized in the urine by the production of alkali urates, or by the retention of carbon dioxide and the precipitation of calcium carbonate. It is noteworthy that although lime occurs in all the leaf-mining Agromyzids, it is absent from the Malpighian tubes of the parasitic Agromyzid *Cryptochaetum*.[140] But, on the other hand, it occurs in the larvae of *Auchmeromyia* (Dipt.) which feed solely on the blood of mammals in which there is only a small excess of base;[123] among Limnobiid larvae it is said to occur more frequently in the predaceous species;[67] and urinary calculi in the rectal caecum of the carnivorous *Dytiscus* have been found to be made up chiefly of calcium oxalate and carbonate.[111] Conversely, where there is an excess of acid-forming matter, as in food rich in sulphur, the urine is most readily neutralized with ammonia. Perhaps that is one factor in the high content of ammonia in the urine of the clothes-moth larvae feeding on keratin, which is rich in cystine.

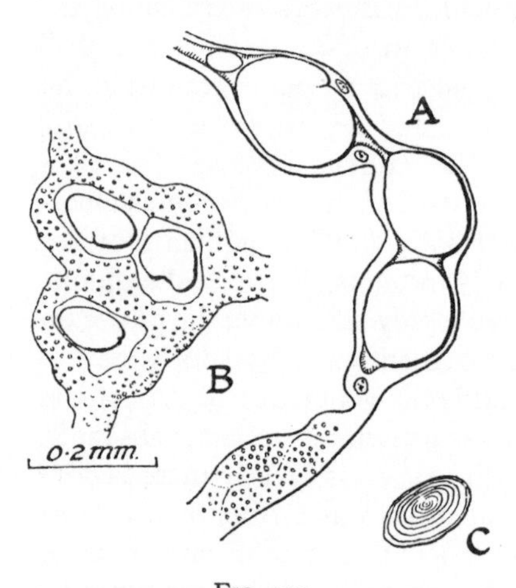

FIG. 323

A, portion of upper segment of Malpighian tubes of larva of *Acidia heraclei* showing calcospherites. *B*, fat body of *Agromyza* larva with calcospherites. *C*, detail of calcospherite showing concentric lamination (*after* KEILIN).

Other constituents of the urine—The urine doubtless contains very many complex waste products which are present in such small amounts that they escape detection. Pigments of unknown nature appear in the urine of many insects;[147, 152] a deep red pigment, which occurs in the meconium of Vanessid butterflies, is an ommochrome (p. 616). Biliverdin is excreted by the Malpighian tubes of *Rhodnius* after the injection of haemoglobin.[157] The glucoside salicin in the leaves of willows and poplars is generally decomposed in the blood of insects feeding upon these plants and excreted by the Malpighian tubes in the form of salicylic acid.[51] In most insects such as *Carausius*, *Ephestia*, *Bombyx mori*, the excreta contain only orthophosphate; but in *Galleria* the phosphorus is for the most part converted into metaphosphate, in part perhaps by micro-organisms in the gut.[162]

INTERMEDIARY NITROGEN METABOLISM

Very little is known of the organs concerned in the intermediary metabolism of the nitrogenous waste products, or of the chemistry of their formation. Extracts of blow-fly larvae contain a deaminase with peculiar properties, being water soluble, and capable of liberating ammonia from the larger breakdown products of proteins, such as peptones, although it is without action upon free amino-acids. The excreta of these larvae do not contain this enzyme, which is presumably intracellular, possibly in the wall of the gut.[15] An important enzyme in this complex system seems to be an adenosine deaminase which occurs not only in the gut but in skeletal muscle and integument; it may be concerned in some way with adenosine in muscle contraction.[81] A deaminase of more usual type, which will liberate ammonia from glycine, is present in the haemolymph of the silkworm.[133]

Since the purines adenine (6-aminopurine), hypoxanthine (6-oxypurine) and xanthine (2, 6-oxypurine) can all be demonstrated in the larva and pupa of *Antheraea pernyi*, a part of the uric acid (2, 6, 8-oxypurine) is believed to arise by oxidation of these products of nucleic acid metabolism.[80] Adenase and guanase (producing hypoxanthine and xanthine respectively from guanine and adenine) are present in *Hydrophilus* and *Dytiscus*,[32] and xanthine oxidase in all the insects studied.[32]

But the greater part of the uric acid must arise by synthesis. In *Periplaneta*, as in mammals, uric acid is synthesized from formate.[188] It may well be that uric acid in the insect is in a dynamic state and may act to some extent as a nitrogen store.[188] Thus, in *Popillia* during metamorphosis uric acid increases and then decreases, although the total nitrogen in the pupa remains constant.[187] Urea formation can be demonstrated *in vitro* in the integument, Malpighian tubules, blood, and mid-gut wall of *Bombyx mori* larvae.[183] [See p. 583.]

Allantoine is produced by the oxidation of uric acid, and the enzyme uricase, effecting this change, has been found very active in the adults of Muscid flies; though absent in *Blattella*, *Melolontha*, *Apis*, and Aphids.[145] Uricase is active also in the pupa of *Antheraea*, all the tissues of which except the Malpighian tubes contain a little allantoine.[80] In *Lucilia*, the pupae of which, as we have seen (p. 502), produce no allantoine, uricase disappears suddenly and entirely when the larvae leave the meat and become 'prepupae'; it reappears abruptly on emergence.[14] Allantoinase, converting allantoine into allantoic acid, has been demonstrated in a long series of adult beetles (Carabids, Dytiscids, &c.) and in *Schistocerca*.[122, 32] In *Popillia japonica* uricase is present in all the larval instars but absent in prepupa, pupa and adult; it disappears suddenly when the larva changes to the prepupa. Xanthine oxidase is present at all stages.[210]

MALPIGHIAN TUBES

The Malpighian tubes were so named by Meckel in 1829 after their discoverer Malpighi, who refers to them in his work on the silkworm (1669) as 'vasa varicosa'. Many of the earlier authors such as Cuvier (1802) regarded them as biliary organs. Their excretory function, which was suggested by Herold in

1815, and supported by the finding of uric acid in them by Brugnatelli in the same year, is now generally accepted.[9, 111, 127]

In number they vary from two in various Coccids, to 150 or more in some Orthoptera and in the honey-bee. Where they are numerous they often tend to be short, long when they are few: in *Periplaneta* with 60 tubules they have a total surface area of some 132 sq. mm. or 412 sq. mm. per gram of insect; in the large moth *Gastropacha*, with 6 tubules, they have a surface area of 209 sq. mm. or 500 sq. mm. per gram.[127] Where they are few in number they may be several times as long as the insect, but in spite of their great length and twisted course their arrangement is extraordinarily constant in a given species.[137]

The Malpighian tubes may open directly into the mid-gut, often in front of the pyloric sphincter, as in the beetles *Necrophorus* and *Gnaptor*;[38] they open in front of the posterior imaginal ring in Muscids;[107] in the Tettigoniid *Stenopelmatus* they discharge by way of six ureters, which clearly agree with the mid-gut in structure, and are separated from the hind-gut by a valve flap;[23] in Coccids they end far in front of the commencement of the hind-gut.[150] Or they may open equally clearly into the hind-gut, as in Cetoniinae (Col.),[10] and in most caterpillars, where the three tubes on each side unite and discharge into a little chitin-lined bladder.[9, 18]

These facts have led to much controversy as to whether the tubes are ectodermal or endodermal in origin. According to some recent authors, they are believed to arise from the neutral zone where the mid-gut and hind-gut meet;[143] or, more specifically, to arise by ingrowth from the undifferentiated or embryonic cells at the inner end of the proctodoeum, which are perhaps homologous with the lips of the blastopore;[46] their final position being determined by subsequent migration. In the Chrysomelid *Galerucella*, the two short anterior tubes open

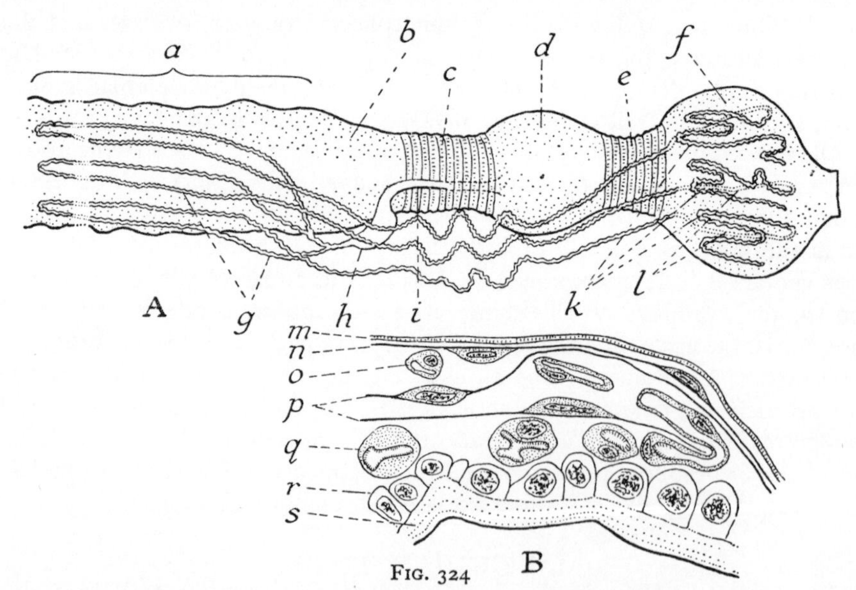

FIG. 324

A, hind part of gut and Malpighian tubes of one side in the larva of *Vanessa urticae*. *a*, mid-gut; *b*, ileum; *c*, anterior sphincter; *d*, colon; *e*, posterior sphincter; *f*, rectum; *g*, free part of Malpighian tubes; *h*, urinary bladder; *i*, free part of common duct; *k*, outer layer of tubes forming the rectal plexus; *l*, inner layer of tubes. *B*, section of rectum and plexus of tubes. *m*, muscular sheet; *n*, outer single membrane; *o*, outer layer of tubes; *p*, inner double membrane; *q*, inner layer of tubes; *r*, rectal epithelium; *s*, lining cuticle of rectum (*after* HENSON).

into the mid-gut, the four posterior tubes are connected by a common chitin-lined stem with the hind-gut.[47]

Anatomical relations of the Malpighian tubes—As a rule the Malpighian tubes lie freely in the body cavity. Occasionally one or more tubes, perhaps merely as an accident,[120] may enter the ventral ostia of the heart and, leaving by the lateral ostia, come to lie in the dorsal or pericardial sinus.[39] In some insects the terminal portions of the tubes, or some of them, are intimately associated with the wall of the rectum. This arrangement occurs in some Tenthredinid larvae[116] and is almost universal among the larvae of Lepidoptera, with the exception of the Hepialids. The typical arrangement in the silkworm and other caterpillars is as follows. The wall of the rectum is lined by a layer of epithelium covered by thin cuticle; outside this is an inner layer of Malpighian tubes; then, separated by a 'double membrane' made up of two layers of flattened cells, is an outer layer of tubes, the whole being enclosed in a single outer membrane and a muscular sheath (Fig. 324). The six tubes, the long lower segments of which lie free in the body cavity, enter at the front end of the rectum, run backwards between the single and double membranes, take a convoluted course forwards between the double membrane and the epithelium, and end blindly.[60]

A similar 'cryptonephridial' arrangement of the Malpighian tubes is common among beetles. It takes two main forms: in *Tenebrio*, Chrysomelids, &c., the tubes lie in a single convoluted layer disposed radially around the rectum; in *Anthrenus*,[98] *Gnaptor*,[38] *Dermestes*,[85] *Araecerus*,[85] &c., they are united in a tangled mass applied to one side of the hind-gut.[91, 139, 212] In Coccinellid beetles there are said to be minute openings through the intima of the hind-gut, leading into the perirectal space.[118] In the Mycetophilid larva *Ceraplatus* (Dipt.) the upper parts of the Malpighian tubes, bearing little diverticula, are bound by a membrane to the surface of the posterior half of the hind-gut.[138]

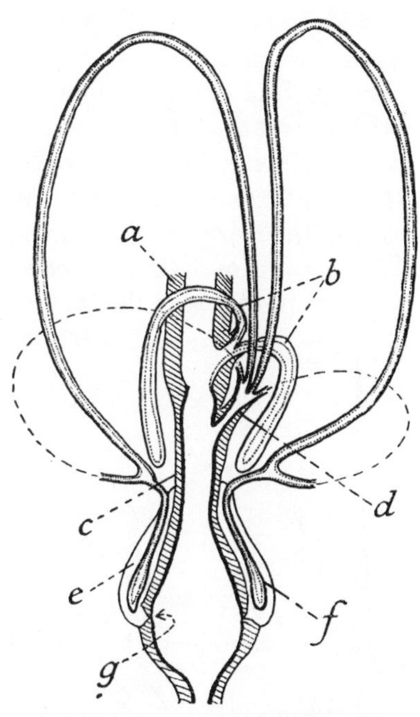

FIG. 325.—Diagram of Malpighian tubes of the Chrysomelid *Galerucella* (*after* HEYMONS and LÜHMANN)

a, mid-gut; *b*, two short Malpighian tubes opening into mid-gut; *c*, hind-gut; *d*, four long Malpighian tubes opening by common chitin-lined duct into hind-gut; *e*, perirectal chamber; *f*, one of the six terminal branches of the Malpighian tubes within the perirectal chamber; *g*, wall of rectum reduced to thin layer of cuticle at this point.

In Myrmeleonid larvae, also, there is a very complex relation between the Malpighian tubes and the intestine.[88] The blind upper parts of the tubes are applied to the hind-gut, and enclosed in a two-layered fold of this consisting of an inner layer of very thin epithelium and an outer of high columnar cells. Between the two is a cavity communicating with the hind-gut.[115]

A somewhat different arrangement exists in some Homoptera, where it is the lower end of the Malpighian tubes which comes into relation with the gut. In the Cercopidae, the lower end of each of the four tubes lies between the muscle wall and the epithelium of the 'filter chamber' (p. 492).[83] In the cicada (*Tibicen septendecim*) the so-called 'internal gland' is a mass of coiled gut and Malpighian tubes, the whole enclosed in muscle fibres.[48]

Sometimes there are obvious anatomical differences between different groups of tubes. In Chrysomelids (*Haltica*,[159] *Galerucella*[47]) the two short anterior tubes, which open into the mid-gut, do not come into intimate relation with the rectum; the four elongated posterior tubes, after a long course in the body cavity, unite on each side into a common stem which then splits up into three terminal branches lying in the chamber on the surface of the rectum. Thus the perirectal chamber contains six terminal segments (Fig. 325). In the Curculionid, *Apion*, there are four long tubes of usual type, and two short tubes forming globular or flask-shaped glands.[58] It is not uncommon for the Malpighian tubes to anastomose with one another to form closed loops, as in *Gnaptor* (Col.),[38] *Tipula*[4] and in Cecidomyids; or even to form crossings through which the lumen is continuous, as in some Homoptera.[83] Sometimes only a single pair of tubes form a closed loop, for example the posterior pair in *Drosophila*.[26]

Before discussing the physiological significance of these anatomical relations we must consider the histological structure of the tubules.

HISTOPHYSIOLOGY OF THE MALPIGHIAN TUBES

The Malpighian tubes are invested by a 'peritoneal coat' carrying abundant tracheoles and perhaps made up of anastomosing tracheal end cells (p. 359).[27] The tracheal branches bind the tubes to adjacent organs, such as the intestine, which carry them passively to and fro. Or they may be connected to other organs or to the body wall by occasional muscular strands, as in the Tachinid larva, *Thrixion*[104] and in *Ptychoptera*;[105] or a delicate muscle may run from the blind end to the alary muscles of the pericardial septum, as in the anterior tubes of *Drosophila*[26] and *Cryptochaetum* (Dipt.) larvae,[140] so that they are constantly agitated in the body fluids.

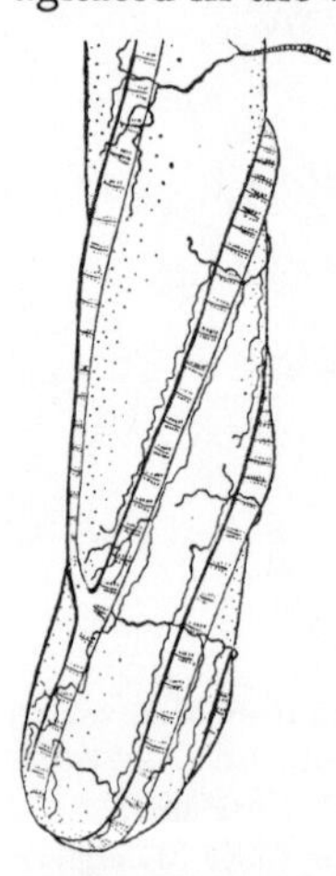

FIG. 326.—End of Malpighian tubule of adult bee, showing tracheoles and spiral muscle fibres (*after* MORISON)

Intrinsic muscles—In addition, the tubes of many insects show active twisting movements brought about by a muscular layer beneath the peritoneal coat. These muscles usually consist of a single layer of striated fibres forming wide spirals around the tube. They have been demonstrated, among other insects, in *Hydrophilus*,[78] *Galerucella*,[117] *Tenebrio*,[84] *Oryctes* larva,[75] Gryllids[78, 127] Phasmids,[136] *Periplaneta*, the honey-bee *Apis* (Fig. 326),[100, 144] the dragon-fly *Epitheca*,[17] and in the larvae of Lepidoptera.[9] In other insects such as Cercopids,[83] *Anthrenus* (Col.),[98] *Dytiscus*[124] and the flea (*Ctenocephalus*),[28, 41] the active contortions, twisting and cramp-like movements of the tubes strongly suggest that a muscular coat must exist. Indeed, muscles seem to be present in all orders of insects except Thysanura, Dermaptera, and Thysanoptera.[103]

A few muscle fibres, circular and longitudinal, often spread over the lower end of the Malpighian tubes where they enter the gut, so that the terminal duct of the tubes may show an active peristalsis, as in *Drosophila*,[26] Syrphid larvae[72] and *Rhodnius*.[152] Or the thin-walled urinary bladder which marks the lower extremity of the tubes in Gryllidae[8] and the larvae of Lepidoptera,[9] may be invested by delicate longitudinal and oblique muscles. Often this is the only musculature on the tubes: *Hyponomeuta* (Lep.),[59] *Rhodnius*,[152] and perhaps Lepidoptera, Diptera and Hemiptera in general.[103]

The function of these movements is probably to secure the maximum amount of contact between the tubes and the body fluid. They seem to be independent of the nervous system, but the movements are accelerated by extracts from the brain, which appears to produce a special secretion not present in other tissues;[65] and dilute aqueous extracts from the corpus cardiacum of *Periplaneta* accelerate the contractions of the Malpighian tubules of *Locusta*.[173] [See p. 583.]

Basement membrane and cells—Beneath the muscle coat is a tough elastic homogeneous membrane covering the entire tube. If the tubes of the mosquito are immersed in dilute sodium hydroxide (N/100) the cells swell up and dissolve and are carried down the lumen by an osmotic stream of water, leaving behind this insoluble basement membrane as an unbroken sheath.

The wall of the tube may be made up of a single cell or several; usually one cell will surround half or two-thirds of the lumen. The cells are anchored to the basement membrane by an apparatus which commonly gives an appearance of vertical or radial striation to the basal region of the cytoplasm.[9, 17] The cyto-

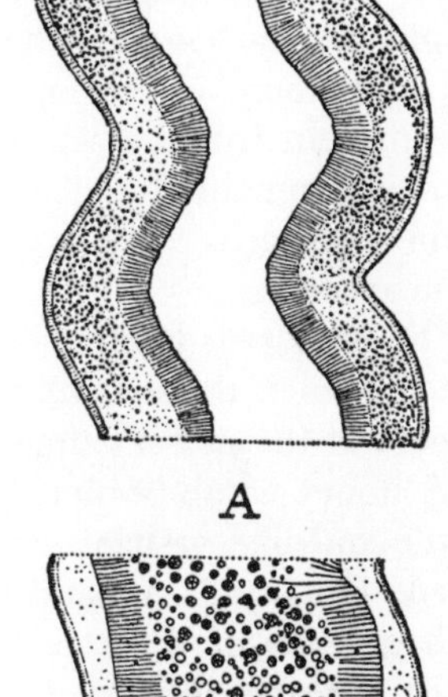

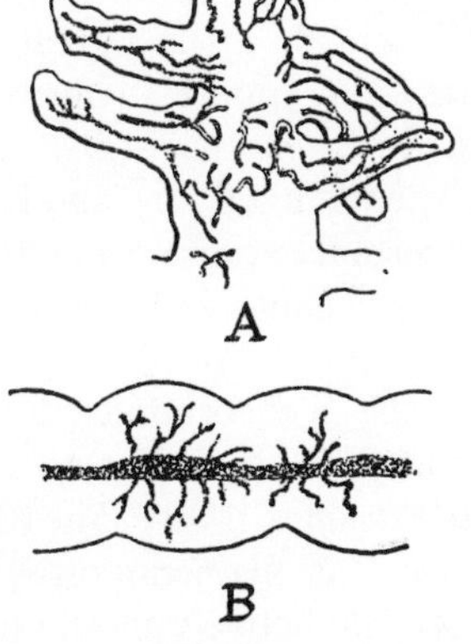

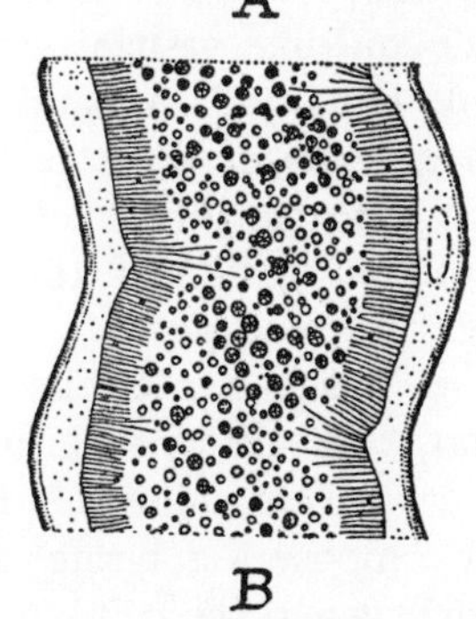

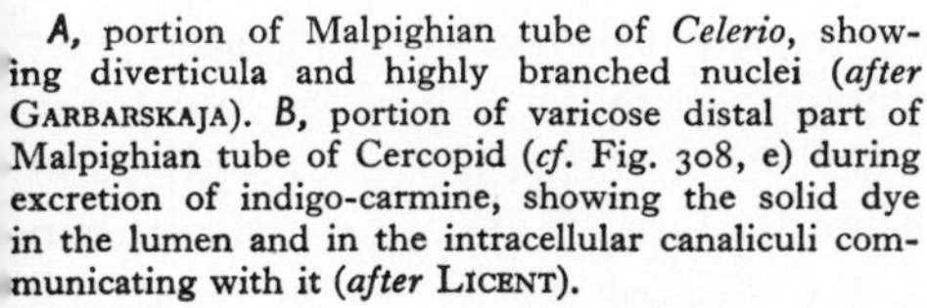

FIG. 327

A, portion of Malpighian tube of *Celerio*, showing diverticula and highly branched nuclei (*after* GARBARSKAJA). *B*, portion of varicose distal part of Malpighian tube of Cercopid (*cf.* Fig. 308, e) during excretion of indigo-carmine, showing the solid dye in the lumen and in the intracellular canaliculi communicating with it (*after* LICENT).

FIG. 328

A, example of Malpighian tube with honeycomb border; upper segment of tube in *Rhodnius* seen in optical section in fresh material. *B*, example of tube with brush border; lower segment of same with uric acid spheres in lumen (*after* WIGGLESWORTH).

plasm of the cells has most varied characters. It may be diffusely tinged with green or yellow; more often it is colourless; sometimes almost clear, but usually more or less filled with refractile or pigmented concretions or droplets, occasionally with needle-like crystals.[17] *Drosophila* and other insects store riboflavine as a yellow pigment in the cells of the Malpighian tubules; and wild type *Drosophila melanogaster*, and certain mutants, store another yellow pigment which may be kynurenine or hydroxykynurenine used for the production of certain of the eye pigments (p. 616).[222]

The cells are generally discrete with well marked boundaries; sometimes there may be small diverticula extending from the lumen into the cells, and these caeca may be so large in parts of the tube that this acquires a feathered appearance, as in *Melolontha*[27] or in Sphingidae (Fig. 327, A).[35] In Cercopidae, the distal varicose segment of the tubes is made up of large cells with intracellular canaliculi communicating with the lumen (Fig. 327, B.)[83]

As seen with the electron microscope the cytoplasm is exceedingly rich in mitochondria; these are particularly evident between the deep invaginations of the plasma membrane at the base of the cells, and in association with the striated border (p. 567). Many mitochondria are commonly breaking down to form amorphous deposits[215, 221, 222, 226] comparable with the 'lysosomes' described in the tissues of vertebrates. All intermediate stages can be seen between these abnormal mitochondria, laminated spheres, and mineralized spheres of calcium phosphate or carbonate with a concentric lamination.[170, 226] The cytoplasm contains also abundant endoplasmic reticulum covered with granules of ribonucleoprotein (ribosomes), and often distended to form vesicles. The tubules commonly contain glycogen along their length, and in *Locusta*[191] and *Gryllus*[170] certain of the cells towards the proximal region of the tubules contain quantities of mucoproteins.

The striated border—The inner surface of the cells is covered by a striated border, the structure of which has excited much controversy. It may be accepted at the present time that this border can be of two kinds: (1) the type called 'honey-comb border' (*Wabensaum*), which is made up of a great number of little rod-like vesicles so closely packed that they appear to form a rigid palisade (Fig. 328, A), and (2) the type called 'brush border' (*Bürstensaum*), which consists of separate filaments quite independent of one another (Fig. 328, B). The distribution of these two types is uncertain because, when the tubes are fixed and sectioned, they are no longer distinguishable with certainty with the light microscope.[152]

The 'brush border' has been described in the Tenebrionids *Blaps* and *Scaurus*,[79] in the mosquito *Culex*,[77] in *Chironomus*[45] and *Drosophila*,[26] in Gryllidae,[8] Tettigoniidae,[23] Lepidoptera[9] and in *Rhodnius*.[152, 160] The filaments which compose this border are of variable length; long filaments often overlie the nucleus and the cell boundaries;[79] and though not actively motile like true cilia, the longer filaments can be seen waving passively to and fro in the lumen. The reality of the distinction between the two types of border can be seen most convincingly in *Rhodnius*, in which the upper 2·8 cm. of each of the four tubes has a border of the honey-comb type; the border then changes abruptly, and in the lower 1·5 cm. it is of the brush border type.[152] The same distinction probably exists in caterpillars, between the striated border of the perirectal tubes and that of the lower free segments.[59]

As seen with the electron microscope the individual filaments or 'microvilli' of the honey-comb type of border are separated by no more than 150–200 Å (the same degree of separation as exists between the plasma membranes of contiguous cells, or between the two membranes of the nucleus). The filaments of the brush

order type are separated by intervals equal to or greater than their own dia-
neter (Fig. 329).[226] The filaments may be as much as 1μ thick in *Melanoplus*,[169]

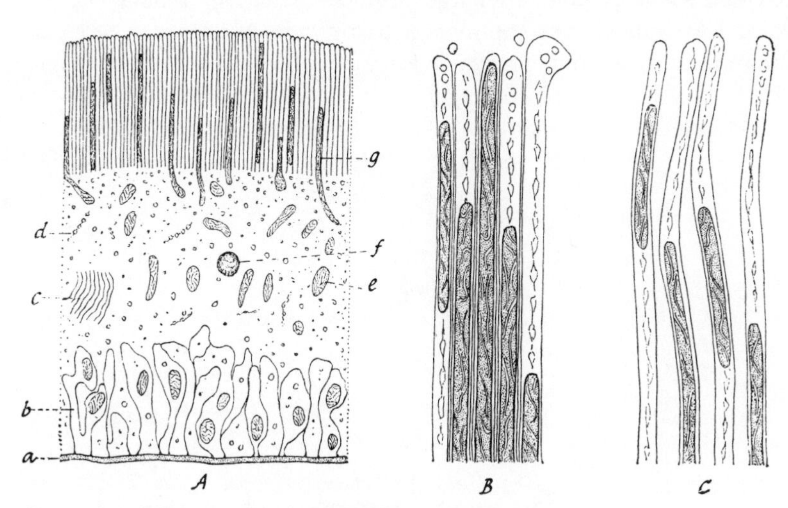

FIG. 329.—Fine structure of the Malpighian tubules in *Rhodnius* (*after* WIGGLESWORTH and SALPETER)

A, from low magnification electron micrograph of the upper segment. *a*, basement membrane: *b*, invaginated plasma membrane with mitochondria between: *c*, laminated endoplasmic reticulum: *d*, vesiculated endoplasmic reticulum: *e*, mitochondria: *f*, mineralized granule: *g*, striated border with mitochondria entering the microvilli. *B*, detail of microvilli of upper segment at higher magnification, showing mitochondria, chains of lozenge shaped vesicles, and discharge of spherical droplets at the apex. *C*, detail of the more widely separated microvilli of the lower segment.

·25μ in *Rhodnius*,[226] $0{\cdot}15\mu$ in the leaf-hopper *Macrosteles*.[215] It was early sug-
ested that the filaments of the striated border might be derived from mito-
hondria;[134] and in *Bombyx* mitochondria from the cytoplasm can be seen ex-
ending far into the lumen of the filaments.[198] This has been confirmed with the
lectron microscope: in *Melanoplus*[169] and other grasshoppers,[221] in *Coccinella*,[167]
hodnius,[226] and other insects, the striated
order appears in places like a palisade of
nitochondria up to 4μ in length. In addition
he filaments contain chains of vesicles con-
inuous with the endoplasmic reticulum in
he body of the cell (Fig. 329).[168, 222, 226]

Visible changes during secretion—
There can be little doubt that the great
najority of the changes described as occur-
ing during secretion in the Malpighian
ubes are artefacts. The cells of the tubes are
xceedingly sensitive to the intense diffusion
urrents set up by fixatives. Such changes as
he distension of the elements of the striated
order to form little vesicles,[95, 134, 144] the
filling of the cells with vacuoles,[38] the rupture
or disappearance of the striated border[95] with the discharge of a foam-like mass
of globules into the lumen[35] are certainly due, at least in most cases, to imperfect

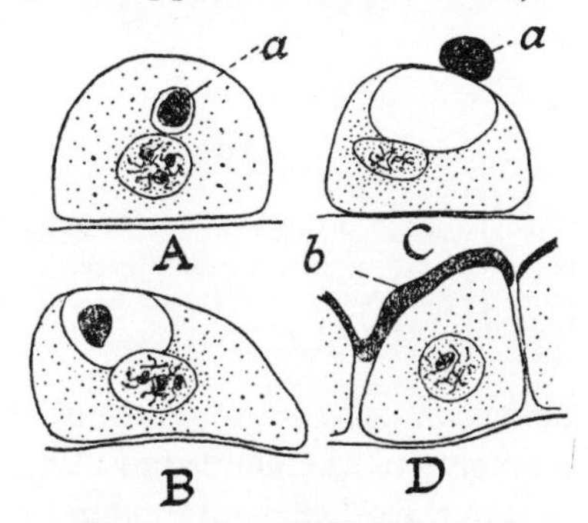

FIG. 330.—Cycle of secretory changes in Malpighian tubules of *Vespa* (*after* GREEN)

a, granule of secretory material being discharged from vacuole; *b*, accumulated material on the surface of the cells.

fixation.[152] The same is true of many of the observations made on isolated tubes in 'physiological solutions'; no artificial fluid is inert towards the Malpighian tubes, which swell up and discharge droplets when so treated.[147, 148] If the tubules of *Rhodnius*, during the most active phases of excretion, are examined, with the minimum of injury, in the insect's own blood, they show only a general

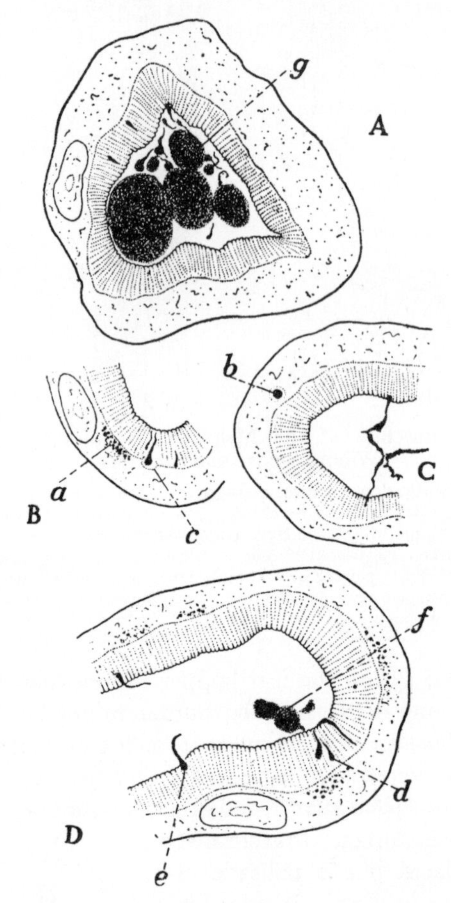

FIG. 331.—Excretion of biliverdin by Malpighian tubes of *Rhodnius* (*after* WIGGLESWORTH)

a, biliverdin granules accumulating in the cell; *b*, droplet of biliverdin in small vacuole; *c*, discharge of biliverdin into the striated border; *d*, droplet in form of filament passing through the border; *e*, discharge of filament to the lumen; *f*, fusion of filaments to form droplets; *g*, masses of large droplets and filaments filling the lumen.

distension of the lumen and, in the region of the brush border, the passive moments of the elongated filaments mentioned above.[152]

On the other hand there may be more than one mechanism of excretion.[27] In *Vespa* a secretory granule staining darkly with haematoxylin appears internal to the nucleus, enlarges as it is moved inwards, and is discharged into the lumen to fuse with other granules and so to form a thick layer over the cell surfaces (Fig. 330).[40] A similar process is described in *Apion* (Col.)[58] and *Dytiscus*;[124] and in some other insects portions of the cells containing solid deposits of excretory substances are said to be pinched off and set free in the lumen.[17] After the injection of haemoglobin into the body cavity of *Rhodnius* it is possible to follow all the stages in the excretion of biliverdin by the Malpighian tubes. The pigment accumulates in small vacuoles and is then discharged through the striated border in the form of tapering filaments which fuse together in the lumen (Fig. 331).[157] Excretory bodies of a remarkable form, termed 'brochosomes', containing protein and lipid components, are produced in quantity within certain cells of the Malpighian tubules of Jassidae and liberated into the lumen.[215]

The study of secretory changes with the electron microscope are still at an early stage; but there is some evidence that vesicles can arise by invagination at the base of the cell, and migration to the lumen.[222] Minute droplets, perhaps derived from such vesicles, are liberated from the tips of the microvilli,[226] and absorptive pinocytosis of vesicles may take place between the bases of adjacent microvilli.[221] [See p. 583.]

Histological differentiation in Malpighian tubules and the formation of urate granules—In most insects the granules present in the cytoplasm of the Malpighian tubes are not the same as the excretory granules in the lumen.[125, 147] The refractile spheres in the cells in mosquitos, which appear superficially very like uric acid, are not blackened by ammoniacal silver nitrate like the granules of uric acid in the lumen.[158] ^{32}P administered to *Culex* larvae accumulates in large amounts in the cells of the Malpighian tubules, being concentrated in the 1–3μ granules.[217] The secretion of the cells is in fact generally fluid; the concretions and crystals separate out later. Precipitation is probably induced by a reabsorption of water by the cells.[17] We have seen that this certainly happens in the hind-gut and rectum; but in many insects, such as Muscid flies during pupal life, and the adults of mosquitos, the tubes themselves contain heavy deposits of uric acid throughout their length. In such insects it is not known whether the cells pass through alternate phases of secretion and reabsorption, or whether different cells perform these two functions constantly. But in some insects a dimorphism has been observed in the cells of the tubes; in the adult *Ptychoptera*, among the pigmented cells there are scattered thin cells devoid of pigment and without a striated border (Fig. 332);[105] similar interstitial cells devoid of granules occur in *Anopheles*,[97] in the larva of *Galleria*,[95] and in the beetle *Dromius*.[127] These cells may possibly be concerned in reabsorption. [See p. 583.]

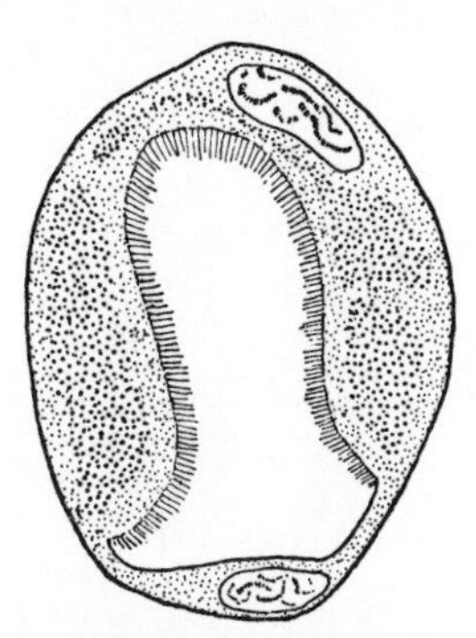

FIG. 332.—Transverse section through Malpighian tubule of *Ptychoptera* larva showing the two types of cell of which it is composed (*after* PANTEL)

A more definite division of labour is seen in *Rhodnius*, in which the granules of uric acid occur only in the lower third of each tube where the cells bear a brush border; the upper two thirds of each tube with the honey-comb border containing only clear fluid (Fig. 333). In *Rhodnius* there seems little doubt that the upper segment is concerned only with secretion, the lower segment with reabsorption. For if the very diffusible dye neutral red is added to the blood of the insect, it can be seen quite clearly to pass through the cells into the lumen in the upper segment, and to be absorbed from the lumen into the cells in the lower segment. Moreover, if two ligatures are applied to the lower segment of the tube some hours after feeding (at a time when all uric acid granules have been swept away by the rapid flow of urine) uric acid will separate out above the upper ligature but between the ligatures no uric acid appears, nor is there any distension of this part of the tube, as would happen if secretion were taking place.

In *Rhodnius*, and this may apply to other insects also, there seems to be another mechanism which favours the precipitation of uric acid. The contents of the upper segment of each tube are faintly alkaline (*p*H 7·2), in the lower segment they are acid (*p*H 6·6). This suggests that the uric acid is secreted in the form of the relatively soluble potassium or sodium acid urate in the upper

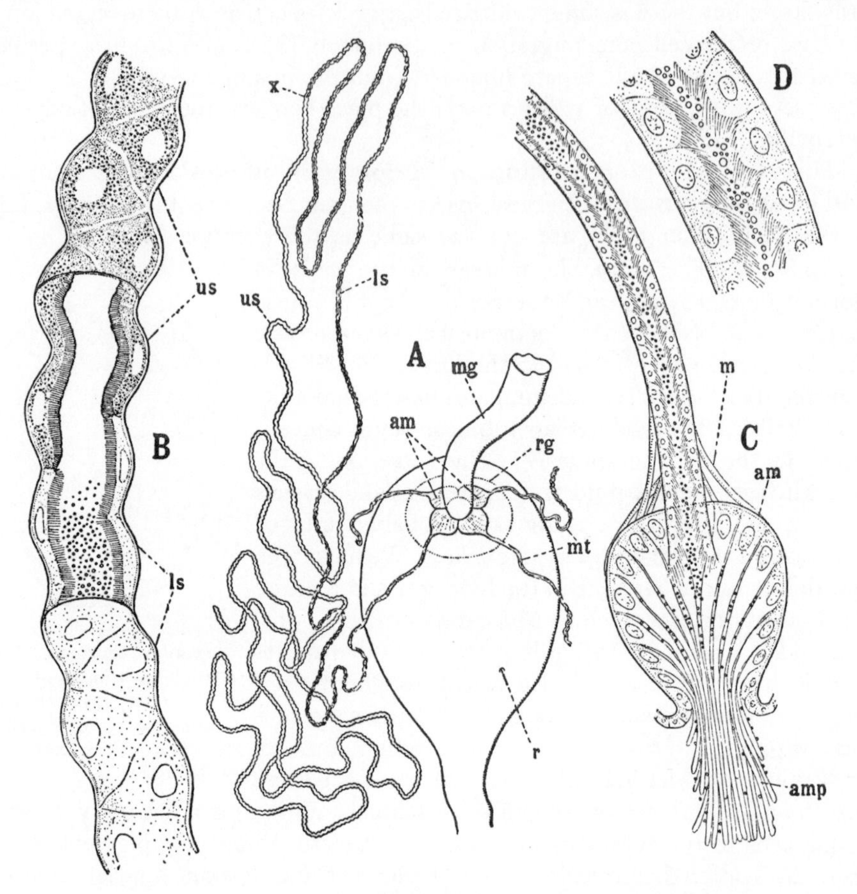

FIG. 333

A, excretory system of *Rhodnius*; only one Malpighian tube shown in full; *B*, junction of upper and lower segments of Malpighian tube, seen in surface view and optical section; *C*, lower end of Malpighian tube and terminal ampulla seen in optical section; *D*, detail of this region of tube (*after* WIGGLESWORTH). *am*, ampulla; *amp*, processes of ampulla cells; *ls*, lower (proximal) segment of Malpighian tube; *m*, muscle fibres; *mg*, mid-gut; *mt*, Malpighian tubes; *r*, rectum; *rg*, rectal gland; *us*, upper (distal) segment of Malpighian tube; *x*, junction of upper and lower segments of tube.

segment, and that the base is reabsorbed along with the water in the lower segment of the tube, leading to the precipitation of free uric acid (p. 558). Thus the same water and base are circulated and used repeatedly in excretion (Fig. 334).[152] Direct estimations of sodium and potassium in the blood and urine of *Rhodnius* show that there is a selective excretion of potassium by the upper segment of the Malpighian tubule and a reabsorption of potassium in the lower segment. Water and sodium, but not potassium, are probably reabsorbed in the rectum during the later stages of excretion.[163]

During this process of reabsorption in the lower segment, the filaments of the brush border may become so elongated as to fill the entire lumen, and hold the uratic spheres among them as in a gelatinous mass. Later, the filaments retract, and the spheres are carried down the free path in the lumen (Fig. 335).

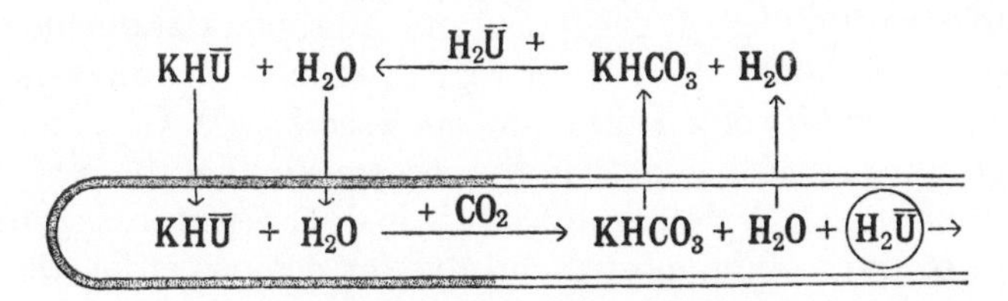

FIG. 334.—Diagram showing possible mechanism by which free uric acid is precipitated in the lower segment of the Malpighian tubes in *Rhodnius* (*after* WIGGLESWORTH)

Shaded part represents upper segment of tube.

Further reabsorption takes place in the peculiar ampullae at the lower ends of the tubes, and in the rectum.[152]

Specialization of parts of the Malpighian system is common in other insects.[125, 126] In Gryllidae some tubes are yellow, others white and loaded with uratic concretions;[66] and in *Gryllotalpa* the dye indigo-carmine is excreted only

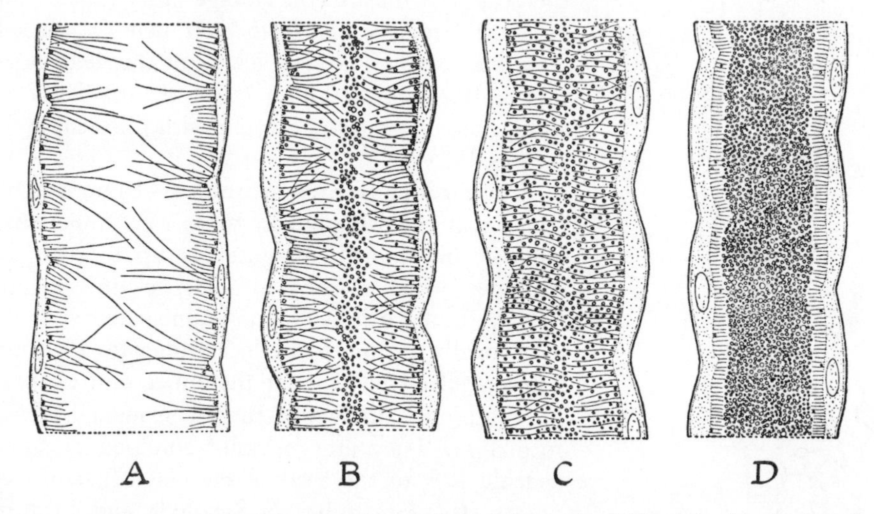

FIG. 335.—Optical sections of lower segment of Malpighian tubes in *Rhodnius* showing variations in the state of the filaments (*after* WIGGLESWORTH)

A, 4 hours after feeding; lumen is distended, many of the filaments are very long, uric acid granules beginning to appear around bases of filaments. *B*, 12 hours after feeding; all filaments elongated, some uric acid granules between filaments, others free in a central channel. *C*, 24 hours after feeding; all uric acid granules entangled among elongated filaments. *D*, 3 days after feeding; filaments contracted, almost all uric acid spheres free in lumen, a few between the filaments.

by the yellow tubes.[22, 68] In nymphs of *Heptagenia* (Ephem.) there are more than 100 tubes, each made up of three very distinct segments.[92] In these Ephemeroptera the Malpighian tubes are highly branched and complex. Each tubule proper, in both larva and adult, consists of an upper clear zone and a lower zone with quantities of excretory material in the lumen.[46] In *Blatta* the total length of the tubule is 16–20 mm. There is a slender clear upper segment of 2–3 mm., and a long lower segment packed with solid material.[46] A similar division occurs in *Forficula*.[46] In the larva of *Dytiscus*, the proximal parts of the four tubes differ histologically from the distal.[3] In the imago of *Borkhausenia pseudospretella* (Lep.), as in *Rhodnius*, uratic spheres occur only in the lower segments of the tubes.[153] [See p. 583.]

The Malpighian tubules of Zygoptera and Anisoptera are made up of five distinct segments. A rather short lower region, termed the 'mucous segment', contains rounded deposits of a neutral mucopolysaccharide. Glycogen occurs in all the segments but is most evident in the receptaculum at the base, which receives several tubules.[178] A similar differentiation is present in *Locusta*.[191]

Excretion of dyes—The most useful dye for demonstrating the excretory function of the Malpighian tubes is indigo-carmine. In *Periplaneta* this is excreted by certain parts only of the Malpighian tubes; and the same regions excrete fluorescein.[36] In *Rhodnius* indigo-carmine is excreted only by the upper segment; neutral red enters the lumen by the same route, but is then taken up by the lower segment, deposited and stored in the cells.[152] Many other dyes are transmitted by the cells to the lumen. In the course of this process they are sometimes retained in part by the cells. But most of the storage of dyes, like that of neutral red, results from their being absorbed from the lumen. The significance of this reabsorption is not known. Often there are gradients of storage along the tube, varying with different substances.[86]

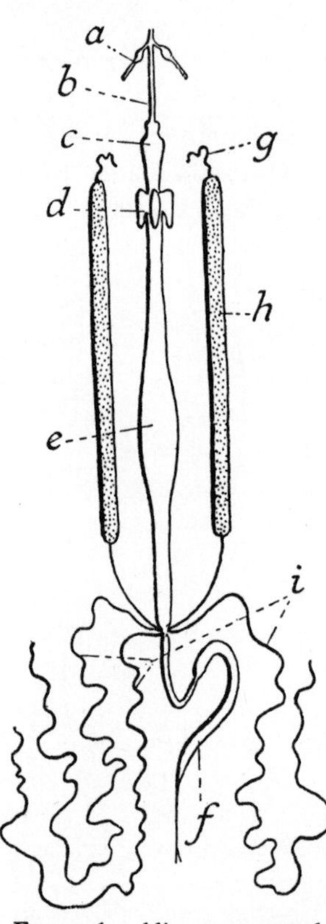

FIG. 336.—Alimentary canal and Malpighian tubes of larva of *Ptychoptera*; schematic (*from* PANTEL *after* VAN GEHUCHTEN)

a, salivary glands; b, fore-gut; c, proventriculus; d, caeca of mid-gut; e, mid-gut; f, hind-gut; g, extremity of anterior Malpighian tubules, of normal appearance; h, middle segment of anterior tubules transformed into a lime sac; i, three posterior tubules of normal structure.

Formation of lime granules—Lime usually appears in the Malpighian tubes as a milky suspension of amorphous granules up to 3·5μ in diameter, which dissolve in acids, leaving behind only a minute particle of some organic material;[63, 105] occasionally the lime forms 'calcospherites' (p. 560). The granules are formed in the lumen of the tubes; but nothing is known of the mechanism of their formation. They are generally confined to some specific part of the system: the distal half of one pair of the four tubes in Syrphids and Stratiomyids;[72, 146] in the distal halves of the anterior pair of tubes in the larvae of *Drosophila* and *Calliphora* and other Diptera;[26] in the dilated tips of the corresponding tubes in the parasitic *Thrixion*;[104] in the dilated middle region of the anterior pair of the five tubes in *Ptychoptera* larvae (Fig. 336);[105] in certain restricted regions of the posterior group of tubes of many female Phasmids;[136] and in the four long tubes of the larva of *Cerambyx* (Col.), not in the two short tubes.[93] The segments where they occur are usually greatly distended, the cells attenuated, with no vestige of a striated border (Fig. 323, A).[10]

Ionic regulation and reabsorption in the hind-gut—In considering digestion (p. 495) and the excretion of water (p. 554) we have already described the part played by the epithelium of the hind-gut, and notably the rectal glands, in the conservation of water. These same cells also serve to recover mineral constituents from the excreta (p. 496). Nymphs of *Libellula* and *Aeschna* take up chloride from fresh water (see p. 515); probably the plaques of argentophil cells at the base of the tracheal gills are responsible.[71] In *Chironomus* and in *Limnephilus* (Trich.) there is certainly an active reabsorption of chloride in the rectum. *Limnephilus* in a medium containing 0·01 per cent. NaCl, had chloride in the blood equivalent to 0·042 per cent. NaCl, in the fluid leaving the Malpighian tubes 0·044 per cent., and in the urine discharged from the rectum 0·005 per cent.[6] A similar reabsorption of salts occurs in the rectum of *Tenebrio*,[106] *Schistocerca*, *Calliphora*, &c. (p. 496). Some of these matters have also been considered in relation to osmoregulation (p. 430) but certain of them will be recapitulated here because they represent important functions of the excretory system.

Besides obtaining chloride from fresh water (p. 514), mosquito larvae actively absorb sodium and potassium by means of the anal papillae.[205, 220] These exchanges take place more actively in well fed larvae.[218] Some indication of the efficiency of the mechanism in larvae of *Aëdes aegypti* is given by the following figures: [205, 218, 224]

Na in haemolymph:	100	mM/l;	in medium		2 mM/l	
K ,, ,,	4·2	,,	,,	,,	0·659 mM/l	
Cl ,, ,,	51·3	,,	,,	,,	3·9	,,

These ions are excreted by the Malpighian tubules and reabsorbed in the rectum. In this way there is a continual circulation of potassium, even when there is an excess in the body, just as there is a continual circulation of potassium in *Rhodnius* (p. 570).[205] Indeed, in all insects examined, the concentration of potassium is higher in the Malpighian tubule than in the haemolymph, whereas the concentration of sodium is always lower. Active transport of potassium seems to be a constant feature in the Malpighian tubules.[206] In *Carausius* it seems to exert an activating effect on the Malpighian tubule and is probably fundamental to urine production in all insects. It has even been suggested that potassium transport is often the primary factor in urine secretion; water and other constituents are then able to follow.[207] It may be that amino acids, sugars, urea, enter the Malpighian tubules passively by diffusion; and sodium, potassium, water, and required metabolites are reabsorbed in the rectum.[203, 207] [See p. 583.]

In the larva of *Sialis*, also, the same system, of potassium secretion by the Malpighian tubules, and subsequent reabsorption in the rectum, seems to operate even though a concentrated excretory fluid is being produced.[214] But *Sialis* seems unable to concentrate chloride from the haemolymph, whereas the caddis larva *Limnephilus affinis*, which can tolerate highly saline waters can concentrate chloride in the rectum until it exceeds that in the haemolymph by a factor of three.[219]

The significance of the cryptonephridial tubes—We have seen that in the larvae of Lepidoptera, in many Coleoptera and in the antlions (Myrmeleonidae), the Malpighian tubes come into intimate relation with the hind-gut (p. 562). In caterpillars, these upper segments of the tubes are so completely cut off from the body cavity that they cannot take up the general waste products; dyes injected into the body cavity are taken up only by the free segments of the tubes.[95] We have seen that the tubes surround that segment of the hind-gut in which water is extracted. It has been suggested that toxic substances of unknown nature may be absorbed at the same time, and that the function of these parts of the Malpighian tubes is to remove all harmful substances before the fluid passes into the body cavity.[95] But it is possible that the water-absorbing power of the gut wall is increased by having another water absorbing mechanism outside it; and that the perirectal tubes are simply part of the mechanism for recovering water from the excreta; this water being returned to the body cavity by the lower segments of the tubes, or used to flush out the solid matter from the lumen.[116] The tubes of the convoluted inner layer have a much thicker epithelium than in the outer layer; perhaps the function of the two layers is different.[60]

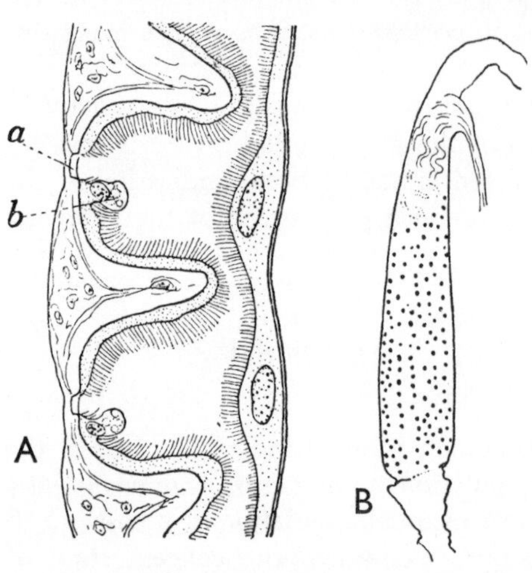

FIG. 337.—*A*, section through the cryptonephridial tubes in the larva of *Tenebrio*

a, 'leptophragma'; *b*, modified cell belonging to the leptophragma. *B*, surface view of the intact rectum of *Tenebrio* larva after treatment with silver nitrate; the leptophragmata blackened (*after* LISON)

In beetles, the cryptonephridial arrangement occurs in phytophagous, carnivorous and omnivorous species.[91] Fluid is certainly taken up from the gut contents and conveyed to the granular blood-filled space in which the tubes lie. Sometimes (in *Tenebrio*, &c.[19]) the gut is here covered with large absorbing cells; sometimes its contents are separated from the Malpighian tubes only by a thin-walled window (as in *Anthrenus*[98]); or there may be several points at which the wall is reduced to a thin membrane (as in Chrysomelids[47, 159]).

The chamber in which the tubes lie is bounded externally by a 'perinephric membrane', and at the points where the convoluted underlying tubes come in contact with this, the wall of the tube is attenuated so as to form a series of little windows closed only by a very thin membrane (in *Adalia*, *Agelastica*, *Gastroidea*, *Melasoma*) or occupied by a cell with large vesicular nucleus, which is clearly differentiated from the other cells that make up the wall of the tube (in *Blaps*, *Bruchus*, *Tenebrio*) (Fig. 337).[19, 114] These little windows in the convoluted cryptonephridial tubes have been termed 'leptophragmata'. They have essentially the same structure in both the symmetrical and lateral arrangements; in *Tenebrio*, and *Dermestes* or *Araecerus*,[85] but they are absent in *Melolontha*.[85]

These segments of the tubes almost certainly take up water reabsorbed from the rectum. They have been regarded as helping in the conservation of water (as in Scolytids,[131] Tenebrionids,[113] Chrysomelids[47]). This idea is supported by the fact that among Lepidopterous larvae (apart from the primitive Hepialidae) the cryptonephridial arrangement is lacking only in the aquatic Pyralids *Cataclysta* and *Nymphula*.[116] It is similar in type in *Lymantria* and

Vanessa which feed on leaves; much more highly developed in *Borkhausenia*, feeding on very dry materials.[116] It is present throughout the Chrysomelids, save in the aquatic genus *Donacia*;[113] but it does occur in vestigial form in some other families of beetles with aquatic habits.[49, 212] The concealed tubes have also been supposed to purify the absorbed fluid before it is returned to the body cavity.[38, 47, 113, 159]

Experimental evidence in support of the belief that the rectal wall and the Malpighian tubules provide two steps in the recovery of water from the rectum has been obtained by measuring the relative concentrations in the different parts of the system. In *Tenebrio* the faecal pellets are dried to the point where they are in equilibrium with an atmosphere of 90 per cent. R.H. or even 75 per cent. R.H. (equivalent to a supersaturated solution of sodium chloride).[208] The haemolymph has a freezing point depression of $\Delta=0{\cdot}75°$ C. increasing to $\Delta=1{\cdot}5°$ C. under dry conditions. The fluid in the perinephric space has an intermediate concentration: $\Delta=8°$ C. at the posterior end of the rectal complex. This high concentration is mainly due to some non-electrolyte. The concentration of fluid within the perirectal tubules is very close to that of the fluid in the perinephric space, but it is due almost entirely to potassium chloride, which may reach a concentration exceeding 2 M. It appears to be this active concentration of potassium chloride in the tubules which creates the concentration gradient that aids water absorption by the rectal epithelium.[208]

The role of the leptophragmata in this process is obscure. Doubt exists as to whether fluid or other substances pass through the thin walled pores from the lumen of the tubes to the body cavity, or in the reverse direction.[19] Most observers have found that indigo-carmine and other dyes, injected into the body cavity, are taken up only by the free portions of the tubes;[38, 47] but in *Tenebrio* they are said sometimes to enter the perirectal segments.[114] Although leptophragmata are present in most beetles with crytonephridial tubules, they are absent in some species and they are absent in the Lepidoptera.[212] It is possible that they merely serve some mechanical function, such as anchoring the tubules to the perinephric sheath.[212] [See p. 584.]

In Myrmeleonid larvae, also, the arrangement is regarded as being concerned with reabsorption from the hind-gut. The curious fold of the hind-gut which invests the upper parts of the Malpighian tubes is perhaps homologous with the rectal glands of the adult, which almost certainly absorb water. Fluids absorbed by the hind-gut enter the perinephric space; part is then believed to be secreted into the lumen of the Malpighian tubes, part to be taken up by the high columnar cells of the investing fold and returned to the haemocele.[115] Injected dyes may reach the upper segments of the tubes by this route.

Another possibility to be considered, in view of the silver staining of the leptophragmata, is that the arrangement is concerned in some way with the recovery of chlorides and perhaps other minerals from the excreta.

ACCESSORY FUNCTIONS OF MALPIGHIAN TUBES

The Malpighian tubes are sometimes regarded as regions of the intestine in which the primitive excretory functions of the gut have become centred.[147] We have discussed elsewhere the extent to which they contribute to digestion (p. 505); here we shall consider new functions they may occasionally acquire.

A sticky secretion, provided apparently by modified cells in the Malpighian tubes, is poured out upon the terminal pseudopod ('Nachschieber') of certain Chrysomelids, notably *Agelastica alni*, and serves to aid progression.[11] In other Chrysomelids the Malpighian tubes furnish the sticky substance with which the female covers her egg chambers.[47] The viscous principle, possibly related to silk, which makes the foam sheltering Cercopid nymphs permanent and resistant, is believed to come from the Malpighian tubes, being mixed with fluid from the gut and whipped into a foam by the repeated closure and separation of the terminal sclerites. This protein in *Philaenus* and *Aphrophora*[227] appears to be a mucoprotein or glycoprotein.[202] The middle segment of the tubes in these nymphs, which is dilated and has branched nuclei, is thought to be the source of this secretion.[83]

Undoubted silk is produced by the Malpighian tubes in several groups. In the larva of *Chrysopa* (Neuropt.), after the 2nd instar, a striking difference appears between the fore-part and hind-part of the tubes. The hind-part is of normal character; in the fore-part the cells become greatly thickened, increasingly so towards the time of spinning the cocoon; the nuclei, round before, become enormously branched and the lumen filled with viscid secretion. There is a sharp transition between these two zones. In the imago, which no longer produces silk, the tubes have the same character throughout their length as the hind-part of the larval tubes.[89] In Myrmeleonid larvae, also, the lower parts of the tubes are concerned in silk secretion, and show similar changes;[88] and the rectal sac serves as a reservoir for silk at the time of pupation.[115] In the larvae of *Phytonomus* (Col.), the Malpighian tubes are massively developed, and are responsible for producing both the silk tissue of the cocoon (from the middle segment of the tubes) and the coating materials which are smeared over its lining (from the larger apical region).[76] In the larva of *Donus crinitus* (Col.) there is some enlargement of the nuclei in the distal segment of the tubule which secretes the silk; the proximal segment secretes a mucoid material.[193] The silk-secreting cells, like those in *Sisyra* (Neuropt.),[172] are rich in acid phosphatase. Other beetles in which the Malpighian tubes secrete silk are *Lebia scapularis*[135] and *Niptus hololeucus*.[91]

In the cocoons of Lepidoptera the silken tissue is produced by the labial glands, but the smeared coating in *Leucoma salicis*, *Eriogaster lanestris*, &c., is provided by the excreta;[24] the lemon yellow powder used for this purpose by *Malacosoma* consists of urates from the Malpighian tubes.[37] [See p. 584.]

Finally, the lime collected in the tubes is often utilized. The Australian Cercopid *Ptyelus*, living on Eucalyptus trees, forms a spiral shell consisting of at least 75 per cent. calcium carbonate.[119] The lime in the tubes of *Cerambyx* larvae passes to the stomach and is ejected through the mouth to form an operculum for the burrow.[93] The great calcospherites in the tubes of *Acidia heraclei*, which are too large to pass down the lumen, dissolve in the blood during the first day of metamorphosis, pass through the newly formed cuticle of the pupa, and are deposited as a hard and brittle layer on the inner surface of the puparium.[63] A similar transference of lime from the Malpighian tubes to calcareous pits in the cuticle takes place at each moult in the larvae of Stratiomyids.[73] And in female Phasmids, lime stored in the tubes is reabsorbed into the blood and incorporated in the chorion of the eggs.[136] [See p. 584.]

Fluorescent substances, notably riboflavine, are of common occurrence in the cells of the Malpighian tubes (p. 613). In the silkworm larva the fluorescent substances present are riboflavine, folic acid and 'fluorescyanine'.[161]

MALPIGHIAN TUBES DURING MOULTING AND METAMORPHOSIS

The Malpighian tubes of caterpillars show a cycle of changes during moulting. Shortly before the skin is cast, there is evidently a copious secretion of fluid by the tubes, which carries down the accumulated crystals of urates and oxalates. These pass between the old cuticle and the new in the ureters and rectum, gush out around the margins of the anus, and spread forwards over the body beneath the old cuticle.[132] They become mixed with the moulting fluid and may be swallowed by the mouth.[39] After moulting the tubes are almost devoid of crystals.

In the embryo of *Blatta* there are four primary tubules; but numerous secondary tubules appear at each instar, so that the adult male has about 130 and the female about 184. These secondaries arise from an embryonic zone at the junction of mid-gut and hind-gut which is homologous with the imaginal ring of Holometabola.[46] A similar process is seen in *Forficula*[46] and *Schistocerca*.[213]

The changes undergone by the Malpighian tubes at metamorphosis differ in different groups. In the Diptera and Coleoptera they suffer little change.[61, 107, 117] In those forms which contain lime, this is generally emptied into the gut at the time of pupation,[93, 105] and the tubes assume the same histological character throughout. In *Ptychoptera* the lime-containing segment is sealed below by a syncytial fusion of cells across the lumen; this obstruction disappears at pupation.[105] In Lepidoptera, the perirectal segments of the tubes break down and are removed by phagocytes;[18, 59] the free segments persist and increase in length, and their epithelium is reconstituted to form the adult tubes. The interstitial ring of embryonic cells at the base of the tubes seems to contribute little to their formation.[46] In Hymenoptera, the four Malpighian tubes of the larva break down completely during histolysis (as in the ant[62, 107] and the bee[7]); the adult organs are a new formation which arise as numerous short finger-like outgrowths from the annular swelling at the junction of the mid-gut and the hind-gut.[7]

The changes in the different orders may be summarized as follows.[46] In Coleoptera imaginal cells occur at the base of the larval cells along the whole length of the tubules and bring about a partial replacement of the larval cells at metamorphosis. In Hymenoptera the larval tubules are destroyed with or without phagocytosis and imaginal tubules grow out from the imaginal ring. In Lepidoptera the larval tubules persist, save in *Tineola* and *Galleria* whose imaginal tubules arise as new growths from the common duct. In Diptera the larval tubules are carried over to the adult. Where new Malpighian tubes are formed they may result from scattered imaginal cells or from the embryonic cells at the anterior end of the hind-gut.

If the spinneret of *Bombyx* larva is ligatured after feeding is completed the insect dies before pupation; but if the same experiment is made on *Galleria*, which produces a less massive cocoon, the pupal moult occurs but it is preceded by an abundant discharge of urates in the excreta.[166] Likewise, after removal of the silk glands in *Bombyx* the excretion of uric acid is almost doubled.[190]

CEPHALIC EXCRETORY ORGANS AND INTESTINAL EXCRETION

'Labial Kidneys'—Malpighian tubes are absent, probably through reduction, in Collembola, in *Japyx* among Thysanura, and in Aphids. Little is known of the excretory organs of Aphids, but in Collembola and Thysanura there are several pairs of glands in the head. Of these, the 'tubular glands', opening by a common duct above the base of the labium, are regarded as excretory organs homologous with the cephalic nephridia of other groups.[16, 110]

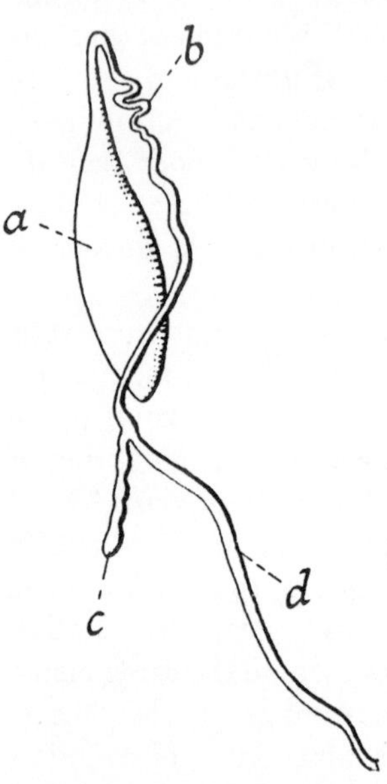

FIG. 338.—Tubular gland of Collembola (*after* MUKERJI) a, saccule; b, labyrinth; c, dorsal loop; d, duct.

In general arrangement the tubular glands are very like the antennal glands of Crustacea. They consist of an upper region or 'saccule' with flattened epithelium, communicating with a long coiled tube or 'labyrinth', the cells of which resemble those of the Malpighian tubes.[16] A third gland discharges into the duct of the tubular gland (Fig. 338). These structures have substantially the same form in all the Apterygota. Nothing is known of the normal products of their excretion; but the saccule takes up ammonia carmine injected into the body cavity and the labyrinth takes up indigo-carmine, passes it through the cells in the reduced or colourless form and deposits it in the form of blue crystals in the lumen, just like the Malpighian tubes.[16] In this these organs resemble the 'green gland' in Crustacea.[68] In *Dolycoris* and other Heteroptera dietary *l*-valine is metabolized, but the unnatural *d*-isomer is excreted from the haemolymph by way of the salivary glands.[20] Perhaps the accumulation of modified haemoglobin as a cherry red pigment in the salivary glands of *Rhodnius*[147] indicates a vestigial excretory function.

Excretion by the gut—There is no satisfactory proof that the gut wall retains in insects the general excretory functions it has in many groups;[27] but there are indications that this is so in Collembola.[177] During metamorphosis, notably in Hymenoptera, quantities of uric acid granules may appear in the mid-gut. This has led to the belief that the mid-gut epithelium may be important in excretion, for instance in Sphegidae.[29, 90, 147] But in many, if not all of these cases the uric acid has passed forwards from the Malpighian tubes. In Collembola the epithelium of the mid-gut is periodically cast off, and this has been regarded as an excretory process;[33] and uric acid has been demonstrated in mid-gut of the beetles *Gnaptor* and *Necrophorus*,[38] though it is absent in the silkworm.[74]

On the other hand there are some special substances which are certainly eliminated by the gut. After the injection of fluorescein into *Periplaneta* the dye is concentrated about tenfold in the mid-gut and is passed on to the hind-gut, and in Aphids it is excreted by the walls of the gut and discharged into the rectal pouch.[36] In *Machilis* injected dyes are discharged into the lumen of the mid-gut and later taken up and stored by special cells in the gut wall as in the Malpighian tubes of other insects (p. 572).[87] In *Rhodnius* biliverdin resulting from the breakdown of haemoglobin in the tissues is discharged by the cells of the mid-gut into the lumen, and the free iron accumulates throughout life in the substance of these cells.[157]

STORAGE EXCRETION

Urate cells—In the Collembola, which lack Malpighian tubes, granules of uric acid collect in certain cells, the so-called 'urate cells', scattered through the substance of the fat body; these cells become increasingly loaded with concretions as the insect grows older.[110] And in the cockroach, the Malpighian tubes, though present, are said never to contain any uric acid;[22] this accumulates throughout life in discrete urate cells which appear in the developing fat body of the embryo[31] and increase in volume, and in the size of the concretions they contain, throughout the life of the insect (Fig. 339).[22, 109]

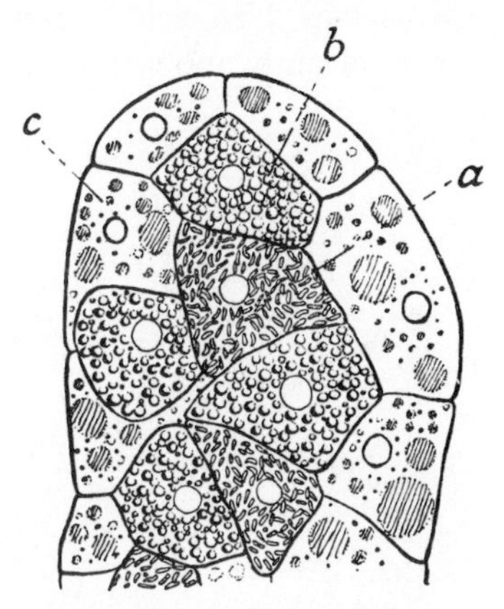

FIG. 339.—Lobe of fat body in *Blatta orientalis*, showing mycetocytes (*a*), urate cells (*b*), and ordinary fat cells (*c*) (*modified after* GIER)

In the larvae of Hymenoptera,[107] the urate cells appear as white points dotted among the cells of the fat body. During the larval stage of some species, such as the honey-bee, *Apis*, the Malpighian tubes end blindly below in the substance of the diaphragm which separates the mid-gut from the hind-gut (p. 478). As the larva grows the tubes become distended, and their walls very thin, but they contain no uric acid; only towards the time of pupation do they establish their connexion with the hind-gut;[102] and only during pupal life, or soon after the emergence of the adult, is the uric acid accumulated in the urate cells transferred to them and voided into the gut.[90, 153] In *Nasonia*, Malpighian tubes are entirely wanting in the larva.[141] [See p. 584.]

In these examples the storage of uric acid within the tissue cells clearly replaces, temporarily or permanently, its elimination by the Malpighian tubes. But there are other insects in which uric acid may crystallize out in the tissues even when the Malpighian tubes are functional. It may appear in the ordinary fat body cells of *Culex*, larvae and adults,[5] of *Dytiscus*[70] and other Coleoptera,[38] and of the larvae of Lepidoptera; or in the epidermal cells of many caterpillars, such as the silkworm,[29, 137] and various Tenthredinids; in certain restricted parts of the epidermis of aquatic Hemiptera (*Notonecta*, &c.) during their nymphal stages,[112] and in the nymphs of *Rhodnius* during moulting.[155] In such cases it seems reasonable to suppose that the uric acid has been produced during the metabolism of the cells in question, and has merely been caused to crystallize out by the conditions, acidity and so forth, existing within them. In the larva of *Aëdes* the fat body cells are largely filled with watery vacuoles some of which contain uric acid. Uric acid increases greatly in amount during starvation and it may sometimes crystallize out even in the living cell (Fig. 340). But if the larva is fed almost all this uric acid is excreted by the Malpighian tubes and its place taken by glycogen and fat.[156] Likewise, uric acid deposits become very conspicuous in the fat body of *Blattella* fed on a high protein diet; but these deposits are greatly diminished when a diet of pure carbohydrate (dextrin) is substituted.[184]

That raises the question whether, in general, uric acid is produced within the urate cells, or collected by them from the blood. In *Vanessa* larvae (Lep.), urates first appear in the fat body, as minute refractile granules around the nucleus, towards the time of pupation. If ammonium urate is injected at this time, urates are excreted by the Malpighian tubes, but the granules in the fat body show no increase.[53] Similar negative results have been obtained with the urate cells of Tettigoniids[56] and Blattids,[109] though in the cockroach the concretions do increase after prolonged starvation, presumably as the result of the utilization of stored protein.[109]

These observations suggest that the uric acid is formed within the cells themselves. Uric acid is certainly synthesized in the fat body of *Periplaneta*.[188] That certainly appears to be the case, also, in Muscid larvae, in which, towards

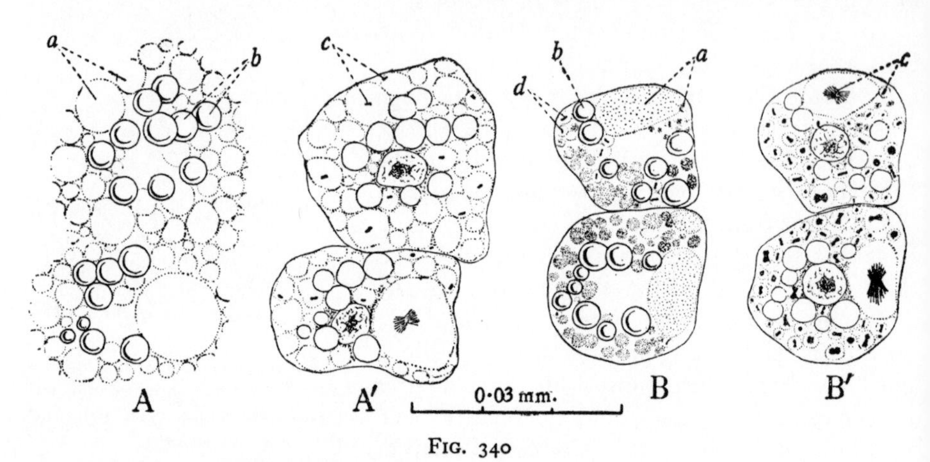

FIG. 340

A, fat body cells of newly moulted living larva of *Aëdes*; *A′*, the same cells after fixation in aqueous Bouin; *B*, fat body cells of living larva starved 7 days; *B′*, the same cells after fixation in aqueous Bouin. *a*, watery vacuoles; *b*, fat droplets; *c*, uric acid crystallized after fixation; *d*, uric acid separated in vacuoles of living cell (*after* WIGGLESWORTH).

the time of pupation, darkly staining 'pseudonuclei' appear within the albuminoid inclusions laid down in the fat body (p. 444)[2] and become transformed into concretions of uric acid (Fig. 285).[107] During the last few days of pupal life these 'pseudonuclei' disappear from the inclusions; the uric acid set free is taken up by the Malpighian tubes, and accumulates as meconium in the rectal pouch.[107, 153] A similar transference of urates from fat body to meconium occurs, as we have seen, in Hymenoptera;[90] and from the fat body and epidermis, in Lepidoptera.[22, 29, 39] Thus, although the uric acid deposited in the cells of the tissues is probably the product of their own metabolism, it often represents a true storage excretion, tiding the insect over the period when the Malpighian tubes are being reconstituted.[108]

In those insects in which the urate cells are distinct from the ordinary cells of the fat body, as in *Lepisma*, Collembola, and the cockroach, &c., they probably arise in the course of development from fat cells.[109] But in the young larva of the ant[107] and bee[129] they can be recognized as distinct rounded cells, without concretions, which later become applied to the surface of the fat lobes. In the

Tenthredinidae they are completely buried in the fat cells,[107] and have been described as arising from leucocytes which penetrate these cells during larval life.[128]

Desert species of Diptera (*Vermileo*) and Neuroptera (Myrmeleonidae and Nemopteridae) have the hind-gut blocked throughout larval development. Stored uric acid is discharged only at pupation or at adult emergence.[180]

Other examples of storage excretion—There are many other waste substances besides uric acid which occasionally accumulate in the tissues. Lime in the form of 'calcospherites' is common in special cells associated with the fat body in the larvae of phytophagous Diptera (Fig. 323).[63] In larvae of *Phytomyza*, mining the leaves of Cineraria, there may be anything up to 60 of these intracellular pellets of lime, each exceeding 200μ in length. During histolysis in the pupa, they are set free and dissolve; they are absent in the adult.[44] The fat body, and special tracts of the epidermis, in old termites become filled with yellow brown granules, perhaps of some waste product.[99] When the larvae of *Tortrix viridana* (Lep.) and other insects feed on the leaves of oak, crystals, believed to be derived from tannin, appear in the cells of the gut wall.[55] The fat body of *Cionus olens* (Col.), feeding on flowers of *Verbascum*, contain granules of purple pigment apparently identical with the anthocyanin in the stamens.[52] In the Vine Phylloxera *Viteus vitifolii*, trypan blue and fluorescein are carried in the blood and stored in the fat body; they are eventually eliminated with the eggs which are produced in enormous numbers.[179] Perhaps this merely signifies that these dyes have become associated with one of the blood proteins and so transmitted to the eggs—like haemoglobin in *Rhodnius*[157] or the yolk proteins of many insects (p. 704). [See p. 584.]

But it is the pericardial cells and the other so-called nephrocytes (p. 441) which are most apt to accumulate waste substances. As we have seen, the readiness with which ammonia carmine and certain other dyes collect in these cells gave rise to the belief that their main function was the segregation of waste products (p. 441).[68] And although that view is probably mistaken, deposits of various kinds may occur naturally in these cells. The brownish granules common in old termites are very abundant in the pericardial cells.[99] The green pigment in these cells in Hemiptera and the carotinoids in some other groups are perhaps excretory.[54] Biliverdin derived from the breakdown of haemoglobin accumulates in the pericardial cells of *Rhodnius* and *Pediculus*.[157] The pericardial cells and the ventral nephrocytes of *Lonchaea chorea* (Dipt. Acalypt.) feeding on decomposed mangel-wurzels, become so laden with brown granules of pigment that they may break down and be consumed by phagocytes.[64]

In *Rhodnius* the iron set free during the breakdown of those small amounts of haemoglobin which are absorbed, accumulates progressively throughout life in the cells of the gut wall and never appears in the excreta.[157] A similar accumulation of iron[149] and of copper[149] occurs in a restricted band of cells in the mid-gut of *Lucilia*, both larva and adult. Riboflavine and other pigmented substances accumulate in the Malpighian tubes (p. 613); and many of the pigments in the wings and other parts of insects are perhaps waste products of metabolism (p. 610).

In most insects relatively small amounts of pteridines are synthesized, but in the developing pupa of *Pieris brassicae* leucopterine and isoxanthopterine account for 14 per cent. of the oxidized nitrogen, purines 80 per cent., urea 3 per cent., and kynurenine 0·5 per cent. Of the purines, all the xanthine is stored in the fat body; 80 per cent. of the uric acid is stored, mainly in the fat body; 99 per cent. of the pteridine is stored, about three quarters in the wings, about one quarter in the fat body and epidermis.[181] [See p. 584.]

ADDENDA TO CHAPTER XII

p. 553. **Role of water in excretion**—*Ilyocoris* and other aquatic bugs are typical freshwater animals in that the excretory system must work continuously to eliminate water gained unavoidably by osmotic uptake through the cuticle.[249]

p. 555. **Diuretic hormones**—The release of the diuretic hormone in *Rhodnius* (p. 205) is induced by stimulation of abdominal stretch receptors, probably located in the vertical tergo-sternal muscles, by the meal of blood distending the mid-gut. 5-Hydroxytryptamine (serotonin) acts like the diuretic hormone on the Malpighian tubules of *Rhodnius* and *Carausius* at a concentration of 5×10^{-8} moles/l., but is not identical with the natural hormone.[243] The tubules of *Rhodnius* and *Carausius* are very sensitive to cyclic AMP (p. 206); perhaps 5-hydroxytryptamine and the natural hormone act by inducing the liberation of cyclic AMP.[243] An important effect of DDT and other insecticides in *Rhodnius* is to cause release of the diuretic hormone with the accumulation of fluid in the rectum.[242] A similar diuretic hormone (and an anti-diuretic hormone), both apparently peptides, are released from the terminal abdominal ganglion of *Periplaneta*. As in *Rhodnius* the diuretic hormone is released after ingestion of liquid.[245] A diuretic hormone, present in the blood of *Dysdercus* after feeding, can be extracted from the neurosecretory cells of the pars intercerebralis and also from the corpus cardiacum and the fused ganglia in the mesothorax.[230] In *Calliphora* a wide variety of substrates can provide energy for the secretory activity of the tubules—presumably by oxidative phosphorylation.[231] A diuretic hormone can be extracted from brain, corpora cardiaca and oesophageal ganglia of *Carausius*.[247] Elimination of the neurosecretory cells of the pars intercerebralis in *Gryllus* leads to retention of water with an increase in blood volume and distension of the abdomen.[236]

p. 557. **Ammonia excretion**—In *Ilyocoris* (Naucoridae) the colourless rectal fluid contains little besides ammonium and bicarbonate ions. The maximum osmotic concentration attained in the fasting insect is about two-thirds that of the haemolymph.[249]

p. 557. **Amino acids in urine**—The mustard Aphid *Lipaphis* utilizes the essential amino acids in the leaf sap and eliminates nitrogen principally as arginine. Urea is present and traces of ammonium salts; amides and carbamyl compounds also occur.[248] Although uric acid makes up more than 60 per cent. of the dry weight of the faeces in *Glossina*, substantial amounts of arginine and histidine are present; these nitrogen-rich amino acids, which make up 10 per cent. of the combined amino acids in the mammalian blood which serves as food, are excreted quantitatively without deamination.[234]

p. 558. **Adenine metabolism**—If labelled adenine is injected into *Leuco-*

phaea it is converted to uric acid and stored semi-permanently in the fat body, strongly bound to protein.[239]

p. 559. **Allantoine excretion**—In *Dysdercus* (Pyrrhocoridae) the rectal contents are always fluid and little water is recycled (p. 527). The urine contains allantoine, urea, amino acids, kynurenine and mucopolysaccharide. Uricase is concentrated in the Malpighian tubules and accounts for the absence of uric acid in the urine and the high concentration of allantoine in the haemolymph. Uric acid, however, is extensively deposited and stored in the epidermal cells.[229]

p. 561. **Uric acid biosynthesis**—By the use of isotopically labelled precursors the origin of all five carbon atoms of uric acid, derived from glycine by way of formic acid and carbon dioxide, has been traced in *Rhodnius*; with nitrogen contributed from glycine, glutamate and aspartate.[228]

p. 565. **Hormones and movements of Malpighian tubules**—In *Carausius*, movements of the Malpighian tubules are strongly stimulated by extracts from brain, corpora cardiaca, suboesophageal ganglion, first thoracic and last abdominal ganglia. The active fraction appears to be a peptide.[247]

p. 569. **Excretion in Drosophila: ultra-structure**—Excretion in *Drosophila* takes place by a process of vesicular flow, or 'cytopempsis'; the basement membrane acts as a molecular sieve; materials of all kinds from the haemolymph are taken up by the detachment of vesicles from the basal invaginations of the plasma membrane; during their intracellular passage, water is absorbed from the vesicles, which may retain contact with the haemolymph by means of an elongated tube, and as they approach the lumen their osmotic concentration increases. The contents are discharged and the resorbed water transported back into the haemolymph.[251] Kynurenine, and injected iron-dextran, are taken up into the ground cytoplasm, collected in the endoplasmic reticulum and conveyed to the lumen, like biliverdin in *Rhodnius* (Fig. 331).

p. 569. **Malpighian tubules of Calliphora**—In *Calliphora* the two cell types in the Malpighian tubules have been called primary and stellate (cf. Fig. 332). It has been suggested that in the primary cells a standing osmotic gradient is set up within the basal folds and between the microvilli which induces a bulk flow of fluid carrying solute into the cell and into the lumen; whereas the stellate cells are reabsorptive and are perhaps responsible for the low sodium concentration in the urine by returning sodium to the blood.[233]

p. 572. **Subdivision of Drosophila tubules**—The Malpighian tubules of *Drosophila* consist of four segments: (i) an initial segment transports and dehydrates vesicles carried to the lumen by 'cytopempsis' and concentrated by intracellular resorption to form the primary urine; (ii) a transitional segment; (iii) a main segment where resorption of useful material occurs and (iv) a ureter controlling transport of the secondary urine to the hind-gut.[252]

p. 573. **Ions and Malpighian tubule activity**—The tubules of *Rhodnius* under the action of the diuretic hormone will secrete at a normal rate in solutions containing K, Na or Cl; there is no special dependence on potassium; there may be an active transport of all three. The tubules secrete a fluid approximately iso-osmotic to the bathing medium.[240] In *Drosophila*, in the dilated first segment of the Malpighian tubules, more K than Na is transferred from the blood to the lumen: the vesicular transport system is selective.[251] The diuretic hormone in *Carausius* appears to act by stimulating active transport of potassium. K^+ is far more effective in inducing urine secretion than Na^+, while NH_4^+, Ca^{++} and

Mg^{++} can support only a very low rate.[247] In *Calliphora* the rate of urine formation is very sensitive to potassium concentration; it may be that there is a coupled sodium-potassium pump at the basal surface of the cell and an electrogenic pump at the apical surface. But K^{+} transport alone will not induce water transport unless accompanied by a transportable anion such as chloride or phosphate.[232]

p. 575. **Leptophragmata in Tenebrio**—Electron microscope studies of the leptophragmata of *Tenebrio* have shown that the cytoplasm contains plentiful mitochrondria, and microvilli are present. It may be that the high osmolarity of the fluids within the rectal complex is due to the inward secretion of potassium chloride, unaccompanied by water, at the leptophragmata.[237]

p. 576. **Calcium oxalate**—The chief component of the yellow powder in the cocoon of *Malacosoma* is calcium oxalate, mostly from dietary oxalic acid.[246]

p. 576. The dwelling tubes of Cercopid larvae consist of a meshwork of microscopic fibrils composed of mucopolysaccharide combined with a resistant protein and impregnated with inorganic salts. The organic material closely resembles the secretion used by Cercopid larvae in 'spittle' formation.[244]

p. 579. **Urate cells**—In dense populations of *Sminthurus* the urate cells become excessively laden with uric acid and this may be a cause of death resulting from cannibalism.[250]

p. 581. **Storage excretion**—The mid-gut cells of *Aphis fabae* become filled with spherical inclusion bodies 2–5μ in diameter consisting of carbonates and phosphates of calcium and magnesium.[235]

p. 582. In *Pieris brassicae* only traces of pteridines are excreted, the rest is stored, mostly in the cuticular scales. If a wing is eliminated pteridine synthesis is correspondingly reduced. In *Vanessa io* which produces far less pteridines, nearly half are excreted in the meconium and there is no storage in the scales; only in some Hemiptera are pteridines formed and stored in quantities comparable to those in Pieridae.[238]

REFERENCES

[1] BABCOCK, S. M. *Univ. Wisconsin Agric. Exp. Sta. Res. Bull.*, No. 22 (1912), 181 pp. (metabolic water: excretion in clothes moth, *Tinea*).

[2] BERLESE, A. *Zool. Anz.*, **23** (1900), 441–9; **24** (1901), 515–21 (fat body during metamorphosis).

[3] BLUNCK, H. *Z. wiss. Zool.*, **121** (1924), 171–391 (physiology, &c., of *Dytiscus*, Col.).

[4] BODENHEIMER, F. *Z. wiss. Zool.*, **121** (1924), 393–441 (histology of Malpighian tubes, &c.: *Tipula*, Dipt.).

[5] de BOISSEZON, P. *Arch. Zool.*, **70** (1930), 281–431 (histophysiology of *Culex*, Dipt.).

[6] BONÉ, G., and KOCH, H. J. *Ann. Soc. Roy. Zool. Belg.*, **73** (1942), 73–87 (reabsorption of chlorides in rectum).

[7] BORDAS, L. *Bull. Sci. Fr. Belg.*, **26** (1894), 402–41 (Malpighian tubes in Hymenoptera).

[8] —— *C.R. Acad. Sci.*, **124** (1897), 46–8 (Malpighian tubes in Orthoptera).

[9] BORDAS, L. *Ann. Sci. Nat., Zool.*, **14** (1911), 191–271 (Malpighian tubes in Lepidoptera larvae, and historical review); *Insecta, Rennes*, **6** (1916), 9–11; *Ann. Sci. Nat., Zool.*, **3** (1920), 175–225 (ditto in adult Lepidoptera).

[10] —— *Insecta, Rennes*, **7** (1917), 25–7 (Malpighian tubes in Cetoninae, Col.).

[11] BRASS, P. *Zool. Jahrb., Syst.*, **37** (1914), 65–122 (tenth abdominal segment as pseudopod in beetle larvae).

[12] BROWN, A. W. A. *J. Exp. Biol.*, **13** (1936), 131–9 (excretion of ammonia and uric acid in Muscid larvae).

[13] —— *J. Exp. Biol.*, **14** (1937), 87–94 (excreta of *Melanoplus*, Orth.).

[14] BROWN, A. W. A. *Biochem. J.*, **32** (1938), 895–911 (nitrogen excretion: *Lucilia*, Dipt.).

[15] BROWN, A. W. A., and FARBER, L. *Biochem. J.*, **30** (1936), 1107–18 (deaminating enzyme in blow-fly larvae).

[16] BRUNTZ, L. *Arch. Zool.*, (sér. 4), **9** (1908), 195–238 (cephalic excretory organs: Thysanura).

[17] BUGAJEW, I. I. *Zool. Anz.*, (1928), 244–55 (structure of Malpighian tubes).

[18] CHOLODKOWSKY, N. *Arch. Biol.*, **6** (1887), 497–514 (morphology of excretory system: Lepidoptera).

[19] CONET, A. *Ann. Soc. Sci. Bruxelles*, **54** (1934), 189–200 (histology of Malpighian tubes in 'cryptonephridial' Coleoptera).

[20] COURTOIS, A. *C. R. Soc. Biol.*, **101** (1929), 365–6 (composition of meconium in Lepidoptera).

[21] CROZIER, W. J. *J. Gen. Physiol.*, **6** (1924), 289–93 (*p*H in gut of *Chironomus* larva, &c.).

[22] CUÉNOT, L. *Arch. Biol.*, **14** (1895), 293–341 (Malpighian tubes, storage excretion, &c.: Orthoptera).

[23] DAVIS, A. C. *Univ. Calif. Publ. Ent.*, **4** (1927), 159–208; *Ann. Ent. Soc. Amer.* **20** (1927), 359–62 (striated border of Malpighian tubes: *Stenopelmatus*, Orth.).

[24] DEEGENER, P. *Schröder's Handb. d. Entomologie*, **1** (1913), 309 (Malpighian tubes).

[25] DELAUNAY, H. *Biol. Rev.*, **6** (1931), 265–301 (excretion in invertebrates).

[26] EASTHAM, L. *Quart. J. Micr. Sci.*, **69** (1925), 385–98 (calcium carbonate and movements in Malpighian tubes: *Drosophila*, *Calliphora*, Dipt.).

[27] EHRENBERG, R. *Winterstein's Handb. d. vergl. Physiol.*, **2** (ii) (1914), 695–759 (excretory system).

[28] FAASCH, W. J. *Z. Morph. Oekol. Tiere*, **29** (1935), 559–84 (muscles on Malpighian tubes: Aphaniptera).

[29] FABRE, J. H. *Ann. Sci. Nat.* (sér. 4), **19** (1863), 351–82 (role of fat body in excretion).

[30] FARKAS, K. *Arch. ges. Physiol.*, **98** (1903), 490–546 (metabolism in egg and pupa: *Bombyx mori*).

[31] FAUSSEK, V. *Z. wiss. Zool.*, **98** (1911), 529–625 (uric acid concretions: *Blattella*, Orth.).

[32] FLORKIN, M., *et al. Bull. Acad. Roy. Belg. Cl. Sci.*, (sér. 5), **27** (1941), 169–78; *Enzymologia*, **11** (1943), 24–5; *Arch. Intern. Physiol.*, **53** (1943), 263–307 (purine metabolism in insects).

[33] FOLSOM, J. W., and WELLES, M. U. *Univ. Illinois Univ. Studies*, **2** (1906), 97 (epithelial degeneration in mid-gut: Collembola).

[34] FRENZEL, J. *Berliner Ent. Zeitschr.*, **26** (1882), 267–316 (alimentary canal of *Tenebrio* larva, Col.).

[35] GARBARSKAJA, M. *Zool. Jahrb., Anat.*, **51** (1929), 63–110 (histology of Malpighian tubes: Sphingidae, Lep.).

[36] GERSCH, M. *Z. vergl. Physiol.*, **29** (1942), 506–31 (excretion of fluorescent substances by insects).

[37] GERSTAECKER, R. *S. B. Ges. Naturf. Freunde Berlin*, 1873, 138–45 (uric acid excretion by insects).

[38] v. GORKA, A. *Zool. Jahrb., Physiol.*, **34** (1914), 233–338 (physiology of Malpighian tubes: *Gnaptor*, *Necrophorus*, Col.).

[39] GRASSÉ, P. P., and LESPERON, L. *C. R. Soc. Biol.*, **122** (1936), 1013–5 (excretion during moulting in the silkworm).

[40] GREEN, T. L. *Proc. Zool. Soc. Lond.*, 1931, 1041–66 (histology of Malpighian tubes: *Vespa*, Hym.).

[41] HARMS, B. *Arch. Mikr. Anat.*, **80** (1912), 167–216 (Malpighian tubes, &c., in larva of *Ctenocephalus*, Aphanipt.).

[42] HELLER, J. *Z. vergl. Physiol.*, **25** (1938), 83–87 (phosphate and potassium in meconium: *Deilephila*, Lep.).

[43] HELLER, J., and AREMOWNA, H. *Z. vergl. Physiol.*, **16** (1932), 362–70 (meconium of *Deilephila*, Lep.).

[44] HENNEGUY, L. F. *Arch. Anat. Micr.*, **1** (1897), 125–8 (calcospherites in fat body of Dipterous larvae).

[45] —— *Les Insectes*, Paris, 1904.

[46] HENSON, H. *Quart. J. Micr. Sci.*, **75** (1932), 283–305 (origin and development of

Malpighian tubes: *Pieris*, Lep.); *Proc. Zool. Soc. Lond.*, B, 1937, 161–71 (ditto: *Vanessa*, Lep.); **109** (1939), 357–72; *Proc. Roy. Ent. Soc. Lond.*, A, **19** (1944), 73–91 (ditto: *Blatta*); *Ibid.*, **21** (1946), 29–39 (ditto: *Forficula*); *Proc. Leeds Phil. Soc.*, **4** (1945), 259–68 (ditto: Ephemeroptera).

[47] HEYMONS, R., and LÜHMANN, M. *Zool. Anz.*, **102** (1933), 78–86 (Malpighian tubes in *Galerucella*, Col.).

[48] HICKERNELL, L. M. *Ann Ent. Soc. Amer.*, **13** (1920), 223–48 (gut and Malpighian tubes in the cicada, *Tibicen*).

[49] HINTON, H. E. *Proc. Roy. Ent. Soc. Lond.*, A, **16** (1941), 39–48 (cryptonephridial tubes: Coleoptera).

[50] HOBSON, R. P. *J. Exp. Biol.*, **9** (1932), 128–38; 366–77 (ammonia production by sterile *Lucilia* larvae, Dipt.).

[51] HOLLANDE, A. C. *Ann. Univ. Grenoble*, **21** (1909), 459–517 (elimination of salicin by insects).

[52] —— *Arch. Zool.*, **51** (1913), 53–8 (storage of anthocyanin pigment in fat body).

[53] —— *Arch Zool.*, **5** (1914), 559–78 (urates in fat body of *Vanessa* larvae, Lep.).

[54] —— *Arch. Anat. Microsc.*, **18** (1923), 85–307 (pericardial cells of insects).

[55] —— *Arch. Anat. Microsc.*, **19** (1923), 349–69 (tannin excretion by mid-gut: *Tortrix viridana*, Lep.).

[56] —— *C.R. Acad. Sci.*, **18** (1925), 1175–6 (urate cells in Orthoptera).

[57] HOLLANDE, A. C., and CORDEBARD, H. *Bull. Soc. Chim. Biol.*, **8** (1926), 631–35 (composition of excreta: *Tineola*, Lep.).

[58] HOLMGREN, N. *Anat. Anz.*, **22** (1902), 225–39 (Malpighian tubes: *Apon*, Col.).

[59] HUFNAGEL, A. *Arch Zool.*, **57** (1918), 47–202 (histology during metamorphosis: *Hyponomeuta*, Lep.).

[60] ISHIMORI, N. *Ann. Ent. Soc. Amer.*, **17** (1924), 75–86 (relation of Malpighian tubes to rectum: Lepidoptera larvae).

[61] ITO, H. *J. Morph.*, **35** (1921), 195–205 (Malpighian tubes during metamorphosis: *Bombyx*, Lep.); *Bull. Biol. Fr. Belg.*, **57** (1923), 152–8 (ditto: *Gnathocerus*, Col.).

[62] KARAWAIEW, W. *Z. wiss. Zool.*, **64** (1898), 385–478 (postembryonic development: *Lasius*, Formicid.).

[63] KEILIN, D. *Quart. J. Micr. Sci.*, **65** (1921), 611–25 (lime in Malpighian tubes and fat body: Diptera larvae).

[64] —— *Ann. Mag. Nat. Hist.* (sér. 9), **13** (1924), 219–23 (nephrocytes in *Lonchaea*, Dipt. Acalypt.).

[65] KOLLER, G. *Biol. Zbl.*, **67** (1948), 201–11 (movement of Malpighian tubes).

[66] KÖLLIKER, A. *Verh. physicalisch-med. Gesell. Würzburg*, **8** (1858), 225–35 (Malpighian tubes and contents).

[67] KÖNNEMANN, R. *Zool. Jahrb., Anat.*, **46** (1924), 343–86 (gut and Malpighian tubes. Limnobiid larvae, Dipt.).

[68] KOWALEWSKY, A. *Congr. Internat. Zool. Moscow*, 1892, **1**, 187–234 (storage excretion).

[69] —— *Arch. Zool.*, (sér. 3), **2** (1894), 485–90 (Malpighian tubes within the heart: Orthoptera).

[70] KREUSCHER, A. *Z. wiss. Zool.*, **119** (1922), 247–84 (fat body and oenocytes: *Dytiscus*, Col.).

[71] KROGH, A. *Osmotic Regulation in Aquatic Animals*, Cambridge, 1939.

[72] KRÜGER, F. *Z. Morph. Oekol. Tiere*, **6** (1926), 83–149 (biology, &c.: Syrphid larvae, Dipt.).

[73] KRÜPER, F. *Arch. Hydrobiol.* **22** (1930), 185–220 (calcium carbonate in larvae of Diptera).

[74] KUWANA, Z. *Japan, J. Zool.*, **7** (1937), 305–9 (uric acid excretion in silkworm).

[75] LAZARENKO, T. *Z. mikr. anat. Forsch.*, **3** (1925), 409–99 (connective tissue of insects).

[76] LEBEDEW, A. *Zool. Anz.*, **44** (1914), 49–56 (silk secretion by Malpighian tubes: *Phytonomus* larva, Col.).

[77] LÉCAILLON, A. *Bull. Soc. Ent. Fr.*, 1899, 353–4 (striated border of Malpighian tubes: *Culex*, Dipt.).

[78] LÉGER, L., and DUBOSQ, O. *C. R. Soc. Biol.*, 1899, 527–9 (muscles of Malpighian tubes: *Gryllus*, &c., Orth.).

[79] LÉGER, L., and HAGENMULLER, P. *Bull. Soc. Ent. Fr.*, 1899, 192–4 (striated border of Malpighian tubes: Tenebrionidae, Col.).

[80] LEIFERT, H. *Zool. Jahrb., Physiol.*, **55** (1935), 131–90 (excretory metabolism: *Antheraea*, Lep.).

[81] LENNOX, F. G. *Nature*, **146** (1940), 268; *Coun. Sci. Ind. Res. Australia*, Pamph. 109 (1941), 1–67 (ammonia formation in *Lucilia*, Dipt.).

[82] LESTER, H. M. O., and LLOYD, L. *Bull. Ent. Res.*, **19** (1928), 39–60 (excretion in *Glossina*, Dipt.).

[83] LICENT, P. E. *La Cellule*, **28** (1912), 7–161 (gut and Malpighian tubes: Homoptera).

[84] LISON, L. *Ann. Soc. Roy. Zool. Belg.*, **67** (1936), 41–9 (muscles of Malpighian tubes: *Tenebrio*, Col.).

[85] —— *Bull Acad. Roy. Belg. Cl. Sci.*, (sér 5), **23** (1937), 317–27 (anatomy and histology of Malpighian tubes: *Tenebrio* and *Dermestes*, Col.); *Ann. Soc. Roy. Zool. Belg.*, **69** (1938), 195–233 (ditto: *Melolontha*, Col.); *C. R. Soc. Biol.*, **129** (1938), 873–5 (ditto: *Leptinotarsa*, Col.).

[86] —— *Arch. Biol.*, **48** (1937), 321–360; 489–512; *Z. Zellforsch. Micr. Anat.*, **28** (1938), 179–209; *Mém. Acad. Roy. Belg. Cl. Sci.*, (sér. 8), **19** (1942), 1–106 (elimination and athrocytosis of dyes in insects).

[87] —— *C. R. Soc. Biol.*, **132** (1939), 309–10 (intestinal excretion: *Machilis*).

[88] LOZINSKI, P. *Zool. Anz.*, **38** (1911), 401–17 (silk secretion by Malpighian tubes: Myrmeleonid larvae).

[89] MCDUNNOUGH, J. *Arch. Naturgesch.*, **55** (1909), 313–60 (gut and Malpighian tubes: *Chrysopa*, Neuropt.).

[90] MARCHAL, P. *Mém. Soc. Zool. Fr.*, **3** (1890), 31–87 (uric acid excretion in insects: Hymenoptera).

[91] MARCUS, B. A. *Z. Morph. Oekol. Tiere*, **19** (1930), 609–77 (Malpighian tubes of Coleoptera).

[92] MARSHALL, W. S. *Ann. Ent. Soc. Amer.*, **20** (1927), 149–55 (Malpighian tubes of *Heptagenia*, Ephem.).

[93] MAYET, V. *Bull. Soc. Ent. Fr.*, 1896, 122–7 (lime in Malpighian tubes: *Cerambyx*, Col.).

[94] MALOEUF, N. S. R. *Physiol. Rev.*, **18** (1938), 28–58 (excretion in Arthropods: review).

[95] METALNIKOV, S. *Arch. Zool.*, **8** (1908), 489-588 (Malpighian tubes and excretion of dyes: *Galleria* larvae, Lep.).

[96] MEYER, P. F. *Z. vergl. Physiol.*, **11** (1930), 173–209 (fate of plant pigments in Lepidoptera).

[97] MISSIROLI, A. *Riv. Malariol.*, **6** (1927), 1–7 (Malpighian tubes in *Anopheles*, Dipt.).

[98] MÖBUSZ, A. *Arch. Naturgesch.*, **63** (1897), 89–128 (gut and Malpighian tubes: *Anthrenus* larva, Col.).

[99] MONTALENTI, G. *Boll. Ist. Zool. R. Univ. Roma*, **6** (1928), 3–15 (storage excretion in epidermis and fat body in termites).

[100] MORISON, G. D. *Quart. J. Micr. Sci.*, **71** (1927), 395–463; 563–631 (muscles of honey-bee).

[101] NEEDHAM, J. *Science Progress*, **23** (1929), 633–46 (uric acid excretion).

[102] NELSON, J. A. *Science*, **46** (1917), 343–5 (relation of Malpighian tubes to hind-gut in honey-bee larva).

[103] PALM, N. B. *Lunds Univ. Årsskr.*, **42** (1946), No. 11, 1–39 (movements of Malpighian tubes).

[104] PANTEL, J. *La Cellule*, **15** (1898), 7–290 (anatomy of *Thrixion* larva, Tachinid.).

[105] —— *La Cellule*, **29** (1914), 393–429 (lime sacs in Malpighian tubes of *Ptychoptera* larva, Dipt.).

[106] PATTON, R. L., and CRAIG, R. *J. Exp. Zool.*, **81** (1939), 437–57 (excretion in *Tenebrio*, Col.).

[107] PÉREZ, C. *Bull. Sci. Fr. Belg.*, **37** (1902), 195–427 (changes during metamorphosis: *Formica*, Hym.); *Arch Zool.*, **4** (1910), 1–274 (ditto: *Calliphora*, Dipt.).

[108] —— *Arch. Zool.*, **59** (1920), 5–10 (inclusions in fat body during metamorphosis).

[109] PHILIPTSCHENKO, J. A. *Rev. Russe Entom.*, **7** (1907), 188–9 (urate deposits in fat body of cockroach).

[110] —— *Z. wiss. Zool.*, **88** (1907), 99–116 (excretory organs of *Ctenolepisma*, Thysanura).

[111] PLATEAU, F. *Mem. Acad. Roy. Belg.*, **41** (1874), 1–124 (digestion and excretion in insects).

[112] POISSON, R. *Bull. Soc. Zool. Fr.*, **50** (1925), 116–24 (epidermal urate cells in aquatic Hemiptera).

[113] POLL, M. *Bull. Ann. Soc. Ent. Belg.*, **72** (1932), 103–9 ('cryptonephridial' arrangement of Malpighian tubes in Coleoptera).

[114] —— *Rec. Inst. Zool. Torley-Rousseau*, **5** (1934), 73–126 (histophysiology of Malpighian tubes: *Tenebrio*, Col.).

[115] —— *Mém. Mus. Roy. Hist. Nat. Belg.*, (sér. 2), **3** (1936), 635–67 (excretory system in Myrmeleonid larvae).

[116] —— *Ann. Soc. Roy. Zool. Belg.*, **69** (1938), 9–52 (Malpighian tubes in caterpillars); *Bull. Ann. Soc. Ent. Belg.*, **77** (1937), 433–42 (ditto: in sawfly larvae).

[117] POYARKOFF, E. *Arch. Anat. Microsc.*, **12** (1910), 333–474 (histology during metamorphosis: *Galerucella*, Col.).

[118] PRADHAN, S. *Ind. J. Ent.*, **4** (1942), 11–21 (Malpighian tubes and hind-gut in Coccinellidae).

[119] RATTE, F. *Proc. Linn. Soc. N. S. Wales*, **9** (1884), 1164–9 (larval cases composed of lime: *Ptyelus*, Cercopidae).

[120] RILEY, W. A. *Ent. News*, **17** (1906), 113–4 (Malpighian tube within the heart: Orthoptera).

[121] ROBINSON, W, *J. Parasitology*, **21** (1935), 354–8 (allantoine excretion: *Lucilia*, Dipt.).

[122] ROCCO, M. L. *C.R. Acad. Sci.*, **202** (1936), 1947–8; **207** (1938), 1006–8 (allantoinase in insects).

[123] ROUBAUD, E. *Bull. Sci. Fr. Belg.*, **47** (1913), 105–202 (biology, &c., of *Auchmeromyia*' Dipt.).

[124] RUNGIUS, H. *Z. wiss. Zool.*, **98** (1911), 179–287 (gut and Malpighian tubes: *Dytiscus*).

[125] SAINT HILAIRE, K. *Zool. Anz.*, **73** (1927), 218–29 (comparative histology of Malpighian tubes).

[126] SAMSON, K. *Zool. Jahrb., Anat.*, **26** (1908), 403–22 (Malpighian tubes and fat body in excretion: *Heterogenea*, Lep.).

[127] SCHINDLER, E. *Z. wiss. Zool.*, **30** (1878), 587–660 (Malpighian tubes: anatomy, histology and chemistry).

[128] SCHMIEDER, R. G. *J. Morph.*, **45** (1928), 121–86 (fat body in excretion: Tenthredinidae, Hym.).

[129] SCHNELLE, H. *Zool. Anz.*, **57** (1923), 172–9 (histology of fat-body in honey-bee).

[130] SCHULZ, F. N. *Biochem. Z.*, **156** (1925), 124–9 (excretion in clothes-moth larva, *Tinea*).

[131] SEDLACZEK, W. *Zbl. ges. Forstw.*, **28** (1902), (gut and Malpighian tubes of Scolytidae, Col.).

[132] SHIMIZU, S. *Proc. Imp. Acad. (Japan)*, **7** (1931), 361–2 (origin of crystals in exuvial fluid of silkworm larva).

[133] SHINODA, O. *Annot. Zool. Japan*, **13** (1931), 117–25 (deaminase in blood of silkworm).

[134] SHIWAGO, P. *Anat. Anz.*, **44** (1913), 365–70; *C. R. Soc. Biol.*, **78** (1915), 180–2 (vesicular secretion in Malpighian tubes of cockroach).

[135] SILVESTRI, F. *Redia*, **2** (1904), 68–84 (silk secretion by Malpighian tubes: *Lebia*, Col.).

[136] de SINÉTY, R. *Bull. Soc. Ent. Fr.*, 1900, 333–5; *La Cellule*, **19** (1901), 119–278 (Malpighian tubes, &c.: Phasmidae).

[137] SIRODOT, S. *Ann. Sci. Nat., Zool.*, (sér. 4), **10** (1858), 141–89; 251–334 (Malpighian tubes: anatomy and chemistry).

[138] STAMMER, H. J. *Z. Morph. Oekol. Tiere*, **26** (1932), 135–46 (biology, &c., of *Ceroplatus* larva, Mycetophil., Dipt.).

[139] —— *Z. Morph. Oekol. Tiere*, **29** (1934), 196–217 (Malpighian tubes in Coleoptera).

[140] THORPE, W. H. *Proc. Zool. Soc. Lond.*, 1930, 929–71 (biology, &c.: *Cryptochaetum*, Agromyzid, Dipt.).

[141] TIEGS, O. W. *Trans. Roy. Soc. S. Australia*, **46** (1922), 319–527 (histology of metamorphosis: *Nasonia*, Hym.).

[142] TIMON-DAVID, J. *Ann. Fac. Sci. Marseille*, **16** (1945), 179–235 (insect excretion: review).

[143] TIRELLI, M. *Rend. Accad. Lincei*, **10** (1929), 278–81 (origin of Malpighian tubes in development).

[144] TRAPMANN, W. *Arch. Bienenk.*, **5** (1923), 177–89 (Malpighian tubes of honey-bee).

[145] TRUSZKOWSKI, R., and CHAJKINÓWNA, S. *Biochem. J.*, **29** (1935), 2361–5 (uricase in Muscid flies, &c.).

[146] VANEY, C. *Bull. Soc. Ent. Fr.*, 1900, 360–1 (Malpighian tubes: *Stratiomys* larva, Dipt.).

[147] VENEZIANI, A. *Redia*, **2** (1904), 177–230 (anatomy and physiology of Malpighian tubes).

[148] VIGNON, P. *Arch. Zool.*, (sér. 3), **7** (1899), xvii–xxv (vesicular secretion in Malpighian tubes, &c.).

[149] WATERHOUSE, D. F. *Coun. Sci. Ind. Res. Australia*, Pamph. 102 (1940), 28–50; *Bull. Coun. Sci. Ind. Res. Australia*, No. 191 (1945), 1–39 (iron and copper in *Lucilia*, Dipt.).

[150] WEBER, H. *Biologie der Hemipteren*, Springer, Berlin, 1930.

[151] WEINLAND, E. *Z. Biol.*, **47** (1906), 232–50; *Biol. Centralbl.*, **29** (1909), 564–77 (ammonia production in *Calliphora* larvae, Dipt.).

[152] WIGGLESWORTH, V. B. *J. Exp. Biol.*, **8** (1931), 411–51 (excretion in *Rhodnius*, Hem.).

[153] —— *Quart. J. Micr. Sci.*, **75** (1932), 131–50 (function of rectal glands).

[154] —— *J. Exp. Biol.*, **10** (1933), 1–37 (water uptake by anal papillae in mosquito larvae).

[155] —— *Quart. J. Micr. Sci.*, **76** (1933), 270–318 (histology during moulting: *Rhodnius*, Hem.).

[156] —— *J. Exp. Biol.*, **19** (1942), 56–77 (Malpighian tubes in intermediary metabolism; storage excretion: *Aëdes* larva, Dipt.).

[157] —— *Proc. Roy. Soc.*, B, **131** (1943), 313–39 (excretion of iron, biliverdin, &c.: *Rhodnius* and other blood-sucking insects).

[158] —— Unpublished observations.

[159] WOODS, W. C. *Ann. Ent. Soc. Amer.*, **9** (1916), 391–407 (Malpighian tubes: *Haltica*, Col.).

[160] ZILCH, A. *Z. wiss. Zool.*, **148** (1936), 89–132 (striated border of Malpighian tubes: *Rhodnius* Hem.).

SUPPLEMENTARY REFERENCES (A)

[161] DRILHON, A., and BUSNEL, R. G. *C.R. Acad. Sci.*, **232** (1951), 182–4 (fluorescent substances in Malpighian tubes of *Bombyx*).

[162] NIEMIERKO, S., and NIEMIERKO, W. *Nature*, **166** (1950), 268–9 (metaphosphate in excreta of *Galleria*).

[163] RAMSAY, J. A. *J. Exp. Biol.*, **29** (1952), 110–26 (excretion of sodium and potassium in *Rhodnius*).

[164] WATERHOUSE, D. F. *Austral. J. Sci. Res.* B. **3** (1950), 76–112 (granules in Malpighian tubes of *Lucilia*).

[165] WOLFRAM, R. *Z. indukt. Abstamm. VererbLehre.*, **83** (1949), 254–98 (formation of excretory pigments in *Ephestia*, etc.).

SUPPLEMENTARY REFERENCES (B)

[166] ALLEGRET, P. *C.R. Acad. Sci.*, **238** (1954), 623–5 (effect of blockage of silk secretion on excretion in *Galleria*).

[167] AMOURIQ, L. *Bull. Soc. Zool. Fr.*, **85** (1960), 21–35 (histology of Malphigian tubules of *Coccinella*).

[168] BACCETTI, B., MAZZI, V., *et al. Redia*, **48** (1963), 47–68; *Z. Zellforsch.*, **59** (1963), 47–70 (fine structure and histochemistry of Malphigian tubules in adult and larva of *Dacus*, Dipt.).

[169] BEAMS, H. W., TAHMISIAN, T. N., and DEVINE, R. L. *J. Biophys. Biochem. Cytol.*, **1** (1955), 197–202.

[170] BERKALOFF, A. *C.R. Acad. Sci.*, **246** (1958), 2807–9; **249** (1958), 1934–6; 2120–1; **250** (1960), 2061–3; *Ann. Sci. Nat., Zool.*, (sér 12) **2** (1960), 869–947 (fine structure of Malphigian tubules).

[171] BERRIDGE, M. J., *The physiology of excretion in the cotton stainer, 'Dysdercus fasciatus'.* Thesis, University of Cambridge, 1964.

[172] BRADFIELD, J. R. G. *Quart. J. Micr. Sci.*, **92** (1950), 87–112 (histochemical characters of silk glands).

[173] CAMERON, M. L. *Nature*, **172** (1953), 349 (movements of Malphigian tubules stimulated by corpus cardiacum extracts).

[174] CRAIG, R. *Ann. Rev. Ent.*, **5** (1960), 53–68 (excretion in insects: review).

[175] DAY, M. F., and BRIGGS, M. *J. Ultrastructure Res.*, **2** (1958), 239–44 (origin and structure of brochosomes).

[176] EGELHAAF, A. *Z. induckt. Abst. Vererbl.*, **87** (1956), 769–83 (excretion in the white-eyed mutant (*a*) of *Ephestia*).

[177] FEUSTEL, H. *Z. wiss. Zool.*, **161** (1958), 209–38 (excretion in Collembola).

[178] GAGNEPAIN, J. *Bull. Soc. Zool. Fr.*, **81** (1956), 395–410 (histochemistry of Malpighian tubules in Odonata).

[179] GERSCH, M. *Zool. Anz.*, **151** (1953), 225–36 (elimination of dyes in the eggs of the vine Phylloxera *Viteus*).

[180] HAFEZ, M., *et al. Bull. Soc. Ent. Egypte*, **40** (1956), 279–99; **43** (1959), 85–8 (storage excretion in *Vermileo*, Dipt., Neuroptera, &c.).

[181] HARMSEN, R. *The storage and excretion of pteridines in 'Pieris brassicae L.'*. Thesis, Cambridge University, 1963.

[182] HARINGTON, J. S. *Nature*, **178** (1956), 268 (amino acids in excreta of *Rhodnius*).

[183] HAYASHI, Y. *J. Sericult. Sci. Japan*, 30 (1961), 13–16 (urea formation in tissues of *Bombyx* larva).

[184] HAYDAK, M. H. *Ann. Ent. Soc. Amer.*, **46** (1953), 547–60 (effect of diet on uric acid in fat body of *Blattella*).

[185] IRREVERRE, F., and TERZIAN, L. A. *Science*, **129** (1959), 1358–9 (nitrogen excretion in mosquitos).

[186] LAMB, K. P. *J. Ins. Physiol.*, **3** (1959), 1–13 (nitrogen excretion in the Aphid *Brevicoryne*).

[187] LUDWIG, D. *Physiol. Zool.*, **27** (1954), 325–34 (excretory nitrogen during metamorphosis in *Popillia*).

[188] MCENROE, W. D., and FORGASH, A. J. *Ann. Ent. Soc. Amer.*, **50** (1957), 429–31; **51** (1958), 126–9 (synthesis of uric acid in *Periplaneta*).

[189] MADDRELL, S. H. P. *Nature*, **194** (1962), 605–6; *J. Exp. Biol.*, **40** (1963), 247–56; **41** (1964), 163–76 (diuretic hormone in *Rhodnius*).

[190] MANUNTA, C. *Genetica ed Entomologia*. **2** (1955), 269–81 (uric acid excretion in *Bombyx* after removal of the silk glands).

[191] MARTOJA, R. *Bull. Soc. Zool. Fr.*, **81** (1956), 172–3; *Acta histochem.*, **6** (1959), 185–217 (mucoid secretion in Malpighian tubules of Orthoptera, &c).

[192] der MAUR, P. A. *Z. Vererb-Lehre*, **92** (1961), 42–62 (uric acid metabolism in *Drosophila* mutants).

[193] MAZZI, V., and BACCETTI, B. *Redia*, **41** (1956), 315–41; **42** (1957), 277–82 (silk secretion in Malpighian tubules of *Donus*, Col.).

[194] MILES, P. W. *Nature*, **182** (1958), 959 (retention of food residues in *Oncopeltus* larvae).

[195] MITCHELL, H. K., and GLASSMAN, E. *Science*, **129** (1959) 268–9 (hypoxanthine in 'rosy' mutant of *Drosophila*).

[196] NATION, J. L., and PATTON, R. L. *J. Ins. Physiol.*, **6** (1961), 299–307; **9** (1963), 195–200 (nitrogen excretion in *Galleria* and other insects).

[197] NELSON, W. A. *Nature*, **182** (1958), 115 (purine excretion in *Melophagus*).

[198] NOËL, R., and TAHIR, E. *Arch. Anat. Microsc.*, **25** (1929), 587–96 (mitochondria in the striated border of *Bombyx* Malpighian tubules).

[199] NUÑEZ, J. A. *Z. vergl. Physiol.*, **38** (1956), 341–54 (neurosecretory control of water excretion: *Anisotarsus, Col.*). *Nature*, **194** (1962), 704; **197** (1963), 313 (ditto: *Rhodnius*).

[200] NUORTEMA, P., and LAUREMA, S. *Ann. Ent. Fenn.*, **27** (1961), 58–100 (excretion of *l*-isomer of valine by the salivary glands of Heteroptera).

[201] PATTON, R. L., GARDNER, J., and ANDERSON, A. D. *J. Ins. Physiol.*, **3** (1959), 256–61 (excretory efficiency of *Periplaneta*).

[202] PESSON, P. *Boll. Lab. Zool. Gen. Agr.* (Portici) **33** (1955), 341–9 (mucoprotein in excreta of Cercopidae).

[203] PHILLIPS, J. E. *J. Exp. Biol.*, **41** (1964), 15–80 (rectal absorption in *Schistocerca*).

[204] POWNING, R. F. *Aust. J. Biol. Sci.*, **6** (1953), 109–17; *J. Ins. Physiol.*, **8** (1962), 92–5 (sulphur excretion in *Tineola* and *Attagenus*).

[205] RAMSAY, J. A. *J. Exp. Biol.*, **30** (1953), 79–89 (exchanges of sodium and potassium in mosquito larvae).

[206] —— *J. Exp. Biol.*, **30** (1953), 358–69 (active transport of potassium by Malpighian tubules).

[207] —— *J. Exp. Biol.*, **31** (1954), 104–13 (excretion in *Carausius*: active transport of water); 183–216; **33** (1956), 697–708 (ditto: inorganic ions); **35** (1958), 871–91 (amino acids, sugars and urea).

[208] —— *Phil. Trans. Roy. Soc.*, B, **248** (1964), 278–314 (physiology of the cryptonephridial system in *Tenebrio*).

[209] RAZET, P. (and POISSON, R.). *C.R. Acad. Sci.*, **234** (1952), 1804–6, 2566–8; **236** (1953), 1304–6; **237** (1953), 1362–3; **239** (1934), 905–7; **243** (1956), 185–7; *Recherches sur L'uricolyse chez les Insectes*, Thesis, Rennes, 1961, 206 pp. (nitrogen excretion in insects with special reference to allantoine, &c.).

[210] ROSS, D. J. *Physiol. Zool.*, **32** (1959), 239–45 (uricase and xanthine oxidase in *Popillia*).

[211] RUSSO-CAIA, S. *Ric. Sci.*, **30** (1960), 1–48 (nitrogen excretion in *Musca* during metamorphosis).

[212] SAINI, R. S. *Trans. Roy. Ent. Soc. Lond.*, **116** (1964), 347–92 (comparative histology of the cryptonephridial system in Coleoptera).

[213] SAVAGE, A. A. *Quart. J. Micr. Sci.*, **97** (1956), 599–615 (development of Malpighian tubules in *Schistocerca*).

[214] SHAW, J. *J. Exp. Biol.*, **32** (1955), 355–82 (ionic regulation in *Sialis* larva).

[215] SMITH, D. S., and LITTAU, V. C. *J. Biophys. Biochem. Cytol.*, **8** (1960), 103–33 (fine structure of Malpighian tubules in *Macrosteles*, Jassidae).

[216] STADDON, B. W. *J. Exp. Biol.*, **32** (1955), 84–94 (nitrogen excretion in aquatic insects: *Sialis*, &c.); **36** (1959), 566–74 (ditto: *Aeschna* larva); **40** (1963), 563–71 (ditto: *Notonecta*).

[217] STICH, H., and GRELL, M. *Nature*, **176** (1955), 930–1 (incorporation of ^{32}P into Malpighian tubules of *Culex*).

[218] STOBBART, R. H. *J. Exp. Biol.*, **36** (1959), 641–53; **37** (1960), 594–608 (ionic exchanges in *Aëdes* larvae).

[219] SUTCLIFFE, D. W. *J. Exp. Biol.*, **38** (1961), 501–19 (osmotic and ionic regulation in *Limnephilus affinis*).

[220] TREHERNE, J. E. *J. Exp. Biol.*, **31** (1954), 386–401 (uptake of sodium and potassium by anal papillae of *Aëdes* larva).

[221] TSUBO, I., *et al. J. Nara Med. Assoc.*, **12** (1961), 301–17; *J. Ultrastructure Res.*, **6** (1962), 28–35 (fine structure of Malpighian tubules in grasshoppers).

[222] WESSING, A. *Verh. Dtsch. Zool. Ges.*, 1961, 167–73 (storage function of Malpighian tubules in *Drosophila*).

[223] —— *Protoplasma*, **55** (1962), 265–93, 295–301 (fine structure of Malpighian tubules in *Drosophila*).

[224] WIGGLESWORTH, V. B. *J. Exp. Biol.*, **15** (1938), 235–47 (osmotic pressure and chloride concentration in haemolymph of *Aëdes aegypti*).

[225] —— unpublished observations.

[226] WIGGLESWORTH, V. B., and SALPETER, M. M. *J. Ins. Physiol.*, **8** (1962), 299–307 (fine structure of the Malpighian tubules in *Rhodnius*).

[227] ZIEGLER, H., and ZIEGLER, I. *Z. vergl. Physiol.*, **40** (1958), 549–55 (composition of the excretory foam of Cercopid larvae).

SUPPLEMENTARY REFERENCES (C)

[228] BARRETT, F. M. and FRIEND, W. G. *J. Insect Physiol.*, **16** (1970), 121–9 (uric acid synthesis in *Rhodnius*).

[229] BERRIDGE, M. J. *J. exp. Biol.*, **43** (1965), 535–52 (nitrogenous excretion in *Dysdercus*).

[230] BERRIDGE, M. J. *J. exp. Biol.*, **44** (1966), 553–66 (diuretic hormone in *Dysdercus*).

[231] BERRIDGE, M. J. *J. Insect Physiol.*, **12** (1966), 1523–38 (energy supply for urine formation in *Calliphora*).

[232] BERRIDGE, M. J. *J. exp. Biol.*, **48** (1968), 159–74; **50** (1969), 15–28 (role of ions in urine formation in *Calliphora*).

[233] BERRIDGE, M. J. and OSCHMAN, J. L. *Tissue & Cell*, **1** (1969), 247–72 (cellular specialization in Malpighian tubules of *Calliphora*).

[234] BURSELL, E. *J. Insect Physiol.*, **11** (1965), 993–1001; *Adv. Insect Physiol.*, 4 (1967), 33–67 (nitrogenous waste products in *Glossina*).

[235] EHRHARDT, P. *Z. vergl. Physiol.*, **50** (1965), 293–312 (storage excretion in mid-gut cells of *Aphis fabae*).

[236] GIRARDIE, A. *C.R. Acad. Sci. Paris*, **262** (1966), 1361–4 (diuretic hormone in *Gryllus*).

[237] GRIMSTONE, A. V., MULLINGER, A. M. and RAMSAY, J. A. *Phil. Trans. Roy. Soc. Lond.* B. **253** (1968), 343–82 (physiology of cryptonephridial system in *Tenebrio*).

[238] HARMSEN, R. *J. Insect Physiol.*, **12** (1966), 9–30 (role of pteridines in nitrogen excretion in *Pieris*).

[239] HOPKINS, T. L. and LOFGREN, P. H. *J. Insect Physiol.*, **14** (1968), 1803–14 (adenine metabolism in *Leucophaea*).

[240] MADDRELL, S. H. P. *J. exp. Biol.*, **51** (1969), 71–97 (movements of ions and water in Malpighian tubules of *Rhodnius*).

[241] MADDRELL, S. H. P. *Adv. Insect Physiol.*, **8** (1971), 200–331 (insect excretion: review).

[242] MADDRELL, S. H. P. and CASIDA, J. E. *Nature*, **231** (1971), 55–6 (mechanism of diuresis induced by insecticides in *Rhodnius*).

[243] MADDRELL, S. H. P., PILCHER, D. E. M. and GARDINER, B. O. C. *Nature*, **222** (1969), 784–5; *J. exp. Biol.*, **54** (1971), 779–804 (5-hydroxytryptamine and cyclic AMP in secretory activity of Malpighian tubules).

[244] MARSHALL, A. T. *Quart. J. micr. Sci.*, **106** (1965), 37–44; *J. Insect Physiol.*, **12** (1966), 635–44, 925–32; **14** (1968), 1435–44 (chemical nature of dwelling tubes of Cercopid larvae).

[245] MILLS, R. R. *J. exp. Biol.*, **46** (1967), 35–41 (diuretic hormone in *Periplaneta*).

[246] OHNISHI, E. *et al. Science*, **160** (1968), 783–4 (calcium oxalate in cocoon of *Malacosoma*).

[247] PILCHER, D. E. M. *J. exp. Biol.*, **52** (1970), 653–65; **53** (1970), 465–84; *J. Insect Physiol.*, **17** (1971), 463–70 (hormonal control of Malpighian tubules of *Carausius*).

[248] SIDHU, H. S. and PATTON, R. L. *J. Insect Physiol.*, **16** (1970), 1339–48 (nitrogenous excretion in *Lipaphis*).

[249] STADDON, B. W. *J. exp. Biol.*, **51** (1969), 643–69 (water balance in *Ilyocoris*, Naucoridae).

[250] WALLACE, M. M. H. *Aust. J. Zool.*, **15** (1967), 1173–206 (excessive uric acid storage in *Sminthurus*, Collembola).

[251] WESSING, A. *Protoplasma*, **56** (1963), 434–65; *Verh. dtsch. zool. Ges.*, 1964, 337–45; 1967, 633–81; *Naturwiss. Rundschau*, **19** (1966), 131–47 (excretion in *Drosphila*, &c.).

[252] WESSING, A. and EICHELBERG, D. *Z. Naturforsch.*, **23** (1968), 376–86 (subdivision of Malpighian tubules of *Drosophila*).

Chapter XIII

Metabolism

METABOLISM COMPRISES all the chemical changes which the constituents of the living body undergo. Broadly speaking, these changes serve two purposes, growth on the one hand and the performance of work on the other. They may be studied by following the chemical transformations or changes in composition of the body, by observing the energy changes, particularly the production of heat, that accompany the chemical phenomena, or by following the consumption of oxygen and the evolution of carbon dioxide, which give an indirect measure of the energy changes that are taking place.

CHEMICAL TRANSFORMATIONS

Fat metabolism—Fat is the chief form in which energy is stored. It is usually present in greatest amounts in the mature larva before metamorphosis. In *Gasterophilus* it rises from 3·5 per cent. of the dry weight in December to 25·9 per cent. in June.[168] In the mature larva of *Calliphora* fat comprises 22 per cent. of the dry weight,[308] of the corpse-fly *Ophyra cadaverina* 41–45 per cent.,[276] of the honey-bee worker 18 per cent., of the drone 21·3 per cent.,[274] of *Bombyx mori* 7·1 per cent.,[89] and of *Galleria* 43·6 per cent.[267] In *Lucilia* it makes up about 11 per cent. of the dry weight until the 3rd instar but rises to 30 per cent. in the full-grown larva.[248]

In composition this fat is exceedingly varied. More than 20 lipids can be recognized in extracts of *Drosophila*, not counting lipids bound to amino acids and proteins.[530] In two Lepidopterous larvae, *Bombyx mori* and *Hyphantria*, the fat body lipids contain small amounts of free fatty acids; 70–80 per cent. of triglycerides; about 10 per cent. of di- and monoglycerides. Eight kinds of triglycerides can be separated and ten fatty acids of chain lengths C_8–C_{18}; more than 80 per cent. are C_{16} or C_{18} fatty acids including the unsaturated acids.[514]

The **unsaponifiable fraction** is usually small (1·5–1·6 per cent. in the 'chrysalis oil' of *B. mori*,[21] 1·56 per cent. in the mealworm larva[19]). In the 'chrysalis oil' a large part of this fraction is composed of hydrocarbons; about one third consists of sterols, of which 85 per cent. is cholesterol, 15 per cent. sitosterol.[21] Whereas there is very little capacity for sterol synthesis from acetic acid in the silkworm larva, sterol is much more actively produced in the pupa, provided the brain is retained.[500] ^{14}C-mevalonic acid introduced into *Musca* is incorporated into various lipid-soluble compounds, but only traces go into sterols;[432] and in *Periplaneta* some 5–8 per cent. of ^{14}C-acetate injected is incorporated into sterols.[453] [See p. 636.]

By far the greater part of the 'total fat' or 'ether extract' consists of **neutral fats.** These usually contain a rather high proportion of unsaturated acids. In the mealworm, for example, the fatty acids are: palmitic and other saturated acids 22·6 per cent., oleic 44·7 per cent., linoleic 32·3 per cent., linolenic 0·35 per cent.;[19] and in 'chrysalis oil', palmitic 20 per cent., stearic 4 per cent., oleic 35 per cent., linoleic 12 per cent., linolenic 28 per cent.[21] Linolenic acid is formed in large amounts before pupation.[227] *Anthonomus grandis* yields a complex mixture of 23 fatty acids of chain length C_6–C_{20}; of these two thirds have at least one double bond.[446] Such high values for linoleic and linolenic acids are quite unlike those found in mammals. Thus the iodine value, which affords a measure of the degree of unsaturation, is generally high: 117 in silkworm oil, 112–159 in other Lepidopterous larvae, 108·6–118 in the phytophagous Chrysomelids, 68·5 in the xylophagous beetle *Ergates faber*, but ranging as low as 37·3 in *Gasterophilus*[288] and 1·5–22 in the butter-like fats of Aphids.[288] The fat of *Melanoplus* is characterized by a great preponderance of stearic acid and by a high proportion of C_{20} acids in the unsaturated fraction. In this and other insect fats sulphur is present in the neutral fat, perhaps as sulphuric esters.[118]

At ordinary temperatures the fat may be a liquid oil, as in *Ergates faber*, or a crystalline solid, as in *Oryctes nasicornis*, both of these being wood-feeding beetles.[288] In the eggs of grasshoppers it seems to be a general rule that those which pass the winter in this stage have fats of lower melting point: in *Melanoplus differentialis*, which spends the winter as an egg, the fat is liquid at room temperature; in *Chortophaga viridifasciata*, which winters as a nymph, the fat is solid. In this case the iodine number is the same in both (135–140);[271] but the low melting point in the winter eggs may possibly be due to a higher proportion of short chain fatty acids. That is the explanation of the liquid fat of the Aphid *Pemphigus*, which contains glycerides of butyric, caprylic, and lauric acids.[288] After prolonged starvation in *Rhodnius* the few remaining fat droplets in the fat body become almost free of unsaturated fatty acids; but these reappear again soon after feeding.[527] [See p. 636.]

The **phosphatides** in the fat body of *Loxostege* (Pyralidae) amount only to 0·2–0·4 per cent. of the total fats, much less than in mammalian adipose tissue.[235] In *Leptinotarsa* the lecithin content varies widely during the life of the insect; it is largely influenced by the diet, being greater in insects feeding on young foliage which is rich in lecithin.[81] *Musca* contains a variety of phospholipids of which the most abundant is phosphatidylethanolamine.[492] Phospholipids are particularly abundant in the haemolymph of *Galleria* (p. 509) which contains many times the amount in other insects; phosphatidylethanolamine again predominates.[449] In the silkworm *B. mori* there is a large increase in the synthesis of phospholipids during moulting. This is probably a reflection of the great increase in the numbers of mitochondria associated with all the intense enzymic synthesis that takes place during growth (p. 69).[468] [See p. 636.]

The neutral fats are doubtless formed in great part from carbohydrates or proteins in the food. *B. mori* at pupation contains twice as much fat as in the leaves it has eaten;[205, 296] and fly larvae will lay down fat when fed on fat-free albumen.[222] The same is seen in mosquito larvae *Aëdes* fed on pure proteins or carbohydrates.[314] In mosquitos there is a sexual difference in lipid metabolism: after feeding on glucose for 7 days the quantity of triglycerides stored in the female is 50 times greater than in the male. Polyunsaturated fatty acids are absent.[415]

When the food is rich in fats, the fat laid down by the insect is influenced by their composition, but it is sufficiently different to prove that the fat is extensively changed in metabolism: there is not a simple storage of ingested fat. This is seen in *Galleria*, in which the pupal fat consists of glycerides of higher fatty acids, quite unlike the waxes of the food (p. 606)[204, 205] and in the mealworm[19] and *Dermestes*.[269] The influence of the food is well seen in the larva of the beetle *Pachymerus*, which contains in its fat some 24 per cent. of lauric acid—far more than in most insect fats, though only about half that in the fat of the oil-palm kernels upon which it feeds;[56] also in *Blattella*, which has an iodine value of 51 when fed on coconut oil (7), 59 when fed on butter fat (27) and 64 when fed with lard (49);[212] and in *Lucilia* larvae in which the fat has an iodine value of 140 when fed on fish fat with a value of 113, and 60 when given a fat with a value of 30.[323] The body fat of the Chironomid *Tanytarsus* resembles that of the algae and plankton on which it feeds.[410]

There is no clear evidence that insects from hot climates have fats of lower iodine value than those from colder regions, such as occurs in plants; systematic differences override any possible effects of temperature.[288] But when a given species is reared at different temperatures there is a fall in the iodine value as the temperature rises. A very small effect was noted in *Lucilia* (15° C.: 75·8; 25° C.: 72·7; 35° C.: 71·0)[248] and in *Heliothis* (Lep.).[74] In the phosphatides in *Calliphora* and *Phormia* there is a very definite effect: the iodine value falls 26 units for a rise of 18° C.[104, 386a]

These fats, which are stored chiefly in the fat body (p. 442) provide most of the energy for growth and metamorphosis (p. 598) and during periods of starvation (p. 600). In many insects they are the chief energy reserve used for flight (p. 602).

Carbohydrate metabolism—Sugar in the blood and tissues is probably the most readily available source of energy. Its importance is most obvious in the bee, in which the average level in the blood of foraging bees is 2·6 per cent. while values up to 4·4 per cent. (even 11·5 per cent. in one instance) may occur! It is a dextro-rotatory sugar, fermentable by yeast, and is presumably glucose. If its concentration in the blood falls below 1 per cent., the bee is unable to fly, but runs along with wings vibrating. Below 0·5 per cent. the bee becomes almost motionless.[24] During flight, the bee, weighing about 100 mg., will utilize sugar at the rate of about 10 mg. per hour.[157] Its normal store is exhausted in about 15 minutes, during which time it may fly some 5½ km. Its sugar disappears completely after a few hours of starvation, so that it must have constant access to the nectar of flowers or honey in the hive.[24] The figures given above refer to the stores of sugar carried in the blood. If the honey sac is filled the bee can fly for hours without fatigue, so that the flight range is greatly increased.[327] In most insects, as we have seen (p. 426), there is very little fermentable sugar in the blood. In the silkworm larva, for example, it averages only 0·022 per cent. This small amount disappears completely after an injection of insulin.[309] The chief sugar in the haemolymph of most insects is the non-reducing disaccharide trehalose (p. 426). [See p. 637.]

In the larva of the bee, some of the carbohydrate ingested with the food is stored in the blood, partly as glucose and partly as a polysaccharide of some kind,[252] and as trehalose. But carbohydrate is stored mainly in the fat body, in the form of glycogen. In the mosquito larva *Aëdes* massive deposits of glycogen in the sarcoplasm around the muscles are at least equal in importance to those in the fat body (Fig. 341, A, B);[314] and in the *Drosophila* adult, there are deposits of glycogen also in the knobs of the halteres, in the flight muscles (Fig. 341, C, D) and in the proventiculus.[316] In the larva of *Bombyx mori* glycogen occurs in greater or less quantities in practically all the tissues: mid-gut, ventral ganglia, epidermis, tracheal epithelium, &c.[40] The same is true in *Melasoma*,[234] in *Aëdes*[314] and *Simulium*[231] larvae.

The importance of glycogen as a reserve substance varies in different groups. In the mature larva of the bee it may comprise 33·5 per cent. of the dry weight;[274] in the larva of *Gasterophilus* at one stage of growth it may reach 31 per cent. of the dry weight.[168] But in *Ophyra*,[276] *Calliphora*,[308] *Lucilia*,[88] *Bombyx mori*,[224] *Malacosoma*,[255] it is present only in small amounts—less than 5 per cent. of the dry weight. In *Heliothis obsoleta* (Lep.) the peak of 7 per cent. of the dry weight

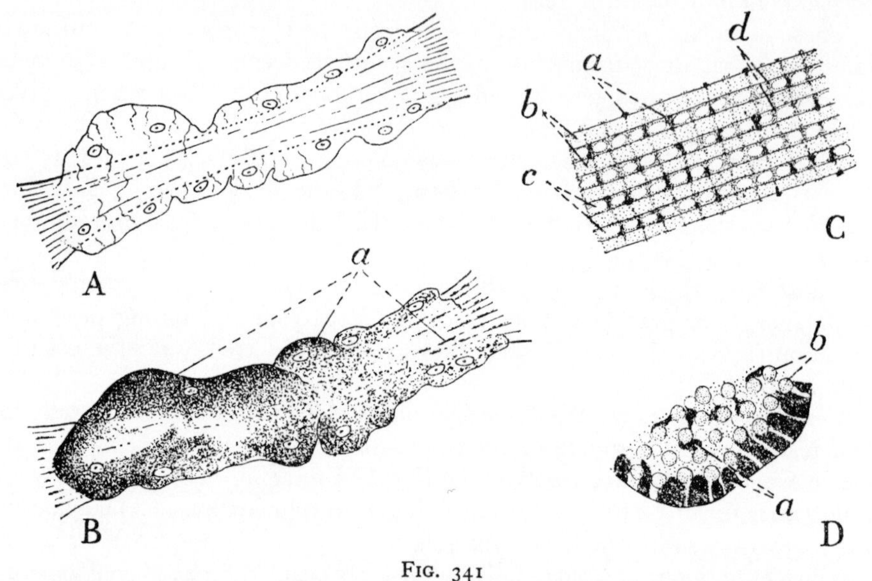

FIG. 341

A, muscle of fully starved *Aëdes* larva; no glycogen can be demonstrated histologically; B, after feeding on casein for 7 days; sarcoplasm on the surface of the muscle and between the fibrils is full of glycogen; C, longitudinal section of indirect flight muscle of *Drosophila*; D, transverse section of the same (*after* WIGGLESWORTH).

a, deposits of glycogen in the sarcoplasm; b, fibrils; c, sarcosomes; d, sarcoplasm.

is reached in the mature larva; it has fallen to 0·5 per cent. in the pupa 12 days old.[74] In the adult bee, glycogen forms from 0·3–0·9 per cent. of the live weight.[24] The total carbohydrate in the honey-bee averages 5·6 mg., of which 36 per cent. (2 mg.) is glycogen; this reaches its maximum in the foraging bee.[428] In many cases this glycogen is doubtless derived from carbohydrate in the food. In the larva of the Noctuid *Prodenia* the blood glucose rises within 15 minutes of feeding; the glycogen in the fat body reaches its maximum 7 hours after a meal.[88] In the larva of *Aëdes* fed upon pure substances, the deposition of glycogen varied with different sugars; it is very evident after feeding on protein or on the amino acids alanine and glycine, but does not occur on a diet of fat.[314] Glycogen is deposited in the fat body of *Rhodnius* in readiness for the formation of chitin, being obviously derived from the proteins in the food.[314] The silkworm may more than double its glycogen reserve in four days at the end of larval life, apparently by transformation of its protein or fat.[12] During spinning, the respiratory quotient in *B. mori* falls to 0·6 suggesting conversion of fat to carbohydrate.[152] But there is no clear evidence for the conversion of fat into stored glycogen (see p. 602).

In the diapausing pupae of *Hyalophora cecropia* and *Samia cynthia* glycerol accumulates in the tissues and haemolymph at the expense of glycogen; the accumulation is exaggerated by anaerobiosis and by chilling (p. 427); eventually, when diapause comes to an end, the glycerol is reconverted into glycogen.[528] After injury to the integument in the diapausing pupa of *Hyalophora* blood trehalose increases about threefold and glycerol falls somewhat.[531]

Other carbohydrates, considered elsewhere, are chitin of the cuticle (p. 31) and the mucopolysaccharides of the connective tissues and basement membranes (p. 438). Mucoid substances are present in the salivary secretion of many insects.[331] Hyaluronic acid is probably important as a cementing substance between the cells; the cells of the mid-gut epithelium fall apart when soaked in a solution of hyaluronidase.[332] [See p. 637.]

Protein metabolism—Proteins provide the chief structural elements of the muscles, glands and other tissues. They comprise some 2·2 per cent. of the fresh substance of the adult bee when newly hatched, 3·2 per cent. in the foraging bee with fully developed flight muscles.[166] Proteins may be transferred from one part of the body to another. In young bees the pharyngeal glands become greatly developed for the feeding of the brood; when these glands atrophy, the protein is transferred to the wax glands and the flight muscles;[23, 137] whereas in the queen termite and in queen ants, the disused flight muscles are utilized for the maturation of the gonads; and after each moult in *Rhodnius* and other insects the intersegmental muscles undergo autolysis and the proteins are utilized elsewhere (p. 45). In addition, a certain amount of protein is stored in the fat body (p. 443) and much is deaminated or converted into carbohydrate or fat and used for energy production. The fate of the waste nitrogen liberated in this process has been considered in Chapter XII; the intermediary metabolism of amino acids and proteins is discussed on p. 603. During growth, and particularly at metamorphosis, there is extensive synthesis of protein. The available evidence suggests that this new protein is always the product of synthesis from free amino acids. [372] The same applies to the silk protein produced by *B. mori* before pupation. The individual amino acids, and the total amino acid concentration in the tissues, go through widely different cycles during the growth and development of different insect species.[386]

If particular amino-acids are utilized for the production of some secretion, the composition of the proteins in the body as a whole may be greatly modified. This is well seen in the silkworm. In the mature larva the proteins consist of: glycine 10 per cent., alanine 8·7 per cent., tyrosine 4·3 per cent., valine 1·7 per cent., leucine 4·8 per cent., aspartic acid 1·6 per cent., glutamic acid 3·5 per cent., phenylalanine 2·4 per cent., proline 1·5 per cent. The silk of the cocoon is composed chiefly of tyrosine, glycine and alanine (p. 604), with the result that the protein of the silkworm moth is greatly altered and consists of: glycine 3·5 per cent., alanine 3·2 per cent., tyrosine 1·6 per cent., valine 1·7 per cent., leucine 8·5 per cent., aspartic acid 2·7 per cent., glutamic acid 5·7 per cent., phenylalanine 2·7 per cent., proline 4·0 per cent.[1] [See p. 637.]

Chemical changes during growth—Little is known of the changes which take place in the separate tissues during development; what information exists is derived from analyses of the body as a whole. By this means only the grossest alterations are detected; if the breakdown of a given substance in one part of the body is compensated by its synthesis in another, analysis of the whole insect may show no change at all.

In the egg the chief reserve substance which provides the energy for growth, the 'work of development',[276] is usually fat. In the silkworm egg, fat yields about two-thirds of the total energy,[89] falling from 8·1 per cent. to 4·4 per cent. of the wet weight.[287] In *Carausius*, as in the silkworm two thirds of the energy consumption of the egg comes from fats: about half the lipid reserves are oxidized during development as against 20–25 per cent. of the proteins.[337] The remaining third is provided by glycogen, which falls from 2 per cent. to 0·74 per cent. of the wet weight,[287] and probably protein.[89] A curious change in the diapausing egg of *B. mori* consists in the conversion of virtually all the glycogen in the egg into sorbitol and glycerol; they are reconverted to glycogen when diapause comes to an end.[387] In *Melanoplus* eggs, higher fatty acids compose 17–22 per cent. of the dry weight at the time of laying; these diminish rapidly during the periods of growth before and after hibernation, and at the time of hatching less than half remains;[271] but in *Oncopeltus* only about 11 per cent. of the total lipid store is utilized.[361] The utilization of glycogen in the egg of *Melanoplus*, as in the insect during moulting (p. 443) seems to be correlated with the formation of chitin in the 'white cuticle' of the egg (p. 3) and the embryonic and first instar cuticles.[147] It provides a comparatively small part of the energy early in development, whereas 72·5 per cent. of the fat is used up[271] and about 6·6 per cent. of the protein, uric acid showing a corresponding increase.[32]

During larval growth there is usually no striking change in composition until shortly before pupation. In *B. mori* the newly hatched larva contains about 4 per cent. of fat; this quickly falls, and remains at a low level; only in the last stage is fat stored extensively.[167] In *Malacosoma* (Lep.), fats compose only 0·66 per cent. of the dry weight at hatching; at the end of larval life they increase rapidly to 28·8 per cent.[255] In *Lucilia*, fat increases in the same manner up to 30 per cent.[323] This great increase in fat may result in a relative fall in other constituents, such as water[323] and protein; in *Malacosoma* larvae the total nitrogen falls from 15·1 per cent. of the dry matter at hatching, to 10·5 per cent. in the full grown caterpillar.[255] In those insects in which it is stored extensively, glycogen seems to be accumulated more gradually during larval life, as in *Apis*[274] and *Gasterophilus*.[73] In *B. mori* there is a sexual difference, accentuated in the adult but already noticeable in the larva; the females storing more glycogen, the males more fat.[297]

Table 3 summarizes the changes in composition which take place in the larva of the worker bee.[274] During the first four days this larva receives a diet consisting of 58 per cent. protein, 8 per cent. fat and 10 per cent. sugar; after the third day the diet consists of 28 per cent. protein, 4 per cent. fat, 45 per cent. sugar.[170] Storage of glycogen takes place in the second period, when carbohydrate predominates in the food.[225] By the end of larval life, about 50 per cent. of the dry weight consists of reserve substances.

The same Table shows the changes in composition in the pupa. This has a wet weight of 150 mg., and the stage lasts 13 days; during this time it consumes 8·4 mg. of glycogen (about 95 per cent. of its store) and 4·9 mg. of fat (about 75 per cent. of its store); the glycogen, as in other insects, being used up first.[274] In the Garden Chafer *Phyllopertha horticola* no food is taken from November,

TABLE 3

Day of larval life		% dry matter	% glycogen	% fat	% nitrogen
1		22·9	—	—	—
2		—	2·5	1·5	2·9
3		17·8	—	—	—
4		—	5·6	—	—
5		20·0	—	3·6	1·5
6		—	6·6	—	—
Day of pupal life	*Weight in mg.*				
1	147–176	23·0	6·2	4·1	1·2
3–4	142	22·0	5·2	3·7	1·6
7	123	19·2	3·0	2·8	1·7
12	113	15·2	0·5	1·5	1·7
13 (hatched bee)	111	14·8	0·5	0·9	2·0

when the full-grown larva goes into hibernation, until June, when the eggs of the adult are matured and laid. Of the store of fat and glycogen present in November, half is used up by the time the adult emerges; the other half is used in the formation of eggs.[447] Most insects consume chiefly fat in the pupal stage; in *Ophyra cadaverina*, over 90 per cent. of the substance consumed is fat;[276] in *Calliphora*, the fat content falls from 7 per cent. of the dry weight at the beginning to 3 per cent. at the end.[308] We shall see later that during pupal development the energy production follows a U-shaped curve; it is high at the beginning and end of development, with a low period between (p. 633). The rate of disappearance of fat follows the same course in *B. mori*,[224] *Ophyra*,[276] *Tenebrio*,[88] &c. In the pupa of *Galleria*, reducing substances follow the same type of curve.[63] Free amino acids fall steadily as histogenesis proceeds (in *Celerio*[139] and other Lepidoptera[60] and in *Lucilia*[88]). In the silkworm pupa about 8·6 per cent. of the dry substance, or 12·1 per cent. of the total energy is used up.[89] [See p. 637.]

In *Pyrausta nubilalis*, in which about 80 per cent. of the total fat is utilized during the pupal period, all the fatty acids, saturated and unsaturated, seem to be affected indifferently and there is no change in the iodine value.[288] The same applies to *Cossus*.[288] In *Pieris brassicae*, on the other hand, there is a selective consumption of the unsaturated acids, especially perhaps linolenic acid. The iodine value in the larva in November is 149·9; in the pupa in April 100·5. The viscous oil extracted in autumn is quite different from the solid fat, melting at 36° C., extracted in the spring.[288] A similar selective consumption of the unsaturated fats is found in *Lucilia*[88] and *Heliothis* (Lep.).[74]

The extent to which protein contributes to the energy metabolism of the pupa varies. *Bombyx mori* and the Sphingid *Celerio* afford a striking contrast in this respect. In *B. mori* about half the protein of the mature larva is used to make the cocoon, half the total dry weight being lost at pupation; whereas *Celerio* loses only one quarter of the dry weight. During pupal development *Celerio* covers 20 per cent. of its energy metabolism from fat, the remaining fraction being chiefly protein; whereas *B. mori* uses fat for nearly 50 per cent. of its total energy, the remainder being covered chiefly by carbohydrate.[139, 224] In all groups there is a great reduction in the water content at pupation. In *B. mori*

100 larvae contain 37·4 gm. of solids, 139·5 gm. of water; 100 cocoons and pupae contain 33·9 gm. of solids and 83 gm. of water.[224]

Chemical changes in starvation—During starvation the relatively small amounts of carbohydrate are first consumed. In the mealworm larva starved at 30° C., the glycogen falls from 2·04 per cent. of the wet weight to 0·68 per cent. in a week;[215] in Odonata it falls from 0·2 per cent. to zero.[272] Proteins may be used extensively—as in *Dytiscus*,[236] adult *Celerio* which use up 41 per cent. of their protein before death,[140] *Melolontha* which uses 22 per cent. and *Geotrupes* which consumes 20 per cent.[272] In starved mosquito larvae the proteins in all the tissues are largely wasted: nuclei and nucleoli are small everywhere, protein is reduced in the fat body cells, oenocytes, &c., the proteins in the muscles are diminished and the epithelium of the mid-gut is attenuated.[314] Or the protein metabolism may be very limited—as in *Periplaneta*,[236] in Odonata, where the protein content falls from 53·4 per cent. to 51·7 per cent. of the dry weight, or *Bombus*, where it falls from 58·3 per cent. to 52·1 per cent.,[272] or in *Apis*, in which, at least at 23° C., there appears to be no utilization of protein at all during fasting.[166] Only under certain conditions does the mature bee seem able to mobilize and break down its protein reserves.[138, 194] The same applies to *Periplaneta*; for if this cockroach is given protein alone, it exhausts its reserves of fat and then starves to death with the fat body still containing masses of protein crystals.[113]

In the larva of *Popillia japonica* starved for 4 weeks, 80 per cent. of the glycogen was consumed (falling from 0·47 to 0·11 per cent. of the dry weight) and 71 per cent. of the lipid (falling from 2·9 to 0·8 per cent.). Glycogen and lipid are used concomitantly, as in *Aëdes* larvae[314] and *Drosophila* adults,[316] and body protein is also used as an energy source. [466] In the starved larva of *Popillia*[456] and of the beetle *Anomala*[480] there is a progressive increase in amino acids and other non-protein nitrogen-containing compounds in the haemolymph. Perhaps this represents the compensation for the loss of chlorides and other inorganic ions during starvation (p. 430).

But fat is always the chief reserve substance that is drawn upon. In the mealworm larva, fat comprises about 14·8 per cent. of the wet weight; and when starved at 30° C. about half of this is used up in one month.[215] In adult *Celerio* 70 per cent. of the fat is broken down by the time death occurs;[140] *Melolontha* adults use up 85·6 per cent. of their fat before death, *Geotrupes* about 75 per cent., and in Odonata the fat content falls from 11·7 per cent. of the dry weight to 5·9 per cent.[272]

The tsetse fly *Glossina* shows a regular hunger cycle, lasting about five days. The hungry fly has an empty gut and a small fat body. Two days after feeding, the gut is empty and the fat body is at its maximum size. Normally the fly becomes hungry and seeks food when the fat in the fat body has been metabolized. Only in very dry air does death occur from desiccation before this happens. In the later days of the hunger cycle the curve for dry matter runs parallel to that for fat; it is mainly fat that is being consumed. The fly does not seem to experience 'thirst' as distinct from 'hunger'; desiccated flies do not become hungry any sooner than those in moist air.[216]

Similarly, during hibernation, fat forms the chief reserve. In *Culex* adults, the mean wet weight fell from 3 mg. in September to 2 mg. in April; the fat fell from 0·91 mg. to 0·13 mg.; solids other than fat remained approximately constant. Meanwhile, chiefly as the result of this loss of fat, the proportion of water rose from 54 per cent. to 65 per cent.[48] In some insects, on the other hand, although much fat is laid in before the winter, very little is consumed until the spring. Full-grown larvae of *Pyrausta nubilalis* contain fat equal to 26·5 per cent. of the wet weight. But during the seven months diapause there is almost no fall. Breakdown occurs shortly before pupation, the pupa containing 9·7 per cent. In this process there is no preferential utilization of unsaturated fats, the iodine value ranges from 76 to 81 throughout.[288] And adults of *Leptinotarsa* increase their fat from 2·6 per cent. of the wet weight in September to 12·6 per cent. in November; by March this has fallen only to 11·6 per cent.; the rapid fall to 3·5 per cent. takes place only with the renewal of activity.[43]

Chemical changes during exertion—During its active stages the unfed insect consumes its reserves in the production of energy. In the larva of *Aëdes* swimming in clean water at 28° C., glycogen and fat are used concomitantly; in ten or twelve days visible fat and glycogen have disappeared, and protein in all

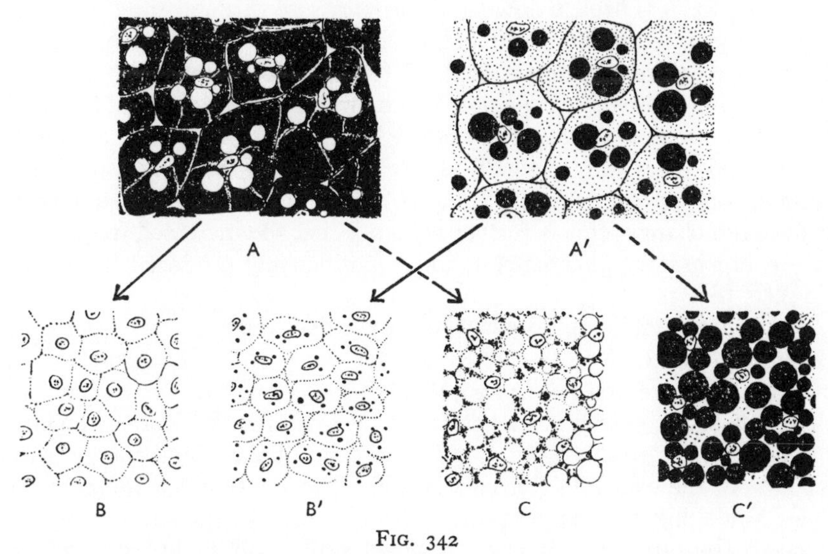

FIG. 342

A, fat body cells in adult *Drosophila* showing glycogen. *A′*, the same showing fat droplets. *B*, the same after starvation until moribund; no glycogen remains. *B′*, the same showing minute fat droplets in the starved fly. *C*, fat body cells in fly which had flown to exhaustion; granules of glycogen between the fat vacuoles. *C′*, the same showing fat droplets (*after* WIGGLESWORTH).

the tissues is greatly reduced.[314] Likewise in the adult *Drosophila* at 25° C.; the insect runs rapidly around and exhausts completely its stores of fat and glycogen in 48 hours (Fig. 342, A, B, A′, B′).[316] And in *Phormia*, in which death occurs in 4–5 days, 69 per cent. of the lipides (the rest being doubtless an integral part of the tissues), 41 per cent. of water soluble extractives (amino acids, sugars, &c.) and 16 per cent. of the non-skeletal protein are used up.[185] [See p. 637.]

But during the exceedingly active metabolism of flight the source of energy may change. Grasshoppers lay down large stores of fat before making migratory flights, and the respiratory quotient of *Schistocerca* during prolonged flight (p. 632) indicates that these fats are being utilized.[182] *Drosophila*, on the other hand, uses only carbohydrate during flight,[51, 317] and the changes which have taken place in the fat body after *Drosophila* has flown to exhaustion are quite different from those in the starved fly (Fig. 342, A, C, A′, C′).[316] Likewise in *Culex*, glycogen disappears during flight without obvious change in the fat deposits.[389] Similar results have been obtained in *Eristalis*.[507] The difference between these insects may lie in the relative speeds of metabolism: only glycogen can be metabolized sufficiently rapidly to meet the intense demands of the indirect flight muscles of *Drosophila*.[316]

Different sugars and other substances vary in their efficiency as sources of energy for flight, depending on the speed with which they can be metabolized; the completely exhausted fly can be restored to continuous flight within 30 seconds of starting to feed on glucose. In *D. melanogaster*, 1 μgm. of glucose will maintain flight for about 6 minutes; the total store of glycogen carried by the mature fly will last about 5 hours.[316, 527a] The failure of protein to prolong life materially in *Calliphora* is likewise attributed to the formation of carbohydrates at such a slow rate that the ordinary muscle metabolism cannot be maintained.[103] The same is seen in *Periplaneta*.[113]

In Diptera, blood trehalose provides the immediate carbohydrate reserve for the flying insect; it is broken down as far as pyruvate through the classic glycolysis pathway (van Emden–Meyerhof pathway) and oxidized by means of the citric acid cycle (Krebs cycle) (p. 623).[529] Trehalose is liberated chiefly from the fat body, and in *Phormia* 'exhaustion' results when trehalose cannot be supplied to the flight muscles at the necessary rate.[388] In the flying bee, sugar is consumed at a rate of 3·9–10·5 (av. 6·7) mg./hr. Partial clipping of the wings causes a large increase in wing-stroke frequency (p. 165), but the rate of sugar consumption per hour does not change; this is further evidence that the increased frequency is a passive compensatory phenomenon and is not actively produced by the insect (p. 165).[506] [See p. 637.]

Aphis fabae, containing 10 per cent. of the dry weight as fat, 3 per cent. as glycogen, can continue in tethered flight for 7–12 hours. They first consume glycogen, during the first hour or so; thereafter they use fat. For a 6-hour flight 90 per cent. of the energy comes from fat.[390] Lipids in *Schistocerca* amount to 10 per cent. of the fresh weight, and 85 per cent. of the stored energy is in the depot fat. All the available glycogen is utilized in the first few hours of flight; during 5 hours' flight 80–85 per cent. of the total energy expenditure comes from fats, mostly from the fat body reserves, which can sustain flight for 12–20 hours. The total metabolic rate is about 75 kcal./kg./hr. A large migrating swarm (say, 15,000 tons) will require as many calories per day as 1½ million men.[522] The hydration of stored glycogen makes it much less profitable than fat as a means of storing energy for flight; weight for weight, fat is about 8 times more profitable than glycogen; honey and nectar are intermediate.[523] The exclusive combustion of carbohydrates in flying Diptera and Hymenoptera is exceptional.

During its long migratory flight the Monarch Butterfly *Danais plexippus* uses only fat as a source of energy;[362] and in all the Lepidoptera studied carbohydrate is converted into fat before utilization by the flight muscles (p. 632). The fuel which enters the citric acid cycle is activated acetic acid; this may come either from pyruvic acid derived from glycolysis or from the breakdown of fatty acids (p. 623). The flight muscles of insects (unlike the skeletal muscles of vertebrates) receive such an abundant supply of oxygen that even during the intense activity of flight the substrates of the citric acid cycle can be completely oxidized, and thus fatty acids are suitable for direct combustion in the flight muscles.[534]

The 'condensing enzyme', by which acetic acid is introduced into the citric acid cycle, shows very high activity in the flight muscles of *Locusta*; it is 130 times more active than in the leg muscles of the rat.[534] Further, all the Lepidoptera examined have lipase in the flight muscles, but this is most active in good fliers; it is present also in other insects capable of prolonged flight activity, such as locusts (*Schistocerca*) and dragon-flies (*Pentala*); it is poorly developed in the bee *Xylocopa*.[406] In *Hyalophora cecropia* there is a large reserve of fat in the male pupa, to provide for the intense flight activity of the adult male. [See p. 637–8.]

In the tsetse fly *Glossina* there is evidence that the large amount of proline in the blood is utilized as a source of energy in the early stage of flight. The liberated nitrogen is conveyed by transamination to glutamic acid.[376] [See p. 638.]

In the honey-bee, flight ability and phosphorus metabolism remain at a high level until old age; in male *Musca* the power of flight is lost by the end of the second week of adult life and at the same time acid glycerophosphatase and ATPase decline to a minimum.[492] These enzymes increase almost 100 per cent. during the first week of adult life in the honey-bee.[493] The corresponding increase in cytochrome, &c., in the young adult has already been described (p. 155.)

Intermediary metabolism—The fat body is perhaps the most important site of intermediary metabolism in insects (p. 442). The oenocytes are probably concerned in lipoprotein metabolism and perhaps in other changes (p. 447). The pericardial cells can break down haemoglobin in *Rhodnius* to biliverdin; perhaps they are concerned in other aspects of protein metabolism (p. 440). The cells of the mid-gut are certainly important. In the larva of *Aëdes*, sugars are converted into glycogen and deposited in these cells before any appears in the fat body; and glycine and alanine are deaminated and converted to glycogen in the mid-gut epithelium.[314] The mid-gut of *Prodenia* has similar functions.[322] Alkaline phosphatase is very abundant in the gut wall of insects but its function is not known.[80] The iron and copper deposited in the mid-gut cells (p. 581) [303, 315] may merely represent an accumulated by-product of metabolism, or it may act as a reserve for future use. The Malpighian tubes retain many of the same metabolic powers as the mid-gut (p. 505). Their high content of riboflavine, which appears to be metabolized during prolonged starvation,[219] and the abundance of acid phosphatase, possibly concerned in the phosphorylation of riboflavin to form the 'yellow enzyme' flavoprotein (p. 622)[80] have suggested that they may have an 'hepatic' as well as a 'renal' function.

Descriptive accounts of the distribution of alkaline and acid phosphatase in many insect tissues have been published[330, 346 352], but no clear evidence of the function of these enzymes in intermediary metabolism has emerged. The abundance of phosphatase in the cell borders of the glandular epithelium of silk glands, and the abundance of ribonucleic acid in the main body of these same cells suggests that phosphatase may form part of the system of enzymes concerned in liberating the finished fibrillar silk protein from a complex with nucleic acids.[328]

In *Prodenia eridania* not only is glucose rapidly converted to trehalose, but trehalose can be formed from acetate; this must indicate that in Lepidoptera fats can be reconverted into carbohydrates. Sugars are converted to fat especially in the fat body.[534] Uridine diphosphate glucose, which in other organisms provides the glucose needed for glucoside synthesis, is present also in the fat body of *Schistocerca* and is utilized in providing the glucose for trehalose formation.[505]

As already pointed out, the tricarboxylic acid cycle (citric acid cycle) has been demonstrated in insects.[448] In the brain of the boll weevil *Anthonomus grandis*, for example, dehydrogenases for all the intermediate compounds of the citric acid cycle are present. Of the glycolytic intermediates α-glycerophosphate is more readily oxidized than lactate.[445]

The general processes of intermediary metabolism in insects are closely similar to those of other animals; they could be surveyed only by writing a text-book of general biochemistry. Thus amino acids suffer extensive transamination (p. 517). This provides for the synthesis of the 'non-essential' amino acids from existing amino acids and sugars (p. 446);[437] and it provides for the oxidative removal of amino groups from aspartate, glutamate, alanine, and other amino acids with the production of α-ketoacids (α-ketoglutarate, for example).[395] These can enter the tricarboxylic acid cycle (citric acid cycle) which operates in insects as one of the chief sources of energy.[365]

A 'balanced amino acid pool' is necessary for protein synthesis, and transamination is the chief mechanism for the regulation of this pool. Transaminase activity in *Hyalophora* pupae follows a U-shaped curve (p. 634), which suggests a close relation between protein synthesis and transamination. The muscles are largely responsible; but in *Periplaneta* the fat body is an important site, maintaining the high amino acid level in the haemolymph.[459]

Amino acids in the blood are doubtless concerned in a great variety of synthetic processes in the body; for example, glutamine is probably concerned in the biosynthesis of glucosamine and therefore of chitin.[412]

SOME CHEMICAL PRODUCTS OF INSECTS

Systematic accounts of the many kinds of glands which open upon the surface of the insect body are given by Weber (1933). Here we shall merely attempt to give some idea of the extraordinary range of the chemical products of these glands, products which in several cases are materials of commerce.[201]

Foremost among these is **silk,** which is secreted not only by the labial spinning glands of Lepidoptera (Fig. 343), Hymenoptera, &c., but by the accessory tubes of the genital organs in *Hydrophilus*, by the tarsal glands of Embioptera (Fig. 344) and Empidae, and by the Malpighian tubes of certain Coleoptera and Neuroptera (p. 576),[191] and by the cerci in the young larva of *Blattella.*[511] In certain Chalcidoids the labial glands become connected with 'ileac glands' and they discharge their secretion jointly into the hind-gut to form the membranous cocoon.[96] Silk secretion is usually confined to the larval stage, but the adult *Psenolus* (Hym.) covers its brood cells with silk tissue, and *Gryllacris* (Orth.) uses a silk secretion to bind leaves together.[290]

In Lepidoptera a silk fibre is formed by each labial gland. As these fibres pass down the common duct they go through a muscular press which squeezes them together and transforms the two cylindrical strands into a flattened ribbon. Accessory glands upon the common duct perhaps secrete a cementing substance which holds the two fibres together.

Chemically, silk consists of 70–75 per cent. of a tough elastic protein, 'fibroin' (probably secreted by the posterior division of the gland) in which the chief amino acids are glycine 36 per cent., alanine 21 per cent. and tyrosine 10 per cent.; and 20–25 per cent. of a gelatinous protein 'sericin' (probably secreted by the middle division[295]) which forms a covering layer. Sericin is readily soluble in warm or soapy water; it is composed mainly of serine 5·8 per cent., alanine 9·2 per cent., and leucine 5·0 per cent.[105, 504] [See p. 638.]

The silk gland of *Bombyx* can synthesize all the amino acids in fibroin from pyruvate, with the exception of tyrosine: isotopically labelled carbon from glucose appears in the alanine, serine, glycine, glutamic acid, and aspartic acid of the silk.[371] But alanine comes also, of course, from the alanine, glycine, threonine, and serine of the haemolymph as well as from aspartic and glutamic acids which are plentiful in mulberry leaves.[370] Different families of Lepidoptera are characterized by the relative quantities of certain amino acids in the silk.[401] If glycine or alanine in which radioactive carbon has been incorporated into the carboxyl group are injected into larvae of *Hyalophora* twenty-four hours before spinning, the radioactive carbon is incorporated in the silk.[353] The part played by phosphatase in silk secretion is discussed above.[328]

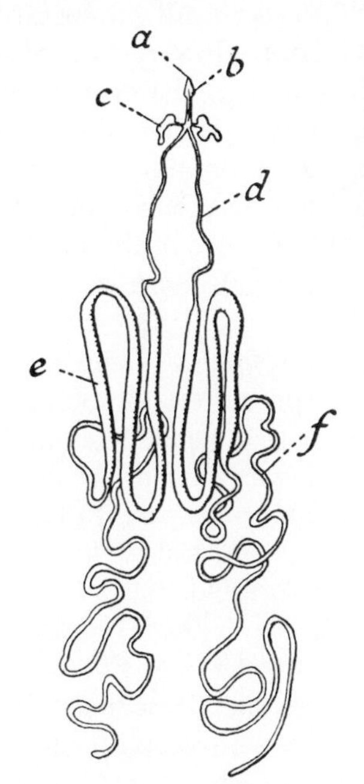

FIG. 343.—Spinning glands of silkworm (*after* LESPERON)

a, spinneret; *b*, press; *c*, Lyonnet's glands *d*, silk ducts; *e*, reservoir; *f*, secreting gland.

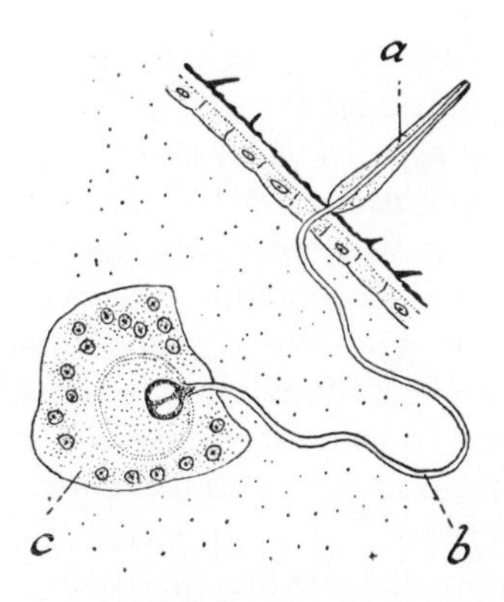

FIG. 344.—Silk gland from tarsus of *Embia* (*after* LESPERON)

a, spinning hair; *b*, duct; *c*, gland with ampulla containing silk substance.

The native protein of the silk gland 'fibroinogen', is water soluble; but on stretching and forcing through the spinneret the coiled long-chain molecules become unfolded and denatured. Mechanical shearing brings together reactive groups normally kept apart by hydration shells. Consequently the chains become aggregated to form the tough insoluble product 'fibroin' in which the molecules assume an orderly crystalline arrangement in the long axis of the fibre.[55, 148, 250, 497a] Traction on the thread is essential to the process of spinning; in *Galleria* and *Bombyx* this induces a continuous flow of secretion which the larva is unable to arrest; a force equivalent to a weight of 200–250 mg. is involved; the role of the 'press' is not important.[357, 358] In *B. mori* the silk filament is about 1,200 metres long.[403] The sericin is dissolved by a specific protease discharged by the adult silkmoth at emergence.[396] [See p. 638.]

Fibrinogen globules appear in the intracisternal spaces of the endoplasmic heticulum of the cells; the protein is elaborated in the rough-surfaced endocytoplasmic membranes in association with the mitochondria.[355] The macromolecules of fibrinogen are stored in the reservoir of the middle division of the silkgland in an ordered succession corresponding to the time of their biosynthesis.[401] [See p. 638.]

The egg-stalks of *Chrysopa* (15–20 μ thick and 1 cm. long) are composed of silk with glycine, alanine, and serine making up 84 per cent. of the total amino acids, as in the silk of *B. mori*, though the proportions are different. The polypeptide chains are normally oriented at right angles to the fibre axis, but become parallel to the fibre axis on stretching.[478] [See p. 638.]

Wax is secreted in the form of powders, threads or thin plates by the dermal glands of many bees,[78] Hemiptera[304] and other insects.[290] All the insect waxes (lac wax, coccerin, psylla wax, beeswax, &c.) are mixtures in varying proportions of (i) even numbered alcohols ranging from C_{24} to C_{36}, (ii) even numbered *n*-fatty acids from C_{24} to C_{34} and (iii) paraffins, always with an odd number of carbon atoms, from C_{23} to C_{37}. The various waxes differ only in the proportions of these products, and in the extent to which the alcohols are free, or combined as true wax-esters with the fatty acids.[22, 53] The waxy fibres of the woolly aphis *Eriosoma lanigera* are said not to be a wax at all, but the glyceride of a fatty acid, that is, a fat.[201]

Waxy or oily secretions are associated with the spiracles of aquatic insects (p. 380). Glands secreting a fatty covering which opposes evaporation occur in *Epiphragma ocellaris* and other Tipulid larvae from dry wood,[164] as well as in the cockroach.[83, 249] We have already discussed the wax layer of the epicuticle, which is secreted by the epidermal cells, and the important part it plays in the waterproofing of the insect cuticle (p. 35), and the similar wax layers in the egg (p. 1). Dermal glands along the articulations of Odonata[265] and other insects[305] perhaps produce an oil which lubricates the joints. The mandibular glands of caterpillars produce an oily secretion; in the larva of *Cossus* they are particularly active, and produce up to 0·4 gm. of a strongly smelling yellow oil, which contains a mixture of hydrocarbons, some with sulphur in organic combination.[144] [See p. 638.]

Lac is a mixture of resinous substances, a hard wax, sugar and pigment secreted by various Coccids, particularly *Tachardia*, *Gascardia*, and *Lecanium*.[198] Its composition varies considerably with the food plant.[193] The pure lac or shellac resin is a mixture of hydroxy acids (*e.g.* aleuritic acid, $C_{13}H_{26}O_4$) united to one another in the form of lactides or lactones; it is soluble in alcohol but not in fat solvents.[294] Stick-lac comprises two fractions: (i) an alcohol-soluble wax containing no fatty acids or paraffins, which is a simple mixture of even-numbered primary alcohols, C_{26}–C_{34}; and (ii) an alcohol-insoluble fraction, soluble in benzene, comprising esters of alcohols C_{30}–C_{36} and *n*-fatty acids C_{30}–C_{34}.[53] [See p. 638.]

Venoms are secreted by the modified accessory glands of female Aculeata. The venom of the bee is produced solely by the so-called 'acid gland' of the sting (Fig. 345).[146] Secretion begins just before emergence, reaches a maximum at 10–16 days and in summer bees it ceases after 20 days.[360a] The chief toxic fraction seems to be a dialysable protein of low molecular weight with an iso-electric point at about *p*H 8·7.[136, 282] Bee venom also contains powerful phosphatases which split oleic acid from lecithin and give rise to a lytic substance 'lysolecithin'. Lysolecithin has the property of dispersing the films of lipoprotein in the protoplasmic structure of the cells. This results in the liberation of histamine which is responsible for many of the symptoms of bee sting.[92] Associated with the sting is a second gland, the 'alkaline' or Dufour's gland, which has nothing to do with the toxic action. It perhaps secretes an oily liquid which serves to clean and lubricate the sting[146, 266] or to line the top of the cell,[266] or to secure the egg to the cell wall.[258, 293]

The chief component of bee venom is named 'melittin', a strongly basic protein with a powerful lytic action on red blood corpuscles. The enzyme phospholipase A (lecithinase) exerts a somewhat weaker haemolytic action. The enzyme hyaluronidase, also present in the venom, favours the spread of the other components by acting on the mucopolysaccharides in the tissues. The pain of the sting is attributed to intense inhibition of dehydrogenases of lactic acid, and of 'citric acid cycle' substrates. Perhaps this is an effect of the lecithinase.[364, 400, 411, 423] The venom of wasps is rich in histamine and 5-hydroxy-tryptamine (serotonin)[501] as well as hyaluronidase and, in *Vespa crabro*, 10 per cent. of acetylcholine.[364]

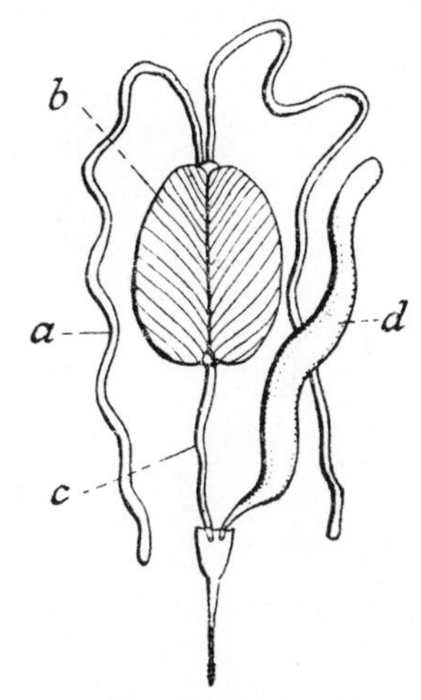

FIG. 345.—Poison glands of the wasp *Vespa germanica* (*after* BORDAS)

a, acid gland; b, poison sac; c, duct of acid gland; d, alkaline or Dufour's gland.

In solitary Hymenoptera (Pompiliids, Sphegids, &c.) the venom is used for paralysing the prey. Spiders or caterpillars when stung may remain motionless for months, although the heart continues to beat. Paralysed spiders and caterpillars show degenerative changes in all the nerve ganglia; not only the ganglion stung. Perhaps the venom contains a specific neurotoxin.[226] The venom of *Habrobracon* will induce paralysis in doses of 1 part of venom in 200 million parts of host blood (*Ephestia*). The site of action seems to be the neuromuscular junction; the action closely resembles that of curare in vertebrates, but there are large species differences in susceptibility.[363] The venom of the hunting wasp *Philanthus* causes a peripheral paralysis. It is equally effective without injection into the nerve ganglia. The haemolymph of *Philanthus* inactivates its own poison.[486] In the invasion of Syrphids by the Ichneumonid *Diplazon*, the young parasite secretes a toxic saliva which arrests growth and differentiation in the imaginal discs of the host long before parasitic histolysis begins.[349] In Cynipids the action of the larva provokes the formation of plant galls.

Myrmicine and Dolichoderine ants produce diverse kinds of venom;[275] 'iridomyrmecin', in the venom of the Argentine ant *Iridomyrmex*, can amount to 1 per cent. of the body weight; it is used in defence and offence against other insects, in which it produces effects resembling those of the insecticide DDT. 'Dendrolasin', the furan of farnesal, is secreted in the mandibular glands of the ant *Lasius*.[479] Among Camponotine ants the chief constituent of the venom is formic acid. In *Formica rufa* from different localities formic acid may range in concentration in the venom from 21 per cent. to 71 per cent. amounting in some cases to 18 per cent. of the total body weight.[275] Insects with soft cuticles are very susceptible to formic acid.[476] *Formica* can project its poison up to 30 cm.; it usually curves the abdomen forwards to spray the wound made by the jaws.[513]

Other well-known poisons produced by insects are: *cantharidin* ($C_{10}H_{12}O_4$), which may form 3–10 per cent. of different Meloid beetles, occurring largely in the blood, but also in the accessory glands of the genital tract;[98] the intensely toxic *arrow poison* obtained by bushmen in the Kalahari Desert from the larva of the beetle *Diamphidia locusta*, which appears to be a toxic saponin;[105] and the *urticating substances* secreted into the hollow cavities of the fragile detachable barbed hairs of many caterpillars;[116] these are of unknown nature, but they are often thought to be related to cantharidin.

Defensive secretions—Many warningly coloured ('aposematic') insects contain not only distasteful substances but potent venoms. High concentrations of histamine are present in *Zygaena*, *Hippocrita*, *Spilosoma menthastri*, and other Lepidoptera. The Garden Tiger moth *Arctia caja* contains a choline ester, perhaps acrylylcholine.[366] Hydrocyanic acid is released from the crushed tissues of Zygaenid moths at all stages of their life cycle.[430]

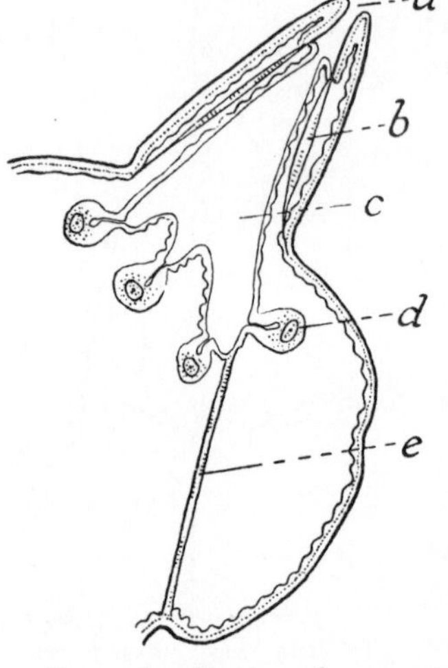

FIG. 346.—Section of segmental gland in *Melasoma* (*after* GARB)

a, pore; b, muscle closing the pore; c, reservoir for secretion; d, unicellular gland with intracellular duct; e, retractor muscle.

In most insects the defensive secretions are produced by glands opening to the exterior; all are toxic in varying degrees; indeed there is no sharp distinction between a venom and a repellent. The following are a few examples. In the cockroach Diploptera a mixture of quinones is stored in glands connected with the second abdominal spiracles and ejected by forcible expiration when the insect is attacked.[496] Many grasshoppers eject venomous froths in various ways (p. 433).[495] Stink glands are widespread in the Heteroptera. The usual components in the glands of Geocorisae are one or more of the unsaturated aldehydes, hexenal, heptenal, decenal, or the unsaturated hydrocarbon *n*-tridecene. These are nerve poisons effective against other insects; they enter the tracheae and cause paralysis; they serve also as repellents against predators.[489, 521] *Scaptocoris* produces a mixture of 7 aldehydes, 3 furans and 2 quinones.[494]

Among Coleoptera, the larvae of the Chrysomelids *Melasoma* and *Phyllodecta* have a series of segmental reservoirs into which they secrete salicylaldehyde, an end-product no doubt of the glucoside salicin in the food (Fig. 346);[110, 300] and the adult of the beetle *Aromia moschata*, feeding on the same group of food plants, eliminates salicylic acid in the form of an ester, through a pair of glands on the hind coxae.[149, 273] The Bombardier beetles *Brachinus* secrete a mixture of hydroquinones into the lumen of the glandular sac; to this is added hydrogen peroxide; and the free oxygen liberated in the explosive reaction that follows provides the pressure which expels and disperses the mixture.[502] *Brachinus* can aim its repellent spray accurately at an assailant and can give off more than twenty consecutive discharges.[397] There are elaborate repugnatorial glands in Tenebrionids, notably in *Tribolium*.[232, 254] Their product is highly toxic to insects; it is a mixture of methyl-, ethyl-, and methoxy-quinone[444] which stains the flour in which they live.[4] In *Eleodes*, besides *p*-benzoquinone, methyl-, and ethyl-quinone there are present tridecene, undecene, and nonene.[398] The pygidial glands of Carabids and other beetles produce a wide variety of chemicals: *m*-cresol in *Chlaenius*;[398] salicylaldehyde in *Calosoma*;[398] methacrylic acid and tiglinic acid in *Abax*, &c.;[502] 75 per cent. formic acid in *Pseudophonus*;[502] benzoic acid in oily solution in *Dytiscus*,[502] ammonia in Sylphidae.[502] [See p. 638.]

Among Lepidoptera, eversible pouches, essentially similar to those of *Melasoma*, form the 'osmeteria' of *Papilio* larvae. These also secrete strongly scented substances; perhaps they arose for the same purpose of eliminating poisonous compounds from the plants (Aristolochiaceae, Umbellifera, &c.) on which these caterpillars feed.[263, 306] The caterpillar of *Notodonta concinnula* emits strong hydrochloric acid.[69] The full-grown caterpillar of *Dicranura vinula* ejects 30–40 per cent. formic acid from a gland behind the head; it also incorporates this with the secretion of the spinning gland to produce the hard and waterproof material of the cocoon.[187] This same species, when it hatches as an imago, softens the cocoon with an alkaline solution, said to contain 1·4 per cent. of potassium hydroxide.[186] Some Mycetophilid larvae secrete on to their web droplets of 0·15 per cent. oxalic acid, which serves to kill the insects, &c., on which they prey, when these come in contact with it.[203]

The anal gland secretion of the ant *Tapinoma* contains methyl heptenone, propyl–isobutyl ketone, and 'iridodial'.[479] [See p. 638.]

Pheromones—A great variety of chemical substances produced by various glands exert an influence upon other members of the same species or upon other species. They may serve as 'warning substances' associated with an evil tasting or poisonous quality of the insect, as 'sexual attractants' secreted by the female (p. 710),[427] as 'aphrodisiac scents' produced by the male (p. 711), communication odours of many kinds, such as are used by honey-bees and ants to influence the behaviour of other members of the community, or substances transmitted by mouth which influence the differentiation of castes among termites (p. 101). This heterogeneous collection of chemical agents has been given the covering name of 'pheromones'.[433] This term implies only the existence of a 'chemical messenger' of some sort which operates between individuals and not necessarily a material having a hormone-like action. The same substance (formic acid, for example) may serve as a venom, as a repellent agent against enemies, or as a means of communication within the society. [See p. 639.]

Scent glands—There is no hard and fast line between defensive secretions and scents used for purposes of communication; but the following are some examples of the second group. One of the best known is the 'queen substance' produced by the mandibular glands of the queen bee and transmitted by 'food-sharing' to the worker bees, on which it has two effects: it prevents them from building the large queen-rearing cells and it suppresses ovary development.[381, 382] This substance does not act alone: other factors, perhaps some volatile odour[477] contribute to ovary inhibition; and removal of the mandibular glands reduces the inhibition of queen-cell building but does not eliminate it completely.[405, 464] The active material is 9-oxodecenoic acid of which 0·13μg per bee will inhibit queen-cell production.[383] It also serves as a sex attractant in mating (p. 710). But the mandibular glands contain other substances; one of these, probably 9-hydroxydecenoic acid, has been recognized as the pheromone which prevents the cluster of the bee swarm from breaking up.[384] The attractive scent from Nassonoff's gland in worker bees is the same in bees of different races or from different colonies;[490] it has been suggested that the active component is geraniol and associated derivatives.[368] The sting glands of bees produce another pheromone, perhaps containing iso-amylacetate, which excites the stinging of other bees.[369] [See p. 639.]

Soldiers of the leaf-cutting ant *Atta sexdens* produce in the mandibular glands the oily substance 'citral' closely related to citronellol; it serves as an alarm and repellent substance.[380] *Acanthomyops* produces a mixture of citronellal and citral.[398] In *Iridomyrmex* the releaser of the alarm reaction is methyl-amyl-ketone from the anal glands.[367] The trail-forming pheromone of *Zootermopsis* and other termites is a species-specific ether-soluble substance, secreted by the sternal glands.[458, 512] All social Hymenoptera produce alarm substances—from mandibular glands, abdominal glands, poison glands, Dufour's gland, anal glands, &c. They are never species specific and even act inter-generically.[460a] In the fire ant *Solenopsis*, mass foraging, trail following, colony emigration, alarm, grooming, and clustering are all largely induced by chemical releasers.[528a]

The warning scents given off by many 'distasteful' insects often have a number of chemical constituents in common. This state of affairs has been compared with 'Müllerian' mimicry, in which a group of distasteful insects show the same colour pattern.[497]

PIGMENT METABOLISM[189, 289]

Attention has always been attracted to the pigmented constituents of insects. In some cases these are, perhaps, substances of physiological importance; but the majority seem to be merely by-products of metabolism. Yet since they furnish the colour patterns they are not always without biological significance.

Melanin pigments are generally incorporated in the substance of the cuticle (p. 34). They range in colour from yellow to black, and are characterized by their complete insolubility in ordinary solvents. Melanins are polymeric in nature, so that it is impossible even in principle to isolate them as definite chemical compounds. Chemically they arise from the polymerization of certain indol ring compounds, which are themselves ultimately derived, by oxidation and ring closure, from the amino acid tyrosine.[9, 286] If larvae of *Aëdes* have insufficient tyrosine or phenylalanine in the diet the cuticle of the respiratory siphon, &c., and of the pupa is wholly unpigmented, although normal pigmentation develops in the adult.[119]

Tyrosine (monoxyphenylalanine) is oxidized in the presence of the enzyme tyrosinase.[59] At least three compounds are recognized in the tyrosinase complex: monophenolase converting tyrosine to 'dopa' or 3-4 dioxyphenylalanine; diphenolase, a copper protein compound, converting 'dopa' to the red substance hallachrome; and enzyme III, apparently a dehydrase, which converts hallachrome to a colourless substance and then to melanin.[65] In the egg of *Melanoplus* the tyrosinase is in an inactive or blocked state (protyrosinase). It is activated by various surface active substances; perhaps these serve to denature the protein and thus expose the 'active groups'.[33]

The immediate precursor of melanin in insects seems usually to be a dioxybenzene derivative; either 'dopa' itself, as in the elytra of *Melolontha*,[259] dioxyphenylacetic acid, as in the elytra of *Tenebrio*,[259] or some related compound. As we have already seen (p. 34) it is not always easy to distinguish between the darkening associated with the phenolic tanning of the cuticle proteins and true melanin formation. These substances occur also in the cocoons of Lasiocampids and Saturniids, being derived from the secretion of the spinning glands and Malpighian tubes. If the cocoons are kept dry they remain pale; but if moistened, the small amount of oxidase present can react with the chromogen, and the cocoons darken.[71, 244]

Thus the deposition of melanins requires the interaction of tyrosinase, chromogen and oxygen. The necessary oxygen is obtained through the tracheal system, not from the air in contact with the surface.[102, 142] The darkening of adult *Tenebrio* is incomplete if the oxygen in the air is reduced.[259] The process is arrested in nitrogen, as in *Leptinotarsa*,[123] *Pyrrhocoris*,[142] the puparium of *Calliphora*;[102] and it is arrested if the tyrosinase is destroyed, as when the elytron of *Leptinotarsa* before pigmentation is heated at 70° C. for 1 minute.[123] Blackening is prevented also if the blood stream is interrupted, as in *Pyrrhocoris*,[142] and the Coccinellid *Epilachna*.[281] In *Tenebrio* the chromogen is almost absent from the cuticle of the newly emerged beetle; it becomes abundant later, reaching a concentration of 5 per cent.[259] In general it seems to be the localization of chromogen which determines the distribution of the pigment patterns; chromogens are absent from white areas of cuticle,[259] whereas tyrosinase is everywhere. The pattern of *Leptinotarsa* is little affected if the unpigmented elytron is immersed in tyrosinase; it is entirely blackened if placed in tyrosine;[123] the same applies to the developing wings of *Pieris*.[229] During the development of the wings in Lepidoptera, the chromogen appears to be deposited in the substance of the scale.[251] But the process as a whole is controlled apparently by the nervous system,[102] perhaps through the medium of hormones (p. 47); and the changes even in the surface layers of the cuticle are under the control of the cells below (p. 47).[312] In *Tenebrio*, darkening occurs only in the living insect.[259]

Tyrosine and tyrosinase co-exist in the blood of insects but in the living insect blackening of the blood does not occur. This has been attributed to the presence of a dehydrogenase, with glucose as substrate, which reduces the redox potential in the blood and keeps the tyrosinase inactive until the time of pupation;[333] but further evidence of this mechanism is required.[350] It seems likely that the absence of tyrosinase activity in undamaged tissue is due to the structure of the cytoplasm, which keeps enzyme and substrate apart.[484] Tyrosinase used for the darkening of the puparium of *Drosophila* is present in the haemolymph of the full-grown larva in an inactive form; it is activated by an agent present in the tissues. The enzyme activity is much greater in the wild-type fly than in 'ebony' which has a paler puparium.[471] The blood of fly larvae blackens most readily immediately before pupation,[70] when the phenolic substances necessary for the hardening of the puparium are being mobilized.

The production of melanin is sometimes regarded as a mechanism for disposing of toxic phenols arising as breakdown products in metabolism.[59] In some cases its distribution seems to be related with the intensity of metabolism in the subjacent tissues: it occurs in the cuticle over the muscle insertions in many insects,[292] and over the fat body and pericardial cells of Vespids (Fig. 347);[15] Xanthopterine will inhibit the final stages in the oxidation of tyrosine or 'dopa' to form melanin. This may provide the basis for the mutual exclusion of pterines and melanin (Fig. 347).[335] In the elytra of Coccinellids tyrosine appears to be generally distributed, but the pattern corresponds with the distribution of 2-amino-4-hydroxy-6-carboxylic pteridine which inhibits the action of tyrosinase.[475]

During the past century, melanic races have arisen among many species of Lepidoptera, particularly in industrial regions.[135] This has sometimes been

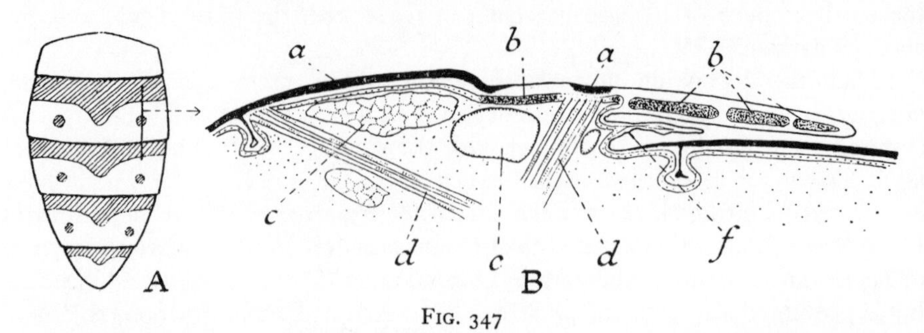

FIG. 347

A, pigment pattern on abdomen of *Vespa*. *B*, longitudinal section of tergite and intersegmental region to show the distribution of melanin and pterine pigments in relation to the underlying structures (*after* BECKER). *a*, melanin in cuticle; *b*, pterine in thickened epidermis; *c*, fat body; *d*, muscles; *e*, tracheal air sac; *f*, intersegmental membrane.

regarded as a direct effect of city fumes upon metabolism;[135] sometimes as a mutation induced by contamination of the larval food plants with such metals as lead and manganese.[133, 283] Probably the influence of industrial conditions is indirect. The genes producing melanism also improve the viability of the insect. With a smoke-grimed background on which to rest, and perhaps with fewer predators, this improved viability may have outweighed the loss of cryptic coloration and led to selection favou ng the black mutant.[100, 120]

Pterine pigments and purines—In his classic work on the pigments of the Pieridae, Hopkins[151] described these coloured substances as 'waste products which function in ornament'. They were long regarded as derivatives of the purines, the white substance being thought to be uric acid itself. It is now known that this group of substances are based on the pteridine ring and are known as pterines.[109] In colour they may be white (leucopterine), yellow (xanthopterine and others) or red (erythropterine). The natural pigments are always mixtures of several pterines.[260, 121] They are particularly characteristic of the Pieridae, but they occur also in Neuroptera (*Ascalaphus*) Syrphidae, Hemiptera (some Cicadidae, *Oncopeltus*[121]), many Hymenoptera such as *Vespa*, and doubtless in other insects.[14, 260] [See p. 640.]

Pterine pigments are widely distributed in nature, being recognized in hay, and in the liver and urine of mammals. They have many possible links with metabolic processes. In the wings of Pieridae, small quantities of uric acid and hypoxanthine,[277] xanthine,[246] isoguanine,[246] and allantoine[261] occur along with leucopterine and xanthopterine. So they may have some connexion with purine metabolism—although they occur only in very small quantities in the excreta.[310] The pterine ring shows some structural relation with the riboflavine molecule; indeed, the biosynthesis of pterines, purines, and riboflavine follows the same path up to a certain point (p. 561).[525] In many insects riboflavine accumulates in quantity in the Malpighian tubes (p. 566) so that there is a possible connexion between the pterines, the flavine-protein system of cellular respiration, and the metabolism of vitamin B_2. Indeed it is claimed that in the egg of *Melanoplus* riboflavine is converted into pterine.[37] Interest in the pterines has been further increased by the discovery that 'folic acid', which is an essential food factor for insects (p. 520), contains the pterine nucleus in the form of pteroylglutamic acid. [See p. 640.]

In the pupa of *Pieris* the pterines occur in haemolymph, wing discs, nervous system, but particularly in the fat body; in the adult wings they are in the free state, but in the pupa they are coupled with protein.[377] In *Pieris*, as we have seen (p. 581) there is certainly a switch over from purine formation to pterine formation during pupal development.[417] Eleven different pteridines have been isolated from *Pieris brassicae* but only two of these, leucopterine and isoxanthopterine, occur in quantitatively significant amounts.[417] Xanthopterine shows a tenfold increase during embryonic development in *Oncopeltus*; in the later stages isoxanthopterine and xanthopterine occur chiefly in the epidermis, but also in the fat body, mid-gut, &.; pteridines account for 0·2 per cent. of the total material in the dried excreta, uric acid of 3·0–5·5 per cent.[425] The normal larva of *Bombyx mori* produces an enzyme which will break down xanthopterine-B, but in the yellow race 'lemon' this enzyme is absent and the yellow pigment accumulates in the integument.[359]

The pterines are deposited in the substance of the epidermal cells, or in the lumen of scales or hairs; they are usually laid down where the metabolism in the adjacent region is low. In *Vespa*, for example, they occur within the rigid fold of cuticle which overlies the intersegmental membrane, over the large tracheal air sacs, and in the distal segments of the limbs (Fig. 347).[15] In the nest-mother of wasps in the latter part of the year concentric rings of dark pigment may be deposited in the yellow regions of the abdominal folds. These appear to be Liesegang rings of precipitation; perhaps the sequel to some injury.[16]

The fluorescent substances in the eyes of *Ephestia* consist of a mixture of riboflavine and isoxanthopterine, along with other pterines.[491, 519] Pterines with blue or yellow fluorescence, which are wide-spread in the eyes, eggs, and luminous organs (where they accompany the closely related substance luciferin (p. 453)) were formerly thought to be a single substance 'fluorescyanine', but they are in fact complex mixtures.[536] A series of three or more have been isolated from the eyes of *Drosophila*.[519] Certain of these pigments are decomposed by illumination; this light sensitivity may prove to be of physiological significance.[536] The *Drosophila* mutant 'rosy' lacks the enzyme xanthine dehydrogenase; this results in an eye-colour change which, like 'vermilion' and 'cinnabar' is non-autonomus (p. 90): the colour depends on a diffusible substance which originates outside the site where the pteridine is finally deposited.[409, 413]

Flavines—The most important natural flavine is riboflavine, which is a vitamin essential for insect growth (p. 519). Riboflavine accumulates in large amounts in the Malpighian tubes where it is deposited in vacuoles as a greenish yellow, fluorescent pigment formerly termed 'entomourochrome'.[44] It appears in greatest concentration in phytophagous insects (250 μgm. per gm. in *Carabus*, 2,150 in *Schistocerca*) being derived probably from the food. On the other hand the content of riboflavine may be doubled in the egg of *Attacus* and increases by 25 per cent. in *Schistocerca* during incubation; and in *Sphinx* it shows a striking increase from 500 to 1,550 μgm. per gm. during metamorphosis.[44] On phosphorylation and combination with protein, riboflavine gives rise to flavoprotein or 'yellow enzyme' which is important in cellular oxidations (p. 622). Its possible conversion to pterine in the *Melanoplus* egg was noted in the previous section.

Riboflavine and pterines occur in the Malpighian tubes of locusts, much of the riboflavine being bound as a flavoprotein.[338]

Carotinoid pigments or lipochromes are almost universal in plants, and they are commonly absorbed by insects and accumulated in the blood or tissues.[68] Carotene itself is most readily absorbed; in the blood of *Leptinotarsa* it occurs at a concentration of 14 mg. per cent., equal to that in many leaves;[174] and it provides the red and yellow colouring of the bug *Perillus* which preys upon this insect.[233] It is deposited in the cellular tissue of the elytra of *Melasoma* and *Haemonia*[179] and in the fat body of *Pyrrhocoris*.[264] (The red and orange pigments in the epidermis of *Pyrrhocoris*,[142] *Rhodnius*[312] and other bugs are not carotinoids; in *Oncopeltus* (Lygaeidae) the red pigment is a pterine (p. 612)). Carotene is absorbed into the blood of many caterpillars, such as Noctuids[220] and Pierids.[111, 205] It is present in the wings of *Melanargia* (Lep.).[284] In *Carausius* it may be transmitted by the female to the eggs and so to the offspring.[291] In *Coccinella* the pigment is lycopene, identical with that of the tomato, combined with carotenes α and β.[188]

Xanthophyll, the dioxide of carotene, on the other hand, is much less readily absorbed. It is absent from *Leptinotarsa*[233] and from Noctuid larvae (*Agrotis*, *Caradrina*[220]); but it is present in the blood and epidermis of *Pieris* larvae.[205] The yellow pigments in the epidermis of the grasshoppers *Melanoplus*,[126] *Schistocerca*, and *Locusta*[52, 122] are multiple in nature: xanthophyll, α- and β-carotene and astaxanthin, identical with that of the lobster, are all present in varying amounts. β-carotene and astaxanthin are transferred to the epidermis in the mature male.[334] Similar carotinoids occur in *Leptinotarsa*.[208] [See p. 640.]

The coloration of the silk cocoons of *Bombyx mori* is also due to carotinoid pigments from the food;[298] and the different silkworm races with distinctive colouring of the cocoon differ in the proportions of xanthophyll and carotene which are transferred from the gut to the spinning gland, and in the rate and time at which this transfer is achieved.[158, 205, 206] Carotinoids (α-carotene and taraxanthine) are likewise taken up from the blood of *Pieris* by the parasitic Braconid *Microgaster* and deposited with the silk to form the yellow cocoons.[207]

Whether carotinoids play a part in insect physiology is not known. In mammals, vitamin A is a derivative of carotene, and a related derivative forms the visual purple of the retina.

Green colours: carotene-albumens and insectoverdins—The green pigments of insects have often been thought to be derived from altered chlorophyll (see p. 506).[242] But chlorophyll cannot be detected in the green blood or epidermal pigment of Noctuid larvae,[220] *Pieris*[205] or *Locusta*[90]; while *Carausius*, *Mantis*,[291] *Locusta*,[90] and *Caradrina*[220] develop their normal green pigment when no chlorophyll is present in the food. The green pigment between the wing membranes of *Pieris* is a 'chromoprotein', the pigmented component of which (pterobilin) is very closely related with the mammalian bile pigment biliverdin.[311]

The colours of insects are frequently due to complex mixtures of pigments, and the same outward appearance may be produced by very different means. This is particularly true of the green coloration which seems to result always from a mixture in varying proportions of blue and yellow components.[244] The chemical nature of the constituent pigments is probably different in different insects. The blue-green component is commonly a bile pigment chromoprotein; the yellow component is also a chromoprotein of which the prosthetic group may be a xanthophyll in *Tettigonia*, *Meconema*, and *Sphinx*, a carotene in *Carausius*. *Colias* and other Pierids have strains with larvae of a blue-green colour. This change depends upon a gene which affects the yellow carotinoid complex in the haemolymph, leaving the blue bile-pigment unchanged.[509] Likewise, *Locusta* and *Schistocerca* hatched from eggs deficient in carotene, and reared on a carotene-free diet, show greatly reduced melanization, and the colour of the insect becomes a greenish-blue and the haemolymph a turquoise-blue. This results from the lack of carotene which normally blends with the blue mesobiliverdin to yield the natural green colour.[334, 393] [See p. 640.]

For these green mixtures of protein pigments the name 'insectoverdins' has been proposed.[160] In the Pentatomid *Nezara* the yellow component consists largely of granules of xanthopterine, while the blue component is a diffuse pigment resembling the anthocyanins of plants.[228] In *Chrysopa perla* the blue pigment is different again.[228] Finally, the Pieridae *Colias*, *Euchlöe*, &c., achieve their green wing markings after the manner of the impressionist painters, by the juxtaposition of black (melanin) and yellow (xanthopterine) scales.[313]

The protein carriers largely determine the properties of these pigments. In the three species of *Oedipoda* (Acridiidae) the same carotenoid astaxanthin is combined with different proteins to give blue hind wings in *O. caerulea*, red in *O. miniata* and yellow in *O. aurea*.[228, 343]

Haemoglobin and chlorophyll derivatives—Small amounts of haemoglobin may be absorbed by some blood-sucking insects and give rise to pigmented derivatives in the tissues. In *Rhodnius*, for example, parahaematin appears in the blood; a pigment resembling haemalbumen gives the salivary glands a cherry-red colour; altered haematin, a green intermediate pigment, and biliverdin appear in the wall of the gut; and the pericardial cells are blue-green with biliverdin deposits; some parahaematin is taken up by the yolk of the eggs giving these a pink tinge.[315] The functional haemoglobin (p. 389) in the blood of the larva of *Chironomus plumosus* is broken down throughout larval life and gives rise to inclusions of bilirubin and biliverdin in the fat body; it is this accumulated biliverdin which is responsible for the green colour of the newly emerged midge.[159, 239]

In the squash bug *Anasa tristis* there is an analogous breakdown of chlorophyll. A red pigment in the epidermis, salivary glands and testis appears to be a phaeophorbide derived from chlorophyll; and a green pigment in the fat body and pericardial cells a tetrapyrrolic derivative analogous to biliverdin.[218] Pigments giving Gmelin's reaction for bile pigments and formerly thought to be derived from the breakdown of chlorophyll are concerned in the red colour of the scales and gut contents in *Vanessa*.[192] It is now known that these pigments are ommochromes (p. 616).

Anthocyanins and anthoxanthins, which are important flower pigments, are not very common among insects. The anthocyanins are sap soluble glucosides giving scarlet, purple and blue colours. The beetle larva *Cionus* contains in its fat body granules of anthocyanin from the *Verbascum* on which it feeds;[150] and the vermilion colour of the Aphid *Tritogenaphis* is perhaps an anthocyanin.[233]

The anthoxanthins or flavones are sap soluble pigments ranging in tint from ivory to deep yellow. The whitish wing pigment of the Satyrine butterfly *Melanargia* is a flavone identical with that present in the grass *Dactylis glomerata* on which the larva feeds;[284] and flavones contribute to the cocoon colours of some silkworm races and are present in the blood and epidermis of *Pieris* larvae.[205] The ability to absorb and deposit these pigments in the wings is capricious. For example, the Satyrine butterflies, *Parage egeria* and *Coenonympha pamphilus*, feed on the same grasses, but only the latter deposits anthoxanthins in the wings. It is only those natural glycosides in the food-plant that resist enzymic attack, which are deposited in the wings; this suggests that in the insect they are waste products.[461] Their distribution may none the less be related to systematics. Among Papilionidae, anthoxanthins are found throughout *Parnassius* and almost all species of *Graphium*; among Pieridae they occur in the South American Dismorphiinae and the related Palaearctic genus *Lepidea*.[99] The yellow pigment deposited in the hairs of *Bombus* is a flavonoid probably derived from the pollen on which the larvae have been fed.[510] The yellow, orange, and pink cocoon colours in *Bombyx*, as we have seen, are due to carotinoids, the green colour is due to flavones.[416]

Anthraquinone pigments and aphins—The pigments of cochineal (carminic acid) and lac insects (laccainic acid) are derivatives of anthraquinone;[189, 289] and the orange pigments obtainable from *Eriosoma* ('lanigerin') and *Adelges* ('strobinin'), formerly believed to be polyhydroxyanthraquinones,[25, 72] are in fact aphins.[391] These red and yellow fat soluble pigments extracted from various Aphid species are probably artefacts arising after death. The natural yellow pigment termed 'protaphin' is water soluble and is a deep magenta colour in alkaline solution; it is readily converted into yellow (xanthaphin), orange (chrysoaphin or lanigerin) and red (erythroaphin or strobinin) derivatives.[82] The stable end product 'erythroaphin' has the empirical formula $C_{30}H_{22}O_8$ and has a symmetrical multi-ringed structure.[391] There is evidence that certain of these 'aphins' are light-sensitive.[488] The pigments of this group are perhaps metabolic products of the vegetable micro-organisms in the mycetomes. The yellow and red-brown pigments of *Aphrophora* species seem to be the products of symbiotic bacteria;[199] and *Psylla mali* contains bacilli in the mycetome which produce a green pigment when the insect feeds on apple, a red pigment when it feeds on cherry.[199]

Ommochromes—These are a group of pigments first recognized among the eye pigments of *Drosophila*, *Ephestia*, and other insects, and in the serosa of *Bombyx mori*.[171, 436] They are yellow, brown or red in colour; several components are recognized in the eye of *Drosophila*[87, 301] with varying distributions in the pigment cells of the eye.[129] They are all derived from tryptophane; this is oxidized to kynurenine and oxykynurenine, which is the precursor of the pigments.[18, 46, 50] The ommochromes fall into two groups: the non-dialysable 'ommines' with large molecules, and the dialysable 'ommatines' with small molecules. Both are insoluble in most organic solvents, soluble in aqueous alkali, and reversibly oxidized and reduced.[450] [See p. 640.]

Four of these pigments have been well characterized. *Xanthommatine* is the sole ommochrome in the eyes of *Calliphora*, &c., and is present in the eyes of *Drosophila*, *Ephestia*, *Rhodnius*, &c.; it occurs in the wings of many butterflies, *Vanessa*, *Argynnis*, &c., as a dermal pigment in *Dicranura* larvae, &c. Xanthommatine is brownish yellow in the normal oxidized state, becoming bright red (hydroxanthommatine) on reduction, and is re-oxidized in air. Under the action of tyrosinase in the presence of 'dopa', hydroxykynurenine is converted to xanthommatine.[378, 379] *Rhodommatine* occurs in the excretory pigment of *Vanessa* and in the wing pigments of *Argynnis*, *Pyrameis*, &c.; it is the glucoside of hydroxanthommatine and in this case it is the reduced form which is stable. *Ommatine D* is a third ommatine in the red meconium of *Vanessa*; it is a sulphur-containing compound intermediate between xanthommatine and rhodommatine. *Ommine A* occurs in the eyes of *B. mori*, *Ephestia*, &c.; it is dark violet in colour.[378, 379] The eye pigments in *Ephestia* appear during imaginal development in the order: xanthommatine, ommatine I, ommatine II, and ommine. This is in agreement with the successive action of known genes.[441] [See p. 640.]

The study of the formation of these pigments has been facilitated by the occurrence of mutations ('vermilion' *Drosophila* and red-eyed *Ephestia* (p. 90)) in which the oxidative breakdown of tryptophane is deficient.[87] If kynurenine is injected into such insects the normal ommochromes are formed and the quantity formed is proportional to the amount of kynurenine injected. If 1 μgm. of kynurenine is introduced, about double that quantity of pigment is produced.[184] Some of the epidermal pigments of insects, such as the 'acridioxanthin' of *Schistocerca*,[52] belong to this same group of substances.[18, 334, 342, 391]

The colour change from green to brown in the fullgrown larva of *Dicranura vinula* and some other Lepidoptera before pupation, under the action of ecdyson (p. 69), results from the successive appearance in the fat body of dihydroxanthommatine, rhodommatine, and ommatine D. The total amount of pigment may reach 0·2 per cent. of the dry weight; its appearance is probably the result of increased protein breakdown.[374, 451]

Fig. 64 shows in simplified form some of the chemical stages, and the genes which control them, during the course of the development of the eye pigments. In *Drosophila* mutants, ommochromes, pterines, and uric acid all seem to be involved in the same genetic chain.[394] In *Ephestia*, if ommochrome synthesis in the head is interrupted by a mutation, there is a simultaneous increase in pteridines; if kynurenine is injected to the *a* mutant (p. 90), pteridines are reduced.[379] In the white-eyed mutant, *wa*, neither pteridines nor ommochromes appear in the eye; both these pigments seem to require the same protein granules; perhaps there is competitive formation at this building site.[413, 441]

Other derivatives of tryptophane may appear as pigments: failure to metabolize tryptophane completely (in the *Corcyra* larva on a diet deficient in pyridoxine, p. 520) leads to the appearance of a yellow pigment in the excreta, consisting of kynurenine and hydroxykynurenine.[516] The yellow pigment of *Papilio xuthus* and other species consists of kynurenine and closely related compounds.[517] In wild type *Drosophila* larvae oxykynurenine is stored as yellow concretions in the cytoplasm of the Malpighian tubules and utilized for ommochrome formation in the adult; in red- and white-eyed mutants oxykynurenine is absent.[524] One strain of *B. mori* ('aka-aka') has a red integument resulting from a large accumulation of 3-hydroykynurenine.[460] Xanthurenic acid, a product of 3-hydroxykynurenine, is present in large amounts in the head of *Drosophila*.[518] [See p. 640.]

There are many **other pigments** in insects the chemical nature of which is quite unknown. A red fluorescent pigment 'lampyrine' occurs in the fat body below the pronotum of Lampyridae.[217] Some of these pigments provide confirmatory evidence of systematic relationships. The types of red pigment in the genus *Delias* (Pieridae) are tied to particular subgenera. The same applies to the red pigments in the Papilionidae; here the red pigment of *Papilio hector*,

P. aristolochiae and their allies (Type A) is different from the red pigment of the forms of *P. polytes* which mimic them (Type B).[99]

Some of the most brilliant colour effects in insects are produced by physical interference (p. 38). The brilliant green in some *Ornithoptera* is due to the modification of a yellow pigment by a structural blue.[230] In the blue forms of Odonata the epidermis is uniformly filled with excessively fine colourless granules, and immediately below the epidermis is a layer of cells containing a dark violet-brown pigment. The incident light is scattered by the fine granules and the blue rays are preponderantly reflected; when viewed against the dark background an intense blue (Tyndall blue) is seen. Sometimes a yellow pigment component is added, and enamel-like yellow-green colours are produced.[17]

Colour change—Some insects can change in colour in response to environmental stimuli. In all such cases, with the exception of the Mantids and the Phasmid *Carausius* to be described below, the change results from alterations in the quantity of pigment formed, and consequently it is not immediately reversible.

Temperature is one such stimulus. We have seen that temperature acting in the early pupal stage of butterflies may determine the seasonal form (p. 105). The effect can be produced by applying the warm or cold stimulus to the head alone. Perhaps the head contains a 'pigmentation centre', which regulates either the rate at which the melanin-forming process proceeds, or the quantity of chromogens capable of yielding this pigment.[115] The formation of melanin in the wasp *Habrobracon* decreases if it is reared at a high temperature;[257] the females of this insect are darkened if the pupa is exposed to extremes of heat or cold; and this effect is transmitted, apparently through the plasma, to the next generation.[161] The same effect, with a partial carry-over to the next generation is seen in the ventral black pigment of *Oncopeltus*.[469] In *Microbracon*, in which there is no transmission to the offspring, the effect has been attributed to the failure of the insect to dispose of breakdown products of metabolism at low temperature.[463] And in the bug *Perillus* both melanin and carotene are at a maximum at low temperature (18–24° C.), at a minimum at high (29–35° C.).[174]

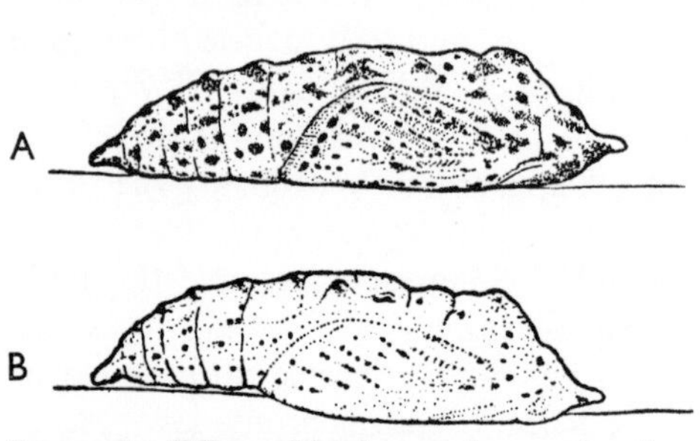

FIG. 348.—Effect of background on coloration of pupae of *Pieris brassicae* (*from photograph by* DÜRKEN)

A, whitish grey pupa with much black pigmentation: black and red background. *B*, pure green pupa with black pigment greatly reduced: green or orange background.

In the alpine grasshopper *Kosciuscola* in Australia the insects change from a bright greenish blue above 25° C. to a dull, almost black, shade below 15° C. The change results from the migration of two types of pigment granule in opposite directions in the epidermal cells. Temperature seems to act directly on the epidermis without nervous or humoral coordination.[435]

Humidity has been regarded as the factor determining the wet- and dry-season forms of the butterfly *Hestina assimilis*; for the application of water to one pupal wing will produce insects with wings of the heavily pigmented wet-season form on that side, and the pale dry-season form on the other;[214] but it is possible that this again may be an effect of temperature. In the highly variable *Locusta migratoria*, bright green hoppers develop only when fed with succulent moist food in a very humid atmosphere.[90]

Illumination and **background colour** are the chief factors controlling colour change. This is well seen in Vanessid and Pierid pupae, which are dark or pale according to their surroundings.[241] The pupal colour of *Pieris* is given by patches of melanin in the superficial layer of the cuticle, by a white pigment in the epidermal cells, and by the green colour of the deeper tissues.[84] The colour varies with the quantity of the black or white pigments, the formation of which is influenced by the illumination of the larva as it rests before pupation. Exposure to green, yellow, and particularly, orange light at this time causes a suppression of the white and black pigments so that the pupae appear plain green (Fig. 348). This influence is transmitted to the offspring, which tend to produce green pupae even under ordinary illumination,[84] an effect which is exaggerated after two generations have been reared in orange light.[134] Ultra-violet light increases the formation of black pigment.[41] Substantially the same effects are obtained with *Vanessa*, but here black pigment is suppressed, and white enhanced, by infra-red rays, which result in whitish pupae.[41] These effects are still produced if the larval eyes are covered with black pigment;[241] but either the eyes or some centre in the head control the response, for it is eliminated if the larva is decapitated or if the eyes are cauterized.[41] Sections of the nerve cord show that the light stimulus is conducted by the nerve cord to the posterior thoracic ganglia and thence apparently by the lateral nerves to the tracheae.[41] There is some evidence that in *Vanessa* it is the incorporation of melanin in the cuticle, rather than melanin formation, which is suppressed.[375] [See p. 640.]

In *Papilio xuthus* there are two main colour types in the pupa: brown and green. The colour type is largely determined by the environment at the time of spinning. Browning is induced by hormone secretion from the prothoracic ganglion, which is controlled by the brain. But superimposed on this is the effect of diapause which is induced by a short photoperiod; and diapause pupae are brown.[421] The same conclusions have been reached in the *Pieris* pupa.[472] The lateral ocelli seem to be the receptors for the photostimuli.[473] In the overwintering caterpillar of *Hestina* the integument changes in colour from green to brown. The appearance of the red pigment seems to be controlled from a centre in the region of the 6th abdominal segment.[474]

A similar adaptation to the background is seen in the nymph or young adult of *Sigara* (*Arctocorisa*). A pale background during the process of moulting inhibits the formation of pigment; but this control disappears if the eyes are covered or if the nerve cord is cut in the neck. The rate of melanin deposition is proportional to the difference between the intensity of light falling on the dorsal surface of the eye and that reflected from the background.[238] Larvae of both Anisoptera and Zygoptera become pale or dark depending on their background. This morphological colour change can take place only after a moult; it is effected by a weaker or stronger diffuse melanization of the outermost cuticular layer.[439] In the louse *Pediculus* there is a different degree of pigmentation in forms from the different races of man; here both heredity and background are involved.[45] In the larvae and pupae of *Thaumalea* and *Simulium* the colour pattern varies with the aggregation of the pigment-containing fat body cells (chromatocytes).[422]

Hoppers of *Locusta migratoria* can also adapt their colours from dirty white to yellow, brown or black according to the background;[90] the changes being dependent on the quantities of orange-yellow and black pigments formed. The amount of black pigment seems to depend on differences in the intensity of the light incident from above and reflected from below, acting upon the eyes.[145] The production of orange-yellow pigment is stimulated by rays with wavelengths within the limits of yellow (5,500–6,000 Å); it is inhibited by blue and violet light (5,000–4,500 Å).[145] The same assimilation to the background occurs in *Schistocerca*.[253] The cuticle of adult Acridiids several months old can become melanized if the insects are kept for a few days on burnt ground.[329] *Acrida turrita* and *Mantis religiosa* likewise show wide powers of colour adaptation to the background, which can occur to some extent even after exclusion of the eyes, though the eyes are necessary for maintaining the green colour. In *Acrida* the green colour is due to background and not to moist food.[399] [See p. 641.]

Crowding—These locusts also show a striking colour change, with conspicuous black and orange markings when they enter the gregarious migratory phase of their existence. This change is associated with intense muscular activity: hoppers of the gregarious type kept singly and undisturbed revert to the solitary phase; whereas if they are continually agitated they retain, at least in part, their 'gregaria' type of colouring. Perhaps under these conditions some diffusible substance is produced which controls the pigmentation. If so, it seems to be passed on by the female to her offspring, for the 1st instar hoppers of 'gregaria' show their characteristic coloration within an hour after hatching.[90, 153] If a mature female of *Locusta* or *Schistocerca* is isolated from a crowd, the successive egg pods laid produce successively paler hatchlings.[356] It is not quite certain, however, whether it is the agitation which induces the change of phase or whether this results from the sight or contact of their neighbours.[52] Whatever its significance, there seems to be a direct relation between the quantity of the pigment 'acridioxanthin' and the degree of gregarism developed.[52] The pigmentary change from *solitaria* to *gregaria* in *Schistocerca* can be produced by injection of haemolymph from *gregaria* hoppers. The active principle appears to be ether soluble.[467] Green pigmentation in locusts seems to be largely brought about by juvenile hormone secretion from the corpus allatum (p. 105).[429]

A similar phase change combining changes in morphology with conspicuous changes in coloration occurs also in Australian stick insects (Phasmatidae). The low density phase is a uniform green. The high density phase is conspicuously patterned with black, yellow, and sometimes white markings. These insects show no kind of agitation or gregariousness; the crowding exerts its effect by sensory stimulation. The change is perhaps a genetic switch-over from a procryptic adaptation to a warning, or aposematic, adaptation in correlation with the population density under which each of these adaptations has survival value.[434]

Phases of the same kind occur in the African army worms *Laphygma exigua* and *L. exempta*[91] and in *Spodoptera*.[211] Reared in a crowd, these caterpillars are active, dark and velvety ('gregaria'); reared in isolation they vary in colour from green to pinkish grey ('solitaria'). In these larvae and in the locusts there are some chemical differences in the blood and tissues of the two phases, notably an increased content of lactic acid and a reduction of uric acid in the gregarious phase.[211] Many other caterpillars become more active and darker in colour if reared under crowded conditions; the change in pigmentation affects both epidermal cells and cuticle.[452] In *Gryllus bimaculatus* there are two genetic colour strains: dark and light; but the population density determines the manifestation of the latent genetic effects. Contrary to the results in most insects, isolation leads to blackening, and crowding (as few as two or three crickets in a cage) produces the light yellow-brown colour. The pigment involved are cuticular melanins and epidermal ommochromes.[404] [See p. 641.]

Physiological colour change—The epidermal cells of *Carausius* contain orange, red, and yellow lipochromes, green 'insectoverdins' and varying amounts of the brown ommochrome pigments ommine and xanthommatine—to provide a 'morphological' colour change.[375a] And each form, except the green variety, has its own range of colours, brought about by the dispersion or clumping of the brown and orange pigments within the cells—a 'physiological' colour change (Fig. 349).[114, 256] Normally the insect is dark at night and pale by day, and this change will continue for several weeks in complete darkness. This diurnal rhythm is absent in insects kept in the dark from the time of hatching; it is induced by periodic illumination. Insects illuminated at night will become pale; and a reversed rhythm, persisting in continuous darkness, can be induced by reversed illumination.[256] High humidity also causes darkening; dry air causes pallor.[114] In Mantids, also, there is a colour change in the epidermal cells which takes place at night, simultaneously with the assumption of the superposition arrangement in the lateral regions of the compound eyes (p. 223).[114]

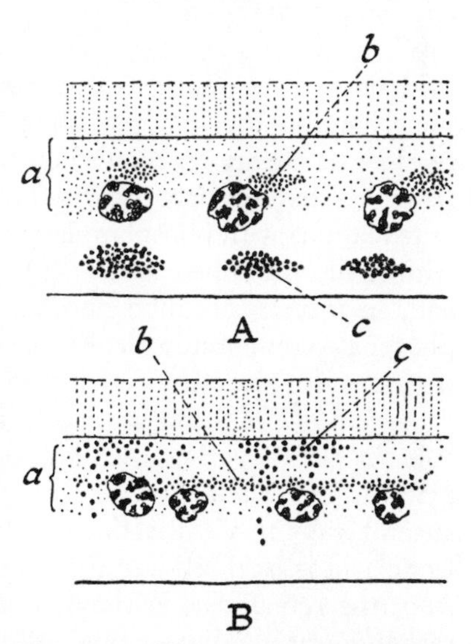

FIG. 349.—Pigment movements in epidermal cells of *Carausius*

A, light position; B, dark position. a, green and yellow pigments; b, orange red pigment; c, brown pigment (*after* GIERSBERG).

These effects are indirect; they apparently depend on some nervous stimulus reaching a centre in the head and leading to the production of a hormone which causes the pigment movement in the cells.[7, 114, 155] The epidermal cells are not innervated; and if pieces of cuticle and epidermis are transplanted from one insect to another they change colour simultaneously with their new host.[155] It is worth noting also that extracts from the head of *Carausius* and other Orthoptera will cause contraction of the chromatophores of Crustacea[130] (p. 199).

The effect of light on *Carausius* is exerted solely through the eyes; section of the eye stem eliminates the response.[7] It is determined by the contrast between the background and the surroundings; if the ventral half of the eye is blackened, darkening takes place; whereas all colour change ceases if the eyes are completely covered. Green is the most effective component of ordinary daylight in inducing the response; but if light of equal energy content is used, ultra-violet is the most effective.[243] The 'morphological' colour change in *Carausius* seems to be brought about by the same stimuli acting over a greater length of time.[7, 114] Several factors seem to be concerned: (*a*) from the corpus cardiacum (probably derived from the deuto- or trito-cerebrum); (*b*) from the tritocerebral region of the brain, where the centre can be localized in many insects.[485] A similar dual control regulates the expansion of the chromatophores on the air sacs of *Corethra*.[407, 414, 438, 485]

The seasonal colour change in *Chrysopa vulgaris* is a much simpler affair. It changes in the autumn from green to yellow or reddish as the result of an accumulation of carotinoids which disappear after hibernation when the insect becomes green again.[156] [See p. 641.]

RESPIRATORY METABOLISM

Oxidative metabolism—The organism derives most of its energy from the oxidation, or combustion, of organic substances; but at the temperature of the body the chief fuels available from the food will not unite directly with molecular oxygen: intermediates or respiratory catalysts of one sort or another are necessary. Furthermore, the energy that is produced by these oxidations is stored in chemical form, in such a state that it is readily available either for doing mechanical work as in muscular contraction, or for providing the energy for endergonic syntheses.

This storage of energy is achieved by utilizing the energy set free in the oxidation of respiratory substrates to synthesize compounds with energy-rich phosphate bonds. The immediate store is adenosine triphosphate (ATP) which releases its energy on conversion to adenosine diphosphate (ADP). ATP can be resynthesized by energy released from further oxidations (the so-called coupling of oxidations with phosphorylation) or by the energy stored in other compounds with labile phosphate bonds (the so-called 'phosphagens', creatine phosphate and, in insects, arginine phosphate). The precise way in which high energy phosphate compounds furnish the energy for contractile processes, for synthetic reactions, for secretory activity, and perhaps for light production, is not known.

It is possible to release a considerable amount of energy, and make it available for ATP synthesis, by the process of 'glycolysis'—the Emden-Meyerhof pathway by which glycogen, or glucose, is phosphorylated and broken down anaerobically to yield lactic acid. When oxygen becomes available a part of the lactic acid is oxidized and the energy liberated is used to resynthesize glycogen from the remainder. Ultimately all energy must come from the oxidation of carbon to carbon dioxide, and of hydrogen to water.

In almost every case the organic substrate must be acted upon by a specific 'dehydrogenase', which 'activates' a part of its hydrogen and transfers it to an 'acceptor'. In a few cases the acceptor may be molecular oxygen itself. Usually oxygen and hydrogen are brought together only through a system of carriers. Glutathione, which can be reduced by active hydrogen and oxidized by molecular oxygen is one such carrier. Co-enzyme I, or diphosphopyridine nucleotide (DPN), and co-enzyme II, or triphosphopyridine nucleotide (TPN), are other hydrogen acceptors, required for different steps in the process of hydrogen transfer, which do not react directly with oxygen but hand on the hydrogen to further carriers.

By far the most important system of carriers is the cytochrome system. A flavoprotein enzyme receives hydrogen from the reduced DPN (or other carrier) and liberates it in the form of hydrogen ions. The electrons are transferred along the series of haemochromogen compounds; cytochrome b, cytochrome c, and 'cytochrome oxidase' ('Atmungsferment') which then activates molecular oxygen so that it combines with the liberated hydrogen ions to form water. This whole system of catalysts is built into the structure of the mitochondria which thus form organized respiratory particles. The energy set free, as we have seen, is largely utilized for the synthesis of ATP and is thus conserved; some, of course, is lost as heat.

A great many substances, produced along a great many different metabolic pathways, may serve as substrates for this oxidative process. We have had occasion to mention the glycolytic pathway to glycero-phosphate and lactic acid,

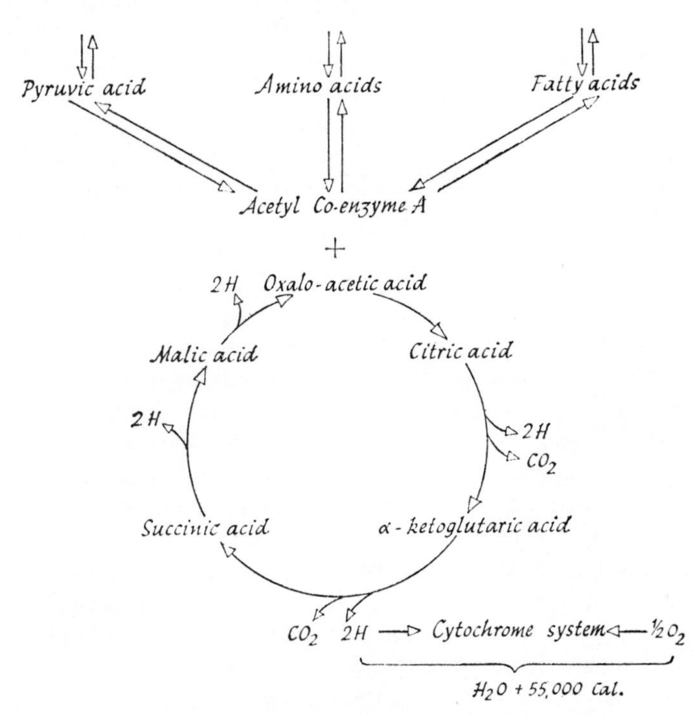

FIG. 350.—Greatly simplified diagram showing the oxidation of metabolites by way of the citric acid cycle and the cytochrome system.

in connexion with muscular activity (p. 155). We have had several occasions to mention the citric acid cycle (p. 602). This is the most important pathway for the production of oxidizable substrates. It is fed with acetic acid (activated by combination with co-enzyme A, thiamine pyrophosphate) which condenses with oxaloacetic acid to form citric acid. This undergoes a series of dehydrogenations and decarboxylations (indicated in greatly simplified form in Fig. 350) until it eventually gives rise once more to oxaloacetic acid, which takes up more acetic acid to repeat the cycle. The hydrogen set free by the dehydrogenation of this succession of substrates provides fuel for the cytochrome system, and energy for the phosphorylation of ADP to ATP. This cycle, likewise, is built into the mitochondria along with the cytochrome system. It can serve for the final oxidation of fats broken down to acetic acid, of carbohydrates broken down by glycolysis to pyruvic acid and then to acetic acid, and of deaminated amino acids.

The details of these metabolic pathways form a large part of general biochemistry, and may be found in text-books of the subject.[408] The cytochrome system, which consists of a series of haemochromogen compounds, was first discovered by David Keilin when he was working with insects.[165] It is present in the muscles in proportion to their activity; there is little in the thorax of wingless insects, much in the bee, and much in the leg muscles of active running forms. It can be seen in living muscles to undergo continual oxidation and reduction: in the resting thoracic muscles of *Galleria* it is oxidized, during activity it is partly reduced.[165] The oxygen storing systems, haemoglobin and myoglobin, are absent from most insects; air is conveyed directly to the tissues. This lack of oxygen storage is compensated for by an increased content of cytochrome. Cytochrome c in the skeletal muscle of the cockroach is much higher than in mammals. At 38° C. metabolism in the muscles of the cockroach is as high as that in the breast muscles of the pigeon.[11] Lacking haemoglobin, insects are not readily poisoned by carbon monoxide; but if the oxygen in the air is reduced to 8 per cent. and carbon monoxide increased to 80 per cent. they are poisoned.[128] This action is reversed in a bright light; probably through a direct effect on the cytochrome-cytochrome oxidase system.[97] At all stages of pupal development in *Drosophila*, the oxygen consumption is strongly inhibited by carbon monoxide. This effect is to some extent prevented by exposure to light. Perhaps the light causes the dissociation of the chemical compound which carbon monoxide forms with the iron-containing respiratory enzyme.[320]

Metal protein compounds of zinc, iron, magnesium and copper play an important part in the catalysis of oxidation and other changes in metabolism. The concentration and distribution of these metals have been studied in the silkworm. The significance of the findings is uncertain, but it may be noted that zinc is most concentrated in the most active tissues, silk glands, ovaries, &c.; manganese is most evident in the cocoon and the cast skin. The order of concentration of the metals is Zn>Fe>Mn, Cu, which is very different from that in mammals.[3] Iron in the *Lucilia* egg and larva is utilized in the formation of iron-containing catalysts. Perhaps some of the free iron appearing in the tissues results from the degradation of such substances.[190]

Glutathione is present in various insect tissues;[94] but probably it is responsible for only a small part of the oxygen transfer. The citric acid cycle occurs in insects as in vertebrates, and the existence of many of the other known metabolic pathways and enzyme systems has been confirmed in insects.[408] The appearance of red pigmentation in certain of the indirect wing muscles and coxal muscles in the adult male *Periplaneta* is associated with a 6·5 fold increase in the activity of succinic dehydrogenase, one of the enzymes of the citric acid cycle.[373] One of the chief end products of glycolysis in insect flight muscle is α-glycerophosphate (p. 155). Instead of being largely reconverted to glycogen (as is lactic acid in the more usual type of glycolysis) it is very actively oxidized by the muscle mitochondria (far more actively even than succinate), and this respiratory activity is coupled with the synthesis of high energy phosphate compounds. It is perhaps the principal substrate for the activation of the respiratory chain during flight.[499] This oxidative system is present also in vertebrate muscle but is relatively insignificant.[535] The classical glycolytic enzyme system is by no means absent in insects. It is present for example in larvae of Chironomids: succinoxidase and cytochrome oxidase systems are at a minimum in the larva; they increase progressively in the pupa and adult.[360] The suggestion has been made that in the tissues of *Celerio* during diapause, tyrosinase and the associated polyphenols and quinones provide a system for hydrogen transfer and thus participate in respiratory processes.[420] [See p. 641.]

Catalase—The transfer of hydrogen to molecular oxygen reduces the latter to hydrogen peroxide, and it is probably the function of the enzyme catalase, which is always present in the tissues, to prevent the accumulation of hydrogen peroxide by decomposing it into oxygen and water. But there seems little correlation between the catalase content of insects and the intensity of their respiration. Catalase increases during feeding and activity, and decreases during starvation (in grasshoppers[27] and *Leptinotarsa*[262]). But it is not markedly reduced in hibernating grasshoppers, when the respiration falls;[27] it does not

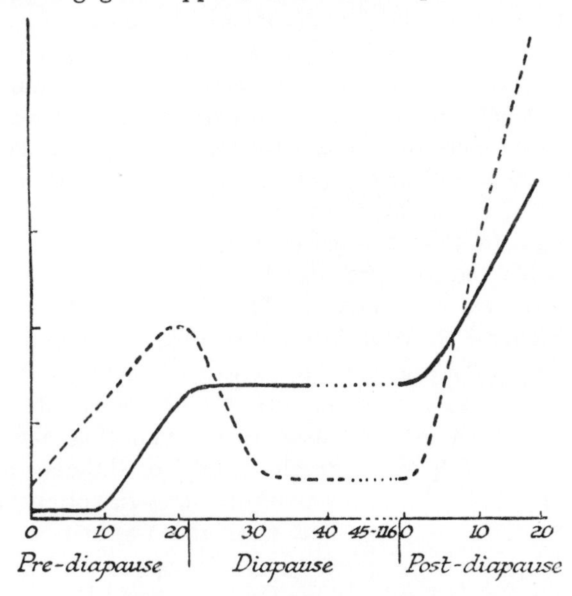

FIG. 351.—Catalase activity (continuous line) and respiratory rate (broken line) in eggs of *Melanoplus* from laying to hatching

Abscissa: days of development (*after* WILLIAMS).

follow the normal U-shaped curve of respiration in the pupa of *Leptinotarsa*;[95] and in the eggs of *Melanoplus* the catalase content continues to rise, and remains at a high level, long after the insect has entered diapause, and the respiratory rate has fallen and mitotic activity has ceased (p. 15) (Fig. 351).[319]

Respiratory metabolism—The rate at which these processes of oxidation take place in the living body, varies within wide limits. Figures are available for many species of insects;[181] but the values recorded depend to so great an extent upon muscular activity, external temperature, tissue growth, and a host of other factors, that such figures do not afford a true measure of the intrinsic differences which must exist between different species. Even values obtained with the insect at rest, at the same temperature, are not strictly comparable; because the tissues of different species, as we shall see, are adapted to different ranges of temperature; some can maintain their metabolism and remain active at a lower temperature than others; and if tested at a low temperature will appear to have a higher metabolic rate.[85] The important effects of temperature on respiration will be considered in the next chapter; here we shall examine the effects of other factors.

The kind of differences which exist are well seen among Ephemerid nymphs. Anaesthetized with urethane at 10° C., these had an oxygen uptake in c.mm. per gm. per hour as follows: *Ephemera vulgata* 278, *Caenis* sp. 290, *Ephemera danica* 370, *Ecdyonurus venosus* 604, *Ephemerella ignita* 950. These figures seem to bear some relation to the habitat of the species, the two first named species, which live in ponds, having a lower rate of metabolism than the last three species, which live in streams.[101] The reason for this difference is not clear; there is no evidence in this case that it depends upon adaptation to different temperatures.

Within the same species there may be a sexual difference. In *Drosophila* adults, under the same conditions, the males use about 2·8 mg. O_2 per gm. per hour, the females about 3·4;[183] whereas in the pupae the metabolism of the males is greater.[240] Larvae and pupae of *Tenebrio* show a diurnal rhythm in metabolism; the oxygen uptake being greatest at about 2 a.m., least about 3 p.m.[221] As in mammals, a protein meal seems to have a 'specific dynamic action' increasing the rate of metabolism more than fats or carbohydrates. Thus cockroaches at 15° C. used oxygen at the rate of 101–113 c.mm. per gm. per hour when fasting, 100–128 when fed with sugar, 117–167 (an average increase of 40 per cent.) when fed with protein.[124]

Metabolism in rest and activity—Metabolism at a given temperature is generally much higher in the adult than in the larva, and higher in the larva than in the pupa: at 30° C. the pupae of flies consumed about 8 c.mm. per gm. per minute, the larvae about 33 c.mm., the imagines 97 c.mm. (Fig. 352).[13] This is the result mainly of differences in muscular activity. In one series of experiments on the bee, when this was at rest at 18° C., the oxygen consumption was 30 c.mm. per gm. per minute. This is equivalent to 1·8 litres per kg. per hour; and since we know that the bee consumes only carbohydrate (p. 595), this will be equivalent

FIG. 352.—Temperature and respiratory exchange of flies in *A*, larval, *B*, pupal and *C*, adult stages (*after* BATELLI and STERN)

Ordinate: oxygen uptake in c.c. per gm. per hour. Abscissa: temperature in ° C.

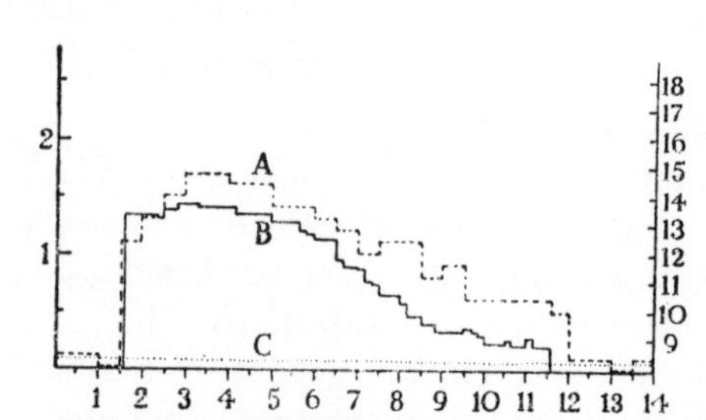

FIG. 353.—Rates of oxygen consumption and wing movement in *Drosophila repleta* (*after* CHADWICK and GILMOUR)

A, oxygen uptake; *B*, wing rate; *C*, average oxygen uptake in 'resting' fly. Ordinates: oxygen uptake in cu. mm. per min. on left; wing rate in thousands of strokes per min. on right.
Abscissa: time in minutes.

to 9·09 calories per kg. per hour.* (The corresponding figure for the 'basal metabolism' in man is about 1·2 calories per kg. per hour.) During flight the oxygen uptake rose to 1,450 c.mm. per gm. per minute, a 48-fold increase.[157] Others have obtained a still higher rate of increase.[175] The Sphingid *Celerio* in one experiment showed a 37-fold increase from 0·33 mg. of CO_2 per gm. per hour during rest, to 12·22 mg. during flight.[162] Various Lepidoptera at rest at 20–25° C. had an average oxygen consumption of about 8 c.mm. per gm. per minute. During the violent vibration of the wings induced by nicotine poisoning, this increased some 200 times.[247]

Detailed studies have been made of the respiratory metabolism of *Drosophila* during flight. In *D. repleta* the average oxygen uptake during rest is 28 c.mm. per gm. per minute. During flight this increases thirteen or fourteen times to 350 c.mm. per gm. per minute. The rate of respiration is closely related with the rate of the wing beat, and as this becomes slower in the exhausted insect the uptake of oxygen falls (Fig. 353). The high oxygen demands are almost wholly met from the supplies available during flight: any 'oxygen debt' resulting from the incomplete oxidation of metabolites is paid off within 1–2 minutes of the cessation of flight.[51] We shall return to the consideration of the metabolism of *Drosophila* during flight in connexion with the respiratory quotient (p. 632). In *Lucilia* the oxygen consumption increases probably about 100 times in going from rest to flight.[67] In *Schistocerca* the oxygen consumption increases 15–50 times on going from rest to flight, the consumption during flight varying from 10–30 litres O_2 per kg. per hour. This figure is of the same order as that in *Drosophila repleta* (21 litres O_2 per kg. per hour).[51] At rest and during the first 30 minutes of flight the respiratory quotient averages 0·82; but during prolonged flight it falls to 0·75, equivalent to 85 per cent. of the energy coming from fats. At the end of a long flight the 'oxygen debt' amounts to no more than the oxygen consumed during 0·5–1·5 minutes of flight; but recovery requires at least one hour.[182] [See p. 641.]

The rate at which the wing muscles of locusts convert energy is between 400 and 800 kcal./kg. muscle/hr. which is the same rate as in the hovering humming-bird or the flying *Drosophila*. Even for very intense muscular work, therefore, fat cannot be regarded as an inferior fuel in well-oxygenated muscles.[522] Male *Periplaneta*, flying for 10–15 minutes without interruption, show a 100-fold increase in oxygen consumption, their resting metabolism being relatively very low. They draw mainly on blood trehalose (which falls 50 per cent.) and glycogen from the thorax and the abdominal fat body. The respiratory quotient, however, averages 0·64, presumably because some substrates are incompletely oxidized.[481] This rate of respiration represents the maximum metabolic capacity of the cytochrome oxidase present.[402]

Metabolism in relation to size—Since each species has its own characteristic intensity of metabolism, it is not possible to establish any law relating one with another; though among similar insects metabolism tends to be less intense in the larger forms.[101, 85] But within a given species, at a given temperature, the metabolic rate shows a fairly definite relation with size. In general, metabolism per gram of body weight falls off as the insect grows, in such a way that respiration is proportional to the mass of the insect with a fractional exponent. This exponent is usually about 2/3 (0·67). Table 4 shows figures for

* In the combustion of carbohydrate 1 litre of oxygen produces 5·047 calories.

Melanoplus.[26] Similar results have been obtained on *Locusta migratoria*.[47] In the three common cockroaches, the exponent varies from 0·75–0·8 instead of 0·67.[127] In *Blatta* the metabolic rate per gram is found to diminish as the insect grows from 80 to 200 mg., but thereafter it seems to remain constant;[66] and in larvae of *Tenebrio*, the rate is high until they reach 5 mg. weight, it falls rapidly to about 45 mg., and then remains roughly constant until they reach 140 mg.[221]

TABLE 4

Weight in gm.	*Ratio of CO_2 output per gm. of body weight.*	*Ratio of CO_2 output per gm. of body weight$^{0·67}$.*
2·16	1,000	1,290
1·35	1,037	1,204
1·11	1,174	1,224
1·08	1,186	1,226
1·05	1,219	1,232
1·01	1,263	1,276
0·94	1,310	1,282

The exponent 0·67 is approximately the same as that which relates body weight with body surface. The two figures agree closely in *Locusta*;[47] in *Aphis rumicis* the surface is estimated to vary with the weight to the power 0·60; in *Blattella germanica* 0·63.[268] The fall in metabolic rate with size is therefore sometimes referred to as the 'surface law'; but there is no known reason why the metabolism of cold-blooded animals should bear a relation to the body surface. In many insects, indeed, the oxygen consumption is not proportional to the body surface;[172] and in the pupae of *Drosophila* in which there is a very close relation between oxygen uptake and surface area, this is regarded merely as a physiological example of the law of allometric growth (p. 62) which often applies equally to morphological and biochemical data.[86]

Effect of oxygen tension on metabolism—So far we have assumed that oxygen consumption is determined solely by the needs of the tissues. We must now consider under what conditions the supply of oxygen may become a limiting factor in oxidation. In general, oxygen uptake is more or less independent of the oxygen tension in the environment, within wide limits. In the mealworm larva the oxygen consumption begins to fall off when the tension drops below 10 per cent. O_2 in the air (Fig. 354, B);[285] in the mealworm pupa at 20° C. it remains

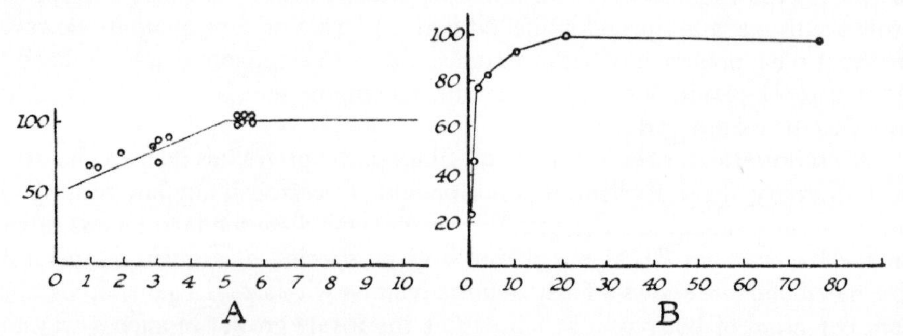

FIG. 354.—Relation between oxygen tension and oxygen uptake in pupa of *Tenebrio molitor* (A) (*after* GAARDER), and larva of *Tenebrio molitor* (B) (*after* THUNBERG)

Ordinates: oxygen uptake as percentage of uptake in air. Abscissae: oxygen concentration in volume per cent.

constant from 50 per cent. O_2 down to 5 per cent. O_2 (Fig. 354, A);[108] in the termite, *Termopsis*, there is no falling off until a concentration of 2 per cent. O_2 is reached;[58] in the actively developing eggs of *Melanoplus*, concentrations between 21 per cent. and 3 per cent. O_2 have no effect, below 3 per cent. the oxygen uptake falls and, unlike most insects, at concentrations above 50 per cent. the respiration is stimulated.[30] The aquatic nymphs of Ephemeroptera show great differences in their ability to maintain their oxygen consumption with a falling oxygen tension. Certain species of *Baëtis*, accustomed to streams with abundant oxygen, are affected far more readily than *Cloëon dipterum*, which lives in ponds. But there is no strict relation of this sort with the type of habitat.[101] The same relation is found in Chironomid larvae. Species such as *Tanytarsus brunnipes* and *Anatopynia nebulosa*, which live in streams, show a fall in oxygen consumption as soon as the oxygen content drops below 100 per cent. air saturation; whereas *Chironomus longistylus* and *A. varia* from stagnant waters maintain their normal rate of oxygen uptake until the oxygen content has fallen to 15·4 per cent. of air saturation. These differences in resistance are not correlated with the presence or absence of haemoglobin (p. 392); perhaps there are differences in the respiratory systems in the cells.[302]

The level of oxygen tension at which consumption is affected, must vary with the rate of metabolism. This can be seen in larvae of *Chironomus thummi*, in which the oxygen uptake of normal insects remained the same as in air, down to a concentration of 3 per cent. O_2; whereas in larvae which were recovering from exposure to nitrogen, and whose respiration was consequently increased 160 per cent. (p. 630), the oxygen uptake began to fall off below about 10 per cent. O_2.[131] Similarly, if an increase in temperature accelerates metabolism more than it favours the supply of oxygen by diffusion, the oxygen consumption will fall off sooner at a high temperature than at a low. Thus in the mealworm pupa at 20° C., oxygen uptake remains constant down to 5 per cent. O_2 in the air; at 32° C. it begins to fall off at about 10 per cent. O_2.[108] In *Lucilia* during flight the oxygen consumption is reduced to one half or less in 10 per cent. oxygen; in 5 per cent. oxygen few insects will fly and the oxygen uptake is still less.[67] In *Drosophila* species exposed at their optimal breeding temperature, mobility ceases at 1·6 per cent. oxygen in *D. melanogaster*, at 2·8 per cent. in *D. obscura*.[64]

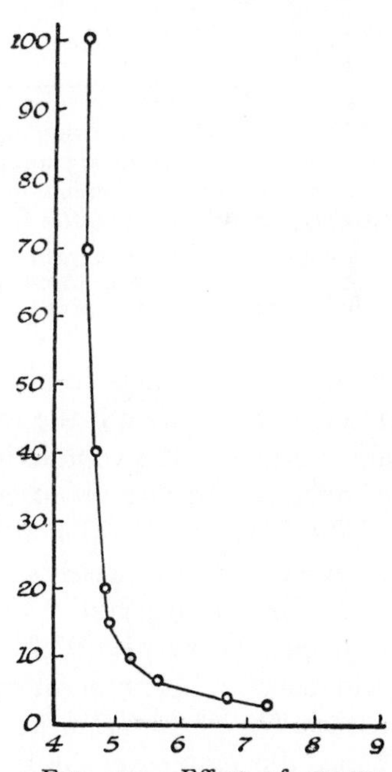

FIG. 355.—Effect of oxygen tension on the duration of the pupal stage of *Drosophila melanogaster* at 18–20° C.

Ordinate: oxygen per cent. (in hydrogen). Abscissa: duration of pupal stage in days (*after* KALMUS).

Probably oxygen tension becomes a limiting factor only when the oxygen supply fails, and the tension in some part of the tissues becomes zero.[181] At 20° C. the oxygen tension in the tissues of the mealworm larva is about 16 per cent. of an atmosphere. Hence there is a gradient of $21-16=5$ per cent. between the tissues and the outside air. The tension in the tissues would therefore be expected to fall to about zero when the oxygen in the air falls to about 5 per cent.; which agrees well with the experimental results (p. 629).[108] But it must be pointed out that this argument is to some extent invalidated by the regulatory mechanisms in the spiracles which have since been discovered (p. 367). In *Chironomus plumosus* in water saturated with oxygen the oxidation-reduction potential in the tissues exceeds 15; on transferring to oxygen free water it falls to 7·7.[5] We have seen that these changes are associated with changes in the extent of air in the tracheoles, possibly as a result of the redox potential affecting swelling processes in the cytoplasm (p. 361).

Oxygen tension may serve as a limiting factor in development. In the pupa of *Drosophila* the time required for development is prolonged from 4·8 to 7·2 days as the oxygen concentration falls from 20 per cent. to 3 per cent. The minimal concentration for development lies between 2–3 per cent.; and up to a concentration of 70 per cent. the rate is increased (Fig. 355).[163]

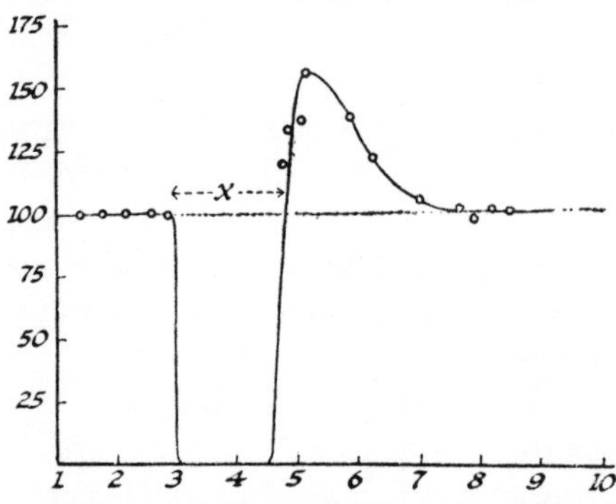

FIG. 356.—Curve showing the effect of immersion in water for 75 minutes on the rate of oxygen consumption in nymph of *Melanoplus differentialis* (*after* BODINE)

Ordinate: percentage oxygen consumption (taken as 100 per cent. before immersion). Abscissa: time in hours. *X*, period of immersion.

Anaerobic metabolism—If the mealworm is kept at a low oxygen tension, such as 3 per cent., for some hours, and then returned to air, its oxygen uptake is raised above the normal level for some time. This is attributed to the removal of intermediary products of metabolism which have been incompletely oxidized.[108] For the initial stages of energy metabolism, notably the production of lactic acid from sugar, takes place anaerobically; the oxidation of such substances may be deferred until oxygen is restored, the insect accumulating an 'oxygen debt'. This process is well seen if the insect is kept in complete absence of oxygen. The cockroach *Blatta orientalis*, kept in nitrogen for ½ hour, contracts a debt which it requires 1½–2 hours to pay off; the extra oxygen consumed during recovery being just about equal to the volume it would have used during the time it was deprived of oxygen, and just about equal to that required for the complete oxidation of the lactic acid which has accumulated.[66, 270] Similar results have been obtained with the grasshoppers *Melanoplus* and *Chortophaga* (Fig. 356)[28] and with *Chironomus* larvae.[131] In the grasshoppers, the blood increases in acidity from *p*H 6·8 to *p*H 5·8;[28] and in the cockroach, and in *Chironomus*, this acid causes a liberation of carbon dioxide from bicarbonates in the blood and tissues.[131, 270] In *Tenebrio*, *Melanoplus* and *Cryptocercus* carbon dioxide is retained during recovery in amounts equal to those given off during anaerobiosis; it has doubtless resulted from the simple buffering of acid by bicarbonates; but in *Zootermopsis*, with or without its intestinal protozoa, carbon dioxide seems to be an actual end product of the anaerobic process.[117]

Metabolism during anaerobiosis doubtless varies in different insects. In the isolated hind femur of *Melanoplus* 125–190 per cent. of the expected oxygen debt is repaid; in the *Tenebrio* larva about 150 per cent.; in *Cryptocercus* the extra oxygen consumed after one hour of anaerobiosis amounts to 3 times the quantity missed. The removal of lactic acid and other metabolites is a more wasteful process than it is during normal respiration.[117] In *Zootermopsis* on the other hand the extra oxygen uptake amounts only to 50 per cent. of that missed during anaerobiosis.[117]

Certain insects show a remarkable capacity for living anaerobically. The Oestrid larva *Gasterophilus* has been kept for 17 days submerged in oil. But it shows almost no accumulation of lactic acid (0·4 mg. per gm. of live weight at the beginning, to 0·53 mg. at 14 days).[73] It seems to convert its store of glycogen into fat and to utilize the oxygen set free for energy production; glycogen disappears, and more fat is produced, in the absence of oxygen. Thus 100 gm. of larvae starved for 24 hours showed an increase of 0·03 gm. fat in the presence of oxygen, 0·06 gm. under anaerobic conditions.[168] [See p. 641.]

The same capacity for anaerobic respiration is shown by some Chironomid larvae. The resistance of these to low oxygen tensions depends not so much on their ability to exploit small amounts of oxygen in the water (p. 349) as on their adaptations to anaerobic respiration. Larvae of *Chironomus thummi* show a tenfold increase in glycogen consumption when transferred to oxygen free water.[132] *C. bathophilus* can withstand 4–6 days of anaerobiosis; it is so highly adapted to low oxygen tensions that it apparently practises anaerobic respiration under all conditions, and after exposure to nitrogen for 48 hours there is no excess oxygen uptake on returning to air.[132] The course of anaerobic metabolism is not known; butyric and caproic acids seem to be the chief end products; lactic acid is formed in very small amounts only. Whether there is a conversion of glycogen to fat remains to be proved.[132]

Carbon dioxide in high concentration acts as a narcotic on most insects.[326] But insects from the deeper layers of the soil, such as Scarabaeid and Elaterid larvae, show a high resistance to carbon dioxide and highly developed powers of anaerobic metabolism.[336] In the larva of *Gasterophilus* from the stomach of the horse a high concentration of carbon dioxide is actually necessary in order to maintain the cells of the tracheal organ in a normal state. This effect is peculiar to the larval stage in this insect.[339]

If the diapausing pupa of *Hyalophora* is kept anaerobically for one day it requires three days to pay off the oxygen debt. After two days anaerobiosis the lactic acid content increased from 0·06 mg. to 1·5 mg. per gm. live weight. This is approximately equivalent to the predicted oxygen debt and suggests that lactic acid is the main end product requiring oxidative removal in this insect,[442] *Musca* adults subjected to anaerobic conditions for some hours show an accumulation of pyruvate and particularly of alanine in the tissues.[483] There is also a large increase in α-glycerophosphate amounting to 8 times the normal level, three quarters of which occurs in the thorax; orthophosphate rises and ATP and arginine phosphate fall. These changes are reversed in air.[487] In the grasshopper and in *Periplaneta*, lactic acid and pyruvic acid accumulate;[426] but in the flight muscles lactic acid formation is insignificant; glycerophosphate and pyruvate replace it.[440] The same result is seen in cell-free preparations of thoracic muscle of *Triatoma*.[354] In *Chironomus thummi* there seems to be no adaptation to prolonged exposure to low oxygen; there is no change in dehydrogenase activity; but this falls during starvation.[482]

Respiratory quotient—When the oxidation of metabolites is complete the respiratory quotient (R.Q.), the ratio of the volume of carbon dioxide evolved to the volume of oxygen consumed, gives some indication of the nature of the substances being utilized. For in the complete combustion of carbohydrate this ratio is 1·0, for fat and protein (when this is oxidized only as far as uric acid) it is about 0·7. Thus the R.Q. of the cockroach fed on starch is 1·01–1·07, on fat it is 0·78–0·83, and after starving for several days (when, as we have seen, fat is chiefly utilized) it falls to 0·65–0·85.[125] In termites the R.Q. is normally 1·0; it falls during starvation.[57] In *Popillia* larvae during starvation it falls from 0·82 to 0·70.[20] In *Zootermopsis* it is 1·09; after elimination of the cellulose digesting protozoa it is 0·76 falling later to 0·57.[117] Defaunation in the cockroach *Cryptocercus* likewise brings about a fall from 1·09 to 0·69;[117] in *Reticulotermes*, from 0·98 to 0·82.[112] In the fasting mealworm it is 0·7.[215] In the bee, which, as we have seen, burns chiefly sugar, it is equal to 1·0 during rest, and during short flights.[157]

Drosophila, as we have seen (p. 601), uses glycogen as the sole source of energy during flight. The respiratory quotient is 1·0. During the resting period after flight, when fat also is being utilized (p. 601), the quotient falls to 0·7. And during the period before flight the R.Q. may average 1·23, suggesting the conversion of carbohydrate to fat.[51] In the locust *Schistocerca*, on the other hand, which utilizes fat, the R.Q. during sustained flight is 0·75–0·76.[182]

At very low oxygen tensions the R.Q. tends to rise, owing to the increasing acidity of the tissues driving off carbon dioxide; for example, in *Tenebrio*[285] and *Chironomus* larvae;[131] it may even go above unity, as in *Anax* larvae.[200] And during the process of recovery from oxygen want, as the carbon dioxide capacity of the blood and tissues increases again, the R.Q. falls for a time below the normal level.[270] Some Lepidoptera, such as *Agrotis segetum*, utilize sugar to synthesize the abundant fat in the egg yolk, and this transformation may send up the R.Q. to an average of 1·5–1·6 or a maximum of 2·09.[177] In the larval stages of the bee also a R.Q. above unity (1·1–1·4) may occur, indicating a synthesis of lipides.[213] In insects exposed to low temperatures, the R.Q. may fall to a very low figure. In ants during the summer at 22° C. the R.Q. is 0·88, at 4° C. it is 0·50;[77] in hibernating nymphs of *Chortophaga* at 27° C. the R.Q. is 0·83, at 15° C. it is 0·62.[173] The low temperature seems to cause some disturbance in metabolism, the nature of which is not known. The values at the low temperature are too low to be explained by an exclusive oxidation of fat, and too prolonged to be explained by the increased solubility of gases in the tissue fluids. In *Phalera bucephala* (Lep.) during diapause the R.Q. is about 0·35, due perhaps to the conversion of fat to carbohydrate; this results, after a period of one year, in an increase in weight of about 10 per cent.[325] But measurement of the R.Q. during diapause must be interpreted with great caution: the cyclic discharge of carbon dioxide (p. 395) may upset the results completely.

The R.Q. of *Galleria*, *Antheraea*, and *Mimas tiliae* at rest is 0·69 through life, and shows no change during flight. These are insects which take little or no food as adults and store mainly fats. But the same results are obtained in *Vanessa* and *Gonepteyx* which feed on nectar: R.Q. at rest 0·65–0·76, and during the 50-fold increase of metabolism in flight: av. 0·73. The sugar has been converted into fat before use. If the resting butterfly is given 40 per cent. glucose solution, the R.Q. rises for several hours to 1·6 or 2·0, with a threefold increase in CO_2 output;

clearly fat synthesis is in progress. If butterflies are induced to fly soon after feeding on sugar, they give an R.Q. of 0·73; but soon after resting it will rise again to 1·4–1·5. These Lepidoptera are apparently unable to utilize sugar directly for flight; but must first convert it into fat (p. 602). Already after 10 minutes' flight the oxygen debt amounts to the equivalent of 1½–3 minutes' flying time.[533]

Metabolism during growth—In the egg, oxygen uptake runs parallel with visible development. In the silkworm, metabolism is active during the initial development up to the blastoderm stage; it then falls and remains at a low level throughout the latent period, to rise rapidly again when renewed development occurs in the spring.[195] In *Melanoplus*, respiration follows the same course, high in early development, low in diapause, and rising steeply during blastokinesis.[42] Whereas in *Chortophaga*, which develops without interruption, the oxygen uptake rises steadily until the time of hatching.[29] In *Melanoplus* the R.Q. is 1·0

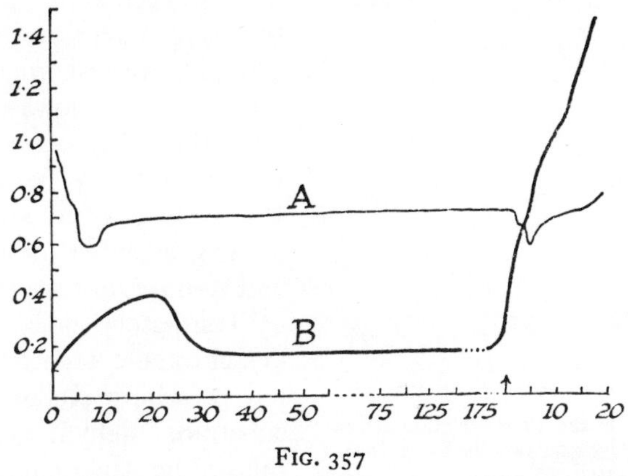

FIG. 357

A, respiratory quotient during entire embryonic development of *Melanoplus* (*after* BOELL). Ordinate: R.Q. Abscissa: days of development; pre-diapause period from 0–20 days; diapause from 21 days to the arrow; post-diapause period from the arrow to 17 or 19 days (hatching). *B*, relative oxygen consumption.

during the first day of embryonic development; thereafter it is generally about 0·71, suggesting that fat is the chief normal source of energy;[39] but throughout development added glucose is readily oxidized by embryos isolated from the egg.[35] There are two periods in the development of the *Melanoplus* egg, before diapause and when blastokinesis is complete, during which the R.Q. sinks very low, suggesting that fat is being utilized for the synthesis of oxygen-rich compounds—possibly the chitin of the serosal membrane (p. 4) and the pre-larval skin (p. 18) (Fig. 357).[39] The period of maximum oxygen uptake in the developing egg of *Rhodnius* coincides with the maximum of cell multiplication;[351] but this does not necessarily signify that mitosis as such is the process requiring an increased supply of oxygen. In the egg of *Tenebrio*[454] and *Popillia*[454] the oxygen consumption is low at first, then rises and remains level for a few days and finally rises to a peak at hatching; it runs more or less parallel with the activity of succinic dehydrogenase and cytochrome oxidase. This relationship is discussed below.

In the pupa, oxygen uptake follows a characteristic U-shaped curve; it is high immediately after pupation, falls to a minimum, rises again shortly before emergence, and finally shows a slight depression before rising to the very high level associated with the escape of the adult insect.[278] Curves of this type (the final drop before emergence has not always been noticed) have been described in *Calliphora*,[106, 307] *Lucilia*[61] and other blow-flies,[278] *Drosophila*,[38] *Galleria* (Fig. 358),[62] *Ephestia*,[278] *Celerio*,[141] and other Lepidoptera,[93] *Tenebrio*,[180] and other Coleoptera.[93] In the pupae of *Lymantria dispar*[10] and *Galleria*[280] the rate of heat production follows the same curve as that of oxygen uptake; but the calorific quotient, the ratio of the heat in gram calories to the oxygen uptake in milligrams, is very low (1·16–2·84 compared with a value of 3·3 for the combustion of fat), perhaps because of the large proportion of endothermic syntheses that are taking place.[10, 280]

When there is a prolonged arrest of growth in the pupa, as in the overwintering pupae of *Celerio*,[141] it is understandable that there should be a long period of low metabolic rate; but the same course is followed when development is apparently continuous.[141] It is certain that there is a real change in oxygen requirements: the fall is not due to failure in the oxygen supply; for in *Tenebrio* the oxygen tension in the tissues is actually greatest when the rate of metabolism is lowest.[108] In *Drosophila* the fall is attributed to changes in the quantity or activity of the oxidizing enzyme system; i.e. this system is either destroyed in part and re-built, or temporarily inhibited by respiratory poisons.[320] Both the cytochrome-cytochrome oxidase system (Warburg-Keilin system) and the substrate dehydrogenase system follow the same curve.[321] The same applies to *Calliphora* where the total activity of the dehydrogenase system follows the usual U-shaped curve; but when the oxidation of particular substances is followed separately they show varying critical periods.[2]

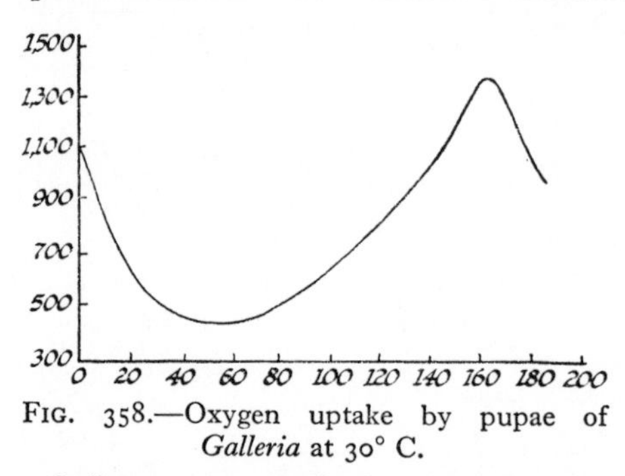

FIG. 358.—Oxygen uptake by pupae of *Galleria* at 30° C.

Ordinate: oxygen uptake in cubic millimetres per gram per hour. Abscissa: time in hours (*after* TAYLOR and STEINBACH).

It has often been suggested that the fall of the curve represents histolysis the rise histogenesis; the intensity of metabolism at any moment being a measure of the amount of organized tissue present.[93, 141, 180] It is not the presence of organized tissue, however, that is important, but its activity. The U-shaped curve may be evident during a period when almost no mechanical work is being done. The oxygen consumed is required for endothermic syntheses and perhaps mainly for protein synthesis.[526, 537] There is often a corresponding curve of activity in the systems of respiratory enzymes;[2, 320, 321, 455, 515] but the enzyme level is commonly adapted secondarily to the metabolic needs.[526, 531, 532] In the larva of *Samia cynthia* there is a precipitous fall in the rate of oxygen consumption two and a half days before spinning begins; this presumably marks the end of the period of active protein synthesis required for silk formation.[431] In *Schistocerca* there is a rapid increase in metabolic rate during the first few days of adult life when sexual maturation begins.[418] The lack of an exact parallel between oxygen consumption and enzyme activity is well seen in the pupa of *Musca* in which the highest values for oxygen consumption are given at the start of the development process, whereas cytochrome oxidase shows its highest activity at the end of pupal development;[470] and a further very large increase takes place after emergence, in readiness for flight (p. 156). In the pupa of *Lucilia* pyrophosphatase activity rises in parallel with protein synthesis,[399a] In *Celerio* (*Deilephila*) respiration in the pupal brei follows the same curve during hibernation as in the intact pupa; tyrosinase seems to form some part in the chain of oxidizing mechanisms.[140] There is no evidence that it plays any part as a terminal oxidase in the intact pupa of Lepidoptera (*Hyalophora*), at any rate during diapause.[350]

The associated changes in the reaction of the blood are variable. In some insects the *p*H is said to fall, for example in *Attacus*,[79] and in *Leptinotarsa*, where it falls from 6·8 to 5·9[93] as the respiratory rate decreases. The same occurs in *Calliphora*.[2] In others, such as *Popillia*, there is no significant change;[197] while in *Galleria* the *p*H is highest (6·5) when metabolism is lowest, and falls to 5·8 when the respiratory rate increases again; here the *p*H is regarded as a reflection of the rate of carbon dioxide production.[279]

We have seen that the chief source of energy during pupal development is fat; and in some insects the respiratory quotient agrees with the expected value for fat combustion: 0·70 in *Tenebrio*,[180] 0·67–0·78 in *Drosophila*,[240] 0·69 in *Galleria*.[62] But in many cases, particularly during the middle period of pupal development, very low values are obtained: less than 0·7 in *Lucilia*,[61] 0·5–0·65 in *Calliphora*,[106, 307] 0·5 in *Attacus* (Lep.),[79] as low as 0·4 in *Popillia*,[196] and 0·42 in *Leptinotarsa*.[93] These values presumably indicate either the synthesis of constituents relatively rich in oxygen, or the temporary incomplete combustion of fats or other substances. Where the pupa is showing a cyclic discharge of carbon dioxide (p. 395) respiratory quotients may be highly misleading.

Metabolism during diapause and quiescence—Depression of metabolism is always a striking characteristic of arrested growth or diapause at whatever stage this may occur (pp. 16,117). In eggs of *Melanoplus* during diapause, respiration is maintained at about one-third or one-quarter of that shown by developing eggs at the same morphological stage.[29] The same depression is seen in overwintering pupae of *Celerio*, in which it has even been suggested that metabolic rate is the causative factor distinguishing strains with and without a diapause.[141] The oxygen uptake by larvae of *Lucilia* increases from 0·34 c.c. per gm. per hour, to 1·10 c.c. when diapause is broken by suitable stimuli.[61] Pyralid larvae in diapause can withstand submergence in water for several days, whereas active larvae have not this resistance.[176] The summer wasp resting at 0° C. has an oxygen uptake nearly 20 times as great as that shown by the hibernating wasp at the same temperature,[223] and the ant *Formica ulkei* shows a similar reduction during hibernation.[77]

In the diapausing adult of *Leptinotarsa* there is an extreme degeneration of the flight muscles with almost complete disappearance of normal sarcosomes. The loss of this respiratory system will account for the 75 per cent. fall in the total respiration of *Leptinotarsa* during diapause. At the termination of diapause, complete regeneration occurs.[508]

We have already discussed the low values for the respiratory quotient which may occur during hibernation (0·6 in *Leptinotarsa*,[93] 0·62 in grasshopper nymphs,[173] 0·52 in ants[77]). These appear to indicate some change in metabolism, perhaps incomplete combustion, which is a direct effect of lowered temperature (p. 632).

There is some evidence that in cold hardy insects, and particularly in insects in diapause, the oxidases dependent on cell structure are largely replaced by anoxybiotic dehydrogenases whose activity is independent of cell structure.[178] In the pupa of *Hyalophora* during diapause cytochrome c is virtually absent. There is a progressive increase in cytochrome oxidase and then of cytochrome c as the internal secretions which terminate diapause come into play (p. 69).[318] Likewise, in the eggs of *Melanoplus* (p. 15), if the actively developing egg is treated with potassium cyanide (which is believed to put out of action the cytochrome-oxidase system), the respiration is reduced to a minimum; whereas respiration during diapause is unaffected.[31] Methylene blue will stimulate the respiration of eggs in diapause, but not active eggs. And respiration of active eggs is depressed by carbon monoxide, while eggs in diapause are quite resistant.[36] That part of the respiration which is sensitive to cyanide is dependent on the intact structure of the embryo, whereas the cyanide insensitive fraction, like the respiration in diapause, is unaffected by breaking up the egg.[34] In the developing egg of *Melanoplus* the activity of cytochrome oxidase increases parallel with the mass of the embryo (which contains at least 90 per cent. of the activity present in the egg); the amount does not fall with the drop in respiration during diapause.[6]

The apparent resistance of the respiratory system during diapause to such inhibitors as cyanide and carbon monoxide is due to the great excess of cytochrome oxidase in relation of the small amount of cytochrome c present.[419, 442, 503] There seems to be no qualitative change in the nature of the respiratory enzymes.

ADDENDA TO CHAPTER XIII

p. 593. **Sterols**—The four sterols most widely distributed in nature, cholesterol, ergosterol, β-sitosterol and stigmasterol are equally effective as growth factors for insect larvae because the last three can all be converted into cholesterol.[544]

p. 594. **Triglycerides**—Radiotracer studies with labelled acetate show that *Hyalophora* synthesizes palmitic, palmitoleic, stearic and oleic acids, but not linoleic or linolenic acids; and this probably applies to all insects.[613] But in some insects body lipids may contain up to 70 per cent. of polyenoic fatty acids, especially linoleic and linolenic, and some tetra-enes and penta-enes. In females of *Formica*, acids with 2–5 double bonds occur;[573] and isolated mitochondria of *Drosophila* will incorporate acetate into even numbered saturated and monoenoic fatty acids, C_{12}–C_{18}.[576]

Changes in environmental temperature during fat synthesis in *Culex* has no influence on the degree of unsaturation of the fatty acids produced.[577] But low temperature combined with a short photoperiod is said to induce more unsaturated fatty acids;[581] and in caterpillars there seem to be higher levels of palmitoleic acid among those which overwinter.[546]

p. 594. **Phosphatides**—Phosphatides are of prime importance in the enzyme activity of mitochondria; the amphipathic character of the molecule allows it to act as a bridge between water and protein; phospholipid can furnish a non-aqueous medium in which specific reactions can take place—such as the electron-transfer chain (p. 622).[573] Phospholipids synthesized in the fat body of *Hyalophora* and *Periplaneta* are liberated into the haemolymph and transported in the form of lipoprotein.[588] A deficiency of choline in the diet of *Musca* causes a reduction in the level of phosphatidylcholine with a corresponding increase in phosphatidylethanolamine in all organs.[548]

p. 595. **Trehalase**—The haemolymph in *Bombyx* larvae contains trehalase; but this is normally inhibited, becoming active only during moulting when trehalose decreases and free glucose rises. Muscles, gut &c. also have an intracellular trehalase and can thus utilize blood trehalose.[570]

p. 597. **Other carbohydrates**—Hyaluronic acid seems to be the sole mucopolysaccharide extractable from *Galleria* mid-gut, that is, from basement membrane, interstitial cements and peritrophic membrane.[567] Radioactive glucose injected into the late pupa (that is, the developing or pharate adult) of *Lucilia* is mainly incorporated into chitin.[560]

p. 597. **Protein**—Collagen extractable from the connective tissue of *Leucophaea* and *Blaberus* contains diagnostic proportions of glycine, proline, hydroxyproline and hydroxylysine.[580] 'Calliphorin' (p. 446) with a molecular weight of 540,000 amounts to 60 per cent. of the total soluble protein in growing larvae of *Calliphora* and 75 per cent. of the blood protein. It falls to 10 per cent. in the young adult fly. It may be a storage protein.[594] A similar protein is present in many insects, notably Lepidoptera;[584] in other Cyclorrhapha the proteins seem to be antigenically identical.[594] *Locusta* injected with labelled blood protein showed extensive uptake by the pericardial cells (p. 440); at certain periods during the instar fat body, epidermis and cuticle, and oenocytes also take up haemolymph proteins.[616] In adult male *Drosophila* about 20 per cent. of the total protein turns over, with a half-life of about 10 days. Among the proteins that are not replaced are the structural and mitochondrial proteins of the flight muscles.[611] There is naturally a great reduction with age. The incorporation of labelled amino acids into body proteins of *Drosophila* 50 days of age is only 40 per cent. of that in 3-day-old flies.[541]

p. 597. **Phenolic substances**—Phenolic substances extractable from the silkworm *Bombyx* are free tyrosine, protocatechuic acid, dihydroxyphenylalanine ('Dopa') and 3-hydroxykynurenine, together with the glucoside of acetyldopamine and other unidentified *o*-diphenols and glucosides of the same.[617]

p. 599. **Synthesis in late pupa of Lucilia**—During the final stages of adult development in the pupa of *Lucilia* oxygen consumption increases greatly; active syntheses are taking place: ATP and arginine phosphate, mitochondrial enzymes, flight muscle sarcosomes, &c. Fatty acid oxidation apparently provides all the energy for these biosynthetic processes; carbohydrate is conserved by synthesis of chitin and for flight.[561]

p. 601. **Energy for flight**—In the flight muscles of *Phormia* the intense rate of glycogenolysis is effected by the conversion of phosphorylase *b* to phosphorylase *a*.[558]

p. 602. **Consumption of trehalose**—In *Musca*, as in *Drosophila*,[316] trehalose is utilized during flight, particularly during the first few minutes; there is a temporary fall in trehalose reserves in the thorax until glycogen consumption is established.[612]

p. 603. **Lipid consumption**—In *Schistocerca*, after 2 hours' flight, the haemolymph lipid, notably lipo-protein, is increased 3–4 times the resting value, to which it has returned by three hours after flight. The main increase is in the diglyceride bound to protein.[591] The same changes occur in *Locusta* which oxidises preferentially palmitic and oleic acids.[542] Male *Dendroctonus* in flight use mainly mono-unsaturated fatty acids.[615]

p. 603. Very rapid oxidation of long chain fatty acids is effected by mitochondria isolated from the flight muscles of Lepidoptera; and very active lipases specific from monoglycerides are present in the myofibrils or the sarcolemma.[614] Carnitine is needed in *Locusta* for the oxidation of fatty acids by the flight muscles: carnitine-coenzyme A transacetylase activity is high in locust muscles, virtually absent in the bee which uses carbohydrate in flight.[543]

p. 603. **Proline**—Besides acting as an energy reserve in *Glossina*, proline may serve to prime the citric acid cycle at the initiation of flight in *Calliphora*.[607, 608]

p. 605. **Sericin**—Sericin is mixed with a large amount of wax, more than 30 per cent., rich in free higher aliphatic alcohols.[586]

p. 605. **'Cocoonase'**—The silk moth *Antheraea*, like *Bombyx*,[396] produces a proteolytic enzyme, termed 'cocoonase', which softens the cocoon at adult emergence (p. 528). It is a general protease not specific for sericin.[583]

p. 606. **Fibroin**—Silk fibroin is elaborated at the rough surface endoplasmic reticulum and transferred to the Golgi apparatus before being discharged.[538]

p. 606. **Other 'silks'**—Some apparent 'silks' are chemically different from the silk of commerce. The silk of *Nematus* and related sawflies is composed of collagen; Blennocampine sawflies secrete a 'polyglycine'; the oötheca of praying mantids and the cocoon of the flea are composed of a protein with an α-helix;[606] the silk-like threads suspending newly hatched mantids are secreted chitin fibrils;[585] in *Ptinus* and some weevils the cocoon fibres are composed of the drawn-out chitinous peritrophic membrane; and in *Chrysopa* the larval silk resembles 'cuticulin' (p. 35) and is perhaps a hydrocarbon polymer with some resemblance to nylon.[606]

p. 606. **Wax from Aphids**—Wax is exuded from the tip of each cornicle of an Aphid if it is touched. The wax is in a liquid or dissolved state; it solidifies instantly on contact and may serve to trap a small predator or parasitoid.[564]

p. 606. **Shellac**—Shellac resin is essentially a polyester, a giant molecule, of which the major component (46 per cent.) is trihydroxypalmitic acid (aleuritic acid); but it contains a number of other hydroxylated fatty acids and a large amount (25 per cent.) of a sesquiterpene, shellolic acid, which carries two carboxyl groups and two hydroxyls, that likewise participate in the polyester system.[559]

p. 609. **Defensive secretions**—Fig. 358*a* shows a diagram of the reaction chambers in the defensive gland of *Brachinus*. The spray mixture of quinones is ejected at 100° C., with a heat content of about 0·2 calorie per milligram.[539] *Tribolium* and *Eleodes*, which produce a mixture of benzoquinones, unsaturated hydrocarbons and caprylic acid, likewise make use of a small 'reaction chamber' with impermeable walls where the final steps in the production of these toxic substances take place.[579] The defensive substance in *Dytiscus* appears to be identical with the vertebrate hormone 'cortexone'.[609] The main components in the secretion from *Papilio* larvae are isobutyric and methylbutyric acids.[565]

p. 609. Besides hydrogen cyanide, salicylaldehyde (p. 608), and alkaloids in *Papilio* spp., there are many other poisonous compounds from plants that are taken up by insects which sometimes use them for their own protection:[549] such as the alkaloids in the cinnabar moth *Callimorpha*;[540] the presence of repellent cardiac glucosides in *Danaus*[598] only when the larvae have fed on *Asclepias* spp. containing such poisons.[550, 600] Cardiac toxins are taken up from milkweeds (Asclepiadaceae) and used in defence also by the grasshopper *Poekilocerus*[568] and in the oleander Aphid.[603] (In *Calosoma* (Carabidae) salicylaldehyde in the venom is synthesized by the insect itself.[556] Likewise hydrocyanic acid can be released from crushed Zygaenids which do not feed on plants containing cyanogenic glucosides.[430])

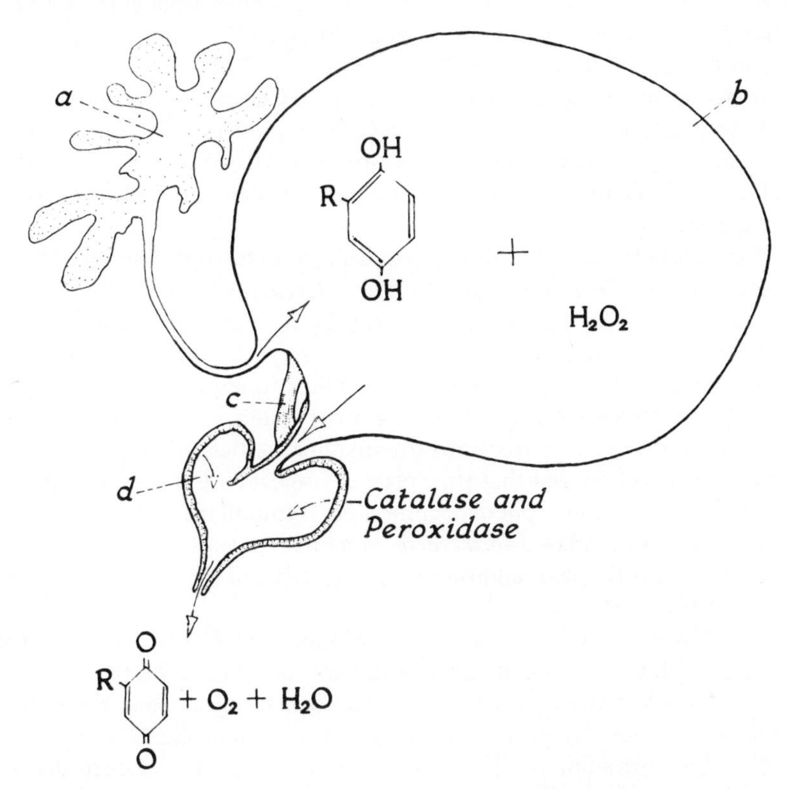

Fig. 358*a*.—Diagram showing the operation of the defensive gland of the bombardier beetle *Brachinus* (after Aneshansley *et al.*, based on Schildknecht).

The gland (*a*) discharges a mixture of hydroquinones and hydrogen peroxide into the reservoir (*b*). The muscle (*c*) allows a little of the mixture to enter the reaction chamber (*d*), the walls of which secrete catalase and peroxidase. The immediate explosive reaction results in a discharge of quinones.

p. 609. **Antibiotic secretions**—The metapleural glands of Myrmecine ants (*Myrmica*, *Atta*, *Tetramorium*, &c.) store in their reservoirs an acid secretion containing phenylacetic acid which serves as a potent antiseptic.[590] The bacteriolytic component in the honey and venom of bees is a lysozyme.[593]

p. 609. **Pheromones**—Pheromones are properly defined as 'chemical messengers which act within the species'—although as noted (p. 609) the same secretion may serve both this purpose and act as a warning stimulus or a 'repellent' directed against other species. Pheromones may be grouped as trail marking pheromones and other olfactory markers (p. 609), sex attractants (p. 710), aphrodisiacs (p. 711), aggregation pheromones, alerting pheromones.[554]

p. 610. **Pheromones and other scents**—In recent years the chemistry of 'pheromones' and other scent gland secretions has been extensively worked out.

These comprise a vast range of saturated and unsaturated hydrocarbons and their derived alcohols and aldehydes (ranging from hexenal and tridecene to pentadecane), fatty acids (such as butyric and caproic acids), a great number of phenolic substances and quinones (such as phenol, *m*-cresol, benzoquinone, toluquinone, &c.) and a variety of terpenoids (such as geraniol, farnesol, citral, citronellol, &c.).[620] For example, the Pentatomid *Nezara viridula* produces at least 20 components in its scent: hexenal, decenal, tridecane, ketohexenal, hexenyl acetate, decenyl acetate, dodecane are major components and 11 other compounds have been identified in trace amounts.[575] The mixtures of substances in the gut of *Dendroctonus* are species specific.[599] The Nassanoff pheromone is a mixture of geraniol, nerolic acid and citral;[610] the most attractive component is citral the two isomers of which amount to only about 3 per cent. of the geraniol.[555] Bumble bee species utilize a wide range of terpenoids and normal aliphatic chain compounds (alcohols, acetates, aldehydes, ethyl esters, &c.) as marking scents.[587]

p. 612. **Pteridines**—Pteridines, particularly erythropterine, are the major pigments in many Hemiptera: *Oncopeltus*,[571] *Rhodnius*.[619]

p. 613. The purine ring of guanine can be a precursor in the biosynthesis of pterines. In normal *Pieris* they are probably synthesized from purines; it is therefore not surprising that a series of purines in small quantities are found along with the abundant pterines in the wing scales of Pieridae.[622] Biopterin and neopterin are incorporated into the imaginal cuticle of *Apis* at the time of sclerotization. It is believed that they play an important part as co-factors in the hydroxylation of tyrosine, phenylalanine and tryptophane.[578]

p. 614. **Carotinoids**—*Leptinotarsa* can also produce keto-derivatives of carotene and of xanthophyll and these are detectable in the larva and accumulate in the adult beetle.[589]

p. 615. **Green colours**—The green coloration of *Carausius* and *Tettigonia* is due to complex mixtures of carotenoids and bile pigments.[621]

p. 616. **Ommochrome**—Ommochrome formation serves in the removal of tryptophane liberated by protein breakdown at metamorphosis.[553]

p. 617. The formation of eye-colour mutants of *Apis* result from the specific pattern of pigments and their precursors, comprising xanthommatine and several ommines and at least five pteridines.[563] If localized oxygen want is produced in *Cerura* by blocking the spiracles, the brown xanthommatine is converted to the red dihydro-xanthommatine in the affected regions.[551]

p. 617. Certain of the yellow pigments ('papiliochromes') of *Papilio xuthus* are compounds of dopamine derivatives and kynurenine.[618]

p. 619. **Colour adaptation in Pieris pupae**—The green type of pupa in *Pieris rapae* results from a combination of a yellow carotenoid (lutein) in the cuticle, overlying mesobiliverdin in the epidermis. In brown pupae, melanin in the cuticle is combined with red-brown xanthommatine in the epidermis.[596] The greatest development of the melanin pattern of *P. brassicae* results on a dark background in bright illumination; somewhat less in total darkness; and the weakest melanization in illumination on a light background. Long-wave light suppresses melanization more than short wave. The light can act both via the eyes and the integument of thorax and abdomen. But this does not happen in the isolated abdomen: if the prepupa is ligatured, only the anterior half will adapt melanization to light conditions; the posterior half is uniformly melanized.

It is concluded that, as in other Rhopalocera, the integument is subject to a 'melanization inhibitory hormone'. But the incorporation of brown ommochrome is influenced by a pigmentation promoting hormone. Ligaturing experiments may therefore give confusing results in different species.[597] In *P. brassicae* the melanizing effects of ligatures and nerve section decrease during the prepupal period and are lost by 12 hours after spinning the girdle. It is suggested that during the critical period a melanization inhibiting factor is secreted by a thoracic centre, which itself is under nervous control from the brain.[552]

p. 620. **Background and colour in Locusta**—The 'homochromic' pigmentation induced in *Locusta* by a black background is of dual origin: increased ommochrome in the epidermis and increased melanin in the cuticle.[572] Labelled tyrosine incorporated into the cuticle before ecdysis is used for sclerotization, tyrosine taken up during and after sclerotization is used for the formation of primary and secondary melanin, as in homochromy of adult insects.[572]

p. 621. **Colour variation in locusts, &c.**—In the variable coloration of locusts and grasshoppers the following systems operate: (i) an epidermal pigment system responsible for a polymorphism between green, brownish green and brown. (ii) In areas with a brown ground-colour, yellow, orange and black granules are responsible for the 'homochrome' response to background. (iii) Superimposed there is often a pattern of cuticular melanin and underlying dark brown ommochrome. (iv) In some species, especially in locusts, there are solitary and gregarious phases of coloration.[605] In *Gryllus bimaculatus* the corpora allata appear to control the pigmentation, perhaps by inhibiting a melanogenic hormone secreted elsewhere. Thus implantation of adult corpora allata into the penultimate instar suppresses melanin formation; whereas extirpation of the corpora allata leads to a complete black pigmentation.[604]

p. 622. **Colour change in Damselflies**—The brilliant colouring of many Zygoptera is due to Tyndall scattering of light in hypodermal cells against a background of dark ommatin pigment. Under conditions of low temperature the dark granules will move to the surface and give a blackish or dull coloration. This appears to be a thermal adaptation to wide diurnal fluctuations in temperature, as in Australia.[595]

p. 625. **α-Glycerophosphate oxidation**—A high rate of oxidation of α-glycerophosphate seems to be a characteristic only of the mitochondria from *Musca* and *Periplaneta*; for other genera, such as *Locusta* and *Leptinotarsa*, the oxidation of α-glycerophosphate relative to the intermediates of the citric acid cycle is no greater and often even lower than in vertebrates.[557]

p. 627. In adult *Musca* there is a progressive fall in the duration of flight with increasing age and this runs parallel with the enzyme and co-enzyme content of the ageing flies.[602]

p. 631. **Anaerobiosis in Musca**—Adult *Musca* can recover from 12–15 hours anaerobiosis; glycogen is markedly reduced; even in oöcytes ready for deposition, glycogen falls by 20 per cent. during 12 hours anoxia.[566]

REFERENCES

1 ABDERHALDEN, E., *et al. Z. physiol. Chem.*, **59** (1909), 170–6 (composition of silk).

2 AGRELL, I. *Acta physiol. Scand.*, **14** (1947), 317–34; **16** (1948), 9–19; **23** (1951) 179–86 (dehydrogenases and *p*H during pupal development).

3 AKAO, A. *J. Biochem.*, **30** (1939), 303–49 (catalytic metals in *Bombyx mori*).

[4] ALEXANDER, P., and BARTON, D. H. R. *Biochem. J.*, **37** (1943), 463–5 (ethylquinone in *Tribolium*, Col.).

[5] ALEXANDROV, W. *Protoplasma*, **17** (1933), 161–217 (oxidation-reduction potential in *Chironomus* larva, Dipt.).

[6] ALLEN, T. H. *J. Cell. Comp. Physiol.*, **16** (1940), 149–63 (cytochrome oxidase in egg of *Melanoplus*, Orth.).

[7] ATZLER, M. *Z. vergl. Physiol.*, **13** (1930), 505–33 (colour change in *Dixippus*, Orth.).

[8] BABERS, P. H. *J. Agric. Res.*, **62** (1941), 509–30 (glycogen in *Prodenia*, Lep.).

[9] BALDWIN, E. *An Introduction to Comparative Biochemistry*, Cambridge, 1937.

[10] BALZAM, N. *Arch. intern. Physiol.*, **37** (1933), 317–28 (heat production and respiration during moulting: *Bombyx mori*).

[11] BARRON, E. S. G., and TAHMISIAN, T. N. *J. Cell. Comp. Physiol.*, **32** (1948), 57–76 (metabolism of muscle: *Periplaneta*).

[12] BATAILLON, E. *Bull. Sci. Fr. Belg.*, **25** (1893), 18–55 (chemical changes during metamorphosis: *Bombyx mori*).

[13] BATTELLI, F., and STERN, L. *Biochem. Z.*, **56** (1913), 35–58 (respiratory metabolism of insects).

[14] BECKER, E., *et al. Liebig's Ann. Chem.*, **524** (1936), 124–44; *Z. physiol. Chem.*, **246** (1937), 177–80 (pterine pigments).

[15] BECKER, E. *Z. Morph. Oekol. Tiere*, **32** (1937), 672–751 (melanin and pterine pigments: *Vespa*, Hym.).

[16] —— *Z. vergl. Physiol.*, **24** (1937), 305–18 (deposition of pigment in Liesegang rings: *Vespa*, Hym.).

[17] —— *Biol. Zbl.*, **61** (1941), 588–602 (colours of Odonata).

[18] —— *Z. indukt. abstam. Vererb.*, **80** (1942), 157–204 (ommochromes in insects).

[19] BECKER, M. *Biochem. Z.*, **272** (1934), 227–34 (changes in fat during metamorphosis: *Tenebrio*, Col.).

[20] BELLUCCI, R. *Physiol. Zool.*, **12** (1939), 50–7 (respiratory metabolism: *Popillia*, Col.).

[21] BERGMANN, W. *J. Biol. Chem.*, **107** (1934), 527–52; **114** (1936), 27–38; **117** (1937), 175–8 (fats and sterols in *Bombyx mori*).

[22] —— *Ann. Ent. Soc. Amer.*, **31** (1938), 315–21 (waxes of cuticle: *Bombyx* larva).

[23] BEUTLER, R. *S. B. Ges. Morph. Physiol. München*, 1934 (protein requirements of adult worker bees).

[24] —— *Z. vergl. Physiol.*, **24** (1936), 71–115 (blood sugar in honey-bee).

[25] BLOUNT, B. K. *J. Chem. Soc.*, 1936, , 1034–6 (pigments of Aphids).

[26] BODINE, J. H. *J. Exp. Zool.*, **32** (1921), 137 ('surface law' in metabolic rate of grasshoppers).

[27] —— *J. Exp. Zool.*, **34** (1921), 143–8 (catalase content of insects).

[28] —— *Biol. Bull.*, **55** (1928), 395–403 (anaerobic metabolism: Orthoptera).

[29] —— *Physiol. Zool.*, **2** (1929), 459–82 (oxygen uptake during development: *Melanoplus* and *Chortophaga*, Orth.).

[30] —— *Physiol. Zool.*, **7** (1934), 459–63 (effect of oxygen tension on oxygen uptake: *Melanoplus* egg, Orth.).

[31] —— *J. Cell. Comp. Physiol.*, **4** (1934), 397–404 (effect of cyanide on oxygen consumption: *Melanoplus* egg, Orth.).

[32] —— *Physiol. Zool.*, **19** (1946), 54–8 (uric acid formation in egg of *Melanoplus*).

[33] BODINE, J. H., *et al. J. Cell. Comp. Physiol.*, **11** (1938), 409–23; **12** (1938), 71–84; **18** (1941), 151–60; *J. Gen. Physiol.*, **24** (1940), 99–103; 423–32 (protyrosinase and tyrosinase in egg of *Melanoplus*).

[34] BODINE, J. H., and BOELL, E. J. *J. Cell. Comp. Physiol.*, **7** (1936), 455–63 (cell structure and oxygen uptake: *Melanoplus* egg, Orth.).

[35] —— *J. Cell. Comp. Physiol.*, **8** (1936), 357–66 (respiratory metabolism of isolated embryos: *Melanoplus*, Orth.).

[36] —— *Proc. Soc. Exp. Biol. Med.*, **36** (1937), 21–3 (effect of carbon monoxide and methylene blue on respiration: *Melanoplus* embryos, Orth.).

[37] BODINE, J. H., and FITZGERALD, L. R. *Physiol. Zool.*, **20** (1947), 146–60; **21** (1948), 93–100; *J. Exp. Zool.*, **104** (1947), 353–63 (riboflavine in egg of *Melanoplus*).

[38] BODINE, J. H., and ORR, P. R. *Biol. Bull.*, **48** (1925), 1–14 (respiratory metabolism: *Drosophila* pupa, Dipt.).

[39] BOELL, E. J. *J. Cell. Comp. Physiol.*, **6** (1935), 369–85 (R. Q. during embryonic development: *Melanoplus*, Orth.).

[40] BOGOJAWLENSKY, K. S. *Z. Zellforsch. mikr. Anat.*, **22** (1935), 207–12 (glycogen in silkworm).

[41] BRECHER, L. *Arch. Entw. Mech.*, **45** (1919), 273–322; **50** (1922), 40–78; 209–308; *Arch. mikr. Anat.*, **102** (1924), 501–48; *Biol. Generalis*, **14** (1938), 212–37 (influence of light on pupal coloration: *Pieris*, *Vanessa*, Lep.).

[42] BURKHOLDER, J. R. *Physiol. Zool.*, **7** (1934), 247–70 (respiratory metabolism: *Melanoplus* eggs, Orth.).

[43] BUSNEL, R. G., and DRILHON, A. *C. R. Soc. Biol.*, **124** (1937), 916–17 (fat and water content during hibernation: *Leptinotarsa*, Col.).

[44] —— —— *Arch. Zool. exp. gén.*, **82** (1942), 321–56 (riboflavine in Malpighian tubes, &c.).

[45] BUSVINE, J. R. *Proc. Roy. Ent. Soc. Lond.*, A, **21** (1946), 98–103 (pigmentation of *Pediculus*).

[46] BUTENANDT, A., *et al. Naturwiss.*, **28** (1940), 447–8; *H.-S. Z. physiol. Chem.*, **279** (1943), 27–37 (kynurenine and eye pigments: *Drosophila* and *Ephestia*).

[47] BUTLER, C. G., and INNES, J. M. *Proc. Roy. Soc.*, B, **119** (1936), 296–304 ('surface law': solitary and migratory phases of *Locusta*).

[48] BUXTON, P. A. *Parasitology*, **27** (1935), 263–5 (changes in composition during hibernation: *Culex*, Dipt.).

[49] CASPARI, E. *Arch. Entw. Mech.*, **130** (1933), 353–81 (eye colours: *Ephestia*, Lep.).

[50] —— *Science*, **98** (1943), 478–9; *Genetics*, **31** (1946), 454–74; *Nature*, **158** (1946), 555; *Quart. Rev. Biol.*, **24** (1949), 185–99 (tryptophane and eye colour: *Ephestia*).

[51] CHADWICK, L. E., *et al. Physiol. Zool.*, **13** (1940), 398–410; *Biol. Bull.*, **93** (1947), 229–39 (respiration of *Drosophila* during flight).

[52] CHAUVIN, R. *Ann. Ent. Soc. France*, **110** (1941), 133–272; *C.R. Acad. Sci.*, **212** (1941), 175–7; *Ann. Sci. Nat. Zool.* (sér 11), **5** (1943), 79–87; *Bull. Soc. Zool. France*, **69** (1944), 154–8 (coloration of locusts; 'acridioxanthin', &c.).

[53] CHIBNALL, A. C., *et al. Biochem. J.*, **28** (1934), 2189–2219 (constitution of insect waxes).

[54] COCKAYNE, E. A. *Trans. Ent. Soc. Lond.*, 1924, 1–19 (fluorescent pigments: Lepidoptera).

[55] COLEMAN, D., and HOWITT, F. O. *Proc. Roy. Soc.*, A, **190** (1947), 145–64 (silk fibroin).

[56] COLLIN, G. *Biochem. J.*, **27** (1933), 1373–4 (fatty acids in *Pachymerus* larva, Col.).

[57] COOK, S. F. *Biol. Bull.*, **63** (1932), 246–57 (respiratory metabolism: *Termopsis*).

[58] COOK, S. F., and SCOTT, K. G. *Biol. Bull.*, **63** (1932), 505–12 (respiratory metabolism and water relations: *Termopsis*).

[59] CORDIER, R. *Ann. Bull. Soc. Roy. Sci. Méd. Nat. Bruxelles*, 1928, 43–57 (melanin pigments).

[60] COURTOIS, A. *C.R. Acad. Sci.*, **186** (1928), 1575–6 (amino acids during metamorphosis: Lepidoptera).

[61] COUSIN, G. *Bull. Biol. Fr. Belg.*, Suppl. **15**, 1932, 341 pp. (diapause: *Lucilia*, Dipt.).

[62] CRESCITELLI, F. *J. Cell. Comp. Physiol.*, **6** (1935), 351–68 (respiratory metabolism: *Galleria* pupa, Lep.).

[63] CRESCITELLI, F., and TAYLOR, I. R. *J. Biol. Chem.*, **108** (1935), 349–53 (changes in reducing substances during metamorphosis: *Galleria*, Lep.).

[64] CSIK, L. *Z. vergl. Physiol.*, **27** (1939), 304–10 (resistance of *Drosophila* to oxygen want).

[65] DANNEEL, R. *Biol. Zbl.*, **63** (1943), 377–94; **65** (1946), 115–19 (melanin formation: *Drosophila*).

[66] DAVIS, J. G., and SLATER, W. K. *Biochem. J.*, **20** (1926), 1167–72; **21** (1927), 198–203; **22** (1928), 331–7 (anaerobic metabolism in cockroach).

[67] DAVIS, R. A., and FRAENKEL, G. *J. Exp. Biol.*, **17** (1940), 402–7 (oxygen consumption during flight: *Lucilia*, Dipt.).

[68] DELAGE, J. *Misc. En tom.*, **33** (1931), 57–88 (insect pigments: review).

[69] DENHAM, C. S. *Insect Life*, **1** (1888), 143 (hydrochloric acid secretion: *Notodonta*, Lep.).

[70] DEWITZ, J. *Arch. Anat. Physiol., Abt. Physiol.*, 1902, 327–40; 425–42; 1905, 389–415 (tyrosinase and pupation in insects).

[71] DEWITZ, J. *Zool. Jahrb., Physiol.*, **38** (1921), 365–404 (pigmentation of cocoons: Lepidoptera).
[72] DHÉRÉ, C. *C. R. Soc. Biol.*, **131** (1939), 672–4 (Aphid pigments).
[73] DINULESCU, G. *Ann. Sci. Nat., Zool.*, (sér. 10), **15** (1932), 1–183 (physiology of *Gasterophilus*, Dipt.).
[74] DITMAN, L. P., *et al. Univ. Maryland Agric. Exp. Sta. Bull.*, No. 414 (1938), 183–206; *Ann. Ent. Soc. Amer.*, **31** (1938), 578–87 (glycogen and fat metabolism: *Heliothis*, Lep.).
[75] DIXON, M. *Perspectives in Biochemistry*, (Ed. Needham & Green), Cambridge, 1937.
[76] DOBZHANSKY, T., and POULSON, D. F. *Z. vergl. Physiol.*, **22** (1935), 473–8 (oxygen consumption: *Drosophila* pupae).
[77] DREYER, W. A. *Physiol. Zool.*, **5** (1932), 301–31 (effect of hibernation and temperature on oxygen consumption of *Formica ulkei*).
[78] DREYLING, L. *Zool. Jahrb., Anat.*, **22** (1905), 289–330 (wax secretion in social bees).
[79] DRILHON, A. *C. R. Soc. Biol.*, **118** (1935), 131–2 (alkaline reserve during metamorphosis: Lepidoptera).
[80] DRILHON, A., *et al. C. R. Soc. Biol.*, **137** (1943), 390–1; *Bull. Soc. Chim. Biol.*, **27** (1945), 415–18 (phosphatases in insects).
[81] DUCET, G., and GRISON, P. *C.R. Acad. Sci.*, **227** (1948), 1272–4 (lecithin in *Leptinotarsa*, Col.).
[82] DUEWELL, H., *et al. Nature*, **162** (1948), 759–61 (Aphid pigments).
[83] DUSHAM, E. H. *J. Morph.*, **31** (1918), 563–81 (wax glands in the cockroach).
[84] DÜRKEN, B. *Z. wiss. Zool.*, **116** (19, 6), 587–626; *Arch. mikr. Anat.*, **99** (1923), 222–389 (effect of surroundings on coloration of pupae: *Pieris* Lep.).
[85] EDWARDS, G. A. *J. Cell. Comp. Physiol.*, **27** (1946), 53–64 (effect of temperature on oxygen consumption).
[86] ELLENBY, C. *J. Exp. Biol.*, **21** (1945), 39–45; *J. Exp. Zool.*, **98** (1945), 23–33 (respiration in relation to surface area: *Drosophila* puparia).
[87] EPHRUSSI, B., *et al. Quart. Rev. Biol.*, **17** (1942), 326–38; *Genetics*, **29** (1944), 148–75; **30** (1945), 62–83 (eye pigments of *Drosophila*).
[88] EVANS, A. C. *J. Exp. Biol.*, **9** (1932), 314–21 (chemical changes during metamorphosis: *Lucilia*, Dipt.); *Ibid.*, **11** (1934), 397–401 (ditto: *Tenebrio*, Col.).
[89] FARKAS, K. *Arch. ges. Physiol.*, **98** (1903), 490–546 (energy exchanges in egg and pupa of *Bombyx mori*).
[90] FAURE, J. C. *Bull. Ent. Res.*, **23** (1932), 293–405 (coloration of different locust phases).
[91] —— *Farming in S. Afr.*, **18** (1943), 69–78; *Union of S. Afr. Dept. Agric. Sci. Bull.* No. 234, 1–17 (phase variation: *Laphygma*, Lep.).
[92] FELDBERG, W., *et al. Austral. J. Exp. Biol. Med.*, **16** (1937), 249; *J. Physiol.*, **99** (1940), 104–18 (bee venom).
[93] FINK, D. E. *J. Gen. Physiol.*, **7** (1925), 527–43 (metabolism in egg and pupa: Coleoptera and Lepidoptera).
[94] —— *Science*, **65** (1927), 143–5 (glutathione in insects).
[95] —— *J. Agric. Res.*, **41** (1930), 691–96 (catalase during metamorphosis: *Leptinotarsa*, Col.).
[96] FLANDERS, S. E. *Ann. Ent. Soc. Amer.*, **31** (1938), 167–80 (cocoon formation in Chalcids).
[97] FLEISCHMANN, W., *et al. Biochem. Z.*, **294** (1937), 280–3 (carbon monoxide poisoning in insects).
[98] FLURY, F. *Arch. exp. Path. Pharm.*, **85** (1920), 319–38; *Oppenheimer's Handb. d. Biochem.*, **5** (1925), 709–14 (bee venom).
[99] FORD, E. B. *Proc. Roy. Ent. Soc. Lond.*, A, **16** (1941), 65–90; **17** (1942), 87–92; **19** (1944), 92–106; **22** (1947), 65–88; *Trans. Roy. Ent. Soc. Lond.*, **94** (1944), 201–23 (pigments in Lepidoptera).
[100] —— *Biol. Rev.*, **20** (1945), 73–88 (polymorphism: Lepidoptera).
[101] FOX, H. M., SIMMONDS, B. G., *et al. J. Exp. Biol.*, **10** (1933), 67–74; **12** (1935), 179–84; **14** (1937), 210–18; *Nature*, **138** (1936), 1015 (oxygen consumption in relation to available oxygen, size of animal, &c.: Ephemerid nymphs).
[102] FRAENKEL, G. *Proc. Zool. Soc. Lond.*, 1935, 893–904 (darkening, &c., in *Calliphora*, Dipt., after emergence from pupa).

[103] FRAENKEL, G. *J. Exp. Biol.*, **17** (1940), 18–29 (utilization of carbohydrates: *Calliphora*).
[104] FRAENKEL, G., and HOPF, H. S. *Biochem. J.*, **34** (1940), 1085–92 (effect of temperature on lipoid chemistry: *Calliphora*, &c., Dipt.).
[105] FREDERICQ, L. *Winterstein's Handb. d. vergl. Physiol.*, **2** (2), (1910), 112–66 (insect secretions).
[106] FREW, J. G. H. *Brit. J. Exp. Biol.*, **6** (1929), 205–18 (metabolism during metamorphosis: *Calliphora*, Dipt.).
[107] FRIZA, F. *Z. vergl. Physiol.*, **8** (1928), 289–336 (colour change in eyes and epidermis: Mantidae).
[108] GAARDER, T. *Biochem. Z.*, **89** (1918), 48–93 (relation between oxygen tension and oxygen uptake: *Tenebrio* pupa, Col.).
[109] GATES, M. *Chem. Rev.*, **41** (1947), 63–95 (pterine chemistry).
[110] GARB, G. *J. Ent. Zool.*, **7** (1915), 88–97 (eversible dermal glands of *Melasoma* larva, Col.).
[111] GEROULD, J. H. *Quart. Rev. Biol.*, **3** (1927), 58–78 (insect pigments: review).
[112] GHIDINI, M. G. *Riv. Biol. Colon. Rome*, **2** (1939), 385–99 (respiratory quotient in termites).
[113] GIER, H. T. *Ann. Ent. Soc. Amer.*, **40** (1947), 303–17 (growth and nutrition: *Periplaneta*).
[114] GIERSBERG, H. *Z. vergl. Physiol.*, **7** (1928), 657–95 (colour change: *Dixippus*, Orth.).
[115] —— *Z. vergl. Physiol.*, **9** (1929), 523–52 (colour pattern formation: *Vanessa*, Lep.).
[116] GILMER, P. M. *Ann. Ent. Soc. Amer.*, **18** (1925), 203–239 (poison apparatus of lepidopterous larvae).
[117] GILMOUR, D. *J. Cell. Comp. Physiol.*, **15** (1940), 331–42 (anaerobic metabolism: *Zootermopsis*); *Biol. Bull.*, **79** (1940), 297–308 (ditto: *Cryptocercus*); *Ibid.*, **80** (1941), 45–9 (ditto: *Melanoplus* muscle); *J. Cell. Comp. Physiol.*, **18** (1941), 93–100 (ditto: *Tenebrio*).
[118] GIRAL, J., *et al. J. Biol. Chem.*, **162** (1946), 55–9 (fat of *Melanoplus*).
[119] GOLBERG, L., and DE MEILLON, B. *Nature*, **160** (1947), 582–5 (tyrosine and pigmentation: *Aëdes*, Dipt.).
[120] GOLDSCHMIDT, R. *Amer. Nat.*, **81** (1947), 474–6 (industrial melanism).
[121] GOOD, P. M., and JOHNSON, A. W. *Nature*, **163** (1949), 31 (pterines).
[122] GOODWIN, T. W., and SRISHUKH, S. *Nature*, **161** (1948), 525–6 (carotenoids in integument of locusts).
[123] GORTNER, R. A. *Amer. Nat.*, **45** (1911), 743–55 (melanin and colour pattern: *Leptinotarsa*, Col.).
[124] GOURÉVITCH, A. *C.R. Acad. Sci.*, **187** (1928), 65–7 (specific dynamic action in cockroach).
[125] —— *C. R. Soc. Biol.*, **98** (1928), 26–7 (R.Q. in relation to diet in cockroach).
[126] GRAYSON, J. M., and TAUBER, O. E. *Iowa State Coll. J. Sci.*, **17** (1943), 191–6 (pigments in *Melanoplus*).
[127] GUNN, D. L. *J. Exp. Biol.*, **12** (1935), 185–90 (oxygen uptake and body size in Blattidae).
[128] HALDANE, J. B. S. *Biochem. J.*, **21** (1927), 1068–75 (carbon monoxide poisoning: *Galleria*).
[129] HANSER, G. *Z. Naturforsch.*, **1** (1946), 396–9; *Z. Vererbungsl.*, **82** (1948), 74–97 (ommochrome pigments).
[130] HANSTRÖM, B. *Kungl. Svenska Vetensk. Handl.* (sér. 3), **16** (1937), No. 3, 1–99; *Lunds Univ. Arsskr. N. F. Avd.* 2, **32** (1937), No. 8, 1–11 (hormones in insects producing colour change in Crustacea).
[131] HARNISCH, O. *Z. vergl. Physiol.*, **12** (1930), 504–23 (effect of reduced oxygen tension on metabolism in fragments of *Chironomus* larvae).
[132] —— *Z. vergl. Physiol.*, **23** (1936), 391–419; **24** (1937), 387–408; **26** (1938), 200–29; 548–69; **27** (1939), 275–303; **28** (1941), 428–56; **30** (1943), 145–66; **33** (1951), 404–6; *Zool. Anz.*, **139** (1942), 1–12; **142** (1943), 240–8; *Arch. Hydrobiol.*, **40** (1943), 184–207; *Biol. Zbl.*, **66** (1947), 179–85; **69** (1950), 449–62 (anaerobic and aerobic metabolism: *Chironomus* larvae, Dipt.).
[133] HARRISON, J. W. H. *Proc. Roy. Soc.*, B, **102** (1928), 338–47 (induction of melanism: *Selenia*, Lep.).

[134] HARRISON, J. W. H. *Proc. Roy. Soc.*, B, **102** (1928), 347–53 (inheritance of pupal coloration induced by orange light: *Pieris*, Lep.).

[135] HASEBROECK, K. *Zool. Jahrb., Physiol.*, **37** (1920), 279–92; **53** (1934), 411–60; *Fermentforsch.*, **5** (1922), 1–40; 297–333 (melanism in industrial regions: Lepidoptera).

[136] HAVEMANN, R., and WOLFF, K. *Biochem. Z.*, **290** (1937), 354–9 (bee venom).

[137] HAYDAK, M. H. *J. Agric. Res.*, **49** (1934), 21–8 (changes in nitrogen content in adult bee).

[138] —— *J. Agric. Res.*, **54** (1937), 791–6 (protein breakdown in adult bee).

[139] HELLER, J. *Biochem. Z.*, **172** (1926), 59–73 (chemical changes during metamorphosis: *Deilephila* and *Bombyx*, Lep.).

[140] —— *Biochem. Z.*, **172** (1926), 71–81 (metabolism during starvation: *Deilephila*, Lep.).

[141] —— *Biochem. Z.*, **169** (1926), 208–34; *Biol. Zentrabl.*, **51** (1931), 259–69; *xvii. Int. Physiol. Congr. Oxford, 1947* (abstracts), 274–6 (respiratory metabolism in pupa of *Deilephila*, Lep.).

[142] HENKE, K. *Z. vergl. Physiol.*, **1** (1924), 297–499 (coloration in *Pyrrhocoris*, Hem.).

[143] HENRICI, H. *Zool. Jahrb.; Abt. Anat.*, **65** (1938), 123–294; **66** (1940), 371–402 (dermal glands; Hemiptera).

[144] HENSEVAL, M. *La Cellule*, **12** (1897), 19–26; 169–81 (essential oil of *Cossus* larva, Lep.).

[145] HERTZ, M., and IMMS, A. D. *Proc. Roy. Soc.* B, **122** (1937), 281–97 (colour change in *Locusta* in response to background).

[146] HESELHAUS, F. *Zool. Jahrb., Anat.*, **43** (1922), 369–464 (dermal glands of Apidae).

[147] HILL, D. L. *J. Cell. Comp. Physiol.*, **25** (1945), 205–16 (carbohydrate metabolism in egg of *Melanoplus*).

[148] HO, C. P., *et al. Physiol. Zool.*, **17** (1944), 78–82 (silk formation).

[149] HOLLANDE, A. C. *Ann. Univ. Grenoble*, **21** (1909), 459–517 (breakdown and elimination of salicin).

[150] HOLLANDE, A. C. *Arch. Zool.*, **51** (1913), 53–8 (anthocyanin in fat body of *Cionus*, Col.).

[151] HOPKINS, F. G. *Phil. Trans. Roy. Soc.*, **186** (1895), 661 (pigments of Pieridae).

[152] HSUEH, T. Y., and TANG, P. S. *Physiol. Zool.*, **17** (1944), 71–8 (growth and respiration: *Bombyx mori*).

[153] HUSAIN, M. A., *et al. Ind. J. Agric. Res.*, **6** (1936), 586–664; 1005–30 (phase changes in *Schistocerca*, Orth.).

[154] JACOBS, W. *Z. Morph. Oekol. Tiere*, **3** (1925), 1–80 (scent organ of *Apis*).

[155] JANDA, V. *Zool. Anz.*, **115** (1936), 177–85; *Trav. Inst. Zool. Univ. Charles IV Prague*, 1934, 1–30 (colour change in transplanted epidermis: *Dixippus*, Orth.).

[156] —— *Trav. Inst. Zool. Univ. Charles IV Prague*, 1936, 1–11 (seasonal colour change: *Chrysopa*, Neuropt.).

[157] JONGBLOED, J., and WEIRMSA, C. A. G. *Z. vergl. Physiol.*, **21** (1934), 519–33 (metabolism of honey-bee during flight).

[158] JUCCI, C. *Arch. Zool. Ital.*, **22** (1935), 259–68 (cocoon colour in *Bombyx mori*: relation with leaf pigments).

[159] JUGA, V. G. *C. R. Soc. Biol.*, **118** (1935), 603–5 (biliverdine in *Chironomus*).

[160] JUNGE, H. *H.-S. Z. physiol. Chem.*, **268** (1941), 178–86 (green pigments in insects).

[161] KAESTNER, H. *Arch. Entw. Mech.*, **124** (1931), 1–16 (effect of temperature on pigmentation: *Habrobracon*, Hym.).

[162] KALMUS, H. *Z. vergl. Physiol.*, **10** (1929), 445–55 (respiratory metabolism during flight: *Deilephila*, Lep.).

[163] KALMUS, H. *Z. vergl. Physiol.*, **24** (1937), 409–12 (oxygen tension and duration of development: *Drosophila* pupae).

[164] KEILIN, D. *Arch. Zool.*, **52** (1913) (Notes et revue), 1–8 (glands of Diptera larvae).

[165] —— *Proc. Roy. Soc.*, B, **98** (1925), 312–39; *Ergebn. Enzymforsch.*, **2** (1933), 239–71 (cytochrome and intracellular respiratory enzymes).

[166] KELLER-KITZINGER, R. *Z. vergl. Physiol.*, **22** (1935), 1–31 (protein metabolism in worker honey-bee).

[167] KELLNER, O. *Landw. Versuchs-Stat.*, **30** (1884), 59; **33** (1887), 381–92 (chemical changes during growth: *Bombyx mori*).

[168] V. KEMNITZ, G. A. *Z. Biol.*, **67** (1916), 129–244 (metabolism in larvae of *Gasterophilus* and *Chironomus*, Dipt.).

[169] KEMPER, H. *Z. Morph. Oekol. Tiere*, **15** (1929), 524–46 (stink glands: *Cimex*, Hem.).

[170] KESTNER, O., and PLAUT, R. *Winterstein's Handb. d. vergl. Physiol.*, **2** (2) (1924), 901–95 (metabolism of insects).

[171] KIKKAWA, H. *Genetics*, **26** (1941), 587–607 (pigment formation: *Bombyx* and *Drosophila*).

[172] KITTEL, A. *Z. vergl. Physiol.*, **28** (1941), 553–62 (oxygen consumption and body size).

[173] KLEINMAN, L. W. *J. Cell. Comp. Physiol.*, **4** (1934), 221–35 (effect of temperature on R.Q.: *Chortophaga*, Orth.; *Popillia*, Col.).

[174] KNIGHT, H. H. *Ann. Ent. Soc. Amer.*, **17** (1924), 258–72 (effect of temperature and humidity on colour patterns: *Leptinotarsa*, Col., *Perillus*, Hem.).

[175] KOSMIN, N. P., *et al. Z. vergl. Physiol.*, **17** (1932), 408–22 (respiratory metabolism during activity: honey-bee).

[176] KOZHANTSCHIKOW, I. V. *C. R. Acad. Sci. U.S.S.R.*, **2** (1935), 322–7 (anoxybiosis during diapause: Pyralid larvae, Lep.).

[177] —— *Bull. Ent. Res.*, **29** (1938), 103–14 (carbohydrate and fat metabolism: adult Lepidoptera).

[178] KOZHANCHIKOV, I. V. *Zool. Zh.* **18** (1939), 97–8 (respiration in diapause).

[179] KREMER, J. *Zool. Jahrb., Anat.*, **40** (1917), 105–54 (carotinoid pigments: *Haemonia*, Col.).

[180] KROGH, A. *Z. all. Physiol.*, **16** (1914), 178–90 (respiratory metabolism: *Tenebrio* pupa, Col.).

[181] —— *The Respiratory Exchange of Animals and Man*, Monographs of Biochemistry 1916.

[182] KROGH, A., and WEIS-FOGH, T. *J. exp. Biol.*, **28** (1951), 344–57 (metabolism in flight: *Schistocerca*).

[183] KUCERA, W. G. *Physiol. Zool.*, **7** (1934), 449–58 (oxygen consumption: *Drosophila* adults).

[184] KÜHN, A., *et al. Nachr. Acad. Wiss. Göttingen, Math. Phys. Kl.*, 1941, 231–61; *Abhandl. Preuss. Akad. Wiss. Math. Naturw., Kl.*, 1942, No. 9, 1–6; *Biol. Zbl.*, **62** (1942), 303–17 (kynurenine and eye pigment formation: *Ephestia*, Lep.).

[185] LAFON, M. *C. R. Soc. Biol.*, **135** (1941), 193–7 (starvation in *Phormia*, Dipt.).

[186] LATTER, O. H. *Trans. Ent. Soc. Lond.*, 1892, 287–92; 1895, 399–412 (secretion of potassium hydroxide: *Dicranura* imago, Lep.).

[187] —— *Trans. Ent. Soc. Lond.*, 1897, 113–25 (secretion of formic acid: *Dicranura* larva, Lep.).

[188] LEDERER, E. *C. R. Soc. Biol.*, **17** (1934), 413–16; *Bull. Soc. Chim. Biol.*, **20** (1938), 568–610 (carotenoids in insects).

[189] —— *Biol. Rev.*, **15** (1940), 273–306 (pigments: review).

[190] LENNOX, F. G. *Coun. Sci. Ind. Res. Australia*, Pamph. No. 102 (1940), 51–67 (iron in *Lucilia*).

[191] LESPERON, L. *Arch. Zool.*, **79** (1937), 1–156 (silk secretion).

[192] v. LINDEN, M. *Arch. ges. Physiol.*, **98** (1903), 1–89; *Ann. Sci. Nat., Zool.*, **20** (1904), 295–363 (pigments of *Vanessa*, Lep.).

[193] LINDSAY, H. A. F., and HARLOW, C. M. *Indian Forest Rec.*, **8** (1921), part 1, 162 pp. (lac and shellac).

[194] LOTMAR, R. *Landwirt. Jarhb. Schweiz.*, **53** (1939), 34–70 (protein metabolism: *Apis*).

[195] LUCIANI, L., and PIUTTI, A. *Arch. ital. Biol.*, **9** (1888), 319–59 (respiration in silkworm egg).

[196] LUDWIG, D. *J. Exp. Zool.*, **60** (1931), 309–23 (respiratory metabolism: *Popillia* larva, pupa and adult, Col.).

[197] —— *Ann. Ent. Soc. Amer.*, **27** (1934), 429–34 (*p*H during metamorphosis: *Popillia*, Col.).

[198] MAHDIHASSAN, S. *Z. Morph. Oekol. Tiere*, **33** (1938), 527–54; *J. Osmania Univ.* (*Sci. Fac.*), **9** (1941), 53–121 (secretions of lac insects).

[199] —— *Nature*, **159** (1947), 749; *Proc. Ind. Acad. Sci.*, B, **25** (1947), 155–62 (symbionts and pigment formation).

[200] MALOEUF, N. S. R. *J. Exp. Zool.*, **74** (1936), 323–51 (respiratory metabolism in aquatic insects).

[201] MALOEUF, N. S. R. *Quart. Rev. Biol.*, **13** (1938), 169–195 (secretion of ectodermal glands: review).

[202] MALUF, N. S. R. *Amer. Nat.*, **73** (1939), 280–5 (starvation of insects: review).

[203] MANSBRIDGE, G. H. *Trans. Ent. Soc. Lond.*, **81** (1933), 75–92 (oxalic acid secretion: Mycetophilid larvae).

[204] MANUNTA, C. *Atti. Accad. Naz. Lincei*, **17** (1933), 309–12 (fat metabolism: *Galleria*, Lep.).

[205] —— *Accad. Naz. Lincei, Mem.*, (sér. 6), **6** (1935), 75–161 (fat and pigment metabolism: *Galleria*, *Bombyx mori*, *Pieris*, Lep.).

[206] —— *Arch. Zool. Ital.*, **24** (1937), 385–401; *Boll. Zool.*, **8** (1937), 167–77 (cocoon colours of *B. mori* in relation to plant pigments).

[207] —— *Atti. R. Accad. Lincei. (ser. vii), Rendic.*, **3** (1941), 151–3 (carotenoids in cocoons of *Microgaster*, Braconidae).

[208] —— *Nature*, **162** (1948), 298 (astaxanthin in insects).

[209] MARCHAL, P. *Richet's 'Dictionaire de Physiologie'*, **9** (1910), 273–386 (insect physiology).

[210] MARCOU, I., and DEREVICI, M. *C. R. Soc. Biol.*, **126** (1937), 726–8 (histamine in *Apis*).

[211] MATTHÉE, J. J. *Bull. Ent. Res.*, **36** (1945), 343–71; *J. Ent. Soc. S. Africa*, **10** (1947), 16–23 (phase characters in locusts and Noctuids).

[212] MELAMPY, R. M., and MAYNARD, L. A. *Physiol. Zool.*, **10** (1937), 36–44 (nutrition in cockroach, *Blattella*).

[213] MELAMPY, R. M., and WILLIS, E. R. *Physiol. Zool.*, **12** (1939), 302–11 (respiratory metabolism: *Apis* larva and pupa).

[214] MELL, R. *Biol. Zbl.*, **51** (1931), 189–94 (determination of dry season form of *Hestina*, Lep.).

[215] MELLANBY, K. *Proc. Roy. Soc.*, B, **111** (1932), 376–90 (metabolism in fasting mealworm, *Tenebrio*).

[216] MELLANBY, K. *Bull. Ent. Res.*, **27** (1936), 611–32 (hunger cycle in *Glossina*, Dipt.).

[217] METCALF, R. L. *Ann. Ent. Soc. Amer.*, **36** (1943), 37–40 (red pigment from Lampyridae).

[218] —— *Ann. Ent. Soc. Amer.*, **38** (1945), 397–402 (metabolism of chlorophyll: *Anasa*, Hem.).

[219] METCALF, R. L., and PATTON, R. L. *J. Cell. Comp. Physiol.*, **19** (1942), 1–4 (riboflavine metabolism: *Periplaneta*).

[220] MEYER, P. F. *Z. vergl. Physiol.*, **11** (1930), 173–209 (uptake of plant pigments by caterpillars).

[221] MICHAL, K. *Zool. Anz.*, **95** (1931), 65–75 (oxygen uptake: *Tenebrio* larva, Col.).

[222] MISHIKATA, T. *J. Biochem. (Tokyo)*, **1** (1922), 261–79 (fat synthesis from protein: Muscid larvae).

[223] NECHELES, H. *Arch. ges. Physiol.*, **204** (1924), 72–93 (temperature regulation, &c., in insects).

[224] NEEDHAM, D. M. *Biol. Rev.*, **4** (1929), 307–26 (chemical changes during metamorphosis).

[225] NELSON, J. A., STURTEVANT, A. P., *et al. U.S. Dept. Agric. Bull.*, No. **1222** (1924), 38 pp. (growth and feeding in honey-bee larvae).

[226] NIELSEN, E. T. *Vidensk. Medd. Dansk Naturhist. Foren.*, **99** (1935), 149–231 (paralysing action by Sphecid wasps).

[227] NIEMIERKO, W. *Acta Biol. Exper.*, **14** (1947), 137–50 (fatty acid metabolism: *Bombyx mori*).

[228] OKAY, S. *Rev. Fac. Sci. Univ. Istanbul*, Ser. B., **12** (1947), 1–8; 89–106 (locust wing pigments: *Oedipoda* spp.; and green insect pigments).

[229] ONSLOW, H. *Biochem. J.*, **10** (1916), 26–30 (development of black markings on wings of *Pieris*, Lep.).

[230] —— *Phil. Trans. Roy. Soc.*, **211** (1921), 1–74 (combined pigments and structural colours).

[231] PACAUD, A. *C. R. Soc. Biol.*, **138** (1944), 264–5; **139** (1945), 471–3; *Ann. Sci. Nat. Zool.*, sér. 11, **12** (1950), 1–13 (glycogen in *Simulium* larvae, Dipt.).

[232] PALM, N. B. *Opusc. Ent.*, 1946, 119–32 (stink glands: *Tribolium*, Col.).

[233] PALMER, L. S., and KNIGHT, H. H. *J. Biol. Chem.*, **59** (1924), 443–55 (carotene in pigmentation of *Perillus*, Hem.).

[234] PARDI, L. *Redia*, **25** (1939), 87–288 (reserves in fat body, &c.).

[235] PEPPER, J. H., and HASTINGS, E. *Montana State Coll. Agr. Exp. Sta. Tech. Bull.*, No. 413 (1943), 1–36 (fatty acid metabolism: *Loxostege*, Lep.).

[236] PILEWICZÓWNA, M. O. *Prace Inst. Nenckiego*, **3** (1926), 1–25 (nitrogen metabolism during starvation: *Periplaneta*, *Dytiscus*).

[237] PLAGGE, E. *Arch. Entw. Mech.*, **132** (1935), 648–70 (eye colours in *Ephestia* larvae and adults, Lep.).

[238] POPHAM, E. J. *Proc. Zool. Soc. Lond.*, A, **111** (1941), 135–72; *J. Animal Ecol.*, **12** (1943), 124–36 (colour of Corixidae in relation to background).

[239] POSSOMPÉS, B. *Mém. Fac. Sci. Univ. Paris*, No. 683 (1937), 1–21 (biliverdine in *Chironomus*, Dipt.).

[240] POULSON, D. F. *Z. vergl. Physiol.*, **22** (1935), 466–72 (oxygen consumption: *Drosophila* pupae).

[241] POULTON, E. B. *The Colours of Animals*, London, 1890.

[242] —— *Proc. Roy. Soc.*, **54** (1894), 417–30 (plant pigments and colours of Lepidoptera larvae).

[243] PRIEBATSCH, I. *Z. vergl. Physiol.*, **19** (1933), 453–88 (effect of light on colour change in *Dixippus*, Orth.).

[244] PRZIBRAM, H. *Arch. mikr. Anat. Entw. Mech.*, **102** (1924), 624–34 (melanin formation in cocoons of Lepidoptera and Tenthredinidae).

[245] PRZIBRAM, H., and LEDERER, E. *Anz. Akad. Wiss. Wien*, **70** (1933), 163–5 (green pigment of Orthoptera).

[246] PURRMANN, R. *H.-S. Z. physiol. Chem.*, **260** (1939), 105–7; *Liebig's Ann. Chem.*, **544** (1940), 182–90 (xanthine and isoguanine in Pierid wing).

[247] RAFFY, A., and PORTIER, P. *C. R. Soc. Biol.*, **108** (1931), 1062–4 (respiration during flight: Lepidoptera).

[248] RAINEY, R. C. *Ann. App. Biol.*, **25** (1938), 822–35 (chemical changes in development: *Lucilia*, Dipt.).

[249] RAMSAY, J. A. *J. Exp. Biol.*, **12** (1935), 373–83 (evaporation from the cockroach).

[250] RAMSDEN, W. *Nature*, **142** (1938), 1120–1 (silk formation).

[251] REICHELT, M. *Z. Morph. Oekol. Tiere*, **3** (1925), 477–525 (development and pigmentation of wing scales: *Lymantria*, Lep.).

[252] RONZONI, E., and BISHOP, G. H. *Trans. 4th internat. Congr. Entom. Ithaca, 1928*, **2** (1929), 361–5 (carbohydrate metabolism in honey-bee larva).

[253] ROONWAL, M. L. *Proc. Nat. Inst. Sci. Ind.*, **13** (1947), 25–8 (colours of *Schistocerca*, Orth. in relation to background).

[254] ROTH, L. M. *Ann. Ent. Soc. Amer.*, **36** (1943), 397–424; **38** (1945), 77–87 (stink glands: *Tribolium*, &c., Col.).

[255] RUDOLFS, W. *J. N.Y. Ent. Soc.*, **34** (1926), 249–56 (chemical changes during growth and metamorphosis in *Malacosoma*, Lep.: moisture and fat); *Ibid.*, 319–30 (ditto: nitrogen); *Ibid.*, **35** (1927), 219–29 (ditto: ash and sulphates); *Ibid.*, **37** (1929), 17–23 (ditto: glycogen).

[256] SCHLIEP, W. *Zool. Jahrb., Physiol.*, **30** (1910), 45–132; **35** (1915), 225–32 (colour change in *Dixippus*, Orth.).

[257] SCHLOTTKE, E. *Z. vergl. Physiol.*, **3** (1926), 692–736; **20** (1934), 370–9 (effect of temperature and oxygen tension on pigmentation: *Habrobracon*, Hym.).

[258] SCHLUSCHE, M. *Zool. Jahrb., Anat.*, **61** (1936), 77–98 (venom glands in Vespidae).

[259] SCHMALFUSS, H., *et al. Biochem. Z.*, **215** (1929), 79–84; **223** (1930), 457–69; **257** (1933), 188–93; *Z. indukt. abstamm. Vererb.*, **58** (1931), 332–71; *Z. physiol. Chem.*, **231** (1935), 161–6; *Z. vergl. Physiol.*, **24** (1937), 493–508; **27** (1939) 434–42 (melanin formation in Coleoptera).

[260] SCHÖPF, C., and BECKER, E. *Liebig's. Ann. Chem.*, **507** (1933), 266–96; **524** (1936), 49–123 (pterine pigments in Lepidoptera, &c.).

[261] SCHÖPF, C., *et al. Liebig's Ann. Chem.*, **539** (1939), 168–78 (allantoine in Pierid wing).

[262] SCHULZ, F. N. *Oppenheimer's Handb. d. Biochemie*, 2nd Edn., **7** (1927), 439–88 (insect metabolism).

[263] SCHULZ, P. *Zool. Jahrb., Anat.*, **32** (1912), 181–224 (osmeterium of Papilionid larvae).

[264] SCHULZ, P. *S. B. Ges. Naturf. Freunde Berlin*, 1913, 1–22; 1914, 398–406 (carotinoid pigments: Chrysomelidae).

[265] SCHULZE, K. *Zool. Jahrb., Anat.*, **58** (1934), 239–74 (dermal glands: Odonata).

[266] SEMICHON, L. *Bull. Sci. Fr. Belg.*, **40** (1906), 281–439 (biology, &c.: solitary bees).

[267] SIEBER, N., and METALNIKOV, S. *Arch. ges. Physiol.*, **102** (1904), 269–86 (nutrition in *Galleria*, Lep.).

[268] SIMANTON, W. A. *Ann. Ent. Soc. Amer.*, **26** (1933), 247–54 (surface area of insects).

[269] SINODA, O., and KURATA, M. *J. Biochem.* (*Tokyo*), **16** (1932), 129–39 (ether extract of Dermestid beetles).

[270] SLATER, W. K. *Biol. Rev.*, **3** (1928), 303–28 (anaerobic life: review).

[271] SLIFER, E. H. *Physiol. Zool.*, **3** (1930), 503–18; **5** (1932), 448–56 (fatty acids of grasshopper eggs).

[272] SLOWTZOFF, B. *Beitr. Chem. Physiol. Path.*, **4** (1904), 23–39; **6** (1905), 162–74; *Biochem. Z.*, **19** (1909), 504 (metabolism of various insects during starvation).

[273] SMIRNOW, D. A. *Trav. Soc. Imp. Natur. St. Petersb. Sect. Zool.*, **40** (1911), pt. 4, 15 pp. (aromatic glands of *Aromia moschata*, Col.).

[274] STRAUSS, J. *Z. Biol.*, **56** (1911), 347–97 (chemical composition during growth and metamorphosis: *Apis*).

[275] STUMPER, R. *Ann. Sci. Nat., Zool.* (sér. 10), **5** (1922), 105–12; *C.R. Acad. Sci.*, **176** (1923), 330–2 (venom of ants).

[276] TANGL, F. *Arch. ges. Physiol.*, **130** (1909), 1–89 (metabolism during metamorphosis: *Ophyra*, Dipt.).

[277] TARTTER, A. *H.-S. Z. physiol. Chem.*, **266** (1940), 130–4 (uric acid and hypoxanthine in Pierid wing).

[278] TAYLOR, I. R. *J. Morph.*, **44** (1927), 313–39 (oxygen consumption during metamorphosis: Dipt. and Lep.).

[279] —— *Physiol. Zool.*, **7** (1934), 593–9 (*p*H during metamorphosis: *Galleria*, Lep.).

[280] TAYLOR, I. R., and CRESCITELLI, F. *J. Cell. Comp. Physiol.*, **10** (1937), 93–112 (heat production during metamorphosis: *Galleria*, Lep.).

[281] TENENBAUM, E. *Arch. Entw. Mech.*, **132** (1934), 42–56 (development of coloration in elytra: *Epilachna* (Col. Coccin.).

[282] TETSCH, C., and WOLFF, K. *Biochem. Z.*, **290** (1937), 354–359 (bee venom).

[283] THOMSEN, M., and LEMCHE, H. *Biol. Zentrabl.*, **53** (1933), 541–60 (melanism in *Selenia*, Lep.).

[284] THOMSON, D. L. *Biochem. J.*, **20** (1926), 73–5; 1026–27 (flavone pigments, &c., in *Melanargia*, Lep.).

[285] THUNBERG, T. *Skand. Arch. Physiol.*, **17** (1905), 133–95 (oxygen tension and oxygen uptake: *Tenebrio* larva, Col.).

[286] —— *Oppenheimer's Handb. Biochem.* (Erganzungsbd.) 1930, 245–81 (oxidative mechanisms).

[287] TICHOMIROFF, A. *Z. physiol. Chem.*, **9** (1885), 518–32 (chemical changes in egg development: *Bombyx mori*).

[288] TIMON-DAVID, J. *C. R. Soc. Biol.*, **96** (1927), 1225–7; **97** (1927), 586–8; **99** (1928), 1797–8; *C.R. Acad. Sci.*, **186** (1928), 104–6; **188** (1928), 1122–4 (insect fats); *Ann. Fac. Sci. Marseille*, 1930, 1–183.

[289] —— *Ann. Fac. Sci. Marseille*, **15** (1941), 49–165; *Année Biol.*, **23** (1947), 236–71 (insect pigments).

[290] —— *Ann. Fac. Sci. Marseille* (sér ii), **16** (1945), 179–235 (insect secretions).

[291] TOUMANOFF, K. *C. R. Soc. Biol.*, **96** (1927), 1392–3; **98** (1928), 198–200 (relation between food and pigmentation: *Dixippus*, Orth.).

[292] TOWER, W. L. *Carnegie Inst. Washington, Publ.*, **48** (1906), 320 pp. (evolution, coloration, &c.: *Leptinotarsa*, Col.).

[293] TROJAN, E. *Z. Morph. Oekol. Tiere*, **19** (1930), 678–85 (Dufour's glands: *Apis mellifica*).

[294] TSCHIRCH, A., and LÜDY, F. *Helv. Chim. Acta*, **6** (1923), 994–1008 (lac chemistry).

[295] UMEYA, Y. *Bull. Sericult. Exp. Sta. Chosen*, No. 1 (1926), 27–48 (silk glands: *B. mori*).

[296] UVAROV, B. P. *Trans. Ent. Soc. Lond.*, 1928, 255–343 (insect nutrition and metabolism: review).

[297] VANEY, C., and MAIGNON, F. *C.R. Acad. Sci.*, **140** (1905), 1192–5 (chemical changes during metamorphosis: *Bombyx mori*).

[298] VANEY, C., and PELOSSE, J. *C.R. Acad. Sci.*, **174** (1922), 1566–8 (cocoon colours: *Bombyx mori*).

[299] VIALLI, M. *Riv. Biol. Colon. Rome*, **2** (1939), 273–7 (secretion of *Pheropsophus*, Col.).

[300] WAIN, R. L. *Ann. Rep. Agric. Hort. Res. Sta., Long Ashton*, 1943, 108–9 (salicylaldehyde secretion: *Phyllodecta*, Col.).

[301] WALD, G., and ALLEN, G. *J. Gen. Physiol.*, **30** (1946), 41–6 (eye pigments: *Drosophila*).

[302] WALSHE, B. M. *J. Exp. Biol.*, **25** (1948), 35–44 (oxygen requirements of Chironomid larvae).

[303] WATERHOUSE, D. F. *Coun. Sci. Ind. Res. Australia*, Pamph. No. 102 (1940), 28–50 (iron in *Lucilia*, Dipt.); *Bull. Coun. Sci. Ind. Res. Australia*, No. 191 (1945), 1–39 (copper ditto).

[304] WEBER, H. *Biologie der Hemipteren*, Springer, Berlin, 1930.

[305] —— *Lehrbuch der Entomologie*, Jena, 1933.

[306] WEGENER, M. *Biol. Zentralbl.*, **43** (1923), 292–301 (osmeterium of Papilionid larvae).

[307] WEINLAND, E. *Z. Biol.*, **47** (1906), 186 (metabolism during metamorphosis: *Calliphora*, Dipt.).

[308] —— *Biol. Centrabl.*, **29** (1909), 564–77 (chemical changes in pupa of *Calliphora*, Dipt.).

[309] WENIG, K., and JOACHIM, J. *Biochem. Z.* **285** (1936), 98–100 (effect of insulin on blood sugar: *Bombyx mori*).

[310] WIELAND, H., and SCHÖPF, *et al. Ber. deutsch. chem. Ges.*, **58** (1926), 2178–83; **59** (1926), 2067–72; *Liebig's Ann. Chem.*, **507** (1933), 226–65; **544** (1940), 163–82; **555** (1944), 146–54 (pterine pigments in Pieridae).

[311] WIELAND, H., and KOTZSCHMAR, A., *et al. Liebig's Ann. Chem.*, **530** (1937), 152–65; **545** (1940), 190–208 (green wing pigment in Pieridae).

[312] WIGGLESWORTH, V. B. *Quart. J. Micr. Sci.*, **76** (1933), 269–318 (melanin formation in the cuticle: *Rhodnius*, Hem.).

[313] WIGGLESWORTH, V. B. *Proc. Ent. Soc. Lond.*, **3** (1928), 4 (impressionist colouring: Lepidoptera).

[314] —— *J. Exp. Biol.*, **19** (1942), 56–77 (storage metabolism: *Aëdes* larva); *Proc. Roy. Soc.*, B, **134** (1947), 163–181 (ditto: *Rhodnius*).

[315] —— *Proc. Roy. Soc.*, B, **131** (1943), 313–39 (haemoglobin metabolism: *Rhodnius* &c.).

[316] —— *J. Exp. Biol.*, **26** (1949), 150–63 (metabolism during flight and starvation: *Drosophila*).

[317] WILLIAMS, C. M., *et al. Biol. Bull.*, **84** (1943), 263–72 (utilization of glycogen: *Drosophila*).

[318] WILLIAMS, C. M., and SANBORN, R. C. *Biol. Bull.*, **95** (1948), 282–3 (cytochrome system during diapause: *Platysamia*, Lep.).

[319] WILLIAMS, M. E. *Physiol. Zool.*, **9** (1936), 231–9 (catalase during embryonic development: *Melanoplus*, Orth.).

[320] WOLSKY, A. *J. Exp. Biol.*, **15** (1938), 225–34 (effect of carbon monoxide on oxygen uptake: pupae of *Drosophila*).

[321] —— *Science*, **94** (1941), 48–9 (dehydrogenase in *Drosophila* pupa).

[322] YEAGER, J. F., and MUNSON, S. C. *J. Agric. Res.*, **63** (1941), 257–94 (glycogen in *Prodenia*, Lep.).

[323] YUILL, J. S., and CRAIG, R. *J. Exp. Zool.*, **75** (1937), 169–78 (fat in *Lucilia* larvae, Dipt.).

SUPPLEMENTARY REFERENCES (A)

[324] AGRELL, I. P. S. *Nature*, **164** (1949), 1039–40; *Acta physiol. Scand.*, **18** (1949), 247–58; 355–60; **23** (1951), 179–86; **28** (1952), 306–35 (dehydrogenase systems and histogenesis during metamorphosis).

[325] —— *Nature*, **167** (1951), 283–4; *Ann. Biol.*, **27** (1951), 287–95 (respiratory quotient during diapause: *Phalera*, Lep.).

[326] BEADLE, L. C., and BEADLE, S. F. *Nature*, **164** (1949), 235 (carbon dioxide narcosis in insects).

[327] BEUTLER, R. *Naturwissenschaften*, **37** (1950), 102; *Bee World*, **32** (1951), 25–27 (range of flight, etc., in the honey bee).

[328] BRADFIELD, J. R. G. *Quart. J. Micr. Sci.*, **92** (1950), 87–112 (phosphatase and silk secretion).

[329] BURTT, E. *Proc. Roy. Ent. Soc. Lond.*, A, **26** (1951), 45–8 (colour change in adult Acridiidae).

[330] DAY, M. F. *Austral. J. Sci. Res.*, B, **2** (1949), 31–41 (alkaline phosphatase in insects).

[331] —— *Austral. J. Sci. Res.*, B, **2** (1949), 421–7 (mucoid substances in insects).

[332] DAY, M. F., and POWNING, R. F. *Austral. J. Sci. Res.*, B, **2** (1949), 175–215 (digestion in insects).

[333] DENNELL, R. *Proc. Roy. Soc.*, B, **136** (1949), 94–109 (control of tyrosinase activity in *Calliphora*).

[334] GOODWIN, T. W., and SRISUKH, S. *Biochem. J.*, **45** (1949), 263–8; 472–9; **46** (1950), xvii; **47** (1950), 549–54; 554–62; **48** (1951), 199–203; **49** (1951), 84–6; 86–7; *Biol. Rev.*, **25** (1950), 391–413; **27** (1952), 439–60 (pigments in locusts).

[335] ISAKA, S. *Nature*, **169** (1952), 74–5 (xanthopterin inhibiting melanin formation).

[336] KUPKA, E., and SCHAERFFENBERG, B. *Österreich. zool. Z.*, **1** (1947), 345–63 (resistance of soil insects to carbon dioxide).

[337] LAFON, M. *Arch. intern. Physiol.*, **57** (1950), 309–42 (utilization of reserves in egg of *Dixippus*).

[338] de LERMA, B. *Ann. Ist. Mus. Zool. Univ. Napoli*, **1** (1949), No. 4, 1–32; *Boll. Zool.*, **18** (1951), 1–2 (fluorescent substances in Orthoptera).

[339] LEVENBOOK, L. *J. Exp. Biol.*, **28** (1951), 181–202 (effect of carbon dioxide on tracheal organ of *Gastrophilus* larva).

[340] —— *J. Exp. Biol.*, **28** (1951), 173–80 (fat and glycogen in larva of *Gastrophilus*, Dipt.).

[341] LUDWIG, D. *Physiol. Zool.*, **23** (1950), 208–13 (metabolism during starvation: *Chortophaga*, Orth.).

[342] OKAY, S. *Comm. Fac. Sci. Univ. Ankara*, **1** (1948), 178–86 (brown pigment of Orthoptera).

[343] —— *Bull. Soc. Zool. France*, **74** (1949), 11–15 (carotenoid pigments in hind-wings of Acridiidae).

[344] O'ROURKE, F. J. *Ann. Ent. Soc. Amer.*, **43** (1950), 437–43 (formic acid secretion in ants).

[345] PAVAN, M. *Mem. Soc. Ent. Ital.*, **30** (1951), 107–32; *IVth. Int. Congr. Biochem.*, **12** (1959), 15–36 ('iridomyrmecin', insecticidal venom from ants).

[346] ROCKSTEIN, M., *et al. Ann. Ent. Soc. Amer.*, **44** (1951), 469–72; *J. cell. comp. Physiol.*, **38** (1951), 451–67 (phosphatase in insects).

[347] SACKTOR, B. *J. Gen. Physiol.*, **35** (1952), 397–407 (cytochrome system in house flies).

[348] SANBORN, R. C., and WILLIAMS, C. M. *J. Gen. Physiol.*, **33** (1950), 579–88 (cytochrome system in *Platysamia*, Lep.).

[349] SCHNEIDER, F. *Z. angew. Ent.*, **33** (1951), 150–62 (arrest of growth in Syrphid larvae by toxic saliva of parasite).

[350] SUSSMANN, A. S. *Quart. Rev. Biol.*, **24** (1949), 328–4; *Bull. Biol.*, **102** (1952), 39–47 (functions of tyrosinase in insects).

[351] TUFT, P. *Proc. 6th Internat. Congr. Exp. Cytology, Stockholm 1947*, 1948, 545–8; *Arch. Néerl. Zool.*, Suppl. 1 (1953), 59–75 (oxygen uptake and mitosis: *Rhodnius* egg).

[352] YAO, T. *Quart, J. Micro. Sci.*, **91** (1950), 79–105 (phosphatase in *Drosophila*).

[353] ZAMECNIK, P. C., *et al. Science*, **109** (1949), 624–6 (amino acids and silk secretion in *Platysamia*, Lep.).

SUPPLEMENTARY REFERENCES (B)

[354] AGOSIN, M., *et al. Comp. Biochem. Physiol.*, **2** (1961), 143–59 (glycolytic pathway in thoracic muscles of *Triatoma*).

[355] AKAI, H. *Bull. Sericult. Exp. Sta.*, **18** (1963), 209–82 (fine structure of active silk gland cells in *Bombyx*).

[356] ALBRECHT, F. O. *Coll. Int. C.N.R.S.* No. 114 (1962), 311–18 (progressive phase transformation in locusts).

[357] ALLEGRET, P. *Etude des glandes séricigènes des larves de Lépidoptères.* Thesis, Paris, 1956.

[358] ALLEGRET, P., and FRAISSE, R. *C.R. Acad. Sci.*, **249** (1959), 165–6 (action of spinneret in *Bombyx*).

[359] ARUGA, H., KAWASE, S., and AKINO, M. *Experientia*, **10** (1954), 336; *J. Sericult. Sci. Jap.*, **23** (1954), 99–107 (xanthopterin metabolism and colouration in *Bombyx* larva).

[360] AUGENFELD, J. M., and NEESS, J. C. *Biol. Bull.*, **120** (1961), 129–39 (respiratory enzymes in *Chironomus* and *Aëdes* during development).

[360a] AUTRUM, H., and KNEITZ, H. *Biol. Zbl.*, **78** (1959), 598–602 (venom secretion in bees of different age).

[361] BABCOCK, K. L., and RUTSCHKY, C. W. *Ann. Ent. Soc. Amer.*, **54** (1961), 156–64 (lipids in the egg of *Oncopeltus*).

[362] BEALL, G. *Ecology*, **29** (1948), 80–94 (fat content of *Danais plexippus*).

[363] BEARD, R. L. *Conn. Agr. Exp. Sta.*, Bull. No. 562 (1952), 26 pp. (*Habrobracon* toxin).

[364] —— *Ann. Rev. Ent.*, **8** (1963), 1–18 (insect toxins and venoms: review).

[365] BHEEMESWAR, B. *Proc. IVth Int. Congr. Biochem.*, **12** (1959), 78–89 (amino acid metabolism in insects: review).

[366] BISSET, G. W., FRAZER, J. F. D., ROTHSCHILD, M., and SCHACHTER, M., *Proc. Roy. Soc.*, B, **152** (1960), 255–62 (toxins in *Arctia caja*).

[367] BLUM, M. S., *et al. J. Ins. Physiol.*, **9** (1963), 881–5 (chemical releasers of social behaviour in *Iridomyrmex*).

[368] BOCH, R., and SHEARER, D. A. *Nature*, **194** (1962), 704–6; **202** (1964), 320–1; *J. Ins. Physiol.*, **9** (1963), 431–4 (active scents produced by Nassonoff's gland in *Apis*).

[369] BOCH, R., SHEARER, D. A., and STONE, B. C. *Nature*, **195** (1962), 1018–20 (sting pheromone in *Apis*).

[370] BRICTEUX-GRÉGOIRE, S., DEWANDRE, A., and FLORKIN, M. *Arch. Int. Physiol. Biochim.*, **68** (1960), 281–4; *Biochem. Z.*, **333** (1960), 370–6 (amino acids and silk formation in *Bombyx*).

[371] BRICTEUX-GRÉGOIRE, S., and FLORKIN, M. *Arch. Int. Physiol. Biochim.*, **70** (1962), 711–717 (use of glucose in amino acid synthesis for silk secretion).

[372] BRICTEUX-GRÉGOIRE, S., VERLY, W. G., and FLORKIN, M. *Nature*, **179** (1957), 678–9 (protein synthesis in *Sphinx ligustri* pupae).

[373] BROOKS, M. A. *Ann. Ent. Soc. Amer.*, **50** (1957), 122–5 (succinoxidase in *Periplaneta*).

[374] BÜCKMANN, D. *Verh. dtsch. zool. Ges.*, 1956, 220–5; 1958, 137–44; *J. Ins. Physiol.*, **3** (1959), 159–89 (pre-pupation colour change in *Dicranura*).

[375] —— *Naturwiss.*, **47** (1960), 610–11 (control of pupal colour in *Vanessa*).

[375a] BÜCKMANN, D., and DUSTMANN, J. H. *Naturwiss.*, **49** (1962), 379 (pigments in the integument of *Carausius*).

[376] BURSELL, E. *J. Ins. Physiol.*, **9** (1963), 439–52 (amino acid metabolism in *Glossina*).

[377] BUSNEL, G., and DRILHON, A. *Bull. Soc. Zool. Fr.*, **74** (1949), 21–3 (pterines in pupa of *Pieris*).

[378] BUTENANDT, A. *Angew. Chem.*, **69** (1957), 16–23; *XVIIth Int. Kongr. rein. angew. Chem.* (Munich, 1959) 11–31; *Hoppe-Seyl. Z.*, **313** (1958), 251–8 (chemistry and distribution of ommochromes: review).

[379] —— *Arb. Forsch. Land. Nordrhein-Westfalen*, No. 62, 1960, 29 pp. (genes and pigment formation: review).

[380] BUTENANDT, A., LINZEN, B., and LINDAUER, M. *Arch. Anat. micr. Morph. exp.*, **48** (1959), 13–20 (mandibular gland scent in *Atta sexdens*).

[381] BUTLER, C. G. *Trans. Roy. Ent. Soc. Lond.*, **105** (1954), 11–29; *Proc. Roy. Ent. Soc. Lond.*, A, **31** (1956), 12–16; *Bee World*, **40** (1959), 269 ('queen substance' in the honey-bee).

[382] BUTLER, C. G., *et al. Proc. Roy. Ent. Soc. Lond.*, A, **33** (1958), 120–2; **37** (1962), 114–16 (source of 'queen substance').

[383] BUTLER, C. G., CALLOW, R. K., and JOHNSTON, N. C. *Nature*, **184** (1959), 1871; *Proc. Roy. Soc.*, B, **155** (1961), 417–32 (isolation and synthesis of 'queen substance').

[384] BUTLER, C. G., CALLOW, R. K., and CHAPMAN, J. R. *Nature*, **201** (1964), 733 (pheromone stabilizing honey-bee swarms).

[385] CHEFURKA, W. *Biochim. Biophys. Acta*, **17** (1955), 294–302 (pentose cycle in flight muscles of *Musca*).

[386] CHEN, P. S. *J. Ins. Physiol.*, **2** (1958), 38–51 (amino acids during development in different insects).

[386a] CHERRY, L. M. *Ent. Exp. et Appl.*, **2** (1959), 68–76 (temperature and composition of fats in Phormia).

[387] CHINO, H. *Nature*, **180** (1957), 606–7; *J. Ins. Physiol.*, **2** (1958), 1–12; **5** (1960), 1–15 (metabolism of glycogen during diapause in egg of *Bombyx*).

[388] CLEGG, J. S., and EVANS, D. R. *J. Exp. Biol.*, **38** (1961), 771–92 (blood trehalose and flight in *Phormia*).

[389] CLEMENTS, A. N. *J. Exp. Biol.*, **32** (1955), 547–51 (energy for flight in mosquitos).

[390] COCKBAIN, A. J. *J. Exp. Biol.*, **38** (1961), 163–74 (full utilization of Aphids in flight).

[391] CROMARTIE, R. I. T. *Ann. Rev. Ent.*, **4** (1959), 59–76 (insect pigments: review).

[392] CRONE, H. D., and BRIDGES, R. G. *Biochem. J.*, **89** (1963), 11–21 (phospholipids in *Musca*).

[393] DADD, R. H. *Bull. Ent. Res.*, **52** (1961), 63–81 (role of carotene in pigmentation of locusts).

[394] DANNEEL, R., and ESCHERICH-ZIMMERMANN, B. *Z. Naturforsch.*, **12b** (1957), 730–2 (genetic relation between chemically different pigments in *Drosophila*).

[395] DESAI, R. M., and KILBY, B. A. *Arch. Int. Physiol. Biochem.*, **66** (1958), 249–59 (transamination in fat body of *Calliphora*).

[396] DUSPIVA, F. *Z. Naturforsch.*, **5b** (1950), 273–81 (sericin-digesting enzyme secreted by emerging *Bombyx*).

[397] EISNER, T. *J. Ins. Physiol.*, **2** (1958), 215–20 (protective spray of the Bombardier beetle *Brachinus*).

[398] EISNER, T., *et al. J. Ins. Physiol.*, **7** (1961), 46–50; *Ann. Ent. Soc. Amer.* **57** (1964), 44–6 (defence mechanisms: *Eleodes*, Tenebrionidae); *J. Ins. Physiol.*, **8** (1962), 175–9 (ditto: *Acanthomyops*, Formicoidea); *Ann. Ent. Soc. Amer.*, **56** (1963), 37–41 (ditto: *Calosoma*, Carabidae); *Psyche*, **70** (1963), 94–116 (*Chlaenius*, Carabidae).

[399] ERGENE, S. *Z. vergl. Physiol.*, **32** (1950), 530–51; **34** (1952), 159–63; **35** (1953), 36–41; **36** (1954), 235–7; **37** (1955), 221–5 (colour adaptation to the background in *Acrida* and *Mantis*).

[399a] FINCH, L. R., and BIRT, L. M. *Comp. Biochem. Physiol.*, **5** (1962), 59–64 (amino acid activation in pupa of *Lucilia*).

[400] FISCHER, F. G., NEUMANN, W. P., and DÖRFEL, H. *Biochem. Z.*, **324** (1954), 447–75 (venom of the honey-bee).

[401] FLORKIN, M., *et al. Arch. Int. Physiol. Biochim.*, **67** (1959), 182–94; 214–21, 586–96, 687–98; **68** (1960), 190–202 (biochemistry of silk production in *Bombyx*).

[402] FOUSTKA, M., and POLACEK, I. *Studie o srovnávací fysiologii metabolismu*, Prague, 1956, 30–2 (metabolic capacity of insects).

[403] FUKUDA, T. *Bull. Sericult. Exp. Sta.* Tokyo, **15** (1960), 595–610 (silk formation in relation to feeding).

[404] FUZEAU-BRAESCH, S. *Bull. Biol. Fr. Belg.*, **94** (1960), 526–657 (biology and biochemistry of pigmentation in *Gryllus*).

[405] GARY, N. E., and MORSE, R. A. *Proc. Roy. Ent. Soc. Lond.*, A, **37** (1962), 76–8 (mandibular glands of the queen and communication in the honey-bee).

[406] GEORGE, J. C., and BHAKTHAN, N. M. G. *J. Exp. Biol.*, **37** (1960), 308–15 (lipase in flight muscles).

[407] GERSCH, M. *Z. vergl. Physiol.*, **39** (1956), 190–208 (hormonal control of melanophores in *Corethra larva*).

[408] GILMOUR, D. *The biochemistry of insects.* Academic Press, New York and London, 1961, 343 pp.

[409] GRAF, G. E., HADORN, E., and URSPRUNG, H. *J. Ins. Physiol.*, **3** (1959), 120–4 (*iso*-xanthopterin metabolism in *Drosophila*).

[410] GRINDLEY, D. N. *J. Exp. Biol.*, **29** (1952), 440–4 (body fat of *Tanytarsus*, Chironomidae).

[411] HABERMANN, E. *Arch. exp. Path. Pharm.*, **225** (1955), 158–60 (bee venom); *Biochem. Z.*, **329** (1957), 1–10 (hyaluronidase in bee venom).

[412] HACKMAN, R. H. *Aust. J. Biol. Sci.*, **9** (1956), 400–5 (metabolism of amino acids in *Calliphora*, &c.).

[413] HADORN, E. *Proc. Xth Int. Congr. Genetics,* **1** (1960), 338–53 (biochemical genetics of pigments formation in *Drosophila*, &c.).

[414] HADORN, E., and FRIZZI, G. *Rev. Suisse Zool.*, **56** (1949), 306–16 (control of melanophores in *Corethra* larva).

[415] van HANDEL, E., and LUM, P. T. M. *Science*, **134** (1961), 1979–80 (sexual differences in lipid metabolism in mosquitos).

[416] HARIZUKA, M., *et al. Bull. Sericult. Exp. Sta.*, **14** (1953), 153–6; **16** (1960), 29–34 (cocoon colours of *Bombyx*).

[417] HARMSEN, R. *The storage and excretion of pteridines in Pieris brassicae* L. Thesis, University of Cambridge, 1963.

[418] HAMILTON, A. G. *Proc. Xth Int. Congr. Ent., 1956*, **2** (1958), 343–7 (variations in metabolic rate in *Schistocerca*).

[419] HARVEY, W. R., and WILLIAMS, C. M. *Biol. Bull.*, **114** (1958), 23–53 (cytochrome system in *Hyalophora* during diapause).

[420] HELLER, J., and SZARKOWSKA, L. *Bull. Acad. Polon. Sci.*, **4** (1956), 331–5; **6** (1958), 413–15 (quinone respiration in *Celerio*).

[421] HIDAKA, T. *Annot. Zool. Japon.*, **29** (1956), 69–74; *J. Fac. Sci. Univ. Tokyo.*, Sect. IV **9** (1961), 223–61 (colour adaptation in *Papilio xuthus* pupae).

[422] HINTON, H. E. *J. Ins. Physiol.*, **2** (1958), 249–60 (changes in colour pattern of *Thaumalea*, Dipt., at metamorphosis).

[423] HODGSON, N. B. *Bee World*, **36** (1955), 217–22 (components of bee venom).

[424] HOSKINS, D. D., CHELDELIN, V. H., and NEWBURGH, R. W. *J. Gen. Physiol.*, **39** (1956), 705–13 (oxidative systems in *Apis*).

[425] HUDSON, B. W., BARTEL, A. H., and CRAIG, R. *J. Ins. Physiol.*, **3** (1959), 63–73 (pteridines in *Oncopeltus*).

[426] HUMPHREY, G. F. *J. Cell. Comp. Physiol.*, **34** (1949), 323–5 (glycolysis in *Periplaneta*, &c.).

[427] JACOBSON, M., and BEROZA, M. *Science*, **140** (1963), 1367–73 (chemical attractants in insects: review).

[428] JOHN, M. *Z. vergl. Physiol.*, **41** (1958), 204–20 (carbohydrate reserves in *Apis*).

[429] JOLY, P. *C. R. Soc. Biol.*, **145** (1951), 1362–4; *C.R. Acad. Sci.*, **235** (1952), 1054–6 (corpus allatum and green colour in locusts).

[430] JONES, D. A., PARSONS, J., and ROTHSCHILD, M. *Nature*, **193** (1962), 52–3 (production of hydrogen cyanide in Zygaenid larvae).

[431] JONES, B. M., and WILSON, R. S. *Biol. Bull.*, **117** (1959), 482–91 (oxygen consumption and silk formation in *Samia*).

[432] KAPLANIS, J. N., DUTKEY, R. C., and ROBBINS, W. E. *Ann. Ent. Soc. Amer.*, **54** (1961), 114–16 (incorporation of mevalonic acid into lipids in *Musca*).

[433] KARLSON, P., and BUTENANDT, A. *Ann. Rev. Ent.*, **4** (1959), 39–58 (pheromones in insects).

[434] KEY, K. H. L. *Aust. J. Zool.*, **5** (1957), 247–84 (phase change in Phasmidae).

[435] KEY, K. H. L., and DAY, M. F. *Aust. J. Zool.*, **2** (1954), 309–63 (colour change in *Kosciuscola* Acridiidae).

[436] KIKKAWA, H. *Advances in Genetics*, **5** (1953), 107–40 (biochemical genetics in *Bombyx*).

[437] KILBY, B. A., and NEVILLE, E. *Biochim. Biophys. Acta*, **19** (1956), 389–90; **34** (1957), 276–89 (amino acid metabolism in locust tissues).

[438] KOPENEC. A. *Z. vergl. Physiol.*, **31** (1949), 490–505 (colour change in *Corethra* larva).

[439] KRIEGER, F. *Z. vergl. Physiol.*, **36** (1954), 352–66 (colour adaptation in Odonata larvae).

[440] KUBIŠTA, V. *Studie o srovnávací fysiologii metabolismu*, Prague, 1960, 101–6, 119–24 (glycolysis in insect muscle).

[441] KÜHN, A., *et al. Naturwiss.*, **43** (1956), 25; *Biol. Zbl.*, **81** (1962), 5–17 (genes and eye pigments in *Ephestia*, &c.).

[442] KURLAND, C. G., and SCHNEIDERMAN, H. A. *Biol. Bull.*, **116** (1959), 136–61 (respiratory enzymes in *Hyalophora* pupa in diapause.

[443] KURLAND, C. G., SCHNEIDERMAN, H. A., and SMITH, R. D. *Anat. Rec.*, **132** (1958), 465–6 (oxygen debt during anaerobiosis in *Hyalophora* pupa).

[444] LACONTI, J. D., and ROTH, L. M. *Ann. Ent. Soc. Amer.*, **46** (1953), 281–9 (protective secretion of *Tribolium*).

[445] LAMBREMONT, E. N. *J. Ins. Physiol.*, **8** (1962), 181–90 (dehydrogenases in brain of *Anthonomus*).

[446] LAMBREMONT, E. N., and BLUM, M. S. *Ann. Ent. Soc. Amer.*, **56** (1963), 612–16 (fatty acids in *Anthonomus*).

[447] LAUGHLIN, R. *J. Exp. Biol.*, **33** (1956), 566–75 (storage and use of reserves in *Phyllopertha*, Scarabaeidae).

[448] LEVENBOOK, L. *Arch. Biochem. Biophys.*, **92** (1961), 114–21 (tricarboxylic acid cycle in *Prodenia*, Lep.).

[449] LENARTOWICZ, E., RUDZISZ, B., and NIEMIERKO, S. *J. Ins. Physiol.*, **10** (1964), 89–96 (phospholipids in *Galleria*).

[450] LINZEN, B. *Verh. Dtsch. zool. Ges.* (Frankfurt) 1958, 154–64 (distribution of ommochromes in insects).

[451] LINZEN, B., and BÜCKMANN, D. *Z. Naturforsch.*, **16b** (1961), 7–19 (chemistry of prepupal colour change in *Dicranura*).

[452] LONG, D. B. *Trans. Roy. Ent. Soc. Lond.*, **104** (1953), 543–85 (crowding and colour change in caterpillars).

[453] LOULOUDES, S. J., *et al.*, *Ann. Ent. Soc. Amer.*, **54** (1961), 99–103 (incorporation of acetate in *Periplaneta* lipids).

[454] LUDWIG, D., *et al. J. Cell. Comp. Physiol.*, **45** (1955), 157–66 (respiratory enzymes in embryogenesis: *Popillia*); *J. Gen. Physiol.*, **38** (1955), 729–834 (ditto: *Tenebrio*).

[455] LUDWIG, D., and BARSA, M. C. *J. N.Y. Ent. Soc.*, **67** (1959), 151–6 (respiratory enzymes in metamorphosis: *Musca*); *Ann. Ent. Soc. Amer.*, **51** (1959), 311–14 (ditto: *Tenebrio*).

[456] LUDWIG, D., and WUGMEISTER, M. *Physiol. Zool.*, **26** (1953), 254–9 (use of reserves in starvation in *Popillia* larva).

[457] LUCAS, F., *et al. Nature*, **179** (1957), 906–7 (composition of silk in the egg stalks of *Chrysopa*).

[458] LÜSCHER, M., and MÜLLER, B. *Naturwiss.*, **47** (1960), 503 (trail-forming pheromone in *Zootermopsis*).

[459] MCALLAN, J. W., and CHEFURKA, W. *Comp. Biochem. Physiol.*, **2** (1961), 290–9 (transamination in *Periplaneta* and *Hyalophora*).

[460] MAKINO, K., TAKAHASHI, H., and SATOH, K. *Nature*, **173** (1954), 586–7 (accumulation of hydroxykynurenine in silkworm mutant).

[460a] MASCHWITZ, U. *Z. vergl. Physiol.*, **47** (1964), 596–655 (alarm substances in social Hymenoptera).

[461] MORRIS, S. J., and THOMPSON, R. H. *J. Ins. Physiol.*, **9** (1963), 391–9 (flavonoid pigments in wings of *Melanargia*, Lep.).

[462] MOTHES, G. *Zool. J. Phys.*, **69** (1960), 133–62 (colour change in *Carausius*).

[463] NARAYANAN, E. S., *et al. Nature*, **173** (1954), 503–4 (low temperature and darkening in *Microbracon*).

[464] NEDEL, J. O. *Z. Morph. Oekol. Tiere*, **49** (1960), 139–83 (structure and function of mandibular glands in Apidae).

[465] NEWBURGH, R. W., and CHELDELIN, V. H. *J. Biol. Chem.*, **214** (1955), 37–41 (pentose cycle in the Aphid *Macrosiphum*).

[466] NEWTON, C. J. *Physiol. Zool.*, **27** (1954), 248–58 (changes during starvation in *Popillia* larva).

[467] NICKERSON, B. *Nature*, **174** (1954), 357–8; *Anti-locust Bull.*, **24** (1956), 1–34 (humoral factors in colour change in *Schistocerca*).

[468] NIEMIERKO, S., WLODAWER, P., and WOJTCZAK, A. F. *Acta Biol. Exper.*, **17** (1956), 255–276 (lipid and phosphorous metabolism during growth in *Bombyx*).

[469] NOVAK, V. *Mém. Soc. Zool. Tchécosl.*, **19** (1955), 233–46 (temperature and melanin formation in *Oncopeltus*).

[470] NOVOTNÝ, I. *Studie o srovnávací fysiologii metabolismu*, Prague, 1960, 79–80 (aerobic metabolism in *Musca* pupae).

[471] OHNISHI, E. *Jap. J. Zool.*, **11** (1953), 69–74; *Annot. Zool. Jap.*, **27** (1954), 34–9, 76–81, 188–93; *J. Ins. Physiol.*, **2** (1959), 219–29 (activation of tyrosinase in *Drosophila* and *Musca*).

[472] OHTAKI, T. *Annot. Zool. Jap.*, **33** (1960), 97–103 (humoral control of pupal colour in *Pieris*).

[473] OKAMOTO, H. *Physiol. Ecol.*, **9** (1960), 84–9 (control of pupal colour in *Pieris*).

[474] OSANAI, M., and ARAI, Y. *Gen. Comp. Endocrin.*, **2** (1962), 311–16 (colour change in overwintering larva of *Hestina*).

[475] OSHIMA, C., SEKI, T., and ISHIZAKI, H. *Genetics*, **41** (1956), 4–20 (development of colour pattern in Coccinellidae).

[476] OSMAN, M. F. H., and KLOFT, W. *Insectes sociaux*, **8** (1961), 384–95 (insecticidal action of formic acid).

[477] PAIN, J., *et al. C.R. Acad. Sci.*, **250** (1960), 3740–2; *Sur la phèromone des reines d'abeilles et ses effets physiologiques*, Thesis, Paris, 1961, 103 pp. (multiple effects of mandibular gland secretion of the queen bee).

[478] PARKER, K. D., and RUDALL, K. M. *Nature*, **179** (1957), 905–6 (egg-stalk silk of *Chrysopa*).

[479] PAVAN, M. *Mem. Soc. Ent. Ital.*, **30** (1951), 107–32 (*Biochemistry of Insects, Proc. IVth Int. Congr. Biochem.*, **12** (1959), 15–36 (biochemistry of insect poisons).

[480] PO-CHEDLEY, D. S. *J. N.Y. Ent. Soc.*, **66** (1958), 171–7 (effects of starvation on amino acids in *Anomala* larva, Col.).

[481] POLACEK, I., and KUBIŠTA, V. *Physiol. Bohem.*, **9** (1960), 228–34 (metabolism in *Periplaneta* during flight).

[482] PRECHT, H., and CARLSEN, H. *Z. vergl. Physiol.*, **35** (1953), 209–18 (effects of starvation, &c., on respiratory enzymes in *Chironomus* larvae).

[483] PRICE, G. M. *Biochem. J.*, **81** (1961), 15–16; **86** (1963), 372–8 (effects of anoxia on *Musca*).

[484] PRYOR, M. G. M. *J. Exp. Biol.*, **32** (1955), 468–84 (activation of tyrosinase in *Drosophila* and *Calliphora* larvae).

[485] RAABE, M. *Ann. Biol.*, **32** (1956), 247–82; *Arch. Zool. Exp. Gén.*, **94** (1957), 63–293; *C.R. Acad. Sci.*, **257** (1963), 1804–6 (colour change in insects).

[486] RATHMAYER, W. *Z. vergl. Physiol.*, **45** (1962), 413–62 (action of *Philanthus* venom).

[487] RAY, J. W., HESLOP, J. P., *et al. Biochem. J.*, **82** (1962), 25P; **87** (1963), 35–8, 39–42 (anaerobic metabolism in *Musca*).

[488] REICHMUTH, W., and KLINK, G. *Z. Naturforsch.*, **15b** (1960), 744–51 (photosensitivity of 'aphin' pigments).

[489] REMOLD, H., *Z. vergl. Physiol.*, **45** (1962), 636–94 (stink glands in Heteroptera).

[490] RENNER, M. *Z. vergl. Physiol.*, **43** (1960), 411–68 (function of Nassonoff's gland in *Apis*).

[491] REISENER-GLASEWALD, E. *Z. indukt. Abstamm. VererbLehre.*, **87** (1956), 668–93 (fluorescent substances in eyes of *Ephestia*).

[492] ROCKSTEIN, M. *J. Gerontol.*, **11** (1956), 282–5 (phosphorus metabolism and ageing in *Musca*).

[493] —— *Smithsonian Misc. Coll.*, **137** (1959), 263–86 (post-emergence maturation of enzyme systems in insects).

[494] ROTH, L. M. *Ann. Ent. Soc. Amer.*, **54** (1961), 900–11 (stink glands of *Scaptocoris*).

[495] ROTH, L. M., and EISNER, T. *Ann. Rev. Ent.*, **7** (1962), 107–36 (chemical defences of insects: review).

[496] ROTH, L. M., and STAY, B. *J. Ins. Physiol.*, **1** (1958), 305–18 (stink glands of *Diploptera*, Blattaria).

[497] ROTHSCHILD, M. *Trans. Roy. Ent. Soc. Lond.*, **113** (1961), 101–21 (Müllerian mimicry of defensive odours).

[497a] RUDALL, K. M. *Lectures on the scientific basis of medicine*, **5** (1956), 217–30 (formation of silk, &c.).

[498] SACKTOR, B. *Arch. Biochem. Biophys.*, **45** (1953), 349–65 (oxidative enzymes in flight muscle of *Musca*).

[499] SACKTOR, B., *et al. J. Gen. Physiol.*, **37** (1954), 343–59; *Arch. Biochem. Biophys.*, **74** (1958), 265–76, 509–31, 532–45 (role of α-glycerophosphate in metabolism of flight muscle).

[500] SAITO, M., YAMAZAKI, M., and KOBAYASHI, M. *Nature*, **198** (1963), 1324 (steroid biosynthesis in *Bombyx*).

[501] SCHACHTER, M., and THAIN, E. M. *Brit. J. Pharmacol.*, **9** (1954), 342–9 (components of wasp venom).

[502] SCHILDKNECHT, H., *et al. Angew. Chem.*, **69** (1957), 62; **73** (1961), 1–17 (repellent secretions of insects: *Brachinus*); *Z. Naturforsch.*, **16b** (1961), 361–3 (ditto: *Pseudophonus*); **17b** (1962), 81–3, 448–52 (ditto: *Dytiscus*); 439–47 (ditto: *Abax ater* and other Carabids); 452–5 (ditto: Silphidae); **18b** (1963), 585–7 (ditto: *Dicranura*).

[503] SHAPPIRIO, D. G., and WILLIAMS, C. M. *Proc. Roy. Soc.*, B, **147** (1957), 218–46 (cytochrome system in *Hyalophora* pupa).

[504] SHIMIZU, M., FUKUDA, T., and KIRIMURA, J. *The silk protein*, Sericult. Exp. Sta., Tokyo, 1957, 145 pp.

[505] SMITH, J. N., and TURBERT, H. B. *Nature*, **189** (1961), 600 (enzymic glucoside synthesis in *Schistocerca*).

[506] SOTAVALTA, O. *Ann. Zool. Soc. 'Vanamo'*, **16** (1954), 1–27 (metabolism and wing-beat frequency in *Apis*).

[507] SOTAVALTA, O., and LAULAJAINEN, E. *Ann. Acad. Sci. Fennicae*, **53** (1961), 6–24 (sugar consumption in *Eristalis* in flight).

[508] STEGWEE, D. *J. Ins. Physiol.*, **10** (1964), 97–102 (breakdown of muscles and sarcosomes during diapause in *Leptinotarsa*).

[509] STEHR, G. *Evolution*, **13** (1959), 537–60 (haemolymph polymorphism in *Colias*).

[510] STEIN, G. *Z. Naturforsch.*, **16b** (1961), 129–34 (yellow flavonoid pigment in hairs of *Bombus*).

[511] STOCK, A., and O'FARRELL, A. F. *Aust. J. Sci.*, **17** (1954), 64–6 (spinning glands in cerci of *Blattella*).

[512] STUART, A. M. *Physiol. Zool.*, **36** (1963), 69–96 (gland producing the scent trail of *Zootermopsis*).

[513] STUMPER, R. *Naturwiss.*, **47** (1960), 457–63 (ant venoms: review).

[514] SUGIYAMA, H. *Bull. Sericult. Exp. Sta.*, **18** (1963), 208 (fat body lipids in the silkworm larva).

[515] SULKOWSKI, E., and WOJTCZAK, L. *Acta Biol. Exp.*, **18** (1958), 239–48 (succinoxidase system during metamorphosis in *Galleria*).

[516] SUNDARAM, T. K., and SARMA, P. S. *Nature*, **172** (1953), 627–8 (tryptophane metabolism in *Corcyra* larva).

[517] UMEBACHI, Y., *et al. J. Biochem.*, **43** (1956), 73–81; *Sci. Rep. Kanazawa Univ.*, **6** (1958), 45–55; 69–75; *Annot. Zool. Japon.*, **32** (1959), 112–16 (kynurenine in the wings of *Papilio* spp.).

[518] UMEBACHI, Y., and TSUCHITANI, K. *J. Biochem.*, **42** (1955), 817–24 (xanthurenic acid in the head of *Drosophila*).

[519] VISCONTINI, M., *et al. Z. Naturforsch.*, **11b** (1956), 501–4; *Helv. Chim. Acta*, **40** (1957), 579–85 (pterines and other fluorescent substances in the eyes of *Ephestia* and *Drosophila*).

[520] WATANABE, M. I., and WILLIAMS, C. M. *J. Gen. Physiol.*, **34** (1951), 675–89 (enzymes in flight muscle mitochondria).

[521] WATERHOUSE, D. F., FORSS, D. A., and HACKMAN, R. H. *J. Ins. Physiol.*, **6** (1961), 113–21 (components in the secretion of stink bugs).

[522] WEIS-FOGH, T., *Phil. Trans. Roy. Soc.*, B, **237** (1952), 1–36 (fat combustion in flying locust).

[523] —— *Trans. IXth Int. Congr. Ent.*, **1** (1952), 344–7 weight economy of flying insects).

[524] WESSING, A., and DANNEEL, R. *Z. Naturforsch.*, **16b** (1961), 389–90 (storage of oxykynurenine in Malpighian tubules of *Drosophila*).

[525] WEYGAND, F., and WALDSCHMIDT, M. *Angew. Chem.*, **67** (1955), 328 (biosynthesis of pterines, purines, and riboflavine).

[526] WIGGLESWORTH, V. B. *Symp. Soc. Exp. Biol.*, **11** (1957), 204–27 (protein synthesis and respiration during moulting and metamorphosis).

[527] —— *Proc. Roy. Soc.*, B, **147** (1957), 185–99 (degree of unsaturation of fat-body lipids).

[527a] —— *New Scientist*, **8** (1960), 101–4 (fuel and power in flying insects).

[528] WILHELM, R. C., SCHNEIDERMAN, H. A., and DANIEL, L. J. *J. Ins. Physiol.*, **7** (1961), 273–288 (effects of anaerobiosis on diapause metabolism of *Hyalophora*, &c.).

[528a] WILSON, E. O. *Anim. Behaviour*, **10** (1962), 134–64 (chemical communication in the fire ant *Solenopsis*).

[529] WINTERINGHAM, F. P. W. *Biochem, J.*, **75** (1960), 38–45 (trehalose metabolism in *Musca*).

[530] WREN, J. J., and MITCHELL, H. K. *J. Biol. Chem.*, **234** (1959), 2823–8 (lipids in *Drosophila*).

[531] WYATT, G. R. in *Insect Physiology*; (23rd Biology Colloquium, Oregon State Univ.) Oregon State Univ. Press, 1962, 23–41 (effect of injury on metabolism in *Hyalophora* pupa).

[532] —— *J. Gen. Physiol.*, **45** (1962), 622; in *Control mechanisms in respiration and fermentation* (B. Wright, Ed.) Ronald Press Co., New York, 1963, 179–88 (metabolic regulation in insect development).

[533] ZEBE, E. *Naturwiss.*, **40** (1953), 298; *Z. vergl. Physiol.*, **36** (1954), 290–317 (respiratory quotient in Lepidoptera at rest and in flight).

[534] —— *Verh. dtsch. zool. Ges. 1958*, (1959), 131–7: *Biochem. Z.*, **332** (1960), 328–32 (fat metabolism in relation to activity in insects).

[535] —— *Umschau*, 1960, 40–3 (metabolism of insect muscles).

[536] ZIEGLER-GUNDER, I. *Biol. Rev.*, **31** (1956), 313–48; *Z. Naturforsch.*, **15b** (1960), 460–5 (pteridine pigments: review, and with reference to photosensitivity).

[537] ZWICKY, K., and WIGGLESWORTH, V. B. *Proc. Roy. Ent. Soc. Lond.*, A, **31** (1956), 153–60 (oxygen consumption during the moulting cycle in *Rhodnius*).

SUPPLEMENTARY REFERENCES (C)

[538] AKAI, H. and KOBAYASHI, M. *Symp. cell. Chem.* **17** (1966), 131–8 (cytology of silk secretion in *Bombyx*).

[539] ANESHANSLEY, D. J., EISNER, T. *et al. Science*, **165** (1969), 61–3 (explosive secretory discharge in *Brachinus*).

[540] APLIN, R. T., BENN, M. H. and ROTHSCHILD, M. *Nature*, **219** (1968), 747–8 (poisonous alkaloids in *Callimorpha*, Arctiidae).

[541] BAUMANN, P. and CHEN, P. S. *Rev. Suisse Zool.*, **75** (1968), 1051–5 (protein synthesis and age in *Drosophila*).

[542] BEENAKKERS, A. M. T. *J. Insect Physiol.*, **11** (1965), 879–88; **15** (1969), 353–61 (lipid consumption during fight in *Locusta*).

[543] BEENAKKERS, A. M. T. and KLINGENBERG, M. *Biochim. biophys. Acta*, **84** (1964), 205–7 (carnitine-coenzyme A transacetylase in mitochondria of *Locusta*).

[544] BERGMANN, E. D. *Bull. Soc. Chim. Fr.*, 1965, 2687–91 (steroids in insects: review).

[545] BEROSA, M. (Ed.) *Chemicals controlling Insect Behaviour*. Academic Press: New York and London, 1970, 170 pp.

[546] BRACKEN, G. K. and HARRIS, P. *Nature*, **224** (1969), 84–5 (high palmitoleic acid in overwintering caterpillars).

[547] BRIDGES, R. G. *J. Insect Physiol.*, **17** (1971), 881–95 (fatty acids in lipids of *Musca*).

[548] BRIDGES, R. G. and PRICE, G. M. *Int. J. Biochem.*, **1** (1970), 483–90 (bases in phospholipids of *Musca*).

[549] BROWER, L. P. and BROWER, J. V. Z. *Zoologica*, **49** (1964), 137–59 (insects obtaining protective poisons from plants).

[550] BROWER, L. P., BROWER, J. V. Z. and CORVINE, J. M. *Proc. nat. Acad. Sci.*, **57** (1967), 893–8 (poisons taken up from *Asclepias* by *Danaus*).

[551] BÜCKMANN, D. *J. Insect Physiol.*, **11** (1965), 1427–62; *Umschau*, **67** (1967), 257 (redox behaviour of ommochromes *in vivo* in *Cerura*).

[552] BÜCKMANN, D. *Wilhelm Roux Arch. Entw. Mech. Organ.*, **166** (1971), 236–53 (control of melanization in *Pieris* pupa).

[553] BÜCKMANN, D., WILLIG, A. and LINZEN, B. *Z. Naturforsch.*, **21** (1966), 1184–95 (ommochrome and disposal of tryptophane).

[554] BUTLER, C. G. *Biol. Rev.* **42** (1967), 42–87 (insect pheromones: review).

[555] BUTLER, C. G. and CALAM, D. H. *J. Insect Physiol.*, **15** (1969), 237–44 (Nassanoff gland pheromones in honey-bee).

[556] CASNATI, G., PAVAN, M. and RICCA, A. *Ann. Soc. ent. Fr.* (N.S.) **1** (1965), 706–10 (venom of *Calosoma*, Carabidae).

[557] CHEFURKA, W. *Ann. Rev. Entom.*, **10** (1965), 345–82 (carbohydrate metabolism in insects).

[558] CHILDRESS, C. C. and SACKTOR, B. *J. biol. Chem.* **245** (1970), 2927–36 (glycogen breakdown in flight muscle of *Phormia*).

[559] COCKERAM, H. S. and LEVINE, S. A. *J. Soc. cosmet. Chem.*, **12** (1961), 316–23 (chemistry of shellac).

[560] CROMPTON, M. and POLAKIS, S. E. *J. Insect Physiol.*, **15** (1969), 1323–9 (incorporation of glucose into chitin in *Lucilia*).

[561] D'COSTA, M. A. and BIRT, L. M. *J. Insect Physiol.*, **15** (1969), 1959–68 (fatty acid oxidation and endergonic syntheses in developing adult of *Lucilia*).

[562] DOMROESE, K. A. and GILBERT, L. I. *J. exp. Biol.*, **41** (1964), 573–90 (lipid metabolism in *Hyalophora*).

[563] DUSTMANN, J. H. *J. Insect Physiol.*, **15** (1969), 2225–38 (eye-colour pigments in *Apis*).

[564] EDWARDS, J. S. *Nature*, **211** (1966), 73–4 (wax secretion from cornicles of Aphids).

[565] EISNER, T. and MEINWALD, Y. C. *Science*, **150** (1965), 1733–5 (defensive secretion of *Papilio* larva).

[566] ENGELS, W. *J. Insect Physiol.*, **14** (1968), 253–60; 869–79 (anaerobiosis in adult *Musca*).

[567] ESTES, Z. E. and FAUST, R. M. *Comp. Biochem. Physiol.*, **13** (1964), 443–52 (mucopolysaccharides in *Galleria* mid-gut).

[568] V. EUW, J. *et al.*, *Nature*, **214** (1967), 35–9 (cardenolides in *Poekilocerus*, Orthopt. feeding on *Asclepias*).

[569] FAST, P. G. *Mem. ent. Soc. Canada*, **37** (1964), 5–50; *Prog. Chem. Fats*, **11** (1970), 181–242 (insect lipids: review).

[570] FLORKIN, M. and JEUNIAUX, C. *Bull. Acad. Roy. Belg.*, **51** (1965), 541–52 (trehalose metabolism in *Bombyx*).

[571] FORREST, H. S. *et al.* *J. Insect Physiol.*, **12** (1966), 1411–21 (pteridines in *Oncopeltus*).

[572] FUZEAU-BRAESCH, S. *J. Insect Physiol.*, **12** (1966), 1363–8 (cuticular blackening in response to background in *Locusta*).

[573] GILBERT, L. I. *Adv. Insect Physiol.*, **4** (1967), 69–211 (lipid metabolism in insects: review).

[574] GILBY, A. R. *Ann. Rev. Entom.*, **10** (1965), 141–60 (lipids and their metabolism in insects).

[575] GILBY, A. R. and WATERHOUSE, D. F. *Proc. R. Soc. Lond. (B)*, **162** (1965), 105–20 (composition of scent of *Nezara*, Hemipt.).

[576] GOLDIN, H. H. and KEITH, A. D. *J. Insect Physiol.*, **14** (1968), 887–99 (fatty acid synthesis in isolated mitochondria of *Drosophila*).

[577] V. HANDEL, E. *J. exp. Biol.*, **46** (1967), 487–90 (temperature and saturation of lipids in *Culex*).

[578] HANSER, G. and REMBOLD, H. *Z. Naturforsch.*, **23** (1968), 666–70 (pteridines in cuticle of *Apis*).

[579] HAPP, G. M. *J. Insect Physiol.*, **14** (1968), 1821–37 (quinone and hydrocarbon production in glands of Tenebrionids).

[580] HARPER, E., SEIFTER, S. and SCHARRER, B. *J. Cell Biol.*, **33** (1967), 385–94 (collagen in connective tissue of cockroaches).

[581] HARWOOD, R. F. and TAKATA, N. *J. Insect Physiol.*, **11** (1965), 711–16 (photoperiod and temperature and fatty acid composition in *Culex*).

[582] HASEGAWA, K. and YAMASHITA, O. *J. Sericult. Sci. Tokyo*, **36** (1967), 297–301 (diapause hormone and ovary metabolism in *Bombyx* pupa).

[583] KAFATOS, F. C., TARTAKOFF, A. M. and LAW, J. H. *J. biol. Chem.*, **7** (1967), 1477–94 ('cocoonase' in *Antheraea*).

[584] KAY, D., ROTHSCHILD, M. and APLIN, R. *J. Cell. Sci.*, **4** (1969), 368–80 (protein particles in blood of Lepidoptera &c.).

[585] KENCHINGTON, W. *J. Morph.*, **129** (1969), 307–16 (chitinous hatching thread in *Mantidae*).

[586] KOMATSU, K. *Bull. sericult. Exp. Sta.*, **23** (1969), 513–14 (wax content of sericin).

[587] KULLENBERG, B., BERSTROM, G. *et al. Acta chem. Scand.*, **24** (1970), 1481–3 (chemistry of marking scents of bumble-bees).

[588] KURIAN, K. T. and GILBERT, L. I. *J. Insect Physiol.*, **13** (1967), 963–80 (release and transport of phospholipids in *Hyalophora* and *Periplaneta*).

[589] LEUENBERGER, F. and THOMMEN, H. *J. Insect Physiol.*, **16** (1970), 1855–8 (ketocarotenoids in *Leptinotarsa*).

[590] MASCHWITZ, U., KOOB, K. and SCHILDKNECHT, H. *J. Insect Physiol.*, **16** (1970), 387–404 (phenylacetic acid as an antibiotic in ants).

[591] MAYER, R. J. and CANDY, D. J. *Nature*, **215** (1967), 987 (lipid consumption by flying *Schistocerca*).

[592] MERLINI, L. and NASINI, G. *J. Insect Physiol.*, **12** (1966), 123–7 (pteridines in Hemiptera).

[593] MOHRIG, W. and MESSNER, B. *Acta biol. med. germ.*, **21** (1968), 85–95 (lysozyme as bactericide in honey and venon of *Apis*).

[594] MUNN, E. A. and GREVILLE, G. D. *J. Insect Physiol.*, **15** (1969), 1935–50 (the protein 'calliphorin' in Cyclorrhapha &c.).

[595] O'FARRELL, A. F. *Proc. R. ent. Soc. Lond.* (*C*), **33** (1968), 21 (colour change and temperature in Zygoptera).

[596] OHTAKI, T. and OHNISHI, E. *J. Insect Physiol.*, **13** (1967), 1569–72 (coloration of pupae of *Pieris*).

[597] OLTMER, A. *Wilhelm Roux Arch. Entw. Mech. Org.*, **160** (1968), 401–27 (control of melanin formation in *Pieris* pupa).

[598] PARSONS, J. A. *J. Physiol.*, **178** (1965), 290–304 (digitalis-like toxin in *Danaus*).

[599] PITMAN, G. B. *et al. J. Insect Physiol.*, **15** (1969), 363–6 (pheromones from gut of *Dendroctonus* spp.).

[600] REICHSTEIN, T. *et al. Science*, **161** (1968), 861–6 (heart poisons in *Danaus* derived from the foodplant).

[601] ROBBINS, W. E. *et al. Ann. Rev. Entom.*, **16** (1971), 53–72 (steroid metabolism in insects: review).

[602] ROCKSTEIN, M. and BHATNAGAR, P. L. *Biol. Bull. mar. biol. Lab., Woods Hole*, **131** (1966), 479–86 (wing-beat frequency and duration in ageing *Musca*).

[603] ROTHSCHILD, M. *et al. J. Insect Physiol.*, **16** (1970), 1141–5 (cardiac glycosides in oleander Aphid).

[604] ROUSSEL, J. P. *J. Insect Physiol.*, **12** (1966), 1085–92 (corpus allatum and colour change in *Gryllus*).

[605] ROWELL, C. H. F. *Adv. Insect Physiol.*, **8** (1971), 146–98 (colour variation in Acrididae: review).

[606] RUDALL, K. M. and KENCHINGTON, W. *Ann. Rev. Entom.*, **16** (1971), 73–9 (chemical varieties of insect silks).

[607] SACKTOR, B. *Adv. Insect Physiol.*, **7** (1970), 267–347 (intermediary metabolism with special reference to insect flight muscle: review).

[608] SACKTOR, B. *et al. J. biol. Chem.* **241** (1966), 624–31. *Archs Biochem. Biophys.*, **120** (1967), 583–8 (proline and flight muscle metabolism).

[609] SCHILDKNECHT, H. *et al. Angew. Chem.*, **5** (1966), 421–2 (defensive secretion of *Dytiscus*).

[610] SHEARER, D. A. and BOCH, R. *J. Insect Physiol.*, **12** (1966), 1513–21 (Nassanoff gland pheromone in *Apis*).

[611] SMITH, J. M. *et al. J. Insect Physiol.*, **16** (1970), 601–13 (protein turnover in adult *Drosophila*).

[612] SRIVASTAVA, P. N. and ROCKSTEIN, M. *J. Insect Physiol.*, **15** (1969), 1181–6 (use of trehalose in flight in *Musca*).

[613] STEPHEN, W. F. and GILBERT, L. I. *J. Insect Physiol.*, **15** (1969), 1833–4; **16** (1970), 851–64 (synthesis and use of fatty acids in *Hyalophora*).

[614] STEVENSON, E. *J. Insect Physiol.*, **15** (1969), 1537–50 (monoglyceride lipase in flight muscle of Lepidoptera).

[615] THOMPSON, S. N. and BENNETT, R. B. *J. Insect Physiol.*, **17** (1971), 1555–63 (use of fat in flight by *Dendroctonus*, &c.).

[616] TOBE, S. S. and LOUGHTON, B. G. *J. Insect Physiol.*, **15** (1969), 1331–46; 1659–72 (movements of protein in tissues of *Locusta*).

[617] TOMINO, S. *J. Insect Physiol.*, **11** (1965), 581–90 (phenolic substances in *Bombyx*).

[618] UMEBACHI, Y. and YOSHIDA, K. *J. Insect Physiol.*, **16** (1970), 1203–28 (wing pigments of *Papilio*).

[619] VISCONTINI, M. and SCHMIDT, G. H. *Helv. chim. Acta*, **46** (1963), 2509–16 (pteridines in *Rhodnius*, Hemipt.).

[620] WEATHERSTON, J. Q. *et al.*, *Q. Rev.* **21** (1967), 287–313; in *Chemicals controlling Insect Behaviour*, Academic Press, New York, 1970, 95–144 (chemistry of defensive secretions in insects: review).

[621] WILLIG, A. *J. Insect Physiol.*, **15** (1969), 1907–27 (green colour of *Carausius*).

[622] ZIEGLER, I. and HARMSEN, R. *Adv. Insect Physiol.*, **6** (1969), 139–203 (pteridines in insects: review).

Chapter XIV

Water and Temperature

THE MOST important factors in the environment, which influence the physiology of insects, are temperature and humidity. In their effects, these are constantly reacting upon one another; it is therefore necessary always to consider them side by side.

WATER RELATIONS

Water content—The quantity of water in different insects ranges from less than 50 per cent. to more than 90 per cent. of the total body weight;[26, 56] and there may be much variation within the same species, even when reared under identical conditions. The percentage tends naturally to be low in insects in which the cuticle contributes largely to the total weight; in the adult of *Calandra granaria* the water content is only 46–47 per cent. It is high in the thin-skinned larvae of Lepidoptera, &c.; thus in the large Saturniid caterpillar *Telea polyphemus* it amounts to 90–92 per cent.[26] The proportion of water is influenced to some extent by the quantity of fat present (p. 593); active larvae of *Phlebotomus papatasii* (Dipt.), in which fat composes 5 per cent. of the dry weight, contain 65–70 per cent. of water, hibernating larvae with 15 per cent. of the dry weight made up of fat, contain 52–56 per cent.[122] But here, as in other insects, the increase in fat will not explain the whole of the fall in water content; there is a second factor, a tendency to lose water before entering hibernation or diapause. Thus growing larvae of *Chortophaga viridifasciata* may have a water content as high as 79 per cent.; on entering hibernation this falls to 65 per cent., rising again to 75 per cent. with the restoration of activity.[16] There is generally a fall in water content on pupation and a further fall in the imago; in *Popillia japonica* the larva contains 78–81 per cent. of water, the pupa 74 per cent., the adult, which has a much higher proportion of skeletal material, 66·6 per cent.[77, 80]

The ability to withstand a reduction in the water content varies considerably. Larvae of the beetles *Leptinotarsa*, *Popillia*,[77, 80] *Tenebrio*,[23] the caterpillar *Ephestia*,[119] and the bug *Rhodnius*[25] succumb when the water content falls from about 75 per cent. to about 60 per cent. or a little less. *Chortophaga* dies at 56–59 per cent.[78] The clothes-moth larva *Tineola*, with a normal water content of about 59 per cent., can survive with a content appreciably less.[83] The termite *Termopsis*, which normally has from 74–80 per cent. of water, dies as soon as this falls much below 68 per cent.[29]

Diminution in the water content usually depresses metabolism and retards development. If eggs of *Melanoplus differentialis* are dehydrated by immersion in hypertonic solutions, the oxygen uptake falls, more or less in proportion to the concentration used.[17] Eggs of the weevil *Sitona* at 20° C. develop in 10½ days in saturated air, in 21 days at a relative humidity of 62 per cent., below which they fail to develop altogether;[3] eggs of *Lucilia* at 22° C. develop in 23 hours at a saturation deficiency of 12 mm. of mercury, in a little over 20 hours at a saturation deficiency of 2 mm.;[40] and the duration of the pupal stage in *Popillia* increases as the water content becomes less.[77] At 70 per cent. relative humidity *Ephestia* required 33 days for larval development and had a water content of 73·5 per cent.; at 33 per cent. relative humidity at the same temperature it required 50 days and had a water content of 57·3 per cent.[4] The extreme cases are those in which eggs of some Collembola and grasshoppers remain dormant for months or even years in a desiccated state (pp. 14, 669). [See p. 690.]

Sometimes the effect of desiccation is purely mechanical; the chorion of the egg may become too hard for the insect to break through (p. 2), or the fully developed insect in the pupa may lack sufficient volume of water in its blood to rupture the pupal sheath, as in *Lucilia*.[40] The size of the body and wings in *Lucilia* is largely determined by the water content; much of this water is stored in the salivary glands which may make up as much as 10 mg. out of a total pupal weight of 45 mg.[89] [See p. 690.]

Sometimes the rate of development is retarded at very high humidities; the pupal stage of *Lucilia* is prolonged by about 5 per cent. in saturated air;[40] pupae of *Sitotroga* (Lep.) develop in 12 days at a relative humidity of 22 per cent., in 17 days in saturated air at the same temperature;[51] and pupae of *Bruchus obtectus* (Col.) which require 14 days at a humidity of 45 per cent., require 22 days in saturated air.[51]

Evaporation—Water is lost by insects chiefly through evaporation. According to Dalton's law, the rate of evaporation from a water surface is proportional to the saturation deficiency of the air; as it is sometimes expressed

$$V = a.E(100 - H) + c$$

where V is the rate of evaporation, E is the aqueous vapour tension when the air is saturated at the temperature in question, H is the relative humidity and a and c are constants.[64] Dalton's law is applicable to the rate of loss of water from insects.[24] But there are many exceptions; and the physiology of evaporation is best appreciated by analysing the factors which cause departures from this rule. Some of these factors are physical, some physiological.

(i) One factor operative in still air is perhaps the increasing rate of diffusion as the temperature rises. At a given saturation deficiency, evaporation may therefore be expected to be greater at high temperatures than at low.[107] Thus in experiments of short duration, eggs of *Lucilia* lose more water in proportion to the saturation deficiency at 22° C. than at 14° C.;[40] and likewise in pupae of *Milionia* (Lep.) the evaporation through the integument per unit of saturation deficiency rises with rising temperature.[64]

(ii) High temperature may increase the permeability of the cuticle to water. That may be the explanation of the increasing loss in *Lucilia* eggs, and through the integument of *Milionia* pupae, with rising temperature, just noted. It is certainly the explanation in the cockroach. Other factors being equal, evaporation from the cockroach is proportional to saturation deficiency until the temperature reaches 30° C. There is then an abrupt increase in evaporation which is due to a change in the properties of the oily film that is responsible for the impermeability of the cuticle in this insect;[107] the same has been found in many other insects[131] (p. 36) and in insect eggs (p. 2); and it may account for the increased rate of loss of water in the eggs of *Calandra* and *Rhizopertha* at a given saturation deficiency as the temperature rises.[11]

(iii) In connexion with respiration we have already discussed some of the evidence which proves that the greater part of evaporation from insects takes place through the spiracles (p. 369). Further evidence to this effect has been obtained by sealing the spiracles with vaseline, or by leaving the spiracles open and covering the remainder of the body with vaseline, and comparing the evaporation under these conditions. By such means it has been concluded that in the pupa of *Bombyx mori* the spiracles are responsible for about 66 per cent. of the total loss, and in the adult grasshopper *Gastrimargus* about 70 per cent., the loss through the cuticle taking place particularly at the intersegmental membranes.[64]

Most of the departures from Dalton's law to be observed in living insects are due to this fact that most of the water vapour escapes from the tracheal system (Fig. 359). Thus increasing the rate of flow of air over the cockroach has little effect upon evaporation from the body surface, but it causes a marked increase in evaporation from the tracheal system, probably by creating eddies within the tracheae.[107] The same thing has been observed in the pupae of *Milionia*.[64]

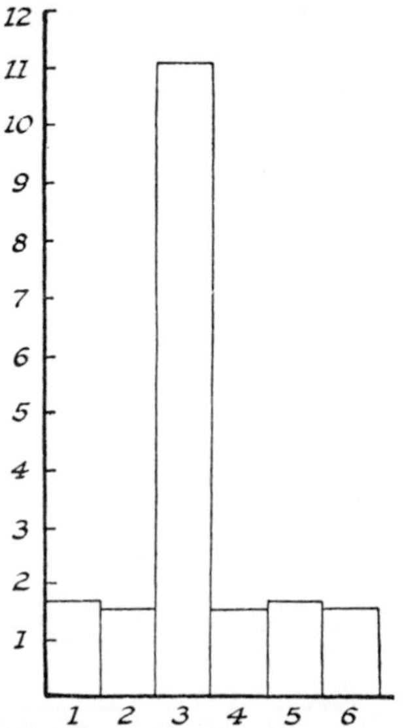

FIG. 359.—Daily loss of watet in adult *Rhodnius* starved a 24° C., 0% Relative Humidity. During the third day the spiracles were kept open by exposure to 5% CO_2 (*after* WIGGLESWORTH and GILLETT)

Ordinate: loss of weight in mg. per day. Abscissa: days.

(iv) If the spiracles are opened more frequently than normal (p. 369), the rate of evaporation will be increased. Thus feeding, exercise, egg production, or any other factor which causes an increase in metabolism, will increase evaporation.[64] At 20° C., under constant conditions of air movement and saturation deficiency, the cockroach *Blatta* lost 3·9 mg. per hour by evaporation while at rest, 6·0 mg. per hour when stimulated.[107] Now increase in temperature is itself a factor which causes an increased ventilation of the tracheal system; a rise in temperature will therefore increase evaporation even when the saturation deficiency remains unchanged.[64] In the tsetse fly *Glossina morsitans* the degree of spiracular regulation is influenced by the physiological state of the insect; when water reserves are low, regulation is more stringent. Spiracular control is markedly influenced also by humidity: in dry air the spiracles are held more completely closed than in air at 90 per cent. relative humidity.[149]

(v) A departure from Dalton's law is also seen sometimes in very dry air. In the mealworm no effect of this kind can be discovered.[83] But in hibernating larvae of *Chortophaga*, the rate of water loss falls off when the saturation deficiency exceeds 20 mm. of mercury.[16] In *Cimex*, also, the rate of loss becomes less than expected in dry air.[83] In the pupa of *Milionia* at 20° C. a similar break occurs at 11 mm. of mercury saturation deficiency.[64] Two explanations are offered for this phenomenon: (*a*) that an active regulation begins at very low humidities, the insect keeping its spiracles even more rigorously closed than usual;[64] (*b*) that in very dry air, water vapour escapes so rapidly from the tracheal system that the air in the tracheae becomes relatively dry, and the rate at which water will diffuse through the tracheal walls becomes the limiting factor in water loss; the level of dryness at which this occurs being peculiar to each species.[86] In the egg of the Australian locust *Austroicetes* the water exchanges are influenced by the vital activity of the protoplasmic layer;[12] we have discussed already the functions of the 'hydropyle' in locust eggs (p. 4).

(vi) Departures from the law may also be encountered at the other end of the humidity scale.[26] *Milionia* pupae lose less water than expected in very moist air, particularly at high temperatures.[64] The clothes-moth larva *Tineola*,[83] the mealworm larva,[83] and the grasshopper *Chortophaga* (Fig. 360)[78] actually increase their water content, they may even increase their total weight by 10 or 15 per

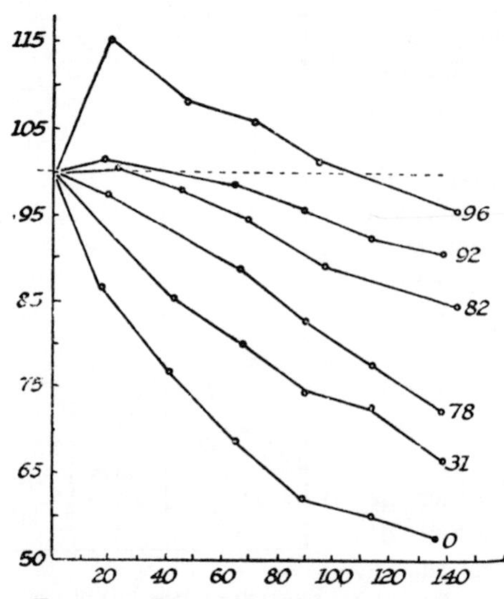

FIG. 360.—Rate of loss of weight in *Chortophaga viridifasciata* at the different relative humidities indicated at the end of each curve (*after* LUDWIG)

Ordinate: weight as percentage of original weight. Abscissa: time in hours.

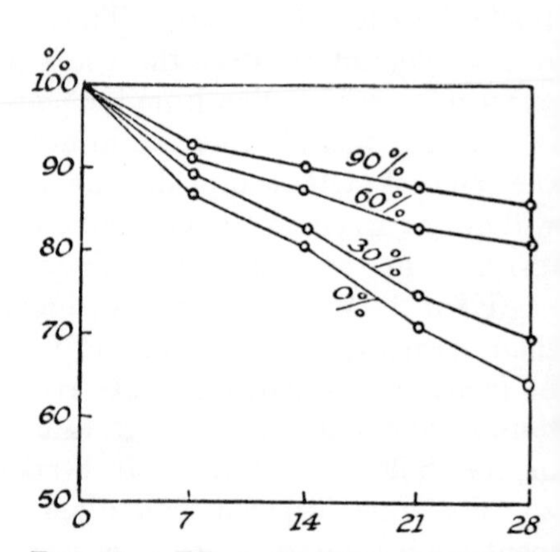

FIG. 361.—Effect of relative humidity on evaporation from the bed-bug *Cimex lectularius* at 8° C. (*after* MELLANBY)

Ordinate: percentage of original weight. Abscissa: days. Figures on the chart show the four relative humidities employed.

cent., when starved at high humidity. This property, like the absorption of water by hygroscopic substances, depends upon relative humidity, not saturation deficiency.[83] In *Chortophaga* this uptake ceases below about 82 per cent. relative humidity;[78] in *Tenebrio* below 88 per cent.[83] In the pupating larva of the flea *Xenopsylla* enclosed in its cocoon, water vapour may be absorbed from the air at a relative humidity as low as 45 per cent. This process continues until pupation and it may result in an increase in weight up to 29 per cent.[39] Water is also taken up from moist air by eggs of the Luna moth *Tropaea* and the Saturniids *Hyalophora*, *Telea*, &c.[81] It can be observed in *Tenebrio* even when there is no possibility of the insect reingesting its hygroscopic excrement;[132] there seems in fact to be an active absorption of water, perhaps in the tracheal endings,[83] perhaps, as occurs in ticks (Acarina), through the general surface of the cuticle. If the *Tenebrio* larva is enclosed in a small air-tight chamber, whatever the initial humidity, it comes into equilibrium at a relative humidity of 90 per cent. or so;

when the insect dies, complete saturation occurs.[63] Not all insects are capable of this; it does not happen in *Cimex* (Fig. 361),[83] *Termopsis*,[29] or larvae of *Popillia*,[80] even in a saturated atmosphere. [See p. 690.]

Water balance—If the water content is to remain constant, the gain of water must equal the loss. *Gain of water* may take place: (i) with the food. This has already been discussed (p. 513); but a further example may be noted. If the bed-bug *Cimex* has been desiccated, it restores its normal body composition by retaining more water than usual at its next meal. It can therefore thrive in any atmosphere provided it is fed sufficiently often.[83] Many insects drink water if it is available.[171]

(ii) By absorption of liquid water through the cuticle (p. 36). In *Onychiurus* (Collembola) the 'ventral tube' on the abdomen seems to be specially concerned in the absorption of liquid water.[97] In *Podura aquatica* water uptake occurs both by mouth and by the ventral tube vesicles; the everted vesicles alone will ensure recovery from desiccation. (Sodium is also taken up by the vesicles, but not potassium.)[175] The same role is played by the eversible vesicles at the hind margin of the sterna of one or more of the abdominal segments in Diplura and Thysanura; in *Campodea* there are six pairs of these vesicles, everted by blood pressure and retracted by muscles.[157]

If desiccated Syrphid larvae are placed in contact with water they bring the anal extremity into contact with this, and extrude the rectal lobes. These absorb the water so actively that the larva may gain more than 50 per cent. in weight in the course of a few hours.[115] The cuticle of larvae in the soil is rendered permeable by abrasion (p. 36). Larvae of *Agriotes* are in osmotic equilibrium with a solution of sucrose of 0·33 M. In fairly dry soils the vapour pressure is depressed because the water bound in the soil pores has a pressure less than that of the atmosphere. This vapour pressure depression or 'suction' is expressed on the *p*F scale.* If wireworms are placed in soils of different water content, they are found to be in equilibrium with soil of *p*F = 3·9 (which is equivalent to 0·33 M sucrose).[41] The largest aggregation of adult beetles and eggs of *Aphodius howitti* is always at a water content giving a *p*F of 2·8 irrespective of soil type.[167]

(iii) By absorption of water vapour from the air as described in the last section.

(iv) By the oxidation of hydrogen in the foodstuffs or reserves, and the retention of the so-called 'metabolic water'. Thus the complete combustion of 100 gm. of fat will yield 107 gm. of water. This is certainly an important source of water in insects such as the mealworm[23] or clothes-moth[5] living in very dry materials. At low humidities more food is eaten to produce a given unit of body weight because part is utilized for the production of water. In *Dermestes* at 30 per cent. relative humidity less than 33 per cent. of the water in the pupa can be derived from water ingested with the food; in *Ephestia* at 1 per cent. R.H., less than 7·6 per cent.[44] It has even been suggested that such insects may increase their rate of metabolism during desiccation in order to make good their loss of water.[23] But it is doubtful if this mechanism could be of service; for increased metabolism necessitates increased respiration, which itself involves an increased loss of water.[87]

Loss of water may take place: (1) by evaporation through the cuticle and the

* *p*F represents the logarithm of the height in cm. of a column of water corresponding to the suction. Since *p*F depends on the texture of the soil it will vary widely in different soils with the same water content.[41]

tracheal system as discussed in the last section. The pupa of *Glossina morsitans* can complete its development at humidities as low as 10 per cent. relative humidity.[150] (ii) By the faeces; insects which retain their excrement, or are efficient at extracting water from it before it is discharged (p. 496), are the most resistant to desiccation. The adult *Glossina morsitans* (like *Cimex*, p. 667) can retain more or less of the water in the blood meal according to the conditions in the environment; in flies maintained at a high relative humidity the water content of the faeces is 75 per cent.; in dry air 35 per cent.[149] [See p. 691.]

When the normal water balance is upset, it is usually in the direction of desiccation; the retention of water being one of the chief problems before all small terrestrial animals. The bed-bug *Cimex*, for example, always loses water unless exposed to an almost saturated atmosphere.[83] But in some insects adapted to very dry conditions the water balance may be upset in the opposite direction. Thus the mealworm, which is able to maintain its water content unchanged even when starved for a month in absolutely dry air,[10, 23] is so well adapted to retain water, that if it is exposed to moderately moist air (relative humidity 70 per cent.), at a temperature of 30–37° C. at which the metabolic rate is high, it is unable to get rid of its water of metabolism, and hence the proportion of water in its body rises excessively.[83] The larva of the clothes-moth *Tineola* occupies an intermediate position; if starved in dry air its proportion of dry matter increases; if starved at a relative humidity of 90 per cent., its water content rises.[83]

Desiccation and survival—The length of time insects are able to survive under adverse conditions may be influenced by many factors, such as temperature, or starvation, as well as desiccation. A flying Aphid *Aphis fabae* (at 25° C. and 70 per cent. R.H.) loses about 1 per cent. of its body weight per hour by evaporation. Indeed, even in dry air, water loss is not a limiting factor in the duration of flight.[153] Larvae and adults of *Chortophaga viridifasciata*, at a given temperature, will survive 5–6½ days regardless of the humidity; here starvation is the cause of death.[78] Whereas survival of *Popillia* larvae varies with the humidity; they will survive less than 4 days at 0 per cent., 23 days at 82 per cent. humidity, death occurring when the water content is reduced from 81 per cent. to 55–59 per cent.[80] When all other factors save loss of water are excluded, the time of survival should be inversely proportional to the saturation deficiency; but since varying temperatures influence both the rate of metabolism, and the rate of water loss through the spiracles, this relation is seldom realized in practice.[86] Where the number of survivals among batches of developing eggs or pupae is being related with saturation deficiency, the duration of development, that is, the time of exposure at each set of conditions, must be taken into consideration. The true measure of evaporative power in such an experiment is the product of time × saturation deficiency. When this time factor is taken into account, it can be shown that the percentage of survivors among eggs of *Habrobracon* or pupating larvae of the flea is in fact inversely proportional to the saturation deficiency.[86]

There is even better agreement if it be assumed that there is a small constant loss of water (a) which is independent of the evaporating power of the air. The formula $x=Ky$, where x is rate of water loss, y is saturation deficiency, and K is a constant, becomes $x=a+Ky$.[61] Now the time necessary for development falls off rapidly with rise of temperature at first, more slowly later (p. 684); while the saturation deficiency increases with temperature more rapidly as the temperature rises. Hence the product of time × saturation deficiency, first falls and then rises; so that with a given water content in the air, the optimal conditions for development will be at some intermediate zone of temperature where this product is smallest (Fig. 362).[75, 86] At low temperatures, the development in the eggs of Saturniid moths may be so slow that they lose more water than at high temperatures.[81] We have already seen that increasing temperatures may cause increased permeability and then the mortality at any particular value of the product saturation deficiency × time becomes higher as the temperature rises.[11]

The humidity conditions to which the insect is exposed are largely influenced by its behaviour; in general, insects avoid a very high humidity or a very low (p. 293). *Schistocerca* hoppers are agitated by low, and even more by very high humidities; they become quiescent within a zone of 60–70 per cent. R.H. where they spend most of their time.[144] There are clear responses to humidity differences in *Drosophila* of both sexes; 3 per cent. R.H. differences are readily perceived at the moist end of the scale; the preferred humidity is 77 per cent. R.H.[178] As desiccation proceeds *Tenebrio* adults show a preference for higher humidities; after allowing access to water the original dry preference is shown.[156]

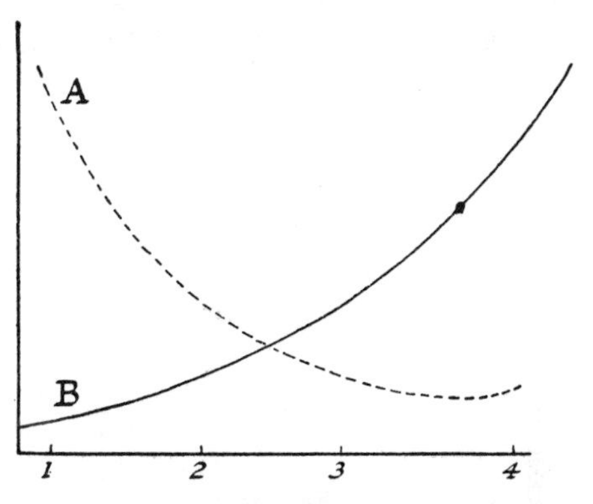

Fig. 362.—Chart to illustrate the effect of temperature on *A*, duration of development, and *B*, saturation deficiency (*after* Mellanby)

At temperature 1, development is stopped by cold, at 4 it is stopped by heat; the optimal zone of temperature is between 2 and 3, where the product of saturation deficiency and duration of development is lowest. (In many insects the relation is complicated by more frequent opening of the spiracles as the temperature rises.)

Cryptobiosis—The foregoing section relates to the majority of insects, which must retain a water content of something like 50 per cent. if they are to survive. But certain insects are able to survive in a state of suspended animation, or 'cryptobiosis', after losing virtually all their water. This happens in certain insect eggs (p. 14). The haemoglobin-containing Chironomid larva, *Polypedilum*, breeds in rock pools in West Africa which are liable to dry up. The larva also dries up in the mud and can survive in the desiccated state, with a water content below 8 per cent., for several years. In contact with liquid water they swell up and are crawling and feeding within half an hour or so. In this desiccated state *Polypedilum* can withstand temperatures of −190° C. (liquid air) for 3 days and 102–104° C. for at least 1 minute.[163] The capacity for cryptobiosis may be more common than is generally supposed; the blood cells in isolated gills of *Sialis* larvae can be desiccated for some months and show some vital activities, such as clotting behaviour, if re-hydrated.[187]

TEMPERATURE RELATIONS

Heat balance and body temperature—In the insect at rest, over a moderate temperature range, heat production in metabolism balances heat loss, and the body temperature is the same as that of the surroundings.[45, 47] In *Periplaneta* that is so between 10–22° C.,[94] in the Scarabaeid *Anomala expansa* between 15–30° C. and in the Libellulid *Orthetrum* between 12–23° C.[64]

Loss of heat is effected almost entirely through evaporation. In *Anomala*, and in the Psychid larva *Clania*, it is estimated that between 10–30° C. evaporation accounts for 80–100 per cent. of the total loss of heat, conduction and radiation for only 0–20 per cent.[64] It follows therefore that any factor which affects evaporation will influence the body temperature. At low temperatures, evaporation is depressed and the temperature of the body is often greater than that of the surroundings; at 5·5° C. the body temperature of the bee averages 4·7° C. above the surrounding air.[102] At high temperatures, evaporation is increased to a greater extent than heat production, and the body temperature falls below that of the air. In the cockroach this happens above 22° C.; it becomes increasingly evident above 32° C.[94] At an air temperature of 23° C. the body temperature of the Noctuid larva *Prodenia* is about 21·5° C.[6] Air movement increases this cooling effect. The humidity of the air, also, greatly influences the body temperature through its effect on evaporation: except at low temperatures the body is always warmer in moist air than in dry;[64] in saturated air the body temperature of the cockroach is invariably above that of the surroundings.[94] Further, the cooling effect of evaporation is much more evident in live individuals, in which the tracheae are ventilated by respiratory movements. The Tenebrionid beetle *Adesmia*, alive on soil with a surface temperature of 38° C., had a body temperature of 36° C.; with the soil at 44° C. the body was at 39·5° C.; but in dead individuals the body was always at least 2° C. hotter than in the living.[22] Enforced opening of the spiracles in *Glossina morsitans* by exposure to 20 per cent. carbon dioxide in air is accompanied by a small depression of body temperature (about 0·6° C.).[159]

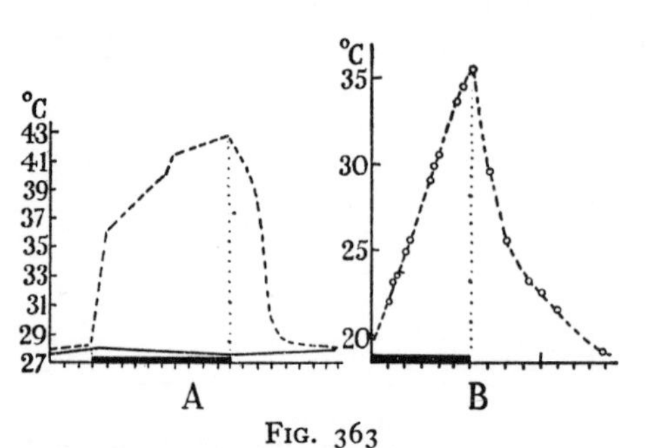

FIG. 363

A, effect of insolation on body temperature of *Locusta* (from UVAROV *after* STRELNIKOV)

Continuous line is the air temperature; broken line is body temperature; thickened base line shows period of exposure to sunlight.

B, effect of fluttering of wings on body temperature of *Vanessa* (*after* KROGH and ZEUTHEN)

Ordinates: temperature in ° C. Abscissae: time in minutes.

Gain of heat is chiefly the result of metabolism, particularly of muscular contraction. Gregarious larvae of *Vanessa* when clustered together may have their temperature raised 1½–2° C. and this is sufficient to shorten the time of larval development by 1–3 days.[93] As was shown by Girard the body temperature of insects is raised above that of the air as soon as they become active; Hymenoptera, it was found, might show an increase of 4° C.; in Sphingids, the temperature in the thorax might be as much as 10° C. above that in the abdomen during vigorous flight.[45] A male of *Saturnia pyri* at an air temperature of 18° C. raised its thoracic temperature to 26° C. during fluttering.[7] Sphingids are said to be unable to take flight until their body temperature has been raised to 32–36° C. by preliminary fluttering, and during flight a temperature as high as 41·5° C. has been recorded in one insect.[37] *Celerio lineata* commonly regulates its temperature during flight between a minimum of 34·8° C. and a maximum of about 38° C.; but it is capable of flight at thoracic temperatures as low as 25° C.[140] Perhaps the thick covering of hairs is of some importance in preventing loss of this heat by conduction.[37] In *Vanessa* and *Bombus* vibratory movements of the wings raise the temperature of the wing muscles above 30° C.; *Vanessa* can fly with the wing muscles at temperatures from 20–42° C., but for high rates the muscles must be heated above 35° C. (Fig. 363).[72, 190] In *Geotrupes*, which requires a temperature above 32° C. before flight can begin, there is no visible movement of the wings during the period of warming up; but the muscles are actively contracting none the less—as may be shown by leading off the action potentials.[72] [See p. 691.]

Study of the regulation of body temperature in flying insects has cast some doubt on the generalization that heat loss in insects is almost all effected by evaporation. Their structure and reactions are adapted for the conservation of water rather than for rapid cooling. In the flying locust, with the thoracic temperature rising towards 40° C., no more than 10 per cent. of the heat loss is effected by evaporation, and only 10–15 per cent. escapes by long-wave radiation in a large insect, flying under a clear sky. 60–80 per cent. of the heat is dissipated by convection. The coats of hair or scales in *Bombus*, *Sphinx*, and *Triphaena* provide improved insulation, and increase the temperature excess in the thorax by 50–100 per cent., perhaps 8–9° C. in a large Sphingid. In Odonata the pterothorax is insulated nearly as effectively by the subcutaneous air sacs.[152]

Radiant energy from the sun is a source of heat in many insects; both visible light and infra-red rays will serve to heat the body above the surroundings.[38] The temperature of *Bombus* rose from 28·7° C. in the shade, to 41·6° C. in the sun, in the course of five minutes, some of this increase being due to increased activity; and a passing cloud caused a rapid fall in temperature.[121] Hoppers of *Locusta* in the shade at 28° C. had a body temperature of 28° C.; on exposure to the sun their temperature rose to 36° C. in the first minute and after 10 minutes it had reached 42·7° C. After shading, the temperature returned to 28° C. in 6 minutes (Fig. 363).[127] Flight begins in *Schistocerca* when the internal temperature reaches 35° C.; it is really active only between 37–44° C.[128] Many butterflies (*Erebia*, &c.) in high altitudes, are dependent on this source of heat, and are quite incapable of flight unless the sun is shining. *Argynnis paphia* controls its body temperature by opening the wings to a varying extent with their surface directed towards the sun. In this way they maintain their temperature between 32° and 37° C. (as much as 17° C. above the surroundings) and they show full activity only within this optimal range. When the temperature rises too high they close the wings above the back, and finally retreat to a shaded spot.[193] [See p. 691.]

The body colour influences the heat absorption; the dark brown race of the Acridid *Calliptamus*, when exposed to the sun, was found to have a body temperature 4–5° C. higher than the buff race of the same species.[22] Differences in the internal temperature of black (gregarious) and green (solitary) hoppers of *Locusta* under identical conditions may be as much as 6·6° C.[127] But the same colours which favour the uptake of heat during the day will favour the loss of heat by radiation during the night.[109] No appreciable differences in temperature response to solar radiation could be observed between the black and buff forms of *Melanoplus differentialis*.[177]

The insect may control its temperature to some extent by changing its position in relation to the sun; locust hoppers in Palestine, with the body perpendicular to the sun's rays, had a body temperature of 41·7° C., with the body parallel to the sun's rays, 38·3° C.[42] They not only turn sideways to the sun but incline over to one side and lower the abdomen below the elytra to expose the maximum surface area; and when facing the sun the fore-part of the body is raised. By these means a ratio of 5 or 6 to 1 is obtained in the surface area exposed in the two postures. In one series of observations on *Schistocerca* the body temperature exceeded that of the air by 6–7° C. when facing the sun, by 16–17° C. in the lateral position. The insect also adopts intermediate positions as the temperature changes.[128] The same response is shown by tsetse flies *Glossina*.[59]

The elytra of many beetles, and the wings of Lepidoptera, show a selective reflection for infra-red rays with a wave length of about 1μ, a region which is of high intensity in solar radiation.[110] This may afford some protection against excessive heating by the sun; and the subelytral air space of beetles is a considerable protection;[73] on the other hand, metallic and interference colours appear to have no influence on the absorption of heat.[38]

Temperature regulation in Hymenoptera—It is evident that in many insects the fluctuations of body temperature are subject to some degree of regulation.[15] A much more definite regulation exists among the social Hymenoptera.[55] Among ants this consists only in opening and closing the entrance to the nest according to the temperature, and in selecting the site for the nest—under stones or with a cap of earth to collect and retain the heat of the sun.[106, 120] The wasp *Polistes* likewise obtains the heat necessary for its brood by solar radiation as well as from the air; but if the temperature becomes excessive, it cools the nest by fanning with its wings (at 31·5–35° C.), and often combines this with bringing water and applying it to the comb (at 34–37·5° C.).[120] *Vespa* also brings water to cool the nest by evaporation, or it may use fluid which it causes the larvae to emit.[130] In *Apis*, the hive is maintained during the summer at a temperature around 34–35° C. If it rises above this it is cooled by fanning and by the carriage of water. During the winter, the temperature of the hive goes through a series of cycles (Lammert's cycles) each lasting about 22–23 hours. The temperature falls to a minimum of 13° C., rises rapidly to a maximum of 24–25° C. and falls very gradually again to 13° C. The bees on the surface of the cluster are stimulated to muscular activity by a temperature of 13° C.; activity ceases again when the temperature has risen. It is estimated that in a hive of average size, 20 gm. of sugar, yielding 80 calories, may be consumed during one Lammert's cycle; if the outside temperature is unduly low the bees draw excessively upon their winter stores.[116] The thoracic temperature of active honey-bees is usually about 10° C. above the surroundings; it is not allowed to rise above 36° C. a figure which agrees closely with the normal thoracic temperature during flight in *Celerio*.[140] Bacterial fermentation in the 'fungus gardens' of *Macrotermes* may play a part in maintaining the high constant temperature of about 30° C. in the nest cavity.[137]

Resistance to high temperature—The resistance of insects to high temperatures is complicated by the interaction of other factors. In nature insects may rest in spots cooled by evaporation, and are thus able to exist in environments far above the lethal temperature; even in laboratory experiments this habit may give an entirely false impression of their resistance.[84] The humidity of the air has a great effect on resistance: the cockroach *Periplaneta* dies at 38° C. in moist air; in dry air it can survive at a temperature of 48° C., being able to cool itself for a time by evaporation.[94] But this ability exists only in insects above a certain size. For the rate at which heat is taken up by the body is proportional to the surface *area*, whereas heat loss, being brought about as we have seen by evaporation, is proportional to the *volume* of water evaporated. As the insect becomes smaller, the ratio of volume to surface becomes less, and below a certain size it could only lower its body temperature appreciably by evaporating more of its water than it could afford to lose.[82] Thus in such small insects as the louse *Pediculus*, the flea *Xenopsylla*, or the blow-fly *Lucilia*, the humidity of the air makes no difference to the lethal temperature when the duration of exposure does not exceed one hour. In larvae of *Tenebrio*, humidity has no influence in the case of small individuals; these die just below 42° C. at all humidities; but larvae

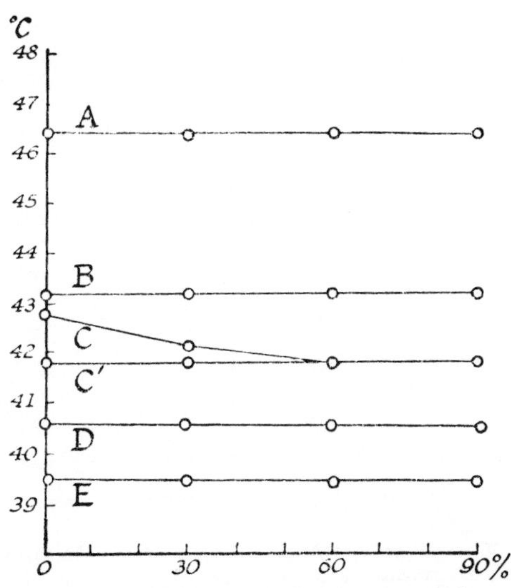

FIG. 364.—Highest temperatures at which certain insects can survive exposure for 1 hour with air of different humidities (*after* MELLANBY)

Ordinate: temperature in ° C. Abscissa: relative humidity. A, *Pediculus humanus*; B, *Lucilia sericata*; C, *Tenebrio molitor* larvae over 100 mg. weight; C′, the same, under 30 mg. weight; D, *Xenopsylla cheopis* adults; E, *X. cheopis* larvae.

weighing over 100 mg. can resist 43° C. in dry air, since they are able to cool themselves to some extent (Fig. 364).[82] [See p. 691.]

In experiments of a longer duration, such as 24 hours, humidity may have the opposite effect; the greater evaporation in dry air may result in the insect dying from desiccation; therefore in long exposures it may withstand a higher temperature in moist air than in dry: *Blatta orientalis* withstands 37–39° C. for 24 hours in moist air, 34–36° C. in dry air.[48] If the insect retains its water efficiently, this effect is not apparent: in experiments lasting 24 hours mealworm larvae can resist 38 5° C. at 90 per cent. and 0 per cent. relative humidity; adults of *Xenopsylla* can resist 38° C. at both humidities; whereas the louse *Pediculus* can withstand 38° C. at 90 per cent., 33° C. at 0 per cent., *Lucilia* 36° C. and 32° C., larvae of *Xenopsylla*, which are quite unable to resist desiccation, 36° C. and 22° C. (Fig. 365).[82] The combined effects of temperature and humidity in affecting survival may be illustrated in the form of a 'thermohygrogram of mortality' as shown in Fig. 366.[2, 65]

Increase in temperature also stimulates metabolism; consequently, in some insects, exhaustion of the food reserves may be the real cause of death at high temperature. In nymphs of *Pediculus* the state of nutrition makes no difference to the lethal temperature in 1 hour experiments, all die at 46·5° C.; but in 24-hour experiments the unfed nymphs die at a lower temperature than those which have been fed.[85] The same is seen in *Trialeurodes*.[129]

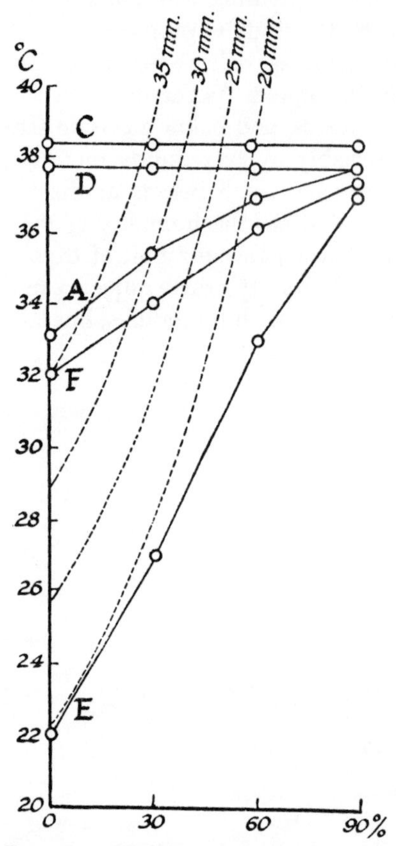

FIG. 365.—Highest temperatures at which certain insects can survive exposure for 24 hours with air of different humidities (*after* MELLANBY)

Ordinate: temperature in ° C. Abscissa: relative humidity. Lines of equal saturation deficiency are also drawn. Lettering as in Fig. 364. *A*, *Pediculus*; *C*, *Tenebrio* larvae; *D*, *Xenopsylla* adults; *E*, *Xenopsylla* larvae; *F*, *Anopheles* adults.

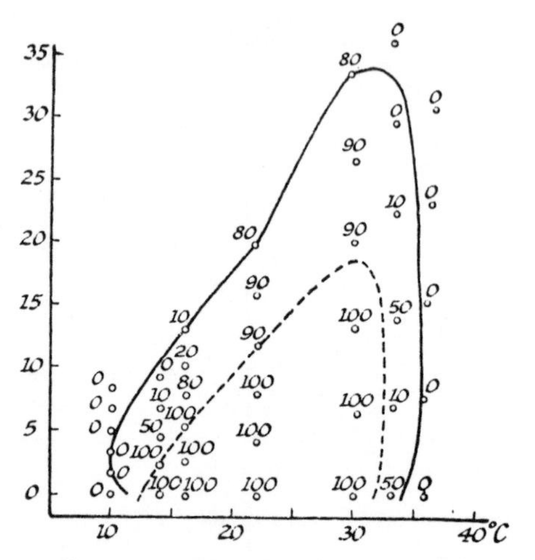

FIG. 366.—Mortality limits of pupae of *Lucilia sericata* at different combinations of temperature and humidity (*after* EVANS)

Ordinate: saturation deficiency in mm. of mercury. Abscissa: temperature in ° C. Figures represent the percentage emergence. The broken line encloses the region of 100% emergence, the continuous line that of 100% development.

Thermal death points—The actual temperatures which the tissues can withstand when these complicating factors have been excluded vary considerably from one species to another. Chironomid larvae have been found breeding in the water of hot springs with a temperature of 49–51° C.;[20] the larva of the cheese-fly *Piophila casei* will resist 52° C. for 1 hour.[118] In 1-hour experiments the thermal death point of *Pediculus* is 46·5° C., *Lucilia* 43° C., *Xenopsylla* adults 40·5° C. and larvae 39·5° C.[82] Heat stupor, which is the prelude to death, occurs in the adult flies of *Musca* from 45–46·5° C., in *Stomoxys* at 43·5° C., in *Fannia* at 39–41° C.[96]

These figures bear little relation to the type of environment in which the insects occur. There is some relation of this kind among Ephemerid nymphs; those from stagnant waters are resistant to higher temperatures. The following figures represent the temperatures at which there was a 50 per cent. mortality after 24 hours' exposure. *Baetis* from swift stream: 21·0° C.; *Ecdyoneurus* from stones at side of stream: 26·6° C.; *Caenis* from slow weed-choked stream: 26·7° C.; *Cloëon* from weedy pond: 28·5–30·2° C.[130a] Similar results have been obtained in Chironomids.[128a] Larvae of the same species of mayfly may be differently affected by temperature, depending on their habitat. After keeping at 40° C. for seven days, 45 per cent. of *Leptophlebia crassa* larvae from a pond were dead; 100 per cent. of the same species from a stream were dead.[135] There is a striking correlation with the environment in *Grylloblatta* and *Thermobia*. *Grylloblatta* is normally active from −2·5° to 11·5° C.; on raising the temperature it shows increased activity up to 18° C., when it becomes stuporous; at 20·5° C. it is irreversibly damaged by heat. *Thermobia* is active in the range 12–50° C.; it is irreversibly injured by cold at 1° C.; at 50° C. activity decreases and heat injury is apparent at 51·3° C.[160]

If *Eristalis* adults are exposed to changing intensities of illumination the mean lethal temperature falls from 250 foot candles to a minimum at 600 foot candles and rises to a maximum at 1,600 foot candles.[134]

The cause of death from high temperature is uncertain. It probably results from an upset in the balance of some metabolic process leading to the accumulation of some product more rapidly than it can be removed.[58, 74] Blow-fly larvae exposed to lethal high temperatures (*Phormia* 45° C., *Calliphora* 39° C.) show an increase in lipoid phosphorus, inorganic phosphorus, and adenyl pyrophosphate in the haemolymph.[58] It has been suggested that the thermodynamic properties of the hydrogen bond may provide an explanation for the upper thermal limits of life.[146] [See p. 692.]

Preferred temperature—In spite of the fact that insects are poikilothermic, their body temperature normally following pretty closely that of the environment, they are to a great extent orientated by temperature. When offered a wide range of temperatures, young feeding larvae of *Musca* show a 'thermopreferendum' of 30–37° C. (which is partly a direct response to temperature, partly a response to products of the fermentation which this temperature has induced in the food medium); by the 6th day of larval life, when they are full grown, these larvae all congregate below 15° C. Larvae of other dung breeding flies have a preferred temperature which accords closely with that of their normal breeding places: *Lyperosia* 27–33° C., *Stomoxys* 23–30° C., *Haematobia* 15–26° C.[123] The louse *Pediculus* has a preferred temperature of about 29° C. which is the temperature of its normal environment (Fig. 367) (p. 327). *Grylloblatta* has a preferred temperature at about 1° C.; these insects can remain active down to at least −5·6° C. and probably lower.[162]

This choice of temperature may be greatly influenced by the conditions to which the insect has been previously exposed: the ant *Formica* showed a preferred temperature of 23–24° C. after exposure to 3–4° C.; but congregated chiefly between 31–32° C. when accustomed to a temperature of 25–27° C.[53] On the other hand, hungry females of the mosquito *Culex fatigans*, when offered two alternative temperatures, always avoid the higher, even down to 15° C. or lower.[124] By selective breeding it has been possible to isolate three strains of the Chalcid *Microplectron* with temperature preferences at 25° C., 15° C., and 9° C.[133]

At low temperatures, the exact level being different for each species, the insect comes to rest and shows no spontaneous activity. When the insect is exposed in a temperature gradient this may result in its becoming trapped at the cold end.[34] As the temperature is raised it becomes normally active, then excessively active, and ultimately it passes into a state of heat stupor followed by death. In *Glossina*, 21° C. is the lowest temperature for spontaneous flight; 14° C. the limit for flying when stimulated; 10° C. for crawling; 8° C. prevents all movement.[88] The physiological differences between species were clearly seen among the insects on a sand dune, the order of their activities during the day being determined largely by the procession of temperature:[28] normal activity began in *Geopinus* at 7° C., *Melanoplus* 10° C., *Sphex* 17° C., *Cicindela* 18° C., *Chlorion*, *Gryllus* 20° C., *Bembex* 26° C., *Microbembex* 30° C.[126] The temperature

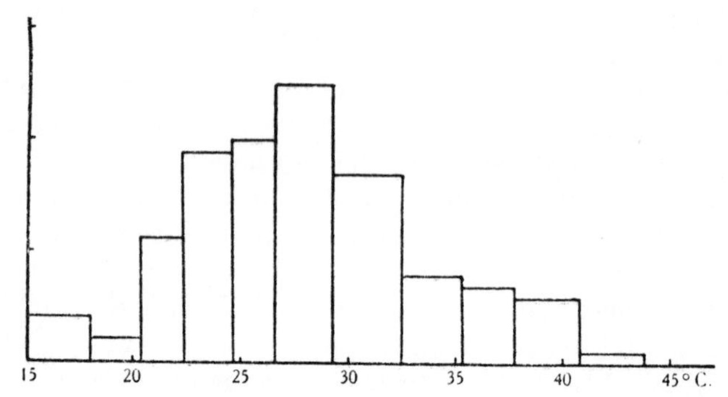

FIG. 367.—Relative numbers of lice *Pediculus* collecting within different sections of a temperature gradient (*after* WIGGLESWORTH)

at which activity begins, and the preferred temperature, seem to bear some relation with the thermal death point. We saw in the last section that the thermal death point in *Musca* was higher than in *Stomoxys*, which was higher than in *Fannia*. The preferred temperature shows the same sequence: *Musca* 33·4° C., *Stomoxys* 27·9° C., *Fannia* 23·7° C.[75, 96] It is interesting to note that *Musca* and *Stomoxys* adults stand in the same relation as do their larvae.[123] But here again, the temperature required for a given level of activity is by no means fixed for a given species: in blow-flies, for example, the effect of any temperature is determined largely by the conditions to which the insect has just been exposed.[95]

Temperature adaptation takes many forms. The activity level of *Ptinus tectus* at a given temperature depends on the temperature at which it has been kept during the previous weeks.[49] The preferred temperature in the wireworm *Limonius* varies with the season, rising to 23° C. in the late summer and falling to 18° C. in the winter and spring.[27] But the most striking examples of adaptation are those which take place in the temperature of immobilization, the 'chill coma' temperature, which is greatly affected by the immediate past history of the insect. The following are the temperatures of activity (that is, 1° C above the chill coma temperature) in insects removed from the temperatures named:

	From 14–17° C.	*From* 30° C.	*From* 36° C.
Blatta	2·0	7·5	9·5
Cimex	4·5	7·0	7·5
Rhodnius . . .	8·6	10·5	12·0

Most species require exposure to a given temperature for about 20 hours before the chill coma temperature becomes fully modified. No acclimatization occurs if the insect is in a state of chill coma; it must be capable of movement. For example, cockroaches from 30° C. remain immobilized for 5 days at 2–3° C., whereas individuals from 15° C. can crawl normally at this temperature. Recovery of movement is very rapid if they have been cooled exactly to the chill coma temperature; much slower if they have been cooled several degrees below. Thus, *Blatta* were immobilized at 1° C. for 24 hours and then transferred to 15° C. They recovered in less than 1 minute if they had previously been acclimatized to 15° C.; they required $1\frac{1}{2}$–$2\frac{1}{2}$ hours if they had been acclimatized at 30° C.[90]

The chill-coma temperature cannot, of course, be lowered below a certain level; in *Blattella* it is the same after acclimatization at 10° C. as at 15° C. (Fig. 368).[154] In intact *Periplaneta* the heart continues to beat well below the chill coma temperature of the animal: to 1·8° C. in cold-adapted and to 4° C. in warm-adapted insects.[181] *Aëdes aegypti* larvae reared at 30° C. failed to recover after exposure to 0·5° C. for 17 hours; but nearly all did so if they were acclimatized for 24 hours at 20° C. before exposure to the low temperature.[170] A similar adaptation is seen in the larva of *Corethra*.[179] Perlid nymphs and *Aëdes punctor* larvae from the Arctic remain active at 0° C. whatever their previous conditions of temperature.[90]

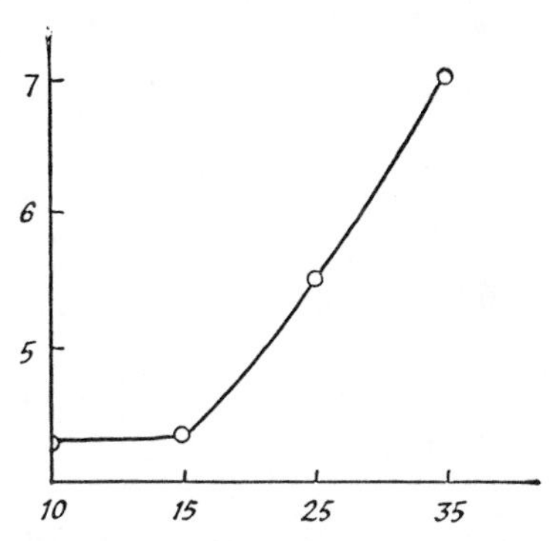

FIG. 368.—The relation between the chill-coma temperature in *Blattella* and the temperature of acclimatization during the preceding 24 hours (*after* COLHOUN)

Ordinate: chill-coma temperature. Abscissa: acclimatization temperature.

If groups of *Periplaneta* are 'cold-adapted' at 10° C. and 'warm-adapted' at 26° C. for 1–3 weeks and the oxygen consumption of each group is then measured at a series of temperatures between 10° C. and 25° C., the 'cold-adapted' insects are found to consume more oxygen per gm. per hour than the 'warm-adapted'.[155] The nature of the adaptation change to low temperature is unknown; in *Periplaneta* and *Tenebrio* the activity of adenosine pyrophosphatase (apyrase) was higher in the cold-adapted than in the warm-adapted insects [173]

A similar acclimatization can occur with respect to the heat-coma point and thermal death point, but the range over which the thermal death point can move is much smaller, and adaptation occurs only when the insect is exposed to temperatures near the top of the temperature range. Thus *Tenebrio* larvae kept at 30° C. have a thermal death point at 42° C.; kept at 37° C. for 24 hours, the thermal death point is 44° C. And in *Aëdes aegypti* larvae the figures are identical: 30° C. (42° C.); 37° C. (44° C.).[169] The ant *Messor* in Palestine is more resistant to heat in July than in March; and the Chalcid *Dahlbominus* and the larvae of sawflies of the genus *Neodiprion* can be readily acclimatized to what would normally be lethal high temperatures.[145]

The resistance of larvae of the Tachinid *Agria* (*Pseudosarcophaga*) *affinis* to high temperature can be increased either by 'thermal conditioning' or by dietary lipids. Larvae reared at 23° C. show a 50 per cent. mortality at 45° C. in an average of 130 minutes; preconditioned for 2 hours at 39° C., the time was increased to 200 minutes. Similarly, in larvae reared on a chemically defined diet with unsaturated fatty acids, the 50 per cent. mortality time at 45° C. was 177 minutes; on a diet with a high proportion of saturated fatty acids 218 minutes.

As in other insects (p. 514) the degree of saturation of the body lipids was directly influenced by that of the lipids in the diet.[164] In *Periplaneta* a rise of temperature causes a similar fall in the iodine number of tissue fats—but no change in thermal death-point is apparent.[172] [See p. 692.]

The effect of atmospheric humidity on the preferred temperature is very difficult to test, because it is impossible to arrange a gradient of temperatures without at the same time creating a gradient in relative humidity. But in *Formica rufa*,[53] in certain beetles,[14] and in *Blatta*,[46] it is said that air humidity does not influence the preferred temperature. Some insects have the same preferred temperature irrespective of their previous treatment, but others collect chiefly at a lower temperature if they have been partially desiccated by exposure to dry conditions. The beetle *Adesmia clathrata* preferred 39·4° C. when in moist conditions, 36·6° C. after keeping in drier air.[14] Cockroaches kept in a graded temperature apparatus tend to settle down at a progressively lower temperature as time goes on,

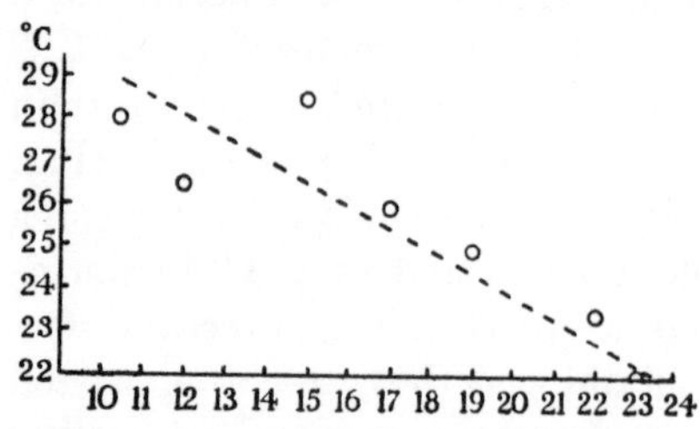

FIG. 369.—Change in the preferred temperature in *Forficula* (*after* VAN HEERDT)

Ordinate: preferred temperature in °C. Abscissa: length of time spent in the apparatus in hours.

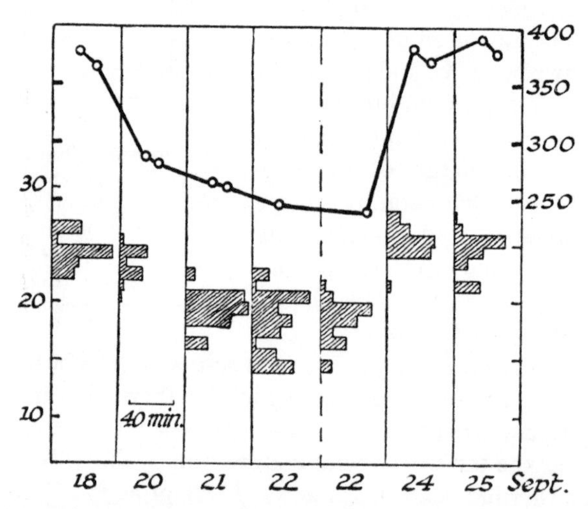

FIG. 370.—The curve shows fall in weight of a cockroach *Blatta orientalis* during desiccation and rise in weight again after drinking (on Sept. 22nd)

Ordinate (to the right): weight in mg. Abscissa: day of month. The blocked figures were arrived at by adding up the periods longer than 2 minutes during which the insect remained at a given temperature. Ordinate (to the left): temperature in ° C. The whole chart shows the fall in the preferred temperature on desiccation, and the return to normal following drinking (*after* GUNN).

especially in dry air.[46] *Blatta* normally shows a preferred temperature or 'indifference zone' at 20–29° C.; individuals desiccated to 70 per cent. of their original weight had this zone shifted to 12–23° C., but returned to the normal range after drinking (Fig. 370).[46] The same is seen in *Forficula* (Fig. 369),[52] *Pediculus*,[53] *Glossina*,[59] and various Acridiidae.[165] In the Spruce Budworm *Choristoneura* it is the rate of evaporation which determines the preferred temperature under a given set of conditions.[129a] For adaptation of the metabolic rate, see p. 683.

Resistance to low temperature—According to their resistance to cold, insects fall into three groups.

(i) Those accustomed to warm surroundings, like many tropical species or many of the insects that infest stored products, soon die even at temperatures well above freezing-point. The cause of death is not understood. It is often attributed to the accumulation of toxic products which at normal temperatures would be eliminated. In the honey-bee, which dies very quickly at temperatures between 1° and 8° C., death is attributed to a differential effect of cold on the various steps in the utilization of sugar: at 1° C. metabolism in the tissues still goes on pretty actively (in Pyralid larvae oxygen consumption continues down to −12° C.[67]) but absorption from the gut is arrested; and since the bee is dependent on sugar in the food it soon dies of starvation at low temperature.[62] This type of effect by cold is sometimes termed the 'quantity factor' because it must act for some time before it causes death.[100]

There is some relation between the chill coma temperature as discussed in the preceding section and the cold death point. Acclimatization of *Blatta* for 20 hours at 15° C. increases the resistance to cold. Insects were exposed at −5·5° C. for 9 hours; of those acclimatized to 30° C. all were dead; whereas all those acclimatized to 15° C. were alive.[90]

(ii) Most insects are killed as soon as their tissues freeze. Since this effect depends only on the level of temperature it is sometimes termed the 'intensity factor' in cold resistance.[100] The actual cause of death from freezing is uncertain. It is attributed sometimes to dehydration of the tissues, sometimes to mechanical injury by ice crystals;[112] it is well known that many oxidation systems in cells are dependent on the integrity of cell structure and are destroyed by freezing.

(iii) A few insects can withstand complete freezing; but they die from some unknown cause when the temperature is lowered still further.[112]

Supercooling—'Cold hardiness' in insects is chiefly a matter of prevention of freezing. When the temperature is lowered, the water in the insect's body behaves like water in capillary tubing; it becomes supercooled, and ice does not form until the temperature has fallen far below the true freezing-point. The temperature of the insect will therefore follow that of the air until it has fallen perhaps to −10° or −15° C. (the 'critical point' or 'under-cooling temperature'); it then suddenly jumps up to say −1·5° C. through the liberation of latent heat, and proceeds to fall once more. It is this 'critical point', where freezing begins, which varies in different species (Fig. 371).[7]

Supercooling is usually exaggerated in hibernating insects, some of which have been said to withstand −50° C.[126] The increased resistance is generally associated with loss of water.[111] Nymphs of *Chortophaga* respond to cold by decreasing their water content from 79 per cent. to 65 per cent. and maintain this low level during hibernation.[16] In the beetle *Popillia* the undercooling temperature and the water content both fall during the winter when the cold hardiness increases greatly; and during the summer this insect may be 'artificially' hardened by desiccation: beetles dehydrated experimentally to half their body weight can resist a temperature of −28° C.[100] The blow-fly larva *Lucilia* can resist only −2° C. when actively feeding, −10° C. in the pre-pupal stage when it has lost much of its water.[30] It is uncertain whether the water eliminated at low temperatures is lost by the excretory or the respiratory systems; but in the case of *Leptinotarsa* there seems to be some connection between this loss and the general mechanisms of water retention. Thus strains of *Leptinotarsa* adapted to life in Arizona are far more efficient at retaining water under dry conditions than the Chicago strain; but they die off for the most part during the winter when brought to the Northern States, since they fail to eliminate their water before hibernation.[125]

But the capacity for supercooling is not necessarily associated with water loss. Full grown larvae of *Ephestia*, feeding, in early prepupal stage, and in the cocoon, have almost the same percentage of water; yet their undercooling points are −5·8° C., −8·0° C. and −21·3° C. respectively. Adult bugs, *Leptocoris*, kept over calcium chloride until they had lost 20 per cent. of their weight, showed no change in their undercooling point. And hibernating bugs, *Chlorochroa*, after feeding for 5 days, increased their water content from 53 per cent. to 67 per cent., but their undercooling point remained stationary. Newly hatched larvae of *Ephestia*, *Sitotroga*, &c., have an undercooling point of about −27° C.; after their first feed it is raised to −6° C.[112] The undercooling temperature in larvae of the codling moth *Cydia pomonella* in winter is −24 to −30° C.; supercooling

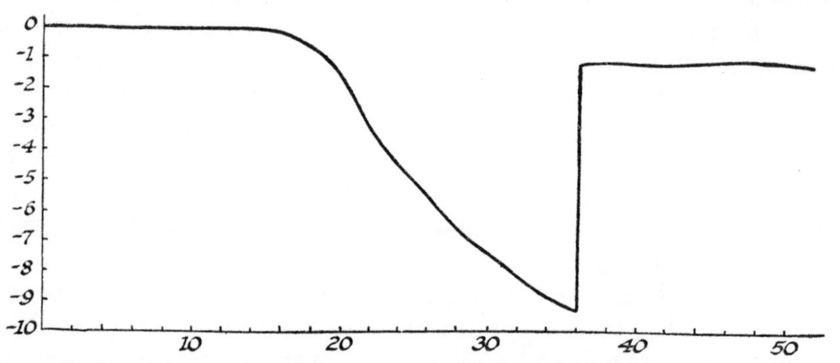

FIG. 371.—Body temperature of a *Saturnia pyri* male exposed to an air temperature of −13·5° C., showing supercooling (*after* BACHMETJEW, 1899)
Ordinate: temperature in ° C. Abscissa: time in minutes.

is favoured by the quiescent state and is reduced by artificial movements.[117] In some insects contact with free moisture prevents undercooling.[57]

The nature of the change which increases the capacity for supercooling is uncertain. This capacity is diminished by injury; for example, in *Bruchus* the undercooling point is normally −15 to −20° C.; if the beetle is pierced it is only −8 to −10° C.;[112] and after repeated freezing and thawing, supercooling is eliminated altogether, and freezing takes place as soon as the freezing-point of the body fluids is reached.[7] It has been shown that ice crystals forming in solutions rich in hydrophilic colloids are liable to become covered with a sheath of dehydrated colloid, and thus fail to 'seed' the entire solution;[112] the quantity of such colloids may be a factor in supercooling. It has been suggested that the water which remains unfrozen at −20° C. is 'bound' to the tissue proteins;[108] but there is at present no way of distinguishing 'bound' water from water which is super-cooled, or which fails to freeze, from some other cause. Complete freezing of the body water does not occur even at temperatures as low as −40° C. But the amount of this unfreezable water is not related to cold hardiness.[35]

One factor in the non-resistance of the feeding insect is the ready freezing of the gut contents which then 'seeds' the rest of the body.[182] The capacity for supercooling is diminished by 'nucleating agents' of an undefined nature, and cold hardening involves the elimination or reduction of these nucleating agents.[185] The supercooling point is not a fixed temperature; it depends largely on chance; and since the probability of nucleation increases with time the undercooling temperature rises as the duration of exposure is prolonged. An insect at −19° C. may take a minute to freeze, at −10° C. it may take a month.[183] Viscous syrups and gels increase supercooling by inhibiting nucleation; they restrict molecular travel by hydrogen bonding. [See p. 692.]

In many insects exposed to low temperatures during diapause the haemolymph comes to contain glycerol (p. 597). This glycerol may be present in sufficiently high concentration to lower the freezing point significantly, and in addition it increases supercooling.[189] In *Campanotus* the glycerol content rises in winter to 2 per cent.; it serves to enhance supercooling but it is not the sole cause of cold hardiness.[192]

In larvae of *Bracon cephi* the concentration of glycerol rises to 5 M during hibernation. By itself this would give a freezing point of −4·8° C. The freezing-melting point of the haemolymph in *Bracon* actually ranges from −7° C. to −17° C.; but the supercooling point may be as low as −47° C.[184] In the slug caterpillar *Cnidocampa* the glycerol content rises to 3–4 per cent. during the winter months and disappears again on warming. Glycogen follows a converse change. The presence of glycerol does seem to increase the frost-resistance, but this resistance is not to be explained by the presence of glycerol alone.[191] The isolated heart of *Popilius disjunctus* (Col.) in insect Ringer's solution failed to recover after exposure to −20° C.; but glycerol at 5 per cent. or 10 per cent. allowed survival after storage for 11 days at −18° to −20° C.[194]

Survival after freezing—In the case of those few insects which are able to withstand freezing of the tissues, it has been suggested that this results from a change in the nature of the cellular respiration; it is suggested that oxidases dependent on the integrity of the cell structure are replaced by dehydrogenase systems which are not associated with structural elements and are therefore not injured by freezing.[69]

That some insects, notably caterpillars, can survive ice formation inside their bodies has been known since Réaumur. Live Chironomid larvae frozen into ice in the Arctic are 10–20 per cent. dehydrated. They spend the winter in this state and can be thawed and refrozen in the laboratory many times without injury. The unfrozen water is not supercooled but is concentrated; as the temperature falls more ice is formed and the residual water increases in concentration.[186] In the fully hydrated larvae of *Polypedilum* (Chironomidae) freezing and thawing causes severe damage and recovery is only temporary. If the larvae are partially dehydrated before freezing, no visible damage occurs and they can complete their development.[166] The undercooling point of the prepupa of the slug moth *Cnidocampa* ranges from −17° C. to −21° C.[141] When this temperature is reached the blood freezes but ice formation does not extend into the tissue cells, which are merely dehydrated and shrunken.[188] They can be kept in this state for weeks at −15° C. to −20° C. But when freezing extends into the tissue cells they no longer survive.[142] It seems that some particular state of the protoplasm is needed in order to tolerate extracellular freezing; only if this state is present does the addition of glycerol, &c., increase frost resistance.[143] Similar changes occur in the larva of the European corn borer, *Pyrausta nubilalis*.[161] [See p. 692.]

Effects of temperature on metabolism and growth—As was first pointed out by Pflüger, whereas the metabolism of warm-blooded animals is

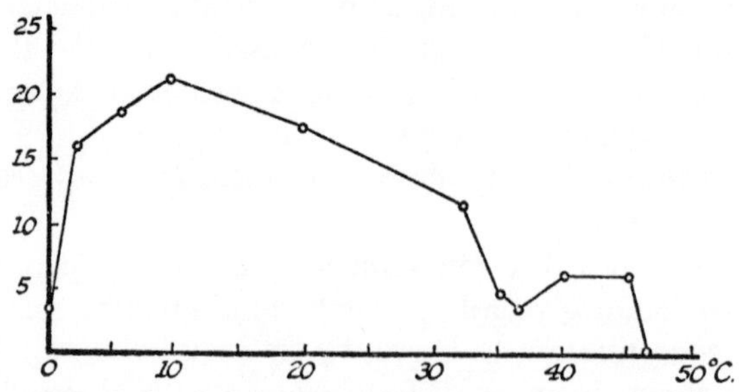

FIG. 372.—Metabolism of the bee at different temperatures (*after* PARHON)

Ordinate: oxygen uptake in litres per kg. per hour. Abscissa: temperature in ° C.

depressed as the external temperature rises, the metabolism of cold-blooded animals increases. The increased activity which the insect shows with rising temperature accounts for most of this increase. What happens to the extra energy produced in the resting or narcotized animal as the result of raising the temperature is not known, but much of it must be expended by the augmented movements of the internal organs. In the developmental stages the extra energy is expended on growth, which is correspondingly accelerated.

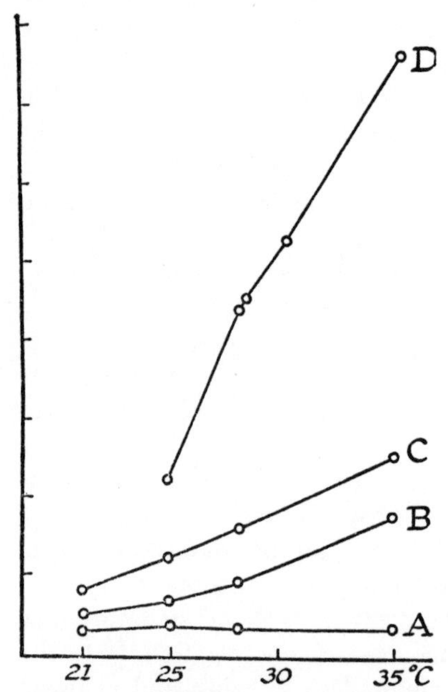

FIG. 373.—Effect of temperature on oxygen consumption in the mud-dauber wasp, *Sceliphron caementarium*, at different stages of development (*after* BODINE and EVANS)

Ordinate: relative oxygen consumption. Abscissa: temperature in ° C. *A*, larva in diapause. *B*, larva coming out of diapause. *C*, pupa. *D*, pupa ready to emerge.

Complicating factors are met with in the honey-bee, which increases its metabolism as the temperature falls in order to maintain the temperature of the hive (p. 672). Its rate of metabolism shows a maximum at 10° C. and then falls to 0° C. (Fig. 372).[66, 98] Diapause also may obscure the relation between temperature and metabolism: the oxygen uptake in eggs of *Hyalophora cecropia* is little affected by raising the temperature from 16° C. to 35° C. during the 'incubation period', but is greatly stimulated during active development;[91] and in the mud-dauber wasp *Sceliphron*, the quantitative response to temperature, which is very slight during diapause, becomes very marked when diapause is broken (Fig. 373).[18]

In the egg of *Oncopeltus* below 20° C., temperature has a very different effect on growth rate and on general energy expenditure; as the temperature falls, more and more energy is required to produce the same amount of growth, and the temperature threshold for hatching is the point below which all the fat reserves in the egg become exhausted before development is complete. In addition, short periods of warming are necessary to provide for some undefined process in metabolism.[180]

There may be some degree of acclimatization to the altered temperature; *Aeschna* nymphs show an increase in oxygen consumption at high temperature and a decrease at low, but after a day or so under either condition metabolism tends to return to the same intermediate level.[114] This adjustment of the rate of metabolism as the environmental temperature is changed seems to be particularly characteristic of aquatic insects (Odonata, Ephemeroptera, Plecoptera, Chironomid larvae, *Dytiscus*, &c.), but there are also aerial insects which show some degree of adaptation, e.g. *Tenebrio* larva, *Blattella*, *Cimex*, *Rhodnius*, *Lucilia*, and *Calliphora*.[176] In a series of insects studied there are all degrees of such adaptation in relation to metamorphosis. In *Tenebrio* larvae and pupae no adaptation occurs; but in *Melasoma* adults, maintained for adaptation for 4 days at 12° C. and 25° C., adaptation is complete: the oxygen uptake at the two adaptation temperatures is approximately the same (horizontal broken line in Fig. 374) and the course of oxygen consumption turns steeply down at 40° C. and 44° C. in the two groups (Fig. 374).[168]

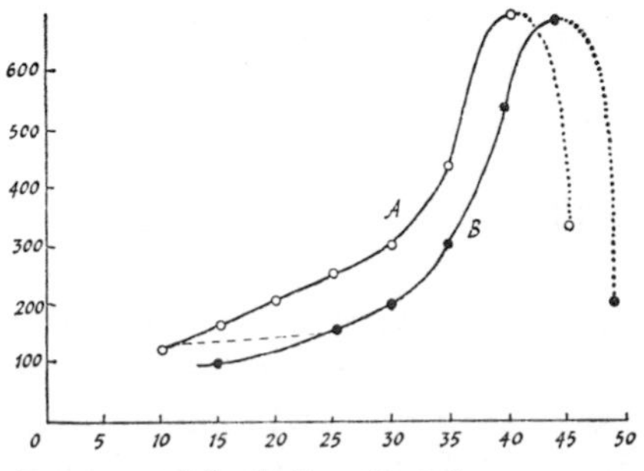

FIG. 374.—Metabolic adaptation to temperature in *Melasoma* adult (*after* MARZUSCH)

A, after keeping for 4 days at 12° C. B, after 4 days at 25° C. The broken line serves to compare the oxygen uptake at the two adaptation temperatures. Ordinate: oxygen uptake in mm^3/g./hr. Abscissa: temperature in °C.

There is one physiological process which seems to be more or less independent of temperature (or in which adaptation is unusually efficient); that is the timing by some types of 'physiological clock' (p. 342). The duration of the endogenous activity rhythm in *Periplaneta* is independent of temperature between 18° C. and 31° C. During a period of chill coma in one or in both of the phases of the daily cycle, the rhythm continues at approximately the normal tempo.[148]

Optimum temperature—Each species has a wide range of temperature between cold stupor and heat stupor within which it can perform its normal functions. Towards the limits of this range, growth or reproduction or some other function may be adversely affected, and some point in the middle of the range is sometimes described as the optimum temperature.[96, 126] This point is often difficult to define, and it is evident from what has been written in earlier sections of this chapter that the optimum may vary with the humidity.

A more definite conception of an optimum temperature is possible in considering development of the egg or pupa. Table 5 shows the effect of temperature

TABLE 5

Temperature	*Duration of pupal stage*	*CO_2 produced in total pupal life in litres per Kg. of pupa*	*CO_2 in c.c. per Kg. of pupa per hour*
32·7	139·9	59·3	427
27·25	172·5	58·0	336
23·65	234·1	59·1	252
20·9	320	59·6	186

on pupal development in *Tenebrio*. As the temperature is raised from 21° C. to 33° C., the duration of pupal life is shortened from 320 hours to 140 hours, but the total carbon dioxide production is unchanged (Fig. 375). There is not

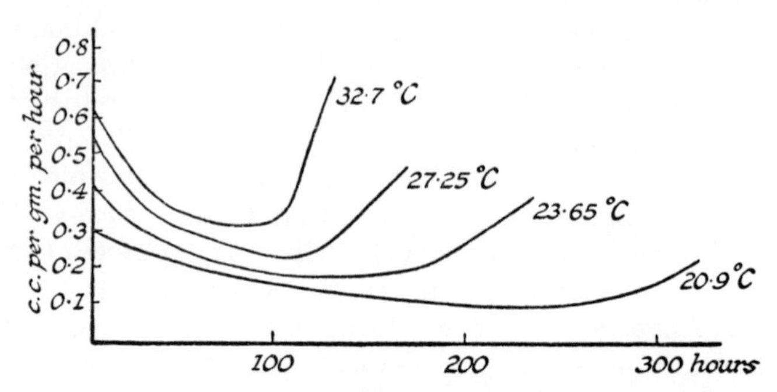

FIG. 375.—Respiratory exchange in pupae of *Tenebrio molitor* kept at four different temperatures throughout the pupal stage (*from* GAARDER *after* KROGH)

Ordinate: CO_2 production in c.c. per gm. per hour. Abscissa: time in hours.

optimum temperature at which metabolism is smallest, and the efficiency of growth therefore greatest.[70] The same applies to the eggs of the beetle *Acilius*,[71] and in the pupa of *Drosophila* between 14° C. and 25° C.[36]

But this is not true of all insects. At abnormal temperatures the pupae of various Lepidoptera (*Agrotis*, *Loxostege*, *Ephestia*) show a higher oxygen uptake, more fat consumption, greater pupal mortality, and shorter imaginal life; the temperature at which oxygen uptake is smallest may be regarded as the optimum.[1, 68] In the pupa of *Galleria* the total metabolism is lowest at 30° C. and increases above and below this temperature; the respiratory quotient being constant at 0·69 throughout.[31] In the pupa of *Glossina morsitans* the consumption of fat is least between 22° C. and 24° C.; this could be regarded as the optimum temperature. At high temperatures the rate of fat consumption is greatly increased without a corresponding reduction in the duration of the pupal period; whereas at low temperatures the pupal period is greatly lengthened without a corresponding decrease in the rate of fat consumption.[151]

Alternatively the optimum temperature may be regarded as that at which the greatest percentage of individuals complete their development; or that at which the time of development is shortest,[13, 43] in other words, where the temperature velocity curve (Fig. 376, A, *b*) has its peak.

Mathematical descriptions of temperature effects—Fig. 376 A, *a*, shows the relation of temperature to the duration of egg development in the weevil *Sitona*, and Fig. 376, A, *b*, shows its relation to the reciprocal of duration which indicates average rate of development.[3] Fig. 376, B, shows the rate of development in eggs of *Phormia* (Dipt.) plotted against temperature,[92] and Fig. 376, C, oxygen consumption in the pupa of *Tenebrio* against temperature.[70] In all cases the curve of intensity or velocity is **S**-shaped. It rises slowly at first, then steeply, then more slowly to a maximum and falls again to end abruptly at the lethal temperature. Sometimes, as in Fig. 376, C, during the steep rise it approximates to a straight line.

This curve represents the sum or resultant of an immense number of chemical and physical reactions, many of which must be differently affected by changes of temperature.[104] But in spite of this, numerous attempts have been made to describe these curves by simple equations. Such formulae can be made to fit particular cases, and to that extent they have some descriptive value, but none

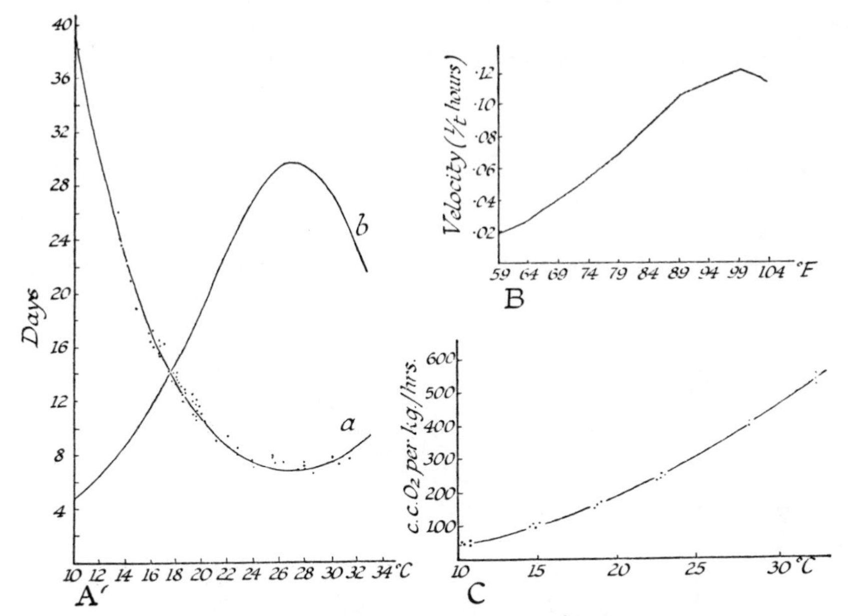

FIG. 376.—Effect of temperature on physiological processes

A, curve a, relation between temperature and duration of egg development in *Sitona lineata*; curve b is the reciprocal of this curve and represents velocity of development (*after* ANDERSEN). B, relation between temperature and velocity of pupal development in *Phormia regina* (*after* MELVIN). C, relation between temperature and rate of oxygen uptake in pupa of *Tenebrio molitor* (*after* KROGH).

of them is of sufficiently general application to be regarded as embodying any rational principle.[9] In view of the vast amount which has been written on this subject we must consider the various formulae that have been proposed.

(i) *That the velocity is proportional to the temperature.*

$$v=k(t-a)$$

where v=velocity, t=temperature, k and a=constants.

Or

$$v_{(t+10)}=v_t+K_{10}$$

where K_{10} is the increase in velocity for a rise of temperature of 10° C. from t to $t+10$.

This relation holds very well over the normal range of temperature at which the egg of *Acilius* develops,[70] and for the pupal development of *Tenebrio* between the temperatures 18·5–28° C., but beyond these temperatures (normal development in *Tenebrio* occurs from 13·5–33° C.) the curve is not straight but bends upwards at lower temperatures and downwards at higher (Fig. 377).[70] With these same departures at the upper and lower limits, this linear relation between rate and temperature holds for the larval and pupal development of many Lepidoptera and Diptera,[101] and the egg and pupal development in *Popillia*.[76] It holds, also, for the rate of metabolism of *Carausius* over most of the range from 5–35° C., provided it remains at rest, and for the respiration of various pupae.[21]

Thermal summation—In the first equation above, the constant a represents the temperature at which the straight line, when prolonged, meets the temperature axis. This temperature is termed the 'developmental zero'. It is the temperature at which development would cease if the curve did not in fact turn upwards in this region.

If y is the time required for complete development at temperature t, the equation may be written

$$y(t-a)=K$$

in other words, the product of time of development in days × the excess of temperature above the developmental zero in °C. is constant. This value K in 'degree-days' is termed the 'thermal constant'. Thus, where the linear relation between velocity and temperature holds, each developmental process will have a characteristic thermal constant and will require a fixed number of 'degree-days'

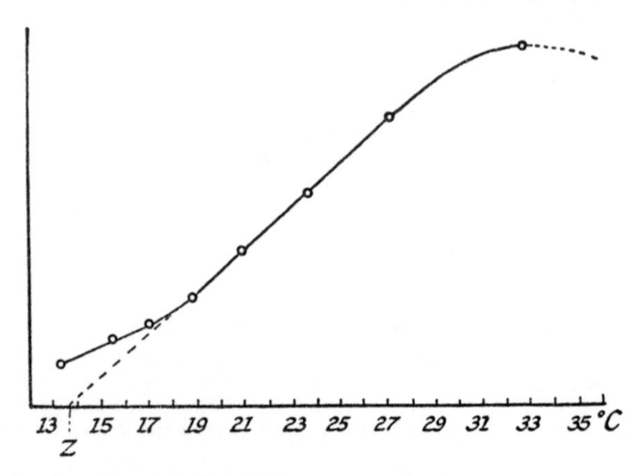

FIG. 377.—Relation between temperature (abscissa) and velocity of development (ordinate) in pupae of *Tenebrio* (*from* IMMS *after* KROGH)

The temperature Z, at which the straight part of the velocity curve meets the temperature axis when extended, represents the theoretical 'developmental zero'.

to bring it to completion. Therefore, even if the temperature is changed in the course of development, it is theoretically possible to predict the time necessary for its completion by adding up the number of 'degree-days' contributed at each temperature. This procedure is called 'thermal summation'.[13, 101] Apart from the linear relation between velocity and temperature, it involves the further assumption that the accelerating effect of temperature is the same at all stages in the developmental process. In the beet leafhopper *Eutettix* and other insects there is some indication that the earlier and later stages of embryonic development are, in fact, differently affected by temperature.[50] Where the relation between rate of development and temperature is not linear, it is possible to apply this summation procedure only after a fairly complete curve has been obtained experimentally, since the thermal constant changes its value at each temperature.[113]

Even where the velocity curve is linear over most of its course, it becomes less steep as it approaches the lowest temperatures; so that the 'developmental zero' is not in fact the true threshold of development; the temperature at which development ceases lies appreciably lower.[101] Thus in *Lucilia* the 'developmental zero' is at 12° C., the lower limit of development at 9–10° C.;[30] in *Rhodnius* these temperatures are 10° C. and 6° C. respectively.[43]

Another disturbing element in the summation of temperatures is the fact that in some cases an alternation of temperatures seems to stimulate development. Thus eggs of *Melanoplus atlanis* collected in the field and exposed at a constant temperature of 32° C. developed in 5 days; if they were exposed each day for 16 hours at 12° C. and 8 hours at 32° C., they developed in 3 days.[99] In the eggs of *Drosophila*, there seems to be no true stimulation of this kind; but development may be more rapid than anticipated from the thermal summation owing to the temperature-velocity curve deviating from the straight line at the lower temperatures.[79]

(ii) *That the logarithm of the velocity is proportional to the temperature.*

$$v = AK^t$$

where v=velocity, t= temperature and K and A=constants. Or

$$v_{(t+10)} = v_t Q_{10}$$

where v_t=velocity a. temperature t, $v_{(t+10)}$=velocity at temperature 10° C. higher, and Q_{10}=the temperature coefficient. Or

$$Q_{10} = \left(\frac{V_1}{V_2}\right)^{\frac{10}{t_1 - t_2}}$$

where V_1 and V_2 are the velocities at temperatures t_1 and t_2.

This is the Q_{10} rule of van't Hoff and Arrhenius, which holds fairly well for chemical reactions. In these the value of Q_{10} is usually between 2 and 3, but even in inorganic reactions some variation in Q_{10} always occurs. In biological reactions the value of Q_{10} usually decreases progressively as the temperature increases. In the development of *Tenebrio* pupae, Q_{10} has the following values:[70]

10–15° C.	5·7
15–20	3·3
20–25	2·6
25–27·5	2·3
27·5–30	2·1
30–32·5	2·0

Therefore this rule rarely holds except over a very limited temperature range. It holds well for the length of the pupal stage in *Aëdes taeniorhynchus* over the lower range of temperatures, but at high temperatures a negative correction has to be applied.[174]

(iii) *That the logarithm of the velocity is proportional to the reciprocal of the absolute temperature.*

$$v = a.k^{\frac{1}{T}}$$

where v=velocity, T=the absolute temperature and k=a constant. Or

$$V_2 = V_1 e^{\frac{\mu}{2}\left(\frac{1}{T_1} - \frac{1}{T_2}\right)}$$

where V_1 and V_2 are velocities at the absolute temperatures T_1 and T_2 respectively, e is the base of natural logarithms, and μ a constant which is termed the 'thermal increment' or 'temperature characteristic'.

This is the law of Arrhenius, which seems to hold good for chemical processes with more accuracy than the Q_{10} rule. But over the narrow range of temperature at which living organisms exist, the reciprocal of the absolute temperature is practically a linear function of the ordinary temperature, so that in biology there is no practical difference between μ and Q_{10}.[9]

When the logarithm of the velocity is plotted against $\frac{1}{T}$ the slope of the curve is determined by the value of μ. In some of the processes of insect physiology a continuous straight line is obtained. But more often the slope changes. It is sometimes claimed that when this happens the change is always abrupt; in

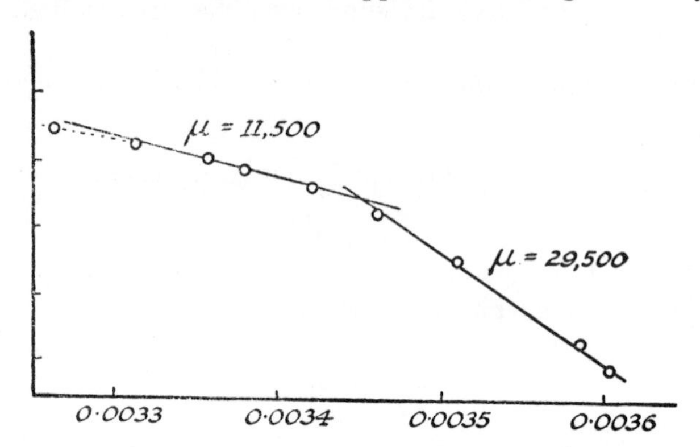

FIG. 378.—Oxygen consumption in *Carausius* at different temperatures (data of v. BUDDENBROCK and ROHR as plotted by CROZIER and STEIR)

Ordinate: logarithm of oxygen consumption in c.c. per hour. Abscissa: reciprocal of the absolute temperature.

other words, that the value of μ is suddenly altered at a given temperature (Fig. 378). The theoretical explanation given, is that over each range of temperature one particular chemical or physical reaction is the 'master reaction' or 'limiting factor', which determines the rate at which the process as a whole can proceed, and that each successive master reaction has a different 'temperature characteristic'.[32] This conclusion rests upon the inspection of the curves obtained experimentally, and some authors consider that the value of μ changes progressively with temperature like the value of Q_{10}.[9, 33, 76]

(iv) *That the logarithm of the velocity is proportional to the logarithm of the temperature.*

$$v = \frac{t^k}{a}$$

where v=velocity, t=temperature, k and a=constants.

In many cases where the value of Q_{10} or μ appears to change with temperature, and breaks occur in curves plotted according to (ii) and (iii) above, a straight line is obtained when the results are plotted in accordance with this empirical equation (as in the development of *Dytiscus*;[13] and the rate of metabolism in the different stages of *B. mori*[8]). When the exponent k is equal to 1, this formula becomes identical with (i) above; it will therefore embrace cases where there is a linear relation between temperature and velocity.[9]

(v) *That the destructive action of temperature at the upper limit of the temperature range also follows the van't Hoff rule.* All the preceding equations apply only to the ascending part of the velocity curve; they take no account of the fall in the curve near the lethal temperature. According to Duclaux's theory of the effect of temperature on enzyme action, both the accelerating and destructive effects of rising temperature follow the van't Hoff rule, and the ordinary velocity-temperature curve represents the sum of these two effects. This same idea has been applied to the effects of temperature on insect metabolism[103] and rate of development.[60] The curve relating time of development with temperature (Fig. 376, A, *a*) would therefore be regarded as a catenary curve, representing the sum of two exponential curves $y=ma^x$ and $y=ma^{-x}$, the complete equation being

$$y=\frac{m}{2}(a^x+a^{-x})$$

where y=time required for development, m=the time at the optimum, x=excess of temperature in °C. above the optimum and a=a constant. The graph of the reciprocal gives the relation between velocity of development and temperature; and by various modifications in the equation, it is possible (after the experimental values have been obtained over the entire range) to obtain a fairly good fit between the curve and the experimental points.[60] Good agreement, for example, can be obtained in the case of egg development in *Sitona*;[3] fairly good agreement with the rate of crawling of *Lymantria* larvae,[19] and the development of Muscid flies (Fig. 379).[75]

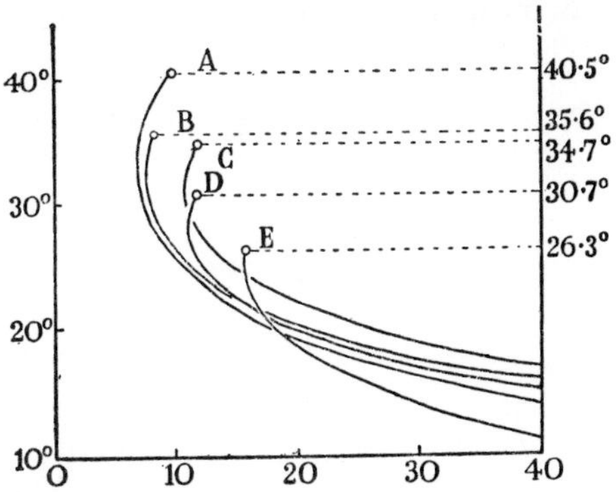

FIG. 379.—Rate of development of Muscid flies at different temperatures (*after* LARSEN and THOMSEN)

Ordinate: temperature in ° C. Abscissa: duration of preimaginal development in days. Figures on the right show upper temperature limits of development.

A, *Musca domestica;* B, *Lyperosia irritans;* C, *Stomoxys calcitrans;* D, *Haematobia stimulans;* E, *Scatophaga stercoraria.*

(vi) *That the relation between temperature and the rate of biological processes is best described by a logistic curve.*[33]

$$\frac{1}{y}=\frac{K}{1+e^{a-bx}}$$

Where $\frac{1}{y}$ represents the reciprocal value of the time required for complete development to be achieved at a given temperature x. K, a and b are constants. K represents the value of the upper asymptote towards which the curve is tending; a indicates the relative position of the origin of the curve on the abscissa; b represents the degree of acceleration of development in relation to temperature and determines the slope of the curve; e is the base of Napierian logarithms.

When plotted this gives an S-shaped curve which affords a faithful representation of the speed of development in many different insects over 85–90 per cent. of the complete temperature range over which development can occur (Fig. 380). Above a certain temperature retarding effects supervene so that the 'peak' temperature is usually well below the temperature at which K is attained. For example, in the egg of *Drosophila* the hatching zero is 14·5° C., the peak temperature is at 29·5° C., growth of the embryo is inhibited at 34° C., whereas K is reached at 42° C.[33]

A modified logistic curve, which also describes the fall in the curve beyond the peak, has the formula

$$y=y_0e^{-ax^2}$$

Where y_0 is the highest value of the 'developmental index' (the reciprocal of the

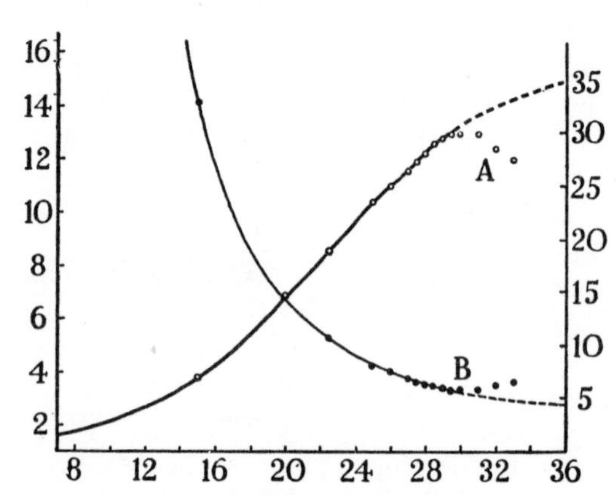

FIG. 380.—Temperature and development of pupae of male *Drosophila melanogaster* (data from LUDWIG and CABLE, *after* DAVIDSON)

A, temperature-velocity data; *B*, temperature time data. The curves are calculated logistic curves. Ordinates: time of development in days, on left; average percentage development in one day, on right. Abscissa: temperature in ° C.

developmental period); y is the developmental index at temperature t. $x=T-t$, where T is the temperature corresponding to y_0. a is a constant and e is the base of natural logarithms. This formula describes both the accelerating and the destructive effects of increasing temperature. Unlike other descriptions, it is claimed that this curve and formula have a rational basis.[105] Although the rate of development in the eggs of *Acheta* and other examples agrees quite well with a logistic curve, when the agreement is critically examined it is found that the departures from the calculated values are highly significant.[147]

ADDENDA TO CHAPTER XIV

p. 664. **Water content**—Respiration of Chironomid larvae is not affected by dehydration until body water has been reduced to below 50 per cent. of normal.[199]

p. 664. In some termites liable to exposure to drought the salivary gland reservoirs serve as 'water-sacs' providing a water reserve.[217]

p. 667. **Active absorption of water from the atmosphere**—This active retention and uptake of water implies that a water pump is associated with the cuticle. In *Periplaneta* and *Schistocerca* there is direct evidence that the water activity within the cuticle is substantially lower than that in the haemolymph: a maintained difference of some 14 atmospheres. Simple physical forces will thus transport water from the atmosphere into the cuticle; but the active mechanism and the expenditure of energy must reside in the epidermal cells.[219] The critical equilibrium humidity for the uptake of water by the third-stage larva of *Xenopsylla* is 65 per cent. relative humidity. The net sorption of water from almost saturated air is about twice as rapid as transpiration into dry air and is regulated by relative vapour pressure. The uptake and retention of water depend upon active processes which operate over the range from 13° C. to about 35° C.[207] Uptake of water from the atmosphere takes place in desiccated, starved *Thermobia* down to a relative humidity of about 60 per cent. over the temperature range 17–37° C. Above this critical level the normal body water content is restored, irrespective of the humidity. Since occlusion of the anus prevents this uptake it has been suggested that absorption takes place in the rectum.[211] *Tenebrio* larvae equilibrate the air spaces in the lumen of the rectum to 88–90 per cent. relative humidity (p. 575) which is the lowest humidity from which they can absorb water from the atmosphere. Here again, blockage of the anus prevents uptake.[211]

p. 668. **Desiccation and water loss**—When the insect is exposed to very dry conditions the drying of the cuticle may reduce its permeability to water. That is seen in *Locusta*;[209] but here most water loss takes place through the spiracles, particularly during respiratory movements: above 42° C. hyperventilation and greatly increased water loss occur; progressive desiccation reduces ventilation and so conserves water.[209]

p. 671. **Heat production in flying insects**—Although oxygen consumption ordinarily rises with an increase in the ambient temperature, the oxygen consumption of *Celerio* in flight shows a 60 per cent. increase when the temperature falls from 31° C. to 26° C.—because of the greater heat required for maintaining the thoracic temperature. Above 31° C. thoracic temperature during flight may become excessive and the moths land periodically and cool off.[203] The same relation is seen in *Hyalophora*. The transition from warming-up behaviour to flight depends on the temperature of the thoracic ganglia: heating the ganglia in a quiescent moth induces flight without warming-up activity.[202] In the Sphingid *Manduca* the high thoracic temperature produced by flight in warm surroundings stimulates the dorsal vessel to increased activity which pumps haemolymph through the thorax and dissipates the heat in the abdomen. If the aorta is ligatured the moths overheat and stop flying.[204] During warm-up the rate of wing beat increases with the thoracic temperature: about 8 per second at 15° C., about 25 per second at 35° C.[204] It was observed by Newport (1838) that honey-bees will raise their body temperature without visible muscular contractions. But although mechanical activity cannot be seen, action potentials of low frequency allow moderate warming of the muscles without vibration.[200]

p. 671. **Solar radiation**—When solar radiation is inadequate *Danaus* can still fly at low ambient temperatures (13–17° C.) through warming up by shivering wing strokes of small amplitude.[206] *Aëdes* mosquitos in the Arctic will bask in the parabolic corolla of *Dryas* and other flowers which always face the sun; a habit which must accelerate the ripening of the germ cells.[205]

p. 673. **Evaporative cooling**—the 'fenestrae' of Acrididae, patches of thin cuticle on various parts of the body, believed at one time to be temperature receptors (p. 292), show a much greater rate of transpiration than the normal cuticle of *Schistocerca* and other locusts. It may be that they provide an emergency cooling system to the overheated locust.[210]

p. 675. **Arrest of growth at high temperature**—Growth in *Rhodnius* if arrested if the temperature rises above 36° C.[218] This is due to inhibition so protein synthesis which is reflected in many aspects of growth (p. 125).[213]

p. 678. **Temperature adaptation**—In mosquitos (*Aëdes sollicitans*) the fatty acid composition was the same from 10 to 35° C.; there was no increase in unsaturation at low temperatures.[200a]

p. 681. **Supercooling**—Insects suffer as a result of cooling even when no freezing has occurred (see the 'quantity factor', p. 679). In *Hyalophora cecropia* this happens over the range 6–0° C.; in *Cephus cinctus* it does not begin until about −20° C.[214] In the ant *Camponotus* there are two supercooling points. The first, at −8·5° C., results from ice formation in the gut contents and can be tolerated; the second, at about −20° C., results from ice seeding from the gut to the other body tissues and is invariably fatal.[212]

Injection of glycerol (3 per cent. of fresh body weight) will increase frost resistance of *Hyalophora* pupae.[197] In the Alaskan Carabid beetle *Pterostichus brevicornis* glycerol may exceed 22 per cent. in the haemolymph during winter; on warming the insects artificially to 20° C. the glycerol content falls to less than 1 per cent. within 36 hours.[198] In the eggs of *Malacosoma* glycerol reaches a peak of 35·2 per cent. of the dry weight in January, compared with 1·0 per cent. in the post-diapause stage just before hatching.[201] The indigenous species *Megachile relativa* which overwinters as a prepupa in southern Alberta produces a glycerol concentration double that of the exotic species *M. rotundata* which cannot overwinter in the open.[208] In the eggs of the Aphid *Hyalopterus* mannitol occurs in high concentration along with glycerol and probably contributes to their cold hardiness.[216] Chilling in *Ephestia* leads to a rise in ninhydrin-positive substances, chiefly alanine, in the haemolymph along with glycerol, glucose and inorganic phosphorus. These same changes may be induced by anoxia for 12 hours.[215] The prepupae of *Trichiocampus* will survive freezing at the temperature of liquid nitrogen; it does not contain glycerol but does contain a high concentration of sugar, mainly trehalose.[196]

p. 681. **Survival after freezing**—In the sawfly *Trichiocampus* the prepupal diapause is so stable that it is unbroken by chilling at 2° C. for 4 months; but freezing of the body below −15° C. always results in termination of diapause. Presumably this is a different process from the chilling effect (p. 118).[195] Muscular tissues, particularly the heart and alary muscles, are highly resistant to freezing. In prepupae of *Cnidocampa* (*Monema*) *flavescens*, even after storage in liquid nitrogen for 810 days, heart beats were resumed after warming but the insects died in a few months.[195]

REFERENCES

[1] AGRELL, I. *Arkiv. Zool.*, **39**A (1947), No. 10, 1–48 (temperature and metabolism in insects).

[2] ALI, M. *Z. angew. Entom.*, **20** (1934), 354–81 (effect of temperature and humidity: *Lymantria*, Lep.).

[3] ANDERSEN, K. T. *Z. Morph. Oekol. Tiere*, **17** (1930), 649–76 (effect of temperature and humidity on egg development: *Sitona*, Col.).

[4] AUBER, L., and RAYMONT, J. E. G. *Nature*, **153** (1944), 314 (water content: *Ephestia*, Lep.).

[5] BABCOCK, S. M. *Univ. Wisconsin Agric. Exp. Sta. Res. Bull.*, No. **22** (1912), 181 pp. (metabolic water: *Tineola* larva, Lep., &c.).

[6] BABERS, F. H. *J. Agric. Res.*, **57** (1938), 697–706 (body temperature of caterpillar: *Prodenia*).

[7] BACHMETJEV, P. *Z. wiss. Zool.*, **66** (1899), 520–604 (body temperature of insects).

[8] BATTELLI, F., and STERN, L. *Biochem. Z.*, **56** (1913), 50–8 (effect of temperature on respiration: Muscidae, &c.).

[9] BĚLEHRÁDEK, J. *Temperature and living matter*, Borntraeger, Berlin, 1935.

[10] BERGER, B. *Arch. ges. Physiol.*, **118** (1907), 607–12 (resistance to desiccation: *Tenebrio* larva, Col.).

[11] BIRCH, L. C., *et al. Austral. J. Exp. Biol. Med. Sci.*, **22** (1944), 265–9; **23** (1945), 37–40 (development of eggs at different temperature and humidity: *Calandra*, *Rhizopertha*, Col.).

[12] BIRCH, L. C., and ANDRAWARTHA, H. G. *Austral. J. Exp. Biol. Med. Sci.*, **20** (1942), 1–8 (water relations of egg: *Austroicetes*, Orth.).

[13] BLUNCK, H. *Z. wiss. Zool.*, **111** (1914), 76–157 (development in *Dytiscus*, Col.: embryo); *Ibid.*, **121** (1924), 171–391 (ditto: larva and pupa).

[14] BODENHEIMER, F. S. *Z. vergl. Physiol.*, **13** (1931), 240–7 (relation between preferred temperature and atmospheric humidity).

[15] —— *Zool. Jahrb., Syst.*, **66** (1934), 113–51 (body temperature: review).

[16] BODINE, J. H. *J. Exp. Zool.*, **32** (1921), 137–64; **37** (1923), 457–76 (water content during hibernation: *Chortophaga*, &c., Orth.).

[17] —— *Physiol. Zool.*, **6** (1933), 150–8 (effect of hypertonic solutions on oxygen uptake: *Melanoplus* egg, Orth.).

[18] BODINE, J. H., and EVANS, T. C. *Biol. Bull.*, **63** (1932), 235–45 (metabolism during diapause: *Sceliphron*, Hym.).

[19] BRANDT, H. *Z. vergl. Physiol.*, **23** (1936), 715–20 (temperature and rate of crawling: *Lymantria*, Lep.).

[20] BRUES, C. T. *Quart. Rev. Biol.*, **2** (1927), 181–207 (animal life in hot springs).

[21] V. BUDDENBROCK, W., and V. ROHR, G. *Arch. ges. Physiol.*, **194** (1922), 468–72 (effect of temperature on metabolism: *Dixippus*, Orth.).

[22] BUXTON, P. A. *Proc. Roy. Soc.*, B, **96** (1924), 123–31 (temperature and humidity relations of desert insects).

[23] —— *Proc. Roy. Soc.*, B, **106** (1930), 560–77 (evaporation from *Tenebrio* larva, Col.).

[24] —— *Proc. Ent. Soc. Lond.*, **6** (1931), 27–31 (Dalton's law applied to insects).

[25] —— *Parasitology*, **24** (1932), 429–39 (humidity relations: *Rhodnius*, Hem.).

[26] —— *Biol. Rev.*, **7** (1932), 275–320 (insects and humidity: review).

[27] CAMPBELL, R. E. *Ecology*, **18** (1937), 479–89 (temperature preference: *Limonius*, Col.).

[28] CHAPMAN, R. N. *Amer. Nat.*, **62** (1928), 298–310 (temperature and activity of insects on sand dunes).

[29] COOK, S. F., and SCOTT, K. G. *Biol. Bull.*, **63** (1932), 505–12 (water relations: *Termopsis*, Isopt.).

[30] COUSIN, G. *Bull. Biol. Fr. Belg.*, Suppl. 15 (1932), 341 pp. (diapause: *Lucilia*, Dipt.).

[31] CRESCITELLI, F. *J. Cell. Comp. Physiol.*, **6** (1935), 351–68 (effect of temperature on respiratory metabolism: *Galleria* pupa, Lep.).

[32] CROZIER, W. J., *et al. J. Gen. Physiol.*, **7** (1924), 123–36; 429–47; 565–70; **10** (1927), 479–500 (effect of temperature on various processes in insects).

[33] DAVIDSON, J. *Austral. J. Exp. Biol. Med. Sci.*, **20** (1942), 233–9; *J. Animal Ecol.*, **13** (1944), 26–38 (temperature and rate of development).

[34] DEAL, J. *J. Animal Ecol.*, **10** (1941), 323–56 (temperature preferendum).

[35] DITMAN, L. P., *et al. J. Econ. Ent.*, **35** (1942), 265–72 (cold hardiness).

[36] DOBZHANSKY, T., and POULSON, D. F. *Z. vergl. Physiol.*, **22** (1935), 473–8 (temperature and oxygen consumption: *Drosophila* pupa).

[37] DOTTERWEICH, H. *Zool. Jahrb., Physiol.*, **44** (1928), 399–425 (body temperature during flight: Sphingidae).

[38] DUSPIVA, F., and CERNY, M. *Z. vergl. Physiol.*, **21** (1934), 267–74 (solar radiation and temperature: Coleoptera).

[39] EDNEY, E. B. *Bull. Ent. Res.*, **35** (1945), 399–416; **38** (1947), 263–80 (water and temperature relations: *Xenopsylla*, Aphanipt.).

[40] EVANS, A. C. *Parasitology*, **26** (1934), 366–77 (effect of temperature and humidity on *Lucilia*, Dipt., eggs); *Ibid.*, **27** (1935), 291–8 (ditto: prepupae and pupae).

[41] —— *Ann. Appl. Biol.*, **31** (1944), 235–50 (water relations: *Agriotes* larva, Col.).

[42] FRAENKEL, G. *Biol. Zentralbl.*, **49** (1929), 657–80 (behaviour in respect to solar radiation: *Schistocerca*, Orth.).

[43] GALLIARD, H. *Recherches morphologiques et biologiques sur la reproduction des réduvidés hématophages* (*Rhodnius et Triatoma*), thesis, Paris, 1935, 160 pp.

[44] FRAENKEL, G., and BLEWETT, M. *Bull. Ent. Res.*, **35** (1944), 127–39 (metabolic water in insects).

[45] GIRARD, M. *Ann. Soc. Ent. Fr.*, (sér. 4), **1** (1861), 503–8; **2** (1862), 345–7; **3** (1863), 92–8; *Ann. Sci. Nat., Zool.*, (sér. 5), **11** (1869), 135–274 (temperature of insects).

[46] GUNN, D. L. *Nature*, **128** (1931), 186–7; *Z. vergl. Physiol.*, **20** (1934), 617–25; *J. Exp. Biol.*, **12** (1935), 185–90; *J. Exp. Biol.*, **15** (1938), 555–63 (effect of desiccation on the temperature preference of the cockroach).

[47] —— *Biol. Rev.*, **17** (1942), 293–314 (body temperature of insects).

[48] GUNN, D. L., and NOTLEY, F. B. *J. Exp. Biol.*, **13** (1936), 28–34 (effect of atmospheric humidity on the thermal death-point of cockroaches).

[49] GUNN, D. L., and HOPF, H. S. *J. Exp. Biol.*, **18** (1942), 278–89 (temperature adaptation: *Ptinus*, Col.).

[50] HARRIES, F. H. *J. Agric. Res.*, **69** (1944), 127–36 (varying temperature and development: *Eutettix*, Hem.).

[51] HEADLEE, T. J. *J. Econ. Entom.*, **10** (1917), 31–8 (effect of humidity on rate of development: *Sitotroga*, Lep., *Bruchus*, Col.).

[52] van HEERDT, P. F. *Thesis, Utrecht*, 1946 (physiology and ecology of *Forficula*).

[53] HERTER, K. *Z. vergl. Physiol.*, **2** (1924), 226–32; *Biol. Zentralbl.*, **43** (1923), 282–5 (preferred temperature: *Acheta*, Orth., *Formica*, Hym., &c.).

[54] —— *Verhandl. vii internat. Kongr. Entom., 1938, Berlin*, **2** (1939), 740–59; *Z. Parasitenk.*, **12** (1942), 552–91 (temperature preferences, &c.).

[55] HIMMER, A. *Biol. Rev.*, **7** (1932), 224–53 (temperature relations of social Hymenoptera: review).

[56] HOCHRAINER, H. *Zool. Jahrb.; Abt. Physiol.*, **60** (1942), 387–436 (water content of insects).

[57] HODSON, A. C. *Ecol. Monogr.*, **7** (1937), 271–315 (water relations during hibernation).

[58] HOPF, H. S. *Biochem. J.*, **34** (1940), 1396–1403 (effect of high temperature: muscid larvae).

[59] JACK, R. W. *Mem. Dept. Agric. No. 1, S. Rhodesia*, 1939, 1–203 (water and temperature relations: *Glossina*, Dipt.).

[60] JANISCH, E. *Z. wiss. Zool.*, **132** (1928), 176–86 (effect of temperature on rate of development in insects).

[61] JOHNSON, C. G. *Parasitology*, **32** (1940), 239–70; *Biol. Rev.*, **17** (1942), 151–77 (insect survival and water loss).

[62] KALABUCHOV, N. I. *Zool. Jahrb., Physiol.*, **53** (1934), 567–602 (effect of low temperature on metabolism: honey-bee).

[63] KALMUS, H. *Z. wiss. Mikr. mikr. Tech.*, **53** (1936), 215–19 (water balance in *Tenebrio*, Col.).

[64] KOIDSUMI, K. *Mem. Fac. Sci. Agric. Taihoku Imp. Univ.*, **12** (1934), 1–380 (temperature and humidity relations of insects).

[65] KOJIMA, T. *Z. angew. Entom.*, **20** (1934), 329–53 (effect of temperature and humidity: *Dendrolimus*, Lep.).

[66] KOSMIN, N. P., *et al. Z. vergl. Physiol.*, **17** (1932), 408–22 (metabolism during activity: honey-bee).

[67] KOZHANTSCHIKOV, I. V. *C. R. Acad. Sci. U.S.S.R.*, **3** (1935), 373–6 (metabolism at temperatures below zero: Pyralidae).

[68] KOZHANTSHIKOV, I., and MASLOWA, E. *Zool. Jahrb., Physiol.*, **55** (1935), 219–30 (optimum temperature for pupal development: Lepidoptera).

[69] KOZHANTSHIKOV, I. W. *Bull. Ent. Res.*, **29** (1938), 253–62 (cold hardiness and respiration).

[70] KROGH, A. *Z. allg. Physiol.*, **16** (1914), 163–90 (effect of temperature on development: *Acilius* eggs, *Tenebrio* pupae, Col.).

[71] KROGH, A. *The Respiratory Exchange of Animals and Man*, Monographs of Biochemistry, London, 1916.

[72] KROGH, A., and ZEUTHEN, E. *J. Exp. Biol.*, **18** (1941), 1–10 (flight preparation and body temperature).

[73] KRÜGER, P., and DUSPIVA, F. *Biol. Generalis*, **9** (1933), 168–88 (effect of solar radiation on body temperature).

[74] LARSEN, E. B. *Kgl. Danske Vidensk. Selsk. Biol. Medd.*, **19** (1943), No. 3, 1–52 (heat death in insects).

[75] LARSEN, E. B., and THOMSEN, M. *Vidensk. Medd. Dansk Naturh. Foren.*, **104** (1940), 1–25 (temperature and development in Muscidae).

[76] LUDWIG, D. *Physiol. Zool.*, **1** (1928), 358–89 (temperature and rate of development: *Popillia*, Col.).

[77] —— *J. Exp. Zool.*, **60** (1931), 309–23; *Physiol. Zool.*, **9** (1936), 27–42 (desiccation and metamorphosis: *Popillia*, Col.).

[78] —— *Physiol. Zool.*, **10** (1937), 342–51 (humidity and survival: *Chortophaga*, Orth.).

[79] LUDWIG, D., and CABLE, R. M. *Physiol. Zool.*, **6** (1933), 493–508 (alternating temperatures and pupal development: *Drosophila*).

[80] LUDWIG, D., and LANDSMAN, H. M. *Physiol. Zool.*, **10** (1937), 171–9 (humidity and survival: *Popillia*, Col.).

[81] LUDWIG, D., *et al. Physiol. Zool.*, **15** (1942), 48–60; **18** (1945), 103–55; *Ecology*, **23** (1942), 259–74 (water and temperature relations in various insects).

[82] MELLANBY, K. *J. Exp. Biol.*, **9** (1932), 222–31 (thermal death point: *Pediculus*, *Lucilia*, *Xenopsylla*, *Tenebrio*).

[83] —— *Proc. Roy. Soc.*, B, **111** (1932), 376–90 (effect of temperature and humidity: *Tenebrio* larva, Col.); *Parasitology*, **24** (1932), 419–28 (ditto: *Cimex*, Hem.); *Bull. Ent. Res.*, **24** (1933), 197–202 (ditto: *Xenopsylla*, Aphanipt. during pupation); *Ann. Appl. Biol.*, **21** (1934), 476–82 (ditto: *Tineola* larva, Lep.).

[84] —— *Proc. Ent. Soc. Lond.*, 1933, 22–4 (evaporation and temperature in the insect's environment).

[85] —— *J. Exp. Biol.*, **11** (1934), 48–53 (effect of nutrition on thermal death-point: *Pediculus*, *Culex*.).

[86] —— *Biol. Rev.*, **10** (1935), 317–33 (evaporation from insects: review).

[87] —— *Nature*, **138** (1936), 124 (atmospheric humidity and metabolism).

[88] —— *Bull. Ent. Res.*, **27** (1936), 611–32 (temperature and activity: *Glossina*, Dipt.).

[89] —— *Parasitology*, **30** (1938), 392–402 (water content and body size: *Lucilia*).

[90] —— *Proc. Roy. Soc. Lond.*, B, **127** (1939), 473–87; *J. Animal Ecol.*, **9** (1940), 296–301 (adaptation of insects to low temperatures).

[91] MELVIN, R. *Biol. Bull.*, **55** (1928), 135–42 (oxygen consumption of eggs: *Samia*, Lep.).

[92] —— *Ann. Ent. Soc. Amer.*, **27** (1934), 406–10 (temperature and rate of development: Muscidae eggs).

[93] MOSEBACH-PUKOWSKI, E. *Z. Morph. Oekol. Tiere*, **33** (1937), 358–80 (temperature in larval colonies: *Vanessa*, Lep.).

[94] NECHELES, H. *Arch. ges. Physiol.*, **204** (1924), 72–93 (temperature regulation: *Periplaneta*, Orth.).

[95] NICHOLSON, A. J. *Bull. Ent. Res.*, **25** (1934), 85–99 (temperature and activity: Muscidae).

[96] NIESCHULZ, O. *Zool. Anz.*, **103** (1933), 21–7; **110** (1935), 225–33; *Z. Parasitenk.*, **6** (1933), 220–42; *Z. angew. Entom.*, **21** (1934), 224–38 (preferred temperature and activity range: *Stomoxys*, *Musca*, *Fannia*, Dipt.).

[97] NUTMAN, S. R. *Nature*, **148** (1941), 168–9 (ventral tube and water uptake: Collembola).

[98] PARHON, M. *Ann. Sci. Nat., Zool.*, (sér. 9), **9** (1909), 1–58 (metabolism of honey-bee in relation to temperature).

[99] PARKER, J. R. *Bull. Univ. Montana Agric. Exp. Sta.*, No. **223** (1930), 132 pp. (effect of temperature on *Melanoplus*, &c., Orth.).

[100] PAYNE, N. M. *J. Morph.*, **43** (1927), 521–46; *Biol. Bull.*, **52** (1927), 449–57; **55** (1928), 163–79; (cold hardiness: *Popillia*, Col., &c.).

[101] PEAIRS, L. M. *West Virginia Univ. Agric. Exp. Sta. Bull.*, No. **208** (1927), 62 pp. (temperature and pupal development).

[102] PIRSCH, G. B. *J. Agric. Res.*, **24** (1923), 275–88 (body temperature: honey-bee).
[103] POTONIÉ, H. W. *Biol. Zbl.*, **44** (1924), 16–57 (temperature and oxygen uptake).
[104] POWSNER, L. *Physiol. Zool.*, **8** (1935), 474–520 (temperature and rate of development: *Drosophila*).
[105] PRADHAN, S. *Proc. Nat. Inst. Sci. India*, **12** (1946), 385–404 (temperature and rate of development).
[106] RAIGNIER, A. *La Cellule*, **51** (1948), 281–368 (temperature in nests of ant: *Formica*).
[107] RAMSAY, J. A. *J. Exp. Biol.*, **12** (1935), 373–83 (evaporation from the cockroach).
[108] ROBINSON, W. *Ann. Ent. Soc. Amer.*, **21** (1928), 407–17 ('bound' water and cold resistance).
[109] RÜCKER, F. *Arch. ges. Physiol.*, **231** (1933), 729–49 (solar radiation and body temperature).
[110] —— *Z. vergl. Physiol.*, **21** (1934), 275–80 (reflection of infra-red rays from elytra of beetles).
[111] SACCHAROV, N. L. *Ecology*, **11** (1930), 505–17 (resistance to cold).
[112] SALT, R. W. *Univ. Minnesota Agric. Exp. Sta. Tech. Bull.*, No. **116** (1936), 41 pp. (freezing in living insects).
[113] SANDERSON, E. D. *J. Econ. Entom.*, **3** (1910), 113–39 (temperature and rate of growth).
[114] SAYLE, M. H. *Biol. Bull.*, **54** (1928), 212–30 (temperature and metabolic rate: *Aeschna* nymphs, Odonata).
[115] SCHNEIDER, F. *Mitt. Schweiz. ent. Ges.*, **21** (1948), 248–85 (water relations in diapause: Syrphidae).
[116] SCHULZ, F. N. *Oppenheimer's Handb. d. Biochemie*, 2nd Edn., **7** (1927), 439–88 (metabolism of insects).
[117] SIEGLER, E. H. *J. Agric. Res.*, **72** (1946), 329–40 (cold hardiness: *Cydia*, Lep.).
[118] SMART, J. *J. Exp. Biol.*, **12** (1935), 384–8 (effects of temperature and humidity: *Piophila*, Dipt.).
[119] SPEICHER, B. R. *Proc. Pa. Acad. Sci.*, **5** (1931), 79 (effect of desiccation on growth: *Ephestia*, Lep.).
[120] STEINER, A. *Z. vergl. Physiol.*, **9** (1929), 1–66 (temperature regulation in nests of ants, *Formica*); *Ibid.*, **11** (1930), 461–502 (ditto in nest of wasp *Polistes*).
[121] STRELNIKOW, I. D. *C.R. Acad. Sci.*, **192** (1931), 1317–19 (solar radiation and body temperature: *Bombus*, Hym.).
[122] THEODOR, O. *Bull. Ent. Res.*, **27** (1936), 653–71 (effect of temperature and humidity: *Phlebotomus*, Dipt. Nemat.).
[123] THOMSEN, E., and THOMSEN, M. *Z. vergl. Physiol.*, **24** (1937), 343–80 (preferred temperature: Muscid larvae).
[124] THOMSON, R. C. M. *Bull. Ent. Res.*, **29** (1938), 125–140 (response to temperature and humidity: *Culex*, Dipt.).
[125] TOWER, W. L. *Biol. Bull.*, **33** (1917), 229–57 (water relations in hibernation: *Leptinotarsa*, Col.).
[126] UVAROV, B. P. *Trans. Ent. Soc. Lond.*, **79** (1931), 1–247 (insects and climate: review).
[127] UVAROV, B. P. *Trans. Roy. Ent. Soc. Lond.*, **99** (1948), 1–75 (temperature relations of locusts: review).
[128] VOLKONSKY, M. *Arch. Inst. Pasteur Algér.*, **17** (1939), 194–220 (insolation and body temperature in locusts).
[128a] WALSHE, B. M. *J. Exp. Biol.*, **25** (1948), 35–44 (thermal resistance: Chironomid larvae).
[129] WEBER, H. *Z. Morph. Oekol. Tiere*, **23** (1931), 575–753 (biology of *Trialeurodes*, Hem.).
[129a] WELLINGTON, W. G. *Sci. Agric.*, **29** (1949) 201–15 (rate of evaporation and temperature preference: *Choristoneura* larva, Lep.).
[130] WEYRAUCH, W. *Z. vergl. Physiol.*, **23** (1936), 51–63 (temperature regulation in nests of social wasps).
[130a] WHITNEY, R. J. *J. Exp. Biol.*, **16** (1939), 374–85 (thermal resistance: Ephemeroptera nymphs).
[131] WIGGLESWORTH, V. B. *J. Exp. Biol.*, **21** (1945), 97–114 (transpiration from insects).
[132] —— Unpublished observations.
[133] WILKES, A. *Proc. Roy. Soc.*, B, **130** (1942), 400–15 (temperature preference in different races of the Chalcid, *Microplectron*).

SUPPLEMENTARY REFERENCES (A)

[134] DOLLEY, W. L., and WHITE, J. D. *Biol. Bull.*, **100** (1951), 84–9 (effect of illumination on lethal temperature: *Eristalis*, Dipt.).

[135] HARKER, J. E. *Proc. Roy. Ent. Soc. London*, A, **25** (1950), 111–14 (temperature adaptation in Ephemeroptera).

[136] HINTON, H. E. *Proc. Zool. Soc. Lond.*, **121** (1951), 371–80 (Chironomid larva surviving dehydration).

[137] LÜSCHER, M. *Nature*, **167** (1951), 34–5 ('fungus gardens' and temperature in termite nests).

[138] SALT, R. W. *Canad. J. Res.*, D, **27** (1949), 236–42; **30** (1952) 55–82 (water uptake in eggs of *Melanoplus bivittatus*).

[139] —— *Canad. J. Res.*, D, **28** (1950), 285–91 (effect of time of exposure on undercooling temperature of insects).

SUPPLEMENTARY REFERENCES (B)

[140] ADAMS, P. A., and HEATH, J. E. *Nature*, **201** (1964), 20–2 (temperature regulation in *Celerio* during flight).

[141] AOKI, K., and SCHINOZAKI, J. *Low Temp. Sci.*, **10** (1953), 103–16 (undercooling of the prepupa of the slug moth *Cnidocampa*).

[142] ASAHINA, E., *et al. Kintyü*, **19** (1951), 13–18; *Bull. Ent. Res.*, **45** (1954), 329–39; *Nature*, **182** (1958), 327–8 (freezing in frosty-hardy caterpillars, *Cnidocampa*).

[143] ASAHINA, E. *Bull. Marine Biol. Sta. Asamushi, Tohoku Univ.*, **10** (1962), 251–6 (mechanism of frost resistance).

[144] AZIZ, S. A. *Bull. Ent. Res.*, **48** (1957), 515–31 (humidity and behaviour in *Schistocerca*).

[145] BALDWIN, W. F., *et al. Canad. J. Zool.*, **32** (1954), 9–15, 157–71; **34** (1956), 565–7 (acclimatization to high temperatures).

[146] BĚLEHRÁDEK, J. *Ann. Rev. Physiol.*, **19** (1957), 59–82 (physiological aspects of heat and cold: review).

[147] BROWNING, T. O. *Aust. J. Sci. Res.*, **5** (1952), 96–111 (temperature and rate of development in eggs of *Gryllulus*).

[148] BÜNNING, E. *Biol. Zbl.*, **77** (1958), 141–52 (temperature and diurnal rhythms in *Periplaneta*).

[149] BURSELL, E. *Proc. Roy. Ent. Soc. Lond.*, A, **32** (1957), 21–9 (water relations in *Glossina*: spiracular control); *J. Exp. Biol.*, **37** (1960), 689–97 (ditto: feeding and excretion).

[150] —— *Phil. Trans. Roy. Soc.*, B, **241** (1958), 179–210 (water balance in *Glossina* pupa).

[151] —— *Bull. Ent. Res.*, **51** (1960), 583–98 (optimum temperature of development in *Glossina*).

[152] CHURCH, N. S. *J. Exp. Biol.*, **37** (1960), 171–85 (heat loss and body temperature in flying insects).

[153] COCKBAIN, A. J. *J. Exp. Biol.*, **38** (1961), 175–80 (water loss of Aphids in flight).

[154] COLHOUN, E. H. *Nature*, **173** (1954), 582; *Ent. Exp. Appl.*, **3** (1960), 27–37 (acclimatization to cold in *Blattella*).

[155] DEHNEL, P. A., and SEGAL, E. *Biol. Bull.*, **111** (1956), 53–61 (acclimatization of oxygen consumption to temperature in *Periplaneta*).

[156] DODDS, S. E., and EWER, D. W. *Nature*, **170** (1952), 758 (desiccation and humidity preference in *Tenebrio*).

[157] DRUMMOND, F. H. *Proc. Roy. Ent. Soc. Lond.*, A, **28** (1953), 145–8 (eversible vesicles of *Campodea* and water absorption).

[158] EDNEY, E. B. *The water relations of terrestrial Arthropods*, Cambridge University Press, 1957, 109 pp.

[159] EDNEY, E. B., and BARASS, R. *J. Ins. Physiol.*, **8** (1962), 469–81 (body temperature in *Glossina*).

[160] EDWARDS, G. A., and NUTTING, W. L. *Psyche*, **57** (1950), 33–44 (temperature range of activity in *Thermobia* and *Grylloblatta*).

[161] HANEC, W., and BECK, S. D. *J. Ins. Physiol.*, **5** (1960), 169–80 (cold hardiness in *Pyrausta*).

[162] HENSON, W. R. *Nature*, **179** (1957), 637 (temperature preference in *Grylloblatta*).

[163] HINTON, H. E. *Nature*, **188** (1960), 336–7; *Proc. Roy. Ent. Soc.*, C, **25** (1960), 7; *J. Ins. Physiol.*, **5** (1960), 286–300 (cryptobiosis in *Polypedilum*).

[164] HOUSE, H. L., RIORDAN, D. F., and BARLOW, J. S. *Canad. J. Zool.*, **36** (1958), 629–32 (dietary lipids and thermal conditioning affecting heat resistance in *Agria* larva, Dipt.).

[165] JAKOVLEV, V., and KRÜGER, F. *Biol. Zbl.*, **73** (1954), 633–50 (desiccation and preferred temperature in grasshoppers).

[166] LEADER, J. P. *J. Ins. Physiol.*, **8** (1962), 155–63 (dehydration and frost resistance in *Polypedilum*).

[167] MAELZER, D. A. *Nature*, **178** (1956), 874 (*p*F scale and behaviour of *Aphodius* in the soil).

[168] MARZUSCH, K. *Z. vergl. Physiol.*, **34** (1952), 75–92 (temperature adaptation of metabolism in various insects).

[169] MELLANBY, K. *Nature*, **173** (1954), 582–3 (adaptation to low temperature in *Tenebrio* and *Aëdes*).

[170] —— *Bull. Ent. Res.*, **50** (1960), 821–3 (adaptation to high temperature in *Aëdes*).

[171] MELLANBY, K., and FRENCH, R. A. *Ent. Exp. et Appl.*, **1** (1958), 116–24 (drinking of water by larval insects).

[172] MUNSON, S. C. *J. Econ. Ent.*, **46** (1953), 657–66 (effect of temperature on the lipids in *Periplaneta*).

[173] MUTCHMOR, J. A., and RICHARDS, A. G. *J. Ins. Physiol.*, **7** (1961), 141–58 (effect of low temperature on apyrase in *Periplaneta*, &c.).

[174] NIELSEN, E. T., and EVANS, D. G. *Oikos*, **11** (1960), 200–22 (temperature and pupal development in *Aëdes taeniorhynchus*).

[175] NOBLE-NESBITT, J. *J. Exp. Biol.*, **40** (1963), 701–11 (water and ion exchange in *Podura*).

[176] PATTÉE, E. *Bull. Biol. Fr. Belg.*, **89** (1955), 369–78; **93** (1959), 320–34 (temperature adaptation in aquatic and terrestrial insects).

[177] PEPPER, J. H., and HASTINGS, E. *Ecology*, **33** (1952), 96–103 (solar radiation and body temperature in *Melanoplus*).

[178] PERTTUNEN, V., and ERKKILÄ, H. *Nature*, **169** (1952), 78 (humidity reactions in *Drosophila*).

[179] PERTTUNEN, V., and LAGERSPETZ, K. *Arch. Zool. Soc. 'Vanamo'*, **11** (1956), 65–70 (temperature adaptation in *Corethra*).

[180] RICHARDS, A. G. In *Influence of temperature on biological systems*, 1957, 145–62; *Proc. Xth Int. Congr. Ent., 1956*, **2** (1958), 67–22; *Biol. Zbl.*, **78** (1959), 308–14 (temperature thresholds in insect development).

[181] —— *J. Ins. Physiol.*, **9** (1963), 597–606 (effect of temperature on heart-beat in *Periplaneta*).

[182] SALT, R. W. *Canad. J. Zool.*, **34** (1956), 1–5, 283–94 (moisture content and cold hardiness).

[183] —— *Proc. Xth. Int. Congr. Ent. 1956*, **2** (1958), 73–7; *J. Ins. Physiol.*, **2** (1958), 178–188 (nucleation theory and freezing of supercooled insects).

[184] —— *Canad. J. Zool.*, **37** (1959), 59–69 (role of glycerol in cold-hardening of *Bracon cephi*).

[185] —— *Ann. Rev. Ent.*, **6** (1961), 55–74 (insect cold-hardiness: review).

[186] SCHOLANDER, P. F., *et al. J. Cell. Comp. Physiol.*, **42** (1953), Suppl. 1–56 (frost resistance in Chironomid larvae).

[187] SELMAN, B. J. *J. Ins. Physiol.*, **6** (1961), 81–3 (tolerance to drying in the haemocytes of *Sialis*).

[188] SHINOZAKI, J. *Low Temp. Sci.*, (Ser. B), **12** (1962), 1–52 (ice formation in the prepupa of *Cnidocampa*).

[189] SØMME, L. *Canad. J. Zool.*, **42** (1964), 87–101 (effects of glycerol on cold-hardiness).

[190] SOTAVALTA, O. *Ann. Zool. Soc. 'Vanamo'*, **16** (1954), 1–22 (thoracic temperature of insects in flight).

[191] TAKEHARA, I., and ASAHINA, E. *Low Temp. Sci.*, (Ser. B), **18** (1960), 52–6, 58–65; **19** (1961), 30–6 (glycerol and frost resistance in the overwintering prepupa of *Cnidocampa*).

[192] TANNO, K. *Low Temp. Sci.*, (Ser. B), **20** (1962), 26–34 (glycerol and frost resistance in *Camponotus*).

[193] VIELMETTER, W. *Naturwiss.*, **41** (1954), 535–6; *J. Ins. Physiol.*, **2** (1958), 13–37 (solar radiation and behaviour of *Argynnis*).

[194] WILBUR, K. M., and MCMAHAN, E. A. *Ann. Ent. Soc. Amer.*, **51** (1958), 27–32 (protective value of glycerol at low temperature in the heart of *Popilius* Col.).

SUPPLEMENTARY REFERENCES (C)

[195] ASAHINA, E. *Adv. Insect Physiol.*, **6** (1969), 1–49 (frost resistance in insects).

[196] ASAHINA, E. and TANNO, K. *Nature*, **204** (1964), 1222 (trehalose and frost resistance in sawfly, *Trichiocampus*).

[197] ASAHINA, E. and TANNO, K. *Low Temp. Sci.*, B., **24** (1966), 25–34 (frost resistance in *Hyalophora*).

[198] BAUST, J. G. and MILLER, L. K. *J. Insect Physiol.*, **16** (1970), 979–90 (glycerol and cold hardiness in *Pterostichus*, Carabidae).

[199] BUCK, J. *J. Insect Physiol.*, **11** (1965), 1503–16 (hydration and respiration in Chironomid larvae).

[200] ESCH, H. *Z. vergl. Physiol.*, **48** (1964), 547–51 (temperature and action potentials in thoracic muscles of *Apis*).

[200a] V. HANDEL, E. *J. exp. Biol.*, **46** (1967), 487–9 (temperature and fatty acid composition in *Aëdes*).

[201] HANEC, W. *J. Insect Physiol.*, **12** (1966), 1443–9 (cold hardiness in *Malacosoma*).

[202] HANEGAN, J. L. and HEATH, J. E. *J. exp. Biol.*, **53** (1970), 349–62; 611–39 (control of body temperature in *Hyalophora*).

[203] HEATH, J. E. and ADAMS, P. A. *Nature*, **205** (1965), 309–10; *J. exp. Biol.*, **47** (1967), 21–33 (temperature regulation during flight in *Celerio*, Sphingidae).

[204] HEINRICH, B. *et al. Science*, **168** (1970), 580–1; **169** (1970), 606–7; *J. exp. Biol.*, **54** (1971), 141–66; **55** (1971), 223–39 (temperature control during flight in *Manduca* Sphingidae).

[205] HOCKING, B. and SHARPLIN, C. D. *Nature*, **206** (1965), 215 (solar radiation and temperature in arctic mosquitos).

[206] KAMMER, A. E. *Z. vergl. Physiol.*, **68** (1970), 334–44 (temperature and flight in *Danaus*).

[207] KNÜLLE, W. *J. Insect Physiol.*, **13** (1967), 333–57 (absorption of water vapour in the flea larva *Xenopsylla*).

[208] KRUNIC, M. D. and SALT, R. W. *Can. J. Zool.*, **49** (1971), 663–6 (glycerol and supercooling in *Magachile* spp.).

[209] LOVERIDGE, J. P. *J. exp. Biol.*, **49** (1968), 1–29 (control of water loss in *Locusta*).

[210] MAKINGS, P. *J. exp. Biol.*, **48** (1968), 247–63 (transpiration through Slifer's patches in Acrididae).

[211] NOBLE-NESBITT, J. *J. exp. Biol.*, **50** (1969), 745–69; **52** (1970), 193–200; *Nature*, **225** (1970), 753–4 (water uptake from subsaturated atmospheres in *Thermobia* and *Tenebrio*).

[212] OHYAMA, Y. and ASAHINA, E. *Low Temp. Sci.*, **27** (1969), 153–60 (ice formation in *Camponotus*).

[213] OKASHA, A. Y. K. *J. exp. Biol.*, **48** (1968), 455–86; *J. Insect Physiol.*, **14** (1968), 1621–34; **16** (1970), 545–53 (high temperature and arrest of growth in *Rhodnius*).

[214] SALT, R. W. *Symp. Soc. exp. Biol.*, **23** (1969), 331–50 (survival of insects at low temperature: review).

[215] SØMME, L. *J. Insect Physiol.*, **12** (1966), 1069–83 (effect of temperature and anoxia on supercooling in *Ephestia*).

[216] SØMME, L. *Norsk ent. Tiddsskr.*, **16** (1969), 107–11 (mannitol and glycerol in overwintering Aphid eggs).

[217] WATSON, J. A. L. *et al. J. insect Physiol.*, **17** (1971), 1705–9 (water sacs in *Hodotermes*).

[218] WIGGLESWORTH, V. B. *J. exp. Biol.*, **32** (1955), 649–63 (high temperature and arrested growth in *Rhodnius*).

[219] WINSTON, P. W. *et al. Nature*, **214** (1967), 383; *J. exp. Biol.*, **50** (1969), 541–6 (water pump in integument of *Periplaneta* and *Schistocerca*).

Chapter XV

Reproductive System

REPRODUCTION IN most insects is bisexual; the egg cell liberated by the female will develop only after fusion with the spermatozoal cell set free by the male. The physiology of reproduction deals with the arrangements for the separation and ripening of these male and female gametes, and with the mechanisms by which they are brought together. The reproductive system consists of paired sexual glands, the ovaries of the female and testes of the male, paired gonoducts of mesodermal origin into which the sexual products are discharged, and a median duct lined with cuticle, derived by invagination from the ventral body wall, forming the vagina in the female and the ejaculatory duct in the male.

FEMALE REPRODUCTIVE SYSTEM

The ovary—Each ovary is made up of a series of egg tubes or ovarioles, varying in number from one in some Aphids, and two in the viviparous Diptera such as *Glossina*,[386] to more than 2,000 in certain termites. The ovarioles consist of a chain of developing ova (Fig. 381). Each is composed of a layer of epithelium resting on a basement membrane, the whole enclosed in a connective tissue coat. They are divided into the following zones.

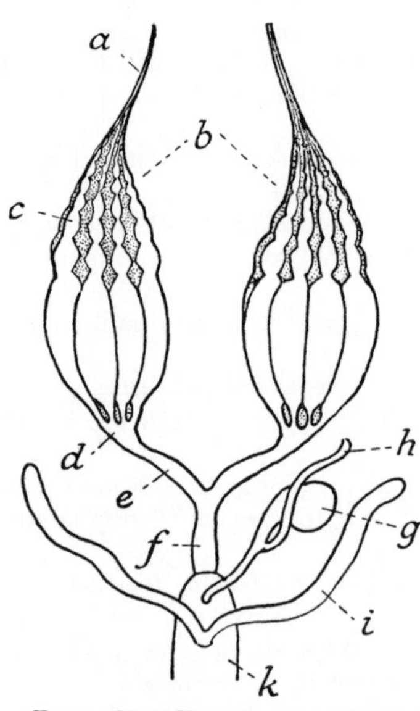

FIG. 381.—Female reproductive system, diagrammatic (*after* SNODGRASS)

a, ovarial ligament; *b*, ovary; *c*, ovariole; *d*, calyx; *e*, lateral oviduct; *f*, common oviduct; *g*, spermatheca; *h*, spermathecal gland; *i*, accessory duct; *k*, vagina.

(i) *Terminal filament*, a thread-like continuation of the connective tissue layer, which often attaches all the ovarioles to the body wall.

(ii) *Germarium*, consisting of densely packed cells from among which the primordial germ cells or oögonia become differentiated into oöcytes and nutritive or nurse cells.

(iii) *Vitellarium*, composing the greater part, which consists of a series of oöcytes, each enclosed in an epithelial sac or follicle, becoming progressively larger towards the lower end.

(iv) *Ovariole stalk*, a thin-walled tube leading to the oviduct. While the leading oöcyte is ripening it is separated from the lumen of this tube by a solid plug of epithelium. This plug is broken down during ovulation (p. 705).

Two types of ovarioles are recognized: (*a*) *Panoistic type*, in which the nutritive cells are wanting, and the yolk of the egg is formed solely by the epithelium of the egg follicles (Fig. 382, A). This is probably the primitive type, but it has been secondarily acquired by many forms such as Ephemeroptera, Aphaniptera and many Orthoptera. (*b*) *Meroistic type*, in which nutritive or nurse cells as well as germ cells arise from the primitive sex cells and contribute to the nourishment of the oöcytes in the early stages of their development. This type is divided into two sub-groups: the *polytrophic group*, in which each oöcyte has a number of nurse cells enclosed with it in its follicle, as in Dermaptera, Lepidoptera, Diptera, Hymenoptera, Coleoptera-Adephaga (Fig. 382, B); and the *acrotrophic* or *telotrophic* group, in which the nurse cells are confined to the apex of each ovariole and are connected to the developing oöcytes in the early stages of their development by means of long nutritive cords, as in Hemiptera, Coleoptera-Polyphaga.

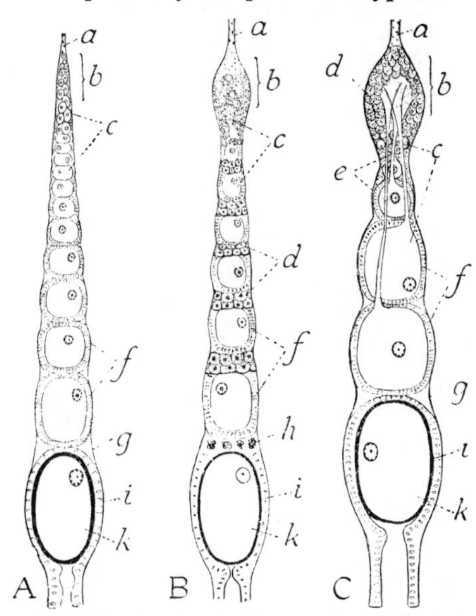

FIG. 382.—The three chief types of ovaries, diagrammatic (*after* WEBER)

A, panoistic type; *B*, polytrophic type; *C*, telotrophic type. *a*, terminal filament; *b*, germarium; *c*, oöcytes; *d*, nurse cells; *e*, nutritive cords; *f*, follicular cells; *g*, peritoneal coat; *h*, degenerated nurse cells; *i*, chorion; *k*, egg.

Genital ducts and accessory structures—The paired oviducts derived from the mesoderm, sometimes, as in Lepidoptera, distended to form pouches for storing eggs, unite with the common oviduct, which is lined with cuticle continuous with the body surface. The common oviduct is generally termed the vagina; in some Diptera its anterior part, which has well-developed muscular walls, is called the uterus. Associated with the vagina are the following:

(i) The *receptacular duct* leading to the *spermatheca* or *receptaculum seminis*. This is usually a fine duct opening on to the dorsum of the vagina. The spermatheca is a pouch lined with cuticle in which the spermatozoa received at copulation are stored. It may contain glands. It is usually single; but there are two in *Blaps* (Col.), *Phlebotomus* and *Dacus* (Dipt.), three in *Culex*, Tabanidae and most higher flies.

(ii) The *bursa copulatrix* into which the sperm are discharged before entering the spermatheca. It is often absent. When present it generally forms a diverticulum from the vagina; and then the bursa or its duct usually receives the receptacular duct. In the higher Lepidoptera, the external opening of the bursa, used for copulation, is quite separate from the vaginal opening through which the eggs are laid (Fig. 383).

(iii) *The accessory glands*, of which there are usually one or two pairs, open into the distal portion of the vagina. They are often termed 'colleterial glands', since they commonly produce an adhesive cement. But they have many other functions (p. 706).

All these structures show great diversity in the details of their arrangements among different insects.[234]

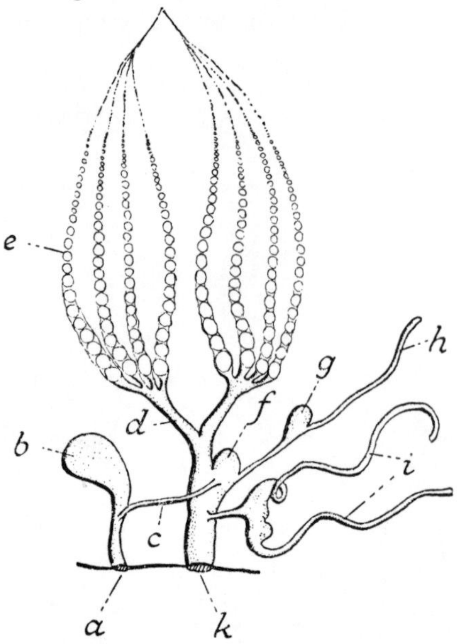

FIG. 383.—Female reproductive system in Lepidoptera; schematic (*after* EIDMANN)

a, ostium of bursa (copulatory opening); *b*, bursa copulatrix; *c*, seminal duct; *d*, paired oviduct; *e* ovary; *f*, vestibulum; *g*, receptaculum seminis; *h*, receptacular gland; *i*, accessory glands; *k*, ovipore.

Oögenesis[284]—The oögonia in the end chamber or germarium become differentiated into oöcytes and nurse cells. As the oöcytes move backwards they become surrounded by the 'follicle cells' of somatic origin, which compose the hind part of the germarium, and each oöcyte in its follicle forms an egg chamber of the vitellarium. At first the follicle is several cells thick; as the oöcyte grows the cells finally arrange themselves in a single layer.

When nurse cells are present there may be a single one to each egg as in Dermaptera, 5 as in Lepidoptera, or up to 48 as in the queen bee.[38] They obtain nourishment from the blood and pass it on to the oöcyte. In Lepidoptera the streaming of secretion from the nurse cells into the oöcyte is clearly visible in histological sections (Fig. 384).[199] In Hemiptera with their telotrophic ovaries, the nutrient cords extend into the oöcytes from a cavity in the germarium surrounded by nurse cells; these cords are probably formed chiefly by the nurse cells, though perhaps in part by the oöcytes themselves.[40] In the polytrophic ovaries the nurse cells ultimately break down completely and are absorbed into the egg. In the telotrophic type the nutrient cords are severed as soon as the oöcyte has grown to a certain size.

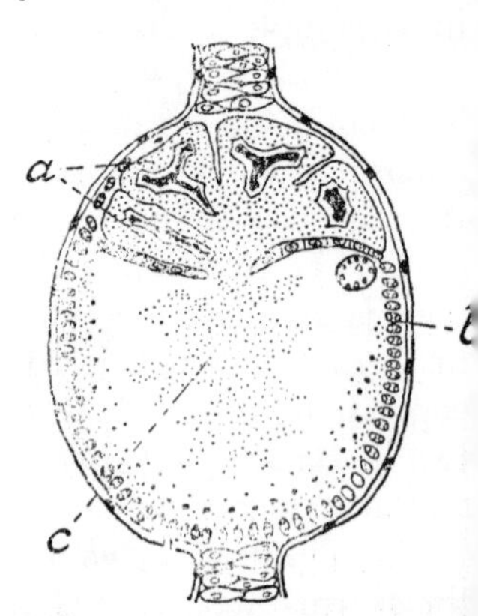

FIG. 384.—Longitudinal section of oöcyte of *Celerio* at the height of the secretory activity of the nurse cells (*after* SCHNEIDER)

a, nurse cells; *b*, follicular cells; *c*, secretion from nurse cells streaming into the egg plasma.

In the later stages of their growth the eggs are nourished by secretion from the follicular cells. At this time the follicular cells become greatly enlarged, and the ratio of plasma to nucleus increases. The nuclei may become constricted in the middle, as in Phasmids[28] and *Pediculus*,[55] or they may divide amitotically into two, in many insects. But this nuclear fission never leads to division of the cells;[117] it is perhaps simply a mechanism for increasing the nuclear and nucleolar surface during the phase of intense secretory activity.[148] Such amitosis is not uncommon among cells which are approaching the

Yolk formation—As soon as the oöcyte begins to enlarge beyond its sister nurse cells, granules and vacuoles of reserve material make their appearance in the cytoplasm. These yolk spheres doubtless have a different composition in different insects; but they consist in the main of proteins, fats and carbohydrates. Droplets or granules of these materials can never be demonstrated in the nurse cells or in the follicular cells; they appear to be formed within the oöcyte from the secretions with which it is provided.[186] The fatty yolk droplets arise by the deposition of fat within the Golgi vacuoles of the oöcytes, in *Periplaneta*,[73, 151] *Luciola* (Col.),[150] Tenthredinidae.[73] The protein spheres arise from basophil granules, which appear in the neighbourhood of the nucleus and subsequently migrate to the periphery and enlarge. These granules are sometimes referred to as 'chromidia'; in Anoplura they are said to arise exclusively from the chromosomal chromatin (though they fail to give the Feulgen reaction for nuclein) and

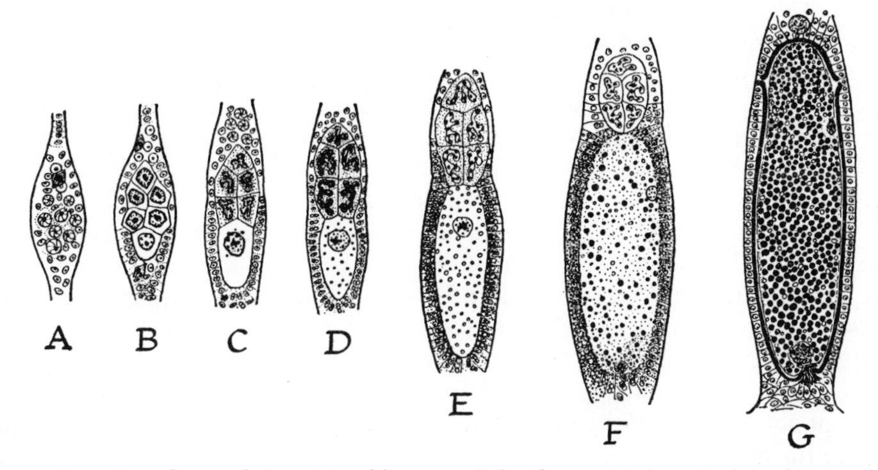

FIG. 385.—Oögenesis in polytrophic ovary of *Anoplura*, semischematic (*after* RIES)

A, stage showing six undifferentiated oögonia. *B*, the six cells differentiated into five nurse cells and one oöcyte; follicular epithelium forming. *C*, the enlarged nucleus of oöcyte giving off chromidia to the plasma; nurse cells beginning to function; next group of oögonia visible above. *D*, nuclei of follicular cells assume dumb-bell form; nurse cells at height of activity; plasma with numerous chromidia; *E*, follicular epithelium active; nurse cells begin to regress; chromidia formation ceases; fat droplets appear in plasma. *F*, follicular epithelium at height of activity; symbionts invade lower pole of egg; nucleus of oöcyte dissolved, chromosomes free in plasma at periphery of egg; yolk formation begins. *G*, ripe egg with reserve materials complete; follicle cells have laid down chorion; mass of symbionts at lower pole; nurse cells degenerated; spindle of first maturation division formed.

to be shed off solely from the nucleus of the oöcyte itself (Fig. 385, C).[186] In *Periplaneta*[92,151] they arise by the extrusion of buds from the nucleolus of the oöcyte. They then migrate to the cytoplasm, and thence to the periphery of the egg. Here they grow and break up into small chromatic bodies, which enlarge to produce the albuminous yolk spheres.[73, 92, 151] A similar process occurs in Tenthredinidae;[73] but here the so-called 'accessory nuclei' are extruded from the nuclei of the nurse cells and follicular cells as well as by budding from the nucleolus of the oöcyte itself.[166] In *Luciola* the nucleolar budding continues throughout oögenesis.[150] In *Bombus* the basophil droplets are discharged from the nurse cells; there is no emission of nucleolar material.[160] [See p. 742.]

In Anoplura the chromidia are formed very early; the fat droplets appear next. The former grow to produce the yolk spheres and the latter appear in abundance during the time when the follicular cells reach the height of their activity and the nurse cells are degenerating. Glycogen is diffused throughout the egg substance; its appearance coincides with the completion of yolk formation when the follicular cells are beginning to secrete the egg-shell (Fig. 385).[186]

Histochemical studies combined with the use of radioactive tracers and electron microscopy have largely confirmed and clarified earlier conclusions. It is evident that there are considerable differences in the different types of ovary. In the panoistic type there is active synthesis and emission of ribonucleic acid (RNA) from the nucleolus, and thence from the nucleus of the oöcyte to the cytoplasm. This is seen in *Periplaneta*,[283] *Blatta*,[317] and *Gryllus*.[393] In the advanced meroistic type, on the other hand, the oöcyte nucleus may apparently play no part in RNA synthesis. In *Bombyx mori* the seven nurse cells become polyploid through endomitosis and generate RNA which flows into the cytoplasm of the oöcyte.[300] In *Drosophila*, of the 16 cells which arise from four consecutive divisions of an oögonium, 15 become nurse cells and the most posterior becomes the oöcyte. Large pores, termed 'fusomes',[360] in the walls separating adjacent nurse cells allow a stream of cytoplasmic material to flow backwards to the oöcyte.[343] Likewise in *Musca* the nurse cells are of primary importance as the source of RNA and other components of the cytoplasm concerned directly with embryonic development;[282] and a similar process is seen in the 12 nurse cells and oöcyte of *Cynips*.[344a] In *Panorpa* an intermediate state of affairs is seen: RNA is supplied mainly from the cytoplasm of the nurse cells, but to a less extent by the oöcyte or by the follicular epithelium.[374] In the telotrophic type of ovary the nurse cells are again the chief source of nucleic acids. In *Oncopeltus* they discharge RNA and DNA into the trophic core.[283] In the nurse cells of *Rhodnius* masses of RNA, apparently derived from the nucleolus, can be seen passing through 400 Å pores in the nuclear membrane to reach the cytoplasm on their way to the oöcyte.[271] [See p. 742.]

Some of the proteins of the egg are synthesized in the nurse cells and follicular cells[406] and in the oöcyte itself; but the bulk of the proteins of the yolk are synthesized elsewhere, chiefly in the fat body (p. 446) and transferred to the yolk by the follicular cells.[401] The blood protein reaches the oöcyte by passing between the follicular cells and appears to be taken up by pinocytosis to form vesicles which later fuse to form the smaller yolk spheres.[394, 402] This process is seen in *Hyalophora*,[394, 402] *Aëdes*,[383] *Periplaneta*,[270] and *Oncopeltus*.[338] [See p. 742.]

The follicular cells certainly play a large part in the synthesis of other components of the yolk. In *Schistocerca* these consist of glycoproteins, phospholipid, triglycerides, and mixtures of these two lipids, associated with β-carotene from the food.[356] In *Drosophila* a phospholipid protein complex, a combination of phospholipids and acid mucopolysaccharides, and triglyceride droplets are recognized;[343] and similar complex mixtures in *Oncopeltus*,[283] *Anisolabis*,[283] *Periplaneta*,[283, 365] *Bombyx*,[300] &c. During yolk deposition in the oöcytes of *Melanoplus* the nuclei oscillate regularly through an angle of 5–20° with a period varying for 5–45 seconds. The significance of this movement is not known.[399] [See p. 742.]

Egg membranes—When the yolk is fully formed, the vitelline membrane (p. 2) arises by the condensation of its outermost layer; and the chorion is secreted by the follicular epithelium (Fig. 386). In *Drosophila* the 'choriovitelline membrane' (p. 2) which contains protein, two different lipids, neutral and acid mucopolysaccharides, is a product of the follicular cells and not of the oöcyte.[344] The endochorion contains protein, mucopolysaccharide and lipid (all distinct from those of the choriovitelline membrane); the exochorion has large amounts of acid mucopolysaccharides. After leaving the ovariole the oöcyte secretes a layer of waterproofing wax between the choriovitelline membrane and the chorion.[344]

The chorion often shows in its surface sculpturing the impress of the follicular cells; it frequently has outgrowths of characteristic form, a detachable cap at the anterior end, or other more or less elaborate structures which indicate a high degree of co-ordination among the epithelial cells by which they are laid down.[11, 8] For example, in some species of Phasmids, after one chorion has been laid down, the follicle is thrown into longitudinal folds; it then smoothes itself out and lays down a second sheath, attached to the first only at the operculum which marks the position of the micropyle.[28]

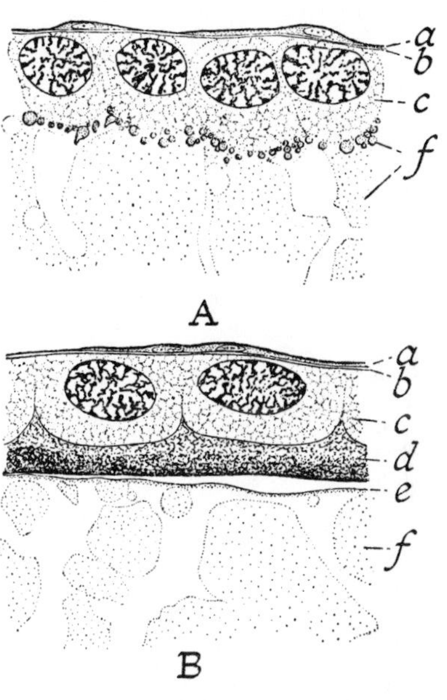

FIG. 386.—Follicular epithelium of *Melanoplus differentialis* (Orthoptera)

A, during the deposition of yolk; *B*, during the secretion of the chorion (*cf.* Fig. 2) (*after* SLIFER). *a*, accessory membrane; *b*, outer limiting membrane; *c*, follicular epithelium; *d*, chorion; *e*, vitelline membrane; *f*, yolk.

Micropyle—At one point the chorion is either very thin and adherent to the vitelline membrane, so that the spermatozoa may enter after the shell is formed, or there may be actual perforations in the chorion. In Muscidae there is an entomicropyle penetrating the vitelline membrane, an ectomicropyle or perforation of the chorion, with an area of modified sculpturing around, and a soft 'mucous' plug projecting to the exterior.[161] The number of openings varies even within a single species. In *Locusta* the apparatus consists of a ring of 35–43 funnel-shaped cuticular canals which taper inwards and open to the interior of the egg by fine apertures in the vitelline membrane.[190] In Tettigoniids, the canaliculi run obliquely through the chorion; they vary in number from 1–23 in different species.[27] In *Rhodnius* there are 10–20 true micropyles scattered irregularly around the rim of the cap; they become reduced in number in the eggs from older females and this may account for the high rate of sterility among such eggs.[8]

Ovulation; 'corpus luteum'—When the leading oöcyte is ripe, it ruptures the epithelial plug which closes the vitellarium below, and passes into the oviduct. This epithelial plug does not regenerate in its original form.[76] The empty follicle collapses, the cells degenerate, undergo autolysis and gradually disappear (Fig. 387). In the early stages of this process the degenerating follicle is a conspicuous object and is known as the 'corpus luteum'; but it shows no progressive development after ovulation like the corresponding organ in mammals. In termites, the cells close to the autolysing follicle may contain deposits of a bright yellow pigment, but this has no relation with the lipochrome lutein of vertebrates.[3] As the eggs above enlarge and extend downwards this body shrinks, and by the time the next egg is ripe it has almost disappeared. When this egg is ovulated it ruptures its follicle and passes through the remains of the preceding corpus luteum.[76]

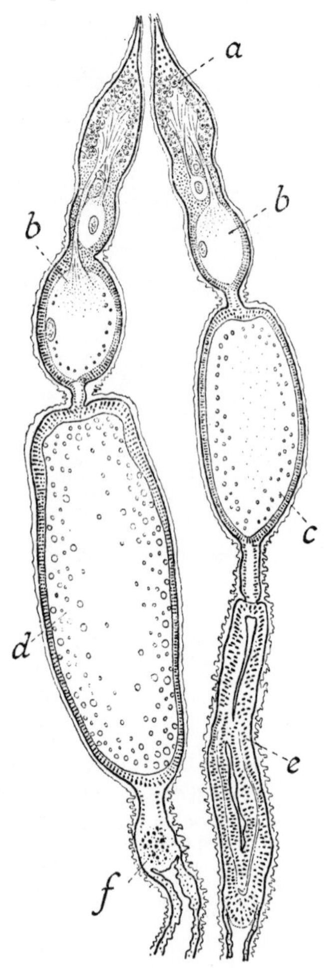

FIG. 387.—Egg development in *Rhodnius* (*after* WIGGLESWORTH)

a, nurse cells; *b*, oöcytes nourished by nutritive cords from nurse cells; *c*, oöcyte nourished solely by follicular epithelium; *d*, egg almost fully developed; *e*, empty follicle, or corpus luteum; *f*, last remains of corpus luteum.

Ovulation results from rhythmical contractions of the lateral oviducts and is initiated by the nervous system.[303] In many insects, for example Lepidoptera, the eggs are set free in rapid succession as they are formed, independently of any external stimulus. In *Glossina*, ovulation seems to take place only after impregnation, the central nervous system perhaps controlling the process; thus, normally one egg is produced at a time; but virgin females 4 or 5 weeks old may contain three or four fully developed eggs in the ovaries; these are ovulated within three days after fecundation.[140] In *Anopheles* females it is possible to judge the number of egg-laying cycles through which the insect has passed by counting the number of dilatations left in the ovaries.[278]

Oviposition—The eggs are carried down the oviducts by waves of peristalsis, and deposited singly or in masses according to the habits of the species. They are generally coated with secretion from the accessory glands; as was suggested by Malpighi, this causes them to adhere to the surface on which they are laid, as in Lepidoptera,[49, 155] Hemiptera,[66] &c. Sometimes these glands provide a gelatinous sheath in which the eggs are embedded, as in the aquatic Trichoptera and Chironomidae. They may be transformed into silk glands by which an elaborate 'egg cocoon' is formed, as in Hydrophilidae.[122] In the oötheca of *Plataspis* (Hem.) the eggs are held together by secretion from a special region of the mid-gut.[30]

In the cockroach, the eggs are retained in the lower end of the vagina, and cemented together by a thick fluid containing calcium oxalate secreted by the accessory glands; this hardens in the air to form the shell of the egg capsule or oötheca.[81, 397] The process of hardening, which results from the tanning of the protein secreted in the left colleterial gland, by means of the quinone formed from the oxidation of protocatechuic acid secreted in the right colleterial gland[177, 286] has already been described (p. 34). Calcium oxalate may form as much as 15 per cent. of the dry weight in the oötheca of *Periplaneta*; 0·3 per cent. in *Blattella*; it is absent from ovoviviparous species.[397] In *Periplaneta* the oötheca contains 6·5 per cent. of calcium; only a third of this is present as oxalate, the remainder perhaps as a calcium-protein complex.[318] As the cement is poured over the eggs to form the oötheca a small process is held in position above the anterior pole of the egg, and serves to mould a pair of ducts which convey oxygen from the atmosphere to the egg shell (Fig. 2).[414]

Gummy material from the accessory glands forms the oötheca of Mantids; it is worked into an elaborate structure, the outermost layers being mixed with air to form a froth which hardens to produce a vacuolated sheath.[245] This is composed of proteinaceous membranes containing small rectangular crystals of calcium citrate secreted by a special gland.[371] In Acridiids the oötheca is contributed by glands in the mesodermal part of the oviducts; but the buried egg mass is covered with a foamy secretion from the accessory glands, which hardens to form a plug. In Hymenoptera the accessory glands are modified to form (i) the poison gland which sometimes serves, in Pompilids and Sphecids, to paralyse the prey on which the larva is to feed, and (ii) the so-called 'alkaline' or Dufour's gland which serves perhaps to lubricate the ovipositor, or sting, perhaps as a 'colleterial gland' (p. 607).

MALE REPRODUCTIVE SYSTEM

Testis and spermatogenesis—The testis is made up of a series of tubular follicles varying greatly in number and arrangement in different insects.[234] Each follicle consists at first of a layer of epithelium resting on a basement membrane. Later, the entire group of follicles is enclosed in a connective tissue sheath, often pigmented, which determines the outward form of the organ. In Lepidoptera the two testes are bound together in a single capsule.

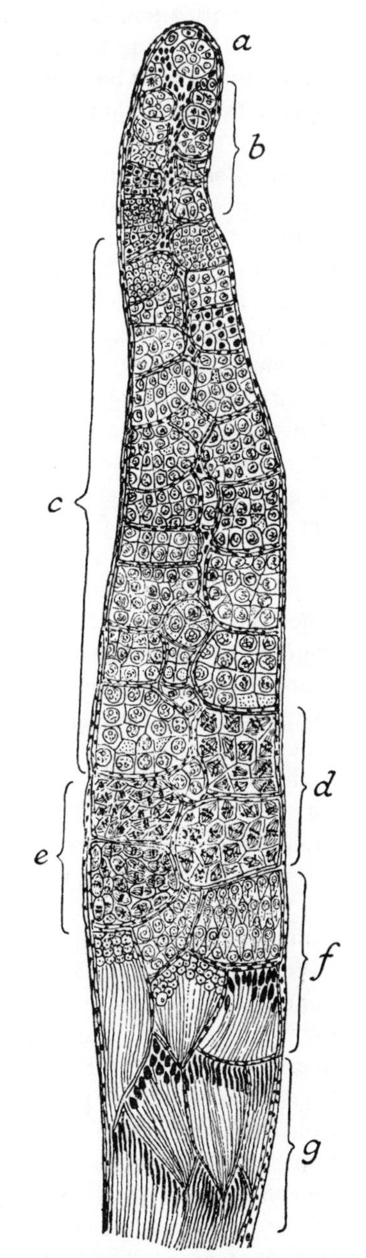

FIG. 388.—Longitudinal section of testis follicle of an Acridiid, semischematic (*from* WEBER *after* DEPDOLLA)

a, apical cells surrounded by spermatogonia; **b**, zone of spermatogonia; **c**, zone of spermatocytes; **d**, cysts with mitoses of first maturation division; **e**, cysts with mitoses of second maturation division; **f**, zone of spermatids; **g**, zone of spermatozoa.

Each follicle contains a succession of zones in which the sex cells are in different stages of development (Fig. 388). These are (i) the *germarium* or *zone of spermatogonia*, consisting of densely packed primordial germ cells or spermatogonia, lying among somatic mesodermal cells. At the upper extremity of the follicles in Lepidoptera and some other insects large 'apical cells' are present which serve as nurse cells in the early stages of sperm development.

(ii) *Zone of spermatocytes*. As each primitive spermatogonium moves backwards, it becomes covered with a mantle of somatic cells; meanwhile it divides repeatedly and distends this sheath to form a cyst. At first the cysts are rounded, but as they increase in size they become polyhedral from mutual pressure. Each contains the 'spermatocytes', from 64 to 256 in number, derived from a single spermatogonium.

(iii) *Zone of maturation and reduction.* The spermatocytes now divide, first into 2 'prespermatids', and then into 4 'spermatids'. In most insects the first of these is the reduction division at which the chromosome number is halved. During the meiotic division in *Melanoplus* the mitochondria enlarge and divide equally between the two 'nebenkern' halves on each side of the tail filament.[400]

(iv) *Zone of transformation*, in which the compressed and rounded spermatids still enclosed in their cysts, are converted into flagellated spermatozoa. The mature spermatozoon of *Drosophila* measures 1·76 mm. in length, which is more than three times the length of the egg (0·5 mm.).[301] In *Orgyia* they are about 335μ long and about 0·4μ thick.[298] The number of spermatozoa produced is commonly very large; but in *Miastor* the male produces exactly 1,024 sperms per individual per life cycle.[240]

The cytology of this process of spermatogenesis has been studied in great detail in many insects.[40] The mature sperm break through their cyst wall to enter the genital ducts by means of the periodic writhing movements of their flagella. At this stage they are often still held together in bundles (spermatodesms) of various forms, their heads being inserted into a cap or rod of gelatinous material. In Scarabaeidae the sperm bundles represent the sperm cells of one cyst still enclosed as in a capsule. They first become activated in the vas deferens, perhaps by oxygen.[408] In the grasshopper *Chortophaga* this hyaline cap remains intact until the bundles reach the seminal receptacle of the female; here it is dissolved, probably by enzyme action, and the spermatozoa set free.[165] In *Popillia japonica* the mature spermatozoa become aligned with their heads embedded in the cyst cell, the sperm tails beating in a co-ordinated manner. The cyst cells contain stores of glycogen which persist until the cells disintegrate in the female tract.[272] In *Pseudococcus* the sperm bundles usually contain 16 sperms so closely bound, with a corkscrew-like head to the bundle, that they are often mistaken for the spermatozoa themselves. The bundle sheath is digested and the sperm liberated shortly before fertilization.[369] In *Bombyx mori* the spermatozoa in the testis become active, perhaps under the influence of some chemical substance from the lower part of the tract, and wriggle in bundles through the basilar membrane which separates the testis lobules from the efferent duct, the bundles being broken down when they come in contact with the secretion of the accessory glands.[157] In most insects, mature spermatozoa have already left the testis at the time of emergence from the pupa. [See p. 742.]

Genital ducts and accessory structures (Fig. 389)—These consist essentially of fine paired ducts, 'vasa deferentia', which are either dilated at one point ('vesicula seminalis') or provided with a tightly coiled middle section (the 'accessory testis' of many Coleoptera) so as to furnish a reservoir for the mature sperm. The vasa deferentia unite to form a common duct continuous with the median 'ejaculatory duct' of ectodermal origin. This is lined with cuticle and provided with a powerful muscular coat made up of outer circular and inner longitudinal fibres. Its terminal part is often enclosed in an evagination of the body wall to form the intromittent penis or 'aedeagus'. 'Accessory glands' of varied number and form discharge into these ducts; in addition, in some insects, parts of the ducts themselves may be glandular in function.

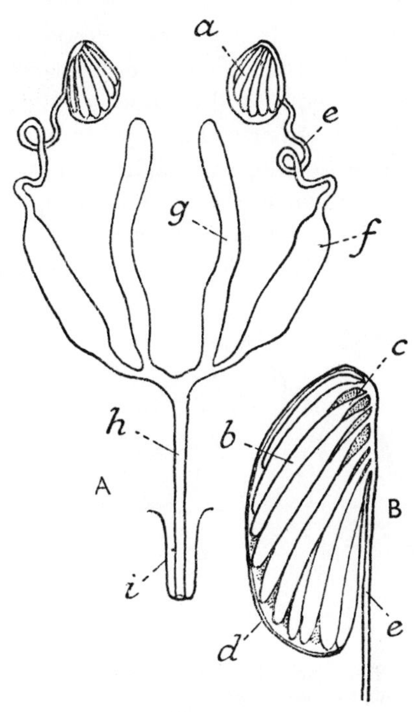

FIG. 389

A, male reproductive system, diagrammatic. *B*, detail of testis (*after* SNODGRASS). *a*, testis; *b*, follicles; *c*, vas efferens; *d*, peritoneal sheath; *e*, vas deferens; *f*, vesicula seminalis; *g*, accessory glands; *h*, ejaculatory duct; *i*, penis.

MATING, IMPREGNATION, AND FERTILIZATION

Mating—The sexual behaviour of insects is apparently not influenced by the gonads. The sexual and social activities in *Polistes* females are not affected by castration.[39] Females of Aleurodidae copulate soon after emergence, long before the ovaries are fully developed.[233] Castrated males of *Lymantria* will copulate normally, although no spermatozoa are passed;[159] and males and females of *Gryllus* will mate in the usual way after castration.[179] In male *Dytiscus*, sexual activity seems to be due to a nervous impulse resulting from distension of the accessory glands.[12] Mating behaviour in *Galleria* is unaffected by removal of the gonads and of the corpora allata.[377] *Chorthippus* and other grasshoppers will stridulate and mate after extirpation of testes and accessory glands.[321] The same is true of the grasshopper *Euthystira*, but this insect fails to show the repulsion response which normally follows copulation; sexual behaviour is not completely independent of the gonads.[375]

Stimuli of many kinds provide the immediate cause of mating. The females may be attracted to the males by their dancing in swarms, as in various Diptera-Nematocera,[310] Trichoptera, Plecoptera, and Ephemeroptera; or by sounds, such as the chirping of crickets and grasshoppers.[179, 288] Communication in *Ephippiger* before mating is effected also by trembling movements that set up vibrations of the vegetation on which the insects rest.[289] Stridulation of the male *Dacus oleae*, which is essential in mating, is effected by combs of hairs on the third abdominal tergites and a modified margin to the wing.[313] In the mating of *Aëdes* and other mosquitos the sounds emitted by the flying female are perceived by the male (p. 270) and elicit a succession of responses leading to clasping and mating. This mating behaviour can be evoked by the sound of a tuning fork of the appropriate frequency.[111] Males may be attracted from a distance by the luminous organs of the female, as in Lampyrid beetles. Among butterflies, the colour of the female may attract the male when he is in the right physiological state. The female may be excited by a complex ceremonial love play.[234] But the most important stimuli are probably scents.[184] In the courtship and mating of *Argynnis*, optic, chemical, and mechanical stimuli are all involved.[261] [See p. 743.]

Scent organs in mating—Scent organs in the female may attract the male from a distance. Such organs are particularly common among Lepidoptera. Here they occur in the neighbourhood of the sexual opening and take the form either of tufts of modified scales, or hairs with gland cells at their base, or a simple fold in the body wall consisting of a glandular epithelium covered by thin cuticle devoid of pores (Fig. 390).[63] Such organs are said to be best developed among those Lepidoptera, such as Lasiocampidae and Bombycidae, in which the eggs of the female are ripe for laying at the time of emergence from the pupa.[49, 111] Males of *Arctias selene* liberated at a distance of 11 km. were able to return to the females.[139] Males of *Callosamia promethea* are readily attracted to the isolated abdomen of the female, or to places where a recently emerged female has rested; they pay no attention to females in a glass container or deprived of the abdomen, and they are not repelled by any abnormalities produced experimentally in the colour or pattern of the wings.[136] Males of *Bombyx mori*,[109] and of *Ephestia* and *Plodia*[45] show the same behaviour. Males of *Orgyia* will endeavour to copulate with pieces of blotting paper on which the droplets of secretion from the everted gland of the female have been imbibed;[63] and males of *Habrobracon* (Hym.) are attracted not only to females, but to males that have recently copulated.[147] [See p. 743.]

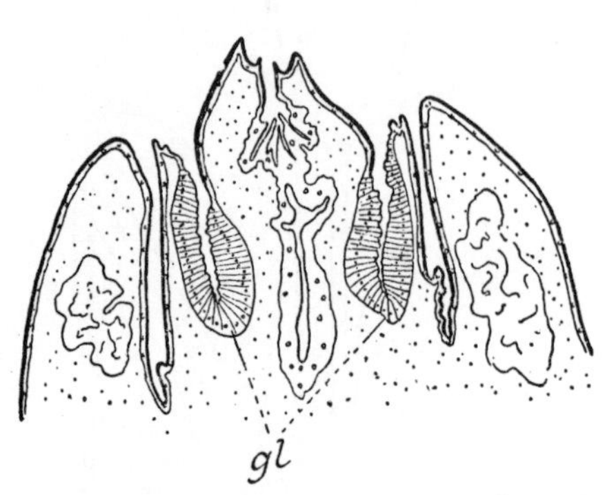

FIG. 390.—Horizontal section through tip of abdomen of *Plodia* female, showing intersegmental scent gland (*gl*) (*after* BARTH)

The attractive sexual scents of a number of female insects have been characterized chemically, notably those of *Bombyx*,[290, 291] *Lymantria*,[332] *Belostoma*,[290] and *Periplaneta*.[333] They are mostly unsaturated, straight-chain aldehydes, alcohols, or ketones, with 8–16 carbon atoms. The attractive unit of *Bombyx* scent is contained in $10^{-10}\mu g$;[290, 291] that for *Lymantria* is attractive in field tests in quantities of less than $10^{-7}\mu g$.[332] In *Periplaneta* the secretion of the female sexual attractant diminishes with age and is depressed by mating.[410] [See p. 743.]

In the honey-bee the female sex attractant is the same 'queen substance' (p. 609) that is secreted by the mandibular glands and is used for communication within the hive.[362] The corresponding attractant in the mandibular glands of *Bombus* is believed to be farnesol.[398]

Females of Lepidoptera awaiting fecundation take up a characteristic 'calling' position, the stimulus to which seems to be provided by a shortage of sperm in the receptaculum.[155] Only during the assumption of this attitude are the scent-producing membranes exposed.[45] Presumably the scent produced by the female in these cases is characteristic of the species. The synthetic sex attractant of *Bombyx* female does not elicit any electrical response in the antenna of any of the Saturniid males tested. The effect of Saturniid glands on male *Bombyx* is far smaller than that of the female *Bombyx* gland.[389] But in confinement the males of the various species of *Ephestia* and *Plodia* are attracted to 'calling' females of other species, and make attempts to copulate with them;[7, 45] and the same is true of the related genera *Galleria* and *Achroea*.[7] There is a similar overlap in the attractiveness of the female scent in *Lymantria monacha* and *L. dispar*.[390] Males of the Trypetid fly *Dacus* are attracted to specific essential oils: *D. zonatus* to methyl-eugenol, *D. diversus* to isoeugenol, *D. ferrugineus* to both, but especially the latter, while eugenol itself is not attractive to any.[96] Perhaps this is a response to an odour approximating to that of the female. [See p. 743.]

The males of many Lepidoptera, also, have scent-producing organs: tufts of glandular scales or hairs on the abdomen, as in Phycitidae;[45] or on the legs;[100] or as 'androconia' scattered or concentrated in patches on the wings.[63, 100] In some cases the pencil of hairs used for dispersing the scent is separated from the glandular area by which it is secreted; as in some Danaine butterflies,[50] or in the abdomen of certain Noctuids and Sphingids.[213] These male scents are generally regarded as having an aphrodisiac function, exciting the female to copulation;[7] in the case of *Ephestia* there is some slight experimental evidence for this belief.[45] In *Eumenis* (*Satyrus*) *semele* there is good evidence that the androconia of the male, which form an elongated patch across the centre of the fore-wing, emit an olfactory stimulus that excites the female to accept the courting male. During an elaborate courtship, the knobs of the antennae of the female come to touch these patches of scent scales.[221] Sexual isolation between *Drosophila pseudo-obscura* and *D. persimilis* is due in part to scent. Females of one species will mate with males of either after the antennae have been removed. It may be that the scent from the male raises the state of readiness to mate in the female.[358] [See p. 743.]

In *Drosophila* the receptivity of the female is inhibited for 5 or 6 days after mating. This has been attributed to the distension of the vagina by a mass of fluid, the source of which is unknown,[353] or to a nervous stimulus from the active sperm, or a chemical stimulus from the sperm.[357] [See p. 743.]

Feeding habits during mating—There are many peculiar habits associated with mating, the physiological significance of which is uncertain. Female *Panorpa* eat secretory globules produced by the salivary glands of the male; the males of certain flies regurgitate a droplet of fluid for the female; the males of Empidae provide the female with food in the form of prey, enclosed by some species in a silk wrapping, while in others the wrapping alone is provided; the females of the tree cricket *Oecanthus* feed on a secretion from the metanotal glands of the male. These habits are sometimes thought to be related to the need of the egg-laying female for a protein diet;[184] in some cases they seem to serve the purpose of protecting the spermatophore (p. 714), or the male himself, from being devoured. Mantid females are particularly liable to eat the male; and it is interesting to recall (p. 194) that removal of the head during such an attack actually stimulates the male to copulation, since the copulatory centre in the last abdominal ganglion is free from the inhibitory influence of the suboesophageal ganglion. Normal pairing, with the transfer of a spermatophore, may take place even when both insects are decapitated.[189]

During courtship the male *Blattella* allows the female to feed on the secretion from his dorsal glands. The sexual stimuli to the male are both mechanical and chemical, but the chief stimulant is the secretion of the female which permeates the grease over her entire cuticle; it acts mainly by contact chemoreception. Contact with the antenna of a female is at once distinguished from contact with a male by receptors on the antennae and mouth parts.[380]

Transfer of sperm—The sexes often remain connected for several hours, held together in some cases by highly complex structures, while the transfer of the sperm takes place. In a few species, such as the butterfly *Parnassius*, the hypertrophied accessory glands of the male produce a secretion which hardens and cements them together during this period.

In some insects, such as the bug *Lygaeus*, the penis after entering the vagina penetrates the duct of the receptaculum, so that the spermatozoa, mixed with the secretion of the accessory glands to form a viscous fluid mass, are ejaculated directly into the spermatheca (Fig. 391);[128] more often, as in *Drosophila*,[154] they are discharged into the vagina; or, as in Lepidoptera, Orthoptera, and many Coleoptera, &c., into the bursa copulatrix. Males of Odonata eject them into a special apparatus below the base of their own abdomen; the ligula of this apparatus is then inserted into the vagina of the female. In *Cimex* the spermatozoa are discharged into a pouch on the lower surface of the female abdomen; they reach the genital tract by wandering through the general body cavity (p. 716).

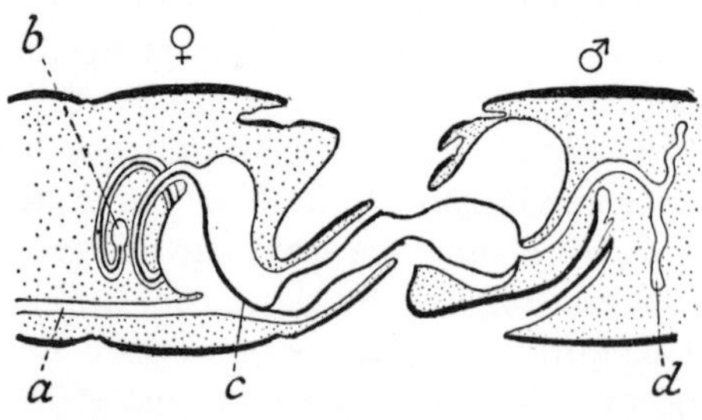

FIG. 391.—Copulation in *Lygaeus equestris*, showing extension of the penis of the male into the receptaculum seminis of the female; schematic (*after* LUDWIG)

a, oviduct; *b*, receptaculum seminis; *c*, penis; *d*, vesicle containing fluid which is driven into the penis in order to extend and uncoil it.

Spermatophores—In Lepidoptera,[155] many Orthoptera (Blattids,[77, 112] Tettigoniids and Gryllids,[13, 67] Acridiids, Mantids, and Phasmids[33]) Neuroptera,[47] Trichoptera,[112] Hymenoptera,[60, 112] and Coleoptera[112] the sperm are not conveyed to the female in a free fluid, but are enclosed in a membranous sac or 'spermatophore', formed by the secretion of the male accessory glands. The spermatophore is rarely transferred to the receptaculum; it merely serves as a provisional sheath which is deposited in the bursa or vagina; but in the cow louse *Linognathus vituli* it is deposited directly in the spermatheca.[363] In *Thermobia* and other Thysanura the spermatophore is simply dropped by the male in the course of a 'love dance'; it is then picked up by the female and inserted in the vagina.[214] In Collembola the spermatophore is just a droplet of sperm, which dries over the surface, on a delicate hyaline stalk. The male deposits a hundred or more of such spermatophores. On contact with the moist vulva of the female the structure bursts and the sperm migrate through the vagina to the eggs.[265] A similar process occurs in *Campodea*.[387]

An early adaptation, in the ancestors of insects, to life on land was the development of internal fertilization brought about by the extension of the spermatophore into the female ducts by means of a long neck. Later a bursa copulatrix was developed as an internal receptacle for the spermaphore.[307]

In Lepidoptera the spermatophore consists of a round sac of cuticular substance with a narrow neck. At the end of the neck are solid transparent horns of the same material, which correspond exactly with the diverticula from the male ejaculatory duct of the species in question. Near the horns is an oval opening in the neck, through which the sperm escape from the sac (Fig. 392, A, B). The formation of this structure in *Plodia* occupies 1–1½ hours, in *Ephestia* 3–4 hours. The male duct has glandular walls throughout much of its length. During the first 15 minutes of mating, in *Plodia*, the secretion from the lower glandular segment flows into the bursal cavity of the female and hardens into a gelatinous mass. The secretion from the middle segment flows into the ejaculatory duct and its diverticula, where it is moulded to form the horns and neck.

The secretion from the uppermost glandular segment is then forced down, pushing the soft core of the previous secretion before it into the bursa. The sac is further distended by the sperm and by the secretion of the accessory glands, which follow. Finally the neck of the spermatophore contracts and is withdrawn from the ejaculatory duct of the male into the bursal cavity of the female. Here it is always orientated so that the aperture of the neck lies near the entrance of the ductus seminalis.[155] In some Lepidoptera several spermatophores (up to nine) may be deposited in the bursa during copulation.[49] They are ultimately digested by a proteolytic secretion from the wall of the bursa; the same applies to the great masses of protein which accompany the spermatophore in the bursa of Trichoptera; in these insects they may be of some importance in nutrition.[112]

In the cricket *Oecanthus*, the spermatophore is produced rather differently. The atrium of the male system is first distended with the mass of sperm; around

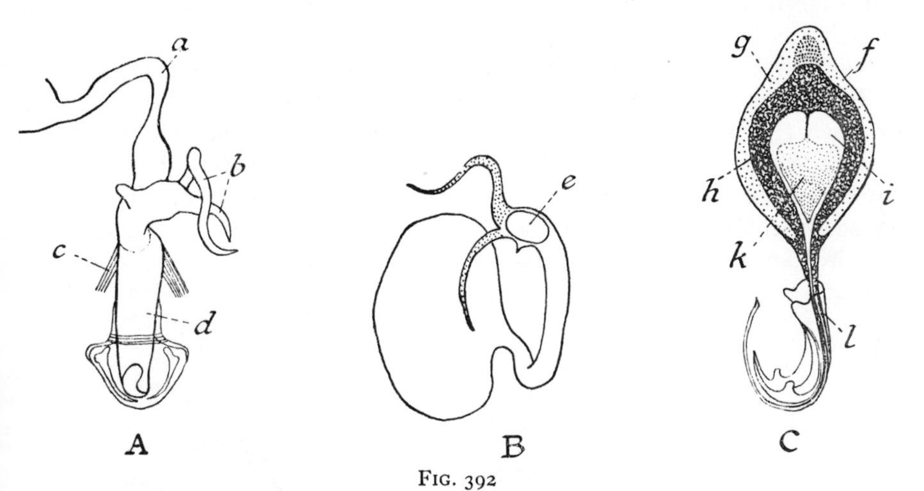

FIG. 392

A, lower part of male reproductive tract in *Ephestia kühniella*. *a*, ejaculatory duct; *b*, diverticula or horns on ejaculatory duct; *c*, protractor penis muscle; *d*, aedeagus. *B*, spermatophore of *Ephestia kühniella*, showing horns moulded in the diverticula of the ejaculatory duct. *e*, aperture in neck of spermatophore through which the sperm escapes (*after* NORRIS). *C*, diagrammatic longitudinal section of spermatophore of *Gryllus campestris*. *f*, outer sheath; *g*, middle layer; *h*, inner layer; *i*, substance which presses out the sperm; *k*, mass of sperm; *l*, sperm tube (*after* REGEN).

this the numerous accessory ducts then pour out their secretion, forming a covering like the shell of an egg. The soft mass is driven by peristalsis into the penis cavity, becoming globular; and finally the terminal filament and the lateral lamellae are formed by the secretion flowing into the penis groove and there rapidly hardening.[93] Only the neck of the completed sac is introduced into the vagina of the female; it is anchored there by the lateral lamellae, and often secured in position by a secretory plug applied by the male. Castrated males of *Gryllus* will produce and transfer empty spermatophores of this type during copulation.[179] [See p. 743.]

The spermatophore of *Rhodnius* is composed of a mucoprotein or neutral mucopolysaccharide originating in the transparent accessory glands of the male. It solidifies in the spermathecal sac of the male intromittent organ as a result of acidification (to *p*H 5·5) by a secretion from the bulbus ejaculatorius, and here it is moulded into shape.[306]

The spermatophore is emptied of sperm soon after it has been inserted. In

Dytiscus, in which it is deposited at the entrance of the vagina, they are perhaps squeezed out by pressure from the terminal sternites of the abdomen; emptying in this case is not complete.[12] The spermatophore of Lepidoptera is also thought by some authors to be emptied by the pressure of the bursa.[49, 155] But this cannot be the explanation in Orthoptera, in which the body of the spermatophore remains outside the genital opening. In *Gryllus* the wall of the capsule is made up of a thin outer, a thicker middle, and a still thicker inner layer; inside this is a gelatinous body which leaves only a relatively small cavity for sperm (Fig. 392, C). During emptying, it is believed that the gelatinous body absorbs water from the middle layer of the capsule and swells to fill the lumen and force out all the spermatozoa.[180] This osmotic process has been described in detail in *Acheta*.[112] In *Oecanthus* the process is completed in about 15 minutes; during this time the female is feeding on the metathoracic secretion of the male; when that is finished she removes the empty spermatophore and devours it.[93] Perhaps emptying is brought about in the same way in Lepidoptera, the inner layers taking up water from the accessory gland secretion which contains the sperm.[86]

Migration of spermatozoa to the receptaculum—Whether the sperm are discharged into the female directly or by way of a spermatophore, they are next transferred to the receptaculum seminis. The mechanism of this transfer is uncertain. According to some authors it is effected mechanically. The duct of the spermatheca of the bee has a complex muscular apparatus arranged in a semicircular manner round the duct; this has been described as a suction pump

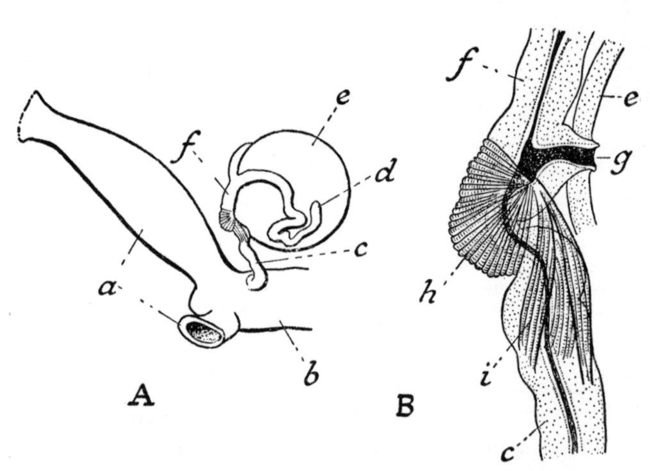

FIG. 393.—Sperm pump of queen bee (*after* BRESSLAU)

A, general anatomy of spermatheca; *B*, detail of entry of spermathecal duct into spermatheca. *a*, oviducts; *b*, vagina; *c*, duct of spermatheca; *d*, spermathecal gland; *e*, spermatheca; *f*, duct of spermathecal glands; *g*, orifice of spermathecal duct into spermatheca; *h*, semicircular muscles of sperm pump; *i*, longitudinal muscles.

which is not only used in fertilization (p. 740) but serves also to transport into the receptaculum the sperm received at copulation (Fig. 393).[2, 19] In Lepidoptera, the sperm have to pass from the bursa copulatrix, through the ductus seminalis to the common oviduct, and thence by another fine duct to the receptaculum seminis (Fig. 383). This movement, also, has been thought to be brought about by muscular contraction, squeezing by the bursa and aspiration by the receptaculum. Alternatively, in *Lymantria*, it has been supposed that contractile movements of the ductus seminalis stimulate the sperm rheotactically to enter the mouth of the ductus seminalis.[277] The bursa is completely emptied of spermatozoa in 1–3 days.[49, 155]

Other authors regard these mechanical theories as untenable, and consider that migration is the result of active movements by the spermatozoa in response to chemical stimuli.[86] In *Drosophila* the spermatozoa are ejaculated into the uterus. Here they remain motionless for some time; but after 2 or 3 minutes they begin to move, activated perhaps by the secretion of the parovaria or accessory glands; and then they swim actively to the two spermathecae and a ventral diverticulum which serve as seminal receptacles.[154] The unicellular glands in the walls of the receptaculum are thought to be the source of the chemotactic stimulus in *Lygaeus*,[128] the receptacular gland in Coleoptera,[187] in *Nematus* (Hym),[191] and Lepidoptera.[86] But at present there is little experimental proof for these ideas, although in Capsids and Nabids the spermatozoa can be seen to aggregate at the mouth of these glands.[121] [See p. 744.]

In *Rhodnius* the spermatophore is deposited in the vagina of the female. The escaping sperm are carried passively forwards to the receptacula by rhythmic contractions in the oviduct. These peristaltic movements are induced by the secretion visible in the form of opaque material in one of the accessory glands of the male, and introduced with the sperm into the female where it acts upon a peripheral nervous system in the oviduct. It has been suggested that the active material may be an *o*-dihydroxy-indolalkyl-amine, perhaps identical with the product which causes acceleration of the heart beat (p. 441). A similar product is detectable in the male accessory glands of many insects.[305]

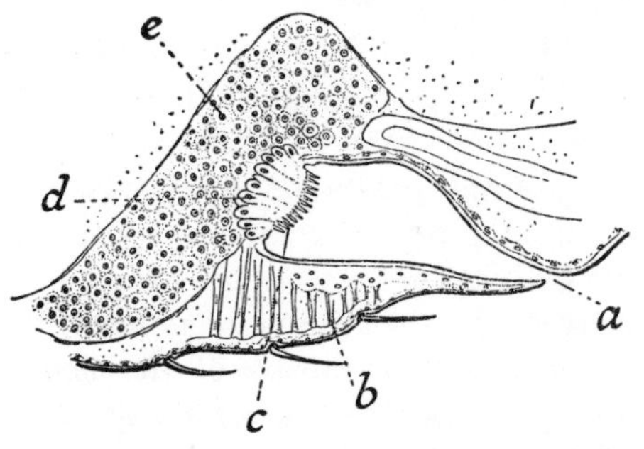

FIG. 394.—Longitudinal section through the fourth abdominal sternite of *Cimex* female showing organ of Ribaga (*after* ABRAHAM)

a, entrance for penis; *b*, muscles; *c*, cuticular processes; *d*, matrix cells of these processes; *e*, cells of organ of Berlese.

In Lepidoptera the receptacular duct has a main lumen, the 'spiral canal' and a very fine subsidiary lumen, the 'fertilization canal' in the substance of the wall.[149, 157, 236] In *Bombyx mori* it is believed that the sperm are activated by the secretion in the lowest part of the male common duct (the 'glandula prostatica') and thus acquire their fertilizing ability;[157] but the opaque material in the accessory glands of *Heliothis* (Lep.) induces peristaltic movements in the female seminal duct, and these movements may serve to transport the sperm.[294] Several hours after the spermatophore has been introduced into the bursa most of the sperm have migrated from it, leaving the fluid contents behind, and the receptaculum seminis is filled with a compact mass of sperms with little seminal fluid.[157] The sperm make their way up along the 'spiral canal' in response perhaps to chemotactic stimuli, perhaps to the downward flow of fluid secretion (rheotaxis), aided probably by peristalsis. The descent from the receptaculum for fertilization of the eggs takes place along the 'fertilization canal' in the wall of the duct.[157, 236] Considerable quantities of pyrophosphate are present in the ejaculatory duct of *Celerio* and are transferred with the sperm to the female. It has been suggested that these high energy compounds may act in some undefined manner in preserving the activity of the spermatozoa.[323]

Migration occurs by a different route in the bed-bug *Cimex*. The spermatozoa are ejaculated into the organ of Ribaga, a sac on the ventral surface of the abdomen (Fig. 394). They make their way through the floor of this pouch into a mass of underlying cellular tissue, sometimes known as the organ of Berlese. This organ appears to produce some internal secretion which it discharges into the blood when the spermatozoa pass through it.[36] Perhaps it serves to activate the sperm,[1] or to prevent bleeding and to protect the internal organs during copulation.[309] The spermatozoa reach the body cavity in 2–3 hours and then migrate, perhaps by chemotaxis, to a pair of evaginations ('spermathecae'[36] or 'absorptive organs'[1]) at the upper end of the common oviduct, which they reach in 2–10 hours.

Haemocoelic fecundation occurs in varied forms in different Anthocorinae, Cimicidae, Nabidae, &c.[30, 296, 297] The vagina may be pierced by a spine and the sperm liberated in the haemocoele (*Alloeorhynchus*); the sperm may be deposited in a pouch in the genital chamber, the wall to the haemocoele being later broken down and the spermatozoa scattered in all parts of the body, including the head and segments of the legs (*Prostemma*); organs like the organ of Ribaga may be present (some Anthocoridae); copulation in Anthocorinae takes place through a tube with an external opening quite separate from the vagina and connected to the genital apparatus (recalling the arrangement in Lepidoptera, Fig. 383); the organ of Ribaga may be dorsal on the abdomen (*Xylocoris*); or the abdominal wall may be pierced without any organ of Ribaga being present (*Primicimex*).[297] Copulation in *Stylops*, also, takes the form of hypodermal injection of sperm to the body cavity.[350]

Migration of spermatozoa into the egg—Fertilization may not take place until long after copulation (p. 648). Usually it occurs immediately before the egg is laid. A few spermatozoa leave the receptaculum and enter the micropyle of the egg just as it passes the receptacular duct. Various mechanisms are concerned in controlling this process. The spermatozoa stored in the spermatheca of the bee are believed to be rendered inactive by self-generated carbon dioxide and to be activated by the alkaline secretion of the spermathecal gland.[385] When the *p*H of the spermathecal capsule of *Mormoniella* changes, the sperm movements change and they begin to spin in a way that suggests that they may be able to screw themselves down the duct.[340]

(i) In the first place the egg is so orientated that the micropyle comes to lie exactly opposite the mouth of the duct. This is clearly seen in *Periplaneta*,[43] Lepidoptera (*Plodia* and *Ephestia*),[155] and *Drosophila* (Fig. 395).[154] In Vespids there is a dilatation of the vagina at this point, in Apids an evagination closed by a flap, and in Formicids a definite 'fertilization pouch', all concerned in directing the cephalic pole of the egg, bearing the micropyle, towards the duct of the spermatheca.[2]

(ii) The discharge of sperm may be under muscular control. Many Coleoptera and Hemiptera have a compressor muscle in the spermatheca (Fig. 396);[86, 232] in the grasshopper *Chortophaga* the sperm are ejected by contraction of the terminal muscular portion of the seminal duct;[165] in some cases perhaps there is a reflex increase in blood pressure when the egg enters the posterior part of the vagina, which may serve to squeeze the spermatheca,[86] or a reflex dilatation of the orifice of the seminal receptacle.[154] In the honey-bee, fertilization is certainly under 'voluntary' control (p. 740). Here the muscle around the duct of the spermatheca serves perhaps to pump or aspirate a few spermatozoa (variously estimated at ten[2] or a hundred[19]) from the sperm pouch to the vagina, or perhaps to control the escape of sperm in response to increased pressure from

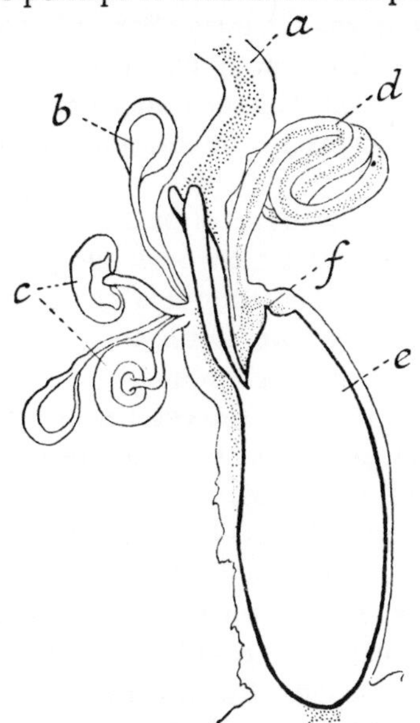

FIG. 395.—Egg of *Drosophila* in vagina at moment of fertilization (*after* NONIDEZ)

a, common oviduct; *b*, accessory gland; *c*, spermathecae; *d*, ventral receptacle; *e*, egg; *f*, micropyle.

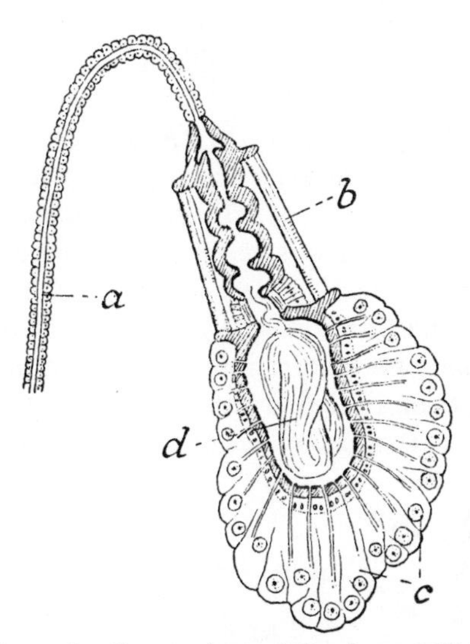

FIG. 396.—Spermatheca of *Graphosoma* (Heteroptera) showing the compressor muscle and the glandular walls (*after* BERLESE)

a, duct of spermatheca; *b*, compressor muscle; *c*, glands with intracellular ducts piercing the cuticular lining of spermatheca; *d*, mass of sperm.

above.[86] (The total number of spermatozoa received at normal mating in *Apis* has been estimated as 5–6 millions.)[316]

(iii) The final entry of sperm into the micropyle is probably a chemotactic response.[2, 232] Eggs of the parasitic Muscids isolated from the uterus show a tangled mass of sperm filaments projecting from the micropyle (Fig. 397).[161] In some cases perhaps this chemical stimulation may explain the discharge of sperm from the seminal receptacle;[154] notably where the receptacle has rigid cuticular walls, as in *Leptocoris*.[165]

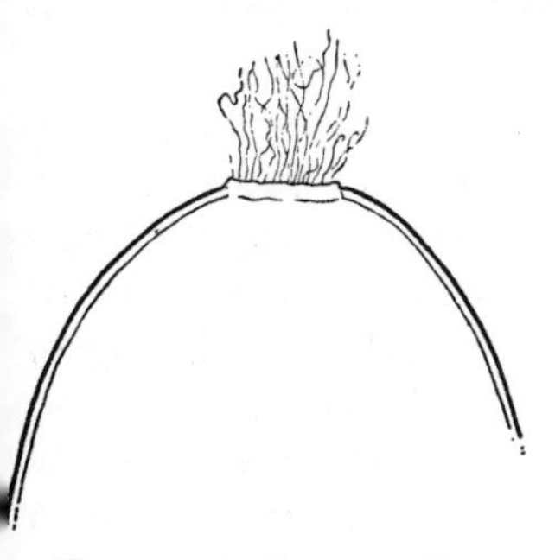

FIG. 397.—Egg of *Fausta radicum* (Tachinidae) extracted from uterus and examined in the fresh state in saline. Micropyle occupied by tangled packets of sperms (*after* PANTEL)

(iv) Spermatozoa of *Periplaneta* always move clockwise in circles and apply themselves to surfaces. These reactions may assist them in finding the micropyle.[43]

In a few insects the sperm may enter the egg while it is still in the ovary. The spermatozoa of *Cimex*, after reaching the receptacula in the common oviduct, migrate in the substance of the walls of the paired oviducts to the ovarioles, where they appear in great numbers in the follicular epithelium. Fertilization takes place in the ovarioles before the chorion is laid down, probably when the egg is still very small.[1, 36] In the Coccid *Aspidiotus* spermatozoa become attached to cells, derived perhaps from the peduncle of the ovariole, by which they are carried up to the ovariole where fertilization takes place.[372] The eggs of viviparous Chrysomelids, also, are fertilized in the egg follicles by sperm travelling up the oviducts.[182] And in the Polyctenid *Hesperoctenes*, spermatozoa occur free in the body cavity and presumably fertilize the egg in the ovary.[79] In the Capsid *Notostira* the sperm are said to enter the vitellaria and penetrate the oöcytes before the chorion is formed.[121]

We have seen that in *Rhodnius* the egg is rendered waterproof by a thin layer of wax, laid down by the oöcyte, which extends across the inner openings of the micropyles (p. 1). The spermatozoa must be able to penetrate this wax layer. As soon as this is accomplished the oöcyte produces a true 'fertilization membrane' which lies between the wax layer and the developing embryo.[8]

Fertilization—At the time of ovulation, the nucleus of the egg has not yet undergone maturation; but the chromatin has already collected into chromosomes and the first maturation spindle has formed. It remains in this state until the spermatozoon has entered the plasma. This seems to provide the stimulus to maturation. As the spermatozoon approaches the egg nucleus and, losing its tail, becomes converted into the male pronucleus, the two maturation divisions of the female follow rapidly upon one another. One of these is the reduction division which halves the chromosome number. The two divisions result in the formation of the female pronucleus and the polar bodies.

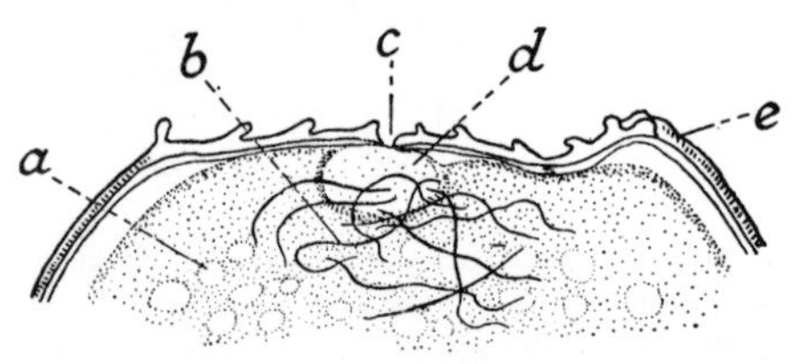

FIG. 398.—Section through upper pole of egg of *Pieris brassicae* (after HENKING)

a, yolk; *b*, spermatozoa; *c*, micropyle; *d*, submicropylar region; *e*, chorion.

Polyspermy seems to be nearly universal among insects; several spermatozoa enter the egg (Fig. 398). Only one of these is normally concerned in fertilization; but where several female nuclei are present, as in the binucleate eggs of Lepidoptera or *Habrobracon* (p. 93) each may be fertilized by a different spermatozoon.[185] In *Drosophila* a few eggs are monospermic; many receive 30 or more spermatozoa. The excess spermatozoa usually degenerate and do not disturb cleavage;[302] but in some cases where hundreds have entered the egg they may disorganize the formation of polar bodies, form mitotic figures themselves, or enter into normal cleavages to form multipolar spindles, and upset the egg to such an extent as to prevent further development.[97] It has been suggested that some mosaics and gynandromorphs may possibly arise by the appearance of extra nuclei from spermatozoa which have conjugated.[97] Some authors claim that polspermy is uncommon in *Drosophila*.[326]

In *Habrobracon* the nuclei of the nurse cells are engulfed in the egg. They normally break down (perhaps under the influence of some agent in the egg cortex), but occasionally one may survive and be fertilized to produce a binucleate egg.[285] In the normal honey-bee a substance seems to be liberated by the newly formed zygote nucleus which prevents excess sperm from dividing. But in mutant stocks sperm nuclei may become equipotent with zygote nuclei.[285]

By strongly heating eggs of the wild silkworm *Bombyx mandarina* which have been inseminated by the domesticated silkworm *B. mori*, at the appropriate time, it has been possible to injure the female nucleus during meiosis and to obtain fully developed adults produced from male nuclei only ('androgenesis'). Their characters, both at the first generation and after back-crossing with normal females of *B. mori*, proved to be indistinguishable from those of the male parent (*B. mori*). The characters of the species had been controlled by the nuclear material of the male sexual cells and were quite uninfluenced by the maternal cytoplasm of the egg.[274]

Fate of excess sperm in the female—The females of many insects mate once only, and the spermatozoa then received fertilize the eggs throughout the oviposition period. In Lepidoptera the sperm remain alive in the receptaculum for several months, nourished and activated perhaps by the secretion from the receptacular gland; even when egg-laying is complete the store of sperm is not exhausted.[49] Females of *Glossina* are impregnated 5–10 days after emergence, and that suffices until their death more than 6 months later.[140] In *Rhodnius* and *Triatoma* a single copulation is sufficient as a rule to ensure fertile ova throughout the entire life of the female; though in some cases the last batches of eggs are sterile in spite of the fact that apparently normal sperm may still be present.[66] And in the honey-bee, the store of sperm received during the nuptial flight lasts for several years. In *Schistocerca* the spermatozoa can fertilize eggs for at least 10 weeks after a single copulation; but a second copulation in some way prevents further use of the previous spermatozoa.[330]

In *Drosophila melanogaster* the male transfers about 4,000 spermatozoa to the female at a single copulation. This number falls rapidly in successive copulations and after a few matings no sperm are transferred. But by next day additional sperm are available. The female *Drosophila* may lay up to 3,000 eggs, so that the sperm transferred at a single copulation are sufficient to ensure fertilization of all the eggs during the first few days only. On subsequent days the percentage of sterile eggs increases, and towards the end of the egg-laying period (4–5 weeks) practically all the eggs are unfertilized.[107]

In a few insects, on the other hand, the sperm which are not immediately used for fertilization are digested and absorbed by the female. This is notably the case in *Cimex*, in which the sperm received at one copulation suffice for egg production for only 4–5 weeks;[1] and at 30° C., if the interval between feeds exceeds 2 weeks, few of the eggs are fertile.[36] Some of the spermatozoa may be absorbed in the cellular organ of Berlese, some by phagocytes in the body cavity, some in the lumen of the common oviduct and in the corpus luteum, but most are digested in the receptacula, which are sometimes described as 'absorptive organs';[1] but these organs still contain many living sperm several weeks after copulation.[142] In the bug *Graphosoma lineatum* one segment of the spermatheca apparently serves to take up excess sperm, which are dissolved by the secretion of the male accessory gland, and the products absorbed.[82] According to Berlese's theory of 'hypergamesis', these excess spermatozoa form an important element in the nutrition of the female. In the case of *Cimex*, which copulates repeatedly, it is estimated that the total bulk of the sperm may amount to as much as one-third of that of the eggs laid.[36] But it has not been possible to obtain any direct evidence of the nutritive value of these sperm.[142]

SOME FACTORS CONTROLLING FERTILITY AND FECUNDITY

Many insects with a short imaginal life, such as Ephemeroptera, Plecoptera, many Lepidoptera, copulate once only and lay their eggs in a single batch; others, such as Acridiids, Mantids, Muscids, mosquitos, &c., lay batches of eggs with intervals between, and may copulate repeatedly; others again, such as *Carausius*, *Tenebrio*, Aleurodids, Aphaniptera, lay single eggs at fairly regular intervals. In this section we shall consider the factors which influence the normal course of egg production in the female, and the factors affecting fertility in the male.

Temperature—The rate of egg production, like other processes of metabolism (p. 449), varies with temperature; it is accelerated up to a point and then falls off rapidly.[233] But the temperature limits between which reproduction can occur are often much narrower than the range of temperature over which the other activities of the same species remain normal. Females of *Locusta migratoria* fail to mature their eggs when the day and night temperatures alternate between 30° and 20° C.;[174] *Pediculus* will not lay eggs below 25° C.;[208] *Anopheles quadrimaculatus* will not lay below 12° C.[137]

The male seems often to be more sensitive than the female to abnormal temperatures. When the Chalcid *Euchalcidia caryobori* was exposed at 16° C. for 10 days, the females still laid the normal number of eggs, but 70 per cent. of the males were sterile.[83] In *Drosophila* kept at 32° C., 50 per cent. of the females were sterile, 96 per cent. of the males; the males were able to copulate but no sperm passed, and the sperm in the male organs lost their motility and subsequently degenerated; if they were returned to the optimum temperature of 24° C. most of the males recovered their fertility.[248] Similarly, when pupating larvae or pupae of *Ephestia kühniella* are kept at temperatures above 27° C. the moths are largely infertile; this is due to an effect on the male; spermatogenesis is retarded and the spermatozoa lose their motility and may disappear.[155, 178] On the other hand, in *Tribolium confusum* reared at 38° C. the males are little affected but nearly all the females are completely sterile; high temperature has the same effect if applied to the adult female.[158]

Exposure of *Drosophila subobscura* adults to a temperature of 30·5° C. causes a partial regression of the ovaries of the female but little reduction in fertility of the male. Exposure of the females to 30·5° C. for a time increases their subsequent longevity at 20° C.: egg-laying accelerates the ageing of females at 20° C. and this prolongation of life after exposure to high temperatures is due to subsequent reduction in the rate of egg-laying.[396] [See p. 744.]

Nutrition—Fertility in the male is comparatively little affected by nutrition, though males of *Rhodnius* appear to be less fertile if they are starved in the adult stage,[24] and well nourished females of *Cimex* mated with unfed males produced an average of 45 normal and 12 sterile eggs, as against 153 normal and 42 sterile eggs when mated with males which had been fed.[36] Egg production in the female, on the other hand, may be profoundly influenced by the food supply.

Lepidoptera fall into three not very sharply defined groups according to the state of the ovaries at the time of emergence. (i) In butterflies, and many moths with a long imaginal life, there are very few fully developed eggs; most are matured after emergence. (ii) In most Heterocera there is perhaps a two- or threefold increase in ripe eggs in the imago. (iii) While in a few Bombycids,

Lymantriids, &c., all the eggs are fully ripe and no more are developed.[49] We have seen that adult Lepidoptera are unable to assimilate proteins (p. 516); all the protein reserves needed for the eggs must therefore be carried over from the larva; consequently, the fat body of the young imago is massive in group (i), small in group (ii), and already quite used up in group (iii).[49] They can, however, assimilate sugar: if *Agrotis segetum* is fed with a 20–40 per cent. solution of glucose, all the eggs are fertile; if given 5 per cent. glucose forty to fifty per cent. are infertile; and if starved, the number of eggs laid is much reduced, and the embryos all die before hatching. On the other hand, starvation in females of *Loxostege sticticalis* causes less reduction, and in *Pyrausta nubilalis* it has a negligible effect.[119] Similarly, *Ephestia* species do not require sugar; although this increases their longevity it has no effect on their fecundity. But they require water. Fecundity in *E. cautella* and *E. elutella* (at 18–27° C. with moderate humidity) is about halved if they receive no drinking water; it is reduced by less than 20 per cent. in *E. kühniella*. This difference is related with the state of the ovaries at emergence; for *E. kühniella* has already a number of ripe eggs, while *E. cautella* has none.[155]

The influence of water and of carbohydrate is seen also in the beetle *Bruchus quadrimaculatus*, in which access to water increases the number of eggs by about 30 per cent., access to sugar and water by about 50 per cent.[123] *Ptinus tectus* requires drinking water for normal egg production;[54] and both the rate of oviposition and the total egg production of *Calandra* are reduced if the food is too dry.[48] At a relative humidity of 20 per cent. *Tenebrio* laid an average of 4 eggs, at 65 per cent. 102 eggs; and *Dermestes* females laid 567 eggs when given water to drink, 30 eggs without water.[44] Partial desiccation of *Platysamia* pupae leads to reduced oviposition by the adult female.[129] *Melanoplus* females produce an average of 140 eggs when fed on fresh alfalfa, 29 eggs on dry alfalfa.[217] The fecundity in *Lucilia* is greater in larger flies. These are produced if the larva is allowed to drink water before pupation; so that larval drinking has an indirect effect on subsequent egg production.[141]

In most insects, proteins are essential for egg production. In the feeding of Aphids two kinds of discrimination are involved; one associated with the kind of host plant, the other with the age of leaf. Most species will feed only if the flavour of the plant is acceptable. Their subsequent readiness to settle down for a long period, and their reproductive rate, vary together: the Aphids prefer to feed, and reproduce faster, on young leaves or early senescent leaves than on healthy mature leaves. Both feeding and fecundity appear to be stimulated by the changes in the contents of the phloem juice that occur during the active translocation of nitrogenous compounds.[258] The same nutritional benefits and preferences are seen on leaves in which gall formation has already been induced by Aphids and on leaves infected with virus disease.[257] In the cabbage aphid *Brevicoryne* the rate of reproduction is positively correlated with the nitrogen and protein content of the host plant.[53]

Pteromalid parasites of the weevil *Phytonomus*, when their ovarial follicles reach a certain stage of development, change from a carbohydrate diet to a protein diet consisting of the body fluids of the host species.[56] The house-fly *Musca* requires both sugar and protein as well as water; it lays no eggs if given protein or sugar alone.[69, 194] *Lucilia* females require at least one meal of protein before eggs are laid, whereas this is not necessary for the fertility of the males.[52, 130]

In most blood-sucking species, *Stomoxys*,[69] *Haematopota*,[26] *Rhodnius*,[24] *Cimex*,[110] most mosquitos,[247] &c., the number of eggs is determined by the quantity of food. In *Culex pipiens* there is a striking difference between the effects of human and avian blood: only half as many eggs are developed after feeding on man;[195] an average of 121 eggs from 3·0 mg. of human blood, 255 eggs from 3·1 mg of canary blood.[247] The stroma of the red blood cells seems to contain some constituent necessary for the development of eggs in *Anopheles sacharovi*.[247a] Casein will satisfy the protein requirements for egg production in *Culex pipiens*, but the need for other growth factors becomes apparent during the development of several batches of eggs.[329] In *Aëdes aegypti* the usual eleven 'essential' amino acids are needed for egg production as for growth (p. 520).[352] In the case of the flea *Xenopsylla* the age of the host is important: if the females are fed on the newborn mouse they start to lay eggs after an unusual delay, and produce fewer eggs than normal although feeding and survival are unaffected.[25] The determining factor in these examples is not known; but it is known that the blood of the new-born rodent is deficient in *p*-amino-benzoic acid. Females of the rabbit flea *Spilopsyllus* will mature their ovaries only on the pregnant rabbit.[359] The factor concerned appears to be one of the hormones secreted by the anterior lobe of the pituitary gland.[384] [See p. 744.]

There are equally striking differences in plant feeding insects. Egg production in *Leptinotarsa* varies greatly with the species of potato on which the larva is reared: *Solanum edinense*, average 35 eggs; *S. tuberosum*, 25; *S. utile demissum*, 12; *S. conmersonii*, 0.[227] In this same insect ageing of the food plant leads to a fall in egg production.[75] Here lecithin seems to be the factor responsible. *Leptinotarsa* produces an average of 0·35 eggs per diem on young leaves, 0 eggs per diem on old leaves. The addition of sucrose to young leaves will increase egg production to 1·0 per diem; but the further addition of lecithin will put up the average number of eggs to 3·1 per diem.[75] In *Loxostege* (Lep.) linoleic acid (p. 519) seems to have some special relation to reproduction, and its deficiency may be connected with the sporadic sterility that occurs in this insect.[167] Polyunsaturated fatty acids and cholesterol are likewise needed for egg production in the cotton boll weevil *Anthonomus grandis*.[407] The ether soluble fraction of royal jelly added to the diet of *Drosophila* caused an increase of about 60 per cent. in the number of eggs produced per day.[226]

Vitamins also are important in egg production. *Lucilia* cannot produce eggs on blood serum alone; it is deficient in phosphate, potassium, and B vitamins; but serum plus 'marmite' (autolysed yeast) is effective.[90] The addition of cholesterol to a sterol-deficient diet doubled the egg-production of *Musca*; β-sitosterol had no effect.[361] *Tribolium* females reared in flour of 85 per cent. extraction show about twice the fecundity of those in 60–75 per cent. extraction flour.[183] The fecundity of *Drosophila* females (and the viability of the eggs) is largely influenced by the quantity and the variety of yeast in the diet. The different strains of yeast probably vary in their vitamin content.[188] During the height of egg production *Drosophila* daily ingests yeasts about equal to its own body weight and produces eggs approximately one third of its weight.[342] Fewer eggs, and more non-viable eggs, are produced if the diet is deficient in folic acid or if the anti-metabolite aminopterin is included in the diet.[342]

Rhodnius prolixus deprived of their intestinal Actinomyces rarely reach the adult stage (p. 524); those few which do so, fail to produce eggs until they are reinfected.[17] It is said that *Cimex* females show reduced egg-laying and many sterile eggs when they are fed on rats which have been deprived of vitamin B_1, in spite of the fact that they contain symbiotic micro-organisms (p. 522).[138]

Among termites, the sterility of the worker and soldier castes, and of the occasional males and females whose gut contains wood particles and protozoa, is attributed to nutrition ('alimentary castration').[239] They do not receive the saliva which seems necessary for the activity of the gonads—perhaps through providing protein, or because it is a source of some specific 'vitamin'.[72] The reverse effect, an activation of the ovaries of the workers, is seen among social Hymenoptera deprived of their queen. In *Vespa* it is sufficient to suppress the queen for half the workers to become fertile and lay eggs. Having no brood to feed, they re-absorb the food and secretions they would have given to the young; it is perhaps their occupation as nurses which determines their normal sterility ('nutricial castration').[134] In colonies of *Polistes* all the females may reproduce actively at first, but later on egg-laying is often maintained by a single female ('functional monogyny'). This limitation of the reproductive function seems to be correlated with socially 'dominant' behaviour. Perhaps these same factors are concerned in *Vespa*, &c.[252]

Larval nutrition—Since many of the reserves which go to form the eggs are laid down in the larval stage, egg production in the adult may be greatly influenced by the nutrition of the larva. Underfeeding in larvae of *Drosophila*,[4] or *Tineola*,[222] reduces the number of eggs laid. The fecundity of *Ephestia* females is much reduced by feeding the larvae on white flour instead of wholemeal, though the average viability of the eggs laid is not affected.[155] In certain races of *Culex pipiens*, the unfed females will develop eggs if their larvae receive a diet rich in proteins, but not after a predominantly carbohydrate diet;[87, 238] and it is noteworthy that those mosquitos with vestigal mouth parts (*Corethra*, *Mochlonyx*) or those which feed only on nectar (*Megarhinus*) are predaceous in the larval stage.[87]

Additional protein received by *Musca* during larval development will not lead to egg production in the adult female receiving sugar and water only;[273] but *Musca* reared on a larval diet supplemented with cholesterol will develop mature ovaries when held on a diet of sucrose and water alone; β-sitosterol had only a very slight effect on the ovaries.[376] The adult females bring over from the larval stage sufficient ribonucleic acid to provide for the first batch of eggs; but for subsequent batches RNA or its precursors must be furnished in the diet.[308]

Larval nutrition provides for the formation of adult structures, including the flight muscles. In the young adult these muscles may be used in mating flights, or migratory flights, and then, when the female settles down to reproduce, the muscles may be autolysed and the proteins utilized for egg production. The classic example is the queen ant;[334] but the same change occurs in Aphids,[336] in the mosquito *Aëdes communis*, where the nitrogen content of the flight muscles is equivalent to about 60 eggs—which agrees with an average egg number of 64,[328] in water beetles,[331] and probably many other insects. [See p. 744.]

Impregnation—Insects differ in the extent to which impregnation affects egg production. In the louse *Haematopinus*, it seems to have no effect; egg laying begins three days after the final moult with or without the presence of the male.[61] But that is exceptional. In *Cimex* no eggs are developed until after impregnation;[36, 142] in *Drosophila*, mating causes an immediate and considerable acceleration in egg production;[124] unmated females of many Acridiids produce notably less eggs than mated females,[99, 113] and the same applies to *Musca*.[69] Among different species of *Anopheles* there are differences in the extent to which the presence of sperm in the spermatheca is necessary for egg formation.[196]

In many cases the effect of impregnation seems to be on oviposition rather than on egg development. In *Thermobia* only one batch of eggs is laid during an instar and impregnation must precede each oviposition.[215] In those Belostomatids in which the eggs are laid on the elytra of the male, copulation is followed by the deposition of a single egg. The process is then repeated until the back of the male is fully laden.[31] Unmated females of Lepidoptera develop the same number of ripe eggs as mated females, but they lay only about one-third of their total store.[49] Females of *Ephestia* lay only a very few of their eggs when mated with males rendered sterile by high temperatures, most of the eggs being retained in the ovaries.[155] (We have seen that in *Glossina*, impregnation seems to provide the stimulus to ovulation (p. 706).) Lack of copulation reduces the number of eggs laid in *Tineola*;[222] and in *Lucilia*, copulation appears to provide some essential stimulus for oviposition.[130] In the Lepidoptera *Bupalus* and *Ephestia*[16] and the beetle *Acanthoscelides*[22] copulation provides a stimulus for both egg development and oviposition. What the nature of this stimulus may be is uncertain. In *Lymantria*, oviposition is incomplete after copulation with a castrated male, and yet the bursa is equally distended with the sterile spermatophore as it is after a normal copulation. Distension of the bursa, or chemical stimulation from the secretion of the accessory glands, cannot therefore be responsible. Perhaps the moving sperms themselves provide a tactile stimulus acting through the nervous system.[114] If this tactile stimulus to the receptacular apparatus ceases, normal egg-laying in *Lymantria* comes to an end and the females behave like unmated individuals. [See p. 744.]

In *Periplaneta* the stimulus of copulation and of the stored sperm are both concerned in increasing egg production. Females associated with males throughout their adult life produced an average of 20·1 egg capsules 7·7 days apart. Females without males produced 9·7 capsules 24·7 days apart. Those females which mated only during the preoviposition period showed a small reduction in the number of capsules (16·7) but these were produced at an average interval of 13·4 days. They showed a normal degree of fertility, 13–15 nymphs emerging from each capsule, for 8 months after removal of the male.[74]

Internal secretions—In many insects the deposition of yolk in the oöcytes is dependent on the presence of a hormone secreted by the corpus allatum.[105, 198, 413] In *Rhodnius* deprived of the corpus allatum the oöcytes continue to grow until the point when yolk should be formed. They then die, and the follicular cells proliferate amitotically in a disorganized manner, invade and absorb the dead oöcyte, and eventually undergo necrosis and absorption themselves like the corpus luteum of the normal ovariole (Fig. 399, 400).[243] The same changes take place in *Schistocerca*.[356] In the male *Rhodnius* the corpus allatum is necessary for the normal activity of the accessory glands.[243] The same changes occur in *Melanoplus* if the corpus allatum is removed; if part is left it hypertrophies and egg development occurs; and in this insect the secretion of the anterior segment of the oviducts, which forms the oötheca, is also dependent on the presence of the gland.[235] In *Leucophaea* the corpora allata are needed for the deposition of the yolk but not for subsequent development in the eggs of this viviparous cockroach.[198] In the adult *Calliphora* the corpus allatum increases greatly in size after emergence. Extirpation prevents the formation of yolk in the oöcytes, but their capacity is restored if the corpus allatum of either sex is implanted.[219] Other insects which need the corpora allata for yolk development include *Leptinotarsa*,[416] *Anisolabis*,[370] *Oncopeltus*,[335] and most species of mosquitos.[346, 351] In *Aëdes* the eggs must reach a certain stage of development before the corpus allatum hormone can influence yolk formation.[349] In *Dytiscus* the corpus allatum is needed at a much earlier stage of ovarian development (p. 728). In locusts it controls the uptake of nutrients and the formation of yolk.[325] The action of the corpus allatum

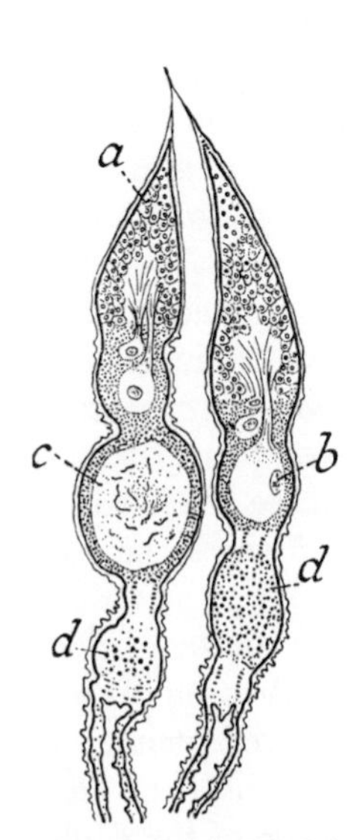

FIG. 399.—Ovarioles of *Rhodnius* deprived of corpus allatum (*cf.* Fig. 387) (*after* WIGGLESWORTH)

a, nurse cells; *b*, live oöcyte still nourished by nutritive cord from the nurse cells; *c*, dead oöcyte; *d*, autolysing residues of dead oöcytes and follicles.

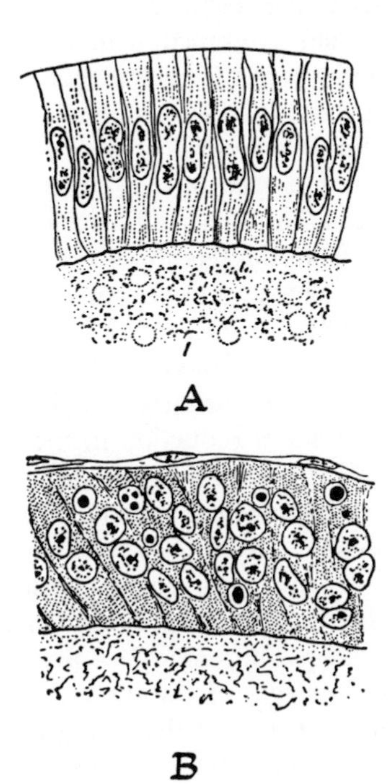

FIG. 400

A, healthy follicular epithelium around normal oöcyte in *Rhodnius*; *B*, degenerating follicular epithelium around dead oöcyte, in *Rhodnius* deprived of corpus allatum (*after* WIGGLESWORTH).

is non-specific as between *Rhodnius* and *Triatoma*,[243] *Periplaneta* and *Oncopeltus*,[156] *Calliphora* and *Rhodnius*,[411] *Tenebrio* and *Galleria*.[373] [See p. 744.]

The activity of the corpus allatum is dependent upon nutrition: the corpus allatum from a well-fed *Oncopeltus* will induce egg production if implanted into a female given water alone.[335] In the ovoviviparous cockroach *Leucophaea*, and in *Blattella* which carries the oötheca until the eggs are ready to hatch, the ripening of the oöcytes is arrested while there is an oötheca in the brood chamber. The secretion of the corpus allatum is inhibited by nervous impulses from the brain; and the brain exerts this action in response to nervous impulses from the brood chamber.[312, 379] The dropping of the oötheca, together with the stimulus of mating and of the introduction of a spermatophore into the bursa copulatrix, reverses these effects, and the corpus allatum begins to secrete its hormone again.[312]

The corpora allata are necessary for maturation in the male of *Schistocerca* and the liberation from the integument of the volatile secretion which excites the female.[354] In the cockroach *Byrsotria* the corpora allata are needed for the production of the sex attractant scent by the female;[276] and in the cockroach *Nauphoeta cinerea* they control mating in the female chiefly by inducing her to feed on the tergum of the displaying male.[378] The development of the ovaries of honey-bee workers is inhibited by acetone washings from the surface of the queen, or by extracts of the head;[409] this 'queen substance' is a product of the mandibular glands (p. 452);[292] it appears to exert its effect by suppressing the activity of the corpus allatum.[355] But the purified principle, 9-oxodecenoic acid, even when combined with the queen scent, is not so effective in inhibiting oögenesis as in the presence of the queen herself.[293]

In some insects, *Calliphora*,[219] *Lucilia*,[37] *Drosophila*,[230] *Melanoplus*,[169] and *Rhodnius*,[243] removal of the ovaries causes hypertrophy of the corpus allatum. The same is seen in the 'female-sterile' mutant of *Drosophila*.[230] Whether this signifies a hormonal influence exerted by the ovaries on the gland,[219] or simply results from the retention of secretion which is not required, is uncertain. Such enlargement does not occur in *Sarcophaga*[37] or *Leucophaea*.[198] Removal of the corpus allatum of the adult insect results in other changes also, such as the very slow digestion of blood in *Rhodnius*,[243] a massive accumulation of fats in the fat body and various changes in the blood in *Melanoplus*,[169] similar changes in the fat body and a reduction in the oenocytes in *Lucilia*, *Sarcophaga*,[37] and *Drosophila*.[231] It is uncertain whether this means that the corpus allatum is secreting a general 'metabolic hormone'[169, 198] or whether all these effects are secondary to the failure of yolk formation in the ovaries. In *Calliphora* the corpus allatum certainly influences oxygen consumption.[219]

The yolk-forming hormone in these insects is quite distinct from the gonadotropic hormones* of mammals.[105, 169] It seems to be present already in the early stages of the insect. *Drosophila hydei* deprived of its corpus allatum will develop eggs if the ring glands (p. 201) from 1st instar larvae of the same insect are implanted.[230] Conversely, the corpus allatum of the adult *Melanoplus* exerts a juvenile effect on the nymphs.[169] And in *Rhodnius* the 'juvenile hormone' (p. 78) of the young nymph and the 'yolk-forming hormone' of the adult appear to be interchangeable and probably identical substances.[243] This conclusion is supported by the fact that farnesol and related compounds (p. 80) likewise induce yolk formation as well as maintaining larval characters;[412, 413] and that the corpora allata of the larva and of the adult *Pyrrhocoris* have the same effects on metamorphosis, on metabolism and on yolk deposition.[395]

There are other insects in which the corpus allatum is not necessary for egg production. *Bombyx mori* and other Lepidoptera develop eggs normally if the corpora allata are excised in the last larval stage[15] or in the pupa.[246] (On the other hand, ovaries of *B. mori* transplanted from an early 5th stage larva into a young pupa are caused to develop ripe eggs prematurely.[64]) *Carausius* (p. 70) from which the corpora allata have been removed in the third or fourth nymphal stage are caused to omit several instars and then produce viable eggs.[171, 172] But most of the Lepidoptera studied in the earlier experiments (*Bombyx*,[15] *Hyalophora*[417]) develop their eggs during the pupal stage; in *Pieris brassicae*, in which the ovaries

* Oestrogens have been obtained from the gonads of female *Attacus* [127] but their significance is unknown.

mature after emergence, removal of the corpora allata suppresses yolk formation.[337] In male *Hyalophora cecropia* much larger amounts of juvenile hormone are accumulated during pupal development and later consumed.[314, 417] It is probably required for the rapid production of spermatophores from the accessory glands.[413]

Although the corpus allatum is of some importance in yolk formation in *Calliphora*, the neurosecretory cells in the dorsum of the brain have a far greater effect: if they are removed the oöcytes never develop beyond a very small size.[403] These cells are probably necessary also in mosquitos.[299, 315] In *Calotermes* adults of both sexes they show maximum activity in the functional sexual pair and in neotenic reproductives.[368] They seem to have relatively little importance in *Rhodnius*,[412] *Oncopeltus*,[335] and *Carausius*.[311] In *Locusta* and *Schistocerca* the secretion of the neurosecretory cells is liberated under the stimulus of copulation, and various artificial stimuli.[324] The male *Schistocerca*, under the action of the corpus allatum, accumulates carotene in the epidermis and becomes bright yellow, and it liberates a volatile secretion[354] which also favours the discharge of the neurosecretory hormone by the female.[324]

The most obvious effect of the brain hormone of the adult insect is to further the synthesis of the protein needed by the developing oöcytes. In *Calliphora* it induces an increased activity of protease in the mid-gut, favouring protein digestion and absorption.[404] In *Schistocerca* the neurosecretory material acts on the fat body and stimulates the uptake of amino acids leading to the synthesis of blood proteins, which are then taken up by the oöcytes.[327] The corpus allatum hormone is concerned in facilitating the absorption of these proteins, and other nutrients, by the oöcytes.[325] In *Rhodnius* the corpus allatum hormone induces the specific synthesis of the yolk protein by the fat body.[299a]

The oenocytes seem to be connected in some way with reproductive activity. The rod-like inclusions in the oenocytes of both sexes of *Tribolium* disappear at the time of sexual maturity.[116] In *Rhodnius* the enlargement of the oenocytes is much greater in the adult female; it has been suggested that they are producing the lipoproteins needed for the egg-shell, as they produce the lipoproteins for the cuticle in the young stages (p. 450).[243] [See p. 744.]

Absorption of oöcytes—If *Rhodnius* females are starved the oöcytes suffer the same fate as in insects deprived of the corpus allatum; they develop until they are ready for the deposition of yolk and are then reabsorbed.[243] The same change is seen in the ovaries of workers among Hymenoptera. In workers of *Formica rufa* and *Camponotus ligniperda* the eggs develop as in the queen up to the stage of yolk formation; they then die and are reabsorbed.[237] The same changes are seen occasionally in the queens; they affect a certain number of the oöcytes in normal mosquitos;[152] and they occur in Pteromalids, either spontaneously (a condition known as 'phasic castration') or if they are prevented from ovipositing.[56] The same applies to *Diadromus* (Ichnumonidae).[345] Whether this regression of the ovaries is an effect of nutrition, of metabolism, or of a specific hormone is uncertain. Probably it results from the interaction of all these factors.

This phenomenon plays an important part in enabling parasitic Hymenoptera to maintain their reproductive capacity during periods when the environmental conditions are unfavourable. Since the eggs of Hymenoptera will develop parthenogenetically (p. 725) they can be stored after ovulation only if they are of the 'hydropic type' which have little yolk and obtain nourishment from the fluids of the host by way of the embryonic membranes. In certain species of *Apanteles* the ovary and oviduct are modified for storing the full quota of eggs of this type.[58] Parasites laying 'anhydropic' eggs, fully provided with reserves at the time of laying, are dependent on the cyclical reabsorption of oöcytes in order to defer their oviposition.[58] In *Mormoniella* the reabsorption of eggs enables the starved female to outlive the male.[341] [See p. 746.]

Hibernation and diapause—Reproduction, like growth, may suffer a periodic arrest; and as with growth (p. 111) this arrest may be a direct effect of an adverse environment, or it may be a true 'diapause' which persists even under favourable conditions. In many parasitic Hymenoptera, under conditions unsuited for oviposition, the eggs are not reabsorbed but enter an embryonic diapause and are retained in the hypertrophied genital tract.[345] In *Dytiscus* and various other beetles, the gonads revert to a resting state after the first reproductive period; they show renewed activity about the same date the following year, sometimes in a third or even a fourth year also. This seems to be a deep-seated rhythm and not a simple effect of warmth following the winter cold; for in *Carabus coriaceus* and *Leistus* spp. the breeding season does not begin until the late summer or autumn,[51] and *Dytiscus marginalis* lays its eggs in March, *D. semisulcatus* in October.[105] The full development of the eggs in *Dytiscus* is dependent on the corpus allatum. Reabsorption of the oöcytes takes place in July as soon as they have grown to 0·8 mm. in length; in February they reach 2 mm. before they break down; and in March they attain the normal length of 7 mm. But if corpora allata are implanted, the full development of the eggs can be produced at any season; and if the corpora allata are removed at the height of egg production, the ovaries regress. The corpus allatum hormone seems to act directly upon the oöcytes, for a localized enlargement can be brought about by implanting the gland within the ovary. But their normal activity seems to depend on a stimulus from the central nervous system to the corpus cardiacum (p. 199) and thence to the corpus allatum.[105] In Carabids, as in *Dytiscus*, the corpus allatum hormone prevents degeneration of the oöcytes and permits ripening of the ovaries.[255]

Mosquitos become more or less immobile in the autumn; and after taking meals of blood they tend to develop an enlarged fat body instead of maturing their ovaries ('gonotrophic dissociation').[216] In *Culex pipiens*, this change seems to be closely related with the temperature: if given blood and kept warm at any time in the winter they will develop eggs.[87] But if *Anopheles maculipennis* race *atroparvus* is fed with blood during the winter and kept at 29° C., although some develop eggs, others lay down fat body.[87, 216] And *A. maculipennis* race *messeae* in the autumn shows complete 'gonotrophic dissociation'; it fails to produce eggs at any temperature.[9] In this state there is no absorption of developing oöcytes as described in the last section; the ovaries show no signs of growth at all.[244] The cessation of blood-feeding in *Culex pipiens*, and the gonotrophic dissociation in *Anopheles maculipennis* during the winter diapause, are the result of exposure to a short (12 hour) photoperiod.[304] The arrested development of the female reproductive system in the Delphacid *Stenocranus minutus*, during the winter, is the result of the long days (18 hours of light) experienced during the larval stage. This diapause is brought to an end by short days (8 hours of light) during 4 weeks in the young adult. If the larva is exposed to these short photoperiods a small adult form (*minutior*) is produced, which develops eggs that enter diapause.[364] In *Leptinotarsa* during diapause the state of the ovaries resembles that in *Rhodnius* deprived of the corpus allatum; it results from absence of the corpus allatum hormone.[416] [See p. 746.]

In the diapausing pupa of male *Papilio xuthus*,[367] as in the pupa of *Hyalophora*,[388] many of the spermatocytes degenerate in a way not seen in non-diapause pupae.

Old Age—Under optimal conditions *Drosophila* females will produce up to 3,000 eggs. In older females the rate of egg-laying diminishes and a smaller percentage of the eggs hatch—partly because there are more unfertilized eggs, partly because the embryos die.[78] By transplantation of the ovaries of ageing females into young flies these effects can be removed; clearly extra-ovarial influences are at work.[181] We have already noted the increasing proportion of sterile eggs laid by ageing *Rhodnius* females which runs parallel with the reduction in the number of micropyles (p. 705).

SPECIAL MODES OF REPRODUCTION

Polyembryony—Some parasitic Hymenoptera show a peculiar form of asexual multiplication in which a single egg divides in the course of development to produce a number of embryos.[126] This condition, described originally by Marchal,[133] has arisen independently among Chalcididae, Proctotrypidae,

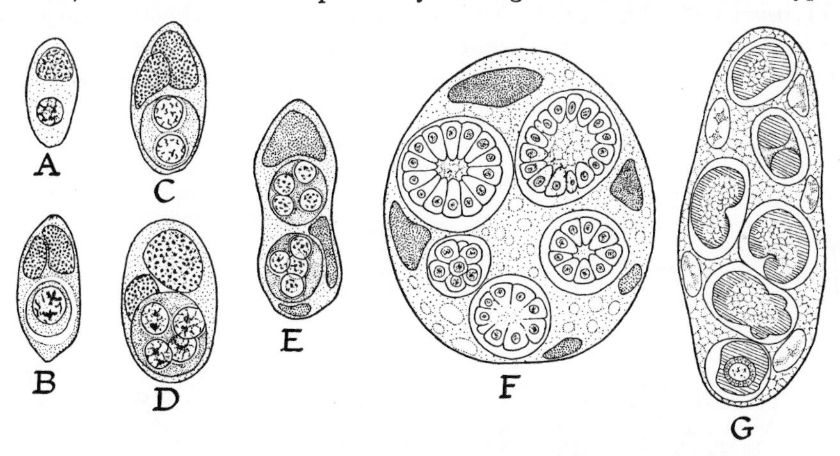

FIG. 401.—Polyembryony in *Platygaster* (*after* LEIBY and HILL)

A–E, *Platygaster hiemalis*; *F*, *G*, *P. vernalis*. *A*, fertilized egg 24 hours old, showing polar nucleus above and segmentation nucleus below. *B*, egg about 2 days old; polar nucleus divided into two paranuclear masses; embryonic region containing segmentation nucleus is now differentiated from the trophamnion. *C*, parasite body about 3 days old; two embryonic nuclei in the embryonic region. *D*, parasite body 4 or 5 days old; four embryonic nuclei present. *E*, embryonic region divided to form twin germs with trophamnion infiltrated between them. *F*, 18-day-old parasite in polyblastula stage; five embryos at early blastula stage; trophamnion vacuolated. *G*, polyembryonal mass at about 26 days, showing portions of six embryos in process of forming their germ layers; trophamnion less dense, paranuclear masses less conspicuous.

Vespoidea, Braconidae and Ichneumonidae.[101, 163] It is seen in its simplest form in *Platygaster hiemalis*, a parasite of the Hessian fly (Fig. 401). After maturation of the egg in this insect, the two polar bodies fuse into a single polar nucleus, which divides amitotically to form a 'paranuclear mass'. This ultimately grows around the germ to produce a nutritive sheath or 'trophamnion', which obtains nourishment from the tissues of the host and gives it up to the embryo. As in all endoparasitic Hymenoptera the egg is very poor in yolk, and cleavage is complete. When division has reached the 4-cell stage, the germ and the paranuclear mass may either split into two and give rise to two embryos, or they may fail to divide and so develop into a single larva.[126] By the same process the egg of *P. vernalis* gives rise to an average of eight individuals.[126]

In *Macrocentrus gifuensis*, a Braconid parasite of the European corn borer, the primary embryonic germ divides by fission to form two secondary germs each accompanied by half the paranucleus; these secondary germs split up into a variable number of morulae, each of which develops into an embryo. Some of the morulae remain grouped in 'polygerms' of 2–6; many become isolated. When the larva is fully formed, the trophamnion is very thin and reduced; the larva then straightens out its body and gnaws its way out.[163] In the Chalcid *Encyrtus fuscicollis* a chain of 100 or so embryos arises; here the blastomeres after breaking up into morulae, each invested by trophamnion, are massed in a single germinal tube.[135] In the Chalcid *Litomastix truncatellus*, developing in the larva of *Plusia gamma*, as many as 2,000 individuals may arise from a single egg.[209] In addition to the trophamnion described above, *Litomastix* has an inner membrane or 'pseudoserosa' formed by delamination of the blastoderm at the morula stage.[209] These membranes usually disintegrate about the time the larva becomes active, but in *Apanteles glomeratus*[71] a fenestrated envelope encloses the larva thoughout its life. In some cases a cellular sheath is also laid down by the host.[101]

Viviparity[319]—The eggs in some insects may be retained within the genital tract of the mother until development is well advanced. In *Cimex* as we have seen (p. 717), fertilization takes place in the ovary, and the embryo has almost

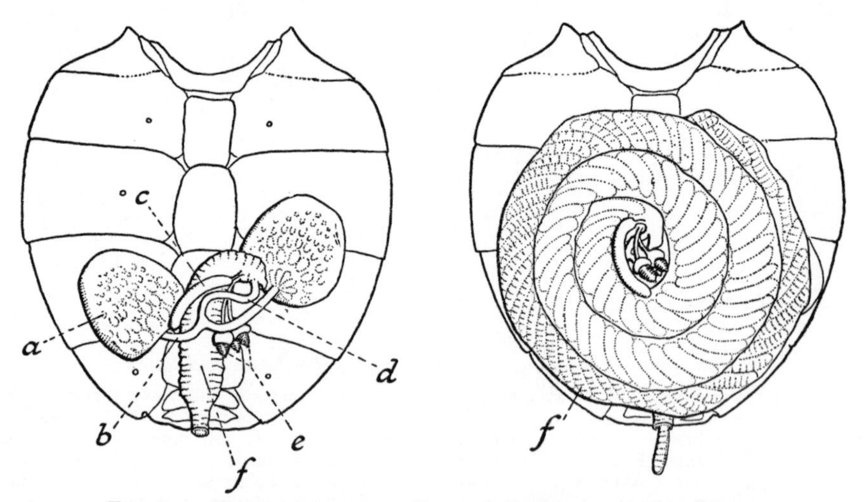

FIG. 402.—Ovoviviparity in the Tachinid *Panzeria rudis* (*after* PRELL)

A, sexual apparatus of newly emerged female; *B*, the same in mature female, showing the vagina hypertrophied to form brood chamber, and eggs containing fully formed larvae being deposited. *a*, ovary *b*, oviduct; *c*, accessory gland; *d*, common oviduct or uterus; *e*, receptacula seminis; *f*, vagina.

reached the stage of blastokinesis by the time the egg is laid;[1, 36] and in *Lecanium hesperidum* and other Coccids, the young nymphs may hatch almost immediately after oviposition.[234] This is sometimes called 'ovoviviparity'. It may be seen occasionally among such oviparous Diptera as *Calliphora*, *Pycnosoma*, *Musca*,

Phora, &c., if the eggs are held back from any cause.[108] It occurs normally among various Tachinids, in which the uterus may increase enormously in extent during the gestation of the eggs.[108] In *Panzeria rudis*, for example, the eggs collect in regular transverse rows, and the uterus becomes so extensive that it is thrown into three or more coils like a snail shell (Fig. 402). Fertilization takes place at the anterior end of the uterus; all intermediate stages of development can be seen, the eggs at the lower end containing fully formed larvae which escape from the chorion during oviposition.[175] In Sarcophagidae and some Oestridae a smaller number of eggs are present, and they may hatch inside the uterus.[108] In other Diptera, such as *Theria muscaria*, *Musca larvipara*, *Dasyphora pratorum*, *Mesembrina meridiana*, &c., only one large egg passes at a time into the dilated uterus; in the last named species the egg measures 4·5–5 mm. in length. In this group each egg hatches at the time of laying; in *Hylemyia strigosa* the cuticle of the 2nd larval instar is already formed by this time, and the larva moults to the 2nd stage before feeding.[108] In the aberrant fly *Termitoxenia*, which lives in the nests of termites, the egg is relatively still larger; it hatches immediately after laying, and the larva enters the pupal rest a few minutes later.[143]

In none of these cases does the embryo receive any nourishment from the mother during its stay in the uterus. But in *Glossina*[193] and the Pupipara[84] the larva, which hatches in the uterus from an egg of normal size, is nourished until it is fully grown by special 'milk' glands, probably modified accessory glands, which ramify throughout the abdomen and open on a

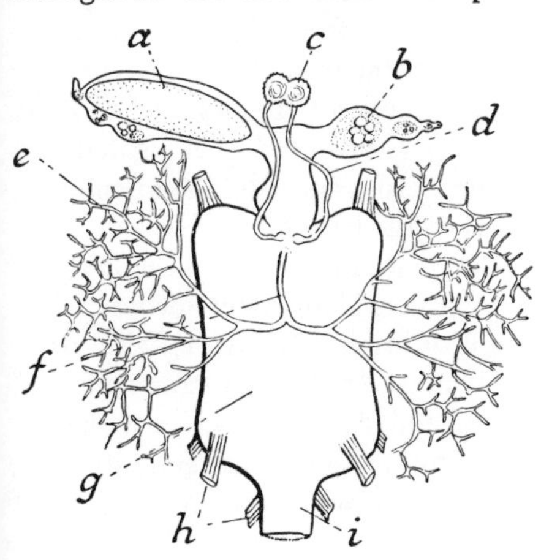

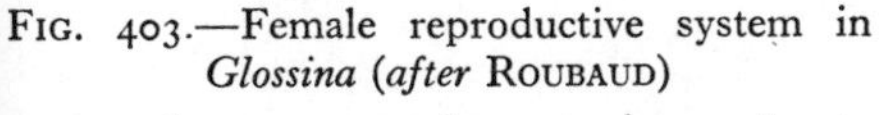

FIG. 403.—Female reproductive system in *Glossina* (*after* ROUBAUD)

a, nearly mature egg in ovary; *b*, egg due to mature next; *c*, spermathecae containing sperm; *d*, duct of spermatheca; *e*, accessory or uterine glands; *f*, common duct of these glands; *g*, uterus; *h*, uterine muscles; *i*, vagina.

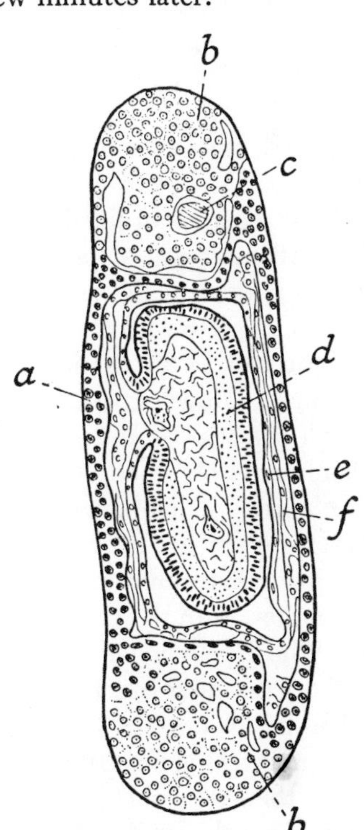

FIG. 404.—Longitudinal section of ovarial follicle in *Hemimerus talpoides* (*after* HEYMONS)

a, follicular epithelium; *b*, follicular epithelium hypertrophied to form a nutritive mass or 'placenta'; *c*, nutritive material produced from the breakdown of these cells; *d*, embryo; *e*, amnion; *f*, serosa.

papilla close to the mouth of the larva (Fig. 403). These larvae are matured singly, breathing by the extrusion of their posterior spiracles through the female opening. They are already immobile 'prepupae' when deposited in the Pupipara; in *Glossina* they become immobile within a few hours without feeding.

In *Cloëon dipterum* the eggs, which have an exceedingly delicate chorion, are not deposited immediately after mating as in most other Ephemeroptera, but are retained in the oviducts. Here they remain for 10–14 days; and then the female seeks the water and deposits the newly hatched nymphs.[10]

In other groups the developing eggs may be retained in the ovarioles, as in some Chrysomelids,[94, 182] Coccids, Aphids. In all such cases the chorion is absent. In Coccids and Aphids the nutritive cords continue to supply nourishment during early segmentation; later, the follicular epithelium takes on this function; the thin amnion and serosa do not interfere.[234] In the parasitic Orthopteron *Hemimerus*, the egg begins to develop as soon as it enters the basal part of the ovariole. The follicular epithelium swells up at each end into a rounded mass. Within these masses, as elsewhere in the epithelium, the cells dissolve and the nuclei break down, and the growing embryo enclosed in serosa and amnion absorbs the nutrient substances so produced (Fig. 404).[88, 234] This enlargement of the follicular epithelium is sometimes described as a sort of 'placenta'. The same term might equally be applied to the general follicular epithelium of viviparous Aphids and Coccids.

A more definite type of 'pseudo-placenta' is seen in the viviparous Polyctenid *Hesperoctenes*. Here the ova are yolk-free and devoid of chorion. Cleavage is complete and begins in the ovariole, the embryonic membranes being formed as usual. Until blastokinesis a 'trophoserosa' surrounds the embryo; the pleuropodia (p. 17) then begin to grow outwards, and fuse over the surface so that the embryo is completely invested by a 'pleuropodial membrane'. It is this membrane which functions as a placenta, collecting nourishment from the egg tube as the embryo passes down it, but it is never attached to the wall.[79] In the viviparous cockroach *Diploptera* the eggs have little yolk and an impervious chorion. Nourishment is obtained during development by means of very elongated persistent pleuropodia which penetrate the amnion, serosa and vitelline membrane, and pass through the micropyle to the outside of the egg.[79, 319]

In a few cases, embryonic development takes place in the general body cavity of the mother, as in the paedogenetic forms to be considered later and in Strepsiptera. The ripe female of Strepsiptera is little more than a sac with a few vestigial organs and a great quantity of fat body in which the immature embryos, nourished by means of a 'trophamnion', are embedded. When these are mature, they escape as triungulin larvae through a series of unpaired ventral genital canals, and make their way out through the 'brood canal', a cleft between the body of the female and the unshed larval and pupal skins.[91]

Neoteny and paedogenesis—We have already noted several abnormal cases of neoteny, that is, the retention of youthful characters beyond the usual period in ontogeny (p. 80).[251] Among Coccids this condition is normal; the females, which always have at least one less moult than the males, retain a wingless form approximating to that of the nymphs.[234] And the females of various Malacoderm beetles, for instance Lampyridae, retain the larval form to a varying extent. An extreme example of this is seen in certain Lycid beetles of Borneo, of which the females are the giant 'trilobite' larvae, the males are normal and much smaller.[144] Among Protermitidae the royal pair may be either perfect insects which have swarmed and shed their wings, or nymphs, with or without wing rudiments, which have become sexually mature.[72]

This last example approximates to *paedogenesis*, that is precocious reproduction in the larval or pupal stages. Paedogenesis is usually associated with parthenogenesis (p. 734); but an exception is afforded by the viviparous Polyctenid *Hesperoctenes* (above) in which some of the female last-stage nymphs may already have spermatozoa in their body cavity, and developing embryos in their ovaries.[79] Most paedogenetic insects are also viviparous: the beetle *Micromalthus debilis* produces larvae with long slender legs which moult into a legless stage; these legless larvae may either transform into pupae and winged adults, or, by viviparous reproduction, give rise again to legged or caraboid larvae.[6, 176, 202] But an exception is seen in the Chironomid *Tanytarsus boiemicus*; this is normally parthenogenetic as an adult, but it may begin to lay its eggs while in the pupal stage.[250] It is probable, however, that it is the fully formed adult within the pupal case which is laying eggs.[89]

The classic examples of paedogenesis occur among Cecidomyidae and Chironomidae. The phenomenon was first observed by Wagner in 1861 in larvae of *Miastor metraloas*. In these larvae,[65, 211] in larvae of *Oligarces* (*Heteropeza*),[85, 228] and in pupae of the Chironomid *Tanytarsus grimmi*,[229] the oöcytes develop parthenogenetically within the ovarioles until a variable number of daughter larvae are produced; these are set free into the body cavity, where they rapidly destroy the tissues of their maternal host and ultimately escape through the body wall. As we shall see later, this cycle occurs alongside a normal cycle of metamorphosis and bisexual reproduction (p. 737). Larvae of *Henria psalliotes* and *Tekomyia populi* develop normally to the third instar; it is then that haemocoelous embryos are produced. At the ensuing moult a pupa is formed which has normal cuticular characters, but has the legs, wings, head capsule, and mouth parts, &c., reduced to distorted vestiges or absent altogether.[419]

Hermaphroditism occurs in a few insects. In the Plecopteron *Perla marginata* the condition is merely rudimentary, the male nymphs having vestigial ovarioles alongside the testis.[200] But the curious fly *Termitoxenia* is a functional hermaphrodite. It has a female system with two ovarioles, paired oviducts, uterus and vagina opening on the 8th segment, and a male system consisting of an unpaired testis, vas deferens and penis opening on the 9th segment. Sperm are present only in young individuals; so presumably the male phase precedes the female.[143] And the Californian race of the Coccid *Icerya purchasi* is both hermaphrodite and self-fertilizing. Two classes of individuals exist: 90–99 per cent. of the population are hermaphrodites, the rest are males; no true females occur, though the hermaphrodites retain the instincts and body form of females, and may copulate with the rare males. In the 1st instar of the hermaphrodite nymphs, certain of the more centrally located cells of the gonads, which are still undifferentiated as to sex, undergo reduction from the diploid to the haploid condition (*cf.* p. 735). These haploid cells form the core of the gonad and give rise to sperm; the more peripheral cells remain diploid and give rise to follicles, nurse cells, and oöcytes. All the eggs show normal maturation, with reduction to the haploid chromosome number. Most of them are fertilized by sperm from the same individual and produce diploid hermaphrodites; occasional eggs escape fertilization and develop parthenogenetically to haploid males. Hermaphroditism is here incomplete, in the sense that the form with both types of gonad can never function as a male and fertilize other individuals.[98] Intersexes and gyandromorphs, which are discussed elsewhere (p. 93), although their gonads may be of mixed sex, seem never to be functional hermaphrodites.

Parthenogenesis—We have already noted several examples of development in unfertilized eggs. This may happen occasionally in a species normally reproducing bisexually (*sporadic parthenogenesis*); or it may constitute the normal mode of reproduction (*constant parthenogenesis*); and all intermediate degrees may exist between these two extremes. Sometimes one or more parthenogenetic or agamic generations may alternate with bisexual, gamic or amphigonic generations (*cyclical parthenogenesis* or *heterogony*). The unfertilized eggs may give rise solely to females (*thelytokous parthenogenesis*) or solely to males (*arrhenotokous parthenogenesis*) or to either sex (*amphitokous parthenogenesis*); and the individuals so developed may contain half the normal number of chromosomes in their somatic cells (*haploid parthenogenesis*) or the normal diploid number (*diploid parthenogenesis*). When parthenogenesis is well established the diploid chromosome number is restored, but by 'automixis' in place of 'amphimixis'. Occasionally, regulation does not cease at diploidy but goes on to polyploidy as the result of repeated nuclear fusion. In some forms, the reduction division becomes rudimentary and the diploid condition is retained throughout.[203]

Sporadic parthenogenesis—In many species of Acridiidae, development begins in nearly all unfertilized eggs; but it is successfully terminated in very few.[99] The haploid female pronucleus divides mitotically giving haploid cleavage nuclei; but in the more successful cases, chromosome division without cytoplasmic division may take place, so that the normal diploid chromosome number is restored. But even when development is complete, many embryos succumb without hatching, or die off as nymphs or young adults. In these eggs, oviposition seems to act as a stimulus for the completion of the maturation divisions; the presence or absence of the sperm nucleus is without effect.[113] A parthenogenetic strain has been selected out from *Schistocerca gregaria* and has been reared in the absence of the male for at least six successive generations. All the progeny are female; sexual maturation and oviposition are delayed; the average length of adult life is increased threefold. The number and size of egg pods are normal, but the average hatch is reduced to one third. Meiosis is complete; the restoration of the diploid number of chromosomes takes place during embryonic development, but some haploid and other abnormal cells persist.[320] Similarly, in *Locusta*[281] and *Periplaneta*[347] viability depends on the degree of chromosomal regulation attained.

A similar condition occurs in Lepidoptera. For example, in the bisexual races of the Psychids *Solenobia triquetrella* and *S. pineti* the diploid chromosome number is restored by nuclear fusion, but hatching never takes place.[204] In many Lepidoptera, however, complete development of unfertilized eggs has been recorded.[35] In *Lymantria*,[70] for example, larvae and adults of both sexes and of normal diploid constitution have occasionally been reared in the absence of males; and some of the females of *Orgyia antiqua* lay viable eggs in nature without mating.[229] Parthenogenesis may occur occasionally in *Periplaneta*[74] and other cockroaches.[382] In certain races of *Culex pipiens* a large proportion of the eggs develop parthenogenetically; this development, however, is initiated only by mating. The zygote is produced, not by fusion with the male gamete but by fusion of cleavage nuclei ('automixis').[348] The converse example of 'androgenic' hybrids in *Bombyx mori* and *B. mandarina* has already been described (p. 718).

As was shown many years ago by Tichomiroff, parthenogenesis may be induced artificially in silkworm eggs by various stimuli; high temperature, acids, alcohol, various histological fixatives and so forth will serve to activate the eggs.[118] Hydrochloric acid combined with heat treatment at 46° C. gives rise to parthenogenetic clones which can be reared for years by repeated heat treatment. There is a single maturation division with no meiosis, so that the cleavage nuclei are diploid from the outset.[274]

Parthenogenesis in Phasmids—Among Phasmids, parthenogenesis is universal.[162] In some species it is constant, and males are excessively rare; other species are ordinarily bisexual, and parthenogenesis occurs only in the absence of the male.[279] In a species like *Carausius morosus*, which is constantly parthenogenetic, when males appear they are generally intersexes, and rarely pure males capable of successful copulation;[29, 225] such males are cytologically sterile;[29] they can be produced experimentally by incubation of the eggs at 30° C. during the first third of their embryonic life.[280, 415] These parthenogenetic Phasmids are of three cytological types. (i) In the first parthenogenetic generation of *Menexenus*, many of the cells are haploid, though some are diploid. (In subsequent generations regulation occurs, all the cells are diploid, mortality is less and development more rapid.) (ii) Most species are always diploid. (iii) In a few species, haploid, diploid, and polyploid cells may occur together.[29]

Sexual races—Geographical or local races of insects may be characterized by their mode of reproduction. The Psychids *Solenobia triquetrella* and *S. pineti* are represented in different parts of Germany by a normal bisexual race in which the unfertilized eggs never develop completely, and a purely thelytokous parthenogenetic race in which males never appear.[204] In the Coccids *Lecanium hesperidum* and *L. hemisphaericum* there are parthenogenetic races producing only females, and bisexual races with a minority of males. The bisexual races show facultative thelytoky: unfertilized eggs develop into females, fertilized eggs into either sex. In the unfertilized eggs of this race the diploid chromosome number is restored at the outset of development by fusion between the second polar body and the egg nucleus, a process very like normal fertilization. In the parthenogenetic races, the reduction division is omitted, so that the oöcytes are always diploid.[220] In *Trialeurodes vaporarium* there is a race showing arrhenotokous parthenogenesis; fertilized eggs producing females, unfertilized producing males; and a race consisting almost exclusively of females (14♂: 4582♀). In the latter race the diploid constitution is restored after the second maturation division by autoregulation—probably through splitting of chromosomes without nuclear division.[220]

Haploid arrhenotokous parthenogenesis—It is evident from the foregoing examples that most insects will develop normally only when the diploid constitution has been regained, but haploid parthenogenesis is a constant occurrence in a few cases. The best known example is the honey-bee and other Hymenoptera, in which the eggs undergo normal maturation, fertilized eggs develop into diploid females, unfertilized into haploid males.[229] The same condition exactly is seen in *Aleurodes proletella*,[220] in one of the local races of *Trialeurodes vaporarium*[220] and in *Icerya purchasi*, *I. littoralis* and related forms—in which the somatic cells of the female contain 4 chromosomes, the males 2.[98]

Cyclical parthenogenesis in Cynipids—Alternation of bisexual and parthenogenetic generations occurs in its most regular form in Cynipids. The two generations often attack different parts of the host plant, and may differ so strikingly that in many cases they have been regarded as separate species. In the oak-gall wasp *Neuroterus lenticularis*, all the fertilized eggs, derived from the bisexual generation, develop in the early spring into females which reproduce parthenogenetically in the early summer, some producing only sexual males, others only sexual females. In the female-producing eggs the maturation divisions are omitted, consequently they have the diploid number of chromosomes; in the male-producing eggs reduction occurs as usual, the males are therefore haploid as in other Hymenoptera. Fertilization of the eggs of this generation furnishes the females of the following spring.[46]

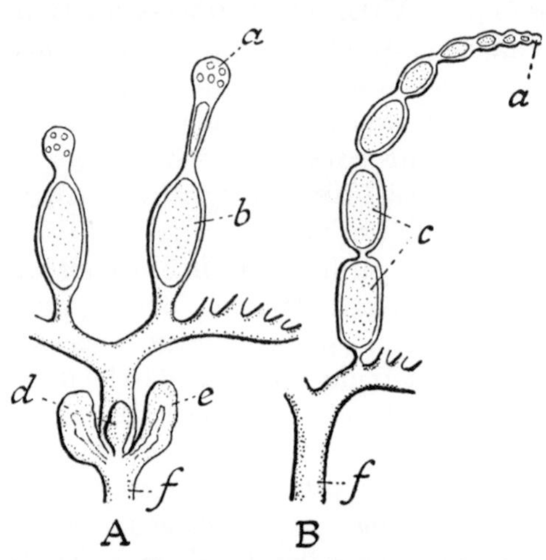

FIG. 405.—Female reproductive system in the Aphid *Macrosiphum solanifolii*

A, gamic type; *B*, parthenogenetic type. (*after* SHULL). *a*, germarium; *b*, egg; *c*, embryos; *d*, seminal receptacle; *e*, colleterial gland; *f*, vagina.

Cyclical parthenogenesis among Aphids—The life cycles of Aphids are complicated by an increase in the number of successive agamic generations, frequently by a periodic change of host plant, by polymorphic wing development, and by the association of parthenogenesis with viviparity. In a simple form such as *Aphis rumicis*, the winter egg gives rise to an agamic, usually wingless, female, the fundatrix, which produces viviparously winged and wingless agamic females (fundatrigenia). The winged forms spread to other specimens of the host plant and again, like their wingless sisters, produce viviparously winged and wingless agamic females. These parthenogenetic viviparous generations continue until the autumn, when, among the agamic females, there appear some (sexuparae) which give rise to winged and wingless males and amphigonic oviparous females (sexuales) which lay a small number of fertilized winter eggs.[232] But the cycle may be far more complicated where the insect changes from one type of host plant to another.[145]

In *Pemphigus*, *Phylloxera*, *Chermes*, viviparity does not occur, the amphigonic generation is always wingless and dwarf, and the amphigonic female produces only one egg. The vine phylloxera (*Viteus vitifolii*) may be noted as an example. The winter egg, laid on the branches of the grape vine, hatches to produce a fundatrix female, forming a gall on the leaf, which gives rise parthenogenetically to a series of gall forming generations ('gallicolae'). In the course of the summer, an increasing proportion of individuals with a longer proboscis are produced, and these migrate to the roots where they continue to reproduce parthenogenetically ('radicolae'). Towards autumn they give rise to individuals which undergo four moults (instead of three as in the 'gallicolae' and 'radicolae') and develop into winged adults. These are 'sexuparae'; they lay eggs which give rise to sexual males or sexual females, a given sexupara usually producing only one sex. The sexual forms are apterous, the proboscis is absent and the gut rudimentary. After mating, the female lays a single large winter egg.[229]

The reproductive system may differ markedly in the different phases of these life cycles. For example, in *Macrosiphum solanifolii* the reproductive system of the amphigonic females consists of a vagina with a pair of colleterial glands and seminal receptacle, a pair of short oviducts and usually 10 ovarioles. There are never more than two oöcytes in each ovariole, and these are surmounted by a large spherical germarium (Fig. 405, A.) In the parthenogenetic females, the accessory glands and seminal receptacle are absent; the ovarioles, which vary in number, each contain 6–9 embryos, eggs, or oöcytes, and bear at the end a very small germarium (Fig. 405, B).[207]

We have discussed in an earlier chapter the mechanisms by which the body form and type of reproduction in Aphids is determined by environmental factors (p. 102). In some Aphids living in the uniform environment of the tropics or of hot-houses, the amphigonic generations have disappeared from the life cycle and they show continuous parthenogenesis.

Alternation of generations in Cecidomyids—The cycle of development in *Miastor* and *Oligarces* (*Heteropeza*) (Fig. 406) shows a degree of instability which is unparalleled elsewhere among insects. Under the conditions of abundant food, high humidity, and constant temperature which obtain in the dark depths of the rotten tree stumps in which *Miastor* larvae occur, paedogenetic reproduction will continue indefinitely.[65, 211] In *Miastor* a moderate standard of feeding results in adults; very good nutrition or starvation conditions cause the larvae to reproduce by paedogenesis.[366] Exposure to light[211] or desiccation[65] may cause these larvae to enter a kind of diapause; they become orange in colour, develop various structural distinguishing characters, acquire a migratory habit, and fail to grow or reproduce. If the ordinary white larvae are starved, these orange larvae appear in the next generation.[65] If the orange larvae are given abundant food, they are 'rejuvenated' and their sexual products mature. If they are ligatured or injured by burning, activation and maturation of the oöcytes is similarly induced.[65]

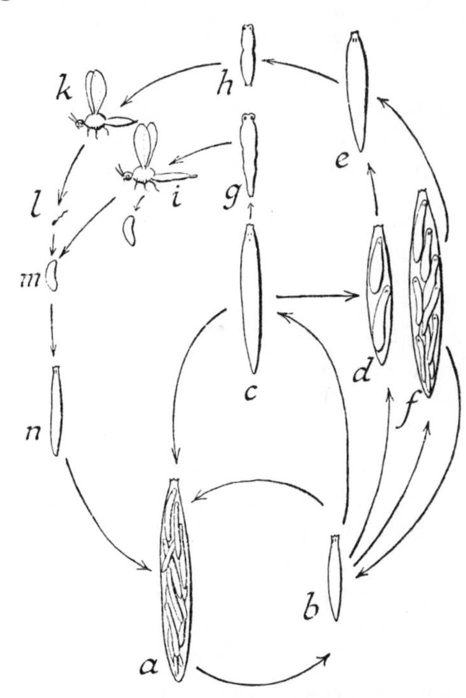

FIG. 406.—Alternative reproductive cycles in *Oligarces paradoxus* (*after* ULRICH)

a, paedogenetic larva giving rise to *b*, undetermined daughter larva which may develop in *a* again or into *c*, female imago-producing larva, or into *d*, which gives rise to male imago-producing larvae (*e*) or into *f*, which gives rise both to undetermined daughter larvae and to male imago-producing larvae; *g*, female pupa; *h*, male pupa; *i*, female imago; *k*, male imago; *l*, sperm; *m*, egg; *n*, young larva from egg.

If white larvae of *Miastor*, growing very slowly at 5–10° C., are suddenly exposed to 30° C., they often give rise to an egg which produces a larva which transforms to a pupa and adult.[65] Sometimes larvae of an intermediate size may produce by paedogenesis 2–5 of these pupating larvae.[211] In *Oligarces*, pupating larvae can be induced at will by overcrowding the previous generation;[85] or, if they are fed on a culture of moulds, more pupating larvae are induced, either directly or in the offspring, as the culture gets older; thus a culture of moulds 2 days old gave 0·5 per cent. of pupae, a 7-day culture 14·4 per cent., a 12-day culture 41·8 per cent.[228] The newly born pupa-forming larvae have imaginal discs (which are always absent from paedogenetic larvae) already visible; yet by transferring them to a fresh culture as soon as they are detected, their type of development may be reversed, the discs vanish and they grow into paedogenetic larvae;[85] in *Miastor*, this reversal may be induced shortly before metamorphosis by burning the head or genital region.[65] *Heteropeza pygmaea* (=*Oligarces paradoxus*) has only one larval instar which, when full grown, becomes the paedogenetic mother. But it actually transforms into an abnormal type of pupa ('hemipupa') enclosed in the larval exuvium before producing the young larvae. The hemipupa exists in two forms, one of which serves for rapid reproduction in favourable conditions, the other for the production of a few young during prolonged adverse conditions.[419] Both bisexual and parthenogenetic lines derived from egg larvae *must* go through one or more paedogenetic generations before their daughter larvae are capable of following all developmental courses.[295]

Since the paedogenetic larvae are all females, while the pupating larvae may be of either sex, it follows that the sex of these larvae is highly indeterminate also. In the parthenogenetic eggs of the paedogenetic larvae, the reduction division is omitted at maturation, so that the egg nucleus has the diploid chromosome number from the outset.[106]

SEX DETERMINATION

Sex is determined by the distribution of the heterochromosomes or sex chromosomes. These may be of several types. (i) There may be an unpaired *X*-chromosome or monosome which occurs singly in the diploid state in one sex. In Orthoptera, Homoptera, some Heteroptera, a few Diptera and some Coleoptera and Neuroptera, the monosome occurs in the male. The female then has two *X*-chromosomes. The monosome occurs in the female of many Lepidoptera, e.g. *Talaeporia*,[203] *Ephestia*.[120] (ii) The *X*-chromosome in one sex may be paired with a chromosome which often differs morphologically from itself. A dissimilar pair of this kind, called *X* and *Y*, and termed idiochromosomes or diplosomes, occurs in the males of most Heteroptera, Diptera and Coleoptera, in the females of which two *X*-chromosomes are present. In many Lepidoptera, e.g. *Lymantria*, the females have the *XY* constitution.

The mode of action of these chromosomes in controlling sex during development has been considered in an earlier chapter (p. 86). Here we are concerned with those disturbances in the sex ratio which result from unusual distribution of the sex chromosomes at maturation and fertilization.

Normally, equal numbers of *X* and *O* or *X* and *Y* gametes will be produced at meiosis in the heterogametic sex; consequently, equal numbers of *XX* and *XO* or *XX* and *XY* zygotes will result at fertilization, and the sex ratio will be equal. But in females of the Psychid *Talaeporia*, which have the *XO* constitution, high temperature and over-ripeness of the eggs cause the *X*-chromosome to migrate more frequently into the female pronucleus. Hence more *XX* individuals, i.e. males, arise at fertilization. At low temperature the *X*-chromosome goes more often to the polar body, and therefore more females are produced. Thus the sex ratio ranges from 100♀: 162♂ at 30–37° C. to 100♀: 65♂ at 3–8·5° C.[203] In *Ephestia*, delayed fertilization has no effect on the sex ratio.[104] In *Drosophila*, an excess of females appears among the latest eggs to be fertilized.[124] In the Coccid *Pseudococcus citri*, delayed fertilization results in a great preponderance of males: the ratio may range from the normal 101♂: 100♀ if fertilization is immediate, to 991♂: 100♀ after 10 weeks delay.[104] In this case, since the male is the heterogametic sex, the cytological explanation must be different from that in *Talaeporia*. Families consisting almost entirely of females in *Abraxas grossulariata* are due to a gene which causes passage of the *X*-chromosome to the second polar body during the maturation of nearly all the ova;[35] in *Lymantria dispar* such families are due to a dominant lethal gene which kills nearly all the males.[35] The cytological basis of the unisexual families which are common in *Pediculus* is not known.[23]

In the amphigonic generations of Aphids, as we have seen, all the offspring are female. That is because during the maturation of the sperm the male-producing spermatids degenerate; all the functional spermatozoa therefore carry the *X*-chromosome and are female-producing.[5, 125, 146]

In the parthenogenetic generations of Aphids, the females of which have the *XX* constitution, only one division occurs at maturation, and there is no reduction. The egg therefore retains the *XX* constitution of the mother, and consequently the offspring are generally female. But occasionally one of the *X*-chromosomes fails to divide (non-disjunction) and passes to the polar body. Consequently the egg acquires the *XO* constitution of the male. In many species a given female of the sexuparous generation is either solely male-producing or solely female-producing;[146, 206] in *Pemphigus spirothecae* a single sexupara produces always males from the two anterior egg tubes, and females from the more caudal egg tubes.[224] In *Miastor* there are no sex chromosomes. The germ line is octaploid. A very strange meiotic division gives rise to haploid sperms from these octaploid spermatogonia. Sex is determined by the number of extra chromosomes shed at maturation, the soma of the female being diploid and of the male haploid.[240] In the paedogenetic larvae of *Heteropeza pygmaea* (*Oligarces paradoxus*) there is a chromosomal sex determination which is somatically controlled. Depending upon nutrition, the oöcytes may become small eggs which do not undergo reduction and so become diploid females; or large eggs which suffer reduction to haploid males; in the males the diploid number is later restored by regulation.[405] As in *Miastor* many chromosomes of both male and female are eliminated from the future somatic nuclei; the final chromosome numbers in somatic nuclei are 10 in the female and 5 in the male.[322] [See p. 746.]

In most Coccids (except the most primitive) the X-chromosome is lost in the heterogametic male and sex determination thus devolves upon the female. In *Stictococcus* this seems to be decided by whether or not the oöcytes come into contact with a mycetocyte. If it does, infection takes place and it develops into a female; uninfected oöcytes develop into males.[287]

In most insects, since the female is homogametic (*XX*), it is easy to see that in parthenogenetic forms, in whatever way the diploid constitution is restored, the offspring will also be *XX* or female (thelytokous parthenogenesis). Whereas in the sporadic parthenogenesis of Lepidoptera, with females *XO* or *XY*, either the

X or the *Y* may predominate in the offspring and, as we have seen, both sexes are produced. But in those races of *Solenobia* (p. 735) which show *constant* parthenogenesis this is purely thelytokous. That is because in these races the *X*-chromosome always passes into the polar body at the first maturation division; consequently the *Y* always predominates and the offspring are always female.[204] An alternative explanation is that in the parthenogenetic races the organism is the product of the fused polar nuclei with the constitution XO or XY and therefore female), and the parthenogenetic egg nucleus degenerates. In the fertilized bisexual egg, fusion of polar nuclei again occurs but their further development is blocked by the amphimictic nucleus. In the unfertilized bisexual egg of some races both nuclei develop and chaotic sexual mosaics are the result.[392] No intermediate forms in the cuticular products of single cells can be seen; these are either male or female. The apparently intermediate structure is due to the mixture of male and female cells.[391]

In the honey-bee, and many other Hymenoptera, all the fertilized eggs become females, the unfertilized become males (Dzierzon's law).[418] All the sperm are in fact female-producing, though the cytological explanation of this must be different from that in the Aphids described above. Thus the control of sex in Hymenoptera resolves itself into control of fertilization.* The 'voluntary' control of the muscular wall of the oviduct orientating the egg towards the receptacular duct, and the control of the so-called sperm pump or sphincter of the duct, regulate the entry of sperm into the egg.[2, 57] In the honey-bee it is uncertain what stimulus determines this response; possibly it is the size of the brood cell; but to some extent the regulation is seasonal, male eggs being produced in the autumn.[229] The female *Osmia* constructs large cells, and deposits fertilized eggs in them, as soon as the seminal receptacle is filled with sperm; whereas the virgin female, or the female whose supply of sperm is exhausted, constructs small cells and lays unfertilized male eggs.[41] The sex of the offspring may be determined by the rapidity of egg-laying; the spermathecal gland may be unable to keep pace with the rate of deposition of the eggs. For example, when oviposition occurs rapidly in *Trichogramma* the first two eggs are fertilized and female, the third egg is unfertilized and therefore male.[253] In the Chalcid *Mormoniella* partially resorbed eggs tend to give rise to males because the micropyle is not fully pressed against the spermathecal duct. A shortage of host puparia leads to partial resorption and so to an increase in the percentage of males.[339]

The parasite *Tiphia*, which oviposits in 2nd and 3rd instar larvae of *Popillia*, has a normal sex ratio with a slight preponderance of females when it develops in 3rd instar larvae; there is a preponderance of males when it develops in 2nd instar larvae; and this difference persists when the eggs after laying are transferred to 3rd instar larvae. Here the sex is clearly determined by the ovipositing female, perhaps in accordance with the size of host[20] (*cf.* p. 97). The same applies to *Pimpla*[32] and other Ichneumonids[205] in which the female deposits fertilized eggs in large hosts; to *Trichogramma*[197] which has a sex ratio of 77·6 per cent. females when many eggs are available as hosts for this parasite, 43·8 per cent. when there is extreme superparasitism; and to the Chalcid *Pleurotopis* which has a sex ratio of 1·66♀: 1·0♂ in the 1st instar larvae of the beetle

* There are exceptions to this rule. For example, in the Cape variety of *Apis mellifica*, fertile workers are common; and they produce males, workers, and even queens, apparently parthenogenetically.[229]

Promecotheca, 4·34♀: 1·0♂ in the 3rd instar of the same host.[34, 218] A similar control exists in the American race of *Trialeurodes vaporarium*, which shows arrhenotokous parthenogenesis: the fertilized eggs becoming female, the unfertilized male; but the mechanism of this control is not known.[201, 220]

In many Hymenoptera both males and females are uniparental; the females are in fact capable of producing two kinds of egg, (i) yielding only uniparental females, (ii) yielding either uniparental males or, if they are fertilized, biparental females. In *Habrobracon* those females which arise occasionally from unfertilized eggs originate from patches of tetraploid tissue in an ovary otherwise diploid. The unfertilized eggs after the reduction process are therefore either haploid (♂) or diploid (♀).[210] It may be that the composition of the ovarian tissue is influenced by the environment. For example, if *Prospaltella perniciosi*, a parasite of the San José Scale, is reared on scales infesting cow-melons (*Citrullus*) the ovaries are entirely tetraploid, producing only females. If this insect is reared on scales infesting peach, the ovaries are either wholly diploid, and produce only haploid males, or there are some patches of tetraploid tissue and both males and females arise parthenogenetically.[59]

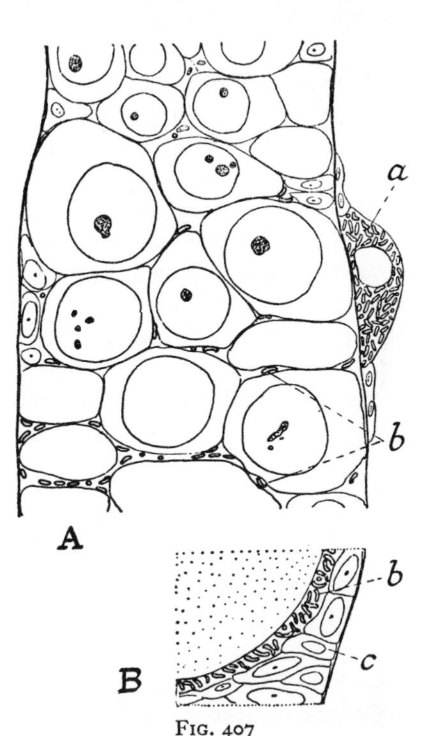

FIG. 407

A, upper end of ovariole of *Periplaneta americana*, showing bacteroids in a mycetocyte and between the oöcytes; B, portion of follicle and oöcyte at later stage, showing bacteroids between oöcyte and follicular cells (*after* GIER). *a*, mycetocyte; *b*, bacteria; *c*, follicular cells.

TRANSMISSION OF SYMBIOTIC MICRO-ORGANISMS

As we saw in an earlier chapter (p. 473), many insects constantly harbour intracellular micro-organisms. The mutual adaptations between these organisms and their host suggests that the relation between them is symbiotic; and one aspect of reproduction in such insects concerns the diverse mechanisms by which these symbionts are transmitted to the offspring.

In the bed-bug *Cimex*,[21] and the weevil *Apion*,[153] the symbionts invade the nurse cells and are conveyed from them to the oöcytes. In the ant *Camponotus* there is a general infection of the young follicle, leading to a generalized infection of the egg plasma,[21] and in Blattidae the bacterial layer forms similarly over the egg surface, later invading the vitellophags and subsequently the fat body and ovary (Fig. 407).[68] In the primitive termite *Mastotermes* there are bacterial rods in the fat body and they are similarly transmitted.[115] In the egg the infected cells are always the extra-embryonic blastoderm or the yolk cells. In *Oryzaephilus* (Col.),[115] *Lyctus* (Col.),[115] Anoplura (Fig. 385, F., G),[186] the posterior pole of the egg is invaded shortly before secretion of the chorion. In parthenogenetic Aphids the embryo is entered while it is in the ovary.[223] In

Rhizopertha (Col.) the micro-organisms invade the testis, mix with the sperm, and after being passed with them into the bursa copulatrix of the female, enter the egg through the micropyle during fertilization.[132] The same thing probably happens in the bug *Nezara*, in which bacterial sacs are associated with the male organs.[131] In Trypetidae the organisms pass from the hind gut into the ovipositor, and enter the micropyle of the egg when it is laid;[212] they also enter the micropyle in Lagriidae (Col.).[212] In Anobiid beetles, the yeasts collect in diverticula in the female tract; from here they contaminate the surface of the egg, and the newly hatched larva is infected by eating the shell.[18] The organisms in the modified Malpighian tubes of some Chrysomelid beetles, are deposited with the mucilaginous secretion in the egg cap, which is eaten by the larva,[212] and a similar transmission of the organisms from the gut occurs in the beetles *Bromius* and *Cassida*.[212] Pentatomidae obtain their symbiotic bacteria by probing the surface of the egg immediately after hatching;[14, 192] and *Rhodnius* pick up their Actinomyces from contaminated surfaces in the same way.[17] In the Plataspid *Coptosoma* the female deposits packets of symbionts among the eggs; these are pierced and sucked by the newly hatched larvae.[267] The fungus which is carried by the wood wasp *Sirex* is maintained in pouches on the first abdominal segment of the female larvae;[164] the hyphae grow into the pouches of the adult female shortly after emergence and while she is still enclosed in the cocoon.[62] Finally, in the Pupipara[249] and *Glossina*,[241] in which the larva is nourished in the uterus until fully grown, the symbionts are transmitted through the 'milk' glands.

ADDENDA TO CHAPTER XV

p. 703. **Yolk formation**—Pre-yolk oöcytes of *Periplaneta* are engaged in the synthesis of RNA and protein; the younger oöcytes are chiefly concerned with the synthesis of ribosomes which accumulate in the cytoplasm, reaching a maximum concentration just prior to the deposition of yolk.[507]

p. 704. **Nurse cells**—Because the divisions of the nurse cells in *Drosophila* and many other insects are incomplete, all the nurse cells and the oöcyte are connected by canals, 'fusomes', which make possible the massive transfer of materials from the nurse cells to the oöcyte.[430, 459] In *Hyalophora* the nurse cells contribute mitochondria, RNA, glycogen particles, lipid droplets and granular vesicles and filaments of unknown composition to the oöplasm.[460]

p. 704. During the period of protein yolk formation, in *Hyalophora*[499] and in *Rhodnius*,[475] the follicle cells separate to form a porous epithelium. The uptake of proteins is selective; partly through the action of the follicle cells, partly through selective pinocytosis into the oöcyte.[468, 499] In *Hyalophora* the specific 'female protein' for incorporation in the eggs appears at the prepupal stage (p. 425). In *Pieris* the corresponding protein first appears at adult emergence.[462]

p. 704. In Lepidoptera the uptake of labelled leucine confirms that protein synthesis occurs both in the yolk itself and in the follicular cells. The substance of the vitelline membrane is synthesized in the cortex of the oöcyte, that of the chorion in the follicle cells.[441]

p. 708. **Spermatodesms**—The chemical nature of the spermatodesm and its mode of formation are quite distinct in the Tettigoniids and Acridids among Orthoptera.[436]

p. 709. **Mating**—In *Pieris rapae* the reflection in the near ultra-violet by the wing scales of the female is an essential agent in the release of the sexual activities of the male.[472] Courtship in male *Drosophila* consists in the production of brief pulses of sound generated by wing vibrations.[427] In *Achroia grisella* the searching female is excited by the male scent but orients chiefly to the sound of his beating wings.[483]

p. 710. **Scent glands**—In *Bombyx* the cytoplasm of the scent-gland cells is rich in smooth endoplasmic reticulum and becomes filled with masses of lipid droplets. The outer surface of the cells in contact with the cuticle is deeply infolded to form lamellar processes which increase the surface 30–60 fold.[496]

p. 710. **Sex pheromones**—Most of the active material in female *Bombyx* is deposited on the gland surface ready for dissipation; there is no large store. It is at a maximum at eclosion and production is suppressed after copulation. The behaviour of the female and the attractiveness of the gland secretion are not affected by allatectomy in the larval stage.[496] Extracts of the female eye gnat *Hippelates* have shown the presence of a male attractant during the period of most active oöcyte development and of a male repellent when development is complete.[421] In *Tenebrio* the same sex hormone is secreted by both sexes, but in much higher concentration by the female.[501] Olfactory receptors responding to a low threshold of the female sex attractant of *Periplaneta* occur in the antennae of male larvae and adults, but are absent from larvae and adults of the female.[432]

The following are some examples of the chemistry of sex pheromones: 'bombycol' of the female *Bombyx* is *trans*, *cis*-10,12-hexadecadienol;[434] 50 per cent. of males will react to 10^{-4} μg of bombycol;[489] in *Trichoplusia ni* it is *cis*-7-dodecenyl acetate;[429] in *Galleria* the male sex attractant is *n*-undecanal;[483] in the leaf-roller *Argyrotaenia*, *cis*-11-tetradecenyl acetate;[481] in the pink bollworm *Pectinophora trans*-5,9-tridecadienyl acetate;[458] whereas in the female wireworm beetle *Limonius* it is valeric acid.[456]

p. 711. In Saturniid moths appropriate environmental signals to the brain stimulate the release of a hormone from the intrinsic cells of the corpora cardiaca and this then acts on the abdominal nervous system to provoke the protrusion of the female genitalia and release of the sex pheromone.[479]

p. 711. **Aphrodisiac secretions in the male**—In some male Noctuid moths the eversible scent brush at the base of the abdomen produces benzaldehyde as a chief component which probably has an aphrodisiac effect during courtship.[423] The hair pencils of male *Danaus* convey a fine 'dust' impregnated with the aphrodisiac secretion to the antennae of the female; a second component probably serves as an adhesive to cause the dust to stick.[477]

p. 711. In *Drosophila*[469] and *Aëdes*[464] the accessory glands, or paragonia, of the male produce a substance which depresses the sexual receptivity of the female. In these Diptera there is commonly a two-day period during which young females cannot be inseminated although mating appears to take place normally.[493]

p. 713. **Spermatophores**—In *Blattella* there are 8 groups of accessory glands, all contributing diverse products, mostly of glycoprotein nature, to the spermatophore;[424] and there is a similar diversity in the mucopolysaccharides and lipotroteins contributed by 9 cell types to the spermatophore of *Schisto-*

cerca.[473] Uric acid present in the utriculi majores of the accessory sex glands of male *Blattella* is poured out over the spermatophore during copulation. It may perhaps serve to protect the spermatophore from being eaten by the female.[484]

p. 715. **Migration of sperm to the receptaculum**—In *Aëdes aegypti* the motile behaviour of the sperm alone will not explain normal filling of the spermathecae. The female also plays some role which is not fully understood.[457]

p. 720. **High temperature and sterility**—Reproductive activity in both sexes of *Rhodnius* is completely inhibited during exposure of adults to 34° C. immediately after feeding. The effects are just like those of starvation or allatectomy (p. 724); they probably result from the arrest of protein synthesis. The effect is reversible but sterility persists for several reproductive cycles.[474]

p. 722. **Reproduction in the rabbit flea**—The male rabbit flea *Spilopsyllus* as well as the female requires some 'factor' provided by the female rabbit in the final stages of pregnancy, or by the new-born rabbit, to make reproductive activity and maturation of sperm possible. This 'factor' may well be the same in the nestling and in the pregnant female; whether it is some nutritional requirement or some sensory signal is still uncertain.[467, 485] (cf. p. 746.)

p. 723. **Flight muscles and ovary development**—Degeneration of indirect flight muscles regularly occurs in adult *Dysdercus* when adequately fed. Breakdown is initiated by hormones from the post-cerebral complex; muscle breakdown and vitellogenesis occur in starved females with the post-cerebral complex implanted.[445]

p. 724. **Impregnation and egg-laying**—In *Rhodnius* the rate of egg laying is greatly increased by mating, but this has little effect on the rate of egg formation.[440] The stimulating action of mating occurs only if sperm are introduced into the spermathecae: it does not happen if the opaque accessory glands of the male are removed so that the spermatozoa fail to reach the receptacula (p. 715). Since implanted spermathecae from normally mated females will induce increased egg-laying by virgin females, it may be that a blood-borne humoral factor from the spermathecae provides the stimulus.[442] If so it probably acts via the brain, since surgical removal of the neurosecretory cells abolishes the effect.[442] But in the cockroach *Pycnoscelus* the sperm exerts this influence on oviposition not by humoral means but by way of the nervous system: innervated spermathecae are required for successful oviposition.[494] In the moth *Zeiraphera* impregnation doubles the number of eggs maturing, and the number is doubled again if oviposition is fully stimulated. Here there appears to be feedback control between oviposition and oögenesis.[428]

p. 725. **Corpus allatum in Cimex**—In *Cimex* the corpus allatum does not become active, even in the fed female, until the sperm, after passage through the body cavity, have reached the spermathecae; and soon after the sperm disappear from the spermathecae the ovary reverts to the virgin condition.[142] This stimulus of the corpus allatum is not hormonal; but activation of the gland is prevented by cutting the nerve cord soon after mating.[443]

p. 727. **Hormonal control of reproduction**—Extensive studies in the past few years have substantiated the conclusion that both neurosecretory cells and corpus allatum are necessary for oögenesis—but there are differences in detail in different species. The corpus allatum hormone, besides inducing yolk deposition, which seems to be an almost universal effect, is often itself responsible for inducing the synthesis of vitellogenic blood proteins, the so-called 'female proteins', usually by the fat body. That happens in *Locusta*;[470] in *Periplaneta*[426] where the two proteins in question, which account for 88 per cent. of the protein content of the yolk, are 100 times more concentrated in the yolk than in the haemolymph; and in the other cockroaches *Leucophaea*[433, 447, 488] and *Nauphoeta*.[466a] In *Leptinotarsa* it induces synthesis of a 'female protein' comprising 75 per cent. of the yolk protein.[466]

The medial neurosecretory cells appear to stimulate directly the synthesis of yolk protein in the fat body of *Pyrrhocoris*: here allatectomy causes a rise in haemolymph protein from a feed-back effect, because the stimulus to oögenesis has been removed.[492] The same effect is reported in *Periplaneta*.[422] The mosquito *Aëdes taeniorhynchus* is an autogenous species: if adequately fed in the last larval stage it produces a batch of eggs without feeding as an adult. Both medial neurosecretory cells and corpora allata are essential for yolk formation; but they seem to regulate different processes, for transplanted neurosecretory cells will not restore egg development arrested by allatectomy; nor will implanted corpora allata restore egg development arrested by ablation of neurosecretory cells.[463] In some insects there is evidence that the neurosecretory cells act indirectly by an effect on the corpora allata. This appears to be a nervous effect by axon pathways in *Adephocoris*[449] and in *Tenebrio*;[471] a hormonal effect in *Locusta*.[453] There is increased secretory activity in the neurosecretory cells of *Leptinotarsa*,[490] *Pterostichus*[455] and *Galeruca tanaceti*[491] when these insects are coming out of diapause; and it may be that their secretion is exerting a direct effect on metabolism in the fat body, oöcytes, or elsewhere. An effect on the production of mid-gut protease occurs alongside the nervous activation of the corpora allata in *Tenebrio*.[471] In *Leucophaea* the neurosecretory cells and the corpus allatum are said to control the synthesis of two different proteins.[487] The extent to which changes in the accumulation of lipids &c. in the fat body and other tissues after excision of the endocrine organs results from direct effects of hormones on the tissues themselves or to indirect, or feed-back, effects from the ovaries, remains uncertain.[503]

Among Acridids there is much variation in the extent to which sexual behaviour is controlled by the corpus allatum.[476] The presence or absence of pheromone production may be correlated with the type of life cycle; such control is absent in the short-lived non-feeding adults of *Antheraea* and *Galleria*.[425] Laying workers of *Apis* produce pheromones (comparable with the 'queen substance') which evoke retinue behaviour in other workers and inhibit ovarial development in them.[502]

The frontal ganglion in *Schistocerca* seems to be involved in the sexual maturation of both sexes. Removal of the frontal ganglion reduces feeding and the failure of maturation may be due to starvation.[454] But there is evidence that removal of the frontal ganglion results in defective liberation of neurosecretory material.[439] It may be that the action of the frontal ganglion is a nervous one via the neurosecretory cells and not a direct endocrine effect.[486]

Attention has been concentrated mainly on oögenesis as a whole. It is often apparent, however, that the presence of the leading oöcyte in the ovariole restrains the maturation of the succeeding oöcytes. The mechanism of this control is not known. It has been attributed to an 'oöstatic hormone' produced by the maturing egg and acting possibly by controlling juvenile hormone secretion, as in *Musca*[420]—but this interpretation is still hypothetical.

p. 727. **Environmental stimuli**—There are many examples of the environment providing sensory stimuli which set in motion the reproductive machinery of particular species. The beet moth *Scrobipalpa* commonly reabsorbs the eggs in the absence of sugar beet; in the presence of the host plant, copulation and oögenesis are stimulated and this leads to rapid infestation of the crop.[480] Effects of this kind are very frequent, though not universal, among Lepidoptera.[444] Reproductive activity in *Schistocerca* appears to be initiated by aromatic scents from terpenoids liberated by desert shrubs at the time of bud-burst.[437] Polyphemus moths require an emanation from oak-leaves which excites the female to release her sex pheromone and thus excite the male.[478]

p. 728. **Absorption of oöcytes**—In *Machilis* resorption consists in digestion of the chorion by a secretion from the follicular cells, which then enter the oöcytes and phagocytose the contents.[431]

p. 728. **Ovulation, oviposition and receptivity**—The normal increase in the rate of oviposition in *Hyalophora* after mating is a reflex response to sperm in the spermatheca, causing the brain to trigger the release of a hormone from the intrinsic cells of the corpora cardiaca.[500] In *Drosophila* the increased egg laying which follows mating depends on a secretion from the male accessory glands (paragonia); for implantation of these glands into virgin females induces ovulation and oviposition;[450] and the isolated peptide from the accessory glands of male *Drosophila* causes a 2–3-fold increase in oviposition when injected into the female.[438] The implanted paragonia also suppress sexual receptivity in the female.[469] Similar effects are found in *Aëdes aegypti*, where the transplanted paragonia from *Drosophila* are equally effective. The extract is inactivated by heat but not by freezing.[464] In the grasshopper *Gomphocerus* there is a sudden change from receptivity to defensive behaviour in the female after copulation. This results from mechanical stimulation of the receptaculum by the spermatozoa; it is abolished if the nerve cord is cut, and the female will then copulate repeatedly.[465]

In the firebrat *Lepismodes* the female undergoes ecdysis every 9–10 days; mating occurs 1–2 days after ecdysis; and oviposition follows on about the 5th day. The oöcytes grow rapidly without deposition of yolk during the 4–5 days before ecdysis, but only if the female mates. Mating is followed by neurosecretory activity; the volume of the corpus allatum increases about 70 per cent.; and these changes initiate yolk deposition.[482]

p. 729. **Reproductive diapause**—Diapause in adult *Leptinotarsa* has been studied in great detail. It is induced mainly by a short-day photoperiod which arrests the secretory activity of the corpora allata. The beetles burrow into the soil and come to rest, and their wing muscles undergo extreme degeneration. Under long-day treatment the corpora allata are activated (probably by hormonal stimuli from the neurosecretory cells), movement is restored and the flight muscles regenerate,[461, 495] and oögenesis proceeds.[505, 506] Diapause also occurs in females fed on aged potato foliage; but this is a result of sensory perception, not of defective nutrition.[505, 506] Similar results have been obtained in *Galeruca tanaceti* in which the male also is affected by arrested development of the accessory glands.[491] Reproductive diapause induced by a short photoperiod occurs in *Anacridium*[451] and in the facefly *Musca autumnalis*.[497]

p. 739. **Sex determination**—Sex determination in *Heteropeza* (*Oligarces*) is not by haploidy–diploidy but by some other, unknown mechanism; for female larvae from unfertilized eggs may have either 5 or 10 chromosomes in their somatic cells.[435]

REFERENCES

[1] ABRAHAM, R. *Z. Parasitenk.*, **6** (1934), 559–91 (behaviour of sperm in female bed-bug *Cimex*, Hem.).

[2] ADAM, A. *Zool. Jahrb., Anat.*, **35** (1912), 1–74 (receptaculum seminis in bees, wasps and ants).

[3] AHRENS, W. *Z. micr. anat. Forsch.*, **37** (1935), 467–500 ('corpus luteum': termites).

[4] ALPATOV, W. W. *J. Exp. Zool.*, **63** (1932), 85–111 (factors in egg production: *Drosophila*, Dipt.).

[5] v. BAEHR, W. B. *Arch. Zellforsch.*, **3** (1909), 269–333 (sex determination in Aphids).

[6] BARBER, H. S. *Proc. Soc. Ent. Wash.*, **15** (1913), 31–8 (paedogenesis in *Micromalthus*, Col.).

[7] BARTH, R. *Zool. Jahrb., Physiol.*, **58** (1937), 297–329; *Z. wiss. Zool.*, **150** (1937), 1–32 (scent glands and mating in male and female Pyralidae, Lep.). *Zool. Jahrb.*; *Abt. Anat.*, **68** (1944), 331–62 (ditto: *Argynnis*, Lep.).

[8] BEAMENT, J. W. L. *Proc. Roy. Soc.*, B, 133 (1946), 407–18; *J. Exp. Biol.*, **23** (1947), 213–33 (micropyles and fertilization membrane: *Rhodnius*, Hem.).

[9] BEKLEMISHEV, V. *et al. Med. Parasit.*, **3** (1934), 460–79 (gonotrophic cycle in *Anopheles*, Dipt.).

[10] BERNHARD, C. *Biol. Zbl.*, **27** (1907), 467–79 (viviparity in *Cloëon*, Ephem.).

[11] BIEDERMANN, W. *Winterstein's Handb. d. vergl. Physiol.*, **3** (i) (1914), 897–908 (formation of chorion in insect eggs).

[12] BLUNCK, H. *Z. wiss. Zool.*, **102** (1912); **104** (1913), *Zool. Anz.*, **41** (1913), (reproduction, &c.: *Dytiscus*, Col.).

[13] BOLDYREV, B. *Horae Soc. Entom. Ross.*, **40** (1912), No. 6, 1–54; **41** (1914), No. 6, 1–244 (spermatophores: Tettigoniidae and Gryllidae).

[14] BONNEMAISON, L. *Bull. Soc. Ent. Fr.*, 1946, 40–2 (symbionts in Pentatomidae).

[15] BOUNHIOL, J. J. *C.R. Acad. Sci.*, **203** (1936), 388–9; *C.R. Acad. Sci.*, **215** (1942), 334–6 (corpora allata and egg production: *Bombyx mori*).

[16] BRANDT, H. *Z. Naturforsch*, **2** (B), (1947), 301–8 (effect of copulation on egg production: *Bupalus* and *Ephestia*, Lep.).

[17] BRECHER, G., and WIGGLESWORTH, V. B. *Parasitology*, **35** (1944), 220–4 (effect of symbiotic actinomyces on reproduction: *Rhodnius*).

[18] BREITSPRECHER, E. *Z. Morph. Oekol. Tiere*, **11** (1928), 495–538 (transmission of symbionts: Adobiidae, Col.).

[19] BRESSLAU, E. *Zool. Anz.*, **29** (1906), 299–323 (spermathecal duct in queen bee).

[20] BRUNSON, M. H. *Science*, **86** (1937), 197 (control of sex in *Tiphia*, Hym.).

[21] BUCHNER, P. *Tier und Pflanze in intracellularer Symbiose*, Berlin, 1930.

[22] BUSHNELL, R. J., and BOUGHTON, D. C. *Ann. Ent. Soc. Amer.*, **33** (1940), 361–70 (mating and egg production: *Acanthoscelides*, Col.).

[23] BUSVINE, J. R. *Proc. Roy. Ent. Soc. Lond.*, **21** (1946), 98–102 (unisexual families: *Pediculus*).

[24] BUXTON, P. A. *Trans. Ent. Soc. Lond.*, **78** (1930), 227–36 (biology and reproduction: *Rhodnius*, Hem.).

[25] —— *Parasitology*, **39** (1948), 119–24 (age of host and egg production: *Xenopsylla*, Aphan.).

[26] CAMERON, A. E. *Trans. Roy. Soc. Edin.*, **58** (1934), 211–50 (biology of *Haematopota*, Dipt.).

[27] CAPPE de BAILLON, P. *La Cellule*, **31** (1920), 1–245 (eggs of Tettigoniidae).

[28] —— *C.R. Acad. Sci.*, **196** (1933), 809–11 (formation of egg-shell: Phasmidae).

[29] —— *C.R. Acad. Sci.*, **199** (1934), 1069–70; **209** (1939), 527–9 (parthenogenesis in Phasmids).

[30] CARAYON, J. *C.R. Acad. Sci.*, **222** (1946), 107–9 (dispersion of spermatozoa in female Nabid); *Bull. Soc. Ent. Fr.*, 1949, 66–9 (oötheca in Plataspididae, Hem.).

[31] CHEN, S. H., and YOUNG, B. *Sinensia*, **14** (1943), 49–53 (carriage of eggs by male Belostomatid).
[32] CHEWYREUX, I. *C. R. Soc. Biol.*, **74** (1913), 695–9 (control of sex by *Pimpla*, Hym.).
[33] CHOPARD, L. *C.R. Acad. Sci.*, **199** (1934), 806–7 (spermatophore: Phasmidae).
[34] CLAUSEN, C. P. *J. N.Y. Ent. Soc.*, **47** (1939), 1–9 (effect of host size on sex ratio in Hymenopterous parasites).
[35] COCKAYNE, E. A. *Biol. Rev.*, **13** (1938), 107–32 (genetics of sex in Lepidoptera).
[36] CRAGG, F. W. *Ind. J. Med. Res.*, **8** (1920), 32–79; **11** (1923), 449–73 (behaviour of spermatozoa in female bed-bug, *Cimex*, Hem.).
[37] DAY, M. F. *Biol. Bull.*, **84** (1943), 127–40 (corpus allatum and ovaries in Diptera).
[38] DEEGENER, P. *Schröder's Handbuch der Entomologie*, **1** (1911), 466–523 (sexual organs).
[39] DELEURANCE, E. P. *C.R. Acad. Sci.*, **226** (1948), 514–16; 601–3; **227** (1948), 866–7 (ovaries and sexual activity: *Polistes*, Hym.).
[40] DEPDOLLA, P. *Schröder's Handbuch der Entomologie*, **1** (1926), 825 (germ cell formation and fertilization).
[41] DESCY, A. *Bull. Biol. Fr. Belg.*, **58** (1924), 1–37 (control of sex in *Osmia*, Hym.).
[42] DETINOVA, T. S. *Zool. Zh.*, **24** (1945), 291–8 (influence of corpora allata on ovaries: *Anopheles*).
[43] DEWITZ, J. *Arch. ges. Physiol.*, **37** (1885), 219–23; **38** (1886), 358–85 (movements of sperm in fertilization in the cockroach).
[44] DICK, J. *Ann. App. Biol.*, **24** (1937), 762–96 (factors in egg production: *Tenebrio*, &c., Col.).
[45] DICKINS, G. R. *Trans. Roy. Ent. Soc. Lond.*, **85** (1936), 331–62 (scent glands and mating: *Ephestia*, &c., Lep.).
[46] DONCASTER, L. *Proc. Roy. Soc.*, B, **82** (1910), 88–113; **83** (1911), 476–89; **89** (1916), 183–200 (sex determination: *Neuroterus*, Cynipidae).
[47] DU BOIS, A. M., and GEIGY, R. *Rev. Suisse Zool.*, **42** (1935), 169–248 (reproduction, &c.: *Sialis*, Neur.).
[48] EASTHAM, L. E. S., and MCCULLY, S. B. *J. Exp. Biol.*, **20** (1943), 35–42 (oviposition: *Calandra*, Col.).
[49] EIDMANN, H. *Z. angew. Entom.*, **15** (1929), 1–66; **18** (1931), 57–112 (female reproductive system: Lepidoptera).
[50] ELTRINGHAM, H. *Trans. Ent. Soc. Lond.*, 1913, 399–406; 1914, 152–76 (scent organs in male butterflies).
[51] van EMDEN, F. *Ve Congr. Internat. Ent. Paris* 1932, 1933, 813–22 (diapause in adult insects).
[52] EVANS, A. C. *Bull. Ent. Res.*, **26** (1935), 115–22 (nutrition and fecundity: *Lucilia*, Dipt.).
[53] —— *Ann. App. Biol.*, **25** (1938), 558–72 (nutrition and reproduction: *Brevicoryne*, Aphididae).
[54] EWER, D. W., and EWER, R. F. *J. Exp. Biol.*, **18** (1942), 290–305 (factors in oviposition: *Ptinus*, Col.).
[55] FISCHER, I. *Z. Zellforsch. mikr. Anat.*, **23** (1935), 219–43 (changes in follicular epithelium: Anoplura).
[56] FLANDERS, S. E. *Ann. Ent. Soc. Amer.*, **28** (1935), 438–44 (feeding habits and egg development: Pteromalidae, Hym.).
[57] —— *J. Econ. Ent.*, **39** (1936), 379–80; *Ann. Ent. Soc. Amer.*, **32** (1939), 11–26 (control of fertilization and sex in Hymenoptera).
[58] —— *Ann. Ent. Soc. Amer.*, **35** (1942), 251–66 (reabsorption of oöcytes in parasitic Hymenoptera).
[59] —— *Science*, **100** (1944), 168–9; *Amer. Nat.*, **79** (1945), 122–41 (bisexuality in uniparental Hymenoptera).
[60] —— *J. Econ. Ent.*, **38** (1945), 323–7 (spermatophore in *Macrocentrus*, Hym.).
[61] FLORENCE, L. *Cornell Univ. Agric. Exp. Stat. Mem.*, **51** (1921), 637–743 (biology of hog louse, *Haematopinus*).
[62] FRANKE-GROSSMANN, H. *Z. angew. Ent.*, **25** (1939), 647–80 (transmission of fungi: Siricidae).

[63] FREILING, H. H. *Z. wiss. Zool.*, **92** (1909), 210–90 (scent organs of male and female Lepidoptera).

[64] FUKUDA, S. *Proc. Imp. Acad. Tokyo*, **15** (1939), 19–21 (accelerated development of ovaries of *Bombyx* larva when transplanted to young pupa).

[65] GABRITSCHEVSKY, E. *Bull. Biol. Fr. Belg.*, **62** (1928), 478–524; *Arch. Entw. Mech.*, **121** (1930), 450–65 (developmental cycles in *Miastor*, Dipt.)

[66] GALLIARD, H. *Recherches morphologiques et biologiques sur la reproduction des réduvidés hématophages (Rhodnius et Triatoma)*. Thèse fac. sci. Univ. Paris, 1935, 160 pp.

[67] GERHARDT, U. *Zool. Jahrb., Syst.*, **35** (1913), 415–532; **37** (1914), 1–64 (copulation and spermatophores: Gryllidae and Tettigoniidae).

[68] GIER, H. T. *Biol. Bull.*, **71** (1936), 433–52 (transmission of bacteroids in cockroach).

[69] GLASER, R. W. *J. Exp. Zool.*, **38** (1923), 383–412 (nutrition and egg production: Muscidae).

[70] GOLDSCHMIDT, R. *Biol. Bull.*, **32** (1917), 35–43 (parthenogenesis in *Lymantria*, Lep.).

[71] GRANDORI, R. *Redia*, **7** (1911), 363–428 (polyembryony: *Apanteles*, Braconidae).

[72] GRASSÉ, P. P., and BONNEVILLE, P. *Bull. Biol. Fr. Belg.*, **69** (1935), 471–91 (alimentary castration, &c.: termites).

[73] GRESSON, R. A. R. *Quart. J. Micr. Sci.*, **73** (1930), 345–64 (yolk formation: Tenthredinidae); *Ibid.*, **74** (1931), 257–74 (ditto: *Periplaneta*, Orth.).

[74] GRIFFITHS, J. T., and TAUBER, O. E. *Physiol. Zool.*, **15** (1942), 196–209 (fecundity and parthenogenesis: *Periplaneta*).

[75] GRISON, P. *C.R. Acad. Sci.*, **219** (1944), 295–6; **227** (1948), 1172–4 (ageing of leaves, and lecithin, affecting egg production: *Leptinotarsa*, Col.).

[76] GROSS, J. *Z. wiss. Zool.*, **69** (1901), 139–201 (ovaries in Hemiptera); *Zool. Jahrb., Anat.* **22** (1905), 347–86 (ovaries in Anoplura).

[77] GUPTA, P. D. *Ind. J. Entom.*, **8** (1946), 79–84 (spermatophore: *Periplaneta*).

[78] HADORN, E., and ZELLER, H. *Arch. Entw. Mech.*, **142** (1943), 276–300 (effect of age on fecundity and egg hatching: *Drosophila*).

[79] HAGAN, H. R. *J. Morph.*, **51** (1931), 1–117; *Ann. Ent. Soc. Amer.*, **32** (1939), 264–6 (ditto: *Diploptera*, Orth.). (viviparity in *Hesperoctenes*, Hem.).

[80] HALLER, H. L., *et al. J. Amer. Chem. Soc.*, **66** (1944), 1659–62 (sex attractant in *Lymantria*, Lep.).

[81] HALLEZ, J. *C.R. Acad. Sci.*, **148** (1909), 317 (oötheca: *Periplaneta*, Orth.).

[82] HANDLIRSCH, A. *Verhandl. Zool.-bot. Ges. Wien*, **50** (1900), 105–12 (fate of excess sperm in insects).

[83] HANNA, A. D. *Bull. Ent. Res.*, **26** (1935), 315–22 (effect of low temperature on fertility: *Euchalcidia*, Hym.).

[84] HARDENBERG, J. D. F. *Zool. Jahrb., Anat.*, **50** (1929), 497–570 (reproduction, &c., in Pupipara).

[85] HARRIS, R. G. *Biol. Bull.*, **48** (1925), 139–44 (reproductive cycles in *Oligarces*, Dipt.).

[86] HEBERDEY, R. F. *Z. Morph. Oekol. Tiere*, **22** (1931), 416–586 (comparative physiology of female reproductive system in insects).

[87] HECHT, O. *Arch. Schiffs. Tropenhyg.*, **37** (Beiheft. 3), 87 pp. (nutrition and egg production in mosquitos).

[88] HEYMONS, R. *Zool. Jahrb., Suppl.*, **15**, 1912 (viviparity in *Hemimerus*, Orth.).

[89] HINTON, H. E. *Nature*, **157** (1946), 552 ('paedogenesis' in *Tanypus* pupa).

[90] HOBSON, R. P. *Ann. App. Biol.*, **25** (1938), 573–82 (nutrition and egg production: *Lucilia*).

[91] HOFFMANN, R. W. *Verhandl. deutsch. zool. Ges.*, **24** (1914), 192–216 (viviparity in Strepsiptera).

[92] HOGBEN, L. T. *Proc. Roy. Soc.*, B, **91** (1920), (oögenesis: Hymenoptera).

[93] HOHORST, W. *Z. Morph. Oekol. Tiere*, **32** (1936), 227–75 (mating in *Oecanthus*, Orth.).

[94] HOLMGREN, N. *Zool. Jahrb., Syst.*, **19** (1903), 421–68 (viviparous insects).

[95] —— *Kungl. Svenska Vetenskap. Handl.*, **44** (1909), No. 3, (corpora allata in queen termites).

[96] HOWLETT, F. M. *Bull. Ent. Res.*, **6** (1915), 297–305 (attraction of *Dacus* spp., Dipt., to chemicals).

[97] HUETTNER, A. F. *J. Morph.*, **39** (1924), 249–65; *Z. Zellforsch. mikr. Anat.*, **4** (1927), 599–610 (polyspermy and early development: *Drosophila*, Dipt.).

[98] HUGHES-SCHRADER, S. *J. Morph.*, **50** (1930), 475–95; *Ann. Ent. Soc. Amer.*, **23** (1930), 359–80 (parthenogenesis and hermaphroditism: *Icerya purchasi*, Hem.).

[99] HUSAIN, M. A., and MATHUR, C. B. *Ind. J. Ent.*, **7** (1945), 89–101 (parthenogenesis in *Schistocerca*, Orth.).

[100] ILLIG, K. G. *Zoologica*, **15** (1902) (scent glands of male Lepidoptera).

[101] IMMS, A. D. *Recent Advances in Entomology*, Churchill, London, 1937.

[102] ITO, H. *Bull. Imp. Tokyo Sericult. Coll.*, **1** (1918), 63–103 (corpora allata in Lepidoptera).

[103] IWANOFF, P. P., and MESTSCHERSKAJA. *Zool. Jahrb., Physiol.*, **55** (1935), 281–348 (factors controlling egg development: *Blattella*, Orth.).

[104] JAMES, H. C. *Bull. Ent. Res.*, **28** (1937), 429–61 (effect of delayed fertilization on sex ratio: *Pseudococcus*, Hem.); *Proc. Roy. Ent. Soc. Lond.*, A, **12** (1937), 92–8 (the same: *Ephestia*, Lep.).

[105] JOLY, P. *Arch. Zool. exp. gén.*, **84** (1945), 49–164 (humoral control of ovarian function: *Dytiscus*, Col.).

[106] KAHLE, W. *Zoologica*, **21** (1908), 1–80 (paedogenesis in Cecidomyidae).

[107] KAUFMANN, B. P., and DEMEREC, M. *Amer. Nat.*, **76** (1942), 445–69 (utilization of sperm: *Drosophila*).

[108] KEILIN, D. *Arch. Zool.*, **55** (1916), 393–415 (viviparity in Diptera).

[109] KELLOGG, V. L. *Biol. Bull.*, **12** (1907), 152–4 (mating behaviour in *Bombyx mori*).

[110] KEMPER, H. *Z. Morph. Oekol. Tiere*, **19** (1930), 160–83 (effect of starvation on *Cimex*, Hem.).

[111] KETTLEWELL, H. B. D. *Entomologist*, **79** (1946), 8–14 (sexual attraction in Lepidoptera).

[112] KHALIFA, A. *Trans. Roy. Ent. Soc. Lond.*, **100** (1949), 449–71; *Quart. J. Micr. Sci.*, **90** (1949), 281–92; *Proc. Roy. Ent. Soc. Lond.*, A, **25** (1950), 33–42; 53–61; *Parasitol.*, **40** (1950), 283–9 (spermatophores in insects).

[113] KING, R. L., and SLIFER, E. H. *J. Morph.*, **56** (1934), 603–19 (development of unfertilized eggs in grasshoppers).

[114] KLATT, B. *Biol. Zbl.* **40** (1920), 539–58 (stimuli to oviposition: *Lymantria*, Lep.).

[115] KOCH, A. *Z. Morph. Oekol. Tiere*, **23** (1931), 389–424 (transmission of symbionts in *Oryzaephilus*, Col.); *Ibid.*, **32** (1936), 92–134 (the same, in *Lyctus*, Col.); *Z. Morph. Oekol. Tiere*, **34** (1938), 584–609 (the same, in termite *Mastotermes*).

[116] —— *Z. Morph. Oekol. Tiere*, **37** (1940), 38–62 (oenocytes in *Tribolium*).

[117] KÖHLER, A. *Z. wiss. Zool.*, **87** (1907), 337–81 (changes in ovary: Hemiptera).

[118] KOLTZOFF, N. K. *Biol. Zbl.*, **52** (1932), 626–42 (artificial parthenogenesis in silkworm).

[119] KOZHANTSHIKOV, I. W. *Bull. Ent. Res.*, **29** (1938), 103–14 (nutrition and egg production: Pyralidae, Lep.).

[120] KÜHN, A., and HENKE, K. *Z. Ges. Wiss. Göttingen Nachr. Biol. N.F.*, **1** (1935), 247–59 (sex linked inheritance in *Ephestia*, Lep.).

[121] KULLENBERG, B. *Zool. Bidr. fr. Uppsala*, **23** (1944), 1–522; **24** (1947), 217–418 (fertilization in Capsids and Nabids).

[122] LAABS, A. *Z. Morph. Oekol. Tiere*, **36** (1939), 123–79 (brood care in *Hydrophilus*, Col.).

[123] LARSON, A. O., and FISCHER, C. K. *J. Agric. Res.*, **29** (1925), 297–305 (nutrition and fecundity: *Bruchus*, Col.).

[124] LAURINAT, K. *Z. indukt. Abstam. Verebungsl.*, **57** (1930), 139–205 (age of gametes and sex ratio: *Drosophila*, Dipt.).

[125] LAWSON, C. A. *Biol. Bull.*, **70** (1936), 288–307 (cytology of sex determination: *Macrosiphum*, Aphid.).

[126] LEIBY, R. W., and HILL, C. C. *J. Agric. Res.*, **25** (1923), 337–50; **28** (1924), 829–40 (polyembryony: *Platygaster* spp.); *Proc. IVth Internat. Congr. Ent. Ithaca*, 1928, **2** (1929), 873–87 (polyembryony: review).

[127] LOEWE, S., *et al. Biochem. Z.*, **244** (1932), 347–56 (oestrogens in *Attacus*, Lep.).

[128] LUDWIG, W. *Z. Morph. Oekol. Tiere*, **5** (1926), 291–380 (copulation in *Lygaeus* and *Pyrrhocoris*, Hem.).

[129] LUDWIG, D. *Physiol. Zool.*, **16** (1943), 381–8 (humidity and fecundity: *Samia*, Lep.).

[130] MACKERRAS, M. J. *Bull. Ent. Res.*, **24** (1933), 353–62 (nutrition and fecundity in blow-flies).

[131] MALOEUF, N. S. R. *Bull. Soc. Roy. Entom. Egypte*, 1933, 96–119 (transmission of intestinal bacteria: *Nezara*, Hem.).

[132] MANSOUR, K. *Quart. J. Micr. Sci.*, **77** (1934), 243–71 (transmission of symbionts: *Rhizopertha*, Col., &c.).

[133] MARCHAL, P. *C.R. Acad. Sci.*, **126** (1898), 662–4; *Arch. Zool.*, **2** (1904), 257–335; **4** (1906), 485–638 (polyembryony).

[134] —— *Richet's 'Dictionaire de Physiologie'*, **9** (1910), 273–386 (physiology of insects).

[135] MARTIN, F. *Z. wiss. Zool.*, **110** (1914), 419–79 (polyembryony in *Encyrtus*, Chalcid.).

[136] MAYER, A. G. *Psyche*, **9** (1900), 14–20 (mating behaviour: *Callosamia*, Lep.).

[137] MAYNE, B. *Rep. U.S. Publ. Health*, **41** (1926), 986–90 (temperature and oviposition: *Anopheles*, Dipt.).

[138] DE MEILLON, B., and GOLDBERG, L. *Nature*, **158** (1946), 269 (egg production in *Cimex* fed on vitamin-deficient host).

[139] MELL, R. *Biologie u. Systematik d. sudchinesischen Sphingiden*, 1922, p. 162.

[140] MELLANBY, H. *Proc. Roy. Ent. Soc. Lond.*, A, **12** (1937), 1–3; *Parasitology*, **29** (1937). 131–41 (reproduction in *Glossina*, Dipt.).

[141] MELLANBY, K. *Parasitology*, **30** (1938), 392–402 (water uptake and body size: *Lucilia*).

[142] —— *Parasitology*, **31** (1939), 193–9 (fertilization and egg production: *Cimex*, Hem.).

[143] MERGELSBERG, O. *Zool. Jahrb., Anat.*, **60** (1935), 345–98 (viviparity and hermaphroditism: *Termitoxenia*, Dipt.).

[144] MJÖBERG, E. *Psyche*, **32** (1925), 119–54 (neoteny: Lycidae, Col.).

[145] MORDWILKO, A. *Biol. Zbl.*, **27** (1907); **28** (1909); **29** (1909) (developmental cycles in Aphididae).

[146] MORGAN, T. H. *J. Exp. Zool.*, **7** (1909), 239–353; **19** (1915), 285–320 (cytology of sex determination in Phylloxerans and Aphids).

[147] MURR, L. *Z. vergl. Physiol.*, **11** (1930), 210–70 (sexual attraction in *Habrobracon*, Hym.).

[148] MURRAY, M. R. *Biol. Bull.*, **50** (1926), 210–34 (changes in follicular cells: *Gryllus*, Orth.).

[149] MUSGRAVE, A. J. *Proc. Zool. Soc. Lond.*, B, 1937, 337–64 (reproductive organs: *Ephestia*, Lep.).

[150] NATH, V., and MEHTA, D. R. *Quart. J. Micr. Sci.*, **73** (1929), 7–24 (yolk formation: *Luciola*, Col.).

[151] NATH, V., and MOHAN, P. *J. Morph.*, **48** (1929), 253–79 (yolk formation: *Periplaneta*, Orth.).

[152] NICHOLSON, A. J. *Quart. J. Micr. Sci.*, **65** (1921), 396–448 (development of ovaries: *Anopheles*, Dipt.).

[153] NOLTE, H. W. *Z. Morph. Oekol. Tiere*, **33** (1937), 165–200 (transmission of symbionts: *Apion*, Col.).

[154] NONIDEZ, J. F. *Biol. Bull.*, **39** (1920), 207–30 (fertilization: *Drosophila*, Dipt.).

[155] NORRIS, M. J. *Proc. Zool. Soc. Lond.*, 1932, 595–611; 1933, 903–34; 1934, 333–60 (reproduction in *Ephestia* and *Plodia*, Lep.).

[156] NOVAK, V. *Nature*, **167** (1951), 132–3 (corpus allatum hormone in *Periplaneta*).

[157] ÔMURA, S. *J. Fac. Agric. Hokkaido Imp. Univ.*, **38** (1936), 151–81; **40** (1938), 111–28; 129–70 (behaviour of sperm: *Bombyx mori*).

[158] OOSTHUIZEN, M. J. *Univ. Minn. Agric. Exp. Sta. Tech. Bull.*, No. 107 (1935), 1–44 (temperature and sterility: *Tribolium*).

[159] OUDEMANS, J. T. *Zool. Jahrb., Syst.*, **12** (1899), 71–85 (behaviour of castrated *Lymantria*, Lep.).

[160] PALM, N. B. *Opusc. Ent.*, Suppl. 7 (1948), 1–101 (corpus allatum and ovaries: *Bombus*, Hym.).

[161] PANTEL, J. *La Cellule*, **29** (1913), 1–289 (egg and fertilization in parasitic Diptera).

[162] PANTEL, J., and de SINÉTY, R. *C.R. Acad. Sci.*, **147** (1908), 1358–60 (parthenogenesis in Phasmids).

[163] PARKER, H. L. *U.S. Dept. Agric. Tech. Bull.*, No. **230** (1931), 66 pp. (polyembryony in *Macrocentrus*, Braconidae).

[164] PARKIN, E. A. *Nature*, **147** (1941), 329; *Ann. Appl. Biol.*, **29** (1942), 268–74 (transmission of fungi in Siricidae).

[165] PAYNE, M. A. *J. Morph.*, **54** (1933), 321–46; *Zool. Jahrb., Anat.*, **61** (1936) 45–50 (behaviour of sperm: *Chortophaga*, Orth. and *Leptocoris*, Hem.).

[166] PEACOCK, A. D., and GRESSON, R. A. R. *Quart. J. Micr. Sci.*, **71** (1928), 541–61 (oögenesis: Tenthredinidae).

[167] PEPPER, J. H., and HASTINGS, E. *Montana State Coll. Agr. Exp. Sta.*, Tech. Bull. No. 413 (1943), 1–36 (linoleic acid and reproduction: *Loxostege*, Lep.).

[168] PESSON, P. *Thesis, Paris*, 1944, 266 pp. (corpus allatum in female Coccids).

[169] PFEIFFER, I. W. *J. Exp. Zool.*, **82** (1939), 439–61; **99** (1945), 183–233 (function of corpus allatum of adult female: *Melanoplus*, Orth.).

[170] PFLUGFELDER, O. *Zoologica*, **93** (1936) (nervous system, corpora allata, &c.: Hemiptera).

[171] —— *Z. wiss. Zool.*, **149** (1937), 477–512 (corpora allata and egg production: *Dixippus*, Orth.).

[172] —— *Z. wiss. Zool.*, **150** (1938), 451–67 (corpora allata in termites).

[173] —— *Biol. Zbl.*, **67** (1948), 223–41 (corpus allatum in *Apis*).

[174] POSPELOV, V. *Bull. Ent. Res.*, **16** (1926), 363–7 (temperature and maturation in locusts).

[175] PRELL, H. *Z. angew. Entom.*, **2** (1915), 57–148 (ovoviviparity: *Panzeria*, Dipt.).

[176] PRINGLE, J. A. *Trans. Roy. Ent. Soc. Lond.*, **87** (1938), 271–90 (reproductive cycles: *Micromalthus*, Col.).

[177] PRYOR, M. G. M. *Proc. Roy. Soc.*, B, **128** (1940), 378–93 (formation of oötheca: *Periplaneta*).

[178] RAICHOUDHURY, D. P. *Proc. Zool. Soc. Lond.*, 1936, 789–805 (high temperature and fertility: *Ephestia*, Lep.).

[179] REGEN, J. *Zool. Anz.*, **35** (1910), 427–32 (behaviour of castrated *Gryllus*, Orth.).

[180] —— *Sitzgsber. Akad. Wien, Math.-Naturwiss. Kl., Abt.* 1, **133** (1924), 347–59 (spermatophore of *Liogryllus*, Orth.).

[181] REIFF, M. *Rev. Suisse Zool.*, **52** (1945), 155–211 (effect of age on fecundity: *Drosophila*).

[182] RETHFELDT, C. *Zool. Jahrb., Anat.*, **46** (1924), 245–302 (viviparity in *Chrysomela*, Col.).

[183] REYNOLDS, J. M. *Ann. App. Biol.*, **31** (1944), 132–42 (effect of food and fertilization on fecundity: *Tribolium*, Col.).

[184] RICHARDS, O. W. *Biol. Rev.*, **2** (1927), 298–364 (sexual behaviour, &c.: review).

[185] RICHARDS, A. G., and MILLER, A. *J. N.Y. Ent. Soc.*, **45** (1937), 1–60 (experimental embryology in insects: review).

[186] RIES, E. *Z. Zellforsch. mikr. Anat.*, **16** (1932), 314–88 (oögenesis: Anoplura).

[187] RITTERHAUS, K. *Z. Morph. Oekol. Tiere*, **8** (1927), 271 (structure and biology: *Phyllopertha* and *Anomala*, Col.).

[188] ROBERTSON, F. W., and SANG, J. H. *Proc. Roy. Soc.*, B, **132** (1944), 258–90 (effect of nutrition on fecundity and egg hatching: *Drosophila*).

[189] ROEDER, K. D. *Biol. Bull.*, **69** (1935), 203–20 (sexual behaviour: *Mantis*, Orth.).

[190] ROONWALL, M. L. *Bull. Ent. Res.*, **27** (1936), 1–14 (structure of egg: *Locusta*).

[191] D'ROSARIO, A. M. *Proc. Roy. Ent. Soc. Lond.*, A, **15** (1940), 69–77 (copulation in *Nematus*, Hym.).

[192] ROSENKRANZ, W. *Z. Morph. Oekol. Tiere*, **36** (1939), 279–309 (symbiont transmission: Pentatomidae).

[193] ROUBAUD, E. *La Glossina palpalis*, Paris, 1909.

[194] —— *Ann. Inst. Pasteur*, **36** (1922), 765–83 (nutrition and fecundity: *Musca*, Dipt.).

[195] ROUBAUD, E., and MEZGER, J. *Bull. Soc. Path. Exot.*, **27** (1934), 666–8 (blood meals and egg production: *Culex pipiens*).

[196] ROY, D. N. *Nature*, **145** (1940), 747–8 (mating and egg production: *Anopheles*).

[197] SALT, G. *J. Exp. Biol.*, **13** (1936), 363–75 (sex and super-parasitism: *Trichogramma*).

[198] SCHARRER, B. *Endocrinology*, **38** (1946), 46–55 (corpus allatum and reproduction: *Leucophaea*, Orth.); *The Hormones* (Pincus and Thimann (Edd.), New York) **1** (1948), 121–50 (ditto: review).

[199] SCHNEIDER, K. *Arch. Zellforsch.*, **14** (1917), 79–143 (development of ovariole: *Deilephila*, Lep.).

[200] SCHOENEMUND, E. *Zool. Jahrb., Anat.*, **34** (1912), 1–56 (rudimentary hermaphroditism: *Perla*, Plecopt.).

[201] SCHRADER, F. *J. Morph.*, **34** (1920), 267–305 (sex determination in *Trialeurodes*, Hem.).

[202] SCOTT, A. C. *Z. Morph. Oekol. Tiere*, **33** (1938), 631–53 (paedogenesis: *Micromalthus*, Col.).

[203] SEILER, J. *Arch. Zellforsch.*, **15** (1920), 249–68; *Arch. J. Klaus-Stift. Vererb.*, **17**

(1942), 513–28 (control of sex ratio in *Talaeporia*, Lep., by environmental conditions).

[204] SEILER, J. *Z. indukt. Abstamm. Vererbgsl.*, **31** (1923), 1–99 (parthenogenesis: *Solenobia*, Lep.).

[205] SEYRIG, A. *Bull. Soc. Ent. France*, **40** (1935), 67–70 (size of host and sex of parasite: Hymenoptera).

[206] SHULL, A. F. *Biol. Rev.*, **4** (1929), 218–48 (determination of sex, &c., in Aphids).

[207] —— *Z. indukt. Abstamm. Verebgsl.*, **55** (1930), 108–26 (reproductive system in gamic and parthenogenetic forms of Aphids).

[208] SIKORA, H. *Centralbl. Bakt.*, **76** (1915), 523–37; *Arch. Schiffs. Tropenhyg. Beheift*, **20** (1916), 76 pp. (biology of *Pediculus*, Anopl.).

[209] SILVESTRI, F. *Redia*, **2** (1904), 68 (polyembryony: *Litomastix*, Chalcid.).

[210] SPEICHER, K. G., and SPEICHER, B. R. *Biol. Bull.*, **74** (1938), 247–52 (females from unfertilized eggs: *Habrobracon*, Hym.).

[211] SPRINGER, F. *Zool. Jahrb., Syst.*, **40** (1915), 57–118 (paedogenesis in *Miastor*, Dipt.).

[212] STAMMER, H. J. *Z. Morph. Oekol. Tiere*, **15** (1929), 1–34 (transmission of symbionts in Lagriidae, Col.); *Ibid.*, 481–523 (the same, in Trypetidae, Dipt.); *Ibid.*, **29** (1935), 585–608 (the same, in Chrysomelidae, Col.); *Ibid.*, **31** (1936), 682–96 (the same, in *Bromius*, Chrysomel. Col.).

[213] STOBBE, R. *Zool. Jahrb., Anat.*, **32** (1912), 493–532 (abdominal scent organs in male Sphingidae and Noctuidae).

[214] SWEETMAN, H. L. *Bull. Brooklyn Ent. Soc.*, **29** (1934), 158–61 (passage of spermatophore in Thysanura).

[215] —— *Ecol. Monog.*, **8** (1938), 285–311 (oviposition: *Thermobia*, Thysan).

[216] SWELLENGREBEL, N. H. *Ann. Inst. Pasteur*, **43** (1929), 1370–80 ('gonotrophic dissociation' in *Anopheles*, Dipt.).

[217] TAUBER, O. E., *et al. Iowa State Coll. J. Sci.*, **19** (1945), 349–59 (food and egg production: *Melanoplus*).

[218] TAYLOR, T. H. C. *The Biological Control of an Insect in Fiji*, London, 1937.

[219] THOMSEN, E. *Dansk Naturhist. Foren.*, **106** (1942), 319–405 (corpus allatum and ovarian function: *Calliphora*); *Nature* **161** (1948), 439 (neurosecretory cells, ditto) (*J. Exp. Biol.*, **26** (1949), 137–49 (corpus allatum and oxygen consumption: *Calliphora*); **29** (1952), 137–72 (corpus cardiacum and ovarian function).

[220] THOMSEN, M. *Z. Zellforsch. mikr. Anat.*, **5** (1927), 1–116 (parthenogenesis in Coccids and Aleurodids, and review).

[221] TINBERGEN, N., *et al. Z. Tierpsychologie*, **5** (1942), 182–226 (androconia and courtship: *Satyrus*, Lep.).

[222] TITSCHACK, E. *Z. wiss. Zool.*, **128** (1926), 509–69 (nutrition and egg production: *Tineola*, Lep.).

[223] TÓTH, L. *Z. Morph. Oekol. Tiere*, **27** (1933), 692–731 (transmission of symbionts in Aphids).

[224] —— *Z. Morph. Oekol. Tiere*, **33** (1937), 412–37 (developmental cycle and symbiosis: *Pemphigus*, Aphid.).

[225] TOUMANOFF, K. *Bull. Soc. Zool. Fr.*, **53** (1928), 528–44; *Bull. Biol. Fr. Belg.*, **62** (1928), 388–411 (gynandromorphism in *Dixippus*, Orth.).

[226] TOWNSEND, G. F., and LUCAS, C. C. *Science*, **92** (1940), 43; *Biochem. J.*, **34** (1940), 1155–62 (chemistry of royal jelly).

[227] TROUVELOT, B., and GRISON, P. *C.R. Acad. Sci.*, **201** (1935), 1053–5 (fecundity of *Leptinotarsa* on different *Solanum* spp.).

[228] ULRICH, H. *Rev. suisse Zool.*, **4** (1934), 423–8; *Biol. Zbl.*, **63** (1943), 109–42 (alternation of generations in *Oligarces*. Dipt.).

[229] VANDEL, A. *La Parthénogenèse*, Encycl. Scientifique, Paris, 1931.

[230] VOGT, M. *Arch. Entw. Mech.*, **140** (1940), 525–46; **141** (1942), 424–54; *Biol. Zbl.* **63** (1943), 467–70 ('gonadotropic' hormone in *Drosophila*).

[231] —— *Z. Zellforsch.*, **34** (1948), 160–4 (fat body and oenocytes after removal of corpus allatum: *Drosophila*).

[232] WEBER, H. *Biologie der Hemipteren*, Springer, Berlin, 1930.

[233] —— *Z. Morph. Oekol. Tiere*, **23** (1931), 575–753 (biology of *Trialeurodes*, Hem.).

[234] WEBER, H. *Lehrbuch der Entomologie*, Jena, 1933.

[235] WEED, I. G. *Proc. Soc. Exp. Biol. Med.*, **34** (1936), 883–5 (corpora allata and egg production: *Melanoplus*, Orth.).

[236] WEIDNER, H. *Z. angew. Ent.*, **21** (1934), 239–90 (fertilization in Lepidoptera).

[237] WEYER, F. *Z. wiss. Zool.*, **131** (1928), 345–501 (ovaries of workers in social Hymenoptera).

[238] —— *Z. Parasitenk.*, **8** (1935), 104–15 (larval feeding and egg production: *Culex pipiens* races, Dipt.).

[239] WHEELER, W. M. *J. Exp. Zool.*, **8** (1910), 377–438 (parasitic castration, &c.).

[240] WHITE, M. J. D. *J. Morph.*, **79** (1946), 323–69; **80** (1947), 1–24; Univ. Texas Publ., No. 5007 (1950) 1–80 (sex determination in *Miastor*, &c., Dipt.).

[241] WIGGLESWORTH, V. B. *Parasitology*, **21** (1929), 288–321 (transmission of symbionts: *Glossina*, Dipt.).

[242] —— *Quart. J. Micr. Sci.*, **76** (1933), 270–318 (oenocytes during reproduction: *Rhodnius*, Hem.).

[243] —— *Quart. J. Micr. Sci.*, **79** (1936), 91–121; *J. Exp. Biol.*, **25** (1948), 1–14 (corpus allatum and reproduction: *Rhodnius*, Hem.).

[244] —— Unpublished work.

[245] WILLIAMS, C. B., and BUXTON, P. A. *Trans. Ent. Soc. Lond.*, 1916, 86–100 (oötheca formation: *Sphodromantis*, Orth.).

[246] WILLIAMS, C. M. *Biol. Bull.*, **90** (1946), 234–43 (corpus allatum and egg production in Lepidoptera).

[247] WOKE, P. A. *J. Parasit.*, **23** (1937), 311–13; *Amer. J. Hyg.*, **25** (1937), 372–80 (nutrition and egg production in mosquitos).

[247a] YEOLI, M., and MER, G. G. *Trans. Roy. Soc. Trop. Med. Hyg.*, **31** (1938), 437–44 (nutrition and maturation of eggs: *Anopheles*).

[248] YOUNG, W. C., and PLOUGH, H. H. *Biol. Bull.*, **51** (1926), 189–98 (sterilization by high temperature: *Drosophila*).

[249] ZACHARIAS, A. *Z. Morph. Oekol. Tiere*, **10** (1928), 676–737 (transmission of symbionts: Pupipara).

[250] ZAVREL, J. *Publ. fac. sci. Univ. Masaryk, Brno*, 1926 (paedogenesis in *Tanytarsus*, Dipt.).

[251] ZIMMER, C. *S. B. Ges. Naturf. Fr. Berlin*, 1934, 304–11 (paedogenesis and neoteny).

SUPPLEMENTARY REFERENCES (A)

[252] DELEURANCE, E. P. *C.R. Acad. Sci.*, **230** (1950), 782–4; **231** (1950), 1565–7 ('functional monogyny' in *Polistes*).

[253] FLANDERS, S. E. *Quart. Rev. Biol.*, **21** (1946), 135–43 (control of sex and sex-limited polymorphism in *Hymenoptera*).

[254] GÖTZ, B. *Experientia*, **7** (1951), 406–18 (sexual scents in Lepidoptera: review).

[255] JOLY, P. *C. R. Soc. Biol.*, **144** (1950), 1217–20 (corpora allata and ovarial function in Carabids).

[256] KAISER, P. *Arch. Entw. Mech.*, **144** (1949), 99–131 (corpora allata in adult Lepidoptera).

[257] KENNEDY, J. S. *Nature*, **168** (1951), 825 (nutrition of aphids on leaf galls and virus-infected plants).

[258] KENNEDY, J. S., and BOOTH, C. O. *Ann. Appl. Biol.*, **38** (1951), 25–64 (feeding preferences and fecundity in relation to age of leaves).

[259] LAFON, M. *Arch. intern. Physiol.*, **57** (1950), 309–42 (nutrition and egg production in *Dixippus*).

[260] de LERMA, B. *Boll. Zool.*, **18** (1951), 1–2 (cholesterol in corpus luteum: *Locusta*).

[261] MAGNUS, D. *Z. Tierpsychol.*, **7** (1950), 439–49; **15** (1958), 397–426 (courting and oviposition in *Argynnis*, Lep.).

[262] NÜESCH, H. *Verh. naturforsch. Ges. Basel*, **61** (1950), 93–113 (sex-determination and intersexuality: review).

[263] POSSOMPÈS, B. *C.R. Acad. Sci.*, **228** (1949), 1527–9 (egg production in *Calliphora* after removal of corpus allatum in larva).

[264] ROTH, L. M. *Amer. Mid. Nat.*, **40** (1948), 265–352 (mating in mosquitos).

[265] SCHALLER, F. *Naturwissenschaften*, **39** (1952), 48; *Z. Morph. Oekol. Tiere*, **41** (1953), 265–77 (spermatophore in Collembola).

[266] SCHMIDT, E. L., and WILLIAMS, C. M. *Anat. Rec.* (suppl.) **105** (1949), No. 70 (growth hormone and differentiation of spermatocytes).

[267] SCHNEIDER, G. *Z. Morph. Oekol. Tiere*, **36** (1940), 595–644 (transmission of symbionts in *Coptosoma*, Hem.).

[268] SCHNEIDERMAN, H. A., *et al. Anat. Rec.* (suppl.) **111** (1951), No. 164 (cytochrome system and spermatogenesis: *Platysamia*, Lep.).

[269] SEILER, J. *Experientia*, **5** (1949), 425–38 (intersexes; review).

SUPPLEMENTARY REFERENCES (B)

[270] ANDERSON, E. *J. Cell. Biol.*, **20** (1964), 131–55 (vitellogenesis in *Periplaneta*).

[271] ANDERSON, E., and BEAMS, H. W. *J. Biophys. Biochem. Cytol.*, **2** (1956), Suppl. 439–43 (fine structure of nurse cells in ovary of *Rhodnius*).

[272] ANDERSON, J. M. *Physiol. Zool.*, **23** (1950), 308–16 (testicular cyst-cells in *Popillia*).

[273] ASCHER, K. R. S., and LEVINSON, Z. H. *Riv. Parassit.*, **17** (1956), 217–22 (larval feeding and egg production in *Musca*).

[274] ASTAUROV, B. L. *Proc. Zool. Soc. Calcutta*, **29** (1957), 30–54 (artificial parthenogenesis in *Bombyx*).

[275] ASTAUROV, B. L., and OSTRIAKOVA-VARSHAVER, V. P. *J. Embryol. Exp. Morph.*, **5** (1957), 449–62 (heterospermic androgenesis in silkworm hybrids).

[276] BARTH, R. H. *Science*, **133** (1961), 1598–9; *Gen. Comp. Endocrin.*, **2** (1962), 53–69 (corpora allata and sex attractant in female *Byrsotria*, Blattaria).

[277] BEHRENZ, W. *Zool. Jahrb., Anat.*, **72** (1952), 147–215 (movements of sperm in genital tract of *Lymantria* female).

[278] BEKLEMISHEV, W. N., DETINOVA, T. S., and POLOVODOVA, V. P. *Bull. Wld. Hlth. Org.*, **21** (1959), 223–32 (age determination in *Anopheles* female).

[279] BERGERARD, J. *Bull. Biol. Fr. Belg.*, **92** (1958), 87–182 (facultative parthenogenesis in *Clitumnus*, Phasmidae).

[280] —— *C.R. Acad. Sci.*, **253** (1961), 2149–51; *Bull. Biol. Fr. Belg.*, **95** (1961), 273–300 (experimental production of males in *Carausius*).

[281] BERGERARD, J., and SENGÉ, J. *Bull. Biol. Fr. Belg.*, **93** (1959), 16–37 (occasional parthenogenesis in *Locusta*).

[282] BIER, K. *Arch. Entw. Mech.*, **154** (1963), 552–75; *J. Cell Biol.*, **16** (1963), 436–9 (nurse cells and follicle cells in yolk formation of *Musca*).

[283] BONHAG, P. F., *et al. J. Morph.*, **93** (1953), 177–229; **96** (1955), 381–411; **97** (1955), 283–312 (histology and histochemistry of ovaries: *Oncopeltus*); *J. Morph.*, **99** (1956), 433–64 (ditto: *Anisolabis*); *Univ. Calif. Publ. Ent.*, **16** (1959), 81–124; *J. Morph.*, **108** (1961), 107–30 (ditto: *Periplaneta*).

[284] BONHAG, P. F. *Ann. Rev. Ent.*, **3** (1958), 137–60 (ovarian structure and vitellogenesis: review).

[285] V. BORSTEL, R. C. in *Beginnings of embryonic development*, 1957, 175–99 (nucleo-cytoplasmic relations in early insect development).

[286] BRUNET, P. C. J. *Quart. J. Micr. Sci.*, **92** (1951), 113–27; **93** (1952), 47–69 (formation of oötheca in *Periplaneta*).

[287] BUCHNER, P. *Z. Morph. Oekol. Tiere*, **43** (1954), 262–312; **43** (1955), 397–428 (symbionts and sex determination in *Stictococcus*).

[288] BUSNEL, R. G., DUMORTIER, B., and BUSNEL, M. C. *Bull. Biol. Fr. Belg.* **90** (1956), 219–86 (acoustic behaviour of *Ephippiger*, &c.).

[289] BUSNEL, R. G., PASQUINELLY, F., and DUMORTIER, B. *Bull. Soc. Zool. Fr.*, **80** (1955), 18–22 (sexual communication in *Ephippiger* by vibrations).

[290] BUTENANDT, A. *Naturwiss. Rundschau*, 1955, 457–64; *Naturwiss.*, **46** (1959), 461–71 (sex attractants, &c.: review).

[291] BUTENANDT, A., and HECKER, E. *Angew Chem.*, **73** (1961), 349–53 (synthesis of 'bombycol', sex attractant of *Bombyx*).

[292] BUTLER, C. G. *Proc. Roy. Ent. Soc. Lond.*, A, **34** (1959), 137–8 (source of the 'queen substance' in *Apis*).

[293] BUTLER, C. G., and FAIREY, E. M. *J. Apicult. Res.*, **2** (1963), 14–18 (role of the queen in preventing oögenesis in worker honey-bees).

[294] CALLAHAN, P. S., and CASCIO, T. *Ann. Ent. Soc. Amer.*, **56** (1963), 535–56 (movements of sperm in female *Heliothis*, Lep.).

[295] CAMENZIND, R. *Rev. Suisse Zool.*, **69** (1962), 378–84 (paedogenesis in *Heteropeza* (*Oligarces*)).

[296] CARAYON, J. *C.R. Acad. Sci.*, **234** (1952), 751–3, 1220–2, 1317–19; *Bull. Mus. Hist. Nat. Paris*, **24** (1952), 89–97 *C.R. Acad. Sci.*, **236** (1953), 1099–1101, 1206–8; **239** (1954), 1542–4 (haemocoelic fecundation in Nabids, Cimicids, Anthocorids).

[297] —— *Rev. Zool. Bot. Africaine*, **60** (1959), 81–104 (extragenital insemination: review).

[298] CHRISTENSEN, P. J. H. *Embryologische und zytologische Studien über die erste und frühe Eientwicklung bei Orgyia antiqua* L. (Lymantriidae, Lep.). Thesis, Copenhagen, 1942.

[299] CLEMENTS, A. N. *J. Exp. Biol.*, **33** (1956), 211–23 (hormonal control of ovary development in mosquitos).

[299a] COLES, G. C. *Nature*, **203** (1964), 323 (corpus allatum and yolk protein synthesis in *Rhodnius*).

[300] COLOMBO, G. *Boll. Zool.*, **23** (1956), 279–88; *Arch. Zool. Ital.*, **42** (1957), 343–4 (histochemistry of oögenesis in *Bombyx*).

[301] COOPER, K. W. in *Biology of Drosophila*, (M. Demerec, Ed.) Wiley, New York, 1950, 1–61 (spermatogenesis in *Drosophila*).

[302] COUNCE, S. J. '*Drosophila*' *Information Service*, **33** (1959), 127–8 (spermatozoa in *Drosophila* eggs).

[303] CURTIS, T. J., and JONES, J. C. *Ann. Ent. Soc. Amer.*, **54** (1961), 298–313 (ovulation in *Aëdes*).

[304] DANILEVSKII, A. S., and GLINYANAYA, E. I. *The Ecology of Insects. Uchen. Zap. Leningr. Gosud. Univ. No. 240* (Ser. biol. Nauk. No. 46) Leningrad, 1958, 34–51 (photoperiod and gonadotropic cycle in *Culex* and *Anopheles*).

[305] DAVEY, K. G. *J. Exp. Biol.*, **35** (1958), 694–701; *Canad. J. Zool.*, **38** (1960), 39–45 (sperm transfer and migration in *Rhodnius*).

[306] —— *Quart. J. Micr. Sci.*, **100** (1959), 221–30 (spermatophore in *Rhodnius*).

[307] —— *Proc. Roy. Ent. Soc. Lond.*, A, **35** (1960), 107–13 (evolution of spermatophores in insects).

[308] DAVIES, D. M. *Nature*, **201** (1964), 948–9 (ribonucleic acid and egg production in *Musca*).

[309] DAVIES, N. T. *Ann. Ent. Soc. Amer.*, **49** (1959), 466–93 (male and female reproductive systems in *Cimex*).

[310] DOWNES, J. A. *Trans. Roy. Ent. Soc. Lond.*, **106** (1955), 213–26 (swarming and mating in *Culicoides*).

[311] DUPONT-RAABE, M. *Arch. Zool. Exp. Gén.*, **89** (1952), 128–38 (neurosecretory cells and egg production in Phasmids).

[312] ENGELMANN, F. *J. Ins. Physiol.*, **1** (1957), 257–78; *Biol. Bull.*, **116** (1959), 406–91; *Ann. N.Y. Acad. Sci.*, **89** (1960), 516–39; *Gen. Comp. Endocrin.*, **2** (1962), 183–92 (hormonal control of reproduction in viviparous cockroaches).

[313] FERON, M. *Bull. Soc. Ent. France*, **65** (1960), 139–43 (male stridulation and mating in *Dacus*, Dipt.).

[314] GILBERT, L. I., and SCHNEIDERMAN, H. A. *Gen. Comp. Endocrin.*, **1** (1961), 453–72 (juvenile hormone in male and female *Hyalophora*).

[315] GILLETT, J. D. *Nature*, **180** (1957), 656–7; *J. Exp. Biol.*, **35** (1958), 685–93 (hormonal control of ovarian development in *Aëdes*).

[316] GOODING, S. J. *Roy. Soc. Arts*, **99** (1951), 597–612 (recent research on bees: review).

[317] GRESSON, R. A. R., and THREADGOLD, L. T. *Quart. J. Micr. Sci.*, **103** (1962), 135–46 (nuclear extrusion by oöcyte in *Blatta*).

[318] HACKMAN, R. H., and GOLDBERG, M. *J. Ins. Physiol.*, **5** (1960), 73–8 (composition of oöthecae in *Periplaneta*).

[319] HAGAN, H. R. *Embryology of the viviparous insects*, 1951, New York: Ronald Press Co. 472 pp.

[320] HAMILTON, A. G. *Proc. Roy. Ent. Soc. Lond.*, A, **30** (1955), 103–14 (parthogenesis in *Schistocerca*).

[321] HASKELL, P. T. *Anim. Behaviour*, **8** (1960), 76–81 (effect of the gonads on mating in grasshoppers).

[322] HAUSCHTECK, E. *Chromosoma*, **13** (1962), 163–82 (cytology of paedogenesis in *Heteropeza*).

[323] HELLER, J., and PIECHOWSKA, M. *Bull. Acad. Polon. Sci.*, **4** (1956), 345–9; **6** (1958), 187–91 (transfer of pyrophosphate during mating in *Celerio*).

[324] HIGHNAM, K. C., *et al. Quart. J. Micr. Sci.*, **103** (1962), 57–72, 73–83; *Mem. Soc. Endocrin.*, **12**, (1962), 379–90 (neurosecretory control of ovarian development in *Schistocerca*).

[325] HIGHNAM, K. C., LUSIS, O., and HILL, L. *J. Ins. Physiol.*, **9** (1963), 587–96, 827–37 (corpus allatum in oögensis in *Schistocerca*).

[326] HILDRETH, P. E., and LUCCHESI, J. C. *Devel. Biol.*, **6** (1963), 262–78 (fertilization in *Drosophila*).

[327] HILL, L. *The endocrine control of oöcyte development in the desert locust, 'Schistocerca gregaria' Forskål.* Thesis, Univ. Sheffield, 1963.

[328] HOCKING, B. *Nature*, **169** (1952), 1101 (autolysis of flight muscles in a mosquito).

[329] HOSOI, T. *Jap. J. Med. Sci. Biol.*, **7** (1954), 57–81, 111–27 (egg production in *Culex pipiens*).

[330] HUNTER-JONES, P. *Nature*, **185** (1960), 336 (fertilization in *Schistocerca* after successive copulations).

[331] JACKSON, D. J. *Proc. Roy. Ent. Soc. Lond.*, A, **27** (1952), 57–70 (break-down of muscles in adult water-beetles).

[332] JACOBSON, M., *et al. Science*, **132** (1958), 1011; *J. Amer. Chem. Soc.*, **83** (1961), 4819–4824 Science, **140** (1963), 1367–73 (sex attractant of *Lymantria*, &c.).

[333] JACOBSON, M., *et al. Science*, **139** (1963), 48–9 (sex attractant of *Periplaneta*).

[334] JANET, C. *C.R. Acad. Sci.*, **144** (1907), 393–6, 1070–3 (involution of flight muscles in queen ants).

[335] JOHANSSON, A. S. *Nature*, **174** (1954), 89; *Nyt. Mag. Zool.*, **7** (1958), 5–132 (corpus allatum and egg production in *Oncopeltus*).

[336] JOHNSON, B. *Nature*, **172** (1953), 813; *J. Ins. Physiol.*, **1** (1957), 248–56; **3** (1959), 367–377 (degeneration of flight muscles in alate Aphids).

[337] KARLINSKY, A. *C.R. Acad. Sci.*, **255** (1962), 191–3; **256** (1963), 4101–3 (corpus allatum and yolk formation in *Pieris*).

[338] KESSEL, R. G., and BEAMS, H. W. *Exp. Cell. Res.*, **30** (1963), 440–3 (yolk formation in *Oncopeltus*).

[339] KING, P. E. *Nature*, **189** (1961), 330–1; *J. Exp. Biol.*, **39** (1962), 161–5 (sex ratio in *Mormoniella*).

[340] —— *Proc. Roy. Ent. Soc. Lond.*, A, **37** (1962), 73–5 (movements of sperm in *Mormoniella*).

[341] KING, P. E., and HOPKINS, C. R. *J. Exp. Biol.*, **40** (1963), 751–61 (egg resorption and longevity in *Mormoniella*).

[342] KING, R. C., *et al. Growth*, **21** (1957), 95–102; **23** (1959), 37–53 (nutrition and oögenesis in *Drosophila*).

[343] —— *Growth*, **24** (1960), 265–323; **26** (1962), 235–55; *Symp. Roy. Ent. Soc. Lond.* No. **2** (1964), 13–25 (histochemistry and fine structure of developing oöcytes in *Drosophila*).

[344] KING, R. C., and KOCH, E. A. *Quart. J. Micr. Sci.*, **104** (1963), 297–320 (fine structure of follicle cells and chorion in *Drosophila*).

[344a] KRAIŃSKA, M. *Quart. J. micr. Sci.*, **102** (1961), 119–29 (histology of oögenesis in *Cynips*).

[345] LABEYRIE, V. *C.R. Acad. Sci.*, **249** (1959), 2115–17 (embryonic diapause in unlaid egg of parasitic Hymenoptera).

[346] LARSEN, J. R., and BODENSTEIN, D. *J. Exp. Zool.*, **140** (1959), 343–81 (hormones and egg production in mosquitos).

[347] LAUGÉ, G. *Bull. Soc. Zool. France*, **85** (1960), 147–57 (parthenogenesis in *Periplaneta*).

[348] LAVEN, H. *Naturwiss.*, **43** (1956), 116–17 (induced parthenogenesis in *Culex pipiens*).

[349] LAURENCE, B. R., and ROSHDY, M. A. *Nature*, **200** (1963), 495–6 (corpus allatum and yolk formation in *Aëdes*).

[350] LAUTERBACH, G. *Z. Parasitenk.*, **16** (1954), 255–97 (mating and parturition in Strepsiptera).

[351] LEA, A. O. *J. Ins. Physiol.*, **9** (1963), 793–809 (corpus allatum and egg maturation in mosquitos).

[352] LEA, A. O., *et al. Science*, **123** (1956), 890–1; *Canad. Ent.*, **88** (1956), 57–62 (nutrition and egg production in *Aëdes aegypti*).

[353] LEE, H. TSUI-YING, *Biol. Bull.*, **98** (1950), 25–33 (insemination reaction in *Drosophila*).

[354] LOHER, W. *Proc. Roy. Soc.*, B, 153 (1960), 380–97 (corpora allata and maturation in male *Schistocerca*).

[355] LÜSCHER, M., and WALKER, I. *Rev. Suisse Zool.*, **70** (1963), 304–11 (inhibition of corpora allata of worker bees by 'queen substance').

[356] LUSIS, O. *Quart. J. Micr. Sci.*, **104** (1963), 57–68 (histology and histochemistry of oöcytes in *Schistocerca*).

[357] MANNING, A. *Nature*, **194** (1962), 252–3 (sperm factor affecting receptivity of *Drosophila* female).

[358] MAHR, E. *Evolution*, **4** (1950), 149–54 (antennae in mating behaviour of *Drosophila*).

[359] MEAD-BRIGGS, A. R., *et al. Nature*, **187** (1960), 1136–7; **201** (1964), 1303–4; *J. Exp. Biol.*, **41** (1964), 371–402 (a factor from the pregnant rabbit, and ovarian maturation in the flea *Spilopsyllus*).

[360] MEYER, G. F. *Z. Zellforsch.*, **54** (1961), 238–51 (intercellular bridges in testis and ovary of *Drosophila*).

[361] MONROE, R. E., KAPLANIS, J. N., and ROBBINS, W. E. *Ann. Ent. Soc. Amer.*, **54** (1961), 537–9 (sterols and egg production in *Musca*).

[362] MORSE, R. A., GARY, N. E., and JOHANSSON, T. S. K. *Nature*, **194** (1962), 605 (mandibular gland secretion and mating in *Apis*).

[363] MUKERJI, D., and SARMA, P. S. *Nature*, **168** (1951), 612 (spermatophore in the sucking louse *Linognathus*).

[364] MÜLLER, H. J. *Beitr. Ent.*, **7** (1957), 203–26; *Zool. Anz.*, **160** (1958), 295–312 (reproductive diapause in *Stenocranus*, Delphacidae).

[365] NATH, V., GUPTA, B. L., and LAL, B. *Quart. J. Micr. Sci.*, **99** (1958), 315–32 (histochemistry of lipids in oögenesis in *Periplaneta*).

[366] NIKOLEI, E. *Z. Morph. Oekol. Tiere*, **50** (1961), 281–329 (nutrition and reproduction in *Miastor*, &c.).

[367] NISHIITSUTSUJI-UWO, J. *Mem. Coll. Sci. Kyoto*, **26** (1959), 9–14 (testis in diapausing pupae of *Papilio xuthus*).

[368] NOIROT, C. *C.R. Acad. Sci.*, **245** (1957), 743–5 (neurosecretion in reproductive forms of *Calotermes*).

[369] NUR, U. *J. Morph.*, **111** (1962), 173–99 (sperm bundles in *Pseudococcus*).

[370] OZEKI, K. *Sci. Papers Coll. Gen. Educ. Univ.* Tokyo, **8** (1958), 85–92 (corpus allatum and yolk formation in *Anisolabis*).

[371] PARKER, K. D., and RUDALL, K. M. *Biochim. Biophys. Acta*, **17** (1955), 287 (calcium citrate in oötheca of *Mantis*).

[372] PESSON, P. *Proc. VIIIth Int. Congr. Ent., Stockholm, 1948*, 1950, 566–70 (fertilization in *Aspidiotus*).

[373] PIEPHO, H. *Biol. Zbl.*, **69** (1950), 1–10 (non-specific character of the corpus allatum hormone).

[374] RAMAMURTY, P. S. *Naturwiss.*, **50** (1963), (RNA supply to the oöcyte in *Panorpa*).

[375] RENNER, M. *Z. Tierpsychol.*, **9** (1952), 122–54 (gonads and mating behaviour in *Euthystira*, Acrididae).

[376] ROBBINS, W. E., and SHORTINO, T. J. *Nature*, **194** (1962) 502–3 (cholesterol and ovary development in *Musca*).

[377] RÖLLER, H., PIEPHO, H., and HOLZ, I., *J. Ins. Physiol.*, **9** (1963), 187–97 (gonads and mating behaviour in *Galleria*).

[378] ROTH, L. M. *Science*, **138** (1962), 1267–9 (corpus allatum and mating in *Nauphoeta*, Blattaria).

[379] ROTH, L. M., and STAY, B. *Science*, **130** (1959), 271–2; *J. Ins. Physiol.*, **7** (1961), 186–202; *Psyche*, **69** (1962), 165–208; *Ann. Ent. Soc. Amer.*, **55** (1962), 633–42 (control of oöcyte development in cockroaches).

[380] ROTH, L. M., and WILLIS, E. R. *Amer. Midl. Nat.*, **47** (1952), 66–129 (mating behaviour of *Blattella*).

[381] —— *Psyche*, **62** (1955), 55–68 (infra-uterine nutrition in *Diploptera*, Blattaria).

[382] —— *Ann. Ent. Soc. Amer.*, **49** (1956), 195–204 (review: parthenogenesis in cockroaches); *Trans. Amer. Ent. Soc.*, **83** (1958), 221–38 (review: viviparity in cockroaches).

[383] ROTH, T. F., and PORTER, K. R. *J. Cell. Biol.*, **20** (1964), 313–32 (yolk protein uptake by oöcyte of *Aëdes*).

[384] ROTHSCHILD, M., and FORD, B. *Nature*, **201** (1964), 103–4 (oögenesis in the rabbit flea *Spilopsyllus* controlled by reproductive hormones from the host).

[385] RUTTNER, F. *Bee World*, **37** (1956), 3–22 (mating in the honey-bee).

[386] SAUNDERS, D. S. *Nature*, **185** (1960), 121–2; *Trans. Roy. Ent. Soc. Lond.*, **112** (1960), 221–38 (ovaries of *Glossina*).

[387] SCHALLER, F. *Naturwiss.*, **41** (1954), 406–7 (spermatophore in *Campodea*).

[388] SCHMIDT, E. L., and WILLIAMS, C. M. *Biol. Bull.*, **105** (1953), 174–87 (spermatocytes in *Hyalophora* pupa in diapause).

[389] SCHNEIDER, D. *J. Ins. Physiol.*, **8** (1962), 15–30 (electro-physiological effect of sex attractant in the antenna).

[390] SCHWINK, I. *Z. angew. Ent.*, **37** (1955), 349–57 (specificity of sexual scent in *Lymantria* spp.).

[391] SEILER, J. *Rev. Suisse Zool.*, **58** (1951), 489–95; *Arch. Entw. Mech.*, **150** (1957), 199–372 (sexual mosaics in *Solenobia*).

[392] SEILER, J., *et al. Chromosoma*, **11** (1960), 29–102 (parthenogenesis and sex determination in *Solenobia*).

[393] SÉRENO, C., and DURAND, M. *C.R. Acad. Sci.*, **251** (1960), 2242–44 (RNA formation in ovary of *Gryllus*).

[394] SKINNER, D. M. *Biol. Bull.*, **125** (1963), 165–76 (incorporation of protein into oöcytes of *Hyalophora*).

[395] SLÁMA, K., and HRUBEŠOVÁ, H. *Zool. Jahrb., Physiol.*, **70** (1963), 291–300 (effect of juvenile hormone on metamorphosis, metabolism, and reproduction in *Pyrrhocoris*).

[396] SMITH, J. M. *J. Exp. Biol.*, **35** (1958), 832–42 (temperature and egg laying in *Drosophila*).

[397] STAY, B., ROTH, L. M., *et al. Ann. Ent. Soc. Amer.*, **53** (1960), 79–86 (oötheca of cockroaches); **55** (1962), 124–30 (colleterial glands of cockroaches)

[398] STEIN, G. *Biol. Zbl.*, **82** (1963), 343–9 (farnesol in mandibular glands of *Bombus*).

[399] TAHMISIAN, T. N., *et al. Exp. Cell. Res.*, **9** (1955), 135–8 (oscillation of oöcyte nuclei in *Melanoplus*).

[400] TAHMISIAN, T. N., *et al. J. Biophys. Biochem. Cytol.*, **2** (1956), 325–30 (mitochondria during spermatogenesis in *Melanoplus*).

[401] TELFER, W. H. *J. Gen. Physiol.*, **37** (1954), 539–58; *Biol. Bull.*, **118** (1960), 338–51 (accumulation of blood proteins in oöcytes of Saturniid moths).

[402] TELFER, W. H., *et al. J. Biophys. Biochem. Cytol.*, **9** (1961), 747–59; *Amer. Zoologist*, **3** (1963), 185–91 (mechanism of protein transfer to the oöcytes).

[403] THOMSEN, E. *J. Exp. Biol.*, **29** (1952), 137–72 (neurosecretory cells and oöcyte growth in *Calliphora*).

[404] THOMSEN, E., and MØLLER, J. *Nature*, **183** (1959), 1401–2; *J. Exp. Biol.*, **40** (1963), 301–21 (neurosecretion and intestinal protease in *Calliphora*).

[405] ULRICH, H. *Verh. dtsch. Zool. Ges., Wien 1961*, 1962, 140–52 (sex determination in *Heteropeza* (*Oligarces*)).

[406] VANDERBERG, J. P. *Biol. Bull.*, **125** (1963), 556–75 (synthesis and transfer of nucleic acids and protein in oöcytes of *Rhodnius*).

[407] VANDERZANT, E. S., and RICHARDSON, C. D. *J. Ins. Physiol.*, **10** (1964), 267–72 (lipid requirements in *Anthonomus grandis*).

[408] VIRKKI, N. *Z. Zellforsch.*, **44** (1956), 644–65 (development of sperm in Scarabaeidae).

[409] VOOGD, S. *Experientia*, **11** (1955), 181–2 (inhibition of ovary development in worker bees by surface extracts from the queen).

[410] WHARTON, D. R. A., *et al. J. Gen. Physiol.*, **37** (1954), 461–9; 471–81; *J. Ins. Physiol.*, **1** (1957), 229–39; *Science*, **137** (1962), 1062–3 (female sex attractant in *Periplaneta*).

[411] WIGGLESWORTH, V. B. *Nature*, **174** (1954), 556 (juvenile hormone in corpus allatum of *Calliphora*).

[412] —— *J. Ins. Physiol.*, **9** (1963), 105–19 (farnesol, &c., and yolk formation in *Rhodnius*)

[413] —— *Advances in Insect Physiol.*, **2** (1964), 243–332 (hormones and reproduction in insects: review).

[414] WIGGLESWORTH, V. B., and BEAMENT, J. W. L. *Quart. J. Micr. Sci.*, **91** (1950), 429–52 (moulding of oötheca in *Blattella*).

[415] WILBERT, H. *Zool. Jahrb., Physiol.*, **64** (1953), 470–95 (experimental production of males in *Carausius*).

[416] de WILDE, J. *Arch. Néerl. Zool.*, **10** (1954), 375–85 (corpus allatum and diapause in *Leptinotarsa* adult).

[417] WILLIAMS, C. M. *Biol. Bull.*, **124** (1963), 355–67 (juvenile hormone in male and female *Hyalophora*).

[418] WILSON, F. *Aust. J. Zool.*, **10** (1962), 349–59 (sex determination in Hymenoptera: review).

[419] WYATT, I. J. *Proc. Roy. Ent. Soc. Lond.*, A, **36** (1961), 133–43; **38** (1963), 136–44 (pupal paedogenesis in Cecidomyiidae).

SUPPLEMENTARY REFERENCES (C)

[420] ADAMS, T. S., HINTZ, A. M. and POMONIS, J. G. *J. Insect Physiol.*, **14** (1968), 983–93 (oöstatic hormone in *Musca*).

[421] ADAMS, T. S. and MULLA, M. S. *J. Insect Physiol.*, **14** (1968), 627–35 (pheromones and mating in the eye gnat *Hippelates*).

[422] ADIYODI, K. G. and NAYAR, K. K. *Biol. Bull. mar. biol. Lab., Woods Hole*, **133** (1967), 271–81 (corpus allatum and haemolymph proteins in *Periplaneta*).

[423] APLIN, R. T. and BIRCH, M. C. *Nature*, **217** (1968), 1167 (aphrodisiac in male *Noctuids*).

[424] BALLAN-DUFRANCAIS, C. *Bull. Soc. zool. Fr.*, **93** (1968), 401–21 (histochemistry of spermatophore in *Blattella*).

[425] BARTH, R. H. *Science*, **149** (1965), 882–3 (endocrine control of sex pheromones).

[426] BELL, W. J. *Biol. Bull. mar. biol. Lab., Woods Hole*, **137** (1969), 239–49; *J. Insect Physiol.*, **16** (1970), 291–9; **17** (1971), 1099–1111 (vitellogenic blood proteins in *Periplaneta*).

[427] BENNET-CLARK, H. C. and EWING, A. W. *J. exp. Biol.*, **49** (1968), 117–28 (courtship in *Drosophila*).

[428] BENZ, G. *J. Insect Physiol.*, **15** (1969), 55–71 (effect of mating &c. on oögenesis in *Zeiraphera*, Lepidopt.).

[429] BERGER, R. S. *Ann. ent. Soc. Am.*, **59** (1966), 767–771 (sex attractant in *Trichoplusia*).

[430] BIER, K. *Naturwissenschaften*, **54** (1967), 189–95 (early stages of oögenesis and nurse cell formation).

[431] BITSCH, J. *Bull. Soc. zool. Fr.*, **93** (1968), 385–95 (resorption of oöcytes in *Machilis*).

[432] BOECKH, J. and SASS, H. *Science*, **168** (1970), 589 (receptors responsive to female sex attractant in *Periplaneta*).

[433] BROOKES, V. J. *Devl. Biol.*, **20** (1969), 459–71 (control of yolk protein synthesis in fat body of *Leucophaea*).

[434] BUTENANDT, A. and HECKER, E. *Angew Chem.*, **73** (1961), 349–53 ('bombycol', sex attractant in *Bombyx*).

[435] CAMENZIND, R. *Chromosoma*, **18** (1966), 123–52 (sex determination in *Heteropeza*).

[436] CANTACUZÈNE, A. M. *Z. Zellforsch. mikrosk. Anat.*, **90** (1968), 113–26 (types of spermatodesm in Orthoptera).

[437] CARLISLE, D. B., ELLIS, P. E. and BETTS, E. *J. Insect Physiol.*, **11** (1965), 1541–58 (odour of aromatic shrubs and sexual maturation in *Schistocerca*).

[438] CHEN, P. S. and BÜHLER, R. *J. Insect Physiol.*, **16** (1970), 615–27 (sex peptide from paragonia of *Drosophila*).

[439] CLARKE, K. U. and GILLOT, C. *J. exp. Biol.*, **46** (1967), 13–34 (metabolic effects of removal of frontal ganglion in *Locusta*).

[440] COLES, G. C. *J. exp. Biol.*, **45** (1965), 425–31; *J. Insect Physiol.*, **11** (1965), 1317–30; **12** (1966), 1029–37 (protein synthesis and yolk formation in *Rhodnius*).

[441] CRUICKSHANK, W. J. *J. Insect Physiol.*, **17** (1971), 217–32 (site of yolk protein synthesis in Lepidoptera).

[442] DAVEY, K. G. *J. exp. Biol.*, **42** (1965), 373–8; *J. Insect Physiol.*, **13** (1967), 1629–36; **14** (1968), 1815–20 (copulation and egg production in *Rhodnius*).

[443] DAVIS, N. T. *J. Insect Physiol.*, **10** (1964), 947–63; **11** (1965), 355–66; 1199–1211 (reproductive physiology in *Cimex*).

[444] DESEÖ, K. V. *Colloqu. int. C.N.R.S.* No. **189**: *L'Influence des stimuli externes sur la gamétogenèse des insectes*, 1970, 163–74 (environmental effects in Lepidoptera).

[445] EDWARDS, F. J. *J. Insect Physiol.*, **16** (1970), 2027–31 (endocrine control of flight muscle histolysis in *Dysdercus*).

[446] EL-IBRASHY, M. T. and BOCTOR, I. Z. *Z. vergl. Physiol.*, **69** (1970), 111–16 (allatectomy and lipid metabolism in *Spodoptera*, Noctuidae).

[447] ENGELMANN, F. *Science*, **165** (1969), 407–9 (corpus allatum and yolk protein synthesis in *Leucophaea*).

[448] ENGELMANN, F. *The Physiology of Insect Reproduction*, Pergamon: Oxford and New York, 1970, 307 pp.

[449] EWEN, A. B. *Can. J. Zool.*, **44** (1966), 719–27 (corpus allatum and oöcyte maturation in *Adephocoris*, Hemipt.).

[450] GARCIA-BELLIDO, A. *Z. Naturforsch.*, **6** (1964), 492–5 (secretion of paragonia and female fecundity in *Drosophila*).

[451] GELDIAY, S. *J. Endocrin.*, **37** (1967), 63–71; *Gen. comp. Endocrin.*, **14** (1970), 35–42 (hormonal control of diapause in adult *Anacridium*).

[452] GILBERT, L. I. *Comp. Biochem. Physiol.*, **21** (1967), 237–54; *Adv. Insect Physiol.*, **4** (1967), 69–211 (lipid content during reproductive cycle in *Leucophaea* and lipid metabolism: review).

[453] GIRARDIE, A. *Bull. Soc. zool. Fr.*, **91** (1966), 423–37 (control of corpus allatum activity in adult *Locusta*).

[454] HILL, L., MORDUE, W. and HIGHNAM, K. C. *J. Insect Physiol.*, **12** (1966), 977–94; 1197–1208 (frontal ganglion and oöcyte growth in *Schistocerca*).

[455] HOFFMANN, H. J. *J. Insect Physiol.*, **16** (1970), 629–42 (neuroendocrine control of diapause in adult *Pterostichus*, Col.).

[456] JACOBSON, M. *et al. Science*, **159** (1968), 208–10 (sex attractant in *Limonius*).

[457] JONES, J. C. and WHEELER, R. E. *Biol. Bull. mar. biol. Lab., Woods Hole*, **129** (1965), 134–50, 532–45 (filling of spermatheca in *Aëdes*).

[458] JONES, W. A. *et al. Science*, **152** (1966), 1516–17 (sex attractant in pink bollworm, *Pectinophora*).

[459] KING, R. C. *Ovarian Development in Drosophila melanogaster*, Academic Press, New York and London, 1970, 227 pp.

[460] KING, R. C. and AGGARWAL, S. K. *Growth*, **29** (1965), 17–83 (oögenesis in *Hyalophora*).

[461] de KORT, C. A. D. *Comm. Agric. Univ. Wageningen*, **69**–2 (1969), 1–63 (hormones and structure and chemistry of flight muscles in *Leptinotarsa*).

[462] LAMY, M. *C.R. Acad. Sci., Paris*, **265** (1967), 990–3 (vitellogenic protein in female *Pieris*).

[463] LEA, A. O. *J. Insect Physiol.*, **13** (1967), 419–29 (neurosecretory cells and egg maturation in mosquitos).

[464] LEAHY, M. G. *J. Insect Physiol.*, **13** (1967), 1283–92 (non-specificity of the paragonial factor in *Diptera*).

[465] LOHER, W. *Z. vergl. Physiol.*, **53** (1966), 277–316 (control of egg ripening and sexual behaviour in *Gomphocerus*).

[466] de LOOF, A. and de WILDE, J. *J. Insect Physiol.*, **16** (1970), 157–69 (hormonal control of yolk protein formation in *Leptinotarsa*).

[466a] LÜSCHER, M. *J. Insect. Physiol.*, **14** (1968), 499–511; 685–8 (control of yolk protein synthesis in *Nauphoeta*).

[467] MEAD-BRIGGS, A. R. and VAUGHAN, J. A. *J. exp. Biol.*, **51** (1969), 495–511 (mating requirements in rabbit flea, *Spilopsyllus*).

[468] MELIUS, M. E. and TELFER, W. H. *J. Morph.*, **129** (1969), 1–15 (yolk deposition in *Hyalophora*).

[469] MERLE, J. *J. Insect Physiol.*, **14** (1968), 1159–68 (effects of male paragonia on egg laying of female *Drosophila*).

[470] MINKS, A. K. *Archs. néerl. Zool.*, **18** (1967), 175–258 (action of corpus allatum in female reproduction in *Locusta*).

[471] MORDUE, W. *J. Insect Physiol.*, **11** (1965), 493–511; 617–29; *Gen. comp. Endocr.*, **9** (1967), 406–17 (corpus allatum and oöcyte development in *Tenebrio*).

[472] OBARA, Y. *Z. vergl. Physiol.*, **69** (1970), 99–116 (ultra-violet reflection and mating behaviour in *Pieris*).

[473] ODHIAMBO, T. R. *Tissue & Cell*, **1** (1969), 155–82; 325–40 (histochemistry of accessory glands in male *Schistocerca*).

[474] OKASHA, A. Y. K. *et al.*, *J. exp. Biol.*, **53** (1970), 25–45 (sublethal high temperature and reproduction in *Rhodnius*).

[475] PATCHIN, S. and DAVEY, K. G. *J. Insect Physiol.*, **14** (1968), 1815–20 (histological changes in vitellogenesis in *Rhodnius*).

[476] PENER, M. P. *Ent. exp. and appl.*, **11** (1968), 94–100 (corpus allatum and male reproductive behaviour in Acrididae).

[477] PLISKE, T. E. and EISNER, T. *Science*, **164** (1969), 1170–2 (sex pheromone in male *Danaus*).

[478] RIDDIFORD, L. M. and WILLIAMS, C. M. *Science*, **155** (1967), 589–90 (volatile principle from oak leaves and reproduction in polyphemus moth).

[479] RIDDIFORD, L. M. and WILLIAMS, C. M. *Biol. Bull. mar. biol. Lab., Woods Hole*, **140** (1971), 1–7 (corpus cardiacum and release of sex pheromone in Saturniid moths).

[480] ROBERT, P. C. *Colloqu. int. C.N.R.S.* No. **189**: (*L'influence des stimuli externes sur la gamétogenèse des insectes*. 1970, 147–62 (host plant and oögenesis in *Scrobipalpa*, Lep., Gelechiidae).

[481] ROELOFS, W. L. and COMEAU, A. *Nature*, **220** (1968), 600–1 (sex pheromone in *Argyrotaenia*, Lep., Tortricidae).

[482] ROHDENDORF, E. B. and WATSON, J. A. L. *J. Insect Physiol.*, **15** (1969), 2085–2101 (control of reproductive cycles in female firebrat *Lepismodes*).

[483] RÖLLER, H. *et al. Acta ent. bohemoslov.*, **65** (1968), 208–11; *Naturwissenschaften*, **58** (1971), 265–6 (sex pheromones and mating in *Galleria* and *Achroia*).

[484] ROTH, L. M. and DATEO, G. P. *J. Insect Physiol.*, **11** (1965), 1023–9 (uric acid storage in accessory sex glands of male *Blattella*).

[485] ROTHSCHILD, M. *et al. Nature*, **221** (1969), 1169–70; *Trans. zool. Soc. Lond.*, **32** (1970), 105–188 (environmental factors and maturation in the rabbit flea *Spilopsyllus*).

[486] ROUSSEL, J. P. *Bull. Soc. zool. Fr.*, **91** (1966), 379–91 (role of frontal ganglion in insects).

[487] SCHEURER, R. *J. Insect Physiol.*, **15** (1969), 1411–19; 1673–82 (endocrine control of protein synthesis and yolk formation in *Leucophaea*).

[488] SCHEURER, R. and LÜSCHER, M. *Rev. Suisse Zool.*, **75** (1968), 715–22 (corpus allatum and yolk protein synthesis in *Leucophaea*).

[489] SCHNEIDER, D. *et al.*, *Z. vergl. Physiol.*, **54** (1967), 192–209 (reaction of male *Bombyx* to 'bombycol').

[490] SCHOONEVELD, H. *Netherl. J. Zool.*, **20** (1970), 151–237 (activity of neurosecretory cells of adult *Leptinotarsa* in relation to diapause).

[491] SIEW, Y. C. *J. Insect Physiol.*, **11** (1965), 1–10; 463–79; 973–81; *Trans. R. ent. Soc. Lond.*, **118** (1966), 359–74 (endocrine control of adult diapause in *Galeruca*).

[492] SLÁMA, K. *J. Insect Physiol.*, **10** (1964), 283–303; 773–82; **11** (1965), 113–22 (effect of hormones on growth, reproduction and metabolism in *Pyrrhocoris*).

[493] SPIELMAN, A., LEAHY, M. G. and SKAFF, V. *J. Insect Physiol.*, **15** (1969), 1471–9 (failure of insemination in young female *Aëdes*).

[494] STAY, B. and GELPERIN, A. *J. Insect Physiol.*, **12** (1966), 1217–26 (control of oviposition in *Pycnoscelus*).

[495] STEGWEE, D. *et al. J. Cell. Biol.*, **19** (1963), 519–27 (corpus allatum and reversible degeneration of flight muscle in *Leptinotarsa*).

[496] STEINBRECHT, R. A. *Z. vergl. Physiol.*, **48** (1964), 342–56; *Z. Zellforsch.*, **64** (1964), 227–61 (sexual scent gland of *Bombyx*).

[497] STOFFOLANO, J. G. and MATTHYSSE, J. G. *Ann. Ent. Soc. Am.*, **60** (1967), 1242–6 (adult diapause in *Musca autumnalis*).

[498] TELFER, W. H. *Ann. Rev. Entom.*, **10** (1965), 161–84 (mechanism and control of yolk formation: review).

[499] TELFER, W. H. and SMITH, D. S. *Symp. Roy. ent. Soc. Lond.*, **5** (1970), 117–34 (egg formation in *Hyalophora*).

[500] TRUMAN, J. W. and RIDDIFORD, L. M. *Biol. Bull. mar. biol. Lab., Woods Hole,* **140** (1971), 8–14 (corpus cardiacum and oviposition in *Hyalophora*).

[501] TSCHINKEL, W., WILLSON, C. and BERN, H. A. *J. exp. Zool.,* **164** (1967), 81–6 (sex pheromone in *Tenebrio*).

[502] VELTHUIS, H. H. W. *et al. Nature,* **207** (1965), 1314 (pheromone secretion by laying workers of *Apis*).

[503] VROMAN, H. E., KAPLANIS, J. N. and ROBBINS, W. E. *J. Insect Physiol.,* **11** (1965), 897–904 (corpus allatum and lipid turnover in *Periplaneta*).

[504] WIGGLESWORTH, V. B. *Insect Hormones,* Oliver and Boyd, Edinburgh, 1970, 159 pp.

[505] de WILDE, J. *et al. J. Insect Physiol.,* **3** (1959), 75–85; **6** (1961), 152–61; *Konink Nederl. Akad. Wetensch.* (C) **71** (1968), 321–6 (photoperiod controlling juvenile hormone secretion in *Leptinotarsa*).

[506] de WILDE, J. and FERKET, P. *Med. Rijksfacult. Landbouw. Wetensch.* Gent. **32** (1967), 387–92 (ageing host plant controlling diapause in *Leptinotarsa*).

[507] ZINSMEISTER, P. P. and DAVENPORT, R. *J. Insect Physiol.,* **17** (1971), 29–34 (RNA and protein synthesis in cockroach ovary).

INDEX OF AUTHORS

GENERAL INDEX